MINING AND THE ENVIRONMENT

FROM ORE TO METAL

KARLHEINZ SPITZ
Environmental Resources Management, Perth, Australia

JOHN TRUDINGER
Trudinger Consultants, St. Kilda, Australia

Supporting website:
www.miningandtheenvironment.com

CRC Press is an imprint of the
Taylor & Francis Group, an **informa** business
A BALKEMA BOOK

Cover illustration: Abandoned Open Pit Copper Mine, Casa Grande, Arizona, USA
Courtesy of Jim Wark, © Jim Wark, www.airphotona.com

Taylor & Francis is an imprint of the Taylor & Francis Group, an informa business

Typeset by Charon Tec Ltd., A Macmillan Company. (www.macmillansolutions.com)
Printed and bound in Great Britain by Cromwell Press Ltd, Trowbridge, Wiltshire.

Published by: CRC Press/Balkema
P.O. Box 447, 2300 AK Leiden, The Netherlands
e-mail: Pub.NL@taylorandfrancis.com
www.crcpress.com – www.taylorandfrancis.co.uk – www.balkema.nl

Library of Congress Cataloguing-in-Publication Data

Mining and the environment : from ore to metal / edited by Karlheinz
Spitz, John Trudinger.
p. cm.
Includes bibliographical references and index.
ISBN 978-0-415-46509-0 (hardback : alk. paper) — ISBN
978-0-415-46510-6 (pbk. : alk. paper) — ISBN 978-0-203-89543-6 (ebook :
alk. paper) 1. Mines and mineral resources—Environmental aspects. 2.
Environmental impact analysis. I. Spitz, Karlheinz, 1955- II.
Trudinger, John. III. Title.

TN275.M477 2008
363.73′1—dc22

2008037125

ISBN 978-0-415-46509-0 (Hardback)
ISBN 978-0-415-46510-6 (Paperback)
ISBN 978-0-203-89543-6 (eBook)

This book is dedicated to **Dibyo Kuntjoro** – Indonesian patriot, mining engineer, environmentalist, public servant and mentor to generations of young Indonesian professionals; and to **Trent R. Dames** and **William M. Moore**, – founders of the pre-eminent international consulting firm – Dames & Moore

Contents

Acknowledgement *xiii*
About the authors *xv*
Color plates *xvii*

1 Minerals, Wealth and Progress **2**
1.1 History of mining 3
1.2 The path of minerals from cradle to grave 7
1.3 Ore – a natural resource blessing? 12
1.4 What makes the mining industry different? 21
1.5 The unique risk profile of mining 25
1.6 Meeting environmental issues head on 29
1.7 Environmental assessment practice – eliminate the negative, accentuate the positive 33
1.8 The equator principles – improved practices for better outcomes 36
1.9 Caring for future generations 41
References 48
Appendix 1.1 Some Useful Websites 51
Appendix 1.2 Some Mines of Interest 59

2 Environmental Impact Assessment **70**
Protection Before Exploitation
2.1 Mining companies are not governments 71
2.2 Environmental assessment in the mining cycle 72
2.3 Managing the environmental impact assessment process 75
2.4 Common themes and core principles 80
2.5 When is an EIA required? 88
2.6 Environmental impact assessment step-by-step 89
2.7 Documenting the findings 96
2.8 Obtaining EIA approval 102
2.9 The costs of delay 103
2.10 Environmental management and monitoring is important 105
2.11 Planning for mine closure 108
2.12 What environmental assessment is not 108
References 110
Appendix 2.1 Data Needs 112

3 Involving the Public **124**
Forging Partnerships and Trust
3.1 Historical perspective 126
3.2 Planning stakeholder involvement 129
3.3 Getting to know your stakeholders 132
3.4 How to identify stakeholders? 142
3.5 Engaging stakeholders 143
3.6 Conflict identification and management 147
3.7 Understanding the benefits and risks of public involvement 149
3.8 Common mistakes 153

References 157
Appendix 3.1 Methods of Public Participation 159

4 The Anatomy of a Mine **164**
This is Mining
4.1 It all begins in the Earth 166
4.2 Exploration – from reviewing data to taking bulk samples 172
4.3 Feasibility – is it worth mining? 183
4.4 Engineering, procurement, and construction 200
4.5 Mining 202
4.6 Ore dressing and thickening 208
4.7 Ancillary facilities 212
4.8 Design for closure 216
References 218

5 Mining Methods Vary Widely **222**
From Excavation to In situ *Leaching*
5.1 The three main categories of commercial minerals 222
5.2 Mining methods 233
5.3 Artisanal mining – mining outside of established law 243
References 252

6 Converting Minerals to Metals **254**
From Ore to Finished Product
6.1 The use of fire – pyrometallurgical mineral processing illustrated 255
6.2 Dissolving ore minerals from gangue – hydrometallurgical mineral processing illustrated 260
6.3 The hydrometallurgical route at industrial scale 266
6.4 Pyrometallurgy and related environmental concerns 281
6.5 Hydrometallurgy and related environmental concerns 286
6.6 Common techniques to estimate emissions 288
References 290

7 Our Environment **294**
A Set of Natural and Man-made Features
7.1 The atmosphere – air, weather, and climate 296
7.2 The lithosphere – geology, landform, and Earth resources 301
7.3 The hydrosphere – storage and movement of water 307
7.4 The biosphere – life on Earth 310
7.5 The social sphere – social and cultural fabric of society 316
7.6 The economic sphere – production, distribution, and consumption of goods and services 321
7.7 Judging the state and value of the environment 324
7.8 What are nature's economic values? 329
7.9 International law pertaining to natural and environmental resources 336
References 344

8 The Baseline **348**
Understanding the Host Environment
8.1 The importance of conflict identification 349

8.2 The use of indicators 351
8.3 Environmental scoping 360
8.4 Conducting baseline surveys – ways and means 364
8.5 Converting data to information 385
8.6 The use of remote sensing techniques and geographic information systems 390
References 391
Appendix 8.1 Common Indicators to Document the State of the Environment and to Detect Change 393
Appendix 8.2 Topics Commonly Included in a Scoping Report 397

9 Identifying and Evaluating Impacts 402
Linking Cause and Effect
9.1 Defining the challenges 402
9.2 Deciding on a direction 408
9.3 Deciding on the methodology 415
9.4 Linking cause and effect 417
9.5 Identifying project impacts 419
9.6 Evaluating project impacts 427
9.7 Cultural heritage sites and mine development 437
9.8 The special nature of community impacts 440
9.9 Environmental justice 444
9.10 Group decision-making in environmental assessment 445
9.11 Reflecting on the objective nature of environmental assessment 448
9.12 Dealing with uncertainties and risks 452
References 461

10 Emphasizing Environmental Management and Monitoring 464
Managing what Matters
10.1 Managing what matters 466
10.2 Management requires measurement 469
10.3 Environmental management systems 475
10.4 Commitment, funds and resources 478
References 479

11 Metals, their Biological Functions and Harmful Impacts 482
Metals are Naturally Occurring Elements
11.1 Persistence, bioaccumulation and toxicity of metals 484
11.2 Some notes on selected metals 488
11.3 Metals, minerals and rock 595
References 501

12 Was the Environmental Assessment Adequate? 504
Identifying Issues, Finding Solutions
12.1 Reviewing the environmental impact statement 504
12.2 Environmental mine audits 509
12.3 Sometimes things go wrong 515
References 516
Appendix 12.1 Comprehensive Mine Audit Checklist (Example) 517

13 The Range of Environmental Concerns **526**
Separating Fact from Fantasy
13.1 Changes in landform 528
13.2 Mine wastes 532
13.3 Mine effluents, acid rock drainage and water balance 537
13.4 Air quality and climate change 542
13.5 Coal – a special case 547
13.6 Biodiversity and habitats 555
13.7 Social and economic change 560
13.8 Surface mining versus underground mining 562
13.9 Accidental environmental impacts 563
13.10 Uranium mining – a special case 564
References 572
Appendix 13.1 An Overview of Environmental and Social Risks and Potential Financial Implications 575
Appendix 13.2 Environmental Impacts at the Exploration Stage 578

14 Land Acquisition and Resettlement **588**
When Property and Development Rights Collide
14.1 Some useful definitions 590
14.2 What determines the severity of resettlement losses? 593
14.3 Resettlement priorities 597
14.4 Compensation for resettlement losses and restoration of livelihood – a right, not a need 597
14.5 Land acquisition and related issues 600
14.6 Livelihood restoration – realizing sustainable value in the compensation of lost assests 604
14.7 The social risks of resettlement 606
14.8 Managing land acquisition and resettlement 610
14.9 Artisanal mining and involuntary resettlement 612
References 613
Appendix 14.1 Full Resettlement Plan: A Recommended Outline 615

15 Community Development **618**
Ensuring Long-term Benefits
15.1 What defines a community? 619
15.2 Pointers to success 621
15.3 Community development process 627
15.4 Preparing for mine closure 638
15.5 Community programmes – what to do? 641
15.6 Local benefits do not always eventuate 650
15.7 Common problems and solutions 651
References 654
Appendix 15.1 Evaluating Community Development Programmes 656

16 Indigenous Peoples Issues **660**
Respecting the Differences
16.1 Who are Indigenous Peoples? 660
16.2 Reason for concern 661
16.3 Important characteristics of indigenous societies 662

16.4 Issues and opportunities 664
16.5 Strategies for interaction with indigenous communities 669
16.6 Rights of Indigenous Peoples 673
16.7 Responsibilities of mining companies in relation to Indigenous Peoples 675
16.8 Preserving or restoring autonomy: partnering for the long-term 678
16.9 Project preparation 684
16.10 In operation and closing down 690
16.11 Conclusions 692
References 692

17 Acid Rock Drainage **696**
A Widespread Problem
17.1 Nature and significance of acid rock drainage 697
17.2 Evaluating the occurrence or risk of ARD 703
References 717

18 Tailings Disposal **720**
Concepts and Practices
18.1 Deciding on the tailings disposal scheme 720
18.2 Alternative approaches to tailings disposal 730
18.3 Surface tailings storage 735
18.4 Submarine tailings placement 759
18.5 Completed and operating STP projects 764
References 769

19 Approaches to Waste Rock Disposal **774**
Issues and Risks
19.1 Nature and characteristics of waste rock 774
19.2 Potential impacts of waste rock disposal 776
19.3 Objectives of waste rock disposal 776
19.4 Site selection for waste rock storages 777
19.5 Alternative design and construction approaches 780
19.6 Landform design 782
19.7 Short-term and long-term erosion control 785
19.8 Monitoring 785
Reference 786

20 Erosion **788**
The Perpetual Disruptive Forces of Water and Wind
20.1 Surface water erosion 789
20.2 Wind erosion 802
References 807

21 Mine Closure **810**
It is Not Over When it is Over
21.1 Reasons for mine closure 811
21.2 Objectives of mine closure 812
21.3 Financing mine closure – the 'polluter pays' principle 812
21.4 Rehabilitation 816

21.5 Pit lakes 839
21.6 Social aspects of mine closure 840
References 840

22 Looking Ahead **844**
22.1 Existing trends in the mining sector 844
22.2 Trends in environmental practice 855
22.3 On and beyond the horizon – global change and challenges 864
22.4 Concluding remarks 879
References 881

Index 883

Acknowledgement

The authors wish to acknowledge the valuable assistance provided by colleagues, friends and peers. Eventually this book summarizes our experience as much as the experience of the many unnamed professionals we have worked with over the years. Particular thanks are extended to Robert McDonough and the late Dibyo Kuntjoro who were involved in formulating the original concept for this book and to Terry Murphy for his editorial guidance in making this text so much more enjoyable to read. Chapter Sixteen "Indigenous Peoples' Issues" had its origins in an unpublished paper prepared in the late 1990s by John Trudinger, John Fargher, Dr Lucy Mitchell and Jim Singleton, at that time all employees of Dames & Moore. Thanks are extended to Greg Guldin of Cross Cultural Consulting Services for his recent review of this Chapter. Thanks also to Robert Barclay for his review and constructive input to Chapter Fourteen "Land Acquisition and Resettlement", to Dr Harvey Van Veldhuizen for reviewing Chapter Eighteen "Tailings Disposal", to Steven Drummond for reviewing Chapter Twenty "Erosion", and (again) to Jim Singleton for his contributions to Chapter Twenty Two "Looking Ahead". Of immense value were the design and text layout provided by graphic designer Diky Halim. We appreciate the support of Jim Wark from Air Photo North America by providing some of his outstanding photographs (including the cover photo). Finally, thanks are extended to Asdora Silalahi who assisted in preparation of the text and in managing the numerous text files through the production process, and to Germaine Seijger, Senior Editor of Taylor & Francis, for her trust in our work and her guidance throughout the process.

About the author

Dr. Karlheinz Spitz is an environmental consultant with more than 20 years of professional experience in Canada, Europe, Asia, and Australia. His main interest is the environmental assessment of large resource development projects in developing countries. He worked on many mines in South East Asia, covering a wide range of minerals and a diverse spectrum of environmental and social settings. Dr. Spitz understands mining as a sustainable economic activity; his focus is on the social, economic and environmental performance of mining. Dr. Spitz provides high level advice to Equator Principles Financial Institutions, and he is regular guest lecturer at various universities.

John Trudinger is an environmental consultant with more than 40 years of professional experience. Qualified as a geologist, his initial experience was on geotechnical investigations for large infrastructure projects. In the early 1970's he became involved in the emerging environmental business, and has since contributed as team member or team leader on environmental assessments for more than 100 resource development and infrastructure projects. He has worked throughout Australia, Asia and North America. His particular interest is the management of mine wastes in the mountainous wet tropics.

Color Plates

1. Environment
2. Exploration
3. Mining
4. Mineral Processing
5. Infrastructure
6. Impact
7. Closure

Wabush Waste Rock Dump in New Foundland - Mine waste management often dominates environmental impacts

In many mining projects, more environmental damage results from waste rock disposal, than any other component of the operations. It is also usually the most visible component.

Photo: courtesy of Jim Wark @airphotona.com

1. Environment
2. Exploration
3. Mining
4. Mineral Processing
5. Infrastructure
6. Impact
7. Closure

Photo 1.1: Karlheinz Spitz

Photo 1.1: Noella and Alina, two sparkling girls in Australia, a well-established mining economy that continues to manage converting natural mineral resources into wealth and sustained development.

Over the ages mining has provided the metals that are the foundation of our civilization. This has not changed. Mining continues to be the nucleus for development. Mining ultimately contributes to a better life for the 6 billion people who depend on the raw materials it produces.

Photo 1.2: Karlheinz Spitz

Photo 1.2

Perhaps because of the large number of bird watchers worldwide, information on the distribution, habitat requirements, behavior and conservation status of birds is readily available for most parts of the world. As a consequence bird sightings (or lack of sightings) often provide a first understanding of the naturalness of a given site.

Photo 1.3: Karlheinz Spitz

Photo 1.3: Rice paddies, Dairi District, North Sumatera, Indonesia.

Few if any mines are developed in inhabited moon-like areas. Land that suddenly has become important with the discovery of valuable minerals has likely been used by humans for generations.

Photo 1.4: Ari Irawan

Photos 1.4: Water sampling, Papua, Indonesia.

For good reasons Earth is known as the water planet within our Solar System. Water, essential to life, is the most controlling resource on our planet, not minerals or oil. Not surprisingly, water is at or near the top end of any mine management's agenda.

Photo 1.5: courtesy of Jim Wark @airphotona.com

Photo 1.5: Haul truck in Arizona, USA.

Host environments vary widely, from flat dry terrains in hot climates such as many parts of Australia or the USA over the wet tropics in South East Asia to the mountainous regions of South America and India.

Photo 1.6: courtesy of Jim Wark @airphotona.com

Photo 1.6: Bingham Kennecott mine, USA.

Large mines often develop their own micro-climate with different climate zones between the bottom of the pit and the pit crest.

1. Environment
2. Exploration
3. Mining
4. Mineral Processing
5. Infrastructure
6. Impact
7. Closure

Photo 2.1: Karlheinz Spitz

Photo 2.1: Exploration track on Halmahera island, Indonesia.

To the extent practical modern exploration activities avoid large access tracks as pictured in this photography. Man-portable rigs or lighter rigs transported by helicopter, which can be transferred from drill site to drill site without the need for vehicular access, are being used increasingly for drilling in rugged forest areas.

Photo 2.2: Road cutting at a nickel mine in Sulawesi, Indonesia.

Road cuttings also provide a picture of the near surface geology.

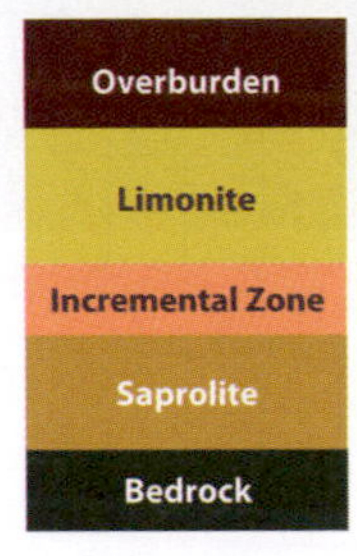

Geological profile of laterite nickel deposits illustrated.

Photo 2.3: Weda Bay Nickel Project, Halmahera Island, Indonesia.

Sometimes test pits are necessary to decide on the most suitable approach to mining or mineral processing, or as in this example, on ways and means to mine rehabilitation.

Photo 2.4: Exploration drilling at the Grasberg mine in Papua, Indonesia.

The only way to test for the possibility of a concealed mineral deposit is to gather data from beneath the surface. This is mainly achieved by drilling. While in the past mining geologists rarely drilled beyond few hundred meters in depths, modern exploration is increasingly a three dimensional search for undiscovered mineral wealth.

Photo 3.1: NASA

Photo 3.1: Freeport's Copper and Gold Mine in Papua, Indonesia.

Iron, copper, stone, or gravel is usually extracted by open pit mining, in which basically a hole is dug in the Earth's crust. It is in this phase that the actual ore body becomes known (see photo 3.1, 3.2 and 3.3).

Photo 3.2: John Trudinger

Photo 3.2: Newmont's Batu Hijau Copper Mine, Indonesia.

1. Environment
2. Exploration
3. Mining
4. Mineral Processing
5. Infrastructure
6. Impact
7. Closure

Photo 3.3: Karlheinz Spitz

Photo 3.3: Tin mining using gravel pumps on Belitung Island, Indonesia.

Photo 3.4: Karlheinz Spitz

Photo 3.4: Small-scale nickel mining on Halmahera Island, Indonesia.

Shovel teams with truck in most mining operations.

Photo 3.5: courtesy of Jim Wark @airphotona.com

Photo 3.6: NASA/GSFC/METI/ERSDAC/JAROS and US/Japan ASTER Science Team

Photo 3.5: (top left) Potash ponds in Utah, USA.

The in-situ leaching process in mining is relatively simple, involving injection of low salinity water with or without pre-heating, the removal of pregnant solution through extraction wells and recovery of salts by means of evaporation, in the same way that common salt and other salts are recovered from sea water by the so called solar salt process.

Photo 3.6: (top right) Three large surface coal mine operations (O) in Germany embedded in a mosaic like pattern of farmland and villages.

Mining is seldom permitted in built-on land or other areas with sensitive land cover such as protected forest, mangrove, or wetlands. Exceptions exist as it is the case of large surface coal mining operations in Northern Germany.

Photo 3.7: Andre Jenny, Painet Inc.

Photo 3.7: Carerra marble mining in Tuscany, Italy.

Mining for industrial minerals easily outweighs metal mining in terms of materials moved.

1. Environment
2. Exploration
3. Mining
4. Mineral Processing
5. Infrastructure
6. Impact
7. Closure

Photo 3.8: Aksel Osterlof

Photo 3.9: Scotty Graham

Photo 3.8, 3.9.

This is big! Modern technologies and equipment allow excavation at a previously unknown scale, exploiting mineral deposits that would have been uneconomical with past mining technologies.

Photo 3.10: Merlina Marra, The Washington Post

Photo 3.10: Oil sand mining in Alberta, Canada.

Nothing illustrates the blurred border between the mineral and oil & gas industry sector better than oil sand mining.

Photo 3.11: Copyright of airphotona.com

Photo 3.11: Junked haul trucks in Arizona, USA.

Eventually even the mightiest machinery comes to a final rest.

Photo 3.12.: Poyry PLC

Photo 3.12: By some estimates peat reserves in Finland alone could, at least in theory, produce twice as much energy as the North Sea oil reserves in Northern Europe.

Formation of peat in a swamp is the first stage in the formation of coal mining. About 325 to 375 million hectares of peat lands exist worldwide. Waste areas are predominantly found in the northern hemisphere: Canada, Russia, the UK and Scandinavia. Predominantly in the northern hemisphere with waste areas in Canada, Russia, UK, and Scandinavia. Equatorial peatland areas (about 20 million hectares) are mostly located at the Sunda shelf in Southeast Asia. Upon drying, peat can be used as a fuel. It provides approximately 7% of Finland's yearly energy production (about 1,500 MW), second only to Ireland (as of 2007). Caloric value.

Photo 3.13: K+S 2007

Photo 3.13.

Underground salt mining in Germany.

Photo 3.14: Armin Kuebelbeck

Photo 3.14.

Placement of waste salt on top of Monte Kali, Germany.

1. Environment
2. Exploration
3. Mining
4. Mineral Processing
5. Infrastructure
6. Impact
7. Closure

Photo 4.1: Ira de Reuver

Photo 4.1: (top left) Search for gemstones in Sri Lanka.

Artisanal mining is defined as small scale mineral extraction, using mainly manual methods, carried out by individuals or small family groups.

Photo 4.2.

The gold rush at Diwalwal in a remote part of Mindanao, Philippines in the 1980's attracted more than 140,000 people. Many died, some when active sites were flood became flooded or collapsed; others were murdered or succumbed to diseases, including those resulting from mercury vapor inhalation to inhalation of mercury vapors.

Photo 4.3: John Trudinger

Photo 4.3: Mill at the Batu Hijau copper mine, Indonesia.

To separate valuable minerals from worthless host rock, crushed ore is commonly ground into fine particles in ball mills or rod-mills which are large, rotating, cylindrical machines.

Photo 4.4: John Trudinger

Photo 4.4: Flotation installation at the Batu Hijau copper mine, Indonesia.

Photo 4.5: (bottom left) Floating heavy minerals in the flotation tank.

Partly physical and partly chemical in its action, flotation is the reverse of gravity concentration (Ch. 4). With its use, heavy minerals can float while light, undesired minerals sink.

Photo 4.5: Karlheinz Spitz

Photo 4.6: John Trudinger

Photo 4.6: Thickener.

Thickening of tailings is a common step prior to pumping the thickened slurry to the tailings pond and ultimately disposing of the thickened slurry. Thickening minimizes the amount of water placed in the pond and the pond size. Thickening is usually accomplished by settling slurries in large tanks, known as thickeners.

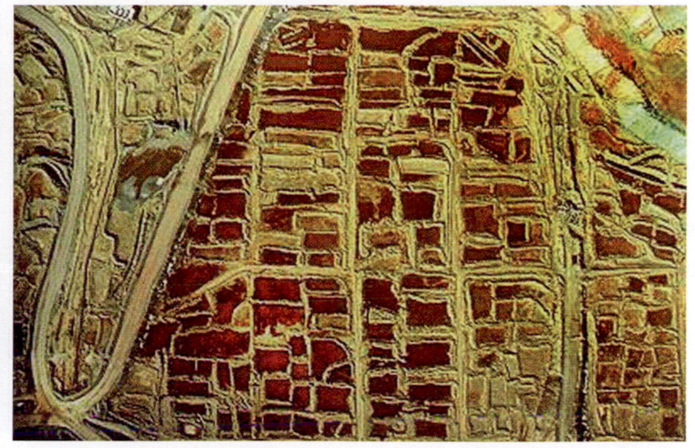

Photo 4.7: courtesy of Jim Wark @airphotona.com

Photo 4.7: Heap leaching operation in Arizona, USA.

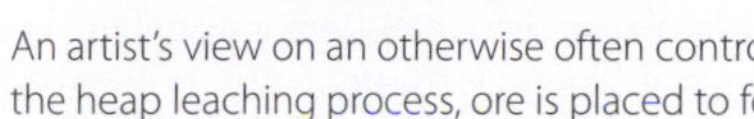

An artist's view on an otherwise often contro-versial mining operation: In the heap leaching process, ore is placed to form heaps on prepared leach pads (Ch. 5). Leach solution applied to the surface of the heaps emerges from the base of the heap as 'pregnant solution', collected in drains for recovery of metals. Heap leaching is controversial because it is difficult to ensure that all leach solution is captured.

Photo 4.8: Tinecke Zelezarny Iron and Steel Works

Photo 4.8: Man at work at the blast furnace.

Photo 4.9: (bottom left) Shaft furnace at Glogow smelter.

The use of fire to extract metals pre-dates recorded history, and it remains the conventional and most commonly used metallurgical process.

Photo 4.11: Stephen Codrington

Photo 4.11: Steel rolling mill.

Iron, the main ferrous metal, is the most important industrial metal, since its alloy with carbon is steel.

Photo 4.9: KGHM

Photo 4.10: Norman Childs

Photo 4.10: Copper electrodes being lifted from sulphuric acid bath at Mantoverdi copper mine, Chile.

Even the best chemical method cannot remove all impurities from copper, but with electro - refining it is possible to produce 99.99% pure copper, the high quality, high purity copper required by end users.

1. Environment
2. Exploration
3. Mining
4. Mineral Processing
5. Infrastructure
6. Impact
7. Closure

Photo 5.1: Karlheinz Spitz

Photo 5.1: Nickel smelter.

Control of gaseous emissions is a continuing challenge to the pyrometallurgical industry. Emission controls focus on stack emissions, but fugitive emissions also contribute considerably to the overall emissions of a smelter.

Photo 5.2: Karlheinz Spitz

Photo 5.2: Amamapare coal-fired power plant, Freeport, Indonesia.

The energy demand of large mines is staggering and may reach several hundred megawatts. The location of the mine, the port, and the type of power generation determine the siting placement of the power plant. Diesel generators are often installed close to the mine; coal-fired power plants close to the port site. In the long run, transporting energy through a transmission line is cheaper transporting coal.

Photo 5.3: John Trudinger

Photo 5.3: Special port, Batu Hijau copper mine, Indonesia.

Mines are often located in remote areas, far from major metropolitan communities and likely end users. Shipment of the final mine product, be it unbeneficiated ore, concentrate or coal, is often via a designated port, which is also usually the main entry point for mine equipment and materials.

Photo 5.4: Karlheinz Spitz

Photo 5.4: Tailings and sea water pipelines, Batu Hijau Copper mine, Indonesia.

The mine operator generally aims to align pipelines and transmission lines along mine roads to avoid the construction and related costs of a dedicated pipeline or transmission line corridor with service road.

Photo 5.5: courtesy of Jim Wark @airphotona.com

Photo 5.5: Iron ore ship on the Great Lakes, USA.

Mines are often located in remote areas, far from major metropolitan communities and likely end users. Shipment of the final mine product, be it unbeneficiated ore, concentrate or coal, is often via a designated port, which is also usually the main entry point for mine equipment and materials.

Photo 5.6: Karlheinz Spitz

Photo 5.6: Ore transport via conveyor belt, Batu Hijau Copper Mine, Indonesia.

Pipelines and conveyors are commonly used for transportation of mine wastes, coal concentrate and commonly also for water and diesel fuel. Similarly, pipelines are the preferred option to transport tailings to the final disposal site.

Photo 5.7: John Trudinger

Photo 5.7: Cross section through enclosed conveyor belt.

Cross-section of a Through Enclosed Conveyor System, developed by ICSI. This in-pit combined crushing-conveyor technique is becoming more common with the availability of more versatile conveyor systems and the construction of increasingly large and deep open pit mines as open pit mines become larger and deeper, and with the availability of far more versatile conveying system.

1. Environment
2. Exploration
3. Mining
4. Mineral Processing
5. Infrastructure
6. Impact
7. Closure

Photo 6.1: Google Earth

Photo 6.1: Mine town in the rainforest of Papua complete with golf course.

Commonly new mine towns with several hundred units are set up for mine workers and their families. Located at a convenient distance from the mine and to the extent possible, a pleasant setting is chosen. High-standard accommodation is provided, supporting the fact that good living conditions contribute to a good working ethos and increased efficiency per staff member. This way the amount of workforce is kept low to keep personnel costs low.

Photo 6.2: John Trudinger

Photo 6.2.

Tailings impoundment at Chatree Gold Mine (Thailand) soon after commissioning. Under-drains have been installed in a 'herringbone' pattern, to facilitate drainage and consolidation of tailings.

Photo 6.3: Carol Stoker, NASA

Photo 6.3: Acid Rock Drainage in Rio Tinto, Spain.

The most serious and pervasive environmental problem related to mine waste management is arguably acid rock drainage (ARD). ARD is a natural process, which takes place wherever sulfide minerals such as pyrite are in contact with oxygen and moisture. Perhaps the oldest incidence/occurence of ARD due to mining is associated with the RioTinto Mine in Spain, where copper has been mined for centuries. The name of the river reflects the characteristic red color caused by iron as it precipitates from acidic solution.

Photo 6.4: courtesy of Jim Wark @airphotona.com

Photo 6.4 & 6.5: The Lady of the Rockies at the Butte Superfund site in Montana, USA, largest superfund site in the US (abandoned mine). The water in the pit lake is toxic with a low pH and high metal content.

Prior to 1985, the issue of mine closure had a low priority for most, as evidenced by the large numbers of abandoned mines that exist in virtually every major mining country. Costs for rehabilitation of these sites can equal or even surpass past financial gains.

Photo 6.6: Ari Irawan

Photo 6.7: Stephen Codrington

Photo 6.6 & 6.7: Indigineous Papuan.

Induced development may easily affect communities that are least prepared for an uncontrolled influx of immigrants and the rapid community changes that may result.

1. Environment
2. Exploration
3. Mining
4. Mineral Processing
5. Infrastructure
6. Impact
7. Closure

Photo 7.1: Karlheinz Spitz

Photo 7.1: Abandoned mine pit with shaft, Belitung Island, Indonesia.

Arguably the most important factor in mine land rehabilitation is the establishment and maintenance of public safety.

Photo 7.2: John Trudinger

Photo 7.2.

To restore visual amenity, rehabilitated areas should blend in with adjacent naturally vegetated areas as is the case in this scene which overlooks the Serujan waste rock dump, Mt Muro Gold Project, Indonesia.

I

Minerals, Wealth and Progress

⊙ GOLD

Gold is the aristocratic metal without equal, and together with copper, among the oldest metals known to man-kind. Gold and copper are also the only two metals that are not grayish in colour, others varying from bluish grey (lead) to white (silver). Unlike copper and most other metals, however, gold is quite inert; it does not tarnish, rust, or corrode – it maintains its elemental condition and yellow colour through geological times. It is however also the most useless metal. Only in recent history has gold found applications beyond its use as currency and for decoration.

1

Minerals, Wealth and Progress

Without the products of mining there would be no civilization as we know it, so a world without mining is unlikely, at least for the foreseeable future. There is, however, a paradox (Mining, Minerals, and Sustainable Development – MMSD – 2002): whereas we enjoy the end products of mining, from simple tools and jewellery to advanced space craft, we are less fond of the 'holes in the ground' needed for their supply. A disconnect between source and product is even reflected in the structure of the metals industry, with some manufacturing companies keen to deny their connection to mining on the basis that large amounts of their raw materials come from secondary sources.

The history of mining is replete with controversy, but in recent decades there has been increasing pressure to improve the environmental performance of mining operations, following from greater awareness of global environmental issues. Although Rachel Carson's 1962 ground breaking text, 'The Silent Spring', focused on pesticide damage, it brought attention for the first time to the worldwide scale of environmental degradation associated with the development of an industrial civilization. This introductory chapter will examine the many facets of the relationship between mining and the environment which follow from the demands of that civilization.

The history of mining and the minerals cycle, of which mining is just one part, reveal the complex linkages between mining and society. Mining operations function within the formal and informal institutional frameworks of the country which hosts the mining project, and therefore inevitably acquire a political dimension, as well as strong links to its economy, ecosystem, and local communities; while these latter almost inevitably come to depend on minerals production for employment, income, and broader development. Unfortunately the perceived divergence between mining based development and environmental conservation, usually focused around the mine site and associated communities, often becomes the subject of controversy in which mining companies find themselves at the centre.

Unfortunately the perceived divergence between mining based development and environmental conservation often becomes the subject of controversy.

Addressing recent history, the chapter touches on the origin and growth of global awareness of environmental issues, and how this has affected regulatory approaches to environmental protection including the now nearly universal environmental impact assessment (EIA) process for new industrial developments. It has become widely recognized that environmental assessment is essential to integrate economic activity with environmental integrity and social concerns. The goal of that integration can be seen as sustainable development.

Finally, there is a discussion of the World Bank's guidance on environmental assessment. First formulated in early 1990, the Bank's approach to environmental assessment of new projects has evolved into a set of standards for industry best practice. Now referred to as the 'Equator Principles', these standards have been adopted by most major international financing institutions.

1.1 HISTORY OF MINING

Mining has been an essential component of social development since prehistoric times. Minerals have met uniquely human needs through the ages, including securing food and shelter, providing defense, enhancing hunting capacities, supplying jewellery and monetary exchange, enabling transport, heat and power systems, and underpinning industry (Hartman 1987). Thus it is no coincidence that we associate most ages of cultural development with minerals or their derivatives: the Stone Age, the Bronze Age, the Iron Age, the Steel Age, and today's Nuclear Age. Gold rushes in recent history contributed to settlements in and development of large areas in Canada, California, South Africa, and Australia.

Minerals have met uniquely human needs through the ages.

Early Mining

Experience of mining varies considerably. Some countries have a long history of mining, either in the form of indigenous small-scale or large-scale, industrial operations, while others show evidence only of recent mining enterprise. There is historical evidence of early mining in Europe, Egypt, and China. In Europe the Iberian peninsula – modern Spain and Portugal – became the focus of the imperial struggle between Rome and Carthage as they fought over its abundance of minerals, including silver, copper, and gold, which in an earlier period had already attracted the interest of the Phoenicians (www.sispain.org).

Mining, of course, has a long history in other parts of the world as well. In the Philippines small-scale mining dates back to the 13th century with the Igorot people, who, for centuries, mined gold and traded it with the Chinese. Historical records show that Southern Africans from Zimbabwe, South Africa, and Tanzania have engaged in mining and smelting for more than a millennium, trading gold with the Arabic world, India, and elsewhere in Asia.

In other areas, mining encouraged the thrust of European colonialism. The invasion of South America, the 'El Dorado' of the 16th century, by Spain and Portugal, is well known. The instructions of the Spanish King Ferdinand to Columbus were plain: 'Get gold, humanely if you can, but at all hazards get gold' (Kettell 1982). Considerations of a shared humanity were to play little part in the early search for and exploitation of mineral wealth. By the end of the 19th century, very few regions remained untouched by the demand for mineral resources to supply the industrialized world.

'Get gold, humanely if you can, but at all hazards get gold'.

However, there are countries where mining commenced relatively recently. In Indonesia, for instance, the first Contract of Work agreement (the legal agreement between the host country and a mining company) was awarded to a US-based company, Freeport McMoran, only in 1967. But even in 'new' mining countries, mining on a small scale may have occurred for centuries.

The first mining was probably done by hand, breaking stones for implements, and working surface deposits of high grade mineral deposits such as copper. This was eventually supplemented by the introduction of simple tools such as picks, shovels, pans, and sluice boxes. Early mining occurred close to human settlements, and the scale was minute compared to today's mining operations.

Because they were small and concentrated on rich and easily extractable deposits, early mining activities tended to have little impact on the environment.

The introduction of new technologies in the past century has changed the nature of mining. The invention of dynamite in 1867 enabled the advent of large-scale mining as practised today.

That has changed with the introduction of new technologies over the past two centuries. The invention of dynamite in 1867, for example, was essential to the large-scale mining of today. Hartman (1987, 1992) summarizes some of the most significant developments that have influenced the mining industry and civilization (**Table 1.1**), but no single chronicle of mining history can be complete. Another good reference, '60 Centuries of Copper' (www.copper.org) provides insight into the history of copper. Along with gold, copper is one of the most important early metals, possibly the first metal used by humans. Another excellent

TABLE 1.1
How Mining Evolved in Human History

Date	Event
450,000 B.C.	First mining (at surface), by Paleolithic man for stone implements
40,000	Surface mining progresses underground, in Swaziland, Africa
30,000	Fired clay pots used in Czechoslovakia
18,000	Possible use of gold and copper in native form
5000	Fire setting, used by Egyptians to break rock
4000	Early use of fabricated metals; start of Bronze Age
3400	First recorded mining, of turquoise by Egyptians in Sinai
3000	Probable first smelting, of copper with coal by Chinese; first use of iron implements by Egyptians
2000	Earliest known gold artifacts in the New World, in Peru
1000	Steel used by Greeks
A.D. 100	Thriving Roman mining industry
122	Coal used by Romans in Great Britain
800	Charlemagne, the first European King, revives mining, contributing to the end of the Dark Ages
1185	Edict by Bishop of Trent gives rights to miners
1550	First use of lift pump, at Joachimstal, Czechoslovakia
1556	First mining technical work, *De Re Metallica*, published in Germany by Georgius Agricola, translated into English in 1912 by Herbert Hoover, the mining engineer who later became President of the United States of America
1600	Mining commences in North America, Era of invasion of South America by Spain and Portugal in search of gold
1627	Explosives first used in European mines, in Hungary (possible prior use in China)

(Continued)

TABLE 1.1
(Continued)

Date	Event
1716	First school of mines established, at Joachimstal, Czechoslovakia
1780	Beginning of Industrial Revolution; pumps are first modern machines used in mining
1800s	American gold rushes help open the West
1815	Sir Humphrey Davy invented miner's safety lamp in England
1843	First mining boom occurs in the US with copper deposits in the Keweenaw Peninsula of Michigan and the Mesabi Iron Range of northern Minnesota
1851	Beginning of the Australian gold rushes by the discovery of gold by Edward Hargraves (resulting in 370,000 new immigrants to Australia in the following year alone)
1866	Discovery of the first diamond, the Eureka, in the Kimberley Diamond Fields in South Africa
1865	Invention of electro refining by Elkington
1867	Dynamite invented by Nobel, applied to mining
1900	Era of mechanization and mass production
1907	Invention of Froth flotation in Australia by Potter
1909	Invention of converting by Pierce and Smith
1940s	First application of solvent extraction technology (solution mining) to uranium purification
1949	Installation of first Flash Smelter (First Outokumpu Furnace)
1963	Solvent Extraction/Electro-winning process developed in Arizona as a means of obtaining copper from oxide ore
1970	NEPA (National Environment Protection Act) promulgated in the USA, signalling the beginning of formal environmental assessment procedures, which rapidly spread to other countries.
1989	Closure of the Bougainville Copper Mine on Bougainville Island, PNG due to civil unrest

Source:
based on Hartman (1987)

reference, 'Gold' by Kettell (1982) provides a thorough history of the one metal that has captivated the imagination more than any other. Further authoritative reviews of mining history can be found in Gregory (1981), Boyns (1997) and Meyerriecks (2003).

Mining Today

Dramatic improvements in mining and mineral processing technologies have effected two main changes in modern mining operations, as compared with mining practices of less

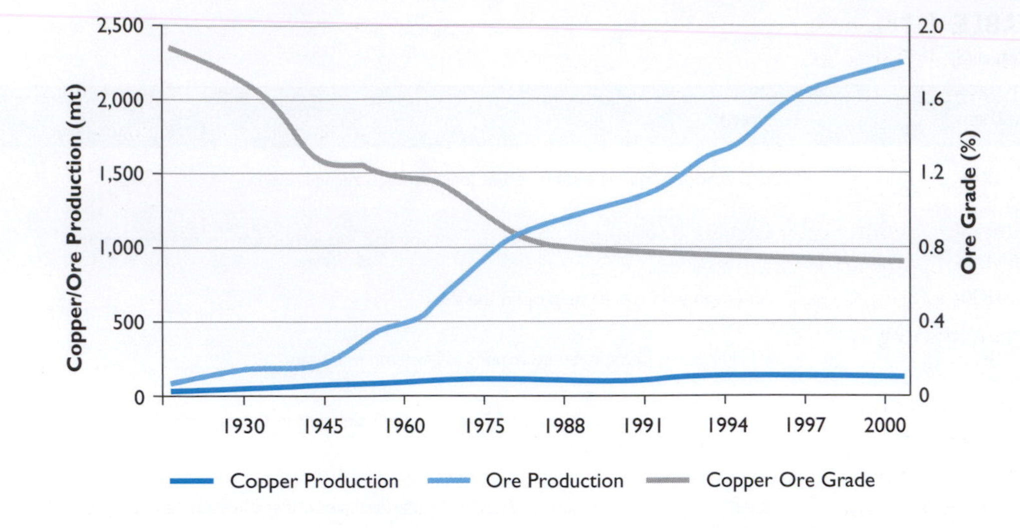

FIGURE 1.1
Shift in Copper/Ore Production and Ore Grade over the Last Century

Modern technologies and equipment permit excavation at a previously unknown scale, exploiting mineral deposits that would have been uneconomical with past mining technologies.

Source:
World Bank 2006

than a century ago: mining now occurs at a much larger scale, and much further from population centres (see Appendix 1.2 for some examples).

The purpose of mining is to excavate from the existing surface down to the mineral deposit. Modern technologies and equipment allow excavation at a previously unknown scale, exploiting mineral deposits that would have been uneconomical with past mining technologies (**Figure 1.1**). Ore production rates can now reach a staggering 100,000 tons per day or more, with total excavation (ore plus waste rock) exceeding 500,000 tons per day at the world's largest mines. As the scale of mining increases, the need for supporting infrastructure also increases. Today's mining operations may require whole mining towns with associated infrastructure, including hospitals, air and seaports, power plants, landfill facilities, and roads.

Given the massive scale of today's developments, a project may become the nucleus of region-wide or even national economic development.

Given the massive scale of today's developments, a project may become the nucleus of region-wide or even national economic development. The consequences of mining at such large scale in remote areas, however, are not all favourable. In fact, few industrial endeavours generate more controversy. In the first instance, land clearing for mine access, excavation of ore, particularly in the case of open Pit mines, and provisions for waste rock and tailings, all change the landscape profoundly. Secondly, there may be community impacts which are difficult to identify and plan for in advance, including social, economic, and political changes which potentially affect the opening, operation, and closing of a mine. It is also true that benefits generated by the mine are almost never equally distributed; although many profit, large numbers may also lose out as mining processes alter the landscape, and disrupt social and economic networks.

The move to larger scale and more remote sites has been accompanied by another challenge for the mining industry, that is, to contain extraction and processing costs in the face of declining ore grades. The resulting increase in the volume of processed ore means that the cost and volume of waste per unit of metal extracted increase, potentially resulting in greater costs of environmental management as well. To date, these costs have been managed by technological advances, specifically in the development of larger, more cost-effective bulk haulage systems.

Mining Terminology

As in any specialized discipline, there are many terms and expressions unique to mining (for a complete glossary of mining terminology see standard references such as Trush 1968;

Gregory 1981; or www.geology.com). Most mining terms in this textbook are introduced selectively, but a few key terms are defined below.

Mining is defined as all activities related to excavating rocks, stones, or minerals that can be sold at a profit. In a more general sense it also includes the subsequent extraction of valuable metals. The extraction of valuable minerals and further refining is referred to as mineral processing, detailed in the following section. Mineral processing covers a wide range of metallurgical processes, ranging from simple gravimetrical separation to complex high-pressure acid leaching to smelting. Primary mineral processing is commonly based on physical processes, referred to as ore beneficiation. Beneficiation is an integral part of many mining operations.

Mining is defined as all activities related to excavating rocks, stones, or minerals that can be sold at a profit.

The excavation made into the Earth-crust to extract minerals is called the mine. The mined mineral itself is a naturally occurring substance, usually inorganic (coal, the most obvious exception, is an organic compound), having a definite chemical composition and distinctive physical characteristics.

Unique mining terms and expressions impose unexpected challenges in working on mining projects in countries with a relatively young mining history. Often no equivalent for a specific mining term exists in the local language, which makes the preparation of documents such as environmental impact statements challenging. It is not uncommon for English mining-specific terms, such as tailings, to be adopted in many languages and even to become part of national mining legislation (e.g. Ind. Government Regulation 19 of 1994).

1.2 THE PATH OF MINERALS FROM CRADLE TO GRAVE

Mining is only the first step in the minerals cycle, that is, the path of any given mineral from cradle to grave. The concept of conducting a detailed examination of the life-cycle of natural resource use, a product, or a process is relatively recent, having emerged in response to increased environmental awareness. The immediate precursors of life-cycle analysis and assessment were the global modelling studies and energy audits of the late 1960s and early 1970s, which were attempts to assess the resource cost and environmental implications of different patterns of human behaviour. Life-cycle analyses were an obvious extension, and they are now vital to the evaluation of mineral use, from mining to the manufacturing processes, the energy consumption in manufacture and use, and the amount and type of waste generated. The study of minerals cycles is instrumental in accurately assessing the total burden placed on the environment by natural resource use. A number of different terms have been coined to describe life-cycle analysis, such as Life Cycle Inventory, Life Cycle Assessment, Cradle to Grave Analysis, Eco-balancing, or Material Flow Analysis. Whichever name is used, life-cycle analysis is a potentially powerful tool, which can improve understanding of the environmental consequences of mineral use.

Minerals Cycle

Minerals are natural resources which are essentially non-renewable, a term applied to resources whose natural regeneration cycle is extremely long. Minerals, metal ores, fossil fuels, and soils do regenerate, but this regeneration takes thousands or millions of years. As a result, non-renewable resources are generally regarded as finite, and their consumption as 'irreversible', a concept underlying the minerals cycle in our economy as depicted in **Figure 1.2**.

Minerals are natural resources which are essentially non-renewable, a term applied to resources whose natural regeneration cycle is extremely long.

Minerals are extracted, transformed into products and goods, transported to other parts of the world, recycled and, sooner or later, released back to the natural environment as waste or

FIGURE 1.2
The Mineral Cycle

The Earth is a closed material system, which means that there are firm limits on natural resource use.

Source:
MMSD 2002

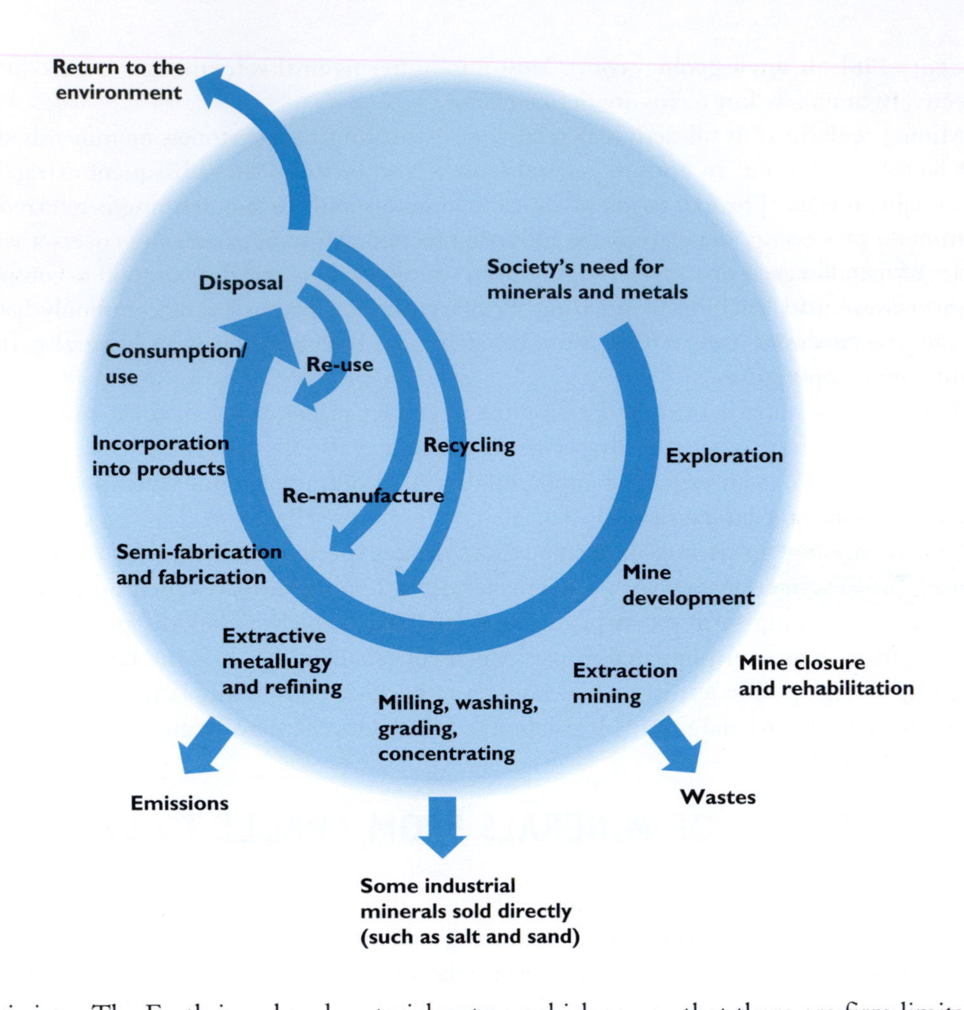

emissions. The Earth is a closed material system, which means that there are firm limits on natural resource use. Some limits are clearly related to availability, which is closely related to price. For some non-renewable resources including many metals and construction minerals, security of supply does not currently give cause for concern; for others, such as oil and land, availability is already a problem that is almost certain to grow. There are also limits related to the ability of the environment to act as a 'sink', that is, to absorb discharges and emissions of pollutants and wastes without serious damage, illustrated most recently by Al Gore's (2006) book on anthropogenic carbon dioxide emission and its relation to global warming.

There are limits related to the ability of the environment to act as a 'sink', that is, to absorb discharges and emissions of pollutants and wastes without serious damage.

Minerals generally have a long life-cycle, and only a small fraction of the minerals put into use each year ends up in the waste stream; those that do include metals in short-life products such as packaging material (e.g. aluminium cans). Most minerals in final products are stocked within the economy for at least several years, since they are used mainly in durable consumer goods, e.g. automobiles, and infrastructure, including the capital stock of industries (e.g. machinery, equipment, and industrial buildings). Gold is an extreme case in that most of the gold that has been extracted throughout history, remains in use or in storage. Similarly, valuable gemstones are seldom discarded.

Recycling – Extending the Life-Cycle of Minerals

Recycling is the most common way of extending the life-cycle of minerals. It saves primary raw material inputs, and reduces the need for new mines with associated environmental impacts. Also, in many cases, processing secondary raw minerals is less environmentally obtrusive and requires less energy than producing primary raw minerals, particularly the case with aluminium. However, mineral recycling has its own set of environmental impacts. For some minerals, high recycling rates have already been achieved (**Table 1.2**). In Europe, the share of the secondary fraction (the share of scrap in the total input to production/smelting) for silver, copper and lead exceeds 50 % and is about 35 to 50 % for steel, aluminium and zinc (EEA 2005). In the US, recycling rates in 1998 were 59% for iron and steel, 39 % for aluminium, 37 % for copper, and about 22% for zinc (Hudson *et al.* 1999).

Mineral recycling has its own set of environmental impacts.

Recycling of any commodity depends on the relative cost of recycling versus the cost of primary production. As the commodity price increases, the economics of recycling become more favourable. However, this simple relativity may be changed by government

TABLE 1.2

Production, Consumption, and Recycling of Metals

	Steel	Aluminium	Copper	Lead	Gold
Cumulative total world production (in tons)	32 billion tons of crude steel	573 million	409 million*	204 million*	128,000 – 140,000
Recent annual world consumption (in tons)	837 million	24.9 million	15.1 million	6.2 million	3,948
Share of total metal consumption derived from recycled material	US 79%, West Europe 55%, East and SE Asia 52%, rest of western world 46%	North America 35%, Western Europe 31%, Asia 25%, world 29%	Western world 35%	US 70%, rest of western world 55%	Western world 35%

*World production from 1900–2000

Source:
MMSD 2002

Because the extraction of aluminium from alumina requires an enormous amount of electrical energy, the aluminium industry initiated processes to recycle used aluminium and was one of the first industries to do so.

CASE 1.1
Battery Lead Recycling in Germany

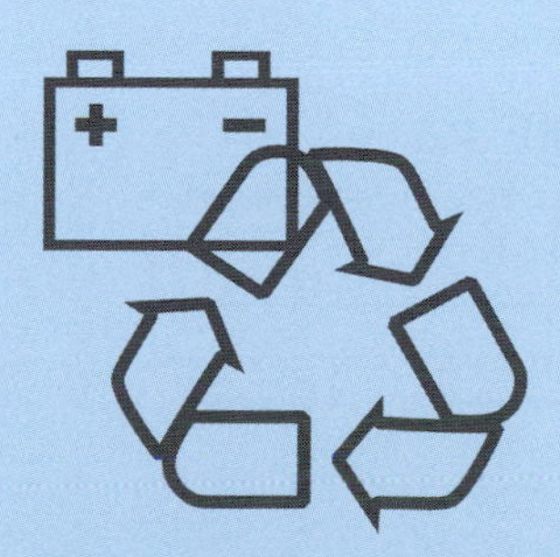

In Germany, as in most countries, discarded automobile batteries constitute the main source of recycled lead. Being environmentally sensible, battery recycling seems a practical and easy solution to extending the life-cycle of lead. On the surface, considering that mining costs do not occur, producing lead from recycled batteries seems also to be a very lucrative business. Reality differs. In the early years there were no incentives for battery owners to prevent uncontrolled dumping of batteries into the environment. Once penalty and reward systems were introduced to support battery recycling, the supporting infrastructure such as means of collection and transport were lacking. Dismantling of batteries proved difficult, and generated a wide range of undesirable hazardous wastes as by products. Lead recyclers expanded into plastic recycling to reduce some of the waste streams now facing the considersable technological challenge of separating various types of plastics. During the 1990s, the lead market became increasingly flooded with primary lead produced by Eastern European mining operations at low cost, being subject to less stringent environmental laws and regulations. Ironically at the same time European governments pressed forward to reduce the use of lead, even considering the extreme step of banning its use. As this example clearly illustrates, a successful recycling scheme requires the commitment of many parties – government, consumers, and industry.

intervention, for example, by the implementation of deposits on cans or bottles which enhance recovery and recycling. Recycling is also strongly influenced by geography, being most economic in large population centres where recyclables are available in large quantities. The variance in recycled minerals occurs primarily because the end uses of some minerals inhibit their effective recovery, and because recycling systems and technologies are less efficient for some minerals. Further increases in mineral recycling will require improved product design that facilitates the dismantling of products after their useful life, increased government commitment, improved recycling infrastructure, as well as a change of consumer habits (**Case 1.1**). Although recycling is important, there is an upper limit to the amount of mineral that it can provide. Mining will still be necessary to meet society's demand for minerals.

Although recycling is important, there is an upper limit to the amount of mineral that it can provide.

Mineral Flows Vary

While **Figure 1.2** is illustrative, mineral flow through most economies differs from the global cycle. Large differences exist between countries, as developed countries make a much greater claim on raw materials than do developing countries. The 20% of the world's population living in rich countries uses, on average, about 50% of the world's mineral reserves. Rich countries are increasingly reliant on minerals extracted abroad. In most European countries, domestic extraction of material resources has decreased while imports have increased as a result of macro-economic restructuring, rising domestic costs of production, availability of cheaper products from abroad, removal of trade barriers, and increased use of recycled materials (EEA 2005, **Figure 1.3**). As global markets open further, this trend is likely to continue.

The scope for opening new mines in developed countries is also decreasing due to the public perception that mining is inherently and unavoidably damaging to the environment. As a result, increasingly large areas in developed countries are now being closed to new mineral development. This manifestation of the 'NIMBY' (Not In My Back Yard) syndrome immediately leads to the question: Well, if not in your back yard, then in whose? Such restrictions raise the question of how environmental and economic responsibilities, including both the responsibility for environmental damage from mining and for providing the world with the raw materials it needs, can be shared equitably. Developed countries have focused on the downstream use of minerals, and most of the ores used in developed countries are imported. Canada and the US are notable exceptions, both continuing to rank high as leading mining countries. The US and Western Europe are generally the highest minerals consumers per

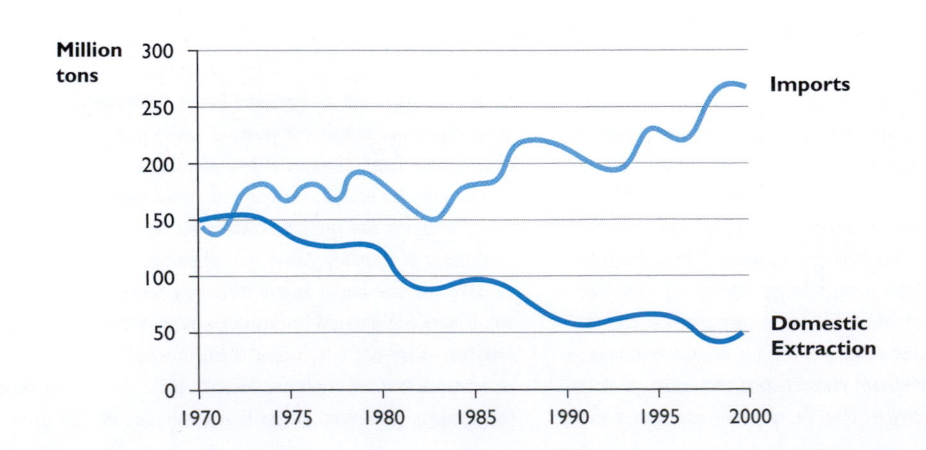

FIGURE 1.3
Metal Ores: Domestic Extraction and Imports, EU-15

Developed countries have focused on the downstream use of minerals, and most of the ores used in developed countries are imported.

Source:
Eurostat/IFF 2004

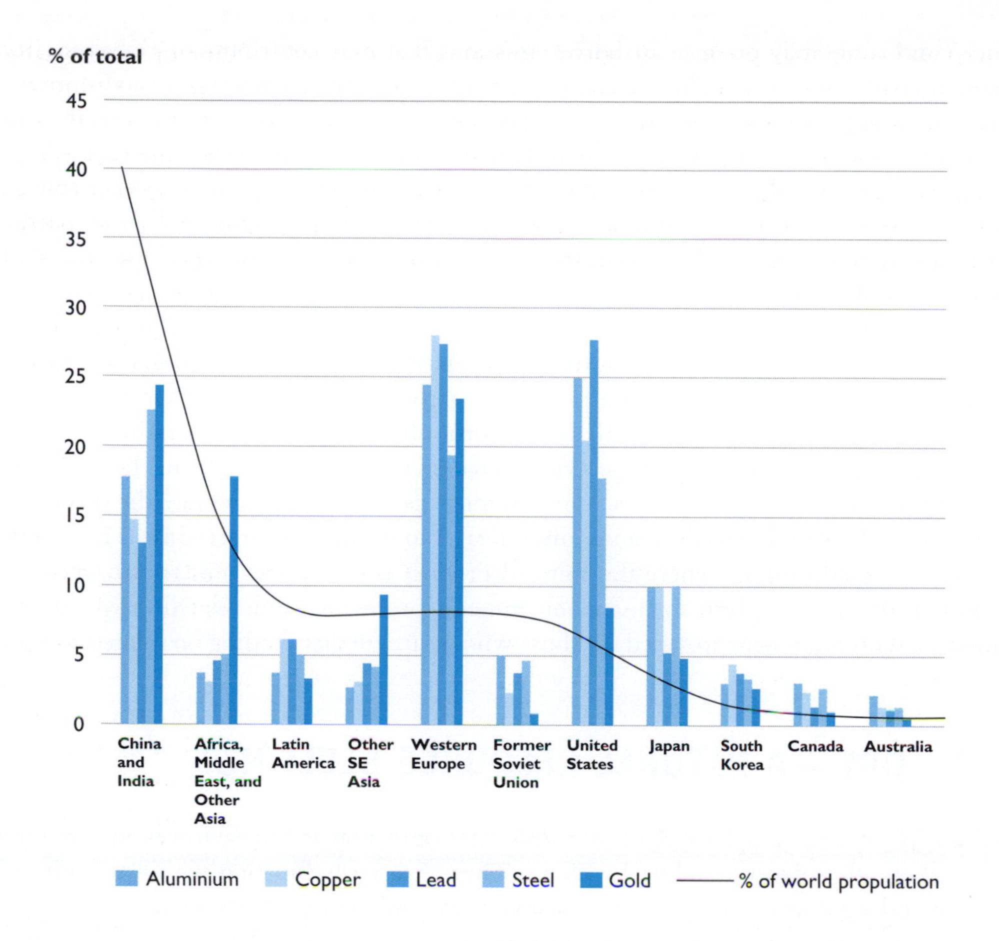

FIGURE 1.4
Consumption of Metals Compared with Population by Region for Selected Countries, 2000

The US and Western Europe are generally the highest minerals consumers per capita.

Source:
MMSD 2002

capita, as shown in **Figure 1.4**. The contrary is found in many developing countries, where the economies often depend on primary production industries such as mining, along with agriculture, fisheries, and forestry.

Path of Minerals and Associated Environmental Impacts

The entire life-cycle of mineral resources can give rise to environmental impacts, from extraction, transportation, through use in the production and consumption of goods and services, to final disposal as waste. Each phase presents its own environmental challenges, affecting different localities. In fact, the consumption of metals and the resources used in their production exemplify the degree to which international trade flows determine the extent and location of environmental pressures. For example, iron ore mined in Western Australia may be converted to steel in Korea using coal mined in Indonesia. Korean steel plate may then be used to build a ship in Japan, which after a lifespan of say 20 years will be beached and disassembled for scrap in India or Bangladesh, with the scrap steel sold to Germany, and so on…

The entire life-cycle of mineral resources can give rise to environmental impacts.

Extraction processes are often very damaging to landscapes. As will be discussed in subsequent chapters of this book, a number of metals, such as gold, nickel, and copper are extracted with environmentally-intrusive mining technologies, resulting in large quantities of mine wastes, potential contamination of soils, landscape destruction, and negative impacts on natural water cycles. Environmental impacts in the later stages of processing will differ, i.e. concentrating and refining crude metal ore, smelting, or forming, but all are energy-intensive activities. They all require other non-renewable resources (e.g. fossil fuels) and some may produce air-borne emissions that may contribute to global environmental challenges such as climate change, air pollution, and acidification. Environmental challenges of the use phase are determined mainly by the final product in which the minerals are embodied, and generally have little to do with the nature of the mineral itself.

The uneven distribution of mineral production and consumption raises another concern.

The uneven distribution of mineral production and consumption raises another concern: the high and increasing consumption of scarce resources and resulting pollution, particularly in the most industrialized countries, is potentially at the expense of the rest of the world, and of generations to come.

A full study of the immediate and long-term impacts of the mineral cycle would cover many volumes. This book, however, focuses primarily on assessing environmental impacts that are related to mineral exploration, mine site development, extraction mining, and primary mineral processing – mostly in the form of milling, washing, grading, and concentrating. As noted above, these activities have increasingly shifted from the industrialized world to developing countries. Many mining operations also integrate the subsequent extraction of valuable metals, commonly referred to as mineral processing. This is often the case in gold mining, where the shipped product is commonly gold, not concentrate. Limited attention is given to discussing mineral processing and related environmental impacts, with references provided for those who desire further reading on related topics.

1.3 ORE – A NATURAL RESOURCE BLESSING?

Mines operate in a complex web of economic, environmental and social forces and are therefore inherently subject to political realities. They necessarily function within the administrative and legal infrastructure of the host country, and are also confronted with local and regional pressures from the communities most directly impacted by mining operations. A brief discussion of these political, economic, and environmental dimensions follows.

The Political Dimension of Mining

Developing country political environments vary widely, but there are two characteristics common to most, if not all: the first is some degree of international economic dependency. The dependency of countries that own mineral resources on international mining companies or, more accurately, their respective home countries as metal consumers, is an important reality of the metal market, with potential consequences for trade and diplomatic relations. The second characteristic is a two level, or 'dual' economic and social structure. That is, one portion of the economy resembles that of a developed country, with access to a modern transportation and communications infrastructure, and participation in a cash based, consumer oriented culture. Generally the middle and upper classes in such a society represent a relatively small portion of the total population. The second level is a much larger percentage that remains embedded in a more traditional agricultural society, with limited access to the goods and services of the modern sector. Good examples are India, China, and Indonesia, but most developing countries display some characteristics of this duality. The existing political and economic distortions of dual economies may be reinforced by large mining projects, which appear to benefit the already rich at the expense of the poor. This can lead to project focused social disruption, which can affect implementation. Exploitation of mineral wealth can also be the basis for broadly based economic development, as in Botswana and Chile, although this is atypical (World Bank 2002 a,b).

The existing political and economic distortions of dual economies may be reinforced by large mining projects, which appear to benefit the already rich at the expense of the poor.

The Investment Climate

Metal mineral reserves are limited, but this does not mean that reserves will be exhausted in the near future. Exploration will go ahead following demand and price movements, and new exploration and mining technologies will allow the mining industry to exploit deposits that previously have been economically unattractive. While the location of each deposit is fixed, international mining companies can often choose between multiple deposits located in different national jurisdictions. Investors will consider not only a proven ore resource but also political, social, economic, and administrative characteristics (the 'investment climate') of the host country prior to committing significant investments (**Case 1.2**). As a consequence, countries that possess mineral reserves are in competition to attract investment dollars. The existence of superior ore bodies will not by itself attract investment.

The existence of superior ore bodies will not by itself attract investment.

Administrative infrastructure differs among developing countries, although some generalizations may be made. Most developing countries suffer from inadequate resources to staff and properly manage administrative organizations, and may lack a tradition of social processes based in law. To those proposing large mining projects, administration in developing countries commonly exhibits one or more of the following deficiencies:

- inefficiency;
- lack of concern for the realities of business and commerce;
- lack of transparency;
- disregard for existing laws and regulations;
- avoidance of accountability and responsibility; and not infrequently, corruption.

Legal Systems and Un-coded Legal Traditions

The legal system of a developing country, including commercial law, tends to reflect its colonial history: British common law in the case of former British colonies, the Napoleonic Codes for France, and Roman law in the case of the Netherlands. Such systems are well understood by investors and provide a degree of comfort, although the existence of a legal system does not necessarily ensure its application in all cases. Of utmost importance to prospective mining investment is "security of tenure", the inherent right of the discoverer to develop the deposit. Development of a new mining operation requires many years of effort and substantial expenditure before there is a financial return to investors. It is therefore

CASE 1.2
Mining in the Philippines

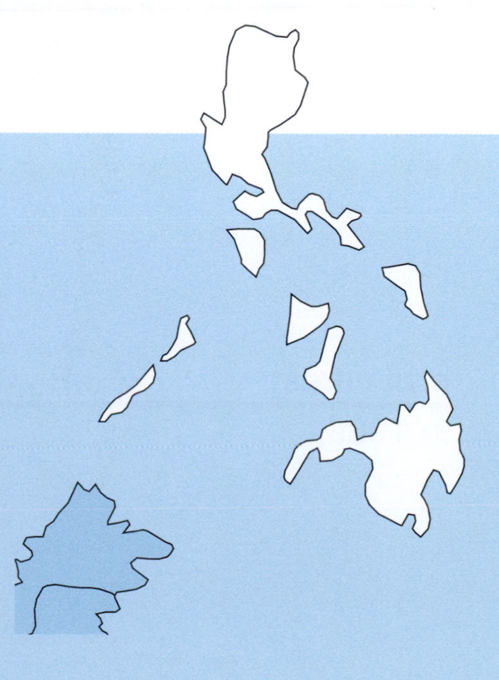

The Philippines, being located on the 'Pacific Ring of Fire', is the home of many mineral enrichments, some of them well known, others yet to be discovered. While mining in the Philippines goes back for centuries, production has declined substantially since 1990, and no modern large scale mine has been developed in the past decade despite surging metal prices and booming exploration and mining activity in less prospective countries. What hinders the interest of foreign mining companies? The Philippines continues to rank high in terms of corruption. Until recently the investment law prevented foreign majority ownership. Land rights remain unclear. Past mining accidents (some involving loss of life) caused by irresponsible mining operations, fostered wide-spread community opposition and resistance to mining. The influential Catholic church has publicly opposed mining on many occasions. As a result, the Philippines has missed out on the investments that could provide jobs and help reduce poverty in the poorest parts of the country.

Those countries where security of tenure is legally guaranteed are the focus of most exploration.

understandable that any doubt relating to security of tenure would be enough to discourage mineral exploration, let alone development. Accordingly, those countries where security of tenure is legally guaranteed are the focus of most exploration. Similarly, any attempt by governments to apply significant retrospective changes after development expenditures have been committed will cause many companies to re-direct their efforts. There is a tendency during periods of high commodity prices for countries to seek higher returns. This has led to various forms of windfall taxes or royalty increases, based purely on the perception that the company can afford to pay the additional impost. Where applied retrospectively, these imposts will certainly curtail future exploration, not only because of the effect on profits, but because they signal that the Government involved does not honour its agreements.

In contrast, social interactions at the community level, including commercial and trading relationships, are often based on un-coded legal traditions, which may be difficult for outsiders to access and understand. Mining companies tend to ignore these local traditions, which are often more important to rural communities than imposed national legal codes. Tension between local communities and mining companies may rise as a direct consequence. The Bougainville Copper and Ok Tedi mines in Papua New Guinea are good examples of what can go wrong as a result of mutual misunderstanding (**Case 1.3**).

The Key Provisions of Mining Codes

There are two apparently conflicting interests in a host country's stance towards mining investment: the need to promote foreign direct investment in order to foster economic growth, and the need to control investment to protect national sovereignty, as well as the interests of national elites. These two interests find expression in country-specific investment laws. Laws that apply to foreign investments govern a wide range of aspects including

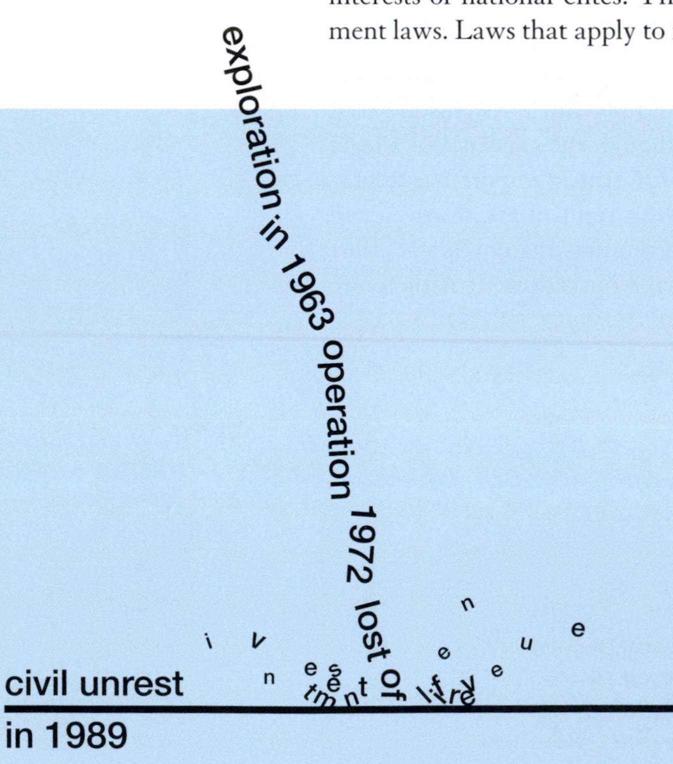

CASE 1.3

The Bougainville Copper Mine in Papua New Guinea

The indefinite closure of the Bougainville Copper Mine on Bougainville Island, PNG, in 1989 is probably the most dramatic event illustrating the complex forces that influence mine development in developing countries. Exploration commenced in 1963; the mine started operation in 1972. Even in hindsight it is difficult to single out any single mistake or misunderstanding that eventually led to the loss of lives and the loss of hundreds of millions of investment dollars and even higher foregone revenue for the host country. The most important issues relate to the unexpected pace of the independence process within PGN, the complex matters of national identity, the conflict between national and provincial government authority, the various view points of ownership of minerals, the at least initially colonial approach to mine development, and the failure to negotiate and to recognize the problems early and to respond appropriately. It is also apparent that the local community had very little conception of the scale of the project and its effects on the landscape and local lifestyles. Over the 25 years mine life (from exploration to closure) intergenerational changes did occur which remained unrecognized. The initial scheme of benefit sharing proved inadequate with an increase in population, together with an increase in education and expectation. It should be noted that the Bougainville Copper Project was implemented prior to the application of environmental impact assessment, and that many of the lessons of Bougainville have been recognized in subsequent project planning.

Source: Clark and Clark 1996

taxation, property protection, labour, social welfare, and foreign exchange. Specialized legislation often complements general law, particularly for the mining industry in the form of mining codes. The provisions of most mining codes can be categorized into the following five areas (Otto 1997a,b):

- *Property and control rights.*
  With few exceptions, ownership of subsurface minerals remains with the host country's national government. Mining companies are essentially contractors who exploit mineral resources on behalf of that government. Exclusive national control of the mineral resources is common, although some governments have moved to share returns with regional or local entities, for example, the Philippines and Indonesia. Exclusive national control of mineral deposits is problematic in that local governments and host communities with traditional land rights feel disadvantaged; they bear the brunt of the development, but don't always share the rewards. In these situations, resentment towards the central government is often channeled towards the mine operator, and environmental assessments need to be sensitive to such underlying tensions.

Resentment towards the central government is often channeled towards the mine operator.

- *Classification of minerals.*
  Most mining laws exclude certain classes of minerals, such as gravel, salt, or uranium, from their application. The management of minerals outside the mining code falls to different agencies. While not the rule, local governments occasionally try to exploit nuances in classification and wording to impose additional taxes or levees on the mine operator (such as classifying overburden removal as rock mining).

- *Qualification of the concessionaire.*
  The qualifications of an applicant for a mining licence are established by applying a set of fitness criteria, such as financial strength, technical capability, no prior violation of national law or trust, and adequate local legal incorporation.

- *Mining licences.*
  Mining licences apply to the prospecting, exploration, and exploitation phases of mine operation. The prospecting and exploration phases are geared to encourage a rapid survey for minerals. The exploitation phase is more intensive, but covers a smaller area since concessionaires are obliged to relinquish concession areas over time. Both the prospecting and the exploration phases have imposed time limits. The production licence generally encompasses mining, processing, and marketing of the mineral. Under most mining codes a production licence is dependant upon presentation of an acceptable feasibility study and environmental impact assessment study. The production licence is limited in time, usually 10 to 20 years, although extensions are usually possible. The project is typically narrowly defined, and all codes describe sanctions for failure to comply with agreed terms and conditions.

Under most mining codes a production licence is dependant upon presentation of an acceptable feasibility study and environmental impact assessment study.

- *Distribution of earnings.*
  Host country participation in mine project earnings takes many forms. Bonuses or agent's fees are sometimes charged at the closing of a contract. In addition to company taxes, royalties are the traditional form of payment. Duties and export taxes play a lesser role than income taxes. Import duties are normally suspended for capital investments to minimize initial investment costs. Mineral codes can also provide incentives, such as tax holidays, or accelerated depreciation. Additionally, mining code provisions may regulate payments to local landowners or local governments, or may

be designed to encourage environmental protection, sustainable development, public health measures, local purchasing, or local labour force content. Such provisions often transfer substantial responsibilities for regional development to the mine investor. In addition, some host countries create state-owned mining companies. Foreign investors are required to enter into a joint venture with these companies to ensure effective state participation in mineral exploitation. An understanding of national mining code provisions that aim to encourage local, sustainable development and environmental protection is essential in the environmental assessment of a new mining project.

The Economic Dimension of Mining – Who Benefits from Mining, and Who Does Not

Mining is first and foremost an economic activity.

Mining is first and foremost an economic activity. As in any other economic sector, mining companies are in business to earn profits, a valid and necessary objective. Their chosen sphere of operation is mining, carried out within a set of constraints put in place to satisfy the interests of various stakeholders. The first and foremost of these is the host country government, which represents the national interest, as well as elite, regional and local powers. Mining companies will try to negotiate the form and nature of those constraints, while host countries may try to shift additional responsibilities onto the industry. But the main objective of mining, to earn profits, remains.

The Concept of Resource Rent

The classic concept of economic rent originated with David Ricardo (1962) in his theory of land rent, and has been subsequently applied (Garnaut and Ross 1975, 1983) to mineral resources development overall and to mining in particular. In the case of mining, Ricardian economic rents can be viewed as the excess of economic return on a project above the total economic cost of the project.

In the case of natural resources, governments often transfer selected property rights to industry, such as the right to mine or to exploit an area in exchange for some amount of economic rent (with mining codes providing the legal vehicle to do so). These economic rents collectively are known as 'resource rents', since they are derived from the utilization of natural resources. Resource rents encompass all direct revenues derived by a nation from a mining project. The most common forms of revenue are direct taxes (corporate income tax, royalty tax, withholding tax, import and export taxes, excess profits tax) and fees (registration, land, water, infrastructure use) for the use and development of the nations' resources (Garnaut and Ross 1975, 1983).

Two additional types of resource rents that are associated with many mining projects are landowner compensation and national/local equity participation in resource development projects (Clark 1994). In the latter case, often the rule rather than the exception in the oil and gas industry, the national government, and occasionally the province, becomes an actual partner in a project, thereby, acquiring a percentage of the profits in addition to taxes and fees. As the equity partner is normally the national government or its agent, the majority of revenues from profit sharing accrue to the national government, which may affect revenue sharing with local governments.

The majority of direct taxes accrues to the national treasury, while the majority of the fees, and often a portion of royalties, accrue to the local government.

As a general rule, albeit with some major exceptions, the majority of direct taxes accrues to the national treasury, while the majority of the fees, and often a portion of royalties, accrue to the local government. These results in a major disparity in revenue distribution, since taxes which accrue to the national government, normally constitute 90 percent or more of all revenues derived from a mining project. Hence the call by local governments in most nations for a more equitable division of resource rents, theoretically a valid request, but one that is difficult to implement in practice.

The economic rents derived from mining may be quite high, accrue on a yearly basis (normally for the life of the development and in some cases beyond), but unfortunately too often are shared by a very small number of people.

National Economic Benefits

Host governments clearly recognize that the people and nation they represent can benefit from mining. It can contribute to the attainment of national development goals, even though it may be accompanied by ecological and social costs. Development goals include increasing gross national product, creating employment, increasing export earnings that can be redirected to national development, promoting import substitution, and facilitating administrative reforms. Mine support facilities such as seaports, airports or roads can complement existing transportation infrastructure, and contribute significantly to linking remote areas with the metropolis. MMSD (2002) argues that mining is important in 51 developing countries, accounting for 15 to 50% of exports in 30 countries, 5 to 15% of exports in a further 18, and being important domestically in three others.

It is argued, whether rightly or wrongly, that local communities too can profit from mining, even though a mining project may be a one-off opportunity for prosperity, and the mineral resource will be exhausted after exploitation. However, given the massive investment required, and the long time-span of implementation and exploitation cycles, the long-term social and environmental costs may appear insignificant and easily discounted.

The ecological and social costs typically associated with mining are not adequately regulated by market mechanisms alone. Developing countries correctly perceive the need to protect their long-term interests through extra-market controls, which are incorporated in mining policies and law, as well as contractual agreements with mine investors. Even the calculation of an optimal mine production rate is not always driven solely by economic factors, but may entail a complex mix of uncertainties.

The ecological and social costs typically associated with mining are not adequately regulated by market mechanisms alone.

Except during times of price depressions in the metal market, investors aim for high production rates in order to transform passive natural capital into financial capital at a maximum rate. Host country governments, however, may want to extend mine life by capping production rates, allowing time to realize regional development, while turning passive natural capital into human capital in the form of, say, a trained and skilled workforce. Thus contractual agreements may aim not only to protect a reasonable return on investment for mine investors, but also maximize the mine's contribution to regional development. According to Carman (1979) and various economists since then, however, a more rapid exploitation of an ore deposit is better for the investor as well as the host government.

There is little dispute that mining can produce wealth. Unfortunately, natural resource extraction has not always led to economic and social development. In well-established mining economies such as Spain, Australia, Canada, US, and South Africa, mining has been of undoubted and significant benefit, although not without some long-term costs, as in the infamous Butte, Montana super-Pit (**Case 13.7**). However, many countries, among them some of the world's poorest, have failed to convert major mining projects into sustained development. The mining industry is often blamed, but where does the responsibility of the industry end and the duty of government begin in ensuring favourable national outcomes from mining?

Mining can produce wealth.

Incorporating Environment Costs into Economic Models

In a simplified linear model of the economy (**Figure 1.5**), the production process of mining and mineral processing results in both raw metals and built capital as outputs (Pearce

FIGURE 1.5
Linear and Circular Economic Models

The amount of outputs in the form of raw materials and wastes will generally equal the amount of natural resources used. The primary reason is the First Law of Thermodynamics, which states that matter can neither be created nor destroyed.

Source:
modified from FEE 2003

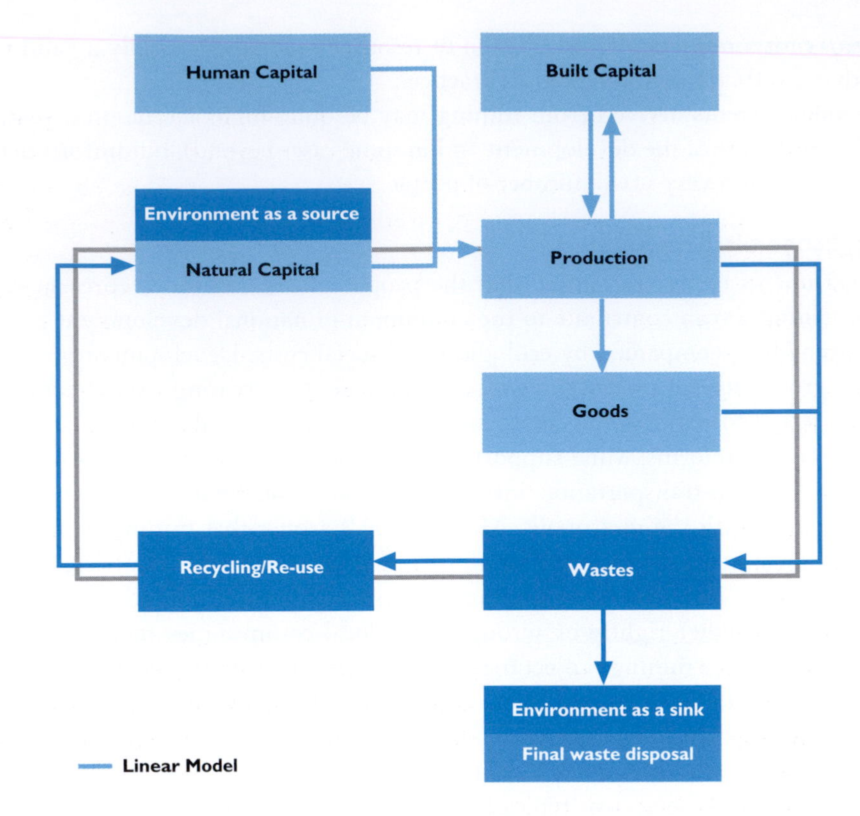

and Turner 1990). The production process itself relies on natural, human, and built capital. Natural capital consists of minerals, fossil fuels, land, topsoil, and water. These can be further divided into renewable resources such as forests, and non-renewables such as minerals. In this model the environment is simply a source of production inputs, and no allowance is made for the waste products that will be generated from all phases of both production and consumption.

A circular economy/environment model (though simplified) is more useful when it is desirable to capture the economic costs of waste streams (FEE 2003, **Figure 1.5**). The First Law of Thermodynamics states that matter can neither be created nor destroyed. Whatever is used up in the production process, however much modified, will eventually end up in the environment. The efficiency of the production process (predominantly a combination of cutoff grade, stripping ratio, and rate of recovery) will influence the rate of consumption of natural resources per unit of target output. Inefficient production processes will waste natural resource inputs. Inevitably, the natural environment becomes a sink for waste products in the form of waste rock, tailings, effluents, and gaseous emissions.

The natural environment becomes a sink for waste products in the form of waste rock, tailings, effluents, and gaseous emissions.

The major concern for efficient production is that the environment has a limit for absorbing or 'assimilating' waste products. Recycling waste can help and is beneficial in two ways: first, it can augment natural resources used in production and thus reduce the raw natural resources required for the same level of output; second, recycling results in reductions in waste volumes. Given technical and economic limits, even with optimal efficiency in recycling, the environment will continue to act as a waste sink. Traditional linear models neglect the externalities of environmental and social costs from waste streams; circular models integrate such externalities into production costs, thereby producing a more accurate picture of the real cost of a mining operation.

The Environmental Dimension of Mining

The most obvious environmental effect of mining is the alteration, sometimes approaching total destruction, of the natural landscape of the mine site. The removal of topsoil can lay vast areas bare for many years. The placement of barren rock (mostly overburden and tailings) creates massive structures with associated risks of failure. Alteration of natural landscape and deposits of rock place a heavy burden on the hydrosphere, especially in areas without ample water supply, as is common in many developing countries. Secondary environmental affects can also arise from induced development. Improved access to previously isolated areas, for example, may impact the region more than the mine itself. Mining roads open areas for illegal logging or hunting, and frequently for illegal mining. Uncontrolled settlements place additional strain on natural resources such as water, wild life, and forest products.

Improved access to previously isolated areas may impact the region more than the mine itself.

Global Benefits at Local Costs

On a global scale, it can be said that the mining industry simply responds to the demands of global society, producing the minerals needed to satisfy that demand. At the local level, broad theory becomes burdened with disruptive practicalities. It is possible for a well-planned and implemented mining project to bring sustainable social benefits to communities in the form of higher levels of education and health care, and other forms of physical and social capital. However, as previously noted, inevitably there are social costs associated with mining projects. They are frequently located in more remote, less developed regions, with little physical or social infrastructure to support industrial operations, and no prior experience with the kinds of environmental and social impacts which follow from a major mining development. Such projects impose a new economic infrastructure with social consequences that may extend well beyond the physical boundaries of the project.

Central to these social concerns is the establishment of settlement areas to support mining operations. Mining towns planned by the mine investors are immigrant settlements. As a result they may suffer from a host of social problems due to the absence of established and familiar social structures. A large number of people may be exposed to ethnic and economic class distinctions not previously encountered. Mining will attract many who expect to benefit from the project: communities or areas in the vicinity of the mining site can become gathering places for migrant workers, extended families of mine personnel, or for unemployed workers who stay after construction is completed. These people often place a strain on supporting infrastructure, which is designed only to cope with personnel directly related to the mine.

Mining may also initiate modernization of the region, which may include, among others, farming methods, transportation, and housing; such rapid change can have social consequences as a result of the destruction of older social and cultural norms. Finally, central to the assessment of social and economic issues is the fact that the mining project will eventually shut down. When this happens, the mine ceases to be a source of employment and a buyer of goods, stranding suppliers of goods and services in the vicinity of the mining operation without their main source of income.

Central to the assessment of social and economic issues is the fact that the mining project will eventually shut down.

The Less Visible but Highly Vulnerable Segments of Population

Given that the distribution of benefits and costs is often inequitable, mining companies have sometimes been caught off-guard by protests from, or on behalf of, less visible but highly vulnerable segments of the population. These may include the poor, the elderly, women, adolescents, the unemployed; also members of groups that are racially, ethnically, or culturally distinctive; and further, occupational, political, or value-based groups for whom a given community, region, or use of a biophysical environment is particularly important. Although comprehensive social assessments might be seen as overly expensive, or simply not possible, good planning in the early stages can anticipate or prevent problems later on.

The World Bank's International Finance Corporation's (IFC) Performance Standard 7 recognizes

> that Indigenous Peoples, as social groups with identities that are distinct from dominant groups in national societies, are often among the most marginalized and vulnerable segments of the population. Their economic, social and legal status often limits their capacity to defend their interests in, and rights to, lands and natural and cultural resources, and may restrict their ability to participate in and benefit from development. They are particularly vulnerable if their lands and resources are transformed, encroached upon by outsiders, or significantly degraded. Their languages, cultures, religions, spiritual beliefs, and institutions may also be under threat. These characteristics expose Indigenous Peoples to different types of risks and severity of impacts, including loss of identity, culture, and natural resource based livelihoods, as well as exposure to impoverishment and disease (IFC 2006).

Perception of Change Differs

Changes in the physical environment may affect all living things in that environment.

Changes in the physical environment may affect all living things in that environment. The human environment can also be profoundly altered in what Berger and Luckman (1966) have termed its social construction of reality, which at its simplest, refers to a community's shared perceptions of reality, based on traditional as well as contemporary beliefs. The community of mining company executives, for example, will share a particular view of the social value of a large mining project, which will probably differ significantly from the particular view of a traditional community within the purview of a proposed mine. It is quite possible, in fact, that both communities will support it, although for different reasons.

A mining project creates opportunities for local communities, including Indigenous Peoples, to participate in and benefit from mine operations. Changes to local lifestyle can be viewed as either positive or negative, depending on one's point of view. Some will view the move to a more money-oriented lifestyle and economy as a sign of positive development, for others, as UNCTAD (2006) observes, it is sign of a destructive erosion of the cultural fabric and heritage.

Due to the external social and environmental costs of mining, some development experts and interest groups see natural mineral resources as a curse, not a blessing. Their view of mining projects is that one cannot assume a country's economy will automatically benefit, over the long-term, from the exploitation of its natural resources. On the contrary, they argue, countries with mining end up with the burden of long-term environmental costs without lasting benefits. The perceived discrepancy between the benefits of industrial development and the needs of environmental conservation has now become the subject of a sharp controversy between industry and protectionists, a dispute in which mining companies often find themselves in the centre.

Whether a curse or blessing will be determined by the commitment of both the host country and the mining company to equitable and sustainable development.

There is no doubt that mining can cause profound environmental and social change. And it also has to be acknowledged that influential interest groups or individuals may limit the potential benefits for the broader public. That said, modern mining practices and appropriate planning when fully and properly implemented can allow not only the mitigation of the negative effects of large mine projects, but also promulgation of initiatives that contribute substantially to regional and national development. Mineral and coal deposits are a passive resource. Whether a curse or blessing will be determined by the commitment of both the host country and the mining company to equitable and sustainable development.

A case in point illustrating the complex inter-connected political, economic, social, and environmental linkages of a large mine to the host region is Freeport's massive copper and gold mine in Papua, East Indonesia (**Case 1.4**).

1.4 WHAT MAKES THE MINING INDUSTRY DIFFERENT?

Clearly the mining industry differs in significant ways from other industrial sectors. For one, mining influences the economic profile of most countries, which are either mineral producers or mineral products consumers, or in many cases both. For another, as discussed in a later section, mining has a unique risk profile. But there are other characteristics that are unique to mining. The pressure-state-impact- response model, illustrated in **Figure 1.6**, helps to elaborate some mining-specific characteristics (partly drawing from the excellent text by Marshall 2001).

The Drivers – Demand and Supply

The demand curve in the mining industry differs in significant ways from other sectors in that it is very long and highly variable. Demand for minerals and thus mining is as old as civilization, and probably much older, and is unlikely to change in the future, irrespective of socio-economic or technological changes. The demand for a core of basic mineral commodities such as iron, copper, gold, silver, and lead is as old as history, although the spectrum of minerals has widened as new technologies required new elements, including uranium for energy production, and silicon for computers and communications infrastructure. Demand for a specific commodity, however, fluctuates greatly with time, as do market prices. Unlike the patterns in other industrial sectors, suspension and resumption of mining activities is common in response to changing demand and price. The recent reprocessing of mine tailings in Romania to extract gold is a good example of response to a large rise in price.

The demand for a core of basic mineral commodities such as iron, copper, gold, silver, and lead is as old as history.

Finding new economic mineral deposits to match increased demand is also difficult. Exploration often lasts five to ten years, with environmental assessment, feasibility study preparation, and ongoing stakeholder consultations leading to necessary government approvals, taking an additional two to three years.

CASE 1.4
Freeport's Massive Copper and Gold Mine in Papua, East Indonesia

Mining the world's richest gold and copper deposit in one of the remotest areas on Earth. The closest most people will ever get to the mining operations of Freeport-McMoRan Copper & Gold in remote Papua is a computer tour using Google Earth.

Royal Dutch Shell first found minerals in the 1930s on an expedition to the nearby Carstenz Glacier, one of the few equatorial glaciers on Earth. In 1959, Freeport Sulphur, now named Freeport- McMoRan, arrived. Systematic exploration began in the 1970s, leading to development of the Ertsberg (the Dutch word for ore mountain) open Pit mine. In 1991, the massive Grasberg deposit was discovered nearby, just as the Ertsberg deposit was depleted. The Grasberg and associated ore bodies have proven reserves of 46 million ounces of gold and about 40 m tons of copper, according to the company's 2004 annual report.

As Freeport prospered into a company with $ 2.3 billion in revenues, it also became among the biggest – in some years the biggest – source of revenue for the Indonesian government. It remains so. Freeport states that it provided Indonesia with about $ 15 billion in direct and indirect benefits between 1992 and 2005, almost 2 percent of the country's gross domestic product (GDP). With a daily ore production rate of well over 200,000 tons and a gold price over US$ 540 an ounce, government payments in form of dividends, royalties, and taxes amount to US $ 1 billion per annum (The Jakarta Post, April 20, 2006).

The Freeport mines contributed about 70% to the GDP of the Province of Papua in 2006, and close to 100% of the Timika' regency in which the mine is located. The company provides additional funds for community development programmes to the amount of US$ 50 m per year.

The original legal agreement between Freeport and the Indonesian Government, signed in 1967, served as a model for all subsequent contract of work agreements. In spite of the enormous economic benefits, however, the mine continues to be the focus of environmental and social controversy. Constrained by unsuitable topography from developing conventional tailings disposal systems, the mine disposes of its tailings into the natural river system. The tailings are contained by a system of levees in the lowlands forming a tailings deposition area covering more than 100 km^2. Fine tailings are also carried into the Arafura Sea.

Many of the decisions during mine development were made by the central government without consultation with local government and local tribes. The central government still holds a 10% share in the mine, while the local government has no share. Freeport has been accused of killing local people, violating the rights of Indigenous Peoples and polluting the environment. Mining, and significant community funds resulting from the mining operation, have attracted large numbers of people, and the population of the town of Timika grew from a few thousand to more than 60,000 over less than a decade.

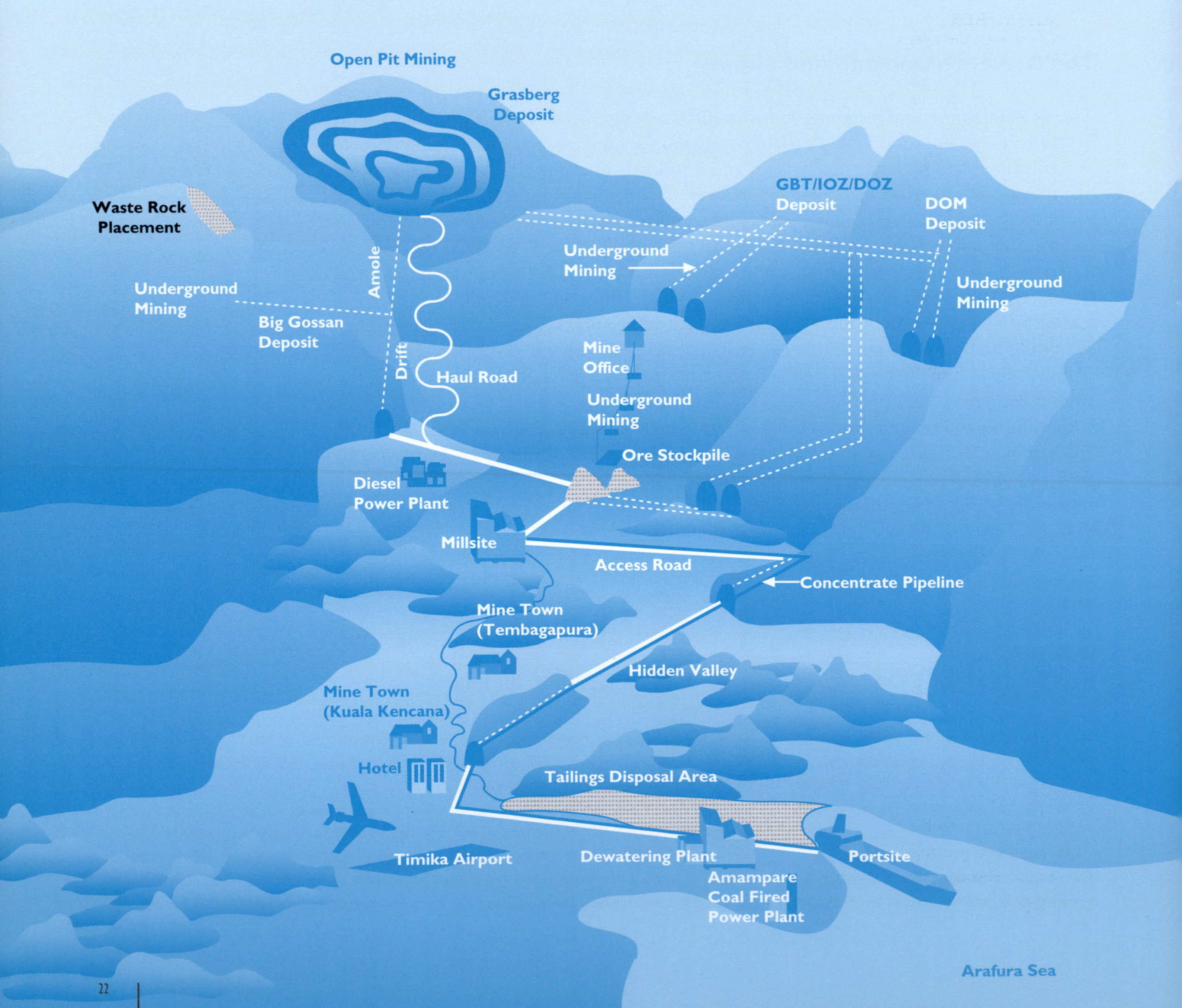

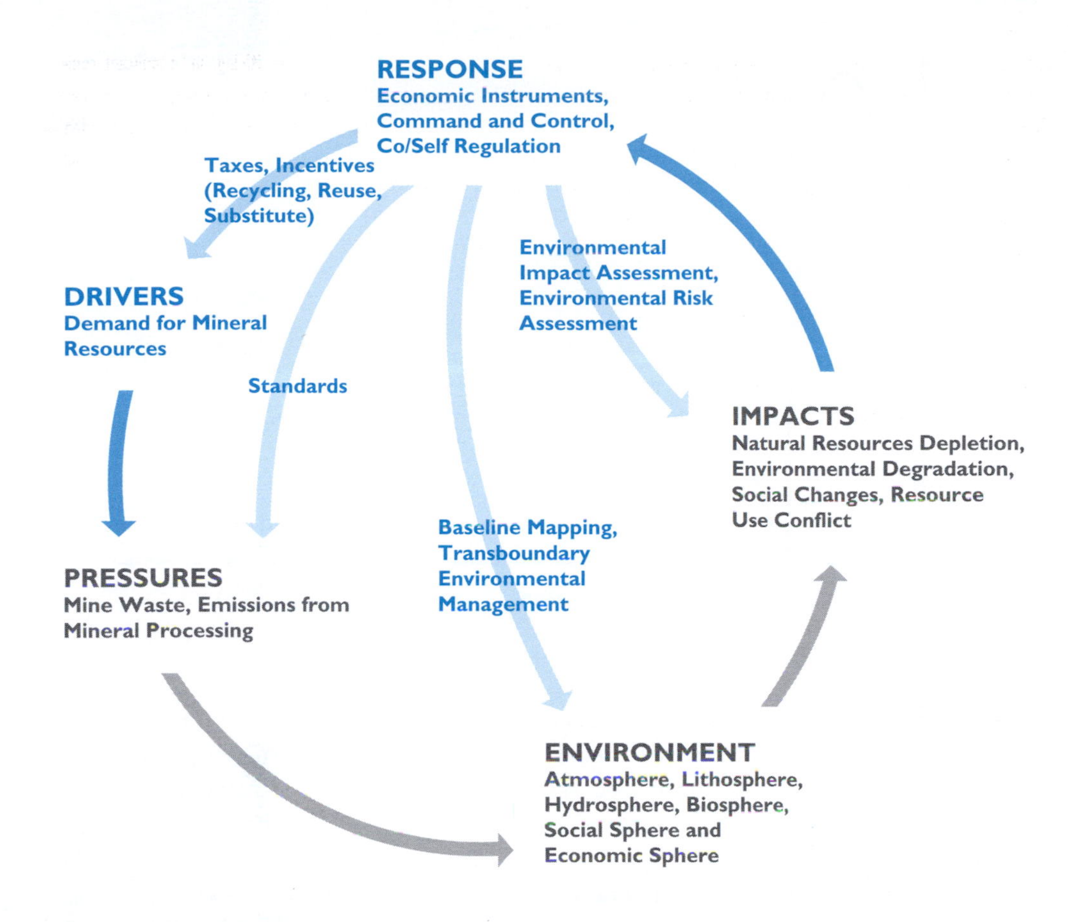

FIGURE 1.6
Schematic DPSIR Framework Applied to Mining

The (Drivers-Pressures- State of Environment- Impact-Response) framework illustrates that societies can respond to mining and mining induced changes in a variety of ways. Environmental impact assessment is one of them.

Construction can take one to three years depending on the size and nature of the mine. In addition, developing a mine usually requires hundreds of millions of dollars in capital investment. Unlike manufacturing, a mine does not usually have the luxury of starting small and, if things go well, expanding (Marshall 2001). To achieve the economies of scale required to generate an adequate return on investment, a modern mine must start large with associated large capital costs. There are almost always extensive upfront development costs incurred before actual ore extraction commences, e.g. to remove overburden in the case of an open Pit mine, or to provide access in the case of an underground mine. All this occurs before the mining company sees any payback of external financing or return on its investment (**Figure 1.7**). Finally, in recent history, the locations of demand and environmental pressures have separated. The main demand resides in the industrial countries but mining has largely moved to areas remote from the markets.

There are almost always extensive upfront development costs incurred before actual ore extraction commences.

Pressures from Mineral Extraction

Pressures on the environment from mineral extraction, mine waste and related emissions have increased because: (1) mining has generally moved from small underground to large surface mining, and (2) the number of mines has increased. Increased amounts of extracted minerals and waste rock, and the liberation of elements such as toxic metals and sulphur have increased global pollutant flows and hence environmental pressures. In terms of emissions, the mining industry is not unlike other industries, but as mining extracts non-renewable resources, easily accessible mineral reserves diminish over time, consequently mining projects often last only ten to twenty years, although occasionally longer. However, it

FIGURE 1.7
A Conceptualized Funding Life-Cycle of a Mine

Mining almost always incurs extensive upfront development costs before actual ore extraction commences and the financial and operational responsibilities of mine owners continue after the mined resource is exhausted.

Source:
Nazari 2000

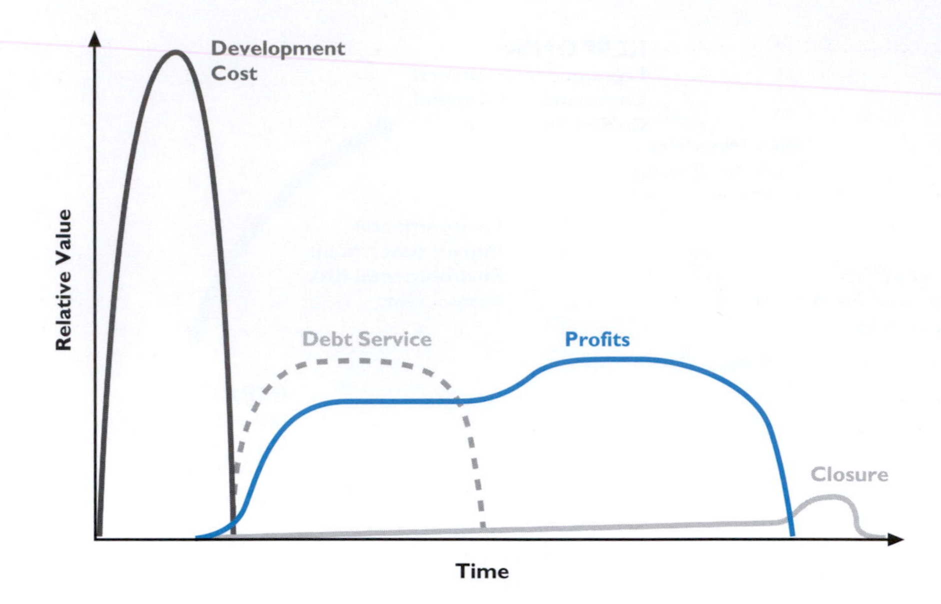

is important to note that pollution problems are not always related to the extracted minerals and waste, but to altered physical-chemical conditions at the mine site. Also, the area of interest to miners, that is, the subsurface, can only be sampled sparsely and accordingly is always subject to a high degree of uncertainty.

The location of environmental pressure is confined, in the main, to the area of mineral deposits.

Another characteristic of mining is that the location of environmental pressure is confined, in the main, to the area of mineral deposits. This means that mining cannot follow and utilize existing infrastructure, but must bring infrastructure and other (traditionally heavy) industry to its location. This results in the creation of secondary pressures or adverse effects, especially in the case of mine closure. Industrial and other supporting infrastructure, as well as mine related settlements suddenly become redundant, giving rise to further socio-economic pressures. However, the financial and operational responsibilities of mine owners continue after the mined resource is exhausted (**Figure 1.7**). The process of mine closure can last up to two years, while rehabilitation of the mine site can continue for another four years or more, with the longer periods being particularly applicable to open Pit mines. Mining company responsibilities may continue even longer when persistent problems, such as neutralization of acid rock drainage, require long-term management.

Mining and the State of Environment

Economically valuable mineral deposits do not necessarily occur below low value surface environments, so that mines are sometimes located in or near ecological reserves and protected areas. They may also be found in conjunction with areas of geologically determined high natural background values or 'natural pollutants' that are associated with mineralized deposits. This means that often the terrestrial environment is already polluted before mining commences, which is, in fact, a useful tool for geochemical mineral exploration. Natural background cannot be remediated to meet limits defined by law. Nevertheless, the state of the environment over mineral deposits and certain geological formations can pose regional scale risk to human and ecosystem health. However, actual risks to health from such pre-existing conditions depend on many factors such as soil organic matter, metal speciation, pH, etc. Additionally, a given site may already be affected by regionally dispersed pollution from historic mining activities. Such pollution makes it difficult to separate given background from pollutants generated by recent human mining activity.

Impacts are Complex and Vary Widely

Mining impacts are many and varied, as discussed in foregoing sections, but tend to be local. However, not all impacts are confined to the immediate vicinity of a mine; regional impacts are commonly related to air pollution (dust, smelter emissions), ground water pollution, naturally elevated background levels, and pollution of down stream water bodies and flood plains. Pollution impacts are often long-term, but also can be delayed, as in long-term acid rock drainage, becoming in effect chemical 'time bombs'. However, the socio-economic impacts of mining and mine closure in the host country are often of a higher significance than the physical and ecological environmental effects, particularly in the short term and in the political sphere.

The socio-economic impacts of mining and mine closure in the host country are often of a higher significance than the physical and ecological environmental effects.

Society's Response

In terms of response, perhaps the most important reality of mining is that 'zero impact' is essentially impossible. However, societies can respond to mining and mining induced changes in a variety of ways (**Figure 1.6**). One is the reduction of demand for minerals through substitution of traditional materials with synthetic ones (recognizing of course that production of synthetic materials involves its own environmental impacts). Demand can be further reduced by product recycling; by reworking of mineral wastes as secondary resources; and by use of material efficient technologies. Although necessarily long-term, ultimately such measures can relieve pressure on the environment.

Mine wastes and emissions can be decreased by improved management, particularly in conjunction with new technology. Overall, the state of the environment can be improved by appropriate environmental management. Environmental Impact Assessment and Risk Assessment studies of mine sites have long been a requirement in identifying and ameliorating environmental degradation and in preparing response strategies for possible accidents. Introduction of environmental management systems as an integral element of project design in all mining projects can further decrease potential impacts, and the 'design for closure' approach can minimize impacts after closure.

1.5 THE UNIQUE RISK PROFILE OF MINING

Mining has a unique risk profile, not only in relation to the environment or applied technology, but financially, politically, and legally. In a legal context, an operator's rights to a mining project (and its ability to generate cash flow and profits) depend on a series of contracts and interpretations of applicable mining laws as well as general law. Political instability may encourage reinterpretation of contracts and legal requirements, or foster social disruptions focused on mining projects. The international economic climate may change, driving costs up and returns down. Even if global economic conditions are favourable for new mining projects, risks remain and a comprehensive risk management plan is essential to a profitable outcome. Table 1.3 lists the main risks to successful implementation according to the level on which they occur: country level, sector level, and project/enterprise level. Lay (2006) and

Mining has a unique risk profile, not only in relation to the environment and technology, but financially, politically, and legally.

TABLE 1.3
Main Risks to Successful Implementation of a Mine

	Key Obstacles to Successful Implementation	Effect on Project Performance
Country Level Factors	Budget deficit Inflation Trade distortions Foreign exchange control Political/ social instability Inadequate legal system Inadequate local capital markets Subsidies rate Inadequate factor endowment Inadequate human capital Inadequate infrastructure	Inflation, crowding out of private investment, cost overruns Increase in cost of local inputs, shift to speculative instead of productive investment, overvalued currency Fluctuations in the real exchange rate; overvalued currency, Limited international competitiveness of exports Inability to make timely decisions on purchase of critical inputs, debts repayment, and repatriation of profits High cost of risk capital Property rights not enforced Difficulty in obtaining long term financing Lack of financial discipline and international competitiveness High cost of doing business
Sector Level Factors	State ownership of enterprises Barriers to entry Barriers to exit Uncompetitive/unstable tax regime Incompetent bureaucracy Shortage of skilled workers Uncompetitive production costs Lax safety procedure Lax environmental control	Political interference, drain on gov. budget; State Owned Enterprise's monopoly over resources; preferential treatment State/favoured company monopoly of production Inability to shut down ailing firms or reduce workforce Inability to control costs Cumbersome and wasteful 'red tape', corruption Low productivity of local labour Low market penetration potential High frequency of accidents; low labour morale Environmental degradation
Project/Enterprise Level Factors	Poor technical design Substandard emission control Price instability of product Poor quality of products High input/output ratio High initial outlay Long gestation period Errors in ore grade and reserves estimates High debt to equity ratio Low profitability ration Low assets turnover Low economic internal rate of return	Low capital utilization and technical efficiency Environmental pollution Earnings instability, high risk of failure High rejection rate: low demand High operating costs Complex financing, high risk and long payback period Capital cost overrun, implementation delays, market risk Potential solvency problems Potential financial problems Inefficient operations Marginal project

Source:
http://www.worldbank.org/html/opr/pmi/industry/industr7.html

Wexler and Lovric (2006) adopt a different approach to risk characterization and broadly categorize risks according to various aspects of exploration and mine operation.

Geological or Reserve Risks

Minerals may not have the quality, quantity or ease, efficiency or cost effectiveness of extraction as originally anticipated. There is no way of seeing directly what variations or discontinuities occur in rock beneath the surface. Geologists depend upon projection of surface formations, plus rock types and structures in combination with interpretation of geophysical responses (such as sonic waves generated by small explosions), and/or widely spaced drill holes. There is no guarantee of continuity of mineralization between adjacent drill holes. Expected continuity varies with the type of mineralized body. For example, there is a greater probability of continuity in base metal ores than in precious metal ores. There is a better chance with silver than with gold. Mining history is replete with examples where a viable ore-grade was encountered in as many as 20 adjacent holes on a grid pattern, only to find that there were no values or only low-grade values between. On the other hand, barren drill holes may have missed a major ore body by a matter of a few metres. Murphy's Law applies – 'If anything can go wrong, it will.'

Operational and Metallurgical Risks

A wide variety of factors and incidents can disrupt operations or make them less economic. In the construction period, delays may follow, for example, from unexpected geological instability, as in the collapse of the Ok Tedi tailings dam, or late arrival of critical production equipment, with consequent cost overruns. In the operational period, poor performance can result from inappropriate metallurgy or technology, equipment failure, unforeseen circumstances, as well as human errors and factors such as lack of skilled labour and/or subcontractors, poor maintenance, labour unrest, failure to identify or obtain the most appropriate equipment, and delays in completing critical infrastructure such as access roads or water supply.

Metallurgical risks can originate from a limited understanding of the ore body (Lay 2006). Mineral deposits are not uniform in the nature of the ore and gangue minerals (i.e. intergrowths, variations in composition, intensity of oxidation and/or alteration processes acting on the ore minerals which influence their metallurgical behaviour). Preliminary understanding of crushing and grinding characteristics, liberation size, and metallurgical recovery, as well as variability of grade of mineralization may be inaccurate. Some portions of the mineralized body may contain deleterious elements (such as arsenic or mercury) that make them unacceptable to smelters, or they may contain unstable pyrite/marcasite that ignites spontaneously within mine workings or during concentrate shipment.

Mineral deposits are not uniform in the nature of the ore and gangue minerals.

Ground instability, specifically rock falls or rock bursts in underground mines and slope failures (land slips) in open Pit mines, are among the most common causes of disruption to mining operations. Again, these risks relate to the difficulties involved in adequately exploring geotechnical conditions in a highly variable, complex subsurface rock mass.

Economic/Market Risks

Mine output may not yield a sufficient return to meet the company's fiscal expectations or obligations either because of a decline in commodity prices, increased cost of fuel or other consumables or because insufficient quantities of output are sold. Mining cash flows are heavily influenced by commodity prices and currency fluctuations, as seen in the closure of many mines during the late 1990s when commodity prices were low. Unlike most other commercial products, prices of metals are generally set internationally. Low-grade products must compete for markets with higher-grade products. Discovery and development of large and/or high-grade deposits may depress the market prices. Substitution of alternate material that will perform the function of a particular metal cheaper or better will diminish the market.

Country Risks

Sometimes referred to as political risk, country risk consists of all factors within a host jurisdiction's economic, political, legal, and social systems that can delay or block the reasonable or scheduled implementation of a mining project (**Table 1.4**). Country risks

TABLE 1.4
Country Risk Factors

Government Stability	A measure of the government's ability to carry out its declared programme(s) and its ability to stay in office. This will depend on the type of governance, the cohesion of the government and the governing party or parties, the proximity of the next election, the government's command of the legislature, popular approval of government policies, and so on.
Investment Profile	This is a measure of the government's attitude to inward investment as determined by the assessment of four sub-components: the risk to operations, taxation, repatriation, and labour costs.
Internal Conflict	This is an assessment of political violence in the country and its actual or potential impact on governance. The highest rating is given to those countries where there is no armed opposition to the government, and the government does not engage in arbitrary violence, direct or indirect, against its own people. The lowest rating is given to a country embroiled in an ongoing civil war. Intermediate ratings take into account kidnapping and terrorist threats.
Corruption	Incorporates the most common form of corruption such as bribes and protection payments, but is more focused on actual or potential corruption in the form of excessive patronage, nepotism and suspiciously close ties between politics and business.
Law and Order	Law and Order are assessed separately. The Law subcomponent is an assessment of the strength and impartiality of the legal system, while the Order subcomponent is an assessment of popular observance of the law.
Ethnic Tensions	This component measures the degree of tension within a country attributable to racial, nationality, religious or language divisions. This may be particularly important where a mine investment may span a particular ethnic enclave, creating potential for disruption due to uprisings or hold-up.
Bureaucratic Quality	The institutional strength and quality of the bureaucracy is another shock absorber that tends to minimize the revisions of policy when governments change. Therefore, high scores are given to countries where the bureaucracy has the strength and expertise to govern without drastic changes in policy or interruptions in government services.

Source:
based on Howell (2001) and Hartley and Medlock (2005)

Until a host jurisdiction develops a reputation for integrity, competence, efficiency, and transparency, large-scale investors will tend to avoid placing themselves in situations where critical outcomes depend on decisions of corrupt and unpredictable politicians and bureaucrats.

are a major factor in attracting mineral exploration and investment. While country risks may sometimes be as dramatic as war or invasion of mining sites by protestors (**Case 1.3**), more common are administrative delays in granting permits, or changes to the legislative and regulatory framework, which can impose additional costs on a project. These issues arise in developed as well as in developing countries. Many governments feel free to apply and increase taxes and royalty assessments at will, and to change permit regulations and restrictions on operations. A change of rules for operations after investment has been made may, for example, result in increased cutoff grade levels, thereby reducing ore reserves or, in some cases, eliminating all profits, leading to closure. Yet another subtle form of country risk includes the use of a mining project by governments, political parties and non-government organizations (NGOs) for their respective political agendas.

Environmental Risks

The costs of environmental and occupational health and safety compliance or obtaining permits may have a negative effect on cash flow or, at its most extreme a catastrophic environmental or public liability incident may lead to a mine shutdown or prevent further development in the affected region (Wexler and Lovric 2006). Environmental legislation differs from country to country but most adopt the 'polluter pays' principle to pollution incidents. In the case of a pollution incident, owners and investors may find that not only have they lost the value of their original investment in a project, but may face substantial clean-up costs, or other liabilities. Closely associated with environmental risk is reputational risk. Mining companies, particularly those that operate internationally, are under intense scrutiny from governments, regulators, NGOs, the public, and the media. Failure to give due consideration to environmental impacts can result in negative publicity for both the respective company and supporting financial institutions.

1.6 MEETING ENVIRONMENTAL ISSUES HEAD ON

Since the 1990s, 'globalization' has become not only the focus of international economic analysis, but the stimulus for an outpouring of literature on the future development of civilization. For some, globalization signifies progress, global environmental awareness, freedom, democracy, and prosperity. For others, it represents exploitation, unemployment, unfettered capitalism, and imperialism. Few, however, would dispute that nations around the world have become more economically interdependent. Anyone who has access to a computer and a telephone line can ascertain the latest information on world markets, finance, technology, political trends, consumer interests, and so on. The same technology has also made it nearly impossible to run away from environmental issues or to 'cover up' incidents involving environmental damage.

Critics of globalization focus their arguments on misuse of power and exploitation of weak economies and political systems. Multinational corporations, including most mining companies, they argue, use political connections in their home countries and economic intervention in others to consolidate their power and maximize their profits. As multinational corporations do business around the globe, critics point out, they become increasingly detached from the interests and values of their countries of origin, with their primary allegiance shifting solely to profits and market share. Any concerns about social benefits or environmental degradation are set aside.

These are real concerns, but globalization also provides an unprecedented opportunity for change for the better; it has brought access to new technologies that give people the potential to learn, communicate, and participate in decision-making as never before. The pace of technological and scientific innovation has brought with it new uncertainties and half-understood risks, but also hope for a better world. Although to capitalize on the opportunities offered by globalization, the wealth and power of the private sector would need to be applied in a manner that recognizes social needs and environmental limits, there are increasing signs that this is happening.

Globalization also provides an unprecedented opportunity for change for the better; it has brought access to new technologies that give people the potential to learn, communicate, and participate in decision-making as never before.

The minerals sector is certainly globalized, but within it the trend toward localization is assuming increased importance. Traditionally, engineers, managers, and political leaders decided how mining would affect local communities. It was assumed that the development of a mine constituted a benefit to those in its vicinity, and few attempts were made to adapt mining infrastructure and practices to suit local interests. As a result most major mining projects looked alike. This point of view is now changing, however slowly, with more emphasis on understanding and accommodating local needs and differences.

As a result of the communications revolution, with low cost and instantaneous point-to-point contact possible almost anywhere on the globe, national borders have become highly permeable. This has not only changed the way mining companies operate internally, but how they relate to and cooperate with host governments and local communities. The companies are aware that the ability to produce, access, adapt and apply information worldwide with relatively little effort has raised international environmental awareness.

Word of environmental incidents, such as an acid spill or a tailings collapse, can spread worldwide within minutes (often with colour photos attached), with possible negative consequences for a company's reputation, as well as its financial stability. At the same time, today's information technology allows mining companies to inform a wide range of stakeholders about environmental performance without delay, making environmental management transparent and environmental performance public.

Evolution of Global Environmental Awareness

Global environmental awareness grew out of widespread dissatisfaction in the late 1960s in North America and Europe with a seemingly insensitive and pointless consumerism. A sustained period of rapid technological change and economic growth led to significant improvements in living standards. But at the same time, with mass communication and mass education, awareness of the drawbacks and social costs of economic development became manifest. Such awareness was a particular source of unrest among the postwar population bulge of young people, generally referred to as the 'baby boom'. Well-educated in unprecedented numbers, and only ever knowing a prosperous life, they had little patience with what they saw as the yawning gap between society's promise and its egregious flaws. Just as far reaching was the message when the crew of Apollo 8 photographed Earth-rise over a lunar horizon on a Christmas Day, 1968: Earth is a fragile pocket of life in a very large and lonely universe (**Figure 1.8**).

FIGURE 1.8
Earth-rise Over a Lunar Horizon

Rather as a foreign country helps a traveller understand his home, so it has taken space flight to understanding Earth. Looking back at a small blue-green planet against black empty space was a strong driving force behind the environmental movement that started in the late 1960s.

Photo: NASA

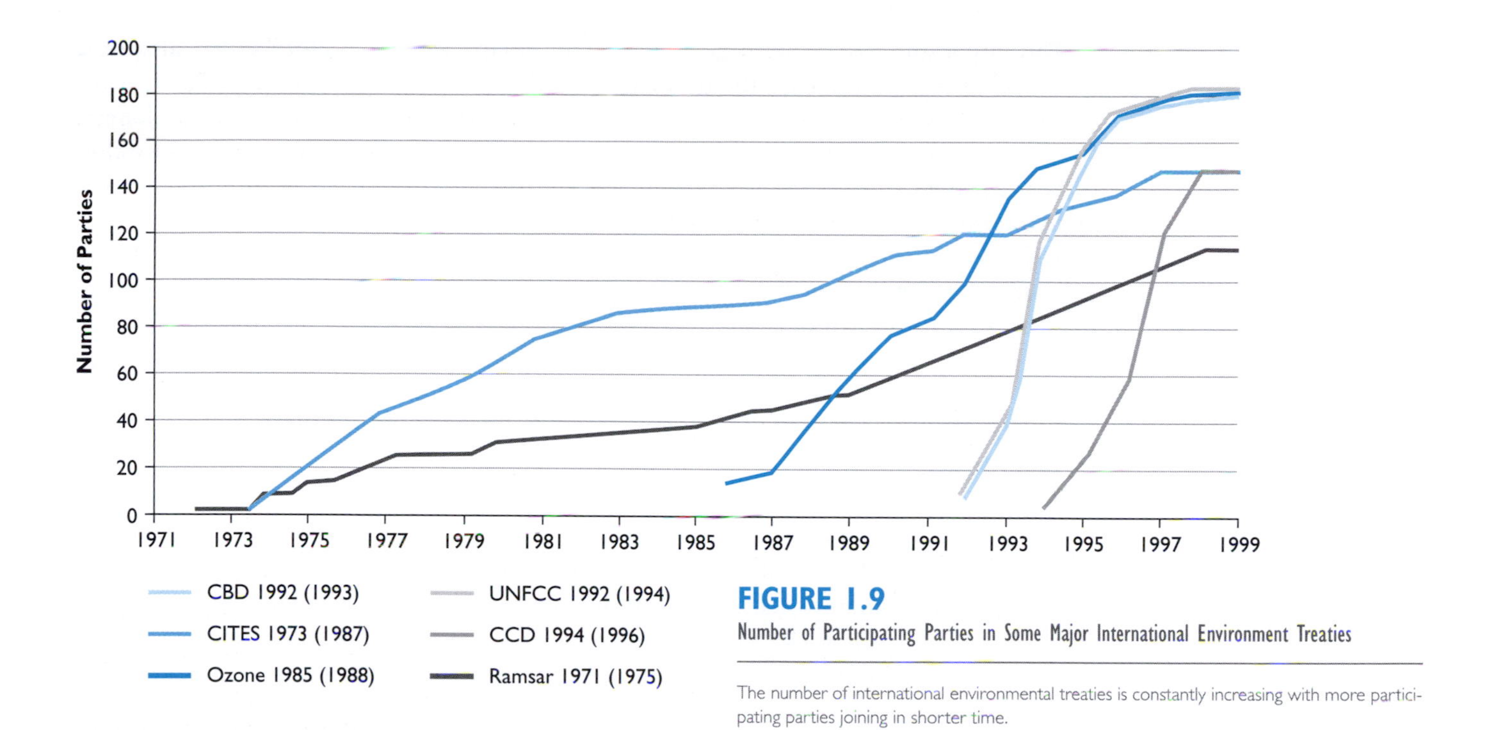

FIGURE 1.9

Number of Participating Parties in Some Major International Environment Treaties

The number of international environmental treaties is constantly increasing with more participating parties joining in shorter time.

By the 1960s understanding of the complex bio-chemical relationships which comprised the natural environment had reached a point where scientists could begin to pinpoint and map the impact of industrial technology on air and water quality, the cycling of toxic chemical compounds such as DDT upward through the food chain (Carson 1962), and impacts on both terrestrial and marine species through development of coastal wetlands, as well as a virtual encyclopedia of other induced environmental changes. Solutions to these problems lay, unfortunately, outside the realm of science, but instead rested in the political sphere.

Prior to the late 1960s, most national legislations had not addressed environmental issues, as there had been little or no public recognition or demand. However, in part stimulated by Carson's book, increasing public pressure generated a spate of environmental legislation. The USA introduced the first comprehensive environmental protection act (NEPA) in 1969. In 1972, the first United Nations Conference on the Human Environment was held in Stockholm, Sweden, and since then global concern for the environment has seldom abated. Today there are over 200 international treaties dealing with an extraordinarily wide range of matters related to environmental and cultural protection (**Figure 1.9**). They govern, inter alia, marine pollution, international transport of hazardous wastes, protection of species, wetland protection, the use of ozone depleting substances, and so on. Many of these treaties apply to mining. For example:

The USA introduced the first comprehensive environmental protection act (NEPA) in 1969.

1. marine pollution that results from dumping of wastes in the sea;
2. slag shipped from smelters for commercial use overseas or trans-boundary movement of automobile batteries covered by agreements on international transport of hazardous waste;
3. large-scale surface mining that requires protection of species and wetlands;
4. coal mining and its contribution to global warming; and
5. human rights and labour issues relating to the employment and interaction with Indigenous Peoples affected by mining projects.

The important role of local governments in environmental protection has also been given due attention through the development of a specific chapter in Agenda 21 of the 1992 United Nations Conference on Environment and Development, in Rio de Janeiro (the so-called 'Earth Summit'). The conference recognized that local governments are closest to the people affected by environmental impacts from mining and other developments, and therefore have a key role to play in responding to environmental threats.

Today many international, regional and national entities include environmental considerations in activities related to mining, and provide useful guidance materials (see **Appendix 1.1** for some examples).

Moving from the Big Boot Mentality to Environmental Protection

Governments worldwide have committed themselves to reviewing their legislative requirements in order to balance environmental protection and economic development. As with other regulatory schemes, governments can choose between the approaches of 'the carrot' or the 'the stick', in this case referred to as the 'big boot'. The initial response was to rely on traditional command and control approaches by setting and enforcing quality standards. The command and control approach is supported by a range of penalties in which the level of penalty reflects the severity of the breach. It is now recognized that this 'big boot' mentality has inherent problems. Implementation of command and control approaches requires adequate government resources to be effective, and constant resort to penalties does little to promote cooperation from industries. On the contrary, it imposes an adversarial 'us and them' mentality, which becomes even more so when lawyers become involved. The environmental protection system of the USA is, perhaps, the most extreme example of such an adversarial system. On the other hand, in most Australian states, the mining industry and government regulators have a far more cooperative relationship.

Implementation of command and control approaches requires adequate government resources to be effective, and constant resort to penalties does little to promote cooperation from industries.

Governments also rely on economic instruments in the form of pollution charges, tax incentives, or tradable permits to alter production and consumption patterns of both industries and individuals. Economic instruments are soft measures that change the business environment in which companies work. Significant reduction in sulphur dioxide emissions in the United States of America, for example, is credited to the introduction of trading schemes in the 1990 Clean Air Act. In emissions trading, companies that have cut pollution more than they are required can sell 'credits' to other companies that still exceed allowed limits.

The Environmental Impact Assessment process, as described in the following section, serves as an illustration of a co-regulatory approach. Environmental management practices are formulated and adopted in consultation with key stakeholders, admittedly within prescribed boundaries. Co-regulation encourages compliance through persuasion and negotiation. Initially, the consultation process was limited to government authorities and business representatives, but modern industry practice makes an effort to reach out to affected people during the EIA process.

A self-regulatory approach has long been promoted by multinational companies, most notably by the petroleum and mining industries. The Australian Minerals Council's voluntary Code for Environmental Performance is an example of a particularly successful self-regulation initiative. Largely as a result of this Code, to which most Australian mining companies subscribe, standards of environmental practice are, in many cases, well ahead of standards prescribed by government regulation. Peer pressure and competition to excel have proved highly effective in raising standards to the extent that some operations have indulged in 'overkill',

The Australian Minerals Council's voluntary Code for Environmental Performance is an example of a particularly successful self-regulation initiative.

TABLE 1.5
Types of Environmental Regulatory Control

Type	Description	Example
Command and Control	Ambient standards setting, using command and control instruments (licences); limited scope for flexibility	• Air and water quality standards • Effluent discharge permit
Economic Instruments	Use of pricing, subsidy, taxes and charges to alter consumption and production patterns of firms and individuals	• Tradeable permits • Pollution charges • Deposit refund • Tax incentives • Resources cost pricing
Co-regulation	Formulation and adoption of rules and regulations in consultation with stakeholders, negotiated within prescribed boundaries	• Environmental Impact Assessment
Self-regulation	Initiatives by firms or industry sectors to regulate themselves through the setting of self-imposed standards, and involving monitoring of member firms to ensure compliance	• Codes of Practice • Self-audit • Pollution reduction targets • Continuous improvement • Global Reporting Initiative 2000

Source:
based on AMC (1993)

going well beyond what is necessary and in some cases beyond what is desirable. The risk here is that the 'bar is raised' so high that others are deterred from subscribing to the code.

Companies argue in support of their ability to adequately manage themselves. The public is not so trusting, asking how it would receive assurance that environmental commitments are met. Companies have responded to such concerns in various forms, more recently in form of the Global Reporting Initiative (GRI). The GRI is a long-term international undertaking whose mission is to develop and disseminate globally applicable sustainability reporting guidelines for voluntary use by organizations reporting on the economic, environmental, and social dimensions of their activities, products and services (GRI 2000). Self-regulation, on its own, is unlikely to be effective as there will always be those who will avoid their responsibilities, either deliberately or through ignorance.

Regulatory controls range from command and control (big boot mentality) to self-regulation

Best practice therefore promotes an integrated management approach, combining the four types of regulatory controls mentioned above and summarized in **Table 1.5**. The introduction of new and flexible regimes is an ongoing process, as governments move to downsize, while corporate responsibility continues to increase.

1.7 ENVIRONMENTAL ASSESSMENT PRACTICE – ELIMINATE THE NEGATIVE, ACCENTUATE THE POSITIVE

The EIA process is designed to adopt management rules agreed upon by the company, the government, and affected communities, negotiated within prescribed boundaries. This section provides some general background on impact assessment at the project level, focusing on the mining industry as much as possible. For detailed and more general reviews on environmental assessment refer to recent texts such as Sadler (1996), Canter (1996), Wood (1995), or IAIA (1999).

The Origin of Environmental Impact Assessment

The National Environmental Policy Act of 1969 (NEPA) was the first significant political response to increasing environmental awareness in the United States, and as such is considered to be the origin of EIA as a distinct discipline.

The National Environmental Policy Act of 1969 (NEPA) was the first significant political response to increasing environmental awareness in the United States, and as such is considered to be the origin of EIA as a distinct discipline. Since available economic appraisal techniques did not take into account environmental and social costs of major developments, it was widely accepted that additional project appraisal instruments were needed. The most prominent provision of this law was the requirement to prepare an 'Environmental Impact Statement (EIS)' for any 'major Federal action' likely to have environmental impacts. Activists welcomed the possibility, if only theoretical, that a project could be stopped on environmental grounds. Pragmatists welcomed a procedure for predicting negative effects, receiving feedback and holding hearings to address controversial proposals, and writing project and site-specific programmes to manage and mitigate negative effects.

The EIA process was quickly adopted internationally, culminating in Principle 17 of the Rio Declaration on Environmental and Development, agreed at the 1992 United Nations Conference on Environmental Development, which states that:

'Environmental impact assessment, as a national instrument, shall be undertaken for proposed activities that are likely to have a significant adverse impact on the environment and are subject to a decision of a competent national authority'.

Provision for EIA began appearing in developing countries' legislation during the 1970s. According to Sadler (1998) EIA provisions now exist in the framework environmental legislation of 55 developing countries. In addition, half of these countries have drawn up specific laws, decrees, or regulations that contain criteria or procedures applicable to EIA guidelines related to specific industries such as mining, energy, or transport. All development banks and most international aid agencies have adopted the EIA process as best practice to protect the environment.

The Early Approach

When introduced in the United States in 1969, it was widely agreed that the environmental impact statement was a great concept. The only problem was that no one knew exactly how to do it. As with many other bureaucratic assignments that filter down from the political system to the civil service, everyone involved believed that the task would be simple and straightforward – for someone else to accomplish. This belief evaporated on the desks of the pioneering individuals who actually had to write the first Environmental Impact Statements.

Government scientists were first given the task of assessing impacts. They soon realized a very basic limitation, that science is a matter of observation, analysis, and explanation. While prediction is an important goal of science, trying to predict all the ramifications of a specific action at a certain location produced results that were, at best, uneven. Once subjected to review, numerous flaws in any approach could be revealed.

The government scientists appealed for assistance from their colleagues in academia, and the confusion and disagreements multiplied. It soon became apparent that while many such predictive tools existed or could be developed, they are generally so data intensive that detailed information on the action to be implemented and intensive studies of the site environment were necessary to arrive at any sort of precision. The costs for achieving this, it soon became apparent, often outweighed the value of the information produced, particularly since the information turned out to not be particularly reliable for prediction in any case.

The impact assessment task shifted to government engineers. The impact statement came to be produced as a pragmatic discussion of the likely effects of an action, based on common sense, professional experience, and logical projections of activities' footprints in space and time. The sequence of agency reviews and hearings, with subsequent incorporations of comments and revisions, gradually improved a specific document as all views were heard and incorporated into the project design. This general tendency has continued to the present – while much scientific inquiry has been directed at improving the precision of impact assessments, the basic approach is a feasibility exercise to improve project planning, decision-making, and management.

Early in the 1970s, private sector consultants began preparing environmental impact statements. While NEPA applied to government 'actions', there was soon the realization that some of the most important government actions involved the granting of permits to private businesses which were given the responsibility of preparing the environmental impact assessments. Private business engaged consulting engineering firms to do this. Private consultants in fact proved more suited to the highly irregular demands of conducting EIAs, and it has become a major niche business for the consulting industry. In a short time, even government projects came to be largely assessed by private consultants.

The use of consultants in the employ of proponents gave rise to questions of objectivity, which have never really been resolved. The question always arises, is 'independent consultant' a contradiction in terms? Subjectivity is part of any individual decision-making or judgment, and a bias in favour of the client may often if not always occur. A bias for or against an action, however, rarely distorts a professional opinion to a degree where right becomes wrong, and wrong becomes right, nor is there any environmental assessor who can claim total objectivity. This dilemma is recognized, and accommodated by EIA guidelines and procedures – objectivity is achieved through team work and through the review and approval process.

The use of consultants in the employ of proponents gave rise to questions of objectivity, which have never really been resolved.

Recent Shift in Emphasis

Difficulties encountered in the assessment of positive or negative, direct or indirect impact interactions for all five environmental components (air, land, water, biota, and people, jointly referred to as 'environment' in this text) very early on focused the EIA process into a simplified technical and scientific discipline. Emphasis was on the more obvious impacts that could be easily assessed by scientific approaches. Only the biophysical impacts of proposed actions were considered, such as impacts on air and water, landscape, or flora and fauna. Over the next 10 years, however, it became apparent that more attention was required on the less obvious impacts, that is the impacts on social, health, and economic dimensions of new projects. There was also a realization that, to be effective, the environmental impact assessment process needed to be more proactive and address practical planning and implementation issues. Emphasis has shifted to include all major stakeholders of a project in the EIA process, underlining the importance of public disclosure and participation. The increased attention on relevance and effectiveness resulted in the emergence of numerous EIA offshoots, such as social impact assessment, cumulative impact assessment, ecological risk assessment, and biodiversity impact assessment, to name a few.

More attention was required on the less obvious impacts, that is the impacts on social, health, and economic dimensions of new projects.

Today the EIA process is widely used to analyze environmental, social, and economic impacts of a proposed action within a single framework. Despite a lack of internationally consistent standards, the impact assessment practice designed by the World Bank (1991) is widely adopted by practitioners in a set of guidelines often referred to as the 'Equator Principles'. Today's environmental assessment has evolved into a planning process that goes well beyond environmental permitting. Best practice EIA identifies environmental risks, lessens conflict by promoting community participation, informs stakeholders, and lays the foundation for sustainable benefits created by mining.

Understanding the impact assessment process requires, above all else, accepting that it is not science per se.

Understanding the impact assessment process requires, above all else, accepting that it is not science per se. While it may apply scientific concepts, impact assessment is in fact a political necessity in a democracy; a bureaucratic tool to negotiate and document acceptance of decision-making; and a planning process to improve project design, construction, and implementation.

1.8 THE EQUATOR PRINCIPLES – IMPROVED PRACTICES FOR BETTER OUTCOMES

The World Bank Operational Directive

The World Bank developed several policies governing environmental assessment (EA) of projects as early as 1989. Operational Directive (OD) 4.01 on Environmental Assessment is the central document that defines the Bank's environmental assessment requirements. The Environmental Assessment Sourcebook (World Bank 1991) and its updates provide technical and authoritative guidance. Subsequently the International Finance Corporation (IFC), the private sector arm of the World Bank, released a similar set of Environmental Safeguard Policies, specifically tailored to private sector financing. Multi-lateral financing institutions such as the Asian Development Bank (ADB), the African Development Bank (AfDB), or the European Bank for Restructuring and Development (EBRD) broadly adopted the World Bank's guidelines and formulated policies and procedures quite similar in nature to the original World Bank Environmental Operational Procedures. While widely praised, the World Bank's approach to environmental assessment remained limited to projects in which the World Bank or IFC had a financial stake. This changed in 2003.

The Buy-in of Private Financial Institutions

The Equator Principles very much reflect the original IFC Environmental and Social Safeguard Policies.

Four private financial institutions, ABN Amro, Barclays, Citigroup, and West LB, drafted a set of voluntary guidelines to ensure environmentally and socially responsible project financing, dubbed the Equator Principles (EP) (**Figure 1.10**). The Equator Principles, drafted with assistance from IFC, very much reflect the original IFC Environmental and Social Safeguard Policies. They represent an attempt by private financial institutions to introduce into lending decisions a more structured and rigorous consideration of social and environmental impacts of the projects banks are being asked to fund. Originally embraced in June 2003 by 10 banks, the Equator Principles have become the accepted standard for environmental assessment, adopted by more than 60 financial institutions (Equator Principles Financial Institutions-EPFI- or in short Equator Banks- EB), by 2007, accounting for about 80% of private project funding. An Equator Bank will only lend to projects whose sponsors the bank considers are able and willing to comply with environmental assessment practice to ensure that projects are conducted in a socially responsible manner and according to sound environmental management practices. While the Equator Principles apply all over the world, they are aimed mainly at projects in developing nations, where government regulators are often ill-equipped, or unwilling, to mitigate the potential side-effects of foreign-backed investments.

The IFC Environmental and Social Performance Standards

In June 2006 IFC released a new set of guidelines for environmental assessment, termed Environmental and Social Performance Standards, or Performance Standards (PS) for

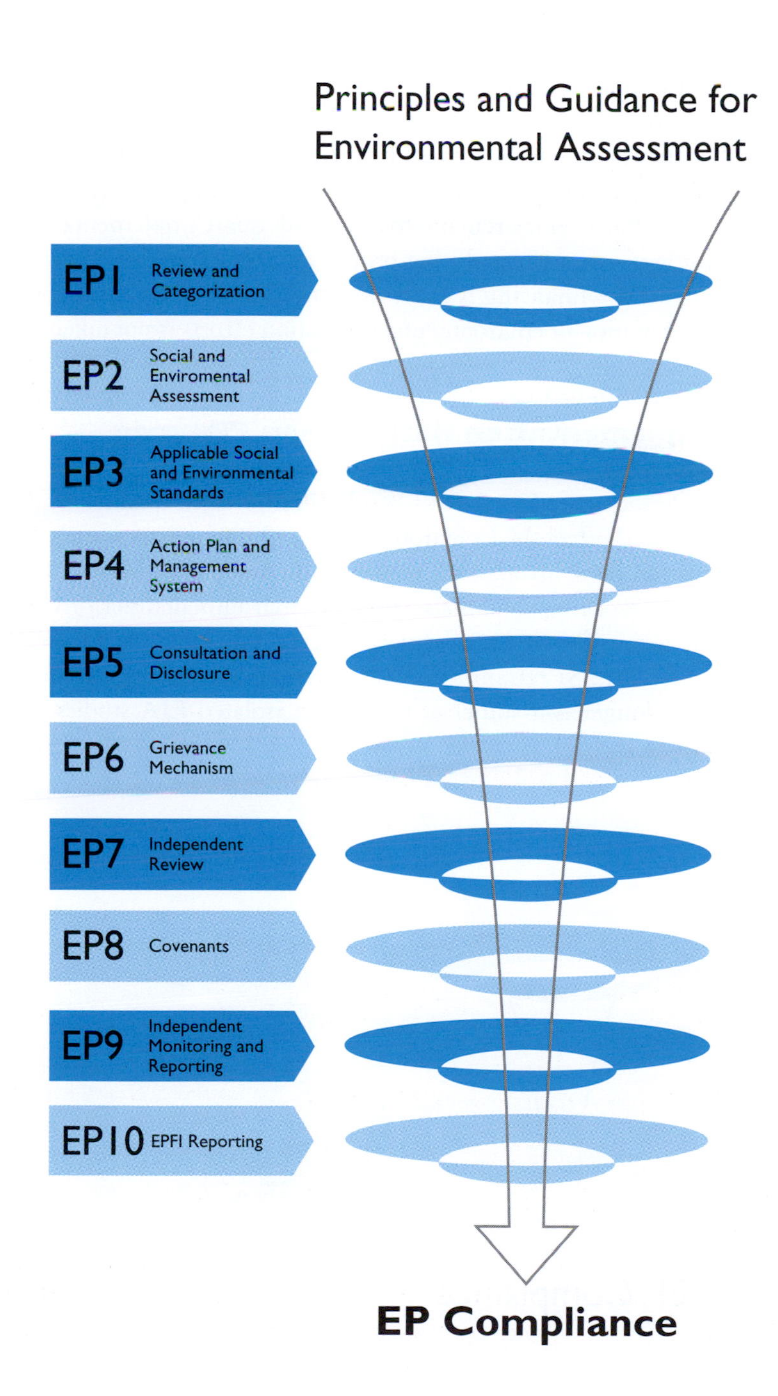

FIGURE 1.10
The Equator Principles

Originally embraced in June 2003 by 10 banks, the Equator Principles have become the accepted standard for environmental assessment, adopted by more than 60 financial institutions by 2007, accounting for about 80% of private project funding.

short, to reflect experiences in past project financing. The Performance Standards are essentially an update of the IFC's environmental Safeguard Policies but with an increased focus on the environmental and social performance during project implementation. Some controversy surrounds the 2006 standards. Certain NGOs have questioned whether the provisions assure that local communities will be adequately consulted, and whether environmental impact assessment reports will be open to full public scrutiny. There are concerns over acceptable levels of pollutants, and the exclusion of 'no-go' zones for projects.

Be that as it may, such commentators are perhaps missing a more substantial point (Warner 2006). For twenty years, environmental regulators and development finance institutions such as the IFC, ADB, and AfDB have required companies to undertake Environmental Impact Assessment (EIA) studies as a mechanism to manage significant environmental and social risks and impacts. The approach works by exploiting 'the moment of maximum leverage'. Securing an environmental clearance certificate from domestic regulators, or closing a project financing deal with international financiers, is made conditional on the applicant committing to a series of environmental and social risk management measures. The problem is that the exercise is dominated by the need for companies to secure a formal licence to operate, or source new financing. Once regulatory approval is obtained and the financial deal is closed, little leverage remains to ensure adequate implementation of environmental and social mitigation measures during project development and operation. The EIA is completed, the hurdle for financing is overcome, the project documentation looks good, but thereafter implementation of environmental mitigation efforts is not taken seriously.

Once regulatory approval is obtained and the financial deal is closed, little leverage remains to ensure adequate implementation of environmental and social mitigation measures during project development and operation.

Focusing on Implementation Rather than Planning

Enter the new IFC Performance Standards – a collection of eight quality standards (**Figure 1.11**), covering some well-established environmental and social issues such as disclosure, biodiversity conservation, and involuntary resettlement, but expanded to encompass new issues such as employee working conditions, supplier environmental performance, community security and Indigenous Peoples' intellectual property. The linchpin of the new standards is Performance Standard 1. PS1 reframes the way in which environmental and social issues are to be handled. No longer is it sufficient to conduct isolated EIA studies, outsourced to

The linchpin of the new standards is Performance Standard 1.

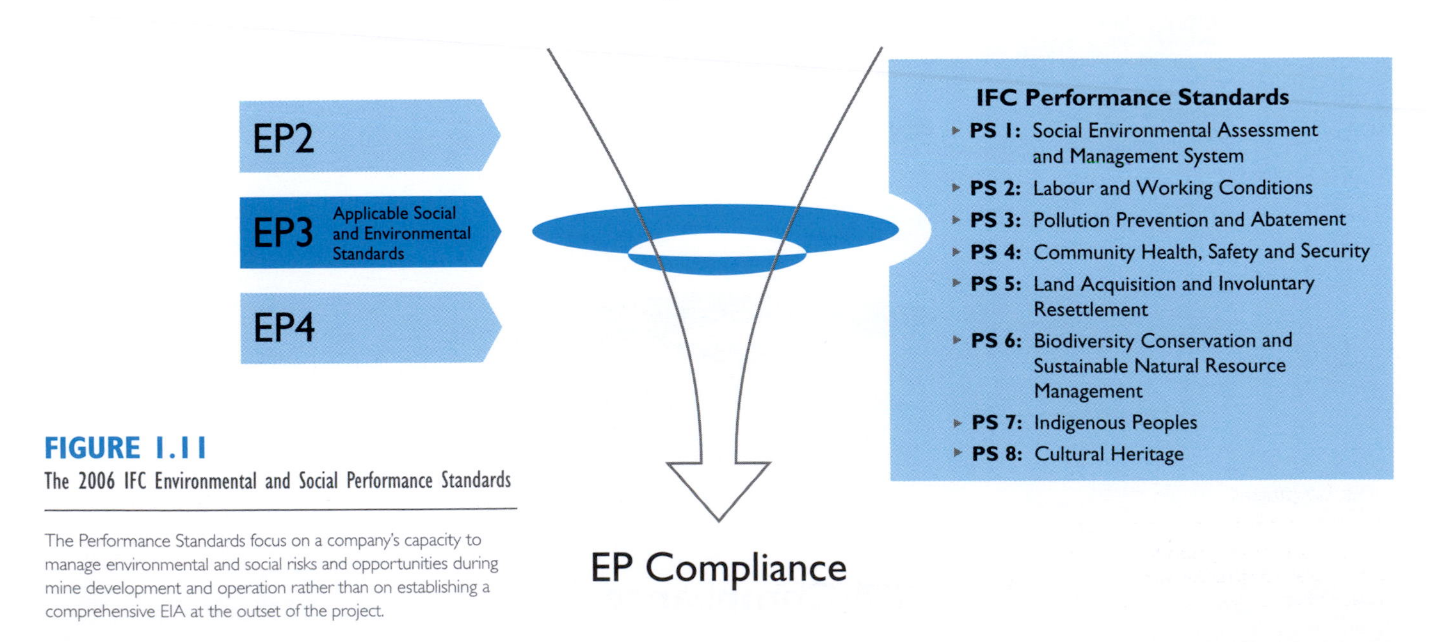

FIGURE 1.11
The 2006 IFC Environmental and Social Performance Standards

The Performance Standards focus on a company's capacity to manage environmental and social risks and opportunities during mine development and operation rather than on establishing a comprehensive EIA at the outset of the project.

external consultants. Instead, PS1 presents the standards as a single, comprehensive, risk and opportunities management framework, fully integrated with the core of the business.

The emphasis on risks and opportunities means that there is the need for project proponents to not only avoid or reduce environmental and social risks, but also to continuously search for opportunities that add environmental and socio-economic value to the investment. PS1 also emphasizes management accountability. Project proponents are asked to audit the adequacy of internal management systems and procedures to implement environmental and social mitigation measures as outlined in the EIA studies. Where found wanting, the proponent may need to develop new business principles, clarify management responsibilities for engagement with workers, local community, local government and regulators, and put in place procedures for long-term monitoring and reporting on the effectiveness of the risk management measures (Warner 2006).

The Equator Banks apply the Equator Principles to all loans for projects with a capital cost of US$ 10 million or more (a decrease compared to the initial threshold of US$ 50 m). They realize, however, that projects differ in their potential environmental impacts, as do the environmental settings.

Project Categorization

As part of its review of a project's expected social and environmental impacts, Equator Banks similar to IFC use a system of social and environmental project categorization (**Figure 1.12**). The scrutiny applied to project financing depends on the project categorization, with Category A Projects attracting the greatest attention to environmental and social due diligence review. The question then arises, 'Who determines the environmental category of a project?' Multi-lateral financing agencies apply banking internal mechanisms and expertise. Equator Banks rely on external expert advice combined with internal assessments based on a set of questions relating to:

1. the sensitivity and vulnerability of environmental resources in project area, and
2. the potential for the project to cause significant adverse environmental impacts.

Who determines the environmental category of a project?

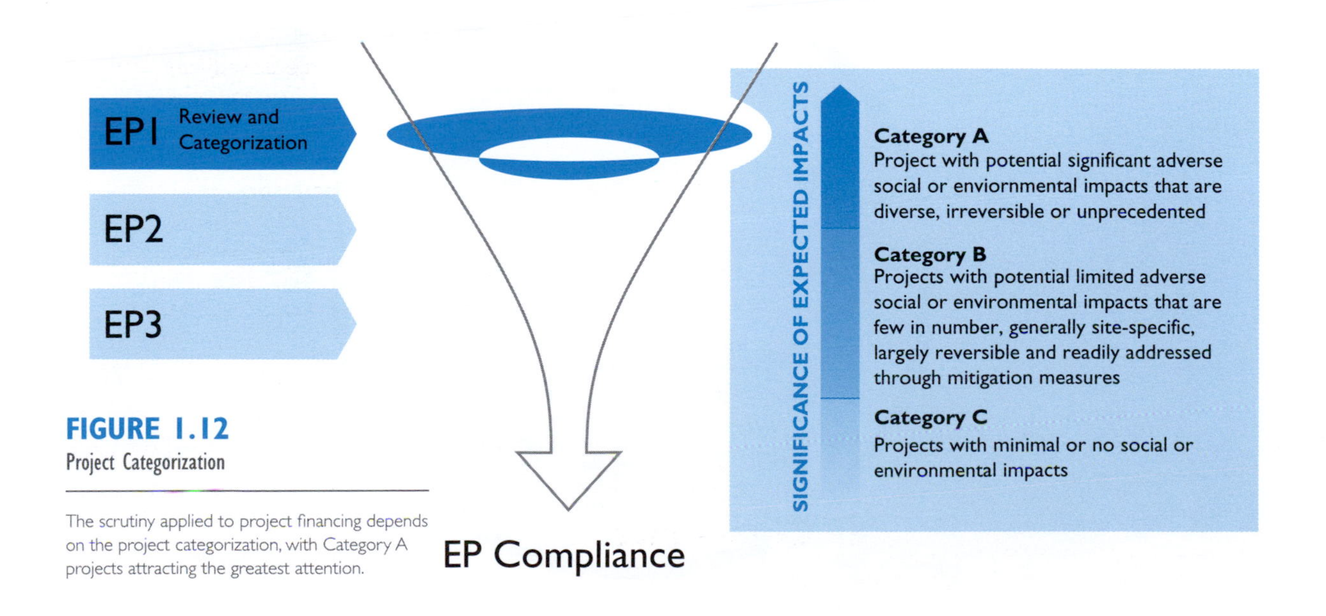

FIGURE 1.12
Project Categorization

The scrutiny applied to project financing depends on the project categorization, with Category A projects attracting the greatest attention.

Categorization is an ongoing process, and the environment category can be changed as more detailed information becomes available during the environmental assessment process.

Eventually the determination of the environment category is to be based on the most environmentally sensitive component of the project. This means that if one part of the project has the potential for significant adverse environmental impacts, then the project is to be classified as Category A regardless of the potential environmental impact of other aspects of the project (ADB 2003). Categorization is an ongoing process, and the environment category can be changed as more detailed information becomes available during the environmental assessment process.

Environmental Assessment of Existing Operations

The new Performance Standards also help to answer the question of how to apply environmental assessment guidelines such as the original Equator Principles to existing operations to which Equator Banks are providing new funds. PS1 focuses on a company's capacity to manage environmental and social risks and opportunities during project development and operation, rather than on establishing a comprehensive EIA at the outset of a new project with subsequent lack of implementation of recommended management efforts. The environmental due diligence then takes on the role of a management systems review (**Figure 1.13**).

Another question arises. Most countries now have their own EIA legislation, so what is the need for additional Performance Standards? There is a realization that challenges still face impact assessment, particularly in countries where governance, ineffective institutional frameworks, and shortages of financial and human resources may render conventional approaches to impact assessment inappropriate. Project proponents may claim that they comply with all relevant national laws or policies, but such laws or policies may be inadequate. International financing institutions consider it their duty, either due to reasons of corporate responsibility, (a recent buzz word) or due to perceived risk to reputation, to ensure that environmental assessment is commensurate with investment exposure and to ensure environmental prudence and due diligence.

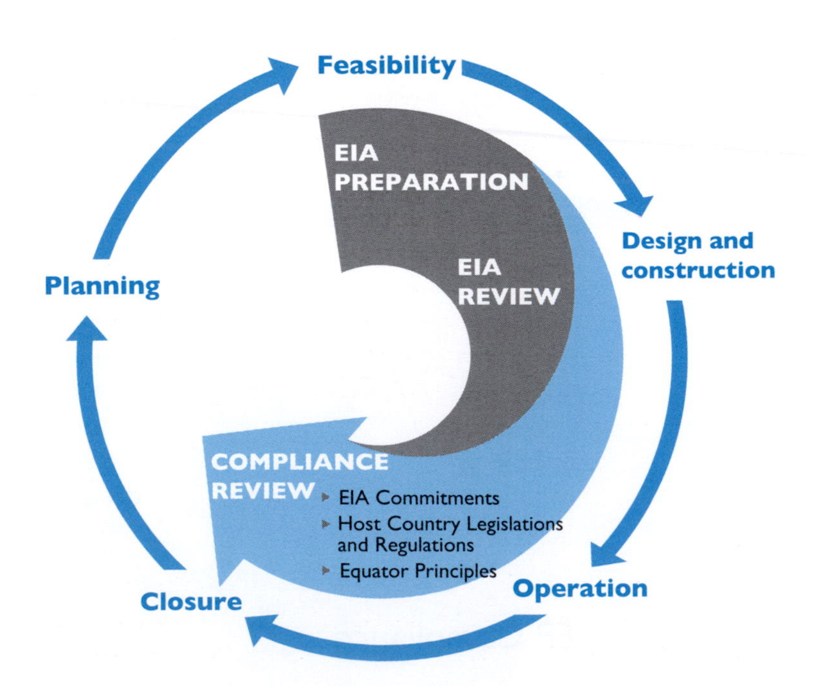

FIGURE 1.13
Equator Principles Due Diligence

The Equator Principles focus on a company's capacity to manage environmental and social risks and opportunities during operation.

1.9 CARING FOR FUTURE GENERATIONS

Our Natural Resources Consumption

A number of factors are important to consider with regard to our natural resource consumption: (1) the continued population growth; (2) the material consumption patterns of developed countries increasingly adopted by developing countries; (3) the imbalance in development, opportunities, and resource allocation between developed and developing countries; (4) the correct pricing of natural resources to account for scarcity and environmental and social costs of natural resource developments; and (5) the efficiency of resource use by industry through adopting best available techniques. These are complex, interlinked issues.

Population Growth

On October 13, 1999 the then UN Secretary-General Kofi Annan welcomed Earth's sixth billionth human being at a maternity hospital in Sarajevo. This rather symbolic threshold signifies a development which carries with it fear of an overpopulated planet that will eventually be unable to bear the burden of hosting human civilization as we know it. Over large areas of Earth's surface we have replaced natural habitats with much simpler ecosystems specialized for agricultural production and human habitation. Logging, clearance by fire, soil cultivation, infrastructure development, industrial production, and other human activities including mining have reduced the richness of ecosystems and species. It is indisputable that human activities have profound local impacts on the natural environment, including local communities. What is now becoming clear is that these activities also have global impacts. These impacts are accepted by most in case of global climate change, or reputational in the case of accidental or operationally adverse environmental impacts, news that today spreads worldwide within hours. There is also the increasing fear that we exploit natural resources at an unsustainable rate, some of them non-renewable in the human time horizon.

It is indisputable that human activities have profound local impacts on the natural environment, including local communities.

Material Consumption

As the global population increases and the average standard of living advances, so does our need for minerals. We now use three times as much copper and four times as much lead and zinc as we did 75 years ago. Not only do we use more traditional minerals, we also exploit new ones. The increasing need for metals in the developed countries is a need shared throughout the world. One of the most important developments in the global mining industry in the last decade has been the rapid emergence of China in the world market. From a small but significant exporter of minor mineral commodities such as tungsten, graphite, and magnesite, China has become a significant influence in virtually all the major mineral markets by virtue of the sheer volumes it is now using, importing, and exporting during its rapid industrialization. Chinese use accounted for one-third of the entire world growth in copper use between 1990 and 2000, and 40% of world growth in aluminium use. China is also the world's largest steel producer and user (MSSD 2002).

The desire to raise global living standards, coupled with a growing world population, will increase worldwide minerals demand in the future. This demand means that the mining industry – responsible for extracting minerals from the Earth for use in our daily lives – will continue to be vital and necessary.

Imbalance in Development

There is concern about disparities in the use of mineral products between rich and poor and the ever-increasing demand, mostly in developed countries. These concerns are heightened by the non-renewable nature of mineral resources and fears of eventual depletion.

There is concern about disparities in the use of mineral products between rich and poor.

Social Costs

Most countries rely on commercial exploitation of natural resources such as minerals, forests, fish, and soils for a significant proportion of national income. Usually natural resources are owned by the State; therefore, formulation and analysis of public policy over natural resources and the environment (so-called 'green' issues) is crucial. At the same time, countries are also trying to encourage greater investment in industrial activities as part of a broad economic development programme. These changes are placing more urgency on 'brown' issues such as air and water pollution. Countries also are increasingly becoming part of the global community. Economic reform programmes being implemented by governments are exposing countries to global competition. International economic, sectoral, and environmental policy is now an important factor shaping national policies for economic and social development, and environmental management.

Most economic arguments for government intervention are based on the idea that the marketplace cannot provide public goods or handle externalities. Public health, education, national and domestic security, and a clean environment, air and water, have all been labelled public goods. Public goods have the distinct aspect of 'non-excludability'. Non-excludability means that non-payers cannot be excluded from the benefits of the good or service. Externalities occur when one person's actions affect another person's well-being and the relevant costs and benefits are not reflected in market prices. A positive externality arises, for example, when non-paying spectators benefit from a fireworks display. (Note that the free-rider problem and positive externalities are two sides of the same coin.) A negative externality arises when one person's actions harm another. When emitting SO_2, mining companies may not consider the costs that SO_2 pollution imposes on others.

Policy debates usually focus on free-rider and externalities problems. Some public good problems can be solved by defining individual property rights in the appropriate economic resource, a less effective solution for environmental problems involving air or water. Property rights to air and water cannot be defined and enforced easily. It is difficult to imagine, for instance, how market mechanisms alone could prevent depletion of the Earth's ozone layer or combat global warming. In such cases economists recognize the likely necessity of a regulatory or governmental solution.

Negative externalities impose costs on society for private gain.

Negative externalities impose costs on society for private gain. This asymmetry means some pay much of the costs and receive few of the benefits. Externalization of social and environmental costs is bad economics, intensifying poverty and hindering sustainability. Externalized costs also too often fall disproportionately on the poor who least can afford to pay them. Consequently, there is also the argument that externalization means developing countries are subsidizing the development in the rest of the world. Minerals extraction and production often incur social and environmental costs, which should be internalized by the industry and accounted for in prices. Good economics, internalizing externalized costs, is the best and fastest way of approaching sustainability. Thus as part of a new mine development, proponents should routinely carry out economic analyses, as well as financial ones, in order to clarify how they contribute to development goals. This is in the company's self-interest as well as the interest of the host country.

Best Available Technologies

Legislative criteria have developed worldwide on the basis of those provided by 'Best Available Technology (BAT)', sometimes modified by economic and managerial factors to 'Best Available Technology Not Entailing Excessive Cost' (BATNEEC). Such criteria can provide a comprehensive system for the control of process emissions to the environment to levels which have a rational basis and are as low as can be achieved with modern technology, as long as the term 'technology' is understood as embracing planning, design, and all relevant managerial systems in addition to the technology and its ancillaries. Today much emphasis is placed on controlling environmental impacts by attacking the problem at the source, termed 'Best Management Practices – BMP' instead of using more costly downstream control technologies (e.g. USEPA 2000). There are two categories of BMP: (1) planning and design with efforts directed at future mining activities, and (2) maintenance and operational practices to minimize impacts from existing mining operations.

It is also important to note that operation criteria based on BAT or BATNEEC become meaningless in time if there is no continuous research in improving existing technologies and practices. World Bank (2006) observed that relatively low levels of research and development expenditures are a long-standing feature of the mining sector, and that there appears to be no role for significant government intervention in these respects. However, World Bank (2006) also states that there may be areas where governments can selectively work with industry and academia to ensure that appropriate research is conducted in respect of long term issues of common interest. These could include the implications of global warming and carbon emission reduction objectives for the mining and metals processing industry (e.g. carbonless steel), and longer-term environmental management issues such as water use in mining and long-term mine closure planning.

The term 'best practice' has been widely used in the mining industry, particularly since publication in the 1990s by Environment Australia, of a series of mainly excellent booklets which are listed in the Bibliography. Undoubtedly, striving to achieve best practice is a worthy ideal. However, the authors of this book have avoided the term, for the following reasons:

- The goal of 'continuous improvement' to which many mining companies subscribe, means that best practice this year should not be so in the future;
- Practices that are highly effective in a particular situation may be ineffective or even deleterious in another. One of very many examples that could be cited is the technique of 'moon-scaping' in mine land rehabilitation (see Chapter Twenty One).

Practices that are highly effective in a particular situation may be ineffective or even deleterious in another.

Notwithstanding the views referred to above from the World Bank (2006), considerable advances and improvements have been made in environmental technology related to mining over the past three decades. Much of the research and many of the advances in environmental technology have been made in Australia, Canada and the USA where strong, profitable mining industries have existed within environmentally conscious societies, and have consequently been subject to relatively strict regulatory control and oversight. Another feature of the mining industry in these countries has been a willingness to cooperate in research and to rapidly disseminate the results. The global supply situation is now changing, with more and more mineral production sourced from developing countries, many of which have little or no history of mining. In most of these countries, effective environmental practices have not been developed or, if developed, have not been widely disseminated. The practices that are appropriate for operations in the established mining countries may be inappropriate in many developing countries where the terrain, climate and social circumstances may be quite different. Indeed, it is particularly important to recognize that, even in the same region, no two projects are identical; subtle differences in the nature of the ore, the treatment process, or characteristics of the local environment, can and do have important ramifications for environmental impacts, requiring different approaches or selection of different technologies to provide the most effective means of environmental management. If a term is needed to describe the ideal, then 'effective practice' or 'appropriate practice' would be preferred to best practice.

No two projects are identical; subtle differences in the nature of the ore, the treatment process, or characteristics of the local environment, can and do have important ramifications for environmental impacts.

Mining and Sustainability

Sustainable development has been proposed as a holistic approach for dealing with these complexities. The term sustainable development was popularized by the World

Commission on Environment and Development in 1987, chaired by Norwegian Prime Minister Gro Harlem Brundtland, and consequently called the Brundtland Commission. In the words of the Brundtland Report (WCED 1987), sustainable development means 'meeting the needs of the present without compromising the ability of future generations to meet their own needs'. This is probably the most widely accepted definition to date.

Sustainability is arguably the most widely used environmental 'buzzword' of the past decade. It has been widely used and misused to denote a variety of concepts. Clearly, the concept of sustainability varies depending on what is being sustained. Sustainable agriculture, for example, refers to agricultural systems that can be continued indefinitely without system failure. At first glance, sustainable mining appears to be a contradiction in terms, as for all practical considerations, minerals are not renewable. Most mining projects have finite operating lives of 5 to 50 years. However, the concept of sustainability can be applied to most aspects of mining and associated activities. A variety of examples are discussed below.

The concept of sustainability varies depending on what is being sustained.

Mining Company View-point

From the view-point of a mining company, sustainability means locating and developing mining projects to provide returns to shareholders, as well as funding to find or acquire replacement projects. Some companies, such as the Benguet Corporation of the Philippines, once a major producer of gold, copper and other minerals, have sustained themselves during downturns in the industry, by alternative revenue generating activities such as real estate development. Profit is clearly the main requirement for sustainability of a commercial enterprise. Most companies aspire to be sustainable; however, many do not achieve this goal, whether due to poor management, lack of profitability, take-over, or inability to find replacement projects.

Community View-point

From the view-point of a community hosting a mining operation it is important that the mining operation itself is not perceived to be sustainable. All mining projects have finite lives. Communities should be well informed so that their expectations are realistic. While there are examples of communities which have been supported by mining operations for more than 100 years, there are many more examples where mining has ceased after much shorter periods. In some cases the associated communities have declined substantially or disappeared totally, as in the case of 'ghost towns', the remnants of abandoned mining communities. This is not to suggest that there is anything intrinsically wrong with temporary communities. Many mines are developed in remote, unpopulated areas with no other potential source of employment and, in such cases, there is usually no reason for the community to be sustained, once mining ceases.

All mining projects have finite lives.

Mining itself, however, can be considered as a sustainable activity as there will always be ores to be mined. This follows because the elements which combine to form ores remain at or close to the Earth's surface, even after they have been used. When the higher grade, readily accessible ores have been mined, lower grade and/or less accessible ores will be mined. And, in the future, particularly if production costs increase, it can be expected that more and more mineral and metal products will be produced by recycling.

In many cases, the communities that have developed in association with mining, have continued long after mining has finished, albeit on a reduced scale. Examples in Australia include many of the larger inland cities, such as Ballarat and Bendigo, which continued to exist and ultimately to thrive following conclusion of mining. What is important to a community considering becoming host to a new mining project, is that the community itself is sustained during and after mining. This usually means that the pre-existing livelihoods and economic bases are maintained and that additional means of income generation are developed to replace mining, once operations cease. These aspects are discussed in more detail in Chapter Fifteen addressing community development.

Environmental View-point

From the view-point of the environment, sustainability means that environmental values should not be lost or permanently degraded. In practice, this is achieved by rehabilitation. In some operations, the total area can be rehabilitated to ultimately reach the same or better environmental condition that existed prior to mining. In other cases, parts of the area can never be restored to their preexisting condition. The walls of large open Pit mines are the most obvious examples. In such cases, environmental compensation such as the use of environmental offsets (see Chapter Seven) can be applied to ensure that there is no net loss of environmental amenity.

From the view-point of the environment, sustainability means that environmental values should not be lost or permanently degraded.

Clearly, environmental sustainability is a complex issue, involving much more than the rehabilitation of surface disturbance. Many aspects of the environment are involved and need to be considered in evaluating sustainability. These include sustainability of:

- community livelihoods;
- indigenous cultures;
- community values;
- water resources – yields and quality;
- air quality;
- ecological functions;
- biodiversity; and
- visual amenity.

The measures used for sustaining or even enhancing these environmental components are the 'building blocks' of environmental management as described throughout this book.

The Three Circles of Sustainable Development

A widely held view of sustainable development is that it refers at once to economic, social, and ecological needs (**Figure 1.14**). According to this view there must be no single focus (or object) of sustainability, but instead all of the economic, social, and ecological systems

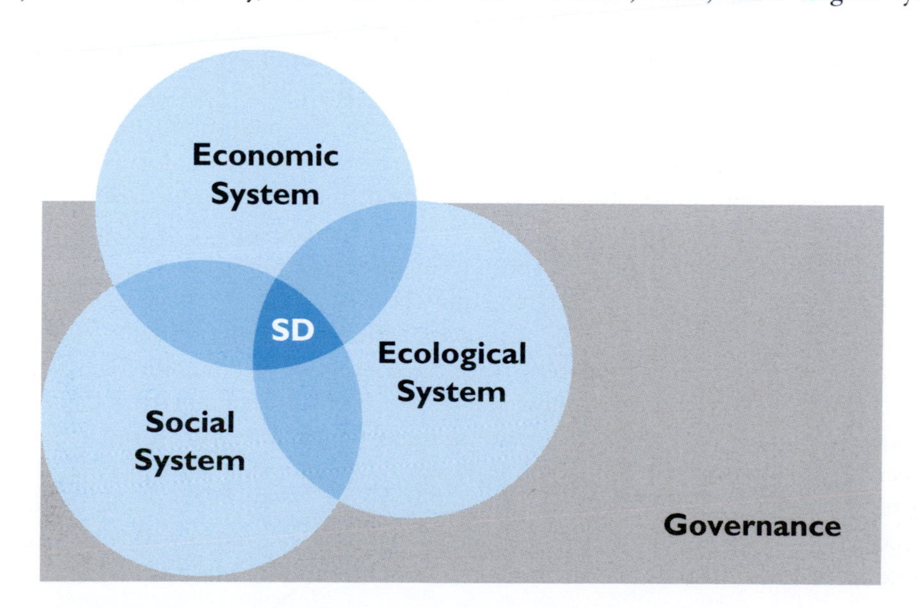

FIGURE 1.14
The Three Circle of Sustainable Development (SD)

For sustainable development to work an environment of adequate governance is needed, expressed e.g. in terms of rule of law, lack of corruption, quality of regulation and political stability.

Source:
Robinson and Tinker 1998

must be simultaneously sustainable in and of themselves. Satisfying any one of these three sustainability circles without also satisfying the others is deemed insufficient. Each of the three circles is independently crucial, but they are interconnected. There is, therefore, a risk of unwittingly causing (or worsening) problems in one system while attempting to correct problems in another. The only sure way to avoid this is to integrate decisions such that effects in all three systems are considered before action is taken (Robinson and Tinker 1998).

Economic Sustainability

The economic circle of sustainability is founded on the concept of maximizing the flow of income from a stock of capital while maintaining the stock yielding this income. The concept encompasses traditional theory on economic growth, that is, determining optimal economic growth with a given capital. The intuition is that future generations can only be better off if they have more capital per capita than we have today. It is immediately obvious that population growth is inimical to sustainable development since it 'dissipates' the capital stock. Technological change, on the other hand, enables a given capital stock to generate more wellbeing per unit of the stock. An easy way to think of it, then, is to say that future generations will be no worse off if capital stocks are 'constant' and for the rate of technological change to just offset the rate of population growth. If technological progress is faster than population change, then future generations could still be as well off as we are today with a lower capital stock, and so on.

Capital, as is now well known, goes well beyond the common idea of financial capital and has five main forms.

Capital, as is now well known, goes well beyond the common idea of financial capital and has five main forms (MSSD 2002):

- *natural (or environmental) capital*, which provides a continuing income of ecosystem benefits, such as biological diversity, mineral resources, forests, wetlands, and clean air and water;
- *built (or productive) capital*, such as machinery, buildings, and infrastructure (roads, housing, health facilities, energy supply, water supply, waste management, etc);
- *human capital*, in the form of knowledge, skills, health, cultural endowment, and economic livelihood (small enterprise development, literacy, health care, inoculation programmes, etc.);
- *social capital*, the institutions and structures that allow individuals and groups to develop collaboratively (training, regional planning, decision sharing culture, etc.); and
- *financial capital*, the value of which is simply representative of the other forms of capital.

The total capital stock could be rising while any one form of capital is declining.

This broadening of the concept of capital is critical to an understanding of sustainable development (**Case 1.5**). It is now easy to see that the total capital stock could be rising while any one form of capital is declining. The idea that forms of capital substitute for each other is embodied in the notion of weak sustainability. If, on the other hand, forms of capital are not substitutable then the requirement that the total stock be constant (rising) has to be supplemented by the requirement than the relevant specific capital stock should also be nondeclining. In the literature, this has been termed strong sustainability.

Links with *social development* centre on equity between different societies, and between the present and future generations. Economic efficiency and optimal use of scarce resources are also underlying principles, since the concept of sustainability is based on the idea that natural resources are somehow scarce, which means that any use today may preclude a use tomorrow and vice versa, that use tomorrow may require a restriction of the use today. Difficulties arise with economic sustainability in terms of identifying the types of capital to be maintained, and substitutability. Linkages with *environmental development* focus on valuation of natural resource capital and the degree to which pricing of the resource accounts for scarcity and the full environmental and social costs.

Social Sustainability

The social circle of sustainability addresses issues such as poverty, health, education, local empowerment, and maintaining culture and heritage. Although social norms change over time, sustaining social and cultural systems is important. *Social sustainability* has obvious links with economic development in terms of addressing poverty and local input into economic decision-making. Links with *environment development* focus on the allocation and distribution of natural resources to future generations as well as local empowerment over natural resource management.

Ecological Sustainability

The ecological circle of sustainability is concerned with maintaining the physical/chemical and biological environment to preserve resilience and the ability of natural systems to adapt to change, protecting from degradation the ecological processes and cycles that are fundamental to life on Earth. An obvious link with economic development is through the supply of raw materials for production and the use of the environment as the final waste sink. One link with social sustainability is through the level of local participation in natural resource management. Closely related to ecological sustainability is the concept of carrying capacity, both in terms of human population that can be supported by Earth and the ability of our environment to assimilate finals waste products. All mine developments intrinsically involve 'trade-offs' between potentially conflicting goals, such as mineral exploitation and respecting traditional land rights, or economic growth and environmental conservation. The challenge is to optimize trade-offs between and across the three spheres basic to sustainable development – the ecological sphere, the economic sphere and the social sphere (Barbier 1987; Holmberg *et al.* 1991).

All mine developments intrinsically involve 'trade-offs' between potentially conflicting goals.

It is easy to appreciate that sustainable development is a challenging concept, and priorities will differ from industry to industry. **Table 1.6** summarizes key elements applicable to the mining sector as suggested by the Conservation Strategy Committee to the Saskatchewan Round Table on Environment and Economy, Regina, 1986.

Patrick James, CEO of Rio Algom Ltd., was among the first to urge that mining, like other forms of development, must contribute not only economic value to stakeholders, but also environmental and social value. In the June 1999 Mining Engineering, he also observed that 'as an industry, we will gradually find ourselves unable to operate anywhere

CASE 1.5

The Almadén Mine in Spain: A Treasure Chest but without Sustained Development

The Almadén mercury mine has been in operation for over 2000 years. Yet it is a typical example of a mining operation that has resulted in non-sustained development. During most of its history, the mine has operated as a state-run enterprise and was used as a treasure chest of the Spanish Crown. Twice it was used to guarantee enormous sovereign loans. Little if any of the rents went back to the local communities and little attention was paid to developing the area or to diversifying the local economy. The community never participated in the decisions that most affected it. Its members seldom had any alternatives to the mine and have seen the state as obliged to give them employment. The very high quality of the deposit took away the need of the parastatal management to innovate and increase the productivity of the operation. The Almadén deposit is the largest, richest mercury deposit known to exist. It contained over 30 percent of the world's known reserves of mercury and its grade has been six times the global average of mercury mines. Since management's main concern has always been to maximize production, the productivity and profitability of the operation were never made priorities.

Source:
World Bank 2002

TABLE 1.6
The Key Elements of Sustainability

• Non-renewable resource development should not threaten the environment and the renewable resources upon which future generations depend.
• Mineral wealth should be maintained from one generation to the next.
• Sustainable mining balances economic growth and protection of the environment through sensible trade-offs that consider all costs and benefits in the decision-making process.
• It is critical to recognize that mining will affect the social structure and culture of local people and to consider these impacts as part of the decision-making process.
• It also must be recognized that the building of local capacities does not happen overnight. Thus, the process should begin as early as possible
• Reducing, reusing, and recycling resources should be encouraged, while avoiding the waste of the resource base by inefficient mining techniques
• Policy and taxation decisions should consider the economic health of the mining industry.

Source:
Conservation Strategy Committee to the Saskatchewan Round Table on Environment and Economy, Regina, 1986.

if we are incapable or reluctant to effectively combine economic, environmental and social goals everywhere we do business'.

In summarizing the major challenges the mining industry faces, Sir Robert Wilson, executive chairman of Rio Tinto, one of the world's largest mining company, writes in the June 2000 Mining Engineering that mining finds itself in increasing disfavour in the United States, Canada, Europe, and many other parts of the world. He adds that industry's traditional responses – to say that criticisms are ill-founded, to remind critics that they depend on mineral products, and to engage in education, advertising and public relations campaigns – have all been to little or no avail. Mining's reputation continues to deteriorate, he concludes. Sir Robert urges the mining industry to change its dialogue with stakeholders, especially with non-government organizations, and supports a new global mining initiative to seek 'independent analysis of issues that will determine the future of mining and that these issues are social and environmental as well as economic' (www.geotimes.com 2006). In summary, however, on a global scale, this book concludes that mining is a sustainable activity which will continue for as long as human civilization itself.

REFERENCES

ADB (2003) Environmental Assessment Guidelines, Asian Development Bank.

AMC (1993) The Environmental Challenge: Best Practice Environmental Regulation, Australian Manufacturing Council, Melbourne.

Barbier EB (1987) The Concept of Sustainable Economic Development, Environmental. Conservation, Vol. 14, No. 2; pp. 101–110.

Bates RL and Jackson JA eds. (1984) Dictionary of Geological Terms, New York: Doubleday.

Berger PL and Luckmann T (1966) The Social Construction of Reality: A Treatise in the Sociology of Knowledge, Garden City, New York: Anchor Books.

Boyns T (1997) The Mining Industry, Tauris Industrial Histories.

Canter L (1996) Environmental Impact Assessment (Second Edition), McGraw Hill Publishing Company. New York, USA.

Carman JS (1979) Obstacles to Mineral Development: A Pragmatic View; Pergamon Press Inc., Elmsford, NY.

Carson R (1962) The Silent Spring, Houghton Mifflin Publisher.

Clark AL (1994) Political, economic and administrative decentralization: Addressing the distributional effects of mineral sector development in Asia. East-West Center Honolulu, Hawaii.

Clark AL and Clark JC (1996) An assessment of social and cultural issues at the Bougainville (Panguna) mine in Papua New Guinea; United Nations Conference on Trade and Development Management of Commodity Resources in the Context of Sustainable Development: Social Impacts of Mining; Papers presented at the Asian/Pacific Workshop on Managing the Social Impacts of Mining Bandung, Indonesia, 14–15 October 1996.

EEA (2005) Sustainable Use and Management of Natural Resources, Report No 9/2005.

Eurostat/IFF (2004) Economy-wide Material Flow Accounts and Indicators of Resource Use for the EU-15: 1970–2001, Series B, (prepared by Weisz H, Amann C, Eisenmenger N and Hausmann F), June 2004.

FEE (2003) Environmental economics and natural resource analysis – Trainings Manual, Forum of Economics and Environment (www.econ4env.co.za).

Garnaut RR and Ross AC (1975) Uncertainty, Risk Aversion and Taxation of Natural Resource Projects, Economic Journal, vol. 85, no. 2, pp. 272–287.

Garnaut RR and Ross AC (1983) Taxation of Mineral Rents, Clarendon Press, Oxford, United Kingdom, 224 p.

Gore A (2006) An Inconvenient Truth: The Planetary Emergency of Global Warming and What We Can Do about It.

Gregory C (1981) A concise history of mining; Pergamon Press.

GRI (2000) Sustainability Reporting Guidelines on Economic, Environmental, and Social Performance, Global Reporting Initiative), Boston.

Hartman HL (1987) Introductory Mining Engineering, New York: John Wiley and Sons.

Hartman HL, ed. (1992) SME Mining Engineering Handbook, Society for Mining, Metallurgy and Exploration, Inc., Littleton, CO, 2260 p.

Hartley P and Medlock K (2005) Political and Economic Influences on the Future World Market for Natural Gas, Prepared for the Geopolitics of Natural Gas Study, a joint project of the Program on Energy and Sustainable Development at Stanford University and the James A. Baker III Institute for Public Policy of Rice University.

Holmberg J, Bass S and Timberlake L (1991) Defending the Future: A Guide to Sustainable Development, IIED/Earthscan, London.

Howell LD (2001) The Handbook of Country and Political Risk Analysis, PRS Group, East Syracuse, N.Y.

Hudson TL, Fox FD and Plumlee GS (1999) Mining and the Environment. AGI Environmental Awareness Series, 3; American Geological Institute Alexandria, Virginia In cooperation with Society of Economic Geologists Society for Mining, Metallurgy, and Exploration, Inc. U.S. Department of the Interior U.S. Geological Survey.

IAIA/IEMA (1999) Principles of Environmental Impact Assessment Best Practice, International Association for Impact Assessment and the Institute of Environmental Management and Assessment, Fargo, North Dakota. (http:/www.iaia.org/publications).

IFC (2006) Environmental and Social Performance Standards, International Finance Cooperation (www.ifc.org).

Indonesian Government Regulation No. 19 (1994) regarding Hazardous and Toxic Waste Management; Government of Indonesia.

Kettell B (1982) Gold; Oxford University Press.

Lacy W (2006) An Introduction to Geology and Hard Rock Mining, Rocky Mountain Mineral Law Foundation Science and Technology Series, http://www.rmmlf.org/SciTech/Lacy/lacy.htm

Marshall IE (2001) A Survey of Corruption Issues in the Mining and Mineral Sector, MMSD Publication.

Meyerriecks W (2003) Drills and Mills: Precious Metal Mining and Milling Methods of the Frontier West.

MMSD (2002) Breaking New Ground; Mining, Minerals and Sustainable Development (MMSD) Project; published by Earthscan for IIED and WBCSD.

Nazari MM (2000) Financial provision in mine closure: developing a policy and regulatory framework in the transition economies. In: Mining and sustainable development II – Challenges and perspectives, UNEP Industry and Environment, Volume 23, Special Edition 2000.

Otto J (1997a) Mineral policy, legislation and regulation; Mining, Environment and Development, UNCTAD.

Otto J (1997b) A national mineral policy as a regulatory tool, Resources Policy, 23 (1/2): 1–8.

Pearce DW and Turner RK (1990) Economics of Natural Resources and the Environment, Harvester Wheatsheaf, Hertfordshire.

Ricardo D (1962) On the Principles of Political Economy and Taxation in Pierro Sraffa, ed.; The Works and Correspondence of David Ricardo, Cambridge University Press, Cambridge, United Kingdom, 187 p.

Robinson J and Tinker J (1998) Reconciling Ecological, Economic, and Social Imperatives; In Schnurr J and S Holtz eds., The Cornerstone of Development: Integrating Environmental, Social and Economic Policies, Ottawa: International Development Research Centre, pp. 9–44.

Sadler B (1996) Environmental Assessment in a Changing World: Evaluating Practice to Improve Performance. (Final Report of the International Study of the Effectiveness of Environmental Assessment). Canadian Environmental Assessment Agency and International Association for Impact Assessment, Ottawa, Canada.

Sadler B (1998) EIA for Industry Status Report on UNEP TIE Initiative to Improve Industrial Project Planning through an international workshop held in Paris, France 30 November – 2 December, 1998.

Trush P (1968) Dictionary of Mining Mineral and related Terms; Mclean Hunter Pub. Co.

USEPA (2000) Coal Remining Best Management Practices Guidance Manual, Washington DC.

Warner M (2006) The New International Benchmark Standard for Environmental and Social Performance of the Private Sector in Developing Countries: Will It Raise or Lower the Bar? Overseas Development Institute, Opinion 66.

WCED (1987) Our Common Future, World Commission on Environment and Development (also referred to as the Brundtland Report).

Wexler T and Lovric K (2006) Buyers, investors and lenders beware: pitfalls in mining projects, Berwin Leighton Paisner LLP; http://www.gettingthedealthrough.com

Wood CM (1995) Environmental Impact Assessment: A Comparative Review, Longman Higher Education, Harlow, UK.

World Bank (1991) Environmental Assessment Sourcebook (three volumes). Technical Papers Nos. 139, 140 and 154, World Bank, Washington D.C.

World Bank (2002a) Mining and Development: Large Mines and Local Communities – Forging Partnerships, Building Sustainability, World Bank and International Finance Corporation, Washington, www.ifc.org/mining

World Bank (2002b) Mining and Development: Treasure or Trouble – Mining in Developing Countries, World Bank and International Finance Corporation, Washington, www.ifc.org/mining

World Bank (2006) Background Paper: The Outlook for Metals Markets; prepared for G20 Deputies Meeting Sydney September 2006. The World Bank Group, Oil, Gas, Mining and Chemicals Department, Washington.

Appendix 1.1
Some Useful Websites

www.miningandtheenvironment.com
The **Mining and the Environment** website supports this book. The website offers study support, and provides access to a wide range of knowledge and environment resources. It has many useful links, and a section targeted to support the environmental consulting industry. Eventually Mining and the Environment aims to portray modern mining industry as an environmentally responsible, international citizen (www.miningandtheenvironment.com).

American Institute of Mining, Metallurgical and Petroleum Engineers (AIME) is organized and operated exclusively to advance, record, and disseminate significant knowledge of engineering and the arts and sciences involved in the production and use of minerals, metals, energy sources and materials for the benefit of humankind. (www.aimehq.org)

Australian Centre for Mining Environmental Research (ACMER) (now University of Queensland: Sustainable Mineral Institute – SMI) was established as a joint industry and research institution initiative. A major focus of its activities is mine closure which is considered as a whole-of-mine-life process which typically results in tenement relinquishment. The Centre pursues its role through research, advisory services and facilitation of stakeholder forums. (www.acmer.uq.edu.au)

Australian Aluminium Council (AAC) is the peak body representing the Australian aluminium industry. (www.aluminium.org.au)

Australian Coal Association (ACA) is an industry body representing the interests of the black coal producers in New South Wales and Queensland, the states that produce 98 per cent of Australia's black coal. (www.australiancoal.com.au)

Australian Coal Association Research Program (ACARP) – Australian black coal producers contribute to a program of collaborative research that is conducted for the benefit of the coal mining industry. (www.acarp.com.au)

Australian Mineral and Energy Foundation (AMEEF) is an independent, not-for-profit body established in 1991 that encourages sustainable development in the resources sector by facilitating dialogue amongst stakeholders, promoting excellence in research, education and training, and recognizing achievements in environmental performance and sustainable development. (www.ameef.com.au)

Australasian Legal Information Institute based at the University of New South Wales provides access to a wide range of databases of legislation in Australia, New Zealand, the South Pacific, Asia, and Africa. (www.austlii.edu.au)

Australian Uranium Industry Framework (UIF) Steering Group established by the Australian Government through its Department of Industry, Tourism and Resources in 2005 aims to propose a uranium strategy. (www.industry.gov.au)

The 1991 **Berlin Guidelines** (http://www.mineralresourcesforum.org/workshops/Berlin) emerged from a United Nations convened roundtable discussion of international mining experts in Berlin to address environmentally sustainable mineral development. The Berlin Guidelines set out mining-environment principles for the mining industry as well as for multilateral and bilateral financing institutions. In 2002, a second roundtable discussion produced a follow-up document referred to as Berlin II.

The **Best Practice Environmental Management** (BPEM) in Mining program is a world-renowned partnership between the mining industry, interested stakeholders and the Australian Government. It aims to help all sectors of the minerals industry – minerals, coal, oil and gas – to protect the environment and to reduce the impact of minerals production. Since the program began in 1994, the Australian Government has worked with industry partners to produce 24 booklets on a range of topics, from community consultation to water management and cleaner production. The booklets present concise, practical information on how to achieve environmental management best practice in the minerals industry anywhere in the world. (www.ea.gov.au/industry/sustainable/mining/bpem.html)

The Colourado School of Mines, USA, website includes information on acid producing potential of mine overburden and acid mine drainage (AMD) chemistry and treatment. (www.mines.edu)

The **Convention on Biological Diversity** (Biodiversity Treaty) – one of the major accomplishments of the 1992 Earth Summit – has major implications for the mining industry. Its core concept is that nations are 'responsible for conserving their biological diversity and for using their biological resources in a sustainable manner'. (www.cbd.int)

The 1961 **Convention on International Trade in Endangered Species of Wild Fauna and Flora** (CITES), in force since 1975, is one of the earlier international environmental agreements. Under this trade agreement, each country is supposed to establish its own system to control movement of wild life exports and imports. Species are separated into three classes according to the degree to which they are endangered. (www.cites.org)

COAL21, initiated by the Australian Coal Industry, is a program aimed at fully realizing the potential of advanced technologies to reduce or eliminate greenhouse gas emissions associated with the use of coal. The program also explores coal's role as a primary source of hydrogen to power the hydrogen-based economy of the future. (www.coal21.com.au)

Coal Institute Advisory Board (CIAB) is a group of high level executives from coal-related industrial enterprises, established by the International Energy Agency (IEA) in July 1979 to provide advice to the IEA on a wide range of issues relating to coal. (www.iea.org)

Cobalt Development Institute is an international organization of a wholly non-profit making character. It is an association of producers, users and traders of cobalt with the objectives of promoting the responsible use of cobalt in all forms; consulting organizations, agencies and governments for research or investigations on all matters concerning cobalt; providing members with topical information on all cobalt matters including health & safety and environmental legislation plus regulatory affairs possibly affecting their interests; promoting co-operation between members; and providing a forum for the exchange of information concerning the resources, production and uses of cobalt. (www.thecdi.com)

Cooperative Research Centre for Greenhouse Gas Technologies (CO2CRC) researches the logistic, technical, financial and environmental issues of storing industrial carbon dioxide emissions in deep geological formations. The CRC also researches the capture and separation of carbon dioxide from industrial systems. (www.co2crc.com.au)

Cooperative Research Centre for Coal in Sustainable Development (CCSD) brings together the majority of Australia's coal research skill base as well as experts in sustainability. The vision of CCSD is to optimize the contribution of coal to a sustainable future, and its research is underpinned by a focus on the three dimensions of sustainability – economic, social and environmental. (www.ccsd.biz)

Copper Development Centre is the peak body for the copper industry in Australia, representing some of the country's most influential companies in mining, manufacturing, production, and recycling. (www.copper.com.au)

Eco-Efficiency and Cleaner Production (EECP) site gives examples, case studies, and tools for reducing environmental impacts and saving money across a wide range of industries including mining. (www.environment.gov.au/settlements/industry/corporate/eecp/index.html)

European Copper Institute (ECI) and its members are committed to strengthening and expanding copper usage in Europe. On behalf of its members, ECI aims to strengthen public awareness of copper's value to society and its role in the environment, based on scientific research. (www.eurocopper.org)

Euromines is a federation whose membership includes many of the larger European mining companies and several national European mining associations. Its interests and activities go beyond the boundaries of the EU, though its primary focus is on European issues. (www.euromines.org)

Council for Responsible Jewellery Practices (CRJP) was founded in May 2005 with 14 members from a cross section of the diamond and gold jewellery supply chain, from mine to retail. Council members believe that a coordinated worldwide approach to addressing ethical, social and environmental challenges will drive continuous improvement throughout the jewellery industry to the benefit of stakeholders. (www.responsiblejewellery.com)

The **Equator Principles** have been a huge step forward for the private financial sector, in terms of having a common framework for assessing and managing environmental and social risk in project financing based on an external and respected benchmark, namely the World Bank and IFC sector-specific pollution prevention and abatement guidelines, and IFC safeguard policies (www. equator-principles.com).

Extractive Industries Transparency Initiative announced by the previous UK Prime Minister Tony Blair at the World Summit on Sustainable Development in Johannesburg, September 2002, aims to increase transparency in transactions between governments and companies within extractive industries. (www.eitransparency.org)

Global Mining Initiative (GMI) is a program by the world's leading mining and metals companies to develop their industry's role in the transition to sustainable development. (www.icmm.com/gmi.php)

Global Reporting Initiative (GRI) is a multi-stakeholder process and independent institution whose mission is to develop and disseminate globally applicable Sustainability Reporting Guidelines. (www.globalreporting.org)

The **Gold Institute** is an international industry association representing companies that mine and refine gold. The Gold Institute provides a central access point for topics like how gold is produced, or answers to frequently asked questions. (www.goldinstitute.org/news/pr2may00.html)

Green LeadTM Project is the use of best practices in all aspects of mining, transport, manufacture, use and reuse of lead in order to minimize people and planet exposure to lead. The concept is based on taking a 'whole of life cycle' approach to lead and its impacts on people and the environment and to analyze all of them. (www.greenlead.com)

Industrial Minerals Association (IMA) provides in-depth coverage of several industrial minerals (namely silica, kaolin, feldspar, perlite, bentonite, borates, talc and calcium carbonate) in Europe. (www.ima-eu.org)

InfoMine Inc. provides access to a fully integrated source of worldwide mining and mineral exploration information. (www.infomine.com)

International Aluminium Institute (IAI) is the global forum of aluminium producers dedicated to the development and wider use of aluminium as a competitive and uniquely valuable material. The IAI in all its activities supports the concept that aluminium is a material that lends itself to improving world living standards and developing a better and sustainable world environment. (www.world-aluminium.org)

International Association for Impact Assessment (IAIA) is a forum for advancing innovation, development, and communication of best practice in impact assessment. The international membership promotes development of local and global capacity for the application of environmental, social, health and other forms of assessment in which sound science and full public participation provide a foundation for equitable and sustainable development (www.iaia.org).

International Commission on Large Dams (ICOLD) promotes improvement in the design, construction operation and maintenance of large dams including tailings containments. The website is available in both French and English. ICOLD has published a large number of technical bulletins. A catalogue is available at the website. (www.icold-cigb.org)

International Copper Association (ICA) is the leading organization for promoting the use of copper worldwide. The Association's 37 member companies represent about 80 per cent of the world's refined copper output and are among the largest copper producers, copper alloy fabricators, and wire and cable companies in the world. (www.copperinfo.com)

International Copper Study Group (ICSG) is an intergovernmental organization that serves to increase copper market transparency and promote international discussions and cooperation on issues related to copper. (www.icsg.org)

International Council on Mining and Metals (ICMM) offers strategic industry leadership towards achieving continuous improvements in sustainable development performance in the mining, minerals and metals industry. ICMM provides a common platform for the industry to share challenges and responsibilities as well as to engage key constituencies on issues of common concern at the international level, based on science and principles of sustainable development. (www.icmm.com)

International Cyanide Management Code of 2005 is a voluntary initiative for the gold mining industry and the producers and transporters of the cyanide used in gold mining. The Code focuses exclusively on the safe management of cyanide that is produced, transported and used for the recovery of gold, and on de-toxification of mill tailings and leach solutions. It includes requirements related to financial assurance, accident prevention, emergency response, training, public reporting, stakeholder involvement and verification procedures. (www.icsg.org)

The **International Finance Cooperation** (IFC) is the private sector arm of the World Bank providing loans, equity, structured finance and risk management products, and advisory services to build the private sector in developing countries. (www.ifc.org)

International Institute for Environment and Development (IIED) is an independent, nonprofit organization promoting sustainable patterns of world development through collaborative research, policy studies, networking and knowledge dissemination. We work to address global issues such as mining, the paper industry and food systems. Its website has links to MMSD sites in Australia, North and South America and southern Africa as well as other resources. (www.iied.org)

International Lead and Zinc Research Organization (ILZRO) was formed in 1958 as a non-profit research foundation. Silver was added to its core group of research metals in 2002 with the launching of the Silver Research Consortium. ILZRO's sponsors include most of the major producers of lead, zinc and silver and significant numbers of end-users of these metals from among the steel, automotive, die casting, battery, galvanizing and other industries. (www.ilzro.org)

International Lead and Zinc Study Group (ILZSG) is an intergovernmental organization that regularly brings together 28 member countries in an international forum to exchange information on lead and zinc. (www.ilzsg.org)

International Lead Management Centre (ILMC) was founded by the lead producing industry and works cooperatively with the lead products applications sectors. Its expertise and advice is available across lead production, applications, recycling and disposal. It is also responsible for working with governments, industries and the international community to manage the risk of lead exposure. (www.ilmc.org)

International Network for Acid Prevention (INAP) is an industry-based initiative that aims to globally coordinate research and development into the management of sulphide mine wastes. (www.inap.com.au)

International Nickel Study Group is an autonomous, intergovernmental organization with nickel producing, consuming and trading countries as members. The objectives of the Group are: (1) to collect and publish improved statistics on nickel markets (including production, consumption, trade, stocks, prices and other statistics such as recycling); (2) to publish other information on nickel, such as data on industry facilities and environmental regulations; and (3) to provide a forum for discussions on nickel issues of interest to nickel producing and consuming countries and their industries, including environmental issues. (www.insg.org)

The **International Union for Conservation of Nature and Natural Resources** (IUCN) lists a total of more than 15,000 endangered and threatened species, including nearly one-quarter of all known bird varieties. (www.iucn.org)

Lead Development Association International is dedicated to encouraging the responsible use of lead and its compounds. (www.ldaint.org)

Martha Mine website has range of useful articles on topics including cyanide and acid mine drainage. (www.marthamine.co.nz)

Millennium Ecosystem Assessment (MEA) is an international work program designed to meet the needs of decision makers and the public for scientific information concerning the consequences of ecosystem change for human wellbeing and options for responding to those changes. (www.millenniumassessment.org)

Mine site Drainage Assessment Group (MDAG) provides a range of useful information about controlling acid drainage. There are a number of publications, including case studies that can be downloaded free of charge from (www.mdag.com)

Mineral Resources Forum (MRF) is an Internet framework on the theme of minerals, metals and sustainable development. The United Nations Environment Program, Division of Technology, Industry and Economics (UNEP DTIE) is a principal partner in the Environment Section of MRF. The website provides information and forums on strategies the mining industry can use to protect the environment. (www.mineralresourcesforum.org)

Minerals Council of Australia (MCA) represents Australia's exploration, mining and minerals processing industry, nationally and internationally, in its contribution to sustainable development and society. It aims to promote the development of a safe, profitable and environmentally responsible minerals industry that is internationally competitive and attuned to community expectations. (www.minerals.org.au)

Mining Minerals and Sustainable Development (MMSD) was an independent two-year project of participatory analysis seeking to understand how the mining and minerals sector can best contribute to the global transition to sustainable development. MMSD was managed by the International Institute for Environment and Development (IIED) in London, UK, under contract to the World Business Council for Sustainable Development (WBCSD). The project was initiated by WBCSD and supported by the Global Mining Initiative (GMI). (www.iied.org/mmsd)

www.mining-technology.com/projects provides its own descriptions of many of the world's larger mining projects, including descriptions of their operations and investments (www.mining-technology.com/projects)

www.mining-journal.com, which provides links direct to the websites of major mining companies (www.mining-journal.com)

www.minesite.com, which also provides links to company sites, but is more geared to gold mining (www.minesite.com).

National Geophysical Data Centre provides extensive information on a range of natural hazards including volcanoes, earthquakes, tsunami, and meteorology. (www.ngdc.noaa.gov)

Nickel Producers Environmental Research Association provides information on the use and properties of nickel and the safe use of nickel in the workplace. (www.nipera.org)

Responsible Gold, Northwest Mining Association, 2005 maintains an interactive site that links amongst others to the National Mining Association's site titled, U.S. Laws and Regulations Governing gold mining on Private and Federal Lands that outlines much of the permitting process that a new mine performs. (www.responsiblegold.org)

The **Shuttle Radar Topography Mission** (SRTM) obtained elevation data on a near-global scale to generate the most complete high-resolution digital topographic database of Earth. SRTM consisted of a specially modified radar system that flew onboard the Space Shuttle Endeavour during an 11-day mission in February of 2000. Topographic data are made available at (www.jpl.nasa.gov/srtm).

Sulphide Solutions aims to increase general knowledge in industry about sulphuric materials and their effective management. (www.ansto.gov.au/sulphide)

South African Chamber of Mines is a prominent industry employers' organization that exists to serve its members and promote their interests in the South African mining industry. (www.bullion.org.za)

UNEP Production and Consumption Unit Environmental Management website provides a number of useful Strategies and Tools for cleaner production, sustainable consumption, APELL, and tools. **APELL** is a strategic program that provides detailed guidelines for the development and implementation of highly effective, integrated, and well-practiced emergency response plans in local communities. The website is part of the Division of Technology, Industry and Economics website. It also has many useful links and a section targeted at the mining industry. (www.uneptie.org)

United Nations Environment Program – The UNEP World Conservation Monitoring Centre provides information for policy and action to conserve the living world. Programs concentrate on species, forests, protected areas, marine, mountains and fresh waters, plus habitats affected by climate change such as polar regions. The relationship between trade and the environment and the wider aspects of biodiversity assessment are also addressed. (www.unep-wcmc.org)

United Nations Convention on the Law of the Sea (UNCLOS) in 1982 is a comprehensive framework for regulating the use, development and preservation of the vast marine areas, including mining and other mineral development in the ocean. (www.un.org/Depts/los/convention_agreements/convention_overview_convention.htm)

United Nations Framework Convention on Climate Change in 1992 and the 1997 Kyoto Protocol provide a comprehensive approach for controlling greenhouse gases (GHG); those chemicals which form a heat-trapping layer in the upper atmosphere and contribute to global warming, chiefly carbon dioxide and methane. (www.unfccc.de)

U.S. Geological Survey (USGS) provides a wealth of mining relevant information including statistics and information on the worldwide supply, demand, and flow of minerals and materials. The Minerals Yearbook: Volume I – Metals and Minerals contains annual statistical data and information on approximately 90 commodities. (www.usgs.gov)

USGS Earthquake Hazards Program provides general information, links, advice on reducing hazards and research. (earthquake.usgs.gov)

World Information Service on Energy (WISE) **Uranium Project** includes information on the safety of tailings dams, current issues, and properties of tailings dams. (www.wise-uranium.org/index.html)

World Bank Operational Directive on Involuntary Resettlement – The World Bank Operational Directive on Involuntary Resettlement states that project planning must avoid and minimize involuntary resettlement and that, if people lose their homes or livelihoods as a result of Bank-financed projects, they should have their standard of living improved or at least restored. (www.ifc.org/ifcext/enviro.nsf/Content/ESRP/$FILE/OD430_Involuntary Resettlement.pdf)

World Business Council for Sustainable Development (WBCSD) is a coalition of 180 international companies united by a shared commitment to sustainable development. The organization pursues this goal via the three pillars of economic growth, environmental protection and social equity. (www.wbcsd.org)

World Coal Institute (WCI) is a non-profit, non-governmental association of coal enterprises and associations, the only international body working on a worldwide basis on behalf of the coal industry. (www.worldcoal.org)

World Conservation Union (ICUN) is the world's largest conservation network bringing together government agencies, non-government organizations, scientists and experts from 181 countries in a worldwide partnership. The Union's mission is to influence, encourage and assist societies throughout the world to conserve the integrity and diversity of nature and to ensure that any use of natural resources is equitable and ecologically sustainable. (www.iucn.org)

World Health Organization is the United Nations' specialized agency for health. Established in 1948, its objective is the attainment by all peoples of the highest possible level of health. Health is defined in WHO's Constitution as a state of complete physical, mental and social wellbeing and not merely the absence of disease or infirmity. (www.who.int)

World Nuclear Association (WNA) is the global organization that seeks to promote the peaceful worldwide use of nuclear power as a sustainable energy resource for the coming centuries. Specifically, the WNA is concerned with nuclear power generation and all aspects of the nuclear fuel cycle, including mining, conversion, enrichment, fuel fabrication, plant manufacture, transport, and the safe disposition of spent fuel. (www.world-nuclear.org)

Appendix 1.2 Some Mines of Interest

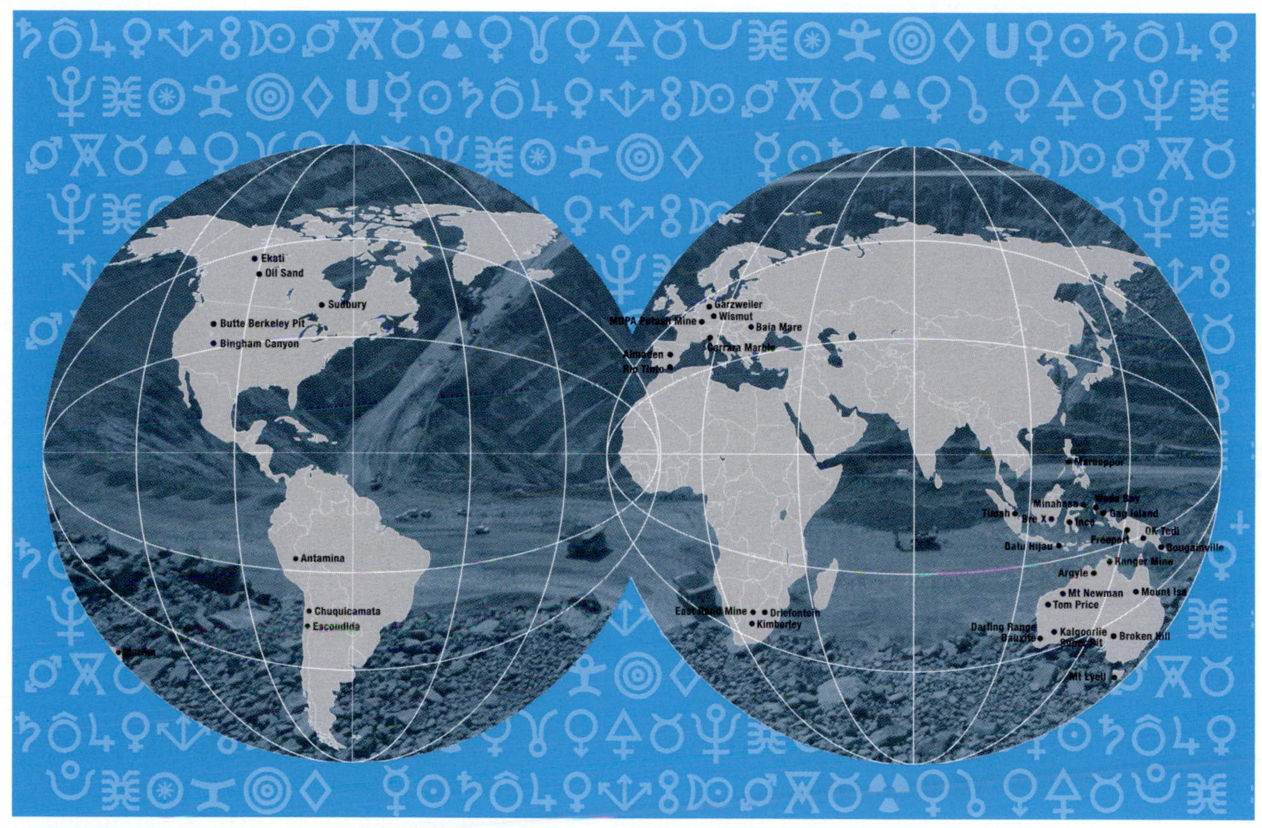

FIGURE A1.2.1
Some mines of interest

ANTAMINA

The Antamina Mine is located in the Antamina valley in the Andes Mountains, north-central Peru. The Antamina deposit is the largest known copper-zinc skarn in the world with zinc, silver, lead, molybdenum and bismuth mineralization. The US$ 2.2 billion construction program was one of the largest new mine development projects ever undertaken. In order to reach the mineral deposits the tops of several mountains had to be removed and a lagoon drained. The mine commenced operation in late 2001 with an estimated mine life of more than 20 years. This project alone increased Peru's gross domestic product by nearly 1 percent and increased Peruvian mining production by 30 percent in 2002.

ALMADEN

The mercury deposits at Almaden, about 200 km south of Madrid, account for the largest mercury production in the world. Its total production from historical times to date totals to about 250,000 metric tons, about one third of the total quantity of mercury produced in recorded history. Almaden is the only known deposit to contain as much as 30 percent of the world's reserves of a single mineral.

ARGYLE

The Argyle Diamond Mine, discovered in 1979 and now the world largest diamond producing mine, is located in the Kimberley Region of Western Australia. It produces over one third of the world's diamonds every year.

BAIA MARE

The tailings storage facility at Baia Mare, Hungary, designed for 'zero discharge', became one of the most publicized tailings spill events to date. In early 2000, a section of the tailings embankment overtopped, washing out a breach 25 m long. An estimated 100,000 $m^3$of mud and wastewater with a 126 mg/liter cyanide load entered into the Lapus River, a tributary to the Szamos River and from there into the Tisza River and the Danube upstream of Belgrade, before finally entering the Black Sea. The Hungarian authorities estimated the total fish kill to have been in excess of one thousand tons.

BATU HIJAU

Mining of this major porphyry copper ore body commenced in 1999. The mine is located in a remote, mountainous area in southwest Sumbawa, Indonesia. Development included construction of major infrastructure including a port and coal-fired power station at Benete Bay. The operations utilize Deep Sea Tailings Placement with tailings depositing on the seabed at depths around 4 km. Also notable is the use of sea water for the process which produces a copper concentrate.

BINGHAM CANYON

The Bingham Canyon mine, near Salt Lake City in the USA which has operated since 1906, is considered to be the world largest excavation. The mine is 4 km in diameter. From bottom to top the mine reaches close to 1,000 vertical meters and, like other large open Pit mines, it has its own climate zones. Operated by the Kennecott Utah Copper Corporation, the mine has produced more than 17 million tons of copper from about 5 billion tons of copper ore (worth around $ 120 billion at today's copper price), and 23 million ounces of gold (worth around $ 21 billion at today's gold price).

BOUGAINVILLE

The indefinite closure of the Bougainville Copper Mine on Bougainville Island, PNG, in 1989 is likely the most dramatic event illustrating the complex forces that influence mine developments in developing countries. The mine started operation in 1972, but violent civil unrest lead to loss of lives and the losses of hundreds of millions of dollars. The abrupt closure of this mine represented a watershed for the industry and has led to the inclusion of social issues among the most important considerations for new projects in developing countries.

Bre-X

Based on reported results of exploration, the Bre-X gold deposit in the head waters of the Busang River in Kalimantan was poised to become the biggest deposit of gold ever discovered. By March 1997 the Canadian company Bre-X Minerals Ltd. reported an estimated 200 million-ounce resource (worth about US$ 100 billion assuming a gold price of US$ 500 per ounce). The deposit immediately attracted the interest of the then President Soeharto of Indonesia plus several large mining companies. The mining company Freeport started its own due-diligence drilling, and on March 26, 1997 Freeport reported that its due-diligence cores, drilled only a meter and a half from Bre-X's, showed 'insignificant amounts of gold.' The next day Bre-X stock lost almost all its value. Industry journalists soon found evidence that the Busang ore samples had been 'salted' with gold dust.

BROKEN HILL

As the era of mining operations at Broken Hill draws to an inevitable close, the name of BHP based on the initials of the mining district, one of the largest mining houses, will be a lasting reminder of the riches it has yielded. The initial discovery of the richly mineralized site occurred in 1883 and, before mining began, Broken Hill was one of the world largest lead-zinc-silver ore bodies. However there is more than mining to Broken Hill: about 380 mineral species are either confirmed, or considered to have occurred in the Broken Hill ore body, making the site one of the top five sources of mineral species in the world.

BUTTE, MONTANA BERKELY PIT

In 1955, Butte's first large-scale open-Pit mine, the Berkeley pit, was constructed. Mining was discontinued in 1982 after the extraction of 1.5 billion tons of copper ore. As soon as the mine dewatering pumps were shut down acidic groundwater with pH less than 3 started to fill the pit in what now has become the Berkeley pit Lake. The water is also toxic because large quantities of heavy metals have leached into the pit lake. For now, the Berkeley pit remains the Environmental Protection Agency's largest Superfund site in the US. That means that the US Government has deemed the Berkeley pit the number one environmental hazard facing the USA today. Many promising new technologies are being trialled at this site.

CARRARA MARBLE

The Carrara marble quarries located in northern Tuscany, have been in operation for more than 2,000 years. Carrara marble is famous for its statuary quality and has been used by most famous stone artists such as Michelangelo, Donatello, Jacopo Della Quercia and Canova. The name Carrara has become synonymous with marble. Carrera marble continues to be requested worldwide for buildings, statues, or objects, and about one million tons of marble is quarried each year.

CHUQUICAMATA (CHUQUI)

The Chuquicamata mine is the world's largest open pit copper mine. The mine is elliptical in form with a surface footprint of almost 8 km^2 and a depth close to 1,000 m. The area has been exploited since pre-Hispanic times, and commercial production in contemporary history commenced in 1915.

DARLING RANGE BAUXITE

Alcoa of Australia mines bauxite from three separate areas and Worsley Alumina has another mining operation, all within the Darling Range of Western Australia. The bauxite is processed in four alumina refineries: Alcoa's Kwinana, Pinjarra and Wagerup refineries and the Worsley refinery. Apart from its bauxite resources, the Darling Range is noted for its Jarrah Forest vegetation. Jarrah, a relatively slow growing eucalypt species, produces high quality timber. The diverse forest vegetation has been threatened by spread of Jarrah Dieback disease caused by a fungus (*Phytophthera cinamomi*). Managing operations to avoid spreading this disease is a major issue for these mining companies. Since the early 1970s, Alcoa has had an active mine site rehabilitation program, and has pioneered many of the techniques that have now become standard practice in the industry.

DRIEFONTEIN

This deep underground mine is located 60 km from Johannesburg in South Africa. It produces more than 1.2 million ounces of gold each year from mines which are more than 3 km in depth.

EAST RAND MINE

Vertical shaft mines hold the record for being the deepest mines in the world. Most are located in South Africa based on an abundance of diamond and gold deposits. As of 2003, the world's deepest mine was the East Rand mine at 3,585 meters, but as technology improves and the search for natural resources continues, many mines are constantly being deepened. In the next few years, the mines may well reach 5 km.

EKATI

'EKATI is a magical and unique place. No other mine in Canada is like it. No other mine in the world is like it' (BHPB's EKATI Diamond Mine Backgrounder). Located 200 kilometers south of the Arctic Circle in the Northwest Territories, Canada, and 100 kilometers north of the treeline, in an area of continuous permafrost with winter temperature of −50°C and below, mine development and operation is at the cutting edge of mining engineering.

ESCONDIDA

The large Escondida open Pit mine located in Chile's Atacama Desert, is currently the world's largest producer of copper and a major producer of gold and silver. It commenced production in 1990, and in 2007 produced nearly 1.5 million tonnes of copper.

FREEPORT

PT Freeport Indonesia (PTFI) operates perhaps the world's most impressive mining complex in the highlands of Papua, Indonesia. The Grasberg and other nearby ore bodies include the world's largest gold reserve and second- largest copper reserve. The mine is located high in the mountains close to one of the few remaining equatorial glaciers. Mine production exceeds 200,000 tons per day, and the mine generated more than two thirds of the GDP of the Province of Papua in 2006. As the Grasberg open pit completes mining in 2015 underground mining will continue producing ore until 2041. However, the mine is also one of the few mines in the world that relies on riverine tailings disposal, cause of continued criticism by environmentalists.

GARZWEILER

The lignite-mining region in the Rhineland covers an area of some 2,500 km^2 to the west of Cologne. Four large opencast mines in the district – Hambach, Garzweiler, Inden and Bergheim – produced between them close to 100 Mt of lignite in 2006. Reserves of surface-mineable lignite are estimated to be 35,000 Mt, of which 4,600 Mt are contained in current areas planned for mining. Garzweiler alone contains 1,300 Mt. Mechanized mining began in the 1890s and the first bucket-wheel excavator was commissioned in 1933. The bucket-wheel excavator remains central to the overburden stripping and lignite mining. The seven largest weigh some 13,000 t making them the largest land vehicles ever built.

KALGOORLIE SUPER PIT

The Super Pit at Boulder, near Kalgoorlie in Western Australia, measuring 3.8 km in length, 1.4 km in width and about 500 m in depth, is Australia's largest open pit gold mine. The gold field was discovered in 1893 and has produced about 1,550 t of gold since then.

KIMBERLEY

In 1871 the discovery of one 83.50 carat diamond on the slopes of the Colesberg Kopje hill in South Africa led to the first diamond rush in the area. As miners arrived in their thousands, the hill disappeared, and became known as the Big Hole. When miners started to work the Kimberley diamond pipe, nobody knew how deep it would go. When mining ceased the Big Hole was the world's biggest mine excavated by hand with a depth of about 800 meters. The Big Hole produced more than 14 million carats of diamonds. Today the mine is gone, but the mine town Kimberley which it created, continues to flourish.

MARCOPPER

In 1996 a major tailings release occurred at the Marcopper Mine on Marinduque Island in the Philippines. The bulk of tailings discharged in the first 3 days, flowing down as a mud flow for most of the 26 km long river to the sea. The incident resulted in closure of the operations, triggered legal proceedings by the Philippine Government, and continues to influence permitting of new mines in the Philippines to the current day.

MARTHA MINE

The Martha Gold Project was the first major hard rock mining operation to be commissioned following the resurgence of the gold mining industry in New Zealand in the late 1970s. The Martha Mine pit is situated right in the town of Waihi. The processing plant and mine waste disposal area have been placed two kilometers away from the pit at the outskirt of the township. The mine serves as a prime example that one can relocate in mine design mine infrastructure but not the ore body. Because of its location, the mining operation has received a high level of public scrutiny, reflected by stringent conditions set for the project.

MINAHASA

Newmont's Minahasa Gold Mine located in North Sulawesi completed its operations in 2004 and was small compared to other gold mines in Indonesia or elsewhere. However, Minahasa was the first mine in Indonesia to dispose of its tailings using submarine tailings placement. The tailings were placed 1 km offshore from Buyat Bay, more than 80 meters below the sea surface. When the mine closed, anti-mining advocates and the then Minister of Environment claimed that Buyat Bay had been polluted by tailings alleging human health problems due to mercury and arsenic contamination. After a lengthy legal battle Newmont was eventually acquitted of all wrong doing. The ill-founded accusations, however, discouraged not only future mining investment, but foreign investment in Indonesia as a whole.

MINES DE POTASSE D'ALSACE

Mines de Potasse d'Alsace (MDPA) has operated the Amelie underground potash mine and processing plant and the Marie-Louise underground mine near Mulhouse in northeast

FIGURE A1.2.2

The Ranger Uranium Mine, near Jabiru, East of Darwin, Australia
Photo credit: Stephen Codrington

France since 1910. The ore deposit is approaching exhaustion, and planning and preparation for this eventuality has been underway for years. The major environmental issue revolves around the waste piles dumped adjacent to the beneficiation plants prior to 1934. These were placed without liners or other containment features. Subsequent dissolution of the salt by rain has contaminated the underlying aquifer giving rise to plumes of salt contamination. The affected groundwater system of the upper Rhine valley is one of the largest sources of drinking water in Europe.

Mt ISA

The Mt Isa Mines produce copper and silver/lead/zinc from separate ore bodies. Located within a metals rich province in Northern Queensland, these major underground mines have been producing since 1931. Built in 1978 with a stack height of 270 m, the Mount Isa lead smelter in Inland Queensland, Australia operates the tallest flue gas stack of any pyrometallurgical operation on Earth. Up to Year 2000, the smelter was operating without a desulphurization unit dispersing significant amounts of SO_2 through the stack. Depending on the prevailing wind direction it was possible to detect the plume in Perth some 2,000 km away.

Mt LYELL

The Mt Lyell Mine, located in western Tasmania has been producing copper since 1896, and has played a major role in development of this rugged, mountainous region. The barren landscape in the vicinity of the mine, due to timer cutting and the effects of sulphur dioxide emissions during the older mining periods, contrasts markedly with the lush vegetation of surrounding areas. For many decades, tailings were discharged to the Queen River, and thence via the King River to Macquarie Harbour where they deposited to form a delta. Acid mine drainage from underground workings has also contributed to contamination of the two rivers and portion of the harbour. Plans have been developed for remediation of the contaminated areas, commencing with treatment of the acid mine drainage.

Mt NEWMAN

Located in the eastern Pilbara region of Western Australia, Mt Newman is one of the world's largest iron ore projects. There are several mines involved, by far the largest being at Mt Whaleback, where the elongated open pit, 5.5 km in length, produces more than 30 million tonnes of iron ore each year. The ore is transported by rail to Port Hedland.

CANADA'S OIL SAND

Nothing demonstrates the blurred boundary between fossil fuels and minerals extraction better than oil sand mining in Canada, home of large oil sand deposits. The vast majority of Canada's oil sand deposits are located in Alberta, and underlie an area larger than Florida. Canada's oil sands represent the second largest hydrocarbon resource on Earth, second only to the vast oil resources in Saudi Arabia.

OK TEDI

At one time the Ok Tedi gold and copper mine which began operating in 1984 provided close to half of the foreign export earnings of Papua New Guinea. Back when the mine first opened a tailings dam had been under construction, but it was destroyed by a landslide. The government subsequently issued a permit to dispose of tailings into the Ok Tedi/Fly River system. Since its beginning the riverine tailings disposal system has attracted world-wide negative publicity as a result of large scale environmental damage in downstream areas. Eventually the mine operator was forced to set aside some US$ 100 million to rehabilitate the affected river system and to compensate affected communities.

RANGER MINE

The Ranger Mine, which has been producing uranium in Northern Australia since 1981, is perhaps the most regulated mining operation in the world. Three major reasons for the level of regulation exist: (1) The mine product, uranium oxide, is classed as a 'prescribed substance'; (2) the mine is surrounded by Kakadu National Park which is listed on the World Heritage register; and (3) the mine is on Aboriginal land.

RIO TINTO

The Rio Tinto ('Coloured River') in Spain, characterized by deep red water that is highly acidic and rich in heavy metals, holds a significant role in history as the birthplace of the Copper Age and Bronze Age. The first Rio Tinto mines were developed in 3000 BC. The Romans made some of the first coins from Rio Tinto's silver and gold. The mines were rediscovered in 1556 and reopened in 1724, only to be sold to the British in 1871, leading to the birth of the British mining giant Rio Tinto. Today the mines are one of the most important sources of copper and sulphur in the world.

SUDBURY

The Sudbury Basin in Northern Canada is a unique geological formation measuring about 60 by 27 kilometers in an oval or elliptical shape. The foundation of the basin is about 10 kilometers deep. Nickel mineralization was first detected in 1856, and in spite of its remote location, the region prospered into an important regional economic center in Northern Ontario. It is home to two of the largest nickel producers in the world, INCO and Falconbridge, and continues to be a leader in high productivity and environmentally sound mining and mineral processing technologies.

TIMAH TIN MINE

Bangka and Belitung Islands in Indonesia are responsible for 85 percent of the tin produced by PT Timah , the world's largest integrated tin producer. Tin is found in placer deposits on-land as well as in the sea surrounding them. Timah operates a fleet of dredges to recover tin from the near coast seabed. Not all tin mining at these islands is legal. By some estimates the number of illegal miners on Bangka Island alone has spiraled to about 30,000 by 2006. Much of the illegal tin ore is believed to be shipped to neighboring Singapore before being sold to Malaysian and Thai smelters.

TOM PRICE

Iron ore was discovered in the Pilbara region of Western Australia in the 1800s, but despite geologists claiming there was enough to supply the whole world, the remoteness and ruggedness of the country stopped people from initially exploiting the deposits. The historic moment in the discovery of one of the world's richest iron ore bodies came in 1962, when two CRA geologists found a large dark outcrop which extended about 6.5 kilometers, later to be known as Mt Tom Price, in tribute to an American engineer who was instrumental in development of the region.

WISMUT

Immediately after the end of World War II, the Soviets started exploration and mining of uranium in the historic mining provinces in the Ore Mountains, Germany. Subsequently, the Wismut Company developed the third-largest uranium mining province in the world. With the political changes in 1989, it was revealed that uranium mining in Eastern Germany had devastated large areas. Uranium production was terminated but huge shut-down uranium mines, hundreds of millions of tons of radiating waste rock and uranium mill tailings remained. This environmental legacy presents an immediate hazard, but also potentially endangers future generations for tens of thousands of years. The German government estimates that clean-up will amount to more than US $ 9 billion.

2

Environmental Impact Assessment

Protection Before Exploitation

♁ COPPER

Copper, along with gold the first metal known to man-kind, was a highly significant metal in ancient civilizations as the first tools, implements and weapons were made from copper. Of all the materials mined, copper is the most versatile and durable – it appears everywhere in our everyday lives. This miraculous metal has a number of unique properties: beside being nonmagnetic, copper is conductive, ductile, malleable, resistant, and biostatic. Copper's durability is legendary. A copper pipe installed 5,400 years ago to carry water to the pyramid of the Egyptian pharaoh Cheops is still operational today. For good reason, copper has been called 'man's eternal metal.'

2 Environmental Impact Assessment

Protection Before Exploitation

Environmental assessment is a planning, decision-making, and management process spanning the entire life-cycle of a mining project. The Environmental Impact Assessment (EIA) study is only one part of the process, usually applied in the feasibility and design stage, or sometimes even in the conceptual stage of mine development. The aim of an EIA is to ensure that potentially adverse environmental impacts are foreseen and addressed at an early stage in the planning cycle so as to pre-empt negative environmental or social issues during mine construction and operation. An equally important aim is to ensure that consequent project opportunities are maximized and project benefits are justly distributed. In brief, the EIA aims to eliminate the negative and to accentuate the positive project impacts.

The EIA aims to eliminate the negative and to accentuate the positive project impacts.

The success of environmental assessment depends on the implementation of agreed-upon environmental and social mitigation measures during mine construction and operation. A lack of continuity between the EIA study and subsequent environmental management practices at the mine operation reduces the environmental impact assessment to a mere permitting exercise designed to obtain regulatory project clearance and/or project financing.

The following chapter illustrates how environmental assessment integrates with the mine life-cycle, and then outlines the purpose and nature of an EIA and its implementation. These topics are dealt with briefly since there are numerous texts on EIA available for those requiring in-depth information. As interest in impact assessment has grown, the volume of impact assessment guidelines has increased. Most OECD countries, and a growing number of developing countries, have prepared guidelines for assessing project and development impacts. In addition, as impact assessment processes have evolved and become increasingly complex, the number of agencies involved in impact assessment within countries has increased, with many of these producing their own guidelines. Almost all bilateral, multilateral, and United Nations agencies have also prepared internal guidelines applicable when they become stakeholders, or providing reference points for mining project developments. For a comprehensive directory of such guidelines see Donnelly *et al.* (1998) and the many useful website links, some of which are listed in the attachment to Chapter One.

Chapter Two poses and answers the following questions:

- How does environmental assessment link with the mining cycle?
- How should environmental impact assessment be managed?
- What are common themes and principles in environmental impact assessment?
- When is an environmental impact assessment required?
- How is an environmental impact assessment done?
- How is an environmental impact assessment documented and distributed?
- How to obtain government approval?
- What are the costs of delay?
- How to incorporate mine closure into an environmental impact assessment?
- What mistakes are commonly made?

Subsequent chapters address these subjects in greater detail.

EIA has now been in use for nearly 40 years, which is sufficient time to appreciate the weaknesses and strengths associated with the process. Experience demonstrates that the role of EIA is more restricted than many environmental practitioners seem willing to admit. There is some resistance to change, on the perfectly rational grounds that it has taken many years for environmentalists to persuade both governments and companies alike to embrace EIA as a planning tool. But to continue with a focus on a limited method, even for good reasons, when there are better ways to do things, seems to be somewhat ostrich-like and, in the long run, counter-productive. The expectation is that the 2006 IFC Performance Standard 1 will change the common approach to environmental assessment. The fundamental question is now: what are the best ways of managing environmental quality throughout the mine life while at the same time encouraging sustainable development of natural resources? This question eventually leads to the conclusion that environmental protection needs to be integrated with regional land-use and development planning, one of the subjects addressed in the concluding chapter of this book.

What are the best ways of managing environmental quality throughout the mine life while at the same time encouraging sustainable development of natural resources?

2.1 MINING COMPANIES ARE NOT GOVERNMENTS

In principle, the responsibilities of mining companies towards the major stakeholders in a mining project are relatively clear. Companies are required to (1) maximize shareholder value; (2) respect human rights and the rights of host communities; (3) conform with all legal requirements stipulated in contractual agreements with government such as those between the mining company and the host country; and (4) obey all other applicable laws and regulations.

In many developing countries the contractual agreement between the mining company and the host country assigns significant responsibilities to the investor including responsibility for capacity building of host communities. This is because mining investment policies try to merge the apparently diverging interests of the various stakeholders (mining company, communities, and government) into one interest: sustainable social transformation as the goal of sustainable human development. It is common that a set of mining policies are in place to support this objective. These mining policies may include policies to address minerals conservation, economic feasibility, employment, community development, taxes and royalties, and industrial development. It follows that the responsibilities of mining companies are multiple, complex and varied; maximizing shareholder value is only one, and this is more than just a technical challenge.

Responsibilities do not include acting as or in place of governments.

It is equally important to establish what is not the responsibility of a mining company. Responsibilities do not include acting as or in place of governments, or taking on government responsibilities towards the development of isolated and underdeveloped regions. Admittedly, mining companies often find it practical and even necessary to undertake actions that should be the responsibility of the local government. In such cases all efforts should be directed towards working these issues through with the local government, and eventually assisting the local government in delivering essential infrastructure and certain community services. Providing such services directly creates a dependence on the mining company that is not sustainable.

2.2 ENVIRONMENTAL ASSESSMENT IN THE MINING CYCLE

Environmental assessment should be conducted as early as possible in the mining cycle to allow environmental mitigation measures to be incorporated into mine planning.

Environmental assessment should be conducted as early as possible in the mining cycle to allow environmental mitigation measures to be incorporated into mine planning. A timely and efficient environmental assessment will result in informed decision-making that supports sustainable development, the ultimate goal of environmental assessment. It should be recognized, however, that there are drawbacks involved in early assessment. These relate to changes in design that commonly occur in the detailed design stage that follows the Feasibility Study. These changes may be sufficiently significant to warrant an Addendum to the approved EIA or even a new EIA. An example is shown in **Case 2.1**. Many environmental regulators are unaware that design changes – usually improvements – commonly occur after EIA approval. Consequently, not all countries have established mechanisms for dealing with such changes and for assessing how the changes affect the various environmental impacts and the related environmental management measures.

Figure 2.1 illustrates how environmental assessment evolves in stages closely linked to the life-cycle of a mine. Site selection, environmental screening, initial assessment, and scoping of significant issues are all activities best done at the project concept stage. Of course the location of the ore body is fixed, but everything else is very much up to mine planning.

CASE 2.1
Newmont's Batu Hijau Copper-Gold Project, Sumbawa, Indonesia

Following approval by the Government of Indonesia of the EIA documents for this major project, ongoing metallurgical studies indicated that the froth flotation process for producing the copper-gold concentrate, could operate at least as effectively using sea water as using fresh water. This represented a major project improvement, obviating the need for a large fresh water reservoir that would otherwise have been required. Removal of the planned reservoir also enabled a more efficient layout of other facilities, and removed the need for a sea water mixing tank in the Deep Sea Tailings Disposal System. These changes were proposed and their impacts evaluated in an Addendum to the EIA, which was subsequently approved.

Detailed assessment of significant impacts, identification of mitigation needs, and inputs to cost/benefit analysis are activities typically aligned with the feasibility study stage of a mining project. These activities combine to constitute the environmental impact assessment study, which is presented as a set of documents which may include an Environmental

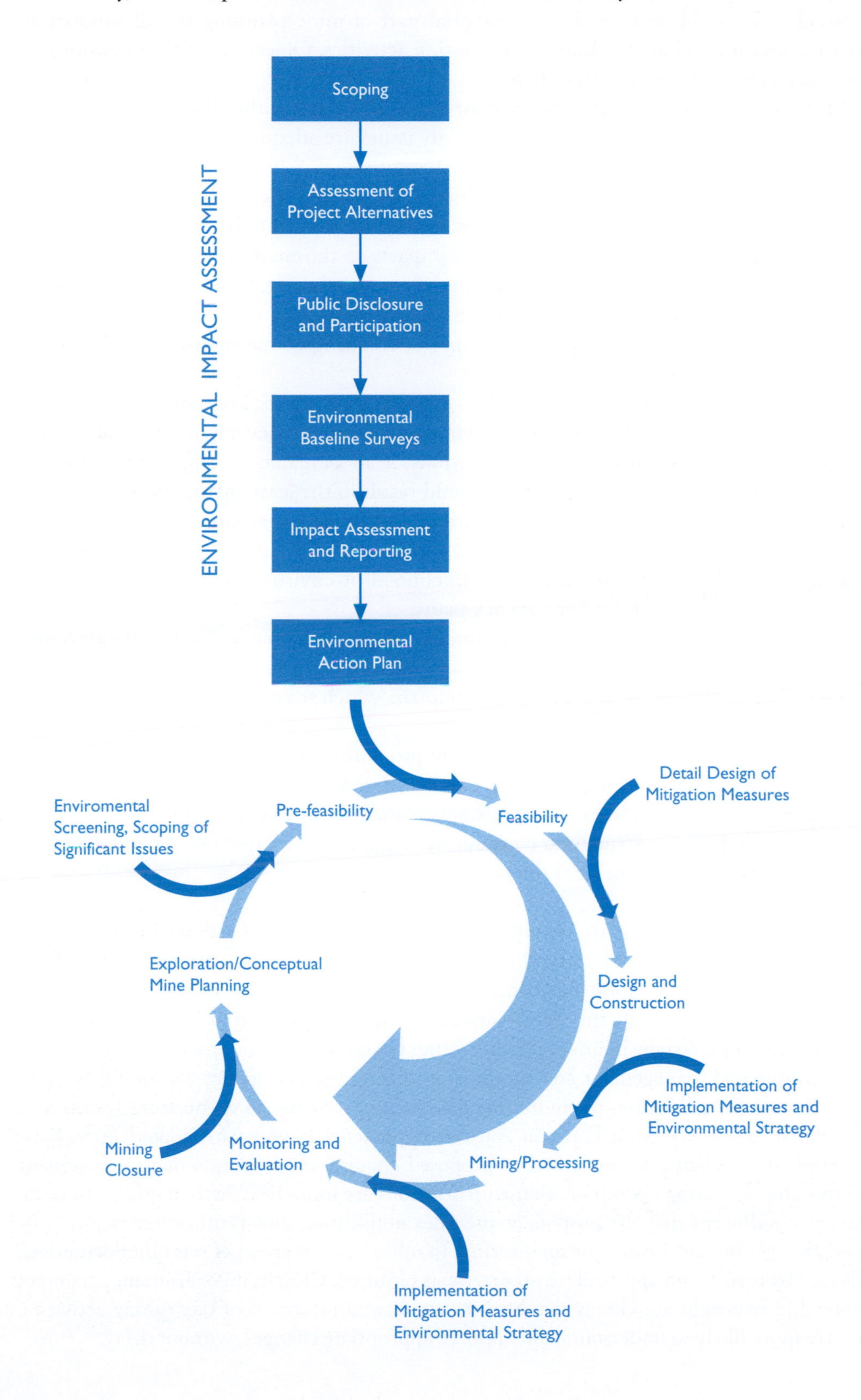

FIGURE 2.1
Links Between Environmental Assessment and Mining Cycle

Environmental assessment is a process that in case of mining starts with an initial environmental scoping and ends with post operation monitoring. The Environmental Impact Assessment study is only one part of the environmental assessment, applied in the feasibility and design stage, or even in the conceptual stage of mine development.

Impact Statement (EIS), an Environmental Action Plan (EAP), and a suite of documents focusing on selected topics such as baseline conditions, land acquisition, resettlement, public consultation, community development, and environmental, health, and safety management. The preparation of these documents is a fundamental part of environmental planning and should be viewed as an essential part of mine planning for all production activities, and for all but very limited exploration activities. Objectives of the environmental impact assessment at this stage are to:

- Identify sensitive components of the existing environment within the mine area and its surroundings, and to ensure that biodiversity issues are adequately addressed;
- Assist mine design and planning by identifying those aspects of location, construction, operation, and decommissioning which may cause environmental and social concerns;
- Enable adverse environmental, economic, and social impacts to be anticipated;
- Develop indicators for measuring mining impacts on the environment;
- Recommend measures during construction, commissioning, and operation to avoid/ameliorate adverse effects and to increase beneficial impacts;
- Provide a forum for developing new ways of thinking about environmental impacts and decision-making;
- Ensure that alternative measures, such as economic incentives, are considered;
- Identify the preferred practicable environmental mine development option. This requires that the environmental implications of all available development options be evaluated. The recommended option should result in the least environmental damage, in balance with other social, health, and economic considerations, and be consistent with prevailing regulations;
- Estimate and describe the nature and likelihood of environmentally damaging incidents to provide a basis for contingency plans;
- Identify existing and expected environmental regulations that will affect the development and advise on standards, consents, and targets;
- Identify any environmental issues and concerns which may, in the future, affect mine development;
- Recommend an environmental management programme for the life of the mine, including compliance, monitoring, auditing, and contingency planning, with focus on implementing environmentally-sensitive business management practice; and
- Provide the basis for structured consultation with and participation of regulatory and non-regulatory authorities and the public.

The completion of the environmental impact assessment study is followed by the implementation of environmental mitigation measures and community development plans. Detailed design of mitigation measures is best undertaken in the detailed engineering phase, in conjunction with the development of the supporting management structure, including defining responsibilities and authorities for environmental management.

Environmental management and monitoring during project operations should be flexible, allowing for adaptation to environmental change, response to monitoring results and operational experience gained.

Environmental management and monitoring during project operations should be flexible, allowing for adaptation to environmental change, response to monitoring results and operational experience gained. Again, various approaches have been adopted by various government jurisdictions, to respond to proposed changes in environmental management and monitoring during operations. Some jurisdictions are more flexible than others. In some cases, it is sufficient that the proponent provides notification and justification of any proposed changes in management or monitoring. In other cases, revision of relevant documents, followed by review and approval (or otherwise) is required. Clearly, if government personnel involved in oversight are closely involved in or are well informed of day-to-day activities, they are more likely to understand and approve appropriate changes, without delay.

Capacity building is central to successful environmental management. It takes time, and requires an open mind. Transparency, openness and respect for interest groups (both those that are directly affected by mining and those that are not, but claim a legitimate interest) are essential to successful environmental management. False or insincere attention to community concerns will be easily detected by the host community, with long-lasting negative consequences on stakeholder relationships. Throughout the life of the mine, ongoing activities include environmental monitoring and auditing (internal and external), rectifying adverse impacts, and understanding lessons learned for future projects.

False or insincere attention to community concerns will be easily detected by the host community, with long-lasting negative consequences on stakeholder relationships.

Planning for mine closure becomes increasingly important as the project approaches the end of its life. However, environmental management does not finish when mining comes to an end. Post-mining environmental management and monitoring may continue for years, well after the last mine product is shipped.

Planning for mine closure becomes increasingly important as the project approaches the end of its life.

In summary, environmental assessment is not only an activity of project preparation, although that is where it starts. Particularly since the advent of the 2006 IFC Performance Standard 1, the focus of environmental assessment is on implementing the Environmental Action Plan. This includes implementing the agreed-upon mitigation and monitoring programme, capacity building, and establishing a management system that can support the implementation of an environmental strategy throughout the mine life.

2.3 MANAGING THE ENVIRONMENTAL IMPACT ASSESSMENT PROCESS

The preparation of an environmental impact assessment is usually the responsibility of the mining company. (In the US, the Lead Agency of the Government has the responsibility for preparing an Environmental Impact Statement, which is normally based on information prepared by the proponent in the form of an EIA). The company owns the EIA documents and the stated environmental commitments. Therefore, although the actual work is often outsourced, the primary responsibility for managing the EIA as well as for preparing the supporting documentation and for implementing management and monitoring recommendations remains with the company.

Over the years, the practice of environmental impact assessment has become more complex and more technically sophisticated. There has also been a tendency towards greater involvement of other parties, the so-called 'stakeholders', in the assessment process. The EIA of large mining projects may be undertaken within a complex multi-jurisdictional legal/regulatory framework; may extend over a number of months or years; often involves a wide range of participants representing donor agencies, governments, proponents, technical experts, affected communities, and NGOs; and incorporates complex scientific data and sophisticated analytical methods. This growing complexity is straining the ability of senior management of mining companies to manage the assessment process effectively. The challenge of managing environmental assessments is now as daunting as the technical complexity of building a mine.

The challenge of managing environmental assessments is now as daunting as the technical complexity of building a mine.

Managing the EIA Process as the Owner

A skilled project manager is the first prerequisite during the environmental permitting stage. It is important for the company to have a cadre of highly qualified and trained professionals assigned to assist in the preparation of the EIA. Good people skills are necessary, especially during the approval period, but traditional management skills are equally important (**Table 2.1**). It is part of a project manager's job to determine resource needs and

TABLE 2.1

Management Tasks of EIA Owner in the Environmental Impact Assessment Process

Task	Description
General	Formulate selection criteria for design consultant and EIA team. Determine roles and responsibilities of involved stakeholders relating to management and administration of the EIA. Ensure agreement on procedures for sharing information and data. Maximize the institutional and technical capacity of the host country to manage the environmental assessment, and maximize local participation. Manage time and cost.
Communication	Establish and communicate the time period for EIA preparation. Ensure that stakeholders, in particular government authorities, are asked to cooperate as early as possible in the planning process. Encourage and manage community consultation and participation. Pre-empt, to the extent feasible, campaigns by anti-mining activists. Establish and maintain communication with key stakeholders, including involved financial institutions. Maintain and circulate a current list of stakeholders including contact details Establish mechanisms whereby mine personnel, the EIA team, and EIA approval authorities can discuss and jointly plan assessment matters in a timely and effective manner. Establish protocols to document all activities related to environmental assessment, in particular activities related to government liaison and public consultation and participation.
Policy Check	Assess the legal requirements (national and local) and the company's internal policy objectives and operational priorities for the proposed mining project. Assess the proposed mining project against the provisions of international agreements to which the host country is a signatory. Assess compliance requirements with the Equator Principles. In consultation with government officials, consider the proposed mining project within the context of regional land use planning and conservation strategies.
Screening	Ensure that all project activities are screened for potential environmental impacts in a manner that meets the procedural requirements of the host country. Should uncertainties in environmental impact assessment requirements occur as a result of environmental screening, resolve in a manner acceptable to all relevant parties. Ensure that, if screening determines that further assessment is required, the government authorities of the host country agree on the need for and scope of such assessment.
Alternatives	Ensure early consideration of alternatives at both project and strategic levels, including consideration of the 'no go' option. Incorporate environmental considerations into selection of preferred option(s). Include approval authorities in the evaluation of controversial or problematic design alternatives.
Terms of Reference	Ensure the development of a single terms of reference acceptable to all parties that defines technical, administrative, and procedural requirements for impact assessment. This should incorporate the results of scoping studies including input from stakeholders.
Project Design	Ensure that environmental measures are integrated into the project design, as far as is practical and feasible. Evaluate and document reasonable alternative project designs and their environmental advantages and disadvantages. Ensure that recommended environmental mitigation and monitoring plans and compensation schemes are included in the project approval documents.
Mitigation Measures	Design management systems and provide budgets to ensure that the Environmental Action Plan (e.g. recommended mitigation, monitoring, relocation and compensation plans) can be adequately implemented.
Review/Quality Control	Ensure adequate review of progress made and contents of the environmental assessment Arrange for publishing and distribution of final assessment reports.

A skilled EIA project manager on the owner's side is the greatest asset during the environmental permitting stage. Good people skills are necessary, especially during the approval period, but other management skills are equally important to request additional resources in a timely manner. The project manager should be directly involved in developing the scope of work and in the consultant selection process, including preparation of the consultant selection criteria. Where a project manager has not previously been through the EIA process it is common for him to be assisted by someone who has – either someone from the company or an individual retained from elsewhere. It is most likely that the bulk of the technical analyses for an EIA will be done by consultants under a third-party contract arrangement. An efficient consultant selection process therefore will avoid delays in initiating an EIA. The selection of consultants who can devote the right combination of skills and resources to the job prevents delays throughout the EIA process.

It is important to establish critical milestones for the completion of EIA tasks and to maintain as tight a schedule as possible. Schedules should be realistic and commensurate with the level of complexity of the EIA. Agreements between companies and governments commonly contain time schedules based on past experience, rather than any foreknowledge of the potential project. For complex projects requiring more time for planning and evaluation, the schedule in the agreement may be inadequate. In such cases, companies have sometimes taken risky short-cuts to meet arbitrary contractual deadlines, when better, more informed decisions could have resulted, had more time been available. Again, flexibility on the part of both government and company is desirable. Schedules are subject to factors beyond the company's control and can change sometimes several times over the course of an EIA. Flexibility in adjusting schedules is essential, both to loosen schedules when needed and to tighten up schedules on remaining tasks to make up for earlier schedule slippages.

A key element of the project manager's responsibility is quality control. If quality control is compromised, there will be delays when analyses and documentation do not pass legal reviews. The best measure of successful EIA management is that the environmental process does not produce conceptual, methodological, or informational surprises towards the end; therefore the choice of an experienced consultant team is essential. The project manager, supported by others assigned to the EIA, needs to look ahead, identify issues and problems as early as possible, and initiate appropriate and timely additional analysis, consultation, or other efforts that will lead to successful resolution and completion of the environmental process.

A key element of the project manager's responsibility is quality control.

Managing the EIA Process as the Consultant

'If I am spending a lot of money I want to have somebody who has more experience than I have.'

'I want advice, not juniors measuring small turtles.'

'You should do more hand-holding – educate us on the regulatory process, on the pitfalls, and how to avoid them.'

'I am not your QA/QC consultant!'

I want advice, not juniors measuring small turtles.

Ask any mining company about their experience with consultants during the EIA process. Some companies will be full of praise; the majority however will air their frustration built up during the EIA process as expressed in the few quotations above. To earn praise, the project manager of the selected consultant needs to know what the company wants, and what it finds counter-productive or annoying.

What Mining Companies Expect of their Environmental Consultants

The client wants full attention.

Firstly, and most importantly, the mining company wants to rely on the consultant as a strategist, as an advisor on both the legal issues and the EIA process, and as a manager and advocate for the project. Secondly, the client wants to access the full expertise of the selected consultant organization and of other specialist expertise where appropriate. Thirdly, the client wants full attention.

As such the main task of the consultant EIA project manager is to facilitate teamwork between the mining company and the consultant. The consultant should facilitate the EIA outcome and should support the mining company communicating with its own internal stakeholders as well as joint venture partners and government authorities. The consultant is part of the team; the consultant EIA project manager should avoid acting as the regulator or the EIA approval authority. The consultant EIA project manager is also responsible for ensuring access to senior expertise and should command enough respect to stand up to the client and communicate what the client has to contribute to make the EIA a success story. The timely delivery of an adequate project description is one example.

What Annoys Mining Companies about their Consultants

People, not companies, are the subject of such embarrassment.

The worst things that a consultant can do to his or her client is to embarrass them or make them appear deficient. When this happens, it is usually a case of poor communications, with responsibility shared by both parties. People, not companies, are the subject of such embarrassment. In the case of a cost overrun, the client representative, usually the Project Manager, must ask upper management for more funding. If there is a project delay, the same individual has to explain the slippage to the next management level. None of this helps the career development of the client representative especially if issues have not been identified and communicated early. Bad news only becomes worse with time. It also does not help the client relationship in submitting draft reports full of typographical and other errors. The mining company should not be responsible for the QA/QC of environmental documents. However, the company is responsible for ensuring that details of the project are adequately described. In many cases, the company or its appointee will also play the role of 'devil's advocate', challenging the consultant to ensure that the conclusions are defendable and that recommendations are cost-effective. Any deliverable will be seen as a final deliverable, and has to be prepared as such. Finally clients hate to bear the cost resulting from a lack of focus and inefficient use of resources. The consultant should understand the key issues, and should avoid labouring on and waffling about minor issues.

The consultant EIA project manager needs to remember a few pointers for success, which are:

- Ensure accountability, management continuity, and safety
- Deliver senior advice
- Act as a mentor for the client during the EIA process
- Manage schedules including the client's schedule – deliver as promised
- Identify and communicate issues early
- Don't use gold plating when writing the initial scope of work
- Ensure the quality of all deliverables (including drafts)
- Use the Equator Principles as a checklist for completeness
- Be professional and act professionally
- Communicate often and positively

Companies should expect that there will be differences of opinion between internal project personnel and external consultants. In particular, company personnel often believe that adverse impacts are overstated in the EIA. This is natural. However, it is the EIA practitioner's responsibility to be objective – to tell things the way they are and not necessarily the way the proponent sees them. This makes for some interesting exchanges between the company and the environmental consultant as the EIA is being reviewed, prior to submittal. This process is beneficial for company and consultant alike. For the company, it raises consciousness and awareness of the environmental issues. For the consultant, it challenges the assessment process and may result in an improved assessment and/or the formulation of better management measures. It is also inevitable that other stakeholders will disagree with a variety of assessments in the EIA. This is a natural outcome of the variety of view-points, perceptions, and prejudices among the stakeholders. The EIA practitioner needs to be aware of all these view-points and to take them into account. However they should also realize that it is impossible to produce an EIA that satisfies all parties. Experienced EIA practitioners feel that they have an appropriately objective balance when they are criticized by the proponent for overstating adverse impacts while also being criticized by NGOs for understating adverse impacts or overstating project benefits.

Experienced EIA practitioners feel that they have an appropriately objective balance when they are criticized by the proponent for overstating adverse impacts while also being criticized by NGOs for understating adverse impacts or overstating project benefits.

Commence Early

It is important to examine environmental factors at an early stage in mine planning. This 'environmental scoping' exercise is discussed in greater detail in Chapter Eight. At a minimum, it is necessary to identify major environmental issues and concerns that have an important influence on the evaluation of the mine. These may range from an issue that is a primary concern of the host community to an issue that poses a legal barrier, such as jeopardy of an endangered species; either case may be of sufficient merit and importance to affect the mine project's basic planning. Late identification of environmental or social constraints is likely to delay the environmental assessment process. Accordingly, it is common practice, particularly in unusual projects or those in previously undeveloped regions, to undertake a 'Fatal Flaw' evaluation, at a very early stage, even before any pre-feasibility engineering studies are conducted. The purpose of such an evaluation is to identify the presence of any 'show stoppers' or 'deal breakers' – issues of such overwhelming importance or concern that they could preclude development. It reduces but does not eliminate, the risk of the project being stopped at a much later stage as a result of an inability to adequately respond to emerging issues. Of course, what is a 'show stopper' to one company may not be a deterrent to another, depending on their internal policies and past experiences. For example, occupation of an ore body by artisanal miners, whether legal or otherwise, may represent a fatal flaw for a large international company, already under fire from NGOs for alleged human rights abuses, while a smaller, more entrepreneurial company may feel confident of its ability to resolve such land use conflicts.

Late identification of environmental or social constraints is likely to delay the environmental assessment process.

Early consideration of environmental issues can be expanded to include a detailed inventory of the existing environmental components in the mine vicinity (e.g. air quality, water quality, and communities), including environmental resources (e.g. wetlands, historic sites, and endangered species). Such an inventory serves the dual purposes of providing improved early environmental information to assist the mining company in selecting from proposed project alternatives, and of providing the existing environmental baseline data set for the subsequent EIA. Environmental inventories become outdated over time, decreasing their usefulness for an EIA. The closer in time that an EIA follows environmental data collection, the less potential there is for problems with data currency and validity.

Mine planning information generated by the company or its appointed engineers is the key to a successful environmental assessment. It defines the proposed project, reasonable alternatives, and the scope and accuracy of the impact analyses. Relevant planning information includes elements such as ore production forecasts, maximum mine capacity, process systems, cutoff grade, tailings disposal system and location, facility requirements, timing and phasing of development, labour requirements, and links with independent planned or existing projects. Mining companies need to ensure that planning information provided to the EIA team is technically sound, reasonably complete and current, and provided in a timely fashion. Inadequacies identified in planning information during an EIA can delay progress. At worst, the proposed mine plan and reasonable alternatives may need to be modified, substantially delaying the environmental work schedule.

2.4 COMMON THEMES AND CORE PRINCIPLES

Environmental assessment is implemented in different ways in different countries, but all have common themes and core principles (Modak and Biswas 1999; Canter 1996; Carroll and Turpin 2002; OECD 1989, 1999; Petts 1999; Goodland and Mercier 1999; Tromans and Fuller 2003; UNEP 1996; Wood 2003).

Environmental Assessment – A Process, Not a Study

Environmental assessment is a process, not a study or a document.

Environmental assessment (EA) is a process, not a study or a document. In the case of a mining project, environmental assessment starts with the initial environmental scoping and ends with post-operation monitoring as shown in **Figure 2.1.** In a wider sense, environmental assessment also includes community development activities during the life of the mine. The EA process consists of planning, executing, monitoring, and redirecting environmental management efforts. The Environmental Impact Assessment (EIA) process is the planning tool, documented in form of the Environmental Impact Statement and its supporting documentation. As such, EIA is only one element of environmental assessment.

Focus on a Definable Geographic Area

One of the requirements of EIA is the determination of boundary conditions in space and time. Where, spatially, does an impact assessment cease? With reference to a mining project for example, should the EIA consider only the mining area and its immediate surroundings, or should it also consider the environmental impacts of mineral exploitation and land clearing; power generation and therefore the impacts of coal mining by the coal supplier; the new seaport constructed to export concentrates; or the larger host region in which economic impacts are felt?

Raff (1997) identifies and substantiates 10 basic principles of quality in EIA, distilled from hundreds of cases, and which he consequently terms the common law of NEPA-style EIA (**Table 2.2**). Of these, the first principle provides some guidance in relation to the EIA study boundary: 'an assessment cannot be restricted to site-specific environmental effects'. This is echoed in subsequent principles. The third principle is that 'greater plans' must be assessed in addition to single phases in their execution (which leads to regional development planning); and the fourth principle is the 'rule' against segmentation of the project. In other words, the EIA must not solely assess individual aspects of the overall mine proposal which on their own will yield negligible impacts; the sum or total effect of the proposal on the environment must be identified and evaluated.

TABLE 2.2
Raff's Ten Principles of Quality in EIA

1. An assessment cannot be restricted to site-specific environmental effects.
2. The assessment must contain a statement of alternative courses of action and their environmental significance even if it is beyond the power of the proponent to implement them.
3. Greater plans must be assessed in addition to single phases in their execution.
4. The project must not be segmented.
5. Time frames must be viable: the range of alternatives that must be discussed in an EIA is limited to realistic alternatives that are reasonably available within the time the decision-maker intends to act.
6. Alternative courses of action must include the option of doing nothing.
7. The assessment must engage in a real inquiry and not merely dispose of alternatives in favour of a decision which has already been arrived at.
8. Environmental effects are not to be disregarded merely because they are difficult to identify or quantify.
9. The EIA must take a 'hard look' at the environmental consequences of the project. While the EIA is not required to be perfect, it must not be superficial, subjective or non-informative; it should be comprehensive in its treatment of subject matter and objective in its approach.
10. The findings of the EIA must be presented in clear language and the methods used to arrive at them must be explained.

Source: Raff 1997

The EIA must not solely assess individual aspects of the overall mining project which on their own will yield negligible impacts; the sum or total effect of the project on the environment must be identified and evaluated.

Although trans-boundary environmental effects such as greenhouse gas emissions may need consideration, environmental assessment is first of all site-based. The obvious study area is the project area itself, which comprises all areas expected to undergo physical changes due to project activities. In the case of mining, these areas include the ore deposit, mine infrastructure, and waste rock and tailings disposal areas. Since impacts extend beyond the areas of physical changes, the geographic boundary of an environmental assessment study is almost always larger than the actual project area. It can be based, among other possibilities, on the natural landscape, such as a watershed area or specific types of ecosystem, downwind 'air-shed' area, or on social and political boundaries.

Although trans-boundary environmental effects such as greenhouse gas emissions may need consideration, environmental assessment is first of all site-based.

Figure 2.2 illustrates the various study boundaries in increasing spatial dimension that typically apply to an environmental assessment. The project boundary includes all areas of direct physical changes – the 'footprint' of the project. The ecological boundary encompasses all areas in which environmental impacts, including indirect impacts, may occur. The socio-economic boundary includes all communities potentially affected by the project. The project area and the area of potential environmental physical and biological impacts always fall within the socioeconomic boundary. The administrative boundary includes the various jurisdictions with influence on the mining project.

For an environmental impact assessment to be fully effective, the study area should include the source of the impacts as well as the area of potential environmental and social concern. At a minimum, the study boundary has to coincide with the socio-economic boundary. The drawing of the study boundary is not always a straightforward exercise. Large mines have regional and national economic impacts. Even the extent of physical and environmental impacts is not always clearly defined, such as in the case of mining activities that employ riverine or marine tailings disposal (see also Chapter Eighteen).

For an environmental impact assessment to be fully effective, the study area should include the source of the impacts as well as the area of potential environmental and social concern.

FIGURE 2.2

Environmental Impact Assessment – Spatial Boundaries

Although trans-boundary environmental effects find consideration, environmental assessment is site based. Since impacts extend beyond the areas of physical changes, the geographic boundary of an environmental assessment study is larger than the actual project area.

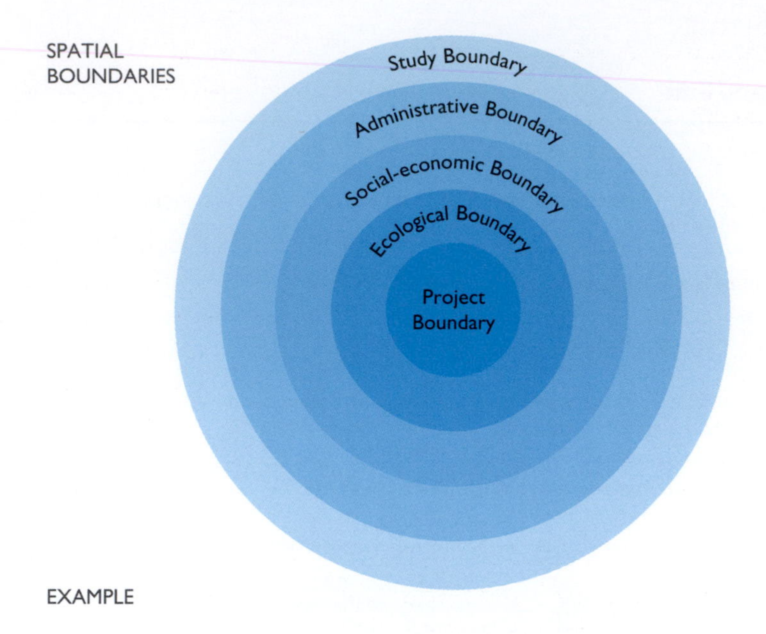

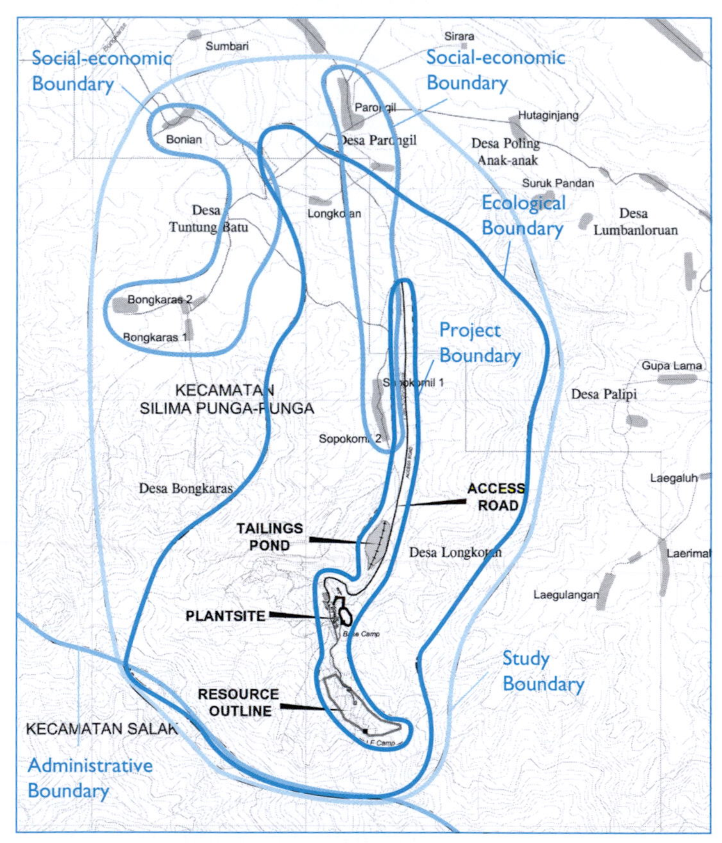

What time period should the EIA cover? In the case of the mining industry, an EIA usually covers the life of the mine, but the actual mine life might be substantially different from that assumed at the EIA stage. Commonly, mining projects are developed when sufficient ore reserves have been defined to support a particular level of investment. Subsequent exploration carried out during the operating stage may identify much larger ore reserves. Accordingly, EIA updates are common for big mines in order to accommodate changes in mine design, operation, or lessons learned. Raff (1997) refers to the 'rules'

against 'science fiction' and 'crystal ball inquiries' in discussing his fifth principle. This states that the time frame should be realistic and 'reasonably' available within the time the decision-maker intends to act (**Table 2.2**).

Evaluate the 'No Action' Scenario for Comparison

To judge mining-induced environmental change an understanding of the 'no action' scenario is important. What would happen to the host environment without the project? This is by no means an easy task. Environments are constantly changing. Natural environments are driven by the forces of nature which include solar energy, wind, rainfall and runoff, and many other factors, none of which is constant over time. Ecosystems do not have simple equilibrium states; they are constantly changing in response to changes in the physical environment and evolution. Individuals, communities, societies and civilizations also change with time. If the mine project were not undertaken, the environment would still exhibit: (1) irreversible trends due to a combination of natural or human-induced factors such as logging, land use changes, or economic development; (2) irreversible trends of natural origin such as soil erosion or eutrophication of lakes; (3) great variability due to daily and annual cycles, variation in weather, or natural ecological cycles; and (4) constant changes in host communities (SCOPE 5, 1975).

Elaborating on the 'no action' scenario provides an insight into environmental changes that would occur in the absence of the project.

Elaborating on the 'no action' scenario provides an insight into environmental changes that would occur in the absence of the project. As mentioned above this is, difficult and necessarily involves a degree of speculation as the scale of future changes can not be accurately predicted.

The environmental impact assessment needs then to identify and evaluate environmental change as a result of the project over time, relative to the dynamic 'no action' baseline, as illustrated schematically in **Figure 2.3**. It is doubly difficult to predict with any pretence of accuracy the effects of proposed project actions on an environment whose attributes and properties are changing continuously.

It is doubly difficult to predict with any pretence of accuracy the effects of proposed project actions on an environment whose attributes and properties are changing continuously.

As monitoring plans are designed to evaluate actual mine-induced environmental changes during mine operation, the environmental action plan needs to incorporate monitoring at environmental and social control locations outside the influence of the mining project, to enable measurement and evaluation of environmental changes that occur which are unrelated to the project.

Focus on the Main Issues, both Negative and Positive

An EIA should not attempt to cover too many topics in too much detail, but should focus on the most likely and most serious potential environmental impacts. It is as important to evaluate positive impacts as to assess negative ones. Otherwise it is not possible to assess the overall or net effects. Large, verbose, complex reports are unnecessary and often counter-productive, as findings are not readily accessible. Mitigation measures should take the form of workable, acceptable solutions to key issues. The EIA should be communicated concisely, preferably including a non-technical summary of information relevant to the needs of decision makers. Supporting data should be provided separately.

It is as important to evaluate positive impacts as to assess negative ones.

Use of Environmental Assessment to Improve Project Design

An EIA supports decisions about project design. If started early enough, it provides information to improve conceptual design such as with the siting of mine infrastructure, including tailings storage facilities and mine access roads. As shown in **Figure 2.1** the EIA evolves concurrently with and is integrated into mine planning. Environmental issues are first considered at the concept stage. During the mine design stage the EIA identifies environmental design standards and impact mitigation measures that should be built into the mine design (**Case 2.2**). The central idea is that environmental assessment and mine planning are iterative processes, with the final outcome being an optimal mine plan.

FIGURE 2.3

Environmental Impact Assessment – Temporal Boundary

The baseline (or benchmark) description should not be that of a static entity, accurately reflecting the 'no action' scenario providing insight into environmental changes that would occur in the absence of the project.

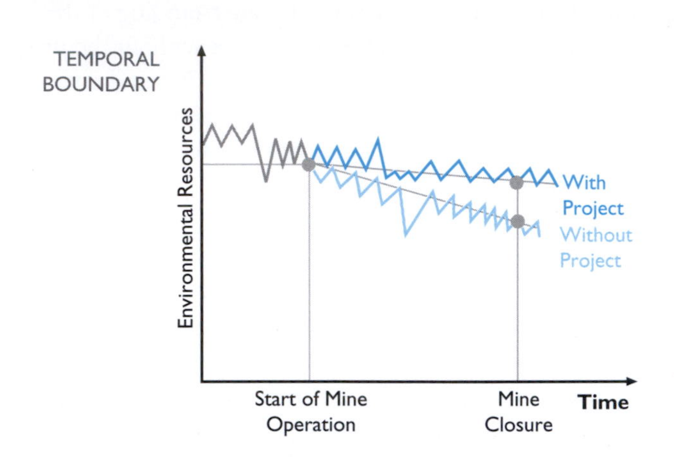

EXAMPLE

Population in Sopokomi Village, Dairi Lead/Zinc Mine, Sumatra, Indonesia		
Year	Without Project	With Project
1980	763	
1985	801	
1990	882	
1995	913	
2000	1,004	
2005	1,153	
Project Start		
2010	?	?
2011	?	?

CASE 2.2
Mitsubishi's Gresik Copper Smelter

To process copper ore from the nearby Grasberg mine, and to help develop downstream metal industries in Indonesia, Mitsubishi, in cooperation with PT. Freeport Indonesia, developed a 200,000 ton per year copper smelter at Gresik, Java. The smelter, based on Mitsubishi's continuous smelting and converting technology, is widely praised as the smelting technology for the 21st century (Prof. Emeritus Herbert H. Kellog 2000: 'I have no hesitation in describing the Gresik plant as the finest copper plant I have ever seen'). All internal solid waste streams are recycled, and final output of trace metals is in the form of inert granulated copper slag. In the original design, slag granulation was based on a once-through sea water cooling system applied at similar smelter operations in Japan. The EIA team successfully proposed the use of a closed water cycle, eliminating potential contamination of sea water in the nearby Madura Strait.

In practice, environmentally driven improvements to project design follow discussions between environmental practitioners and key members of the project design team. To encourage and facilitate these interactions, it is valuable to schedule 'environmental workshops' involving 'brainstorming' among these key personnel, at key intervals in the project planning process (see also Chapter Nine). For a large project, appropriate occasions for such workshops would include:

- At the start of the Pre-feasibility Study;
- Shortly prior to completion of the Pre-feasibility Study; and
- Shortly prior to completion of the EIA and/or the Feasibility Study, depending on whether or not these studies are synchronized.

Present Clear Options for Mitigation Measures

The main outcome of an EIA is a set of clear choices on the planning and implementation of environmental mitigation measures. For instance the EIA can propose: (1) pollution control technologies or design features; (2) reduction, treatment, and/or criteria for disposal of wastes; (3) water recycling schemes; (4) recommendations for tailings storage site selection; (5) compensation or concessions to affected people; (6) community development initiatives; (7) limitations to the initial size and/or growth of the mine (albeit imposing a challenge in convincing the mine developer); (8) separate programmes to contribute in a positive way to protecting local resources or to enhancing quality of life; and (9) involvement of the local community in later decisions about the mining project.

CASE 2.3
Internalizing Externalities

Until recently, mining companies would not dream of calculating how much greenhouse gases (GHG) they would emit, not even from integrated coal-fired power plants, although this has been best practice at least in the USA since 1991 (US EPA 1992). Now, since the Kyoto Protocol in 1997, the publication of the World Bank's OD 10.04 in 1994 and the recent 2006 IFC PS, an increasing number of mining projects calculate GHG emissions. Such calculations are the first step towards internalizing externalities. The valuation of environmental costs and benefits has progressed greatly in recent years, and should become standard methodology where practicable. Environmental costs and benefits, including aspects of scarcity and externalities, should be included in the overall financial and economic evaluation of any new mining project as a matter of routine.

Integrate Economic, Environmental, and Social Objectives

Environmental assessment aims to ensure that the development of a mining project is consistent with the goals of sustainable development. Although no single definition of sustainable development has emerged, there is consensus that sustainable development should result in a stable, more efficient economy, a clean and healthy environment, and an improved quality of life for all community members. As such, environmental assessment is concerned with the economic, environmental, and social resources within the study area. The basic assumption is that all resources in a place are interconnected parts of one system. Taking an integrated and holistic approach, rather than focusing on one particular resource or concern, more appropriately addresses environmental problems (**Case 2.3**). However, as uncertainties and limits always attend a holistic approach, environmental assessment will always be based on simplifications.

Increase the Overall Capital Stock

Environmental assessment seeks to bring benefits to affected communities, along with natural resource planning and management efforts.

Economists differentiate various types of capital that exist within the region of a prospective mine (e.g. natural resource capital, productive capital, human capital, social capital, and financial capital). Productive, human, social, and some forms of environmental capital are sometimes collectively referred to as renewable capital, as the level of this capital is influenced by human activities and investment. During the operating stage of a mining project, environmental assessment is concerned with maximizing the productive, human, social, and environmental benefits of the operation through a wide range of community development activities, commonly referred to as private sector social investment. Environmental assessment can and should be the planning vehicle for social investment. Environmental assessment seeks to bring benefits to affected communities, along with natural resource planning and management efforts. Environmental assessment is a process that enables communities to participate through training and education, assistance agreements, information sharing, and technical assistance.

As mineral resources are exploited, renewable capital is created.

Social investment aims to offset any negative impacts associated with the project operations and seeks to increase the levels of renewable capital in areas of mining activities and beyond. In an ideal world, the increase of renewable capital through social investment would balance or outweigh the loss of non-renewable mineral resource capital over the mine life as illustrated in **Figure 2.4**. As mineral resources are exploited, renewable capital is created. This is probably the single most important challenge in ensuring sustainable development due to mining, and it will be discussed in more detail in Chapter Fifteen.

A second challenge exists. Environmental assessment should not only aim to maintain or to increase the level of capital stocks, but should also seek an equitable distribution of environmental costs and economic benefits. Social tension is unavoidable if resources and capital in the region of the mine activity decline (depletion of mineral resources, environmental degradation, disruption of social patterns, etc.) while capital stock in regions beyond the mining area increases (tax income to central government, increase of foreign shareholder wealth, etc.).

Throughout history, mineral resources have provided income for national governments, but only a portion of created wealth and capital will flow back to local communities.

Equitable distribution of renewable capital at the local, national, and international scale is challenging. Throughout history, mineral resources have provided income for national governments, but only a portion of the created wealth and capital flows back to local communities. Mining companies are often blamed for the real or perceived unequal distribution of financial benefits at the local and national levels, but much of the responsibility rests with the host government. International mining companies are also seen as exploiting national resources to the benefit of distant foreign beneficiaries. This argument has been successfully used by anti-mining advocates to exploit existing nationalist feelings to create or increase negative community perceptions.

Work Collaboratively with Stakeholders

Strong local opposition to a proposed mine slows the environmental process. An EIA proceeds more rapidly and smoothly when mining companies are able to build a broad base of local consensus in support of the project and to maintain a sense of trust and fair treatment among concerned communities.

Apart from avoiding project delays, environmental assessment is designed as an open, inclusive decision-making process. Environmental assessment necessarily seeks collaboration with all stakeholders, but particularly project-affected people and host communities. In a wider sense, however, stakeholders are all the people that perceive themselves to be affected by the project, including voluntary groups, NGOs, and issue-based pressure groups.

No aspect of environmental assessment is more important than encouraging the participation of stakeholders, the subject of Chapter Three. A process that is open to public scrutiny is more likely to reflect diverse views, and to ensure an outcome that is satisfactory for both the mining company and the affected communities.

A public outreach programme is a responsibility the mining company should always consider, since it can generate a positive effect on the environmental review process. When a proposed project is highly controversial, the mining company is well advised to assign a skilled community relations specialist to assist in providing the optimal interface with the local community during project planning and environmental review. The specialist can facilitate the two-way flow of information so critical to avoiding unnecessary conflicts. It is important that such communications on the part of the company be frank and accurate, presenting a balanced picture of the project, what it will involve and what impacts it will cause. This should not be a public relations exercise in the sense of selling the project to a

Any attempt to 'gild the lily' is likely to generate expectations which can not be realized, providing seeds of future problems.

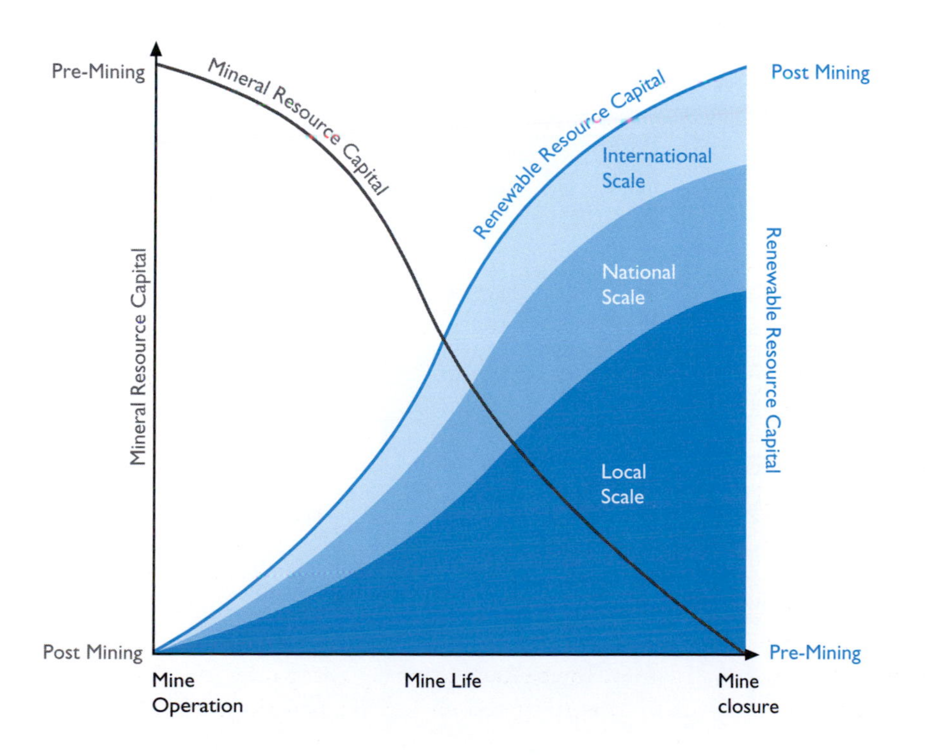

FIGURE 2.4
Principle of Resource Conservation

In an ideal world, the increase of renewable capital through social investment would balance or outweigh the loss of non-renewable mineral resource capital over mine life. There would also be an adequate distribution of environmental cost and economic benefits.

doubting public. Any attempt to 'gild the lily' is likely to generate expectations which can not be realized, providing seeds of future problems.

In an area with no previous experience of mining, one of the impediments to effective communication is the lack of awareness on the part of the community to the nature and scale of the proposed project. To overcome this problem, some companies arrange for representatives of local communities to visit existing mining operations. This has proved to be particularly valuable. However, it is important that the mine or mines that are visited are of a similar nature and scale to the proposed project. Otherwise, false expectations will be generated.

Unfortunately, not all the stakeholders in a mining project have genuine interest. Radical anti-mining activists, often grouped as an NGO, tend to insert themselves in new mine developments to block them, regardless of their benefits. False stakeholders have become a major concern for both governments and investors.

2.5 WHEN IS AN EIA REQUIRED?

In the first instance the answer to this question depends on the host country's jurisdiction. A screening process is used in most countries in order to identify those projects for which a full-scale environmental impact assessment must be made. In effect, screening clears all projects with minor environmental impacts for development. Most jurisdictions and international lenders have established exclusive and inclusive thresholds to simplify the screening process (**Figure 2.5**). For projects within the inclusive threshold, the preparation of the environmental impact assessment is mandatory. Projects with minor environmental impacts fall within the exclusive threshold and environmental assessment of these projects can be kept very simple. The World Bank and the Equator Banks define projects that require a full environmental impact assessment as Category A projects (IFC 2006). Mine projects are commonly classed as Category A projects. For projects that fall between the inclusive and exclusive thresholds, the screening process, together with consultation of the relevant authorities, leads to a decision on the extent of the required environmental assessment.

The World Bank and the Equator Banks define projects that require a full environmental impact assessment as Category A projects

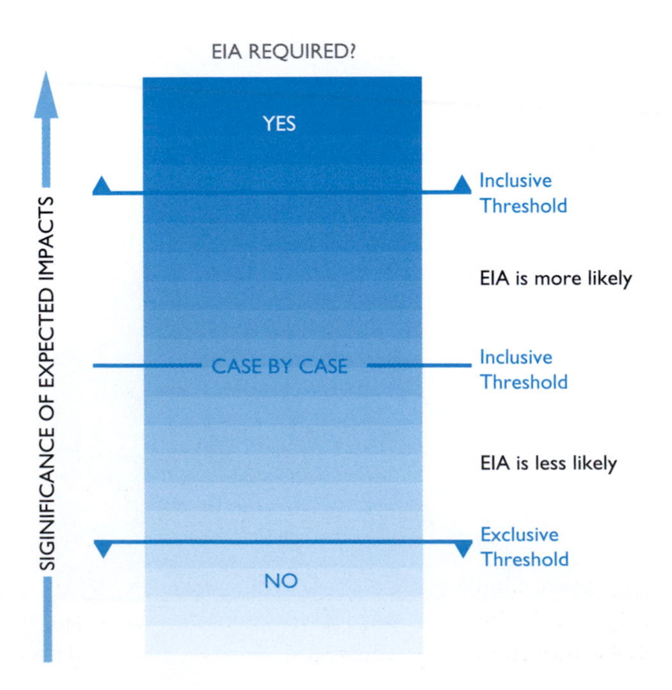

FIGURE 2.5
Threshold for EIA Requirement

A Screening process is used in most countries to identify those project for which a full-scale environmental impact assessment must be made. As such screening clears all projects with minor environmental impacts for development.

In the event that a full environmental impact assessment is required, there are some key points to remember (Pallen 1997):

- Keep environmental assessment in perspective. Recognize it as a tool to enhance the decision-making process, not the decision-making process itself.
- Keep the assessment simple and concentrate on pertinent factors and data. It should be rigorous but not necessarily laborious.
- Focus time and effort on the most relevant matters.
- Invest neither too much nor too little time on an assessment. Stay flexible throughout the assessment in order to meet new challenges as they arise.
- Tailor each assessment to the particular project needs. Each project has a unique set of environmental, economic, and social characteristics.
- Be inventive. There is no standard format available for interpreting the information gathered during an environmental assessment.
- Be prepared for inexact and suggestive data. Typically data is imperfect, and assumptions open to challenge. Quantification may be difficult (and in some circumstances, impossible). Exposing the limits and inadequacies of knowledge, data, and interpretation can help stimulate improvements in the understanding of environmental issues and accelerate the provision of reliable information to support informed decision-making.
- Avoid secrecy. Open communication among all stakeholders throughout the assessment process not only produces better results, but also increases the project's credibility and builds trust and acceptance within the wider community.
- Seek external help and advice in situations that require more expertise than is available in the project management and environmental assessment teams.

2.6 ENVIRONMENTAL IMPACT ASSESSMENT STEP-BY-STEP

An EIA is generally conducted step-by-step. The tiered system can have variants in different countries. The EIA team typically complements the regulatory tiered system with its own step-by-step methodology. Usually, environmental impact assessment evolves along the following lines (see also **Table 2.3**): (1) screening as discussed in the previous section; (2) scoping; (3) assessment of project and project alternatives; (4) public disclosure, consultation and participation; (5) establishing the institutional setting; (6) establishing the environmental baseline; (7) identifying and quantifying impacts; and (8) designing environmental management and monitoring measures.

Much information is needed during the EIA process, and much information is generated. A disciplined collection and organization of data is helpful. An example of data needs and data organization is presented in **Appendix 2.1**.

Scoping

Screening is carried out to determine whether an environmental impact assessment is necessary. Scoping aims to define the focus. More specifically, scoping allows one to

- Specify what issues and impacts the EIA study shall focus on;
- Specify regulatory design criteria in form of applicable standards;
- Specify methods that should be used in impact identification and quantification;
- Identify stakeholders and their need for information and participation; and
- Define the study boundaries in time and space.

Scoping aims to define the focus.

TABLE 2.3
Steps that should be included in any EIA

1.	Screening the proposal to determine whether an EIA should be conducted;
2.	Scoping the proposal to determine the scope of environmental assessment, the scope of factors to be considered, the parties involved and their interests and concerns, the appropriate level of effort and analysis, and to prepare guidelines for the conduct of the EIA;
3.	Description of the proposal and of relevant alternatives, including doing nothing at all;
4.	Description of the environmental baseline, including changes without the proposed project;
5.	Identification and prediction of the nature and magnitude of key environmental effects, both positive and negative, for each of the alternatives studied, over both short and longer term periods;
6.	Assessment of how such effects are valued by representative sections of society;
7.	Testing of impact indicators as well as the methods used to determine their scales of magnitude and relative weights;
8.	Prediction of the magnitude of these impact indicators and of the total impact of the proposed project and relevant alternatives;
9.	Proposal of mitigation measures including identification of critical thresholds;
10.	EIA presentation, public consultation and participation, including feedback;
11.	Recommendations for project rejection, or acceptance plus advice on how to reduce or remove the most serious impacts as measured and as socially valued, or adoption of the most suitable alternative scheme;
12.	Decision-making;
13.	Post-decision monitoring and auditing both during project construction and after project completion to ensure that environmental effects are minimized, and to compare actual with predicted impacts;
14.	Amelioration if thresholds are exceeded or unacceptable impacts are identified, and
15.	Longer period reviews (5 years, 10 years, 20–25 years), requiring comprehensive assessment and adaptive management. A period of 25 years is a common limit for licence periods – 'in perpetuity' arrangements should be avoided.

Source: Conacher, 2000, drawn mainly from O'Riordan and Turner 1983, Wood 2003, and Harvey 1998

Scoping should be regarded as an early part of the decision process. Two aspects of scoping are particularly important. Scoping provides an excellent opportunity to identify project alternatives before mine planning has proceeded too far to allow radical changes. Secondly, scoping offers an opportunity to identify the key stakeholders in the mine project and to initiate their participation in the environmental assessment process at an early stage.

A scoping study is incomplete without a site visit.

To a certain extent, scoping can be done as a desk study through a critical review of existing data. However, a scoping study is incomplete without a site visit (see Chapter Eight). The site visit provides important contacts with the people affected by the mine at an early stage of planning, enabling the incorporation of feedback from the local community into mine development. Also, the site visit frequently reveals important environmental issues that would not necessarily be apparent from a desk study.

Review the Project Design

It seems odd if the EIA team for a mining project does not include a mining specialist. And yet this is often the norm, not the exception. In spite of all the complexities of mining,

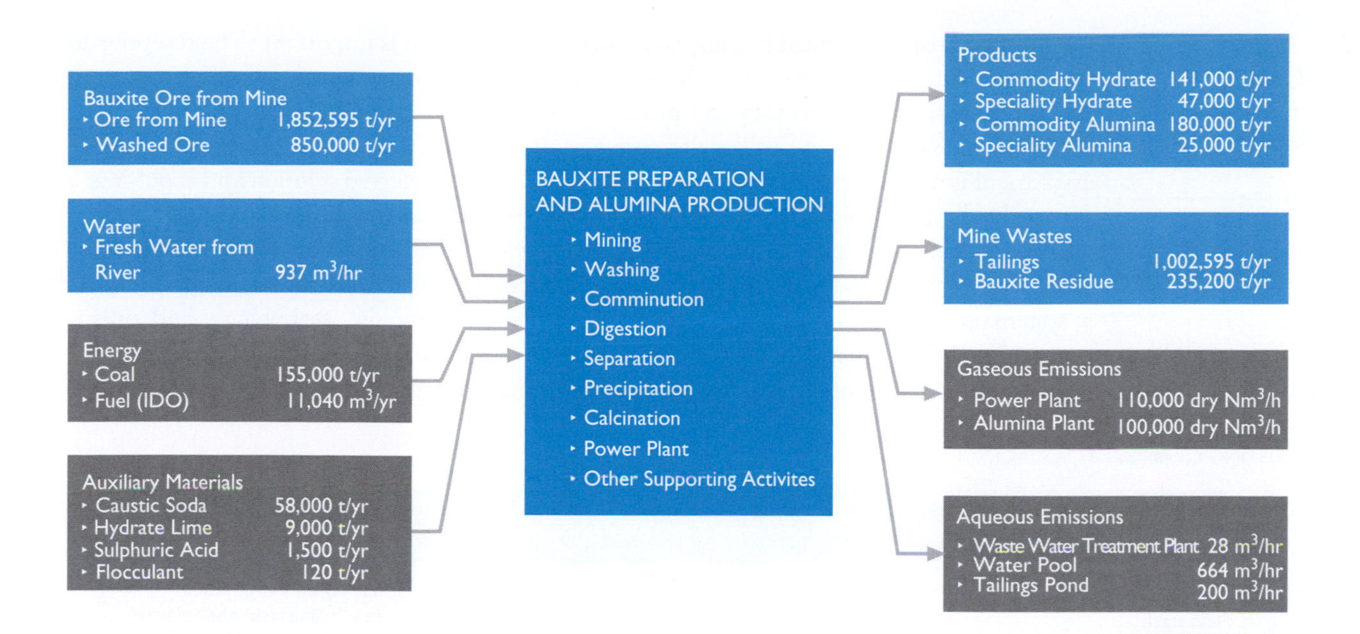

FIGURE 2.6
Indicative Mine Inputs and Outputs for a Hypothetical Maximum Mine Operation Illustrated

a biologist, an environmental engineer, or a physicist reviews the mining project (a trend often repeated by the approval authorities).

The review of the project design is critical. Is acid generation likely, and if so when and where? Which metals are likely to remain in the mine waste, and at what concentrations? Does the process produce acidic or caustic effluents? Does an alternate tailings detoxification scheme exist? What are the technical implications of relocating the mill site? Is backfilling of tailings a realistic option? Admittedly the answer to most technical questions is best left to the project proponent. After all, the mining company owns the EIA, and the preparation of the EIA is a team effort: the mining company is providing technical expertise; environmental input is outsourced to environmental specialists. However, for the biologist, the environmental engineer, or the physicist to meaningfully participate in the impact assessment, and to ask the right questions, experience with and knowledge of the mining sector is essential.

For the biologist, the environmental engineer, or the physicist to meaningfully participate in the impact assessment, and to ask the right questions, experience with and knowledge of the mining sector is essential.

A methodological approach to the project review will help. Firstly, what are the main mine inputs and outputs (see **Figure 2.6** as an example)?

The next question is: What characterizes emissions? For each emission stream (that is, emissions to air, emissions to water and waste emissions), list source, quantity, quality or characteristic, and discharge point and/or management option as illustrated in **Figure 2.7**. Once this exercise is completed, there is a good understanding of the various waste streams and emissions associated with the mine development.

Assess Project Alternatives

The section in which project alternatives are addressed is often called the heart of an environmental impact assessment because it organizes and clarifies the choices available to the decision makers. The project proponent, the approval authority, or the public can generate alternatives, including the 'no action' alternative (see the section below on establishing the environmental baseline). While generally the assessment of alternatives can relate to both the design and the location of a project, alternative siting of a mine is understandably not an option. However, alternative methods exist to accomplish the proposed action, such as different mining and mineral processing methods, alternative size or production rate, different timing in mine development, different layouts of supporting mine infrastructure, and different

The credibility of the environmental impact assessment is at risk if the mine proposal is identified as the preferred solution and is put into a favourable light by comparing it to poor or 'straw man' alternative solutions.

FIGURE 2.7

Specifying Individual Emissions – Source, Quantity, Quality and Emission Point

A set of information helps illustrate mine outputs a tabular listing of related group of emissions, together with an illustration showing where emissions are generated (using the operational flowchart) and discharged to the environment (using the location map).

human resource deployment in mine construction. Note that it is important to be receptive to completely different solutions than those presented in the mine proposal. It is also important that alternatives be real, so they can provide a basis for meaningful comparison. The credibility of an assessment of project alternatives and, ultimately, of the environmental impact assessment is at risk if the mine proposal is identified as the preferred solution and is put into a favourable light by comparing it to poor or 'straw man' alternative solutions (**Case 2.4**).

The assessment of project alternatives is not always mandatory, but it should always be part of an EIA as good practice. International lenders such as the World Bank often require a systematic comparison of the proposed investment design, site, technology, and operational alternatives (World Bank OD 4.01, Annex B 1989; IFC 2006). Alternatives are to be considered in terms of their potential environmental impacts, capital and recurrent costs,

LISTING OF EMISSION SOURCES

Emission Source	Estimated Quantity	Estimated Quality	Emission Point
List emission sources e.g. ▸ Main stack ▸ Uncondensated gas ▸ Coal fired power plant	Specify total annual emission per source (e.g. tons per annum)	List emission concentration or refer to table detailing flue gas composition including relevant emission standards (see table below)	Specify discharge point (e.g. main stack) and refer to figure 'Location Map' and/or 'Process Flow Diagram' (see graphic below)

ESTIMATED EMISSION CONCENTRATIONS (e.g. Main Stack)

Parameter	Unit	Estimated Concentration	Host Country Standards	IFC Guidelines	Other Guidelines Used in the Assessment
List of relevant Parameters, e.g ▸ Sulphur dioxide	$\mu g/m^3$				

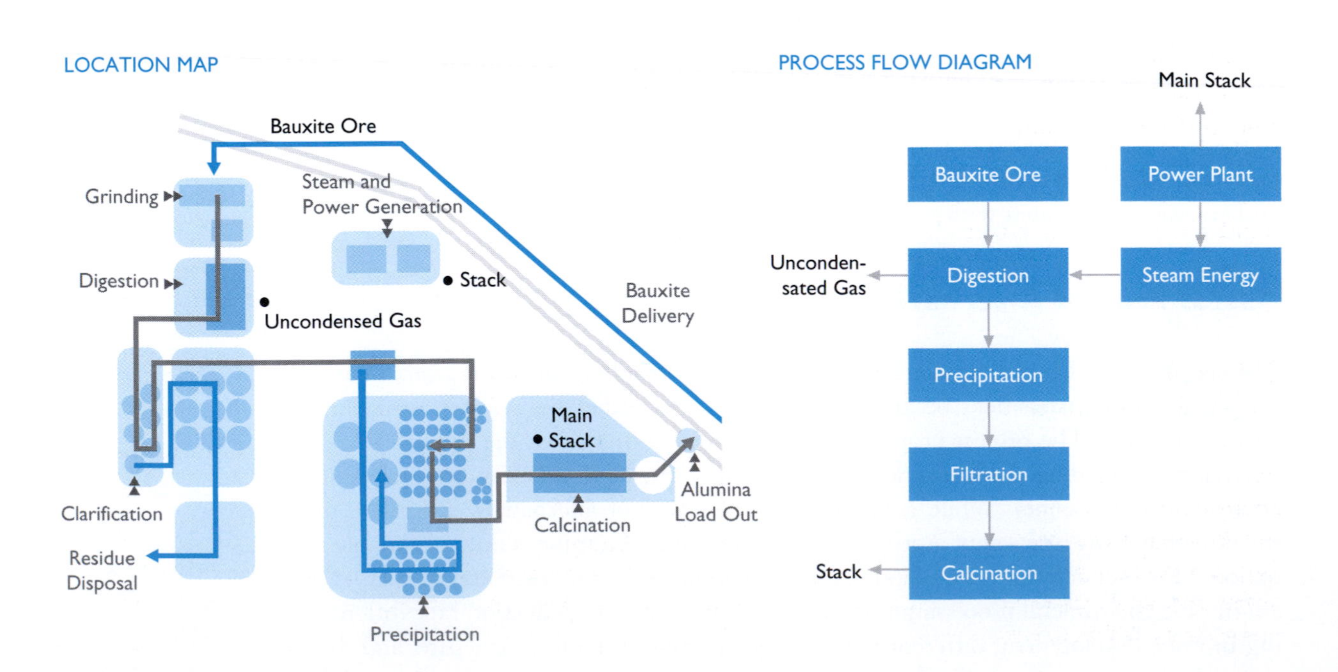

suitability under local conditions, and institutional, training, and monitoring requirements. For each alternative, the environmental costs and benefits are to be quantified to the extent possible. The basis for the selection of the preferred mine development scheme must be stated.

Embrace Participation

The Equator Principles are clear and firm on the need for public disclosure and participation. The people affected by the mine development at all levels must be consulted and given the opportunity to participate in the environmental assessment process. Public disclosure and participation must be open, transparent, and voluntary (IFC 2006). It is recognized, however, that traditions in different countries vary where participation and consultation are concerned. In consultation with local groups, a clear gender perspective must also be maintained. People also differ. As such, consideration should be given to the ways and extent to which women, men, and children may be affected by the proposed mine development.

Traditions in different countries vary where participation and consultation are concerned.

Few principles guide the process of participation of stakeholders. An initial stakeholder mapping should be carried out to identify project-affected people and other stakeholders, and to understand their expectations and interests. Consultation should start at the initial phase of the mine proposal and should include the opportunity for stakeholders to actively participate in the environmental assessment of the project. Affected people should be informed of the nature of the project and its likely impacts in an honest and unbiased matter. To avoid the creation of unrealistic expectations, no promises should be made that may later be revoked.

The importance of public disclosure and participation cannot be overstated, and we discuss this subject in detail in Chapter Three.

Understand the Legislative and Regulatory Framework

Attention to the legislative and regulatory framework as it applies to a given mining project is essential for understanding the administrative procedures required to support the environmental assessment process. Most countries have developed specific EIA guidelines, and most of them are similar in their general nature and their intent. OECD (1998, 1999) provide a helpful comparison of various EIA schemes.

Environmental assessment must always comply with applicable environmental laws and regulations of the host country. Developing countries, however, sometimes lack specific guidance or standards. Regulatory standards may be incomplete, or due to lack of experience, may be too stringent to be practical for some industries (**Case 2.5**).

If international lenders are involved, established international industry practices and standards often complement the applicable national legislative and regulatory framework.

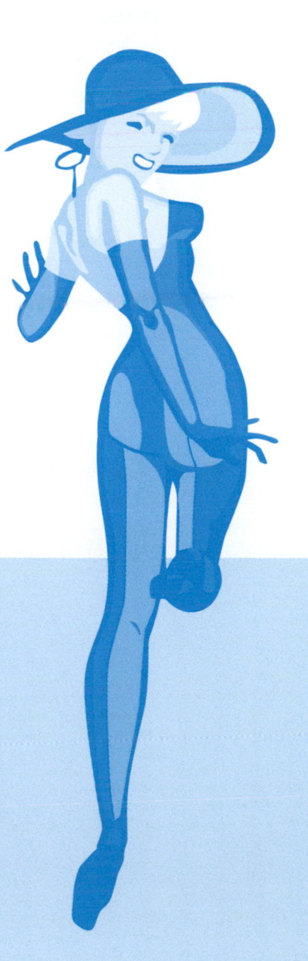

CASE 2.4
Exporting Oil from Sakhalin Island

The Russian Government has argued for turning natural gas from the Sakhalin Island into liquid form and shipping it on tankers to broaden the range of potential customers. Exxon says this option is too expensive. But Ivan Malakhov, the current governor of Sakhalin, says Exxon should show Russia all the potential options, not just its preferred one. 'If you're telling us this girl is the prettiest one, then show us the photos of the others we haven't seen. So far, we've seen only one lady.'

Source: The Wall Street Journal, May 8, 2007)

Pinup Girl: adopted from Art by Andy Nortnik (www.andynortnik.com)

The Equator Banks require compliance with Equator Principles as well as local regulations.

Mining projects that are partly funded by IFC, the private investment arm of the World Bank, must be implemented according to the administrative and legislative procedures of the host country as well as in accordance with applicable IFC directives. In the same way, the Equator Banks require compliance with Equator Principles as well as local regulations.

Finally, most international mining companies have developed their own corporate policies and codes of practice with which all mining activities within the group have to comply. In some cases these go beyond the requirements of regulatory and financial agencies.

Establish the Environmental Baseline

An environmental impact assessment study starts with the collection of background information on relevant physical/chemical, ecological, demographic, socio-cultural, and economic conditions of the host region (often simply termed 'the baseline'). Baseline data form the basis of describing the existing environmental setting with which the mine will interact. As its derivation suggests, the term environment is always a relative one, meaning 'surroundings'. An environment is the environment of 'somebody' or of 'something'. Hence there is no such thing as a single environment. The environment of the mine is the area in which it operates, and the features of that area that affect the mine in some way, providing sources of mineral resources and environmental and social challenges. Strictly speaking the mine and the host community do not have identical environments because each is part of the other's. Less strictly speaking they can be said to share an environment because most of the features of the area affect each of them in similar ways.

This, however, is the cause of much of the controversy that surrounds mining. To speak of an environment in this way is to speak of it as a set of resources. That is to emphasize one side of a two-way relationship. The other side is the impact that the mine has on the environments of people and other living things in the host region. Hence, which the environmental baseline do we establish during the EIA process – the environment of the mine or the environment of the communities living in the mine area? This differentiation is important since what improves the locality in terms of the environment of the mine, may spoil the environment of the host community, or any other living thing in the host region. The answer is to address both environments. Firstly, the EIA practitioner has to be concerned with the environment of the people and the living things in the host regions, and the effects of mining upon their environment. Secondly, she has to consider the environmental features or resources that have the potential to affect mine development, both negatively as well as positively.

What improves the locality in terms of the environment of the mine, may spoil the environment of the host community.

Once established, the baseline describes the no-action or zero alternative, against which the mining project is assessed. As mentioned earlier there must be the consideration of

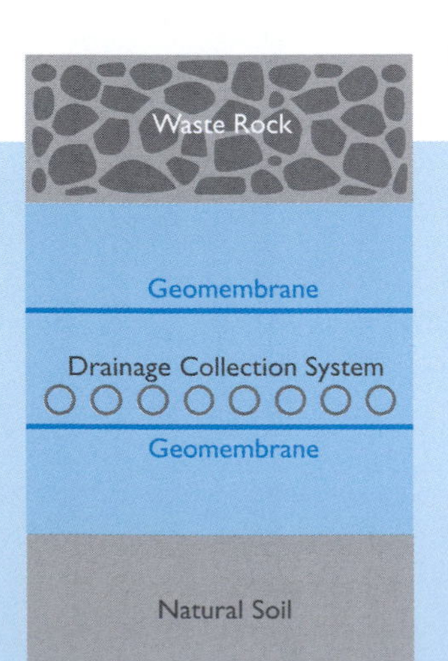

CASE 2.5
Indonesian Government Regulation No. 19 of 1994

Indonesian Government Regulation No. 19 of 1994, concerning hazardous and toxic waste management and now superseded with new regulations, serves as a case in point for too-stringent national regulation. The regulation categorized overburden and tailings indiscriminately as hazardous waste, requiring overburden and tailings placement areas to be designed to the same stringent standards as a hazardous and toxic waste landfill. The regulation, if strictly enforced, would have basically precluded mining.

future changes that are expected to occur in the absence of the project. Establishing the environmental baseline commands considerable commitment in term of human resources, time, and finances, and is further discussed in Chapter Eight.

Identify Potential Impacts

The variety of environments is matched by the variety of impacts that can be termed 'environmental'. The environmental impact assessment study attempts to identify changes to the environments of host communities and existing fauna and flora that can arise as direct or indirect consequences of the mining project, considering economic, social, and environmental aspects. Chapter Nine elaborates on identifying and evaluating environmental impacts.

A common mistake in environmental impact assessment is focusing on negative effects rather than elaborating on unbiased positive and negative project consequences. The second common mistake is to emphasize physical chemical and biological changes (say change of total dissolved solids in streams, dust concentrations, or how many hectares are cleared), and to address social and economic effects only briefly (say changes to quality of life of host communities, cultural changes, induced development, economic growth opportunities or the distribution of economic benefits).

A common mistake in environmental impact assessment is focusing on negative effects rather than elaborating on unbiased positive and negative project consequences.

Effects of course depend on the place where they occur and the people who are affected. Environmental effects can be quite local, like water erosion at the mine site. They can also be far reaching, such as with emissions from a roaster or smelter. And, at the extreme, they can be global in extent like the effects on the Earth's atmosphere of the emissions of carbon dioxide due to say, power generation to operate a mill, or due to the release of coal bed methane.

Some negative effects on communities and fauna and flora in the host region are unintentional and may be difficult to predict. The act of land clearing and mining is intentional but the indirect consequences are not intended, or even foreseen. The effects on the environments of men, women, children, and minority groups of the society differ and, therefore, should be presented separately, whenever relevant. Special consideration should be given to vulnerable groups and to future generations. Similarly, special attention should be given to sensitive receptors such as protected or pristine ecosystems, and endangered species.

Most positive effects on host communities and fauna and flora, however, are intentional. In the past there has been the tendency to concentrate on whether mine-induced changes result in harmful effects on humans (or non-human environmental components), and whether the mine developer can do something about these effects. Today, the focus of environmental impact assessment has shifted to include the opportunities for betterment that a mining project offers. Both mining companies and governments have become increasingly aware and supportive of the many ways mine development can support improving the quality of life of host communities, and improvement of other environmental values. The adoption of the Equator Principles by financial institutions and the mining sector in recent years is proof of this increasing awareness.

Today, the focus of environmental impact assessment has shifted to include the opportunities for betterment that a mining project offers.

Quantifying Significant Impacts

Quantifying significant impacts – both positive and negative – is a central element of the environmental impact assessment process. In a sense, the process of quantifying impacts is similar to the mine project proposal screening process, in that impacts with minor effects are cleared, enabling greater focus of management efforts on impacts that are significant. In quantifying impacts, it is important that stakeholders are permitted to give their views; this is a main objective of public consultation and disclosure.

The quality of an environmental impact assessment depends largely on the quality of the team leader.

Chapter Nine details the various methods commonly applied to quantifying impacts. Since potential impacts of a large mine development are complex and varied, it is common to simplify the impact analysis by dividing the mining project into various activities and the environment into different components. Activities and their potential interrelationships to individual environmental components are studied separately. It becomes clear that senior stewardship is needed to understand individual sub-studies as parts of an overall picture, and to eventually integrate these efforts in a coherent manner. It is fair to say that the quality of an environmental impact assessment depends largely on the quality of the team leader.

Develop an Environmental Action Plan

The identification of significant positive or negative impacts is meaningless without the development of measures to avoid or minimize damage and to optimize the contribution of the mining project to sustainable development. In this respect it is important that the Environmental Action Plan (EAP) is operational and covers the physical and biological environment, as well as the social environment, as elaborated in more detail in Chapter Ten.

Having a plan, however, is not sufficient. The Equator Principles require identification of the people who will be responsible for the implementation of impact avoidance or mitigation measures and that sufficient funding be allocated for environmental management. Companies need to establish an environmental management system that supports successful implementation of the EAP (IFC 2006).

Management is incomplete without a mechanism to monitor management success and redirecting management efforts if necessary. Environmental monitoring is necessary to support effective environmental management. Monitoring is intended to evaluate the effectiveness of adopted operations and management. Monitoring also provides an early warning if mitigation measures are unsuccessful or, alternatively, monitoring helps to verify that expected environmental targets are met. Data collected in an on-going monitoring programme facilitates the detection of trends and the determination of whether a particular value is within the normal range for a specific site.

The establishment of appropriate indicators for gauging environmental and social impacts is essential; indicators that are measured will be considered important, while those that are not measured are likely to remain relatively unnoticed.

Monitoring should also include reference or control areas and communities unaffected by mine activities.

Environmental and social impacts require monitoring not only at the project site, but in the downstream environment, and in adjacent areas and communities. Monitoring should also include reference or control areas and communities unaffected by mine activities. Monitoring the development of a nearby community that has a similar social and economic setting, but is unaffected by the mine enables the documentation of changes over time that occur naturally in the absence of the project. This facilitates evaluation of the success of community development programmes for communities that are directly affected by and benefit from the mine activity. Monitoring should be carried out at an agreed-upon frequency until mine closure is completed. Reduced post-mining monitoring will demonstrate the effectiveness of mine closure measures. Similar to the management plan, the monitoring plan is a dynamic document and parameters, frequency, and locations will change over the life of the mine.

2.7 DOCUMENTING THE FINDINGS

Much can be said about the quality of EIA documentation, or the lack of it. Why bother? Firstly, the EIA documentation is an important instrument for mine approval. In most jurisdictions, approval of the environmental impact assessment is a prerequisite for environmental approval of the proposed project. Secondly, the EIA documentation becomes a legal document. Stated management and monitoring measures become compliance requirements for mine operation. Thirdly, the EIA documentation is an important reference for future reviews of the mining operation. If unforeseen environmental impacts do surface, government authorities will first refer to the original EIA documentation. Have these impacts been addressed in the EIA? If so, has the mining company stringently followed 'all' commitments made in the EIA to mitigate impacts? Disputes caused by environmental impacts do end up in courts, in which case lawyers dissect documents word by word. And so will anti-mining activists. Most major mining companies understand the importance of clearly written documents and apply significant efforts including legal reviews of each document before submittal to government authorities or other third parties such as banks.

The EIA documentation becomes a legal document.

Major mining companies understand the importance of clearly written documents.

Lastly, some government officials will eventually read the entire documentation, irrespective of length. This will not necessarily happen during the EIA approval process, but at later mine reviews, or mine inspections. Commitments made, even vaguely, may suddenly become important, and costly to implement (**Case 2.6**).

Format and Content

Most EIA references listed in this chapter will provide useful guidance on the content of the EIA documentation. In some cases the legislation of the host country will provide the mandatory format. Formats may differ from country to country, or organization to organization, but most have similar requirements in respect to the minimal content of an EIA.

There is the need to describe the existing environment to provide baseline or benchmark data. This baseline, when compared to the results of monitoring, facilitates future identification of changes that have occurred following mine development. This information can then also be used to assess the accuracy of the EIA predictions (Beanlands 1983, 1988).

There is also the need to elaborate on the full range of proposed actions and activities that may adversely affect the environment. It is perhaps this step more than any other that requires the involvement of the mining company in the EIA preparation, since no other institution or group has all of the necessary information.

The discussion of the predicted affects is, of course, the key section of the EIA document, and should comprise the bulk of the text. Often it does not. The EIA should discuss the range of environmental effects, considering simple, complex, direct and indirect

CASE 2.6
The Moment of Maximum Leverage

There is a common perception that once the EIA is completed and the project is approved and financed, the project documentation finds its way into the archives and is never read again.

Wrong. In most jurisdictions, the EIA represents a legally binding document. One example is the construction of a second transverse levee for the Freeport tailings deposition area in Papua, Indonesia, based on a brief reference as an optional measure in an otherwise voluminous EIA, that was picked up by government officials during a routine mine audit. Another example is the word-by-word review of the original EIA for the Minahasa Gold Mine in Indonesia during the Buyat Bay court case (and the questioning of the EIA team leader) 10 years after EIA approval.

It is for this reason that bigger companies increasingly prefer to produce two EIAs – one, reviewed (and often controlled) by lawyers line by line, to fulfil Government requirements with minimum commitments and a second more comprehensive document (sometimes termed EHSIA – Environmental, Health, and Social Impact Assessment) to fulfil the company's policy or other requirements. At least one US. based company has prepared separate EIAs on the same project – one to conform with host country requirements, the other using the format of the US. NEPA regulations. The jury on the benefits of producing two separate EIAs is still out.

impacts, and unavoidable impacts. The probability of the impacts occurring and the significance of those impacts should also be outlined. Depending on study constraints, evaluations of probability and significance of impact occurrence may be subjective or objective, qualitative or quantitative, vague or precise.

The EIA would be incomplete without presenting environmental mitigation and monitoring measures, together with financial and human resource requirements.

The EIA would be incomplete without presenting environmental mitigation and monitoring measures, together with financial and human resource requirements.

While the format of documentation varies from country to country, the EIA documentation commonly comprises the following documents:

- Terms of Reference (TOR) or Scoping Evaluation for the environmental impact assessment study
- Environmental impact assessment report (referred to as Environmental Impact Statement or EIS in the US) (for an example Table of Contents see **Table 2.4**)
- Environmental management and monitoring plan, often termed Environmental Action Plan (EAP)
- Executive Summary in non-technical language
- Appendices containing the data used in impact analysis
- Supporting specialized studies (for an example list see **Figure 2.8**).

These are not always separate documents. For example, in many jurisdictions, the EIA document also includes environmental management and monitoring plans.

The Language

Theoretically at least, writing an EIA report should be relatively straightforward. The format is likely to be prescribed by the host jurisdiction, and there are ample other EIAs that can serve as a guidance. And, as far as possible, an EIA should be written in simple language that can be understood by the general public. Real life however differs for many reasons.

Contrary to common belief, writing in simple language is often more difficult than writing in complex technical jargon.

The first is a general one. It cannot be assumed that all team members involved in the EIA preparation are decent writers. The second reason springs directly from the directive to write in plain, simple language. Contrary to common belief, writing in simple language is often more difficult than writing in complex technical jargon. The third reason is a result of teamwork. Each contributing team member has his or her own style and way of writing. All points considered, the preparation of a consistent, well-written, readable EIA report is a daunting task.

It is even more painful, however, to prepare an EIA in two languages, the language of the host jurisdiction as well as in English for the international stakeholders such as non-native mine executives, shareholders, or involved financial institutions. It is a mistake to think that this is only a matter of translation. **Case 2.7**, which provides some sample translations of well-written Indonesian EIA paragraphs by an official translator into English, speaks for itself.

The Presentation

An environmental impact assessment for a major mine is a substantial effort, and documentation of the study can comprise several thousands of pages. This leads to the question: 'How to manage the scope and size of EIA documents?'

The EIA documentation is supposed to be 'brief but thorough'. It is not supposed to be encyclopedic (though many are), nor is it supposed to be a 'mini-EIA'. As early as 1978, a

TABLE 2.4
Framework Table of Contents for Environmental Impact Assessment Report

Introduction	Background Project definition Cooperation among jurisdictions Objectives of the assessment Project proponent Study compiler
Description of Mining Project	Construction phase Mine operation Mine closure and post-closure activities
Institutional Setting	Legislative framework Applicable regulations and standards Company environmental and social policy International agreements and conventions
Project Alternatives	Alternatives within the project Zero-alternative (no project)
Public Involvement	Institutional cooperation Project disclosure Public consultation
Environmental Setting	Air Water Land Fauna and flora People
Analysis of Impacts	Natural resources Human resources Relocation and compensation Direct, indirect, and cumulative impacts Trans-boundary impacts Impact significance
Environmental Management Plan	Management system Budgeting Time line Management efforts per significant impact
Environmental Monitoring Plan	Monitoring efforts Indicators Location and time line Reporting requirements

The EIA would be incomplete without presenting environmental mitigation and monitoring measures, together with financial and human resource requirements.

regulation promulgated by the US. Council on Environmental Quality (CEQ) established the target size for an EIS as 'normally not to exceed 150 pages in length and for proposals of unusual scope or complexity 300 pages' (40 CFR 1502.7). Regulators and companies routinely fail to meet or even approach such page limits, probably driven by the hope that 'beefing up' the EIS will deter potential opponents and litigants, or, in the event that deterrence fails, support the argument that 'it's in there somewhere'. This tends to result in long, complicated, costly documents that are, in essence, EIAs with little or no public participation, and that aren't particularly clear about which impacts are or are not significant.

Several techniques for reducing the size of EIA documents exist, but companies are somewhat restricted by local regulations and practice. EIA formats prescribed in many jurisdictions lead to unnecessary repetition, throughout the document(s). Attempts on the part of the EIA preparer to streamline the text may be rejected by the regulator as incompatible with the format. As such rejection can lead to costly delays, EIA preparers tend to err on the side of slavish adherence to prescribed formats.

The scoping process provides the first and generally one of the best opportunities to keep the document from excessive growth later.

When preparing an EIA, the scoping process provides the first and generally one of the best opportunities to keep the document from excessive growth later. A proper analysis of the scope of the project allows limits to be set for what has to be analyzed later. It is particularly important at this stage to understand the nature of the decision that is to be supported by the contents of the environmental document. Documents not directly used in an EIA

FIGURE 2.8
List of Supporting Specialized Studies

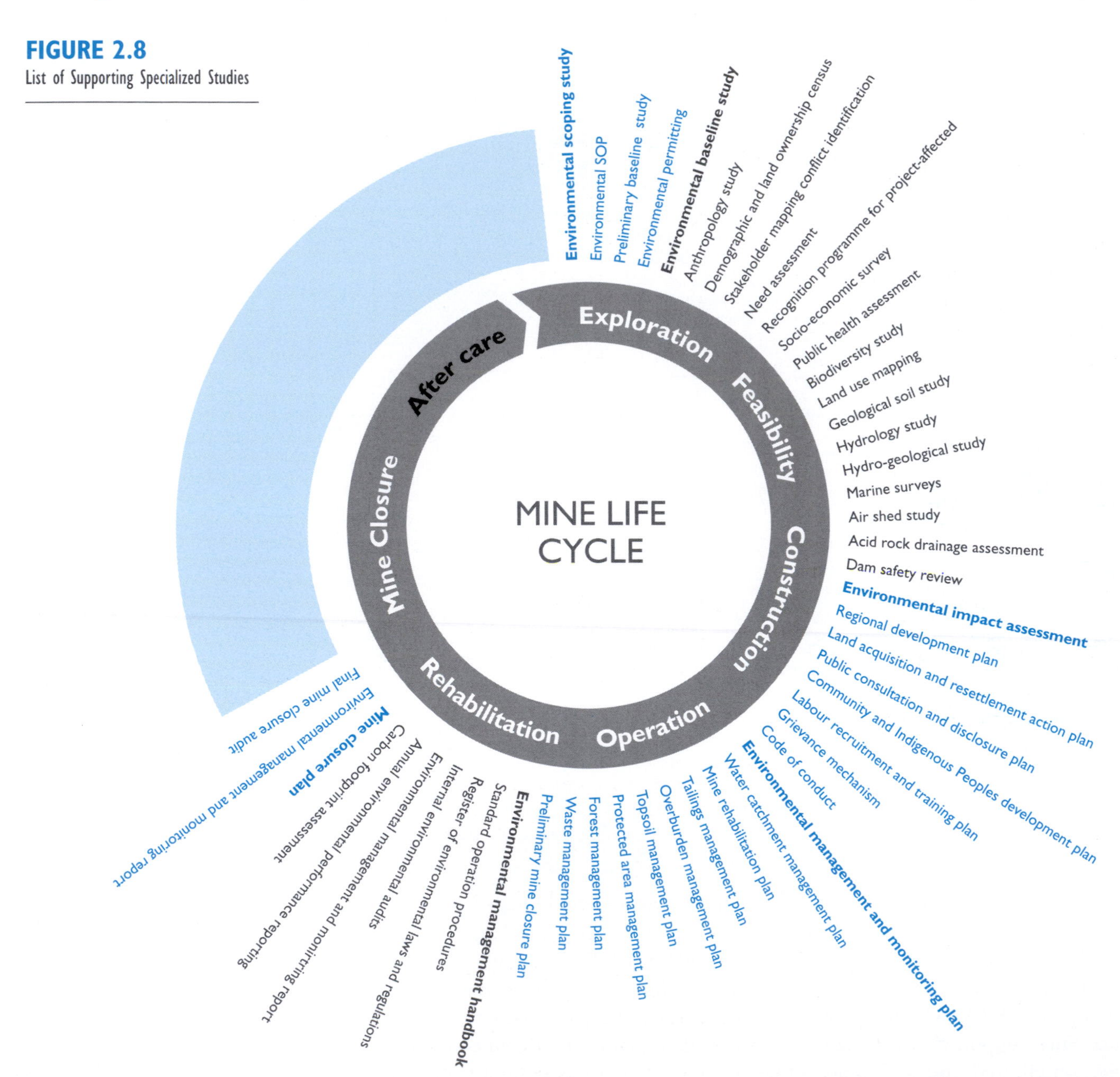

can and should be incorporated by reference. If this is done, care should be taken that the documents referenced are reasonably available to any reviewer who wants to review them.

A well-written statement of the purpose and need for the project (not why the document is prepared) lays the groundwork for a well-written, disciplined EIA document. It is often possible to reduce the size of an EIA by taking special care in describing the alternatives in this initial statement of purpose and needs. It is normally the practice to compare the impacts of the various alternatives in detail in the environmental impact section, and so detailed comparisons of impacts may be avoided earlier in the report. One technique is to use a summary table comparing the proposed project and its alternatives, referring to the detailed discussions in the subsequent environmental impact section.

Because significant amounts of data are generally available on current conditions, there is a tendency to 'load up' an EIA with such data simply because it is there. For example, most weather stations measure a variety of climatic parameters, including temperature, rainfall, wind speed and direction, relative, humidity, and solar radiation, all of which are usually presented and discussed in the baseline section of the EIA. Yet, rarely if ever are the data on solar radiation used in any of the subsequent impact assessments. There is a case therefore to include only that information which is used for impact assessment. One method that seems to help is to limit the description of the affected environment to a relatively minor discussion of where the proposed project is located plus general conditions in the area, and to include specific detailed information in the environmental impact section. In doing this, care must be taken to not simply transfer the problem from one chapter to another.

The environmental impact section should focus attention on significant effects. Elaborating on all mining-related effects with equal enthusiasm demonstrates poor judgement, and loads the documentation with unnecessary verbal baggage.

Elaborating on all mining-related effects with equal enthusiasm demonstrates poor judgement.

CASE 2.7

The Agony of Preparing EIA Documents in Two Languages: Translation from Original Text into English – Some Examples

'Mathematic model results can be modified into either a 1 or 2 dimensional graphic. This graphic results from overlay into a thematic map so that dimensions of impact distribution can be determined with concentration or intensity contour degradation. The limit of significant impacts could be determined if Environmental Standards, or environmental tolerance limits according to similar types of references results or literature references, have been set-up.'

'The natural forest is vegetated with mother plants that grow naturally. Vegetation is located throughout the study area, so its condition can be evaluated using a sample of vegetation representing the class. On-site observation and sampling shows that the forest seems to be already exploited by people there. The level of disturbance to the forest depends on people's activities or pressure on the forest. The forest areas of cultivated dryland were recorded planted with cash crops. This secondary/natural forest can be preserved as a conservation area.'

'Thus, the impact occurs could be categorized as unimportant negative impact in small category of impact (−2).'

'During operations of the alumina plant, the machines could be considered to unite with operation of power plant.'

'Climate and meteorological data including: air temperature, relative humidity, wind (direction and speed), duration of solar radiation and rainfall is generated from the climatologic station of The Meteorology and Geophysics Agency of Supadio Airport, Pontianak. The data on climate will be collected and analyzed to determine:

...

The range and average of annual global warming;

...'

'Level of impact importance symbolized in notation P (Important) and TP (unimportant) as follows:

Table VI-2: Notation of Level of impact importance

Notation of	Level of impact importance
TP	Unimportant
P	Important

Admittedly, the later examples highlight another challenge – a lack of national expertise in the host country, a common problem in developing countries.

In summary, as environmental documentation continues to increase in volume and complexity, methods must be developed to ensure that EIAs and other environmental documents are distributed as widely as needed, while reducing document bulk. Although it is probably not practical to think of the 'paperless EIA', there are some efforts that may be undertaken. EIA documentation can be made available on compact disk, which allows wide dissemination at a much lower cost than otherwise could have been accomplished. Likewise, most multilateral banking financing institutions upload a significant part of the environmental information, at least the Executive Summary, to a site on the World Wide Web.

Our closing comment addresses the need to produce and distribute EIA documentation for the approval process, typically involving 40 to 50 sets of documents for each round of EIA presentation to the approval authorities, depending on the host government EIA legislation. The effort involved in terms of time and cost is considerable.

2.8 OBTAINING EIA APPROVAL

Review of the environmental impact assessment documentation as part of the EIA approval process is one of the main checks built into most national EIA guidelines. Depending on the national, state or provincial legislation, the review process is carried out by an environmental agency, inter-departmental committee, commission of independent experts, or a review panel that may combine several of the aforementioned groups.

The review panel is a group of experts selected on the basis of their knowledge and expertise, appointed by the approving authority to review and assess, in an impartial and objective manner, a project with probable adverse environmental effects. The review panel submits its recommendations to the responsible approval authority. Review panels have the unique capacity to encourage an open discussion and exchange of views. They often inform and involve large numbers of interested groups and members of the public by allowing individuals to present evidence, concerns, and recommendations. The panel allows the mining company to present the proposed mining project and to explain projected environmental effects, and provides opportunities for the company to hear the views of government experts about the project, as well as the views of the public if it is invited to EIA presentations.

The review panel often recommends whether the mine proposal should proceed, or requires additional study and revision, or if it is environmentally unsatisfactory or not feasible. In the worst case, the mine project proposal must be withdrawn, and may or may not be acceptable as a resubmitted, redesigned project.

A mine proposal that is environmentally unacceptable would normally be recognized as such well before the formal submittal of the EIA documentation to the review panel.

In practice, environmental impact assessment reports are rarely rejected. Environmental assessment is a process and reviewers of the final EIA are commonly consulted during the EIA preparation. A mine proposal that is environmentally unacceptable would normally be recognized as such well before the formal submittal of the EIA documentation to the review panel.

The duty of the EIA compiler is not to persuade the decision-maker to cast a vote in favour of the project. The EIA compiler has the responsibility to bring environmental and social issues into focus, and to provide significant input into the development of alternative management strategies, and thus into the decision-making process. The EIA compiler assigns somewhat subjective weights to conclusions on environmental and social impacts. The presentation of findings should be clear enough to allow decision-makers to change these weightings to accommodate other considerations or their own experience.

The project proponent cannot always assume that the review panel has the resources or the experience to reach educated decisions. Mining as a whole, or specific mining mineral processing technologies, may be new to the members of the review panels. The review authority may lack the funds to invite all review members to site visits or study tours and to shoulder the associated costs for transportation, accommodation, and presentation. often, the mine proponent provides logistical assistance to the review panel to facilitate the review process. Fear of perceived corruption (and corporate rules) can limit the proponent's ability to provide such assistance that may include: (1) organizing and financing study tours to existing mines where proposed mining and mineral processing technologies are applied; (2) providing logistic and financial assistance in arranging workshops and EIA presentations; and (3) inviting recognized industry experts to provide seminars on selected topics.

Fear of perceived corruption (and corporate rules) can limit the proponent's ability to provide such assistance.

There is a common misperception that environmental permitting by means of the EIA process is the sole environmental permit that is required to commence mine construction and operation. Bureaucratic reality differs. Operating a large mine may require obtaining and maintaining several hundred operating permits, some at local, others at central government level, with quite a few directly related to the larger aspects of environment (see Chapter Four for an example of environmental permit requirements for a relatively small underground gold mine in Indonesia).

Operating a large mine may require obtaining and maintaining several hundred operating permits.

2.9 THE COSTS OF DELAY

Delays in EIA approval are common. Among other reasons, they may be caused by issues raised by anti-mining activists. However, there are many other contributing factors. The mining company may have yet to decide on final mine planning. Review and approval may be delayed to accommodate different points of view within or among government agencies (**Case 2.8**). The host government may assign its best civil servants who are most likely overloaded with a variety of duties, and hence will be slow in their study review. In developing countries, a delay in EIA approval may also occur due to the simple fact that the distribution of EIA documents to review panel members proves to be too time and cost intensive for the review panel. Members of the review panel are likely to ask for study clarification, even when the issues are well covered in the report. They may do so simply because they have not read the EIA reports.

Underpaid or venal bureaucrats who have the power to prevent, delay or halt the approval or construction process know that time is money for mine developers.

There is also the issue of corruption. Underpaid or venal bureaucrats who have the power to prevent, delay or halt the approval or construction process know that time is money for

CASE 2.8
Getting Trapped in the Red Tape

Although the mine proponent received official government approval for the environmental impact assessment, the decision was subsequently challenged by the Ministry of Forestry. Ironically, representatives of the Ministry of Forestry were members of the inter-departmental EIA approval commission. As of 2007, the issue is still unresolved causing a delay of mine development of more than two years, and the early retirement of the Senior Project Manager. His retirement notice says it all:

'Dear All

Please note that I am retiring from fulltime work for the ... group as of today in order to stop waking up in the middle of the night in a lather – agonising about when we will ever get the ... Project up and running. The Forestry bastards beat me in the end but the company will prevail!

Thanks to all of you in making life more bearable up in .... I will miss my trips up there (but you might not have seen the last of me yet). I will retire until I have written my memoirs and a few papers or get bored shitless (maybe 1–2 weeks) but hope to be doing a bit of consulting for ... and whomever else.

Kindest regards and best wishes'

mine developers. 'If construction is financed by loans, interest costs begin to grow immediately as soon as the first money is drawn from the lender and will continue through the mine's development and well into the production phase until the loan is repaid. Even if development funds come from the shareholders or joint venture partners, there is a need to bring the investment to production, establish a cash flow and commence dividends as rapidly as possible' (Marshall 2001). It is no coincidence that, particularly in developing countries, international companies who refuse to pay bribes experience more and longer delays in securing project approvals than 'local' companies who follow local corrupt practices.

Is project delay really significant?

Is project delay really significant? A positive feasibility report, showing a clear and reasonable expectation of a satisfactory return on the investment, is the basis on which financing is usually obtained for the construction of the mine. Since mine construction is so capital intensive, some sort of outside financing is usually required or desirable. Such financing must not only cover most of the costs of building the mine, but provide the initial working capital to allow the mine to operate until cash flow from the sale of minerals is established on a regular basis. The loan should also provide funds to cover initial payments on the interest incurred on borrowings made during the construction period. Financing often requires guarantees to the lenders that the project will be built by a certain date and will perform in accordance with certain specified standards consistent with the feasibility study. Consequently, for a company constructing a mine, maintaining the schedule becomes a critical factor (Marshall 2001).

'Much more significant, however, are costs associated with deferred or cancelled production, foregone earnings, disruption to customers, lost market opportunities, and interest on borrowed money.'

A study by Jorgensen *et al.* (1996) indicates that the risk associated with 'project delays is the primary concern of respondents in the areas of project approvals and environmental assessments. Delay costs may involve out of pocket expenses if construction schedules cannot be met. Much more significant, however, are costs associated with deferred or cancelled production, foregone earnings, disruption to customers, lost market opportunities, and interest on borrowed money.' The project may also be tied against long-term sales contracts and penalties may apply if contractual sales commitments are not met (**Table 2.5**).

TABLE 2.5
The Costs of Delay

Loss for Owners and Investor
Interest
Special Damages
Lost Profits
Loss of Use
Increased Financing
Extended Maintenance and Operations Expenses
Extended Field Office Overhead
1. Labour Costs
2. Equipment Costs
3. Material Costs
4. Bonding Costs
5. Subcontractor Costs
Unabsorbed Home Office Overhead
Attorney's Fees
Loss for Host Country/Region
Forgone Income
Forgone Taxes, Royalties, and Dividends
Delayed Social Investments

The costs of delay can amount to up to US$ 110 million for a US$ 240 million project for delays between 6 and 18 months. This cost does not account for the loss of economic benefits to the host country. Foregone income to project employees, contractors, and suppliers could amount to more than US$ 100 million for one year of delay, not accounting for loss in royalties, taxes, and dividends

Hence, contrary to the prevailing thinking in project management or in the management of most commercial endeavours, the real costs of delay are not related to capital cost increases in developing the mine but in the loss of mine value (and sales). Only if the value of the mining project greatly exceeds its cost at the planning stage it is worth pursuing. It has been estimated by Jorgensen *et al.* (1996) that the costs of delay range from US$ 38 million to US$ 110 million for a US$ 240 million project, when financed with a debt/equity ratio of 50/50 at 9% on debt and a 15% rate of return on equity for delays between 6 months and 1.5 years.

The biggest financial impact will be felt in the early months of a project delay. Investors look at actual cash flows discounted to their present value. The value of money over time is a key consideration, as are interest payments on drawn loans. But there are other costs of regulatory delay. Equipment may have been ordered (an important consideration given the long delivery times for some items) and delivered, but cannot be installed. Financial guarantees may have been paid to suppliers to ensure that equipment is available when needed. In addition the estimated costs of external consultants and lawyers involved in a company's regulatory application can exceed US$ 100,000 per month.

The biggest financial impact will be felt in the early months of a project delay.

There is, of course, the argument that costs associated with project delay are factored into the project financing of a mining company that has already made the decision to proceed with a project. This assumes that approval timelines can be reasonably predicted. If uncertainty is too great a factor, or if timelines are felt to be unreasonably long, mining companies may well decide to invest elsewhere.

The above costs do not account for the loss of economic benefits to the host country for the duration of the regulatory delay. Foregone income to project employees, contractors, and suppliers could amount, for some projects, to more than US$ 100 million for one year of delay, not accounting for loss in royalties, taxes, and dividends. A loss of benefits of that magnitude is a significant cost to any host country and community.

2.10 ENVIRONMENTAL MANAGEMENT AND MONITORING IS IMPORTANT

Accurate predictions of impacts and well-planned design of mitigation measures are important, but alone they are not sufficient. Until it is implemented, the EIA is merely a paper report. Implementation is fostered by establishing an adequate environmental management system and by integrating environmental budgets, including allocation of social investment funds, into the mine development budget. Then, specifically trained staff are needed to implement programmes and put to use allocated environmental budgets. Monitoring and reporting requirements, and milestones and spot checks all support implementation and help ensure that the EIA recommendations are fully implemented (Goodland and Mercier, 1999; IFC 2006).

Until it is implemented, the EIA is merely a paper report.

Environmental Management – More a Question of Company Culture than Systems

A systematic approach to environmental management from the outset of the mine project helps to avoid or to minimize key negative impacts, while maximizing positive contributions to sustainable development. Formal environmental management begins at the exploration stage, but management efforts increase to a much higher level when mine operation commences.

The mining company has overall responsibility for environmental performance even when the activities are carried out by contractors.

Numerous environmental and social impacts occur during construction. Mining companies at times hide behind the Engineering Procurement and Construction (EPC) system, and argue that contractually, the project is only 'handed over' to them after commissioning. Regulators and people in local communities who may be affected by the construction phase are not sensitive to or impressed by this 'handover' argument. The mining company has overall responsibility for environmental performance even when the activities are carried out by contractors. Accordingly, the mining company needs to ensure that contract documents include the requirement for appropriate environmental management, and also to maintain sufficient oversight of construction operations to ensure that contractual obligations are met. Negative perceptions towards the mining project developed through poor environmental management during mine construction are difficult to reverse, even if subsequent environmental management of the project achieves the highest standards.

Environmental management is greatly assisted by a formal environmental policy and management system that guides mine personnel. The environmental management structure should clearly define formal lines of responsibilities for maintaining legislative and regulatory compliance, and for achieving environmental goals. Virtually all project staff have environmental responsibilities. These responsibilities need to be well communicated, documented in individual job descriptions and should be reviewed regularly. Where appropriate, training should be provided to ensure that all staff are adequately trained to carry out their environmental responsibilities.

Overall environmental responsibility should rest with senior management, separated from day-to-day supervision of production.

Overall environmental responsibility should rest with senior management, separated from day-to-day supervision of production. Environmental site personnel need to have environmental expertise and skills. Dedicated personnel (in the sense of full-time, as well as personally committed) with relevant expertise should be assigned to social programmes.

There should be adequate financial and personnel support to carry out the many tasks related to environmental and social management. Regular reporting to top management is essential, with direct access in situations that require immediate management decisions. Periodic external audits, submitted to top management, should be carried out to evaluate the effectiveness of management policies and practices.

Effective collaboration with production personnel is essential to successful environmental management. Environmental and social performance policies need to be known to and respected by all mine personnel; production personnel need to develop a feeling of ownership for environmental targets. Ideally, as with safety and quality control, environmental management should be a production responsibility, not classified merely as a support function.

Environmental management should be designed to allow for effective and transparent communication with the public and with government agencies. Public environmental reporting has become a key element of major mining companies and is guided by the Global Reporting Initiative (2000).

While mining operations rarely generate major accidents that endanger local communities, this has happened, most notably in the failure of tailings storages. Environmental management should include the preparation of emergency plans for potentially hazardous activities related to the mining operation. To the extent that they could be affected, communities need to be apprised of such emergency preparedness plans. Tailings disposal in particular stands out as hazardous, especially in regions with rugged terrain, high seismicity and extreme rainfall events. The UNEP Industry and Environment Program Activity Center has developed a Handbook on Awareness and Preparedness for Emergencies at the Local Level (UNEP 2001) that can assist in the preparation of such plans.

Social Investment – More than a Buzzword

Social investments encompass the range of community development activities undertaken by the mining company to maximize the benefit of the operation to the local communities. To some degree, they are designed to offset negative impacts caused by the mine. More generally they are also the means by which the company builds renewable resource capital to offset the loss of the mineral resource capital in the region.

No single definition of social investments has emerged, nor is there a single approach that can apply to all mining projects. Social investment is very much dependent on context, and community development programmes differ from mine to mine. Social investments are typically undertaken in partnership with local governmental organizations and community based NGOs. Social investment includes programmes that improve the social infrastructure of the region, encourage environmental protection, build community capacity, and lead to improvements in education and livelihoods. In developing countries and elsewhere where Indigenous Peoples are involved social investments are essential to the success of the project. They may also bring considerable benefit to the reputation of the company. To the extent possible, programmes should be designed to avoid long-term dependency on the mining company.

Social investment is very much dependent on context, and community development programmes differ from mine to mine.

Monitoring and Auditing

Environmental monitoring is necessary for several reasons. Monitoring prior to commencing operation of a mine provides the benchmark against which environmental impacts are measured. It provides information on the environmental quality before, during, and after mine operation. Monitoring needs to address all five environmental components: air, water, land, biota, and people.

It is recommended that monitoring programmes include reference locations similar to those in the mine environment but which are unaffected by the mine (also called control points). Monitoring a reference location provides helpful insight into environmental, social, and economic changes that occur in a similar setting over time without the mine. Only by comparison can the impacts of the project be distinguished from natural changes.

It is recommended that monitoring programmes include reference locations similar to those in the mine environment but which are unaffected by the mine.

Ongoing monitoring during operation will help to pinpoint vulnerable aspects of the operation and occasionally will highlight general operating deficiencies. It demonstrates the effectiveness of environmental controls and protection measures. Environmental monitoring is also a regulatory requirement to demonstrate compliance with applicable national standards and project specific requirements.

Environmental auditing is essential since it includes the review of administrative and managerial practices. Audits can be usefully conducted at all stages of the project. For example, a Construction Audit may be scheduled to evaluate whether or not construction is being carried out in accordance with project commitments and applicable regulations. During operations, audits are usually carried out at regular intervals – usually yearly. However, surprise audits may also be undertaken. Mine Closure Audits are carried out to assess closure progress and achievement of closure goals and acceptance criteria.

Environmental audits critically review the adequacy of the company's environmental policy and objectives, the environmental management structure, and the financial and human resources allocated to environmental management. Audits can be carried out by the company (internal) or by third parties, usually consultants (external). An environmental

Environmental audits critically review the adequacy of the company's environmental policy and objectives, the environmental management structure, and the financial and human resources allocated to environmental management.

audit gives an overall view of the mine company's ability to provide adequate environmental and social safeguards. The host country's government can request independent audits as part of their appraisal of mine projects (Freeport 1996, 2000, 2005); financial institutions investing in a mine often appoint their own independent environmental and social auditor(s).

2.11 PLANNING FOR MINE CLOSURE

A legacy of environmental and social issues at many mines world-wide underlines the importance of careful planning of a mining operation, from mine construction to mine closure.

For the non-mining specialist, it may seem unnecessary to plan for mine closure before mining activities have even started. Actual mine operations differs from the initial mine planning, and obviously so does mine closure. Early planning for mine closure, however, is good practice, and most national jurisdictions have recently added the requirement for a preliminary mine closure plan as part of the feasibility study and mine approval. A legacy of environmental and social issues at many mines world-wide underlines the importance of careful planning of a mining operation, from mine construction to mine closure.

Mining, especially opencast mining, is very much a temporary use of land and its resources. Landscape architects introduced the term 'landscape on loan' to describe potential impacts on land. Because of the potential long-term nature of mining impacts, the restoration of disturbed land is as important as the management of impacts during mine operation. Since the exploitation of minerals is only temporary, impact mitigation should focus on restoration planning from the outset of mine development. Even where there is no practical post-mining land use for surface mine pits, it is essential to ensure that post-closure impacts are not propagated to adjacent or downstream lands.

The socio-economic dimensions of a mine operation provide the most compelling reasons for early mine closure planning. Mining project operations provide a major stimulant for local and regional development, including employment and business opportunities, and tax incomes which can change the social patterns of nearby communities. Once the minerals are depleted, the mine as an engine for economic development is gone. Without careful early planning by both the mining company and the local government, local communities will be unprepared to cope with mine closure. Without mine closure planning, productive human, social, and environmental capital created by the mining operation may prove to be unsustainable, and local communities as well as the region may well end up worse off than had the project never been developed. On the other hand, with foresight and commitment, many positive outcomes can be achieved that will strengthen the capacity of local communities to cope with mine closure and to embrace replacement livelihoods.

Without careful early planning by both the mining company and the local government, local communities will be unprepared to cope with mine closure.

2.12 WHAT ENVIRONMENTAL ASSESSMENT IS NOT

Common misperceptions of the term environmental assessment exist. Environmental assessment is…

Not a Permitting Effort. It is fair to say that today's senior mine management does understand that environmental assessment is an important element of mine planning. This understanding is often based on past experience where an initial lack of attention to environmental and social issues troubled mine operations throughout the mine life, or even contributed to unscheduled mine closure (see **Case 1.3** as one example). This broad view, however, is not always shared by all technical personnel in charge of preparing the mine feasibility study and obtaining mine approval. They may see environmental assessment as an environmental permitting exercise. As a result, efforts are kept to a minimum, and the opportunity to elevate management of environmental and social aspects of mine operation from the very beginning to a level of importance equivalent to technical mine management is lost. Similarly, compilers of the EIA often lack the experience or the interest to view an environmental impact assessment study as anything more than an academic exercise to support a permitting effort.

Not an Afterthought. Considering the many challenges involved in developing a mine prospect to the bankable feasibility stage, initiation of environmental assessment is frequently postponed as long as possible without jeopardizing overall mine development. Again the importance of environmental planning at an early stage and its benefits for future mine operation are poorly understood.

In emerging countries, where mining companies are challenged to include local communities and experts as much as possible in mine planning and operation, the appointment of local universities to prepare the EIA is a convenient tool to accommodate nationalistic pressures.

Not a Scientific or Academic Exercise. Environmental assessment studies are often considered to be the domain of scientific institutions and universities. In emerging countries, where mining companies are challenged to include local communities and experts as much as possible in mine planning and operation, the appointment of local universities to prepare the EIA is a convenient tool to accommodate nationalistic pressures. To a large extent the preparation of EIA documents follow boilerplate formats, and environmental action plans in particular often reflect the lack of practical experience. A practical compromise in merging international mining experience and local expertise is to appoint an international consultancy as the lead in preparing the EIA with a strong mandate to include local expertise as much as practical and feasible.

Not Concerned with a Single Resource. Environmental assessment is a truly multidisciplinary undertaking that must consider a wide range of environmental, social, and economic issues all of which may be important. Most EIAs fall short of realizing their full potential. More often than not the focus of environmental assessment is on the physical, chemical, and biological environment. Social aspects are addressed only to a limited degree. This is particularly the case in the USA, where, although included in NEPA requirements, social impacts have typically received only superficial attention (Projects on Indian Reservations are exceptions). A consequence of this is that US-based firms venturing outside the US for the first time are unprepared for the social issues that need to be addressed and the breadth and depth of studies required. Local and regional economic impacts of the mine development are rarely adequately studied or understood.

Not a Top-down Management Exercise. The social dimension of mine development commands a participatory approach to environmental assessment. Affected people and other stakeholders need to develop ownership in the identification and assessment of impacts, and even more, in development of the management and monitoring plans that will affect their lives in a fundamental way. Environmental assessment provides a unique opportunity for mine management to begin building productive human and social resource capital. To be sustainable, a bottom-up management approach is needed, contrary to the conventional and much more convenient top-down management style that is usually applied. Environmental assessment requires taking a collaborative approach to address community concerns. It is about getting useful public participation to develop a shared responsibility to improve environmental decision-making.

Environmental assessment requires taking a collaborative approach to address community concerns.

Social patterns are highly dynamic and community development efforts need to be designed to be responsive to trends and issues that have not been anticipated.

Not a One-Off Exercise. The feasibility study outlines the mine planning and operation in as much detail as is possible with the information available at the time of document compilation. Actual mine operation will differ as actual conditions are encountered and new information is generated during mine development and operation. As such, mining operations are never static, flowing exactly as predicted by an initial mine blueprint.This is also true for environmental management. Environmental assessment is not a one-off exercise as part of the feasibility study. Environmental assessment and its findings need to be revisited as new information emerges during mine operation. Management and monitoring efforts need to be readjusted in response to changes in impact prediction. This is nowhere more evident than in social management. Social patterns are highly dynamic and community development efforts need to be designed to be responsive to trends and issues that have not been anticipated. Priorities will change over time. For example, in the early stages of a project, major issues may include the management of incoming workers, particularly single men, who may comprise a high proportion of the workforce. However, as operations continue, locally trained people from local communities will replace most of the non-local workers. As circumstances change some practices can be de-emphasized or discontinued as new measures are implemented

Not an Environmental Public Relations/Media Strategy. It is important for a mining operation to publicly demonstrate good stewardship and to maintain effective public awareness programmes. This is necessary to offset the negative and often unjustified criticism that a mining operation will receive. A publicly available environmental impact assessment document that reflects honestly and without bias the feedback from a large range of stakeholders will serve as a formidable defence against unfair criticism. However, environmental assessment is not intended to be a public relations tool, and it becomes very obvious when it is being used as such. The point is to inform the public and to involve the public, not to influence public opinion.

The point is to inform the public and to involve the public, not to influence public opinion.

Not a Study, but a Way of Doing Business. Corporate culture is reflected in the way environmental assessment is conducted. For mining companies with a well developed corporate culture and sense of environmental and social responsibility, environmental assessment is not a paper exercise but a way of doing business. Success is reflected in mine projects that receive public acknowledgement of their contribution to local and national welfare (e.g. Newmont Batu Hijau Project, Jakarta Post, 05 June 2004).

Not a Panacea to Solving all Issues. Environmental assessment is not the one solution to all of a mining operation's environmental and social challenges. As a mining operation learns from its experience, it will adopt additional approaches and actions to respond to environmental problems as they arise.

REFERENCES

Beanlands GE and Duinker PN (1983) An Ecological Framework for Environmental Impact Assessment in Canada, Institute for Resource and Environmental Studies, Dalhousie University.

Beanlands GE (1988) Scoping Methods and Baseline Studies in EIA, Environmental Impact Assessment: Theory and Practice. P. Wathern, ed. London: Unwin Hyman.

Canter LW (1996) Environmental Impact Assessment. 2nd edn. New York: McGraw Hill.

Carroll B and Turpin T (2002) Environmental Impact Assessment Handbook ISBN 0727727818 London: Thomas Telford.

Conacher A and Conacher J (2000) Environmental Planning and Management in Australia, Oxford University Press, Australia.

Donnelly A, Dalal-Clayton B and Hughes R (1998) A Directory of Impact Assessment Guidelines, (Second Edition), International Institute for Environment and Development (IIED). Russell Press, Nottingham.

Goodland R and Mercier J (1999) The Evolution of Environmental Assessment in the World Bank: from 'Approval' to Results. Paper No. 67, Environmental Management Series.

GRI (2000) Sustainability Reporting Guidelines on Economic, Environmental, and Social Performance, Global Reporting Initiative, Boston.

Harvey N (1998) Environmental Impact Assessment, procedures, practice and prospects in Australia, Oxford University Press, Melbourne.

IFC (2006) Environmental and Social Performance Standards, International Finance Cooperation, www.ifc.org

Jorgensen J, Mokkelbost P, Smith A, Butler Wutzke C, McCoy E, and Yarranton GA (1996). Overlapping Environmental Jurisdictions – Estimation of Economic Costs Associated with Regulatory Delay. Macleod Institute for Environmental Analysis, University of Calgary.

Marshall IE (2001) A Survey of Corruption Issues in the Mining and Mineral Sector; joint MMSD and IIED Publication.

OECD (1989) Environmental Assessment in Developing Countries, Note by the Delegation of the Netherlands to the DAC, Working Party on Development Assistance and Environment, October 9, 1989.

OECD (1999) Coherence in Environmental Assessment Practical Guidance on Environmental Assessment For Development Co-operation Projects.

O'Riordan T and Turner RK (1983) An Annotated Reader in Environmental Planning and Management. Oxford: Pergamon Press.

Pallen D (1997) Environmental Sourcebook for Micro-Finance Institutions. Asia Branch, Canadian International Development Agency.

Petts J (1999) Handbook of Environmental Impact Assessment, Vol. 1 EIA: Process, Methods and Potential ISBN 0632047720 Oxford: Blackwell Science.

Petts J (1999) Handbook of Environmental Impact Assessment, Vol. 2 EIA In Practice: Impacts and Limitations ISBN 0632047712 Oxford: Blackwell Science.

Prasad Modak and Asit K Biswas (1999) Conducting Environmental Impact Assessment in Developing Countries, 375 pp., United Nations University Press.

Raff M (1997) 'Ten principles of quality in environmental impact assessment', Environmental and Planning Law Journal, vol. 14, no. 3, pp. 201–221.

SCOPE 5 (1975) Environmental Impact Assessment: Principles and Procedures; Scientific Committee on Problems of the Environment (SCOPE), established by the International Council of Scientific Unions (ICSU) in 1969 http://www.icsuscope.org/downloadpubs/scope5 3/16/2006

Tromans S and Fuller K (2003) Environmental Impact Assessment – Law and Practice ISBN 0406959544 London: Reed Elsevier.

UNEP (2002) UNEP Environmental Impact Assessment Training Resource Manual, Second Edition, www.unep.ch/etu/publications/EIAMan_2edition_toc.htm

UNEP (2001) APELL for Mining: Guidance for the mining industry in raising awareness and preparedness for emergencies at local level.

Wood C (2003) Environmental impact assessment: a comparative review (2nd edition). ISBN 058236969X Harlow, UK: Pearson Education (Prentice Hall).

World Bank (1989) Operational Directive 4.01 on Environmental Assessment, converted in 1999 into a new format: Operational Policy (OP) 4.01 and Bank Procedures (BP) 4.01.

● ● ● ●

Appendix 2.1 Data Needs

	Data Collection		Data Exist?	
	Proponent	Consultant	Yes	No
1. INSTITUTIONAL SETTING				
1.1 Project proponent, specifying				
• Shareholders	☐	☐	☐	☐
• Address(es)	☐	☐	☐	☐
• Person(s) responsible for EIA/environmental compliance	☐	☐	☐	☐
1.2 Tentative organization chart of project illustrating				
• Environmental Management Unit	☐	☐	☐	☐
• Community Development Unit	☐	☐	☐	☐
• Health & Safety Unit	☐	☐	☐	☐
1.3 Environmental assessor(s) specifying				
• Company(ies)	☐	☐	☐	☐
• Address(es)	☐	☐	☐	☐
• Person(s) responsible (EIA project manager)	☐	☐	☐	☐
• Team members specifying area of expertise/responsibility	☐	☐	☐	☐
• Resumes of team members	☐	☐	☐	☐
1.4 Public authorities including contact details				
• Legislative authorities	☐	☐	☐	☐
• National authorities	☐	☐	☐	☐
• Regional authorities	☐	☐	☐	☐
• Local authorities	☐	☐	☐	☐
• Administrative boundaries	☐	☐	☐	☐
• EIA review committee	☐	☐	☐	☐
• EIA approval authority	☐	☐	☐	☐
1.5 Relevant NGO's involvement including				
• Local NGOs/interest groups	☐	☐	☐	☐
• Scientific NGOs	☐	☐	☐	☐
• Anti-mining advocacies	☐	☐	☐	☐
• International NGO	☐	☐	☐	☐

	Responsibility		Data Source	
	Project Proponent	Consultant	Exist	To be generated
2. LEGISLATIVE AND REGULATORY SETTING				
2.1 Project proponent's environmental and social policy(ies)	☐	☐	☐	☐
2.2 Applicable international conventions and treaties	☐	☐	☐	☐

	Responsibility		**Data Source**	
	Project Proponent	Consultant	Exist	To be generated
2.3 Other relevant international guidelines (e.g. Ramsar Site, Unesco Listing, 'The Equator Principles', 'The Australian Best Mining Practices', or the IFC Environmental and Social Performance Standards)	☐	☐	☐	☐
2.4 Environmental impact assessment guidelines of the host country	☐	☐	☐	☐
2.5 Applicable standards such as				
Land				
• Land use classification (such as protected forest or nature reserves)	☐	☐	☐	☐
• Soil quality standards	☐	☐	☐	☐
• Sediment quality standards				
Water				
• Effluent discharge standards	☐	☐	☐	☐
• Water quality standards	☐	☐	☐	☐
• Groundwater quality standards	☐	☐	☐	☐
• Drinking water standards	☐	☐	☐	☐
Air				
• Air emission standards	☐	☐	☐	☐
• Ambient air quality standards	☐	☐	☐	☐
• Ambient noise standards	☐	☐	☐	☐
• Work place noise standards	☐	☐	☐	☐
Fauna and Flora				
• Endangered species (Red list)	☐	☐	☐	☐
• Indigenous species	☐	☐	☐	☐
Waste				
• Sanitary and non-toxic wastes	☐	☐	☐	☐
• Hazardous and toxic wastes	☐	☐	☐	☐
• Medical wastes	☐	☐	☐	☐
2.6 Current spatial planning by host government	☐	☐	☐	☐

	Responsibility		**Data Source**	
	Client	Consultant	Exist	To be generated
3. PROJECT DEFINITION				
3.1 Project brief				
• Type of project, including purposes and objectives	☐	☐	☐	☐
• Project benefits, including need for project to proceed	☐	☐	☐	☐
• Project alternatives	☐	☐	☐	☐
3.2 Project location				
• Location map at regional scale	☐	☐	☐	☐
• Location map at local scale	☐	☐	☐	☐
• Site/plant layout, including supporting infrastructure	☐	☐	☐	☐

	Responsibility		Data Source	
	Client	Consultant	Exist	To be generated
3.3 Project time line				
• Exploration phase	☐	☐	☐	☐
• Bankable feasibility phase	☐	☐	☐	☐
• EIA approval	☐	☐	☐	☐
• Construction phase	☐	☐	☐	☐
• Expected mine life	☐	☐	☐	☐
• Mine closure	☐	☐	☐	☐
• Post-mining environmental care period	☐	☐	☐	☐
3.4 Ore reserves/resources in terms of				
• Cutoff grade	☐	☐	☐	☐
• Proven reserves (spatial distribution, depth, volume)	☐	☐	☐	☐
• Grade and ore composition (associated metals)	☐	☐	☐	☐
3.5 Mining schedule, including				
• Average ore production	☐	☐	☐	☐
• Maximum mine production	☐	☐	☐	☐
3.6 Mining methods including simplified flow chart covering mining up to product shipping	☐	☐	☐	☐
3.7 Details of processing facilities, such as				
• Type and size of mill	☐	☐	☐	☐
• Type of beneficiation (hydro-, pyro-metallurgical process details)	☐	☐	☐	☐
• Chemicals/input materials used	☐	☐	☐	☐
• Energy demand	☐	☐	☐	☐
• Gaseous emissions (gases, noise, and particulates)	☐	☐	☐	☐
• Hydrocarbon wastes	☐	☐	☐	☐
• Toxic and hazardous wastes	☐	☐	☐	☐
3.8 Details of supporting infrastructure including				
• Roads and bridges	☐	☐	☐	☐
• Haul road	☐	☐	☐	☐
• Rail lines and bridges	☐	☐	☐	☐
• Conveyor belts	☐	☐	☐	☐
• Harbour facilities (wharves/jetties, breakwaters, anchorages, aids to navigation)	☐	☐	☐	☐
• Airfield/heliport	☐	☐	☐	☐
• Conveyor	☐	☐	☐	☐
• Housing/Community complexes	☐	☐	☐	☐
• Health facility	☐	☐	☐	☐
• Ware house, maintenance yard	☐	☐	☐	☐
• Energy (source, generation and distribution)	☐	☐	☐	☐
• Waste water treatment facility	☐	☐	☐	☐
• Fuel pipelines	☐	☐	☐	☐
• Quarries, borrow pits	☐	☐	☐	☐
• Dams, dikes, levees, causeways, canals	☐	☐	☐	☐
• Sanitary land fill	☐	☐	☐	☐
• Medical waste incinerator	☐	☐	☐	☐

	Responsibility		**Data Source**	
	Client	Consultant	Exist	To be generated
3.9 Water demand by project				
• Use of fresh water (users and demand, quality requirements, intake points, storage, treatment process, and distribution)	☐	☐	☐	☐
• Use of sea water	☐	☐	☐	☐
• Use of recycled water	☐	☐	☐	☐
• Water balance diagram for project, including rain water flow and runoff/drainage	☐	☐	☐	☐

	Data Collection		**Data Source**	
	Project Proponent	Consultant	Exist	To be generated
3.10 Summary of all effluent discharges by project (source, quantity, quality, and discharge point)				
• Waste water treatment plants	☐	☐	☐	☐
• Cooling water	☐	☐	☐	☐
• Storm water drainage	☐	☐	☐	☐
• Mine dewatering	☐	☐	☐	☐
• Potentially acid mine runoff	☐	☐	☐	☐
• Discharge points	☐	☐	☐	☐
3.11 Summary of all gaseous and related emissions by project (source, quantity, quality, and discharge point)				
• Mining (especially mine mobiles)	☐	☐	☐	☐
• Mineral processing	☐	☐	☐	☐
• Transportation	☐	☐	☐	☐
• Greenhouse gases	☐	☐	☐	☐
• Diffuse emissions	☐	☐	☐	☐
• Noise	☐	☐	☐	☐
• Vibration	☐	☐	☐	☐
• Light	☐	☐	☐	☐
3.12 Summary of all solid/waste emissions (source, quantity, and quality)				
• Mine wastes (waste rock and tailings)	☐	☐	☐	☐
• Sanitary wastes	☐	☐	☐	☐
• Hydrocarbon wastes	☐	☐	☐	☐
• Non-toxic industrial wastes (e.g. fly ash)	☐	☐	☐	☐
• Hazardous and toxic wastes	☐	☐	☐	☐
• Medical wastes	☐	☐	☐	☐
• Biomass	☐	☐	☐	☐
3.13 Areas required/affected by the project including				
• Construction camps	☐	☐	☐	☐
• Roads and other support infrastructure	☐	☐	☐	☐
• Run on mine (ROM)	☐	☐	☐	☐
• Open pits	☐	☐	☐	☐
• Ore stockpiles	☐	☐	☐	☐
• Topsoil storage	☐	☐	☐	☐
• Overburden/waste rock stockpiles	☐	☐	☐	☐

	Responsibility		Data Source	
	Project Proponent	Consultant	Exist	To be generated
• Tailings storage facilities	☐	☐	☐	☐
• Groundwater depression zones due to underground mining or pit dewatering	☐	☐	☐	☐

	Responsibility		Data Source	
	Client	Consultant	Exist	To be generated
3.14 Land preparation and top soil management				
• Amount of topsoil available, removed, and stored	☐	☐	☐	☐
• Top soil management	☐	☐	☐	☐
• Types of vegetation removed and biomass management methods	☐	☐	☐	☐
3.15 Overburden/waste rock management plan:				
• Amount of overburden produced over time	☐	☐	☐	☐
• Location, size and design of overburden stockpiles	☐	☐	☐	☐
• Geochemical characteristics (including grade)	☐	☐	☐	☐
• Water management	☐	☐	☐	☐
3.16 Tailings management plan, including:				
• Amount of tailings produced over time	☐	☐	☐	☐
• Location, size and design of tailings ponds/deposition areas	☐	☐	☐	☐
• Geochemical, physical and toxicological characteristics of tailings	☐	☐	☐	☐
• Water management	☐	☐	☐	☐
3.17 Labour force requirements for exploration, construction, and operation, including:				
• Total number (maximum, average)	☐	☐	☐	☐
• Origin (including number of expatriate work force)	☐	☐	☐	☐
• Educational qualifications and skills needed	☐	☐	☐	☐
• Management efforts for workforce mobilisation and demobilisation	☐	☐	☐	☐
3.18 Mobilisation or construction of unusual equipment or facilities (unusual due to type, size, amount, etc. particularly items rarely or never used in host country)	☐	☐	☐	☐
3.19 Installation and maintenance of environmental protection measures (such as sediment traps, etc)	☐	☐	☐	☐
3.20 Installation and maintenance of environmental monitoring equipment (such as meteorological or stream gauging stations)	☐	☐	☐	☐
3.21 Mine closure and post operation phase				
• Removal of temporary structures at the end of exploration and construction activities and related rehabilitation measures	☐	☐	☐	☐
• Plans to restore/rehabilitate site after removal of mining and support facilities	☐	☐	☐	☐
• List of infrastructure elements that could be used in place or relocated after mine closure	☐	☐	☐	☐
• Potential land uses for all areas used in project, after mine closure	☐	☐	☐	☐

	Responsibility		Data Source	
	Client	Consultant	Exist	To be generated
• Reclamation schedule for open pit areas, overburden stockpiles, and tailings disposal sites	☐	☐	☐	☐
• Post closure plans for dams, reservoir, intake/diversion, and other surface water structures	☐	☐	☐	☐
• Pit void's/lake's hydraulic behaviour and outflow water quality	☐	☐	☐	☐
• Number of employees to be demobilized or employed elsewhere	☐	☐	☐	☐
• Post mining care (monitoring and management)	☐	☐	☐	☐
ENVIRONMENTAL SETTING				
4. LAND (PHYSIOGRAPHY, GEOLOGY, AND SOIL DATA)				
4.1 Sources of geological and related data for the project site	☐	☐	☐	☐
4.2 Topography	☐	☐	☐	☐
4.3 Geology	☐	☐	☐	☐
4.4 Geochemistry				
• NAG (Net Acid Generating Materials)	☐	☐	☐	☐
• ANC (Acid Neutralization Capacity)	☐	☐	☐	☐
4.4 Soil units	☐	☐	☐	☐
4.5 Seismicivity				
• Faults	☐	☐	☐	☐
• Past earthquakes	☐	☐	☐	☐
• Active volcanoes	☐	☐	☐	☐
4.6 Unique, special, and sensitive landforms and geological formations	☐	☐	☐	☐
4.7 Existing illegal mining activities	☐	☐	☐	☐
4.8 Magnitudes probabilities for geo hazards in the project area, including:				
• Landslide	☐	☐	☐	☐
• Earthquake	☐	☐	☐	☐
• Volcanic activity	☐	☐	☐	☐
5. WATER (HYDROLOGY, HYDROGEOLOGY, AND OCEANOGRAPHY)				
5.1 Existing and planned sources of hydrological, hydrogeological, and oceanographic data for the project site, including location, parameters, and period of record for each station	☐	☐	☐	☐
5.2 Surface water				
• Identify existing surface water bodies and their drainage basins	☐	☐	☐	☐
• Water quality (physical, chemical, and microbiological characteristics)	☐	☐	☐	☐
• Flows (instantaneous, monthly, annual, maximum and minimum flows)	☐	☐	☐	☐

	Responsibility		**Data Source**	
	Client	Consultant	Exist	To be generated
• Patterns of erosion and sedimentation	☐	☐	☐	☐
• Bed/suspensa load (quantity and quality)	☐	☐	☐	☐
5.3 Ground water				
• Aquifer system	☐	☐	☐	☐
• Potentially exploitable aquifers	☐	☐	☐	☐
• Groundwater recharge and discharge areas	☐	☐	☐	☐
• Groundwater levels	☐	☐	☐	☐
• Groundwater quality (physical, chemical, and microbiological characteristics)	☐	☐	☐	☐
• List of known groundwater wells and springs	☐	☐	☐	☐
5.4 Oceanography				
• Bathymetry	☐	☐	☐	☐
• Natural hydrodynamic patterns, including tides, currents, waves, scour, accretion, and sediment transport	☐	☐	☐	☐
• Salinity, temperature, and density	☐	☐	☐	☐
• Stratification and thermocline	☐	☐	☐	☐
• Dissolved oxygen and concentrations of major ions and metals	☐	☐	☐	☐
• Sediment physical and chemical characteristics, including metals content	☐	☐	☐	☐
5.5 Water availability				
• Availability and exploitation of water for drinking and domestic use	☐	☐	☐	☐
• Availability and exploitation of water for agriculture, industry, and other uses	☐	☐	☐	☐
• Existing water usage	☐	☐	☐	☐
5.6 Unique, special, and sensitive water bodies and/or wetlands	☐	☐	☐	☐
5.7 Magnitudes probabilities for geo hazards in the project area, including:				
• Floods including flash floods	☐	☐	☐	☐
• Droughts	☐	☐	☐	☐
• Tsunami	☐	☐	☐	☐
• Storm surges	☐	☐	☐	☐
6. AIR (METEOROLOGY AND AIR QUALITY)				
6.1 Existing and planned sources of meteorological data for the project site, including location, parameters, and period of record for each station	☐	☐	☐	☐
6.2 Existing meteorological data on:				
• Climatic type				
• Temperature regime	☐	☐	☐	☐
• Humidity	☐	☐	☐	☐
• Rainfall and rainfall pattern	☐	☐	☐	☐
• Wind direction and speed	☐	☐	☐	☐
• Solar radiation	☐	☐	☐	☐
• Evaporation	☐	☐	☐	☐

	Responsibility		**Data Source**	
	Client	Consultant	Exist	To be generated
6.3 Information on ambient air quality in the project area				
• Selected air quality parameters in accordance to host country legislation	☐	☐	☐	☐
• Noise	☐	☐	☐	☐
• Vibration	☐	☐	☐	☐
6.4 Existing sources of air emissions already operating in project area, together with emissions hey are producing (measured or estimated)	☐	☐	☐	☐
6.5 Existing sources of noise and vibration already operating in the project area, together with their noise levels and the time periods they are generated (measured or estimated)	☐	☐	☐	☐
6.6 Magnitudes probabilities for geo hazards in the project area, including:				
• Extreme temperatures	☐	☐	☐	☐
• Maximum precipitation, for all available return periods	☐	☐	☐	☐
• Maximum wind speeds	☐	☐	☐	☐
• Unusual storms (such as cyclones)	☐	☐	☐	☐
7. BIOLOGY				
7.1 Existing and planned sources of habitat data for the project site, including aerial and satellite imagery	☐	☐	☐	☐
7.2 Typical vegetation communities/associations and their functions as habitats	☐	☐	☐	☐
7.3 Internationally or nationally recognized critical habitats, natural reserves, national parks, or other protected areas	☐	☐	☐	☐
7.4 Common, characteristic, and economically important fauna in project area, including:	☐	☐	☐	☐
• Terrestrial	☐	☐	☐	☐
• Aquatic (fresh water, marine, estuarine)	☐	☐	☐	☐
7.5 Common, characteristic, and economically important flora in project area, in cluding:	☐	☐	☐	☐
• Terrestrial	☐	☐	☐	☐
• Aquatic (freshwater, marine, estuarine)	☐	☐	☐	☐
7.6 Threatened, endangered, protected, special status, endemic or culturally significant fauna and flora in project area	☐	☐	☐	☐
7.7 Sensitive habitats, including:				
• Wetlands	☐	☐	☐	☐
• Mangrove habitats	☐	☐	☐	☐
• Pristine forest	☐	☐	☐	☐

	Responsibility		Data Source	
	Client	Consultant	Exist	To be generated
• Coral reefs	☐	☐	☐	☐
• Sea grass habitats	☐	☐	☐	☐
• Turtle nestling grounds	☐	☐	☐	☐
• Migratory birds nesting grounds	☐	☐	☐	☐
7.8 Patterns of hunting, fishing, and trade in flora and fauna	☐	☐	☐	☐
7.9 Existing illegal logging activities	☐	☐	☐	☐
8. SOCIAL SETTING (PEOPLE; SOCIAL, ECONOMIC, CULTURAL, AND HEALTH ASPECTS)				
8.1 Existing and planned sources of demographic data for the project site, including location, parameters, and period of record for each community	☐	☐	☐	☐
8.2 Information on spatial organization and land use, including:	☐	☐	☐	☐
• Inventory of land and other resources at time of project and likely future developments	☐	☐	☐	☐
• Regional development plans, spatial plans, land use plans, and other natural resource plans applicable to the project area that have been officially adopted or are in preparation by the government at the local, provincial and national levels	☐	☐	☐	☐
• Probability of conflict or limitations arising between the project and plans for land and resource use	☐	☐	☐	☐
• Inventory of aesthetic and natural beauty values together with recreation areas in the project area	☐	☐	☐	☐
8.3 Demographic data such as:				
• Population numbers	☐	☐	☐	☐
• Age/gender composition/structure	☐	☐	☐	☐
• Population growth and trends (fertility, and natural increase rates)	☐	☐	☐	☐
• Mobility (in – and out – migration)	☐	☐	☐	☐
• Settlements and population density	☐	☐	☐	☐
• Circulation patterns	☐	☐	☐	☐
8.4 Information on economy, including:				
• Infrastructure, including market centres, roads, transportation facilities, educational facilities	☐	☐	☐	☐
• Economic activities (sectoral composition) and sources of livelihood	☐	☐	☐	☐
• Patterns of whole sale and retail/trade	☐	☐	☐	☐
• Traditional economies, including agriculture, fisheries, forest gathering	☐	☐	☐	☐
• Patterns of subsistence and monetized economic activity	☐	☐	☐	☐
• Workforce participation	☐	☐	☐	☐
• Average household income	☐	☐	☐	☐
• Local government budgets, major expenditures, and sources of revenue	☐	☐	☐	☐
• Crime rate	☐	☐	☐	☐

	Responsibility		Data Source	
	Client	Consultant	Exist	To be generated
8.5 Information on cultures and religion, including:				
• Ethnic origins, in particular Indigenous Peoples	☐	☐	☐	☐
• Traditions and beliefs regarding use and ownership of land and other natural resources	☐	☐	☐	☐
• Education participation and trends	☐	☐	☐	☐
• Division of labour (by age, gender, ethnicity)	☐	☐	☐	☐
• Arts and unique cultural values	☐	☐	☐	☐
• Religious practices, places of worship, sectarian, divisions, religious organizations	☐	☐	☐	☐
• Patterns of formal and informal leadership	☐	☐	☐	☐
• Government sponsored and community to used institutions, mass organizations, youth and women organizations	☐	☐	☐	☐
• Patterns of social interaction	☐	☐	☐	☐
• Sources of social conflict (interethnic, intraregional, intergenerational)	☐	☐	☐	☐
• Prevailing nutrition	☐	☐	☐	☐
8.6 Public health information on:				
• Existing medical facilities	☐	☐	☐	☐
• Number and skill levels of medical/paramedical personnel	☐	☐	☐	☐
• Prevalent diseases including sexual transmitted diseases	☐	☐	☐	☐
• Commonly prescribed medications	☐	☐	☐	☐
• Health trends (changing disease patterns)	☐	☐	☐	☐
• Morbidity and mortality (indicators, life expectancy, maternal/infant mortality rate, percentage of population with malaria, or hepatitis, etc.)	☐	☐	☐	☐
• Environmental health factors	☐	☐	☐	☐
• Disease vectors	☐	☐	☐	☐
• Roles of traditional practitioners and treatments	☐	☐	☐	☐

3

Involving the Public

Forging Partnerships and Trust

SILVER

Silver, like gold, is a rare and noble metal. It is the most chemically active of the noble metals, is harder than gold but softer than copper. Silver occurs in the metallic state, commonly associated with gold, copper, lead, and zinc. The beauty, weight and lack of corrosion made silver valuable, and hence one of the earliest of metals to be used as a medium of exchange. Today, demand for silver is built on three main pillars – industrial uses, photography, and jewellery and silverware. Silver is the best electric conductor of all metals; it can also achieve the most brilliant polish of any metal.

3 Involving the Public

Forging Partnerships and Trust

Agenda 21 and Principle 10, which emerged from the Earth Summit in 1992, both called for increased public involvement in environmental decision-making. In Europe, this led to the adoption of the Aarhus Convention in 1998, which, to date, is the most comprehensive international agreement on the right and importance of public participation in environmental decision-making. Public consultation and, as a mining project evolves, public participation is now accepted as an integrated part of any mine development (WSSD 2002), as it should be. The importance of public consultation and participation is also reflected in the World Bank's Policy on Environmental Assessment OP/BP 4.01 and the more recent IFC Performance Standards (IFC 2006), part of what is now referred to as 'The Equator Principles' (EP), the industry standard for best environmental and social assessment practice.

Public participation is now accepted as an integrated part of any mine development.

Equator Principle 5 (Consultation and Disclosure) states (www.equatorprinciples.com July 2006):

> 'For all Category A and, as appropriate, Category B projects located in non-OECD countries, and those located in OECD countries not designated as High-Income, as defined by the World Bank Development Indicators Database, the government, borrower or third party expert has consulted with project-affected communities in a structured and culturally appropriate manner. For projects with significant adverse impacts on affected communities, the process will ensure their free, prior and informed consultation and facilitate their informed participation as a means to establish, to the satisfaction of the Equator Principles Financial Institutions (EPFI), whether a project has adequately incorporated affected communities' concerns. In order to accomplish this, the Assessment documentation and Action Plan, or non-technical summaries thereof, will be made available to the public by the borrower for a reasonable minimum period in the relevant local language and in a culturally appropriate manner. The borrower will take account of and document the process and results of the consultation, including any actions agreed resulting from the consultation. For projects with adverse social or environmental impacts, disclosure should occur early in the assessment process and in any event before the project construction commences, and on an ongoing basis.'

The principle of participation, defined by the World Bank (1999) as 'a process through which stakeholders' influence and share control over development initiatives and the decisions and resources which affect them', derives from an acceptance that people are at the heart of development. At the broader, societal level, recent research has demonstrated that governments are often most effective when they operate within a robust civil society. At the project level, a growing body of empirical evidence demonstrates that project implementations tend to be more successful when stakeholders and beneficiaries are integrated into the planning process. The principle of participation also contains a normative component, in the belief that people have a right to be consulted about initiatives or activities that will have a major impact upon their welfare and lifestyle. Participation implies that management structures are flexible enough to offer beneficiaries and other interested and affected peoples the opportunity to improve the design and implementation of public policies or private sector investments.

Project implementations tend to be more successful when stakeholders and beneficiaries are integrated into the planning process.

The importance of public involvement in mine planning is generally acknowledged, but there is some disparity in the terminology used to describe this involvement, since there are no consistent definitions for the terms 'public' and 'involvement'.

The use of the general term public has given concern in the past. Mining companies prefer to see a direct causal link between the mining project and parties having a genuine stake in the project. For this reason mining companies prefer to clearly identify all interested and affected parties, commonly referred to as stakeholders, in the early stage of a mine proposal.

Mining companies prefer to see a direct causal link between the mining project and parties having a genuine stake in the project.

The use of the term 'involvement' also has different meanings for different people. It is commonly accepted, however, that involvement includes disclosure, consultation, and participation. The differences between these facets of involvement are significant (**Figure 3.1**). Disclosure represents a one-way flow of information from the mining company to the public. No opportunity for questions and discussion exists. Consultation is a two-way flow of information allowing the public to express their opinions on the project and to provide feedback. The mining company commits itself to considering stakeholder input into making its decision. Finally, participation involves shared analysis of the mining proposal, and provides the public with some level of control over the project.

More often than not in the past, mining companies have limited public involvement to a type of communication represented in the lower left-hand corner of **Figure 3.1**. Experience, however, demonstrates that mining projects with a history of local empowerment and

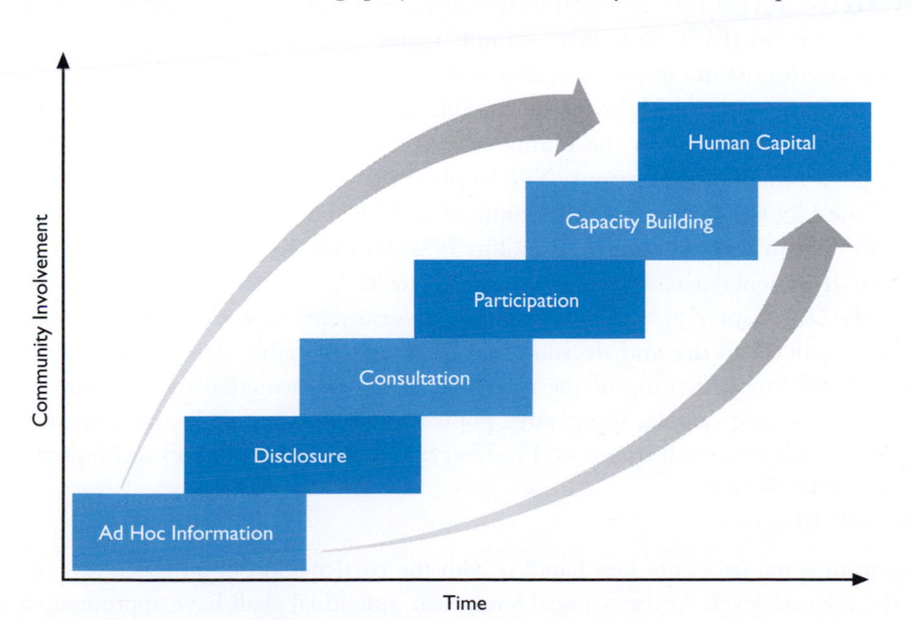

FIGURE 3.1
Community Involvement over Time

Participation involves shared analysis of the mining proposal, and provides the public some level of control over the project.

Mining projects with a history of local empowerment and dialogue generally have lower levels of social risk.

dialogue generally have lower levels of social risk. True two-way communication leads to greater benefits for both the company and the host community. What is sought in environmental assessment is that proponents go beyond the phase of informing. Proponents are asked to clearly and honestly explain the mining project to affected people, to actively listen to responses received, and to take their views into account in mine development. Truly participatory practice in public involvement – shared decision-making – is usually not appropriate or practical initially, even when it might be desirable. It requires a level of skills that need to be developed over time through capacity building and community development.

This chapter presents a brief historical perspective on public involvement, capturing the main aspects that will lead to successful stakeholder engagement: (1) planning of stakeholder involvement; (2) capacity building in the company; (3) identifying and engaging stakeholders; (4) capacity building in the communities; (5) understanding benefits and risks of public involvement; and (6) common mistakes.

3.1 HISTORICAL PERSPECTIVE

In democratic societies we, the public, have at least in theory the right and the power to decide on our future, and how we use natural resources to best suit our interests and needs.

The Right of Access to Information, Participation, and Justice

General provisions for public access to information, participation, and justice go back as much as 14 centuries in Islamic Law (Ahmad 2002). The International Boundary Waters Treaty in 1909 however was the first global treaty to include strong provisions for public access to information and participation (Bruch and Filbey 2002). The Universal Declaration of Human Rights in 1948 provided the kernels for generalized access to information and justice as well as the right to participate.

In the USA, as in much of the West, the 1960s and 1970s were marked by a participation movement (Breggin and Hallman 1999). The Freedom of Information Act in 1966 gave access to data and information held by government agencies. The National Environmental Policy Act (NEPA) of 1969 amongst its measures required public review of Environmental Impact Statements (EIS). As mentioned in Chapter One NEPA has largely been the basis on which environmental impact assessment world-wide has been based.

The UN Conference on Environment and Development in 1992 laid the foundation for public consultation in environmental assessment as we now know it.

The voice of environmental concern was internationalized at the Stockholm Conference in 1972 (UN Conference on the Human Environment). For the first time environmental non-government organizations (NGOs) played an important part. The follow-up conference, the UN Conference on Environment and Development (UNCED or 'The Earth Summit') held in Rio De Janeiro, Brazil in 1992, laid the foundation for public consultation in environmental assessment as we now know it.

Agenda 21, Chapter 8, Section 1 calls on governments to seek broader public participation in policy making and decision-making for sustainable development. Section 3 is devoted to the strengthening of the roles played by the principal social groups: women, children and young people, indigenous populations; NGOs; local government; workers associations and unions; business and industry; and the scientific and technological community (NCEDR 2003).

Principle 10 states:

> 'Environmental issues are best handled with the participation of all concerned citizens, at the relevant level. At the national level, each individual shall have appropriate access

to information concerning the environment that is held by public authorities, including information on hazardous materials and activities in their communities, and the opportunity to participate in decision-making processes. States shall facilitate and encourage public awareness and participation by making information widely available. Effective access to judicial and administrative proceedings, including redress and remedy, shall be provided'.

It would be difficult to overstate the significance of Principle 10 in providing a mandate to regional and international governance development in the decade following the Rio Declaration, as well as to an underlying commitment motivating national law making. Since 1992 international organizations, in particular the World Bank, have made steps towards making public consultation and involvement an integral constituent of their development loans (World Bank 1999). With the advent of the Equator Principles in 2003 public involvement has now also become mandatory in private sector financing.

From a European perspective, the most important initiative is the 1998 UN/ECE Convention on Access to Information, Public Participation in Decision-Making and Access to Justice in Environmental Matters. The ECE (Economic Commission for Europe) is a regional body of the United Nations and includes 55 countries. The European Community plus 39 other nations signed the convention (also known as the Aarhus Convention) in 1998 and it came into force on 30 October 2001. Described by the then UN Secretary General, Kofi Annan, as 'the most ambitious venture in environmental democracy so far undertaken under the United Nations,' the Aarhus Convention is based on three pillars: (1) access to information; (2) public participation in decision-making; and (3) access to justice.

The first pillar of the Aarhus Convention addresses concerns over the difficulties faced by the public in gaining access to environmental information. It aims to ensure that individuals are informed about their environment and that their role in decision-making is on an informed basis. It includes requirements that public authorities supply certain information on request and within time limits, and that they collect, possess, and disseminate certain information.

The second pillar aims to improve public participation in decisions relating to the environment. It accords individuals the right to participate in decision-making that may have significant environmental impacts. It requires the establishment of a transparent and fair framework, within which the public will participate in the preparation of plans and programmes relating to the environment, in other words, that there should be effective public participation in the preparation of law and rules with environmental impacts.

The second pillar aims to improve public participation in decisions relating to the environment. It accords individuals the right to participate in decision-making that may have significant environmental impacts.

The third pillar promotes access to justice. It requires that individuals have the right to challenge decisions concerning their access to information or public participation, by means of an independent review by a court of law or other independent body. Individuals should also have access to these review procedures to challenge violations of national law on the environment. While the Aarhus Convention is relatively restricted both in its procedural and in its geographical reach, it is path-breaking in the depth to which it seeks to democratize environmental debate and protection.

The Extent of Information Disclosure and Public Consultation at the Project Level

At the project level information and environmental assessment reports for EPFI projects are intended to be accessible to interested parties and the general public. The extent of disclosure and the need for consultation depends on the project categorization (EP 1), and the adequacy of public involvement is one of the criteria used to determine the project's compliance with the Equator Principles.

Public consultation is mandatory for all Category A projects as defined by the Equator Principles.

Category A Projects

Public consultation is mandatory for all Category A projects as defined by the Equator Principles (see Chapter One). To facilitate the required consultations with the affected people and local NGOs, the information about the project's environmental issues as well as technical data needs to be transferred into a form and language(s) accessible to those being consulted. The full environmental impact statement is also made available to interested parties upon request. Public consultation needs to be carried out during the early stage of Environmental Assessment (EA) preparation and throughout the project implementation to address any environmental issues that affect local communities, NGOs, governments, and other interested parties. Consultation should be 'free' (free of external manipulation, interference or coercion, and intimidation), 'prior' (timely disclosure of information) and 'informed' (relevant, understandable and accessible information), and apply to the entire project process and not to the early stages of the project alone. The consultation process needs to be tailored to the language preferences of the affected communities, their decision-making processes, and the needs of disadvantaged or vulnerable groups. Consultation with Indigenous Peoples must conform to specific and detailed requirements as found in IFC Performance Standard 7 (IFC 2006). Furthermore, the special rights of Indigenous Peoples as recognized by host-country legislation will need to be addressed.

Consultation should be 'free' 'prior' and 'informed'.

Category B Projects

It is recommended but not mandatory that public consultation be carried out during the early stages of the EA process and throughout the project implementation to address any environmental issues that affect the local communities, NGOs, governments, and other interested parties. To facilitate the required consultations, information about the project's environmental issues needs to be transferred into a form and language accessible to those being consulted.

Category C Projects

Information disclosure and consultation is not required for the Category C projects. However, consultation requirements may be recommended on a case-by-case basis, depending on the nature of the project and the relevant environmental and social issues, and interest level of the public.

The Increasing Voice of Non Government Organizations (NGOs)

NGOs are not an invention of recent history. Historically, associations of private individuals have gathered for public purposes, usually to provide a service not available from the state, well before the establishment of democratic governments. They have preceded, and now complement, the growth of government services. Many have been, and continue to be church-based, concentrating on the needs of individuals for assistance, typically in welfare, health and education.

In more recent times, a new class of NGOs has evolved, which focuses directly on changing public policy.

In more recent times, however, a new class of NGOs has evolved, which focuses directly on changing public policy. Though membership-based, they are unlike the representative interest groups of employers and employees that provide both, service to membership and public advocacy on behalf of their members. Many such NGOs consist, typically, of middle-class activists who want government to reallocate resources or change laws according to their views on the good society. Few matters of public policy or industrial development, not to mention new mine ventures, pass without an NGO spokesperson advocating a position. They have, in some regards, become the official opposition to government policies and private sector investments, a reality painfully known to all senior mine management (Johns 2000).

3.2 PLANNING STAKEHOLDER INVOLVEMENT

A number of authoritative guidelines on the public consultation process exist that can guide the formulation of a formal Public Consultation and Disclosure Plan, commonly abbreviated PCDP (e.g. Doing Better Business Through Effective Public Consultation and Disclosure – A Good Practice Manual, IFC 1998). The IFC standards are widely acknowledged as industry best-practice, and they became the 'role model' for subsequent guidelines published by other multi-lateral funding agencies. IFC requires consultation to be undertaken throughout the various phases of project activity. Consultation with stakeholders should commence during the environmental assessment preparation stage, and be maintained throughout mine life. Stakeholders include those directly and indirectly affected by the project, as well as other interested parties including members of the public, local authorities, NGOs, and businesses.

Consultation with stakeholders should commence during the environmental assessment preparation stage, and be maintained throughout mine life.

A 'consultation and disclosure' plan documents objectives, management structure and skill requirements, stakeholder identification, strategies for communication and involvement, conflict management, time lines, and funding. An example outline of a public consultation and participation plan is given in **Table 3.1**. A well-developed plan will help to answer the following questions:

- Why involve stakeholders?
- Who should be involved?
- What should be communicated?
- What feedback is desired and expected?
- When to start involvement?
- How to communicate with and involve stakeholders?
- What are the internal resource requirements?

The Three Levels of Public Involvement

Roberts (2003) distinguishes between disclosure (which he termed involvement), consultation, and full participation as the three progressively more inclusive levels of participation. In these three levels, a range of activities takes place, from education and information sharing, to consultation and community advisory groups, to the final level of public participation in consensus-based decision-making plans.

Step 1 Stakeholder mapping. The project proponent needs to determine who the stakeholders are and engage them in the project. Conducting the assessment itself can be a valuable way to involve stakeholders. The analysis is also important for starting to collect information on which aspects of peoples' lives might be affected, and to identify any groups that might be affected, but who had not been identified previously.

Step 2 Education and awareness creation (one-way information transfer). Stakeholders cannot participate meaningfully in the absence of information. Information transfer and capacity building are ongoing activities that underpin the entire process of public participation and consensus building. Providing information by itself is not a form of public participation, but rather an initializing step that begins to involve people, and should underpin any kind of further involvement, consultation, and process.

Providing information by itself is not a form of public participation.

TABLE 3.1
Example Outline for a Public Consultation and Disclosure Plan (PCDP)

Introduction • Overall purpose of consultation and disclosure • Structure of PCDP
Mine Proposal • Key elements of mine proposal • Time line
Stakeholder Mapping • Project affected people • Other affected people • Stakeholder profile • Key issues that need to be managed
Principles and Policy Objectives • Legislative requirements • Emergence of international standards • Company's internal requirements
Review of Past Disclosure and Consultation
Strategies for Continuous Communication and Stakeholder Involvement • Applied methods • Risks and opportunities • Grievance mechanisms
Resource Allocation • Funding • Organizational structure
Time Line
Monitoring • Indicators • Reporting

Step 3 Consultation and information sharing (gathering diverse opinions and contributions). Tens or even hundreds of stakeholders may initially participate in one way or another in the project. They are likely to present a diversity of perspectives e.g. different concerns, different suggestions, different local expectations and needs, and different priorities. In many cases this step includes mechanisms to integrate these perspectives into project planning, supported by technical evaluation.

Step 4 Option building, (possibly) converging in opinions, and a decision. Eventually, convergence in opinions and perspectives should start to emerge, and fewer stakeholders may be directly involved, until a decision is reached. The final decision may not necessarily develop as a consensus of all the participants, but it may be simply a formal decision by the mining company based on the various inputs received.

Step 5 Consensus-building for the long term (during project implementation). The highest level of public participation is collaborative decision-making which involves consensus building procedures. There are many reasons why collaborative decision-making, while desirable, is restricted to selected project aspects such as designing community development programmes. This level of participation will only be reached at the later stage of mine development.

A large number of public participation methods exist on which a consultation and disclosure plan can be built (see, for example, Glasson, Therivel and Chadwick 1994; Canter 1996). Each has its own advantages and disadvantages (see Appendix 3.1).

The common thread to all methods is that communication and public involvement is not a one-off activity. It is an interactive process, one that begins with informing stakeholders and leads from consultation to capacity building and eventually to shared decision-making. The main barriers to effective participation are: (1) attitudinal/proponent – proponents do not want to share decision-making with other stakeholders; (2) attitudinal/public – participation in highly complex and technically oriented environmental/engineering studies are intimidating; (3) financial – activist groups/NGOs operate on shoe-string budgets with overburdened volunteers, and therefore, cannot afford technical consultants or to put too much effort into one EIA; (4) procedural – in many planning/regulatory processes the many stages of review, notification, and submission are complex and aggravated by obscure and legalistic public notification procedures.

Creating an Enabling Environment

Not surprisingly theory and practice often differ in public engagement. A good understanding of the principles, the various phases, and the key elements of public involvement, 'the theory', is of course helpful and in fact essential to prepare for engaging in a successful campaign. Given the complexity and variety of individuals and human groups, however, it would be an unrealistic expectation to assume that most projects, in particular mining projects, could be implemented without controversy. Practice demonstrates that probably more than for all other aspects of a mining development, the key to success is having the right people involved in public engagement.

It would be an unrealistic expectation to assume that mining projects could be implemented without controversy.

A high quality consultation and disclosure plan at the outset of a mine proposal forms the foundation of effective public involvement. As with all plans, good practice also includes having the right management structures and skills in place to ensure long-term and consistent implementation of the plan.

A senior management champion is required to develop the communication policy and to serve as the final advocate in funding and final decision-making. The success of public involvement will depend on the credentials and attitude of the management champion.

CASE 3.1
Same Book, Different Title

The primary role of mining companies is, and always will be, to extract the maximum long-term value from the ore deposits in their stewardship. However, this by no means requires that a mining company should compromise the environmental or social integrity of its host region and the communities that live in it. On the contrary, past experience shows, and most mining companies recognize, that mining projects are more successful if environmental and social concerns are integrated into mine planning.

Now nearly forty years ago, for example, the RTZ Corporation, as it then was, set up a Commission on Mining and the Environment, in conjunction with other leading companies. Environmental and social initiatives of mining companies, individual and combined, have snowballed since then. Clearly, Corporate Social Responsibility was associated with mining long before it rose to prominence as it has today. It may have come under a different name, such as community development, community support, social investment, or social stewardship. Supporting the community that supports the mine, however, is a concept that goes back centuries, easily forgotten in the contemporary fashionable discussion of Corporate Social Responsibility.

Technical minded management often understands public involvement as an unavoidable hassle.

Technical minded management often understands public involvement as an unavoidable hassle, an attitude that will ultimately be reflected in the consultation process. Cultural sensitivity, previous relevant experience, and an appreciation of the holistic nature of public involvement are essential qualification criteria for the selected champion. The management champion must truly believe in the value of public involvement.

It is also necessary to appoint a lead person as a primary contact for all communication activities. The lead person is responsible for designing and implementing public consultation and for the day-to-day management. In the best circumstances the lead person is a recruit from the region in which the proposed mine is located, who understands the local language, local values and traditions, and is seen as part of the community.

Finally, the company needs to establish a communication team to participate in the development and implementation of the plan. Team members may include a diverse group of staff with different expertise and external advisors. Individual team members may be assigned to liaise with selected stakeholders, to track issues, to provide technical information, and to liaise with mass media. But just having people involved is not enough. It is important to have the right people in the right jobs with the appropriate knowledge and abilities.

3.3 GETTING TO KNOW YOUR STAKEHOLDERS

There are as many stakeholders as there are different people who care, positively or negatively, about a mining project.

There are as many stakeholders as there are different people who care, positively or negatively, about a mining project. Who they are may depend on their ethical, moral, interest, welfare or viewpoints and on how they may be affected by the project in regard to: (1) proximity (important in regard of say, pollution or the need for resettlement); (2) economics (that is changes to land ownership, property value, and creation of employment and business opportunities); (3) use (say amenity value, rights of way, or impact on vista); (4) social and environmental issues (environmental justice to risks being the two most likely key issues); and (5) values (ranging from religion and ecological values to animal rights) (Ortolano 1997).

Often these members of the public are termed 'stakeholders' alongside the mining company, regulatory bodies, involved industry sectors and NGOs. Wates (2000) defines stakeholders as 'persons or organizations with an interest, because they will be affected and may have some influence.' Different stakeholders are affected to varying degrees, and the greater their interest in the project, or the greater the potential impact on them, then the more involvement they should have. There are of course also those who could affect or impact on the project without being affected by the project.

A stakeholder's interest may change over the course of participation, and so they may opt in or out at various stages, while others may want to be involved throughout mine development. Some may purposefully reject participation (**Case 3.2**). It is essential, however, to ensure that no group is excluded, as this, whether intentional or not, will be resented, and may lead to action being taken in the political or legal spheres (e.g. demonstrations, press releases, or law suits).

Not everyone can, or needs to, sit around the same table.

The number of potential public stakeholders on some issues is practically limitless; however, not everyone can, or needs to, sit around the same table. The challenge then is to balance the need to consider the many view-points of all the stakeholders, with the practical considerations of convening a group of individuals who have a role in making, or directly influencing, decisions. 'Tiered' stakeholder involvement is one approach used to strike that balance, within this there are three categories: (1) those who want to be directly involved in the process; (2) those affected who just wish to be kept informed; and (3) anyone with an interest in the project.

Some stakeholders are not able to effectively participate in decision-making even when they are the most potentially affected. This may be due to poor language skills, poor self-esteem, lower educational achievement, lack of participation culture in the host country, physical handicap or intellectual disability and so on. Many remain unaware of their opportunity to participate. The Environmental Justice Program of the US EPA attempts to address these issues requiring 'the decision makers to seek out and facilitate the involvement of those potentially affected' (USEPA 17/11/02). The Equator Principles require the process will ensure their free, prior and informed consultation (see also IFC PS 1 2006).

It becomes appropriate to ask: Who are the stakeholders in a mining project? They generally fall in one of the following categories (Zemek 2002, **Figure 3.2**).

Project-Affected People Come First

Local communities near the proposed mine are people whose daily life is likely to be most affected by the new mining operation. Project-affected people (PAP) deserve the highest attention in public involvement. They generally want to know what is proposed and what changes will occur. They are concerned that their values are known and respected, and that their suggestions are carefully considered. Individuals or groups in the local community expect benefits from a mining project, usually monetary benefits. When involved in an open and honest manner local communities are often the best allies against unfounded opposition. If local communities are not well briefed about a proposal at an early stage, they may learn about the project at second or third hand, often obtaining distorted information from anti-mining activists. Since the education level in remote areas is often low misconceptions can easily develop. Once local communities are manipulated by outsiders it is challenging for a mining company to regain their trust.

Project-affected people deserve the highest attention in public involvement.

Minority and indigenous groups (IP) need special consideration (see also Chapter Sixteen). They often represent the weakest members of society such as transitory or nomadic groups, refugees, old people, or the poor and homeless. Any negative changes to their environment, even those considered minor, have the potential to cause significant distress and hardship. If multilateral lenders or EPFIs are involved in project financing, consultation with Indigenous Peoples must conform to specific and detailed requirements as for example found in IFC Performance Standard 7.

CASE 3.2
Opposing Mining as a Matter of Principle

When anti-mining activists talk about the 'mining mafia' they generally refer to mine owners, politicians and law enforcement officials, united as they see it in a desire to keep profits flowing, all too often for their personal benefit. Mistrust has led some anti-mining activists to believe that mining is bad, no matter what the circumstances. They fail to recognize that the mining industry has progressed a long way towards the goal of mining being a truly sustainable economic activity. They also fail to realize that each mine is different. Both authors have worked on projects that operate in a manner that should never have been allowed. On balance, however, there are many more successes than failures. There have also been, and will always be, genuine efforts by responsible mining company directors and mine managers to share the benefits with host communities. It is people, not companies that plan, develop, and operate a mining project.

FIGURE 3.2
Types of Stakeholders

Project-affected people deserve highest attention in public involvement. They generally want to know what is proposed and what change will occur. They are concerned that their values are known and respected, and their suggestion are carefully considered.

Source:
Zemex 2002

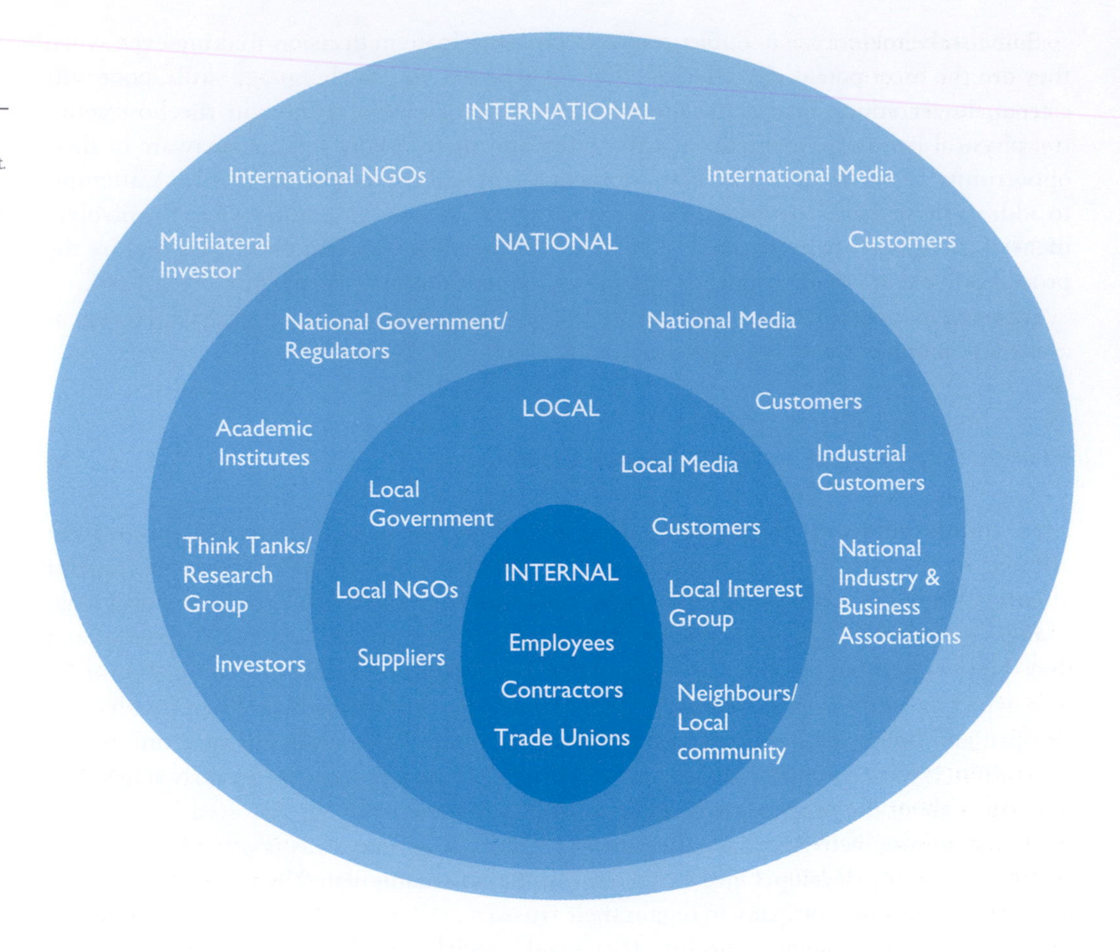

Strict disclosure requirements of institutions like IFC or EPFIs make consultation with local communities compulsory. The level of consultation may vary considerably if these organizations are not involved. Where such consultation needs to be undertaken in a structured manner, EPFIs may require the preparation of a Public Consultation and Disclosure Plan (PCDP). Some provision for local community consultation is also made in EIA regulations in most countries.

Mining projects often have indirect impacts that reach far beyond the adjacent mine areas.

Mining projects often have indirect impacts that reach far beyond the adjacent mine areas. If resettlement occurs, communities which receive migrants are also affected. Employment opportunities attracted people from communities far away from the actual mine site. Downstream communities may be affected from proposed work due to actual or perceived changes in water quality or quantity. Transportation corridors have immediate impacts on adjacent communities and make previously inaccessible areas open to migrants.

The Government which Owns the Mineral Resource

The government of the host country has significant interest in a mining project, not only because in most jurisdictions the ore body remains the property of the state. Governments facilitate mining investment and they set the rules of engagement, two areas that overlap and form the mining policy of a country. On the national level a mining project secures foreign investment and a long-term stream of tax revenue. On the regional level it will significantly stimulate economic development. On the local level, the mining project often represents the main source of income and employment opportunities. From the government's point of view public involvement may help to clarify legal and regulatory requirements and help to avoid the project becoming controversial in the later stages. In most jurisdictions governments form official commissions to evaluate environmental impact assessment studies. Commission members are intimate stakeholders in the environmental assessment process.

The Company's Shareholders

Various names are common to describe the owners of a mine. Owners and sponsors are the most common ones but other names – project proponent or juniors and majors in particular, depending on the size of the mining company – are also used. Owners are key movers of a mining project. Owners and their shareholders have a vested business interest. Their focus is (quite naturally) on the economic viability of the project and, ultimately, its profitability. Their mandate is to maximize shareholder value.

While environmental and social issues are 'among many' issues to be addressed, owners realize the need to shape the mining project to ensure long-term success. Today few if any owners deliberately avoid environmental or social issues. Problems, if they occur, are due to a combination of lack of experience, optimism ('things won't go wrong'), a lack of urgency ('we'll do that task tomorrow'), the difficulties in working with different cultures, a lack of trust on both sides, and lack of local reward because taxes are paid to central coffers, to give just a few reasons (Zemek 2002).

A common view is that mining projects with substantial external funding are usually more environmentally sound than those funded mostly by the owners themselves, as bank guidelines (particularly in a consortium with a multilateral institution or a Equator Principles Financial Institution) are often much stricter than the in-house ones of mining companies. Equally, majors are more likely to impose rigorous environmental safeguards on project development than juniors.

The extent of public involvement in a mining project more often than not depends on the mining company involved. National companies, especially when government owned, often enjoy strong national and administrative support, and few international stakeholders such as international NGOs raise opposition to a proposed mine development. In developing countries with a strong central government, local public involvement is often also limited. Similarly, junior mining companies do not fall immediately on the radar screen of international NGOs, and neither are they seen as cash cows by local interest groups. In contrast international mining corporations attract, from the very outset of mine exploration, national and international attention. Public interest and hence public involvement is large, both at home and in the host country.

The extent of public involvement in a mining project more often than not depends on the mining company involved.

Multilateral Financial Institutions

Multilateral financial institutions (or multilateral lenders) are created and funded by governments. Their role is broader than just providing equity or debt for profit, as they are supposed to contribute to the development of lesser-developed countries, promote environmentally and socially responsible investment, or help with education and awareness. In strictly financial terms the key word is 'additionality'; they are not supposed to replace the commercial banks, but to provide 'additional' services where commercial financial institutions are unable or unwilling to provide finance. Multilateral lenders active in the mining sector include the International Finance Corporation (IFC), the private sector arm of the World Bank Group, the European Bank for Reconstruction and Development (EBRD) and regional development banks such as the African Development Bank (AfDB), the Inter-American Development Bank (IADB), or the Asian Development Bank (ADB).

Given their mandate, multilateral financial institutions are generally reluctant to become involved in projects that are commercially viable and hence, whose financial needs could be met through the ordinary tools of financial markets. Exceptions include mine proposals by state-owned mining companies with limited access to private capital markets, because of a country's low credit standing, and proposals where international agency funding for project infrastructure would provide identifiable benefits to the host region. Multilateral lenders, however, see additional roles of project funding that serve as a catalyst for the involvement of other participants, providing initial technical and environmental assistance in project development, and to a lesser extent, assisting in identifying viable project opportunities through worldwide surveys. The involvement of the IFC is often seen as giving a stamp of respectability to projects. An environmental impact assessment that meets IFC requirements, as is mandatory for IFC participation in a mine project, is considered meeting best industry practice.

The involvement of the IFC is often seen as giving a stamp of respectability to projects.

Multilateral lenders involved in project financing fully recognize the environmental and social risks of mining projects. In the past, ill-planned mine developments have caused social unrest, loss of reputation, and eventually loss of money. Many large mining projects continue to be controversial (such as Ok Tedi, Freeport, or Lihir, all discussed elsewhere in this book). Today information on mining projects and related issues are easily available to all parties, including to groups or individuals that oppose mining as a matter of principle (**Case 3.3**). In response, multilateral lenders have adopted an open and transparent approach to investment and, in many respects, set the benchmark for best industry practice for public involvement (such as World Bank 1999). It is now common practice for multilateral lenders to use the Internet to publicize their involvement in a project and to make environmentally related information public knowledge.

Multilateral lenders have adopted an open and transparent approach to investment and, in many respects, set the benchmark for best industry practice for public involvement.

Commercial Financial Institutions – Increasingly Embracing Sound Environmental Practices

Commercial financial institutions have always been involved in foreign direct investments in developing countries. They have expanded along with their multinational clients, and have established institutional frameworks through which large and risky ventures can be financed. Commercial financial institutions, however, are notoriously conservative and

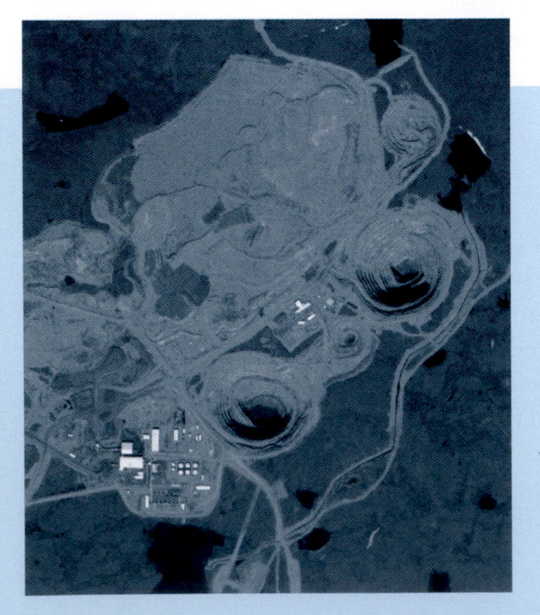

CASE 3.3
Taking a Tour on Google Earth

'EKATI is a magical and unique place. No other mine in Canada is like it. No other mine in the world is like it' (BHPB's EKATI Diamond Mine Backgrounder). And yet, the mine attracts few if any passers-by. Located 200 kilometres south of the Arctic Circle in the Northwest Territories, Canada, and 100 kilometres north of the treeline, in an area of continuous permafrost with winter temperature of −50°C and below, the mine is hardly a place that invites visitors, be they tourists or anti-mining activists. However, the mine is only a few clicks away on Google Earth. Today high resolution satellite images and their easy access thanks to Google Earth (and other image providers or image libraries) bring even the remotest areas within comfortable reach.

Photo Credit: Google Earth

they survive on their reputations. Also beside the test of commerciality, a mining project must pass a rigorous environmental assessment to demonstrate that its development is without severe negative environmental and social consequences to the host region.

Commercial financial institutions survive on their reputation.

As mentioned in the introductory chapter of this text, leading commercial banks have now subscribed to World Bank standards for environmental assessment as a prerequisite for participation in project financing in excess to 10 million US dollars:

> The Equator Principles Financial Institutions (EPFIs) have consequently adopted these Principles (the Equator Principles) in order to ensure that the projects we finance are developed in a manner that is socially responsible and reflect sound environmental management practices. By doing so, negative impacts on project-affected ecosystems and communities should be avoided where possible, and if these impacts are unavoidable, they should be reduced, mitigated and/or compensated for appropriately. We believe that adoption of and adherence to these Principles offers significant benefits to ourselves, our borrowers and local stakeholders through our borrowers' engagement with locally affected communities. We therefore recognize that our role as financiers affords us opportunities to promote responsible environmental stewardship and socially responsible development. As such, EPFIs will consider reviewing these Principles from time-to-time based on implementation experience, and in order to reflect ongoing learning and emerging good practice (www.equator-principles.com).

In case of a large consortium of lenders, environmental matters are usually delegated to one of the consortium members. This bank appoints a specialized independent environmental advisor (often a well-established independent environmental consultancy) to advise on environmental matters and to liaise with the mine owners.

Export Credit Agencies – The Quiet Giants of Mining Financing

Related to the role of financial institutions are investment guarantee schemes to facilitate the assembly of finance packages for mining projects in developing countries. Export Credit Agencies, Government Development Agencies, and Investment Insurance Agencies, commonly known as ECAs, are public agencies that provide government-backed loans, guarantees, credits, and insurance to private corporations from their home country to do business abroad, particularly in the financially and politically risky developing world. Most industrialized nations have at least one ECA, usually an official or quasi-official branch of government.

The oldest insurance system is the Overseas Private Investment Corporation (OPIC), an agency established by the US government in 1948. Since then many countries have established similar schemes such as Hermes (Germany), DANIDA (Denmark), PFCE and COFACE (France), ECGD (United Kingdom), MITI and JBIC (Japan), Finnvera (Finland) – one of the biggest lenders to the mining sector, EDC (Canada) and MIGA, established by the World Bank in 1990 to complement the IFC, the World Bank's private investment arm. The official OECD website now lists 49 such agencies as of October 2006 (www.oecd.org).

Today, ECAs are collectively among the largest sources of public financial support for foreign corporate involvement in industrial projects in the developing world. For example, ECAs are estimated to support four times as many oil, gas, and mining projects as all the multilateral development banks such as the World Bank Group combined (NGO Eca-Watch 2002). Typically, ECAs provide part of the finance required for a mining project's purchases of goods and services in the agency's home country. For example, if Komatsu Earth moving equipment is to be used, the Japanese agencies would finance its purchase; if Caterpillar equipment was planned, the American agencies would step in. Bearing in mind that most of the heavy equipment used in mining operations is manufactured in developed countries it is understandable that the involvement of ECAs from those countries can be quite substantial (**Case 3.4**). ECAs are the 'quiet giants' of mining finance (to quote Zemek 2002). Their aggregated lending is probably much higher than all the commercial banks and multilateral institutions combined.

ECAs are the 'quiet giants' of mining finance .

Up to the year 2000, ECAs had a poor environmental record. In 1999 the French COFACE, the Japanese JBIC, and the Canadian EDC, involved in two of the most controversial mining projects to date – the Ok Tedi copper project in PNG, and the Omai gold mine in Guyana, published environmental guidelines for the first time. The Environmental Guidelines for JBIC International Financial Operations, a comprehensive 54-page document, is available at: http://www.jbic.go.jp/english/environ/guide/finance.

Insurers

The following discussion of insurers as stakeholders draws heavily on text from Zemek 2002. In the discussion of insurers as stakeholders there is the need to differentiate between insurance of the project and insurance of the investment. While, as a rule, a project is insured by its owners, the equity and debt provided are usually not insured, except for Political Risk Insurance in some cases. The rationale behind not insuring bank investment is that, for project finance, banks either have recourse to the sponsors (before completion) or to the project (after completion).

Banks avoid taking possession of troublesome assets.

In reality, banks avoid taking possession of troublesome assets. If the project defaults (i.e. there is something wrong with it) most banks would rather extend extra funding to it to 'get things right' than rush into taking it over. In some cases, the lenders may decide to write-off the debt rather than become administrators of a project. This was the case at Baia Mare, Romania (Case 18.3), where banks did not wish to be associated with one of the worst environmental disasters in Europe in the last decade.

For base metals and coal, banks also require the project to be 'insured' against price variation through long-term off-take contracts. For precious metals, hedging is often required and may become part of the structured finance package offered by the bank.

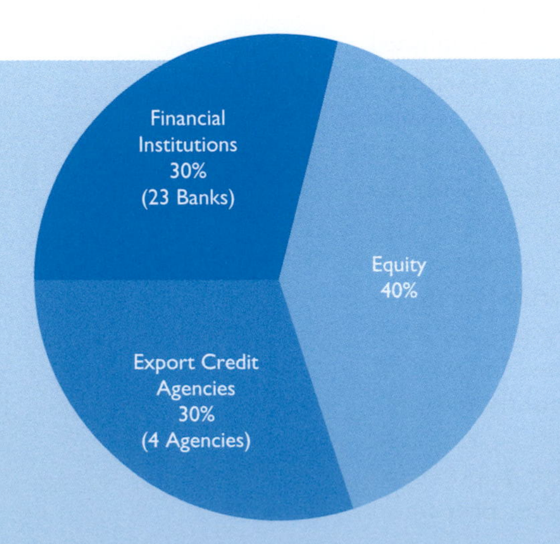

CASE 3.4
Financing scheme for the Antamina Zinc and Copper Mine in Peru

For example, 51 per cent of finance required for the largest mining project ever – the Antamina zinc and copper mine, in Peru (US$ 2.3 billion) was provided by ECAs (Mining Finance Magazine 1999). They also played a significant part in the insurance cover for the remaining part. Japan's JEXIM (predecessor of JBIC) alone provided US$ 245 million (18 per cent of the total finance) followed by the German KfW's US$ 200 million and Canada's EDC's US$ 135 million. For comparison, the largest commitment from any of the participating commercial banks was less than US$ 50 million.

Project Insurance

The standard Project Insurance at the construction stage would cover: (1) loss or damage to the project; (2) third party liability (this would also cover environmental liabilities to certain limits); and (3) marine cover (particularly if large items of equipment needed for the project are to be transported by sea). The cover would normally extend to anything mentioned in the project's detailed description. This may include housing if this is a part of the project. Project insurance is usually bought via insurance brokers, and banks may employ a special advisor to help them in finding the best balance between cost and cover.

While the insured party is the owner of the mine, the lending banks are heavily involved in drafting the policy and liaising with the insurers. No policy will come into effect if it is not approved by the lending banks. The banks are also named as parties to the insurance policy and the policy is assigned to them. In case of a claim, the proceeds go into a special bank account and the lending bank(s) decide how they should be distributed between the interested parties. A minimum insurance schedule is written into the loan agreement. Banks also frequently insist on a sponsor's guarantee – a pledge by the project owners that in case of the project not proceeding to completion, the loan will be paid back by the project owners. Therefore, technically speaking, banks are not insured, but from the practical point of view they are well covered.

Operational Insurance Programme

When a project moves to the operational stage, it is covered by an Operational Insurance Programme which typically covers: (1) property damage; (2) theft; (3) business interruption (7 to 14 days of lost production might be covered under 'business interruption'); (4) mechanical breakdown; and (5) third party liability (property damage and bodily damage). It is not clear if events like a 'business interruption due to social disturbance caused by environmental concerns' would be covered. In general, most insurers would require that any damage suffered be 'sudden' and 'accidental'. Gradual events (like increasing pollution of a river by tailings) would not be covered. In theory, specialized Environmental Cover is available covering a broader range of situations including gradual events, but it is very rarely (if ever) contracted.

It is not clear if events like a 'business interruption due to social disturbance caused by environmental concerns' would be covered.

Political Risk Insurance (PRI)

Political Risk Insurance (PRI) is taken directly by banks, either through private sector insurers, through ECAs or multilateral institutions like MIGA. The cost is normally recovered in fees from the sponsors. PRI would usually cover: (1) war; (2) terrorism (not clear if environmental terrorism would be covered); (3) confiscation; and (4) local currency non-conversion risk. There is also a possibility of insuring against licence suspension by the host government but, in most cases, this is an optional extra.

In case of a large consortium of lenders, insurance matters are usually delegated to one of the consortium members (known as an 'insurance bank' or 'technical bank' or 'facility agent'). Such a bank appoints a specialized insurance adviser to liaise with insurance brokers and sponsors. The insurance adviser also makes sure that any compulsory insurance requirements of the host country are met. The insurance cost for an average project is estimated to be around 5 per cent of the project value.

Many mining projects are developed without PRI schemes. National schemes are too small to cope with the massive investments required for a mining project. Another difficulty is the increasingly typical multinational consortium involved in large mining projects. The consortium approach may make it more difficult to arrange political risk insurance, since some members may be entitled to obtain guarantees while others may not.

Many mining projects are developed without PRI schemes.

Private sector political risk insurance providers are useful in these conditions, but are limited in the amount of risk that they will underwrite.

Police and Military – A Fact of Life

In some countries, it is common to find anti-government insurgents and/or high levels of lawlessness. The Philippines, for example, is host to various rebel groups, while in Papua New Guinea and the Solomon Islands there is widespread small and medium scale violence. In these and many other countries, it would be impossible for a mining project to be developed and operated without the presence of a security force to protect life and property. Depending on the situation the security force may be private or it may be provided by the host government.

Unfortunately, security forces associated with mining projects have been involved in incidents which have reflected poorly on the mine operator despite the fact that, in most cases, the mining company was not directly involved and could not have prevented the incidents. Relations between the company and the community have been damaged in some of these cases, while in other cases relations between the company and host government have suffered.

National police and military possess near monopoly of force, and they may propose or impose military assistance on the mine operator, usually financed by the project.

National police and military frequently play a major role in the political and social structure of developing countries. Both often possess near monopoly of force, and they may propose or impose military assistance on the mine operator, usually financed by the project (**Case 3.5**). Police or military detachments may be assigned by the host government with or without any request from the project operators. Dealing with these security forces involves a delicate balance between up-holding human rights while acknowledging and respecting the dominant role of these security forces in the host country.

CASE 3.5
Mining, Police and Military

'*Freeport is contributing to the suffering in West Papua because it funds the Indonesian government and military*', a West Papuan refugee told a mining meeting in 2006. '*You can't separate what the mine is doing from the political situation in West Papua. It is directly linked to the human rights problems.*' (www.mpi.org.au)

'*In August 1997, a farming village of 45 houses, Nkwatakrom, near Tarkwa was completely demolished by some policemen and a group of thugs allegedly hired by the Ghana Australia Goldfields Limited, now a subsidiary of AGC.*' (Speaker at a seminar on Human Rights Violations in the Extractive Sector, Nigeria, November 2000, www.rainforest.org.au)

Mining companies may become caught in the middle of conflicts between national security forces and local rebel or opposition groups. The problem is exacerbated if company equipment is commandeered by the security forces. Companies also often face another all too familiar dilemma. Government authorities, military, and police offer their services to provide security to the mine (expecting, of course, to be compensated). The government will argue, quite correctly, that it is their right and imperative to protect national assets and to provide security and stability. Mining companies on the other hand recognize that community based security (or a 'social fence') will prevent the development of security issues, and that prevention is better than cure.

Many mine operators find it useful to engage private security advisors, to interface between the company and the police/military personnel, while at the same time advising on any threats to security. Such advisors are usually retired senior military people who, accordingly, communicate well with the security forces. The authors are aware of several instances where involvement of such security advisors has been effective in avoiding conflict between security forces and local communities.

Most mine operators, if given the choice, will opt to engage a private security company, rather than government police or the military. The reason is that this provides the mine operator with more control over the actions of the security force – in particular relating to the use of force; any excessive use of force inevitably leads to adverse publicity or worse, for the mine operator.

NGOs – The 'New' Government Organizations

In recent years societies have experienced a rapid growth of NGOs both internationally and domestically. By some estimates, there are now about 40,000 internationally operating NGOs (Kovach *et al.* 2003). Domestically the USA alone may have roughly 2 million NGOs, while India accounts for about 1 million (Zadek 2003). Even the emerging NGO scene in China comprises between 1.4 and 2 million non-registered NGOs (Edele 2005). 'Suitcase NGOs' have emerged (Lee 2004) – NGOs made up of one person travelling from event to event – as well as imitation NGOs who use the NGO model for other interests (what Cohen (2004) refers to as 'Astroturf NGOs').

NGOs have become a force in many societies, or are at least so perceived. NGOs argue that national and international policies as well as commercial market forces often undermine sustainable development efforts and limit the ability of people at the grassroots level from participating in public or private policy decisions that will affect them. Informed by the needs and experiences of the poorer or disadvantaged sectors in their or other societies, NGOs have come to mobilize, articulate, and represent peoples' interests or concerns (or so they like to argue) at different levels of decision-making: locally, nationally, and internationally. This advocacy work is increasingly seen by NGOs as an integral part of the role they play in civil society. Using information as a key tool, they seek to change the course of human development by promoting equal power relationships in national and international arenas. There is a consensus that NGOs play a vital role in society and that this role will probably expand in the future. The strength of the NGO community lies in its variability and the degree to which individual organizations reflect local concerns and aspirations. It is not always immediately obvious, however, how the involvement of NGOs could lead to improving environmental assessment practices in the private investment sector. To the contrary, their capability or their genuine interest in the betterment of society is often over-estimated. Also, as elaborated later in this chapter, they are not held accountable for their actions.

NGOs have become a force in many societies, or are at least so perceived.

It is not always immediately obvious, however, how the involvement of NGOs could lead to improving environmental assessment practices in the private investment sector.

The NGO community is very diverse. However, although it can be misleading to generalize, the environmental NGO community can be characterized as follows (Edele 2005): (1) the majority of NGOs are locally based, poorly organized and have somewhat uncertain futures; (2) the degree to which NGOs are independent of governments is highly variable; some of them are 'agents' of government; (3) although environmental NGOs have a common interest in the environment, there is often a wide divergence in the objectives and agendas of individual organizations; (4) environmental NGOs within a country are no more likely to work collaboratively than any other group of organizations; and (5) large, successful NGOs may lose touch with their own 'grass roots' and be mistrusted by their local counterparts.

Others

Other interested parties to a project include influential individuals, employees, customers, and consultants or engineers involved in some parts of the project. Academics may also demonstrate interest in particular aspects of a mining project.

Influential individuals may include large landowners, business owners, local government officials, and formal or informal community leaders, including religious leaders. Communication with all these parties is essential to ensure that they are properly informed about the project, and that they are afforded the opportunity to participate in the community involvement process.

The role of consultants and engineers must not be underestimated. The owner will appoint a project team to evaluate the deposit, to identify the key issues (technical, economic, and environmental) and to determine the best way of addressing them. The project team usually consists of a core team with consultants or engineering companies appointed to address specific issues as required.

The employment of local academics also helps to establish the mining company as a member of the local or regional community, reducing the impression of newcomers and intruders.

Members of the environmental study team are often recruited from local universities to benefit from local knowledge. The employment of local academics also helps to establish the mining company as a member of the local or regional community, reducing the impression of newcomers and intruders. It demonstrates the commitment of the proponent to channel benefits to the local community from the very beginning of a mine project. However, if environmental impact assessment studies are written for an international audience, local academics often lack the relevant experience, and assistance from international consultancies will become desirable if not essential.

3.4 HOW TO IDENTIFY STAKEHOLDERS?

It is important for the mining company to be comprehensive in identifying and prioritizing stakeholders that are genuinely directly or indirectly affected by the project.

Stakeholder identification is context-specific and what works for one project may not be appropriate for another. Stakeholders are initially identified during the screening stage of environmental assessment. The scoping study provides an opportunity to ensure that all relevant stakeholders are identified. It is important for the mining company to be comprehensive in identifying and prioritizing stakeholders that are genuinely directly or indirectly affected by the project. Subsequently it is important to augment this list with all other stakeholders.

Participation specialists distinguish three ways of identifying different interest groups (Ortolano 1997): self identification, staff identification, and third party identification.

In self identification, individuals and groups come forward as a result of publicity by the proponent conducting the participation programme. These are 'active' responses and individuals who pro-actively open a dialogue should be kept continually informed of the mine's progress. In staff identification, government or company personnel learn from past experience, supplemented by local inquiry, the names of individuals and groups that might be interested in becoming involved. International anti-mining advocacies and their local counterparts are included. Third party identification can be achieved by asking those already involved about other individuals or groups that should be sought out. As each person becomes involved, they are asked to suggest others, until the response becomes minimal.

Another good way to identify stakeholders who should be involved in the consultation processes is to start by asking questions (World Bank 1999):

Who owns the land?

- Who might be affected by the development?
- Who owns the land?
- Who are the 'voiceless' for whom special efforts may have to be made?

- Who are the representatives for those who are likely to be affected?
- Who is responsible for what is intended?
- Who is likely to mobilize for or against what is intended?
- Who can make what is intended more effective through their participation, or less effective by their non-participation or outright opposition?
- Who can contribute financial and technical resources?
- Whose behaviour has to change for the effort to succeed?
- Who gets things done in affected communities?
- Whom do others listen to?
- Who manages the leading businesses in the project area?
- Are there any environmental or social groups with interests in the project areas?

Answers to these questions will identify stakeholder representatives. Communicating with these representatives is an effective way to disseminate information to a large group of people and to channel community feedback.

Another technique for stakeholder identification, at least initially, is to draw a sketch map of the mining project and potential impact areas. Through subsequent consultation with local people, project-affected groups can be identified (**Figure 3.3**).

Identification of stakeholders is incomplete without understanding the stakeholders' concerns. How will the mine proposal affect stakeholders and how will the mine proposal be affected by these stakeholder effects? Stakeholder mapping will provide answers to both questions. A stakeholder matrix helps to summarize key information in a concise manner (**Table 3.2**).

Defining stakeholders' concerns is not always a straight-forward exercise. Again the best way is by asking direct questions. Understanding all issues early on and factoring them into planning is critical in avoiding problems, thereby saving time and money. A stakeholder map, however, will evolve over time and a structured database is an efficient way to order and maintain information on stakeholders. There is no standard format for such database but it should include information on stakeholder group, stakeholder representatives, main concerns and interests, past contacts, and any other relevant information.

3.5 ENGAGING STAKEHOLDERS

The goal of engaging stakeholders is to forge relationships, partnerships and trust, allowing directly affected individuals and groups to participate in the mining project. The more solid these relationships, the fewer community issues will develop. Since participation is strongly influenced by culture, educational level, and the political system in the jurisdiction concerned, engagement of stakeholders in developing countries differs from engaging stakeholders in the western world. Indeed, Western-style public participation in decision-making can be alien to developing country communities.

Western-style public participation in decision-making can be alien to developing country communities.

Arnstein (1969) represents the levels of participation as eight rungs of a ladder, then groups the rungs into three categories (**Figure 3.4**), ranging from going through the empty ritual of non-participation to having the real power needed to affect the outcome of decisions. Arnstein describes his first category 'non-participation' as tactics whose real objective is 'to enable power holders to educate, or cure the participants'. In the second category, 'degrees of tokenism', Arnstein argues that when these are 'proffered by power holders as the total extent of participation, citizens may indeed hear and be heard. But they lack the power to ensure that they are heeded'. Arnstein groups both informing, or respectively disclosure, and consultation into the category of tokenism. While informing is important, it is a one-way

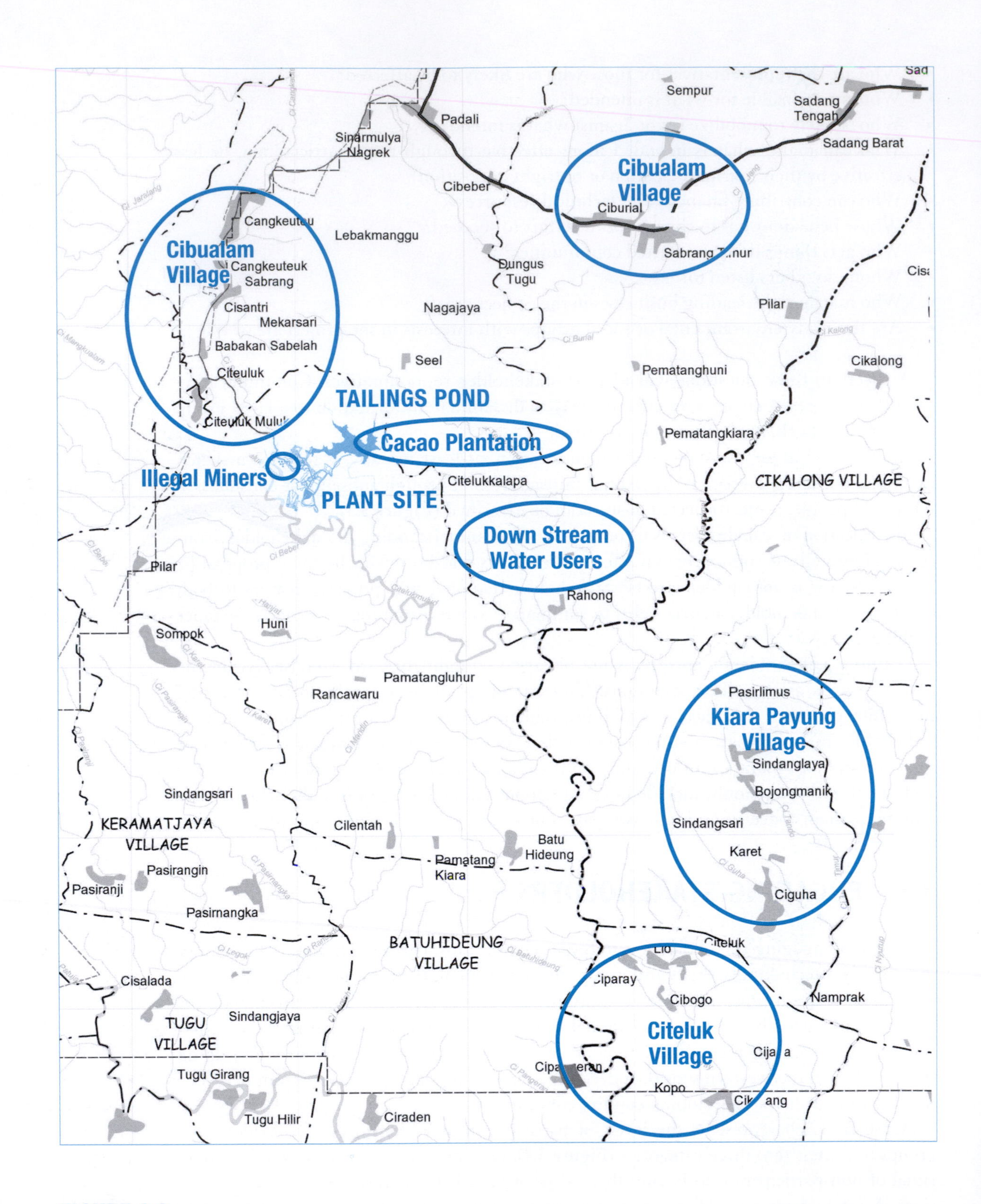

FIGURE 3.3
The Use of Sketch Maps to Identify Stakeholders Illustrated

Consultation with local people is always necessary to identify project-affected groups.

TABLE 3.2

Example Stakeholder Matrix

Group	Priority	Primary Impact	Secondary Impact	Power Assessment	Perceived Attitude Re: Mine Proposal	Mine Counterpart	Communication Consideration	Disclosure Method(s)	Consultation Method(s)
Project-Affected People									
Sangsang Village	High	Dust, Noise	None	Very High	(+)	Eko Tutuk	Banjarnese Language	Meeting	Community Board
People living along haul roads	High	Dust, Noise	None	Very High	(−)	Ahmad	Bahasa Indonesia	Meeting	Community Board
People whose land has been or will be acquired	Very High	Disrupted incomes, social issues, jealousy	Home business	Very High	(−) (+)	Nandang	Banjarnese language	Meeting	Direct Meeting
Indigenous Peoples	Very High	Cultural Change	Social Change	Medium	(+)	Amin	High illiteracy Banjarnese language	Meeting	Direct Meeting, Participatory Rural Appraisal
Other Stakeholders									
National level government agencies (Min of Mines, etc)	Very High	Taxes Royalties	National/local economic growth	Very High	(+)	Budi, John Smith	English Indonesia	Letters, Direct meeting, Newspaper	Direct meeting, Presentations Letters Personal communication
Governor	Very High	Taxes Royalties	National/local economic growth	Very High	(+)	John Smith, Budi	English Indonesia	Letters, Direct meeting, Newspaper	Direct meeting, Presentations Letters Personal communication
Religious leaders, informal leaders	High	S, I	P/M	High	(+)	Achmad F	Banjarnese language Bahasa Indonesia	Letters; direct meeting	Direct meeting
Local NGO_S	High	–	P/M	High	(+)/(−)	Budi		Letters; direct meeting	Direct Meeting

FIGURE 3.4
Arnstein's Ladder of Participation

The continuum stretches from the empty ritual of non-participation, to having the real power needed to affect outcomes of the processes.

Source:
Arnstein 1969

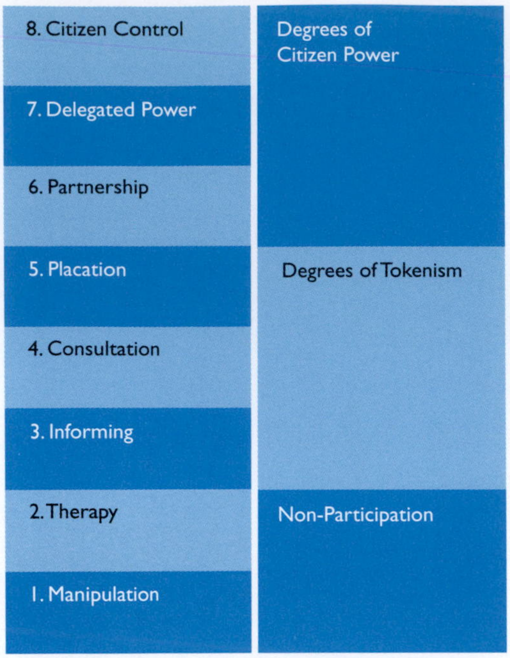

Rung	Category
8. Citizen Control	Degrees of Citizen Power
7. Delegated Power	
6. Partnership	
5. Placation	Degrees of Tokenism
4. Consultation	
3. Informing	
2. Therapy	Non-Participation
1. Manipulation	

1. Manipulation and 2. Therapy. Both are non-participative. This aim to cure or educate the participants. The proposed is best and the job of participation is to achieve public support by public relations.

3. Informing. Citizens holding a clear majority of seats on comittees with delegate powers to make decisions. Public now has the power to assure accountability of the programme to them.

4. Consultation. Legitimate step – attitude surveys, neighbourhood meetings and public enquiries. Arnstein felt this is just window dressing.

5. Placation. For example co-option of hand-picked 'worthies' onto comittees. it allows citizens to advise or plan ad infinitum but retains for power holders the right to judge the legitimacy or feseabilities of the advice.

6. Partnership. Power in fact redistributed through negotiation between citizens and power holders. Planning and desicion-making responsibilities are shared e.g. joint commitees.

7. Delegated Power. Citizens holding a clear majority of seats on comittees with delegate powers to make decisions. Public now has the power to assure accountability of the programme to them.

8. Citizen Control. Disenfranchised and stakeholders handle the entire job of planning, policy making and managing programme e.g. neighbourhood corporation with no intermediaries between it and the source of funds.

flow of information. The final category, citizen power, involves citizens-power holder partnerships, then finally programmes in which citizens are in control, or can veto decisions.

Without clearly understanding the objectives of communicating and sharing information, the activity is empty and will raise more questions than answers.

Understanding the intent of informing and communicating is critical to success. Without clearly understanding the objectives of communicating and sharing information, the activity is empty and will raise more questions than answers. Communication should also be kept clear, simple, honest, and consistent. Ask for feedback only if the response will be heeded. If the mine management does not intend to use stakeholder feedback, a public relations campaign is probably more appropriate.

When informing evolves into a two-way communication it takes the form of consultation for which VicRoads (1997) suggests five underlying principles: (1) The community should be able to understand the decision-making process; (2) it should be easy to participate; (3) consultation should be fair; (4) consultation should follow a logical sequence; and (5) consultation should deal with conflict resolution. While Arnstein (1969) felt that consultation is merely window dressing, it is the first step toward legitimate decision-sharing.

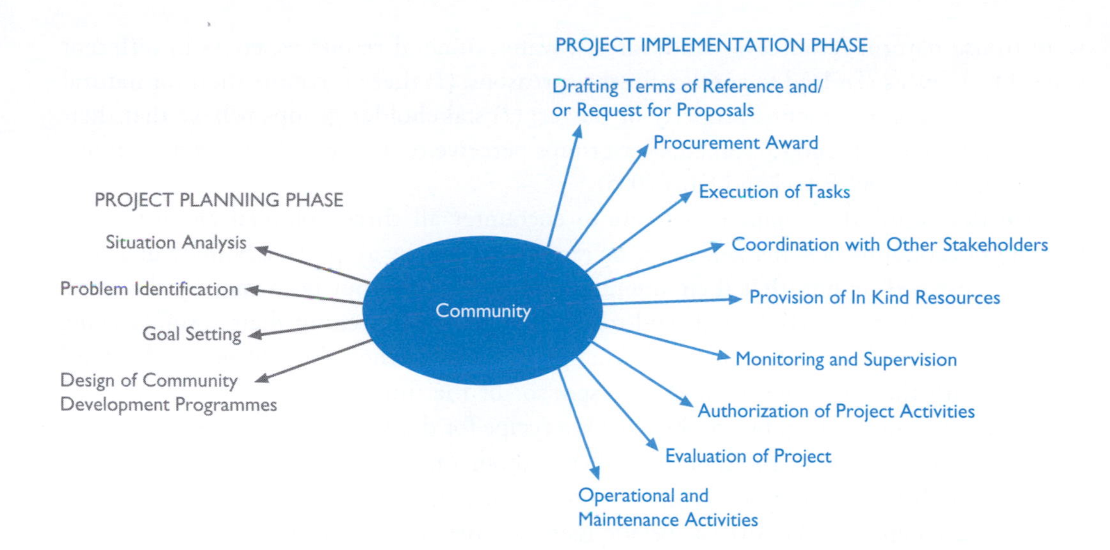

FIGURE 3.5
Level of Community Decision-Making

As the community capacity grows, a number of decisions can (and should) be shifted to the communities.

Arnstein terms the highest level of participation 'Citizen Power' which is about inclusiveness. The mining company is asking host communities to 'join in'. The difference between 'joining in' and 'buying in' is important. Participation is about taking part in decision-making, not necessarily about consensus in all matters. The mine management, however, cannot assume that it has included host communities in decision-making until such a time as the host communities feel able to contribute as fully and equally as they would wish. To be included is to have an equitable opportunity to have one's individual perspectives and needs accommodated, absorbed, and integrated by decision-makers.

Participation is about taking part in decision-making, not necessarily about consensus in all matters.

The mine management needs to actively provide host communities with the information and means needed for them to be full and effective participants. Host communities should be enabled to take part and to contribute meaningfully by being armed with an awareness of the purpose, opportunities, and risks associated with mine development. Without this there can be no effective dialogue or participation. Having the appropriate information to know how, why and what to do is one of the ways in which host communities are empowered.

Empowerment is a word open to a range of interpretations, and used (sometimes misused) to support a range of agendas. The idea that we can 'empower' others is a contradiction in terms. The mine management can and should create an environment through community capacity-building initiatives in which community members can empower themselves, since empowerment cannot be imposed. As the community capacity grows, a number of decisions can (and should) be shifted to the communities (**Figure 3.5**).

The idea that we can 'empower' others is a contradiction in terms.

While it is important to engage all stakeholders, it is also important to recognize that three groups of stakeholders, broadly defined, are likely to exist: (1) the group of people that are in favour of mine development; (2) the group of people without a clear position (either pro or con mine development); and (3) the group of people who oppose mine development. Efforts are best directed towards engaging the group of people without a clear position, without neglecting mine supporters. To engage with self-proclaimed anti-mining activists is also necessary, but it should not be where most effort and time is spent.

3.6 CONFLICT IDENTIFICATION AND MANAGEMENT

It would be great if a mining project could be developed without conflicting interests and disputes. Public involvement certainly helps to minimize conflicts and to build positive relationships. However, different people have different needs and interests. Some want to use natural resources – land, forests, water, mineral resources, etc. – in different ways. Differences can lead to conflict for three reasons: (1) there is competition for natural resources, economic benefits, property or power; (2) stakeholder groups believe that their needs cannot be met; and (3) stakeholder groups perceive that their values, needs or interests are under threat (Engel and Korf 2005).

Each new mine development is likely to encounter all three potential conflict areas. Sharing of economic benefits will never be received as fair by all parties. Some will argue, whether justified or not, that their interests and needs have not been met by the mine developer, while most people will tend to see change as threatening their comfort zone, at least initially. People generally resist change, especially when they do not understand or agree with the goals, the methods, the sponsor or the timing of the proposed change. Accordingly, to keep the public in the dark is a recipe for disaster (Connor 2001).

Some will argue, whether justified or not, that their interests and needs have not been met by the mine developer, while most people will tend to see change as threatening their comfort zone, at least initially.

Sometimes it is best to monitor a conflict without intervening. At other times, however, if a conflict is ignored or attempts to manage it fail, it can grow into dispute. A dispute occurs when a conflict over a specific issue becomes public. A dispute can be a fight, an appeal to authorities, or a legal action. The difference between dispute and conflict is important. All disputes reflect conflict, but not all conflicts develop into disputes (see also Chapter Fifteen on community development).

Some conflicts develop into disputes quickly. As one example a disagreement over land compensation will quickly turn into a dispute if not managed and resolved in a timely fashion. Then there are other conflicts that remain latent for a long time until triggered or aggravated by something new. Dissatisfaction with the way the mine operates may build up over time before anger is released, sometimes violently, triggered by say a mining accident. Conflicts may occur only locally, but in the mining sector they often involve regional, national or even global parties. Conflicts range from disagreement over land use, to conflicts between host community and the mine over shared economic benefits. Another common example is disagreement by community members concerning workforce recruitment policies.

Dissatisfaction with the way the mine operates may build up over time before anger is released, sometimes violently, triggered by say a mining accident.

Mining conflicts often have many causes that are closely linked – some form the core, others are underlying or contributing. Engel and Korf (2005) divide the various factors causing conflict into four principle types: (1) growing competition for natural resources; (2) structural causes; (3) development pressures; and (4) natural resource management policies, programmes, and projects.

Competition for natural resources, of course, can often be accommodated by compensation. Land acquisition is an obvious example. Compensation can take many forms. Experiences demonstrate that financial compensation, often the preferred option by affected people, is the least effective choice. Multi-lateral development agencies favour land for land compensation, and encourage non-monetary assistance to affected communities rather than cash payments. This is further discussed in Chapter Fifteen.

Experiences demonstrate that financial compensation, often the preferred option by affected people, is the least effective choice.

Structural causes of conflicts are more difficult to address. Established organizations and patterns of host communities govern how traditional law works. Natural resource conflicts are often underpinned by these traditional or customary structures. A conflict becomes more complicated when customary law and State law are organized differently; one is local and the other national. Broader social, political, economic or legal frameworks at national level may be perceived as unjust, ineffective or exclusionary by local affected communities. This makes it difficult to solve structural conflicts, which often lie dormant until awakened by development of a new project. Customary land rights often remain unclear, even when they are acknowledged legally. Wider inequalities (real or perceived) may also lead to conflicts over use or control of natural resources, and the distribution of economic benefits in terms of taxes, royalties, and dividends. Struggling for resource access then often becomes linked with a search for recognition of identity, status, and political

Customary land rights often remain unclear, even when they are acknowledged legally.

rights. The resolution of structural economic and legal inequalities, of course, is out of the control of mining companies and may require intervention at the national level through, for example, land reform, legal recognition of land rights, devolution of authority and responsibility, or improved governance and accountability (see also Chapter Fourteen). Public involvement in mine development will not resolve these broader issues, but will help to identify and understand these underpinning conflict areas.

The resolution of structural economic and legal inequalities, is out of the control of mining companies.

Development pressures and associated conflicts occur when society and the economy undergo change, inevitably as a mine development moves forward: (1) introduction of new technologies, new ways of communication, and improved access and infrastructure may have positive and negative effects on affected communities; (2) commercialization of common property resources may invite powerful external groups to monopolize benefits, often excluding marginalized Indigenous Peoples from using the resources; and (3) migration caused by mine development may change the ways in which host communities and local resource use are organized.

National resource management policies and programmes themselves can serve as sources or arenas of conflict, although their intention is to reduce conflicts or improve livelihoods. FAO (2000) provides the following reasons: (1) policies imposed without local participation; (2) poor stakeholder identification and consultation; (3) uncoordinated planning as many government and other agencies still rely on sectoral approaches with limited cross-sectoral planning and coordination; (4) inadequate or poor information sharing, leading to suspicion and mistrust; (5) limited institutional capacity – conflicts arise when government and other organizations lack the capacity to engage in sustainable natural resource management; (6) inadequate monitoring and evaluation of programmes concerning natural resource management; and, most importantly, (6) lack of effective mechanisms for conflict management.

For natural resource management to be effective, mechanisms for participatory conflict management need to be incorporated into mine management. These should ensure that open or latent conflicts are constructively dealt with to reduce the chances of conflict escalation. IFC Performance Standard 1, Article 23 summarizes the key requirements of an effective Grievance Mechanism:

> The client will respond to communities' concerns related to the project. If the client anticipates ongoing risks to or adverse impacts on affected communities, the client will establish a grievance mechanism to receive and facilitate resolution of the affected communities' concerns and grievances about the client's environmental and social performance. The grievance mechanism should be scaled to the risks and adverse impacts of the project. It should address concerns promptly, using an understandable and transparent process that is culturally appropriate and readily accessible to all segments of the affected communities, and at no cost and without retribution. The mechanism should not impede access to judicial or administrative remedies. The client will inform the affected communities about the mechanism in the course of its community engagement process.

3.7 UNDERSTANDING THE BENEFITS AND RISKS OF PUBLIC INVOLVEMENT

Benefits

Direct benefits for the mining company in public involvement include reducing financial risks and costs, and creating positive community perceptions and corporate image. Hence, consultation and collaboration with the public makes common sense as well as good business sense, considering that most mining investments are long-term investments. Lack

Consultation and collaboration with the public makes common sense as well as good business sense.

of support from local communities will eventually turn into outright opposition to the project, and may lead to legal disputes, delays, social unrest, and to long-term security risks and negative publicity. Effective public involvement reduces these project risks and consequently reduces project costs. In addition to these project benefits, public involvement will enhance social benefits for the community. This in return will help to build and expand local support for the project, further reducing financial risks.

The Committee on the Challenges of Modern Society (1995) has summarized the benefits of public involvement as: (1) increased quality of decisions; (2) reduction in costs and delays; (3) achievement of transparency of decisions and commitment to decisions; and (4) avoidance of public controversy and confrontation.

Informed host communities will better understand the tradeoffs between project benefits and disadvantages; be able to contribute meaningfully to the mine design; and will develop greater trust and support.

ADB (2005) argues that the information gained through public consultation on the stakeholders' concerns, interests, and their ability to influence decision-making, helps identify key causes of environmental problems. This can be used in the environmental impact assessment to evaluate direct and indirect environmental impacts, and assess short-term and long-term resource use implications. According to the ADB, input from local communities and NGOs can also help the evaluation of alternatives and strengthen environmental mitigation measures by incorporating local knowledge. Informed host communities will better understand the tradeoffs between project benefits and disadvantages; be able to contribute meaningfully to the mine design; and will develop greater trust and support.

The Dangers of Public Involvement

The way in which public participation is conducted is often a source of conflict in itself.

While there are clear benefits in public involvement, there are also risks. The way in which public participation is conducted is often a source of conflict in itself. Stakeholder groups, when the outcomes of the process do not correspond with their desired outcomes, are likely to attack the project and the public participation process itself. The process was either too long or too short; there was too little time to comment, or too much; the process provided too little or too much confusing information; the public participation was biased; the consultants and engineers cannot claim to be independent because they are being paid for by the company; and so on. It is for these reasons that it is important to build checks and balances into the public participation process. Everybody involved should be encouraged to thoughtfully analyze the importance of disputed issues, the interests of other parties, and the alternatives that other parties offer. This, of course, is easier said than done. For instance, anti-mining activists often flatly refuse to even attend meetings to discuss controversial issues. And 'compromise' is not part of their agenda.

However, the main risk of early public involvement is in creating unrealistic demands and expectation. Clear and consistent communication in describing what the project can deliver in benefits (and what not) helps to avoid this pitfall. It is particularly dangerous to overstate benefits to gain consensus or to make promises that later cannot be fulfilled. If the company employs external consultants for public consultation or public relations, it must maintain overall responsibility throughout the consultation process. The community will not differentiate between consultant and mining company, and promises made by consultants are essentially perceived as promises made by the company. In short, the mining company needs to manage external consultants and contractors carefully.

There is the real risk that information relating to a mine proposal will lead to land speculation.

Finally, there is the real risk that information relating to a mine proposal will lead to land speculation. In many developing countries, clear and documented land ownership does not exist, allowing false or unjustified land claims. Entrepreneurs are attracted and speculate on increasing land value. The best safeguard against land speculation is a detailed census of land ownership at the very beginning of the project. This includes an accurate assessment of people living in the area and of current land use, an assessment that needs to be well-documented and best witnessed by local community members (see Chapter Eighteen).

NGOs and Mining

Because of the massive and highly visible impacts large mining projects have on both the physical and human environment around a mine site, mining projects are an easy target for environmentalists and community development NGOs. Photographs of deforested landscapes, and videos of poor village people who somehow got lost in the shuffle make for excellent copy in both national and international media, while the whole story of carefully planned community development programmes, not to mention massive economic benefits to the host country gets completely overlooked in the rush to publish. Some national NGOs, sometimes with funding and other support from international NGOs, have used the obvious, and acknowledged, impacts from mining to mount effective anti-mining campaigns locally and internationally. Given the reach of the Internet, and the skill that some anti-mining organizations have acquired in using it, international campaigns to block or significantly modify major mining enterprises can be quite successful in slowing project development, adding costs through delays and litigation.

Mining projects are an easy target for environmentalists and community development NGOs.

On the other hand, NGO involvement can be productive. NGOs often combine strong social commitment with sound knowledge of their areas of interest. Local NGOs tend to have a genuine interest in local development, and companies have much to gain from working in partnership with these not-for-profit organizations in spurring development. Such partnerships can arise in the context of a company's philanthropic activities, or as a part of its core business operations. The motivation for and method of collaboration may differ, but it is grounded in the fact that the private sector and development organizations have many long-term goals in common: both have an interest in a stable society; both want to foster income-generating activities and build the capacity of local entrepreneurs; both want to ensure that people are healthy and educated. These shared goals can provide a basis for partnership (WBCSD and IBLF 2004).

Local NGOs tend to have a genuine interest in local development, and companies have much to gain from working in partnership with these not-for-profit organizations in spurring development.

Among the wide variety of roles that NGOs play, the following nine can be identified as important, at the risk of generalization, although no one organization performs all of them (Cousin 1991; OECD 1999):

- Provision of expertise – NGOs can contribute to specific aspects in environmental assessments;
- Training – competent NGOs can be involved in training for government, the private sector and the general public;
- Community links – one of the reasons given for the growing involvement of NGOs in environmental assessment is related to the fact that environmental assessment is the only environmental management tool accepted around the world that carries an obligation to consult with local people. NGOs have been instrumental in achieving such 'grass roots' consultation links.
- Education and awareness – either directly, or through their networks, NGOs are able to reach a wide spectrum of the public on environmental issues;
- Sensitizing private sector and politicians – NGOs may be in a unique position to sway development decisions through publicizing environmental issues;
- Advocacy for and with the poor – in some cases, NGOs become spokespersons for the poor and attempt to influence government policies or private sector developments on their behalf. This may be done through a variety of means ranging from demonstration

and pilot projects to participation in public forums and the formulation of policies and plans, to publicizing research results and case studies on the poor. Thus NGOs play roles from advocates for the poor to implementers of programmes; from agitators and critics to partners and advisors; from sponsors of pilot projects to mediators;

- Development and operation of infrastructure – community-based organizations and cooperatives can acquire, subdivide and develop land, construct housing, provide infrastructure, and operate and maintain infrastructure such as wells or public toilets and solid waste collection services. They can also develop building material supply centres and other community-based economic enterprises, although in many cases, they will need technical assistance or advice from the private sector, governmental agencies or higher-level NGOs;
- Supporting innovation, demonstration and pilot projects – NGOs have the advantage of selecting particular places for innovative projects and specifying in advance the length of time which they will be supporting the project – overcoming some of the shortcomings that governments face in this respect. NGOs can also be pilots for larger government projects by virtue of their ability to act more quickly than the government bureaucracy;
- Research, monitoring and evaluation – innovative activities need to be carefully documented and shared – effective participatory monitoring permits sharing of results with the people themselves as well as with project staff.

Be Aware

The inclusion of NGOs as ancillaries to the mining itself is not without risk, as greater familiarity with the project may lead to more nuanced opposition.

Nevertheless, the inclusion of NGOs as ancillaries to the mining itself is not without risk, as greater familiarity with the project may lead to more nuanced opposition. If companies make concessions, NGOs may come back for more. Through opposition, an NGO may derive power that is inconsistent with its knowledge or constituency. The notion that 'power corrupts' can apply to NGOs as well as other groups (**Case 3.6**).

It is imperative, therefore, that those responsible for planning the response to the environmental and social impacts of a project need to consider the full potential of the NGO sector, both in what it can contribute, and the degree to which it can be a source of frustration, delay and added costs. The experience of other major projects will provide invaluable guidelines for maximizing the utility and minimizing the pain of NGOs in the arena of mining and the environment.

CASE 3.6
Manila warns it will arrest Canadian anti-mining activists.

Thu, May 05 2005. Back off or you'll be arrested. This is the warning issued by the Philippines Environment Secretary Michael Defensor to Canadian anti-mining activists who are helping some members of an ancient tribe save their ancestral land. '*They have no right to meddle in the affairs of the country. They should respect the policies of the Philippines, especially the economic policy to do mining*,' Defensor said after a recent meeting with pro-mining tribal elders from Mindanao who claimed that they were being used by foreign groups. Defensor noted that international non-government organizations were becoming invasive in their efforts to rally support against mining activities not only in the Philippines but worldwide. '*I don't mind that they have an advocacy, but for them to agitate, finance people [to go against mining], that's a different thing*'

3.8 COMMON MISTAKES

Promises That Are Not Kept

Refer to any text on educating children. Almost all will advise you never to make promises that you can't keep. This holds equally true for adults. Mining companies that promise employment opportunities will be held accountable for that promise. Promise to provide better water supply. You will be reminded as soon as mine development commences. It does not matter whether the promise was made by the mine owner or another party perceived to speak on behalf of the mine owner. It is important to choose words carefully. Once a promise is made, the mine management will be held accountable.

Too Little – Too Late

Participation should not be so late in the life of an issue that it is tokenistic, or merely confirms decisions already made. The timing should occur when affected people have the best chance of influencing outcomes. People should be provided with sufficient time to express their views and to become involved in subsequent decision-making.

People should be provided with sufficient time to express their views and to become involved in subsequent decision-making.

Top-down

In the past, the mine management or the government could simply inform the public on decisions that had already been made. It then became necessary to explain why the decision had been made which, in turn, led the management to consult in advance of decision-making. Good practice now goes even further and involves the public and key stakeholders at all stages of decision-making. Participation offers an additional and complementary means of channelling the energies of host communities. Enhancing participation at the community level fosters a 'bottom-up' approach to economic and social development.

It is important to ensure that any recommendations which emerge from the consultative process will have a strong likelihood of being adopted. If they are not, it is important that a public explanation is provided. Trust in the process is important for both the power holders and the participants. Although the decision-making process can aim for consensus, complete agreement need not be the outcome. The main point is that all parties should be clear on how decisions will be made so that participants know and understand the impact of their involvement.

Not Sincere or Transparent

Consultation makes sense only if all participants are sincere and concerned about a balanced outcome. Use of the consultation process to manipulate other parties will eventually prove counter-productive since manipulation destroys trust. Information should not be hidden, especially on investments. Lack of full disclosure of environmental impacts, or emphasizing mainly positive aspects is also all too common and reduces credibility.

The consultation method should also be appropriate to the target group. Evaluation questions should be formulated in advance, with consideration of how the 'success' of the

Feedback to the community after consultation is essential.

consultation will be measured. Factors beyond the adoption of recommendations should be included. Feedback to the community after consultation is essential.

Buying Support

Paying for support may appear attractive in the short term, but this is a short-sighted approach. The fact that paying money for support is accepted in many developing countries and used by politicians at all levels, does not legitimize its use. As with any other form of corruption, it leads to expectations and the development of a repetitive pattern that is difficult to end.

Many stakeholders may be unable to participate in community involvement activities unless their legitimate expenses are met.

On the other hand, many stakeholders, including government representatives and NGOs, may be unable to participate in community involvement activities unless their legitimate expenses are met. Accordingly, it is common practice and even mandatory in some countries, for the project proponent to pay for transport, accommodation *per diems* to government (including academic) attendees at public forums, presentations and hearings. This can induce reactions of shock and horror from US-based proponents who are highly sensitive to any suggestion of corruption, as well they might be considering the penalties that apply in the USA. Another common reaction is 'Why should we pay for Dr X to attend when we know she is spreading misinformation about our project?'

Offering payment for participation in decision-making by other stakeholders should also be considered, for those with low incomes. Although occasionally resulting in accusations by anti-mining activists of buying support, this is only fair, and reduces the barriers of financial inequality. Expenses should be paid as standard, and it should be borne in mind that for a 'vulnerable' person the same expense may be more significant than for other participants. Experience shows that payment of reasonable attendance expenses have absolutely no effect on support for a project from those involved. However, failure to pay attendance costs in situations where such payments are customary, will inevitably lead to resentment and distrust.

Trying to Please Everyone

It is unrealistic to assume that all stakeholders' interests can be met. At the very least there will be intractable anti-mining activists who will maintain their opposition regardless of the merits of a project. Proponents need to recognize that conflicts are sometimes inevitable and that some stakeholders may be unreasonable and therefore plan for conflict management and dispute resolution. As mentioned previously, while issues raised by all stakeholders should be considered, most of the effort should be devoted to those with legitimate concerns.

Diversity, Interest Groups are not Recognized

A common failure is to overlook disadvantaged people, minorities and women, who may be less accessible for cultural or linguistic reasons.

It is important to focus attention on stakeholder identification, especially of affected people and communities, local authorities and decision makers, the media, the scientific community, NGOs, and other interested groups or parties. A common failure is to overlook disadvantaged people, minorities and women, who may be less accessible for cultural or linguistic reasons. Often the spectrum of participants is kept very narrow (e.g. only government and business). Participants should be selected in a way that is not open to manipulation, and should include a cross-section of the population – as individuals and as groups. Some degree of random selection offers the best chance of achieving this. However, it **is** important that community leaders **also** be included as their cooperation is usually critical to any community interaction. The problem arises when these leaders advise that 'It is not necessary to involve X, Y and Z because they are not important'. This is only one of many occasions when the proponent needs to tactfully but firmly explain its own viewpoint and the reasons behind it.

Consultation Stops with Disclosure

A common failure is to engage in disclosure without consultation, or to complete consultation prior to making a decision. All participants should have time to become well-informed and to understand material that is unfamiliar to them. Over-simplification of complex issues should be avoided. The 'big picture' should be presented, so people can really become engaged.

Mixed Messages

Mixed signals undermine public trust. Despite this obvious concern, mixed messages are surprisingly common in mining developments. One of the reasons is the variety of spokespersons that may be involved during the 5 to 10 years that it takes to move from exploration to production. Early exploration staff can have no real idea of what, if anything may develop, and so should emphasize the uncertainties involved. In all stages of the project, spokespersons should avoid speculative comments. On the other hand, whenever things do become clear, they should be communicated unequivocally. All company personnel interacting with local communities should be well briefed so as to provide consistent information.

Mixed signals undermine public trust.

Mixed message are not confined to the project planning stages. They may continue through the entire project and are particularly common toward closure when spokespersons wish to soften the blow by de-emphasizing the bad news. It is most unwise for employees to receive a different version of a situation from the one provided to external stakeholders as inevitably the conflicting messages will cause confusion or worse. An example is where one company spokesperson is reassuring the community that the tailings system is 100% safe, while another is training employees on contingency plans for initiation in the event of a tailings spill.

Lack of Cultural Sensitivity

Information should be disseminated early and in a culturally meaningful fashion, including using local languages, visual methods and, where appropriate, specialist communication expertise. It is important to recognize that language can be used, intentionally or otherwise in ways which can either be powerfully marginalizing or powerfully inclusive. Clearly problems arise when language is developed and used to tackle complex specialized issues, such as in mining. While it is most often used with the intentions to be both accurate and helpful, specialized language including technical terms and industry jargon

While it is most often used with the intentions to be both accurate and helpful, specialized language including technical terms and industry jargon is marginalizing for the non-specialist.

is marginalizing for the non-specialist. Where specific mining terms must be used, as in the case of 'tailings' it is important that the meaning be fully explained and that feedback be obtained so that it can be assessed whether or not the true meaning has been conveyed and understood.

Lack of Documentation

The entire process of consultation should be well documented, including all participants, issues raised, responses provided by project proponents and the impact upon subsequent decisions. Minutes of meetings signed by all participants are helpful (if appropriate for the type of meeting). Some companies opt to electronically record all important meetings. This again is not appropriate for all meetings, but when done should be clearly stated at the outset.

Lack of Capacity Building

Instruction and training to enhance cultural awareness (in both directions) as well as training on the objectives and methods of public involvement may be needed for employees, project managers, central and local government authorities, affected communities, and NGOs.

Capacity building is an external as well as an internal concern. Instruction and training to enhance cultural awareness (in both directions) as well as training on the objectives and methods of public involvement may be needed for employees, project managers, central and local government authorities, affected communities, and NGOs. It can be very valuable for all concerned if affected people are involved in the design and implementation of project monitoring. However, this is only effective if those involved are sufficiently familiar with the issues to understand what is being done and why. This may require extensive and extended training. A common deficiency is for the company to devote insufficient resources and time to the process of capacity building so that meaningful public involvement can be achieved.

Make it Flexible

A variety of consultation mechanisms exist. Project planners should select the one which best suits the local circumstances. A variety of mechanisms could be tried over time, and specific feedback should be sought so as to gauge relative effectiveness. Particular consideration will be required in the case of stakeholders with special needs (e.g. language, disabilities, the elderly, and the young). Different communities and different questions will produce better responses with different forms of consultation. A mix of qualitative and quantitative research methods generally provides the best results.

Make it Community-focused

Mine development is not about the interests of particular individuals. Mining is about providing benefits for society as a whole.

Individuals should be asked not what they want personally or what is in their self-interest, but what they consider appropriate in their role as community members. Mine development is not about the interests of particular individuals. On the contrary, to be sustainable, mining is about providing benefits for society as a whole. Discussions and participation should not be centred on individuals' needs (although often it is) but about community benefits and the needs of the entire society.

Failure to Recognize the Crucial Difference between Releasing Information and Informing the Public

The wholesale release of vast amounts of data does not necessarily inform anyone. There should of course be no question of hiding or distorting information, but care should be taken to ensure that the overall effect of the release of information is to improve recipients' understanding of the issues (and the uncertainties) rather than simply providing data, which may add to confusion.

To be most effective, the consultation process should follow the customary and legal requirements of the host country. Anything else will be uncomfortable to the participants and could lead to distrust. Although difficult for project planners to avoid, to write off a particular option as 'too expensive' should be avoided. Who has the authority to say that? The alternative is to canvass the advantages and disadvantages of all options, including costs. Also, statements that action to protect the environment or the public could harm the project, are counterproductive as they will reinforce any concern that a risk is being transferred from those who are benefiting to those who are not. When dealing with risks to health and safety, it is important to recognize that nothing is entirely 'safe'. The Government's role is to ensure that everything is 'safe enough'.

By avoiding these mistakes and pitfalls, consultation is more likely to be all-embracing, meaningful, useful and effective. Avoiding common pitfalls is the first step towards achieving community consultation that works.

REFERENCES

ADB (2005) The Public Communications Policy of the Asian Development Bank, Disclosure and Exchange of Information, March 2005.

Ahmad A (2002) Righting Public Wrongs and Enforcing Private Rights; The New Public. Environmental Law Institute.

Arnstein SR (1969) A Ladder of Citizen Participation; Journal of the American Institute of Planners, Vol. 34, pp. 216–224.

Bruch C and Filbey M (2002) Emerging Global Norms of Public Involvement; The New Public Environmental Law Institute.

Canter L (1996) Environmental Impact Assessment (Second Edition); McGraw Hill Publishing Company. New York, USA.

Cohen J (2004) Governance by and of NGOs. London, Accountability.

Committee on the Challenges of Modern Society (1995) Evaluation of Public Participation in EIA; Report 207, North Atlantic Treaty Organization, Brussels.

Connor DM (2001) Constructive Citizen Participation: A Resource Book; 8th ed. Victoria, BC: Development Press, 5096 Catalina Tce., Victoria, B.C., Canada, V8Y 2A5. www.connor.bc.ca/connor.

Cousins W (1991) 'Non-Governmental Initiatives' in ADB, The Urban Poor and Basic Infrastructure Services in Asia and the Pacific; Asian Development Bank, Manila.

Edele A (2005) Non-Governmental Organizations in China; Geneva, The Centre for Applied Studies in International Negotiations (CASIN).

Engel A and Korf B (2005) Negotiation and mediation techniques for natural resource management; FAO Rome.

FAO (2000) Natural resource conflict management, by V. Matiru. Rome.

Glasson J, Therivel R and Chadwick A (1994) Introduction to Environmental Impact Assessment, UCL Press, London.

Johns G (2000) NGO way to go: Political Accountability of Non-government Organizations in a Democratic Society.

Kovach H, Neligan C and Burall S (2003) The Global Accountability Report: Power without Accountability? London, One World Trust.

Lee J (2004) NGO Accountability: Rights and Responsibilities, Programme on NGOs and Civil Society, CASIN Geneva, Switzerland.

NCEDR (2003) Tools-Information Gathering and Analysis; National Centre for Environmental Decision-Making Research.

NGO Eca-Watch (2002) (www.eca-watch.org)

OECD (1999) Coherence in Environmental Assessment Practical Guidance on Environmental Assessment for Development Co-operation Projects.

Ortolano L (1997) Environmental Regulation and Impact Assessment, New York: John Wiley and Sons. pp. 72–92.

Roberts (2003) Involving the public, in H. Becker and F. Vanclay (eds) International Handbook of Social Impact Assessment. Cheltenham: Edward Elgar. pp. 259–260.

University of Manchester EIA Centre (1995) Consultation and public participation within EIA. Manchester EIA Centre leaflet series No 10.

VicRoads (1997) Community Participation Strategies and Guidelines (Victoria), www.vicroads.vic.gov.au

Wates N (2000) The Community Planning Handbook: How people can shape their cities, towns and villages in any part of the world; Earthscan, London http://www.earthscan.co.uk 230 pp book (English & Chinese) and e-book (English) ISBN 1 85383 654 0 Website: www.communityplanning.net

WBCSD and IBLF (2004) A business guide to development actors: Introducing company managers to the development community.

World Bank (1999) Public Consultation in the EA Process: A Strategic Approach, Environment Department.

Zadek S (2003) In Defense of Non-Profit Accountability; Ethical Corporation Magazine. September. pp. 34–36.

Zemek A (2002) The Role of Financial Institutions in Sustainable Mineral Development, United Nations Environment Program (UNEP). Mineral Resources Forum (www.mineralresourcesforum.org)

● ● ● ●

Appendix 3.1 Methods of Public Participation

ANNOUNCEMENT

Use of the media. Newspaper and radio coverage are relatively inexpensive and effective methods of reaching a large audience. Using television as a medium is increasingly common. This method is restricted by the amount of information that can be delivered to the public. Priority should be given to local newspapers in local language. Providing contact details allows one to collect some feedback.

Advertisements. Paid public notices describing the details of the project and the issues involved may also sue print media, and also can placed in accessible locations (for example: libraries, community centers). They provide a quick, easy and low-cost method of informing a wide section of the community and can often advertise other participation activities. However, the quantity of information that can be effectively communicated is limited and it does not provide the opportunity for feedback from the public.

Leafleting. Brochures, leaflets, and information packs can be distributed to the general public as a quick and easy method of providing general information. This information should be kept brief but should include a summary of the project, the issues involved and details of other participation activities (telephone numbers and addresses). The one-way flow of information from this method may be subject to bias or open to misinterpretation by the general public.

Displays and exhibitions. These methods employ visual communication of information to educate and inform the public. When set up at an accessible location they provide a low-cost method of reaching a large number of people who may otherwise fail to participate in the public participation program. The displays should rely upon photographs, maps and diagrams rather than text to allow easy communication. A two-way dialogue is only facilitated by staffing the display or providing other means of participation.

INFORMAL DISCUSSIONS

Telephone 'hot lines'. Telephone numbers which are allocated for use by the general public allow them to pose questions or to make comments on the different aspects of the project. Its merits lie in its accessibility to all individuals who possess a telephone, making it an appropriate method where the public are widely distributed in remote areas. However, they rely upon knowledgeable staff with effective communication skills and also incur higher costs than alternative methods of receiving feedback.

Open houses. These should be held at an accessible location in the community where the public may view displays, ask questions and discuss issues with the project staff. The informal nature of such participation may attract individuals who may be dissuaded from attending formal public meetings. Problems associated with this method include high costs of providing staff to receive feedback and to discuss the relevant issues.

Personal contact. This involves direct discussions between project staff and individuals who are particularly interested in the effects of the project. Close personal interaction allows detailed and complex information to be explained and discussed. However, it is time-consuming and therefore many interested persons may fail to participate.

Community liaison staff. Staff employees of the proponent are asked with liaising with members of the community to provide them with information and to reduce the misinterpretation of commitments made by the proponent. Such staff needs to possess good communication skills and have a sound understanding of the project details. Problems of using liaison staff may arise when the community is widely distributed or is composed of various ethnic/cultural groups.

Community advisory committees. Consist of members of the public who represent the interests of the community. They advise the proponent on the project and communicate with the rest of the community. An effective two-way dialogue can be set up early in the planning process and should demonstrate the proponent's willingness to work with the community. Effective participation relies upon the group being representative of the full range of interests and having regular communication with the public.

Online consultation. Online consultation refers to a process by which the project proponent and/or the government provides opportunities for citizens to submit feedback and input on a particular project/issue, or participate in discussions with the project proponent/government using the Internet. Within the definition of online consultation two subcategories or distinctions deserve attention: online feedback and online discussion based consultations. Online consultation is a relatively new form of consultation, and requires a well established Internet infrastructure as well as a well educated audience. While online consultation may well become the preferred consultation method in the future, it is rarely used in developing countries.

Online feedback. Online feedback refers to a process by which the project proponent and/or government creates opportunities for citizens to provide feedback and input on a particular project/issue, using the Internet. Examples of online feedback could include, using web-based forms, responding to questions in an online poll, or submitting qualitative responses to a EIS/policy document via email. These activities can be directed toward the general public or take the form of a targeted consultation with specific stakeholders invited to submit expertise and opinion on an issue or policy. Participants are generally provided with feedback at the end of the process in the form of a final report or consultation summary. Increasingly these feedback-based consultations are becoming more interactive through the use of online workbooks, where citizens work through a survey activity complete with supporting information and interactive decision trees. In this latter example, individuals make deliberative choices as they work through specific policy questions.

Online discussion. Online discussion refers to a process by which the project proponent and/or the government provide opportunities for citizens to interact with government and one another in a discussion, debate or online meeting. Feedback and information is generally provided to participants on an ongoing basis throughout the consultation activity. Unlike the online feedback process of consultation, online discussion based consultations provide opportunities for the public, or selected representatives from particular interests or communities, to join a conversation, share ideas, collaborate on projects and build relationships through electronic communication. Online discussion based consultations utilize Internet-based discussion tools, including Email Lists, Live Chat, Web-based discussion boards, and/or Group Collaboration applications.

FORMAL GROUP DISCUSSIONS

Questionnaires and surveys. These methods aim to determine public attitudes and perceptions on various issues, through a structured questioning process carried out by the proponent. To avoid potential misinterpretation of the findings, the survey must be carefully designed to ensure that the surveyed individuals accurately represent the community as a whole. The disadvantages of this method are its high demand upon personnel resources and the lack of a two-way dialogue.

Public meetings. Such meetings require the proponent to make a presentation describing the project and the relevant issues, which is then followed by a question and answer session. These should be held at a central facility, readily accessible to the community. Public meetings are relatively inexpensive. The information provided by the proponent should be kept simple in order to communicate with a wide section of society; otherwise the discussion may fail to be beneficial to the proponent or the public.

Public hearings and inquiries. These are more formal than public meetings and are viewed as the traditional method of public participation in many jurisdictions. However, many individuals are unwilling to voice their opinions in such a formal setting where the discussion is highly structured. They are also time-consuming and tend to be expensive for all parties wishing to present their case. Nevertheless, they provide an opportunity for two-way communication to occur.

Group presentations. These involve presentations given by the proponent to specific community groups. They allow the proponent to focus upon specific issues relevant to the audience and to select an appropriate level of detail. Discussion after the presentations allows feedback and prevents **thuds** misinterpretation of the information. The restricted coverage of all of the issues at these meetings may preclude the audience from receiving a realistic overview of the project.

Workshops. These are most effective for discussing and identifying solutions to problems, scoping of potential impacts and creating other plans of action. However, they are limited to using small groups of selected community representatives and require extensive organization by the proponents, which make them expensive.

Source: based on University of Manchester EIA Center 1995.

4

The Anatomy of a Mine

This is Mining

♄ LEAD

Lead, a bluish-grey metal, has no taste or smell. Lead is readily produced by reduction from Galena (lead sulphide). With a melting point of only 327°C, lead flows easily and collects in the lowest point of a fireplace or furnace. Lead is highly malleable, ductile and non-corrosive, making it an excellent piping material. Lead pipes bearing the insignia of Roman emperors can still be found. Lead is also toxic. Lead has been shown to affect virtually every organ and system in the body in both humans and animals. It has been hypothesized that lead poisoning contributed to the fall of the Roman Empire. Today, the USA is the largest producer and consumer of lead metal.

4 The Anatomy of a Mine

This is Mining

The goal of this chapter is to foster a better understanding of mining, from the initial prospecting to final mine product, raising awareness of common mining methods as well as environmental, social, and economic issues associated with the mining industry. Specialized textbooks such as Lacy (2006), Hartman (1992), Hartman and Mutmansky (2002) or USDA (1995) (from which this chapter title is borrowed) are available for those who might wish to study particular aspects in detail. Nevertheless, the discussion that follows will be most useful for those with some working knowledge of geology, mining, environmental and social sciences.

A wide range of mining methods are available to extract ore from the Earth, and there is more than one way to extract metals and other valuable components from the host rock.

Ore is an economic term used to distinguish potentially profitable from non-profitable mineral deposits.

As noted in Chapter One, the need for environmental impact analysis of mining developments has long been established. While this may be achieved for existing mining operations by means of a mine audit as described in Chapter Twelve, a new mining development generally requires a full environmental impact assessment (EIA). This can involve a complex and extended process since mining and mineral processing cover many disciplines, mining-related professions are highly specialized, and mines differ widely in their layouts and products. A wide range of mining methods are available to extract ore from the Earth, and there is more than one way to extract metals and other valuable components from the host rock. In summary, each mine development is unique, involving many different issues and presenting an array of different demands to those conducting the EIA.

Conceptually, mining begins with ore, which is defined as an accumulation of minerals from which one or more valuable substances can be extracted at a profit (Bateman 1950). In effect, ore is an economic term used to distinguish potentially profitable from non-profitable mineral deposits, while mineral deposit is a more general geological term which does not necessarily mean that the deposit can be mined at a profit. It follows that all ores are mineral deposits, but not all mineral deposits are ores. Ore genesis is defined as the origin or mode of formation of ore; in other words, how, when, and where it was formed. Ore bodies are found in the Earth's crust, and can be on the surface, near the surface or at depth. Both economic and geological factors will combine to determine the exploration and mining techniques employed.

The subject matters of this chapter are arranged along the mining life-cycle (as shown in **Figure 4.1**), addressing topics in the sequence typical of mining activities. Modern mining

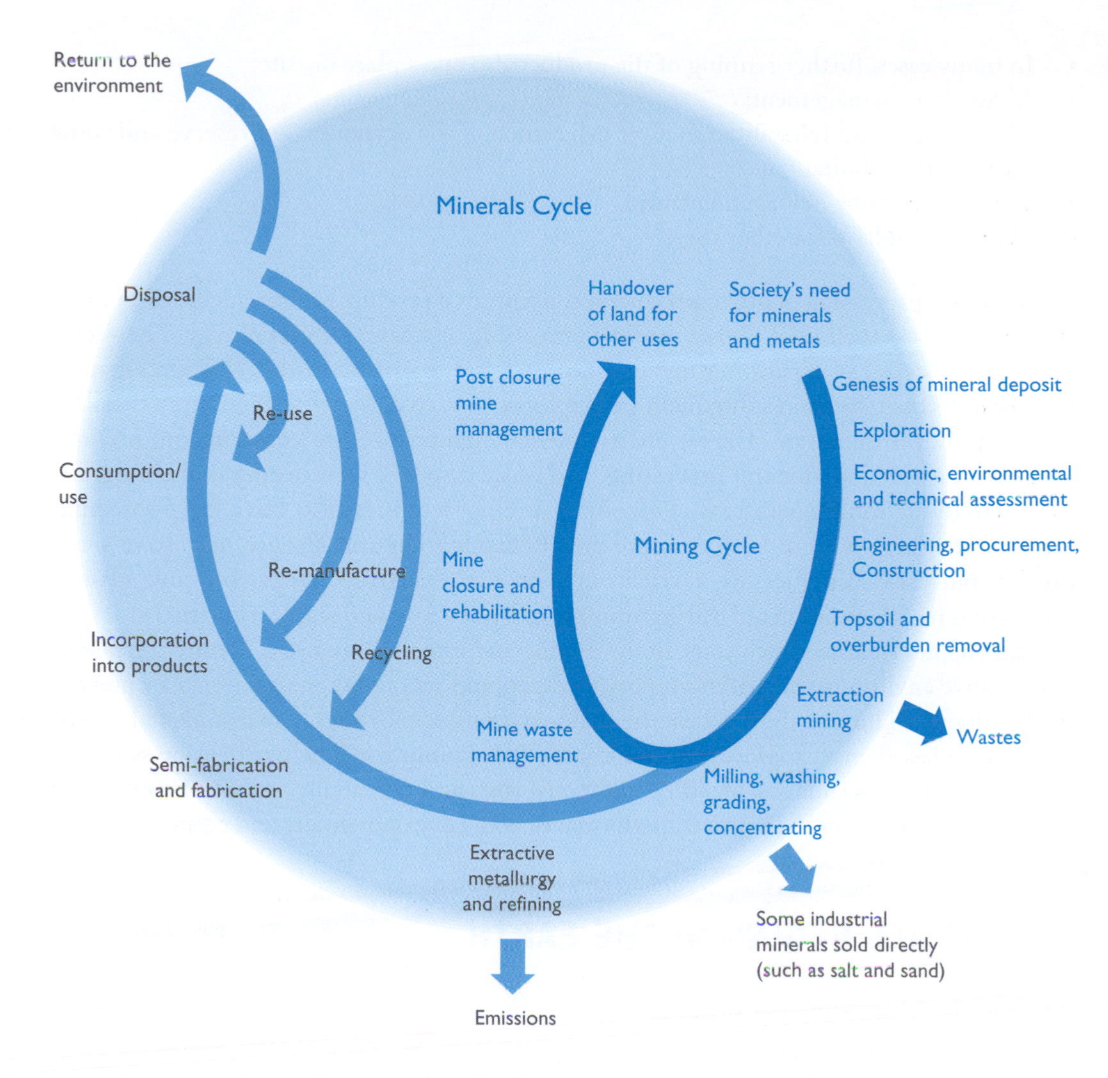

FIGURE 4.1
The Mining Life-cycle

The three main phases of environmental concerns are exploration (site access, drilling, and sampling), production (mining and mine waste management), and mine closure (decommissioning and rehabilitation).

Source:
Concept for the schematic is drawn from MMSD 2002

operations place emphasis on closing this circle, with the final activity being rehabilitation of the mine site. As a consequence, rehabilitation is a serious consideration in the initial operations planning.

The mining cycle begins with our need for minerals and metals, and nature's distribution of these elements in the Earth's crust, the ore genesis. Human activities in the mining life-cycle then typically include (modified from UNEP and IFA 2001):

- Exploration – prospecting and exploration to identify potential economic mineral deposits. By definition exploration is a forerunner to mining, but it often continues throughout the mine operation stage and even beyond, usually at a smaller scale;
- Economic, environmental, and technical assessment – assessing the mineral deposit to determine whether it can be economically extracted and processed under current and predicted future market conditions at an acceptable economic and environmental cost;
- Engineering, procurement, and construction – design, planning and construction of the mine, ore handling and processing plant, together with associated infrastructure such as roads, power generation facilities, workforce accommodations and ports;
- Mining itself – involving removal of overburden, and waste rock in open pits and excavation of underground declines, shafts and tunnels to access deeper ore bodies; extracting ore; and transportation of ore from mine to beneficiation plant;
- Milling, washing, grading, and concentrating – beneficiation and primary processing of the ore to produce a concentrated product;

- In many cases, further refining of the product also takes place on-site;
- Mine waste management;
- Mine closure and rehabilitation after exhaustion of the economic ore reserve and completion of rehabilitation;
- Post closure mine management; and
- Handover of land for other uses.

The primary processing of ore to produce a concentrated product is often included in mine design, since it is the initial step of mineral processing or extractive metallurgy. Therefore, the basic processes of separating metals from rock are discussed in this chapter. Chapter Six discusses the two main subsequent principles of extractive metallurgy in more detail.

Generally the activities of most environmental significance are construction, ore and waste rock extraction, mineral processing, and waste disposal. Dewatering is also a significant activity at some operations. Processing mining activities also have environmental impacts but these tend to be less important. Rehabilitation and closure may have some impact, but these activities are carried out with the objective of repairing any adverse effects that may have occurred during mining, to leave a safe and stable mine site.

This chapter examines each stage of the mining and minerals life-cycle, identifying potential negative environmental effects and briefly describing actions taken to mitigate or prevent them. Chapter Thirteen then details the main environmental issues. Environmental assessment that is based on an appreciation and knowledge of mining will help to develop synergies and to identify opportunities, avoiding additional costs and potentially creating new streams of revenue, resulting in an improved performance both environmentally and economically.

4.1 IT ALL BEGINS IN THE EARTH

The ores which provide the substance of the mining industry are often complex compositions which require sophisticated understanding to locate, process, and utilize. A more detailed discussion of the properties of metals and minerals for those who might wish to review the basics is included in Chapter Eleven.

Types of Ores

As with arsenic, cadmium is a by-product which is recovered in smelters; no dedicated cadmium or arsenic mine exists.

Ore may yield a single metal (simple ores) or several metals (complex ores) as shown in **Table 4.1**. Ores that are generally exploited for only a single metal are those of iron, aluminium, chromium, tin, mercury, manganese, tungsten, and some ores of copper. Gold ores may yield only gold, but silver is commonly associated. Much gold, however, is extracted as a by-product from other ores, predominantly copper. Ores that commonly yield two or three metals are those of gold, silver, lead, zinc, nickel, cobalt, platinum, and manganese. Arsenic is commonly associated with many gold deposits as well as copper and lead deposits, and in fact copper ores are the primary sources of arsenic. Cadmium, which is often found in zinc-concentrate after mineral processing, is usually removed at the smelter. Lead and copper ores may also contain small amounts of cadmium (EIPPCB 2001). As with arsenic, cadmium is a by-product which is recovered in smelters; no dedicated cadmium or arsenic mine exists, although artisanal mining to extract arsenic compounds used as a cosmetic, has occurred in the Middle East.

The knowledge of primary and associated constituents in an ore body, is of prime importance for the environmental assessment of a mine proposal. Fragmentation and removal of rock materials and tailings expose rock surfaces that previously have been

TABLE 4.1

Common Ore Types – Minerals can consist of a single element such as carbon (as graphite or diamond) or native gold. More often minerals are compounds of two or more elements, and about 3,000 mineral species are known.

Metal	Ore Mineral	Composition	Percent Metal
Gold	Native gold	Au	100
	Calaverite	$AuTe_2$	39
	Sylvanite	$(Au, Ag)Te_2$	
Silver	Native silver	Ag	100
	Argentite	Ag_2S	87
	Cerangyrite	$AgCl$	75
Iron	Magnetite	$FeOFe_2O_3$	72
	Hematite	Fe_2O_3	70
	Limonite & Goethite	$Fe_2O_3H_2O$ (variable)	60 or less
	Siderite	$FeCO_3$	48
Copper	Native copper	Cu	100
	Bornite	Cu_5FeS_4	63
	Brochantite	$CuSO_4\ 3Cu(OH)_2$	62
	Chalcocite	Cu_2S	80
	Chalcopyrite	$CuFeS_2$	34
	Covelite	CuS	66
	Cuprite	Cu_2O	89
	Enargite	$3Cu_2S\ As_2S_2$	48
	Malachite	$CuCO_3\ Cu(OH)_2$	57
	Azurite	$2CuCo_3\ 2H_2O$	55
	Chrysocolla	$CuSiO_3\ 2H_2O$	36
Lead	Galena	PbS	86
	Cerussite	$PbCO_3$	77
	Anglesite	$PbSO_4$	68
Zinc	Sphalerite	ZnS	67
	Smithsonite	$ZnCO_3$	52
	Calamine	$H_2Zn_2SiO_5$	54
	Zincite	ZnO	80
Tin	Cassiterite	SnO_2	78
	Stannite	$Cu_2S\ FeS\ SnS_2$	27
Nickel	Pentlandite	$(Fe, Ni)S$	22
	Garnierite	$H_2(Ni, Mg)SiO_3\ H_2O$	
Chromium	Chromite	$FeO\ Cr_2O_3$	68
Manganese	Pyrolusite	MnO_2	63
	Psilomelane	$Mn_2O_3\ (xH_2O)$	39
	Braunite	$3Mn_2O_3\ MnSiO_3$	71
	Manganite	$Mn_2O_3\ H_2O$	81
Aluminium	Gibbsite (found in lateritic ore termed Bauxite)	$Al_2O_3\ 2H_2O$	39

(Continued)

TABLE 4.1
(Continued)

Metal	Ore Mineral	Composition	Percent Metal
Antimony	Stibnite	Sb_2S_3	71
Bismoth	Bismuthianite	Bi_2S_3	81
Cobalt	Smaltite	$CoAs_2$	28
	Cobaltite	CoAsS	35
Mercury	Cinnabar	HgS	86
Molybdenum	Molybdenite	MoS_2	60
	Wulfenite	$PbMoO_4$	39
Tungsten	Wolframite	$(Fe\ Mn)WO_4$	76

Source:
Lacy (2006)

isolated from the surface environment. Natural forces of oxidation and erosion will eventually release metals that have not been extracted by mining, into the environment. While beneficial to life at low concentrations, many metals at increased concentrations become detrimental (see also Chapter Eleven).

How Mineral Deposits are Formed

For a mineral deposit to be formed, some process or combination of processes must produce localized enrichment of one or more elements. Minerals can become concentrated in a number of ways. Concentration of elements or minerals may result from (Robb 2005):

- the flow of hot magmatic fluids through fractures in crustal rock producing hydrothermal mineral deposits;
- igneous processes (the term igneous means 'origin by fire', derived from the Latin word *ignis* meaning fire) within a magma body producing magmatic mineral deposits;
- metamorphic banding or re-crystallization producing metamorphic mineral deposits;
- precipitation from lake water or seawater producing sedimentary mineral deposits;
- hydraulic forces such as waves or currents in flowing surface water producing placer deposits;
- weathering processes including lateritization, producing both residual and enriched mineral deposits.

A mineral deposit is always a geological anomaly with an unusual concentration of formerly diffuse metals.

A mineral deposit is always a geological anomaly with an unusual concentration of formerly diffuse metals. A pure concentration of elements or minerals is rare and minerals are commonly interspersed with non-metallic rock, which are referred to as gangue minerals. Gangue minerals are of no commercial value. Familiar minerals that commonly occur as gangue are silicates, such as quartz, feldspar, and mica, and the carbonate minerals, calcite and dolomite. The mixture of minerals and gangue constitute the mineral deposit that is generally enclosed in surrounding host rock of no economic value (**Figure 4.2**). The transition from the zone of unusually high concentration of elements to host rock with average concentration may be sharply defined but is more often gradual.

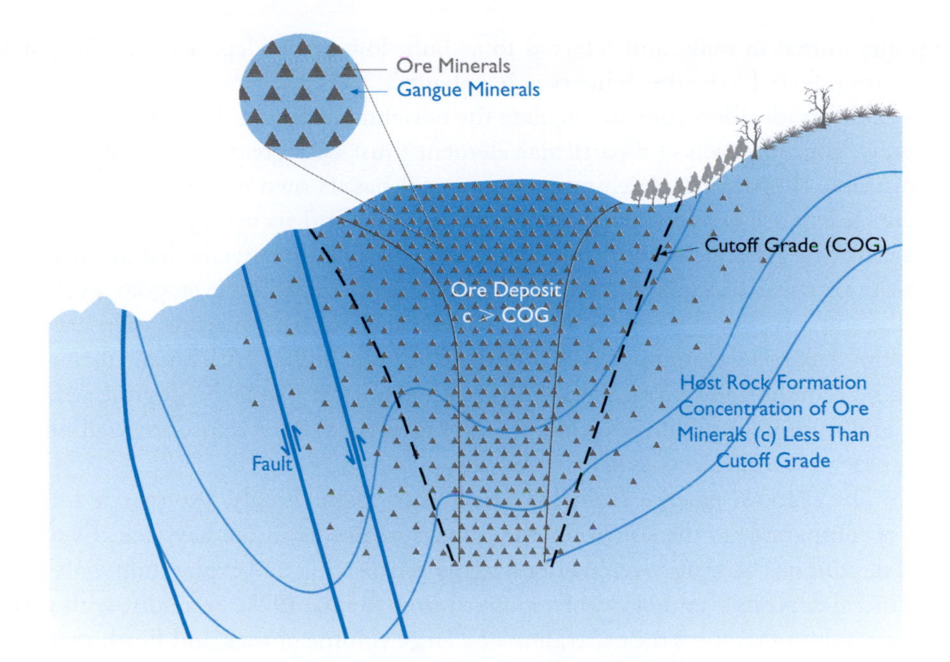

FIGURE 4.2
Ore Body, Host Rock and Cutoff Grade

The transition from the zone of unusual high concentration of metals to host rock with average concentration is often gradual or blurred. The cutoff grade is an economic term and marks the transition from ore to the geological envelope of 'no commercial value'.

Cutoff Grade – The Economic Definition of Ore

Only a few mineral deposits are commercially viable. For a mineral deposit to become an ore deposit requires that the commercially valuable minerals can be extracted at a profit. The metal content of a mineral deposit is called the grade. This is generally expressed as a percentage by weight, or in the case of precious metals, as ounces per ton. Higher mineral prices allow lower metal concentrations to become economically attractive to mining. Most iron ores need a grade of at least 50 percent of iron to attract interest in mining, whereas economically attractive concentrations of gold are as low as 1/1000 of 1 percent (**Case 4.1**). Of course, the more concentrated the desired minerals, the more valuable the deposit. The lower limit, however, is fixed primarily by economic considerations, but other aspects are also important. The lower limit of grade is called the cutoff grade (COG) and it varies according to metal prices, available mining technology, and the nature, size and location of the mineral deposits. The cutoff grade is an economic term, and it marks the transition from ore to waste rock in a given ore deposit (as illustrated in **Figure 4.2**). Waste rock is defined as rock which does not contain an adequate percentage of ore minerals to be economically valuable as a source of these minerals. The cutoff grade represents this threshold between economic and uneconomic ore. Mine plans are drawn up with some prior assumptions about cutoff grades. As the nature of the actual ore body is revealed during mining, and in response to changes in mineral prices and available mining technology, the cutoff grade of an ore will probably change over the life of the mine. The cutoff grade is an operational criterion which is applied at the point of mining (Lane 1988). An increase in metal prices may well turn past mine waste dumps into valuable ore resources.

Most iron ores need a grade of at least 50 percent of iron to attract interest in mining, whereas economically attractive concentrations of gold are as low as 1/1000 of 1 percent.

The cutoff grade is an economic term, and it marks the transition from ore to waste rock in a given ore deposit.

It is not always possible to say exactly what the grade must be, or how much of the given mineral must be present, for a substance in a given deposit to be considered an ore. Two deposits may have the same size and identical grade, but one may be identified as an ore and the other may not. Much depends on location, as illustrated in **Case 4.2**.

Small, selectively mined deposits with a high gold–silver content and a high profit margin are often referred to as bonanza deposits, the focus of mining in the early days. Bonanza deposits are few in number, and in the past their discoveries were the cause of gold rushes (see the Chapter Five). Today's deposits are usually of large tonnage and

Today's deposits are usually of large tonnage and low grade, mined in bulk, and referred to as bulk low-grade deposits.

low grade, mined in bulk, and referred to as bulk low-grade deposits. The largest such deposits are referred to by the industry as 'elephants'.

The cutoff grade allows one to calculate the enrichment factor, the multiplier by which the average concentration of a particular element must be increased to become economically mineable (**Figure 4.3**). Each mineral or metal has its own enrichment factor, which represents a balance between the price of the material and its average abundance in the Earth's crust. In general, the enrichment factor is greatest for metals that are least abundant in the crust, such as gold or mercury. Rather surprisingly however, gold, even though one of the scarcest mined elements, does not rank at the top of the list – mercury does. If the price of a particular metal falls, it requires an even higher enrichment factor for the metal to be economically extractable. On the other side, due to technological advances in finding and mining ores, some deposits that are now considered as ore are well below the lowest grade ores some decades ago.

The search for an ore deposit is comparable to the search for the proverbial needle in the haystack.

The cutoff grade is one of the most important numbers in the mining industry.

Three observations emerge from the above discussion. Firstly, the search for an ore deposit is comparable to the search for the proverbial needle in the haystack. Even after a mineral deposit has been discovered after lengthy prospecting and exploration, only 1 out of 1,000 mineral deposits eventually evolves into a mine (Roscoe 1971). Secondly, with very few exceptions, mining requires the excavation of a large volume of rock and Earth materials of which only a small percentage of volume is extracted as mined minerals. The remaining materials have to be discarded as waste close to the mine. In fact the placement of unwanted materials imposes the biggest direct environmental impact of any mine. Thirdly, the cutoff grade is one of the most important numbers in the mining industry. Although

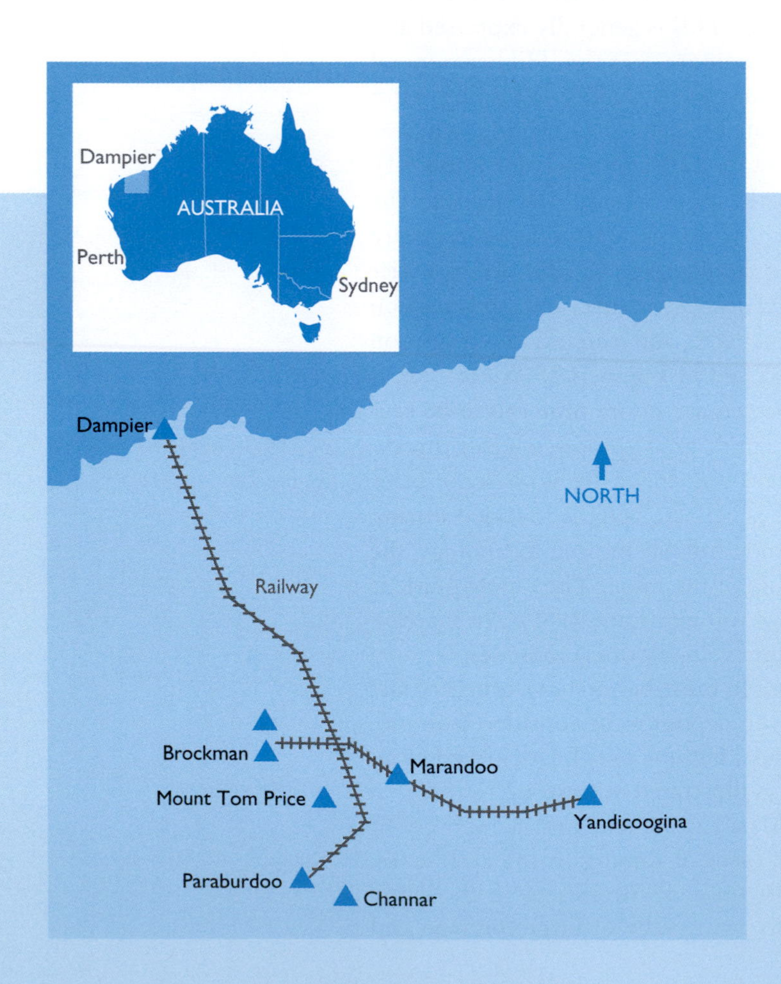

CASE 4.1
Tom Price Iron Ore Mine in Australia

Iron ore was discovered in the Pilbara region of Western Australia in the 1800s, but despite geologists claiming there was enough to supply the whole world, the remoteness and ruggedness of the country stopped people from initially exploiting the deposits. The historic moment in the discovery of one of the world's richest iron ore bodies came in 1962, when two CRA geologists found a large dark outcrop which extended about 6.5 kilometres, later to be known as Mount Tom Price, in tribute to an American engineer who was instrumental in the Pilbara's development. Production at Mt Tom Price began in 1966, four years after the discovery of this huge hematite deposit. Today ore from this mine is still an important part of the Hamersley Range mining: the chemical and physical quality of Mt Tom Price lump and fines is outstanding with iron content well above 60 percent.

Source: Hamersley Iron – Technical Fact Book.

simple in concept, the definition of the cutoff grade for a particular mine is challenging and open to discussion and interpretation.

From the perspective of the host country, which usually retains ownership of an ore deposit, the cutoff grade should be as low as possible to maximize the amount of minerals mined or to achieve maximum resource utilization. From the perspective of the mine investor, the best cutoff grade is the one that produces the highest return on investment. This can and does in many cases mean that much of the mineral deposit remains unmined, either temporarily until prices improve, or permanently if prices never increase sufficiently to justify extending or re-opening the mine. Discrepancies in defining an appropriate cutoff grade exist and, to some extent, the environmental assessment of a mine proposal needs to elaborate the different factors that have influenced the definition of a particular cutoff grade. A critical review of a chosen cutoff grade, of course, is well beyond the scope of the environmental assessment, as well as the expertise of environmental specialists. Nevertheless, since the cutoff grade will influence the footprint of the mine, the volume and type of waste produced, and the life of the project, it represents a major factor in determining the environmental impacts of a mining project.

From the perspective of the mine investor, the best cutoff grade is the one that produces the highest return on investment.

In general, the cutoff grade will fall when project borrowings have been repaid, and near the end of a project when many of the project operations have been curtailed. Recognizing this likelihood, many operators segregate and stockpile low grade mined material with the intention of processing it after mining has been completed, providing that this can be done economically. This practice introduces its own uncertainties in environmental management. Should the low grade material be managed as waste rock, subject to whatever management strategies apply, or should it be assumed that the low grade stockpile will ultimately be processed? This is of particular importance in many sulfide deposits as the low grade stockpile may contain much higher concentrations of potentially acid-generating materials than the waste rock.

CASE 4.2
Effects of Location on Cutoff Grade

Bauxite, the major ore of aluminium, was discovered in the Darling Range region of Western Australia more than 50 years ago. However, the average grade of Darling Range bauxites proved to be much lower than in most other bauxites mined at the time. Despite aluminium contents generally up to 20% lower than those in other bauxite mining areas such as Jamaica, the Darling Range bauxite mines and associated alumina refineries emerged as some of the world's lowest cost producers of alumina. The main reason for this is the location of the Darling Range bauxite deposits in a region close to infrastructure including ports and towns, and with a skilled workforce located nearby. The three alumina projects developed by Alcoa (Kwinana, Pinjarra and Wagerup) together with the Worsley Alumina Project, established Australia as the world's leading source of alumina (AAC 2006).

FIGURE 4.3
Enrichment Factors for Selected Metals

Each mineral or metal has its own enrichment factor, which represents a balance between the price of the material and its average abundance in the Earth's crust. If the price of a particular metal falls, it requires an even higher enrichment factor for the metal to be economically extractable.

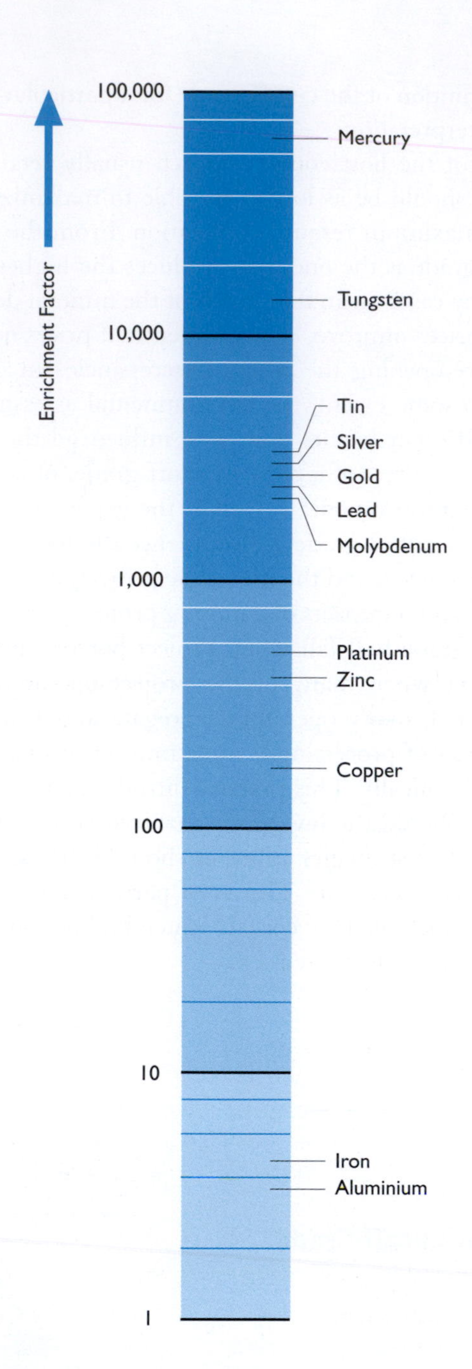

4.2 EXPLORATION – FROM REVIEWING DATA TO TAKING BULK SAMPLES

Minerals cannot be considered as wealth unless known, and the discovery of mineral wealth comes with great difficulty. One Canadian research study relates exploration efforts to the number of actual mines which result (Roscoe 1971). From around 1,000 mineral prospects identified, only 100 are drilled for reconnaissance. From these, only 10 progress to intensive exploration drilling. From these 10, eventually only one mineral deposit will become a mine (**Figure 4.4**). Modern exploration, loosely defined as all activities that eventually lead to the discovery of an ore deposit, may have altered these ratios, but not much. Significant economic deposits are difficult to find. They have to be discovered.

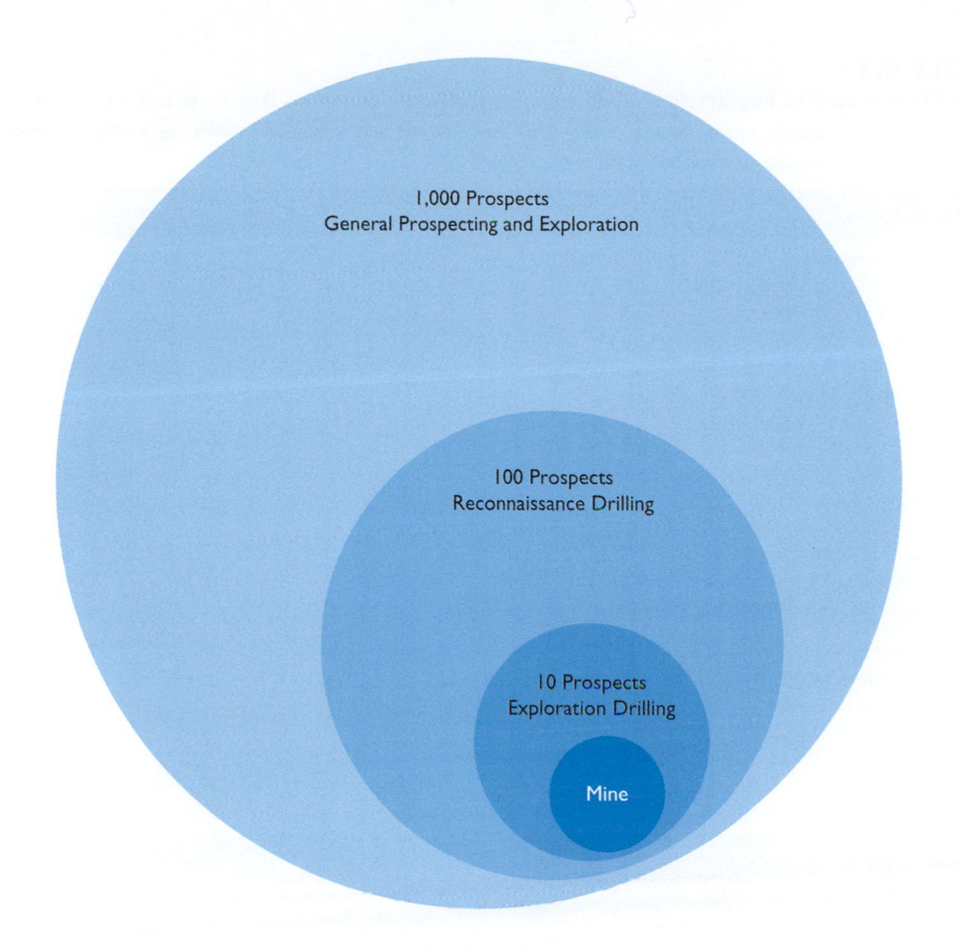

FIGURE 4.4
From Exploration to Exploitation

How hard is it to find a mineral deposit? The art and science of finding new mineral deposits is much better than pure luck, but is still far from perfect. Successful mining projects can be spectacularly profitable, but overall, mining has one of the lowest returns on investment of all major industries.

Source:
CGER 1999

Even the largest deposit is very small compared to the area that has to be explored to discover the deposit. The exploration tools of the past such as picks, shovels, and more recently, simple drilling rigs, were used to locate near surface deposits, often found by surface outcrops or gossans. Freeport's giant copper and gold deposit in Papua was first mentioned in a journey diary in 1937 by the Dutch explorer, Jean Jacques Dozy:

> Close to our right was the Ertsberg. This exposure of ore was an impressive sight....
> as far as I could determine it was all copper ore.

It is still possible to find new deposits near the surface by exploring close to previously discovered mineralization, or old mines and workings (sometimes referred to as Brownfield exploration as opposed to Greenfield exploration, which occurs in previously unexplored areas). However, there is an increasing trend towards exploration for concealed mineral deposits located at considerable depth (also termed 'blind' deposits). In effect, exploration has shifted from a two-dimensional to a three-dimensional search.

Exploration has shifted from a two-dimensional to a three-dimensional search.

This presents major challenges. Every exploration programme is different and there are no hard and fast rules governing the process and the outcome. However most programmes are structured in three stages (**Table 4.2**): area selection, exploration (field data gathering), and data evaluation.

Area selection is usually an office task with minimum fieldwork. In most countries governments produce regional geological and geophysical maps outlining the country's geological framework. In addition, records of previous exploration data may be available. Once an area has been selected and the right to explore is granted, data gathering or exploration can begin.

Much of the initial exploration data is from remote sensing methods including satellite imagery, conventional and infra-red aerial photography, and airborne geophysical surveys

TABLE 4.2

The Three Stages of Exploration – Every exploration programme is different and there are no hard and fast rules governing the process and the outcome. However, drilling is the main technique used to eventually confirm the presence of concealed mineral deposit, and to outline its extent and grade.

Area Selection
Preliminary preparation
Selection of programme criteria
Commodities to be sought
Guidelines for identifying deposit location
Familiarization
Literature and data review
Analysis of regional geology
Review of known areas of mineral occurrence
Reconnaissance exploration
Geological reconnaissance
Airborne reconnaissance
Reconnaissance geological mapping
Airborne sensed geo-physical reconnaissance
Land acquisition/right to explore
Determination of land status
Ensure right to explore
By government lease
By lease from private owner
By staking
By prospector's permit
Exploration – Gathering Field Data
Geological mapping
Geochemical exploration
Stream sediment sampling
Rock sampling
Soil sampling
Other geo-chemical methods
Geo-physical exploration
Gravimetric and magnetic methods
Electrical methods
Forced current coil methods
Seismic surveys
Radiometric surveys
Down hole methods
Exploration drilling
Auger drilling
Rotary-percussion drilling
Diamond core drilling
Bulk sampling
Test pitting and costeaning
Adits and shafts
Data Evaluation

from helicopters or fixed-wing aircraft. Such reconnaissance exploration has virtually no environmental impact as it does not require contact with the land surface.

Ground surveys include geological mapping, geo-chemical sampling, and geophysical exploration, which are initiated if the results of remote sensing are encouraging. Ground surveys aim to define small areas of heightened interest that warrant further testing by drilling. Drilling, supplemented in some cases by test pitting, is the main technique to

REGION	2001	2005
Latin America	29	23
Canada	17	19
Africa	14	16
Australia	17	13
United States	8	8
Southeast Asia	6	4
Rest of World	9	17
Total	100	100

Rest of World Exploration

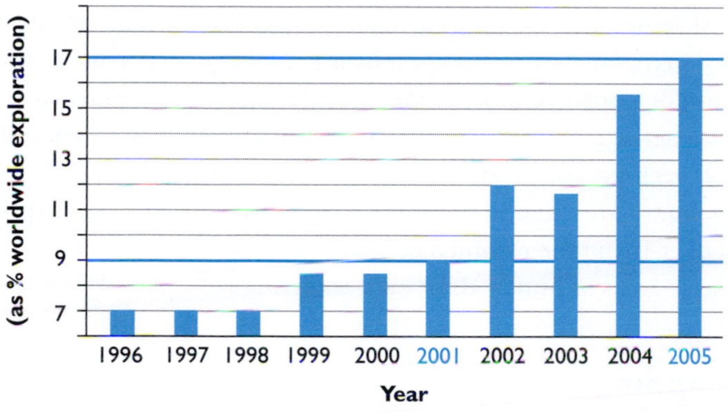

(as % worldwide exploration)

7 9 11 13 15 17

1996 1997 1998 1999 2000 2001 2002 2003 2004 2005

Year

FIGURE 4.5

Worldwide Exploration Spending by Region as Percentage of Worldwide Exploration

While Canada, Australia, and the USA remain the countries of preferred exploration investment, a large number of mining companies from these countries are increasingly exploring outside – in over 100 countries.

Source:
Metals Economic Group 2006 and World Bank 2006

eventually confirm the presence of a concealed mineral deposit, and to outline its extent and grade. All other techniques are merely indicative.

Where is most of the exploration money spent? The Metals Economic Group identified the 10 countries that accounted for roughly 70 per cent of the total US$ 4.9 billion global exploration investment in 2005 (**Figure 4.5**). Not surprisingly the traditional big three – Canada, Australia, and the United States – head the list, with Canada hosting 19 per cent of the total global exploration budget, Australia 13 per cent and the United States 8 per cent. Russia moved to fourth in the 2005 ranking, up from ninth in 2003 and fifteenth in 2001. In 2006, China accounted for a meager 2 per cent, a reflection, perhaps, of its Byzantine matrix of bureaucratic controls. Given China's vast area, much of which remains unexplored, future intensive search for resources there is inevitable, and China will attract a much higher share of the exploration dollar.

China serves as a prime example of the understanding that to attract mining investment, a favourable investment climate is as important as favourable geological conditions. Almost half of worldwide exploration spending in 2005 was attributed to gold, 30 per cent to base metals, 13 per cent to diamonds and 3 per cent to platinum-group metals.

To attract mining investment, a favourable investment climate is as important as favourable geological conditions.

There is uncertainty regarding the current state of mining in major producing regions such as China and India and it is likely that investments in these areas are not fully captured by existing surveys (World Bank 2006). Given their production bases and land masses, this group's share of long-term investment is likely to be larger than implied in

Figure 4.5. In the short term, planned exploration spending for the rest of the world, particular China and India, has grown rapidly.

At any stage of exploration, results are assessed to decide whether to proceed with further work. A decision to abandon the search is not proof that ore deposits do not exist in the area. There are many examples of major mineral deposits being discovered in areas explored on previous occasions. Re-evaluation of previously explored areas is very important in the search for minerals.

Exploration geology is a well-researched topic and ample literature is available for detailed subject coverage (Sheriff and Geldart 1995, Robb 2005, Moon *et al.* 2005, and McDonald 2007). The following text is limited to a brief illustration of reconnaissance exploration, geological mapping, geo-chemical and geo-physical exploration, and drilling. These activities normally do not require the preparation of environmental impact assessment studies, but good industry environmental practices and standard operating procedures apply (see also Chapter Thirteen).

Reconnaissance Exploration

In the reconnaissance or early exploration phase, remote sensing methods such as satellite imagery and various types of aerial photography are used as a 'first-pass' to home-in on possible mineralization sites worthy of more detailed, and hence more expensive ground appraisal. Digital satellite and aerial imageries are commercially available, but generally require computer enhancement techniques for effective use. Selected geophysical data such as magnetic intensity and radiometry are often in the form of remotely sensed data, acquired by specially equipped aircraft over-flying the reconnaissance area in a grid pattern. Reconnaissance exploration can cover large areas of hundreds or even thousands of square kilometres.

Geological Mapping

The next step in exploration data gathering is normally the production of suitable geological maps of target areas identified in the reconnaissance phase. Field geologists explore a target area, typically tens of square kilometres in size, accurately recording the nature, location, and structure of various rock units. Small hand-sized samples may be collected for mineralogical and textural studies by microscope or chemical analysis in the laboratory. It is normal for an area to be mapped by different geologists at different times. As geological concepts change over time, repeated mapping improves the understanding of an area. The final product of geological mapping is a map which accurately documents rock types, alteration mineralogy, and structural data such as faults, folds, and stress patterns. Geological mapping is of vital importance in deducing the location of hidden ore deposits.

Geo-chemical Exploration

Geochemical exploration is the scientific process of locating geo-chemical haloes by systematic sampling and chemical analysis of rock, soil, water, or vegetation.

When an ore deposit is formed, elevated concentrations of metals and other elements are often dispersed in the surrounding host rock. Further dispersion of metals and elements may occur when the deposit weathers in a near surface environment. This dispersion feature is known as a geo-chemical halo, which eventually may lead towards the origin of the anomalies; in concentration these may indicate a concealed mineral deposit. Geochemical exploration is the scientific process of locating geo-chemical haloes by systematic sampling and chemical analysis of rock, soil, water, or vegetation. The survey typically follows a series of steps from examination of large areas to progressively smaller and smaller areas, until further exploration locates the deposit. This type of survey depends on the current state of geo-chemical and geological knowledge, the nature of the area investigated, and the exploration philosophy of the organization involved.

Stream Sediment Sampling

Stream sediment sampling involves the systematic collection of small samples of sediment from the active beds of rivers and creeks. In combination with geological mapping and reconnaissance remote sensing methods, it can provide insight into the mineral potential of a large area. In rugged areas, locating sampling sites can be difficult, but exploration personnel are aided by using high quality aerial photos, satellite-based global positioning systems and experienced ground crews. Sediment at a sample point may be sieved to obtain the most sensitive size fraction, or panned in the traditional way to determine if gold or other heavy minerals are present. It is not advisable to rely on panning alone as it is easy to miss very fine-grained gold. Sample size ranges from 100 grams to five kilograms. The number of stream sediment samples taken can vary substantially but one to four samples per square kilometre is typical. The environmental impact of stream sediment surveys is minor and temporary. No special access developments are required. The small sample holes in stream beds are self-repairing through normal river action. Flagging tape used to temporarily mark sample sites is often biodegradable.

Rock Sampling

The geochemical halo associated with some mineral deposits may be detected by chemical analysis of rocks. Such halos are typically larger than the actual ore deposit, making them an easier target to locate. Rock sampling involves the taking of small hand-sized samples generally of less that one kilogram. Once an anomalous concentration of the target elements is detected, more detailed sampling and careful geological mapping is required to locate the actual deposit.

The geochemical halo associated with some mineral deposits are typically larger than the actual ore deposit.

Soil Sampling

A geochemical indication of a nearby mineral deposit may be obtained by sampling and analyzing the soil. Under some weathering conditions, elements present in the ore body may be widely dispersed in a geochemical halo that is much larger than the source deposit. It is this enlarged halo that is the target for many soil sampling programmes. Geochemical soil sampling programmes are usually conducted over relatively small areas and the location of the sample sites is controlled by a surveyed grid established on the ground or by temporary location techniques. Soil samples are typically collected by hand from a small hole dug to a depth of about ten centimetres. They are generally less than one kilogram in weight. In some flat, open environments, a small hand or vehicle mounted auger drill may be used to obtain a deeper sample. Appropriate rehabilitation of sample sites is carried out during the survey. Soil samples are transferred directly to a laboratory for geochemical analysis.

Other Geo-chemical Methods

A variety of other geochemical methods have been developed in recent years, often in response to specific needs. Minerals deposits may emit a variety of gases that leak to the surface and either accumulate in soils or are emitted to the atmosphere in very low concentrations. Radon emanating from uranium deposits is an example. Attempts have been made to collect and analyze these gases, both with ground and airborne collection systems. Results to date vary. Techniques employing the direct analysis of water samples have similarly been developed with mixed success. Certain plant species can take up metals from the soil through their root systems and concentrate them in various parts of the plants, such as the leaves or bark. Geo-botanical surveys based on direct sampling of plants or ground litter have been undertaken with some success around the world, notably in the Copper Belt of Africa.

Geo-physical Exploration

Geophysical methods offer means of seeing into the Earth itself, something few other exploration techniques can do.

Mineral deposits usually possess physical properties that are different from those of the surrounding host rocks. Mining geologists use scientific techniques called geophysical methods to measure these variations in the physical properties of rocks (such as density, magnetism, electrical conductivity, natural radioactivity, or heat capacity) as a guide to the possible locations of mineral deposits. Geophysical methods offer a means of seeing into the Earth itself, something few other exploration techniques can do. They are a very important component of most modern exploration programmes.

Geophysical observations are not new. However, the systematic measurement of geophysical properties as a means of discovering deposits did not commence until the late 1940s (Kearey *et al.* 2002). Technology development was rapid and actual geophysical surveying commenced in the 1950s and continues to be an integral part of today's exploration activity. When conducted from the air, geophysical exploration is called remote sensing. It may be undertaken from fixed-wing aircraft or helicopters normally flying 60 to 200 metres above the surface or from satellites up to 1,000 kilometres above the Earth. Ground geophysical surveys are expensive and are generally only undertaken over relatively small areas of established interest.

Gravimetric and Magnetic Methods

The Earth acts as a giant magnet, generating a field about itself that influences or captures other objects that are either magnetic or may be magnetized, particularly objects (or respectively, ore deposits) containing iron. Magnetometers, that allow magnetic surveys, are simple but highly sensitive. Magnetic surveys may be undertaken from the air or on the ground. Survey data are processed in the office and presented as a magnetic map. There are three common gravimetric and magnetic methods.

- Electromagnetic methods are electrical exploration methods. They determine the magnetic field that is associated with an electrical current induced through the ground.
- Gravity methods map the force of gravity at different locations with a gravimeter determining differences in specific gravity of rock masses and, through this, the distribution of masses of different specific gravity. Often ore minerals have a higher density than the surrounding rocks. By measuring variations in the Earth's gravity field over an area, an indication of the nature of the underlying geology and the likely presence of deposits can be obtained. Surveys are often undertaken over large areas, and provide regional information on the nature of rocks often at considerable depth. As the gravity survey instruments have no surface impact and no special access development is required, this type of surveying has no environmental impact on the area being surveyed.
- Magnetic methods are commonly used in geophysical prospecting to map variations in the magnetic field of the Earth attributable to changes in structure or magnetic susceptibility in certain near-surface rocks. Most magnetic prospecting is done with airborne instruments. Detailed magnetic surveys designed for closer studies of the geology may be undertaken on the ground. Ground surveys normally require a grid of pegged lines over relatively small areas. Stringent conditions apply to any land clearing for such grid lines.

Electrical Methods

Mineral deposits and geological structures display a wide variety of electrical properties, including electrical conductivity, and capacity to hold an electric charge. Electrical surveys are normally conducted along surveyed grid lines and require electrodes, usually in the form of either short metal stakes driven into the ground or porous ceramic pots filled with copper sulphate, to be placed in shallow holes dug with a spade. Both stakes and pots are withdrawn on completion of a survey.

In the resistivity method, a current is introduced into the ground by two contact electrodes and potential differences are measured between two or more other electrodes. Electrodes are removed on completion of the survey. Telluric methods measure variations in electric fields caused by such diverse mechanisms as electrical storms and solar activity, and are generally used to define large-scale variations in conductivity of the Earth's crust. The forced current electrode methods take advantage of the fact that a forced electric current circulated by two electrodes results in voltages measured at two others. The electrodes are usually arranged in fixed linear patterns and are systematically moved along survey lines. The flow of the induced current is affected by the properties of the rocks, including mineral deposits that occur along the grid line. There are two basic electrode methods: (1) Induced Polarization (IP), the production of a double layer of charge at a mineral interface; or production of changes in double-layer density of charge, brought about by the application of an electric or magnetic field; and (2) Spontaneous Polarization (SP), the electrochemical reactions of certain ore bodies causing spontaneous electrical potentials.

Forced Current Coil Methods

Collectively known as electromagnetic or EM methods, these depend upon electric current being transmitted through a coil or loop of wire laid upon the ground. Rocks or mineral deposits that are electrical conductors deform the resultant magnetic field. If the area to be explored is large or remote, reconnaissance stage EM surveying may be achieved by remote sensing. This technique is known as airborne EM surveying.

Seismic Surveys

These exploration techniques utilize the variation in the rate of propagation of shock waves in different media. Seismic methods rely on studying the ways in which sound or equivalent wave forms produced on the surface travel through the underlying rock. Different rock formations and geological structures affect these energy waves in specific ways and, by studying the results obtained, it is possible to predict the nature of the concealed geology.

Seismic methods rely on studying the ways in which sound or equivalent wave forms produced on the surface travel through the underlying rock.

There are two basic types of seismic surveying – refraction and reflection. The refraction method is used to study ground conditions, such as depth of weathering or faulting, within 50 metres of the surface. Therefore, it is extensively used in quarrying and construction foundation studies. It is not generally used in exploration for mineral deposits. The reflection method is normally used for deep penetration and the understanding of geological stratigraphy and structure.

The energy source is critical in seismic surveying. It determines the depth of penetration. A vibrator or explosives are usually used in reflection surveys. Vibrator systems are non-damaging to the surface, but explosives, which are inserted in shallow shot holes, can affect small areas of four to ten square metres, which require rehabilitation on completion of the survey.

Seismic methods are commonly used in exploration for oil, gas, and coal but are less common in exploration for metalliferous deposits in geologically complex mineralized areas. Large three-dimensional seismic exploration is a major activity with potentially significant environmental and social impacts.

Radiometric Surveys

Radiometric methods rely on the use of portable Geiger-Muller apparatus for field detection of emission counts in the search for radioactive minerals. Many rocks and minerals are naturally radioactive. Radiometric surveys measure variations in the natural radioactivity of an area. Modern spectrometers allow the detection of radioactivity at very low levels that were not previously detectable. Surveys of this type are undertaken from the air with follow-up ground work wherever anomalies are identified.

Down-hole Methods

Down-hole geophysical surveying is usually a secondary exploration method used to maximize the value of a drill hole. It increases the effective search area of a hole and may detect an ore deposit narrowly missed by drilling. Rock properties such as natural radioactivity, density, conductivity or other electrical or magnetic properties may be measured by lowering a probe down the hole. Alternatively, a receiver may be placed in the hole, and a transmitting loop on the surface. This configuration can be used to measure the rock properties between the surface and the drill hole. Holes are rarely drilled purely for geophysical exploration, but as technology improves this is likely to become more common, as will measurement between drill holes. Down-hole methods are also extensively used in oil and gas exploration and in some groundwater investigation programmes.

Trenching and Pitting

Trenching or costeaning and pitting are traditional methods of exposing concealed bedrock or taking samples for analysis. In the early days excavations were usually undertaken using hand tools. The practice is still carried out today, particularly in hard to reach areas, but to a much lesser extent. Costeaning to map and sample the bedrock is normally undertaken with an excavator. Larger-scale excavations may be used to provide bulk samples of known deposits for detailed testing prior to mining. Prior approval is usually required for excavations of this nature, with strict requirements governing the back-filling and surface rehabilitation of any such approved works after sampling has been completed.

Exploration Drilling

The only way to test for the possibility of a concealed mineral deposit is to gather data from beneath the surface. This is mainly achieved by drilling.

The exploration methods described so far are generally applied over relatively large areas of land and have only a very minor impact, if any, on the surface. They are designed to indicate the possible presence of a mineral deposit, but cannot physically define it. Once all the data from these surveys have been collected, they are plotted on various maps, with the assistance of sophisticated computers, and then interpreted by geoscientists. The only way to test for the possibility of a concealed mineral deposit is to gather data from beneath the surface. This is mainly achieved by drilling.

The first stage of drilling is usually termed reconnaissance drilling. If the results from this drilling are encouraging, a further programme of follow-up drilling may take place. If the results of this additional drilling are also encouraging, geologists may embark upon a programme of detailed or deposit delineation drilling. The selection of the drilling method and equipment is determined by such factors as the nature and depth of the target deposit, location and access factors associated with the drill site, and cost. Only a few exploration holes will intersect an ore body. Within gold deposits, a discovery hole may be one in one thousand and with some base metals commodities, strike rates range from one in fifty to one in one hundred. However, once the deposit has been located, most of the delineation drill holes will intersect the deposit.

Drilling can result in significant environmental impacts. Although rare, exploration drilling can encounter artesian water which, if not prevented can flow for many years, wasting groundwater resources. Also, drill cuttings containing pyrite can oxidize. Where these cuttings are left on the surface adjacent to the drill hole, the resulting oxidation can leave a patch of contaminated ground which resists attempts at re-establishment of vegetation.

Auger Drilling

If disturbed samples are required from relatively shallow depths in unconsolidated sediments, soils or soft rocks, an auger rig may be used. The auger may be truck or tractor mounted, similar to a farm post hole drill, or it may be portable. The main limitations of auger drilling are boulders in the soil, excessive moisture, and inability to penetrate far into the bedrock.

Rotary-percussion Drilling

This includes a wide range of drilling techniques. In most cases the energy source is compressed air, and the sample obtained is in the form of chips. The chip samples produced at the bit face are continuously blown to the surface by compressed air and collected at regular intervals by the crew for examination by the geoscientists. In the reverse circulation technique the chips are blown to the surface through the centre of the drill rods, reducing contamination of the sample and danger of hole collapse. If the ground is soft, special bits may be used with rotary percussion rigs to produce small pieces of rock similar to diamond cores. This is known as air-core drilling. Any form of percussion drilling requires quite large volumes of compressed air to operate the rig and to lift the samples to surface. The compressors required are usually mounted on wheeled or tracked carriers. Truck-mounted rigs generally require a reasonable quality road to access a drill area, although lightweight rigs are increasingly being used at the reconnaissance stage.

Core or Diamond Drilling

Samples are obtained with this method by rotating a string of drill rods into the ground under hydraulic or mechanical pressure. The drill bit, which is usually faced with diamonds, cuts a cylindrical core of rock that continuously passes up into the drill barrel as the bit penetrates. Diamond drilling rigs are truck, track, or skid mounted and an access track is usually, but not always, required. The standard of such tracks is determined by a number of factors, including the location of the area, terrain, type of drill rig and the estimated duration of the drilling programme. Lightweight rigs can also be used for this type of drilling for transport to rough or remote mountainous areas by helicopter. This approach requires that drill sites be prepared in advance by ground crews. **Figure 4.6** shows the distribution of drill sites accessed by helicopter at the Batu Hijau copper deposit in Sumbawa, Indonesia. Man-portable rigs, which can be transferred from drill site to drill site without the need for vehicular access, are being used increasingly for drilling in rugged forest areas. Most of the drilling at the Tampakan site in the Philippines used this minimally invasive approach.

Man-portable rigs, which can be transferred from drill site to drill site without the need for vehicular access, are being used increasingly for drilling in rugged forest areas.

Diamond drilling usually involves directing the cuttings to a sump where they are deposited, enabling the water with any drilling additives to be recycled. These sumps require careful filling after drilling operations are completed. Otherwise they can pose a hazard for people, livestock and wildlife.

FIGURE 4.6
Distribution of Drill Sites Accessed by Helicopter at the Batu Hijau Copper Deposit in Sumbawa, Indonesia

Recurrent Drilling

In 50 years time, mining geologists will undoubtedly be probing routinely to depths of 2,000 or 3,000 metres, finding deep deposits missed today.

As new exploration and mining concepts, methods and targets become known and available, previously explored areas are often re-explored and re-drilled. Mining geologists in the 1930s rarely drilled beyond 100 metres in depth. Thus, they only found shallow deposits. Today, drill holes beyond 1,000 metres are rare and we only find deposits above that depth. In 50 years time, mining geologists will undoubtedly be probing routinely to depths of 2,000 or 3,000 metres, finding deep deposits missed today.

Rigs and their associated down hole equipment are becoming lighter and more efficient. This results in greater opportunity to undertake drilling programmes, with drilling equipment carried into the field manually or by helicopter. The use of lighter machinery for preparing access roads and sites for large skid and wheel-mounted rigs has minimized environmental impacts, allowing a high standard of post-drilling rehabilitation to be achieved. Chemical additives used to improve drilling performance are biodegradable, and simple procedures have been developed to prevent drilling fluids from contaminating adjacent areas. Drill crews and supervisors are industry-trained in all aspects of environmental awareness and in the strict regulations designed to minimize environmental impacts that apply to drilling operations. This ensures that modern exploration drilling is a low-impact activity, involving temporary effects to very small areas.

Geological survey is thus a systematic investigation of an area determining the distribution, structure, composition, history, and interrelations of rock units. Its purpose may be either purely scientific or economic with special attention to the distribution, reserves, and potential recovery of mineral resources.

Bulk Sampling

In most exploration work, there is eventually a need for large, representative samples of the ore deposit (USDA 1995). A final cross-check of the grade of the deposit must be made, as well as testing to determine the best choice of metallurgical method. Bulk sampling may also yield other valuable data of use in planning mine and haulage facilities, the treatment method, or disposal of waste. The mining characteristics concern such factors as the way the rock in an open pit mine may be expected to break during blasting and to support itself on a bench face, and the manner in which the rock will cave in an underground block caving operation or support itself in an underground mine. Metallurgical treatment methods can be most effectively researched by pilot testing techniques, and disposal of waste can be carefully researched using the waste from these original testing programmes, with the crushed bulk sample fed into a pilot plant.

Typically, large-scale bulk sampling is undertaken in the last stages of exploration of a low grade ore deposit to be developed by open pit methods. In a major project, bulk sampling provides sufficient ore to operate a pilot plant with a capacity of 50 to 100 tons per day for several months. The pilot plant may be established on-site, but more commonly, a pre-existing off-site plant will be used, with the bulk sample shipped to that facility. Bulk sampling takes on the character of a small-scale mining operation with all its related environmental impacts, requiring its own environmental permit.

Typically, large-scale bulk sampling is undertaken in the last stages of exploration of a low grade ore deposit to be developed by open pit methods.

The pilot plant is a miniature version of the full-scale plant to be built to concentrate the ore from the mine. The design of the pilot testing plant is based on knowledge of the type of ore in the deposit and the details of bench testing of ore from exploration core drilling. Details of crushing, grinding, concentration characteristics, and waste disposal can be studied over a period of time in a pilot plant. Also considered are the effects that a change in one part of the process will have on another, as well as the overall efficiency of the process. Depending on the results, alterations are made in the design of the full-scale plant. Costs of construction, operating costs, and waste disposal problems can be determined for use in broad planning and in the final feasibility study.

Greenfield and Brownfield Exploration

Exploration is termed either Greenfield or Brownfield, depending on the extent to which previous exploration has been conducted on the target area in question. While loosely defined, the general meaning of Brownfield exploration is that which is conducted within geological terrains within close proximity to known ore deposits or within a known mineral province. Greenfields are the remainder.

Greenfield exploration is highly conceptual, relying on the predictive power of ore genesis models to search for mineralization in unexplored virgin ground. This may be territory which has been drilled for other commodities, but with a new exploration concept is considered prospective for commodities not sought there before.

Greenfield exploration has a lower strike rate, because the geology is poorly understood at the conception of an exploration programme, but the rewards are greater because it is easier to find the biggest deposit in an area earlier, and it is only with more effort that the smaller satellite deposits are found. Also, the explorer making the first discovery in a Greenfield area may be able to secure exploration rights to most or all of the prospective ground, whereas, in Brownfield areas exploration and mining areas are commonly fragmented and new discoveries may extend into areas controlled by competitors. Brownfield exploration is less risky, as the geology is better understood and exploration methodology is well known, but since most large deposits are already found, the rewards may be incrementally less.

4.3 FEASIBILITY – IS IT WORTH MINING?

In mineral exploration, something is always better than nothing (**Figure 4.7**) and, when successful, exploration delivers the discovery of a mineral deposit. But is the discovery worth mining? The evaluation of a mineral deposit from exploration through development and production is a lengthy and complicated process. Determining the technical and

The evaluation of a mineral deposit from exploration through development and production is a lengthy and complicated process.

FIGURE 4.7
Something is Always Better than Nothing

Credit: www.CartoonStock.com.

environmental feasibility and the economic viability of each project is a continuing process, elaborated during each phase of mine development, with more detailed environmental information and engineering data required at each stage.

Integrated evaluation and planning, which considers economic, environmental, and technical aspects in parallel, helps to minimize the environmental impact of the operation. It helps to avoid or reduce adverse environmental impacts over the life of the mine and after closure, and to minimize costs. Defining the final objectives of mine closure from the outset allows an optimum balance between operational, rehabilitation, and closure goals to be selected, thus minimizing the cost of these activities.

Each mining company has a minimum acceptable rate of return on investment, which considers the cost of borrowing capital for developing a new mine, or of generating the needed capital internally within the company. If a company has a number of attractive investment opportunities, the rate of return from the proposed mine venture is compared with the rate expected on a different mining venture elsewhere, or with some other business opportunity unrelated to mining. Eventually the project with the best rate of return is selected. The size of an ore deposit is equally important. Large mining companies concentrate their efforts on developing large ore deposits, and may divest rich but small mineral discoveries.

As a general rule of thumb, a project must have better than a 15 percent rate of return to be considered by a major mining company.

As a general rule of thumb, a project must have better than a 15 percent rate of return to be considered by a major mining company. An individual or a junior mining company usually expects a 30 to 50 percent rate of return to consider investing in a mining venture. Among other uses of the cash flow generated by the mine, these funds must finance continuing exploration elsewhere, pay for past failures, and contribute to the mine's portion of main office and general overheads (USDA 1995).

Once the existence of a potentially valuable ore deposit is demonstrated, the project is normally turned over to the mining and metallurgical engineers. They decide on the basis of engineering

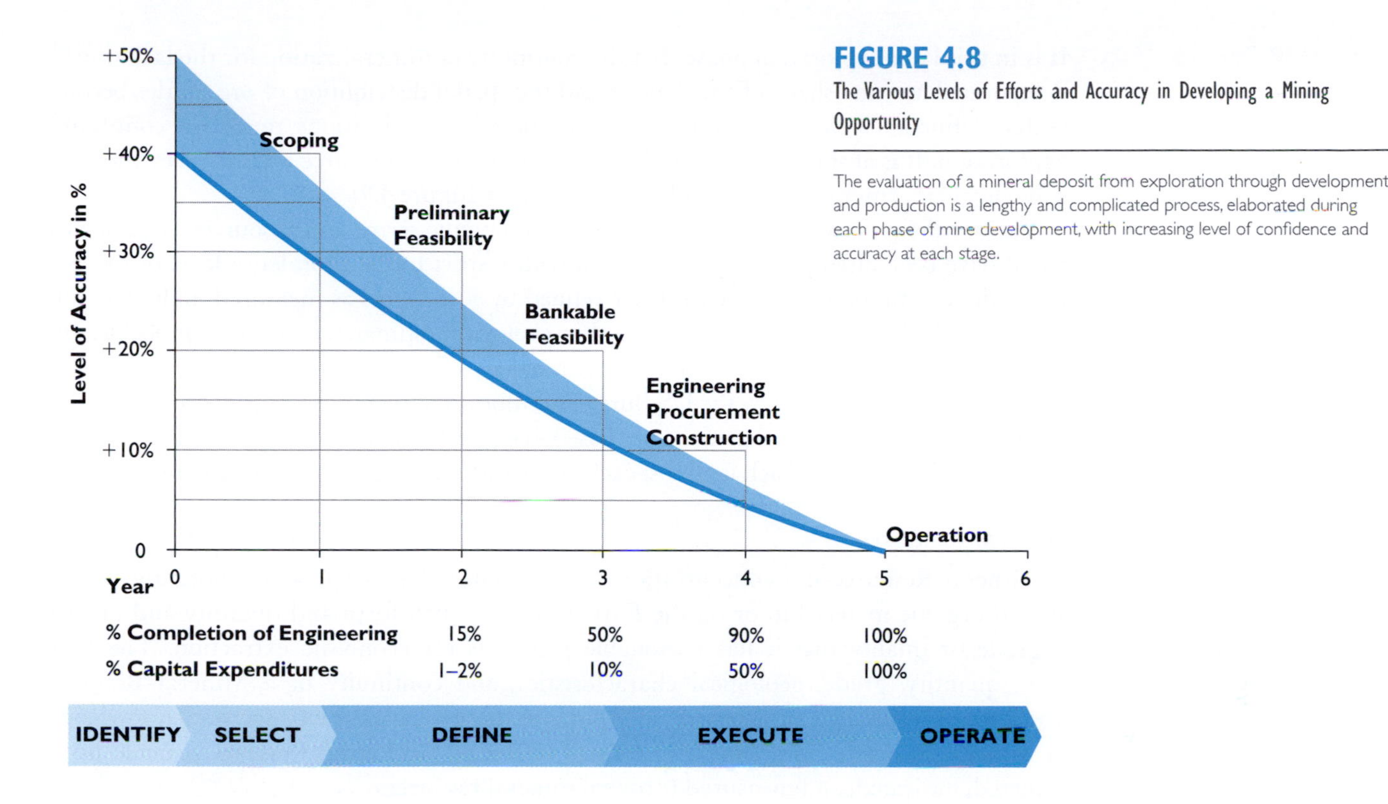

FIGURE 4.8

The Various Levels of Efforts and Accuracy in Developing a Mining Opportunity

The evaluation of a mineral deposit from exploration through development and production is a lengthy and complicated process, elaborated during each phase of mine development, with increasing level of confidence and accuracy at each stage.

studies if, how, and when the deposit should and can be mined. There are two levels of engineering studies during development that are commonly acknowledged by the mining industry (**Figure 4.8**): conceptual and feasibility, the latter further subdivided into pre-feasibility and bankable feasibility (Pincock *et al.* 2005 from which the following discussion borrows).

Figure 4.8 illustrates the sequential activities and related studies in evaluating and developing a mining opportunity:

- Identify – determining the potential value of the opportunity through exploration;
- Select – selecting the preferred opportunity realization concept based on conceptual studies;
- Define – finalizing scope, cost, schedule, and financing based on feasibility studies;
- Execute – producing an operating asset using EPC contractors (engineering, procurement, and construction); and
- Operate – producing return to shareholders through ore extraction.

A brief discussion of each step follows.

Evaluation of the Exploration Programme – Differentiating Resources from Reserves

Should we? The evaluation of an exploration programme is a continuing process applied to every step in the exploration programme. At each stage the exploration manager has to decide whether to continue exploration or abandon the site. If exploration is terminated too early, an ore body may be missed. On the other hand, continued exploration where no ore body is to be found wastes time and money.

As elaborated above, exploration is a progressive process. Initial exploration drilling and testing at selected locations shows the presence and extent of ore-grade mineralization. Defining the mineral deposit is then achieved with more and more closely spaced holes.

Exploration is a progressive process.

It is in this later exploration phase that the continuity of mineralization (or the lack of it) is determined and the shape of the deposit and the spatial distribution of ore grades become better defined. Geological resources are computed, and decisions made on continuing exploration. It is at this stage where the difference between resource and reserves and their sub-categories emerges (www.insidemetals.com, see **Figure 4.9**).

Various definitions of the word reserves (an economic term) and resources (a geological term) have been attempted in the past, and, with respect to the confidence level of the estimate, these definitions have been further refined by such terms as measured, indicated, and inferred, which have commonly been used in geological estimates; or proven, probable, and possible, which is the more readily used terminology of industry for economic evaluations. The definitions introduced by the US Bureau of Mines and the US Geological Survey in 1974, on which the following explanations are based, have achieved wide acceptance and form the basis of the JORC code which is almost exclusively used in Australia and elsewhere in Asia.

Mineral Resource

A Mineral Resource is a concentration or occurrence of natural, solid, inorganic or fossilized organic material in or on the Earth's crust in such form and quantity and of such a grade or quality that it has reasonable prospects for economic extraction. The location, quantity, grade, geological characteristics, and continuity of a Mineral Resource are known, estimated or interpreted from specific geological evidence and knowledge. A Mineral Resource, sometime also termed *in situ* or Geological Reserve, is made up of inferred, indicated, and measured (proven) mineral resources.

Inferred Mineral Resource is that part of a Mineral Resource for which quantity and grade or quality can be estimated on the basis of geological evidence and limited sampling and reasonably assumed, but not verified, geological and grade continuity. The estimate is based on limited information and sampling gathered through appropriate techniques from locations such as outcrops, trenches, pits, workings and drill holes.

Indicated Mineral Resource is that part of a Mineral Resource for which quantity, grade or quality, densities, shape and physical characteristics, can be estimated with a level of confidence sufficient to allow the appropriate application of technical and economic

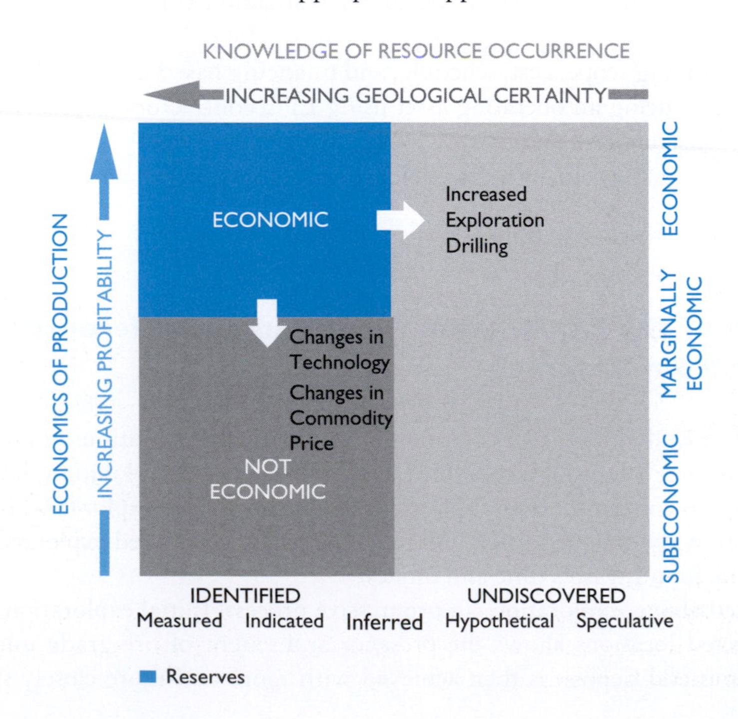

FIGURE 4.9
The Relationship Between The Various Categories of Mineral Resources and Mineral Reserves

Mineral resource is a geological term, while mineral reserve is an economic term. A mineral reserve is the economically mineable part of a mineral resource demonstrated by at least a preliminary feasibility study.

parameters to support mine planning and evaluation of the economic viability of the deposit. The estimate is based on detailed and reliable exploration and testing information gathered through appropriate techniques from locations such as outcrops, trenches, pits, workings, and drill holes that are spaced closely enough for geological and grade continuity to be reasonably assumed.

Measured Mineral Resource is that part of a Mineral Resource for which quantity, grade or quality, densities, shape, and physical characteristics are well established (**Case 4.3**).

Mineral Reserve

Mineral Reserve (also termed Mineable Reserve) is the economically mineable part of a Measured or Indicated Mineral Resource demonstrated by at least a preliminary feasibility study. This study must include adequate information on mining, processing, metallurgical, economic, and other relevant factors that demonstrate, at the time of reporting, that economic extraction can be justified. A Mineral Reserve allows for dilution of ore by nearby non-ore materials and allowances for losses that may occur when the material is mined and processed, including unrecovered mineral that will be included in the tailings.

Statistical studies have shown that less than 5 percent of the total number of a specific type of mineral deposit produce more than 90 percent of that commodity or group of related commodities (McLemore 2005). These deposits, so-called 'world-class' deposits, are the largest of the known deposits in the world in terms of size and grade.

Less than 5 percent of the total number of a specific type of mineral deposit produces more than 90 percent of that commodity or group of related commodities.

Probable Mineral Reserve is the economically mineable part of an Indicated, and in some circumstances a Measured Mineral Resource, demonstrated by at least a preliminary feasibility study. This study must include adequate information on mining, processing, metallurgical, economic, and other relevant factors that demonstrate, at the time of reporting, that economic extraction can be justified. Proven Mineral Reserve is the economically mineable part of a Measured Mineral Resource demonstrated by at least a preliminary feasibility study. This study must include adequate information on mining, processing, metallurgical, economic, and other relevant factors that demonstrate, at the time of reporting, that economic extraction is justified.

CASE 4.3
Watch Out for Salting

This old and dishonourable method of making a mining prospect seem better than it is, has unfortunately not vanished. While as old as mining itself, the most recent and probably most remembered 'salting' case is the Bre-X case. In 1996. Bre-X, a little known Canadian mining company, convinced financial markets, as well as the then President Soeharto in Indonesia of the discovery of a gold resource in East Kalimantan, with a value well in excess of several tens of billion US dollars. Only subsequent independent third party drilling suggested salting, whereby rich ore or native gold is mixed before (or even after) it has reached the assayer. The bonanza mine turned into a hoax with the loss of hundreds of millions of US dollars for the unfortunate shareholders, not to mention the embarrassment for the Indonesian Government. Salting can be achieved by adding metallic gold in the material to be sampled or the sample itself. Not all salting, however, is due to dishonesty; carelessness in sample preparation or in the laboratory may result in accidental salting.

Exploration activities normally continue throughout the feasibility study, mine development and operation.

From the environmental point of view two considerations deserve special attention. Firstly, exploration activities normally continue throughout the feasibility study, mine development and operation. The chances are that proven mineral reserves will increase as more exploration data are generated, although the reverse may also occur. Generated volumes of mine wastes may be well above the volumes estimated during the feasibility and the EIA studies. Secondly, measured mineral resources may turn into proven mineral reserves in time if mineral prices increase during the mine life, lowering the initially selected cutoff grade. Mine operations may be adjusted accordingly with the need to re-visit the original EIA study. Similarly, low grade ore temporally stockpiled for future ore processing might turn into permanent waste rock dumps if mineral prices were to fall. The EIA practitioner needs to appreciate these potential changes in future mine planning to adequately evaluate environmental protection measures that are built into mine design.

Selecting the Preferred Opportunity Realization Concept

The environmental scoping study is part of the conceptual design phase with the main objective of identifying 'fatal flaws' or 'show stoppers'.

What objective? The conceptual study, also commonly referred to as a scoping study, is the first level study and the preliminary evaluation of the mining project. The environmental scoping study is part of the conceptual design phase with the main objective of identifying 'fatal flaws' or 'show stoppers'. The principal parameters for a conceptual engineering study are mostly based on broad assumptions. Accordingly, the level of accuracy is low at about 50 percent. Although the level of drilling and sampling must be sufficient to define a resource, flow sheet development, cost estimation, and production scheduling are often based on limited data, test work, and engineering design. The results of a conceptual study typically identify: technical parameters requiring additional examination or test work; general features and parameters of the proposed project; magnitude of capital and operating cost estimates; and level of effort for project development. A conceptual study is useful as a tool to determine if subsequent engineering studies are warranted. However, it is not valid for economic decision-making, nor is it sufficient for reserve reporting.

The Pre-feasibility Stage

The level of accuracy is higher than the scoping study at about 25 percent.

What needs to be done? The pre-feasibility study represents an intermediate step in the engineering process to evaluate a mining project. The principal parameters for a pre-feasibility study are based on various engineering data. The level of accuracy is higher than the scoping study at about 25 percent. At the pre-feasibility study stage adequate geology and mine engineering work has been conducted to define a resource and a reserve. Sufficient test work has been completed to develop mining and processing parameters for equipment selection, flow sheet development, and production and development scheduling. Capital and operating cost estimates are derived from preliminary test work, assumed factors, and some vendor quotes. The economic analysis of a pre-feasibility study is of sufficient accuracy to assess various development options and the overall project viability. While cost estimates and engineering parameters are typically not considered of sufficient accuracy for final decision-making or bank financing, the pre-feasibility study helps to define the limits of a deposit and its development. Typically the pre-feasibility study considers the following categories of limitations:

- Geological limits of the mineralization correspond to the distribution zone of known mineral resources. These limits are liable to change in time as exploration activities continue and the general degree of knowledge of the actual mineral accumulation and its geological environment increases. Advanced computerized resource modelling will further influence the definition of the deposit.

- Economic limits differentiate between mineral accumulation and ore. Economic concepts are used to define zones of rich ore(s) and zones of lean ore(s) within the deposit calculated on a balance between general production costs and net income, including profit margin. Market fluctuations could lead to the consideration in turn of one part or another of the mineralization as a rich ore, or as a lean ore, or as mining waste.
- Technological limits are concerned with the optimal conditions for mining the ore. Depending on a number of morphological criteria (depth, dissemination, or segregation in a formation or datum level, dip, type of substance), the pre-feasibility study makes an initial recommendation on the mining method designed for optimal recovery in terms of quality and cost from the many mining methods available for surface or underground workings. For the miner, only the part of the ore accessible by the available mining techniques actually defines the deposit.
- Environmental limits recognize the importance of safeguarding the mine environment. Miners, during the early years of commercial mining, wanted only one thing: valuable minerals. They didn't care about elegance, craft, aesthetics, or the environment. In most cases in these days before environmental consciousness, they developed a mineral deposit without considering the long-term environmental consequences. Today's mine developments differ. They are, from the outset, subject to intense scrutiny from the public, including anti-mining advocacies. Environmental concerns are addressed and accommodated, and may even prevent a mine from being developed. On the other hand, shareholders and their Board representatives may demand that the NPV be maximized. Showstoppers are manifold: mineral deposits may be located in or near protected areas; endangered species may be present in the mining areas; there may be a lack of safe on-land tailings disposal sites; traditional land ownership issues may be unresolved; mine development would result in significant resettlement; or widespread community opposition etc.

Miners, during the early years of commercial mining, didn't care about elegance, craft, aesthetics, or the environment.

- Political limits are a reflection of the large capital investment that a new mine development commands. Is the host jurisdiction stable and investor friendly? Are there security concerns? The overall political instability of some countries can be a great deterrent to the development of mines. Mining companies working in many developing countries can also encounter problems such as high tax and tariff costs, and the corruption of civil servants such as customs officials, without whose help they would have difficulty implementing their project. A favourable, stable investment climate is required to attract the amount of money required to develop a mine. Country risk also influences interest rates charged by the international financial markets. Higher country risk premiums may cause a mine investor to allocate funds to countries with lower risk rating. Somewhat perversely, however, the existence of any combination of negative factors leads to less exploration in that country or region, which, for a more adventurous company, can increase the chances of discovering a large ore body.

A favourable, stable investment climate is required to attract the amount of money required to develop a mine.

Preparing the Bankable Feasibility Stage

A feasibility study represents the last and most detailed step for evaluating a mining project for a 'go/no-go' decision and financing purposes. The principal parameters for a feasibility study (FS) are based on sound and complete engineering and test work. Accuracy is higher than the pre-feasibility study and is typically around 10 to 15 percent. Feasibility study objectives are the same as for the pre-feasibility study, but the level of detail and accuracy for each objective are more stringent. The level of detail is typically also dictated by whether the project is to be financed by the company or bank financed. Often the term 'bankable' is used to describe a feasibility study (bankable feasibility study or BFS). This term simply defines that the level of detail of the study is sufficient to secure financing, provided that the results are positive. In a BFS, geological and mine engineering work has been conducted in sufficient detail to define proven resources and reserves. Detailed test work has been completed to develop all mining and processing parameters for pit slope design, hydrology, geotechnical, flow sheet development, equipment selection and sizing, consumables and power consumption, material balance, general arrangement drawings, production and development schedules, and capital and operating cost estimates. Capital and operating cost estimates are derived from take-offs and vendor quotes. Economic analysis with sensitivities is based on annual cash flow calculations for the mine life. Provided that the project is feasible, a proven and probable reserve statement can be made.

The feasibility study including the EIA documentation demonstrates the economic, environmental, and technical viability of the proposed mine development.

The feasibility study including the EIA documentation demonstrates the economic, environmental, and technical viability of the proposed mine development. It is often also a prerequisite of obtaining government approval for mining. The preparation of a BFS, however, does not automatically lead to mine development. Farrell (1997) estimated that only about one of ten mine proposals in Australia for which a feasibility study is prepared will eventually progress to mining. This may be a premature observation. It may be that most of these proposals will eventually become mining projects. In fact, the current boom in most mineral commodities is leading to the development of many projects for which previous feasibility studies were less than compelling.

In sum the feasibility study aims to determine and judge all aspects that are important in assessing the worth of a potential mining project such as resource estimation, suitable mining methods and beneficiation, mine waste management options, required mine infrastructure, cost/benefit analyses, environmental and project risks, and financing.

Resource Estimation

As discussed above, mining geologists differ between geological and mining resource estimates. The geological resource estimate is based on modelling the size, shape, and grade distribution of the ore deposit, and as such is based on the prediction of geometry and continuity of ore body, distribution and variability of grades, and recoverability of metal values based on bulk sampling and metallurgical testing. The mining resource estimate predicts the part of the estimated geological resources that can be mined at a profit. The mining resource estimate is developed with an understanding of the most likely mining method, and it considers mining factors such as the range of likely cutoff grades, selected mining method and production rates, and characteristics of the ore deposit that may affect the ability to mine and process the ore.

Mining Method

A preliminary screening of suitable mining methods provides estimates on mining resources and production rates, and associated investment requirements. The simple aim in selecting and implementing a particular mine plan is always to mine a mineral deposit so that profit is maximized given the unique characteristics of the deposit and its location, current market prices for the mined mineral, and the limits imposed by safety, economy, and environment.

The EIA practitioner needs to be aware that ore production rates during the mine operation may change.

The EIA practitioner needs to be aware that ore production rates during the mine operation may change for two main reasons (**Figure 4.10**). Firstly, an increase in mineral prices may lead to an increase in ore production in existing mines rather than in the opening of new mines: less money is involved in expanding than opening a mine; project risks in an operating mine are well understood; and increases in production rates are

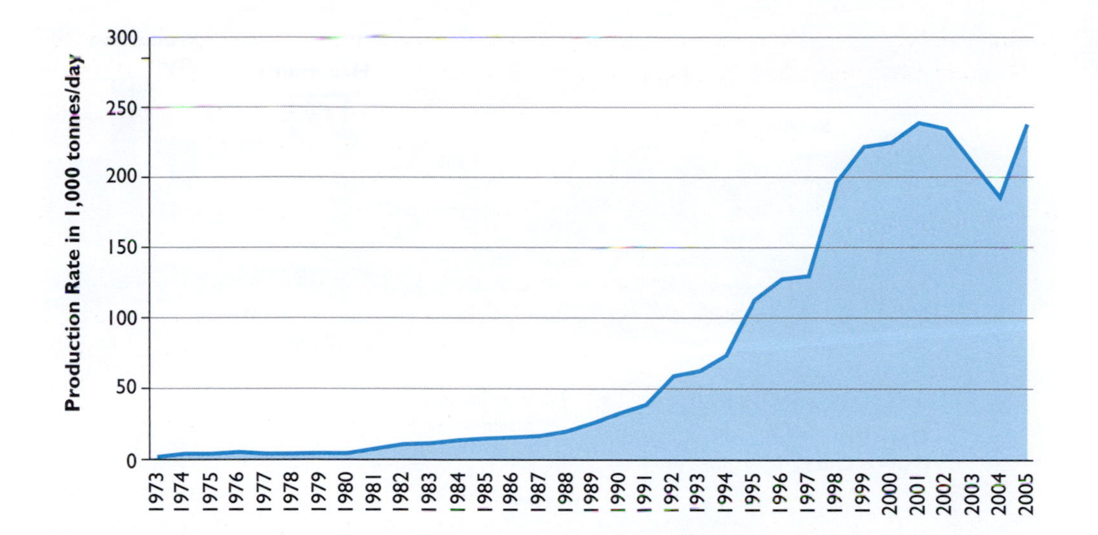

FIGURE 4.10

Increase of Production Rate over Mine Life Illustrated

From an initial production rate of less than 20,000 tons per day of ore in early 1970, Freeport's mining operations in the Papua Province of Indonesia now extract copper and gold bearing ore at a staggering rate of more than 250,000 tons per day, which equates to more than 90 million tons each year. In comparison, most mid-sized gold mines process about 1 million tons of ore per year.

faster to achieve. Secondly, exploration continues during the mine operation and additional mineable ore reserves may be discovered. For the purpose of environmental permits, it is often advisable to obtain the permit for a realistic but upper limit of ore production rates. The environmental permit remains valid if actual ore production rates are below the approved ore production rate; a new environmental approval however may be required if actual rates significantly exceed the approved design rate.

A new environmental approval may be required if actual rates significantly exceed the approved design rate.

Ore production rates largely depend on the mining method employed. The depth of a particular deposit determines whether it will be extracted by surface mining, in which minerals or coal are extracted directly at the surface, or underground mining, in which deposits are too deep to be economically removed by surface mining (**Figure 4.11**). Surface mining is more common, since it is less expensive than underground mining. Hence, only high ore grades normally warrant underground mining.

At this point some semantic difficulties relating to the size of a mine operation need to be examined. Depending on the point of view, various means of referring to the size of operations are common, as illustrated in **Table 4.3** for a hypothetical copper mine operation. For example, the mine superintendent thinks in terms of material moved, while the mill superintendent cares about the amount of ore processed per day. Top management thinks in terms of copper equivalent per year. The figures in **Table 4.3** are based on a mine producing 100,000 tons of copper per year from copper ore with an average grade of 0.5 percentage, and with an average stripping ratio 2 to 1. In most circumstances, however, mine size is expressed as tons per day of processed ore.

In most circumstances mine size is expressed as tons per day of processed ore.

Beneficiation

Excavating ore from the Earth is only half the battle since the required products are rarely found in their pure form. More often they are mixed with rock and gangue minerals, and usually as compounds of several elements. Some minerals yield their elemental constituents more readily than others. Therefore the economic viability of an ore deposit depends not only on the quantity and quality of accessible ores, but also on the ease with which valuable metals can be extracted from gangue materials (schematically illustrated in **Figure 4.12** for pyro-metallurgical copper extraction). Mineral extraction from ore conventionally takes place in mills (ore dressing or beneficiation), smelters (converting concentrates to metals), and refineries (producing the final high grade products as required by the market).

Ore beneficiation is the processing of ores to regulate the size of the product, to remove unwanted constituents, or to improve the quality, purity, or grade of a desired product.

FIGURE 4.11
Transition from Surface to Underground Mining

Surface mining is more common, since it is less expensive than underground mining. Only high ore grades normally warrant underground mining.

Source:
Adapted from Atlas Copco 2002

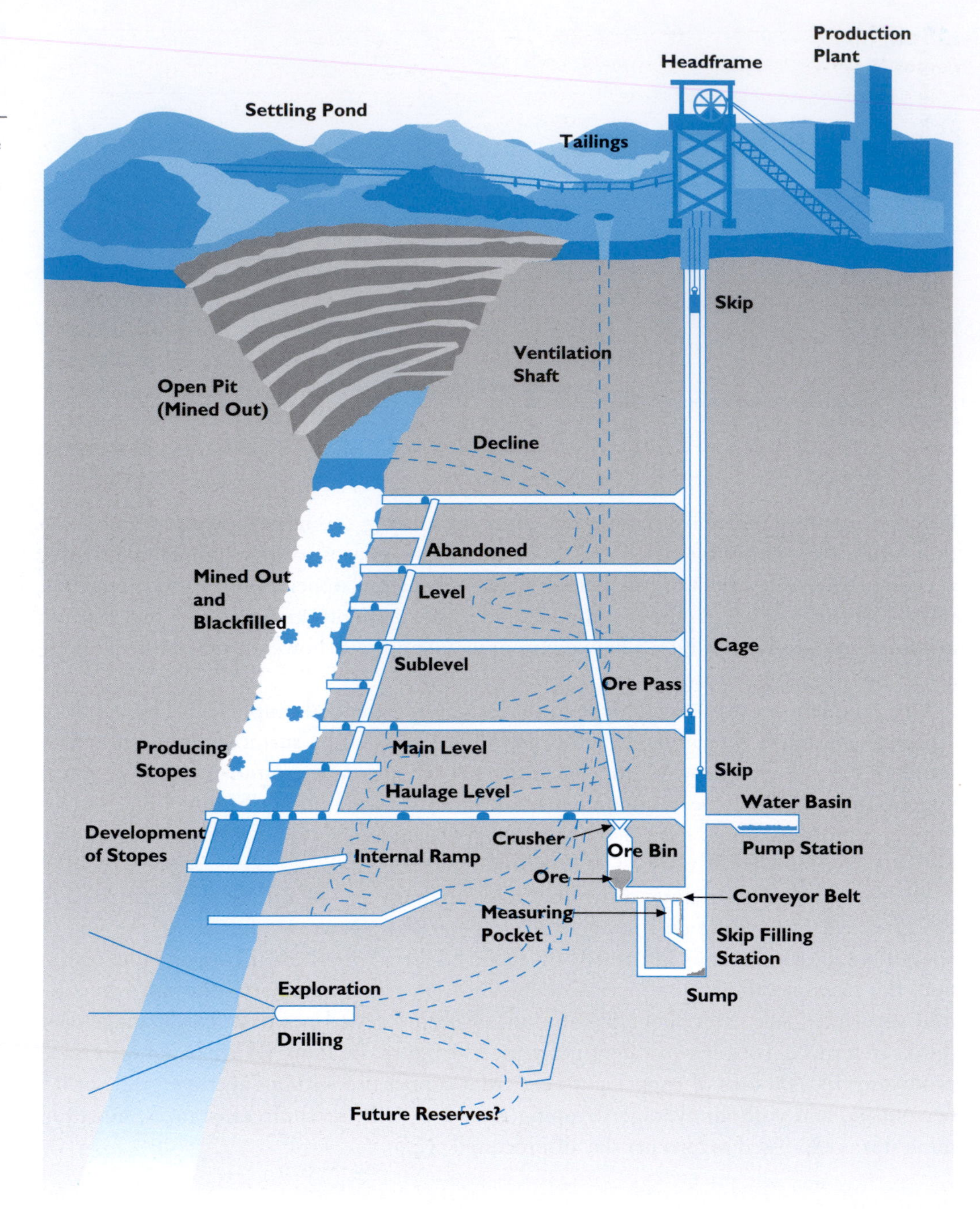

Ore beneficiation conventionally separates ore minerals from gangue minerals to the degree of concentration required for downstream mineral processing industries. Under regulations pursuant to the US. Resource Conservation and Recovery Act (40 CFR §261.4), beneficiation is restricted to the following activities: crushing; grinding; washing; dissolution; crystallization; filtration; sorting; sizing; drying; sintering; pelletizing, briquetting; calcining to remove water and/or carbon dioxide; roasting, autoclaving, and/or chlorination in preparation for leaching; gravity concentration; magnetic separation; electrostatic separation; flotation; ion exchange; solvent extraction; electrowinning; precipitation; amalgamation; and heap, dump, vat, tank, and *in situ* leaching (USEPA 1995a,b).

TABLE 4.3

Various Means of Referring to the Size of Mine Operations – All of these capacities are interrelated and any change in one quantity will affect others

Annual Capacity	Daily Capacity (Tons per Day)
100,000 tons per year of copper	
400,000 tons of concentrate (25% copper content)	1,100 TPD (shipped to smelter)
20 million tons of ore (0.5% grade ore)	55,000 TPD (mill capacity)
60 million tons of material moved (2-to-1 stripping ratio)	165,000 TPD (mine capacity)

Source:
Based on Navin 1978

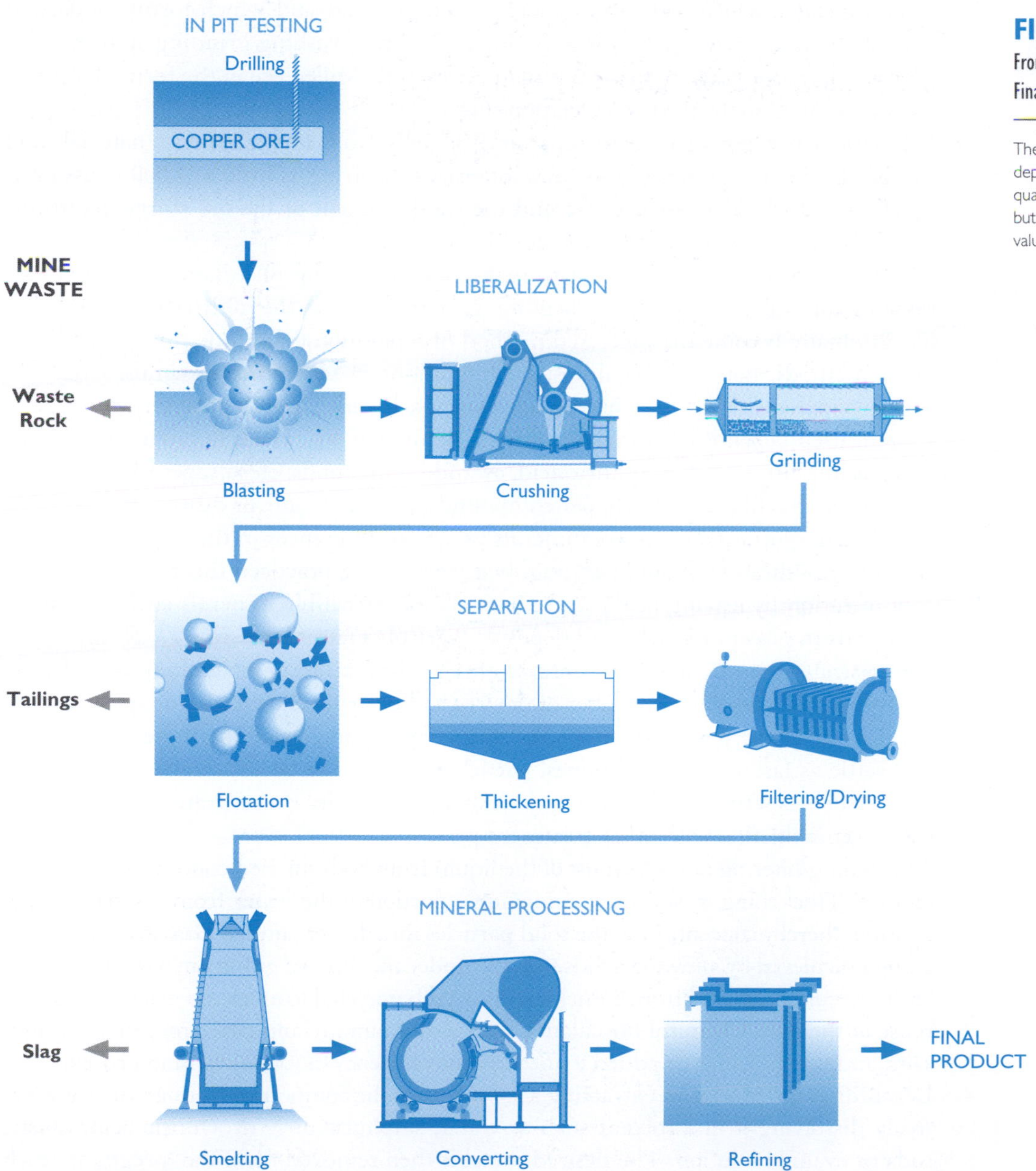

FIGURE 4.12

From Liberalization to Separation to Final Product

The economic viability of an ore deposit depends not only on the quantity and quality of accessible ores, but also on the ease with which valuable products can be extracted.

The 'mill' in a mining operation comprises all activities related to ore beneficiation.

Conventionally the 'mill' in a mining operation comprises all activities related to ore beneficiation. While the interior of a mill may seem to the visitor to be a baffling maze of tanks, pipes, pumps, conveyors, motors, chemicals, pulps, and solutions, this seeming confusion is actually a carefully designed system constructed for one objective – to recover the valuable minerals locked up in the ore. The end product from a mill is called a concentrate, or in the case of gold and silver, a doré bar of the metal itself.

The most common beneficiation processes include gravity concentration; milling and flotation (used for base metal ores); leaching (used for tank and heap leaching); dump leaching (used for low-grade copper and low grade gold ores); heavy media separation and magnetic separation. Typical beneficiation steps include one or more of the following (USEPA 1995a,b): milling; washing; filtration; sorting; sizing; magnetic separation; pressure oxidation; flotation; leaching; gravity concentration; and agglomeration (pelletizing, sintering, briquetting, or nodulizing).

All milling and concentrating processes begin with a crushing and grinding stage, which usually represents most of the total cost of processing the ore.

- Milling processes. All milling and concentrating processes begin with a crushing and grinding stage, which is the most energy-intensive stage and which forms the largest part of the costs of ore processing. As many as three crushing/grinding steps may be required to reduce the ore to the desired particle size. Milled ore in the form of a slurry is then pumped to the next beneficiation stage.
- Magnetic separation is used to separate iron ores from less magnetic material, and can be classified as either high-or low-intensity (requiring as little as 1,000 gauss or as much as 20,000). The particle size and the solids content of the ore slurry determine which type of magnetic separator system is used.
- Flotation uses a chemical reagent to make one or a group of minerals adhere to air bubbles for collection. From its beginning at the turn of the 20th century, flotation has gradually become the preferred method of separation for sulfide minerals. Unlike other methods, flotation has almost no limitations in separating minerals. Chemical reagents include collectors, frothers, antifoams, activators, and depressants; the type of reagent used depends on the characteristics of a given ore. These flotation agents may contain sulphur dioxide, sulphuric acid, cyanide compounds, cresols, petroleum hydrocarbons, hydrochloric acids, copper compounds, and zinc fume or dust.

Flotation has gradually become the preferred method of separation for sulfide minerals.

- Gravity concentration separates minerals based on differences in their specific gravity, and is a modification of gold panning. Nature has long practiced this form of mineral concentration by leaving heavy and relatively indestructible minerals such as gold or diamonds in placer or beach sand deposits. Agricola's famous treatise '*De Re Metallica*' demonstrates that, by the 16th century, gravity concentration technology was already well advanced and a large number of devices had been developed. Beside specific gravity, the size of the particles being separated is important. Since small, heavy minerals will settle as fast as large, light ones, particle sizes need to be kept uniform with classifiers (such as screens and hydro-cyclones). Today gravity concentration is most often used in combination with other treatment processes.
- Thickening/filtering removes most of the liquid from both slurried concentrates and mill tailings. Thickening is defined as removing a portion of the water from a slurry or suspension, thereby concentrating the solid particles into the remainder. Gravitational thickening is achieved by allowing solids to settle under the force of gravity in a sedimentation basin. Liquid recovered from a thickener is usually recycled to a process water storage for reuse at the mill. Chemical flocculants, such as aluminium sulphate, lime, iron, calcium salts, and starches, may be added to increase the efficiency of the thickening process.

Leaching is the process of extracting a soluble metallic compound from an ore by selectively dissolving it in a solvent.

- Leaching is the process of extracting a soluble metallic compound from an ore by selectively dissolving it in a solvent such as water, sulphuric or hydrochloric acid, caustic soda or cyanide solution. The desired metal is then removed from the 'pregnant' leach solution by chemical precipitation or another chemical or electrochemical process. Leaching methods include dump, heap, and tank operations. Heap leaching is widely used in the gold industry, and dump leaching in the copper industry (see Chapters Five and Six for a detailed discussion on leaching).

During the exploration phase, laboratory testing indicates the feasibility of concentration and the expected percentage recovery of metals. The physical properties of a mineral often lead directly to a process for separating that mineral from others. Some well-known examples of these are: magnetic separation for minerals such as magnetite, rutile, ilmenite; gravity separation for gold, silver, galena, cassiterite; flotation separation for most sulfides; distillation for mercury from cinnabar; and selective dissolution in specific solvents such as halite or sylvite in water, gold and silver in cyanide solutions, copper oxides, carbonates, sulphates, and silicates in dilute sulphuric acid (Bolles 1985). During the feasibility evaluation, testing at a larger scale using bulk samples with processing in a pilot concentrator is required to confirm the degree of metal recovery in beneficiation, on which project planning can be based.

Mine Waste Management Options

In the last few decades, advances in bulk haulage methods and equipment have favoured larger bulk mining in open pits, over underground operations, particularly for base metal and coal operations. As a consequence, larger amounts of wastes are produced from these operations, mainly because large quantities of waste rock have to be removed to gain access to the ore. In many cases, the amount of waste rock to be mined, transported and disposed of is many times more than the tonnage of ore that is extracted (see Chapter Nineteen for a detailed discussion of waste rock management). Waste to ore ratios of more than 20:1 are common in high grade vein gold deposits, and are also sometimes found in high rank coal mines. Larger bulk mining operations also produce more tailings. The amount of tailings generated depends on the content of the desirable mineral in the ore, its grade, and the efficiency of the mineral processing stage in recovering it. Another factor is the duration of an operation. In most mining operations, the total amount of tailings is very large in comparison to the amount of mine product, and roughly equals the amount of ore mined. Tailings management, discussed in more detail in Chapter Eighteen, is arguably the most challenging aspect of any mining operation. Favourable site conditions for mine waste disposal, or the lack of them, may be the decisive factor in determining if mine development is feasible or not.

Waste to ore ratios of more than 20:1 are common in high grade vein gold deposits, and are also sometimes found in high rank coal mines.

Favourable site conditions for mine waste disposal, or the lack of them, may be the decisive factor in determining if mine development is feasible or not.

Required Mine Infrastructure

The infrastructure requirements for mining projects are project specific. The capital cost for infrastructure can vary substantially from site to site as a percentage of the total capital cost, and is more often a function of the location than the mining or processing methods. Thus, the capital cost estimate in engineering studies must be based on a proper identification and assessment of the infrastructure requirements, such as access to the ore deposit, availability of local resources, and access to port location. Only massive ore deposits, such as the Freeport copper and gold deposits in Irian Jaya, are attractive when the deposit is sited in unexplored frontier land, far from potential seaport locations and without any existing supporting infrastructure whatsoever. Large coal deposits are known in Central Kalimantan, Indonesia but the costs of coal transportation to seaports continues to prevent their development.

Cost/Benefit Analyses

Benefits are all of the positive factors associated with a mining project. The return on investment is important, but the time value of money also plays an important role. Simply put, the annual profits generated by a mine must be sufficient to pay back (within a reasonable time) the money invested in the mine. It is the job of mining engineers to estimate the 'payback period' as well as the Net Present Value of a mining project. While financial benefits are of prime concern, non-financial benefits find equal consideration in decision-making. Costs represent all of the negative factors, primarily investment and operational costs, but also political, environmental, and social costs. At the pre-feasibility stage, costs are often based on comparative costs of similar operations, and on rules of thumb. Cost estimates are unlikely to be more than 30 to 40 percent accurate, and of course this accuracy is insufficient to secure project financing and commence mine development. The cost/benefit analysis in a feasibility study is more detailed, reducing the margin of error in cost estimates to less than 15 percent. The most important factor affecting costs is the size of the mine in terms of ore mined and milled per day of operation. The mine size in turn depends on the geological resource estimates. Consequently, the cost estimates for mine development are sensitive to the accuracy of the resource estimates.

Cost estimates for mine development are sensitive to the accuracy of resource estimates.

Each cost/benefit analysis, even when based on sophisticated financial models, includes uncertainties. Assumptions must be made, but the cost/benefit analysis can only provide indicative answers, together with an estimated degree of uncertainty. Mining ventures may be the subject of distorted cost/benefit analyses from both their proponents and opponents. A mining proponent may downplay the external environmental costs of the mine development to encourage better financing arrangements. Opponents on the other hand may exaggerate the magnitude of the social costs in attempts to kill a mine proposal before it gets off the ground. Regrettably, both kind of distortions do occur.

Opponents on the other hand may exaggerate the magnitude of the social costs in attempts to kill a mine proposal before it gets off the ground.

Project Risks

The most serious risks in any mining project are those associated with geology (the actual size and grade of the mineable portion of the ore body), metallurgy (how much of the metal can be recovered), and economics (metal markets, interest rates, transportation costs). But there are many others, such as problems arising from unforeseen political developments, new restrictive mining regulations or the scarcity of skilled labour, to name a few.

Geological risks always remain. No matter how sophisticated the exploration programme and subsequent resource modelling, geological risks in terms of uncertainty of the actual ore body remain. The only way to be certain of the extent and enrichment of an ore body is by mining the deposit. Other geological risks include the presence of highly permeable groundwater aquifers, or challenging geotechnical conditions.

Technological risks are in the first instance related to estimated recoveries of metal contained in the processed ore. Actual crushing and grinding characteristics, liberation size, metallurgical recovery, or variability of ore grade may fail to meet expectations. Some portions of the ore body may contain deleterious elements such as arsenic or mercury that make the concentrate unacceptable to smelters or other downstream processing. There are risks in other areas as well. For example the tailings disposal pipeline of Bougainville, even though it was never used, was the first slurry pipeline of such length at the time of mine development, and hence no comparable industry experience could give certainty on its performance.

Political risks are real, particularly as more mines are located in less developed countries with unstable political regimes or less developed legal investment frameworks. Mining history is full of examples where political factors have determined the fate of a mine development, either for better or worse. The issuance of the Indigeneous Peoples Rights Act in the Philippines in 1997 almost brought mining activities to an end. A similar effect was felt by the Indonesian mining industry with the proclamation of Forestry Law 20 of 1999, which prohibited mining in protected forest areas. Both laws were based on good intentions, but were detrimental to mining interests in implementation. In many parts of Africa, political instability, tribal conflicts, legal uncertainties, and lawlessness have combined to ensure

Mining history is full of examples where political factors have determined the fate of a mine development, either for better or worse.

that many known mineral deposits, some of large size and high grade, remain unmined to this day.

In most jurisdictions, mineral deposits are national assets and remain in the ownership of the government. Mining companies act more or less as contractors. This tempts government to apply and increase taxes or royalties or to change permit requirements arbitrarily. Changes in ground rules after an investment has been made may result in 'high grading', reducing ore reserves or even entire ore bodies. Political risks, however, are not restricted to developing countries. A change in government can change the investment climate in any country, and there is no way of assuring absolute safety.

In most jurisdictions, mineral deposits are national assets and remain in the ownership of the government.

Environmental risks are almost always a reality in mine development, but proper planning and reasonable precautions can minimize them (see Chapter Thirteen for an overview of mining related environmental change processes). In the pre-feasibility stage the environmental scoping study aims to identify key environmental constraints and possibilities. The best safeguard against environmental and social risks, however, is a comprehensive and unbiased environmental impact assessment study, backed up with the resources and commitments to implement its recommendations. The EIA complements the BFS, and provides documentation on whether or not the project is acceptable on environmental and social grounds. The EIA aims to demonstrate the environmental feasibility of the mining project and covers all activities from mine construction up to mine closure and post closure activities as detailed in Chapter Two of this book. To some the EIA forms part of the BFS; others understand the preparation of the EIA as a separate and stand alone exercise, a difference that matters little in practice. However mining projects remain controversial and social tensions have caused more than one mine to cease operation.

The best safeguard against environmental and social risks is a comprehensive and unbiased environmental impact assessment study.

Eventually all risks contribute to economic risks. The cost/benefit analysis is based on assumptions and forecasts. Metal prices are known to be highly volatile. Unlike most other commercial products, metal prices are set internationally. Development of new mines elsewhere in the world may depress market prices. On the other hand given the long lead time of mine development, an increase in demand for a particular metal will be soon followed by a price increase. In the feasibility evaluation, however, an assumption must be made about the price at which the final product can be sold. The forecast must cover the time period of mine operation until the initial investment is recouped with a typical market profit. Economic risks are also related to investment costs. Given the size of today's mines, investment costs are huge and any percentage increase in actual costs compared to initial estimates will have a notable impact on mine profitability. Other examples of economic risks include fluctuation in exchange rates or an increase in the costs of construction materials such as steel.

Today, most mines and virtually all downstream processing, smelting, and refining plants make use of a feasibility study, usually prepared by one of the leading international geological and technical consultants. It is almost inconceivable that a major mining project could be financed today without a bankable feasibility study. What constitutes 'bankable' depends on the circumstances, but as an overall definition, a document is considered bankable if it is in the standard form as seen in similar transactions, and as to matters of fact and projection, analyses the required scope of data using accepted methodology, subject to customary limitations (Berwin Leighton Paisner 2006).

Given the complexity of large-scale mining projects, a feasibility study is itself a complex undertaking. It requires careful analysis of many variables to insure the economic development of an ore body. It may take one to two years to complete, and cost up to several million US dollars. And it is often the biggest item of capital expenditure during a project's early stages, the period of highest risk. Unfortunately, considerations for financing are sometimes left to be completed in the last few weeks, and by those unfamiliar with the implications of technical decisions already cast in stone. In light of their significance,

Given the complexity of large-scale mining projects, a feasibility study is itself a complex undertaking.

financing considerations should be highlighted and carefully considered throughout the feasibility study. It is the proper marriage of both technical and financial considerations that gives rise to a viable project.

Financing

Adequate financing may be the difference between success and failure.

One of the features that distinguishes a mining project from most other businesses is that during production, the company's asset, the ore, is progressively consumed. At some point in the future the project's main asset will be gone, which is the reason a mine is referred to as a wasting asset. This has important implications for financing and for the justification of allocating capital to any new mining project. Financing, always an important consideration in resource development, has reached the point where it may well be the controlling element in bringing an ore body into production. Adequate financing may be the difference between success and failure.

The following text on mine financing is based in part on Ulatowski (1978), Ulatowski *et al.* (1977), and Vickers (2006). Due to the large scale and the high upfront capital investment of modern mining projects, the role of the mining company has shifted from its classic position of organizer and provider of funds to that of a sponsor, providing upfront risk money and managing the project. In this environment, new partnerships are forged: involvement of lenders; joining together of more than one mining company; offering host country participation; use of contract mining and infrastructure outsourcing arrangements; and, due to the heavy reliance on sales contracts, often involving the purchaser of the product.

The partnership character is partly the reason why a new mining project emerges as a separate entity, a new mining company in other words, relying heavily on its intrinsic worth to attract the capital needed to bring it into production. Financing costs of this new venture are significant, often equal to or exceeding all of the other operating costs put together (Vickers 2006). These costs are real and place high demands on the future cash flow of the mine.

Financing and technical considerations are not independent. Mine and finance planners must work closely every step of the way, from the beginning to the completion of a feasibility study, to ensure financial success. Mine planning must not only focus on the characteristics of the ore body in the development of an optimal exploitation system, but must equally consider the implications of mine design on the financing of the ore body. Mine design characteristics include scale of operation, cutoff grades, and timing of both the construction and production. Each of these factors significantly affects the financing requirements, and the credit strength that the deposit can convey to the suppliers of capital.

Environmental approval of the mining project can become a barrier on the critical path to project financing, with wide-reaching financial implications if approval deadlines are not met.

Environmental approval of the mining project can, at this point, become a barrier on the critical path to project financing, with wide-reaching financial implications if approval deadlines are not met (see Chapter Two). Depending on the partners involved in the new venture, additional environmental conditions may apply that exceed the requirements of the host country. For example, lenders may choose to apply the Equator Principles, or joint venture partners may have adopted selected environmental performance standards of specific industry organizations, thereby potentially increasing costs.

Postponing Mine Development

Sometimes, after the excitement of the original exploration drilling and feasibility studies, work on mining development is halted. The reasons why a company cannot put the property into operation immediately are many and varied, and some of the more critical ones are completely beyond the control of the company (USDA 1995): (1) drop in price of mine product or no rise in price if this had been anticipated; (2) increase in labour costs; (3) unfavourable legislation or regulations; (4) change in tax laws or assessment procedures; (5) threat of litigation; (6) action of private conservation groups; (7) lack of smelter or refinery capacity; (8) lack of capital; (9) delay in obtaining delivery of major equipment; and (10) lack of transportation facilities. Although one or two of these considerations may be paramount in deciding to postpone development, there are usually multiple factors involved. The 'go' or 'no go' decision is carefully weighed against a list of favourable and unfavourable factors, some of which may be changing as the decision is being made.

Sometimes, after the excitement of the original exploration drilling and feasibility studies, work on mining development is halted.

From the standpoint of surface damage to the environment, a delay of mine development is often unfortunate. Often, considerable damage has already been done during the exploration and early development stages, and it will remain until the decision to mine is eventually made. For example, an ore deposit near the surface that is to be mined by open pit methods will have been drilled in a close-spaced pattern and the close network of access roads over the property will usually have made a mess of the surface, particularly from the visual standpoint. There are many people who find nothing particularly ugly about a well-engineered and smoothly running open pit mining operation; however, no one would view the drill roads as anything but an eyesore (**Case 4.4**).

A postponement of mine development is particularly perplexing to local people, some of whom may have begun to make changes in their personal and business lives in anticipation of the new mining operation. Maintaining good community relations and participation may now become even more important. Rumours abound whenever there is a delay in mine development, and community leaders sometimes call for a clear statement of intent, so that everyone will know what to expect. Sometimes this is possible, and periodic updates may be issued in the interests of community relations. Just as often, company management has been so taken aback by an unforeseen or uncontrollable event, or series of events, that they do not know what the best plan for the future might be. Rather than issue a false statement, or speak in misleading generalizations, the company may choose, or be advised, to remain silent.

A postponement of mine development is particularly perplexing to local people, some of whom may have begun to make changes in their personal and business lives in anticipation of the new mining operation.

CASE 4.4
Nickel Exploration

A geological region rich in surface nickel laterite deposits that stretches from Papua, over Sulawesi, to the southern part of the Philippines. Various lateritic nickel deposits have been systematically explored by drilling, trenching and bulk sampling. To a mining novice much of the land area may resemble an active mine. Lateritic nickel pits are difficult to rehabilitate as illustrated in photograph taken at a nickel mine in the Philippines. Depending on their grade and accessibility, some of these deposits will remain unmined, at least for the forseeable future. Unless difficult and expensive rehabilitation is carried out, the adverse effects of exploration will remain.

A delay in mine development may cause the EIA to become outdated and invalid.

Postponement in mine development may also have implications for environmental approvals. Environmental impact statements are often only valid for a defined period of time, within which the sponsor needs to commence mine construction and operation. A delay in mine development may cause the EIA to become outdated and invalid.

4.4 ENGINEERING, PROCUREMENT, AND CONSTRUCTION

How do we do it? Engineering, procurement, and construction are major undertakings. More people are employed during the construction phase than during mine operation. Access roads are built, heavy equipment and machinery are mobilized, and the project area turns into a huge construction site with all the environmental impacts that conventionally occur during construction. Before construction commences, however, the mining company needs to secure a wide range of permits and access to the mine site.

Approved feasibility and environmental impact studies are pre-requisites for mine development, but they need to be complemented with a wide range of additional permits and approvals prior to mine construction and before mine operations can commence. **Table 4.4** illustrates the various permits that were required for a relatively small-scale underground gold mine in West Java, Indonesia. Depending on the jurisdiction and size of the mine, the list of required permits may be even longer. Some permits are dependent on EIA approval. To avoid lengthy and costly delays, the project owner needs to secure some of these permits immediately following EIA approval before mine construction can begin (**Case 4.5**).

Securing access to the mine site, that is, acquisition of land, and if required, resettlement of inhabitants, are the first activities in mine development that have potentially major social impacts.

Securing access to the mine site, that is, acquisition of land, and if required, resettlement of inhabitants, are the first activities in mine development that have potentially major social impacts. Land acquisition is always sensitive and there are many associated issues (see Chapter Fourteen for a detailed discussion of this subject matter). Land speculators may seek financial benefit by buying land from local people and selling it to the mining company at a large profit. Such speculation not only inflates the cost of mining, it may leave previous land owners feeling cheated, with such feelings ultimately directed against the mining operator. Furthermore, a formal land title, supported by accurate land surveys and documentation, is problematic in most developing countries. Land acquisition procedures need to respect and comply with traditional communal customs and laws, which outsiders often find difficult to grasp. This is particularly the case where local tribes claim undocumented ancestral land rights, which in some cases may be in dispute between two or more tribes.

Construction activities that follow after land access is secured have the potential for significant adverse environmental and social impacts. Movement of material and equipment can be severely intrusive. Major earthworks create noise, dust, and heavy traffic on access roads. Large numbers of transient workers may encourage prostitution and alcoholism, and bring sexually transmitted diseases, along with a general increase in crime rates.

In many cases, specialized third party companies (the engineering, procurement, and construction contractor, dubbed EPC contractor) will construct a mine's operating infrastructure, often under 'Turn-key' contracts, that is, completed through commissioning to the operational stage. Mining companies are then left to focus their environmental management efforts on the actual mine operation, while overlooking environmental and social impacts that occurred during construction.

The construction contractor, normally a large engineering company or a consortium of engineering companies, has a different outlook to the mine operator. Its aim is to get the job done within budget and on time, particularly if as is customary, contractual penalties apply to delays. The contractor has no incentive to develop a working rapport with local

TABLE 4.4

Permits, Certificates, and Report Requirements for a Small Gold Mine in Indonesia (as of 2004) – The Bankable Feasibility Study together with the Environmental Impact. Assessment are very much umbrella approvals. The mine operator has to secure a wide range of additional specific permits and certificates before mine construction can begin

Company

1. Foreign Investment Approval Model IPMA (SPPP)
2. Articles of Association (Establishment Deed)
3. Establishment Deed Legalization
4. Company Domicile Permit (Surat Izin Tempat Usaha/SITU)
5. Taxpayer Registration Code Number (NPWP)
6. Affirmation of Taxable Company (PKP)

Manpower

7. Registration of Company at Manpower Department
8. Approval for Deviation from Standard Working Hours
9. Legalization of Company Regulations (two yearly renewal)
10. Establishment Worker's Union (SPSI – Kesepakatan Kerja Bersama)
11. Collective Labour Agreement (CLA)
12. Manpower Insurance and Social Benefit (JAMSOSTEK)
13. Occupational Health
14. Expatriate Utilization Plan Approval (RPTKA-Rencana Penggunaan Tenaga Kerja Asing)
15. TA-01 Recommendation for Expatriate Visa – VBS
16. Travelling Permit (SKJ - Surat Keterangan Jalan)
17. Temporary Resident Visa (VBS – Visa Berdiam Sementara)
18. Temporary Entrance Regional Immigration Permit (KITAS – Kartu Izin Tinggal Terbatas)
19. Work Permit for Expatriates (IKTA – Izin Kerja Tenaga Kerja Asing)
20. Tax Registration for Foreigners (PBA)
21. Certification of Registration for Temporary Resident for Expatriates (SKPPS – Surat Keterangan Pendaftaran Penduduk Sementara)
22. SKTTS – Surat Keterangan Tempat Tinggal Penduduk Sementara
23. Multiple Exit Reentry Permit (MERP)

Environmental Assessment/General Permits

24. Environmental Impact Assessment (AMDAL) Approval
25. Permit to Conduct General Mining Activities in Forest Areas
26. Appointment of Kepala Teknik & Wakil Kepala Teknik
27. Location Permit (Izin Lokasi)
28. Building Utilization Right Certificate, permission to use the land for project (HGB)
29. Building Construction Permit (IMB)
30. Deep-Well Drilling Permit for industrial purpose (SIPA)
31. Limited Importer Identification Number (APTI)
32. Master List/Approval to import the materials and equipment and obtain the duty exemption
33. Import Duty and VAT exemption (postponement to pay value added tax of M/L importation)
34. Material Import Identification (RIB)
35. Permits for Temporary Import of Construction Equipment
36. Pressurized Equipment Certification (included in this category gas cylinders, transport vessel, air conditioning, vessel or storing gas, mixture of gas, liquid, etc.) (Reg. No. PER 01/MEN/1982)
37. Inspection and Testing of Power Machine and Production Machines (included in this category motors, turbines, working tool machines, genset, transmission, casting, pressing, piping, crushing, rolling, cutting, furnaces, trolley, etc.) (Reg. No. PER01/MEN/1985)
38. Inspection and Testing of Lifts and Transport Equipment (included in this tackles, pneumatics, gondolas, cranes, escalator, conveyors, towing, truck, tractor, forklifts, locomotive, etc) (Reg. No. PER02/MEN/89)
39. Lighting Conductor Circuit Installation Approval (Reg. No. PER02/MEN/89)
40. Automatic Conductor Circuit Installation Approval (Reg. No. PER02/MEN/89)
41. Portable Fire Extinguisher Installation and Maintenance Approval
42. Quarry Permit (S. I. P. D) Izin Bahan Galian C

(Continued)

TABLE 4.4
(Continued)

43. Access/Road/Infrastructure Construction
44. Power Station Permit (Izin Usaha Kelistrikan Sendiri/U.K.S) (to construct, to operate)
45. Explosive Permit (Approval on construction of explosive storage building, Permit to possess, to take charge and to store explosive P3, Permit to purchase and to use explosive P2)
46. Permits to Build and to Operate a Port/Wharf
47. Lease of Sea Surface to Port Site
48. Fuel Storage Tank Construction Permit
49. Tailings Dam Construction Permit
50. Agreement to Borrow and to Use Land for Mining Activities in Forest Area
51. Timber Utilization Permit
52. Shot Firer Certificate of 2nd class, License to Shot Firer Card/KIM
53. Fuel Storage Permit
54. Fuel Allocation Permit
55. Approval of the Use of Public Road for Earthwork Equipment Mobilization (commencement of construction)
56. Approval for Mine Bench/Terrace higher than 6 metres
57. VSAT Telephone System Permit
58. Radio Communication Permit
59. Permit to Possess, Store and Utilize Hazardous Substances
60. Permit to Import, Install, Export, and/or Assemble Radioactive Substances
61. Permit to Use Radioactive Substances (Periodically checked or spot check if needed)
62. Permit to Transport Radioactive Substances
63. Radioactive Waste Storage Permit (to be stored at Radioactive Waste Research Centre)
64. Toxic and other Hazardous Waste Storage Permit (included in this category: explosives, inflammable, reactive, toxic, infectious, corrosive and others that, after toxicity testing, are classified hazardous waste)
65. Permit to Dispose of Waste Water (based on the Effluent Quality Standards stipulated by the Central and Local Government)

The contractor has no incentive to develop a working rapport with local communities, since the construction stage is relatively short compared to the operations stage.

communities, since the construction stage is relatively short compared to the operations stage. Consequently, the implementation of environmental and social programmes during construction is rare, and the mine operator is left to cope with any environmental damage and social tension created during construction. Local communities, of course, will not differentiate between the mine constructor and the mine operator, so poor environmental performance or social tensions generated during construction will be blamed on the mine operator. For this reason, there is a strong incentive for the company or consortium which owns the mine to adopt a strong set of policies to pre-empt adverse construction impacts and to incorporate appropriate environmental management procedures in the construction tender documents and contracts.

4.5 MINING

In its general sense, mining typically includes the following activities (see also **Figures 4.13** to **4.15**): land clearing, excavation and management of topsoil and waste rock, excavation and transportation to the mill of ore, ore processing, tailings management, and shipment of mine product. While mining activities follow a sequential order, they may also occur in parallel. For widely distributed ores such as many laterite ores, exploration, land clearing, and topsoil removal may be underway in some areas while ore mining and rehabilitation proceed in others. Mining commonly takes place in a continuous operation, 24 hours a day, 7 days per week.

Mining commonly takes place in a continuous operation, 24 hours a day, 7 days per week.

Each mining activity in isolation may seem rather simple, but to optimize the many contemporaneous activities requires sophisticated planning and scheduling.

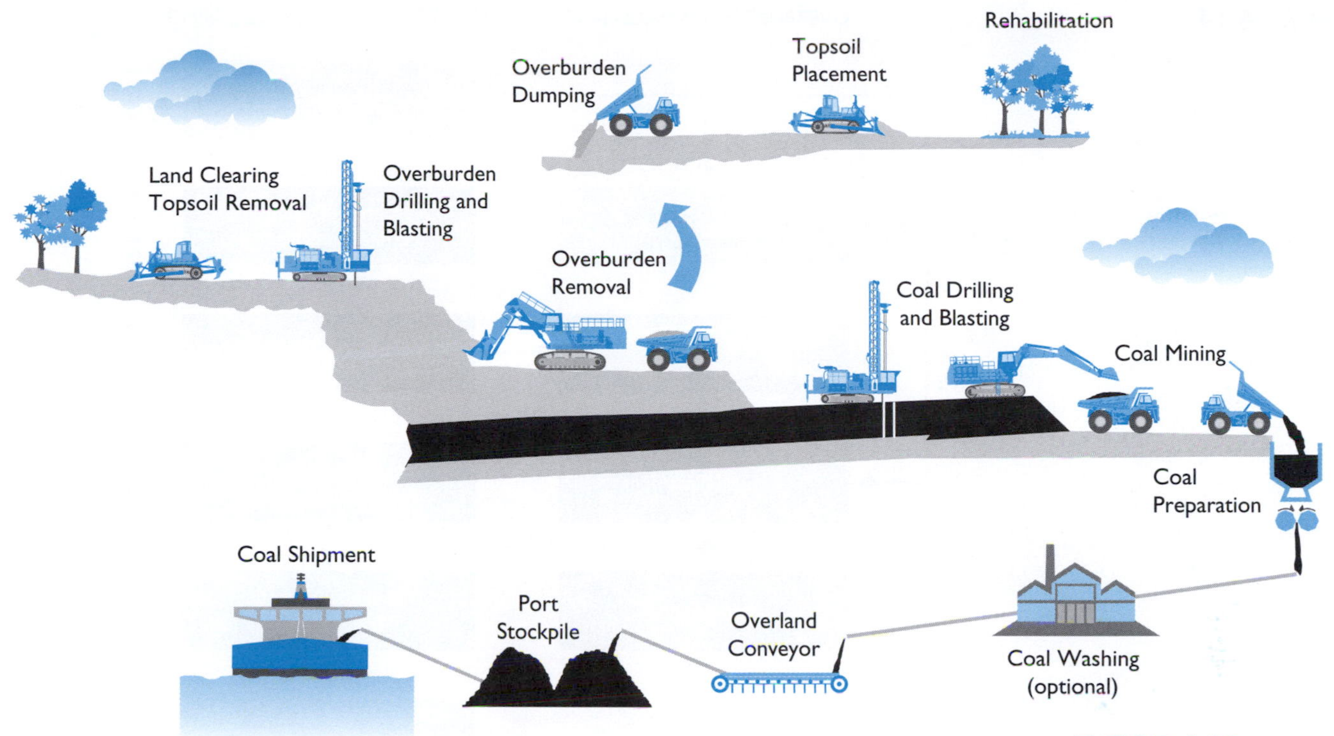

FIGURE 4.13

Common Activities in Surface Coal Mining

Open cast mining suits topsoil management uniquely. The rate of rehabilitation is similar to the rate of mining. To a large extent, direct return of topsoil is possible eliminating the need for extensive topsoil stockpiling.

Once the mine is in full operation, as new information about the ore body emerges, assumptions made at the feasibility stage may prove to be inaccurate or irrelevant, so that actual operations differ from planned operations; also market conditions may vary significantly from original assumptions, forcing variations from the original plan. Therefore, while good developmental planning is essential, operational management has to be well

CASE 4.5

Expanding INCO's Nickel Mining Operation in Sulawesi, Indonesia

INCO has operated this large scale nickel mining operation for decades. Since traditional nickel ore processing using smelting is very energy intensive, INCO is fortunate that its operation in Indonesia can rely to a large extent on its own hydro-electric power generating scheme. As the mine expanded INCO commenced planning of a third hydropower scheme as early as 2000. The environmental permitting of the new scheme was obtained in 2002. In December 2004 the Department of Forestry notified the company that an additional forestry permit is required before dam construction can begin. Obtaining the permit delayed the project by more than two years with a financial loss to the company and eventually to the Government of Indonesia amounting to millions of US dollars.

Source:
The Jakarta Post, 07 November 2006.

FIGURE 4.14
Illustrating Sequential Activities in Surface Coal Mining

Coal mining produces small amounts of tailings compared to metal mining, if at all. Tailings in coal mining consist of reject materials such as stones and clay which are present in the coal seams.

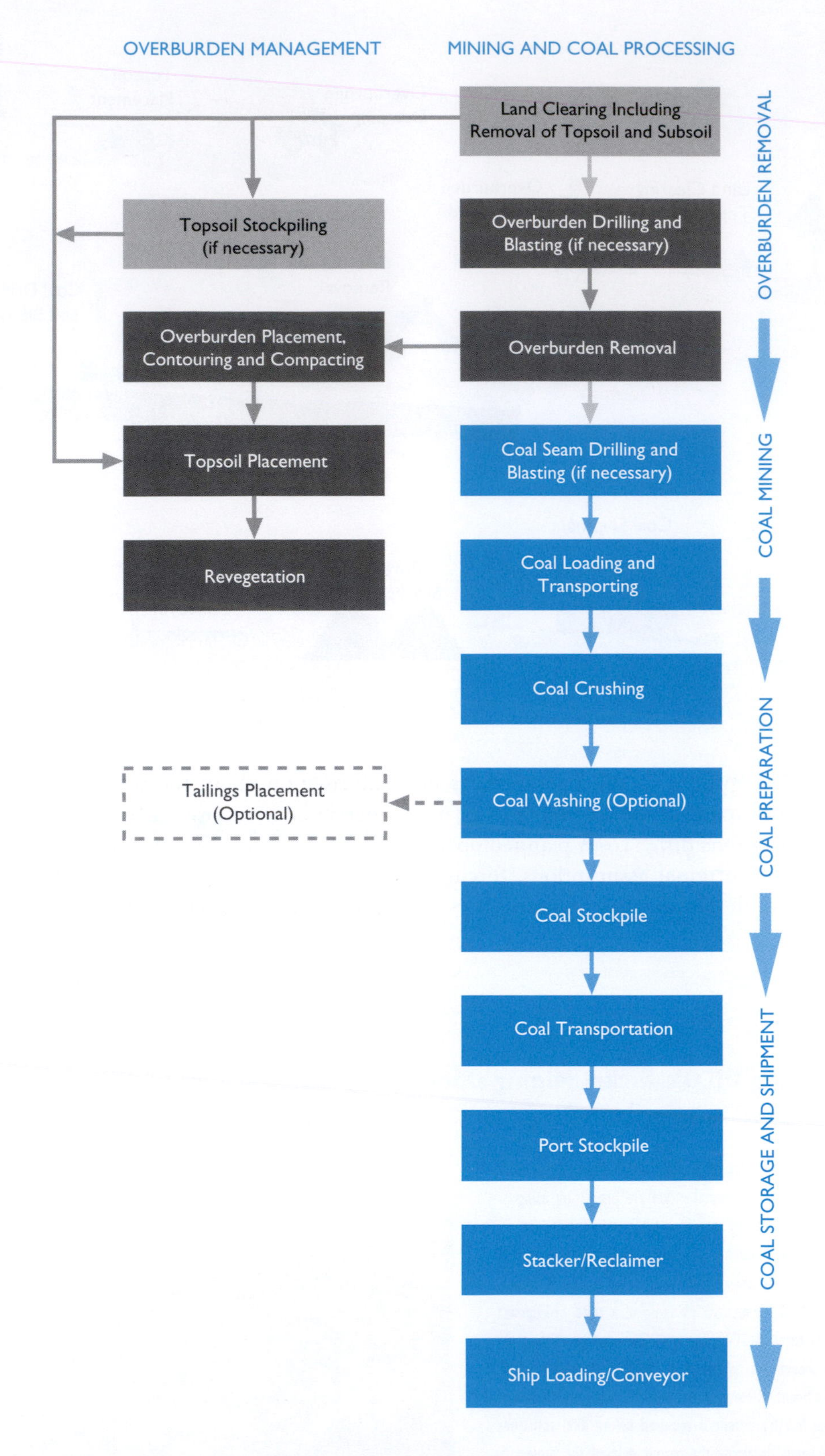

informed and flexible. In the following discussion open pit mining has been chosen for an illustration of organizational complexities (see Chapter Five for a broader discussion of mining methods). Differences in environmental impacts between open cast and underground mining are highlighted.

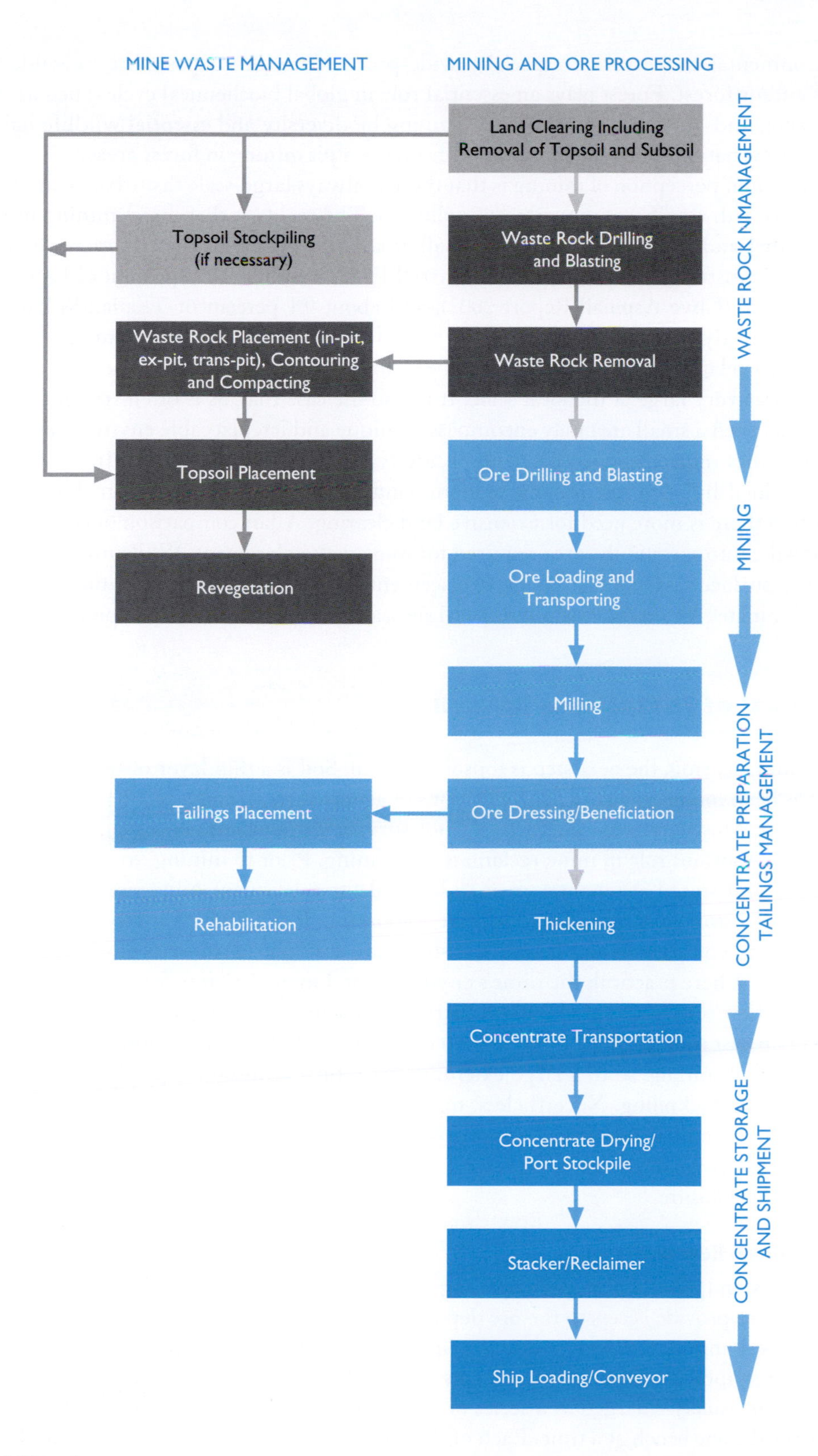

FIGURE 4.15
Sequential Activities in Surface Hard Rock Mining

Each mining activity considered alone is rather simple, but to optimize the many contemporaneous activities requires sophisticated planning and scheduling.

Land Clearing

Land clearing is the first step in the physical preparation of the mine site. It leads to the first significant visual impact, removes wildlife habitat, and predisposes the surface to accelerated erosion by wind and water (discussed in detail in Chapter Twenty).

Most national jurisdictions put limits on open pit mining in forest areas.

Environmental concerns are especially widespread if the ore deposit lies in wilderness areas and/or forest. Forest plays an essential role in global biochemical cycles such as those for carbon and nitrogen, as well as maintaining biodiversity and essential wildlife habitat. Today most national jurisdictions put limits on open pit mining in forest areas.

A common perception of mining is that there is always large-scale disturbance in wilderness areas with significant downstream pollution. The reality is that direct mining impacts are isolated and confined to relatively small areas. It is estimated that no more than 0.02 percent of Australia's land surface area (Farrell 1996), less than 0.001 percent of Indonesia's land surface (Clive Aspinall Report 2002), and about 0.1 percent of Thailand's land surface are directly affected by mining. Similar figures hold true for most countries. Small is, however, a relative term. What seems to be a small impact on a national or regional scale may appear very large at the local scale. It is also the case that, as is often true in Western Australia, even a small area may encompass a unique and irreplaceable environment.

Ore bodies mined by open pit methods are typically three-dimensional. In contrast, ore bodies mined by open cast mining predominantly spread in two-dimensions horizontally, and often there is more need for extensive land clearing. A fair comparison of course, needs also to take into account the area required for mine waste placement. While mined areas are larger in surface cast mining, the in-pit placement of waste rock and, sometimes, tailings as well, eliminates the need for off-pit disposal sites, a key requirement of open pit mining.

Removal of Topsoil and Subsoil

The mine's environmental manager aims for an early return of the topsoil for rehabilitation of the reconstructed landforms.

After land clearing, the next step is topsoil removal. Soil is a thin layer of the Earth's crust that has been modified by the natural actions of agents such as water, plant-life, and microorganisms. Most terrestrial plants could not survive without soil, so topsoil management plays an important role in mine reclamation planning. Prior to mining, topsoil is carefully removed and stored at separate areas to be used for subsequent mine site rehabilitation. Timing and storage of topsoil removal are important. Topsoil which is too wet or too dry can result in compaction and loss of soil structure, and long-term stockpiling will reduce its quality. Where practical, the mine's environmental manager aims for an early return of the topsoil for rehabilitation of the reconstructed landforms.

In open cast mining (most often seen in coal mining) the rate of rehabilitation is similar to the rate of mining, so to a large extent, direct return of topsoil is possible, eliminating the need for stockpiling. Nevertheless, topsoil management can be a demanding process, requiring substantial resources in terms of expertise, staffing, equipment, and finances. Further discussion of the value of topsoil in rehabilitation is provided in Chapter Twenty One – Mine Closure.

Waste Rock Removal

From the standpoint of physically opening a mine, the major purpose of initial mine development is to provide access to the ore deposit. In a surface mine, this means removing sufficient overlying material to expose the uppermost part of the ore body. This is referred to as 'pre-stripping' and sometimes forms part of the construction contract. Subsequently, the mine is usually enlarged as a series of 'cut-backs' each of which extends outwards and downwards, one bench at a time. Each of these cut-backs requires mining of ore and waste rock, but the ratio between them will vary depending on the shape of the ore body. Apart from tailings management, handling and relocation of waste rock continues to remain one of the main environmental challenges in the mining industry, and is considered in detail in Chapter Nineteen.

Waste rock comprises all material removed from the mine other than topsoil, subsoil, and ore. The extent of waste rock removal depends on pit geometry, which is determined by the geometry and grade of the ore deposit, the equipment to be used and the planned production rate of the mine. Government regulations may stipulate maximum overall slope or batter angles which, in turn will effect the quantity of waste produced. It is the one of the first tasks in surface mine design to estimate the final pit boundary. The pit limit is typically established by the maximum allowable stripping ratio, a simple break-even ratio based solely on economics. It is defined as the ratio of waste rock to ore at the ultimate boundary of the pit, where the profit margin becomes zero. In other words, the stripping ratio indicates the point where the costs of removing and disposing of the waste rock increase mining costs to the point where costs exceed revenue. Deposits generally decline in grade outwards, and cutoff and pit limits may vary with economic parameters. This means that the final pit size may differ from the initial estimate, depending on the information which subsequently becomes available on the actual ore grade distribution.

The pit limit is typically established by the maximum allowable stripping ratio, a simple break-even ratio based solely on economics.

The allowable stripping ratio has also a direct environmental significance. Since it locates the ultimate pit boundary, it determines the amount of waste rock that needs to be removed during mining.

Ore extraction

Ore Drilling and Blasting

Ore breakage historically has been a drill and blast cycle. Continuous breaking is restricted to soft rock and coal, using mechanical rotating cutters or chain-saw type devices. In-pit drilling and sampling is commonly required to accurately define the limits of the ore and to delineate ores of different grades which may require blending to maintain a reasonably uniform feed to the mill, or to distinguish different ores for separate stockpiling and treatment. This is known as 'grade control' and is the responsibility of the mine geologists. Low grade ores may also be directed to leach dumps or pads instead of to the mill.

Environmental impacts due to drilling and blasting include dust emissions, noise, and vibration. Visible dust plumes from blasting are classed more as an aesthetic impact or a source of neighbourhood annoyance, rather than a health risk. For most mining operations, however, the major sources of dust emission are the haul roads rather than the more transient dust clouds caused by blasting. Peak noise is used to assess overpressure from blasting. Planners limit blast noise by using good stemming in drill holes, and by limiting the Maximum Instantaneous Charge detonated simultaneously, through the incorporation of delays. Noise and vibration due to blasting are among the most noticeable effects of mining and hence figure prominently among complaints from neighbouring communities.

For most mining operations the major sources of dust emission are the haul roads rather than the more transient dust clouds caused by blasting.

Ore Loading and Transporting

Rock and ore materials in mining are transported by haulage or hoisting (primary vertical movement). In open pit mining, truck teams with shovels are used/operate in most applications (**Figure 4.16**). In open cast mining with few benches, belt conveyors are preferred for haulage in spite of the high investment costs for their continuous high output and low operating costs. The main environmental impacts associated with ore loading and transportation are dust and noise. At least half of the transport-related dust emissions are eliminated by careful watering of the haul roads. Water sprays on ore dump stations and conveyor belt transfer points, further reduce dust emissions. The use of covered conveyor belts practically eliminates dust emission during haulage. Interestingly, the main source of transport noise perceived as a nuisance is the warning noise of trucks when backing up. Night time noise is of particular concern.

In open pit mining, truck teams with shovels operate in most applications.

FIGURE 4.16
Truck Teams with Shovel operate in Most Open Pit Applications

4.6 ORE DRESSING AND THICKENING

The separation from valuable minerals and worthless gangue requires the following two steps, size reduction (milling and grinding) and separation of ore minerals from gangue materials (concentrating).

At most modern mining operations, whether surface or underground, the ores are not of sufficiently high grade to ship long distances to smelters, and they are subjected to milling, mineral dressing, or beneficiation, usually at or near the mine site. All of these terms are sometimes referred to as ore dressing. Ore dressing is the mechanical separation of the grains of ore minerals from the worthless gangue. The resulting concentrate contains most of the ore minerals, and the waste is called tailings. The separation from valuable minerals and worthless gangue requires the following two steps, size reduction (milling and grinding) and separation of ore minerals from gangue materials (concentrating).

Milling and Grinding

The mining, crushing, and grinding portions of the processing are extremely energy intensive.

The reduction of ores to small particles is generally the most expensive phase of mineral beneficiation. The first ore breakage occurs during blasting. Subsequently, the crushing plant reduces the ore to fist-sized rocks. Two stages of crushing using different equipment may be required. Crushed ore is then ground into fine particles in ball mills or rod-mills which are large, rotating, cylindrical machines (**Figure 4.17**). The mining, crushing, and grinding portions of the processing are extremely energy intensive since the rock must be reduced essentially to fine sand and silt-sized particles in order to liberate the valuable minerals for subsequent extraction. The manipulation of particle size by crushing and grinding the ore, combined with particle size classification is also termed comminution (**Table 4.5**).

It is advantageous to remove ores from the crushing and grinding circuit as soon as the desired grain size is achieved. Further crushing or grinding is energy intensive, and over-grinding may in fact hinder subsequent metal recovery. Apart from operational safety and health issues, and potential dust emissions at first stage crushing, milling and grinding have no significant direct environmental impacts. Indirect impacts are associated with the high energy consumption.

Concentrating

Milling aims to liberate commercially valuable minerals from gangue minerals. Concentrating refers to the subsequent separation and removal of the valuable constituents

FIGURE 4.17
Ball Mill Illustrated

The reduction of ores to small particles is generally the most expensive phase of mineral beneficiation.

TABLE 4.5
Classification of Basic Size-reduction Steps – Comminution is a general term for size reduction that may be applied without regard to the actual breakage mechanism involved

Size reduction step	Upper size	Lower size
Explosive shattering	Unlimited	1 m
Primary crushing	1 m	100 mm
Secondary crushing	100 mm	10 mm
Coarse grinding	10 mm	1 mm
Fine grinding	1 mm	100 μm
Very fine grinding	100 μm	10 μm
Superfine grinding	10 μm	1 μm

Source:
Weiss 1985

using gravity, flotation, or other physical processes. Chemicals may be used for mineral separation or ore concentration. A variety of gravity concentration techniques are used to separate targeted minerals of high specific gravity from the less dense gangue minerals. Partly physical and partly chemical in its action, flotation is the reverse of gravity concentration. With its use, heavy minerals can float while light, undesired minerals sink. Flotation involves the use of various chemical flotation agents and pH regulators such as lime. Physical processes include magnetic separation for magnetic ores.

Flotation is the reverse of gravity concentration.

The method of concentrating depends on the ore type. For example ground copper sulfide ore is commonly enriched (concentrated) by Froth Flotation, in which the powdered ore is mixed with special paraffin oil which makes the copper mineral particles water repellent. It is then fed into a bath of water containing a foaming agent which produces a kind of bubble bath. When jets of air are forced up through the bath, the water repellent copper mineral particles are picked up by the bubbles of foam. They float to the surface as froth, hence the name flotation. The unwanted gangue (host rock) falls to the bottom and is removed as tailings. The froth is skimmed off the surface and the enriched ore (mainly the copper mineral) is dewatered and stored or shipped as concentrate for further processing. The mixture of water, foaming agent and paraffin is recycled. In the concentrating operation, minerals are separated from the gangue material of the ore that might contain less than 1% metals to form a concentrate containing more than 30% metals.

Flotation has become the principal means of separating most metallic minerals as well as some non-metallic ones.

Flotation has become the principal means of separating most metallic minerals as well as some non-metallic ones. By selective flotation, whereby appropriate reagents are used, and the alkalinity (pH) of the solution is controlled, almost any sulfide mineral can be separated from its associates, both ore and gangue, no matter how complex. Native gold, tungsten, and certain non-sulfide minerals can also be isolated by flotation, but sulfide minerals, especially those containing copper, lead, and zinc, are most amenable to this treatment, with silicates the least amenable (Pearl 1973).

From an environmental viewpoint the main issue of concentrating minerals is the generation of large amounts of fine ground waste materials of no commercial value- the tailings.

From an environmental viewpoint the main issue of concentrating minerals is the generation of large amounts of fine ground waste materials of no commercial value – the tailings. These often pose the biggest environmental challenge associated with a mine operation. How to dispose of tailings material in a safe manner? Safe refers to the physical integrity of the tailings containment structures as well as the long-term quality of water emerging from the tailings containment over time. Finding beneficial uses for tailings storage sites after mine closure also poses challenges. Tailings management is discussed in detail in Chapter Eighteen.

A second issue of particular public concern is the environmental impact of chemicals used in concentration. Most of these are harmless to the environment and most concentration processes, including flotation, basically rely on physics to liberate ore minerals from gangue. Notable exceptions are amalgamation and cyanidation of gold, which attracts public attention because the two processes rely on the use of chemicals associated in the public mind with, respectively, environmental disaster and murder, that is, mercury and cyanide. The use of mercury is discussed in Chapter Five in the section on artisanal mining; the use of cyanide to leach gold from rock is discussed in Chapter Six.

Thickening

Thickening is usually accomplished by settling slurries in large tanks, known as thickeners (**Figure 4.18**). Thickeners are used to remove liquids from concentrate slurries and from tailings slurries. Concentrates exiting the flotation circuit typically with water contents of 60 to 80 percent, are thickened to reduce the moisture content and reclaim the process water and flotation reagents before the final dewatering. Slurried tailings may be thickened to reclaim water and reagents and to facilitate final tailings disposal. Mills usually employ a number of thickeners concurrently. Typically, mills use continuous thickeners equipped with a raking mechanism to remove solids. Gravity causes the solids to settle to the bottom of the thickener, where they are scraped to a discharge outlet by a slowly rotating rake. Several variations of rakes are commonly used in thickener. The thickened solids, whether mineral concentrates or tailings, exit from the thickener as underflow.

Dewatering the concentrate in a thickener, then in disc, drum, or vacuum filters for final dewatering, produces a relatively dry product, the final ore concentrate, ready for further shipping and processing (Weiss 1985). The liquid component removed during the thickening process may contain flotation reagents, and/or dissolved and suspended mineral products. The liquid is usually recycled for reuse at the mill.

Thickening of tailings is a common step prior to pumping the thickened slurry to the tailings pond and ultimately disposing of the thickened slurry. Thickening minimizes the

FIGURE 4.18
Thickener Illustrated

Gravity causes the solids to settle to the bottom of the thickener, where they are scraped to a discharge outlet by a slowly rotating rake.

amount of water placed in the pond and the pond size. The thickened tailings retain sufficient water to allow them to flow in the tailings pipeline without undue wear on the transport system.

The settling of solids in the thickeners is often enhanced by chemical reagents known as flocculants. Filter cake moisture is regulated by reagents known as filtering agents. Typical flocculants and filtering agents used are polymers, nonionic surfactants, polyacrylate, and anionic and nonionic polyacrylamides (ASARCO 1991).

Concentrate Storage and Shipping

The final mine product is shipped to downstream mineral processing for further treatment, or as in coal mining to the end user. Concentrate transportation is by truck, conveyor belt, pipeline, railway, ship, or a combination of those. In view of environmental concerns, concentrate and coal transportation to the downstream industry or the end user differ in one particular aspect to movement of waste rock or tailings – the transportation corridor is often outside the actual mine site, may stretch over tens of kilometres, and may interfere with public transportation or land uses (**Case 4.6**). If the existing public transportation infrastructure is used, additional traffic due to ore transport will increase noise, dust, and gaseous emission, and will decrease road safety. Newly established transportation corridors require land acquisition with related environmental and social issues. Concentrate transportation in pipelines is a preferred solution on environmental grounds, but requires high initial capital commitment. Concentrates are transported in slurry and dewatering at the port site is required. Effluent discharge becomes an environmental issue. Independent of the mode of transportation, there is always the need for storage for concentrate or coal. If not protected against rain by shelters, runoff water or leachate from such storage areas may cause water pollution.

The transportation corridor is often outside the actual mine site, may stretch over tens of kilometres, and may interfere with public transportation or land uses.

Precious metals and precious stones such as diamonds provide a special case. Here, the product is highly valuable but occupies relatively little space. Security against theft is a major consideration. As a result, precious commodities are usually transported by air rather than by land or over water.

4.7 ANCILLARY FACILITIES

While environmental impacts of ancillary facilities are generally less significant than impacts related to mining activities, they are important nevertheless.

Ancillary facilities are all facilities that are required to support mining, but do not directly contribute to extraction, beneficiation, and concentrate shipping. Depending on the mine size, ancillary facilities such as power stations, water supply reservoirs, and ports may be sufficiently large to warrant stand-alone environmental impact assessments, and in some cases this is required. More commonly, however, the impacts of construction and operation of ancillary facilities are included in the overall EIA for the mining project. While environmental impacts of ancillary facilities are generally less significant than impacts related to mining activities, they are important nevertheless. It is risky to attempt to summarize the wide range of potential environmental impacts that could result from ancillary operations, but some indications of their levels of importance follow.

Port

Mines are often located in remote areas, far from major metropolitan communities and likely end users. Shipment of the final mine product, be it unbeneficiated ore, concentrate or coal, is often via a designated port, which is also usually the main entry point for mine equipment and materials (**Figure 4.19**). The port may be located on the coast or on the bank of a navigable river. Sometimes as in PT Arutmin's coal operations in East Kalimantan, Indonesia, barge-loading facilities are constructed on rivers close to the mines, with subsequent transfer to ships at a major port facility.

Ship movements need to be coordinated with existing local sea traffic, even though local traffic may be a few small crafts.

A number of environmental issues are related to construction and operation of the seaport. Ship movements need to be coordinated with existing local sea traffic, even though local traffic may be a few small crafts. Dredging, if required, will generate its own impacts, which may be trivial or significant, depending on the nature of the local aquatic or marine environment. Port design needs to accommodate secure storage areas for equipment and material. Waste management and concentrate dewatering facilities, if required, are critical to effective environmental management, as is control of potential spillage from material

CASE 4.6
Transporting Lead Concentrate Stirs Public Fear

In 2000 Magellan Metals Pty Ltd (Magellan) obtained approval to develop an open-cut lead carbonate mine and processing facility approximately 30 kilometres west of the Wiluna townsite in Western Australia to produce a lead concentrate and to export the concentrate through the Port of Geraldton. In 2004 the then Minister for the Environment approved a variation to export through the Port of Esperance rather than the Port of Geraldton.

Mining commenced in November 2004 and Magellan exported concentrate in bulk from June 2005 until March 2007. The export of bulk lead carbonate concentrate through the Port of Esperance was halted by the Department of Environment and Conservation due to fugitive lead and nickel dust pollution from the Port of Esperance. Subsequently, the Magellan Mine was put on care and maintenance.

In a revised shipment proposal the mine operator suggested to export containerized lead concentrate through the Port of Fremantle, currently the only container terminal in Western Australia capable of handling the volume and proposed method of transport of lead carbonate. At the mine site it is proposed to seal the lead concentrate in approved bulk bags to be loaded into shipping containers, which are then closed with a steel bolt until arrival overseas. An accredited, independent, inspector shall ensure that there is no fugitive lead dust on the outside surfaces of both the bulk bags and the shipping containers prior to leaving the mine site.

The population of Fremantle, however, is heavily opposing the proposal. They argue that a highly toxic, volatile product would be transported over many years through the city with a too great inherent risk of an accident.

handling, for example, when loading concentrates. Oil spills are another potential hazard and require a contingency plan, and appropriate barriers and cleanup equipment.

Airport

Transport of employees, contractors, government officials, and visitors is often by aircraft. If a mine-site requires an airstrip, then significant land clearing will be required. As is the case with seaports, airport construction and operation must comply with relevant national and international standards and requirements. Safety is always the first concern, but noise pollution is likely to be the most immediate environmental issue once an airport is in operation; this needs to be addressed in the initial planning. More often than not, the airport will also be open to public transportation, imposing further environmental and planning obligations.

Access and Haulage Roads

Newly constructed roads as part of mine development may create access to previously inaccessible forest or wilderness areas, facilitating hunting, illegal logging, informal housing, and farming. Road construction itself has a number of adverse environmental impacts. Road cutting is a major cause of accelerated water erosion, a particularly sensitive issue at surface water crossings. Depending on their geochemistry, exposed rock faces in road cuttings may become acid generating. In steep unstable terrain, road cuttings can trigger landslides. Traffic on unpaved roads is a major source of dust in most mining operations, and requires adequate management.

Newly constructed roads as part of mine development may create access to previously inaccessible forest or wilderness areas.

Pipelines and Conveyors

Pipelines and conveyors used for transportation of mine wastes, concentrate or coal have been discussed in previous sections, but other pipelines or conveyors may exist. Pipelines to transport water or diesel fuel are common. Similarly, pipelines are the preferred option to transport tailings to the final disposal site. The mine operator generally aims to align pipelines along mine roads to avoid the construction and related costs of a dedicated pipeline corridor with a service road.

FIGURE 4.19
The Special Port of the Batu Hijau Copper Mine in Indonesia

Shipment of the final mine product is often via a designated port, which is also usually the main entry point for mine equipment and materials.

FIGURE 4.20
The Chatree Gold Mine in Thailand

The Chatree Gold Mine has a particularly small footprint with all facilities – open pit mine, processing plant, waste rock dumps, and tailing storage – all situated in close proximity.

Storage

A mine operation consumes a large amount of diesel fuel, oil, and other materials. Large diesel fuel storage facilities are required if the mine operates diesel generator power plants. Depending on the beneficiation process, a wide range of chemicals are stored at the port and mill site. Effluent treatment facilities may require storage of additional chemicals. Coal storage areas not only exist in coal mining, but also in metal mining where there is an integrated coal-fired power plant.

Environmental concerns related to material storage are accidental release of stored materials, and contaminated runoff or leachate.

Environmental concerns related to material storage are accidental release of stored materials, and contaminated runoff or leachate. Environmental controls include the provision of sealed ground surface, shelter, secondary containment, leak detection equipment, runoff collection drains with associated settlement ponds, or ground water monitoring wells.

Mine Town

Whether or not to construct a new town to accommodate the project workforce is often a difficult decision for proponents of new mine developments. Many factors are involved, including:

- the wishes of local landholders and other stakeholders;
- the presence of existing towns in reasonable proximity to the planned operations, and the capacity of these towns to accommodate that part of the workforce which cannot be recruited locally (**Figure 4.20**);
- the likely scale and duration of the project; and
- government policies; often, governments use resource development projects in remote areas to contribute to the overall development of the regions involved.

Where existing towns are available within reasonable proximity, using these to accommodate mine employees should be given serious consideration.

Where existing towns are available within reasonable proximity, using these to accommodate mine employees should be given serious consideration. Reasonable proximity will depend on circumstances but would typically mean within about 40 km. The workforce could be domiciled by and obtain services from the existing communities which, if well planned, should bring benefits to both the workforce and the communities. The reality is, however, that the mine is often located in a physically hostile setting too far from major settlements to commute to and from work on a daily basis. Furthermore if housing facilities do exist, they are rarely adequate or sufficient in number. This, of course, can be overcome by planning and investment, carried out by the company or sponsored by the company.

Often new mine towns with several hundred units are set up for the mine workers and their families at a convenient distance from the mine. Where possible a pleasant setting is chosen and accommodation is of a high standard, in recognition of the fact that good living conditions contribute to working morale, thereby reducing turnover, which is a major concern and cost for remote operations. To complete the picture, mine towns boast facilities such as kindergartens and schools, places of worship, mess houses, hotels, supermarkets, department stores, recreation facilities (possibly including a gymnasium and golf course), libraries, coffee shops, administrative buildings, and banks most of which may be absent in the region.

A mine town has the same environmental issues as any other town. To begin with, a new mine town requires suitable land in terms of topography and soil conditions, land that may serve other competing land use purposes. Sewage and liquid wastes are generated, so the town planning must allow for collection and treatment of liquid waste streams and their final disposal. Finally, solid municipal waste is generated that requires proper waste collection and disposal systems, often via an engineered sanitary landfill.

A mine town has the same environmental issues as any other town.

Medical Facilities

Without a nearby hospital, mines often need to provide and operate their own medical facilities. They are commonly designed and staffed to also provide services to existing communities as part of community development programmes. As with all hospitals and clinics, medical waste will be generated and, to avoid spread of contaminants, provision must be made for its treatment, most commonly through the installation of a high temperature medical waste incinerator.

Power Generation and Distribution

The energy demand of large mines is staggering and may reach several hundred megawatts. The main energy consumer is the mill where ore is physically broken down into fine-grained particles. But even when milling is not required, such as in coal mining, the energy demand of all other mine activities remains high.

The energy demand of large mines is staggering and may reach several hundred megawatts.

In remote locations mines operate their own power plants. Frequently these are huge diesel generators. The main environmental issues relate to gaseous emissions, and to diesel fuel and lubricant handling. Other mines install dedicated coal-fired power plants, so coal handling, flue gases, and ash management are the main areas of concern.

The location of the mine, the port site, and the type of power generation determine the siting of the power plant. Diesel generators are often installed close to the mine; coal-fired power plants close to the port site (see Figure 4.19). Transmission lines, if necessary, need special corridors where adjacent tall trees, a potential hazard to transmission lines, are cut.

Supporting Operations and Maintenance Activities

The operation and maintenance of the mine and its supporting infrastructure require additional activities that have not yet been discussed. Operation and maintenance activities include warehouses, workshops, administration, security, fire fighting, water supply, waste management, plant nurseries, quarries, and borrow pits. Most of these activities present environmental issues of their own, some major, others less so. Of the former, for example, the construction of a mine and mine operations require a substantial amount of sand, gravel, or rock as building materials. Such material is not always easily available at the mine site, and quarries and borrow pits need to be created outside the actual mining area.

Quarrying may continue during mine operation to provide limestone for acid rock drainage prevention measures, to acquire material for staged tailings dam construction, to cater for ongoing road maintenance or to acquire materials needed for primary mineral processing (such as limestone in acid leaching).

4.8 DESIGN FOR CLOSURE

The aim of environmentally effective mine closure is to leave all areas affected by the project, in a safe and stable condition.

Mine closure activities occur on cessation of mining and beneficiation activities, following the exhaustion of the ore reserve. The aim of environmentally effective mine closure is to leave all areas affected by the project, in a safe and stable condition. This usually involves: (1) decommissioning and removal of mine equipment and facilities; (2) other activities to ensure safe conditions after mine closure such as sealing shafts, stabilizing slopes, or long-term acid rock drainage treatment measures; (3) managing the complex social impacts on the workforce and on the community associated with the closure of the operation; (4) finalizing rehabilitation that commenced earlier in the mine life; and (5) monitoring the success of rehabilitation and closure activities over the long term.

Planning for mine closure starts with mine planning.

Ideally, planning for mine closure starts with mine planning. Modern EIA regulation requires the mine proponent to present a preliminary mine closure plan as part of the environmental permitting documents. Mine closure planning is then periodically reviewed and updated during mine operation with an increasing level of detail and accuracy as the mine reaches the final years of operation. In opencast mining, however, mine closure is a continuous effort: mined out areas are rehabilitated as soon as practical and in parallel with the mining.

The formation of acid drainage with its associated contaminants has been described as the largest environmental problem facing the US. mining industry.

Environmental aspects of mine closure are many and varied. Decommissioning of the physical mine infrastructure that is of no use for local communities is achieved relatively easily. One of the main social challenges is to plan for social changes that will occur due to the absence of the mine. Employment opportunities are gone, and secondary businesses that may have depended solely on the mine will lose all custom. Community development programmes run out of funds and cease to exist, unless ongoing funding is generated from previous initiatives. One of the main environmental challenges is long-term water management (**Case 4.7**). Water can be a long-term destructive force on the mine landscape, particularly open pits, waste rock storages, and tailings disposal areas. Long-term water quality also is a concern. Large volumes of rocks have been relocated and are exposed to a new environment with different natural chemical reactions. The formation of acid drainage with its associated contaminants has been described as the largest environmental problem facing the US. mining industry (USEPA 2000). Commonly referred to as acid rock drainage (ARD) or acid mine drainage (AMD), acid drainage may be generated from mine waste rock or tailings (i.e. ARD) or mine structures such as pits and underground workings (i.e. AMD).

Finally there is the issue of the long-term integrity of structures left behind by the mine. Mine openings, both horizontal and vertical, can be a significant physical hazard at an abandoned mine site. In many cases the openings are well known, but are still a potential threat to the adventurous who may decide to enter and explore. Those who do face a number of hazards that can injure or kill, including unstable ground that can collapse, or foul air, either insufficient in oxygen or containing poisonous gases such as carbon monoxide. Mine openings, particularly vertical shafts, where the opening has been covered either by a collapsed building or by overgrown vegetation, pose a threat to persons or wildlife of falling, sometimes hundreds of metres (USEPA 2000).

Mine closure is not always permanent. Some circumstances lead to temporary mine closure. Metal prices may drop to a level that makes it attractive to temporarily stop mining earlier than originally planned. A future increase in metal prices may lead to the re-opening of the mine. Some mines such as the Climax Molybdenum Mine, near Leadville in Colorado, USA, have closed and re-opened several times over a period of decades. Advances in mining and mineral processing technologies may also make it financially attractive to re-open past mining areas. On the other hand, temporary or permanent mine closure may result from unsafe conditions, a major mine accident and/or public opposition. The tailings spill in 1989 at the Marcopper Mine in the Philippines led to mine closure. Civil unrest in Bougainville involving opposition to mining, resulted in the closure of all mining activities at the Bougainville Copper mine in 1989, and it is unlikely that the mine will re-open in the near future, if ever. A detailed examination of the environmental issues related to mine closure follows in Chapter Twenty One.

Mine closure is not always permanent.

CASE 4.7
Remediation at Mines de Potasse d'Alsace, France

DEPOLLUTION WELL
Rain
Runoff
Towards Rhine
Percolation
Ground Water
Impervious Basement
Depollution Well

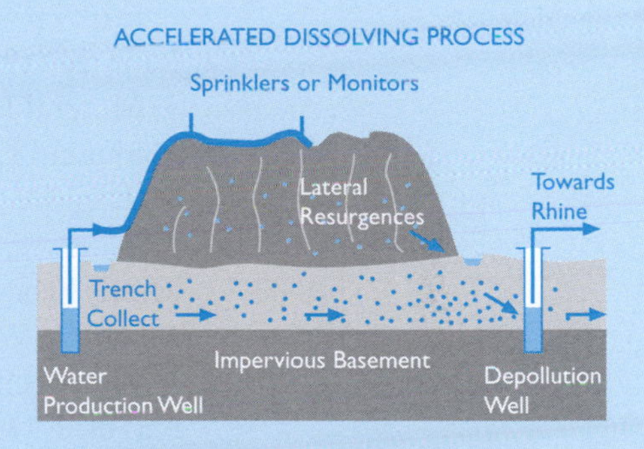

Mines de Potasse d'Alsace (MDPA) operates the Amelie underground potash mine and processing plant and the Marie-Louise underground mine near Mulhouse in north-east France. Mining of potash ore bodies in the region began in 1910. The workforce peaked in 1965 at around 14,000 and declined to 1,300 by mid-2000. The number of operating mines and plants has similarly reduced. The ore deposit is approaching exhaustion, and planning and preparation for this eventuality began a number of years ago. The long potash mining history of the region has left a legacy of challenges. Closure planning and preparation has focused on: (1) rehabilitation of the surface waste piles (referred to as terrils); (2) remediation of the salt-contaminated surface aquifer; (3) removal of decommissioned plant, equipment and materials; and (4) creation of alternative employment for the workforce and community.

The major environmental issue revolves around the waste piles dumped adjacent to the beneficiation plants prior to 1934. These were placed without liners or other containment features. Subsequent dissolution of the salt by rain has contaminated the underlying aquifer giving rise to plumes of salt contamination down-flow from the piles. Rehabilitation activities have focused on both removing the source of the pollution (the waste piles) as well as reducing the level of salt contamination in the groundwater.

Seventeen salt piles covering an area of 220 hectares are to be rehabilitated. Two methods are currently used to rehabilitate them. One involves reshaping the piles, capping them with plastic, clay or bituminous liners to isolate the waste from the rain, then covering with a growing medium and revegetation with grasses. The other method involves accelerating the dissolution of the salt component of the pile using fresh water. The resulting brines are discharged in a controlled manner to the adjacent Rhine River, while the insoluble clay residue piles are reshaped and revegetated. Both these methods are intended to prevent any further contamination of the groundwater. The existing salt contaminated groundwater is contained and remediated by a pumping programme. A series of wells have been drilled down-flow from each of the piles. Pumping from these wells both restricts further contamination of the aquifer by salt, while providing the processing operations with a source of industrial water.

The former mine land is being converted mainly to industrial use following removal of unwanted plant, equipment and infrastructure. Incentives have been provided for the creation of new businesses with an emphasis on small and medium size enterprises. A number of the old mine buildings are being refurbished to provide space to house some of these new enterprises. Training is provided to interested employees to allow them to develop their own businesses or prepare for work in a new industry.

In addition, a hazardous waste management company has begun operation, using underground openings created by mining rock salt for long-term storage of hazardous waste. Communication with employees and the community on the closure plans and activities has been important in alleviating concerns. An interesting method that has been used to assist this process is the publishing of a comic book story on the MDPA operation. The story conveys information about the forthcoming closure in a manner and style that is easily understood by all.

Source:
UNEP and IFA (2001)

REFERENCES

ASARCO (1991) Ray Unit Tailing Impoundment Alternative Analysis. Appendix 11.19. Submitted to EPA Water Management Division Region IX Wetlands Program; February 4.

Bateman A (1950) Economic Mineral Deposits, John Wiley, 2nd. edition New York.

CGER (1999) Hardrock Mining on Federal Lands; Committee on Hardrock Mining on Federal Lands, Committee on Earth Resources, Board on Earth Sciences and Resources, USA.

Clive Aspinall Report (2002). The Clive Aspinall Report, Fax: 62-(0) 21-750-2874, gsm@jakarta3.wasantara.net.id

EIPPCB (2001) IPPC Reference Document on Best Available Techniques in the Non-Ferrous Metals Processes. Seville: European Integrated Pollution Prevention and Control Bureau. Available at http://eippcb.jrc.es/pages/FActivities.htm

Hartman HL (ed.) (1992) SME Mining Engineering Handbook, Society for Mining, Metallurgy and Exploration, Inc., Littleton, CO, 2260 p.

Hartman HL and Mutmansky JM (2002) Introductory Mining Engineering, John Wiley & Son, 2nd edition.

Kearey P, Brooks M and Hill I (2002) An introduction to geophysical exploration, Blackwell Publishing.

Lacy W (2006) An Introduction to Geology and Hard Rock Mining, ROCKY MOUNTAIN MINERAL LAW FOUNDATION Science and Technology Series, http://www.rmmlf.org/SciTech/Lacy/lacy.htm

Lane K (1988) The economic definition of ore – cutoff grades in theory and practice. Mineral Journal Books Ltd.

Mcdonald EH (2007) Handbook of gold exploration and evaluation, CRC.

McLemore VT (2005) Significant Metal Deposits in New Mexico – Resources and Reserves, in Decision Makers Field Guide.

Metals Economics Group (2006) Corporate Exploration Strategies, www.metalseconomics.com

Moon C, Whateley M and Evans AM (eds) (2005) Introduction to mineral exploration. Blackwell Publishing.

Navin T (1978) Copper Mining and Management, University of Arizona Press. OECD 1983. Coal – Environmental Issues and Remedies.

Pearl R (1973) Handbook for Prospectors, McGraw-Hill, Inc.

Pincock, Allen and Holt (2005) Pincock Perspectives Issue No. 70, September 2005.

Robb L (2005) Introduction to ore-forming processes; Blackwell Publishing.

Roscoe WE (1971) Probability of an exploration discovery in Canada. Canadian Mining and Metallurgical Bulletin, Vol. 64, (707):134–137.

Sheriff RE and Geldart LP (1995) Exploration seismology; Cambridge University Press.

Ulatowski T (1978) 'Importance of Financing in Project Planning,' Mining Engineering, June 1978.

Ulatowski T, Frohling E, and Lewis FM (1977) Mine Financing: When Is the Best Time to Borrow?; Engineering and Mining Journal, May 1977.

UNEP and IFA (2001) Environmental Aspects of Phosphate and Potash Mining. First edition. UNITED NATIONS PUBLICATION (www.fertilizer.org)

USDA (1995) Anatomy of a mine – from prospect to production. Gen. Tech. Rep. INT-GTR-35 revised. Ogden, UT: US Department of Agriculture, Forest Service, Intermountain Research Station. 69 p.

USEPA (1995a) EPA Office of Compliance Sector Notebook Project-Profile of the Non-Metal, Non-Fuel Mining Industry. Office of Compliance Office of Enforcement and

Compliance Assurance U.S. Environmental Protection Agency 401 M St., SW (MC 2221-A) Washington, DC 20460.

USEPA (1995b) Sector Notebook Project Profile of the Metal Mining Industry, Office of Compliance, Office of Enforcement and Compliance Assurance U.S. Environmental Protection Agency 401 M St., SW (MC 2221-A) Washington, DC 20460, EPA/ 310-R-95-008

USEPA (2000) Abandoned mine site characterization and cleanup handbook, EPA 910-B-00-001 August 2000.

Vickers E (2006) in: Feasibility Studies and Project Financing (Guerdon E. Jackson, Editor), Surface Mining, 2nd edition, SME Online Digital Library: http://books.smenet.org/Surf_Min_2ndEd/sm-ch04-sc04-ss00-bod.cfm 10/28/2006

Weiss NL (Editor) (1985) SME Mineral Processing Handbook, Kingsport Press.

Wilson F (1981) The conquest of copper mountain. McClelland and Stewart Ltd.

World Bank (2006) Background Paper: The Outlook for Metals Markets; prepared for G20 Deputies Meeting Sydney September 2006. The World Bank Group, Oil, Gas, Mining and Chemicals Department, Washington.

• • • •

5

Mining Methods Vary Widely

From Excavation to *In situ* Leaching

♃ TIN

Native tin is not found in nature. The first tin artefacts date from 2000 B.C. However, it was not until 1800 B.C. that tin smelting became common in western Asia. Tin was rarely used on its own and was most commonly alloyed to copper forming bronze. Bronze flowed more easily than pure copper, was stronger after forming and was easy to cast. Tin is highly crystalline and during deformation is subject to mechanical twinning producing an audible 'tin cry.'

5 Mining Methods Vary Widely

From Excavation to *In situ* Leaching

This chapter has several purposes. First, it illustrates that mining is not restricted to extracting metals, coal, or diamonds; mining also includes the wide range of operations to extract dimension stone; sand and gravel; clay, ceramic, and refractory minerals; chemical deposits and fertilizer minerals; and other non-fuel, non-metallic minerals. The mining method will depend on the type of material and its mode of occurrence.

Second, it outlines common mining methods. Surface and underground mining as discussed in Chapter Four are revisited, before the concepts of solution mining and heap leaching are introduced. Heap leaching is a relatively common process used for the recovery of base metals and precious metals (gold and silver) from amenable low grade ores or, occasionally, from previously processed tailings or from past waste rock dumps.

Third, the chapter discusses artisanal mining which, while it represents a traditional way for many people, also has the potential to cause severe environmental damage. Artisanal mining is defined as small-scale mineral extraction, using mainly manual methods, carried out by individuals or small family groups. Other terms in common use are 'small-scale mining' and 'illegal mining', the latter when artisanal mining is carried out on a tenement granting exclusive mining rights to others. Occasionally, artisanal mining can be at the scale of commercial mining when backed by wealthy businessmen and supported by government.

Mining ultimately contributes to a better life for the 6 billion people who depend on the raw materials it produces.

Finally, this chapter is intended to reinforce the fact that mining ultimately contributes to a better life for the 6 billion people who depend on the raw materials it produces.

5.1 THE THREE MAIN CATEGORIES OF COMMERCIAL MINERALS

A common classification of commercial minerals into three main categories is based on the primary elements in the ore (**Table 5.1**): metallic ores, non-metallic ores including gemstones, and mineral (fossil) fuels.

TABLE 5.1

The Three Main Categories of Commercial Minerals – Deposits of fossil fuels do not fit the rigorous definition of minerals of chemists, but are frequently referred to as fuel minerals in the mining industry.

Metallic Ores	
Ferrous and ferroalloys	Iron (Fe), manganese (Mn), nickel (Ni), chromium (Cr), molybdenum (Mo), tungsten, vanadium, and cobalt (Co)
Non-ferrous	
Base metals	Polymetallic metals copper (Cu), lead (Pb), zinc (Zn), and the metal tin
Precious metals	Gold (Au), silver (Ag), and the platinum family
Lightweight metals	Aluminium (Al), magnesium (Mg), beryllium, and titanium
Radioactive metals	Uranium (U), Thorium (Th), Radium (Ra)
Rare Earth metals	Lanthanide series
Mercury	
Other minor metals	Arsenic, bismuth, gallium, selenium
Non-Metallic Ores (Industrial Materials)	
Dimension stones	Marble, granite, limestone, sandstone, slate
Crushed and broken stones	All principal types of stones
Sand and gravel	Alluvial sediments
Clay, ceramic, and refractory minerals	Kaolin, bentonite, clay, shale
Chemical and fertilizer minerals	Potash, salt, phosphate rock, Guano
Other non-metallic minerals	Sulphur (S)
Gemstones	Diamonds, rubies, sapphires, emeralds, jade, opals, topaz, etc
Mineral Fuels (Fossil Fuels)	
Peat	
Coal	Brown coal, lignite, coking coal, bituminous coal
Oil	Oil, oil sand, oil shale
Gas	Methane, coal bed methane

Metallic Ores

Metallic ores encompass ores of ferrous and ferroalloy metals (iron, manganese, nickel, chromium, molybdenum, tungsten, cobalt, and vanadium); base metals (copper, lead, tin, and zinc); precious metals (gold, silver, and the platinum family); light metals (aluminium, magnesium, beryllium, and titanium); radioactive metals (uranium, thorium, and radium); rare Earth metals (lanthanide series); mercury and a wide range of minor metals.

Metallic minerals are mined specifically for the metals that can be extracted from them. We are all familiar with the most common metals but there are a host of others that play vital roles in society yet remain virtually unknown.

Ferrous and Ferroalloy Metals

Iron bearing minerals are found practically everywhere, and a profitable discovery of iron ore is likely to be one of very large size.

Iron, the main ferrous metal, is the most important industrial metal, since its alloy with carbon is steel. Iron bearing minerals are found practically everywhere, and a profitable discovery of iron ore is likely to be one of very large size. Manganese is an essential ingredient in the manufacture of carbon steel. Curiously, manganese metal has no use of its own as a metal, but as a metallurgical material, it has no substitute. Beside manganese, nickel is the most important ferroalloy, being an essential ingredient in stainless steel. Sulphide nickel ore often contains chromium, platinum, and cobalt. Serpentine and lateritic nickel deposits are derived from weathering of ultrabasic rocks and are found in New Caledonia, Cuba, Australia, Papua New Guinea, the Philippines, and Indonesia. Chromium, like manganese, is found mostly in non-industrial countries which have little use for it. Chromium mineralogy is simple: one mineral, chromite, is the only ore (Pearl 1974). Cobalt and vanadium are often produced as by-products of mining for other metals (uranium in case of vanadium; nickel, silver, or copper mining in case of cobalt).

Base Metals

The term base metal is used informally to refer to a non-ferrous metal that oxidizes or corrodes relatively easily, and reacts variably with dilute hydrochloric acid (HCl) to form hydrogen. Copper is considered a base metal as it oxidizes relatively easily, although it does not react with HCl. In alchemy, a base metal was a common and inexpensive metal, as opposed to precious metals. A long time goal of the alchemists, of course never achieved, was to transform base metals into precious metals, gold and silver. Like ferrous metals, base metal mining is capital intensive. Copper is a metal with many applications, and apart from gold, was the first metal known to humans. Characteristic signs of copper are the bright green and blue minerals that stain host rocks. Copper minerals vary widely and lead and zinc are common associates. The main zinc mineral of economic value is zinc sulphide or sphalerite (Pearl 1974), containing 57 to 67 percent of zinc. Tin has a restricted distribution, with a 2000 km belt of tin deposits in Southeast Asia extending from Belitung Island East of Sumatra, Indonesia through Malaysia, Thailand, and Burma and into China, produces the majority of the world's tin output.

Base metal mining is capital intensive.

Precious Metals

The term precious metal is used informally to refer to rare metallic chemical elements of high economic value. Chemically, precious metals are less reactive than most other elements, have high lustre, and high electrical conductivity. Historically, precious metals were important as currency and jewellery, but are now regarded mainly as investment and industrial commodities. The best-known precious metals are gold (Aurum, Au, in Latin, or 'shining dawn') and silver (argentum, Ag, in Latin) the brightest of all known metals, known for their uses in jewellery and coinage (**Case 5.1**). Other precious metals include the platinum group metals (platinum – which means 'little silver', ruthenium, rhodium, palladium, and osmium and iridium – or its alloy osmiridium). Platinum was not known in Europe until Spanish explorers acquired some platinum objects from South America (Pearl 1974). Interestingly certain other metals – for example, gallium – although not termed a precious metal, command higher prices.

Chemically, precious metals are less reactive than most other elements, have high lustre, and high electrical conductivity.

Light Metals

Four metals that are exceptionally light but strong for their weight, have become highly important for structural use since the beginning of the industrial revolution. Aluminium, the oldest and best-known of the light metals, is the most common metal in the Earth's crust. It can only be produced commercially from bauxite deposits, the result of lateritic weathering of granitic rocks. Bauxite mining requires large-scale operations, and the processing facilities which produce alumina are expensive (see Chapter Six for a brief discussion of the Bayer process, the common method of extracting alumina from ore). Smelting of alumina to produce aluminium metal is energy intensive. Aluminium metal is used extensively for the manufacture of aircraft and motor vehicles. Magnesium is the lightest metal known. It is being used increasingly in new alloys for motor vehicles where lighter weight means increased fuel efficiency. Compounds of magnesium are used in agriculture and in water treatment. It is produced from sea water and also from hard rock deposits of magnesite (Magnesium carbonate). Beryllium is a newcomer in the world of light metals, used to some extent in the nuclear power industry and as an alloy ingredient. Titanium, the fourth light weight metal, is produced mainly from ilmenite and rutile, two heavy minerals found in mineral sand deposits, in which heavy mineral grains have been concentrated by the winnowing action of waves, currents or winds. Commercial production of these titanium minerals occurs in Australia, Mozambique, Eastern USA, and, in the future, India. Most titanium is used as a pigment in paints. However, increasingly titanium metal is being used in applications where a very strong but light weight metal is required; golf clubs and tennis racquets are two examples, rockets are another.

Aluminium, the oldest and best known of the light metals, is the most common metal in the Earth's crust.

Radioactive Metals

For radioactive metals, their metallic nature is generally irrelevant. Uranium is used as nuclear fuel to produce energy and its use is further discussed in Chapter Eleven. Thorium,

For radioactive metals, their metallic nature is generally irrelevant.

CASE 5.1

'It's not how much ground you search, but how well you search the ground.'

Electronic prospecting (better known as metal detecting) for gold is a relatively new development in gold prospecting. Improvements in this technology mean that metal objects can be detected at depths up to 1 m and more, depending on their size. It was the advent of these electronic devices that started the detector gold rush in Australia. As the name implies, a metal detector will locate all types of metal whether in native form like gold nuggets or man-made metallic objects such as coins, cans, nails or bolts. Most gold nugget hunters work the old gold-rush grounds hoping to strike it lucky.

Is there a prospect of a 21st century gold rush based on this type of prospecting? Probably not. Gold nuggets account for very little of the world's gold production. Today the real money in gold belongs to serious miners, guided by serious geologists. The gold ore being mined today shows no sign of its valuable fraction, yielding a few grams in every ton of rock.

However, gold has a call unique to itself, a call that only a few can resist. Nuggets can still be found – the luck is in how big they are. There is always the hope to strike it as lucky as German prospector Bernard Otto Holtermann in 1872 who found a 630 lb mass of almost solid gold at Hill End, New South Wales, Australia.

discussed below, is a potential future source of nuclear fuel, but it finds some application outside the nuclear energy field, such as alloyed with magnesium. The radioactive metal radium is used in medical technology as are many radioactive isotopes produced in nuclear reactors.

Rare Earths

Rare Earth metals, shown at the bottom of the Periodic Table (see Chapter Eleven) as a block of two rows of elements but occupying only two positions in the periodic table of elements, are neither rare nor Earths. The thirty rare Earth elements comprise the lanthanide and actinide series, the latter comprising only radioactive elements, including uranium and thorium (**Case 5.2**). Apart from uranium, thorium, actinium, and protactinium, the other elements in the actinide series and one element in the lanthanide series, promethium, are synthetic elements (also called trans-uranic elements). The name rare Earth originates from difficulties experienced prior to 1945 in purifying the metals from their oxides, due to their complex chemistry. Ion-exchange and solvent extraction processes are used today to quickly produce high purity, low-cost rare Earths, but the old name remains. All of the rare Earth metals are found in group 3 of the periodic table, and the 6th and 7th periods. The lanthanides series comprises the 15 elements in the periodic table with atomic numbers 57 to 71. Often included with the lanthanides are scandium (atomic number 21) and yttrium (atomic number 39). Because of their similar physical and chemical characteristics, the lanthanide elements and yttrium often occur together in nature. While scandium may occur with the rare Earth elements, it is also found in a range of other minerals (Spooner 2005).

The rare Earth metal thorium is an alternative to uranium as fuel for nuclear reactors. The most common ore of thorium, the phosphate mineral monazite, which contains up to 12% of thorium oxide, also contains other economically extractable rare Earth metals such as cerium, lanthanum, and neodymium, together with yttrium and iridium.

Industry uses rare Earth metals in a wide range of specialized applications, but in small amounts compared to most other mined metals.

Industry uses rare Earth metals in a wide range of specialized applications, but in small amounts compared to most other mined metals. A few examples follow, which illustrate the diverse uses of these metals: (1) chemical catalysis, especially in petroleum refining (cerium used in the petroleum industry and emissions control catalysts for gasoline and diesel fueled vehicles); (2) battery manufacturing (lanthanum used in the production of

CASE 5.2
The Case of Bukit Merah in Malaysia

In the early 1980s, the Asian Rare Earth (ARE) Company commenced extraction of the rare Earth yttrium oxide from monazite in the town of Bukit Merah in Malaysia. Yttrium finds application as chemical catalysts or pigments which can be used for colour TV screens. The extraction process leaves a sludge containing concentrated amounts of NORM – meaning 'naturally occurring radioactive materials'.

ARE commenced operation without having a permanent and safe waste disposal option for produced NORM. A few years later ARE made front page news when it was alledged that the radioactive waste had contributed to the deterioration in children's health. In 1983, eight residents filed a legal complaint against ARE. While neither established facts nor scientific proof was produced to support the claim that the alleged health problems were related to the operation of ARE, the company was forced to cease operations in 1994 due to mounting public pressure in Malaysia and Japan, combined with a shortage of locally mined monazite and increasing competition from newly emerging rare Earth producers in China.

It took almost ten years and tens of millions of US$ to develop a closure plan for the operation. Eventually a secure land fill was built for the final disposal of NORM and other radioactive materials at the ARE site.

nickel metal hydride rechargeable batteries for hybrid vehicles and portable electronic devices); (3) small powerful magnets (neodymium used to make the magnets used in miniature electric motors such as about 40 of these used in a typical automobile), and; (4) high temperature corrosion resistant alloys (iridium used in surgical devices, thermocouples and automotive emission catalysts).

China holds the leading position among producers of the rare Earth elements. Processing of rare Earth concentrates, and partially processed or intermediate products, is carried out at several locations in Europe, the USA and Japan, as well as in China. World production of rare Earths is around 100,000 t/y rare Earth oxide (REO). Output of yttrium is significantly less (about 2,500 t/y yttrium oxide – Y_2O_3). Global production data for scandium are not easily publicly available (Spooner 2005).

Mercury

Mercury is one of the most interesting of all natural substances because it is the only metal which occurs as a liquid at ordinary temperatures. It is also one of the easiest metals for prospectors to recognize, since most mercury comes from the bright red mineral cinnabar. Traditionally mercury has been of great importance for its ability to combine with gold, and hence to extract gold from placer or oxidized deposits by amalgamation. Until recently, mercury was widely used in scientific instruments such as thermometers and barometers. However, new technologies have replaced most of these uses. Mercury is notoriously toxic, particularly in its organic compound methyl mercury. It readily volatilizes and re-condenses so that, in an environment where mercury is present, the metal can be absorbed by dermal contact or inhaled as vapour, or ingested on food. In the marine environment it often bio-magnifies so that long-lived predators such as sharks and barracuda may have relatively high concentrations of mercury in their tissue. Like gold, in its metallic form it does not oxidize or react readily and, although it can be transformed into a less problematic compound (mercury sulphide – the compound found in nature as cinnabar) the chemical process is costly. This means that there is a 'global pool' of metallic mercury circulating in the environment – between air, water, sediments, soil and living things.

There is a 'global pool' of metallic mercury circulating in the environment – between air, water, sediments, soil and living things.

Other minor metals

A number of metals are recovered often only as by-products of smelting or other mineral processing operations (Pearl 1974): arsenic (primarily a by-product of processing copper ore), bismuth (a by-product of processing lead ore), cadmium (usually a by-product of processing zinc ore), gallium (solely a by-product of bauxite – aluminium – or sphalerite – zinc), or selenium (obtained in copper refining from electrolytic slime).

Non-metallic Ores – Industrial Minerals Mining

Mining is not restricted to extracting metals or fossil fuels such as coal. The industrial mining industry comprises the wide range of mining operations to extract dimension stone; sand and gravel; clay, ceramic, and refractory minerals; chemical and fertilizer minerals, and other non-fuel, non-metallic minerals. Non-metallic ores also include rocks in which gems such as diamonds are found. With the notable exception of gemstones and diamonds, non-metallic mineral resources are predominantly mined as a source of chemical feedstocks or building materials. They are mined for a wide range of chemical and physical properties. Although metals such as gold or silver may be more glamorous, the value of non-metallic minerals mined worldwide far exceeds that of metals (**Figure 5.1**).

Although metals such as gold or silver may be more glamorous, the value of non-metallic minerals mined worldwide far exceeds that of metals .

FIGURE 5.1
Metals and Industrial Materials Consumed in the USA, 1900 to 1995

Construction materials (crushed stone, sand and gravel) are separated from the remainder of the industrial minerals to illustrate the upsurge in construction following the end of World War II.

Source:
David McConnell 1998–2001

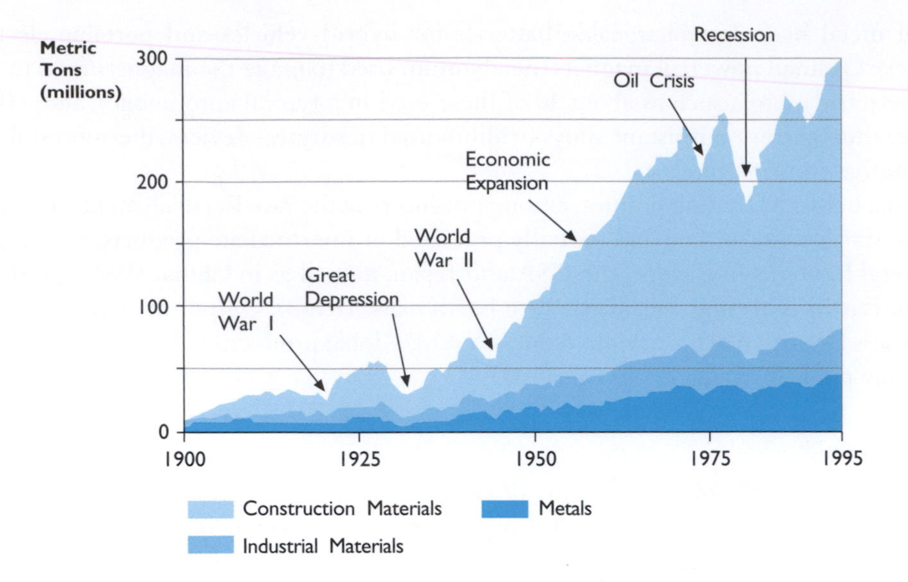

Dimension stone refers to rock that is cut to a certain shape and size, commonly used as building material: limestone, granite, dolomite, sandstone, marble, and slate (**Case 5.3**). Processing the stone begins with sawing the excavated rock into slabs using a rotating diamond or circular saw. Water is used to cool the saws and to remove particles. After the stone has been cut to the desired size, it is finished using natural and synthetic abrasives. Natural abrasives include iron oxide, silica, garnet, and diamond dust, themselves usually products of mining. Synthetic abrasives include silicon carbide, boron carbide, and fused alumina (USEPA 1995).

Most crushed and broken stone is mined from open quarries.

Nearly all the principal types of stone, including granite, limestone, sandstone, dolomite, and marble may be used as sources of commercial **crushed and broken stone or aggregate**. Stone that breaks in chunky, cubical fragments and is free of surface alteration from weathering is preferred for crushed stone. Most crushed and broken stone is mined from open quarries. Processing activities include blasting, conveying, screening, secondary and

CASE 5.3
Mining Carrara Marble

The Carrara marble quarries have been in operation for more than 2,000 years. Carrara marble, found in northern Tuscany, is famous for its statuary quality and it has been used by the most famous sculptors including Michelangelo, Donatello, Jacopo Della Quercia and Canova. The name of Carrara, Tuscany became synonymous with marble. Carrera marble continues to be requested worldwide for buildings, statues, and other ornamental objects, and about one million tons of marble is quarried each year.

tertiary crushing, and sizing. Screening is the single most important part of the processing cycle of crushed stone particles. Stone washing is something performed in order to remove unwanted material.

Sand and gravel deposits are commonly found in or adjacent to rivers or in areas with glaciated or weathered rock. There are two main types of sand and gravel. Construction sand and gravel are used mainly in concrete aggregates, road-base, asphaltic aggregates, and construction fill. Generally, the physical characteristics of construction sand and gravel and their proximity to construction sites are more important than their chemical characteristics. Construction sand and gravel represent some of the most accessible natural resources and are major basic raw materials. Despite the low unit value of these products, the construction sand and gravel industry is often a major contributor to and an indicator of the economic wellbeing of a nation. While most countries have adequate sand and gravel resources, this is not always the case. Singapore, for example, imports large quantities of sand and gravel from Indonesia, which has led to some sensitive political issues.

The construction sand and gravel industry is often a major contributor to and an indicator of the economic wellbeing of a nation.

Industrial sand, gravel, and clay are used in manufacture of glass, ceramics, and chemicals. The chemical and physical characteristics of these materials are very important to their end uses, and are therefore subject to stricter chemical and physical characterization than construction sand and gravel. Loose sand and gravel deposits are usually mined without the necessity of drilling and blasting. Material mined below water, in rivers, estuaries, lakes, and oceans must be removed using specialized equipment such as dredges, draglines, and floating cranes.

Common types of **clay, ceramic, and refractory minerals** include kaolin, ball clay, bentonite, fuller's Earth, fire clay, common clay, and shale. Processing of minerals in this category usually entails a combination of crushing, grinding, screening, and shredding to reduce particle size. The cement industry consumes large quantities of clay and limestone, and the occurrence of both raw materials in close proximity is an important criterion for siting of cement plants. A cement producing operation is as much about mining as it is about chemical processing (**Figure 5.2**).

Chemical and fertilizer minerals include potash and phosphate rock. UNEP (2001) provides an excellent summary on the chemical and fertilizer industry sector, and its environmental challenges. Potash, a term that describes minerals containing potassium compounds, is used in fertilizers. Processing potash involves mixing crushed potash ore

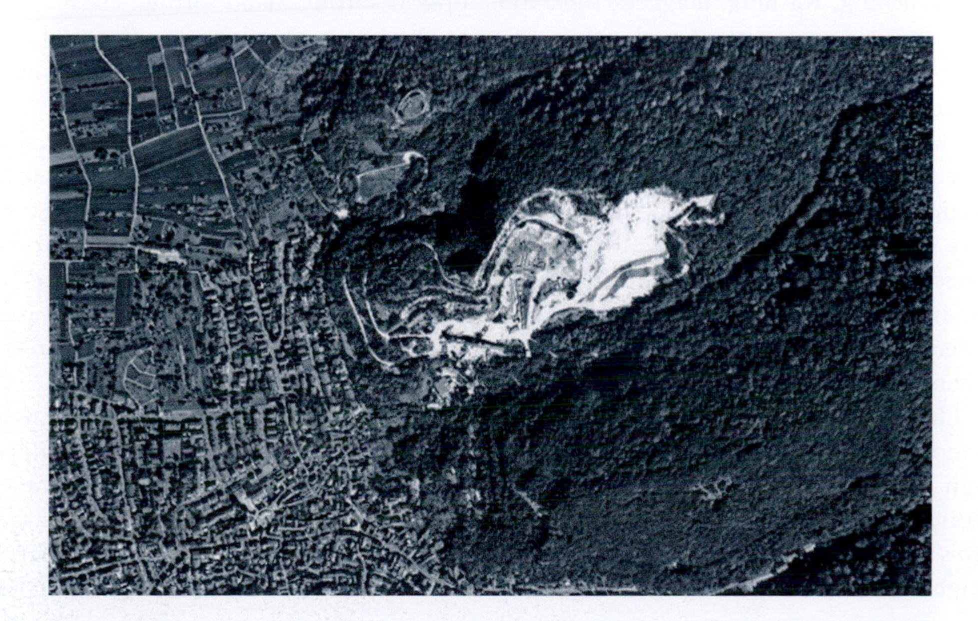

FIGURE 5.2
Lime Stone Quarries Close to Heidelberg, Germany

The cement industry consumes large quantities of clay and limestone, and the occurrence of both raw materials in close proximity is an important criterion for siting of cement plants.

Source:
Google Earth 2008

with brine which is saturated with potassium chloride and sodium chloride. Flotation, crystallization, or heavy media separation methods are then used to recover potassium-bearing compounds from the saturated solution (UNEP 2001).

Worldwide potash production is dominated by Canada, Russia, and Germany, which together account for about 76% of the total production. About one fifth of the world potash production comes from European mines in France, Germany, Spain, and the UK (EC 2004). Potassium chloride (KCl), commonly referred to as muriate of potash (MOP), is the most common and least expensive source of potash. Potassium chloride accounts for about 95% of world potash production (EC 2004). Potash tailings are composed of table salt (sodium chloride) together with a few per cent of other salts (e.g. chlorides and sulphates of potassium, magnesium and calcium) and insoluble materials such as clay and anhydrite. Dry stacking of large volumes of potash tailings (salt) on un-lined ground was historically common practice in France and Germany and has led to large-scale ground water contamination by salt (Spitz 1981, Rinaudo 2003, and Lottermoser 2007). Some potash heaps reached 240 m in height, containing about 130 million tons of salt tailings. The tailings heaps themselves generate saline solutions when atmospheric precipitation dissolves salt from the tailings material (see also Case 4.7). Today, five methods are common for managing potash tailings, each of them having their own specific environmental issues. These tailings management options are: (1) storing solid tailings on tailings heaps; (2) backfilling solid tailings into mined out voids of underground workings; (3) discharging solid and liquid tailings into the ocean/sea (e.g. marine tailings management); (4) discharging liquid tailings into deep wells; and (5) discharging liquid tailings into natural flowing waters (e.g. rivers) (EC 2004; UNEP 2001).

The tailings heaps generate saline solutions when atmospheric precipitation dissolves salt from the tailings material.

Environmental Impacts of Mining Non-metallic Minerals

In general, mining of industrial minerals is of less environmental concern than mining of metals or fossil fuels (see Chapter Thirteen for an overview of environmental concerns relating to metals and coal mining). It is recognized, however, that industrial minerals are recovered in much larger volumes, at far more locations, and in many different ways. Some industrial minerals are sold as mined, i.e. without being processed. In other cases a variety of mineral processing methods are applied to achieve a highly concentrated product. The majority of mines in the industrial minerals sector use only physical treatment (e.g. crushing, washing, magnetic separation, optical sorting, hand sorting, classification, flotation, and evaporation), with only a minority of projects carrying out a chemical treatment of the mineral (e.g. leaching). Industrial minerals mining operations are generally small compared to most metal mines. The value of one unit of industrial mineral is generally lower than one unit of product from metals mining, and most industrial minerals are only mined if they can be extracted with minimum movement of waste materials (overburden and waste rock). While the amounts and characteristics of tailings and waste rock vary significantly, in most cases the amounts of waste rock and tailings, the areas of most concern in metals mining, are much smaller compared to metals mining. Acid rock drainage is seldom an issue in the industrial minerals sector. Many other environmental issues remain the same: land disturbance, vegetation clearing, topsoil removal, dust, erosion, rehabilitation, and mine closure.

Most industrial minerals are only mined if they can be extracted with minimum movement of waste materials.

The building sector in particular reflects the size of industrial minerals extraction, while demonstrating the low value per ton of product. Most industrial minerals do not support high transportation costs and mining of industrial minerals often occurs close to the end users; hence most people are familiar and comfortable with this industry sector having grown up with the familiar sight of at least one industrial mine or quarry in their extended neighbourhood. Social issues if they occur tend to be localized, often related to land compensation, aesthetic visual impacts, increased transportation, noise, vibrations and dust. Finally with very few exceptions mining of industrial minerals does not attract the attention of international anti-mining activists. One notable exception is potash mining which has different processing techniques and tailings than most others industrial mineral mining operations. Adverse aesthetic impacts are also commonly associated with industrial mineral extraction, most cities having examples of unsightly scars from clay pits, gravel pits or aggregate quarries.

Gemstones and Diamonds

Jewels or gemstones are sought for their beauty, durability, and rarity. Beauty may be due to colour, brilliancy when properly faceted, or other special optical effects. Rarity of course is a psychological necessity. Much literature on gemstones is available (Ali 2003; Burke 2006; www.casmsite.org) and the subject is only briefly covered in this text. Diamonds are undoubtedly the king of all jewels. The primary source of diamonds is kimberlite, a strong igneous rock that occurs as funnel-shaped pipes that extend deep into the Earth's crust (**Case 5.4**). Diamond crystals are scattered sparsely through the rock. The high price of diamonds makes it economic to follow kimberlite pipes deep into the Earth. Once eroded from kimberlite, diamond is highly durable, remaining in the sediment in which it may be to some extent concentrated. It is commonly mined from placer deposits in stream and beach gravels. Mining of these placer deposits may be by conventional excavation equipment or dredging. Vacuum devices are operated off the coast of South-West Africa to extract diamonds from sea bed placer deposits (**Figure 5.3**). Diamond grains are separated from other minerals by various physical processes, with the final sorting carried out manually.

The high price of diamonds makes it economic to follow kimberlite pipes deep into the Earth.

Fossil Fuels

The predominant fossil fuels are coal, petroleum, and natural gas. Coal mining is addressed in more detail elsewhere in this book; the environmental impacts of coal mining are similar to those associated with the mining of metalliferous ores. Peat mining, still

CASE 5.4
The Kimberly Diamond Mine in South Africa – The Largest Excavation by Hand

In 1866, Erasmus Jacobs found a small white pebble on the banks of the Orange River near Hopetown. The pebble turned out to be a 21.25 carat diamond. In 1871, an even larger 83.50 carat diamond was found on the slopes of a nearby hill, Colesberg Kopje, and led to the first diamond rush into the area. As miners arrived in their thousands, the hill disappeared, and became known as the Big Hole. A town, New Rush, was formed in the area, and was renamed Kimberley on 5 June 1873, after the British Secretary of State for the Colonies at the time, John Wodehouse, 1st Earl of Kimberley.

When the miners started to work the Kimberley diamond pipe, nobody knew how deep it would go. There seemed to be no end to the diamonds and people became unbelievably rich. The deeper the workings the more it resembled the inside of an ant heap. Up to 30,000 men were working day and night to clear the rubble and rock. When mining stopped the Big Hole was about 800 metres deep and more than 14 million carats of diamonds have been extracted.

Source: Based on Wikipedia.

FIGURE 5.3
Seabed Mining near the South-West Coast of Africa, and Traditional Mining by Hand

Eroded from kimberlite, diamond becomes an ideal placer mineral in streambeds and in beach and coastal sands and gravels. Vacuum devices are used to excavate diamonds-bearing sediment from sea bed placer deposits.

Photo Credits:
www.debeersgroup.com
Ira de Reuter 2007 (photo to the right)

extensively applied in Northern European countries to generate fuel for more than 100 peat fired power plants, is closely related to coal mining.

Although coal and petroleum would not be classified by chemists as minerals, they are frequently referred to as fuel minerals in the mining industry (**Case 5.5**). It may also be argued that groundwater should be considered as a mineral, and groundwater abstraction should consequently be subject to mining legislation. However, the main distinction here is that groundwater is (usually but not invariably) a renewable resource, which is harvested, rather than mined. The exploitation of petroleum and gas, while originating from mining, has evolved into a separate highly specialized discipline involving technologies quite different from those used in mining. Oil and gas extraction will not be further discussed in this book. However, it will become apparent that many of the environmental and social issues associated with mining are similar to issues faced by the oil and gas industry.

Many of the environmental and social issues associated with mining are similar to issues faced by the oil and gas industry.

CASE 5.5
Oil Sand Mining in Canada

Nothing demonstrates the blurred boundary between fossil fuels and minerals extraction better than oil sand mining in Canada, home of large oil sand deposits. The vast majority of Canada's oil sands are located in Alberta, where they underlie an area larger than Florida. Canada's oil sands are the second single largest hydrocarbon deposit on Earth, second only to the vast oil resources in Saudi Arabia. While this resource is next door to the biggest market for oil products, the USA, large-scale development started less than 10 years ago. Why? The costs of extracting oil from sand are tremendous. Depending on the depth of the reserves, oil sands are either surface mined from open pits (strip mining) or heated so the bitumen can flow to a well and be pumped to the surface (*in situ* extraction). On average it requires two tonnes of oil sand to produce one barrel of oil. Crude bitumen is extracted from the mined oil sands through a process that essentially mixes the oil sands with hot water to wash the bitumen from the sand. Producing the final synthetic crude oil from bitumen requires two further stages of upgrading, hydro-cracking and hydro-treating. Synthetic crude oil is then refined to the final petroleum products.

Source: Woynillowicz *et al.* 2005; Photo Credit: Melina Mara 2005, The Washington Post

5.2 MINING METHODS

The most common mining methods are surface, underground and solution mining subdivided into various classes and subclasses as illustrated in **Figure 5.4**. Since the range of environmental concerns associated with mining is discussed in detail in Chapter Thirteen, the following sections address only those environmental impacts associated with a particular mining method.

Surface Mining

Surface mining has been discussed in detail in Chapter Four and we will revisit the two main types of surface mining, open pit and strip mining, only briefly.

Iron ore, stone, gravel and many other minerals and ores are extracted by open pit mining, in which a hole is excavated into the Earth's surface. Overburden, if present, is stripped and placed outside the pit area to uncover the mineral deposit. The pit excavation usually produces additional large volumes of waste rock that need to be removed to provide access to the ore and to create stable slopes. The ore is usually separated from waste materials (either overburden or adjacent host rock) within the pit during excavation. Rock fragmentation (that is drilling and blasting) and laboratory testing of samples recovered during the drilling programme prior to blasting allows the mining engineer to classify waste rock and ore type for selective placement or, respectively, treatment. It is at this stage that the detailed characteristics of the ore body become apparent.

Strip or open cast mining, used for near-surface laterally extensive deposits such as coal seams, stratified ores or lateritic deposits such as bauxite or nickel laterites, resembles open pit mining but differs in one unique aspect: overburden is hind-cast or transported a short distance into adjacent mined-out panels. Mine rehabilitation can be carried out progressively at the same rate as mining.

Most mineral products and coal are extracted by surface mining. Surface mines are large-scale operations moving very large volumes of materials. By contrast, underground mining produces much lower volumes. Climate, topography, and existing land cover and use are more critical to surface than underground mining operations. In surface mining, most physical and chemical environmental impacts are directly related to extracting and relocating large amounts of rock materials.

Most mineral products and coal are extracted by surface mining.

Underground mining

Underground mining is more complex than surface mining. While surface mining can be characterized as a mass production of ore and associated mine waste at minimal cost, underground mining aims to target the main ore body with minimal waste rock removal. Overall mining costs may be similar; however, the unit excavation cost in underground mining is much higher.

Underground mining aims to target the main ore body with minimal waste rock removal.

Access to underground mines may be by means of vertical shafts, horizontal drives also known as adits, or by declines which are inclined tunnels. In a shaft mine, one or more vertical shafts connect the deposit to the surface. The ore is fragmented underground and then hoisted in buckets or skips, up through the shaft to the surface. Where adits or declines are used the broken ore can be hauled out of the mine in rail trucks or conventional trucks. If further crushing takes place underground, conveyors may be used to transport the ore along adits or declines. Each underground mine is different in the way ore is accessed and fragmented, but underground mines differ little in their environmental impacts.

FIGURE 5.4
Common Mining Methods

Most of the metals, coal, and industrial minerals that we use have been extracted by surface mining, predominantly by open pit and open cast mining. Surface mining is a large-scale operation, moving a staggering volume of materials, far more materials than underground mining.

Open Pit Mining

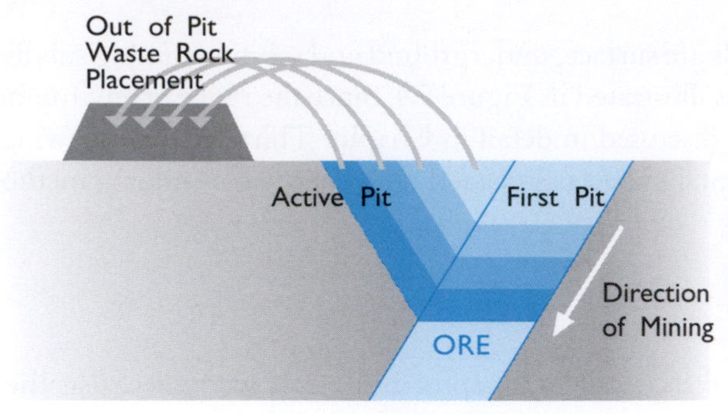

Open Pit Mining involves digging a hole in the Earth's crust to access and extract ore. Although the basic concept of open pit is quite simple, the planning required to develop a large deposit for surface mining is a very complex and costly undertaking.

Open Cast Mining

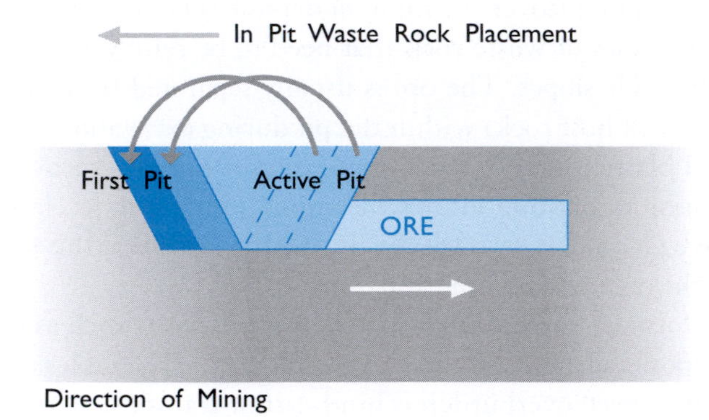

Strip Mining or Open Cast Mining is used for near surface deposits of predominantly lateral extent such as coal seams and lateritic deposits. Open cast mining resembles open pit mining but differs in two aspects: overburden is cast or hauled directly into adjacent mined out panels; Mine rehabilitation can progress at the same rate as mining.

Underground Mining

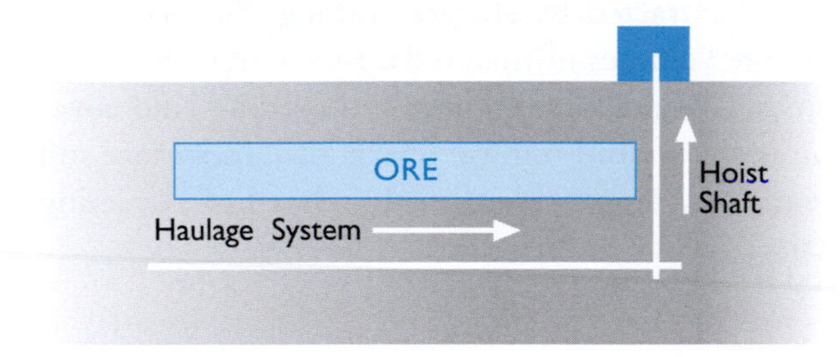

Underground Mining involves a mine opening at the surface, from which ore is removed that has been broken apart underground. Underground mining aims to target the main ore body with minimal waste rock removal.

Solution Mining

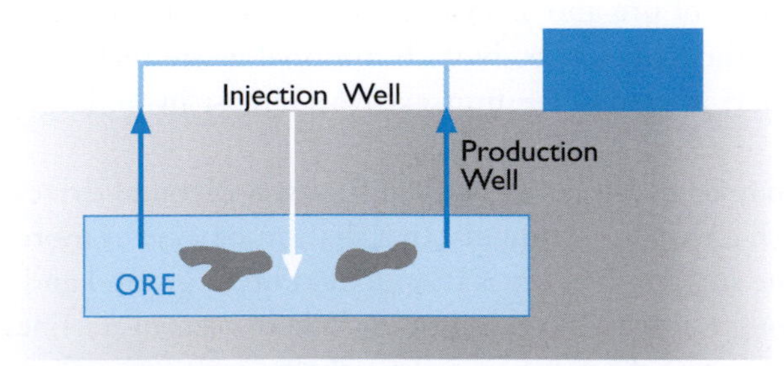

In Situ Leaching originally involved solution mining techniques used for extracting soluble ores such as uranium, potash and salt in situations where conventional mining methods would not be economic.

Heap Leaching

Heap leaching is a relatively common process used for the recovery of base metals and precious metals (gold and silver) from amenable low grade ores or, occasionally, from previously processed tailings. Heap leaching is a single stage process in which the ore is in a stationary heap and the solvent percolates through the heap. Some gold mining projects use heap leaching for low grade ores with conventional vat leaching for higher grade ores. At other projects where overall grades are lower, all ore may be treated by heap leaching. Oxidizing bacteria may first be used to decompose sulphide minerals, thereby releasing metals and facilitating the leaching process.

In the heap leaching process, ore is placed to form heaps on prepared leach pads. Leach solution applied to the surface of the heaps by drip irrigation devices or sprinklers, percolates through the ore and emerges from the base of the heap as 'pregnant solution'; this solution is collected in drains which discharge to a pregnant solution pond, from where it is pumped to recover the metals. After processing, the 'barren' solution is stored in a separate barren solution pond from where it is pumped for re-use. Amenable ores are usually oxidized. For gold and silver ores, weak cyanide solutions with a cyanide (CN-) concentration typically in the range of 0.02% to 0.05% NaCN are used. The pH value is also important. Most cyanide leaching is carried out at an alkaline pH of between 10 and 11, depending upon lab testing of individual ores and the optimum leaching/chemical use rates.

Critical requirements in the leaching process are that the ore is crushed sufficiently that the process solution is able to contact all the target minerals; that percolation proceeds evenly through the heap; and that percolation rates result in partially saturated conditions. In general, these requirements are achieved by crushing to produce a range of particles in the gravel and coarse sand sizes which provide a permeability of 10^{-4} cm/sec, or more. Ores containing excessive fines (silt and clay sized particles) may be pre-prepared by agglomeration to form larger particles that enable appropriate permeability to be achieved.

Following completion of leaching, the heap is rinsed by applying water or in areas of higher precipitation, may be allowed to rinse naturally. Collection and treatment continue until the quality of drainage from the heap achieves discharge standards. The rinsed heap may be left in place for rehabilitation or removed to a separate final disposal site.

The main features of a heap leaching operation are:

- Ore preparation, involving one or more stages of crushing, and/or agglomeration, if appropriate;
- Leach pads, which provide the foundation for the heap, and contain the solution by means of a 'liner system' which may use clay, concrete, asphalt or a geomembrane such as polyethylene or polyvinyl chloride (PVC);
- The heap itself, which can be placed by end-dumping using trucks, front end loaders, or conveyors, the main considerations being to avoid damage to the liner and to avoid segregation of particles;
- The solution application system which may involve sprinklers, or drippers to provide application rates that produce partially saturated conditions, and avoid surface ponding;
- Solution collection system involving under-drains installed immediately above the pad liner, either as a gravel layer or a series of closely spaced perforated pipes, draining to the pregnant solution pond;
- Pregnant solution storage, comprising a pond, lined by clay or geo-membrane to minimize seepage;
- Processing circuit for metal recovery, and
- Barren solution storage, also comprising a lined pond, together with reagent dosing facilities.

Further details of the design and operation of heap leaching systems are provided in Hutchison and Ellison (1992), from which much of this section has been summarized.

Three approaches to leach pad design have been utilized, namely: reusable pads, expanding pads and valley leach dumps. These are depicted schematically in **Figures 5.5** and **5.6**.

The features and requirements of each system are compared in **Table 5.2**. The reusable pad system requires that the leached and rinsed ore be relocated to a separate spent ore dump.

The most critical design considerations for heap leach systems relate to the liners, which provide containment for leach pads, solution ponds and any drainage channels used for process solutions.

Environmentally, the most critical design considerations for heap leach systems relate to the liners, which provide containment for leach pads, solution ponds and any drainage channels used for process solutions. Common liner types include single liners composed of high density polyethylene (HDPE) or polyvinyl chloride (PVC), compacted clay liners, and composite liners involving compacted clay and geomembranes (Hutchison and Ellison 1992). Very low permeabilities may be achieved using geomembranes; however, the use of suitable clay soils, appropriately conditioned and compacted avoids the risks of rupture, puncture or tearing that apply to geomembranes. Selection of liner design depends on many factors including leach system, local environmental conditions, local availability of suitable soils, and availability of skilled geomembrane installation personnel.

Impacts

Construction and operation of heap leaching structures involve most of the same short-term and long-term impacts as other mining facilities, including loss of vegetation and faunal habitat, modification of landforms, soil profile changes, land use changes, and modifications to surface and subsurface drainage. These impacts can be largely reversed by rehabilitation during and following operations.

Heap leaching does involve additional risks of serious environmental damage including: (1) leakage of pregnant solution through the pad liner, or through solution pond or drain liners, causing contamination of underlying groundwater; and (2) discharge from or overtopping of solution ponds due to excess water, pump failure or damage to structures, causing contamination of downstream surface water.

These risks are minimized by: careful calculation of water balances under all conceivable operating scenarios; sizing of solution ponds to accommodate all predicted inflows plus an appropriate freeboard to provide additional security; selection of appropriate liner systems; close attention to construction control (particularly compaction of embankments and clay liners, and welding of geomembranes); and installation of appropriate instrumentation to monitor key parameters and provide early warning and/or automatic shutdown in the case of malfunction (**Case 5.6**).

Heap leaching operations are less common in areas of high rainfall because of the difficulty of storing and treating the large volumes of water entering the system from precipitation.

Heap leaching operations are less common in areas of high rainfall because of the difficulty of storing and treating the large volumes of water entering the system from precipitation. However, successful heap leaching operations have been conducted in both Indonesia and the Philippines.

Arguably heap leaching also offers a number of environmental and social benefits. Smith (2004) uses *The Seven Questions to Sustainability: How to Access the Contribution of Mining and Minerals Activities* (**Table 5.3**) as framework for evaluating mineral development involving heap leaching in terms of the goal of sustainable development and provides the following seven answers (for full text see original publication).

1. *Engagement* – Since the construction technologies are more accessible to local contractors, there are more opportunities for partnering between the owner and contractors, and less reliance on imported technologies and equipment. Thus, a heap leach operation is more accessible to the local community, improving opportunities for engagement.
2. *People* – Heap leaching is an inherently more 'hands-on' process then milling. It is the 'low technology' solution for low grade ores, and as such requires more people doing things

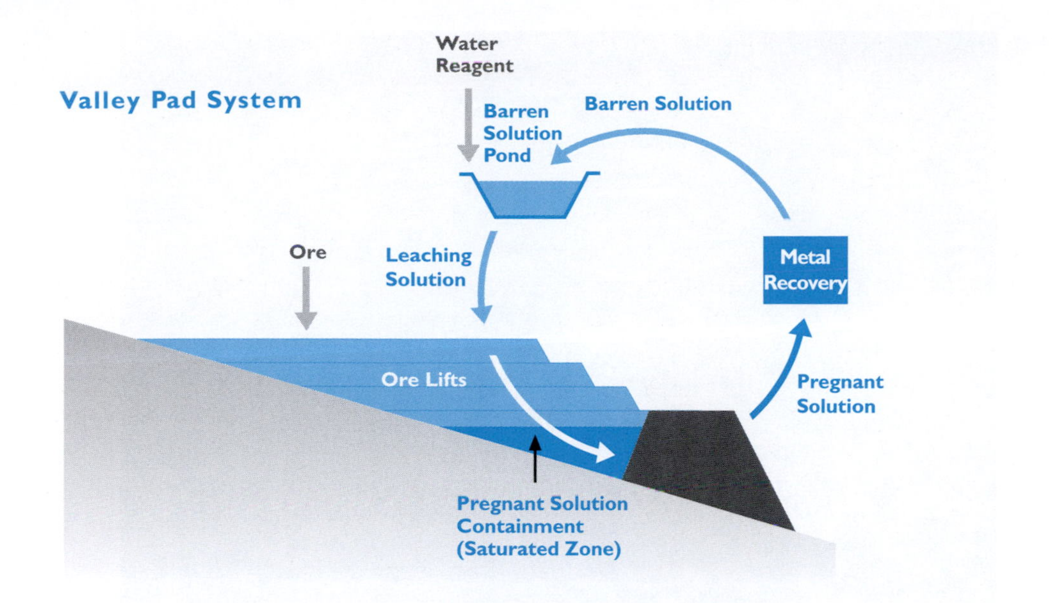

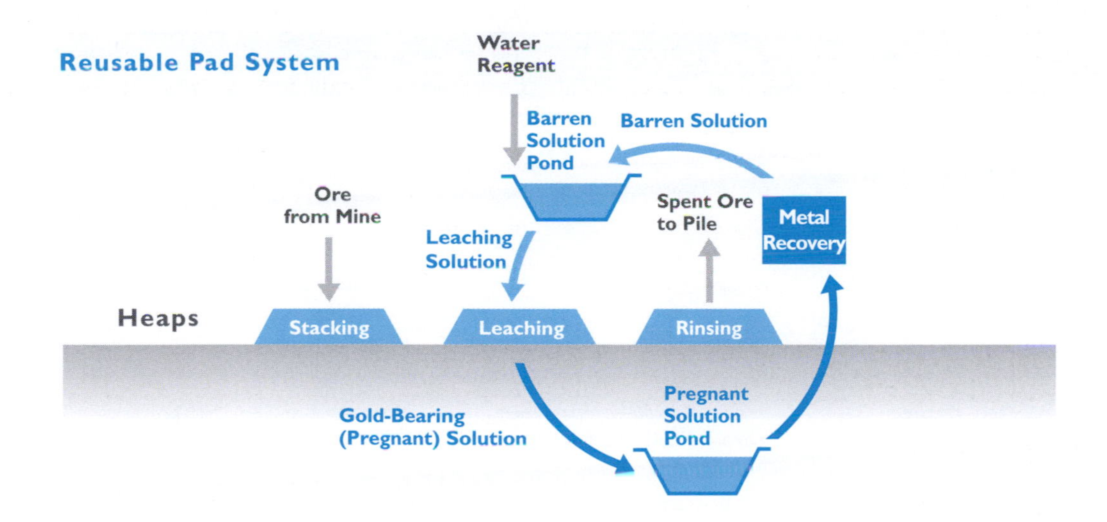

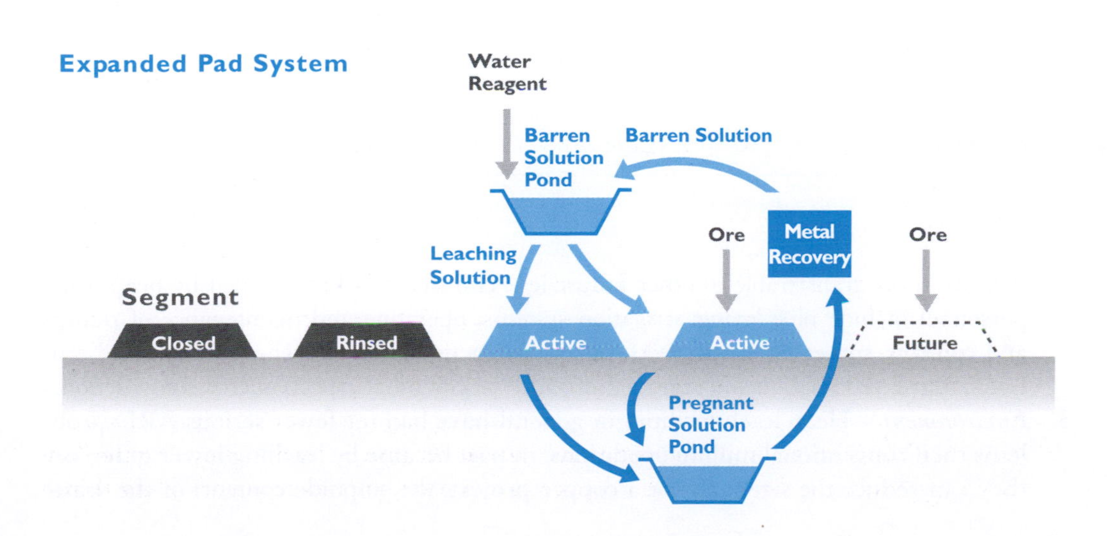

FIGURE 5.5
Three Approaches to Leach Pad Design

Heap leaching is a relatively common process used for the recovery of base metals and precious metals (gold and silver) from amenable low grade ores or, occasionally, from previously processed tailings.

FIGURE 5.6
A Bird's Eye View on a Field of Expanding Heap Pads

Photo credit:
www.airphotona.com

TABLE 5.2
Comparison of Leach Pad Approaches

Consideration	Reusable Pad	Expanding Pad	Valley Leach Method
Land area required	Small	Large	Moderate
Topography required	Flat or gently sloping	Flat or gently sloping	Incised valley including steep terrain
Climatic factors	Not suited to high rainfall situations[1]	Requires dry climate	Suitable for wet climates
Pond sizes	Small	Large	Ponds not usually required
Liner requirements	High impact resistance required due to repeated exposure/traffic	Relatively low stress environment may allow lower resistance liner	High resistance due to high pressures
Ore variability	Requires ore of consistent leachability	Can be adapted to ores of varying leachability	Can accommodate ores of varying leachability including ores requiring extended leach times
Rehabilitation	Not possible until operations cease (except for spent ore storage)	Segments can be rehabilitated progressively	Outslope can be rehabilitated progressively
Other	Requires separate spent ore storage	Low initial capital cost	Requires retaining impoundment

[1] Installation of roof or other cover above heap can overcome this limitation.

that are more transferable to other industries. Transferable skills learned by heap leach personnel include pipe laying, irrigation systems, operating and maintenance of pumps and controls, surveying, earthworks, liner construction and maintenance, slope and erosion control, reclamation and revegetation, and various other aspects of civil construction.

3. *Environment* – Heap leach facilities in general have had far fewer serious ARD problems then conventional milling operations, in part because by leaching lower grade ore they can reduce the size (and for a copper project, the sulphide content) of the waste

dumps. Spent heap leach ore from gold operations is usually strongly alkaline; mixing waste types can help compensate for acidic waste rock (MMSD 2002). In terms of catastrophic failures, the history of heap and dump leaching is, in comparison to tailings dams, very good. There have been no significant leach heap or dump slope failures and no failure-related fatalities. Spent heaps are also more stable and easier to reclaim than old tailings deposits, if for no other reason than their self-draining characteristics. Thus, simply reducing reliance on conventional tailings disposal is of itself a move towards more sustainable development (Leduc *et al.* 2004).

4. *Economy* – The obvious points here are that heap leach technology allows more ore to be processed since a lower cut off grade results, allowing a longer life or a larger operation, or both. It is also less capital intensive and thus less sensitive to commodity price fluctuations and generally a lower risk investment. Beyond these obvious answers lie some other important considerations. Heap leach technology shifts the balance of investments, de-emphasizing capital in favour of operating expenses. Payments for operations are something that the communities can generally share in to a greater extent than initial capitalization. Heap leaching also increases employment beyond the process circuit.
5. *Traditional and non-market activities* – By expanding employment in areas with transferable skills, a more sustainable workforce results. While the tools and techniques used at a modern mine are not directly applicable to traditional activities, certainly they have more in common in the case of heap leaching then mill operations and maintenance.
6. *Institutional arrangements and governance* – The types of problems inherent to a heap leach project tend to be more manageable at a local level. The typical threat from a leach heap is excessive leakage from the leach pad during its operating life. At closure, heaps are usually self-draining and thus the potential for post-closure leakage is significantly reduced.
7. *Synthesis and continuous learning* – Heap leaching has the potential for reducing all types of impacts, improving economic benefits in the local community and reducing economic risk. It also leaves a more secure site after operations and thus reduces long-term environmental liability. At the same time, it uses technologies that are both more locally available and have more applications outside mining. Since the projects are less capital intensive and typically subject to expansions or revisions in the leach pad and stacking operations annually or bi-annually, project reevaluation is a deeply engrained part of the heap leach culture. Expanding this to include the local community should be an easy step.

CASE 5.6
Heap Leaching Applied to Barrick Gold's Pierina Mine in Peru

Barrick's Pierina Mine in Peru uses heap leaching to extract gold. Pierina was expected to produce more than 800,000 ounces of gold at a total cash cost of less than $ 50 per ounce in 1999, making it the world's lowest-cost major gold mine. The ore is stacked in a lined containment area behind a retention dam as depicted right. A leach solution is applied to the top of the ore and allowed to percolate through the heap. As the solution migrates through the ore, it leaches the gold and silver from the rock and holds it in a solution. The gold-bearing solution is collected at the base of the leach pad in the pore space within the heap. The pregnant solution is pumped to the gold recovery plant where suspended solids are removed and the solution is then treated in a conventional Merrill-Crowe precious metal circuit. The same valley-fill system was successfully used at Barrick's Mercur Mine in Utah. (Article and photo borrowed from METSOC 2006)

TABLE 5.3

Seven Questions to Sustainability (from MMSD North America) – In 1999 the International Institute for Environment and Development was commissioned by the World Business Council for Sustainable Development to undertake the Mining, Minerals and Sustainable Development (MMSD) project. MMSD North America was then formed as a partnership of the International Institute for Sustainable Development and the Mining Life-Cycle Center at Mackay School of Mines, which produced *The Seven Questions to Sustainability: How to Access the Contribution of Mining and Minerals Activities*

Engagement	Are engagement processes in place and working effectively?
People	Will people's well-being be maintained or improved?
Environment	Is the integrity of the environment assured over the long term?
Economy	Is the economic viability of the project or operation assured, and will the economy of the community and beyond be better off as a result?
Traditional and Non-market Activities	Are traditional and non-market activities in the community and surrounding area accounted for in a way that is acceptable to the local people?
Institutional Arrangements and Governance	Are rules, incentives, programmes and capacities in place to address project or operational consequences?
Synthesis and Continuous Learning	Does a full synthesis show that the net result will be positive or negative in the long term, and will there be periodic reassessments?

The above arguments perhaps over-state the virtues of heap leaching while understating the potential problems. Heap leaching does have a significant role in the processing of oxidized low grade gold and copper ores, but is unlikely to find further applications.

Closure

Rehabilitation of spent ore from heap leaching operations is relatively straightforward. Rinsing is the first step as it utilizes the same application and containment facilities used for leaching. Generally, up to eight pore volume displacements (Hutchison and Ellison 1992) will remove all but the smallest trace of reagents. Oxidizing agents such as hypochlorite, peroxide or specially bred strains of reagent destroying bacteria may be added to the rinse solution to reduce the required volume of rinse water. Intermittent rinsing is more effective than continuous rinsing.

Following rinsing, the piping and other surface installations are removed as are the liners from solution ponds and drains. The surface is then re-graded to construct a landform suited to the required land use. Consideration of drainage from the heap is required, given that the pad is usually retained, preventing downward drainage beneath the heap. This results in seepages from the base of the heap, unless measures are implemented to collect and discharge this drainage water. Generally, drains and ponds are back-filled using the same materials removed in their original construction, with further filling using spent ore, if appropriate. The heap is usually graded to produce a rounded, natural-looking profile. Following application of a layer of topsoil, if available, the entire surface can then be re-vegetated.

Dump Leaching

Dump leaching refers to leaching that takes place on an unlined surface. The term 'dump leaching' derives from the practice of leaching materials that were initially deposited as waste rock; however, it is now also applied to run-of-mine, low-grade sulphide or mixed grade sulphide and oxide rock dumped specifically for leaching. Copper dump leaches are typically large, with low-grade rock piled into heaps ranging from 10 m to over 30 m in height. These may cover hundreds of hectares and contain millions of tons of waste rock and low-grade ore (Biswas and Davenport 1976).

Dump leaching may involve the application and recovery of leach solutions as for heap leaching, or it may involve recovery and processing of leachate from rainfall percolating through the dump. Generally, the waste rock will not have been crushed, so that particle sizes will be relatively large and metal recovery will be low compared to heap leach operations. Copper is the main metal recovered in this way, with sulphuric acid the main leach solution. The principles that apply are the same as for heap leaching. Groundwater and surface water contamination are the main potential impacts of concern.

Vat Leaching

Vat leaching, a hydro-metallurgical rather than a mining process which is further discussed in Chapter Six, works on the same principles as the dump and heap leaching operations except that it is a high-production-rate method conducted in a system of vats or tanks using concentrated leaching solutions (solvent). Vat leaching is typically used to extract copper from oxide ores by exposing the crushed ore to solvents (concentrated sulphuric acid) in a series of large tanks or vats. Similarly, gold is extracted in vats, using cyanide solutions as discussed elsewhere. The vats are usually designed in a series configuration, which acts to concentrate the copper content of the solutions as a function of ore-solvent contact time (USEPA 1989). In contrast to heap and dump leaching, vat leaching operations may be conducted under a number of environments, including slightly sub-atmospheric, atmospheric, or super-atmospheric pressures, and under ambient or elevated temperatures (Weiss 1985). Industrial leaching processes applying countercurrent stage-wise leaching can deliver the highest possible concentration in the pregnant solution (also termed extract), but multi-stage hydrometallurgical processes are often expensive, technically challenging, and only applicable to large-scale mining operations. Agitation accelerates leaching, but due the high energy consumption for ore preparation and leaching, the method is only warranted for high-grade ore (see **Figure 6.7**).

In contrast to heap and dump leaching, vat leaching operations may be conducted under a number of environments.

In situ Leaching and Solution Mining

In situ leaching (ISL) involves dissolving minerals from the ore without its removal from the ground. *In situ* leaching includes the leaching of either disturbed or undisturbed ore. In either case, *in situ* leaching allows only limited control of the solution compared to a lined heap leach type operation. Theoretically, this approach to mineral extraction could be applied to a range of minerals. However, to the present time, its commercial use has been confined mainly to various uranium minerals which are soluble in acid or alkaline solutions. To a lesser extent ISL has been applied to recover copper from low-grade oxidized ore. In each of these cases solution is pumped into the ore body through a network of injection bores or wells and then pumped out together with salts dissolved from the ore body through a series of extraction bores (production wells) as shown in **Figure 5.4**. Further processing of the pregnant solution is usually carried out at the surface to precipitate the mineral product (Yellowcake -U_3O_8- in the case of uranium ISL operations) from the solution which is subsequently re-used.

The main issues of ISL relate to ensuring that the process solutions do not migrate beyond the immediate area of leaching.

The main issues of ISL relate to ensuring that the process solutions do not migrate beyond the immediate area of leaching.

Solution mining has been used for many decades to extract soluble evaporite salts such as halite ($NaCl$), trona ($3Na_2O \cdot 4CO_2$), nahcolite ($NaHCO_3$), epsomite ($MgSO_4 \cdot 7H_2O$), carnallite ($KMgCl_3 \cdot 6H_2O$), and borax ($Na_2B_4O_7 \cdot 10H_2O$) from buried evaporite deposits in various parts of the world including the UK, Russia, Germany, Turkey, Thailand, and the USA. The process is relatively simple, involving the injection of low salinity water with or without pre-heating, the removal of pregnant solution through extraction wells, and the recovery of salts by means of evaporation, in the same way that common salt and other salts are recovered from sea water by the so called solar salt process (**Figure 5.7**). Where there are difficulties in establishing a connection between injection and recovery bores, a different technique can be used, involving concentric pipes, with the solution being pumped up through the inner pipe. Horizontal bores are commonly used, drilled within horizontal or gently dipping evaporite strata. Areas where evaporites have previously been mined by conventional room and pillar underground methods may be further mined by introducing water which, after contact with pillars in the mine, is pumped out for recovery of the dissolved salts.

There are generally no solid wastes from this form of mining. Low volumes of liquid wastes may be produced which are brines containing salts which are uneconomic to separate or for which there is no market. Such liquid wastes are usually re-injected, either into the same stratum which is being leached or into a separate aquifer.

The main environmental issues associated with solution mining of evaporites relate to surface or shallow groundwater contamination in and around the evaporation ponds. The evaporite solutions may be highly corrosive and phyto-toxic. They can react with natural soil materials used to construct evaporation basins and may migrate to surrounding areas through seepage, overflow or windblown spray. These same impacts are common in solar salt production operations. Surface subsidence including formation of 'sink holes' is another potential impact after prolonged solution mining, particularly if the solution mining follows underground mining.

FIGURE 5.7
Evaporite Mining in Arizona

Photo Credit:
Jim Wark, www.airphotona.com

Seabed Mining

Other mining methods have been subjects of interest for a number of research groups for several decades, with primary attention being to deep seabed mining, due to its potential to recover large reserves of minerals that could provide a future alternative source of metals. A deep-sea mining operation would offer a variety of challenges, owing to distant locations (thousands of kilometres from the coast), deep-sea mineral occurrences (5 to 6 km of water depths), extreme physical and chemical conditions (high pressure, low temperature) and unknown environmental settings. Deep-sea mining and other non-commercial mining concepts are not covered in this text, as they are still in their infancy. The interested reader may find these specialized references useful for further study (e.g. ISOPE 2002; Cronan 2000).

5.3 ARTISANAL MINING – MINING OUTSIDE OF ESTABLISHED LAW

Artisanal mining is defined as small-scale mineral extraction, using mainly manual methods, carried out by individuals or small family groups. Other terms in common usage are 'small-scale mining' and 'illegal mining', the latter when artisanal mining is carried out on a tenement granting exclusive mining rights to others, usually a mining company. The main products of artisanal mining are gold, silver (usually as a by-product), tin and gemstones, although coal is also produced by small-scale miners in China and Indonesia.

Two contrasting categories of artisanal mining are recognized: (1) the 'Gold rush', which is a relatively short term phenomenon, characterized by large influxes of mainly unskilled, transient people hoping to 'strike it lucky'; and (2) traditional small-scale mining by skilled practitioners, using methods adapted to local conditions, usually operating within an established mineral field, sometimes for generations.

Gold Rushes

Famous gold rushes occurred in California in the late 1840s, followed by a succession of gold rushes in central Victoria, Australia in the 1850s, and the Klondike gold rush in the Canadian Yukon at the end of the 19th century. In the late 20th century, gold rushes have occurred in Brazil, Papua New Guinea, the Philippines, and Indonesia. Remarkably, the nature of gold rushes has changed little since the 1840s, while corporate mining has been completely transformed.

The typical gold rush phenomenon follows a familiar pattern:

- Gold (or precious stones) are discovered at or near the surface in an area of no previously known occurrence; often, the initial discovery will be an alluvial or placer deposit, as these are readily identified by prospectors (**Case 5.4**);
- As the news spreads, thousands of people flock to the area; the people may come from diverse backgrounds and include adventurers, veterans of previous gold rushes and the extremely poor, together with large numbers of opportunists, including criminals;
- Only the first arrivals will normally be successful in securing land for mining. Most people arrive after all available, prospective land is taken; late arrivals may return to their original homes but, more commonly, they stay in the hope that opportunities will arise;
- Initial mining techniques are simple and include excavation by pick and shovel, panning, dry blowing or cradling to collect free gold or gemstones;

Often, the initial discovery will be an alluvial or placer deposit, as these are readily identified by prospectors.

- Socially, gold rushes are characterized by lawlessness, unsafe work practices and poor hygiene;
- Only a very small proportion of miners earn significant wealth; more successful are the money lenders and traders, while the majority of people make very little money;
- Once near-surface deposits have been mined, the population declines rapidly;
- As mines are deepened, miners tend to organize into groups or co-operatives, with division of labour into miners, carriers, rock breakers, equipment operators, and processors. At this stage new equipment is introduced for rock fragmentation, such as small grinding mills known as trommels, larger stamp mills, and various jigs and sluice boxes to recover free gold. Mines usually comprise adits or shafts that follow the lode. Support and ventilation are minimal;
- Population declines further as mines encounter stronger rock, groundwater or other constraints, which reduce profitability; typically the end of small-scale mining corresponds to the base of the oxidized zone;
- For major ore bodies continuing at depth, small-scale mining may give way to corporate mining, with one or more companies floated to raise funds for larger-scale development required to access the ore bodies at depth.

The above scenario may take place over five to ten years. A gold rush at Diwalwal in a remote part of Mindanao, Philippines in the 1980s attracted more than 140,000 people (see also Case 7.6). Many died, some when mine workings became flooded or collapsed; others were murdered or succumbed to diseases including those relating to inhalation of mercury vapours.

A characteristic of gold rushes is the low recovery of gold that is achieved.

A characteristic of gold rushes is the low recovery of gold that is achieved. Although the ores mined may have relatively high grade (25 g/T or more), recoveries achieved are typically 50% or less. This means that the residues or tailings contain significant remnant grades, often justifying re-treatment. In most modern gold rushes, the mercury amalgam process has been used to improve the recovery of fine gold and silver particles (**Figure 5.8**). However, even with mercury extraction, considerable gold, silver, and mercury remain in the tailings. During the 1980s in the Philippines, gold grades in tailings from artisanal mining were sufficiently high to attract the attention of operating mining companies, who routinely purchased tailings from small-scale miners operating in many parts of the country.

Gold rushes on mining tenements have led to conflict between companies with legal mining entitlements and illegal miners. At Mt Kare in Papua New Guinea, the mining company RioTinto, abandoned its plans for mining after a large influx of people to the area. Similarly, PT Tambang Tondano Nusajaya, a subsidiary of Aurora Gold, from Australia, was unable to prevent an 'invasion' of small-scale miners at its high grade Talawaan prospect near Manado in Indonesia (**Case 5.7**). Other operations have been able to co-exist with small-scale miners. For example, Newmont at its Minahasa Gold mine in Sulawesi, Indonesia, allowed small-scale miners to exploit several of its smaller ore bodies, while the company concentrated its attentions on the main Mesel ore body. Similarly PT Aneka Tambang at its Pongkor gold project in West Java conducts mining in underground stopes, while small-scale miners actively exploit near-surface ores. Similarly, local people have extracted gold from mine wastes produced by mining companies, such practices usually being tolerated if not encouraged by the companies involved. Examples include people fossicking and hand-picking ore from waste rock dumps at Mt Muro in Indonesia, and small enterprises using simple gravity separation techniques to extract gold particles from tailings at the Atlas Copper Project in the Philippines and Freeport's Grasberg Project in Indonesia.

Narrow, gold-bearing veins may prove unattractive to operators using large-scale methods, but can prove lucrative to artisanal miners. This can attract small-scale miners to mined out

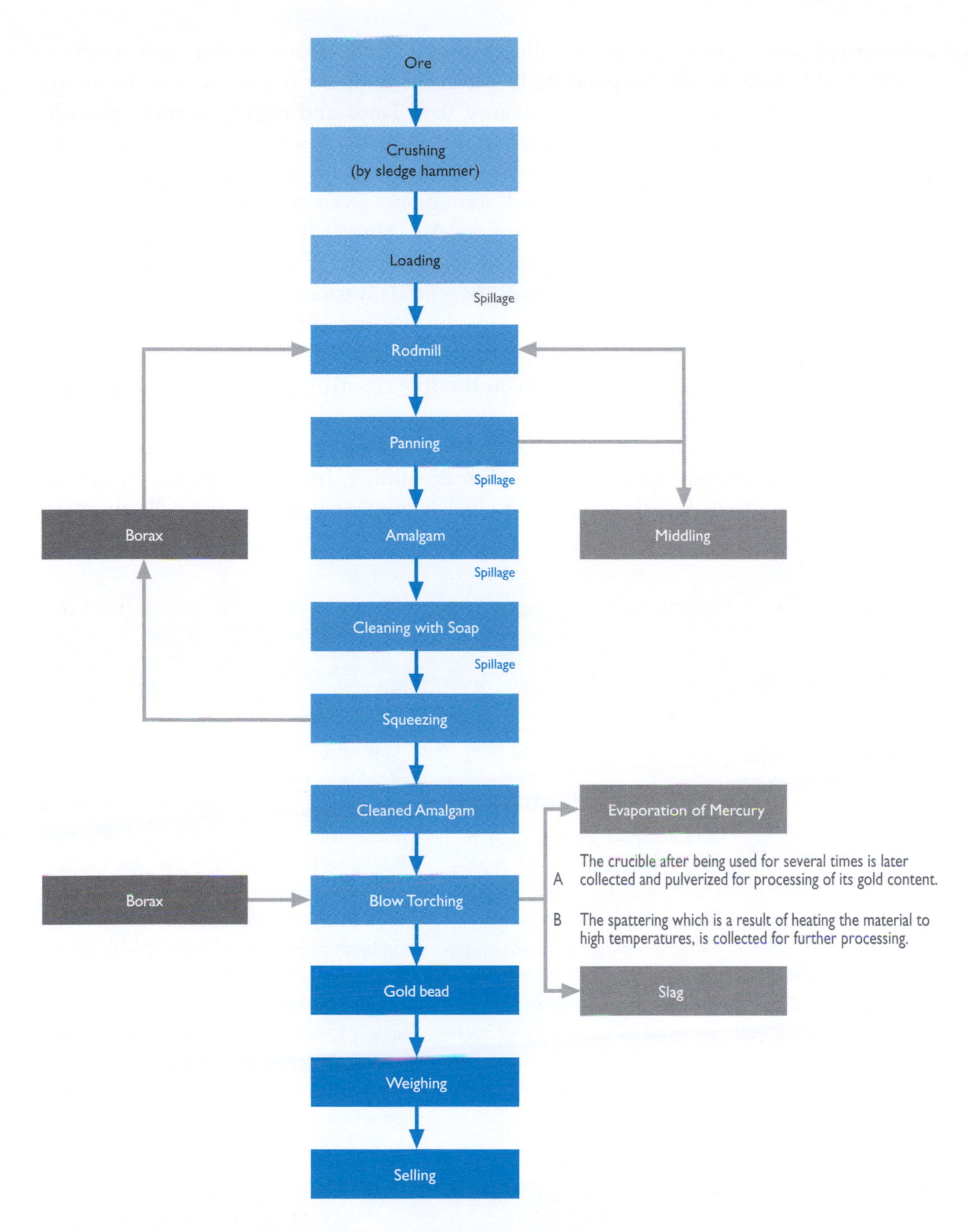

FIGURE 5.8
Recovery Process Flowchart: Amalgamation Gold Rush Areas

areas, even close to areas of active large-scale mining. During the final year of the Mt Muro operation, small-scale miners were extracting ore from the upper benches of the Kerikil pit, under cover of darkness, while the company was still mining in the base of the pit. When mining ceased in 2003, the small-scale miners stayed and are still active at this time (2008).

In Indonesia and some other countries, mining companies have had difficulties enforcing their legal rights against influxes of small-scale miners. Sometimes the small-scale miners are local people with a history of mining in the area. Despite the strict legal situation, the historical connection provides a strong moral argument for tolerating, if not encouraging, their activities. More often, however, the influx comprises outsiders who may or may not be welcomed by the local communities. Companies are usually reluctant to seek

Companies are usually reluctant to seek government intervention to remove 'illegal miners'.

government intervention to remove 'illegal miners'. One reason is that such intervention invariably leads to adverse publicity in which the company is portrayed as the greedy capitalist, exploiting the poor by occupation of their lands and depriving them of traditional livelihoods. NGOs are quick to seize on these situations, sometimes visiting the sites with attempts to inflame the situation and invariably producing a biased story including half-truths, exaggerations, and un-verified accusations. Even well-respected aid organizations such as the Australia-based Community Aid Abroad have been involved in such campaigns, perpetuating erroneous allegations and ignoring the less newsworthy efforts of a company to reach an amicable and mutually beneficial outcome.

Even worse publicity follows when attempts by government forces to remove 'illegal miners', lead to violence and bloodshed. Again, such incidents are often exaggerated in press reports based on hearsay, rather than on-the-spot reporting.

In efforts to avoid conflicts between mining companies and small-scale miners, governments in several countries, including Thailand and Indonesia, have established 'Peoples' Mining Areas' specifically set aside for small-scale mining. In these cases, the operations may be closely controlled, with government officials supplying permits and collecting fees.

CASE 5.7
Illegal Gold Mining at Talawaan near Manado, North Sulawesi, Indonesia

The Talawaan gold ore body, located only 10 km from the Provincial capital Manado in North Sulawesi, Indonesia, was discovered by PT Tambang Tondano Nusajaya, an Indonesian subsidiary of Australia's Aurora Gold in 1997. Subsequent drilling delineated a medium-sized, relatively high grade ore body amenable to mining by open pit methods.

While exploration drilling operations were underway, in 1998, the site was invaded and occupied by large numbers of illegal miners. Efforts by the company to have the illegal miners evicted, were unsuccessful and the company abandoned its activities, essentially leaving the area to the illegal miners. The situation then rapidly became 'legitimized' in the sense that local government officials arrived on the scene, establishing order, setting aside areas for mining, mineral processing and retail services, issuing permits and collecting fees.

At its peak in 2001, about 3000 people were working at the site – mining, hauling, and processing ore. Up to 400 processing units involving multiple trommels (small electric powered rotating mills) were active at this time. Mercury was used to extract gold and silver from the milled ore. Solid wastes were dumped on-site with liquid effluents discharged to the nearby river. Subsequent investigations indicated widespread mercury contamination of surface water and river-bed sediment, at levels well above human health risk thresholds. Activities have since declined markedly as most of the high grade ore above the water table has been extracted.

Although low grade ore remains above the water table and high grade ore below it, the overall resource has been substantially depleted, to the extent that it may not be economically feasible, in the foreseeable future, to extract the remaining ore. In the meantime, the legacy from mercury contamination will persist for decades. The parent company (Aurora Gold) which no longer exists, received no revenue or compensation from Talawaan, despite having discovered the ore body, spending millions of dollars on its delineation, and owning the legal rights for its exploitation. Furthermore, largely as a result of this situation, confidence in Indonesia's Contract-of-Work system was severely diminished, and mineral exploration in Indonesia suffered a severe downturn from which it has not completely recovered.

FIGURE 5.9
Dredging for Gold on the Barito River in Central Kalimantan, Indonesia

In at least one case, the government actually carries out the mining, providing ore for treatment by local people. Similarly, as mentioned previously, companies such as Newmont, Aneka Tambang, and Aurora Gold have voluntarily allowed small-scale miners to exploit deposits within their tenements. It is relatively common for systematic exploration to identify mineral deposits, which because of their small size or remoteness from infrastructure, do not justify mining. Some of these deposits may be ideally suited to small-scale mining, providing a potential 'win – win' situation. An extension of this situation would be if tailings from small scale mining were then collected by the company for re-treatment. However, many companies would be reluctant to treat wastes from small-scale miners on the basis that they would be exposed to the associated risks and liabilities.

Modern technology has increased the range of methods available to small-scale miners. One environmentally damaging example is hydraulic mining or sluicing, in which high pressure water jets are used to excavate alluvial or oxidized ores. Another is dredging (**Figure 5.9**), where river bed or river bank sediments are dredged from a floating vessel. In both cases, the excavated sediment is passed over riffle boxes or similar sluicing arrangements to concentrate the heavy minerals – usually gold. Panning and mercury amalgamation may also be used in some of these operations.

Traditional Small-scale Mining

Traditional small-scale mining activities are characterized by community co-operation, a skilled workforce, relatively high resource recovery, and stability – even in circumstances of fluctuating prices. In fact, the best examples of traditional small-scale mining have proved to be 'sustainable' over many generations; few corporate mining operations have achieved comparable sustainability.

One of the best examples is the *Kankana-ey* artisanal mining community in Benguet Province, near Baguio in the Philippines, a region of extensive small-scale and corporate mining activity over a long period. There is evidence that these traditional gold mining activities date back more than 800 years (Caballero 1996). A flow chart, illustrating the steps followed by the *Kankana-ey* in mining and processing, is shown in **Figure 5.10**.

According to Caballero, a husband and wife and their extended kin work together in both placer and lode mining. While both men and women participate in mining, some

FIGURE 5.10

Flowchart of Traditional Small-scale Gold Recovery Process

Source:
CABALLERO 1981, 1984, 1989;
CABALLERO, J. 1991;
How (1991)

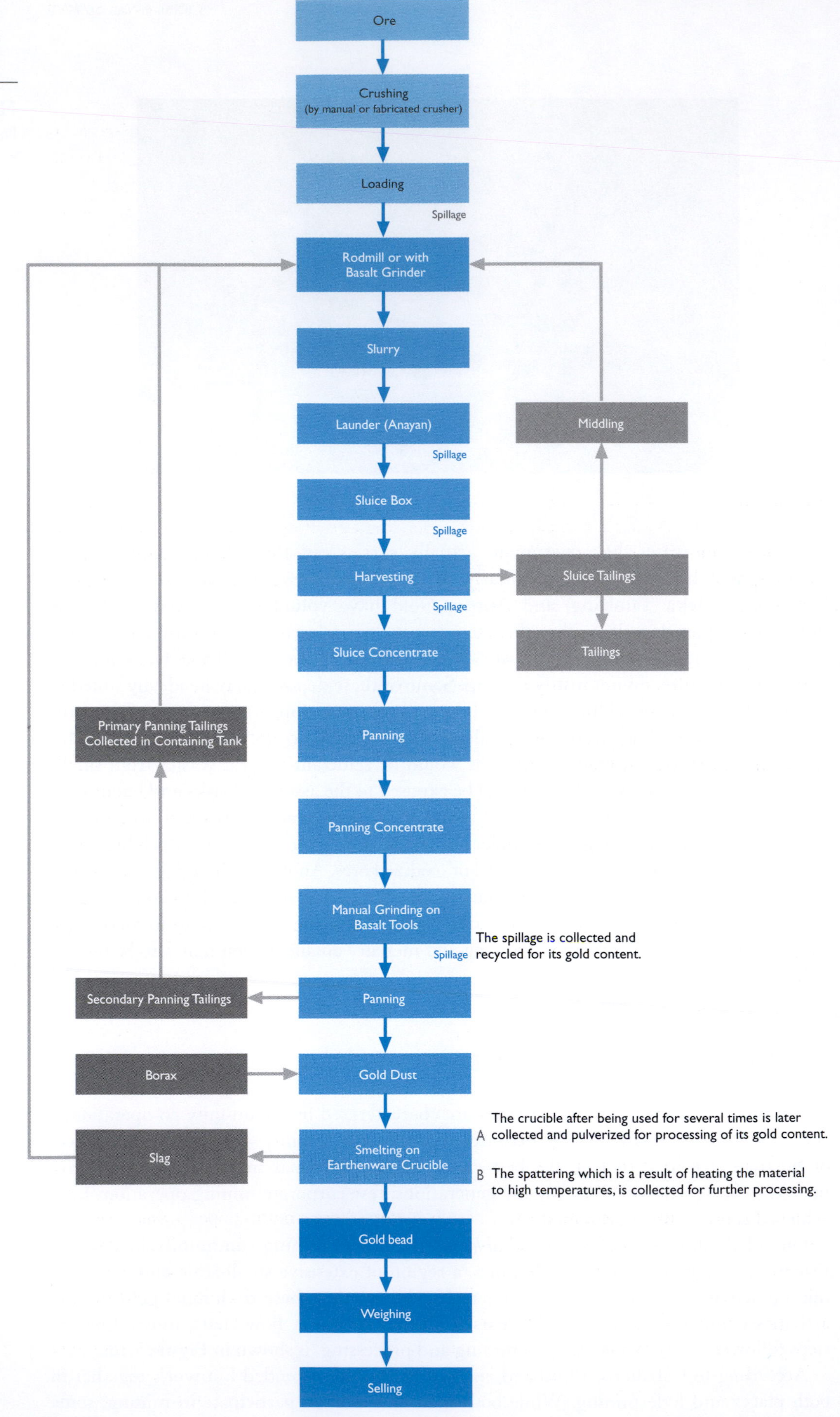

tasks are performed more by women. The processing of gold is a woman's job. Processing areas are located close to homes, facilitating integration of processing activities with child-rearing and household tasks. The man or woman who owns a tunnel is the 'supplier' who finances the operations including providing food for the workers. One feature of the *Kankana-ey* tradition is the sharing of ore and concentrate among community members. Older women invariably receive priority in distribution of both ore and concentrate.

The *Kankana-ey* have also been involved in co-operative ventures with mining companies, both as contract miners and as suppliers of tailings for re-treatment in Carbon-In-Pulp plants.

Environmental Impacts of Artisanal Mining

Clearly, with the wide range of techniques used by small-scale miners, resulting environmental impacts are also variable in type, extent, and severity. The worst excesses of the gold rushes cause the most severe adverse environmental effects with widespread surface disturbance leading to accelerated erosion and sedimentation of local streams. Rainfall on the typically 'pock-marked' surface creates numerous small pools which provide ideal breeding sites for mosquitoes. The situation is usually aggravated by uncontrolled disposal of wastes including sewage, which in the worst cases, leads to diseases such as typhoid and cholera.

With the wide range of techniques used by small-scale miners, resulting environmental impacts are also variable in type, extent and severity.

The use of mercury introduces a range of other problems, some due to occupational exposure, and others due to environmental contamination. Partly chemical and partly physical, gold concentration by amalgamation takes advantage of the fact that free gold particles amalgamate with mercury. Amalgamation consists of alloying mercury to gold and silver. Gold in pyrite or other minerals cannot be recovered by amalgamation. Today only a very few commercial gold production facilities use amalgamation, but the use of mercury continues to be the preferred method in artisanal gold mining operations. Mercury is cheap, and its application is simple. However, it is also environmentally damaging, both to workers' health and the surrounding ecosystem. Aritisanal miners mix finely ground gold-bearing ore with mercury. Free gold particles amalgamate with mercury and gold is separated by simply heating to evaporate the mercury. Two sources of mercury release occur: (1) much mercury escapes in the often rudimentary heating/evaporation process; (2) remaining mercury is discharged with tailings directly to the terrestrial or aquatic environments (**Figure 5.11**).

Today only a very few commercial gold production facilities use amalgamation, but the use of mercury continues to be the preferred method in artisanal gold mining operations.

Smelting of mercury amalgam to recover the contained gold and silver involves heating to volatilize the mercury. While it is a relatively simple process to recover the volatilized mercury by condensation, this seldom occurs. Where smelting takes place indoors without any vapour extraction system, as in many small-scale mining situations, the mercury condenses on interior surfaces such as ceilings, floors, walls, and furnishings, where it is indistinguishable from the general grime from other combustion deposits. Accordingly, workers involved in gold recovery receive exposure by breathing mercury vapour, and by dermal contact with contaminated surfaces. As discussed later, they may also be exposed to mercury in drinking water and in food.

In terms of worker exposure, the situation is better where smelting operations take place outdoors or in open-sided buildings, as was the case at Talawaan. However, in such cases, the mercury vapour is more widely dispersed in the surrounding environment, condensing on vegetation and the exteriors of buildings, from where it will be washed by rainfall into the soils, and ultimately into the surface drainage system.

Another pathway for surface water contamination by mercury, concerns the disposal of tailings and runoff from the grinding and smelting activities. Often, these wastes are discharged to the nearest stream or river (**Figure 5.11**), which may be used by local people for

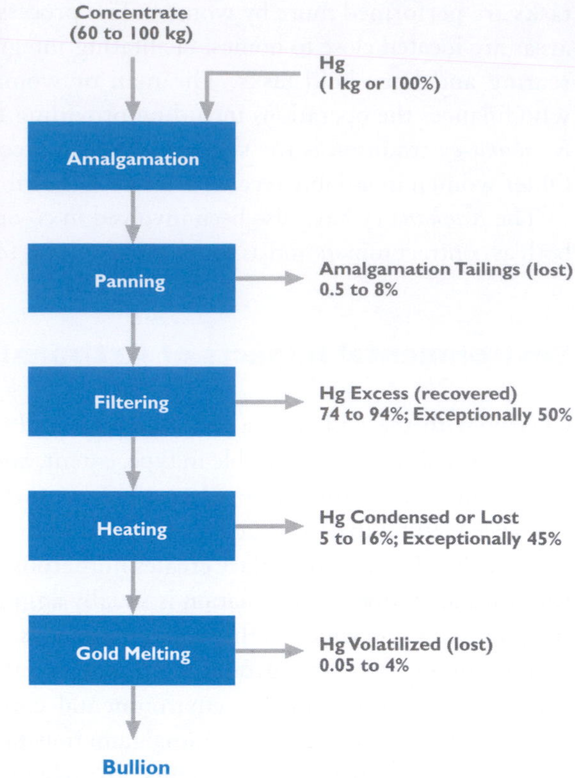

FIGURE 5.11
Mercury Balance for Artisanal Gold Mining in Brazil

In Latin America, up to 200 t/y gold are produced by artisanal mining, releasing more than 200 t/y of mercury into the environment.

Source:
Fharid *et al.* 1992

drinking and bathing purposes. Monitoring of situations where there has been widespread use of mercury, for example near the Talawaan small-scale mining complex, shows elevated concentrations of dissolved mercury in surface water supplies used for bathing and drinking. High concentrations of mercury are also found in stream bed sediments, ensuring a continuing source of mercury long after gold recovery activities end. The mercury may also accumulate in filter-feeding organisms such as shellfish, and may bio-magnify along the food chain. Consumption of aquatic organisms therefore provides an additional potential exposure pathway for local people and those living downstream.

Health effects of mercury vary, depending on the chemical form in which the mercury occurs. Organic forms of mercury are more toxic, with methyl mercury being the most toxic. Methyl mercury was implicated in causing Minimata disease, which was first discovered in 1956, and which was responsible for killing more than 1,000 people in Japan. Allegations of Minimata disease from small-scale mining operations have been made but have not been substantiated, and it appears that most of the mercury contamination remains in an inorganic form. Ingestion or dermal contact with inorganic mercury may adversely affect the central nervous system and endocrine system and damage the mouth, gums, and teeth. High exposure over long periods of time results in brain damage causing mental illness ('mad hatters' disease') and ultimately death. Early symptoms may be subtle and difficult to distinguish from those of other maladies. Serious health effects may take many years to appear. In typical gold rush situations in remote parts of developing countries with poor sanitation and low life expectancy, the effects of mercury poisoning may never be recognized. This may explain why the results of monitoring indicating dangerously high exposure to mercury, do not seem to generate the levels of concern that would be expected among the people involved. A campaign by the Bureau of Mines and Geosciences in the Philippines to persuade small-scale miners to use simple condensers to enable mercury to be recovered and re-used, proved unsuccessful, despite the equipment

In typical gold rush situations in remote parts of developing countries with poor sanitation and low life expectancy, the effects of mercury poisoning may never be recognized.

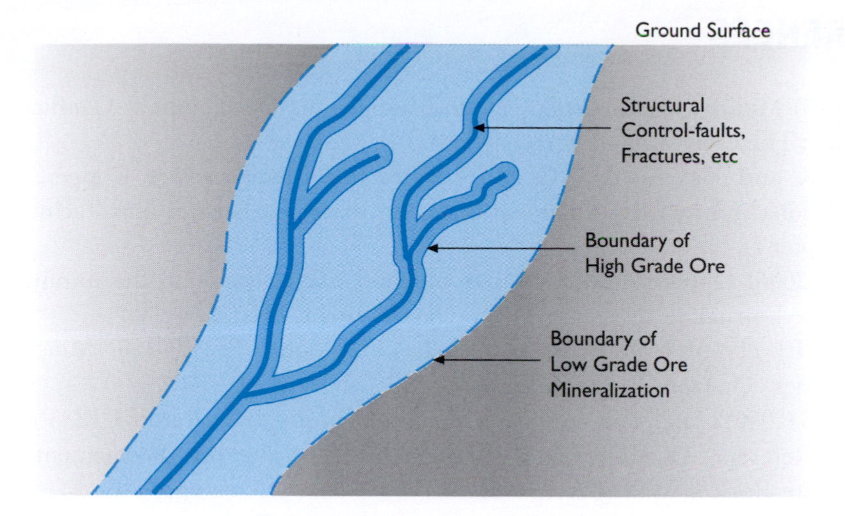

FIGURE 5.12
Typical Distribution of High Grade and Low Grade Ore Types

While small-scale miners only mine high grade zones, large-scale enterprises mine the high grade ores together with low grade materials which generally occur nearby.

being offered at a low price, and the prospect of significantly reduced expenditure through the recovery of mercury.

In addition to these health effects, gold rushes can cause serious social problems, particularly if a large influx of people takes place in an area occupied by a small population of vulnerable, Indigenous People. Population pressure, environmental damage, crime, loss of traditional resources, prostitution, alcohol, drugs, and the effects of HIV/AIDS can cause a complete breakdown of traditional societies. Even in the absence of vulnerable groups, gold rushes may cause a variety of social ills, and it could be argued that, in most cases, the economic benefits do not justify the social and environmental costs.

Another significant impact of much small-scale mining is the loss of mineral resources. The low recovery rate has already been mentioned. Of even more concern is the 'high-grading' involved in virtually all small-scale mining. While small-scale miners only mine high grade zones, large-scale enterprises mine the high grade ores together with low grade materials which generally occur nearby. This is illustrated in **Figure 5.12**.

Removal of the high grade ore can reduce the overall grade to the extent that the remaining resources are uneconomic to mine. Even where this is not the case, the presence of numerous small excavations throughout the ore body can lead to instability and flooding of subsequent mining operations. It is therefore not surprising that many mining companies deliberately avoid areas and situations where small-scale miners have been active or are likely to appear in the event of a discovery.

In contrast to the impacts of gold rush mining, traditional small-scale mining may be undertaken with minimal impact, as shown by the activities of the *Kankana-ey* people. The underground mines of the *Kankana-ey* cause little surface disturbance and are relatively safe compared to those in gold rush areas, because these highly skilled miners recognize signs of instability and understand how to provide ground support. As the *Kankana-ey* live close to their mines and have the expectation that they and their descendants will continue to do so, they have a vested interest in protecting their environment and maintaining pleasant living conditions. Waste rock removed from underground workings is used for construction of buildings, stairways, retaining walls, and terraces in the local villages. The *Kankana-ey* do not use mercury, but manage to achieve gold recoveries of more than 80%, using a range of traditional methods. Ores are, in fact, processed and reprocessed using a variety of physical separation techniques, before being discarded. Similarly, water used in processing is conserved and re-used by the community.

In contrast to the impacts of gold rush mining, traditional small-scale mining may be undertaken with minimal impact

REFERENCES

Ali S (2003) Mining, the Environment, and Indigenous Development Conflicts. Tucson, Arizona: The University of Arizona Press.

Biswas AK and Davenport WG (1976) Extractive Metallurgy of Copper, Pergamon International Library, International Series on Materials Science and Technology, Vol. 20, Chapter 2.

Burke G (2006) Opportunities for environmental management in the mining sector in Asia, The Journal of Environment and Development, 15(2), 224–235.

Caballero EJ (1996) Gold from the Gods: Traditional Small-scale Miners In The Philippines. Quezon City. Giraffe Books

Cronan DS (2000) Handbook of Marine Mineral Deposits, Boca Raton: CRC Press.

EC (2004) Reference Document on Best Available Techniques for Management of Tailings and Waste-Rock in Mining Activities

ISOPE (2002) Deep Seabed Mining Environment: Preliminary Engineering and Environmental Assessment, International Society of Offshore and Polar Engineers, www.isope.org

Hutchison IPG and Ellison (eds) (1992) Mine Waste Management, sponsored by California Mining Association, Lewis Publishers Inc., Michigan, USA

Leduc M, Bachens M, and Smith ME (2004) Tailings Co-disposal in Sustainable Development, proceedings of the annual meeting of the Society for Mining, Metals and Exploration, SME, Denver, February, 2004 and 'Safer, Cleaner, and Potentially Cost-Effective,' Mining Environmental Management, March, 2004.

Lottermoser B (2007) Waste of Phosphate and Potash Ores, Springer Berlin Heidelberg

MMSD (2002) Breaking New Ground. pp420. Earthscan

Pearl R (1974) Handbook for Prospectors, McGraw-Hill, Inc.

Rinaudo JD (2003b) Economic assessment of Groundwater Protection: impact of groundwater diffuse pollution of the upper Rhine valley aquifer. Case study report n° 2 – BRGM/RC: 52325-FR. Orléans: BRGM

Spitz K (1981) Salzwasserausbreitung im Oberrheintal, University Stuttgart, IFW Report

Spooner J (2005) Rare Earth, USEPA Fact sheets, Micon International Ltd

UNEP (2001) Environmental Aspects of Phosphate and Potash Mining; United Nations Environment Programme – International Fertilizer Industry Association

USEPA (1995) EPA Office of Compliance Sector Notebook Project – Profile of the Non-Metal, Non-Fuel Mining Industry. Office of Compliance Office of Enforcement and Compliance Assurance U.S. Environmental Protection Agency 401 M St., SW (MC 2221-A) Washington, DC 20460

6

Converting Minerals to Metals

From Ore to Finished Product

IRON

While rarely found in its native state, iron was available to the ancients in small amounts from meteorites. This native iron is easily distinguished as it contains 6 to 8% nickel. In the early times, iron was five times more expensive than gold and its first uses were ornamental. Later in history iron weapons revolutionized warfare and iron implements did the same for farming. Iron and steel became a building block for civilization, and remain so to this day. In 2004, annual world production of crude steel exceeded one billion tons for the first time. No other metal is used in such quantity.

6 Converting Minerals to Metals

From Ore to Finished Product

This chapter is concerned with mineral processing. Most mining projects involve some processing operations, which are generally integrated with the mining operations. Furthermore, many projects integrate processing activities, including smelting and refining, further downstream.

Mineral processing or extractive metallurgy includes all activities required to extract valuable minerals from their ores to produce pure metals. Metals are rarely found in their native metal form, gold being a notable exception. Most other metals exist as oxide, sulphide or silicate minerals, and these compounds must be reduced to extract metals. Mineral processing plants usually combine a number of methods to extract and to purify metals.

Most metals exist as oxide, sulphide or silicate minerals, and these compounds must be reduced to extract metals.

It is beyond the scope of this book to provide more than a brief overview of mineral processing technologies and associated environmental impacts. Ample literature covers the wide field of metallurgy, and the interested reader is referred to Weiss (1985).

This said the chapter is selective in its emphasis. The copper industry has been selected to illustrate the scale and range of environmental challenges that this industry sector faces. Subsequent discussions focus on the Bayer Process to leach alumina from bauxite ore, and cyanide leaching of gold, which together identify additional issues associated with mineral processing.

From its original location deep underground to its use in a finished product such as wire or pipe, copper passes through a number of stages to convert copper-bearing rock to pure metal (**Figure 6.1**). Once copper ore has been mined, it can be converted to pure copper metal in two different ways: smelting and electrolytic refining (pyrometallurgical processing used for concentrated sulphide ores) and leaching and electrowinning (hydrometallurgical processing, the preferred option for oxide ores).

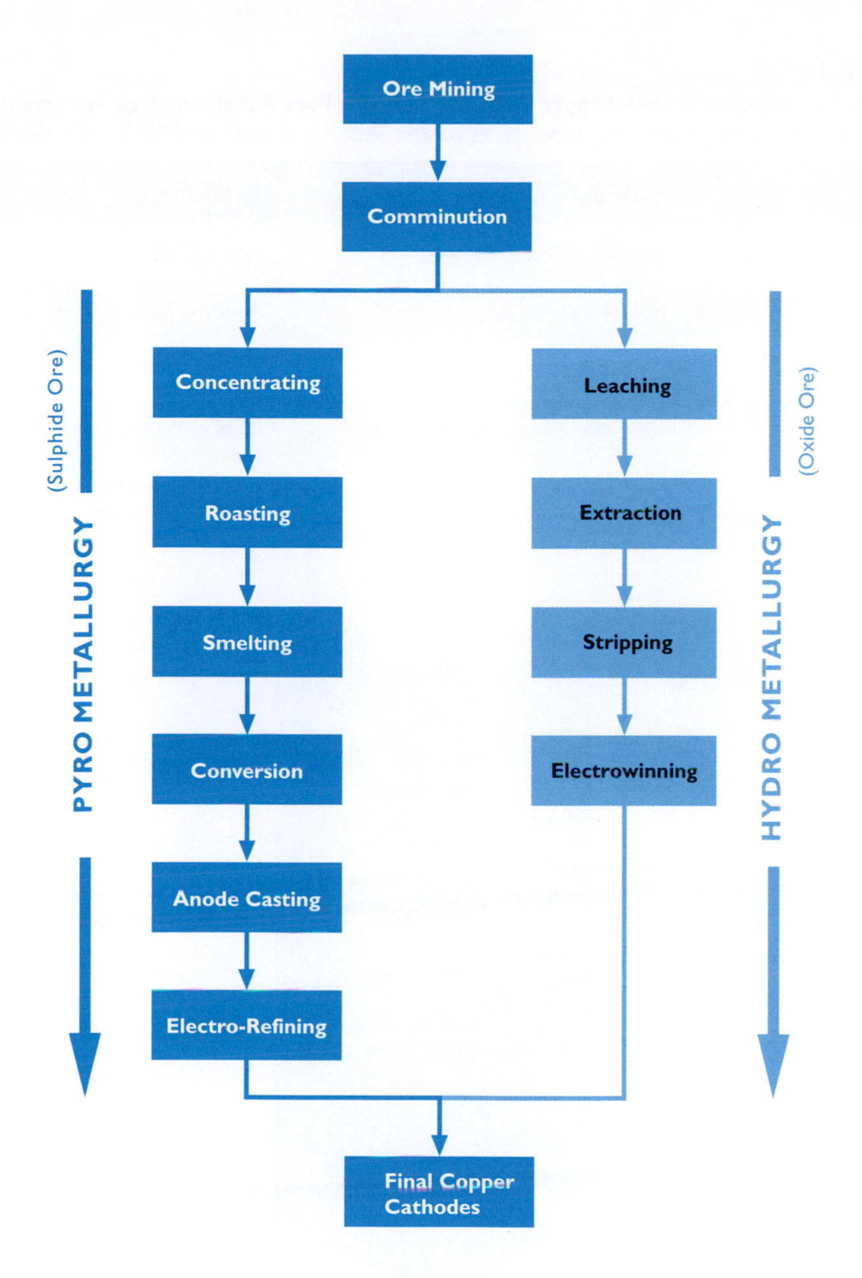

FIGURE 6.1
The Path of Copper from Mineral to Pure Metal

It is important to differentiate between sulphide ore and leachable ore. Generally sulphide ore is composed of minerals containing copper, iron and sulphur, hence the name sulphide. Leachable ore is composed primarily of minerals containing copper oxide.

6.1 THE USE OF FIRE – PYROMETALLURGICAL MINERAL PROCESSING ILLUSTRATED

In pyrometallurgy, concentrated copper ore is smelted and refined. The use of fire (pyro in Greek) to extract metals pre-dates recorded history, and it remains the conventional and most commonly used metallurgical process: reactions are faster at higher temperatures; molten metals are easy to handle and alloy; reagents (fluxes) are cheap; the metal and impurities separate easily; the precious metals stay with the primary metal; and the concentration of copper in a smelter is 20 times that in a hydrometallurgical process (Dresher 2001). Due to the high recovery rate the process remains especially attractive for high grade ores when even a small increase in recovery results in a substantial increase in profit. It would surprise most people, even those involved in the industry, that smelting and refining technologies were introduced on an industrial scale as early as 1865 (**Table 6.1**).

The use of fire to extract metals pre-dates recorded history, and it remains the conventional and most commonly used metallurgical process.

TABLE 6.1

The Main Technologies to Turn Copper Concentrates into Pure Metal – In an age when some of today's most powerful industries scarcely existed three decades ago, the extractive metal industry is the 'Grand Old Man' of modern economy

Year	Method	Technology
1865	Electrolytic (Electro) Refining	Patent (Elkington)
1907	Froth Flotation	Australian Patent (Potter)
1909	Converting	Patent (Pierce & Smith)
1940s	Solvent Extraction – SX/EW	Uranium Purification
1949	Flash Smelting	First Outokumpu Furnace

Source:
Dresher 2001

FIGURE 6.2

Increase of Copper Content through Smelting and Electro-Refining

To be amenable directly for smelting ores must contain copper minerals in sulphide the form. Only high-grade ores in form of concentrates can be economically smelted.

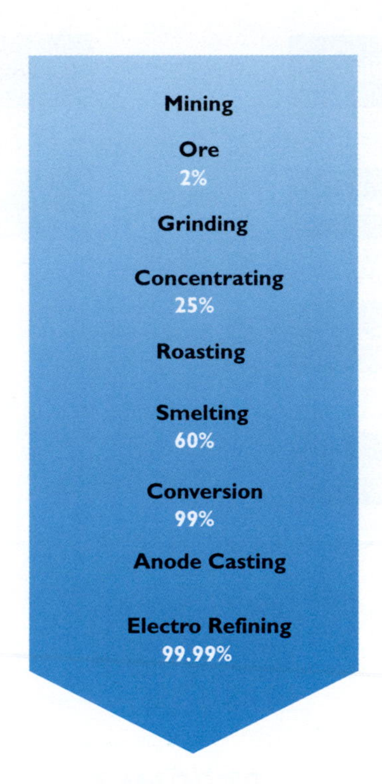

We will illustrate the pyrometallurgical process using the example of converting chalcopyrite ($CuFeS_2$) to copper. From the formula, it is clear that iron and sulphur have to be removed in order to produce copper which is the purpose of smelting ore. The iron is removed from the furnace as slag, the sulphur in the form of sulphur dioxide is processed into sulphuric acid. The copper leaves the smelter in anode form – 99 percent pure. Anode copper is further processed in an electrolytic refinery where impurities are removed, upgrading the purity to 99.99 percent (**Figure 6.2**).

Flotation

To be amenable for smelting, copper minerals must be sulphides.

To be amenable for smelting, copper minerals must be sulphides, such as chalcocite (Cu_2S), chalcopyrite ($CuFeS_2$) and covellite (CuS). Because of the high energy requirements only high-grade ores in form of concentrates can be economically smelted. Froth flotation is the usual method of producing copper concentrate, prior to smelting. In most cases the concentration processes take place at a plant located adjacent to the mine. After crushing and milling, one or more flotation agents are added to the slurry. The slurry is agitated to maintain the solids in suspension and air is pumped into the tank. The flotation reagent binds to the sulphide minerals which float to the surface as bubbles, which are then skimmed from the surface of the tank, leaving the unwanted gangue minerals. For ores containing different sulphides, more than one flotation stage can be used to produce concentrates of different minerals.

Roasting

Sometimes roasting precedes smelting. Roasting of the concentrate at temperatures between 500°C and 700°C in air converts some of the $CuFeS_2$ to copper sulphide, and removes some of the sulphur as sulphur dioxide. The roasting temperature is too low to actually melt the concentrate. The product from the roaster is called calcine, which is a solid mixture of oxides, sulphides and sulphates. The roasting reaction may be expressed typically as

$$2MS + 3O_2 \rightarrow 2MO + 2SO_2 \qquad (6.1)$$

where M represents a metal. One such reaction when chalcopyrite ($CuFeS_2$) is roasted is:

$$2CuFeS_2 + 3O_2 \rightarrow 2FeO + 2CuS + 2SO_2 \qquad (6.2)$$

Smelting with Fluxes-producing Matte

The initial letter 's' of the word 'smelting' implies that, in addition to the physical process of melting, a chemical change also occurs. The calcine is heated to over 1,200°C with fluxes such as silica and limestone. In all smelting operations the kind of flux depends on the ore: limestone flux for iron; iron flux for siliceous ore; and siliceous flux for sulphide concentrates. The calcine melts and reacts with silica and limestone. Some impurities combine with the flux to form a slag which floats on the surface of the molten metal (like oil on water) and is easily removed (**Figure 6.3**). The reaction is:

In all smelting operations the kind of flux depends on the ore: limestone flux for iron; iron flux for siliceous ore; and siliceous flux for sulphide concentrates.

$$\begin{gathered} 2CuFeS_2 + \text{heat} + O_2 \rightarrow \text{Copper-iron matte} + SO_2 \\ CaCO_3 + Al_2O_3 + SiO_2 + \text{iron} \rightarrow \text{slag} \end{gathered} \qquad (6.3)$$

The remaining liquid is a mixture of copper sulphides and iron sulphides called matte.

Conversion of Matte to Copper Blister

The liquid matte is oxidized with air in a converter to form blister copper. The reactions are twofold. First is the elimination of iron sulphide by oxidation to iron oxide which forms a slag:

$$2FeS + 5O_2 \rightarrow 2Fe_2O_3 + 2SO_2 \qquad (6.4)$$

Second is the formation of blister copper by reduction of copper sulphide:

$$CuS + O_2 \rightarrow Cu + SO_2 \qquad (6.5)$$

FIGURE 6.3
Transfer from Matte at the Glogow Smelter

Some impurities form a slag which floats on the surface of the molten liquid (like oil on water) and is easily removed.

Photo Credit: www.kghm.com

FIGURE 6.4
Anode Casting at the Glogow Smelter

While already virtually pure (in excess of 99% copper), it is not really pure enough for the international metal market.

Photo Credit: www.kghm.com

The name 'blister' copper comes from the fact that this final process produces bubbles of sulphur dioxide on the surface of the copper.

The name 'blister' copper comes from the fact that this final process produces bubbles of sulphur dioxide on the surface of the copper. The blister copper is cast into large slabs to be used as the anodes in the next step – electro-refining (**Figure 6.4**). While blister copper is more than 99% pure copper, it is not sufficiently pure for the international metal market. Electro-refining of copper produces the high quality, high purity copper required by end users.

Electro-refining

Even the best chemical method cannot remove all impurities from copper, but with electro-refining (or electrolytic-refining) it is possible to produce 99.99% pure copper. In industry, electrolysis is carried out on a massive scale in tank houses. Blister copper anodes are

FIGURE 6.5
Final Cathode Copper in the Tank House

Even the best chemical method cannot remove all the impurities from the copper, but with electro-refining it is possible to produce 99.99% pure copper.

Photo Credit: Greenshoots Communications

immersed in an electrolyte containing copper sulphate and sulphuric acid. Pure copper cathodes are arranged between the blister copper anodes and a current of over 200A is applied. Gradually, the anode erodes while the cathode grows. Pure copper ions migrate from the blister copper anodes to the starter sheets or 'mother blanks' where they deposit and accumulate over several days to form 100 to 150 kg copper cathodes (**Figure 6.5**).
At the blister copper anode:

$$Cu(s) \rightarrow Cu_2 + (aq) + 2e^- \qquad (6.6)$$

At the copper cathode:

$$Cu_2 + (aq) + 2e^- \rightarrow Cu(s) \qquad (6.7)$$

Over the past decade, starting sheets of stainless steel or titanium have replaced the traditional pure copper starter sheets in a number of refineries. This blank and the techniques developed to optimize its use are known as the 'ISA' method, which was first used in electroplating (Engineering and Mining Journal 1990). If the cathode copper is plated onto a stainless steel 'blank', the copper plate is peeled off the blank prior to shipment and the blank is reused. Deposited copper is separated from the stainless steel or titanium starting sheet by use of a guillotine and/or by flexing with an air blast. The elimination of sheet production and reduced inspection means the work force is up to 60 percent smaller than it is at a conventional plant (Engineering and Mining Journal 1990). Further labour-cost reductions are achieved by automated handling of anodes and cathodes in the tank house.

The anode slime is a valuable by-product that is further processed in specialized smelters to recover valuable metals.

Insoluble impurities in the blister copper anode fall to the bottom forming a sludge that collects under the anodes called anode slime. What happens to the anode slime? The slime contains gold, silver, platinum, and tin all elements that are insoluble in the electrolytic refining and so do not deposit on the cathode. The anode slime is a valuable by-product that is further processed in specialized smelters to recover these valuable metals (**Case 6.1**).

6.2 DISSOLVING ORE MINERALS FROM GANGUE – HYDROMETALLURGICAL MINERAL PROCESSING ILLUSTRATED

Since the mid-1980s, ore leaching has been developed on a larger scale to recover copper from an entirely different set of ores and mining by-products than is possible by smelting; namely, oxidized materials. These may be mined copper minerals that are in an oxidized form – minerals such as Azurite ($2CuCO_3 \cdot Cu(OH)_3$), Brochantite ($CuSO_4$), Chrysocolla ($CuSiO_3 \cdot 2H_2O$), and Cuprite (Cu_2O), residual copper in old mine waste dumps whose sulphide minerals have been oxidized by exposure to the air or sulphide copper minerals that have been oxidized by bacterial leaching, a technology discussed below.

Leaching of Oxide Ores

Leaching is the process by which metals are liberated from minerals by dissolving them away from solids.

Leaching is the process by which metals are liberated from minerals by dissolving them away from solids; it constitutes the hydrometallurgical branch in extractive metallurgy. Leaching operations resemble chemical plants. The chemical process industries also use leaching but the process is usually called extraction, and organic solvents are often used.

An easily understood example of liberating materials by dissolving them away from solids is removing salt (the solute) from salty sand (the ore) by extraction with water (the solvent or the extractant). Using again the example of copper extraction, **Figure 6.6** illustrates the two steps involved for single stage leaching: (1) contact of solid (copper ore) and solvent (dilute sulphuric acid) for transfer of solute (copper) to solvent; and (2) separation of resulting solution from the residual solid. The extract is the solvent phase, now copper bearing (the pregnant leach solution or PLS, also termed pregnant liquor); the raffinate is the solid material, leached ore (also termed spent ore), and its adhering solution. While this process is designed for copper, other metals can be extracted simultaneously.

CASE 6.1
The Unknown Fort Knoxes of this Earth

Gold is a common associate of copper, and eventually accumulates in the anode slime during electro-refining of copper, together with silver and other valuable metals. Some copper smelters incorporate special smelting and refining units to process produced anode slimes and to extract these valuables metals. These specialized smelters also often process anode slimes from other copper smelters that lack the anode slime processing step.

Assuming that copper concentrates contain gold at a concentration of say 1/100 of one percent a copper smelter with a production capacity of 100,000t/y may produce about 1 ton of gold per month. It is alleged that gold produced at the PASAR copper smelter in the Philippines contributed to the wealth of the former President Marcos.

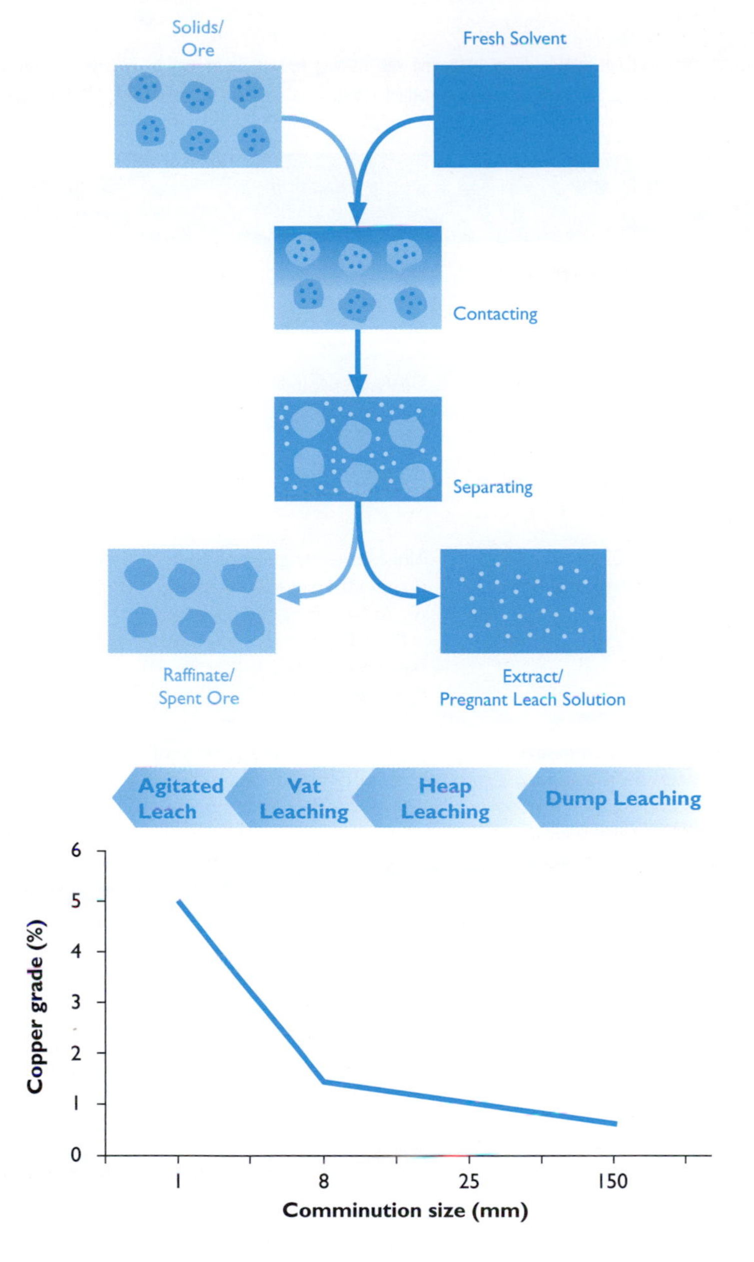

FIGURE 6.6
Single Stage Leaching

Removal of materials by dissolving them away from solids is called leaching. An example is removing salt from salty sand by extraction with water.

FIGURE 6.7
Leaching Process Versus Copper Ore Grade and Comminution Size

Source: US Congress 1988

In hydrometallurgical processing, two routes are common for removing metals from ores. The first route is removing metals directly from ore. As discussed in Chapter Five this is the case when selecting heap, dump, and underground (or *in situ*) leaching. The selection of the leaching methods very much depends on the ore type, ore grade, and comminution size (**Figure 6.7**). Characteristics of these leaching operations applied to copper processing are presented in **Table 6.2**. The leach cycle depends on the ore type, and applied leaching conditions.

Oxidized copper minerals such as azurite, malachite, tenorite, and chrysocolla, are completely soluble in sulphuric acid at room temperature. Extracting copper from these oxide ores typically involves spraying the crushed ore with a sulphuric acid solution to dissolve the copper. In heap leaching, large piles of oxide ore are placed on impermeable pads and are sprayed with leach solution by a sprinkler system. The leach solution percolates down through the heap over time, picking up copper as it travels along. A central drainage system collects the copper bearing runoff at the bottom of the heap.

Oxidized copper minerals are completely soluble in sulphuric acid at room temperature.

TABLE 6.2

Characteristics of Copper Leaching – Hydrometallurgical methods include dump, heap, and vat leaching techniques, as well as underground (or *in situ*) leaching methods. The leach cycle depends on ore type and applied leaching conditions. Each copper leaching method results in a pregnant leach solution (PLS). Copper is recovered from the PLS through precipitation or by solvent extraction/electrowinning

	Vat Leaching	Heap Leaching	Dump Leaching	Underground and *in situ* Leaching
Ore grade	Moderate to high	Moderate to high	Low	Low to high (dependent upon mine conditions and layout)
Types of ore	Oxide, silicates, and some sulphides	Oxides, silicates, and some sulphides	Sulphides, silicates, and oxides	Oxides, silicates, and some sulphides
Ore preparation	May be crushed to optimize copper recovery	May be crushed to optimize copper recovery	Blasting	None
Container or pad	Large impervious vat	Impervious barrier of clay, synthetic material, or both	None for existing dumps; new dumps intended to be leached would be graded, and covered with an impermeable polyethylene membrane, or bedrock, protected by a layer of select fill	None
Solution	Sulphuric acid for oxides; acid cure and acid-ferric cure provide oxidant needed for mixed oxide/sulphide ores	Sulphuric acid for oxides; acid cure and acid-ferric cure provide oxidant needed for mixed oxide/sulphide ores	Acid ferric-sulphate solutions with good air circulation and bacterial activity for sulphides	Sulphuric acid, acid cure, acid-ferric cure, or acid ferric-sulphate, depending on the ore type
Length of leach cycle	Days to months	Days to months	Months to years	Months
Solution application method	Spraying, flooding, and circulation	Spraying or sprinkling	Ponding/flooding, spraying, sprinkling, and trickle systems	Injection holes, recovery holes
Metal recovery method	SX/EW for oxides and mixed oxide/sulphide ores; iron precipitation for mixed ores	SX/EW for oxides and mixed oxide/sulphide ores; iron precipitation for mixed ores	SX/EW for oxides and mixed oxide/sulphide ores; iron precipitation for mixed ores	SX/EW for oxides and mixed oxide/sulphide ores; iron precipitation for mixed ores

Source:
US Congress, Office of Technology Assessment 1988

VAT leaching of copper concentrates is only viable for high-grade ore when even a slight increase in copper recovery adds substantial profit.

The second route is removing copper from copper concentrate. This is done in tanks, vessels or vats (VAT leaching). Considering the cost of ore grinding and concentrating, VAT leaching of copper concentrates is only viable for high-grade ore when even a slight increase in copper recovery adds substantial profit. VAT leaching is also applied to leaching ores that are not amenable to simple atmospheric leaching as illustrated in a later section using the Bayer Process to separate alumina from bauxite ore as an example.

In its simplest form, hydrometallurgical mineral processing requires low capital investment relative to smelting, and the process can be operated economically at a small scale. Especially where applied to ore *in situ* rather than as concentrate the processing of low-grade ores can be achieved at much lower cost. The net result is that metals can be produced from low grade ore that in the past would have gone untouched or would have been discarded as waste rock. As such, it is amenable to use by small 'mom and pop' operators. In China, for example, there are about 40 to 50 small-scale leaching operations as of 2000 (Dresher 2001).

Bioleaching of Sulphide Ores

Only oxidized ores are amenable to atmospheric leaching using sulphuric acid. Other, less oxidized, cuprite and sulphide ores, such as chalcocite, bornite, covelite, and chalcopyrite, require the addition of ferric sulphate and oxygen (as oxidants) to accomplish leaching. Leaching ores containing bornite and chalcopyrite with ferric sulphate is very slow, even at elevated temperatures (Weiss 1985).

Bacterial leaching, otherwise also known as bioleaching, is the extraction of a metal from these less oxidized or sulphide ores using materials found native to the environment: water, air and microorganisms. In other words, bioleaching is the commercialization of the ability of certain bacteria, found in nature, to catalyze the oxidation of sulphide minerals (Brierley 2000). Leaching sulphide materials, whether in ore or concentrate, requires a chemical oxidizing agent – ferric ions (Fe^{3+}). These can be generated by reactions with air assisted by bacteria. The oxidation can also be assisted by pressure as in an autoclave.

Bacterial leaching is the extraction of a metal from these less oxidized or sulphide ores using materials found native to the environment: water, air and micro-organisms.

The results of natural microbial leaching, not the cause, have been known since ancient times. Iron-rich acidic waters draining from abandoned coal and metal mines as well as from unmined mineralized areas provide evidence of microbial leaching. In fact, history records that mine water problems began at the same time that mining activities began. At Rio Tinto in Spain, seventeenth century records describe the occurrence of copper-bearing waters. The UK based mining company, Rio Tinto, which was formed in 1873, owes its name to these copper-bearing waters, although it is iron that imparts the red colouration.

Commercial application of bacterial leaching, however, only began in the 1950s at Kennecott's Bingham mine near Salt Lake City, Utah. It was noticed that blue copper-containing solutions were produced in waste piles that contained copper sulphide minerals – a condition that should not have happened in the absence of powerful oxidizing agents and acid. On investigation it was found that naturally occurring bacteria were oxidizing iron sulphides and the resulting ferric sulphate was acting as an oxidizer and leachant for copper sulphides. These bacteria, thriving at pH 1.5 to 3.0 (Blesing *et al.* 1975, Hawley 1977), were given the name *ferro-oxidans* for their action in oxidizing iron sulphides. A second set of bacteria were also identified and given the name *thio-oxidans* for their action in oxidizing sulphur to yield sulphuric acid.

Bacterial leaching offers a method of exploiting small ore bodies with a minimum of capital investment. Today most commercial operations leaching copper from ore dumps are located in the Southern Hemisphere: in Australia, Chile, Myanmar and Peru (Dresher 2001). The process consists of injecting the material to be leached with cultivated strains of appropriate bacteria and maintaining conditions that are conducive to their effective operation and propagation. Air, for instance, is blown into the heap through air lines situated under the leach pad.

Bacterial leaching offers a method of exploiting small ore bodies with a minimum of capital investment.

Copper bacterial leaching is still confined to the leaching of ore, but pilot plant tests are underway for the leaching of chalcopyrite concentrates that would normally be processed by flotation and smelting. Unlike heap leaching, industrial leaching processes are not subject to climatic conditions at the mine site. Leaching of concentrate using vats allows the metallurgist to apply various combinations of temperature and bacteria or temperature and pressure.

Leaching with cyanide has been applied almost exclusively to gold and silver as discussed later in this chapter, but cyanide has also been applied to copper for both oxidized and low-grade sulphide ores. The effectiveness of cyanide in leaching depends on the ability of the cyanide ion to form stable complexes with the majority of transition metals. These complexes are strong enough to overcome the relative inertness of gold and silver and the insolubility of copper minerals, such as chalcocite, to form copper-cyanide complexes (Weiss 1985).

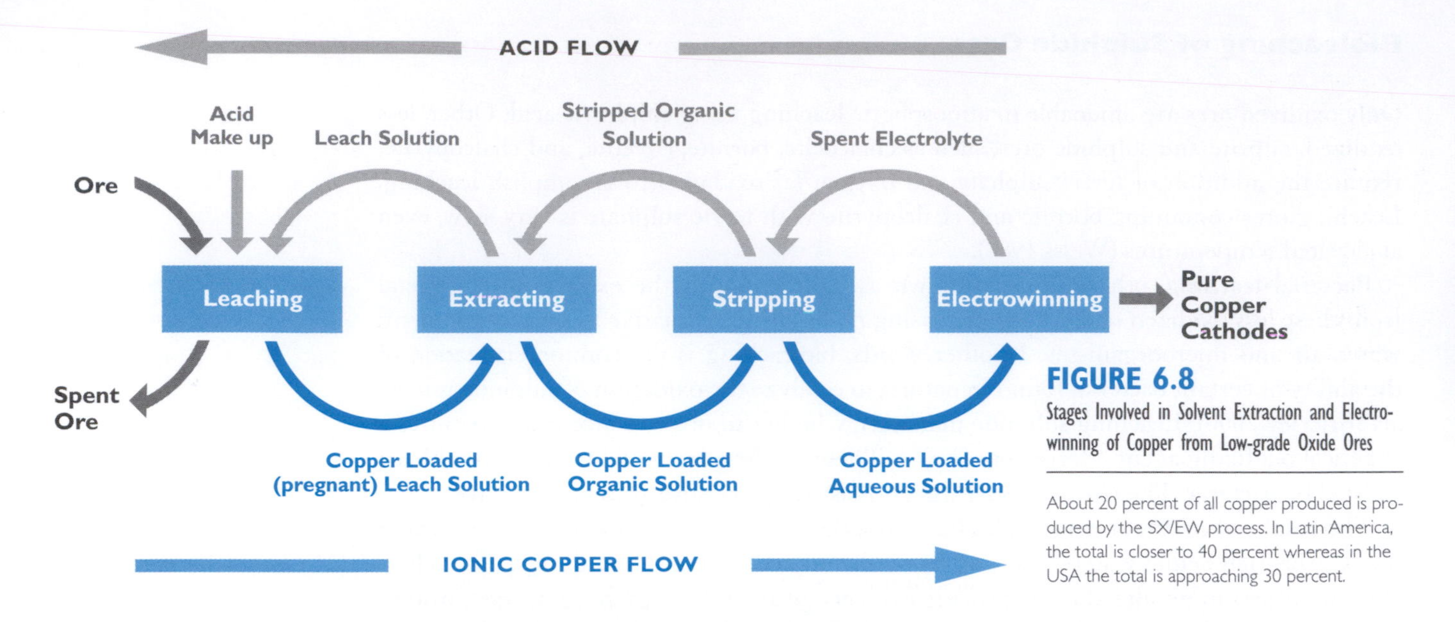

FIGURE 6.8

Stages Involved in Solvent Extraction and Electrowinning of Copper from Low-grade Oxide Ores

About 20 percent of all copper produced is produced by the SX/EW process. In Latin America, the total is closer to 40 percent whereas in the USA the total is approaching 30 percent.

Recovering Copper from the Pregnant Leach Solution

Each copper leaching method results in a pregnant leach solution (PLS). The method used for recovering copper from the leach solution is the solvent extraction-electrowinning process (**Figure 6.8**; U.S. Congress, Office of Technology Assessment 1988). The widely applied leach-solvent extraction-electrowinning process (or in short SX/EW) process has its roots in analytical chemistry where it is used to separate one metallic ion from another. It was first used as a large-scale process during World War II for the recovery of uranium from its ores. The key to the process is the development of organic extractants that are specific to the metal to be extracted. The first extractant that was specific for copper and used at a commercial scale was developed by General Mills Corporation and sold under the name LIX 64® (LIX for Liquid Ion Exchange and Roman for 1959 – the year of the first patent). Ranchers Exploration and Development Corporation at its Bluebird Mine in Arizona in 1968 first demonstrated it on a large scale. Today, worldwide, about 20 percent of all copper is produced by the SX/EW process. In Latin America, the total is closer to 40 percent, whereas in the USA the total is approaching 30 percent (Tilton and Lansberg 1997).

Today, worldwide, about 20 percent of all copper is produced by the SX/EW process.

Solvent extraction stage – The solvent extraction (SX) part of the process occurs in two steps and each step includes a process of ion exchange between two solutions. In the first step the pregnant leach solution is mixed vigorously with a kerosene-based solvent containing an organic chemical specifically designed to extract copper. Here the copper is extracted away from the aqueous phase leaving behind most of the impurities that were in the pregnant leach solution. Since in the solvent extraction stage, copper ions are exchanged for hydrogen ions, the aqueous phase is returned to its original acidity and recycled to the leaching step of the process. During the ion exchange, the lighter organic copper bearing solution floats on top of the heavier leach solution, much like oil floats on water, and can be easily separated and directed to the next processing step.

In the second step the copper exchange is reversed in another tank that strips the copper from the organic solution. The copper-laden organic solution is mixed with a strong sulphuric acid solution known as the electrolyte or aqueous solution. In the mixing and settling stage, the copper is transferred from the organic to the aqueous solution, filtered and then pumped to the electrowinning (EW) tank house. During the process, the organic phase is reconstituted in its hydrogen form and, in its barren organic phase, is returned to the extraction stage of the process.

The SX/EW process operates at ambient temperatures. Copper is kept in either an aqueous environment or an organic environment during its processing until it is reduced to its metallic form. Because of its dependence on sulphuric acid, the SX/EW process is often not a substitute for, but rather an adjunct to conventional smelting, using sulphuric acid produced from smelter gases. It is also applicable in locations where smelter acid is not available by importation of sulphuric acid or through importation of sulphur or pyrite, which are used to manufacture sulphuric acid on-site. Such acid plants are found associated with many hydrometallurgical plants including lateritic nickel plants and uranium plants.

Because of its dependence on sulphuric acid, the SX/EW process is often not a substitute for, but rather an adjunct to conventional smelting, using sulphuric acid produced from smelter gases.

Electrowinning – In electrowinning, copper is reduced electrochemically from copper sulphate in the aqueous solution to a metallic copper cathode. When electrically charged, pure copper ions migrate directly from the aqueous solution produced by solvent extraction to starter cathodes made from pure copper foil. The inert (non-dissolving) anodes are made of lead (alloyed with calcium and tin) or stainless steel, referred to as sheets (US Congress, Office of Technology Assessment 1988). Electrowon copper cathodes are as pure as electro-refined cathodes from the smelting process.

The electrochemical reaction at the lead-based anodes produces oxygen gas and sulphuric acid by electrolysis. Spent acid is either pumped to the solvent extraction process via electrolyte heat exchangers (to balance the tank house temperature by heating up incoming copper bearing electrolyte from the solvent extraction process), or is recycled and pumped back to the leaching operation (US Congress, Office of Technology Assessment 1988; Engineering and Mining Journal 1990).

While smelting is more energy consuming than applying SX/EW technology, electrowinning consumes more energy than electro-refining. The electrowinning of copper requires considerably more electrical energy than does the electro-refining process – an average of about 8 MJ/kg for electrowinning compared to about 1.5 MJ/kg for electro-refining: in electrowinning the copper must be reduced from the cupric form to metal; whereas, in electro-refining the copper is already in metallic form and is merely transported from the anode to the cathode to purify it.

While smelting is more energy consuming than applying SX/EW technology, electrowinning consumes more energy than electro-refining.

In recent years hydrometallurgical routes have become more popular in metallurgical research because: total energy costs are lower; increased environmental awareness favours 'zero discharge' conditions; hydrometallurgical processes can separate impurities better; operating temperatures are lower making plants easier to operate; while pyrometallurgical processes generate sulphur dioxide, a gas that has to be converted into sulphuric acid, hydrometallurgical processes use and even form sulphuric acid. However, apart from the SX/WE process, industrial leaching plants for metals such as copper or nickel at the scale required for modern mineral processing have yet to establish a track record comparable with that of smelters. The hydrometallurgical route at industrial plants has proved to be challenging and in many respects is still in its infancy (**Case 6.2**).

CASE 6.2
Leaching Nickel using Pressure, Temperature, and Acid

Most of today's nickel is produced in ferro-nickel smelters that process nickel sulphide ore. Advances in leaching technology make it increasingly attractive to process laterite nickel ore, which result from preferential leaching and enrichment of nickel under prolonged tropical weathering. Lateritic nickel ores are found in abundance along a geological belt stretching from New Caledonia to Southern Philippines, and also in Western Australia and Cuba.

As the name suggests, in the High Pressure Acid Leaching (HPAL) process, nickel and cobalt, its common companion, are dissolved from ore under high pressure, and high temperature, using concentrated sulphuric acid solution. At present, only a few HPAL plants are in commercial production, and even fewer plants operate without technical problems. The HPAL operation at Coral Bay in Southern Philippines demonstrates that the technology can work successfully, the Goro HPAL plant in New Caledonia and its past challenges highlights the perils associated with this technology.

FIGURE 6.9
Countercurrent Stage-Wise Leaching

Industrial leaching processes such as High Preassure Acid Leaching (HPAL) apply countercurrent stage wise leaching because the process can deliver the highest possible concentration in the extract and can minimize the amount of solvent needed

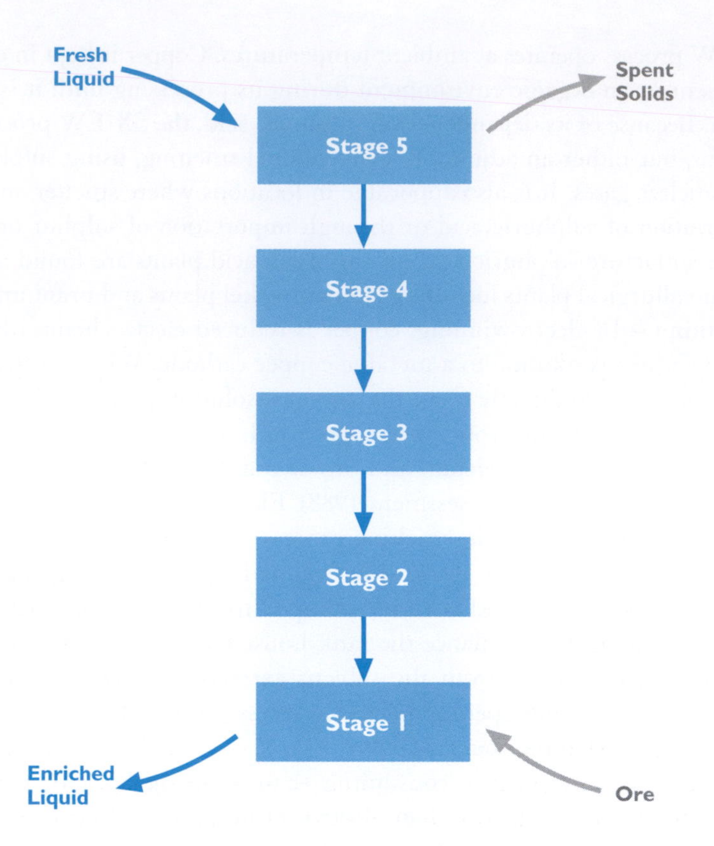

6.3 THE HYDROMETALLURGICAL ROUTE AT INDUSTRIAL SCALE

Atmospheric leaching is an operation with relatively low capital investment requirement. This cannot be said for large industrial leaching processing plants such as the use of the Bayer Process in alumina refining or the use of High Pressure Acid Leaching (HPAL) processes, an emerging technology to recover nickel from laterite ore. These hydrometallurgical mineral processing plants are chemical processing plants, expensive to develop and often difficult to operate.

A countercurrent stage-wise process delivers the highest possible concentration in the extract with a minimum amount of solvent.

Rather than the single stage of atmospheric leaching, industrial tank or VAT leaching allows countercurrent stage-wise leaching at higher pressure and/or temperature. A countercurrent stage-wise process delivers the highest possible concentration in the extract with a minimum amount of solvent. Solvent and solid are mixed, allowed to approach equilibrium, and the two phases are separated. Liquid and solids move counter-currently to adjacent stages (**Figure 6.9**). The solvent phase (or extract) becomes more concentrated as it contacts in a stage-wise fashion the increasingly solute-rich solid. The raffinate (leached ore) becomes less concentrated in soluble material as it moves toward the fresh solvent phase.

The design equations comprise the overall and component material balances for the entire process and for each separation stage. It is assumed that the solute-free solid is insoluble in the extracting solvent.

The total system material balance is:

$$VB + LA = VA + LB \quad (6.8)$$

The material balance for each component calculates as:

$$VB\ YB + LA\ XA = VA\ YA + LB\ XB \quad (6.9)$$

where

V = mass of liquid extractant (B entering, A leaving)
YB = concentration entering with solvent (zero if fresh solvent)
YA = concentration in solvent leaving
L = mass of retained liquid on inert solids (A entering, B leaving)
XA = concentration of solute in adhering entering liquid
XB = concentration in adhering liquid leaving process

Not only does VAT leaching allow several stages of leaching, it enables the metallurgist to modify other parameters such as pressure or temperature to obtain an optimum environment to leach metals from minerals. Two examples of hydrometallurgical plants follow.

The Use of Cyanide Solution to Dissolve Gold

In the late nineteenth century amalgamation with mercury was the main method for separating gold from its ore. By the 1880s, however, problems were being experienced in the South African gold mines as workings deepened and the sulphide content in the ore increased, markedly reducing the effectiveness of amalgamation. The cyanide process, introduced in South Africa in 1880 (Kettel 1982), represented a vast improvement over amalgamation and other earlier methods, and has been extensively used ever since. The use of cyanide leaching of gold is a hydrometallurgical process, and is the most effective way of extracting fine gold particles from ores. Cyanide (CN) is a compound of carbon and nitrogen, two of the most common elements in the Earth's crust. It is produced naturally in a number of micro-organisms, insects, and plants. Today it is a chemical manufactured for use in a number of important industries. About 20% of all manufactured cyanide is used in the form of sodium cyanide for mineral processing (TRI/Right-To-Know Communications Handbook, Section 5).

The reason that cyanide is so widely used in gold mining is that it is one of the very few chemical reagents that will dissolve gold in water, using only oxygen from the air as an oxidizing agent. Other chemicals will work but only in much higher concentrations using much stronger oxidizing agents, e.g. chlorine, nitric acid or hypochloride. At these higher concentrations they can be more dangerous to handle than cyanide. In commercial mining, cyanide has thus become the chemical of choice for the recovery of gold from ores. Used in metal extraction since 1887 it is safely managed in gold recovery around the world. In 2000, there were about 875 gold or gold and silver mining operations in the world. This number does not include the contribution from base metal mines where some gold is recovered as a by-product at the mine or the smelter. Of those 875 sites, 460 (i.e. 52%) used cyanide, of which 15% were heap leaches and 37% used cyanidation in tank leaching. The remaining 48% used a variety of processes, primarily gravity separation and flotation to form a concentrate. These concentrates were then sent to a smelter for final processing (Mudder 2000).

The reason that cyanide is so widely used in gold mining is that it is one of the very few chemical reagents that will dissolve gold in water.

In general, there are two basic types of cyanidation operations, tank leaching and heap leaching. Tank leaching involves one of three distinct types of operations, Carbon-in-Pulp (CIP), Carbon-in-Leach (CIL), and the Merrill Crowe Process. In Carbon-in-Pulp operations, the ore pulp is leached in an initial set of tanks with carbon adsorption occurring in a second set of tanks. In Carbon-in-Leach operations, leaching and carbon recovery of the gold values occur simultaneously in the same set of tanks. The Merrill Crowe process uses zinc to remove the gold from solution and is generally used for ores that have high silver to gold ratio.

The discussion that follows includes material from Environment Australia (1998), which provides an excellent synthesis of information, and to which the reader is referred for further details.

Prior to dissolving or leaching of the gold, most types of ore must first be crushed and ground to liberate or expose the individual grains of gold.

Prior to dissolving or leaching of the gold, most types of ore must first be crushed and ground to liberate or expose the individual grains of gold. The ground ore is mixed with a solution of sodium cyanide in an alkaline solution. Typically, between 0.5 and 2 kg of sodium cyanide is used for each tonne of gold-bearing ore. Both high alkalinity (typically pH 10.3) and salinity are essential to ensure that most of the cyanide is present as CN^- ion, minimizing the formation of hydrogen cyanide (HCN) gas, which is highly toxic.

The process in which gold and other metals are dissolved in cyanide solution is known as cyanidation and the resulting solution is known as 'pregnant liquor'. Gold is dissolved according to the following reactions:

$$2Au + 4CN + O_2 + 2H_2O \leftrightarrow 2Au(CN)_2^- + H_2O_2 + 2OH^- \text{ (Bodlander's Equation)} \quad (6.10)$$

$$4Au + 8CN^- + O_2 + 2H_2O \leftrightarrow 4Au(CN)_2^- + 4OH^- \text{ (Elsener's Equation)} \quad (6.11)$$

Apart from the gold cyanide ions, cyanide forms a variety of complex ions with other metals – for example:

$$Fe^{2+} + 6CN^- \leftrightarrow Fe(CN)_6^{4-} \quad (6.12)$$

The optimization of gold extraction is a highly complex process, requiring close attention to cyanide concentration, pH, salinity, and sometimes oxygen.

The cyanide-metal complex ions differ considerably in their toxicities and stabilities, factors that are highly important in determining detoxification requirements.

In practice, the optimization of gold extraction is a highly complex process, requiring close attention to cyanide concentration, pH, salinity, and sometimes oxygen. For optimum leaching conditions, a pH value of somewhere between 9.5 and 11.0 should be maintained, depending on the requirements of the ore being treated at the time. Generally the pH is maintained at 10.5 by the addition of lime or sodium hydroxide to the slurry. This is essential to prevent the loss of sodium cyanide in solution to gaseous hydrogen cyanide (this would obviously have safety implications) which would result in high cyanide consumption.

The time required to complete the cyanidation reaction varies depending on the gold particle size, presence of other metals, cyanide concentration, oxygen concentration, and amount of agitation provided. Optimum concentrations of reagents will vary, not only from ore body to ore body, but within the same ore body in accordance with variations in gold grades and the presence of other cyanide complexing metals. It is relatively common that 'spikes' in concentration of one or more base metals or the occurrence of some carbonaceous substances, cause rapid depletion of cyanide, requiring addition of more cyanide solution to maintain gold extraction. Subsequently, when base metal concentrations decline to normal levels, there may be an excess of cyanide in the pregnant leach solution. For this reason, it is common practice to blend ores of differing composition, prior to the crushing stage, in an attempt to maintain a feedstock of uniform grade so that variation in cyanide consumption is minimized. In modern CIP and CIL plants, the concentration of cyanide, pH, and salinity are monitored continually so that frequent adjustments can be made, based on the reactions as they occur. However, it is unavoidable that variations in the ore will cause variations in the consumption of cyanide which will be reflected in the chemical quality of the pregnant liquor and, ultimately, the tailings liquor.

This text uses the example of the CIL process to illustrate ore processing and refining using cyanide. In this process gold is extracted from the ore, after crushing and grinding,

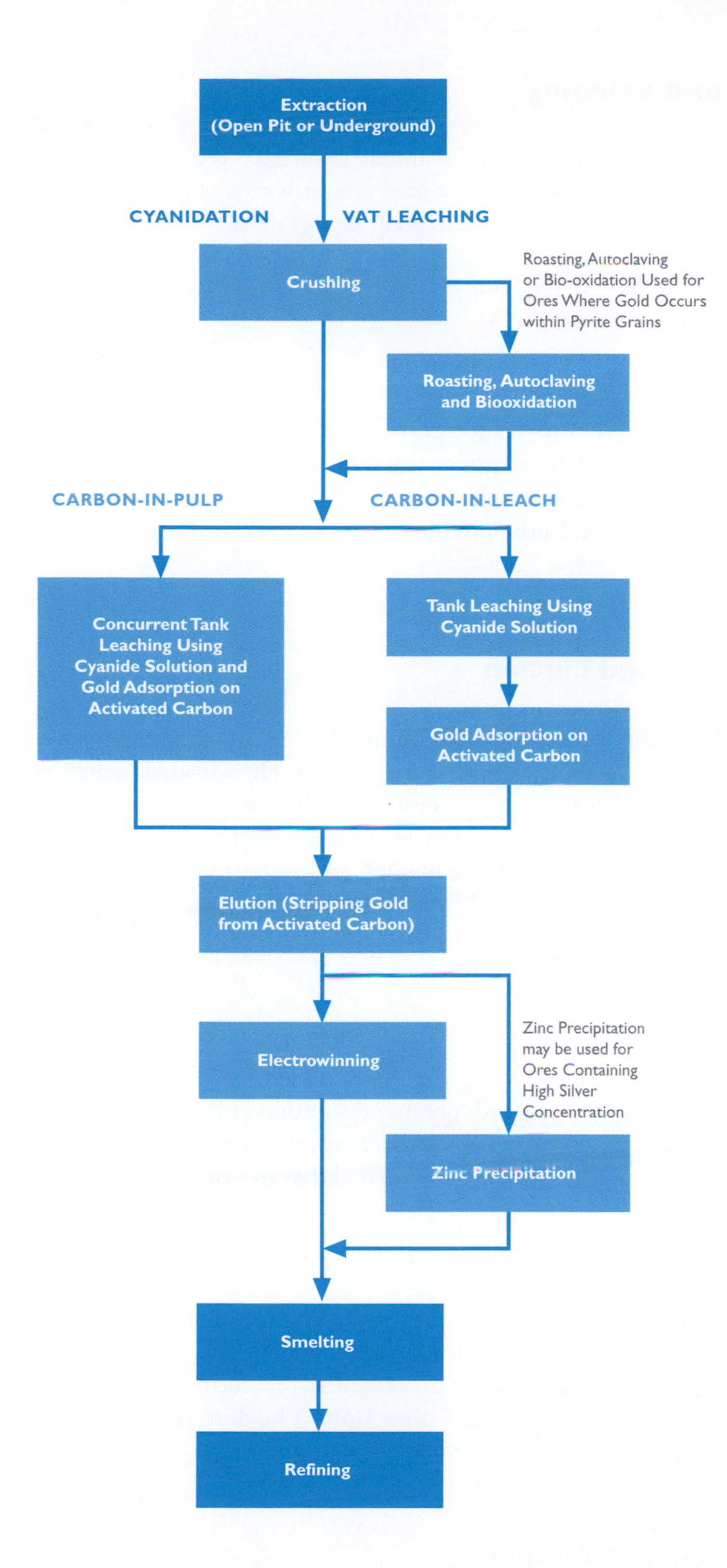

FIGURE 6.10
Process Involved in Gold Extraction and Beneficiation

applying a sodium cyanide solution to dissolve the gold which is then adsorbed on to carbon. Gold is then stripped from the carbon, using sodium hydroxide solution, and recovered by electrowinning, the product of which is smelted to produce doré bars containing gold and silver metal. A simplified flowchart for the process is provided as **Figure 6.10**. Common steps are as follows.

Crushing and Grinding

Ore is commonly fed to a crusher for initial grain size reduction. Crushed ore is then directed to the mill which operates in closed circuit with a cluster of cyclones to grind the ore so that the product passes a pre-defined design grain diameter (typically in the order of 100 microns).

Leach and CIL

Milled ore from the cyclone overflow gravitates through a vibrating trash screen into the leach and CIL circuit, where it passes through leach tanks and CIL tanks over a total period of about 48 hours. Each tank is agitated. Screens are installed at the outlets of each CIL tank to retain carbon. Carbon in the CIL circuit is transferred counter-current to the flow, by air lifts.

Acid Washing and Elution

From the CIL circuit, loaded carbon is pumped to a vibrating screen where it is washed in water before gravitating to an acid wash column. Here loaded carbon is soaked for 30 to 60 minutes in a hydrochloric acid solution, washed in water again to remove the acid, and transferred to the elution column. The elution circuit process uses staged additions of different solutions, including $NaCN_3$, NaOH, and water to strip gold from carbon. After stripping, the carbon is regenerated in a rotary kiln, for reuse. The pregnant solution is directed to the electrowinning process.

Electrowinning

In electrowinning, the stripped and 'pregnant' solution circulates through electrowinning cells. Sludge containing gold and silver collects at cathodes from which it is washed into a collection vessel. The sludge is then dewatered and oven dried.

Smelting

The oven-dried material is smelted in a furnace, located within a secured gold room. It is only then that gold emerges from the process in visible form. Molten metal from the furnace solidifies to produce bars of gold-silver doré. The doré is stored in a safe within the gold room until it is transported off-site for final refining.

Detoxification of Tailings

Detoxification needs to be carried out to decompose cyanide, which may be present in the thickener underflow at concentrations up to 300 mg/L.

Tailings emerge from the CIL circuit as fine-grinded cyanide-containing material. Detoxification needs to be carried out to decompose cyanide, which may be present in the thickener underflow at concentrations up to 300 mg/L. Detoxification commonly uses the well-proven INCO SO_2 Air System, in which sodium metabisulphite is added, together with oxygen (in air) and small amounts of copper sulphate catalyst, to oxidize the cyanide to cyanate. Free cyanide is oxidized to cyanate in accordance with the following reaction:

$$CN^{-}free + SO_2 + O_2 + H_2O \rightarrow CNO^{2-} + H_2SO_4 \quad (6.13)$$

Metal (Me) cyanide complexes are also decomposed, according to reactions such as the following:

$$Me(CN)_4^{2-} + 4SO_2 + 4O_2 + 4H_2O \rightarrow CNO^{-} + 4H_2SO_4 + Me^{2+} \quad (6.14)$$

These processes take place in a single reaction tank, fitted with an agitator, air sparging system and pH control involving lime addition.

Various environmental and occupational health concerns are associated with the use of cyanide. At high concentration cyanide is acutely toxic. At mines, however, cyanide is used in low concentrations in water, typically 0.01% to 0.05% cyanide. Its use is tightly controlled and monitored. Mining employees are trained to handle cyanide with care, while the manufacture, transport, storage, use, and disposal of cyanide are strictly regulated; it has been used safely for decades (The International Cyanide Management Code 2005). At gold mines, all tanks, pipes, ponds, and other areas that contain cyanide are required by law to have appropriately designed and engineered containment facilities (USEPA 1981). This means that if there is any problem, there is a back-up system to contain the material. Cyanide in mining solutions is collected either to be recycled or effectively destroyed to permitted levels on-site after gold is removed. Common well-proven detoxification technologies are the above mentioned INCO SO_2 Air System or the use of hydrogen peroxide (H_2O_2). Cyanide can be destroyed quickly when required using several commonly available chemicals kept at all mines.

In the natural environment, cyanide is naturally decomposed, generally by oxidation (see also Chapter Eighteen). It is not persistent in the environment, neither does it bioaccumulate. It is not carcinogenic, or mutagenic. However it does present three potential environmental hazards: (1) cyanide-containing ponds and ditches can represent an acute hazard to wildlife and birds. Tailings ponds may pose similar hazards, although cyanide concentrations are typically much lower; (2) spills can result in cyanide reaching surface water or groundwater and causing short-term (e.g. fish kills) or long-term (e.g. contamination of drinking water) impacts; (3) cyanide in active heaps and ponds and in mining wastes (e.g. heaps and dumps of spent ore, tailings impoundments) may be released and present a hazard in surface water or groundwater. Cyanide may also increase the potential for metals to go into solution and, therefore, be transported to other locations.

In the natural environment, cyanide is naturally decomposed, generally by oxidation.

Some basic knowledge of the different forms of cyanide is necessary to understand regulatory standards. Cyanide concentrations are generally measured as one of the following forms:

- Free Cyanide – the sum of the free cyanide (CN^{-}) and hydrocyanic acid (HCN) and includes cyanide-bonded sodium, potassium, calcium, or magnesium. Free cyanide is very difficult to measure except at high concentrations and its results are often unreliable, difficult to duplicate, or inaccurate.
- Titratable Cyanide – the cyanide concentration measured by titration with silver nitrate ($AgNO_3$); may include cyanide from dissociation of some complex forms in addition to free cyanide.
- Simple cyanides – containing only one type of metal ion which dissociates in water to release free cyanide.
- Complex cyanides – contain more than one metal and dissociate in water to release a metal ion and a cyanide-metal ion complex, which may subsequently dissociate further to release free cyanide.

- Total cyanide – the sum of all forms of cyanide present in the sample, including iron, cobalt, and gold complexes. 'Total cyanide' is a toxicologically meaningless term since its measurement requires harsh treatment to disintegrate intractable complex cyanides before free cyanide can be liberated and measured.
- Weak acid dissociable (WAD) cyanide – cyanide that is readily released from cyanide-containing complexes when the pH is lowered. Any free cyanide already present plus cyanide released from nickel, zinc, copper, and cadmium complexes (but not iron or cobalt complexes) is measured. WAD cyanide (CN_{WAD}) is measured by treating the sample with a weak acid buffer solution such as sodium acetate/acetic acid mixture at pH 4.5 to 6.
- Cyanide Amenable to Chlorination (CATC) which refers to the cyanide that is destroyed by chlorination. CATC is an analytical quantity that requires similar sample treatment to WAD, but is much less reliable. CATC is commonly used at water treatment plants.

Typically, residual cyanide concentrations of total cyanide in tailings will be in the range of 80 to 400 mg/L, of which about 50% may be CN_{WAD}. As cyanide solutions at such concentrations are toxic to many organisms, treatment is required before tailings solutions can be safely released to the environment.

Unfortunately some jurisdictions do not explain which form of cyanide is being regulated. Total cyanide is not particularly relevant from an environmental perspective and free cyanide is difficult to measure and does not include all potentially damaging forms. Accordingly, WAD cyanide is considered to be the best measure for assessing human and animal toxicity.

WAD cyanide is considered to be the best measure for assessing human and animal toxicity.

Many different processes and procedures have been used to remove cyanide from process wastes. These include both destruction processes and recovery processes and are listed in **Table 6.3**.

The most common means of reducing cyanide levels to achieve safe disposal include:

- Natural degradation by volatilization, enhanced by exposure to sunlight;
- Dilution of cyanide-bearing solutions by mixing with non-cyanide-bearing water;
- Oxidation, which can be accomplished using a variety of oxidants such as hydrogen peroxide (H_2O_2), chlorine (in the form of calcium or sodium hypochlorite), Caro's acid (H_2SO_5), or sulphur dioxide (in the form of sodium metabisulphite);
- Cyanide recovery and reuse, which may be achieved by lowering pH to form HCN, which is then volatilized and re-dissolved for addition to the leach solution.

Natural degradation was widely used in the past and can be effective and reliable, particularly in the flat arid goldfields of Australia where large, shallow tailings ponds facilitate rapid degradation of cyanide to form HCN gas. Natural oxidation of cyanide ions from alkaline solutions may also generate ammonia (NH_3), cyanate ion (CNO^-) and cyanogen gas $(CN)_2$ according to the following equations:

$$
\begin{aligned}
CN^- + \tfrac{1}{2}O_2 + 2H_2O &\leftrightarrow NH_3 + HCO_3^- ;\\
CN^- + 2OH^- &\leftrightarrow CNO^- + H_2O + 2e^-;\\
2CN^- &\leftrightarrow (CN)_2\uparrow + 2e^-.
\end{aligned}
\tag{6.15}
$$

Degradation of certain dissolved cyanide complexes may also result from precipitation of stable metallo-cyanides, which in some cases involve the release of HCN gas.

TABLE 6.3

Cyanide Removal Technologies

Technology (and Type*)	Short Description	Basic Reagents	Basic Products
A. OXIDATIVE			
Alkaline Chlorination (C)	Oxidation to CNO^- and then N_2 and CO_3^{2-} with Cl_2 or ClO^- at pH >11	Cl_2/ClO^-, NaOH	CNO^-, CO_3^{2-}, N_2
SO_2/Air (C)	Oxidation to CNO^- with SO_2/Air and soluble Cu catalyst; INCO process	SO_2, air, Cu catalyst	CNO^-
Hydrogen Peroxide (C)	Oxidation to CNO^- with H_2O_2 and Cu^{2+} catalyst; Degussa Process	H_2O_2	CNO^-, CO_3^{2-}, NH_4^+
Caro's Acid (C) (C)	Oxidation to CNO^- with H_2SO_5	H_2SO_5	CNO^-
Activated Carbon (C & P)	Oxidation to CNO^- and then partially to CO_3^{2-} and NH_4^+ with activated carbon and Cu catalyst	Activated carbon, air/O_2, Cu catalyst	CNO^-, CO_3^{2-}, NH_4^+
Biodegration (B)	Oxidation to CO_3^{2-} and NH_4^+ and then NO_3^- using indigenous micro-organisms	Na_2CO_3, H_3PO_4	CO_3^{2-}, NH_4^+, NO_3^-, PO_x
UOP Catalytic Oxidation (C)	Oxidation to CO_2, N_2 and NH_4^+ with air at mild temperatures (<130°C) and pressures (550 kPa) with a catalyst	Catalyst	CO_2, N_2, and NH_4^+
Ozonation (C)	Oxidation to CO_3^{2-} and N_2 with O_3	O_3	CO_3^{2-}, N_2
Wet Air Oxidation (C)	Oxidation to CO_2 and N_2 at high temperatures (175 – 320°C) and high pressures (2,100 – 20,700 kPa)	none	CO_2, N_2
Photocatalytic Oxidation (C & P)	Oxidation to CNO^- and then NO_3^- and CO_3^{2-} using uv/visible light and semiconductor type substrate, e.g. TiO_2, ZnO or CdS		
B. NON-OXIDATIVE			
Natural Degradation (B, C & P)	Mainly volatilization of HCN from tailing dams	none	Mainly HCN
AVR (C & P)	Acidification-Volatilization-Reneutralization. After acidification to pH <3, HCN(g) is volatilized and absorbed in NaOH and recycled. Metals are precipitated after reneutralization	H_2SO_4, NaOH	HCN, SCN^-
CYANISORB® (C & P)	Similar to AVR bur HCN(g) stripped at higher pH values (5.5–7.5)	H_2SO_4, NaOH	HCN, SCN^-
CRP (C & P)	Cyanide Regenaration Process; similar to AVR but with better HCN(g) stripping and metal precipitation	H_2SO_4, NaOH	HCN, SCN^-
Thermal Hydrolysis (C)	Hydrolysis to NH_4^+ and formate at high temperatures	none	NH_4^+, $HCOO^-$
Alkaline Hydrolysis (C)	Hydrolysis to NH_4^+ and formate at high temperatures (100–250°C) and high pH	NaOH	NH_4^+, $HCOO^-$
GM-IX (C & P)	Gas Membrane-Ion Exchange; Ion Exchange concentrates CN. After regeneration the Gas Membrane recovers pure CN.	Resin	CN^-
Prussion Blue Precipitation (C)	Precipitation of $Fe_4[Fe(CN)_6]_3$ on addition of $FeSO_4$	$FeSO_4$	$Fe_4[Fe(CN)_6]_3$

(Continued)

TABLE 6.3
(Continued)

Technology (and Type*)	Short Description	Basic Reagents	Basic Products
Pregnant Pulp Air Stripping (P)	Air stripping from pregnant pulps	Air	HCN
Reverse Osmosis (P)	Physical removal of cyanide and its complexes by a semipermeable membrane process under pressure	H_2SO_4	CN^-
Flotation (P)	Adsorption of precipitated CN particles onto fine air bubbles	$FeSO_4$, Surfactant	$Fe_4[Fe(CN_6)]_3$
High Rate Thickeners (P)	Fast thickening and recycling of CIP tailings	none	CN^-

Source:
Environment Australia 1998

There are several reasons why natural degradation may not be appropriate as the sole means of cyanide control. These include:

- An excess of water, requiring discharge without sufficient residence time to achieve the required degradation;
- The risk that the tailings storage could be overtopped, discharging cyanide-bearing solution, before sufficient degradation is achieved; and
- The risk that animals, including livestock or wildlife including water birds are attracted to the tailings ponds and are poisoned as a result. This risk is particularly high in arid areas where surface water occurrences are usually rare.

Away from sunlight, cyanide may persist for many years if the solution remains alkaline and there are no other reactions to form complex metal precipitates.

Cyanide, being relatively unstable when exposed to sunlight, is generally not persistent in the environment; however, there are some situations in which cyanide can persist. One example is in groundwater; away from sunlight, cyanide may persist for many years if the solution remains alkaline and there are no other reactions to form complex metal precipitates. A case of persistent cyanide is described in **Case 6.3.**

Dilution is not commonly used on its own but may be used in combination with other processes, or to provide an additional factor of safety. In the event of an accidental spill of

CASE 6.3
Cyanide and Groundwater

Montana is the only state in the USA to have implemented a ban on cyanide leaching of gold ores. Three primary reasons have been given for the decision to phase-out existing projects and prohibit new developments involving cyanide leach mining in Montana:

1. Open pit, cyanide leach mines threaten the private property rights of neighbouring landowners (e.g. Landowners downstream of the Golden Sunlight mine were forced to sell their properties to Placer Dome Corp. after their drinking water well was contaminated with cyanide)

2. Open pit, cyanide leach mines expose Montana taxpayers to the costs of reclamation and leave liabilities for future generations (e.g. Pegasus Gold Corp. declared bankruptcy in 1997 leaving the State with insufficient funds to reclaim the Zortman/Landusky mine, Montana's largest gold mine. The State has estimated that water treatment will have to occur at the mine site in perpetuity).

3. Open pit, cyanide-leach mines consistently contaminate Montana's water resources with cyanide and other pollutants placing human and environmental health at risk. Since 1982, there have been 50 cyanide releases at Montana mines, releasing millions of gallons of cyanide solution into Montana's soil, surface and groundwater resources. Cyanide can persist for very long periods of time in groundwater because the sunlight and oxygen needed to break it down to less harmful substances are largely absent. Groundwater contamination is the most prevalent form of cyanide contamination at Montana's open pit cyanide leach mines because the liner systems designed to prevent this type of occurrence are not impermeable and are prone to structural damage (punctures or tears).

Source: Montana Environmental Information Center (MEIC), meic@meic.org cyanide into surface water, dilution may be the most important determinant of the fate of cyanide and the environmental damage that results (**Case 6.4**).

The most common method of cyanide destruction from process waste streams is the INCO SO_2/Air™ detoxification process in which sodium metabisulphite solution is used in the presence of a copper sulphate catalyst to oxidize the cyanide in accordance with the following reaction:

$$CN^- + SO_2 + H_2O + O_2 \rightarrow CNO^- + H_2SO_4 \qquad (6.16)$$

This process can oxidize weak and moderately strong metal cyanide complexes as well as free cyanide (CN^-). Depending on a variety of considerations including water balance, likely presence of birds, and applicable regulations, the process may be applied to the total tailings stream before discharge to the tailings storage facility (TSF), or it may be applied only to excess supernatant water removed from the TSF, for subsequent release to the environment. Clearly, cyanide destruction prior to discharge to the TSF removes any risk to wildlife or to downstream communities or ecosystems in the event of overflow from the TSF. However, this represents a much more expensive approach as it decreases the extent to which the cyanide-bearing solution can be recycled as well as requiring higher consumption of detoxifying reagents.

As shown in **Table 6.3**, there are several processes to recover cyanide for reuse. These non-oxidative processes involve acidification to form HCN, volatilization of the HCN and absorption into solution. One such process is CYANISORB®, which is used at the Golden Cross Gold Mine in New Zealand and at a mine in Argentina. In this process, the tailings are first acidified using sulphuric acid to lower the pH to 7.5, following which the tailings are contacted with high volumes of turbulent air which strips the HCN. The HCN vapour is then redissolved in caustic solution which flows to the leach circuit. Subsequently, lime is added to restore alkaline conditions, thereby precipitating de-complexed cations. At Golden Cross, this process reportedly recovers 80% to 90% of cyanide. The process is reported to be cost-effective compared to other cyanide control technologies. In view of its reported efficacy and cost benefits, it is somewhat surprising that this technology has not been more widely applied.

The mining industry has been using cyanide for gold extraction for more than 100 years without human fatalities.

As an environmental threat, the use of cyanide to leach gold has been much discussed in recent years, but that cyanide leaching is a problem is perception rather than fact. The mining industry has been using cyanide for gold extraction for more than 100 years without

CASE 6.4
Comparing Two Cyanide Spills

A notorious example of a serious cyanide spill was the case of the tailings overflow from the Baia Mare gold mining operation in Romania, further detailed in Case 18.3. The tailings overflow resulted in the sudden discharge of an estimated 100,000 m^3 of mud and wastewater with a 126 mg/L cyanide load into the Lapus River, a tributary to the Szamos River, which both were largely frozen at the time. As a result, the plume or 'slug' of contaminated water flowed downstream along the bed of this river and, subsequently, the Danube River with relatively low rates of dilution. Cyanide decomposition was minimized by the very low temperature and the absence of sunlight beneath the river ice. Undoubtedly, free cyanide would have been generated progressively from complex cyanide ions as mixing lowered the pH. As a result of this spill event, toxic effects, notably fish kills, extended many kilometres downstream of the spill site.

This contrasts with another notorious incident at the Omai Gold Mine in Guiana in 1995, where a tailings dam failure caused a spill of 3 million m^3 containing cyanide at 25 – 30 mg/L. However, in this case, fish kills were limited to the Omai River River and, possibly, a short portion of the larger Essequibo River, into which it flowed. The much more serious impact of the Romanian event can partly be explained by the cold temperature, the presence of ice shielding the contaminated water from sunlight, and the low rate of mixing with uncontaminated water (dilution).

human fatalities. Other relevant facts that help to put the toxicity of cyanide into context include:

- There are well-established threshold concentrations, below which cyanide exposure is completely harmless. (This is in contrast to some toxins which can be cumulative, or others which are considered to pose risks at even extremely low concentrations);
- Forms of cyanide occur in nature, particularly in many plants and their fruits, and are commonly consumed in small amounts without adverse effects;
- Cigarette smoke contains cyanide at concentrations above those that would be permitted in a working environment;
- Toxic forms of cyanide tend to be reactive and unstable and hence are subject to natural degradation by a variety of reactions, and generally do not persist in the environment.

Residual cyanide after gold adsorption occurs as free cyanide (CN^-) plus a variety of metal cyanide complexes of differing toxicities, some of which are weak and readily dissociated. **Table 6.4** lists the cyanide species found in gold mine tailings with information on stability and toxicity.

There remains a need for wider public education, information, and communication, but what counts most is overall industry performance, since the public judges the industry by its worst performers, not its best. Recent accidental cyanide tailings and effluent spills have tended to affect public confidence in the mining sector (see also Chapter Eighteen). In January 2000, the accidental release of large amounts of cyanide effluent from the Aurul mine in Romania resulted in major media exposure of the cyanide leaching process and the resulting river pollution (see also Case 18.3). In March 2000, a second accident at another nearby mine released heavy metal-containing effluent and sludge into the same river system. These unfortunate spills tend to mask the overwhelming safe performance of hundreds of mines over many decades.

The Cyanide Code has been developed as a joint initiative between groups of gold mining companies and cyanide producers, with the objective of encouraging and facilitating the responsible management of cyanide.

The Cyanide Code has been developed as a joint initiative between groups of gold mining companies and cyanide producers, with the objective of encouraging and facilitating the responsible management of cyanide. It is a voluntary code for producers, transporters and users of cyanide, with specific verification protocols for each of these groups. The Code contains a series of nine Principles, each of which is supported by one or more Standards of Practice, as shown in **Table 6.5**. The Code is administered by the International Cyanide Management Institute, a non-profit, industry sponsored corporation whose responsibilities are to:

- Promote adoption of and compliance with the Code, and to monitor its effectiveness and implementation within the world gold mining industry;
- Develop funding sources and support for Institute activities;
- Work with governments, NGOs, financial interests and others to foster widespread adoption and support of the Code;
- Identify technical or administrative problems or deficiencies that may exist with Code implementation; and
- Determine when and how the Code should be revised and updated.

Signatories must submit the relevant operations to rigorous certification and verification procedures including an external audit every three years. Transparency is assured through publication of Summary Audit Reports and Action Plans on the Cyanide Code website (www.cyanidecode.org). The Code was first introduced in 2005, and by the end of 2006, had 27 signatories throughout the world including 14 mining companies, 8 cyanide producers and 5 cyanide transporters.

TABLE 6.4

Nomenclature, Stability and Toxicity of Some Important Cyanide Species in Gold Mining Tailings

Term	Analytical type [a]	Species or compound	Log equilibrium constant [b] and solubility data	Toxicity to fish [c] (LC_{50} in mg/L)
1. Free cyanide	free cyanide	CN-HCN	not applicable 9.2	−0.1 0.05 to 0.18
2. Simple compounds				
a) readily soluble	free cyanide	$KCN(s)$*	sol = 71.6 g/100 g H_2O (25°C)	0.02 to 0.08
		$NaCN \cdot H_2O(s)$	sol = 34.2 g/100 g H_2O (15°C)	0.4 to 0.7
		$Ca(CN)_2(s)$		–
b) relatively insoluble	WAD/CATC/total	$CuCN(s)$	−19.5	–
		$Zn(CN)_2(s)$	−15.9	–
		$Ni(CN)_2(s)$	sol = 9.1×10^{-4} g/100 g H_2O (25°C)	–
3. Weak complexes	WAD/CATC/total	$Cd(CN)_4^{2-}$	17.9	–
		$Zn(CN)_4^{2-}$	19.6	0.18
4. Moderately strong complexes	WAD/CATC/total	$Ni(CN)_4^{2-}$	30.2	0.42
		$Cu(CN)_2^{-}$	16.3	–
		$Cu(CN)_3^{2-}$	21.6	0.71 (24 hours)
		$Cu(CN)_4^{3-}$	23.1	–
		$Ag(CN)_2^{-}$	20.5	–
5. Strong complexes	total	$Fe(CN)_6^{4-}$	35.4	35.0 (light); 860 to 940 (dark)
		$Fe(CN)_6^{3-}$	43.6	35.2 (light); 860 to 1210 (dark)
		$Au(CN)_2^{-}$	38.3	–
Thiocynate		SCN^-	not relevant	50–200
Cyanate		CNO^-	not relevant	34–54

(a) WAD = weak acid dissociable; CATC = cyanide amendable to chloride; total = total cyanide following acid distillation.
(b) Measure of stability, given only for comparative purposes; values quoted in the literature are quite variable.
(c) Toxicity data is given only for guidance and should be used for comparative purposes only, a dash indicates 'no relevant data found'.
(Beck, 1987, Hagelstein, 1997, Minerals Council of Australia, 1996, Richardson, 1992)
* (s) = solid

Source:
Environment Australia 1998

TABLE 6.5

Cyanide Code Principles and Standards of Practice

Principle	Standards of Practice
1. PRODUCTION. Encourage responsible cyanide manufacturing, by purchasing from manufacturers who operate in a safe and environmentally protective manner.	1.1 Purchase cyanide from manufacturers employing appropriate practices and procedures to limit exposure of their workforce to cyanide and to prevent releases of cyanide to the environment.
2. TRANSPORTATION. Protect communities and the environment during cyanide transport.	2.1 Establish clear lines of responsibility for safety, security, release prevention, training, and emergency response, in written agreements with producers, distributors and transporters. 2.2 Require that transporters implement appropriate emergency response plans and capabilities, and employ adequate measures for cyanide management.
3. HANDLING and STORAGE. Protect workers and the environment during cyanide handling and storage.	3.1 Design and construct unloading, storage and mixing facilities consistent with sound accepted engineering practices, and quality control and quality assurance procedures, spill prevention and spill containment measures.

(Continued)

TABLE 6.5
(Continued)

Principle	Standards of Practice
	3.2 Operate unloading, storage and mixing facilities using inspections, preventive maintenance and contingency plans to prevent or contain releases, and control and respond to worker exposures.
4. OPERATIONS. Manage cyanide process solutions and waste streams to protect human health and the environment.	4.1 Implement management and operating systems designed to protect human health and the environment, including contingency planning and inspection and preventive maintenance procedures. 4.2 Introduce management and operating systems to minimize cyanide use, thereby limiting concentrations of cyanide in mill tailings. 4.3 Implement a comprehensive water management programme to protect against unintentional releases. 4.4 Implement measures to protect birds, other wildlife and livestock from adverse effects of cyanide process solutions. 4.5 Implement measures to protect fish and wildlife from direct and indirect discharge of cyanide process solutions to surface water. 4.6 Implement measures designed to manage seepage from cyanide facilities to protect the beneficial uses of groundwater. 4.7 Provide spill prevention and containment measures for process tanks and pipelines. 4.8 Implement quality control/quality assurance procedures to confirm that cyanide facilities are constructed according to accepted engineering standards and specifications. 4.9 Implement monitoring programmes to evaluate the effects of cyanide use on wildlife, surface and groundwater quality.
5. DECOMMISSIONING. Protect communities and the environment from cyanide through development and implementation of decommissioning plans for cyanide facilities.	5.1 Plan and implement procedures for effective decommissioning of cyanide facilities to protect human health, wildlife, and livestock. 5.2 Establish an assurance mechanism capable of fully funding cyanide-related decommissioning activities.
6. WORKER SAFETY. Protect workers' health and safety from exposure to cyanide.	6.1 Identify potential cyanide exposure scenarios and take measures, as necessary, to eliminate, reduce and control them. 6.2 Operate and monitor cyanide facilities to protect worker health and safety, and periodically evaluate the effectiveness of health and safety measures. 6.3 Develop and implement emergency response plans and procedures to respond to worker exposure to cyanide.
7. EMERGENCY RESPONSE. Protect communities and the environment through the development of emergency response strategies and capabilities.	7.1 Prepare detailed emergency response plans for potential cyanide releases. 7.2 Involve site personnel and stakeholders in the planning process. 7.3 Designate appropriate personnel and commit necessary equipment and resources for emergency response. 7.4 Develop procedures for internal and external emergency notification and reporting. 7.5 Incorporate into response plans, monitoring elements and remediation measures that account for the additional hazards of using cyanide treatment chemicals. 7.6 Periodically evaluate response procedures and capabilities, and revise them as needed.
8. TRAINING. Train workers and emergency response personnel to manage cyanide in a safe and environmentally protective manner.	8.1 Train workers to understand the hazards associated with cyanide use. 8.2 Train appropriate personnel to operate the facility according to systems and procedures that protect human health, the community and the environment. 8.3 Train appropriate workers and personnel to respond to worker exposures and environmental releases of cyanide.
9. DIALOGUE. Engage in public consultation and disclosure.	9.1 Provide stakeholders the opportunity to communicate issues of concern. 9.2 Initiate dialogue describing cyanide management procedures and responsively address identified concerns. 9.3 Make appropriate operational and environmental information regarding cyanide available to stakeholders.

Source:
Cyanide Code (International Cyanide Management Institute)

The Bayer Process in Alumina Refining

Alumina (aluminium oxide and aluminium trihydroxide) is the material from which all of the world's aluminium is produced. Transforming bauxite ore into alumina is based on the Bayer process and the principles of this process are essentially the same today as when it was first introduced about 100 years ago. Bauxite ore is ground and mixed with caustic soda. On average 3 tons of ore is needed to produce 1 ton of alumina (**Figure 6.11**). The mixed solution is pumped into high-pressure, heated vessels in which the aluminium trihydroxide in the bauxite is dissolved. The hydrate ($Al(OH)_3$) is precipitated from the solution after mud separation. Hydrate solids are washed using filters. Calcination converts the hydrate into alumina oxide, a sugar-like white powder.

Figure 6.12 illustrates these steps. The four basic steps are (1) digestion, (2) clarification, (3) precipitation and filtration, and (4) calcination.

Alumina (aluminium oxide and aluminium trihydroxide) is the material from which all of the world's aluminium is produced.

Digestion

Mined bauxite is crushed and ground to produce a uniform fine grained material with a maximum grain diameter of 2 mm to ensure sufficient solid-liquid contact during the digestion phase. After grinding, the ore is mixed with spent liquor (typically a solution of sodium hydroxide from the recycled caustic soda liquor circuit) and pumped into a series of digestion vessels. The prepared slurry ranges from 40 to 50% solids content.

The controlling factors for the digestion step are the concentrations of caustic soda and bauxite, grain size, pressure, temperature, and retention time. Pressure, temperature, and retention time vary from plant to plant, but typical figures are as follows: a digester pressure equivalent of about 4 kg/cm^2, a temperature of about 140°C, and a retention time of 30 to 45 minutes.

The controlling factors for the digestion step are the concentrations of caustic soda and bauxite, grain size, pressure, temperature, and retention time.

At the digestion process, solid hydrate alumina in the bauxite (mineralogically gibbsite) reacts with sodium hydroxide:

$$Al_2O_3\,3H_2O + 2NaOH \rightarrow 2NaAlO_2 + 4H_2O \tag{6.17}$$

The slurry leaving the digestion tanks contains the aluminium ion in solution (in the form of a sodium aluminate solution), together with undissolved ore solids.

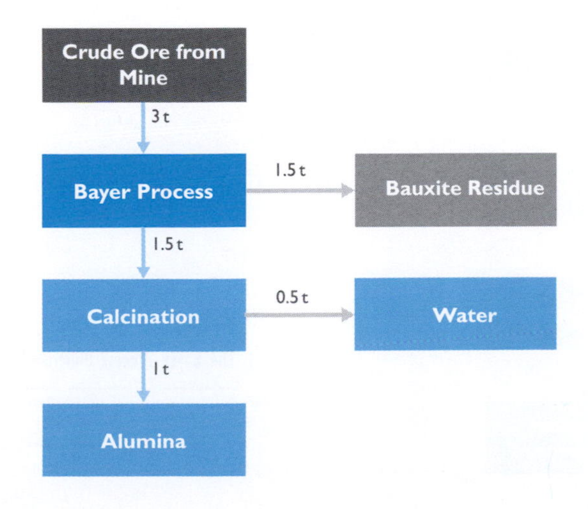

FIGURE 6.11
From Bauxite Ore to Alumina: A Simplified Mass Balance

On average 3 tons of ore is needed to produce 1 ton of alumina.

Clarification

Clarification refers to the separation of undissolved solids from the aluminium-bearing digestion solution. The hot slurry from digestion is first directed through a series of flash tanks gradually reducing the pressure and recovering heat that can be reused in the digestion process.

Separation is then achieved using large clarification vessels (mud thickeners), which allow the solids to settle out. A flocculant is added to improve separation by gravitation and subsequent filtration. Separated solids are passed through a counter-current washing train (mud washers) using water to recover caustic soda which then is directed to the caustic liquor circuit (step 1).

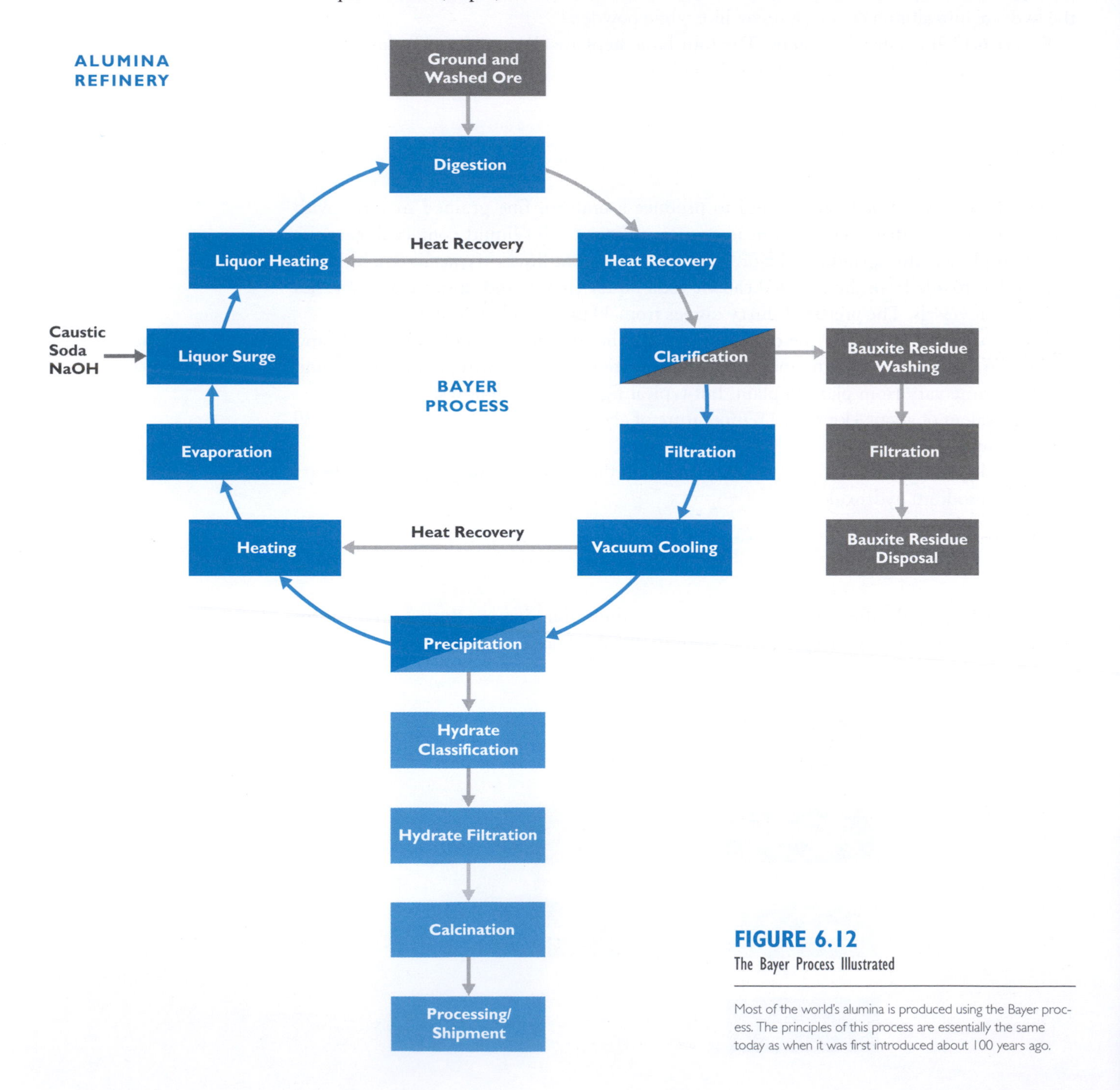

FIGURE 6.12
The Bayer Process Illustrated

Most of the world's alumina is produced using the Bayer process. The principles of this process are essentially the same today as when it was first introduced about 100 years ago.

In conventional wet tailings disposal schemes, washed solids (termed red mud or bauxite residues) are pumped to tailings ponds for final disposal (see Chapter Eighteen for a discussion of tailings management practices). In recent years dry tailings disposal schemes emerged as the preferred tailings management practices in which washed solids are dewatered using filter units to produce bauxite residue cake. Red mud consists of iron oxide, sodium aluminium silicate, titanium oxide, and some other metals, and is alkaline with a pH of about 13.

Red mud consists of iron oxide, sodium aluminium silicate, titanium oxide, and some other metals, and is alkali with a pH of about 13.

Precipitation and Filtration

After separation of solids, the remaining liquor (or pregnant liquor) is passed to precipitation via heat exchangers where the pregnant liquor is cooled to 60 to 70°C. Recovered heat is transferred to the spent liquor (liquor from which aluminium ions have been separated) that is returned to the digestion process.

The cooled pregnant liquor is seeded with hydrate crystals, which act as a nucleus (or seed crystal) for precipitation. Seed crystals grow as they settle through the liquor by attracting dissolved aluminium ions to them. The chemical reaction during precipitation that removes dissolved aluminium ions from the pregnant liquor is as follows:

$$2NaAlO_2 + 4H_2O \rightarrow Al_2O_3 \cdot 3H_2O + 2NaOH \qquad (6.18)$$

The reaction is slow with a large fraction of soluble alumina remaining in the spent liquor. While the retention time in precipitation tanks is about 35 to 40 hours, most of the yield and particle growth is obtained in the first few hours.

Calcination

Calcination transforms precipitated hydrates into alumina. After filtration, alumina hydrates are fed into a rotating, cylindrical kiln that is tilted to allow gravity to move the material through it, heating it to a temperature of about 1,100°C to 1,300°C. After leaving the kiln, the crystals pass through a cooler. The final product is in the form of white alumina powder.

6.4 PYROMETALLURGY AND RELATED ENVIRONMENTAL CONCERNS

The main environmental challenges in pyrometallurgical processes are the production of large quantities of gaseous emissions (e.g. sulphur dioxide, vaporized metals, and dust), solid emissions (e.g. slag, gypsum, and wastewater treatment sludge), and liquid emissions (e.g. effluents and cooling water).

Gaseous Emissions

Control of gaseous emissions is a continuing challenge to the pyrometallurgical industry. Emission controls focus mainly on stack emissions, but fugitive emissions also contribute to the overall emissions of a smelter. Fugitive emissions are emissions that are not released through a vent or stack. Examples of fugitive emissions include volatilization of vapour from vats or open vessels during material movement or dust from stockpiles and conveyors.

Control of gaseous emissions is a continuing challenge to the pyrometallurgical industry.

Atmospheric emissions may contain only gases, only particulate matter, or a combination of both. Considering the use of sulphide ores, copper smelters, lead smelters, and zinc roasters produce huge amounts of sulphur dioxide. In copper smelting, for example, typically more sulphur dioxide is produced by weight than copper. Rather than discharging sulphur dioxide into the air, as was once the practice (e.g. The Mount Isa Mine in Australia released all of its sulphur dioxide directly to the atmosphere as recently as 1999), more than 99 percent of the sulphur dioxide is now captured. The standard technique to limit sulphur dioxide emission is to convert sulphur dioxide to sulphuric acid, a valuable by-product. Common technologies to capture particulate matter include electrostatic precipitators and bag houses.

In the USA, smelter-produced sulphuric acid amounts to approximately 10% of the total acid production from all sources. Prior to the mid-1980s, this by-product sulphuric acid was mostly sold as a raw material to other industries such as the fertilizer industry, often at a loss due to the long shipping distances involved. Today mining operations themselves increasingly use sulphuric acid for hydrometallurgical processing of oxide ores and sometimes of tailings as discussed in the earlier sections of this chapter.

Schlesinger (1991) established the global sulphur cycle, illustrating how liberation of elements from the Earth by the extractive industries has resulted in regional and global impacts, primarily the generation of acid rain and its subsequent impacts on natural ecosystems (**Figure 6.13, Case 6.5**). The numbers may have changed since then but the underlying message remains the same (Jordan and D'Alessandro 2004). The major pool of sulphur in the global sulphur cycle is found in crustal minerals, gypsum, and pyrite (Schlesinger 1991). Many metals are mined from sulphide minerals. Sulphur is also an important constituent of coal and oil, and large amounts of SO_2 are emitted during the combustion of fossil fuel and the smelting of metal ores.

Large amounts of SO_2 are emitted during the combustion of fossil fuel and the smelting of metal ores.

Coal and petroleum extraction mobilizes about 149 million tons per year (Mt/y) of sulphur, more than double the amount produced 100 years ago (Brimblecombe *et al.* 1989), and almost twice the amount naturally liberated from the Earth's crust. Of this, about 93 Mt/y is released to the atmosphere, about twice the natural (volcanic, dust and biogenic gases) emissions (**Figure 6.13**). Human activities contribute the largest atmospheric sulphur

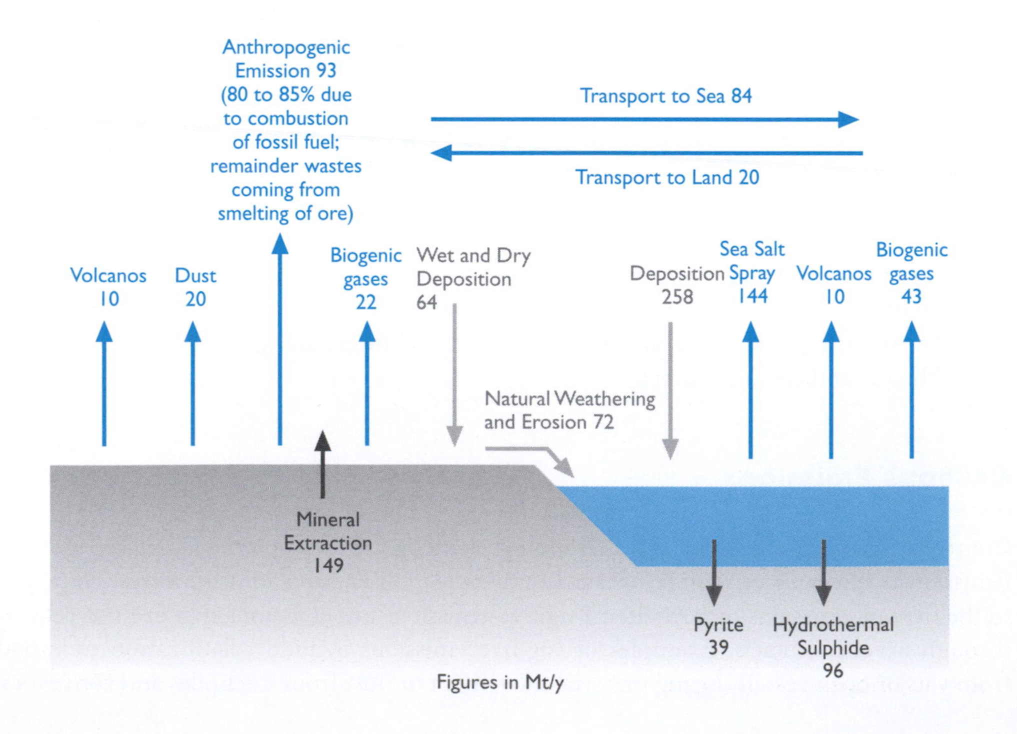

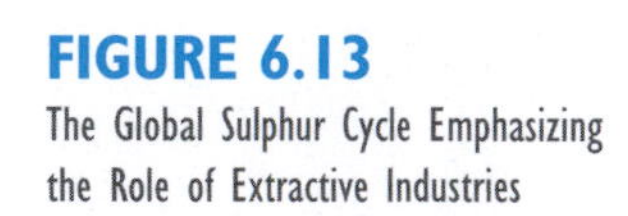

FIGURE 6.13

The Global Sulphur Cycle Emphasizing the Role of Extractive Industries

Despite the relative insignificance of human inputs to the global sulphur cycle, anthropogenic atmospheric emissions due to smelting, fossil fuel combustion and other activities are more than 10 times the emissions from natural sources.

Source: Jordan and D'Alessandro 2004, Schlesinger 1991

releases. Ivanov *et al.* (1983) suggested that the present river sulphur transport (from air pollution, mining, erosion, and other sources) of 200 Mt/y is about double that of pre-industrial conditions. Excess acidic deposition is likely to cause changes in rock weathering, and in mobilization of metals in soils and sediments. Thus, global cycles of heavy metals are coupled to the sulphur cycle primarily by liberation from metal sulphide ores by mining, metal-sulphide precipitation in anoxic conditions, and metal mobilization due to sulphate-induced acidification in terrestrial environments (Benjamin and Honeyman 1992). Human-induced emissions have led to a net flux from land to sea through the atmosphere, where 100 years ago the net flux was the reverse. The global cycle of sulphur is not at a natural steady state at present and probably never has been. Schlesinger (1992) concludes that although human activities have caused only a minor change in global pools of sulphur, very large changes have occurred to the annual flux through the atmosphere.

On the other hand, there are large areas of the land surface where sulphur is deficient. Much of Australia is in this category. The argument can be made that such areas benefit from deposition of sulphur from the atmosphere.

Solid Emissions

The production of metals results in the generation of a wide variety of wastes and residues. They are a result of the metals separation that is necessary for the production of pure metals from complex sources. These wastes and residues arise from the different stages of processing as well as from the off-gas and water treatment systems. The key solid wastes and residues which result from production of non-ferrous metals are: (1) slag; (2) drosses and skimmings; (3) spent linings and refractories; (4) wastes/residues/by-products of air and water pollution abatement systems; and (5) wastes and residues from hydrometallurgical processes.

Slag is produced by the reaction of slag-forming elements (e.g. iron) in the ore with added fluxes. In the smelting process, the slag is liquid and has a different density than the melted metal and separates on top of the metal-rich matte. The slag can, therefore, be tapped-off separately. It is either rapidly quenched with water to form granules that can be

Slag is produced by the reaction of slag-forming elements (e.g. iron) in the ore with added fluxes.

CASE 6.5
Coal Combustion and Acid Rain

In Europe acid rain started to make the front page in the 1970s when large forest areas started to die. While much of the damage was concentrated around the coal-fired power plants in Eastern Europe burning low quality, high sulphur coal, conifer trees also started to shed their needles in the picturesque Black Forest, a forest close to the heart of Germans and tourists alike. Forest damage due to acid rain, of course, also occurred elsewhere. Maple trees in Quebec started to produce less syrup, largely attributed by the maple syrup producers to the sulphur dioxide emissions generated in the US by the extensive coal and steel industry in the Appalachian region. The issue of trans-boundary pollution, and its political fallouts emerged. As with the current debate on global warming, the controversy about acid rain was emotionally charged and highly polarized. Nevertheless, the severe environmental consequences of acid rain were soon recognized, and, among other measures, more stringent SO_2 emission standards were implemented.

reused or discarded, or alternatively, the slag is transported in a liquid state to a cleaning operation or slag dump (**Case 6.6**).

Drosses and skimmings result from the oxidation of metals at the bath surface or by reactions with fireproof material used as furnace linings.

Spent linings and refractories result when refractory material falls out of the furnace linings or when the furnace lining has to be replaced completely (**Figure 6.14**).

Pollution abatement systems wastes and residues include flue gas dust and sludge recovered from the air pollution control equipment as well as other solids such as spent filter materials. Sulphuric acid and liquid sulphur dioxide are also wastes/residues from pollution abatement systems as discussed above. In addition, wastewater treatment plants generate metal-containing sludge.

Hydrometallurgical wastes and residues include sludge generated in the leaching process. Purification and electrolysis processes can generate metal rich solids such as anode slime. Other wastes and residues typically produced are oils and greases and industrial scrap such as steel and wood wastes.

In terms of quantity, slag as the final product of unwanted materials, contributes to most of the solid waste outputs of a smelter. Not all slag is waste. Depending on the smelter type and mineral process, slag can be a valuable by-product to be used as construction material or for sand blasting (Borell 2005) (**Case 6.7**). When slag is rapidly cooled by instant granulation in water, a stable granular product is formed. The granulated slag is often virtually completely amorphous (glassy). Remaining trace metals incorporated in glass phases show a low solubility in comparison with those existing mainly in crystalline phases (Konradsson 2003), a characteristic that holds metals immobile in granulated copper slag and in most other slag.

In terms of quantity, slag as the final product of unwanted materials, contributes to most of the solid waste outputs of a smelter.

Alternative materials are often labelled as waste which puts focus on their assumed environmental properties. Concerns about assumed environmental risks are sometimes a barrier to a wider use of alternative materials. A far too common approach is to classify a material based only on its total chemical analysis (say metal content) without further analysis of

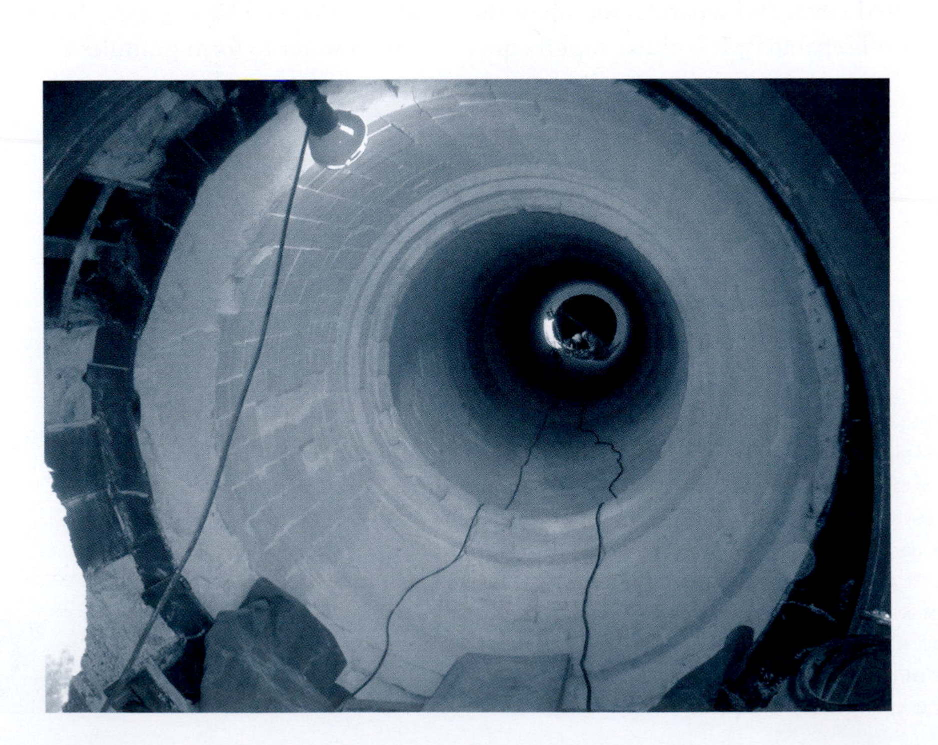

FIGURE 6.14
Renewing the Furnace Linings of a Rotary Furnace

TABLE 6.6

Laboratory Leaching Test Methods – A far too common approach is to classify a material based only on the chemical assays. The appropriate approach for assessing environmental characteristics of solid materials is by laboratory leaching testing

Type of Leaching Test	Test Standard
Column-leaching test	(EN 14405)
Two-stage agitated leaching test	(EN 12457-3)
Availability-leaching test	(NT Envir 003)
Oxidized availability leaching test	(NT Envir 006)
pH-static leaching test	(CEN/TC 292/WG6)
The US EPA Toxicity Characteristic Leaching Procedure (TCLP)	Test method 1311

how metals are bound in the matrix. The appropriate approach for assessing environmental characteristics of slag is in form of laboratory leach testing. A variety of factors influence the leaching characteristics of different elements. A variety of different standard tests are used to characterize the leaching properties of a material (**Table 6.6**).

Column tests and two-stage serial batch tests are frequently used for the characterization of by-products, including materials to be characterized before land filling. A leachate test predicts whether a material is likely to leach contaminants, specifically metals, at levels of concern. The material is subjected to an extraction fluid (e.g. a basic or buffered solution) and then the resulting extract is analyzed and compared to a leachate criteria list. If the resulting concentration is equal to or in excess of the concentration specified for that contaminant in the applicable schedules, the material is considered leachate toxic and may be considered potentially toxic or hazardous. In addition, the material may require treatment to eliminate or reduce the leaching potential prior to disposal or may require disposal in a secure, hazardous waste landfill.

A leachate test predicts whether a material is likely to leach contaminants, specifically metals, at levels of concern.

The second main category of solid waste streams originates from air pollution controls (e.g. gypsum from scrubbers) and water treatment facilities. These wastes are usually hazardous and in the form of sludge. Final disposal is difficult and generally costly. Modern smelting technologies aim to reduce gaseous and solid waste emissions. Wastes destined for disposal from metal production are kept to a minimum. Flue dusts are typically recycled to

CASE 6.6

Slag Disposal into the Environment – An Environmental Eyesore or Reason for Concern?

The nickel mining and smelting operation of INCO in Sulawesi, Indonesia commenced about four decades ago and has been in operation since then. The ferro-nickel smelter produces slag at a rate of several tons per hour, which is dumped at valleys nearby the smelter. While leachate testing indicates that the slag is inert, the practice continually invites mine opponents to criticize the mining operation in general and the slag disposal practice in particular. Claims continue to be made that leachate from slag disposal sites cause environmental damage to groundwater and surface water runoff.

the smelting furnaces. Dust from the bag filters and metals precipitated in the wastewater treatment plant are also fed to flash smelting furnaces as secondary materials. A fine example of internal waste material management is the Mitsubishi Continuous Smelting technology that practically eliminates the production of hazardous solid waste in copper production (**Case 6.8**).

Liquid Emissions

It is practically impossible to operate a closed circuit system without allowing for a bleed-off stream to avoid the build up of dissolved solids and/or a consistent decrease or increase in pH.

Liquid wastes in pyrometallurgical operations range from wash water to effluent from wet scrubbers in air emission controls. Efforts are made to recycle water to the maximum possible extent. However, it is practically impossible to operate a closed circuit system without allowing for a bleed-off stream to avoid the build up of dissolved solids and/or a consistent decrease or increase in pH. Liquid effluent treatment wastes and residues also result from treatment of wastewater streams. Process water from processing usually requires cleaning in a wastewater treatment plant. The cleaning takes place by neutralization and precipitation of specific ions. The main wastes/residues from these effluent treatment systems are gypsum ($CaSO_4$), and metal hydroxides and sulphides.

6.5 HYDROMETALLURGY AND RELATED ENVIRONMENTAL CONCERNS

Hydrometallurgical operations are chemical operations, with their own set of environmental challenges. The main environmental challenges are the production of various waste streams, including spent ore, and the management of leaching solution and water, including maintaining the integrity of liners.

CASE 6.7
Concerns about Assumed Risks as Barrier to Wider Use of Alternative Materials

Copper smelter slag is used worldwide as an alternative material. The Minerals and Metals Policy of the Government of Canada, 1996 promotes the use of slag as alternative materials. In Germany copper slag from the NA smelter in Hamburg is a preferred material for dyke construction. In Sweden granulated slag from the Rönnskär smelter is widely used as a fill material and sub-base in thickness between 30 and 70 cm for road construction as shown the photograph. The granulated slag marketed as Iron Sand reveals good heat insulating and draining qualities, which makes it particularly suitable for road- and groundconstructions in cold climates. Iron Sand has been used for over 30 years for this purpose in the Skellefteå region (Borell 2005). It is estimated that in Skellefteå alone at least 100 km of road and about 80,000 m^2 of industrial areas are built on copper slag.

In other parts of the world granulated copper slag is labelled as waste, a regulatory barrier to wider use. As one example, the Indonesian Government continues to be one of the most out-spoken critics of the use of copper slag, and continues to label copper slag as hazardous and toxic material.

Photo Credit: Borell 2005

Management of Waste Streams

Spent ore consists of the material remaining in either dump or heap leach piles when leaching ceases. Spent ore from heap or dump leaching may contain residual lixiviant and other constituents of the ore (a lixiviant is a solution used in hydrometallurgy to selectively extract the desired metal from the ore). Wastes from vat leaching operations are commonly referred to as tailings. Another consideration is that heap leach piles may periodically be used as a disposal area since most spills that occur at the mine site may be ore excavated and transported to the heap for leaching. Also, sludge generated at the solvent extraction facility may be deposited on the heap for leaching of residual metals.

SX/EW sludge is the semi-solid gelatinous material (i.e. soft mud, slime, slush, or mire) that can accumulate in SX/EW tanks. Sludge is in the form of colloids of suspended material (mostly silt or clay, usually less than 5 angstroms in size) that cannot be easily settled or filtered. Sludge accumulated on the bottom of the tanks over time needs to be removed. Sludge volume and its method for disposal, is relevant; it may be deposited on the heap for additional leaching.

The solvent extraction process specifically generates a sludge termed 'crud' or 'gunk' in the copper industry. This sludge consists of a solid stabilized emulsion of organic and aqueous solutions from solvent extraction. It is located at the organic/aqueous interface in the settlers and is periodically removed from the system, and centrifuged or otherwise treated to remove the organic constituents. The aqueous solutions and the solids are disposed of and the organics are returned to the solvent extraction circuit for reuse. Depending on the characteristics of the ore, SX/EW sludge may contain base or precious metals in quantities sufficient for recovery.

Spent electrolyte is generated during electrowinning activities. Historically, the electrolyte went through a stripping step and was subsequently discharged to a tailings pond. Today, this effluent is recycled to reduce the costs associated with the electrolytic acids used in these operations.

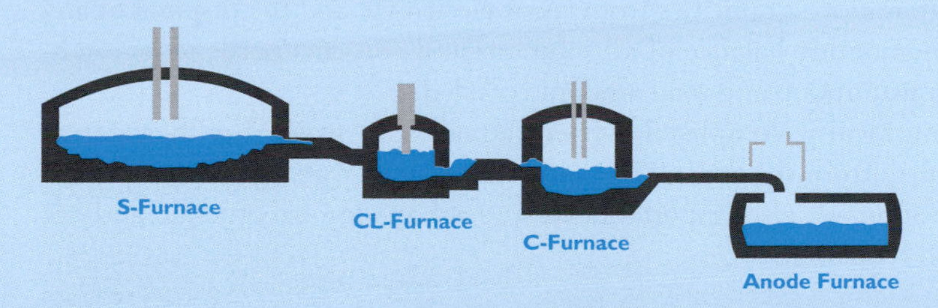

CASE 6.8
Smelting Technology for the Twenty-first Century

The graphic for this case schematically illustrates the Mitsubishi Process, showing the sequential smelting, slag cleaning, and converting stages and the continuous delivery of blister copper to the anode furnaces.

Three launder-connected furnaces are used: a circular Smelting (S) Furnace, an elliptical Slag Cleaning (CL) Furnace, and a circular Converting (C) Furnace. The mixture of matte and slag formed in the S-Furnace flows continuously to the CL-Furnace, where the denser copper matte separates from the discard slag. The matte is then siphoned to the C-Furnace, to be continuously converted to blister copper and C-slag. The latter is water granulated, dried, and recycled to the S-Furnace, while blister copper is siphoned continuously from the C-Furnace to the Anode Furnace.

Material flow from one furnace to another is continuous and enclosed greatly reducing fugitive emissions. Sludge from the wastewater treatment plant is recycled to the S-Furnace with granulated slag being the final outlet of most gangue material and impurities.

Spent leaching solution or barren solution (raffinate) is an acidic aqueous solution that has been stripped of copper but still has some carryover of the organic extraction/diluent used in the solvent extraction operation. The raffinate generated at hydrometallurgical plants is typically stored in ponds and recycled to the dump leaching operation. As a result, it does not become a waste until closure of the plant. Following mine closure, spent leaching solutions must be treated prior to disposal.

Other wastes and materials are typical of industrial operations, and include spent solvents, refuse, and used oil. As is true for all waste streams, accurate recording of wastes generated and documentation of waste disposal operations are important.

Management of Leach Solution and Water

Although heap leach operations are designed for zero discharge solution losses from the heap to undetermined sources will occur.

It is often argued that hydrometallurgical processing has very little environmental impact because its liquid streams are contained. In the SX/EW process, all impurities are returned to the site where they originated and the sulphuric acid is eventually neutralized by lime or limestone to form calcium sulphate (gypsum) – a relatively insoluble substance. Although heap leach operations are designed for zero discharge, (with the heap, its collection ditches, the pregnant solution pond, and plant feed ponds all lined to prevent solution loss), solution losses from the heap to undetermined sources (e.g. evaporation or seepage) will occur. Released solution potentially impacts ambient groundwater.

As for mine waste storages there is concern for the long-term integrity of the remaining leached ore materials or mine wastes. At heap closure the leached ore will remain on the pad. Closure will entail rinsing of the heap by application of water or by natural rainfall, often with no cover or reclamation planned for the heap pad. It is often necessary to continue operation of an SX/EW plant to treat leachate resulting from rinsing operations. This may be required for several years. The life of pond liners becomes important, as are anticipated flow rates from the heap after application of lixiviant ceases. There is also the need to establish 'trigger levels' for determining the point at which leachate collection will no longer be required.

All solutions (leaching, organic, and electrolyte) are reused in the SX/EW operation. Electrolytic solutions decrease in effectiveness over time due to an accumulation of impurities. The method used to remove impurities from spent electrolyte and the disposal of any wastes are important. A monthly balance of all solutions and reagents needs to be established indicating total consumption and total amount recycled.

At the electrowinning facility, completed copper cathodes are washed with water to remove residual electrolyte from the sheets. The amount of wash water generated in this operation and its disposition are also important. Waste lead anodes are of value and are sent to a smelter for lead recycling.

6.6 COMMON TECHNIQUES TO ESTIMATE EMISSIONS

To assess potential environmental impacts of mineral processing activities, it is necessary to estimate the quantity and quality of the various emissions that are associated with the project. Four techniques are commonly used to estimate these emissions: (1) sampling or direct measurement; (2) mass balance; (3) engineering calculations; and (4) application of emission factors. While particular estimation methods are generally more suited to particular applications, the final choice will depend on factors such as costs, level of required accuracy, nature of emission and substance, and data availability.

Direct Measurements

Direct measurement is one of the more accurate techniques but is only applicable to existing and active operations. Measurements should take into account standard and non-standard operating conditions. Direct measurement data allow calculation of actual loads to the environment, by multiplying concentration times emission volume. For hot gaseous emissions it is necessary to account for temperature differences. It is also essential that sampling procedures conform to established measurement protocols. Collection and analysis of samples is expensive and complicated, especially where various substances are emitted, and emission sources are fugitive in nature. Sampling may not be representative for the entire process, and may provide only one example of actual emissions.

Measurements should take into account standard and non-standard operating conditions.

Mass Balances

Mass balances identify the quantity of a substance going in and out of the control volume, which may be a process, a piece of equipment, or a facility as such as a smelter (see **Figure 2.6** as one example). While common practice, the validity of a stationary mass balance that does not account for accumulation or depletion of the substance within the control volume, cannot be assumed. There will always be errors inherent in the estimation of inputs and losses from the control volume, and in the fates of the substance under consideration, increasing the error bands of the mass balance, especially for minor emissions streams. A rigid inventory of chemical use and emissions is essential to establish accurate mass balances. Any mass balance should consider chemical or biological degradation that may occur during process or treatment. It is also important to realize that even small errors in any one step of the mass flow can significantly skew emission estimates. While there may be many variables in a mass balance, errors may be minimized through continued development and refinement of the mass balance equation.

A rigid inventory of chemical use and emissions is essential to establish accurate mass balances.

Engineering Calculations

Engineering calculations may be used to estimate emissions subject to rigid controls, when there is a thorough understanding of the substance's fate in relation to process conditions and chemistry. Engineering calculations commonly use standard physical and chemical laws and constants to allow estimate emissions. They are based on known performance standards of processes and equipments, and of the physical and chemicals properties and reactions of the substance under consideration. Engineering calculations may be very simple (such as a simple black box model) or extremely complex with factors for weather influences, chemical speciation, and the many site specific factors. As in all calculations the old rule applies: 'garbage in garbage out'. The complexity of a calculation will influence the accuracy of the emission estimate. Whenever possible, the calculations should be verified by direct measurements.

As in all calculations the old rule applies: 'garbage in garbage out'.

Emission Factors

Emission factors are tools used to estimate emissions to the environment based on standard formulae or experience derived from similar operations. They are useful tools for estimating emissions where the relationship between emission and substance use is well known. Emission factors are widely used in estimating emissions from combustion sources, and as such are at the centre of the current greenhouse gas emission debate. Carbon credits are commonly based on differences in greenhouse gas emissions factors for different fuels.

Emission factors are industry-averaged data, and as such are less accurate than direct measurements.

Emission factors are derived from direct measurements of actual emissions from a range of similar plants. While emission factors may be empirical, they are based on standard equipment and operating practices. Emission factors are also supported by engineering calculations. Eventually, however, emission factors are industry-averaged data, and as such are less accurate than direct measurements.

REFERENCES

Benjamin MM and Honeyman BD (1992) Trace metals. In: Butcher SS, Charlson RJ.

Blesing N, Lackey J, and Spry A (1975) in 'Minerals and the Environment', ed. Jones MJ, Institute of Mining and Metallurgy, London.

Borell M (2005) Slag – a resource in the sustainable society, Boliden Mineral AB, Rönnskär Smelter, S 932 81 Skelleftehamn, Sweden.

Brimblecombe C, Hammer H, Rodhe A, Ryaboshapko and Boutron CF (1989) Human influence on the sulphur cycle. In: Brimblecombe P, Lein AY (eds), Evolution of the Global Biogeochemical Sulphur Cycle, pp. 77–121. Wiley, New York.

Dresher W (2001) How Hydrometallurgy and the SX/EW Process Made Copper the 'Green' Metal, Copper Applications in Mining & Metallurgy.

Engineering and Mining Journal. (1990). Technology Turns Southwest Waste into Ore. Vol. 191, January, pp. 41–44.

Environment Australia (1998) Cyanide Management; Booklet in the series 'Best Practice Environmental Management In Mining'. Principal authors: Dr John Duffield and Prof Peter May.

Hawley JR (1977) The Problem of Acid Mine in the Province of Ontario', Ministry of the Environment, Ontario.

International Cyanide Management Institute (2005) The International Cyanide Management Code (www.cyanidecode.org).

Ivanov MV, Grinenko YA and Rabinovich AP (1983) Sulphur flux from continents to oceans. In: Ivanov MV and Frenet JR (eds), The Global Biogeochemical Sulphur Cycle, pp. 331–356. Wiley, New York.

Jordan G and D'Alessandro M (eds) (2004) Mining, mining waste and related environmental issues: problems and solutions in Central and Eastern European Candidate Countries – Joint Research Centre of the European Commission, Ispra, EUR 20868 EN, ISBN 92-894-4935-7, 208 p.

Kettell B (1982) Gold; Oxford University Press.

Konradsson A (2003) Leaching characteristics of Boliden Iron Sand – Environmental aspects of product certification. Degree project in Engineering Biology 20p, Institute of technology, Umeå University.

Mudder T (editor) (2000) The Cyanide Monograph, collection of 25 full length papers on all aspects of cyanide in mining, published by Mining Journal Books, London, England, United Kingdom, 650 p.

Schlesinger WH (1991) Biogeochemistry. An Analysis of Global Change. Academic Press, San Diego.

Tilton JE and Lansberg HH (1997) Innovation, Productivity Growth and the Survival of the U.S. Copper Industry, Discussion Paper No. 97-41, Resources for the Future, Washington, D.C.

US Congress, Office of Technology Assessment (1988) Copper: Technology and Competitiveness, OTA-E-367, US Government Printing Office. Washington DC., September 1988.

USEPA (1981) Cyanide risk assessment, An Exposure and Risk Assessment for Cyanide (EPA-440/4-85-008)

Weiss NL (1985) SME Mineral Processing Handbook, Editor-in-Chief, Vols. 1 and 2, Society of Mining Engineers of AIME, Seeley W. Mudd Memorial Fund of AIME, New York, New York.

• • • •

7

Our Environment

A Set of Natural and Man-made Features

Mercury, one of the seven metals known to the ancients, was in the past widely used because of its ability to dissolve silver and gold. Also known as quicksilver, it is the only metal which is liquid at room temperature. Extraction from ore is most simply carried out by distillation as mercury compounds decompose and volatilize at moderate temperatures, to be readily recovered by condensation. Most of the mercury known to man originated from one mine. The Almadén deposit in Spain, exploited over 2000 years, contained over 30 percent of the world's known reserves of mercury with a grade six times the global average for mercury mines.

7 Our Environment

A Set of Natural and Man-made Features

'Nature, natural, and the group of words derived from them, or allied to them in etymology, have at all times filled a great place in the thoughts and taken a strong hold on the feelings of mankind. That they should have done so is not surprising, when we consider what words, in their primitive and most obvious signification, represent; but it is unfortunate that a set of terms which play so great a part in a moral and metaphysical speculation, should have acquired many meanings different from the primary one, yet sufficiently allied to it to admit confusion. The words have thus become entangled in so many foreign associations, mostly of a very powerful and tenacious character, that they have come to excite, and to be the symbols of, feelings which their original meaning will by no means justify; and which have made them one of the most copious sources of false taste, false philosophy, false morality, and even bad law'

Mill's essay, 'Nature', is one of the few undoubted classics of environmental philosophy (Mill 1963–77). The beautifully written quote illustrates the challenges when discussing nature, environment, and impacts. First, what is the meaning of nature or, respectively, environment? Second, is an unbiased discussion of the environment possible?

When discussing the meaning of environment the legislated definition of environment in the host country becomes important.

When discussing the meaning of environment the legislated definition of environment in the host country becomes important. Most jurisdictions now define the term broadly to include not only physical/chemical and biological but also social, cultural, and economic aspects, and by implication refer to the environment of people. The general characteristics of the physical/chemical (air, water, and land), biological (fauna and flora), and human environment (social and economic systems) are discussed in this chapter, together with a discussion of judging and evaluating the state of the environment once baseline data become available.

By the time Mill wrote his essay, environment was already an overworked word. Webster's 9th Collegiate Dictionary defines environment as 'the complex of physical, chemical, and biotic factors (such as climate, soil, and living things) that act upon an organism or an ecological community and ultimately determines its form and survival', a set of natural and man-made features which exist in a given place and point of time. While it means the totality of things that in any way affect an organism, we associate with the word 'environment' the environment of people. In it purest sense environment comprises all living and non-living things that occur naturally on Earth. Natural sciences and physical geography (the 'hard' sciences) differentiate four environmental spheres, namely atmosphere ('air'), hydrosphere ('water'), lithosphere ('land'), and biosphere ('fauna and flora'), in an intricately interlocked natural system. On this natural system we have erected our own human environment, modifying, altering, or destroying natural conditions that existed before human impact was expressed (sometimes coined 'built environment' as compared to natural environment). The influence of human activities on nature became very apparent in the middle of the last century, and has since been studied by many researchers. Some of the dominant study themes include environmental degradation and natural resource use, natural hazards and impact assessment, and the effects of urbanization and land use. Apart from human influences on the natural environment, are all the complex aspects of human society and human systems, typically covered in social sciences and human geography (the 'soft' sciences). Hence our environment is also defined by demographics, by man-made systems and structures and by cultural and heritage resources, in this text conveniently grouped into social and economic systems, or the social and economic spheres.

In brief, environment is often thought of as consisting of the physical/chemical and the biological environment, jointly referred to as natural environment, and the human or man-made environment comprising the social and economic spheres, closely interlinked as illustrated in **Figure 7.1.** The link to the three circles of sustainability – economic, social and ecological systems – is immediately apparent. The environmental assessment of a new mining project as understood in this text, will cover all six environmental spheres, the overlapping areas of the three circles of sustainability. Each mining project will influence all these spheres, and all of them will influence the mine.

Environment is often thought of as consisting of the physical/chemical and the biological environment, jointly referred to as natural environment, and the human or man-made environment comprising the social and economic spheres.

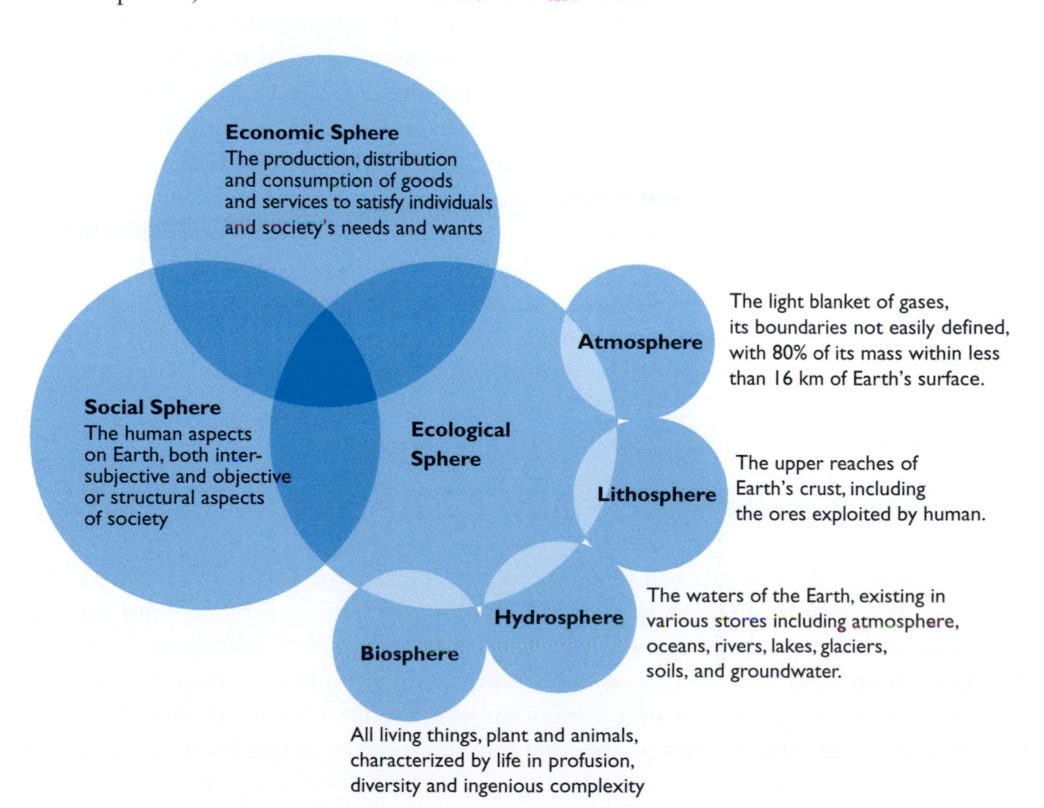

FIGURE 7.1
Environmental Components Illustrated

Environment is often thought of as consisting of the physical/chemical and the biological environment, jointly referred to as natural environment, and the human environment comprising the social and economic spheres, closely interlinked.

As mentioned above, the natural environment consists of two distinct, but closely interrelated parts, the physical/chemical and the biological (or living) environment.

The physical/chemical environment, that is air, land, and water, also known as the abiotic or inorganic environment, comprises everything in nature that is not living (climate, weather, noise, topography, hydrogeology, non-organic soil matrix, minerals, etc.). The physical/chemical environment has two sub-components, abiotic inputs (e.g. energy, climate, atmospheric, aquatic, and terrestrial conditions) and abiotic matter (e.g. soil matrix, sediments, particulate matter, dissolved organic matter, nutrients in aquatic systems and dead or inactive organic matter in terrestrial systems).

The biological environment is all that in nature which is living (the fauna and flora) and includes all levels of life from genes to species and ecosystems. Thus, the biological environment is strongly linked with the concept of biodiversity. Biodiversity is, essentially, all life on Earth and the systems that support it.

The human environment also consists of two parts: the social sphere defined by human aspects such as demographics, community health, religion, norms and values, tradition, education, community cohesion, and cultural and heritage resources; and the economic sphere, defined by aspects such as economic and institutional structures, public and private infrastructures, production and consumption patterns, and the many facets of natural, and renewable resource availability and use.

Nearly all natural environments have been directly or indirectly influenced by humans at some point in time.

It is important to recognize that dividing the environment into a natural and built environment is somewhat misleading. Nearly all natural environments have been directly or indirectly influenced by humans at some point in time. However it is this subdivision and the perception of the relation between humans and other living beings that often defines people and their attitude towards nature and development.

Some people argue that humans, because of culture and technology, have separated significantly from nature (Angermeier 2000). This separation is accompanied by a massive human occupation of land and over-exploitation of natural resources. Humans, they say, are responsible for the current biodiversity crisis, and consequently the natural imperative of economic development is to protect and conserve environmental integrity. This group advocates human humility and respect towards non-human nature. It is this thought of separation from nature that seems to guide and motivate most anti-mining advocacies.

Other people defy a separation of humans and nature. Humans are seen as participants in ecosystems, and from this point of view, the imperative of economic development is to guide a sensible and sustainable human use of natural resources (Callicott and Mumford 1997; Povilitis 2001). Not surprisingly, government and mining companies adopt the second school of thinking. However, despite the obvious differences in thinking, both schools agree that without a deep respect for nature, neither view of the human-nature relationship can justify human behaviour that diminishes the Earth's integrity and diversity.

7.1 THE ATMOSPHERE – AIR, WEATHER, AND CLIMATE

Mining projects are very much influenced by the climatic conditions experienced at the mine site.

Mining projects are very much influenced by the climatic conditions experienced at the mine site. Temperature, precipitation (or lack of it), temperature, visibility, wind, and their variations are all important factors in the design and operation of mining and processing facilities. Mining and mineral processing, however, can also influence climatic conditions, at local and global scales, but more importantly can impact local air quality affecting human health and comfort. Hence, the importance of understanding local climate and air quality.

The atmosphere contains four different layers, but it is the first layer, called the troposphere, where the majority of our weather, the climate, occurs. This layer, which varies in thickness from about 8 to 16 kilometres, contains about 80% of the total mass of the atmosphere.

Temperature

The maximum air temperature occurs near the Earth's surface in the troposphere. Air temperature declines uniformly with altitude at a rate of approximately 6.5 Celsius degree per 1,000 metres, a phenomenon termed environmental lapse rate.

Much of the Earth's climatic variation results from uneven heating of its surface by solar radiation or, respectively, by uneven lateral distribution of temperature, caused by the spherical shape of the Earth and the angle at which the Earth rotates on its axis as it orbits the sun. About half of the energy received from the sun in the form of solar radiation is absorbed by the land and the oceans. To maintain the Earth's long-term mean surface temperature of 16°C, however, the Earth must lose heat as well as gain it. The energy gained by the surface must be transferred to the atmosphere, with the result that on average, our atmosphere is primarily heated from below by heat given off from the Earth. This means that the atmosphere is set into convective motion.

Measurements made over the Earth's surface also show that on average, more heat is gained than lost at equatorial latitudes, while the opposite is observed at higher latitudes. Winds and ocean currents remove accumulated excess heat in the tropics and release it at higher latitudes, where a heat deficit exists. Closer inspection also shows that land and ocean areas respond differently to solar radiation. Land has a lower heat capacity; it changes temperature more rapidly than oceans as it gains or loses heat between day and night, or summer and winter.

The long-term average distribution of gained and lost radiant energy with latitude does not include the annual cycle of radiant changes related to the seasonal north-south migration of the sun. The intensity of solar radiation and thus temperature remains fairly constant at equatorial latitudes over the year, but seasonal changes increase with increasing latitudes.

Understanding the climate at a specific location on the Earth, such as a mine site, usually requires long-term meteorological measurements at the particular location.

The complex pattern of average solar radiation determines the Earth's average climate pattern, but daily weather is influenced by countless others parameters, some better understood than others. Understanding the climate at a specific location on the Earth, such as a mine site, usually requires long-term meteorological measurements at the particular location.

Fallen and Deposed Precipitation

Precipitation drives the land surface hydrology in the same way that incoming solar radiation drives the surface thermal regimes.

Precipitation is a critical variable for establishing water balances and their variability. Precipitation includes fallen (liquid or solid, i.e. rain, snow, and hail) as well as deposed (dew or frost) forms. The observed variable is precipitation depth defined as the depth of liquid water accumulated during a defined time interval on a horizontal surface. Hence, precipitation is measured in mm per time interval. Precipitation drives the land surface hydrology in the same way that incoming solar radiation drives the surface thermal regimes. It is therefore the key variable in the terrestrial hydrological cycle (surface water budget), and is essential for all vegetation growth. Reliable, high-resolution records of precipitation are critical inputs for the design of mining projects, for understanding and monitoring regional effects of project development on hydrology, and for estimating water availability (or respectively water scarcity) for project consumption. For most applications (water cycle and budget), area averaged information on precipitation is sufficient.

Evaporation and Evapotranspiration

Water is removed from oceans, lakes, and other surface bodies by evaporation. Evaporation is the process where a liquid, in this case water, changes from a liquid to a gaseous state. It is a misconception that at a pressure of 1 atm, water vapour only exists at 100°C. Water molecules are in a constant state of evaporation and condensation flux near the surface of any water body. If a surface molecule receives enough energy, that is solar radiation, it will leave the liquid and turn into vapour. At mine sites located in high temperature regimes with low precipitation, most of the water used in mining will probably be lost to evaporation, rather than returned to surface water or ground water systems.

At mine sites located in high temperature regimes with low precipitation, most of the water used in mining will probably be lost to evaporation.

Evapotranspiration is the loss of water to the atmosphere by evaporation from the surface as well as by transpiration from vegetation growing on the soil. As for evaporation, transpiration is measured in kg of H_2O/m^2/day, the loss of water to the atmosphere per square metre of land surface per day. Evaporation and transpiration are critical hydrological links between the Earth's surface and the atmosphere. Both are therefore important for issues involving climate change and ecosystem response. Changes in vegetation may have large impacts on transpiration and evaporation rates, with consequent effects on the water balance. Conversely, if changes in the water balance are significant, major shifts in vegetation patterns and composition are a likely outcome. Equally, changes in transpiration are likely to impact atmospheric composition of greenhouse gases, and climate, as the hydrological cycle increases in intensity with warming.

The Atmosphere in Motion

Air moves because at one place less dense (warmer) air rises, while in another place denser (cooler air) sinks towards the Earth. Between these two areas, the air flows horizontally along the Earth's surface as wind. Simplified, air circulation is in the form of giant convection cells. Less-dense, warmer air rises at the equator where temperatures are high, and cooler, denser air sinks at both poles, where temperatures are very low. The Earth's rotation, and the associated *Coriolis effect*, complicates this basic concept. The result is three convection cells in each hemisphere wrapped around the rotating Earth (**Figure 7.2**). Moving from this conceptual model to real air movement requires consideration of three factors: (1) seasonal changes in temperature due to solar heating, (2) large continental blocks with differences in elevations, and (3) the difference in heat capacity of land and water, cause of the monsoon effect. While the general wind and pressure belts are still identifiable over the Earth, actual air movement can fluctuate widely at any given time and place, making forecasting wind and weather notoriously difficult. Local wind data become especially important for mining projects with integrated power or pyrometallurgical processing plants. The surface footprints of gaseous emissions from these operations will very much depend on the prevailing wind strength and direction.

Local wind data become especially important for mining projects with integrated power or pyrometallurgical processing plants.

Ambient Air Quality

The atmosphere has a finite ability to absorb anthropogenic emissions before those emissions inflict harm on human health. In many parts of the world air pollution has already become the environmental factor with the greatest impact on health and is responsible for the largest burden of environment-related disease. As one example, recent estimates indicate that 20 million Europeans suffer from respiratory problems every day (EEA 2005).

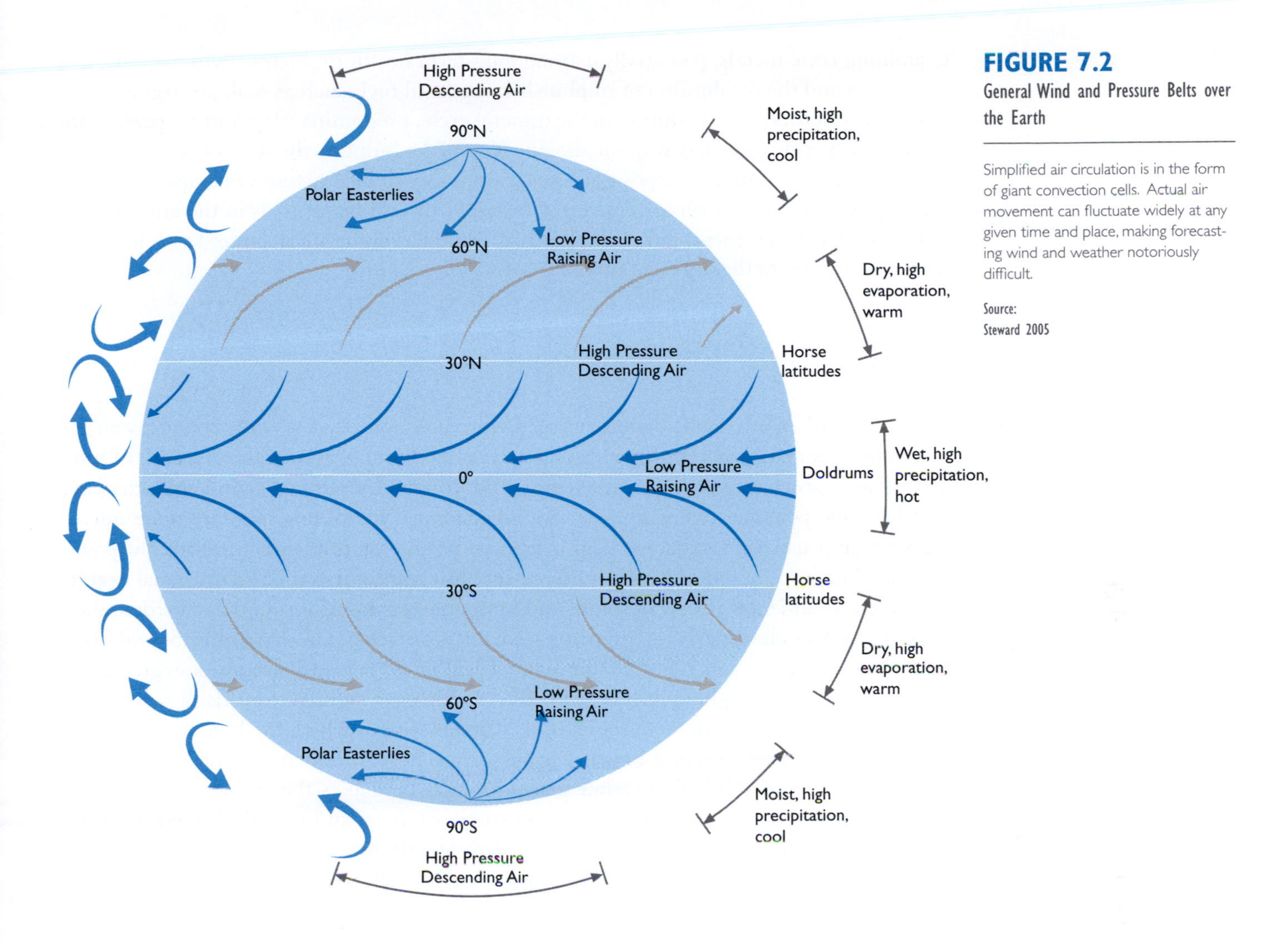

FIGURE 7.2
General Wind and Pressure Belts over the Earth

Simplified air circulation is in the form of giant convection cells. Actual air movement can fluctuate widely at any given time and place, making forecasting wind and weather notoriously difficult.

Source:
Steward 2005

Nobel laureate Paul Crutzen estimates that at least 3 million people die each year from diseases triggered by air pollution.

Most jurisdictions now recognize that people have a right to breathe air that is not harmful to their health and that sustainable economic vitality is tied to proper air resource protection. Ambient air quality standards are designed to protect health, environment, and the quality of life from the pervasive threat of air pollution. The six most common pollutants are ozone, particulate matter (PM), carbon dioxide, nitrogen dioxide, sulphur dioxide, and lead.

Particulates matter (PM) may well be the air pollutant that most commonly affects people's health.

Mining and mineral processing have the potential to contribute to air pollution in various ways. Particulate matter (dust) is usually the most obvious air pollutant associated with mining. Particulate matter (PM) may well be the air pollutant that most commonly affects people's health.

Airborne particles come in different shapes or sizes, and may be solid particles or liquid droplets.

Airborne particles come in different shapes or sizes, and may be solid particles or liquid droplets. Larger particles (PM_{10}) range between 2.5 and 10 micrometres (from about 25 to 100 times thinner than a human hair). These cause less severe health effects than smaller particles. Small particles ($PM_{2.5}$), less than 2.5 micrometres in size, travel over longer distances, and penetrate deeper in the respiratory system. But size is not the only difference; particle composition also counts. Pyrometallurgical processes may emit small particles containing trace metals, potentially causing long-term health problems. Pyrometallurgical processing and the combustion of sulphur-bearing fossil fuels such as coal, are arguably the main contributors of air pollution in the mineral cycle, predominantly emitting particulate matter, carbon dioxide and sulphur dioxide (**Case 7.1**). Sulphur dioxide, of course, is one of the main sources of acid deposition. Acid rain is caused by emissions of sulphur dioxide and nitrogen oxides, which are converted to sulphuric and nitric acids in the atmosphere. Dilute forms of these acids fall to the Earth as precipitation or are deposited as acid gas or particles, leading to the degradation of forests.

Noise

Noise, defined as any loud, discordant or disagreeable sound or sounds, counts as an air pollutant.

The environmental effects of sound and human perceptions of sound can be described in terms of four characteristics, pressure, frequency, duration, and pure tone.

Noise, defined as any loud, discordant or disagreeable sound or sounds, counts as an air pollutant. In an environmental context, noise is defined simply as unwanted sound. Mining and mineral processing activities inherently produce sound levels and sound characteristics that have the potential to create noise. Sound generated by mining becomes noise due to adjacent land use, i.e. if adjacent land is used by people for residential or other purposes. Close to the source of sound itself, sound pressure levels are a matter of occupational health.

The environmental effects of sound and human perceptions of sound can be described in terms of four characteristics, pressure, frequency, duration, and pure tone. Sound pressure level (SPL, also identified by the symbol Lp) or perceived loudness is expressed in decibels (dB) or A-weighted decibel scale dB(A) which is weighted towards those portions of the frequency spectrum, between 20 and 20,000 Hertz, to which the human ear is most sensitive. Both measure sound pressure in the atmosphere. Frequency (perceived as pitch) represents the rate at which a sound source vibrates or makes the air vibrate. Duration represents recurring fluctuation in sound pressure or tone at an interval; sharp or startling noise at recurring interval; the temporal nature (continuous vs. intermittent) of sound. Pure tone is sound comprising a single frequency, relatively rare in nature but, if pure tone does occur, it can be extremely annoying.

Another term, related to the average of the sound energy over time, is the Equivalent Sound Level or Leq. The Leq integrates fluctuating sound levels over a period of time to express them as a steady state sound level. As an example, if two sounds are measured and

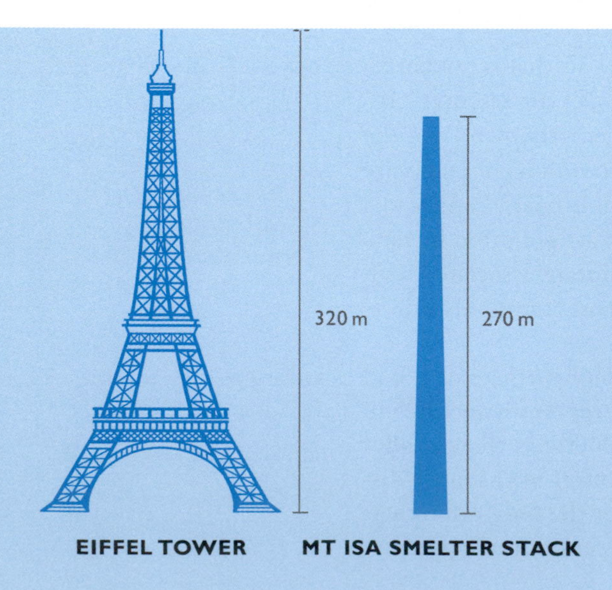

CASE 7.1
Lead Smelting at Mount Isa

Built in 1978 with a stack height of 270 m, the Mount Isa lead smelter in Inland Queensland, Australia operates the tallest flue gas stack of any pyrometallurgical operation on Earth. Mining in Mount Isa commenced in 1923 following the discovery of substantial deposits of lead. Today Mount Isa is one of the top ten producers of copper, silver, lead, and zinc. Until 2000, the lead smelter operated without a desulphurization unit dispersing significant amounts of SO_2 through the stack. Depending on the prevailing wind direction it was possible to detect the plume in Perth some 2,000 km away.

TABLE 7.1

Human Reaction to Increases in Sound Pressure Levels – Absolute sound level is not the only determinant of effect; increases in sound pressure level above ambient levels at the chosen point of sound reception are equally important

Increase in Sound Pressure (dB)	Human Reaction
Under 5	Unnoticed to tolerable
5–10	Intrusive
10–15	Very noticeable
15–20	Objectionable
Over 20	Very objectionable to intolerable

Source:
Down and Stocks (1978)

one sound has twice the energy but lasts half as long, the two sounds would be characterized as having the same equivalent sound level. Equivalent Sound Level is considered to be directly related to the effects of sound on people since it expresses the equivalent magnitude of the sound as a function of frequency of occurrence and time. By its derivation Leq does not express the maximum or minimum sound pressure levels that may occur in a given time period. These maximum and minimum sound pressure levels should be given in the noise analysis. The time interval over which the Leq is measured should always be given. It is generally shown as a parenthetic; Leq (8) would indicate that the sound had been measured for a period of eight hours.

It is important to have an understanding of the way noise decreases with distance. The decrease in sound level from any single noise source normally follows the 'inverse square law'. That is, sound pressure level changes in inverse proportion to the square of the distance from the sound source. Moisture, temperature, and wind all influence the propagation of sound levels, but their influence is minor and time dependent, usually neglected in environmental assessment.

Most humans find a sound level of 60 to 70 dB(A) as beginning to create a condition of significant noise effect (USEPA 550/9-79-100, November 1978). However, absolute sound level is not the only determinant of effect; increases in sound pressure level above ambient levels at the chosen point of sound reception are equally important. Increases ranging from 0 to 6 dB should have no appreciable effect on receptors (**Table 7.1**). Sound pressure increases of more than 6 dB may require a closer analysis of impact potential, depending on existing sound pressure levels and the character of surrounding land use and receptors. Sound pressure level (SPL) increases approaching 10 dB result in a perceived doubling of SPL. The perceived doubling of the SPL results from the fact that SPLs are measured on a logarithmic scale.

Most humans find a sound level of 60 to 70 dB(A) as beginning to create a condition of significant noise effect.

7.2 THE LITHOSPHERE – GEOLOGY, LANDFORM, AND EARTH RESOURCES

The ores that we exploit are all located in the upper part of the Earth's crust – the lithosphere. Accordingly the nature of the land is of fundamental importance to any mining project. Geology, landform, soils, land quality, and land use all represent important characteristics which influence mine development and may be affected, temporally or permanently, by project operations.

The nature of the land is of fundamental importance to any mining project.

Geology

The science of geology is concerned with the history, behaviour, and nature of the Earth. It deals with the physical and chemical composition, distribution and structure of rocks, soils, and their mineral constituents; and with the processes of magmatic differentiation, crystallization, metamorphism, chemical alteration and precipitation, weathering, sedimentation and induration by which soils and rocks are formed and subsequently decomposed. It also deals with movements of tectonic plates which form the Earth's crust; with the associated stresses (tectonic forces) which lead to earthquakes (sometimes causing tsunamis); with ruptures of the Earth (faults) caused by these stresses; and with volcanos and geysers that periodically erupt ejecting large volumes of solids and/or liquids, commonly accompanied by toxic gases.

Economic geology is the branch of geology that deals with the genesis and nature of ores, and the range of techniques used to discover and evaluate these ores.

Rocks are classified into three categories, based on their origin:

- Igneous rocks are those which solidify by cooling of magma, either within the Earth's crust (plutonic rocks) or extruded on to the Earth's surface (volcanic rocks). These rocks are categorized and classified by their chemical and mineralogical composition and texture. Common examples include granite which is a plutonic rock of acidic composition, and basalt which is a volcanic rock of basic (alkaline) composition.
- Metamorphic rocks result from the transformation of pre-existing rocks under conditions of high temperature and/or high pressure, which change the mineral assemblages and also cause textural changes, particularly the development of mineral banding and foliation. Classification of metamorphic rocks is based partly on the characteristics of the pre-existing rock, as well as the type and grade of metamorphism as reflected in the mineral assemblage and the texture. Examples are slate (which results from low grade metamorphism of shale), schist which results from higher grade metamorphism, and marble, which results from re-crystallization of limestone at high temperature.
- Sedimentary rocks which form by deposition and accumulation of particles from weathering of other rocks; particles ejected into the atmosphere by volcanic eruptions; the solid remains of organisms; and crystals and colloids precipitating from surface waters. Deep burial, consolidation over time and in some cases inter-granular reactions cause the deposited sediments to lithify, i.e. become rock. Classification of sedimentary rocks is based mainly on composition, and grain size. The most common sedimentary rocks are sandstone, shale, and limestone. Peat and coal are also classified as sedimentary.

Soils are derived from rocks initially by chemical weathering, which involves oxidation and leaching, preferentially dissolving some mineral components. This weakens the rock fabric, eventually leading to physical disintegration under the forces of temperature change, freezing and thawing, and erosion by wind and water.

Residual soils are those which form from the parent rock *in situ*. The development of soil from parent rock is a slow process. In some situations, any material weakened by weathering, is removed by erosion almost immediately, so that little or no residual soil accumulates. Elsewhere, in the absence of significant erosion, soil profiles may reach depths of 50 m or more. Most older residual soils exhibit profile development, in which several more or less distinct horizons can be distinguished, each with a different texture and colour.

Transported soils are more or less unconsolidated sediments removed from the parent material and subsequently re-deposited elsewhere. They include:

- Fluvial soils deposited in rivers;
- Alluvium – soils deposited by moving water in flood plains and deltas;

- Till – comprising sediment from glaciers;
- Colluvium – soils transported by gravity, usually accumulating towards the base of a slope;
- Eolian or windblown soils.

It should be recognized that the term 'soil' has different meanings in different branches of science and engineering. The definitions presented above are those used by geologists. However, unconsolidated marine or lake bed sediments are not included by geologists in the definition of soils, despite having similar origins, compositions, and properties to other transported sediments. Geotechnical engineers, however, use the term soil for unlithified natural materials, whatever their origin, and classify soils based on physical properties of strength (cohesion and internal friction) as well as grain size. The agronomists have a slightly different view of soils based on their suitability for agriculture, leading to the development of completely different classification systems. These are particularly concerned with profiles that develop as the parent material interacts with the climatic, hydrological, and biological processes and the textural, drainage, moisture retention, and nutrient status of different soil horizons.

The term 'soil' has different meanings in different branches of science and engineering.

Ores occur in igneous, metamorphic, and sedimentary rocks, although some rock types are far more prospective for economic mineralization than others. Base metal and precious metal ores are commonly associated with volcanic activity, although this may have occurred in the distant geological past, with only very subtle evidence of its origin. Residual soils may also contain economic mineral deposits, the most notable examples being lateritic ores of iron, aluminium, and nickel, which result from preferential leaching and enrichment of particular metals under prolonged tropical weathering.

Base metal and precious metal ores are commonly associated with volcanic activity

Residual soils may also contain economic mineral deposits, the most notable examples being lateritic ores of iron, aluminium, and nickel, which result from preferential leaching and enrichment of particular metals under prolonged tropical weathering.

Clearly, geology influences most mining project activities and operations. Apart from the composition and characteristics of the ore itself which influence the selection of excavation, crushing, milling, and processing methods and equipment, the mine design is very much influenced by geological structure, while the need for mine drainage or dewatering will depend on the hydrogeology.

Landform

Landform is any physical feature of Earth's surface having a characteristic, recognizable shape, influenced by geology and produced by natural causes as illustrated in **Figure 7.3**. Landforms can be broadly grouped into four categories (Pidwirny 2006):

- Structural Landforms are created by massive Earth movements due to plate tectonics (e.g. fold mountains, rift valleys, or volcanos).
- Weathering Landforms are created by the physical or chemical weathering which removes susceptible materials while leaving more resistant rock. Weathering produces landforms such as mesas, hogback ridges 'breakaways' and karst features in limestone such as pinnacles and sink holes.
- Erosional Landforms are formed from the removal of weathered and eroded surface materials by wind, water, glaciers, and gravity (e.g. river valleys, glacial valleys, and coastal cliffs).
- Depositional Landforms are formed from the deposition of sediment. These deposits may eventually be compressed, altered by pressure, heat, and chemical processes to become sedimentary rocks (e.g. dunes, beaches, deltas, alluvial terraces, flood plains, and glacial moraines).

FIGURE 7.3
Simplified Model of Landform Development

Landforms are influenced by geology (such as rock type), and natural forces (such as tectonic processes, weathering, erosion, and sedimentation).

Source:
Pidwirny 2006

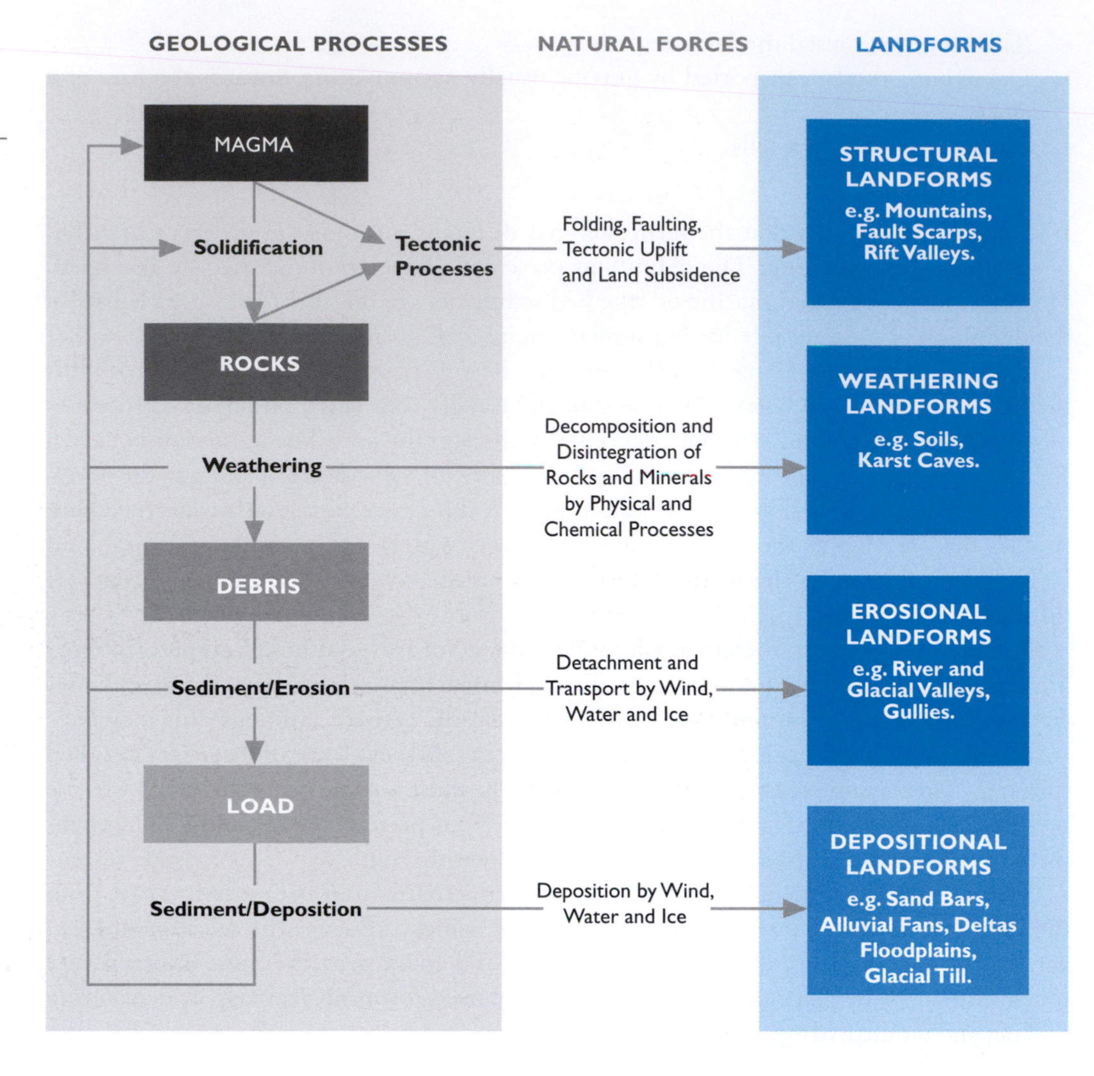

Many landforms show the influence of several of the above processes (termed polygenetic landforms). Processes acting on landforms can also change over time, and a single landscape can undergo several cycles of development (also termed polycyclic development).

Landform and landscapes should be treated the same as any other resource subject to the effects of mining.

Landscape is how people perceive landforms and land cover. Landform and landscapes should be treated the same as any other resource subject to the effects of mining. While a study of landform is based on geology, and geological processes, a study of landscapes has to be based on more subjective criteria:

- Landscape values are the sum of the values of individual components, predominantly water, rocks, soil, and vegetation, and their spatial relationships.
- Landscape encompasses the aesthetic and emotional values of the natural environment, underlining the importance of visual qualities and values such as visibility, colour, contrast, and fragility.
- Visibility involves elements of scale and distance as well as possible observation points.
- Fragility in this context refers to those landscape characteristics relating to the capacity to withstand or respond to changes in the landscape. It is used as a guide for locating the possible facilities or their elements in such a way as to minimize visual impact. Biophysical, perceptive, and historical-cultural factors usually affect fragility. Proximity and visual exposure can also be considered.

- Quality or beauty of the landscape, sometimes termed 'visual amenity' also deserves to be assessed in terms comparable to those used for other resources. Perception of the landscape depends on the sensitivity, conditioning and attitudes of the observer, educational or cultural conditions, and the relationship between the observer and the observed object. A mine, for example, will be perceived as beautiful by a mining engineer but as an eyesore by some environmentalists. Also, although the landscape components and their relations with the surrounding environment can be described in terms of design, size, shape, colour, and space, many different considerations may be involved in assessing the relative value of each and its weight in the total composition.

Perception of the landscape depends on the sensitivity, conditioning and attitudes of the observer, educational or cultural conditions, and the relationship between the observer and the observed object.

For most of us, of course, topography is the most intuitive feature of landform. Topography influences landscape, but also land stability, land cover, erosion, and sedimentation, as well as the interactions between the solar radiation, water, and the heterogeneous land surface. The most complex interaction between the biosphere and the climate system is through the elevational and orographic effects of topography. Topographic structure strongly influences the regional climate, which controls soil development, water balance, and vegetation distribution. For hydrological modelling, mine design, and a host of related activities, detailed mapping of topography is required. The results may be presented as contour plans, cross-sectional profiles, slope maps, 3-dimensional models, and various computer-generated images designed to enhance particular topographic features.

For hydrological modeling, mine design, and a host of related activities, detailed mapping of topography is required.

In 2000, the US Government sponsored a satellite altimetry of the Earth's surface. The Shuttle Radar Topography Mission (SRTM) obtained elevation data on a near-global scale to generate the most complete high-resolution digital topographic database of the Earth. SRTM consisted of a specially modified radar system that flew onboard the Space Shuttle Endeavour during an 11-day mission in February of 2000. SRTM is an international project spearheaded by the US National Geospatial-Intelligence Agency (NGA) and the US National Aeronautics and Space Administration (NASA). Topographic data are made available at www2.jpl.nasa.gov/srtm/index.hltm and can be used for regional analysis, with a horizontal resolution of 90 metres and a vertical accuracy of 15 metres. Higher accuracy data require stereo-photographic mapping, aircraft laser altimetry, or field surveying.

Land Cover

Land cover is defined as that which overlies or currently covers the ground including vegetation, permanent snow and ice fields, water bodies, or man-made structures. Barren land is also considered a land cover although technically it is lack of cover. Classes of land cover include built-on land, cultivated land, grassland, wetland, forest, water areas, and barren land. Each class can be further sub-divided. Forest as an example is sub-divided according to type of forest (rainforest, wetland forest, mangrove forest, etc.) and/or to its designated use (production forest, conservation forest, protected forest, etc.). Mining is seldom permitted in built-on land or other areas with sensitive land cover such as protected forest, mangrove, or wetlands. As with any rule, however, exceptions do exist (**Case 7.2**).

Mining is seldom permitted in built-on land or other areas with sensitive land cover such as protected forest, mangrove, or wetlands.

An important factor influencing land cover and the productivity of the Earth's various ecosystems is the nature of their soils, the very upper part of the Earth's crust. Soils are vital for the existence of many forms of life that have evolved on Earth. Soil itself is very complex and is much more than merely a combination of fine mineral particles. Soil also contains air, water, dead organic matter, and various types of living organisms. The formation of a soil is influenced by organisms, climate, topography, parent material, and time.

Land Quality

The holistic concept of land as a natural resource was recognized in the Framework for Land Evaluation (FAO 1976), repeated implicitly in Chapter 12 of the UN Conference on the Environment and Development (UNCED) in 1992, and formally described in FAO 1995. It reads: 'Land is a delineable area of the Earth's terrestrial surface, encompassing all attributes of the biosphere immediately above or below this surface, including those of the near-surface climate, the soil and terrain forms, the surface hydrology (including shallow lakes, rivers, marshes and swamps), the near-surface sedimentary layers and associated groundwater reserve, the plant and animal populations, the human settlement pattern and physical results of past and present human activity (terracing, water storage or drainage structures, roads, buildings, etc.).'

Land quality (or its lack) is a complex land attribute. It is not an absolute value, but has to be assessed in relation to land functions and the specific land use that one has in mind. The various land functions are detailed in FAO (1997) and they are summarized in **Table 7.2.**

Land Use

As a space for any human activity, land is a resource *sui generis*. It is literally the foundation for all economic activity and for all life, limited only by the extent of the Earth's surface. This means that alternative patterns of land use, the predominant purpose for which an area is utilized, are mutually exclusive. Intended alternative patterns of land use are reflected in the spatial planning of governments. Accordingly, an understanding of land use planning of the area in which the ore body is located is crucial at the very outset of mine planning. In some jurisdictions such as Indonesia, if mining conflicts with existing official land use plans (termed Spatial Plans), mine development can only commence after the spatial plans have been adjusted to accommodate mining activities.

An understanding of land use planning of the area in which the ore body is located is crucial at the very outset of mine planning.

From an ecological point of view, we distinguish protective use (areas left to nature, e.g. primary forest or wetlands), productive use (e.g. agricultural land or mining), and sterile use (e.g. urban areas, housing, buildings, or utility/transportation corridors such as roads or pipeline corridors). Environmental assessments sometimes overlook land required for human activity, raising the potential for major conflict associated with alternative land uses.

CASE 7.2
Coal Mining in Germany

Mining in Germany goes back a long time. This rich history is reflected in the names of entire mountain ranges such as Erzgebirge (ore mountains) and townships such as Stolberg (tunnel mountain). Today, the German mining industry has disappeared with the notable exception of large-scale surface coal mining in the State of North Rhine Westfalen, and a few government subsidized hard coal under ground mines. The photograph shows a simulated natural colour ASTER image in the German state of North Rhine Westphalia, acquired on August 26, 2000, and depicts one of the enormous opencast coal mines. To enable these mines to be expanded, entire villages are relocated. One mine, the Hambach opencast coal mine, operates the No. 293 giant bucket wheel excavator, the largest machine in the world, twice as long as a soccer field and as tall as a building with 30 floors. To uncover the 2.4 billion tons of brown coal (lignite) found at Hambach, five years were required to remove a 200 m thick layer of waste sand and to redeposit it off site. The mine currently yields 30 million tons of lignite annually, with annual capacity scheduled to increase to 40 million tons in coming years.

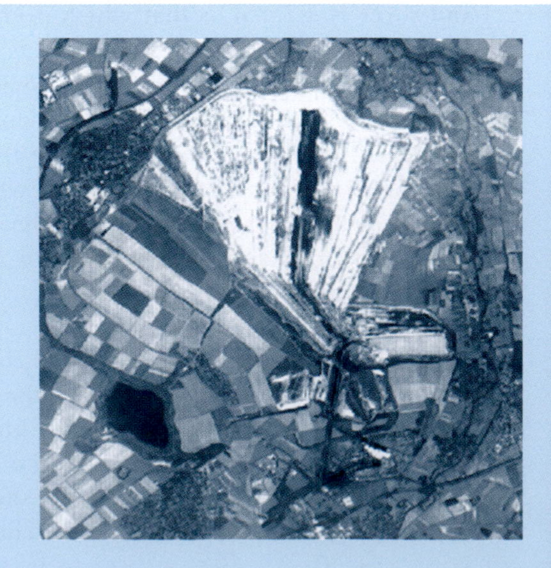

TABLE 7.2

Functions of Land – Land quality (or its lack) is not an absolute value, but has to be assessed in relation to land functions and the specific land use that one has in mind

Function	Description
Production	Land is the basis for many life support systems, through production of biomass that provides food, fodder, fibre, fuel, timber and other biotic materials for human use, either directly or through animal husbandry including aquaculture and inland and coastal fishery
Biotic environmental	Land is the basis of terrestrial biodiversity by providing the biological habitats and gene reserves for plants, animals, and micro-organisms, above and below ground
Climate regulative	Land and its use are a source and sink of greenhouse gases and form a co-determinant of the global energy balance – reflection, absorption, and transformation of radiative energy of the sun, and of the global hydrological cycle
Hydrologic	Land regulates the storage and flow of surface and groundwater resources, and influences their quality
Storage	Land is a storehouse of raw materials and minerals for human use
Waste and pollution control	Land has a receptive, filtering, buffering, and transforming function for hazardous compounds
Living space	Land provides the physical basis for human settlements, industrial plants, and social activities such as sports and recreation
Archive or heritage	Land is a medium to store and protect the evidence of the cultural history of humankind, and source of information on past climatic conditions and past land uses
Connective space	Land provides space for the transport of people, inputs and produce, and for the movement of plants and animals between discrete areas of natural ecosystems

Source:
FAO (1995)

7.3 THE HYDROSPHERE – STORAGE AND MOVEMENT OF WATER

For a mining project water may be both friend and foe. Mining project operations invariably require water, sometimes in very large quantities. The uses vary widely: dust suppression, milling and processing, transport of tailings and/or concentrates, recovery of metals by leaching, and reclamation of mined lands. Waste waters may be generated from dewatering, runoff from disturbed land, or excess water from tailings impoundments or water storages. Impacts on the local hydrologic cycle in terms of quantity, quality, or both, often constitute the most prominent or the most serious environmental impacts associated with mining.

For a mining project water may be both friend and foe.

Water in Motion

The hydrologic cycle is a conceptual model that describes the storage and movement of water between the atmosphere, lithosphere, biosphere, and the hydrosphere. Water on Earth can be stored in any one of the following media (**Figure 7.4**): atmosphere, oceans, surface waters, soils, glaciers, snowfields, and groundwater. Water moves from one medium to another by way of natural processes such as evaporation, condensation,

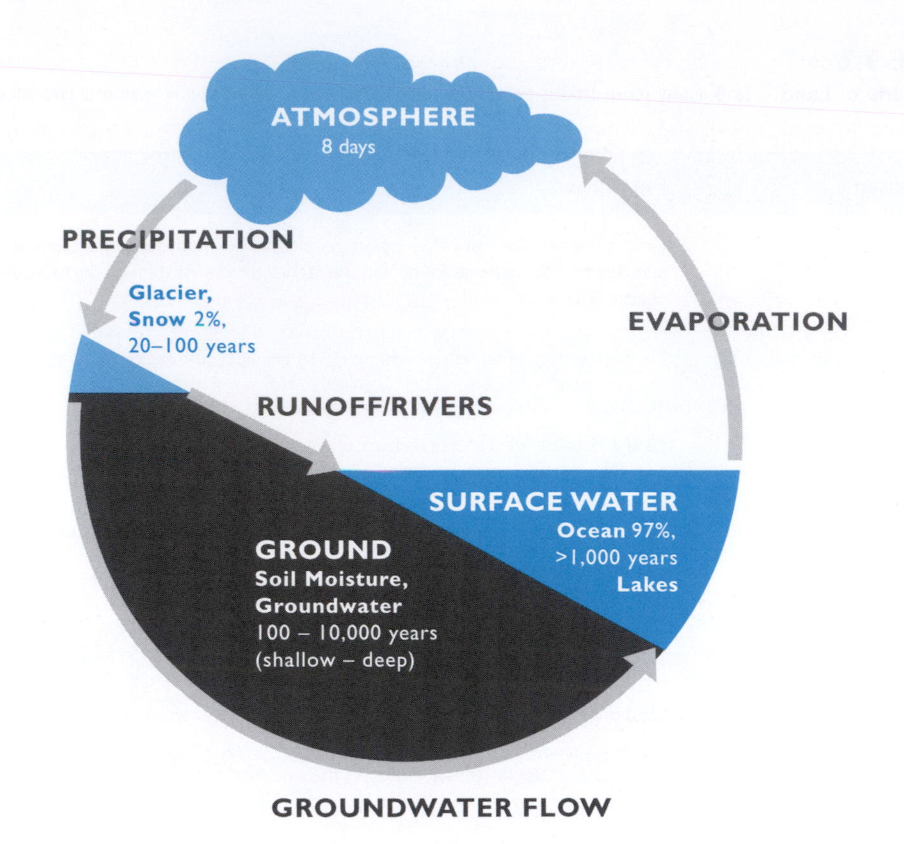

FIGURE 7.4
The Hydrologic Cycle and Typical Residence Time of Water in Natural Reservoirs

Water moves from one reservoir to another by way of natural processes like evaporation, condensation, precipitation, runoff, infiltration, transpiration, melting, and groundwater flow.

precipitation, runoff, infiltration, transpiration, melting, and groundwater flow. In areas where water is scarce or human population is high, human activities can change the local natural hydrologic cycle. It is now also widely accepted that industrialization including mining has contributed to global warming, which is associated with notable shifts in the global hydrologic cycle.

Water is continually cycled between the various storage media. The time of residence in any particular medium is an important factor in any assessment of water pollution (**Figure 7.4**). On average water is renewed in rivers once every 16 days. Water in the atmosphere is completely replaced every eight days. Slower rates of replacement occur in large lakes, glaciers, ocean bodies, and groundwater. Replacement in these reservoirs can take from hundreds to thousands of years. Some of these resources (especially groundwater) are being used by humans at rates that far exceed their renewal times and water becomes effectively nonrenewable. Some groundwater, referred to as 'fossil water' is water which was stored in the distant past and is no longer subject to renewal.

Water as a Scarce Resource

Water is essential to life. It is also the most widely occurring substance on Earth; however, fresh water represents less than 3 percent, the remainder being salt water, as illustrated in **Figure 7.4**. Some two-thirds of freshwater is locked up in glaciers and permanent snow cover. Availability of the remaining one-third differs widely in different parts of the world, with wide variation in seasonal and annual precipitation. Industrial, human, and agricultural wastes are discharged to receiving waters, further polluting much of the fresh water that exists. At the same time, the human population is growing, as is per capita water use. The consequence is that unpolluted fresh water is becoming increasingly scarce.

This is particularly so in many parts of Asia where particularly high rates of population growth and industrialization are occurring. Not surprisingly, mining activities with notable impacts on water quantity and quality have and always will generate controversy, be it the industry as whole as in Mexico where mining contributes about 2% to the nation-wide water depletion (Shomaker 2005), or individual mines such as the Ok Tedi mine in PNG or the Freeport Mine in Papua, both of which utilize riverine tailings disposal.

Mining activities with notable impacts on water quantity and quality have and always will generate controversy.

The UN Conference on the Environment and Development (UNCED) in 1992 (Earth Summit), was instrumental in placing water at the centre of the sustainable development debate. Public authorities now recognize the vital importance of water as a natural resource. It is also understood that water has not only economic value, but social, religious, cultural, and environmental values as well. Water management now sits at or near the top end of any environmental authority's agenda.

Water has not only economic value, but social, religious, cultural, and environmental values as well.

The International Conference on Water and the Environment in Dublin, also held in 1992, sets out four principles – the Dublin Principles – to guide water management:

- Principle 1: Fresh water is a finite and vulnerable resource, essential to sustain life, development and the environment.
- Principle 2: Water development and management should be based on a participatory approach, involving users, planners, and policymakers at all levels.
- Principle 3: Women play a central part in the provision, management, and safeguarding of water.
- Principle 4: Water has an economic value in all its competing uses and should be recognized as an economic good.

The WHO drinking water guidelines (WHO 2006) often serve as the main reference in developing national water quality standards. The guidelines also provide information on water quality and health, and on effective management approaches.

What is the Local Water Resource?

Local water resources are highly variable, both spatially and temporally. Accommodating competing water demands against variable supply is difficult, and the need for management increases as more of the resource is relied on for extraction and use. The available total water resource can be defined as the difference between total inflows and outflows to surface and groundwater in a given time period, for a defined area, plus the net volume of water in storage at the start of the time period under consideration. Total inflows include total surface water flows and deep drainage to groundwater (groundwater recharge) and transfers into the water system (both surface and groundwater), for a defined area. Total outflows are the opposites; they represent water losses for a defined area.

The water balance will only provide insights into local water resources if sufficient data are available. More often than not the water balance will only highlight the known data and help identify information gaps. From the water balance, a picture of water availability (or scarcity) in the host region will emerge. Water balances for different time intervals will be required for this understanding. For example, the average annual balance may be insufficient to reflect daily water availability.

More often than not the water balance will only highlight the known data and help identify information gaps.

Equitable and sustainable management of shared water resources requires responsible contributions from all beneficiaries, particularly true for a new mining venture becoming a new and significant player in the host region's hydrologic cycle. The appropriate management response is to quantify existing water cycles in the project area and its usage in sufficient detail to evaluate any changes imposed by the mining project (**Case 7.3**).

7.4 THE BIOSPHERE – LIFE ON EARTH

Today's mining often takes place in remote areas, largely unaffected by human presence and activities, and often home to unique ecosystems. Such natural areas are becoming increasingly rare, and their relative importance to human society is constantly increasing. Large-scale mining by its very nature alters these ecosystems, sometimes permanently.

The biosphere comprises all in nature which is living (separated into flora and fauna) and includes all levels of life from genes to species to ecosystems. The levels of life are strongly linked with the concept of biological diversity (also termed biodiversity). Biodiversity is decreasing at an increasing rate due to the impact of the growing human population and increasing rates of resource consumption (**Case 7.4**). Recognition of the worldwide impact of this decline prompted the global community to negotiate, in 1992, the United Nations Convention on Biological Diversity (CBD). The Biodiversity Convention, as it is commonly known, is a legally binding international treaty. The Convention obliges signatory countries to assess the adequacy of current efforts to conserve biodiversity and to use biological resources in a sustainable manner. Environmental assessment has been recognized as a key element in meeting the obligations of the Biodiversity Convention and Strategy, manifested in Article 14 of the CBD which recognizes environmental assessment as an important decision-making process toward the protection of biodiversity.

The biosphere comprises all in nature which is living (separated into flora and fauna) and includes all levels of life from genes to species to ecosystems.

What is Biodiversity?

The United Nations Convention on Biodiversity (1992) defines biodiversity as: 'The variability among living organisms from all sources including inter alia, terrestrial, marine and other aquatic ecosystems and the ecological complexes of which they are part; this includes diversity within species, between species, and of ecosystems'.

CASE 7.3
Water Management at Olympic Dam Mine, South Australia

The Great Artesian Basin (GAB) provides much needed water to large parts of the eastern and southern part of Australia. In the past water abstraction was primarily for pastoral water use, estimated to be around 132 Mega Litre/day (Ml/d) in SA.

WMC (Olympic Dam) Pty. Ltd. commenced groundwater extraction in 1988 to support its mining operation including a mine town (copper and associated products of uranium, gold, and silver). Groundwater abstraction initially at 15 Ml/d for a production rate of 85,000 tons copper per annum (tpa) in 1996, increased to 42 Ml/d due to an increase of mine production to 350,000 tpa. Currently an increase of production to 500,000 tpa is under study.

In its effort to limit water consumption, WMC looked beyond the fence. The company sponsored a programme to assist pastoralists to control bore flows and pipe water to stock drinking troughs rather them allowing it to flow many kilometres along open drains where most evaporates or soaks into the ground. It is estimated that this programme will eventually conserve about 37 Ml/d, close to today's water consumption of the mine.

Source: Mineral Council of Australia (www.minerals.org.au/education); Power (2002)

It follows from this definition that biodiversity applies at different levels, namely:

- Ecosystems, which include a diversity of
- Habitats, each of which supports an assemblage or community of plant and animal
- Species, each of which includes individuals, each with their own unique set of
- Genes.

Any significant change in conditions, whether natural or man-made, has the potential to reduce biodiversity at one or more of these levels. The geological record contains numerous examples of events that caused sudden or progressive loss of biodiversity. Examples of the former include the impact of large meteors; the extinction of the dinosaurs is commonly attributed to such an event. Examples of the latter include climate change, which among other effects has caused a succession of 'ice ages' during which glaciers have advanced over major portions of the Earth's surface. However, although these events may have been catastrophic at the time, the palaeontological record also indicates that each drastic loss of biodiversity is followed by a major and, in geological terms, rapid evolution of new species. It is as if the creation of new areas available for colonization provides the impetus for renewed competition, involving rapid adaptation and a proliferation of new species.

In recent years, there have been numerous predictions of impending loss of biodiversity as a result of man's activities. **Case 7.4** is typical of the more extreme predictions. Many practising biologists consider these predictions to be extreme as they are based on questionable assumptions.

Biodiversity in Relation to Biogeography

To evaluate biodiversity in perspective, it is also instructive to consider bio-geography, the branch of biology that deals with the geographic distribution of plants and animals. Accordingly, the Earth's surface can be divided into bio-geographic provinces, representing areas defined by the endemic species present. The world's continents include many large biological provinces of relatively uniform biological conditions. This is particularly the

To evaluate biodiversity in perspective, it is also instructive to consider bio-geography.

CASE 7.4
Human Threats to Biodiversity

In 1995 the United Nations Environmental Program (UNEP) estimated the number of known species to be 1,750,000. Annually about 12,000 new species are added. Only a fraction of all species is currently known. Estimates say that the total number of species may be as high as 10 or even 100 million.

The global variety of life varies according to the different biospheres. Tropical rainforests which cover only 7% of the world's land surface host up to 90% of all terrestrial species. More recent studies indicate that there are areas on Earth which have a special significance because of their great biodiversity. These 'Hot Spots' represent only 1.4% of the Earth's surface, but are especially important because they host 44% of all plant species and 35% of all vertebrates.

Some estimates predict that the loss of species in the next 50 years will be between 10% and 50% of the total number of species that ever existed on Earth and that between 70 and 300 species disappear every day. The current rate of extinction is said to be 50 to 100 times higher than extinction due to natural factors.

Source: The Lutheran World Federation 2003

case in low relief, mid-continental areas, where climatic conditions differ little over large distances. Plant and animal species living in these environments, which are usually quite uniform and lack the variety of special habitats or niches that encourage the evolution of new species, are generally widely distributed. An example is the Mulga Woodland which occupies much of the arid zone of central and Western Australia.

By contrast, in the adjacent high mountains, elevation and aspect combine to produce marked variations in temperature and rainfall which, with differences in geological substrate that are also characteristic of mountains, result in numerous habitat differences, supporting a much greater diversity of species, some of which may have relatively limited distributions.

Tropical rainforests generally support very high diversities of plants and animals.

Tropical rainforests generally support very high diversities of plants and animals, largely because biological productivity is very high and the structure of the forest, in itself, provides a wide variety of habitats and micro-habitats, each of which supports a different assemblage of species. While most tropical rainforest species are common with wide distributions, this is not always the case. In particular, some species have become rare or endangered as a result of exploitation by man. Examples are the great apes of Africa, the Indian tiger, many species of commercially valuable timber trees, parrots and cockatoos popular as caged birds, and numerous orchid species. And, of course, some previously common species are now uncommon because the forest habitat on which they depend, has been largely removed. High diversity is, however, not restricted to tropical rainforests. For example, the heathlands of southwest Australia support a very high diversity of plants, many of which have very limited distributions, based on adaptations to subtle differences in substrate conditions. Several large mineral sand mines are located in this environment which poses particular challenges to rehabilitation.

Islands, particularly remote oceanic islands, are at the other extreme from mid-continental areas. Individual islands commonly have highly distinctive assemblages of flora and fauna, with relatively little overlap of species between islands, except where they are in close proximity. Species and sub-species may be endemic to a single island or group of islands. This high degree of endemism is particularly pronounced in the central and eastern islands of Indonesia. Plants and animals endemic to islands are also particularly vulnerable to invasion by introduced species. Accordingly, the potential for species extinction as a direct or indirect result of mining, is relatively high on small or medium sized islands.

The potential for species extinction as a direct or indirect result of mining, is relatively high on small or medium sized islands.

Finally, uncommon or unusual habitats, particularly permanently moist areas in otherwise arid environments, may support remnant populations of plants and animals that were widespread during different, usually wetter periods, but which have contracted to one or more refuge areas, as the climate has changed. Examples are ferns which are confined to permanently shaded areas at the bases of cliffs in central Australia. These relict species may expand their distributions again when cooler or wetter conditions return. The geological record contains many examples of extensive expansion and contraction in areas occupied by species as a result of fluctuating climatic conditions. In fact, it appears that the climate has seldom, if ever, been static for long and accordingly, the distributions of species are constantly changing in response to climate change, and these changes can be quite rapid.

It follows from the foregoing, admittedly oversimplified bio-geographic summary that the risks to biodiversity from mining, or any other invasive activity, will depend on the bio-geographic situation as much as on the nature of the activity itself.

The risks to biodiversity from mining, or any other invasive activity, will depend on the biogeographic situation as much as on the nature of the activity itself.

To understand how biodiversity is likely to respond to mining, impacts at each level of diversity are best assessed in terms of composition (what biological units are present and how abundant they are), structure (or pattern) (how biological units are organized in time and space) and function (the role different biological units play in maintaining natural processes and dynamics) as illustrated in **Table 7.3**.

TABLE 7.3

Biodiversity Checklist of the Three Components of Biodiversity – Composition, Structure, and Function

	Components of Biological Diversity			
Level of biological diversity	**Composition**	**Structure (temporal)**	**Structure (spatial: horizontal and vertical)**	**Key processes**
Genetic diversity	Minimal viable population (avoid destruction by inbreeding/gene erosion)	Cycles with high and low genetic diversity within a population	Dispersal of natural genetic variability	Exchange of genetic material between populations (gene flow)
	Local cultivars		Dispersal of agricultural cultivars	Mutagenic influences
	Living modified organisms			Intraspecific competition
Species diversity	Species composition, genera, families etc., rarity/abundance, endemism/exotics	Seasonal, lunar, tidal, diurnal rhythms (migration, breeding, flowering, leaf development, etc.)	Minimal areas of species to survive	Regulation mechanisms such as predation, herbivory, parasitism
	Population size and trends	Reproductive rate, fertility, mortality, growth rate.	Essential areas (stepping stones) for migrating species.	Interactions between species
	Know key species (essential role)	Reproduction strategy	Niche requirements within ecosystem (substrate preference, layer within ecosystem)	Ecological function of a species
	Conservation status		Relative or absolute isolation	
Ecosystem diversity	Types and surface area of ecosystems	Adaptations to/dependency on regular seasonal rhythms	Spatial relations between landscape elements (local and remote)	Structuring process(es) of key importance for the maintenance of the ecosystem itself or for other ecosystems
	Uniqueness/abundance	Adaptations to/dependency of on irregular events; droughts, floods, frost, fire, wind	Spatial distribution (continu -ous or discontinuous/patchy)	
	Succession stage, existing disturbances and trends (= autonomous development)	Succession (rate)	Minimal area for ecosystem to survive	
			Vertical structure (layered horizons, stratified)	

Source:
Convention on Biological Diversity, COP 6 Decision VI/7 (www.unep.org)

Why is Biodiversity Important?

Biodiversity values, conveniently categorized as economic, social, and intrinsic values, are often underestimated. *Economic values* comprise biodiversity as a source of harvestable goods (food, medicines, building materials, and products sold for income or used as inputs to other economic activities) and its regulative function of natural processes and the

Earth's life support systems, e.g. carbon sequestration, soil formation, purification of water, or coastal protection by dunes or mangroves.

Another example of economic values is the importance of biodiversity in agriculture in which biodiversity is often essential for pollination of commercially valuable crops and biological control of pests and diseases.

Social values relate to the importance of biodiversity for employment, health, quality of life, social security, and appreciation. In many cultures and societies, all or some components of biodiversity have 'intrinsic' value in their own right, irrespective of any material contribution to human wellbeing. Biodiversity is then a source of spiritual and religious enrichment and well-being, the *intrinsic values*. Perhaps most important of all, biodiversity is the basis for evolution and adaptation to changing environments, making it essential for survival of life.

Some habitats, ecosystems and bio-geographic provinces are far more diverse (and valuable) than others.

As is evident from the discussion of bio-geography, some habitats, ecosystems and bio-geographic provinces are far more diverse (and valuable) than others. Areas of important biodiversity are defined by IAIA (2005) as those that:

- Support endemic, rare, declining habitats/species/genotypes.
- Support genotypes and species whose presence is a prerequisite for the persistence of other species.
- Act as a buffer, linking habitat or ecological corridor, or play an important part in maintaining environmental quality.
- Have important seasonal uses or are critical for migration.
- Support habitats, species populations, ecosystems that are vulnerable, threatened throughout their range and slow to recover.
- Support particularly large or continuous areas of previously undisturbed habitat.
- Act as refugia for biodiversity during climate change, enabling persistence and continuation of evolutionary processes.
- Support biodiversity for which mitigation is difficult or its effectiveness unproven including habitats that take a long time to develop characteristic biodiversity.
- Are currently poor in biodiversity but have potential to develop high biodiversity with appropriate intervention.

CASE 7.5
Mining and Forest Destruction – Fact or Fantasy?

The ten most forest-rich countries account for two-thirds of the Earth's forest area: Australia, Brazil, Canada, China, the Democratic Republic of the Congo, India, Indonesia, Peru, the Russian Federation, and the United States of America. The total natural forest area is annually reduced by about 14.6 million hectares, an area of the size of England (House of Common 2006). From the remaining forest areas only 12% lies in protected areas. More than half of all logging takes place in vulnerable forests in SE Asia (in particular Indonesia), Central and South America (in particular Brazil), Russia, and Africa. Logging at current rates in some of the world's most pristine habitats, such as Indonesian Papua and Papua New Guinea, is predicted to destroy their forests and habitats in as little as ten years from now.

Mining is often singled out as the main cause of forest loss, an accusation that contradicts public data. Forest clearing to allow agriculture by far outweighs all other activities combined that contribute to rainforest loss. That said, environmental impacts are site-specific and a particular mine may in fact destroy valuable forest areas, even so, on average, this is not the case.

Source of Graphic: European Aluminium Association (www.eaa.net)

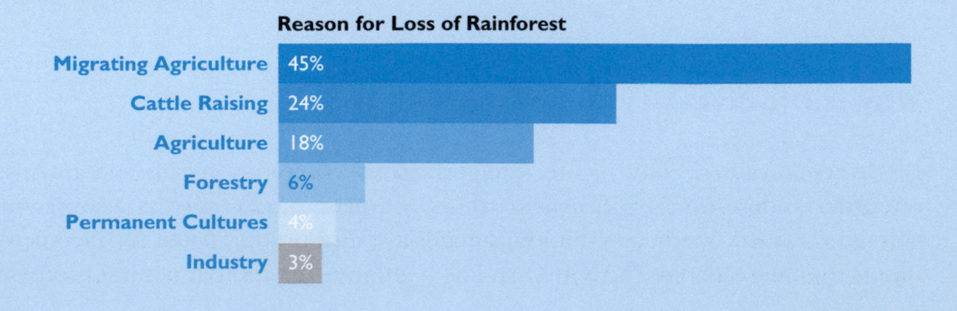

Forests, particularly tropical rainforests, meet most of these criteria and as such, are a vital part of the Earth's ecosystems, and of biodiversity. Tropical forests contain some of the most bio-diverse habitats on Earth, representing a vast natural resource, containing the majority of the Earth's terrestrial species, many as yet unknown, and some of great potential benefit to society should they survive to be discovered. It is therefore no surprise that mining in forest areas is becoming increasingly controversial (**Case 7.5**). Loss of natural forests, which once covered 48% of the land on Earth, has profound economic, social, and environmental impacts. Forest cover has now been reduced to 29% (3,900 million hectares) and continues to decrease every year at an alarming rate. Sixty million Indigenous Peoples are almost wholly dependent on forests for their livelihood, and forests support up to 1.6 billion people to greater or lesser extent (House of Common 2006). Forests also act as an enormous carbon sink. The total carbon stored in forest biomass, deadwood, litter, and soil represents roughly 50% more than the amount of carbon in the atmosphere. The destruction of forests is estimated to contribute almost two billion tons of carbon to the atmosphere every year. This represents a quarter of annual man-made carbon emissions (FAO 2005). Furthermore there are concerns that the impacts of climate change may further increase carbon emissions, as a result of forest die-off in tropical and temperate areas caused by the predicted increase in temperatures and droughts.

Mining in forest areas is becoming increasingly controversial.

Biodiversity and Indigenous Peoples

Indigenous Peoples living traditional lifestyles may use many different plant and animal species as food, and for a variety of other purposes including healing, birth control, shelter, and ritual practices. Involvement of local Indigenous Peoples in baseline surveys is highly beneficial as such people have a wealth of knowledge relating to the occurrence and behaviours of plants and animals inhabiting their lands. In situations where mining takes place in areas inhabited by Indigenous Peoples, the use of local plants and animals should be researched and documented so that the relevant species can be included in the rehabilitation programme.

Involvement of local Indigenous Peoples in baseline surveys is highly beneficial as such people have a wealth of knowledge relating to the occurrence and behaviours of plants and animals inhabiting their lands.

Conservation Reserves

Conservation reserves, including National Parks, provide the first line of defence against species extinction, the aim being to protect representative areas of all habitat types. However, while in some countries this aim has been largely achieved, this is not the case in many other countries, including most of the developing world. Moreover, even where conservation reserves have been established, many countries lack the capacity to implement the necessary conservation procedures, so that the reserves become degraded despite their conservation status. Nevertheless, such reserves provide the only hope for some species which are already at low population levels. Proponents of mining projects in the vicinity of conservation reserves may assist the conservation of biodiversity within and around these reserves by sponsoring scientific research or contributing to management initiatives and capacity building.

Occurrence of a species within a conservation reserve does not necessarily guarantee the continued survival of that species. Also, the absence of conservation status does not indicate that an area has no valuable ecological attributes or that it does not support populations of rare or endangered species. Each situation needs to be evaluated on its own merits.

The absence of conservation status does not indicate that an area has no valuable ecological attributes.

7.5 THE SOCIAL SPHERE – SOCIAL AND CULTURAL FABRIC OF SOCIETY

The essential feature of a group is that its members have something in common and that they believe that what they have in common is important.

In studying social impacts we are concerned about changes in the social and cultural fabric of human groups. Groups are the essence of life in society since we become who we are because of our membership in human groups. The essential feature of a group is that its members have something in common and that they believe that what they have in common is important. The largest and most complex group is society, people who share a culture and a territory.

Anthropologists and sociologists have identified five types of society that have developed in the course of human history: hunting and gathering, pastoral and horticultural, agricultural, industrial, and postindustrial, each characterized by distinct forms of social division, social labour, and social inequality. These societies and their attributes are described in the following drawn from the excellent textbook by Hensin (2005).

The hunting and gathering society has the fewest social divisions and is the most egalitarian. In this society, the men hunt large animals and the women usually gather edible plants, fruits, and other food found growing in the wild. The first social revolution was based on the domestication of plants and animals. This resulted in the development of the pastoral society that concentrated on the herding of animals and the horticultural society that specialized in planting and harvesting crops. The horticultural society made it possible for permanent settlements to be established since it was no longer necessary for people to follow the food supply. In the third social revolution, the invention of the plough made it possible for large areas of land to be cultivated and harvested. The society that developed, known as the agricultural society, made large cities possible because it freed some members of society from being dedicated to producing their own food. In the fourth social revolution, the invention of the steam engine introduced the industrial society which concentrated on the manufacturing and consumption of goods. The fifth social revolution commenced with the invention of the microchip in 1973. In our current society, the emphasis is on the development and transfer of technology, information, and knowledge, today central to the ever-growing service industry.

Dysfunctional social change may be associated with large-scale developments such as mining when too much happens too suddenly.

Human groups and societies are characterized by a number of closely interrelated attributes that combine to form the human environment in which we live. These attributes relate to demographics, type of organization, race and ethnicity, gender, culture, norms, values, and religion. Our human environment is constantly changing; some changes are too subtle to be noticed; some changes are too violent to cope with. Dysfunctional social change may be associated with large-scale developments such as mining when too much happens too suddenly.

Demography – Size, Composition, Growth, and Distribution of Population

The study of the size, composition, growth, and distribution of human population is called demography. Three demographic variables are used to project population trends: fertility, mortality, and migration. Fertility refers to the number of children the average woman bears while mortality is measured by the crude death rate – the annual number of deaths per 1,000 population. Known as the basic demographic equation, the growth rate equals births minus deaths, plus net migration. The growth rate is affected by anticipated variables, such as the percentage of women who are in their childbearing years, and unanticipated variables like wars, famines, and changing economic and political conditions. Development of a new mining project would introduce unanticipated economic variables. At a local scale, mines may drastically influence the basic demographic equation by increasing net migration, which is often accompanied by undesirable social consequences (**Case 7.6**).

At a local scale, mines may drastically influence the basic demographic equation by increasing net migration, which is often accompanied by undesirable social consequences

People in most developing countries continue to have large families, because parenthood elevates their status and large numbers of children are economic assets. The implications of these different growth rates can be illustrated using population pyramids, graphic representations of a population divided into age and sex classes. Industrialization is the primary factor influencing growth rate. In every country that industrializes, the growth rate declines and the standard of living, including life expectancy, improves. The primary problems of uncontrolled growth, sometimes associated with mining projects, are the development of pockets of poverty, environmental decay, and a decline in the quality of life for some people. To be truly successful, managing growth needs to follow three guiding principles: scale (regional and local planning), livability (respect of human needs), and social justice (how growth affects people). The social management plan for a mining project must reflect these three principles.

The primary problems of uncontrolled growth, sometimes associated with mining projects, are the development of pockets of poverty, environmental decay, and a decline in the quality of life for some people.

Social Organization and Structure

Society is organized 'to get its job done'. Nations do so through formal organizations. The same system that can be frustrating and impersonal is also the one on which we rely for our personal welfare and to fulfil our daily needs. Formal organizations, secondary groups designed to achieve specific objectives, have become a central feature of contemporary society. Today, we take their existence for granted.

The larger the formal organization, the more likely that it will become a bureaucracy. Bureaucracies are defined as formal organizations characterized by five features that help

CASE 7.6
Shanty Town Housing of Illegal Miners in Diwalwal, Philippines

Diwalwal, the gold rush site in Southern Mindanao is said to be the home of the biggest undeveloped gold deposit in the Philippines. It is also the site of armed conflicts and tragedies. Tens of thousands of people, mostly small miners, rely on it for their livelihood. Several mining companies have tried to take control, some using military force. The Philippine Government continues to search for a balanced approach that allows of the opening the area for large-scale modern mining while at the same time providing an income and livelihood to the thousands of current small-scale miners.

Dysfunctions of bureaucracies can include 'red tape' and lack of communication between individual units (such as between ministries, or between units within one ministry), dysfunctions all too familiar to mining companies operating in developing countries.

them reach their goals: to grow, and to endure. These five features are (1) clear levels, with assignments flowing downward and accountability flowing upward; (2) a division of labour; (3) written rules; (4) written communications with records; and (5) impartiality. Although bureaucracies are the most efficient forms of social organization, they can also be dysfunctional. Dysfunctions of bureaucracies can include 'red tape' and lack of communication between individual units (such as between ministries, or between units within one ministry), dysfunctions all too familiar to mining companies operating in developing countries (**Case 7.7**).

At the community level, social structures complement formal systems. The major components of social structure include culture, social class, social status, roles, groups, and social institutions. Social structure guides peoples' behaviours. A person's location in the social structure (her social class and social status, the roles he plays, and the culture, groups, and social institutions to which she belongs) underlies the person's perceptions, attitudes, and behaviours. People develop these perceptions, attitudes, and behaviours from their place in the social structure, and they act accordingly. Social structure is not static. It responds to changes in culture, technology, economic conditions, group relationships, and societal needs and priorities. Structural changes can, sometimes, fundamentally and permanently alter the way a society organizes itself.

Race and Ethnicity

The Earth offers a fascinating array of human characteristics. Race refers to the inherited physical characteristics that distinguish one group from another. These distinguishing characteristics include a variety of complexions, colours, and shapes. Although there have been significant strides in the understanding of race and racial equality, two myths of race are still common. One is the perception that some races are superior to others, and the other is that 'pure' races exist.

Sociologically, a minority group is a group of people who are singled out for unequal treatment and who regard themselves as objects of collective discrimination.

While race refers to biological characteristics that distinguish one group of people from another, ethnicity refers to cultural characteristics that distinguish one group of people from another. Derived from the Greek word 'ethnos', meaning 'people' or 'nation', ethnicity may centre on nation of origin, distinctive foods, dress, language, music, religion, or family names and relationships. It is common for people to confuse the terms race and ethnic group. Along with race and ethnicity, the concept of minority group is often misunderstood. Sociologically, a minority group is a group of people who

CASE 7.7
The Many Facets of Environmental Concern

Australian based Herald Resources Ltd. obtained government approval at provincial level for its mine development in Sumatra, Indonesia in 2005. The company raised funds for mine development, commissioned detailed engineering work, and ordered mining equipment. Twelve months later the Department of Forestry at the Central Government level questioned the validity of the local approval, requesting Herald Resources Ltd. to redo the environmental permitting at the central level. Eventually the dispute was resolved with support from the Ministry of Environment upholding the legitimacy of the original approval. However, the Department of Forestry continued to refuse issuing a forestry permit needed for mine development. The associated financial loss to the company caused by the delay in mine development was substantial, as was the emotional stress on those involved. As at the end of 2007, the company still has to receive the official forest permit allowing the mine to go ahead. Coincidentally, in early 2008 an Indonesian-owned resource company with well-established government relationships at the highest level eventually stepped forward to take over the ore deposit.

are singled out for unequal treatment and who regard themselves as objects of collective discrimination.

Gender

Gender stratification refers to males' and females' unequal access to power, prestige, and property on the basis of their sex. Gender is especially significant because it is a master status that cuts across all aspects of social life. Sex refers to the biological characteristics that distinguish males from females; gender refers to the social characteristics that a society considers proper for its males and females. Although human beings are born male or female, they learn how to be masculine or feminine. This process of gender socialization begins at birth and continues through life. In short, we inherit our sex, but learn our gender.

In practically every society, greater prestige is given to male activities, regardless of the nature of these activities. There are many forms of gender inequality in various aspects of everyday life that continue to persist. Among these are a devaluation of things feminine, violence against women, and sexual harassment.

Women are not usually associated with mining, and very few professions attract fewer women than mining. Modern mining equipment has improved working conditions, and female equipment operators have become a common sight at mines, but mining remains 'a man's world'. The influx of thousands of male mine worker into a new mining area causes social changes that extend well beyond the borders of the mine. Gender issues become important, within the families of mine workers and in adjacent communities.

Modern mining equipment have improved working conditions, and female equipment operators have become a common sight at mines, but mining remains 'a man's world'.

Culture, Norms and Values

The concept of culture is sometimes easier to grasp by description rather than definition. All human groups possess material and non-material culture. Material culture consists of art, buildings, weapons, jewellery, and all other man-made objects that are passed from one generation to the next. Non-material culture includes a group's ways of thinking (beliefs, values, and other assumptions about the world) and patterns of behaviour (language, gestures, and other forms of social interaction). Sociologists sometimes refer to non-material culture as symbolic culture, since the central component of non-material culture is symbols. Symbols include gestures, language, values, norms, sanctions, folkways, and mores. Gestures involve the ways in which people use their bodies to communicate with one another. Although people in every culture use gestures, the gestures people use and the meanings they associate with those gestures vary greatly from one culture to another.

Language is the basis of culture. It is critical to human life and essential for cultural development. Among other things, language allows people to communicate with each other and for human experience to be accumulated. It also provides people with the capacity to share understandings about the past and develop common perceptions about the future. Like gestures, all human groups have language. And like gestures, the meanings that people associate with different sounds and symbols can vary greatly from one culture to another.

Culture's effects are profound and pervasive, touching almost every aspect of people's lives. Cultures may contain numerous subcultures, a group whose values and related behaviours set it apart from the larger culture. All groups have values (beliefs regarding what is desirable or undesirable, good or bad, beautiful or ugly), which they channel into norms (expectations, or rules of behaviour, that develop out of values). Human groups need norms to exist. By making behaviour predictable, norms make social life possible.

People may become more aware of their own culture when their cultural assumptions are challenged by exposure to other people's culture, particularly those with fundamentally different ways of believing and doing.

Most people are generally unaware of their own culture; it is so ingrained into their beings that it is often taken for granted. People may become more aware of their own culture when their cultural assumptions are challenged by exposure to other people's culture, particularly those with fundamentally different ways of believing and doing. When people come into contact with cultures that significantly differ from their own, they often experience cultural shock – a condition of disorientation that requires them to question the cultural assumptions that they previously took for granted.

When technology changes, society also changes.

Mining projects do interfere with local culture. There is an influx of outsiders with different cultural backgrounds into the project area, in high numbers during the relatively short construction period, and in reduced numbers during the longer mine operation period. Mining projects also eventually influence local technology, which is central to a group's material culture, and also provides the framework for its non-material culture. When technology changes, society also changes. William Ogburn (1922) understands technology as the basic cause of social change. According to Ogburn, technology changes society through invention, discovery, and diffusion. Ogburn also coined the term, 'cultural lag', to describe how some elements of a culture typically lag behind the changes that come from invention, discovery, and diffusion. Cultural lag develops if a group's non-material culture lags behind its material culture. Changes in technology often significantly change social organization, ideology, values, and social relationships. Eventually however, social change, the alteration of culture and societies over time, is a vital part of social life. The society of today is not the society of yesterday, nor will it be the society of tomorrow.

Not all societal changes are necessarily good. As developing countries rush to industrialize, they are also damaging the environment, often causing even greater damage to the environment than occurs in developed countries. This is because they lack the pollution controls, anti-pollution laws, and experience in dealing with environmental issues. Concerns about the environment have sparked a worldwide environmental movement. Today, there is a mutual concern for a sense of harmony to be developed between technology and the natural environment.

Religion

Religion is a unified system of beliefs and practices relative to sacred things. A strong relationship exists between society and religion with religion commonly playing an important role in people's lives. Durkheim states that religion is defined by three elements: beliefs, practices, and a moral community (Giddens 1972). He also argues that all religions separate the profane (common elements of everyday life) from the sacred (things set apart or forbidden, that inspire fear, awe, reverence, or deep respect). Through the use of religious symbols, rituals, and beliefs, people build and maintain a community of similar-minded people. All religions use symbols to provide identity and social solidarity for its members.

Religion meets basic human needs by providing answers to questions about ultimate meaning, emotional comfort, social solidarity, guidelines to everyday life, social control, adaptation to a new environment, support for government, and occasionally an impetus for social change. Many of the functions that religion provides can also be fulfilled by functional equivalents (other components of society that serve the same functions as religion).

Throughout history religious organizations have also exercised political power. Religion can support or oppose industrial development. In the Philippines as one example, the Catholic Church remains an influential political stakeholder in new mine ventures, quite often opposing mining.

Religion can support or oppose industrial development. In the Philippines as one example, the Catholic Church remains an influential political stakeholder in new mine ventures, quite often opposing mining.

Community Health

Not least because of IFC Performance Standard (PS) 4, community health has now emerged as an important consideration in environmental assessment. Health concerns exist at the individual, family, and community level. As is true for the individual, community health is influenced by a complex web of factors ranging from nutrition, education level, sexual behaviour and attitudes, and societal factors, to biological risk and genetic predisposition. Apart from physical health it also encompasses mental and sexual health, including the problems of HIV and other sexually transmitted diseases (STD), unintended pregnancy and abortion. Physical, sexual, and mental community health issues are always closely interrelated.

IFC PS 4 is primarily concerned about protecting the health of host communities from potential harm associated with mine development. There is however also the case in which community health may impact on mine operations. For example, average life expectancy relates directly to community health. In some countries average life expectancy can be well below 40 years (such as in many African countries due to problems associated with HIV). Accordingly in such circumstances, ensuring a stable and healthy work force becomes a formidable challenge. Community development programmes to improve community health serve as a prime example of environmental management that enhances rather than merely protecting existing environmental resources.

Community development programmes to improve community health serve as a prime example of environmental management that enhances rather than merely protecting existing environmental resources.

7.6 THE ECONOMIC SPHERE – PRODUCTION, DISTRIBUTION, AND CONSUMPTION OF GOODS AND SERVICES

Mining converts rock into wealth; its relation to economics is intuitive. Economy is about how wealth is created, distributed, and consumed. It concerns the ways in which a group, society, or country produces, distributes, and consumes the tangible, material commodities of life. It is also about how the proceeds or income from these activities are distributed between those that contribute toward them: businesses, workers, government, and the whole of society. Every person affects the economy in some way and we are all affected by it.

Hunting and gathering societies have a subsistence economy. Composed of groups of 25 to 40 people, they live off the land, produce little or no surplus, and have little, if any, social inequality. Of all societies, the hunting and gathering society is the most egalitarian. As societies become more complex, they consistently produce greater surpluses. With the increase in surplus, trade also increases. This, in turn, creates social inequality as some people begin to accumulate more than others. As societies continue to become more complex, their division of labour, surpluses, trade, and social inequality all continue to increase.

To understand how a new mining venture impacts the local economy, it is necessary to understand how the local economy functions, specifically: (1) What is currently produced and how? (2) Why does a particular human group produce particular goods and services? (3) How are the natural resources used? (4) How do communities in the area earn and spend their money? (5) How are local people employed, and what technology do they use

To understand how a new mining venture impacts the local economy, it is necessary to understand how the local economy functions.

in their work? (6) What is the relationship between these things and the wealth and poverty of different human groups or communities in the host region?

How to Measure the Size of an Economy?

The way we usually measure the size of an economy is by Gross Domestic Product (GDP). GDP is the value of all the goods and services produced within a country's borders in one year. This value is equal to the economic wealth of the country, all the things of economic value, which can be bought or sold, that have been produced in one year. This includes all goods such as food, merchandise, or minerals, as well as all services such as labour, transportation, or accommodation. The total prices of these things combined is the GDP. The GDP concept also applies to local economies. A mining project will change the local and regional economy, often with notable effects on the host country's GDP. The mine creates a surplus of goods (e.g. produced ore) and creates a wide range of service and business opportunities (e.g. labour). Economic impacts may be small at the national level, but they are always significant at the local and regional levels (**Case 7.8**).

A mining project will change the local and regional economy, often with notable effects on the host country's GDP.

One of the drawbacks of the GDP number is that it excludes unrecorded economic activity. Unrecorded transactions include the informal sector or subsistence production (i.e. growing rice for own consumption). Also unrecorded in total GDP is work carried out by members of society that contributes to production but which is not sold for money. The classic example is housework performed mainly by women. Cooking, raising children and housekeeping are all economic activities essential for producing wealth, but are not included in GDP, unless they are remunerated and registered with government, such as formal domestic work.

The GDP figure provides a snapshot of how large an economy is in any particular year, but the economy is changing all the time. There are a huge variety of factors that influence economic growth over the long run. These include economic factors (such as developing a mine), but also all kinds of social, political, and even cultural conditions.

CASE 7.8
Mining in Sumbawa

In 1996, Newmont commenced operation at its Batu Hijau copper mine in Sumbawa, an island in the Province of NTT, Indonesia. In 2006, ore production averaged 120,000 tons per day, a production rate that ranks the Batu Hijau mine amongst the biggest copper mines on Earth. All economic sectors combined, the mine contributes most of the GDP of Sumbawa. The mine is expected to continue operation until 2021. At the end of the mine life the mine will have dominated the provincial economy over a quarter of a century, or over one generation.

Economic Sectors

The economy can be divided or classified, in various ways. One common approach is to define economic sectors. Some examples follow.

- Extractive industry (oil and gas, mining, and quarrying), the extraction of our natural mineral wealth such as oil, gas, coal, gold, diamonds, or sand and gravel. In many countries (e.g. Peru, Bolivia, Chile, South Africa, and Australia) the extractive industry was the first industrial sector in the economy and the whole subsequent economic structure has grown up around this sector. In other regions the extractive industry is a new sector, often becoming the dominant economic activity.
- Agriculture to produce foods such as rice, wheat, fruit, vegetables, fish, and meat. In many developing countries the agricultural sector continues to outweigh the contribution of the industrial sector.
- Manufacturing, the production of goods in factories. This includes a wide range of products such as food, beverages, textiles and clothing, wood and paper and their products, petroleum products, metals like steel as well as machinery and metal equipment, electrical goods, transport equipment, and furniture.
- Utilities including electricity, gas, and water supplies to the community and to industry, infrastructure and services usually provided by the state.
- Construction, the building of houses, roads, bridges, factories, and other buildings.
- Finance, the institutions that hold the savings, or surplus wealth, such as banks, pension funds, and the insurance industry. The products these institutions sell to the public (such as life insurance and bank accounts) account for a significant proportion of the goods and services produced in developed countries, but the financial sector is small in most developing countries.
- Government, all government taxation and spending contributes to economic interactions in society. Government contributions to the economy can be divided into three types of economic interaction (1) transfers, where the government takes money from one sector the economy and gives it to another; (2) consumption, the paying of salaries and other recurring costs that keeps the system of government moving; and (3) investment, where the government builds new infrastructure.
- Retail, comprising activities such as wholesale, retail, catering, and accommodation. It includes the shops, the restaurants, the hotels, and other services that we use every day.
- Transport (transport and communications), the moving of goods and people by road, rail, ship, and air, as well as communications via telephone.

In many countries (e.g. Peru, Bolivia, Chile, South Africa, and Australia) the extractive industry was the first industrial sector in the economy and the whole subsequent economic structure has grown up around this sector.

Another way of looking at the economy is to define it according to primary, secondary, and tertiary sectors. The primary sector is the closest to natural resources and includes mining, agriculture, forestry, and fishing. Both mining and agriculture work directly on the products of nature found on, or under, our soil. The secondary sector is a step away from this: it consists of activities that process raw materials into manufactured products or material goods that are used by consumers (such as electricity and water). While mining and mineral processing is part of the primary sector, the production of steel or copper pipes is a manufacturing process. The tertiary sector is even further removed from natural resources. This is where the steel or copper pipes are sold to the consumer and marketed under a particular brand identity. Other activities belong to the service sector (such as finance, government, accommodation, and transport).

Both mining and agriculture work directly on the products of nature found on, or under, our soil.

Having defined economic sectors and grouped them into broader economic categories, we can assess the economic change associated with a new mining project. Since each economic sector is interlinked with other sectors in various ways, the economic impacts of a mine development affect most sectors. A mine could not operate without electricity, which may be generated from coal, while from its profits the government takes taxes and may for example, invest in new infrastructure. The mine generates employment, attracting newcomers to the project area. Increased population combined with an increase in purchasing power stimulates growth in the services sector.

7.7 JUDGING THE STATE AND VALUE OF THE ENVIRONMENT

If the environment is defined as a set of natural and man-made features at a given place and time, the question arises as to how to judge the state and value of a natural resource, once baseline surveys have helped to identify and describe these features. Three approaches help. The straightforward approach is to apply quantitative metrics such as regulatory standards or established scientific criteria, if they exist. The second, admittedly more challenging approach, is to apply subjective criteria to judging environmental values as are commonly used in nature conservation. In the third approach, economic considerations are applied to evaluate the benefits and services that nature provides.

Applying Regulatory Standards and Scientific Criteria

Ambient environmental standards, set by public authorities, are helpful in judging the state of our environment. They are numerical concentrations of environmental parameters established to support or protect a designated use of a resource in an ecosystem. They represent concentrations of parameters in an environmental medium (such as air, soil, water, or fish tissue) that must not be exceeded, or levels of environmental quality that must be maintained. They provide lower or upper limits within which the concentration of a given parameter (such as concentration of trace metals) must fall to protect human health, or to ensure long-term sustainability of an environmental resource (ADB 1998). Thus, the comparison of observed concentrations with ambient quality standards provides a qualitative valuation of the environment under consideration.

The comparison of observed concentrations with ambient quality standards provides a qualitative valuation of the environment under consideration.

Ambient quality standards are normally expressed in terms of average concentration levels over time. Some standards such as those for air quality may have two criteria defined as averages over two time periods, often daily and annual averages. The reason for taking averages is to recognize natural daily and seasonal fluctuations, and fluctuations in pollutant emissions. Averaging also allows for the limits to be exceeded for a short time, provided that this situation does not persist.

Of course, ambient environmental standards are not directly enforceable. What can be enforced are the various emissions that lead to ambient quality levels, the prime driver of emission standards. Emission standards are not-to-exceed levels applied to the quantities of emissions coming from pollution sources, normally expressed in terms of the mass of material per some unit of time. As with ambient quality standards continuous emission streams such as flue gas emissions may be subject to two averages, a short-term and a long-term average. It is important to realize that meeting emission standards does not necessarily result in meeting ambient environmental quality standards. Between emission and ambient quality stands nature, in particular meteorological and hydrological mechanisms that link the two.

Meeting emission standards does not necessarily result in meeting ambient environmental quality standards.

Applying Conservation Values

Subjective criteria to judge the state and value of the environment, common in nature conservation, are more subtle. In an assessment of criteria used in geological conservation, Erikstad (1991) discriminates between primary and secondary criteria, recognizing 'naturalness, diversity, rarity, and function' ('part of a system') as primary value criteria, as they are more connected to the intrinsic properties of ecosystems. TemaNord (2005) suggests including 'vulnerability' as a primary criterion, as a means for securing cost-effectiveness in environmental management. Secondary criteria are more connected to how people experience the site from a scientific, educational, or recreational point of view. Secondary criteria, such as importance to research, education, recreation, and intrinsic value, are relevant, especially in defining objectives and priorities in conservation. In the following we will discuss the primary criteria in more detail, drawing on the description in TemaNord (2005).

Naturalness

The concept of naturalness is often used in a sense that implies freedom from human influence (Margules and Usher 1981). In the traditional nature protection movement in western civilization, and in the rationale of some anti-mining activists this seems to be one of the most important aspects: the steadily increasing use of land for economic purposes triggered the need for conserving pristine areas from such activities. The dominating underlying reasons for using the criterion probably originated from aesthetic, ethical, and recreational considerations. A more fundamental scientific rationale lies in the need for maintaining intact ecosystems for comparison with areas being more influenced by human activity. The view point in developing countries, of course, often differs. Developing countries understandably resent influences seeking to protect their environment at the expense of their own economic development, to offset the past destruction of nature in the western world.

Developing countries understandably resent influences seeking to protect their environment at the expense of their own economic development, to offset the past destruction of nature in the western world.

The rationale behind naturalness as a value criterion is that most of us appreciate the experience of ecosystems unaffected by humans more than artificial ones. As truly natural areas are rare, they may be valued for this reason. However, totally unmodified nature is a rare condition, and it is more useful to regard naturalness as a continuous variable, ranging from completely natural (100% natural) to completely artificial (0% natural).

As truly natural areas are rare, they may be valued for this reason.

The naturalness criterion can also be applied to species. It is generally considered that introduced species have limited conservation value, compared to native or endemic species. Native species may be defined as species occurring naturally within a region, contrary to alien species which may be defined as species occurring outside their known natural range as a result of intentional or accidental dispersal by human activities (UNEP 1995). Of note is that many species now regarded as native have been introduced by humans in historic times, and the immigration time is often unknown. Thus, some subjective judgement is unavoidable in the application of the IUCN red list criteria.

Many species now regarded as native have been introduced by humans in historic times.

The naturalness criterion can also be applied to humans. Indigenous (or aboriginal) Peoples are those people regarded as the original inhabitants of an area. For example the Indians of the American continents are Indigenous Peoples in those lands, the Balinese are indigenous to the island of Bali, and Australian Aborigines are the Indigenous Peoples of Australia. The term Indigenous Peoples can be used for all original people of a host region, but Indigenous Peoples in different parts of the host region have individual tribal and area names. The term Asmat is such a name in Papua, arguably the best known local tribe outside of Papua, but it is not applicable to all Papuan Indigenous Peoples. On the

Although the concept of naturalness is easy to understand, and its relevance to environmental assessment process and conservation is intuitive, it is often more difficult to define and to quantify.

contrary, the island of Papua is considered as the place on Earth with the highest concentration of distinct tribes and languages.

The nature quality concept is usually closely related to naturalness. Although the concept of naturalness is easy to understand, and its relevance to environmental assessment process and conservation is intuitive, it is often more difficult to define and to quantify.

Diversity

The use of the term 'diversity' has increased tremendously over the last years, especially since the Rio Earth Summit, which made politicians and their audiences aware of the term. Diversity however is an older concept in ecology, and it had been used as a value criterion as well as in ecological studies long before the Rio summit. It has been referred to as community, habitat, or species diversity by Margules and Usher (1981). Diversity may appear to be a straightforward and easily measured concept. Most people have an intuitive idea of what is meant by diversity, but nevertheless there is no consensus of how it should be defined and measured. Perhaps the reason is that diversity has two components. Diversity may refer to the number of different species, the species richness, but also to their relative abundance (Magurran 1988). Whittaker (1972) introduced the concept of alpha, beta, and gamma-diversity, thereby relating diversity to geographical scale. Alpha-diversity is the number of species found in a small homogeneous area, while beta-diversity is the difference in species composition between different sites (or along gradients), the so-called species turnover. Gamma diversity is the total diversity in the area. These concepts are obviously scale dependent. Noss (1990) offered a hierarchical approach in which structural, compositional, and functional diversity is recognized at multiple levels of organization: genes, species, community, and landscapes. His concept has many features in common with the definition of biodiversity from the UN Convention of Biological Diversity. It was suggested as a guideline for the monitoring of biodiversity. By choosing good indicators, the topic of the next section, the approach may also offer some opportunities for site assessment.

Rarity

Rarity is often assumed to be related to vulnerability – the rarer a species, the more vulnerable it will be, and the more likely to become extinct.

As with diversity defining rarity as a value criterion is not easy. Both man-induced and natural factors may contribute to the rarity of a species. Rarity is often assumed to be related to vulnerability – the rarer a species, the more vulnerable it will be, and the more likely to become extinct. Therefore information on rarity is important for classification of species in terms of their conservation status. Rabinowitz *et al.* (1986) introduced a clarifying concept of rarity by partitioning the species distribution and abundance into three levels. The species are subdivided according to (1) geographic range (extensive versus restricted), (2) habitat tolerance (broad versus narrow), and (3) local population (large versus small) as shown in **Figure 7.5**. By combining these three categories, one class of common species and seven classes of rarity emerge. Species with a wide geographical distribution, wide habitat demands and large population sizes, are common species not particularly vulnerable. Species with a small population size and/or narrow habitat demands are more or less vulnerable depending on the intensity and extent of disturbance. Species both with small population size, small geographical distribution, and narrow habitat demands are on the other hand very vulnerable to disturbances. A somewhat similar approach is discussed by Smith and Theberge (1986), with emphasis on geographic and demographic criteria. They discern five types of rarity. Widespread rare species are species with a wide geographic distribution, but they are scarce wherever they occur. Tigers are an example of widespread rare species. Endemic species have a restricted geographic range, while disjunct species have populations separated from the main range of the

FIGURE 7.5
Commonness, Rarity, and Vulnerability to Extinction

Species can be subdivided according to:
(1) geographic range (extensive versus restricted),
(2) habitat tolerance (broad versus narrow), and
(3) local population (large versus small).

By combining these three categories, one class of common species and seven classes of rarity emerge.

Source:
Molles 2005

MOST COMMON

	SPECIES	
Extensive geographic range Broad habitat tolerance Large local population	 House Sparrow Passer Domesticus	*Species such as these show no aspect of rarity; they are among the most common in the biosphere.*
Restricted geographic range Broad habitat tolerance Large local population	 Galapagos Medium Ground Finch	*Each of these species show one aspect of rarity, which gives them some vulnerability to extinction.*
Extensive geographic range **Narrow habitat tolerance** Large local population	 California Grey Whale	*Each of these species show one aspect of rarity, which gives them some vulnerability to extinction.*
Extensive geographic range Broad habitat tolerance **Small local population**	 Tiger	*Each of these species show one aspect of rarity, which gives them some vulnerability to extinction.*
Restricted geographic range **Narrow habitat tolerance** Large local population	 Fish Crow	*With two aspects of rarity, these species are even more vulnerable to extinction.*
Restricted geographic range Broad habitat tolerance **Small local population**	 Tasmanian Devil	*With two aspects of rarity, these species are even more vulnerable to extinction.*
Extensive geographic range **Narrow habitat tolerance** **Small local population**	 Northern Spotted Owl	*With two aspects of rarity, these species are even more vulnerable to extinction.*
Restricted geographic range **Narrow habitat tolerance** **Small local population**	 Mountain Gorilla	*Species such as these are the rarest in the biosphere and are the most vulnerable to extinction.*

RAREST — *text on white highlights aspect of rarity*

species. Peripheral populations are at the edge of the species geographical range, and a declining species is one which was previously abundant but now with declining populations. The orangutan which has only a few remaining populations, is an example of a declining species.

Threat or vulnerability is often used as a criterion for environmental protection (Margules and Usher 1981; Smith and Theberge 1986). If a specific area or environmental component is especially prone to negative project-induced changes or activities, it deserves special attention with respect to environmental protection. Vulnerability is not a measure of the value of a site (Erikstad 1991), but understanding threat and vulnerability is important when designing protection measures. This makes vulnerability a useful concept in environmental impact assessment, in which vulnerability can be analyzed with respect to specific project-induced impacts. Not all project activities are necessarily detrimental to the natural area in which the activity is performed. A given nature type can have different

Vulnerability is not a measure of the value of a site, but understanding threat and vulnerability is important when designing protection measures.

vulnerability with respect to different types of project activities or human influence. The vulnerability may also change over time.

Considering Economic Values

From scenic beauty and recreational opportunities to direct use, environmental resources provide a complex set of values and benefits to society. Coastal areas, for example, offer scenic panoramas and radiant sunsets, and, as the ever-increasing number of sun-searching tourists demonstrates, provide excellent recreational areas. Fish and other edible sea life provide a rich and nutritious source of food. Other coastal values not directly tied to use, include climate modulation. Both use and non-use benefits are relevant to economic valuation of environmental resources.

Many people argue that environmental resources are priceless. Thus, they say, it is not appropriate to consider them within the context of economic valuation. In a sense of absolute purism, this stance equates to environmental conservation at any cost. Human history, however, proves that we are unwilling to forgo all future economic activity to eliminate all factors that cause environmental damage. We commonly sacrifice environmental assets to generate economic gains. The value placed on environmental assets is illustrated in our choices made for other competing economic needs.

We commonly sacrifice environmental assets to generate economic gains.

Willingness to Pay

Value in economics means exchange value. Since money is the medium of exchange, the value of a benefit is determined by its price – that is, the quantity of money for which it will be exchanged. However, the value of a benefit is not simply the price of that product on the open market. It is, rather, the worth of that benefit to a potential buyer. The worth can be quantified in terms of willingness to pay (WTP) for goods and services, including environmental resources. True, in many cases we do not pay for the environmental benefits we receive. However, our willingness to pay for these benefits can be derived from surveys, observed behaviour, or through other methods. The WTP concept is central to environmental valuation, providing a framework on which to examine and measure individual preferences. Positive preferences for environmental resources translate into an expressed or observed willingness to pay for them. Conversely, individuals are not willing to pay for environmental resources that they do not value.

Fortunately for future generations, there is growing evidence of consumers' willingness to pay for ecological benefits. Trends such as the growing demand for ecologically renewable energy, certified wood products, organic foods, shade-grown coffee, non-toxic cleaners, and other goods and services with a real or perceived environmental advantage, as well as the increasing concern about global warming and society's willingness to pay for carbon offsets, suggest that there is increasing market recognition of the economic value of preserving natural areas and processes.

The increasing concern about global warming and society's willingness to pay for carbon offsets, suggest that there is increasing market recognition of the economic value of preserving natural areas and processes.

Willingness to Accept

In any real-world world situation, willingness to pay implies ability to pay. Especially in developing countries, it is necessary to be sensitive to differences in income levels. Thus, an alternative approach of valuing environmental resources is to ask people how much they would be willing to accept (WTA) to give up some environmental amenity. To value better water quality, for instance, one could ask either how much people would be willing to pay for a small improvement or how much they would have to receive to compensate them for a small reduction in water quality. WTA is not constrained by one's income, as is WTP. Environmental conservation and non-tangible environmental benefits, however, do not stir the imagination of households in developing countries to the same extent as quality of life, household income, and the prospects of children. It should therefore be no surprise that willingness to accept some degree of environmental degradation is higher in developing countries than in developed countries.

Environmental conservation and non-tangible environmental benefits, however, do not stir the imagination of households in developing countries to the same extent as quality of life, household income, and the prospects of children.

7.8 WHAT ARE NATURE'S ECONOMIC VALUES?

Economists posit that society's valuation of ecosystems is based on the benefits and services that they provide (Pearce and Turner 1990). They differentiate various values. Total economic value encompasses both use value and non-use value (**Figure 7.6**).

Use and Non-use Values

Use values, not surprisingly, are those derived from the benefits people gain from using the resource and environment. These are classified into direct and indirect use values.

Direct use values refer to ecosystem goods and services that are used directly by human beings. They include the value of consumptive uses such as mineral exploitation, timber harvesting, fish harvesting, or hunting of animals for consumption. Direct use values also include the value of non-consumptive uses such as the enjoyment of recreational and cultural activities that do not require harvesting of products. Direct use values are most often enjoyed by people visiting or residing in the ecosystem itself.

Indirect use values refer to the functional or ecological service benefits generated by the environment, benefits often felt outside the ecosystem itself. People benefit from these but do not directly consume them. Examples include natural water filtration, which often benefits people far downstream; the storm protection function of mangrove forests, which benefits coastal properties and infrastructure; and carbon sequestration in forests, which benefits the entire global community by abating climate change.

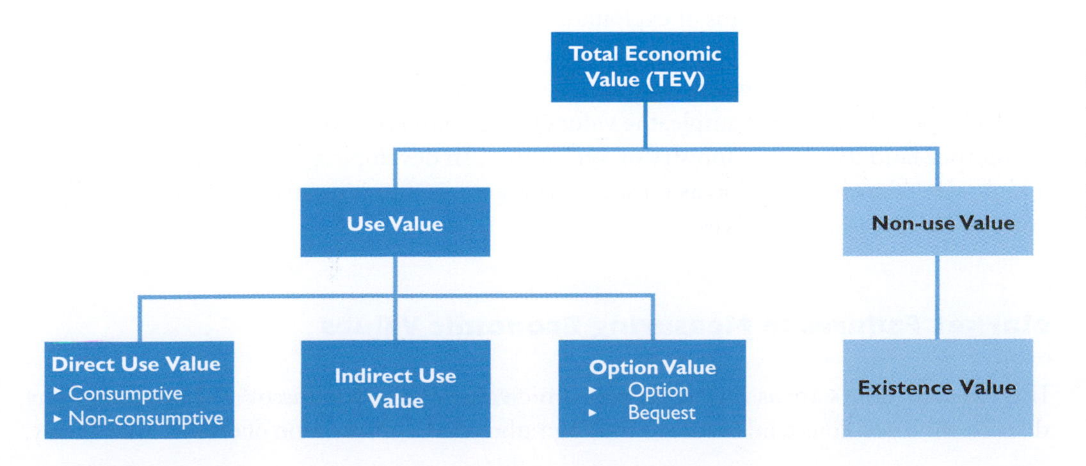

FIGURE 7.6
Categories of Ecosystem Services

Values represented by use value on the left-hand side are more easily derived and tangible. Direct use values in particular are fairly easy to estimate.

Option values are derived from preserving the option to use in the future ecosystem goods and services that may not be used at present, either by oneself (option value) or by others/heirs (bequest value). Provisioning, regulating, and cultural services may all form part of option value to the extent that they are not used now but may be used in the future.

Non-use values (sometimes called passive use values) refer to the enjoyment people may experience simply by knowing that a resource exists even if they themselves never expect to use that resource (termed existence value). Non-use values reflect the intrinsic significance of ecosystems, e.g. cultural, aesthetic, heritage, existence, or bequest values. Non-use values are important to people even when the resource is not directly or indirectly used. A good example is the campaign to raise money for Black Rhino protection in Southern Africa. People donate money to save the species although they may only see it in books or on television (www.blackrhino.org).

Option values cut across both use and non-use values alike. A mining project could destroy a unique habitat that has high potential tourism and ecological service functions (use values) that can be passed on to future generations (non-use values). A decision not to approve the project could be based on the option values at risk. By not proceeding with the mine, society maintains the option of generating other use and non-use values in the future.

Values represented by use value in the left-hand side of **Figure 7.6** are more easily derived and tangible. Direct use values in particular are also fairly easy to estimate. Society understands the value of mining. Moving towards the right-hand side of **Figure 7.6**, values become more difficult to grasp and measure. How do we put a dollar value on existence value? In spite of these difficulties, valuation of a mining project, at least in theory, should strive for total economic value, including exploitation of ore body and associated loss of ecological values. In the environmental assessment process, we have to settle for a partial measure of economic value.

Denying approval to mine may conserve the environment and the goods and services which nature provides, but also precludes the benefits that would have been provided.

Denying approval to mine may conserve the environment and the goods and services which nature provides, but also precludes the benefits that would have been provided. As any anti-mining activist will argue, this can be offset by the many indirect use values that are derived from conserving this area. However maintaining the existing environment through protected areas or through some other mechanisms, also requires expenditure of resources, and there are often many competing claims on these resources. Devoting more effort to nature conservation may mean having fewer resources to address other pressing needs, such as improving education, health, or infrastructure. Accordingly, to assess the consequences of different courses of action, it is not enough to know that ecosystems are valuable; we also need to know how valuable they are, valuable to whom, and how values are affected by different forms of exploitation and management.

To deny mining in the interests of conservation would be pointless if, for example, the values being conserved were subsequently destroyed by another land use such as forestry or agriculture.

In assessing these issues of use and non-use, it is also important to assess all potential uses and their relative costs and benefits. To deny mining in the interests of conservation would be pointless if, for example, the values being conserved were subsequently destroyed by another land use such as forestry or agriculture. In developing countries, there is always a high risk of such a situation as most governments in these countries lack the capacity to enforce conservation initiatives.

Market Failures in Measuring Economic Values

Why do economic decisions tend to favour mining rather than conservation of ecosystems?

If ecosystem values are as real as the economic value of a newly discovered ore body, why do economic decisions tend to favour mining rather than conservation of ecosystems? Partly, this is because the importance of ecosystem processes is not traditionally captured in financial markets. Partly also this is because many costs associated with changes in levels of ecosystem functions may take some time to become apparent or may be apparent only at some geographical distance from where the change occurred. Partly again, this may be because some changes involve thresholds or changes in stability that are difficult to measure. This further hinders the assessment of the real and complete value of an ecosystem and its services. Finally, this is because private and social values of conserving biodiversity and ecosystem services often differ widely. The private use value of biodiversity and ecosystem services will typically ignore the external benefits of conservation that accrue to society in general. Consequently, if private decision-makers such as mining companies are not given the incentive to value the larger social benefits of conservation, their decisions will often result in inadequate conservation. To summarize, while we routinely measure ecological direct use values, the economic value of most supporting, cultural, and regulating values are rarely considered. Consequently, many decisions continue to be made in the absence of a detailed analysis of the full costs, risks, and benefits resulting from associated changes within ecosystems.

While we routinely measure ecological direct use values, the economic value of most supporting, cultural, and regulating values are rarely considered.

Economists trace the underlying problem of missing full costs and benefits to something called market failure – the failure of markets to reflect the full or true cost of goods or services. The calculation of the economic value of mining does not, in most cases, include environmental costs such as reduced water quality, loss of biodiversity, or loss of option values because these values and services do not have readily available dollar values such as those available for minerals. In fact, ecosystem services are provided for free – they do not need to be purchased. Only when these services are lost, are actual monetary costs incurred. Paradoxically, the zero price for nature services is of very high value to society. Various factors contribute to market failure when it comes to the environment.

Paradoxically, the zero price for nature services is of very high value to society.

Distribution of Benefits between Owners and Non-owners

Stating that ecosystems are valuable begs the question 'Valuable to whom?' The benefits provided by a given ecosystem often fall unequally across different groups and so do the costs of resource development. Ecosystem uses which seem highly valuable to one group (e.g. mineral use directly benefiting mining company and host government), may cause losses to another (e.g. host communities that need to be relocated to allow mine access). Answering the question about the economic value of an ecosystem from the aggregate perspective of all groups would thus give very different answers to answering it from the perspective of a particular group. The mining company upholds the economic value of the ore body and the development right. Anti-mining advocacies uphold the property rights of host communities and the intrinsic values of ecosystems.

The benefits provided by a given ecosystem often fall unequally across different groups and so do the costs of resource development.

There is yet another consideration. Unlike most assets, an area supporting a valuable ecosystem may deliver more benefits to society than to individual land owners. Individual owners may receive only a small proportion of benefits, and therefore will tend to undervalue these benefits. A case in point is carbon sequestration, a benefit for the entire world population rather than an exclusive benefit for owners of forest land. In fact, individual owners may even feel economically penalized for preserving the ecosystem for the good of society and may see more immediate value in developing land, for which society as a whole will bear most of the costs in terms of lost benefits. Since it is difficult for an individual owner to receive direct monetary benefit for those benefits which nature provides to others, the true value of such benefits is generally not taken into account in land use decisions.

Economists term these 'externalities'. An externality is a benefit (positive externality) or cost (negative externality) that is experienced by someone who is not a party to the transaction that produced it. Externalities are important because they can create incentives to engage in too much or too little of an activity, from an efficiency perspective. Considering all internal costs and benefits of a mine development, meaning all costs and benefits directly experienced by the mining company, we expect mine development to take place only if the benefits are greater than the costs.

Distribution of Costs between Owners and Non-owners

The market does normally not include negative externalities, the lost economic value of the environment, in the company's production costs.

The market price of mine products may not reflect all of the production costs. For example, if mining companies discharge effluents into a stream, the economic damage done to the downstream aquatic environment, whether fewer fish are produced or water quality is impaired, may not be reflected in the market price of the mine product. In other words, the market does normally not include negative externalities, the lost economic value of the environment, in the company's production costs.

Unaccounted environmental costs are often substantial. In 1990 the USA introduced a nationwide SO_2 trading system similar to the CO_2 trading mechanism embedded in the Kyoto Protocol in 1997. The current trading price of 1 ton of SO_2 (as of end of 2007) averages to about 400 US dollars. If this is applied to the average SO_2 emissions of the Mount Isa Lead Smelting operation, which are about 500 kt/y (Case 7.1), then there would have been an additional operation cost of about 200 million US dollars. Applying a price of 5 to 10 US dollars to 1 ton of CO_2 emissions to the annual methane emissions caused by coal mining sector worldwide, detailed in a later chapter of this text, would amount to a staggering annual bill to the coal industry well exceeding 1 billion US dollars.

Underestimating Cumulative Effects

Even in large-scale mining, the affected land area is usually very small compared to the total area of the host region affected by human activities.

Even in large-scale mining, the affected land area is usually very small compared to the total area of the host region affected by human activities. When taken together, however, a combination of relatively small incremental changes to an ecosystem may have more dramatic effects than those recognized when individual changes are made. These cumulative effects result from past, planned and future changes, often unrelated, and are difficult to recognize and assess physically as well as economically, in part because of the dynamic nature of ecosystems.

Our Limited Understanding of Science

The ability to measure the economic value of an ecosystem is also limited by our scientific understanding of ecological functioning in nature. Societies do not yet fully appreciate or understand all the benefits that the environment provides to their wellbeing and to the Earth's ecological stability, and the odds are that society never will. This lack of scientific understanding will always undervalue environmental benefits contributing to market failure.

Deriving Economic Values for Natural Assets and Services

A comparison of the full economic costs of mine development versus generated benefits, for example, could prompt governments to put resources into the conservation of an ecosystem rather than in the exploitation of minerals within the ecosystem.

To counteract the problem of market failure, we need to find ways to calculate the total economic value of an ecosystem in a way that the market understands – in dollars. This helps both individuals and societies more easily compare alternative uses and policy options. A comparison of the full economic costs of mine development versus generated benefits, for example, could prompt governments to put resources into the conservation of

FIGURE 7.7
Environmental Valuation Methods

The choice of the valuation method will depend on each situation, the information readily available, the time and budget available, and level of expertise.

Source: based on Pagiola *et al.* 2004

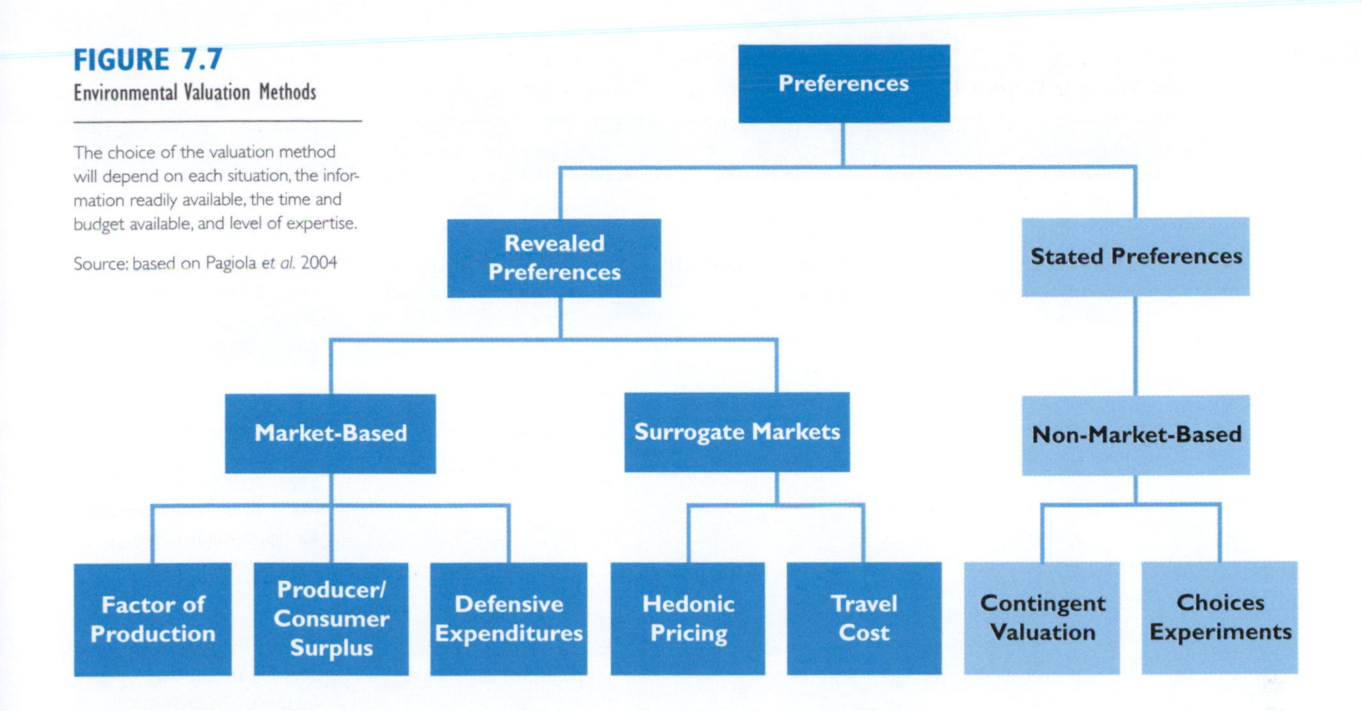

an ecosystem rather than in the exploitation of minerals within the ecosystem. Similarly, such evaluations could lead to a better understanding of tax incentives, rebates or subsidies that could give individuals an economic incentive to retain an ecosystem. Thus, society could purchase the services provided by nature from individual land owners.

Methods of valuing ecosystem benefits can be divided into three categories as illustrated in **Figure 7.7**: market-based (direct proxies), surrogate markets (indirect proxies), and non-market based (no proxies). In measuring value, it is important to remember that net value is desired – the gross value of a benefit less the costs that must be incurred to receive that benefit. The choice of the valuation method will depend on each situation, the information readily available, the time and budget available, and level of expertise (**Table 7.4**).

Direct Proxies

Direct proxies involve cost or price information, which approximate values of environmental assets. The advantage of using costs or market prices as proxies for willingness to pay/willingness to accept is that they are easily observable.

Market equilibrium prices are an acceptable base for applying these techniques if markets of the goods and services involved are competitive and apply to small changes in their demand and supply. The actual costs incurred as a result of environmental degradation can represent the minimum benefits from avoiding environmental impacts. Examples of direct proxies include (Pagiola *et al.* 2004):

- The loss of agriculture productivity from soil erosion (productivity loss);
- Medical expenditure due to air pollution (cost of illness);
- An ill person or premature death from air pollution (human capital cost);
- Averting or mitigating environmental impacts (response/preventative cost);
- Replacing environmental goods or services (replacement costs);
- An environmental aid project such as grants and donations (aid costs) ;
- The reconstruction of an environmental good (shadow project);
- Cost per unit of output (cost-effectiveness); and
- A close substitute (substitute costs).

The actual costs incurred as a result of environmental degradation can represent the minimum benefits from avoiding environmental impacts.

TABLE 7.4

Ways to Calculate the Total Economic Value of an Ecosystem

Methodology	Approach	Application	Data requirement	Limitation
Revealed preferences				
Production function (also know as 'change in productivity')	Trace impact of change in ecosystem services on produced goods	Any impact that affects produced goods	Change in services; impact on production; net value of produced goods	Data on change in services and consequent impact on production often lacking
Cost of illness, human capital	Trace impact of change in ecosystem services on morbidity and mortality	Any impact that affects health (e.g. air or water pollution)	Change in services impact on health (dose-response function); cost of illness or value of life	Dose-response function linking environmental conditions to health often lacking; underestimates, as omits preferences for health; value of life cannot be estimated easily
Replacement cost (and variants, such as relocation cost)	Use cost of replacing the lost good or services	Any loss of goods or services	Extent of loss of goods or services, cost of replacing them	Tends to overestimate actual value; should be used with extreme caution
Travel cost (TCM)	Drive demand curve from data on actual travel costs	Any loss of goods or services	Extent loss of goods or services, cost of replacing them	Limited to recreational benefits; hard to use when trips are to multiple destinations
Hedonic pricing	Extract effect of environmental factors on price of goods that include those factors	Air quality, scenic beauty, cultural benefits	Prices and characteristic of goods	Requires vast quantities of data; very sensitive to specification
Stated preference methods				
Contingent valuation (CV)	Ask respondents directly their WTP for a specified services	Any services	Survey that presents scenario and elicits WTP for specified service	Many potential sources of bias in responses; guidelines exist for reliable application
Choice modelling	Ask respondents to choose their preferred option from a set alternatives with particular attribute	Any services	Survey of respondents	Similar to those of CV; analysis of the data generated is complex
Other methods				
Benefits transfer	Use results obtained in one context in a different context	Any for which suitable comparison studies are available	Valuation exercises at another; similar site	Can be very inaccurate, as many factors vary even when contexts seem 'similar'; should be used with extreme caution

Source:
Pagiola *et al.* 2004

All these proxies can be considered opportunity costs of environmental assets. Furthermore, different costs of the same impact can also value different functions of an environmental good or service. For example, the cost of illness, human capital and productivity loss, are often complementary in that each reflects a specific aspects of a healthy life. If more than one approach is used to value different facets of a human health impact, one has to be cautious not to double count.

Indirect Proxies

Indirect proxies for environmental goods and services are based on the observed behaviour of individuals with respect to related markets. The Travel Cost Method derives the value of a recreational site from the revealed information on the time and costs people spent to get there. Hedonic prices infer the value of an environmental attribute from the price of a related market good. For example, the noise associated with a particular residential area will be reflected in lower land and real estate value, everything else being constant. Wages paid in a safe and quiet factory are expected to be lower than in a dangerous and noisy factory producing the same quantity and quality products. The residual value method derives the net price of a natural resource by deducting all the costs from the finished product price. An implicit value is obtained from a reverse analysis (bottom-up analysis) similar to the residual approach but for a project instead of a specific good or service. While indirect proxy methods involve more calculation they will not necessarily provide a better estimate of willingness to pay for environmental goods and services than the direct proxies. They have the advantages however, of relying on observed behaviour and existing market prices directly related to the environmental attribute being valued. These valuation methods are more costly, time consuming, and require skilled analysis. Yet, if the information is available and the analysis is done properly, these methods should provide a better approximation of willingness to pay than the direct proxy approaches.

No Proxies

Where no proxies are available, the Contingent Valuation Method (CVM) is often the best valuation approach. It consists of asking people directly, via questionnaires or experimental techniques, what they would be willing to pay for a benefit or what they would be willing to receive as compensation for a deterioration of their environment. The questionnaire simulates a hypothetical (contingent) market of a particular environmental good in which individuals (demand side) are asked to state their willingness to pay / willingness to accept for a change (improvement or deterioration) in the provision (supply side) of the good in question. The questionnaire has to provide the institutional context in which the good would be provided and on the payment vehicle. CVM may apply equally to changes in public goods such as air quality, landscape, or the existence values of wildlife, as to goods and services sold to individuals, like water supply and sanitation. It may apply to both use and non-use values which is not the case for the 'proxy' techniques. A challenge in developing countries is to modify the technique for cases where people have little income on which to base a willingness to pay judgement.

Challenges and Limitations in Attaching a Dollar Value to Nature

Arguably the two main reasons for society's failure to conserve nature are that (1) society does not realize how valuable nature is, and (2) if society does, it fails to translate the various ecological values in dollars.

The notion that intact ecosystems deliver multiple benefits simultaneously, is often overlooked in comparison with mineral production and usage, which constitute a more traditional economic benefit.

GDP often identifies activities that destroy ecosystems as benefits.

A supporting argument for the first point is that when both mining companies and governments alike decide whether to clear land for mining, they focus on the minerals produced, and pay little attention to lost ecological services. The notion that intact ecosystems deliver multiple benefits simultaneously, is often overlooked in comparison with mineral production and usage, which constitute a more traditional economic benefit. Host governments also base their budget decisions solely on the basis of indicators such as tax receipts, GDP, or foreign exchange balances, in which ecosystem values either do not appear or are not recognized as such – indeed, perversely, GDP often identifies activities that destroy ecosystems as benefits. Extensive legalized timber harvesting in many countries is a case in point.

A supporting argument for the second point is that there is rarely an existing market value for a given ecosystem. Different approaches that discern value through more intuitive means must be examined. Efforts to put an accurate dollar value on ecological benefits are limited by the scientific understanding of natural systems, current economic methods for establishing the values of the non-market benefits produced by ecosystems, and, not surprisingly, by time and resources. The science of calculating economic values for ecological services is still relatively new and evolving. New methods are being developed, refined, and enhanced. Putting values on naturally occurring services such as water filtration, erosion control, sediment trapping, or carbon sequestration is a much newer concept than valuing traditional consumptive or extractive uses such as mining, logging, fishing, or hunting. It is, however, now widely accepted that natural areas possess substantial economic value.

It is fair to say that in almost every effort to estimate value, additional challenges surface. For example economic valuations are typically undertaken for the total of a specific ecosystem benefit; estimating a small or marginal change in an ecosystem is more difficult. As such a change in the benefit within one corner of a forest area is more difficult to value than loss of the entire benefit associated with the total forest area, as the change may not be proportional to the area lost or degraded. For these reasons, instead of trying to judge the total value of the ecosystem, the focus is often on calculating the net value of specific ecosystem benefits that will be affected by change (such as mining). So rather than attempting to calculate the total value of all ecosystem benefits now and in the future, a few specific existing benefits may show a comparable value to a proposed alternative use. Consequently, valuation efforts often determine only a fraction of an ecosystem's total value, a partial valuation which is not always adequate for balanced decision-making.

Economic values can not and should not be the sole reason for making decisions about natural resources and the environment.

Economic values can not and should not be the sole reason for making decisions about natural resources and the environment. The most economically efficient choice is not necessarily the most socially acceptable or environmentally beneficial choice. Ecosystems simultaneously produce a number of benefits and unlike many traditional economic benefits, a number of people can enjoy these benefits without the value to the individual being in any way diminished. For example, society gains from clean water or air produced by a forest, but only a few prosper from timber harvesting or forest clearing to allow mine access. Economics however equip decision-makers with information that helps to make better decisions and to plan appropriate action.

7.9 INTERNATIONAL LAW PERTAINING TO NATURAL AND ENVIRONMENTAL RESOURCES

People throughout the world are struggling to understand their local environment. However, over the last few decades people have broadened their outlook to recognize that many apparently local environmental issues have global implications, and that protection of the global environment is critical to human welfare. There is also the recognition that ignoring environmental problems is no longer workable.

Ignoring environmental problems is no longer workable.

Since environmental problems cause impacts beyond national borders the willingness of nations to enter into protective covenants and treaties is constantly growing. The different designations of such agreements – treaty (an agreement in which all the details have presumably been worked out), convention (an agreement in which parties agree on a general framework that is expected to be supplemented in the future by one or more protocols that work out the details), or protocol – all mean essentially the same thing and have no special significance for the purpose of the following discussion, which is based largely on relevant text from Pring (1999) and Zillman *et al.* (2002).

By signing multilateral agreements, a State surrenders portions of its sovereign powers, and becomes bound to new standards of conduct along with other States; this is what is termed international law. International law – the growing body of laws and policies, not at the single nation level but at the multi-nation level – now forms a significant part of the continuously changing environmental regulatory framework that applies to mining. Since laws are rules set by authority, society, or custom, a variety of other codes of behaviour can be considered laws if they are backed by some enforcement power. Mining industry codes and guidelines established by influential NGOs fall under this category.

Since laws are rules set by authority, society, or custom, a variety of other codes of behaviour can be considered laws if they are backed by some enforcement power.

There are now over 1,000 international treaties and other legal authorities focussed on the environment, most developed since 1970 (Pring and Joeris 1993). There is not, as yet, a comprehensive international law of mining. In part, this is because States are naturally reluctant to surrender their sovereign control over such an important part of their economies. In part, it is because of the still primitive state of international law, which lacks conventional law-making bodies (there is an international court but it has no enforcement power). It is partly because most international law lacks conventional command-control enforcement, leaving compliance largely to the political will of the individual sovereign States. It is also partly because international laws are not created equal; some international agreements are classed as legally binding or 'hard law' (chiefly treaties and litigation), while others are viewed as non-binding, aspirational, or 'soft law' (such as United Nations and other international bodies' declarations, resolutions, statements of principles, guidelines, etc.) (Guruswamy *et al.* 1994, Wälde 1993, Armstrong 1996).

While there is no comprehensive international mining law, a number of treaties or conventions have provisions regulating the industry. Typically, these mining related treaties use very general language, lack adequate enforcement regimes, and focus on only a small portion of mineral development that is ore access, ore processing, and product controls.

While there is no comprehensive international mining law, a number of treaties or conventions have provisions regulating the industry.

The mining industry of course took notice of the global trend to environmental awareness, and was one of the first sectors to respond to this trend. Once at the margins of mine planning and operation, environment protection is now a mainstream issue that lies at the heart of new mine developments. The mining industry, often working together with government and community organizations, has established a series of codes of practice to offer innovative and far-reaching solutions to mining induced environmental problems, from minimizing waste and monitoring operations, to returning mined land as close as practicable to its original condition. While not yet mandatory, applying sound environmental management practices is becoming the industry norm and can be considered *de facto* part of international law.

Once at the margins of mine planning and operation, environment protection is now a mainstream issue that lies at the heart of new mine developments.

While requirements based on international law are not mandatory, new mine developments can not afford to ignore them. A description of the main mining related multilateral environmental agreements and codes of practice follows.

International Law and Mine Access

Various international treaties have been developed since the 1940s to protect outstanding natural areas and resources. A listing under one of these international treaties practically places an area off-limits to development including mining. Nature preservation laws at national and international levels often become battlegrounds between mining interests and environmentalists **(Case 7.9)**

The World Heritage Convention

The 1972 United Nations Educational, Scientific, and Cultural Organization (UNESCO) Convention for the Protection of the World Cultural and Natural Heritage (the World Heritage Convention) (www.unesco.org/whc) is the foremost example of international nature protection initiatives. It provides for the preservation of outstanding natural and cultural sites by listing them as part of 'the world heritage'. States nominate their own sites for inclusion, a 21-State elected committee of the treaty parties (the World Heritage Committee) decides which to list, and then States are obligated to protect their sites in perpetuity. As of October 2006, 184 states are parties to the treaty, and 851 natural and mixed natural-cultural sites have been established. Two important international NGOs assist the Committee in evaluating newly proposed world heritage sites – the World Conservation Union (IUCN) and the International Council on Monuments and Sites. In general, for inclusion, natural areas must be outstanding examples of major stages of the Earth's history; significant ongoing ecological and biological processes; superlative natural phenomena; exceptional natural beauty and aesthetic importance; or important natural habitats for *in situ* conservation of biological diversity. Once a site is listed, enforcement is left to each State, the only express sanction being delisting a site that a State has failed to preserve adequately. Threats of delisting have, in fact, been made to focus attention on management of particular heritage areas. Any developments in world heritage sites are practically off-limits (**Case 7.10**).

Any developments in world heritage sites are practically off-limits.

The Ramsar Convention

The 1971 Ramsar Convention on Wetlands of International Importance (www.iucn.org/themes/ramsar), aims to protects wetlands, including marshes, peat lands, and marine water less than 6 metres deep at low tide, with particular emphasis on wildfowl habitat.

CASE 7.9
Forestry Law 41/1999 in Indonesia

In 1999 the Government of Indonesia issued Law No. 41 prohibiting open Pit mining in protected forest. While good in its intent Law No. 41 was controversial from the outset. In part, this is because the law was applied retrospectively and violated the mining rights of 11 companies granted earlier. In part, this is because the definition of protected forest turned out to be as much a legal as an environmental protection issue. Large land areas of Indonesia have been and are being declared as protected forest even though the original forest has been harvested long ago. After years of lobbying the Government of Indonesia eventually upheld the existing legal rights of all 11 affected companies much to the despair of anti-mining advocacies which in turn filed a law suit against the government.

As of May 2006, 1,604 wetlands were listed from 152 participating countries, covering 134,722,002 hectares. Unlike the World Heritage Convention, individual States have unilateral power to list their wetlands of 'international importance', which, once listed, must be preserved and protected. Here too, enforcement is left to the discretion of individual States, with delisting again the only overt sanction. In addition, there are a number of similar regional nature treaties for the Americas, Africa, Europe, and Asia. These include the 1940 Convention on Nature Protection and Wildlife Preservation in the Western Hemisphere, the 1968 African Convention on the Conservation of Nature and Natural Resources, the 1979 Berne Convention (Europe) and subsequent EU Council Directives, and the 1985 ASEAN Agreement on the Conservation of Nature and Natural Resources (not yet in force).

Protection of sites under these nature treaties (even plans to study sites for possible protection) can provide significant leverage for NGOs and others in dealing with mining proposals. Site protection under these treaties can be used to block or redirect mining access and development (**Case 7.10**).

Protection of sites under these nature treaties (even plans to study sites for possible protection) can provide significant leverage for NGOs and others in dealing with mining proposals.

Agenda 21

The international scope of many environmental problems was first highlighted by the 1972 United Nations Conference on the Human Environment (the Stockholm Conference or the first Earth summit), which led to the United Nations Environmental Program (UNEP) and the Earth summits of Rio de Janeiro in 1992 and Johannesburg in 2002.

The 1992 Rio de Janeiro was attended by 172 nations making it the greatest international summit on any subject in history (**Table 7.5**). The Assembly set out an ambitious agenda, calling for an Earth Charter that would have the status of international constitutional law, the Rio Declaration on Environment and Development, consisting of 27 revised environmental principles, an action plan for the 21st-century accomplishment of these goals (called 'Agenda 21'), and the ceremonial signing of three treaties on biodiversity, climate change, and forestry. All documents have much to say about global mining operations, and nearly half the Declaration's principles have relevance to mineral development as illustrated in **Table 7.6.**

CASE 7.10
Mineral Resources in World Heritage Areas

The Lorentz National Park (2.5 million ha) is the largest protected area in South-East Asia and a World Heritage Site since 1999. It is the only protected area in the world to incorporate a continuous, intact transect from snowcap to tropical marine environment, including extensive lowland wetlands. Located at the meeting-point of two colliding continental plates, the area has a complex geology with ongoing mountain formation as well as major sculpting by glaciation. The area also contains fossil sites which provide evidence of the evolution of life on New Guinea, a high level of endemism and the highest level of biodiversity in the region.

Airborne geological surveys have confirmed the existence of potential mineralization zones that rival the nearby Grasberg Copper and Gold deposit. Developing of these potential mineral resources has been explicitly stopped by the Government of Indonesia.

A second example is the defeat of the Windy Craggy mine proposal by the listing of the Tatshenshini-Alsek Region, British Columbia, Canada, as a World Heritage Site.

The Coronation Hill mine in the Northern Territory of Australia is yet another example being turned down primarily because of Aborigines' claims; significantly, however, concerns were also raised that, while the mine itself would not have negative environmental effects on downstream Kakadu National Park, the cumulative environmental impact if other mines were allowed in the area 'would threaten... the World Heritage listing' of the Park.

TABLE 7.5

The Earth Summit (United Nations 1997) – The Earth Summit in Rio de Janeiro was unprecedented for a UN Conference, in terms of both its size and the scope of its concerns. The Earth Summit influenced all subsequent UN conferences, which have examined the relationship between human rights, population, social development, women and human settlements – and the need for environmentally sustainable development (www.un.org)

Conference	United Nations Conference on Environment and Development (UNCED), Rio de Janeiro, 3 to 14 June 1992
Informal name	The Earth Summit
Host Government	Brazil
Number of Governments participating	172, 108 at level of heads of State or Government
Conference Secretary-General	Maurice F. Strong, Canada
Organizers	UNCED secretariat
Principal themes	Environment and sustainable development
NGO presence	Some 2,400 representatives of non-governmental organizations (NGOs); 17,000 people attended the parallel NGO Forum
Resulting document	Agenda 21, the Rio Declaration on Environment and Development, the Statement of Forest Principles, the United Nations Framework Convention on Climate Change (UNFCCC) and the United Nations Convention on Biological Diversity (UNCBD)
Follow-up mechanisms	Follow-up mechanisms: Commission on Sustainable Development; Inter-agency Committee on Sustainable Development; High level Advisory Board on Sustainable Development
Previous/subsequent conferences	UN Conference on the Human Environment, Stockholm (1972) and Johannesburg (2002)

The Biodiversity Treaty

One of the major accomplishments of the 1992 Earth Summit – the Convention on Biological Diversity (Biodiversity Treaty) (www.biodiv.org) – has major implications for the mining industry. Its core concept is that nations are '*responsible for conserving their biological diversity and for using their biological resources in a sustainable manner*'. It is this preservation provision that has the most immediate relevance to mining. It requires State governments to develop and implement national biodiversity plans, which include the establishment of protected areas of *in situ* biodiversity, and the prevention of significant impacts on biological diversity. The Biodiversity Treaty encourages States to place new areas off-limits to mining and development and to urge more support from international companies; that is greater financial responsibilities from mining companies operating in developing countries with substantial biodiversity resources. This in turn offers the opportunity to mining companies to contribute to the establishment of protected areas containing valuable wildlife in the host region, to offset potential mining induced impacts on wildlife (**Case 7.11**).

This in turn offers the opportunity to mining companies to contribute to the establishment of protected areas containing valuable wildlife in the host region, to offset potential mining induced impacts on wildlife.

Currently, both onsite and offsite opportunities are being pursued by leading companies to enhance their contributions to biodiversity conservation. These include assessments and conservation of unique flora and fauna, research and development, support for protected area site management programmes, and proactive community development programmes to provide sustainable economic and social benefits, even after mine closure. A number

TABLE 7.6

The Rio Principles and their Relevance to Mining

Rio Principles	Relevance of Mining
Rio Principles 1 and 3	Stating, for the first time, a 'right to development' for each state
Rio Principle 2	(1) Re-affirming state sovereignty over resources and the prohibition against transboundary harms (adopting the Stockholm Principle 21 from the first Earth Summit)
	(2) Affirming that states may 'exploit their own natural resources pursuant to their own environmental *and developmental* policies'
Rio Principles 4 and 8	(1) Adopting the concept of 'sustainable development' as the guiding paradigm for the future
	(2) Requiring environmental protection to be 'an integral part' of development
	(3) Limiting the development right 'so as to equitably meet developmental and environmental needs of present and future generations'
Rio Principle 15	Emphasizing the important 'preventive' or 'precautionary approach', which states that, 'where there are threats of serious or irreversible damage, lack of full scientific certainty shall not be used as a reason for postponing cost-effective measures to prevent environmental degradation'.
Rio Principle 16	Adopting the 'polluter pays principle'
Rio Principle 10	Calling for increased public participation in environmental issues: citizen access to environmental information from their Governments, opportunity to participate in environmental decision-making, and effective access to courts and agencies for redress and remedies.
Rio Principles 11 and 13	Calling on States to enact 'effective environmental legislation' and laws of 'liability and compensation' for victims of environmental damage. 'Indigenous Peoples' and 'local communities' are positively singled out; States should support their 'identity, culture and interests' and enable their effective participation in sustainable development
Rio Principle 14	Calling for an end to the 'export' of toxic substances and harmful activities from one State to another.
Rio Principle 17	Making environmental impact assessment a generally accepted procedure for all activities likely to have significant adverse environmental impacts.
Rio Principles 18 and 19	Re-stating the longstanding international environmental laws of 'timely notification' and 'good faith consultation' for environmental emergencies and trans-boundary impacts.

CASE 7.11

Support in Establishing Nature Conservation Areas

Eramat is currently studying the development of a large nickel laterite deposit in Halmahera Island, North Indonesia. To date Halmahera Island remains largely unaffected by industrial developments and tourism, and large land areas are covered by pristine vegetation, with landforms of stunning beauty. In collaboration with BirdLife International, Eramet is pursuing the establishment of a nature conservation area near the mine site to offset the impact of proposed mining activities. Financial support to the long-term management of the nature park is embedded in the mine's environmental management plan.

of companies have also established partnerships with conservation groups, and these are beginning to deliver real 'on-the-ground' conservation outcomes.

Convention on the Law of the Sea (UNCLOS)

Years in negotiation, the 1982 United Nations Convention on the Law of the Sea (www.un.org/Depts/los/IYO/UNCLOS) is a comprehensive framework for regulating our use, development, and preservation of the vast marine areas, including mining and other mineral development in the ocean. The treaty establishes two different mining regimes. Mineral resources generally within 200 miles of shore are under the exclusive sovereignty of the coastal States. In these areas (about one-third of all oceans) national laws control mining access, environmental protection, and other requirements. Environmental protection provisions require States to adopt laws and regulations to control all forms of pollution within their areas of national jurisdiction, as well as monitoring and environmental assessment.

The other two-thirds of the ocean (termed the 'International Seabed Area') are beyond this national jurisdiction and are governed by this treaty under a unique 'global commons' regime, the benefits of mining and other development to be shared among all nations. To accomplish this, the Convention established the 'International Sea Bed Authority' (ISA) (www.isa.org.jm) with the power to permit and control all mining exploration and activity in the International Seabed Area. While ratified by 100 countries as of 1996, an almost equal number (including the United States and other developed nations) have so far not become parties, chiefly because of the Convention's deep seabed mining provisions.

The Convention on Climate Change

The 1992 United Nations Framework Convention on Climate Change (www.unfccc.de) and the 1997 Kyoto Protocol provide a comprehensive approach for controlling greenhouse gases (GHG), those chemicals, chiefly carbon dioxide and methane, which form a heat-trapping layer in the upper atmosphere and are commonly believed to contribute to global warming. In force since 2005, the GHG reduction rules will have a profound effect on the mining industry to the extent that its processes release CO_2, methane, and other GHGs.

The Convention on International Trade in Endangered Species of Wild Fauna and Flora (CITES)

The 1961 Convention on International Trade in Endangered Species of Wild Fauna and Flora (CITES), in force since 1975, is one of the earlier international environmental agreements. Under this trade agreement, each country is supposed to establish its own system to control movement of wildlife exports and imports. Species are separated into three classes: Class I are species threatened with extinction, in which commercial trade is banned and non-commercial trade regulated; Class II are species that become threatened if trade is not held to levels consistent with biological processes; and Class III are species that are not currently threatened but for which international cooperation is appropriate, and for which trade requires permits. Beside CITES, the International Union for Conservation of Nature and Natural Resources (IUCN) lists a total of more than 15,000 endangered and threatened species, including nearly one-quarter of all known bird varieties. While the mining industry is not involved in wildlife trade, a CITES or IUCN listing of an endangered or threatened species present at a proposed mine site, would provide significant leverage to stakeholders opposing the mine proposal. It is also widely recognized that improved access to wilderness areas due to any development including mining, can contribute to increased wildlife trade, unless specific steps are taken to avoid such an outcome.

All international agreements, once consent has been reached, become enforceable only after certain criteria are fulfilled, the prominent criteria being the minimum number of nations that have ratified the agreement. The rate at which parties are signing on and the speed at which agreements take force have increased rapidly. The Biodiversity Convention was enforceable after just one year, and had 160 signatories only four years after its introduction.

One of the principal problems with most international agreements is the tradition that they must be by unanimous consent. A single recalcitrant nation effectively has veto power over the wishes of the vast majority. For instance, to reach unanimous consent more than 100 countries at the 1992 Earth Summit agreed to reword the climate convention, after persistent representations by US negotiators, so that it only urges, not requires, nations to stabilize their greenhouse gas emissions.

It is also widely recognized that improved access to wilderness areas due to any development including mining, can contribute to increased wildlife trade, unless specific steps are taken to avoid such an outcome.

Guidelines and Codes by Private Sectors, Financial Institutions, and Other Stakeholder Organizations

The mining industry itself and NGOs are producing an expanding body of guidelines, standards, best practices, codes of conduct, technical and management procedures and intra-company rules – both of general application and mining-industry-specific. It is not surprising that anti-mining advocacies now treat the industry's own pronouncements as the best evidence of the international standards to which the industry should be held accountable.

It is not surprising that anti-mining advocacies now treat the industry's own pronouncements as the best evidence of the international standards to which the industry should be held accountable.

The 1991 Berlin Guidelines

The 1991 Berlin Guidelines emerged from a United Nations convened roundtable discussion of international mining experts in Berlin to address environmentally sustainable mineral development. The Berlin Guidelines set out mining-environment principles for the mining industry as well as for multilateral and bilateral financing institutions. In 2002, a second roundtable discussion produced a follow-up document referred to as Berlin II.

The Best Practice Environmental Management (BPEM) in Mining

The Best Practice Environmental Management (BPEM) in Mining programme is a world-renowned partnership between the mining industry, interested stakeholders, and the Australian Government (www.ea.gov.au/industry/sustainable/mining/bpem.html). It aims to help all sectors of the minerals industry – minerals, coal, oil, and gas – to protect the environment and to reduce the impact of minerals production. Since the programme began in 1994, the Australian Government has worked with industry partners to produce 24 booklets on a range of topics, from community consultation to water management and cleaner production. The booklets present concise, practical information on how to achieve environmental management best practice in the minerals industry anywhere in the world.

The 2005 International Cyanide Management Code

The 2005 International Cyanide Management Code (www.cyanidecode.org) is a voluntary initiative for the gold mining industry and the producers and transporters of the cyanide used in gold mining. It is intended to complement an operation's existing regulatory

The Code focusses exclusively on the safe management of cyanide that is produced, transported, and used for the recovery of gold, and on de-toxification of mill tailings and leach solutions.

requirements. Compliance with the rules, regulations, and laws of the host country's jurisdiction is necessary; this Code is not intended to contravene such laws. The Code focusses exclusively on the safe management of cyanide that is produced, transported, and used for the recovery of gold, and on de-toxification of mill tailings and leach solutions. It includes requirements related to financial assurance, accident prevention, emergency response, training, public reporting, stakeholder involvement, and verification procedures.

REFERENCES

ADB (1998) Environmental assessment requirements of the Asian Development Bank. Environment Division, Office of Environment and Social Development. Asian Development Bank.

Angermeier PL (2000) The natural imperative for biological conservation. Conservation Biology 14: 373–381.

Armstrong K (1996) The Green Challenge – Managing Environmental Issues in Natural Resources Projects in Developing Countries, 42 ROCKY MOUNTAIN MINERAL LAW INSTITUTE 3–1.

Callicott JB and Mumford K (1997) Ecological sustainability as a conservation concept. Conservation Biology 11: 32–40.

Down CG and Stocks J (1978) Environmental Impact of Mining. Applied Science Publishers Ltd., ISBN 0853347166.

EEA (2005) Environment and health, Report No 10/2005.

Erikstad L (1991) Østfold. Kvartærgeologisk verneverdige områder. Norsk Institut Naturforsk. Utredn. 26: 1–61.

FAO (1976) A framework for land evaluation. Soils Bulletin 32, FAO, Rome. 72 p. Also, Publication 22, (R. Brinkman and A. Young (eds.)), ILRI, Wageningen, The Netherlands.

FAO (1995) Planning for sustainable use of land resources: a new approach, W.G. Sombroek and D. Sims. Land and Water Bulletin 2, FAO, Rome.

FAO (1997) Land quality indicators and their use in sustainable agriculture and rural development, Proceedings of the Workshop organized by the Land and Water Development Division FAO Agriculture Department and the Research, Extension and Training Division FAO Sustainable Development Department 25–26 January 1996.

FAO (2005) Global Forest Resources Assessment 2005, http://www.fao.org/newsroom/en/news/2005/1000127/index.html.

FAO (2006) Land resources evaluation and the role of land-related indicators, by W. G. Sombroek, Land and Water Development Division, FAO, Rome, Italy http://www.fao.org/docrep/W4745E/w4745e05.htm 5/28/2006.

Giddens A (ed) (1972) Emile Durkheim: Selected Writings. London: Cambridge University Press.

Guruswamy L, Sir GWR Palmer and HW Burns (1994) International Environmental Law and World Order: A problem-oriented course book.

Hensin JM (2005) Sociology – A down to Earth approach, Pearson Allyn & Beacon Publisher.

House of Commons (2006) Environmental Audit – Second Report. http://www.publications.parliament.uk/pa/cm200506/cmselect/cmenvaud/6 07/60705.htm 5/6/2006.

IAIA (2005) Biodiversity in impact assessment. Special publication series no. 3.

Mill John Stuart (1963–77) Nature, Collected Works, University of Toronto Press.

Magurran AE (1988) Ecological diversity and its measurements. Croom Helm Limited, London.

Margules C and Usher MB (1981) Criteria used in assessing wildlife conservation potential: a review. Biological Conservation 21: 70–109.

Molles (2005) Ecology: Concepts and Applications, McGraw Hill.

Noss RF (1990) Indicators for monitoring biodiversity: a hierarchical approach. Conserv. Biol. 4: 355–364.

Ogburn WF (1922) Social Change with Respect to Culture and Original Nature.

Pagiola S, Ritter KV, and Bishop J (2004) Assessing the Economic Value of Ecosystem Conservation; THE WORLD BANK ENVIRONMENT DEPARTMENT ENVIRONMENT DEPARTMENT PAPER No.101 in collaboration with The Nature Conservancy and IUCN—The World Conservation Union, October 2004.

Pearce DW and Turner RK (1990) Economics of Natural Resources and the Environment. 2nd Ed. The Johns Hopkins University Press. Baltimore, MD.

Pidwirny M (2006) University of British Columbia Okanagan, http://www.physicalgeography.net/fundamentals/10q.html(1of 3)7/23/2006

Povilitis T (2001) Toward a robust natural imperative for conservation. Conservation Biology 15: 533–535.

Power N (2002) Roxby Downs Indenture Arrangements for Water Management, GABCC Paper.

Pring WR (1999) 2. International law and mineral resources; University of Denver College of Law; in MINING, ENVIRONMENT AND DEVELOPMENT A series of papers prepared for the United Nations Conference on Trade and Development (UNCTAD).

Pring GW and Joeris D (1993) Various International Environmental Law Collections, 4 COLORADO JOURNAL OF INTERNATIONAL ENVIRONMENTAL LAW & POLICY 422.

Rabinowitz D, Cairns S and Dillon T (1986) Seven forms of rarity and their frequency of the flora of the British Isles. In M.E. Soulé (Ed.), Conservation Biology, pp. 182–204. Sinauer Associates Inc., Sunderland, Massachusetts.

Shomaker JW (2005) Will there be water to support mining's future in New Mexico? In: Policy, economics, and the regulatory framework, Decision-makers Field Conference, Taos Region.

Smith PGR and Theberge JB (1986) A review of criteria for evaluating natural areas. Environmental Management 10: 715–734.

Steward RH (2005) Introduction to physical oceanography, Texas A&M University, www.Oceanworld.tamu.edu

TemaNord (2005) The Valuation of Habitats for Conservation – Concepts, methods and applications TemaNord 2005:519; Rasmus Ejrnæs *et al.*; Nordisk Ministerråd, København 2005 ISBN 92-893-1132-0 www. norden.org; www.norden.org/publikationer.

The Lutheran World Federation, Department for World Service, 2003. Environmental Reporting, Monitoring and Evaluation in LWF/DWS, A toolbox for LWF/DWS' country programs, Geneva.

UNEP (1995) Global Biodiversity Assessment. Cambridge University Press. Cambridge.

USEPA (1978) United States Environmental Protection Agency, Protective Noise Levels, Condensed Version of EPA Levels document, EPA 550/9-79-100, November 1978, Office of Noise Abatement & Control, Washington, D.C.

Wälde T (1993) Environmental Policies towards Mining in Developing Countries, 10 JOURNAL OF ENERGY & NATURAL RESOURCES LAW 327 (1992); reprinted in 30 PUBLIC LAND & RESOURCES LAW DIGEST 41 (1993).

Whittaker RH (1972) Evolution and measurement of species diversity. Taxon 21: 213–251.

WHO (World Health Organization) 2006 Guidelines for drinking-water quality [electronic resource]: incorporating first addendum. Vol. 1, Recommendations. – 3rd ed. Electronic version for the Web.

Zillman DN, Lucas AR and Pring G (2002) Human Rights in Natural Resource Development. Public Participation in the Sustainable Development of Mining and Energy Resources. Oxford University Press Inc.: New York. 710p. ISBN 0–19–925378–1.

8

The Baseline

Understanding the Host Environment

ARSENIC

Arsenic is steel grey metalloid, very brittle and crystalline; it tarnishes in air and when heated rapidly forms arsenious oxide with the odor of garlic. Many arsenic compounds are poisonous with the result that, historically, arsenic has been the 'poison of preference' for would be murderers. Arsenic in solution is particularly toxic to freshwater aquatic organisms; however, most marine organisms are less susceptible, having adapted to the significant concentrations that occur in sea water. Arsenic is associated with volcanic activity and its most commonly occurring mineral – arsenopyrite – is commonly present in gold-bearing ore bodies. The main uses of arsenic are as a pesticide and fungicide, particularly for the treatment of timber.

8

The Baseline

Understanding the Host Environment

In order to assess changes that will result from a project, it is necessary to understand the existing environmental conditions, and the likely changes that would occur in the absence of the project. This understanding is commonly referred to as the 'baseline'. The baseline also provides benchmark data for identifying and measuring actual changes based on monitoring data generated during mine operation, which then can be used to assess the accuracy of impact predictions made at the mine planning stage. But how much baseline data is needed? There is no fixed number; rather the spatial and temporal extent, numbers of sampling or observation sites and parameters to be studied will depend on the environmental situation, the nature of the proposed project, and the concerns of stakeholders as identified through the scoping process. In the documentation of the environmental impact assessment the baseline description should be relatively brief, and not used to inflate the document with largely irrelevant information. Excessive data collection with little relevance to potential change is a waste of time and money, and so is data presentation without data interpretation. A good measure of relevance is the 'so what?' test. If an item can be deleted without any detriment to the interpretations and assessments, then it should not be included, providing of course that the item is not in response to regulation or stakeholder request. Solar radiation is an example of an environmental parameter that is included in most baseline descriptions but which is rarely, if ever, used in impact assessment. The reason that solar radiation receives this undeserved attention is that most commercially available weather stations measure this parameter, along with many others, and hence the results are included.

Excessive data collection with little relevance to potential change is a waste of time and money, and so is data presentation without data interpretation.

Environmental scientists prefer to focus attention on data that serve as indicators of environmental state, change, and trend. Environmental indicators, as discussed later in this chapter, help to identify, quantify, and monitor the occurrence of environmental impacts. The use of indicators is discussed later in this chapter. Scoping, which is also discussed in this chapter, determines how environmental features, project design, and stakeholder concerns shape the extent and detail of environmental baseline surveys. Baseline studies initially involve a review and compilation of data available from the scientific literature and relevant government sources, collectively termed 'secondary data'. Subsequent site surveys are conducted to generate primary or site specific data. Since most environmental data and information have spatial attributes lending themselves to be managed and analyzed using geographic information systems (GIS), this chapter also addresses GIS applications.

Scoping, determines how environmental features, project design, and stakeholder concerns shape the extent and detail of environmental baseline surveys.

One main challenge remains, to differentiate data from information. What do data say? Environmental data collected in the field, the primary data, or data from existing sources, the secondary data, can only capture a narrow aspect of the structure, state, and functioning of natural and built components in environment systems at a given time and space. Data on water composition reflect water chemistry. But what does water chemistry tell us? Demographic data reflect the composition of the population in such terms as gender or age. But what does the age pyramid indicate? Accordingly, we attempt to assess the data to derive information on the state or condition of the environment and on environmental processes taking place. This is not always simple. Converting data to information is explored in the concluding section of this chapter.

8.1 THE IMPORTANCE OF CONFLICT IDENTIFICATION

Almost invariably, development of a new mining project will compete or interfere with other land uses, even in those rare situations where the land is not occupied. In most cases, the land will have been long inhabited by other groups, utilizing local natural resources in other ways. Competition for resources potentially leads to conflict. WCD (2000) defines conflicts as 'interactions of interdependent people who see their goals as incompatible'. In a conflict situation some people believe 'other' people are interfering with their efforts to satisfy their interests or values. Conflicts are not necessarily bad. They can be useful as a means of stimulating engagement and creativity. Handled ineffectively, however, conflicts become destructive.

Competition for resources potentially leads to conflict.

In recent years, mining companies have paid increasing attention to the controversies, disputes, and occasionally violent confrontations around mining developments. Why are large mining projects prone to such conflicts? First, there is often a mismatch of benefits and costs. Benefits are widely dispersed among shareholders, employees, and governments while costs in terms of adverse environmental impacts are locally concentrated. This real or perceived mismatch of benefits and costs at these different scales creates a structural challenge to dialogue and, thus, easily leads to confrontational attitudes. Second, large mines may affect critical local needs, such as quality and allocation of freshwater, an increasingly scarce and coveted resource. Third, insufficient recognition of or response to the social and environmental costs of large mine developments in the eyes of project-affected people often results in social mobilization. Other causes of conflict include disparities in social and economic power, different roles of different institutions, unmet expectations (which may include misunderstandings or commitments not met), an unpleasant surprise or simply the specifics of mine location and design

One or more of the above are usually involved in cases where conflicts emerge. Conflicts may be exacerbated by cultural and social differences between different stakeholders. And,

Conscious or sub-conscious distrust of governments may make conflicts more difficult to resolve.

of course, conscious or sub-conscious distrust of governments may make conflicts more difficult to resolve.

Conflicts may develop from differences in values and interests, shaped by the history of relationships and existing social structures. Conflicts may also emerge through lack of information or different interpretations of available information. Coleman (1957) reasons that conflicts are more likely to emerge when (1) people perceive that a decision or event significantly affects them, (2) the distribution of risks, benefits, and costs is unclear, and (3) people perceive that they can take some action, that change is a political decision rather than fate.

When individual or group interests are thwarted or unacknowledged, people look for ways to assert or achieve them. The types of issues and the relationships of the parties affect how the individual or group chooses to pursue its interests or values. Options for satisfying interests and values are also shaped by social and cultural factors, and the institutions and contexts in which they arise. Given the diversity of individuals, groups, and institutions involved in mining projects it is important to understand how social and cultural factors will influence public involvement and conflict dynamics. Gender, ethnicity, class, land tenure, and religion all influence the options stakeholders have for pursuing their interests and values in conflict situations. Most societies, even those committed to equal opportunities, are stratified. Women, religious and ethnic minorities, poor economic classes, and Indigenous Peoples face different social opportunities. This will affect conflict dynamics in many ways.

First, groups who have traditionally had little voice in society may be alienated, apathetic, and passive. Since they have had little experience with being heard or taken into account, when confronted with issues that affect their lives, they do not participate or voice their interests because they have no experience or expectation that it will be meaningful (Gaventa 1980). Second, groups who have little voice and input into decisions may quickly escalate the tactics they use to seek a hearing or involvement. Since they have little experience of being heard or taken account of with traditional non-violent mechanisms, they may resort to extreme demands and/or violent tactics faster than would a group whose experience involves being included, heard, and taken into account. Third, individuals and groups with little voice or input may be led (or influenced) by individuals and groups from outside their group. Since they have had little experience with normal democratic participation and conflict handling, they may lack internal spokespersons or leaders. This third avenue is commonly pursued by anti-mining activists when engaging with host communities to oppose a mining project. Anti-mining activists often try to influence, to lead, and to speak for the communities.

Anti-mining activists often try to influence, to lead, and to speak for the communities.

Different social institutions also affect involvement and conflict dynamics. Laws and social regulations vary considerably across societies and countries. These variations will affect the availability of different ways of participating and handling conflict. Family and community norms, education, government regulations, and bureaucratic organization will all affect the tactics and processes that groups use to pursue their interests. Individuals and groups in conflict usually have disparate access to resources and different capacities. Resources may include money, time, and information; capacities may be organizational, linguistic, cultural, and informational. These differences in legal or cultural rights, abilities and capacities to speak and be heard, and resources to sustain activities affect involvement and conflict dynamics.

All of the above illustrate why involving host communities is so important in the design of public participation and dispute resolution processes. Even people from the same country, but working for the government or the mining company may not be able to fully appreciate and incorporate into the consultation process the best approaches for involving local people.

8.2 THE USE OF INDICATORS

We are used to economic indicators, either at private level such as cash at the bank or at national level such as GDP or market interest rates. Economic indicators are designed to provide information on the state of the economy in a simple and clear manner. The use of indicators, however, is not restricted to the economic sphere; they exist for all parts of the environment. An indicator may be as simple as the concentration of a chemical element in an environmental medium or a much more complex index that combines various data in one number. The following introduces a number of site and company-level environmental indicators designed to facilitate assessing the current state of the environment, and to monitor future environmental change. It is not intended to provide a prescriptive listing of indicators to be used in every circumstance. The diverse nature of the environment at any given mine site, in particular in regard to biodiversity and social fabric, makes this an unrealistic expectation.

Why Indicators?

Environmental indicators are a way of presenting complex information in a simple and clear manner (**Case 8.1**). It is impractical if not impossible to measure all environmental parameters, such as for example the concentration in water of all elements in the periodic table. Accordingly, the practice that has developed is to select environmental indicators that will represent the situation. These are physical, chemical, biological, or socio-economic measures that best represent the key elements of a complex ecosystem or environmental issue. An indicator's defining characteristic is that it quantifies and simplifies information in a manner that facilitates understanding of environmental issues by both decision-makers and the public. Indicators are superior data as an analytical tool, since they commonly present several data in one number.

Environmental indicators are a way of presenting complex information in a simple and clear manner.

An indicator is not only important for establishing the environmental baseline but is embedded in a well-developed interpretive framework that has meaning beyond the

CASE 8.1
The Gini Coefficient as a Measure of Inequality of a Distribution

The Gini coefficient is a measure of inequality of a distribution, defined as the ratio of area between the Lorenz curve of the distribution and the curve of the uniform distribution, to the area under the uniform distribution. It is often used to measure income inequality. It is a number between 0 and 1, where 0 corresponds to perfect equality (i.e. everyone has the same income) and 1 corresponds to perfect inequality (e.g. one person has all the income, while everyone else has zero income). It was developed by the Italian statistician Corrado Gini and published in his 1912 paper 'Variabilità e mutabilità' ('Variability and Mutability'). The Gini coefficient is equal to half of the relative mean difference. The *Gini index* is the Gini coefficient expressed as a percentage, and is equal to the Gini coefficient multiplied by 100.

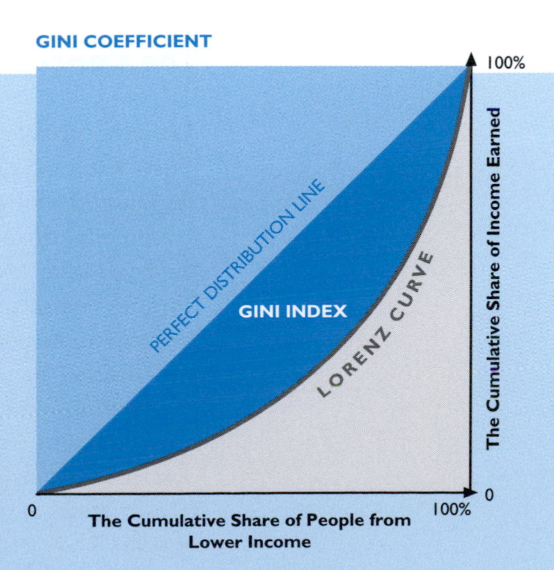

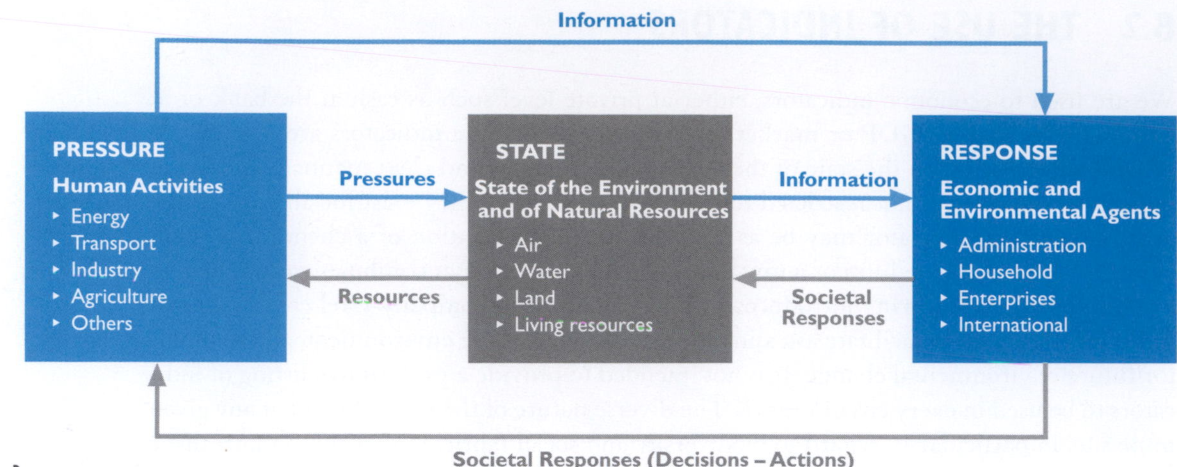

FIGURE 8.1
The Pressure – State – Response Framework

The PSR approach examines human activities which impact on components of the environment and the subsequent human and natural responses to these impacts.

Source:
OECD 1994

measure that it represents. Repeated measurements of the variables that comprise the indicator in various places and times, and in a defined way, comprise the monitoring programme for that indicator during mine operation. Comparison of this repeated set of measurements with a benchmark set (the control site) or condition prior to mining provides the basis for detecting and quantifying change. Thus, environmental indicators are measurable features that provide managerially and scientifically useful evidence of environmental quality and reliable evidence of trends in quality.

In selecting indicators the goal is not only to define the environmental baseline, but also to be able to assess how mining activities affect the direction of change in environmental performance, and to measure the magnitude of that change.

In selecting indicators, the goal is not only to define the environmental baseline, but also to be able to assess how mining activities affect the direction of change in environmental performance, and to measure the magnitude of that change. Indicators that allow a quantitative evaluation of mining impacts are particularly useful, since they provide more information than just whether mining is improving or degrading the state of the environment. Information on the magnitude of a benefit is required to determine whether it is worth the resources being expended to achieve it. Similarly, information on the magnitude of adverse impacts might indicate whether the harm is justified given the other benefits of the activities in question (Segnestam 1999, 2002).

The Concept of Indicator Frameworks

Two interpretive indicator frameworks are widely used to select environmental indicators. The OECD (1994) based on work of Adriaanse (1993) introduced the pressure-state-response (PSR) framework to select national-level indicator sets. The PSR approach examines human activities which impact on components of the environment (usually at the ecosystem level) and the subsequent human and natural responses to these impacts. Typical stress factors are pollution loadings, land-use changes, or resource exploitation. Ecosystem responses (**Figure 8.1**) include changes in productivity, species composition, and disease incidence. The framework provides the means to structure sets of indicators in a manner that facilitates their interpretation; it also aids the understanding of how different issues are interrelated. The PSR framework differentiates indicators that describe pressures exerted on the environment, indicators of the environmental condition or state, and indicators of responses to the pressures, or to changes in the condition or state.

Segnestam (1999 and 2002) developed an indicator framework based on project cycles to derive project-level indicators, classifying indicators quite similar in nature to the PSR framework as follows:

- Input indicators – monitor the project-specific resources provided;
- Output indicators – measure goods and services provided by the project; and
- Impact indicators.

While the input-output-outcome-impact approach distinguishes between project outcomes (measuring immediate, or short-term, results of project implementation) and project impacts (monitoring longer-term or more pervasive results of the project) outcome and impact indicators are not easily differentiated in practice; hence they are often bundled together and jointly referred to as impact indicators.

Environmental indicators are commonly divided into five categories.

1. Input indicators measure resources in terms of people, equipment, and materials that go into environmental and social management. Examples of input indicators include funding of various environmental management activities, establishment of the environmental management organization, installation of environmental controls, or the provision of monitoring hardware and software. In addition there are project input indicators such as materials used, money spent, or people employed in the mining project or, at a more detailed level, for any activity within the mining project, including environmental protection measures.
2. Process indicators measure the change in quality and quantity of access and coverage of the activities and services. Examples of process indicators include Standard Operating Procedures, information campaigns for communities or local government, and the creation of emergency response plans.
3. Output indicators measure the results of activities and services that are produced with the inputs. Examples of output indicators include environmental monitoring results or the creation of a database for tracking employment or numbers of people trained.
4. Outcome indicators measure environmental changes. Examples of outcome indicators include the changes of biodiversity, fish populations, family and individual income levels, and overall employment rates.
5. Impact indicators measure medium and long-term changes in selected environmental attributes. Examples of impact indicators are demographic and ethnographic changes compared against the baseline.

Indicators of both outputs (e.g. gaseous project emissions) and impacts (e.g. ambient air quality at project area) are typically required to properly evaluate project impacts.

Output indicators alone are often insufficient because the link between output and consequent environmental impact may be ambiguous or of unknown magnitude. This point is important because it is the end result that is of most concern to people. People care about emissions primarily because emissions increase air pollution and hence affect people's health.

Impact indicators alone are also often insufficient because changes in environmental conditions depend on the combined effect of multiple pressures (and on random natural factors). Unless the project's contribution to change is measured, the project might be blamed for problems it did not cause or credited for improvements to which it did not contribute.

Table 8.1 summarizes these and other commonly used indicators. One common approach, for example, is to develop one set of alarm indicators, and one set of diagnostic indicators. In this approach, the first priority of monitoring is to provide sufficiently early warning about impending adverse environmental change so that corrective action can be

TABLE 8.1
Definition of Selected Environmental Indicators

Indicator	Definition
Input Indicator	Monitors the project-specific resources provided
Output Indicator	Measures goods and services provided by the project
Impact Indicator	Monitors longer-term environmental impacts of the project
Pressure Indicator	Measures human activity with potential environmental effects
State Indicator	Measures current environmental conditions
Response Indicator	Measures human and natural response to environmental pressures
Bio-Indicator	Plant /animal signalling environmental condition
Performance Indicator	Indicates environmental performance
Alarm Indicator	Provides timely warning about impending adverse changes
Diagnostic Indicator	Enables in-depth analysis of changes
Leading Indicator	Measures the implementation of environmental practices
Lagging Indicator	Measures the result of environmental practices

The difference in nomenclature largely depends on the underlying framework used to derive indicators.

Alarm indicators are a small set of indicators, easily monitored at low cost, hence allowing frequent or continuous monitoring.

taken. Alarm indicators are a small set of indicators, easily monitored at low cost, hence allowing frequent or continuous monitoring. Examples are monitoring of water levels in tailings ponds or of pH values in discharge water.

Diagnostic indicators, in comparison, are a second set of indicators activated if the value of the alarm indicators crosses a predetermined trigger value (**Figure 8.2**). Diagnostic indicators are designed to allow a more in-depth analysis (or diagnosis) of the causes that triggered the alarm indicator. The diagnostic indicators give more detailed information about the issues at hand, but data collection and analysis is generally more time consuming and costly.

Environmental indicators can also be broadly classed into two types. Lagging indicators, the type of metrics most commonly reported, 'lag' or measure the results of environmental practices or operations. Types of data include tons of waste generated, number of penalties and violations, number of lost work days, or tons of shipped materials. Frequently these types of data are collected because environmental laws require that they be reported. Lagging indicators are usually readily quantifiable and understandable, and include statistical data collected for other business purposes. The main disadvantage is that, as the name implies, they lag or reflect situations where corrective action can only be taken after the fact, and often after incurring some type of cost, whether it be in penalties or decreased credibility with regulators or the public.

Leading or in-process indicators measure the implementation of practices or measures which are expected to lead to improved environmental performance.

Leading or in-process indicators measure the implementation of practices or measures which are expected to lead to improved environmental performance. For example, instead

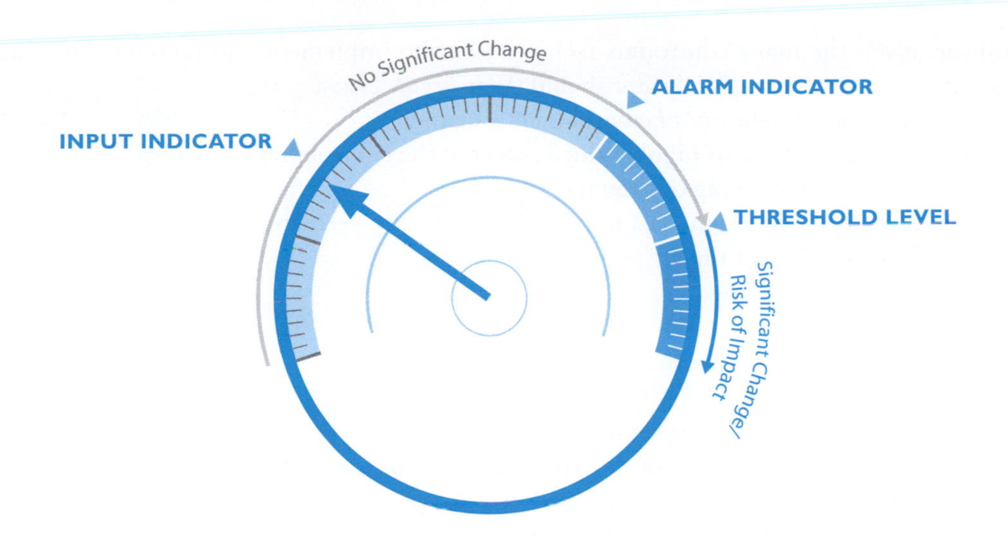

FIGURE 8.2
Illustrating the Significance of Alarm and Diagnostic Indicators, and Threshold Level in Environmental Management

of the numbers of non-compliances, a leading indicator would be the number of internal environmental compliance audits conducted per year. The major advantage of leading indicators is that corrective actions can often be taken before deficiencies show up. Unfortunately, leading indicators can be difficult to quantify (some may be qualitative rather than quantitative), and the results may not address the concerns of some stakeholders (such as the public), who may still want to know the quantities of waste released into the environment, or the number of health and safety violations.

Which Indicator to Choose

Much research has been done in developing environmental indicators (Bakkes *et al.* 1994, Segnestam 1999, Saunders *et al.* 1998). At the environmental scoping stage, since all indicators are derived from data, the environmental assessment can only rely on indicators based on data that can be easily acquired. In the scoping process, the focus is on indicators describing the current state of the environment. As the mine moves into the feasibility stage, detailed surveys are carried out which increase the quantity of environmental data providing an expanded set of indicators for future monitoring and evaluation of impacts associated with mine operation. During mine operation, environmental indicators will eventually evolve into environmental performance indicators (EPI). As such, indicators are very much dynamic tools. Clearly, an understanding of mining in general and of the proposed mining project in particular, is beneficial and in fact necessary in the design of field surveys, and the selection of meaningful environmental indicators.

An understanding of mining in general and of the proposed mining project in particular, is beneficial and in fact necessary in the design of field surveys, and the selection of meaningful environmental indicators.

Priority environmental issues identified on the basis of community concerns also influence the selection of environmental indicators. They are driven by the current environmental debate and may not always imply an obvious cause of pressure on the environment. In practice, however, categories of stresses or pressures on the environment (e.g. vegetation clearing, illegal logging, disease patterns, community unrest) are often themselves priority issues. Because priority issues and stress response approaches overlap, stress-response indicators can be considered as a subset of indicators within priority issues. This avoids the problem with a pure priority issues approach that, while immediate community concerns will be addressed, underlying causal processes may be ignored.

Standard selection criteria for environmental indicators centre around scientific validity, practical, and programmatic considerations. Above all, an indicator must be practical and

An indicator must be practical and realistic, given the many constraints facing those who implement and monitor a mining project.

realistic, given the many constraints facing those who implement and monitor a mining project. An environmental indicator should (Ward *et al.* 1998):

- serve as a robust indicator of environmental change;
- reflect a fundamental or highly valued aspect of the environment;
- provide an early warning of potential problems;
- be capable of being monitored to provide statistically verifiable and reproducible data that show trends over time;
- be scientifically credible;
- be easy to understand;
- be monitored regularly with relative ease;
- be cost-effective;
- have relevance to management needs;
- contribute to monitoring of progress towards implementing stated management commitments;
- where possible and appropriate, facilitate community involvement;
- contribute to the fulfilment of reporting obligations; and
- where possible and appropriate, use existing commercial and managerial indicators.

It is most effective to be selective and use smaller sets of well-chosen indicators.

It is most effective to be selective and use smaller sets of well-chosen indicators. Using too many indicators risks diluting their usefulness. Priorities may become confused and the details may seem overwhelming for both the developers and the users (Segnestam 1999). It is also recommended that a concise indicator profile as illustrated in **Table 8.2** be established at the outset of a long-term environmental monitoring programme. Establishing such profiles helps in understanding and documenting the rationale of monitoring efforts.

Biological systems are organized hierarchically from the molecular through the ecosystem to the landscape level. Logical classes such as genotypes, populations, species, communities, and ecosystems are heterogeneous; all members of each class can be distinguished from one another. The variety of biological configurations at all levels is extremely large, currently unknown and probably not measurable. Yet for monitoring and reporting on the condition of biological diversity there has to be some acceptable baseline against which change can be measured, and some related biological indicators.

Figure 8.3 shows a biological hierarchy with precision of measurement increasing from the higher more heterogeneous levels down to the molecular level, while practicality (including effort and cost) increases in the opposite direction. Higher levels of organization also integrate ecological processes and functions such as nutrient and energy cycling, which result partly from components of biological diversity. A decision on which level of surrogacy to use, depends on the scale of measurement and reporting and the resources available: the greater the level of precision, the more useful the result. For project-level environment reporting it is possible, in some cases, to use sub-sets of taxa as surrogates for biological diversity, but vegetation classes and environmental domains are also commonly adopted. **Appendix 8.1** lists some common indicators to measure the current state of the biological environment, and to detect future change. The listing can not be exhaustive, and actual indicators will change from site to site.

Using Wildlife as Early Warning Signals for Human Impacts

Using wildlife (or sometimes plants) as early warning for environmental and human impacts, differentiation is made between indicator species (indicating human impacts on the environment) and sentinel species (indicating potential health impacts on humans).

TABLE 8.2

Example of an Indicator Profile, Adopted from Ward *et al.* (1998)

Indicator – Sediment Quality (Contaminants)

Description – This indicator documents levels of, and changes in, major contaminants of surface sediments in surface water bodies including estuaries and coastal areas.

Rationale – Contaminants commonly accumulate in sediments, and measurement of sediment concentrations of contaminants is a useful way to track long-term trends in concentrations of most contaminants in aquatic systems. These concentrations indicate the extent and magnitude of pressure imposed by contaminants on the flora and fauna of the shallow water ecosystems.

Analysis and interpretation – Contaminants are sometimes anthropogenic materials (such as cyanide) introduced by mining, while others are naturally occurring materials (such as trace metals) and become contaminants when they occur in higher than usual concentrations. Some find their way into the surface sediments of waterways after various periods (sometimes brief) in the water column. Naturally occurring materials are often in the natural make up of the sediment material. For both anthropogenic and natural materials, the precise level at which an effect can be expressed in the biological system is difficult to define. Hence, besides using absolute concentration criteria to determine when levels are acceptable, trend assessment becomes important in evaluating level of stress imposed by contaminants. For anthropogenic materials, levels should be trending downwards, hopefully to near-zero, while for natural materials they should be close to natural background levels and not trending upwards. Monitoring results that do not fit these objectives may indicate the need for environmental mitigation and remedial actions. Change can only be detected against a baseline of existing or historic data, and then only with many caveats about collection and analysis techniques. Since laboratory techniques become increasingly sophisticated and a project may become controversial with time, data from earlier times may become questionable. Full documentation of procedures, quality assurance and controls is critical if currently collected data are to be useful in the future.

Monitoring design and strategies – Sediment quality as indicator is measured using traditional field sampling and laboratory analysis protocols documented in form of an SOP. It includes analysis of major chemical constituents (trace metals relevant for the specific mining project and/or as specified in relevant regulatory standards), organic carbon, and of physical characteristics such as grain size and size distribution. The indicator is monitored annually (or as otherwise specified in the SOP) in a small number of carefully selected downstream locations and reference areas. Monitoring has three purposes: to track changes in contaminants that are probably influencing waters downstream of the mine site; to provide the basis for control/reference conditions so that local-scale effects of contamination can be determined more robustly; and to enable rehabilitation programmes to have a relatively undisturbed condition as a target for the restoration of degraded habitats.

Reporting scale – The data for each year is summarized annually, together with year-to-year changes for each site, and with estimates of uncertainty (say 95% confidence limits). The difference between these estimates and any previous (or baseline) estimates will be expressed as an estimate of change, together with an estimate of the size of change that could be statistically detected with the methods used.

Outputs – The outputs are presented in the form of maps annotated with tables summarizing site-specific levels of each contaminant. Changes are summarized by tables setting out the percentages of significant change (positive or negative change of statistical significance).

Data sources – Often extensive sediment data are collected during exploration. Such data are collected with a specific purpose, and are not viewed as objective data in case of a controversy. Also realistic data on background concentrations over regional areas are seldom available.

Linkages to other indicators – This indicator is closely allied to water quality, fish tissue quality, and seabird eggs (contamination).

Using too many indicators risks diluting their usefulness. Priorities may become confused and the details may seem overwhelming for users.

Indicator Species

An indicator species (or bio-indicator) is a plant or animal that indicates, by its presence in a given area, the existence of certain environmental conditions. Indicator species are unique environmental indicators as they offer a signal of the biological condition. Using bio-indicators as an early warning of pollution or degradation in an ecosystem can help with directing environmental mitigation efforts. While indicator species is a term that is often used, it is somewhat inaccurate. Indicators are actually often groups or types of biological resources. Within each group, individual species can be used to calculate metrics such as percent of species or groups of species or individual orders in an effort to assess environmental conditions.

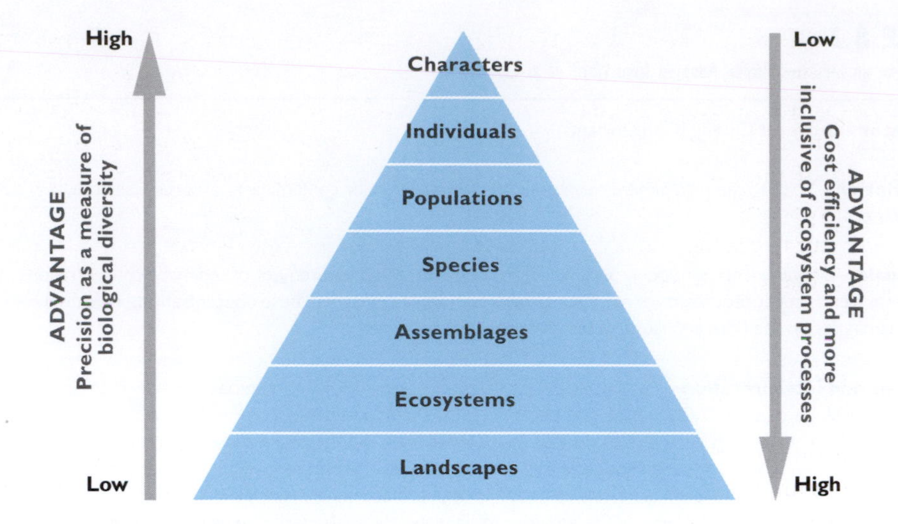

FIGURE 8.3
Level of Biological Diversity Surrogacy

Precision increases from the landscape to the character level, but cost decreases and ease of measurement increases in the opposite direction

Source:
Adapted from Williams 1996

Fish have been used for many years to indicate whether waters are clean or polluted.

Fish have been used for many years to indicate whether waters are clean or polluted, doing better or getting worse. Knowing just whether fish live in the waters is not enough – we need to know what kinds of fish are there, how many, and their health. Fish are excellent indicators of water quality because they (1) live in the water all of their life, (2) differ in their tolerance to amount and types of pollution, (3) are easy to collect with the right equipment, (4) live for several years, and (5) are easy to identify in the field (EPA 2006).

Sentinel Species

One of the more difficult aspects is the realistic assessment of the health impacts, once environmental degradation is evident. Humans are exposed to low levels of contaminants in their daily lives, but exposures may be over long, often lifelong, periods. The exposure situation is usually complex, with environmental contaminants being present in mixtures or coming from different sources.

Certain plants and animals can serve as sentinel species, indicators and early warners of the impacts of environmental exposure.

Certain plants and animals can serve as sentinel species, indicators and early warners of the impacts of environmental exposure. High trophic-level organisms (at the top of the food chain) occupy a similar niche to many human consumers and are often exposed to similar types of environmental contaminants through daily dietary intake. Many fundamental physiological functions in wild animals and humans are similar. If a chemical at a certain concentration is causing a certain effect in a wild animal, it could be taken as an indication of the same effect occurring in humans. A substance that is neurotoxic to birds is very likely to be so to humans and a substance that is estrogenic in fish is likely to be so in humans. Trace metals in the environment and their effects on wildlife provide a good example of this. The persistent and bioaccumulative nature of some metals causes them to bio-magnify to high concentrations from water in aquatic systems to the tissues of top predators. Evaluating the patterns, levels, trends, and effects of metals in higher trophic-level consumers contributes to the understanding of both the contamination of ecosystems and the risks posed to the health of humans and wildlife. Even when the concentrations are below the limit of detection in water they may easily be measured in animals. Wildlife examples have shown that some human groups may be at increased risk because of their preferential habit of consuming fish and other aquatic food.

A sentinel species can be selected for its ability to reflect different types of environmental disturbances. The best known example is perhaps the canary, which is sensitive to carbon monoxide in the air and was used in underground coal mines as an indicator of potentially toxic concentrations of this gas. Similarly, lichens are sensitive to sulphur dioxide and have been used as indicators of SO_2 pollution around smelters and coal-fired power stations.

Based on its life history and physical characteristics, a selected species can provide insights about environmental changes at various scales, for example, over time or space. Each sentinel species is specific to particular environmental conditions. In selecting a sentinel species we attempt to answer the same question: Can any of these species tell us something about the impacts of the mining project, and their relationship to human health and biological integrity? **Appendix 8.1** contains examples of environmental indicators covering the full range of natural resources. Other indicators may be more appropriate taking into account the specific characteristics of a particular mine site.

Indicators Applied to the Human Environment

Evaluating the human environment is more difficult, particularly in assessing how well communities are doing, and progress towards sustainability. Statistical indicators as commonly used in environmental assessment can only portray a narrow view of a particular sector or issue, rather than a holistic picture of the community: economic progress (growth rates, unemployment, incomes); social well-being (infant mortality, years of schooling, number of people per dwelling); and environmental monitoring (air and water quality, emission of pollutants, hectares of protected areas). As such, designing appropriate indicators for a given community is crucial, and their interpretation has always to be in a larger context, preferably including comparison to one or more control communities unaffected by the project. An important aspect of developing indicators at community level is that they be defined, developed, and used jointly with the community. The community then becomes stewards of the indicators, and communicating community benefits due to mine development becomes easier and more transparent.

Designing appropriate indicators for a given community is crucial, and their interpretation has always to be in a larger context.

Social indicators are designed to point to measurable change in social and economic systems among communities in the host region resulting from a mining development. **Appendix 8.1** suggests some social indicators. A novel, readily measured indicator of disposable income is illustrated in **Case 8.2**.

The indicator grouping depends on the framework used to present baseline data, and it can vary widely. One example of the many alternative groupings follows (NOAA 1992):

- Population characteristics including: present population and expected change, ethnic and racial diversity, and influxes and outflows of temporary residents as well as the arrival of seasonal or leisure residents.
- Community and institutional structures including: the size, structure, and level of organization of local government including links to the larger political systems. They also include historical and present patterns of employment and industrial diversification, the size and level of activity of voluntary associations, religious organizations, and interest groups, and finally, how these institutions relate to each other.
- Political and social resources including: the distribution of power authority, the interested and affected publics, and leadership capability and capacity within the community or region.
- Individual and family changes refer to factors which influence daily life including: attitudes, perceptions, family characteristics, and friendship networks. These changes range from alteration in family and friendship networks to perceptions of risk, health, and safety.
- Community resources include: patterns of natural resource and land use; availability of housing and community services such as health, welfare, police protection, and sanitation facilities. The continuity and survival of human communities depends partly on their historical and cultural resources.

8.3 ENVIRONMENTAL SCOPING

Environmental scoping, also referred to as rapid environmental and social appraisal or briefly, rapid assessment, is the process of determining the context and extent of matters which should be covered in the environmental assessment. While scoping is not always a mandatory requirement, it is important. The outcome of the scoping process defines the 'scope' of environmental information to be prepared for the environmental impact assessment study, and outlines the extent and detail of baseline surveys to be undertaken to compile that information. Since scoping is designed to provide a concise but brief picture of the environment as well as an initial understanding of expected interactions between mine and environment it must fulfil two apparently contradictory requirements: it must be comprehensive and complete, but, as the alternative name rapid assessment suggests, scoping is conducted within a short time frame and often with limited resources. As such environmental scoping requires considerable expertise.

Environmental scoping requires considerable expertise.

Environmental Scoping Guidance

While there is no single universally accepted methodology for scoping, there is a general understanding about the various approaches that can be used, required expertise, importance of planning, inputs, need for secondary data collection, benefits of a site visit, and content of the scoping report.

CASE 8.2
Satellite Dishes as Indicators of Disposable Income

A village close to a gold mining project in Indonesia, was studied as part of the environmental assessment process and subsequently monitored throughout the life of the operation. Most of the indicators used to assess economic impacts were statistics obtained from village records and national census data. However, the mine staff selected some additional indicators which were easily measured. One of these was the number of satellite dishes installed on village houses. This provided an indication of increased disposable income and its distribution throughout the village.

The results over a period of seven years are summarized below:

	Number of Dwellings	Number of Satellite Dishes
Preconstruction	311	4
End of Construction	325	11
After 2 Years of Operations	340	69
After 5 Years of Operations	342	95

The survey involved in producing this indicator was particularly simple and involved only 30 minutes of observations. Further information on the distribution of satellite dishes indicated that these installations were distributed throughout the village. In many cases it was clear that neighbours had cooperated through joint ownership of a satellite dish.

Unfortunately, there was no control village that would have provided a comparison of the normal (non-project related) growth in satellite dish ownership over the same period. Similarly simple to monitor indicators include home extensions and motor vehicles.

Two approaches are used for the scoping process. The first is based on applicable requirements of the regulators, prospective financial institutions, or the proponent. Regulatory and financial institutions often apply pre-defined scoping criteria; however, no universally accepted scoping methodology or format exists in the mining sector. Environments in which mines are developed vary widely and may be highly complex, meaning there can be no single scoping method applicable to every mine, anywhere in the world. Furthermore, the extent of what is possible in any particular case will depend on the resources and capacities available. Environmental scoping may include desk studies, satellite imagery analysis, interviews and discussions with relevant experts, and limited field surveys. It usually includes compilation of existing data and information, including traditional knowledge (jointly termed secondary data). Most commonly, environmental scoping comprises both secondary data collection and a site visit, but without actual field sampling and laboratory analyses. However, many mining companies initiate some primary data collection activities, such as climate and water quality parameters during the early exploration stage. In these cases, the primary data would be included in the scoping process.

Environmental scoping comprises both secondary data collection and a site visit.

The second approach to scoping is public participation and the involvement of experts. Ideally, effective scoping involves dialogue between the mining company and public authorities. This will be supplemented by consultation with formal and informal public leaders at the proposed mine site and the general public. Public involvement, however, is the exception rather than the rule in most developing countries, at least at the early stage of a mine development. Environmental scoping then follows a management approach, whereby issues of importance are addressed based on professional judgements.

It is evident that successful scoping requires appropriate expertise. Whoever is undertaking an environmental scoping study, it is important that involved professionals have sufficient information about the project and the area which will be affected, to enable potential project related concerns and opportunities to be identified. An understanding of relevant legislation and regulations of the host country and its implications for the project are also important. A site visit is extremely useful; experienced professionals will derive much useful information from even brief observations, which will provide important input to the subsequent design of data collection programmes.

An understanding of relevant legislation and regulation of the host country and its implications for the project are important.

Whether undertaken as part of a legal process or as good practice in environmental assessment process, scoping brings a number of benefits. When scoping is carried out well, the EIA process has a good chance of proceeding smoothly and efficiently; when carried out poorly, it is likely that important issues will be overlooked or underestimated, leading to unnecessary delays and costs. Information generated during scoping is useful in planning the extent of baseline surveys, ensuring that field surveys address natural resources of concern that may be affected by the project (or that will affect the project) while at the same time avoiding lengthy research and investigation of topics with little relevance to the project. Scoping also helps to identify 'fatal flaws' or 'show stoppers' early in mine design. One of the main benefits of scoping is that it provides environmental input in mine design at an early stage where fundamental changes can be made to the project without excessive cost implications and without causing project delays (**Figure 8.4**). Figure 8.4 illustrates that the space available to resolve problems diminishes as project planning advances. Of course, the attention level and capacity available to resolve problems are higher during the design phase.

One of the main benefits of scoping is that it provides environmental input in mine design at an early stage where fundamental changes can be made to the project without excessive cost implications and without causing project delays.

Scoping is particularly important in addressing social issues, since its purpose is to develop a basic understanding of a project's social setting, potential stakeholders, stakeholder issues, and the range of probable social costs and benefits to be addressed.

Scoping also provides the best opportunity to minimize the quantity of EIA documentation. As well as identifying the issues of most importance during the EIA process, scoping also identifies issues of little concern, enabling them to be eliminated from further

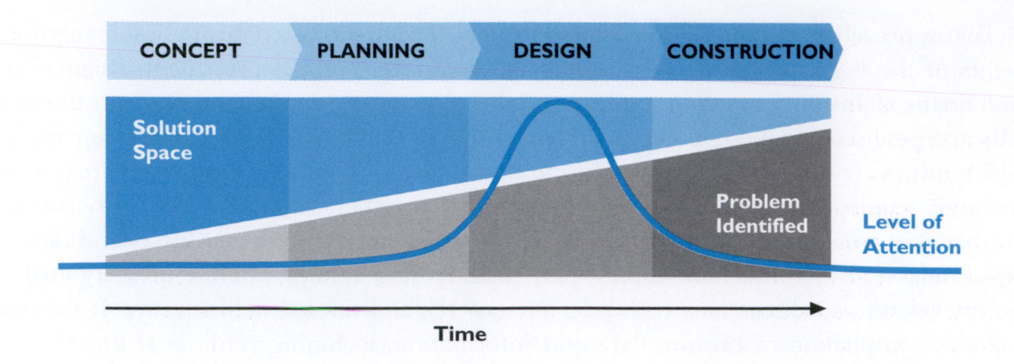

FIGURE 8.4
Scoping in the Project Design Cycle

Solution space decreases over time.

consideration. Thus, a proper analysis of the scope of the project will allow limitations on what has to be analyzed later. In this way, scoping ensures that EIA studies are focussed on significant issues and the environmental components involved in their evaluation, so that time and money are not wasted on unnecessary studies.

While environmental scoping usually requires speed, it can be expensive. Costs will increase when visiting remote areas or large areas, involving difficult terrain, and/or a large number of sensitive or important environmental features. Undertaking scoping can also mean a higher cost owing to the need, for example, to mobilize a qualified team rapidly and to work long hours. Use of helicopters, although expensive, can considerably reduce field time and is therefore likely to be cost effective for large sites with limited accessibility.

Scoping Planning and Assessment Context

In practice the actual scoping process is determined by project type and its context, more specifically by the institutional setting of the mine and by available resources. Therefore, there is the need in scoping to elaborate on regulatory and project-specific requirements for planning and executing the project environmental assessment. These contextual considerations are fundamental, and should be the subject of a shared understanding between EIA compiler and mine proponent as to what constitutes a satisfactory assessment for the project. Lee (2006) groups the contextual considerations into the following four categories:

- Regulatory and institutional context including (1) the key requirements and constraints which the regulatory framework imposes on the project and assessment (e.g. formal stages and procedures in project implementation, provisions for public consultation, time schedules, and applicable standards); (2) the main authorities, other agencies and stakeholders likely to be involved in the preparation and approval of the EIA; and (3) environmental approval process of similar projects within the same jurisdiction which may provide useful information on environmental approval process.
- Environmental context including (1) the environmental setting of the project and access to the project site; (2) affected communities and Indigenous Peoples likely to be affected, positively or negatively, by mine development; (3) sensitive ecosystems and endangered or threatened species encountered at the project site; and (4) natural hazards.
- Characteristics of the mining project including (1) size of mine and key project components (e.g. mineral processing, power generation, and harbour, deciding whether the project is a single or multi-sector project); (2) project and environmental assessment timelines; and (3) geographic extent of the area likely to be impacted and the extent of expected environmental changes.

- Resource requirements and constraints including (1) staffing resources, other expertise and financial support likely to be required/available for undertaking the assessment; (2) the extent of information needs and what is readily available; and (3) time constraints within which the assessment must be completed.

The information in the first three categories can be used to form a first impression of the likely nature, duration, scope, level of detail, and complexity of the environmental assessment. The information provided in the fourth category should indicate the types and scale of resources likely to be available to complete the assessment, and whether additional resources may be required.

Inputs to Scoping

Two inputs into scoping are important. First, there is the need to understand the mine activities which are likely to take place, and their likely impacts. Scoping provides an initial indication of activities that are likely to generate significant impacts during various stages through mine development and operation. Second, there is the need to understand the nature of the environment in which the mine is located. Scoping initiates the process of compiling relevant environment information required to assess the potential impacts of various project activities on various environmental components. Scoping can help to identify 'hot' issues.

Information on a specific project area is available from a myriad of sources. Some of the common sources of information include government agencies (to collect information on geology, topography, meteorology, land use planning, administrative boundaries, population census, protected areas and species, heritage sites, and so on), the internet, environmental and scientific NGOs such as BirdLife, WWF, or The Nature Conservancy (TNC), and environmental documents prepared for nearby projects. Areas of particular ecological interest have usually been studied, to some extent, by academic researchers, although in many cases these studies are focussed on a narrow biological feature rather than a broad ecological assessment. Perhaps because of the large number of bird watchers worldwide, information on the distribution, habitat requirements, behaviour, and conservation status of birds is readily available for most parts of the world.

Areas of particular ecological interest have usually been studied, to some extent, by academic researchers.

The internet in particular has developed into a source of information at a scale never experienced before the advent of the web. No listing can do justice to the vast amount of information sources at the internet, but some useful links follow:

- Satellite imagery – Google Earth (www.googleearth.com)
- Digital elevation data – Shuttle Radar Topographical Mission (SRTM) (www2.jpl.nasa.gov/srtm)
- International conventions and treaties – Environmental Treaties and Resource Indicators (ENTRI) (www.sedac.ciesin.org)
- NGO – NGO Global Network (www.ngo.org)
- Mining – International Council on Mining and Metals (ICMM) (www.icmm.com)
- Environmental assessment – International Association For Impact Assessment (IAIA) (www.iaia.org)
- Libraries
- International Institutions – World Bank (www.worldbank.org)

Google Earth in particular has revolutionized the way we view the Earth. Any place on Earth is only a few clicks away, and no mine site should be studied without an initial free

The background research should include a critical review of legislation, regulations, and standards of the host region that apply to the mining project.

tour on the web. There are many other sources of information, and a more comprehensive listing of useful websites is given in Appendix 1.2. The background research should also include a critical review of legislation, regulations, and standards of the host region that apply to the mining project, sometimes referred to as the institutional setting.

In spite of all the data that may be readily available, a visit to the mine site and its surrounding is invaluable. Observing the various types of natural habitats and communities and their condition at the mine site and collecting additional background information will help with focussing subsequent baseline surveys. It is important to realize that scoping is not generally designed to take into account temporal variance, such as seasonality, in ecosystems. Only when scoping is used in repeat surveys as elements of an integrated monitoring programme, can temporal variance be addressed. It is also important to note that assessments at the genetic level of biodiversity do not generally lend themselves to rapid approaches. However, a site visit, in particular when combined with remote sensing techniques, is suited to assess ecosystems at the habitat level.

Indigenous Peoples have depended for their survival, on their understanding of the nature of a wide variety of plants and animals, the functioning of ecosystems and the particular techniques required for harvesting and managing these biological resources.

The site visit may also allow the study team to access the vast local environmental knowledge found in local communities. In particular, Indigenous Peoples have depended for their survival, on their understanding of the nature of a wide variety of plants and animals, the functioning of ecosystems and the particular techniques required for harvesting and managing these biological resources. In rural communities in developing countries, locally occurring species are relied on for many – sometimes all – foods, medicines, fuel, building materials, and other products. Traditional knowledge has value and validity. Local non-indigenous communities of long standing also have traditional knowledge of the local conditions, environment, and wildlife. This knowledge may be as profound as indigenous traditional knowledge in certain areas. A useful reference document regarding traditional knowledge has been compiled by Emery (2000).

Issuing the Scoping Report

A key output from the scoping process is the scoping report, frequently used to prepare the Terms of Reference for the EIA. In many jurisdictions, this is the first public document in the environmental assessment process. The scoping report presents the relevant environmental information obtained through the scoping process, including input received from community consultation. It should also lay out the approach for subsequent stakeholder consultation, data requirements, methodology to be used in the environmental assessment, timing, and resources required.

No generally accepted standard defining the content of environmental scoping exists, neither is a 'cook book' approach recommended. Scoping reports should contain information, results, and recommendations for further action to guide mining companies or authorities in the environmental assessment. Reporting should describe the institutional setting of the project, illustrate the environmental and social setting, summarize site-specific project opportunities and challenges, and should provide a listing of recommended environmental studies. **Appendix 8.2** provides a listing of topics commonly covered in a scoping report.

8.4 CONDUCTING BASELINE SURVEYS – WAYS AND MEANS

A fundamental decision in selecting methods for acquiring information is whether the quantitative information is required or whether the programme only needs to illustrate general characteristics. Quantitative methods are often used when uncertainty exists or

TABLE 8.3
Examples of Qualitative and Quantitative Evaluation Techniques

Qualitative Measures	Quantitative Measures
Plan view map showing observation points and estimated vegetation coverage	Scaled vegetation map quantifying coverage areas
Vegetation (species list and relative abundance estimates)	Vegetative diversity, dominance, density, cover, and biomass
Fixed point panoramic photographs, including satellite/ aerial photography	Topographic mapping
General observations on climate, hydrology, soils	Rainfall and water level data, Stream gauging, Water quality and soil properties data
Wildlife observations	Wildlife counts
Fish and macro-invertebrate (species list and relative abundance estimates)	Fish and macro-invertebrates (diversity, density and distribution)

when baseline data relate to specific thresholds. Conversely, in situations where survey results are not tied to quantitative criteria, qualitative evaluations are often appropriate. In practice, a combination of quantitative and qualitative methods is usually employed in the same survey. **Table 8.3** lists examples of qualitative and quantitative evaluation techniques.

Except in those rare cases where relevant, detailed data have previously been collected from a proposed mine site, desktop studies can only indicate the likely range of environmental conditions at the site. Fieldwork undertaken as part of the baseline survey is necessary to establish quantitative data. There are many approaches to establishing the environmental baseline, considering that the host environment may be already impacted by human activities, and that any environment varies over time, even in the absence of a project. As a general rule, the longer and the more comprehensive the scope of baseline surveys, the more representative the outcome. Baseline studies, however, always represent snapshots in time. Consideration of relevant long-term data such as climate records and historical information about local communities, is required to provide a context for the results of fieldwork. For example, it is important to know whether the field surveys were undertaken during or following normal or extreme climatic conditions.

The longer and the more comprehensive the scope of baseline surveys, the more representative the outcome.

Planning is the first step to successful fieldwork. The planning process is to baseline surveys what a foundation is to a building. Poor planning, like a poor foundation, often produces shaky results. Planning will also save time and money but more importantly planning outlines what are required to accomplish the objectives of field surveys. Planning, whether viewed as a process or an analytical framework, is not linear but iterative. That is, the results of early steps often need to be revised after later steps are completed. A good reason for documenting the fieldwork plan is so that it can be reviewed by data users. Some jurisdictions require that work plans for field surveys and for EIA preparation be reviewed and approved by the regulating authority prior to initiating actual fieldwork.

Poor planning, like a poor foundation, often produces shaky results.

While this text may assist in the design of baseline surveys it does not offer detailed descriptions of field and laboratory methods. These issues are well documented in the literature (e.g. Clark, editor, 2003). We also do not attempt to provide a concise synopsis of the wide range of field investigations, which indeed would be too ambitious for a single text book. A well designed fieldwork plan, however, will consider the following, in addition to other site and task specific issues:

- Objectives: Why collect field data?
- Health and safety of study team: Creating a climate of care
- Logistics: Accessing hard to reach areas
- Environmental setting: What do we know?
- Data collection: Measuring what matters
- Quality control and quality assurance: What it is required to make data credible?

A discussion of these issues follows.

What are the Objectives?

Fieldwork requires prior consideration of the why, who, what, when, where, and how of data collection.

Fieldwork requires prior consideration of the why, who, what, when, where, and how of data collection. The why is particularly important; field personnel commonly commence data collection before developing a clear work plan that includes purpose and desired use of the data. While the main objective of any baseline survey is to describe the current state of the environment, data collection often serves a number of related purposes. Some follow.

Understanding the Affected Environment and its Components

Baseline surveys provide an understanding of the people and environmental components that may be affected by the project, positively or negatively. It is often desirable to carry out surveys in consultation with the affected indigenous and local people so that natural resources of particular significance to local communities can be identified. An understanding of natural resources (such as ecosystems, species, and genetic diversity) and their existing usage, including valuation of these resources, is essential to the protection of both biodiversity and cultural values, and their potential enhancement. Baseline surveys typically cover a wide spectrum of the host environment as illustrated in **Table 8.4.**

Biodiversity Baseline Studies

Not only do the baseline studies provide data on biodiversity, they provide the basis for the design of rehabilitation programmes.

Baseline studies are of paramount importance in assessing the biodiversity of a project area. Not only do the baseline studies provide data on biodiversity, they provide the basis for the design of rehabilitation programmes. However, it is neither possible nor necessary to identify every species occurring in the area. In particular, the micro-flora and micro-fauna are seldom targeted in baseline studies unless there is reason to believe that rare or endangered species are present. The most important need is to identify and characterize each habitat in the area of interest, and for each habitat to evaluate its extent, not only locally, but regionally. Of particular interest and concern would be the presence of small, uncommon habitats within the potential mining project footprint.

It is important that baseline studies are designed and implemented so as to identify the presence or absence of rare, endangered species that may be affected by project activities (see Chapter Nine). Each country maintains its own register of protected species, while the International Union for Nature Conservation (IUCN) in its Red Data Book and the Convention on International Trade in Endangered Species (CITES) in its Appendices, maintain lists of endangered, vulnerable, threatened, and near-threatened species. Compiling a list of such species from these sources is a straightforward process. The next step is to review the relevant biological literature to evaluate the distribution and habitat data for each species so that its potential presence in the area of interest can be assessed. The biological field programme is then designed so as to specifically target each species identified

TABLE 8.4

Overview of Information Typically Collected as Part of Baseline Surveys

Air

- Meteorological data (wind speed and direction, precipitation, temperature, humidity, extreme weather conditions, etc)
- Air quality (measured parameters commonly defined by relevant regulations by the host country; common parameters are particulate matter, SO_x, NO_x, lead, and ozone)
- Noise (at day and night)
- Inventory of existing air emission sources

Land

- Topography (preliminary using existing topographical data and SRTM data)
- Geology (data that is often provided by the mining company, but expanded to include information on seismicity, geohydrology, geomorphology, etc.)
- Soil types, their distribution and characteristics (also serving as initial input data in mine rehabilitation planning)
- Geo-hazards (such as landslide potential, volcanism, earthquakes, tsunamis, collapsing or swelling soils, etc)

Water

- Surface water (inventory of existing various surface water bodies, their discharge characteristics; water quality considering average, seasonal, and daily fluctuations; water usage; flood hazard identification)
- Groundwater (differentiating shallow and deep aquifer systems, recharge and discharge areas, groundwater level distribution, and groundwater quality)
- Inventory of water users (grouped according to water usage; identification of particular important types of water usages for local communities; comparison of water availability and water demand)
- Coastal and marine waters (coverage and extent very much dependent on mine location and design; e.g. potential subsea tailings disposal or other marine discharge, and/or development of port facilities)

Fauna and Flora

- Species inventories (including identification of particular species important to the affected indigenous or local community as food, medicine, fuel, fodder, construction, artifact, production, clothing, and for religious and ceremonial purposes, etc);
- Identification of endangered species, species at risk, etc (possibly referenced to the World Conservation Union (IUCN) Red Data Book, the Convention on International Trade in Endangered Species of Wild Fauna and Flora (CITES), and national lists);
- Identification of particularly significant habitats (as breeding/spawning grounds, remnant native vegetation, wildlife refuge areas including buffer zones and corridors, habitats, and routes for migratory species) and crucial breeding seasons for endangered and critical species;
- Identification of areas of particular economic significance (as hunting areas and trapping sites, fishing grounds, gathering areas, grazing lands, timber harvesting sites, and other harvesting areas);
- Identification of particularly significant physical features and other natural factors which provide for biodiversity and ecosystems (e.g. watercourses, springs, lakes);
- Mines/quarries that supply local needs; and
- Sites or routes of religious, spiritual, ceremonial, or sacred significance.

People

- Demographic factors (number and age structure of population, ethnic grouping, population distribution and movement – including seasonal movements);
- Housing and human settlements, including involuntary resettlement, expulsion of Indigenous Peoples from lands and involuntary sedentarization of mobile peoples;
- Health status of the community (particularly health problems/issues – availability of clean water – infectious and endemic diseases, nutritional deficiencies, life expectancy, use of traditional medicine, etc);
- Levels of employment, areas of employment, skills (particularly traditional skills), education levels (including levels attained through informal and formal education processes), training, capacity building requirements;
- Level of infrastructure and services (medical services, transport, waste disposal, water supply, social amenities or lack thereof for recreation, etc);
- Level and distribution of income (including traditional systems of distribution of goods and services based on reciprocity, barter and exchange);
- Asset distribution (e.g. land tenure arrangements, natural resource rights, ownership of other assets in terms of who has rights to income and other benefits);
- Traditional systems of production (food, medicine, artifacts), including gender roles in such systems;

(Continued)

TABLE 8.4
(Continued)

- Views of indigenous and local communities regarding their future and ways to bring about future aspirations;
- Traditional non-monetary systems of exchange such as hunting, barter, and other forms of trade, including labour exchange;
- Related economic and social relations, including cooperative assistance;
- Importance of gender roles and relations;
- Traditional responsibilities and concepts of equity and equality in society; and
- Traditional systems of sharing natural resources, including resources that have been hunted, collected, or harvested.

as potentially present. It helps if the biologists involved already have experience in similar environments as they will know what to expect and the most effective techniques for establishing the presence or absence of particular species. It should be realized that failure of field programmes to identify the presence of a particular species does not necessarily prove that the species is absent.

When an endangered, vulnerable, threatened, or near-threatened species is located in the field, it is necessary to further evaluate the status of the species so that the threat to the species from project activities can be assessed (**Case 13.6**). Where species in any of these categories are found living within the potential footprint of the project, it is important to obtain detailed distribution data, both locally and regionally, so that the potential reduction in population can be seen in perspective. It is also advantageous to understand specific requirements of each species, including food sources and breeding habits so that these can subsequently be incorporated in rehabilitation programmes.

Documenting Traditional Knowledge

Traditional knowledge, innovations, and practices should be considered an important and integral component of baseline studies.

Consistent with Principle 11 of the ecosystem approach endorsed by the CBD, traditional knowledge, innovations, and practices should be considered an important and integral component of baseline studies, particularly the traditional knowledge, innovations, and practices of those who have a long association with the particular area for which the mine development is proposed. Traditional knowledge, innovations, and practices can be cross-referenced by old photographs, newspaper articles, known historical events, archaeological records, anthropological reports, and other records contained in archival collections and in the academic literature.

Identifying Existing Natural Hazards and Project Constraints

Baseline surveys provide an understanding of site characteristics that have the potential to affect the mine, or its design. Examples are intense rainfall, large upstream catchment areas, and steep terrain which combine to impose significant constraints on the siting of tailings ponds; lack of suitable on-site construction materials; unstable slopes hampering the development of access roads; flood-prone areas, areas prone to earthquake damage, protected or sensitive areas, the presence of threatened or endangered species, or traditional land rights, or practices restricting access.

Some of these characteristics have the potential to be 'show stoppers' while others may be managed, but at a cost. It is therefore most important that the key environmental issues be identified at an early stage. Modifying a conceptual design requires relatively little effort and money; changes at a later stage in mine design are difficult and expensive.

Providing the Basis for Impact Evaluation

Impacts can only be assessed with the knowledge of present baseline conditions. All sources of baseline information should be fully referenced and any gaps and uncertainties acknowledged, with measures proposed to deal with them. Baseline information should be as accurate and up-to-date as practicable and consequently, some jurisdictions impose time limits on the validity of an environmental data. If the proposed mine development does not commence within a predefined timeframe, the environmental impact assessment itself may become invalid.

Defining Reference Points for Future Monitoring

Baseline data provide reference points against which change is monitored. Thus, baseline data collection should already consider future environmental monitoring needs. Involving the mining company early in the design of baseline surveys is advisable since the mining company will eventually be responsible for environmental monitoring during mine operation. To differentiate between project and natural or third party induced changes, it is important to include control sites in the baseline survey programme. A control site (also termed reference site) is a site that is similar to the project area in terms of its environmental characteristics, but which remains unaffected by mine development. Since environmental impact assessment is concerned with all natural resources, control sites for air, land, water, fauna and flora, and people are required.

A control site is a site that is similar to the project area in terms of its environmental characteristics, but which remains unaffected by mine development.

Environmental conditions at the mining area during mine operation can be compared with the control sites. Control sites provide information about the natural range of values for the parameters used in the monitoring programme and show the annual variation in these parameters. Without a control site, mining impacts on the natural environment, including human communities, may be difficult to ascertain. To maximize comparability and to allow evaluation of natural variations within the system, control sites are located, wherever possible within the same geographic region where mining takes place. It is best if control sites have similar climates, soils, plant and animal assemblages; similar human influences; and similar functions. Including several control sites in the long-term monitoring programme can further delineate the natural range of values and annual variation in the parameters that are monitored. It is often difficult to find appropriate control sites, particularly for evaluating mine-induced changes on communities.

Without a control site, mining impacts on the natural environment, including human communities, may be difficult to ascertain.

There is one more challenge in using reference communities to evaluate mine-induced change on host communities. In spite of best efforts, local average education or income levels in host communities may decline over the life of a mining project. How can this happen? Simple statistics such as averages, do not separate between people that were genuine inhabitants of the mining-affected area when mining commenced, and newcomers. Mining may attract many more in-migrants than are needed. The host government may, in fact, encourage in-migration since it will view mining as a stimulus for economic development. In statistical terms the effect of a surplus in newcomers may easily outweigh the benefits generated for the traditional inhabitants.

There is one more challenge in using reference communities to evaluate mine-induced change on host communities.

Providing Data Used For Project Design

Environmental data generated from baseline studies, are used by project design personnel in designing various project components. Common examples are listed in **Table 8.5.**

Identifying Existing Environmental Liabilities

Baseline surveys also help to identify any liabilities, if present, that may be inherited from past or current activities. Relevant activities include past mining activities, existing artisan mining, legal and illegal logging, transmigration programmes, or illegal hunting (**Case 8.3**).

Identifying Data Gaps

Baseline surveys help to identify information gaps and areas requiring further study. Gaps in knowledge may not necessarily obstruct impact evaluation, but effective long-term

TABLE 8.5
Environmental Data used in Project Planning and Design

Environmental Parameter	Use in Planning & Design
Climatic data – wind speed and direction	• Siting of air emissions facilities in relation to local communities, • Design of elevated structures, • Planning of blasting operations.
Climatic data – temperature and relative humidity	• Selection of appropriate air conditioning systems.
Landscape analysis – visibility	• Location of facilities and infrastructure, • Need for screening of facilities.
Geological data – seismicity, earthquake risk, potential loadings, etc.	• Incorporated in mine design to minimize the risk of slope failure or collapse; • Earthquake resistant design of buildings, embankments, and other structures.
Geological data – slope stability, landslide hazard	• Siting of facilities including roads, waste rock dumps and tailings storages.
Geological data – surficial deposits	• Design of leach fields from septic tank systems.
Groundwater levels, aquifer characteristics	• Design of mine dewatering systems, • Water balance as input to project water supply decisions.
Surface water flows, including seasonal and extreme variations	• Water balance as input to project water supply decisions, • Need to supplement downstream users, • Design of culverts and water storage structures.
Groundwater and surface water quality	• Suitability for process water requirements, or • Treatment required to ensure suitability for process water requirements.
Wetland habitats, presence of water birds	• Presence of nearby wetlands with large water-bird populations may be important in siting of an airstrip, • Lack of nearby wetlands may mean that a tailings storage proves attractive to bird-life, • Migratory birds may provide vectors for diseases such as malaria, requiring control programmes.
Vegetation	• Risk of wildfires, design of fire protection and fire control systems, • Invasive aquatic plants can occupy water storages, clogging intakes etc.
Demographics	• Availability of labour, • Requirement for project accommodation facilities.
Education and skill levels	• Availability of skilled, semi-skilled and unskilled labour, • Recruitment policies, • Design of training programmes.
Existing community services, law enforcement, health and educational facilities and staffing	• Need for project to supply its own services or augment existing services.
Community attitudes	• Selection among workforce accommodation options, • Siting of facilities, • Also relevant to many other components of project design, operation and closure.
Land tenure, land use, resources usage, value etc.	• Approach to land acquisition, compensation etc.

(Continued)

TABLE 8.5
(Continued)

Environmental Parameter	Use in Planning & Design
Existing road networks – capacities, traffic characteristics etc	• Need for new roads and/or upgrades; • Management of project-related traffic
Local agricultural and fisheries production	• Potential to supply project food requirements, • Policy to encourage local purchases • Input to land acquisition and compensation approach.
Local business enterprises	• Ability to supply goods and services, • Policy to encourage local purchases
Local customs – religious, cultural etc.	• Working hours, rosters, holidays, • Scheduling of blasting to avoid interference with religious observances.

management may benefit from further studies during mine development. A common example is in rehabilitation. Baseline surveys may identify candidate plant species for use in rehabilitation. However, frequently, there will be a lack of available information concerning the most effective means of germination, propagation, and cultivation. Research, including field trials, will typically commence prior to or during project construction and continue into the operations stage.

Creating a Climate of Care

Health and safety is a black and white affair. Fieldwork is either safe, or it is not. Fieldwork is a professional activity and should be conducted in a climate of care in which safety of

Fieldwork is either safe, or it is not.

CASE 8.3
Environmental Due Diligence to Establish the Legacy of Past Mining Activities

Chromite mining has been undertaken in the Co Dinh Chromite Placer area, Vietnam, in varying capacities, since the early 1900s. A variety of mining methods have been used, to a varying degree of success. In addition to mining chromite ore, there was a period of bichromate processing in the late 1990s; this was stopped however due to environmental and health issues. There is no available information on this process, other than the fact that both hazardous liquid and solid waste were generated as a result of the process and dumped on-site. Faced with the environmental degradation and uncontrolled export of chromite ore to China, in 2005 the Thanh Hoa People' Committee withdrew all mining licences in this area. However, illegal mining continues to occur.

At present, various international mining companies have expressed interest in developing the chromite deposit. Clearly any investor is well advised to establish a comprehensive environmental baseline to document the current state of the environment. The baseline is probably the best safeguard against any potential future allegation of environmental pollution due to new mining activity.

Photo Credit: Toby Whincup

Personal safety must, ultimately, be a personal responsibility.

personnel has priority. There is no successful method of guaranteeing safety of personnel at all times and circumstances, but safe practices reduce risks and prepare for emergencies if they occur. In professional planning it is customary and normal that all possible precautions have been taken and all proper responsibilities met. The major requirements – assessment of risk, planning, training, equipment, communication, responsibilities – are addressed in the following paragraphs. Eventually, however, personal safety must, ultimately, be a personal responsibility. Complacency and inattention, lack of preparation or lack of training may all cause problems. No amount of regulation or planning can replace personal vigilance.

Planning and Assessment of Risk

Planning for fieldwork starts with assessing the potential risks. This is essentially a four-step process: (1) identifying risks, i.e. those things that may impede the successful outcome of the fieldwork; (2) assessing their likelihood and potential consequences; (3) implementing controls to minimize risks; and (4) monitoring and reviewing the effectiveness of the risk control measures, and implementing improvements as needed. The leader of the field team is responsible for ensuring that there is adequate advance planning and assessment of risks. The same level of risk assessment should be used whether fieldwork involves a large number of team members or a party of two or three travelling to the site. Checklists of possible risks and how to identify them are helpful. Examples of typical risks include extreme weather conditions, encounters with wild animals, exposure to tropical diseases, disorientation in remote areas, hiking in difficult terrain, driving vehicles, and the use of boats. Issues to be considered in assessment and reduction of risk during fieldwork are summarized in **Table 8.6.**

TABLE 8.6
Some Important Issues in Assessing and Reducing Risks Associated with Fieldwork

Chain of command	Determine explicitly in advance who is in charge of the group so that there is reduced confusion in the event of accidents or other untoward circumstances
Number of people involved	Preferably at least two people with experience of the type of fieldwork being carried out
Fitness, health, and competency	Knowing medical conditions, swimming ability, etc. assists in deciding whether risks are acceptable or not, and planning alternatives. For example, exclude non-swimmers from offshore work
Nature of fieldwork	Nature of the work and the area where it is done, including its remoteness, terrain, likely weather conditions including possible weather extremes, possibility of encountering dangerous animals or plants (or people); work in or near water; working at height or below ground
Transportation	Methods and availability of transport and assistance in case of breakdown or accident
Communication	Availability of reliable channels of communication
First aid kits	Inclusion in the party of people who have training in first aid and provision of adequate first aid kits
Time and length of travel	Adequacy of water and other provisions such as food, fuel, and shelter
Site map and location equipment	Suitable maps should be available for routes to and from the fieldwork site, and of the fieldwork area, including information about relevant support services etc.; GPS equipment, compasses; consider satellite imagery analysis prior to fieldwork in remote areas
Personal preparation	Insect repellant, sun block, long sleeved shirts, proper walking shoes, rain coat, cap, mosquito net, knife, personal amenities such as soap, towel, toothbrush, and toilet paper

Coastal and offshore work commands special attention. Information about tides, currents, weather, and other factors affecting safety should be considered. Work on rock-platforms can be particularly hazardous and adequate precautions must be taken to prevent fieldworkers being swept from rocks or injured by unexpected waves. Training, experience of team leaders, and adequate personnel to ensure continuous vigilance are required. Ensure that appropriate clothing, including footwear is worn by all personnel (this is particularly important if someone has to go to the aid of someone else who is in difficulty).

Coastal and offshore work commands special attention.

Precautions required for terrestrial fieldwork vary according to the type of environment and likely weather conditions, including possible weather extremes which may be encountered. Rainforest, desert, or mountain environments present different hazards, and the health & safety plan should always consider project specific potential natural hazards.

Safety and First Aid

Field teams must have adequate first aid training and supplies, as appropriate to the type of work and the hazards that may be encountered, and the size of the field trip party. Precautions should be taken to minimize the potential for accidents of any kind; such preparations are essential for an adequate response to accidents. Where appropriate, portable survival kits should be provided. Standard vehicle first aid kits are almost certainly inadequate for fieldwork. Survival in difficult circumstances in remote areas, at sea, or in deserts requires specialist knowledge, skills, preparation, and training. For extended fieldwork in remote areas the inclusion of a paramedic in the field party is worth considering (**Case 8.4**).

Standard vehicle first aid kits are almost certainly inadequate for fieldwork.

Communication

Reliable means of communication are essential for all fieldwork. What is needed will obviously vary according to the circumstances and, in some cases (e.g. boating), statutory requirements apply. It is essential that, if something does go wrong, assistance can be summoned and emergency services notified. Mobile telephones are one convenient form of communication, but they are not suitable for all circumstances or areas. Satellite telephones are available for use in areas without mobile telephone coverage. For remote operations, it is important to establish a schedule for contacting the home base or other appropriate base. Such contacts not directly involved in the work, must have the means to initiate a rescue if regular contact breaks down and there is evidence that something has gone wrong. Global positioning systems (GPS) should be available, preferably with back-up units in case of loss or damage to the primary unit. GPS use is essential in all travel by boat or in remote areas.

Use of Vehicles

For most of us driving is part of daily life. So why bother about special safety consideration when working in the field? There is a simple answer to this. Driving is a dangerous activity

Driving is a dangerous activity.

CASE 8.4
Health & Safety Planning – Preparing for the Unexpected

Some plant species release a poisonous substance when in contact to exposed skin. These plants are found in low land forests in Kalimantan, and other parts of South East Asia. Skin contact causes a painful rash, and in rare cases can trigger allergies leading to death. Such an allergic reaction did happen to a party member during the biological baseline study of the Juloi coal deposit in remote Central Kalimantan. Fieldwork was professionally planned, with satellite communication in place, and emergency helicopters within reach. The party also included a paramedic who provided the initial medical support. Within one day the affected team member was hospitalized, and she fully recovered within a few days.

in urban cities, even more so in areas where road conditions may be hazardous and traffic unpredictable. Accordingly only appropriately trained and experienced drivers should be in charge of field vehicles. Not surprisingly access to most mine sites requires specific training and induction and a special driving licence. Modern mines typically allow access to vehicles fitted with specific equipment to maximize visibility, and safety equipment such as air bags.

Diesel vehicles are preferred for reasons of greater range and fuel safety, particularly for off-road use. Vehicles used for field trips should be well-maintained and equipped with adequate spare parts and tools, according to the area and length of trip. Care is required when loading vehicles to maintain a low centre of gravity and to adequately secure items adequately in the cabin or on roof racks. Luggage should always be securely stowed. Netting, mesh or solid barriers between the rear luggage compartment and cabin protect occupants from loose objects, which may be propelled through the cabin if the vehicle stops suddenly. Vehicles should be driven with caution and attention to the prevailing road and weather conditions.

Only vehicles designed and/or equipped for the purpose should travel on unsealed roads. The vehicle should be selected for the type of terrain likely to be encountered. Drivers should be familiar with the vehicle before embarking on the trip. Drivers intending to use four wheel drive (4WD) vehicles should have received training in 4WD or be able to demonstrate experience in driving such vehicles. Drivers should be familiar with routine maintenance procedures such as checking oil, water, tyre pressure, coolant, and battery, and changing tyres. Drivers should also be aware of the fuel capacity and range of the vehicle. Prior to setting out, the driver should check the vehicle to ensure it has been adequately maintained, and is equipped with all necessary tools, spare parts, and special equipment for the trip. A check should be made that the luggage is secure. Rest stops and fuel stops should be used to check that the vehicle is operating normally with respect to tyre pressure, engine leaks, etc, and that the luggage remains secured. Every day, before setting out, check the oil, water, fuel, battery fluid, coolant, brake fluid, and tyre pressures, and ensure that the controls are working.

Driving times and distances should be planned to prevent fatigue. Usually a driver should not drive for more than about two hours before changing over or taking a short break that incorporates some light physical activity such as walking. Driving at night is more hazardous than during daytime and should be minimized. Drivers should always heed applicable road rules, including those pertaining to consumption of alcohol. Driving should always be done at safe and legal speeds. Safe speeds depend upon the road and weather conditions, experience of the driver, time of day, alertness of the driver and the vehicle itself. Unfamiliarity with the road or conditions and the possible presence of animals contribute to driving hazards. Occupants should wear seat belts when travelling in vehicles.

Particular care is required when driving through human settlements. In many developing countries, roads through towns and villages are commonly rudimentary; blind turns and crests, and general unfamiliarity with traffic may add to the hazards of driving. Slow speeds and great care are required to safely negotiate such areas.

Use of Boats

Boating field trips must comply with the requirements of maritime legislation. Personnel in charge of boats are responsible for ensuring they have the appropriate licences and any appropriate boat registrations are obtained. Boats should be well-maintained and equipped with adequate spare parts and tools, according to the area worked and the length of the trip. Care must be taken when loading boats. The maximum capacity that the boat can carry must be displayed on the boat and must not be exceeded. Boats must contain adequate safety devices such as distress flares and personal flotation devices. Only boats designed and equipped for the purpose may be taken out to the open sea. A radio transceiver should be provided for vessels going more than two nautical miles offshore. Only licensed and appropriately trained personnel should be in charge of boats. Boats must be driven with caution and attention to prevailing conditions. Navigation skills may also be required. Only those personnel necessary and trained for the fieldwork may be carried in boats. The minimum size of a boating fieldwork party is two and at least one must be a competent swimmer.

Before setting out on boating trips, check prevailing and predicted weather conditions. Boat trips should not be undertaken in poor weather (e.g. high winds or rough seas) or when poor weather is predicted over the period of the planned trip. Even when good weather is predicted, changing weather should be anticipated in planning the trip. Prior to setting out, the vessel should be checked for safety equipment, personal flotation devices, fully charged battery, correct fuel mix, spare plugs, cotter pins, anchor, and a small bucket for bailing.

Partly due to safety considerations, off-shore surveys tend to be expensive (**Case 8.5**). Delays to fieldwork due to bad weather will cost money. Standby costs of vessels are substantial and can accumulate to large amounts. Nevertheless, safety comes first. Proper planning and costing should allow for delays due to adverse weather conditions. In fact, proper planning will take into account anticipated climate patterns in the survey area.

Off-shore surveys tend to be expensive.

Security

Elementary security rules apply to safe travel and fieldwork, the most important ones being not to appear wealthy or to flaunt valuables, not to travel at night unless absolutely necessary, and not to work alone. A minimum of two persons should jointly perform all fieldwork, plus, if appropriate, a local guide who speaks the local language, and is familiar with the project site. It is also recommended, and in most jurisdictions mandatory, to coordinate fieldwork well in advance with local public authorities including police and informal public leaders such as the head of the village (**Case 8.6**). Discussing fieldwork with local public authorities prior to commencing work is common sense and represents simple courtesy. Debriefing at the conclusion of each field campaign is similarly a matter of courtesy. Other basic safeguards apply to avoid theft or harm, such as locking rooms and vehicles to secure their contents. A state of awareness is a valuable protection.

Elementary security rules apply to safe travel and fieldwork.

Discussing fieldwork with local public authorities prior to commencing work is common sense and represents simple courtesy.

There is almost no paper – except cash – that cannot be replaced, eventually, if a copy or a record of it exists. Before commencing fieldwork, important papers (such travel and work permit, passport photo page, vaccination certificate, air ticket, and driver's licence) should be copied, with one copy left at the home base and at least two more copies carried in various parts of the luggage. A list should be made and carried, recording important information such as insurance policies, bank accounts, social security or national identity numbers, credit card numbers, and camera serial number.

CASE 8.5
SCUBA Diving and Other Dangerous Fieldwork

Some service companies flatly exclude SCUBA diving from their range of services. It is too risky. Unsurprisingly SCUBA diving should only be authorized when it is done in accordance with established safe practices. Only competent, certificated divers who are medically fit to dive may participate in diving operations.

Physical attacks causing bodily harm during fieldwork are rare, but not unknown. A team member unlucky enough to have a bandit or thief threaten him or her with a knife or gun and demand money, is well advised to hand over money and valuables without demurring. Thieves will usually leave immediately. Most thieves are not interested in harming people, or even in confronting people. Considering the small probability of being attacked, arming team members cannot be recommended, especially since most of us are unlikely to be able to use any weapon to much advantage.

The best protection is not to work alone; safety is in numbers.

There is sometimes a problem of sexual harassment of women in a field team. The best protection is not to work alone; safety is in numbers. Female fieldworkers should obtain information about local customs before they arrive, and choose clothing that is not provocative and is in line with dress codes of local women.

What Do We Know?

When planning fieldwork the region in which the project site is located, should be considered as well as the site itself. Collection and compilation of relevant secondary data prior to fieldwork, as part of the scoping process, should have generated background information about the host region, to an extent adequate for the objectives of planning the survey. Information on geology and soils, topography, seismicity, land use, demographics, local customs, norms and values, administrative boundaries, watershed boundaries and drainage patterns, water quality and biota, or rainfall records are usually available, although extensive background research may be required. After reviewing available background information, survey objectives and field programmes should be reassessed to take advantage of the available information.

Google Earth often provides amazingly clear imagery of the site to be visited.

Google Earth may be highly beneficial. Depending on the location Google Earth often provides amazingly clear imagery of the site to be visited. Overlaying digital topographical data, a virtual flight around is an enlightening experience. Settlement areas are readily identified, and largely undisturbed wilderness areas are also usually recognizable as such.

CASE 8.6
Environmental Scoping and Security Risks

From 1999 to 2000 violent religious unrest raged through the northern part of Indonesia, causing the deaths of several thousand people. Extremists travelled from village to village, provoking civil war. In 2001 your authors were engaged in environmental fieldwork at the Weda Bay Nickel mine project area in Halmahera Island at the northern tip of Indonesia. Travel was still restricted by the military although the unrest had ceased during the previous year. Security arrangements were central to fieldwork planning, and military escorts accompanied all field teams.

Measuring What Matters

It is not possible or desirable to provide a comprehensive and definitive list of environmental baseline information needs applicable to each mine, everywhere in the world. Environmental information needs, by definition, are always site and project specific. However, **Table 8.4** lists information commonly required. In practice, baseline data collection will be designed to commence with basic information, adding increasing complexity as appropriate. For example, the faunal element of terrestrial biodiversity baseline information is further split into invertebrates, birds, and mammals. For mammals, the required information may then encompass home range size, population density and dynamics, social organization, seasonal patterns of use or activity, mobility, resource dependence and habitat specificity, and inter-dependencies. A well defined survey plan will always list data which are proposed to be collected and the justification for the data, given the significant costs involved in data collection and laboratory analyses (**Case 8.7**).

Environmental information needs, by definition, are always site and project specific.

Ideally, field surveys should be designed to yield information about ecological cycles, ecosystems or species functioning, as well as recording the habitats and species that are present. In relation to the biological environment this could include, for example, watershed dynamics, extent of habitat intactness, seasonality, migration and breeding patterns, and predator-prey relationships. In practice however, defining the habitats and the main floral and faunal communities in each, usually represents the major objective of field surveys, the more complex information being derived from the scientific literature. There will be situations where little relevant literature exists; in these cases much more intensive research may be warranted.

In the social context, field surveys should be designed to yield information about the origin and composition of the local population, particular Indigenous Peoples, local norms and values, economic activities at local and regional scales, and community perceptions towards the project. Such elements will be important in developing an understanding of how local communities and the ecosystem and its component species will react to changes caused by mine development.

Coverage versus Intensity

The sampling programme depends on factors such as available resources, the geographic and temporal scope of the assessment, and confidence levels. Often the main difficulty associated with baseline surveys is limited available time. This means that issues associated with seasonal and year-to-year variation may not be adequately addressed. Selecting the number of sample locations is also difficult. In general the greater the number of sampling locations the greater is the coverage of the project area. However, selection of sites to be representative may be a more important consideration. It follows that the more diverse the site, the more sampling sites will be required to ensure representative coverage. On

The more diverse the site, the more sampling sites will be required to ensure representative coverage.

CASE 8.7
The Case for Air Quality Monitoring

Most jurisdictions specify standard physical/chemical air quality parameters to be analyzed as part of the EIA baseline data collection. As with all standards and 'guidelines', they are indiscriminate in relation to actual site characteristics. Regulatory requirements may specify sampling for a set of air quality parameters appropriate to an urban or industrial location but irrelevant for a mine site located in the deep forests of Papua New Guinea. Permission to deviate from regulatory requirements may be sought in advance from the relevant regulatory authority but in the experience of your authors, such permission is rarely granted. Accordingly, considerable meaningless or irrelevant data are obtained.

the other hand, given limited resources, selecting fewer sites allows for a more in-depth survey at each site. The number of sampling locations depends on what is measured. A baseline inventory for fauna and flora, for example, requires a relatively broad assessment of the biodiversity at several sites with variable habitats. In contrast, a species-specific assessment would concentrate on habitats used by the target species and may forego several sampling sites to provide greater depth of study in fewer sites. In selecting a greater or a smaller number of sites the choice is not either/or, but to reach the best compromise between coverage and intensity. Finally, and often most importantly, the number of sample locations depends on the nature and size of the project. Ideally, data obtained from a well designed set of sampling locations serve as output indicators (characterizing the source of impacts), as impact and pressure indicators (providing information on long-term mine-induced environmental change), and as a state indicator (or control location representing environmental change without the mine), as well as providing design data sought by project planners. Sets of sampling locations need to be developed for each environmental component. As illustrated in **Figure 8.5** for water, while the number and function of sampling locations may vary throughout the life of mine, they are interlinked by the indicator framework used to design the baseline and long-term monitoring programme. **Figure 8.5** also illustrates that the baseline surveys are an integral part of designing a meaningful long-term environmental management and monitoring programme for the mine. Environmental management and monitoring is further discussed in Chapter Ten.

The number of sample locations depends on the nature and size of the project.

It follows from this discussion that the sampling programme should be formulated carefully and judiciously by experienced professionals, and that all parties including the field team, the project proponent and regulators should be aware of the trade-offs involved and the limitations of the data.

Distinct versus Representative Sample Locations

The selection of sample locations depends on geography, on what natural resource is sampled and with what purpose, and whether sampling locations are chosen by virtue of being characteristic or distinct. Choosing between distinct versus representative sampling sites depends on the purpose of data collection.

Representative or characteristic sites are representative of the typical habitats present within a given area. They are selected to provide a general description of the natural resources at the mine site. Biological sampling locations are often chosen to be representative of each specific habitat. Characteristic sites are often selected as control sites. Control sites are sampling sites that will stay unaffected by mine development, and hence serve as reference points in long-term monitoring. A comprehensive monitoring programme will include control sites for all aspects of the host environment, including the human environment.

A comprehensive monitoring programme will include control sites for all aspects of the host environment, including the human environment

Distinct sites are representatives of a distinct habitat of a specific area. Usually habitat is not continuous, and localized gradations in habitat create a mosaic of related but distinct communities that grade into one another. Distinct sites are established to provide accurate baseline information at specific sites, independent of whether or not the sampling site is representative of a larger project area. Distinct sites are often monitoring sites that are intended to become part of long-term monitoring during mine operation, such as downstream water monitoring sites.

Available Expertise

Data collection is best carried out by experts who, through practice and education in a particular field, are accorded authority and status by the public or their peers. An expert in terrestrial botany is someone who, for example, is familiar with current sampling and

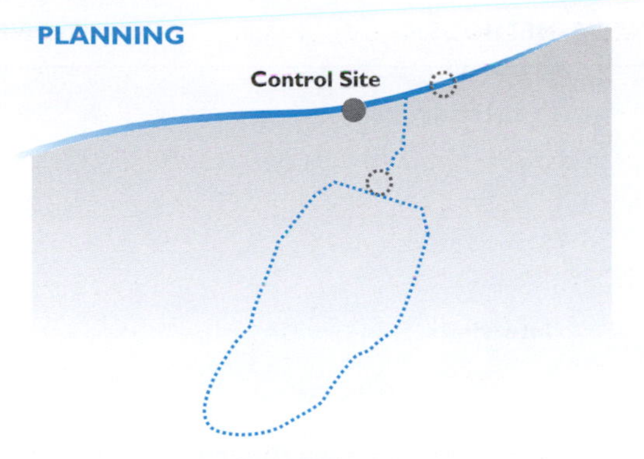

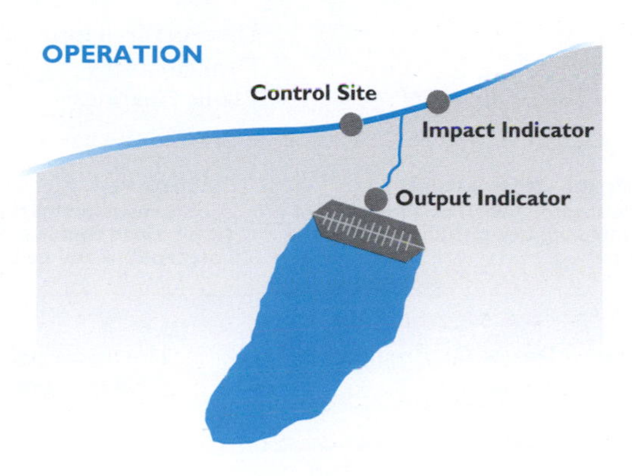

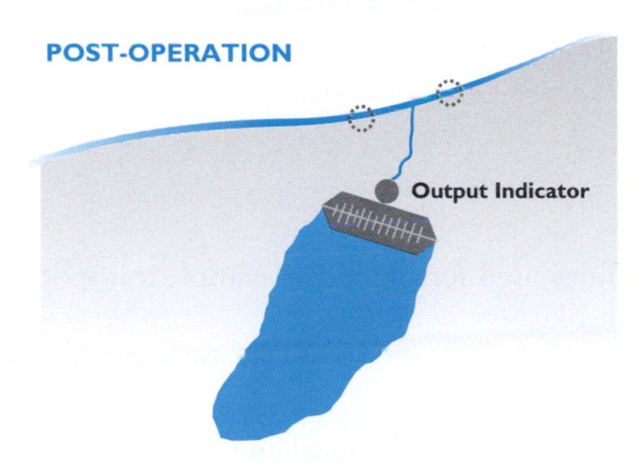

FIGURE 8.5

Example of Water Monitoring over of the Life of Mine

Sets of sampling locations need to be developed for each environmental component. While the number and function of sampling locations may vary throughout the life of mine, they are interlinked by the indicator framework used to design the baseline and long-term environmental monitoring programme.

collection methods, can analyze data, and is familiar with the taxonomic plant groups within a larger biological and ecological context. He or she will be able, in the field, to identify species at least to family level, and, subsequently, to genus and species level, with the aid of herbarium facilities and in consultation with appropriate specialists.

Using the skills of experts expedites both fieldwork and the interpretation of results. Consideration should also be given to the availability of experts on a local, regional, and international level. Local expertise is highly beneficial whenever it is available. Often local experts will have a good understanding of local geography, ecology, and community issues. However, if there is no local expert, there is no choice but to use outside expertise. In highly specialized fields there may only be a small number of people, or even just one person, who can be considered an expert in a narrow specialty. Consideration of available expertise should be complemented with consideration of available logistical support, as this may present a limitation to the capacity and scope of any baseline survey. Logistical

Local expertise is highly beneficial whenever it is available.

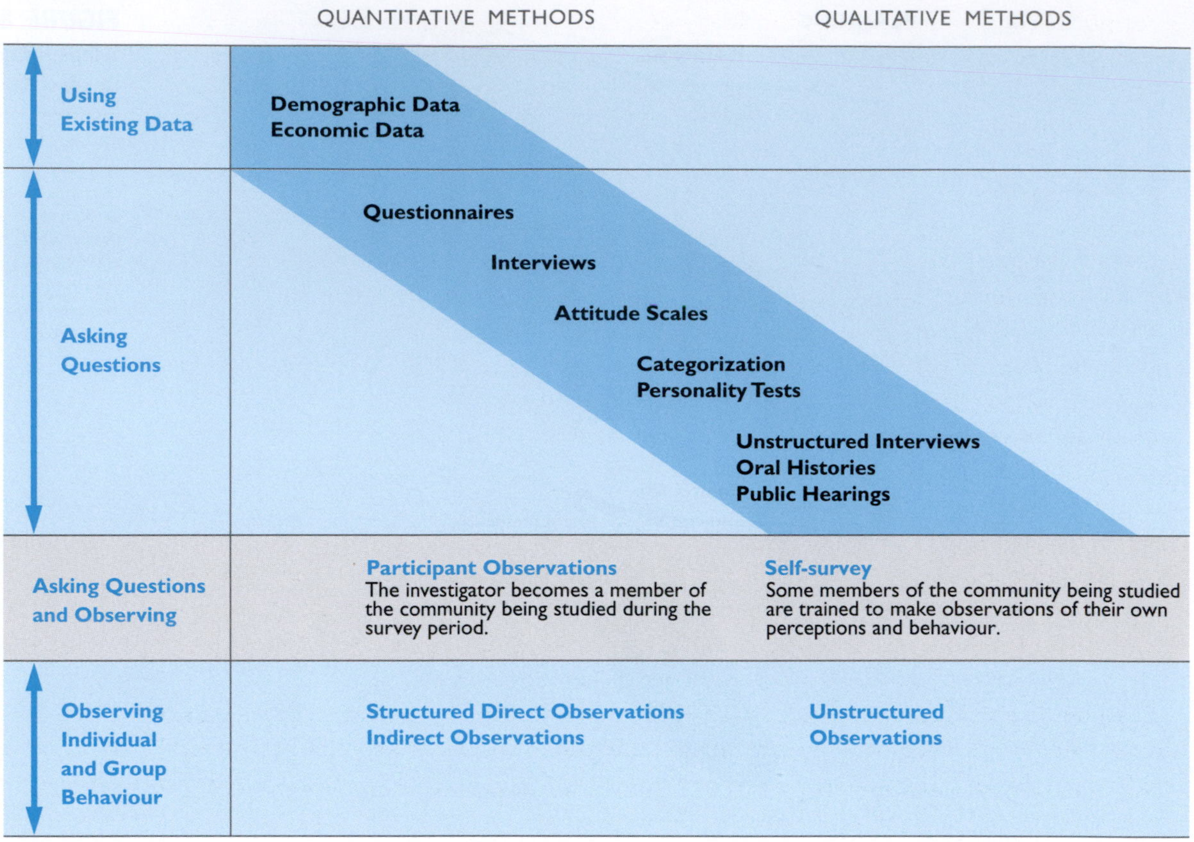

QUANTITATIVE METHODS

QUALITATIVE METHODS

Using Existing Data

Demographic Data
Economic Data

Asking Questions

Questionnaires
Interviews
Attitude Scales
Categorization
Personality Tests
Unstructured Interviews
Oral Histories
Public Hearings

Asking Questions and Observing

Participant Observations
The investigator becomes a member of the community being studied during the survey period.

Self-survey
Some members of the community being studied are trained to make observations of their own perceptions and behaviour.

Observing Individual and Group Behaviour

Structured Direct Observations
Indirect Observations

Unstructured Observations

FIGURE 8.6

Socio-economic Data Collection Methods

Oral histories – events are described to interviewers by persons who directly experienced them. The technique is of great value when no documentary materials exist.

Source:
After Whyte 1997

support refers to the facilities used for site access, sample transportation, analysis, storage of samples or data, and the like.

Sample Collection Methods

Sample collection methods for physical/chemical and biological parameters are well documented (e.g. Clark, editor, 2003). Data collection in social sciences, however, differs, but generally involves three basic techniques: (1) using existing data, (2) asking questions, and (3) observing people (**Figure 8.6**).

- Using existing data – Examples of existing data include statistics on age, sex, and income distributions; ethnic origin; mortality; housing type and occupancy; and education.
- Asking questions – Information about attitudes, feelings, and beliefs cannot be easily obtained except by asking questions. The many techniques that have been developed range from highly structured, randomized pre-coded questionnaires to informal interviews. The less structured the interview, the more likely it is that the interviewer can probe more deeply if he receives an unexpected reply or comment. On the other hand, the responses will appear less systematic and may be harder to interpret.
- Observations – Observations can be direct or indirect. Human behaviour can be observed directly by watching people in public places, in unusual situations (say accident or festival), or by watching the response of a community to a public announcement (say as part

of public consultation during the EIA process for a mining project). Human behaviour can also be observed indirectly, for example, through measurements of the width and wear of footpaths, studying the extent of littering, obtaining data on purchases, or performing traffic surveys.

Baseline surveys as well as subsequent monitoring programmes should be carried out according to a systematic schedule. Ideally, the operations stage monitoring programme should be designed, at least conceptually, prior to conducting any baseline studies so that pre- and post-mining sampling and analysis sites and methods will be strictly comparable.

Baseline surveys as well as subsequent monitoring programmes should be carried out according to a systematic schedule.

The field survey plan should include a start date, the time of the year during which field studies will take place, the frequency of field studies, and the end date for the baseline data collection. Timing, frequency, and duration are dependent on the type of ecological systems, their complexity, and existing uncertainties. In addition, controversy over the project can force a higher degree of scrutiny and may increase the level of monitoring effort. Seasonality must be taken into consideration. Seasonality usually affects surface water discharge and quality, groundwater levels, air quality, and fauna and flora assemblages. Well timed sampling minimizes the number of sampling efforts and thereby reduces the cost of the programme. Because weather varies from year to year, it is wise to 'bracket' the season. For example, tropical countries with distinct dry and wet seasons are generally surveyed over two field surveys, one for the dry season and one for the wet season.

Seasonality must be taken into consideration.

Baseline data collection can be performed in two ways: (1) by concentrating all tasks during a single site visit, or (2) by carrying out one task or a similar set of tasks at several sites in a single day. The latter strategy is often preferable, because it minimizes seasonal effects and variability in conditions from day to day, and repeating the same task on the same day may be more efficient. However, it is not always practical if sampling sites are distant or difficult to access. Sampling of specific parameters in control sites should take place during the same time of year as sampling in the mine area.

What Does It Take To Make Data Credible?

Ensuring credibility of data depends primarily on good documentation. Pre-prepared field sheets are used to record sampling details and results of field tests and observations. Standard forms are available from numerous manuals (USGS 1998; EQB 1994). Field sheets typically include simple instructions and examples for calculation and provide ample space for the following:

Ensuring credibility of data depends primarily on good documentation.

- Site designation and exact location (including sample depth).
- Time and date of sampling.
- Data collector's name and phone number.
- Weather conditions (recent as well as current).
- Name and model number of equipment or test kits used.
- Actual readings, including duplicate readings (not only the average or final reading). For example, for dissolved oxygen titrations, the actual volume titrated should be recorded as well as the final concentration of dissolved oxygen (**Case 8.8**).
- Site conditions, in particular unusual conditions at the site, and any factor that may affect results.
- Space for comments to record unusual observations (construction, land clearing, dead animals, etc.) as well as any problems encountered during sampling. Field sheets should also include observations on habitats, and field procedures, including calibration and documentation procedures.

Once sampling is completed, field sheets are the sole monitoring records, and it is good advice to always maintain a copy of field data collection sheets in case originals become lost.

Baseline data will eventually be used for decision-making, evaluating impacts and/or to fulfil regulatory requirements. Thus, data will need to meet quality objectives set by those who will ultimately use the data. Developing a Quality Assurance and Control Plan (QA/QC Plan) is good practice, and is required by some jurisdictions. A QA/QC Plan is a written document that outlines the procedures adopted to ensure that the samples collected and analyzed, the analyses themselves, the data that is stored and managed, and the reports reported are of sufficient quality to meet the required data needs. Developing a QA/QC Plan requires developing Standard Operating Procedures (SOPs) for all field sampling and field/laboratory analytical methods. SOPs are step-by-step directions, including calibration and maintenance procedures for field and laboratory analytical instrumentation. A number of manuals provide detailed method descriptions, SOPs and even field sheets (e.g. USGS 1998 and EQB 1994). Most laboratories maintain SOPs for all their analytical protocols.

The Quality Assurance and Control Plan needs to consider the full spectrum of activities related to sampling and analysis, from the maintenance of sampling equipment at the office to the critical review of laboratory data.

The Quality Assurance and Control Plan needs to consider the full spectrum of activities related to sampling and analysis, from the maintenance of sampling equipment at the office to the critical review of laboratory data (**Figure 8.7**).

Some parameters are measured *in situ* (such as climatic parameters, sound levels, pH and conductivity of water, and bird sightings) but many samples are sent to a laboratory for analyses. Using field meters for some *in situ* measurements is acceptable, in fact necessary for some parameters such as temperature, dissolved oxygen (DO), and pH, which can change during the period between sampling and laboratory testing. Instruments can also change, so that it is important that instruments or meters used for *in situ* measurements are calibrated each time they are used and that all calibration results are recorded. Parameters for which meters are frequently used include: DO, temperature, pH, conductivity, and turbidity.

Field kits are also available for a wide variety of other analyses, including metals. However, laboratory analyses are generally used for cations and anions as much of the work is automated, detection limits are generally much lower, and quality control is easier to achieve in a laboratory than in the field. Most field practitioners would agree that there are enough problems involved in field sampling, that they would not wish to cope with the multitude of additional issues involved in chemical analyses.

Reproducibility of results between laboratories is notoriously difficult to achieve.

Notwithstanding the above, laboratory analyses probably cause more confusion, uncertainty, and disputation than any other part of the baseline data collection process. Reproducibility of results between laboratories is notoriously difficult to achieve for a wide variety of reasons including differences in the type and age of equipment, operating and quality control procedures. In response to this issue which may sometimes have serious legal implications, some countries such as Australia have introduced laboratory accreditation schemes. Australia's National Association of Testing Authorities (NATA – see www.nata.asn.au) provides this service which involves inspection of laboratories, conducting examinations for laboratory staff, and sending reference samples for analysis by member laboratories.

CASE 8.8
Reporting Results of Numerical Calculations

$c = 52.35273489$ ppr

When reporting results of calculations, excess decimal places should be avoided. The following rule of thumb may be used: scan all the values used for the calculation, and identify the measured value with the fewest decimal places. The final answer should have that same number of decimal places.

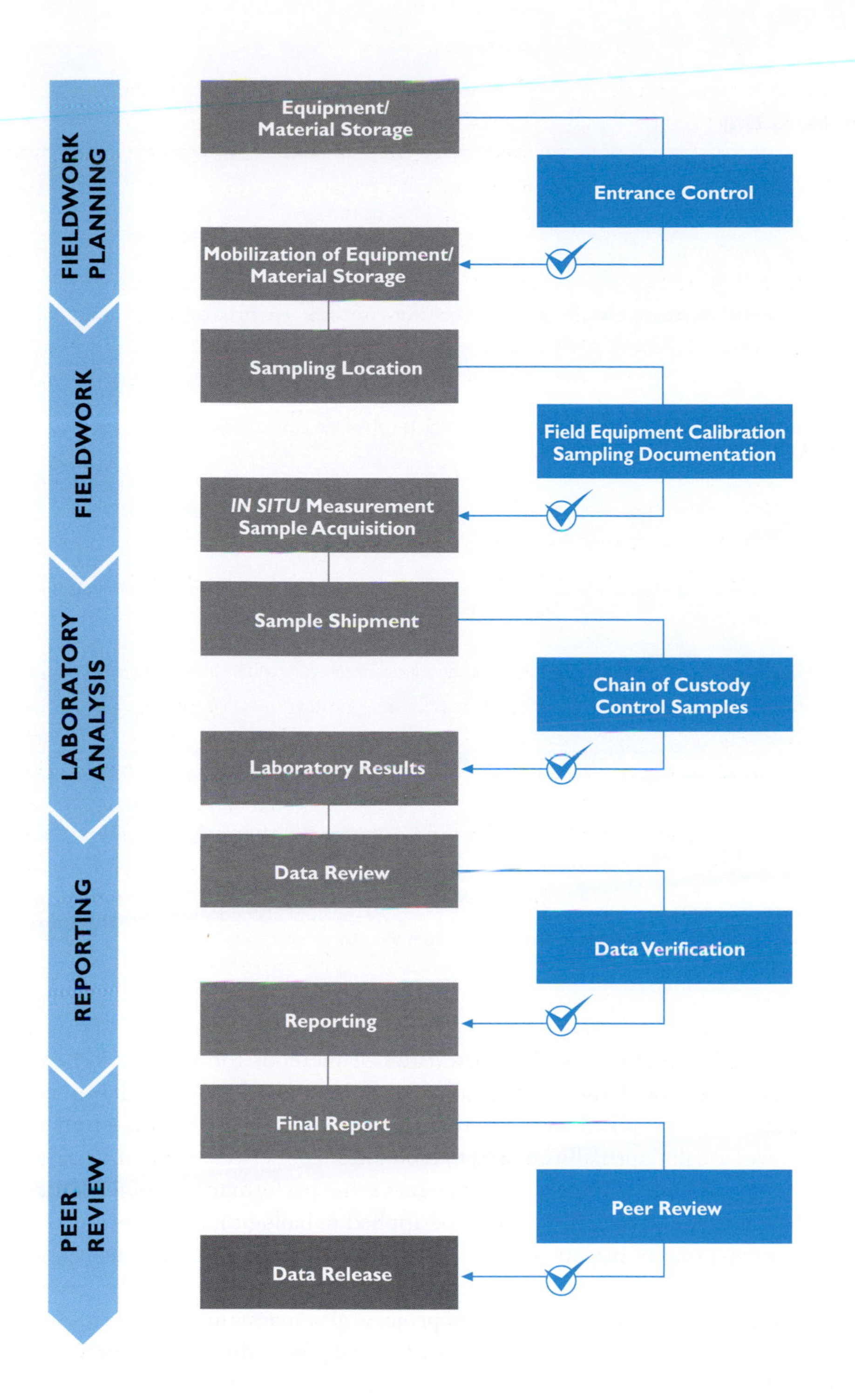

FIGURE 8.7
Elements of a QA/QC Plan for Fieldwork

Laboratory analyses probably cause more confusion, uncertainty, and disputation than any other part of the baseline data collection process.

Table 8.7 provides guidance for selecting and using a contract laboratory. When sending samples to a laboratory, holding times and sample preservation methods become important. Holding times, acceptable bottle types, and sample preservation methods should be included in SOPs/fieldwork plans; information that can be easily requested from the laboratory based on the parameter to be analyzed and the method used. Many laboratories provide a service whereby they supply pre-prepared sample containers suitably labelled for the analyses to be conducted. The chain of custody should be clearly documented for all external sample analyses.

TABLE 8.7

Laboratory Considerations (Minnesota Pollution Control Agency 2003)

Bottles and preservatives Ask if the laboratory will provide appropriate bottle types with preservatives.

Cost Consider cost not only of the sample analyses, but also of shipping. Inform the laboratory of the number of samples anticipated. Better unit prices are usually available for large orders.

Certification In countries where laboratories can be certified, request evidence of certification for the particular tests to be conducted.

Chain of custody Request a description of their chain of custody procedures and copies of their chain of custody forms.

Consistency If you have a long-term project, consider a laboratory you expect will survive for the length of the project. Staying with the same laboratory for the duration of the project will help minimize variability between laboratories/analysts.

Delivery and shipping Inquire about the delivery service and whether or not this cost is included in the cost per sample analysis. Some labs allow sample drop-off to satellite locations, which avoids shipping costs.

Detection limits Make sure that the laboratory can achieve the detection limits you need for your project. Ask if they have the necessary equipment to achieve these limits.

Hours of operation Make sure the laboratory will be able to receive samples at the times you anticipate collecting and be able to complete analyses within specified holding times.

Methods Specify the methods you want used. Ask if the laboratory has experience with these methods and if they have Standard Operating Procedures already prepared for these methods. If so, you may want to ask for copies.

QA/QC Ask for a copy of the laboratory's Quality Assurance/Quality Control Manual. Check to make sure that the laboratory's data quality objectives are consistent with your project objectives and needs.

Reporting Suggest the format for the reported results (i.e. paper report, electronic). Ask them to include results of laboratory QA/QC efforts for precision and accuracy and to note if data quality objectives were met in the reports. Direct data transfer from the laboratory reports into the mining project database is preferred as it avoids the risk of transcription errors.

Data quality and performance characteristics of methods for analytical chemistry are typically validated through the use of quality control samples including blanks, calibration standards, and samples spiked with a known quantity of the analyte of interest. Such QA/QC procedures are not normally applied to biological laboratory determinations which are therefore more subjective. **Table 8.8** summarizes some performance characteristics used in analytical chemistry and how these might be applied to biological methods.

Most mining projects maintain their own metallurgical laboratory, used for routine assays and process research. This laboratory is usually situated within the general vicinity of the mill and processing plant. Some projects also maintain laboratories for analysis of samples from exploration programmes, which may be combined with the metallurgical laboratory. Some mining operations operate an environmental laboratory to carry out water quality testing, avoiding the need to send samples to an offsite laboratory (**Case 8.9**). It is essential that such an environmental laboratory be located well away from the metallurgical and/or exploration laboratories, and also well away from the mill/process plant area. This is because environmental laboratories usually need to measure metal concentrations to very low levels (parts per billion); metallurgical laboratories deal with metal concentrations in percentage levels. Cross-contamination would be extremely difficult to avoid if the laboratories were situated in close proximity. This applies not only to project laboratories, but to contract laboratories that carry out both environmental and exploration/metallurgical analyses.

Environmental laboratories usually need to measure metal concentrations to very low levels (parts per billion); metallurgical laboratories deal with metal concentrations in percentage levels.

TABLE 8.8

Performance Characteristics Used in Analytical Chemistry and How These Might be Translated to Biological Methods

Performance Characteristic	Analytical Chemical Methods	Biological Methods
Precision	Replicate samples	Multiple taxonomists identifying 1 sample; split sample for sorting, identification, enumeration; replication samples within sites; duplicate reaches
Bias	Matrix spiked samples; standard reference materials; performance evaluation samples	Taxonomic reference samples: 'spiked' organism samples
Performance range	Standard reference materials at various concentrations; evaluation of spiked samples by using different matrices	Efficiency of filed sorting procedures under different sample conditions (mud detritus, sand, low light)
Interferences	Occurrence of chemical reactions involved in procedure; spiked samples; procedural blanks; contamination	Excessive detrital material or mud in sample; identification of young life stages; taxonomic uncertainty
Sensitivity	Standards; instrument calibration	Organism-spiked samples; standards level of identification
Accuracy	Performance standards, procedural blanks	Confirmation of identification percentage of 'missed' specimens

Source: Diamond *et al.* 1996

Data quality and performance characteristics of methods for analytical chemistry are typically validated through the use of quality control samples including blanks, calibration standards, and samples spiked with a known quantity of the analyte of interest.

8.5 CONVERTING DATA TO INFORMATION

The goal of collecting baseline data is to make the results available for decision-making, and to others – the mine proponent, interested community members, or regulatory agencies. However, raw data has little meaning until transferred into a format and context which can be readily understood. Clearly, some data is readily interpreted, while other data may be very difficult to interpret. Deductions concerning the data and its meaning are interpreted based on scientific principles. This section provides some guidance in converting data to information, based on relevant text in the 'Volunteer Surface Water Monitoring Guide', Minnesota Pollution Control Agency (2003).

Raw data has little meaning until transferred into a format and context which can be readily understood.

CASE 8.9
Freeport's Environmental Laboratory

At the time that PT Freeport Indonesia (PTFI) was developing its Grasberg Mine in the Papua Province of Indonesia, there were no laboratories in Indonesia capable of accurately analyzing dissolved metals at concentrations in the parts per billion range. Accordingly, PTFI established its own state-of-the-art laboratory which is located at Timika, more than 100 km from the mine and process plant areas.

The 1,000 square metre lab building was completed in January 1994 at a total cost of more than $US 3 million, and has been used for environmental laboratory operations since then. Hundreds of samples and thousands of measurements are conducted at the lab each month. The environmental laboratory provides analyses of samples from mine water, river water, groundwater, tailing, soil, plant tissues, and fish for heavy metals (both dissolved and total), suspended solids, pH, conductivity, alkalinity, hardness and other parameters which are of environmental concern.

Interpreting Data

Data interpretation is the process of asking questions to arrive at findings and conclusions. Findings are objective observations about data. Conclusions are how we explain why data look the way they do. For example, consider water quality monitoring conducted downstream from a proposed mine. Findings indicate high turbidity and limited water clarity. Based on the spatial and temporal variations in these parameters, we can draw conclusions as to whether or not turbidity is caused by exploration activities, logging, high rainfall event, or other natural or human events, and whether or not turbidity alone is the sole cause of decreased water clarity.

Questions like the following help us to arrive at findings:

- How do data compare with reference data (standards, control sites, quality goals, etc.)?
- Are there seasonal differences in results?
- Did natural events such as rainfall or wind affect results?
- Do results change in a consistent manner upstream or downstream?
- Do changes in one parameter coincide with changes in another?

It is helpful to consider data as part of the bigger scientific picture by considering how data fit the established and accepted ecological models (**Table 8.9**). This approach converts data into information by analyzing data in context. Data are analyzed holistically using one of many conceptual scientific models, such as hydrological, biochemical or demographic cycles. Accordingly, data and environmental components are not only analyzed according to their isolated attributes, but also in the context of their interactions.

It is helpful to consider data as part of the bigger scientific picture by considering how data fit the established and accepted ecological models.

TABLE 8.9
Data and Ecological Models

Ecological model/cycle	Description
Atmospheric energy cycle	Balance of radiation, reflection, and dispersion in the atmosphere, emission of long-wave energy, transfer of sensitive heat (temperature) and latent heat (evaporation) to the atmosphere
Hydrological cycle	Evaporation and evapotranspiration, atmospheric moisture (relative and absolute moisture, points and nuclei of concentration), condensation (clouds, fogs, mists), precipitation (rainfall and solid forms), infiltration, runoff, and water storage (groundwater, snow, and glaciers)
Pollution cycle	Emission of particles (total and breathable), transportation and diffusion by air-water or through the soil, suspension in dry and liquid media, dry precipitation and rainfall (acid rain), emission or deposit (concentration in the air, water, and soil)
Biogeochemical cycles or element-transformation cycles	Carbon, phosphate, nitrogen, or sulphates when passing through the atmosphere, hydrosphere, lithosphere, and biosphere
Trophic chains	Primary, secondary, and tertiary productivity levels; trophic levels; prey-predator or producer-consumer-reducer relationships
Demographic cycles	Population dynamics, relationship between birth and mortality rates, levels of morbidity and risk for populations, growth curves and rates, migratory balances
Economic cycles	Economic growth rates, composition and evolution of the gross domestic product, main activities, occupation, and labour productivity
Social components indicating accumulation in socioeconomic cycles	Society's values regarding the environment, educational levels, quality of housing, saving and investment rates, added value levels, population per services

Dealing With Variability and Uncertainty

Variability happens. In spite of rigorous data collection and QA/QC protocols, field data will exhibit variability. Natural systems are inherently variable, and through sampling and sample handling and analysis, we introduce additional variability. Uncertainty, in turn, compounds variability. Uncertainty arises because there is no such thing as a truly exact measurement, and samples cannot be collected continuously, forever. Instead, we periodically collect samples to represent an environment that is continuously changing over time and space. And we analyze these periodic samples using methods that have limits in resolution, precision, and accuracy. Much of the following information on variability is based on Rector 1995.

Natural systems are inherently variable, and through sampling and sample handling and analysis, we introduce additional variability.

Statistics is the science of reaching conclusions in the face of variability and uncertainty. We cannot eliminate uncertainty and variability, but we can use statistics to estimate their contribution to our observed results and make informed decisions based on the data. The most frequently used descriptive statistics are those that describe central tendency (i.e. arithmetic means, geometric means, and medians) and those that describe the distribution or variability (range, quartiles, and confidence interval/standard deviation). In general, it is appropriate to use the average when data sets are normally distributed around a mean value with no outliers. It is better to use the median if the data is skewed and/or if there are outliers. The geometric mean is rarely used except for bacteria monitoring.

Other statistical techniques, such as trend analysis, can also be conducted. However, they may require years of data and/or more advanced statistical techniques. Deciding which statistical measure to use depends upon the type of data to be summarized. **Table 8.10** recommends common summaries for different indicators.

TABLE 8.10

Suggested Statistical Summaries for General Chemical and Physical Water Parameters

(Table 8.10 is adapted from Data to Information for Coastal Volunteer Water Quality Monitoring Groups in New Hampshire and Maine, by Dates and Schloss (University of Maine Cooperation Extension and University of New Hampshire/Maine Seas Grant Extension, 1998))

Parameter	Statistical summary	Parameter	Statistical summary
Total suspended solids	Average Median Flow-weighted average Range Quartiles Confidence intervals or standard deviation	pH Alkalinity	Median or average Quartiles Minimum Median Quartiles Minimum
Temperature (water or air)	Seasonal average Seasonal median Maximum Range Quartiles	Chlorophyll-a	Seasonal average Range Maximum and minimum Median Quartiles
Dissolved oxygen (as mg/l)	Seasonal median Minimum Quartiles		Confidence intervals or standard deviation
Turbidity	Median Maximum Quartiles	Flow	Average Maximum and minimum Median Quartiles

(Continued)

TABLE 8.10
(Continued)

Parameter	Statistical summary	Parameter	Statistical summary
Nutrients (e.g. nitrite plus nitrate or total phosphorus)	Seasonal average Flow-weighted average Median Quartiles Confidence intervals or standard deviation	Water clarity/transparency	Seasonal average Seasonal median Maximum and minimum Range Quartiles Confidence intervals or standard deviation
Conductivity	Average Median Quartiles	Bacteria (water contact safety)	Geometric mean Quartiles

TABLE 8.11
Graphs and Comparisons to Consider when Assessing Aquatic Data (Minnesota Pollution Control Agency 2003)

Graph	Comment
Flow vs. any parameter	May show non-point source pollution effects or dilution of dissolved parameters at high flows
Date vs. observed values/concentrations	May show trends or seasonal variation
Precipitation vs. any parameter	May show how parameters respond to rainfall and/or non-point source pollution effects
Parameters vs. numerical standards/criteria	May indicate problem areas
Dissolved oxygen and temperature depth profiles	May show stratification or mixing status in lakes
Bacteria vs. total suspended solids or turbidity	May indicate that bacteria are associated with solids, and reductions in bacteria could be achieved with technologies that trap solids
Observed values or biometrics vs. river station	May show trends by location or points/locations where major changes are noticeable

To assess findings, graphs are useful to visually display results. They are also helpful in comparing parameters. **Table 8.11** lists some common forms of graphic data presentation.

Reaching Conclusions

Once data are organized into findings, conclusions can be made. Clear presentation of conclusions is essential to effectively communicate and gain credibility for study results. In reducing data to usable information, the key is to make only those conclusions that are supported by the data. One conclusion may be that additional data is needed. Sometimes conclusions are controversial. This is acceptable providing that controversial conclusions follow a logical process with scientific basis, and assumptions are well documented. Often controversial findings and conclusions can be explained by natural conditions, human alterations, and/or errors in sampling and analysis. Independent professional input always helps to verify and to understand controversial findings and conclusions (**Case 8.10**).

Independent professional input always helps to verify and to understand controversial findings and conclusions.

Various questions help to decide if human alterations or natural conditions affect findings and conclusions.

- Might natural upstream-to-downstream changes account for monitoring results? As one example benthic macro-invertebrate results might be explained by natural shifts in macro-invertebrate community composition from headwaters to mouth.
- Does weather appear to influence monitoring results? For example, do problem levels coincide with intense rainstorms? Might elevated water temperature levels be caused by unusually hot weather?
- Does the presence of specific sources explain results? As one example, can increased water turbidity be attributed to logging activities? Can mercury content in sediment be related to small-scale gold mining?
- Do changes in one parameter appear to explain or to relate to changes in another? As one example, could low dissolved oxygen be explained by high temperature? Do problem levels coincide with rising flow?
- Do visual observations support findings? For example, observed land clearing upstream of the monitoring location may explain high turbidity.
- For multiple years of data, are there overall trends?

Sometimes, monitoring results may be simply explained by the way samples were collected and analyzed. Flaws in field and/or laboratory techniques do occur. For example, high concentrations at one location may be the result of cross contamination, due to a sampling error, or simply the result of erroneous laboratory analysis. Field equipment should

CASE 8.10
Sometimes Conclusions are Controversial

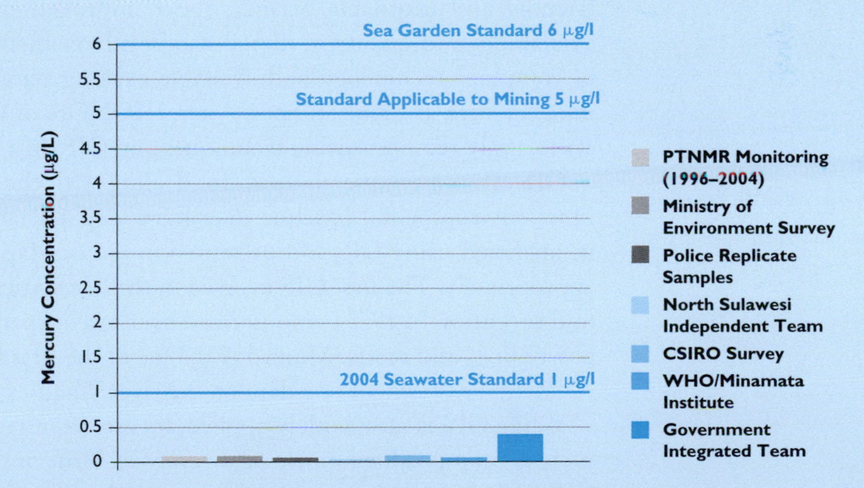

Newmont Mining Corporation's 80%-owned subsidiary, PT Newmont Minahasa Raya (PTNMR), operated the Minahasa Gold Mine at Mesel in North Sulawesi, Indonesia from 1996 to August 2004. PTNMR was authorized by the Indonesian government to use submarine placement of tailings (tailings are finely ground rock from which gold has been recovered). The tailings were placed on the seabed at a depth of 82 metres via a pipeline that rested on the seabed. The pipeline extended 900 metres from the coast with the outfall well outside 'Buyat Bay'. As known and analyzed in the environmental impact assessment completed prior to operations, the Mesel ore contained small amounts of naturally occurring, insoluble mercury and arsenic compounds. Roasting of the ore prior to gold extraction removed the mercury, which was recovered and sent to a government approved hazardous waste storage site. Arsenic remained in the tailings as insoluble compounds. In July 2004, shortly before the eight year project operations were scheduled to cease due to completion of the mine, non-governmental organizations (NGOs) instigated a campaign alleging that PTNMR's use of submarine tailings placement had polluted Buyat Bay with mercury and arsenic causing adverse health effects to area residents. In subsequent months, the Government of Indonesia initiated civil and criminal proceedings against PTNMR and its senior management. This conclusion was based on samples collected by the Indonesian police and tested in government laboratory. Subsequently the company and government regulators carried out additional sampling and analyses, the results of which failed to support the results obtained in the police investigation. In fact these results indicated that mercury and arsenic concentrations in the sea water were well within the range of natural concentrations for these elements. Despite such findings from respected independent authorities such as the CSIRO from Australia and the Minimata Institute from Japan, the case against PTNMR continued until in 2007, the mining company and its management were eventually acquitted of the charges.

Source:
Newmont's letter to the stakeholders, 31 May 2006

be carefully calibrated prior to use, and laboratory results should be checked for consistency. QA/QC sample results should also be compared to confirm data quality.

Another challenge is to attempt to sample the range of conditions that may be present or may occur. For example, is sampling primarily conducted when river flows are low? Did sampling catch storm-related runoff or merely base flow? Do soil tests reflect the entire range of variation in soil properties? It is important to recognize that samples represent a tiny fraction of the whole and, accordingly, it is highly unlikely that maximum or minimum values of any parameter would be found.

Yet another aspect to consider is whether the analytical method is sufficiently sensitive to detect levels of concern. Time of sampling may also be important. Could the time of day the sample was taken affect results? Is sampling always being carried out at the same time of the day? Streams and rivers fed by glaciers or snow melt vary diurnally in their discharge and quality. Also, dissolved oxygen in surface water is typically lowest in the morning and highest in the late afternoon.

8.6 THE USE OF REMOTE SENSING TECHNIQUES AND GEOGRAPHIC INFORMATION SYSTEMS

Remote sensing is the interpretation of airborne or satellite imagery to provide a picture of selected aspects of physical geography, such as topography, habitat distribution, settlements, or mangrove forest areas. Geographic information systems (GIS) are computer systems that can store, integrate, analyze, and display spatial data. Two important developments have helped to reduce the complexity of spatial analysis and to allow remote sensing techniques and GIS to become standard applications in environmental assessment. First, due to advances in computer and software technology, GIS have become powerful, user-friendly, and affordable. Second, due to improvements in remote sensing technologies, the availability and quality of digital data sets have increased exponentially. The current state of computer technology and available existing data sets (commercially available satellite imagery such as Landsat, Spot, Aster, IKONOS, or Quickbird) allows the EIA practitioners to apply remote sensing techniques and GIS on a routine basis.

GIS have become powerful, user-friendly, and affordable.

GIS offers a unique platform for dealing with the spatial properties of a mining project. Most environmental baseline data have a spatial attribute, and they lend themselves to be analyzed using GIS and presented in maps. Maps are powerful analytical tools if used appropriately. The first GIS evolved in the late sixties, and entered EIA applications in the mid seventies. In 1972 computerized overlays of spatial data were first used in the siting of power lines and roads (Munn 1975). One of the first full-scale GIS was applied to the environmental assessment of a dam on the river Thames (Griffith 1980).

Most environmental baseline data have a spatial attribute, and they lend themselves to be analyzed using GIS.

While GIS is now widely applied, its use often remains limited to some basic functions such as map production, classic overlay or buffering (Joao 1996). Standard application of GIS analysis include mapping of topography (elevation and slopes), habitats, in particular mangrove areas, surface water bodies (rivers, lakes, and wetlands), coral reefs and other shallow marine habitats, settlement areas, and road networks. Implementation of GIS, however, does not always fully utilize the analytical capacities that GIS offers, possibly because the spatial analysis and modelling requires a higher level of expertise, than is normally found among EIA practitioners.

Although use of GIS and the resulting maps can be very powerful both in their presentation as well as an analytical tool, EIA practitioners need to recognize potential constraints and pitfalls associated with GIS application, examples of which are illustrated in

TABLE 8.12

The Good and Bad of Geographic Information Systems (Segnestam 2000)

Advantages
• GIS compile and analyze large amount of data from different sources
• GIS provides the only good and simple tool for analyzing interrelated aspects through overlaying data or indicators
• It provides an unrivalled visual tool for communicating spatial data
• GIS visualize data and information at different analytical scales
• GIS is a dynamic tool and can be easily expanded as new data are acquired
Disadvantages
• GIS presentations may imply a quality of data that does not exist
• Original data used in GIS are not always accessible to the reviewer
• Interpolation programs may distort data sets
• Maps may lead the user to think there are causal links that in reality may not exist
• Working with GIS demands resources in terms of money and competent personnel

Table 8.12. As with most computer analysis the main concern with GIS application is that generated information easily creates a perception of accuracy that does not reflect reality. GIS maps need to be interpreted with an understanding of the input data used in the analysis, and their accuracy.

GIS-based analyses help to identify site-specific constraints and opportunities in the regular scoping effort often prior to the actual site visit. At a minimum, GIS will equip the EIA practitioner with a set of environmental baseline maps prior to the actual site visit. The site visit can then also be used to verify data (ground truthing), and to improve environmental mapping.

GIS is a highly useful tool for designing field investigations, and for presenting environmental data during the EIA process. As the mining project evolves, the initial GIS can easily be expanded to assist environmental monitoring during operation. More importantly, remote sensing combined with GIS is a powerful tool in documenting environmental change over time. GIS helps to differentiate land use changes as a direct consequence of mining, and environmental changes that occur due to other human, or natural, pressures. In essence, the GIS application in a mining project can be designed as a central knowledge base that concisely stores, analyses, and communicates environmentally related spatial data throughout the life of the mine.

The GIS application in a mining project can be designed as a central knowledge base that concisely stores, analyses, and communicates environmentally related spatial data throughout the life of the mine.

REFERENCES

Adriaanse A (1993) Environmental Policy Performance Indicators. The Hague: Ministry of Housing, Physical Planning and the Environment.

Bakkes JA, van den Born GJ, Helder JC, Swart RJ, Hope CW, and Parker JDE (1994) An Overview of Environmental Indicators: State of the Art and Perspectives. UNEP/EATR.94–01, RIVM/402001001. Environmental Assessment Sub-Programme, UNEP, Nairobi.

Clark MJR ed. (2003) British Columbia Field Sampling Manual. Water, Air and Climate Change Branch, Ministry of Water, Land and Air Protection, Victoria, BC, Canada; 312 pp.

Coleman JS (1957) Community Conflict. New York: Free Press.

Diamond J, Barbour MT, and Stribling JB (1996) Characterizing and comparing bio-assessment approaches and their results: A perspective. J. North Am. Benthol. Soc. 15:713–727.

Emery AR (2000) Guidelines: Integrating Indigenous Knowledge in Project Planning and Implementation. Prepared by KIVU Inc. for the World Bank and the Canadian International Development Agency; www.worldbank.org/afr/ik/guidelines

EQB (1994) Minnesota Lake and Watershed Data Collection Manual, Lakes Task Force (EQB); written by several agencies; http://www.shorelandmanagement.org/depth/index.html

Gaventa J (1980) Power and Powerlessness: Quiescence and Rebellion in an Appalachian Valley. Urbana: University of Illinois Press.

Griffith C (1980) Geographic Information Systems and Environmental Impact Assessment. Environmental Management, 4 (1):21–25.

João EM (1996) Use of Geographic Information Systems in Impact Assessment. In: Environmental Methods Review: Retooling Impact Assessment for the New Century, edited by: Porter A and J Fittipaldi.

Lee N (2006) Bridging the gap between theory and practice in integrated assessment, Environmental Impact Assessment Review 26 (2006) pp. 57–78.

Minnesota Pollution Control Agency (2003) Volunteer Surface Water Monitoring Guide.

Munn RE ed. (1975) Environmental Impact Assessment: Principles and Procedures. SCOPE report 5: Toronto.

NOAA (1992) Technical Memorandum NMFS-F/SPO-16May – Guidelines and Principles for Social Impact Assessment. The Interorganizational Committee on Guidelines and Principles for Social Impact Assessment.

OECD (1994) Environmental Indicators. OECD Core Set. Organisation for Economic Co-operation and Development: Paris, France.

Rector J (1995) Variability Happens: Basic Descriptive Statistics for Volunteer Programs, The Volunteer Monitor, Vol. 7, No., 1, Spring 1995.

Segnestam L (1999) Environmental Performance Indicators, A Second Edition Note, Environmental Economics Series, Paper No. 71, World Bank.

Segnestam L (2002) Indicators of Environment and Sustainable Development: Theories and Practical Experience, Environmental Economics Series, Paper No. 89, World Bank.

Saunders D, Margules C and Hill B (1998) Environmental indicators for national state of the environment reporting – Biodiversity, Australia: State of the Environment (Environmental Indicator Reports), Department of the Environment, Canberra.

USGS (1998) National Field Manual for the Collection of Water-Quality Data; http://water.usgs.gov/owq/pubs.html

Ward T, Butler E and Hill B (1998) Environmental indicators for national state of the environment reporting – Estuaries and the Sea, Australia: State of the Environment (Environmental Indicator Reports), Department of the Environment, Canberra.

WCD (2000) Participation, Negotiation and Conflict Management in Large Dams Projects; World Commission on Dams Secretariat; P.O. Box 16002, Vlaeberg, Cape Town 8018, South Africa.

Williams P (1996) Measuring Biodiversity Value. World Conservation, IUCN, Switzerland, 1/96, 12–14.

Whyte AVT (1977) Guidelines for field studies in environmental perception. MAB Tech. Note 5, UNESCO, Paris. 117 pp.

Appendix 8.1 Common Indicators to Document the State of the Environment and to Detect Change

PHYSICAL/CHEMICAL ENVIRONMENT

Air

- Emission/discharge outputs (quantity and quality)
- Consumption of Ozone Depleting Substances (ODS)

Air quality indices
- Particulate matter
- Carbon monoxide
- Carbon dioxide
- Sulphur dioxide
- Nitrogen dioxide
- Lead
- Number of complaints about air quality/noise per year
- Number of respiratory diseases per 1,000 habitants

Land

Land requirement
- Rate of change of wilderness area
- Agricultural land loss
- Use of fertilizers, pesticides, and herbicides

Land form
- Topography and slope
- Geology
- Areas of geohazards
- Area of land affected by soil erosion

Land use
- Forest area as percentage of total land area
- Area of urban formal and informal settlement
- Agricultural areas

Land quality
- Soil distribution
- Soil/sediment quality (contamination)
- Soil organic matter content

Water

Emission/discharge outputs
- Effluent discharges (quantity and quality)
- Solid waste emission (sanitary, industrial, hazardous, medical, etc)

- Total water consumption (maximum, minimum, average)
- Water consumption as percent of available water

Water quality indices
- BOD in freshwater
- Concentration of Faecal Coliform
- Turbidity, temperature, and pH
- Concentrations of chemical constitutes (such as trace metals)

Water consumption
- Bernie Fowler's Sneaker Index – water clarity
- Quality of river water entering mine area / quality leaving mine area

Water quantity indices
- Precipitation/evaporation
- Drainage system and surface water bodies
- Water levels and water level variation
- Flow velocities
- Groundwater levels and aquifer characteristics
- Compliance with dissolved oxygen standards
- Compliance with water quality standards

BIOLOGICAL ENVIRONMENT

Vegetation degradation/Habitat indicators
- Extent and rate of clearing or modification
- Location and configuration of remnant land areas with sensitive habitats (mangrove forest, ancient forest, wetlands, coral reefs, etc)
- Portion of land area covered by forest

Species diversity
- Species diversity, conservation status, economic importance and extent of knowledge
- Number, distribution and abundance of migratory species
- Demographic characteristics of target taxa

Birds
- Restricted-range species
- Number of species at risk

Genetic diversity
Alien and exotic species

Endangered or threatened species in area
- Pest numbers

Harvesting/Hunting (Species used by local people).
- Species trafficked
- Permits requested and issued for harvesting
- Fish catch/Animal hunted
- By-catch to target species

Ecologic diversity
- Ecosystem diversity
- Number and extent/area of ecological communities of high conservation potential
- Protected area compared to total project area

Contamination
- Concentration of chemicals/trace metals in tissue

(Continued)

HUMAN ENVIRONMENT

Social Systems

Population

Growth indicators

- Total population
- Population growth

Migration indicators

- Population density
- Net migration gain and loss
- Community stability index

Structure indicators

- Age trends
- Gender distribution

Institutional capacity

Origin (Indigenous Peoples)

- Ethnic groups and Indigenous Peoples
- Unemployment rate by ethnicity
- Occupational distribution of women and minorities

Norms and values

Religion

Equity

- Gini index of income inequity
- Percentage of population living below poverty line
- Unemployment rate
- Ratio of average female wage to male wage
- Income of top 10 percent/ income of bottom 10 percent
- Percentage of time contributed to causes other than household earning

Health

- Nutritional status of children
- Mortality rate under 5 years old
- Life expectancy at birth
- Percentage of population with adequate sewage disposal/ access to safe drinking water/access to primary health facilities
- Number of water borne diseases

Education/Awareness

- Children reaching Grade 5 of primary education
- Adult literacy rate
- Percent population perceive pollution a priority

Housing

- Floor area per person
- Secure tenure – Proportion of population with access to secure tenure (owned or rented)

Security

- Number of recorded crimes per 1,000 people
- Number of security personnel per 1,000 people

Economic System

Economic structure

- GDP per capita
- Investment share in GDP
- Balance of trade and services

Consumption and production patterns

- Distance traveled per capita by mode of transportation
- Energy sources/consumption per household/capita
- Intensity of energy usage
- Intensity of material use
- Generation of municipal/industrial/hazardous wastes
- Waste recycling and reuse

Institutional framework

- Host region sustainable development strategy
- Regulatory enforcement

Institutional capacity

- Expenditure in research and development
- Economic and human losses due to natural disasters

Natural resource capital

- Proven mineral/coal reserves
- Cutoff grade

Productive capital

- Road infrastructure
- Public health facilities
- Sources of energy/water supply
- Waste treatment facilities

Human capital

- Population
- Public health
- Education levels
- Literacy
- Economic livelihood

Social capital

- Training
- Regional development planning
- Participatory decision-making

Environmental capital

- Extent of areas of particular value (biodiversity, forest, wet lands, coral reefs, etc)
- Cultural heritage sites
- Recreation areas
- Landscape of recognized beauty

Appendix 8.2 Topics Commonly Included in a Scoping Report

Describe the Institutional Setting of the Project	
Public authorities	Briefly outline the legislative authorities of the host country, together with its administrative structure (central government, provincial government, district head, village head, or equivalent authorities). Describe administrative boundaries within the project area. Identify environmental authorities likely to be involved in the project. Provide contact details.
Main environmental legislative and regulatory requirements	Identify the main environmental legislations and regulations that apply to the project. Detail the environmental approval process, including time line, approval process, EIA review and approval authorities, and involved parties. Give recognition to informal traditional laws in the project area.
Applicable ambient environmental and emission standards	As part of environmental scoping, EIA practitioners compile a listing of ambient and emission standards applicable to mining project at national and local level. Where neither ambient nor discharge standards exist, initial minimum standards should be provided based on past relevant practice in the host country for similar projects, or in reference to international accepted standards such as laid out in the World Bank's Pollution Prevention and Abatement Handbook or the IFC EHS Guidelines. Discharge standards represent the design criteria for abatement technologies at pollution sources, and hence directly influence project costs.
Relevant international law	Environmental international agreements and treaties to which the host country has subscribed potentially represent an additional layer of requirements that affect mine planning and development. While a long list of treaties applies to most projects, the most relevant treaties are probably the Basler Convention, Ramsar, Biodiversity Treaty, CITES, and the listing of World Cultural Heritage sites.
Company environmental and social policy	Identify particular requirements in regard of environmental and social performance of the project due to the project proponent internal environmental and social policies (e.g. zero- discharge, commitment to reduce greenhouse gas emissions, adoption of The Equator Principles or the Australian Best Practices in Mining)
Describe the Environmental Setting of the Project	
Land resources	Describe topography, geology, and soil types. Provide information on seimicivity in the project area. Identify unique, special, and sensitive landforms and geologic formations. Note existing illegal mining activities. Make reference to magnitudes and probabilities for geo-hazards in the project area such as landslides or earthquakes.
Water resources	Describe hydrology, hydro-geology, and oceanography at the project area (if relevant). In regard to surface water identify existing surface water bodies and their drainage basins, water quality, flows, and patterns of erosion and sedimentation. In regard to ground water, identify aquifer systems, groundwater recharge and discharge areas, groundwater levels and groundwater quality. List known groundwater wells and springs. Identify existing water usage, effluent discharge points, and unique, special, and sensitive water bodies and/or wetlands. In particular, identify sources of drinking water. Note magnitudes and probabilities for natural geo-hazards such as floods, droughts, or tsunamis. Identify and list existing sources of hydrological, hydro-geological, and oceanographic data for the project site, including location, parameters, and period of record for each station.

Air	Identify and list existing sources of meteorological data for the project site, including location, parameters, and period of record for each station. Provide information on meteorology such as climatic type, temperature regime, rainfall and rainfall pattern, and prevailing wind direction and speed. If available provide information on ambient air quality in the project area, including noise. Identify existing sources of air emissions already operating in project area, together with emissions they are producing (measured or estimated). Make reference to the magnitudes and probabilities for geo-hazards in the project area, including extreme temperatures and precipitation events, maximum wind speeds, or unusual storms (such as cyclones).
Fauna and flora	Identify typical vegetation communities/associations and their functions as habitats. Highlight recognized critical habitats, natural reserves, national parks, or other protected areas in or close the project area. Evaluate common, characteristic, and economically important fauna and flora in project area. Focus on threatened, endangered, protected, special status, endemic or culturally significant fauna and flora. Identify sensitive habitats (wetlands, mangrove forest, coral reefs, and the like). Describe patterns of hunting, fishing, and trade in flora and fauna, including existing illegal logging activities.
Describe Social and Economic Systems	
Land use	Provide inventory of land and other resources at time of project and likely future developments. Identify regional development plans, spatial plans, land use plans, and other natural resource plans applicable to the project area that have been officially adopted or are in preparation by the government at the local, provincial and national levels, together with potential conflicts or limitations arising between the project and plans for land and resource use. Provide inventory of aesthetic and natural beauty values together with recreation areas in the project area.
Demography	Provide basic demographic data such settlements, population numbers, and age/gender composition/structure. Consider seasonal workers and transmigrants. Comment on ease of transportation/ movement between settlements.
Local economy	Gather information on local economy, such as existing public infrastructure, including market centers, roads, transportation facilities, educational facilities. Detail existing economic activities (including traditional economies, including agriculture, fisheries, and logging) and sources of livelihood. Estimate average household income. Do economic activities exist that can support/ conflict with the proposed mine?
Cultures and religious beliefs	Provide information on cultures and religion, including ethnic groups, traditions and beliefs, and traditional use and ownership of land and other natural resources. Identify formal and informal leadership, and religious practices and places of worship. Appreciate existing sources of conflicts, and education levels and trends. Discuss the role of women in households.
Public health	Discuss public health aspects such as existing medical facilities, number and skill levels of medical/paramedical personnel, and prevalent diseases including sexual transmitted diseases. Identify environmental related health factors. Describe public sanitation. Make reference to roles of traditional practitioners and treatments.
Identify Project Opportunities and Challenges	
Project Definition	Provide a synopsis of the proposed mine. A methodological approach to the project definition will help: First, what are the main project components and activities? Second, what are the main mine inputs and outputs? The next question is: What characterizes emissions? For each emission stream (that is, emissions to air, emissions to water and waste emissions), list source, expected quantity, quality or characteristic, and discharge point and/or management option (if available).
Community perception and conflict identification	Describe if and how the development of the new mining project will compete or interfere with existing land uses, even in those situations where the land is not occupied. Is there a prevailing community perception towards the project? Are there outspoken opponents?

(Continued)

Potential show stoppers	Raise red flags. Environmental input in mine design at an early stage is essential – fundamental changes can be made to the project without excessive cost implications and without causing project delays. Potential show stoppers are many and varied. They may relate to the environmental setting (e.g. coral reef habitats close to a proposed port site) over stringent host government standards (e.g. stringent NO_x standards) to complex land ownerships issues.
Project opportunities	Identify potential project benefits (besides the usual suspects of providing employment opportunities and improving water supply). Is there a potential environmental offset component? Does rehabilitation allow developing a long-term biofuel program?
Recommended future studies	Scoping provides one of the best opportunities to minimize the quantity of baseline work. Identify issues of little concern to enable them to be eliminated from detailed consideration. Ensure that EIA studies are focused on significant issues and the environmental components involved in their evaluation

9

Identifying and Evaluating Impacts

Linking Cause and Effect

PLATINUM

A heavy gray-white, malleable metal, platinum is extremely rare, with the result that it is significantly more valuable than gold. Like gold, platinum is resistant to tarnish and corrosion. Used in a variety of technological processes and applications, platinum is a highly effective catalyst; demand for this metal increased markedly with the advent of catalytic converters, installed to reduce pollution from internal combustion engines.

9 Identifying and Evaluating Impacts

Linking Cause and Effect

'How does society weigh the benefits of jewellery, which in most cases is a luxury item like a fur coat, against the potential environmental, community and landscape impacts of new large-scale gold mines?'

Steve D'Esposito
President of the Minerals Policy Center
May 2002

This chapter introduces some common methodologies for analyzing impacts and designing mitigation measures. Analyzing impacts combines two activities, identifying impacts and evaluating their significance. Mitigation measures also involve two activities, managing impacts and monitoring management success. The term 'methodology' refers to a structured approach to accomplish these activities. Methodologies are not meant to be 'cook books' in which successful outcomes are guaranteed by adhering to defined approaches. Most, if not all, methodologies require assumptions to be made, introducing elements of uncertainty, potential errors, and risk.

Additionally, this chapter discusses the various ramifications of these assumptions, and embedded uncertainties and risks. Differentiation is made in discussing impacts on the physical-chemical and biological environment, and on the human and economic environment of mining projects.

9.1 DEFINING THE CHALLENGES

Change is sometimes beneficial to some people while at the same time harmful to others.

In this text, impact analysis, or impact prediction, is defined as identifying project-induced environmental changes and evaluating their significance. A number of terms are used to distinguish between natural and project-induced environmental changes; and between changes and their harmful (negative) and beneficial (positive) consequences, referred to as the direction or nature of change. Change is sometimes beneficial to some people while at the same time harmful to others. One approach is to refer to changes as 'effects', and consequences as 'impacts'. Another convention is to use the term 'impacts' to denote only harmful effects. In still other texts, the words 'effects' and 'impacts' are synonymous and harmful effects are termed 'damage'. No matter how the words are defined, a change, effect, or impact is usually described in terms of its nature and its significance. In this text the terms 'effect', 'impact', and 'change' are used synonymously, without any value judgement.

Table 9.1 identifies mining-induced environmental changes and human concerns, changes that may be 'good' or 'bad' depending upon the point of view. These change processes

TABLE 9.1

Environmental Attributes for Projects, Environmental Changes and Areas of Human Concern in Industrial Developments – Value judgements ('good' or 'bad') should be avoided at the early stage of an environmental assessment, since the assessment of change significance follows once changes and their interrelationships are identified.

LAND

Soil stability; natural hazard; land use patterns; landscape changes

WATER

Aquifer safe yield; flow variations; change in runoff; oil; radioactivity; suspended solids; acid rock drainage; trace metals; biochemical oxygen demand (BOD); dissolved oxygen (DO); dissolved solids; nutrients; toxic compounds; aquatic life; fecal coliforms

AIR

Air quality – Particulates; sulphur oxides; hydrocarbons; nitrogen oxides; carbon monoxide; carbon dioxide; photochemical oxidants; hazardous toxicants; diffusion

Noise – Physical effects; psychological effects; communication effects; performance effects; social behaviour effects; effects on wildlife.

BIOTA

Abundance/scarcity of species or genetic resources; extent of crops, ecosystems, vegetation, and forests; diversity of species; extent of provision of nesting grounds, etc., for migratory species; abundance/scarcity of pests and disease organisms; large animals (wild and domestic); predatory birds; small game; fish, shellfish, and waterfowl; threatened species; endemic species; natural habitat and vegetation; aquatic plants

PEOPLE

Economic and occupational status – Displacement of population; relocation of population in response to employment opportunities; services and distribution patterns; property values; regional economic stability; public sector review; per capita consumption; disposable income; renewable resources; non-renewable resources; aesthetics

Social amenities and relationships

- Family life styles; schools; transportation; community feelings; participation vs alienation; local and national pride vs. regret; stability; disruptions; language; hospitals; clubs; recreation; neighbourliness; community needs
- Physical amenities (intellectual, cultural, aesthetic, and sensual)
- National parks; wildlife; historic and archaeological monuments; beauty of landscape; wilderness; quiet; clean air and water; visual physical changes; moral conduct; sentimental values

Religion and traditional belief/ Culture – Belief; symbols; taboos; values; leisure, new values; heritage; traditional and religious rites

Social pattern or life style – Resettlement; rural depopulation; change in population density; food; housing; material goods; nomadic; settled; pastoral agricultural; rural; urban

Psychological features – Involvement; expectations; stress; frustrations; commitment; challenges; work satisfaction; national or community pride; freedom of choice; stability and continuity; self-expression; company or solitude; mobility; physiological needs

(Continued)

TABLE 9.1
(Continued)

Health – Changes in health; medical services; medical standards; waste management

Personal security – Freedom from molestation; freedom from natural disasters/hazards; safety measures; security

Political – Authority; level and degree of involvement; priorities; structure of decision-making; responsibility and responsiveness; resource allocation; local and minority interests; defence needs; contributing or limiting factors; tolerances

Legal – Restructuring of administrative management; changes in taxes; public policy; statutory laws and acts; air and water quality standards; safety standards; national building acts; noise abatement by-laws

Source:
Based on SCOPE 5 (1975)

are further discussed in Chapter Thirteen. Value judgements should be avoided at the early stage of an environmental assessment, since the assessment of change significance (including its nature) follows once changes and their interrelationships are identified. Note that directly related to change is rate of change. A slow change may be acceptable, especially if it leads to a new stability, whereas rapid change or large fluctuations may place an intolerable burden on humans and/or ecosystems.

A slow change may be acceptable, especially if it leads to a new stability.

Analyzing environmental change is simple in theory but not an easy task in practice. The main challenges are differentiating between natural and project-induced change, appreciating different value systems, acknowledging resource limitations, and understanding the available methodologies to assess environmental changes.

Natural versus Project-Induced Change

'Would change occur in the absence of the project?' Even in the absence of humans, the natural environment undergoes continual change (see also related discussion in SCOPE 5, 1975 on which the following is partly based). This may be on a time-scale of hundreds of millions of years, as with continental drift and mountain-building, or over a period of a few years or less, as with siltation of lakes or changes associated with natural geo-hazards such as floods, tsunamis, forest fires, or landslides. Some natural changes are irreversible (e.g. destruction of ancient rain forest), while others are cyclic (e.g. seasonal climate variation or El Nino events) or transient (e.g. droughts).

Even in the absence of humans, the natural environment undergoes continual change.

The natural environment appears static except where the results of human intervention are evident. This is because most changes are imperceptible. However, changes do occur and they do so at surprisingly rapid rates. Plant communities advance and retreat with fluctuations in climate. Individual species of plants and animals invade new territories or are displaced from existing territories due to evolutionary adaptations or to changes in climate and habitat that favour some species over others. Pollen preserved in lake bed sediments show that numerous major changes in vegetation have occurred in Southwest Australia within the last 1,000 years, in response to climatic changes. Over a time-scale of a century, extensions or contractions of 40 km or more, have occurred repeatedly.

Superimposed on natural environmental changes are changes caused by human activities, such as developing a mine. Because the natural environment changes with time, it is not always easy to distinguish incremental project-induced changes (**Case 9.1**). But in order to understand project-induced changes, it is necessary to know what environmental conditions would have occurred had mine development not taken place. It is not an easy

It is not an easy task to measure present environmental conditions, the baseline, far less to assess the significance of past trends and to extrapolate these accurately into the future.

task to measure present environmental conditions, the baseline, far less to assess the significance of past trends and to extrapolate these accurately into the future. The assessment of mining-induced changes is even more difficult when human activities unrelated to mining take place in the project area.

Development of new settlements or land conversion from forest areas to agricultural land may occur in the absence of the project, and the challenge is to isolate incremental changes that are direct consequences of mine development. A good example of a geographic area which is undergoing dramatic changes is the eastern part of Kalimantan in Indonesia, as discussed in **Case 9.2**.

Environmental Changes and the Stage of Development

'Is change good or bad?' Perceptions about environmental changes can be different in different countries. Where poverty is widespread and large numbers of people do not have adequate food, shelter, health care, or education, the lack of development may constitute a greater collective degradation to quality of life than the negative environmental impacts of development. The imperative for development to remedy these defects may be so great that the consequent environmental degradation may be tolerated. The grinding and pervasive poverty in developing countries has been spoken of as the 'pollution of poverty', while the widespread social and environmental degradation in developed countries has been characterized in its advanced state as the 'pollution of affluence' (Scope 5 1975). Decisions on mine developments should be amenable to different value judgements concerning the net cost-benefit assessment of project-induced environmental, economic, and social changes. This does not mean accepting environmental degradation in the name of development; on the contrary, it is now widely accepted that mine development can be planned to make the best use of natural resources without jeopardizing environmental quality for the future generations.

Where poverty is widespread and large numbers of people do not have adequate food, shelter, health care, or education, the lack of development may constitute a greater collective degradation to quality of life than the negative environmental impacts of development.

Sadly, over-dominant advocacy groups often impose limits on a balanced cost-benefit assessment. Anti-mining advocacy groups tend to retreat to emotionally charged view-points, disregarding project type and/or project location. They refuse to be open to the argument that mining can contribute substantially to the Gross Domestic Product (GDP) of developing countries, hence contributing to the reduction of poverty and improvement in the quality of life. Ironically, many of these advocacy groups are headquartered in countries with a long history of mining, such as the USA, Canada or Australia, and so enjoy the benefits that mining has brought and continues to bring to these nations. It is also true that mining companies

CASE 9.1
Impact of Mining on the Melting of Glaciers

During the environmental impact assessment of the Freeport Gold and Copper Mine Expansion in Papua, East Indonesia (1997), the argument was brought forward that mining will contribute to the decline of the nearby picturesque Carstenz Glacier, one of the few remaining equatorial glaciers. However, it is well documented that equatorial glaciers have been melting since the end of the Little Ice Age in the nineteenth century and this melting has accelerated with the generally higher temperatures experienced over the past 15 years. It is predicted that most equatorial glaciers, including the Carstenz Glacier will disappear by the middle of the twenty-first century. How much, if at all, will mining-induced changes in micro climate contribute to the retreat of the Carstenz Glacier, and are these incremental changes significant?

Environmental assessment should avoid being guided by one-sided arguments.

sometimes overemphasize the project benefits to help ensure that the project finds public acceptance. Both these view-points are equally unhelpful, and environmental assessment should avoid being guided by one-sided arguments.

Available Resources

Of pragmatic importance is the question of resources available to assess environmental impacts. Any discussion of resource commitments must recognize two very real dangers: *over-complexity* and *over-simplicity*. Between these undesirable extremes is arguably always a useful working range of options for selecting an approach that matches resource needs with project complexity. Obvious resource constraints are money, time, and available expertise.

The cost of preparing an Environmental Impact Assessment (EIA) rarely exceeds one percent of the total project cost.

The cost of environmental assessment should not be viewed as a lost investment. Costs for an environmental impact assessment study differ from country to country, and from project to project. A compounding factor is that there is not always a clear distinction as to whether a particular study is related to environmental assessment or to mine planning, as many studies contribute information to both. However, we can look to the experience with large projects for some indicative costs. According to Goodland and Mercier (1999), the cost of preparing an Environmental Impact Assessment (EIA) rarely exceeds one percent of the total project cost. Capital investments for new mining projects are in the order of several hundreds of million of US dollars, and of course, it is much easier to keep environmental assessment costs down to one percent on a project whose total budget is $500 million or more than on one whose budget is $10 million. Our experience on a wide range of

CASE 9.2
Eastern Kalimantan – a Region Subject to Dramatic Change

Until recently, most of Eastern Kalimantan was covered in dense equatorial rainforest in a patchwork of swamps and low hills. First came the loggers, clear-felling the forest, removing the valuable timber and burning the remaining vegetation. During the *El Nino* drought years of 1982–83, and again in 1997–98, fires spread into uncleared areas, including peat swamps where the drought had caused peat to dry sufficiently to ignite and where fires continued to burn for many months. The results in 1998 included more than 200,000 hectares of forest destroyed in East Kalimantan, smoke causing health problems over much of Southeast Asia as far away as Singapore and Bangkok, and carbon dioxide emissions estimated at 17% of total world-wide carbon dioxide emissions in that year.

Eastern Kalimantan is also the location of most of Indonesia's lucrative coal mines which supply much of Indonesia's domestic power generation needs. In 2005, about 152 million tons of coal were mined in Indonesia, including exports of about 107 million tons valued at about US$ 4 billion, which represented 21% of world exports.

Recently, particularly in the past five years, there has been a vast land clearing programme associated with the establishment of oil palm plantations and the associated infrastructure. One source (Leslie Potter, Australian National University) has estimated that 3.1 million hectares of land have been cleared in East Kalimantan under the guise of oil palm plantation development. While many of the oil palm plantation areas have been established on land that had been left in a seriously degraded condition as a result of logging and fires, other plantations have been and are being established on forested areas, requiring new campaigns of clear-felling and burning, prior to planting of oil palms. It is ironic that the push for renewable energy which has increased the demand for palm oil as a renewable bio-fuel, is causing widespread environmental damage and contributing to increased carbon dioxide emissions.

Many species of plants and animals are considered to be at risk due to these widespread reductions in natural habitat, the most notable being the orangutan, whose population in Kalimantan has declined substantially in recent years. The population of orangutans declined by one third as a result of forest fires during 1997–98. Several coal mining projects including the large Kaltim Prima Coal Project are also located in orangutan territory. The areas affected by mining are very small in comparison with areas affected by logging, fires and oil palm plantations. However, mining companies provide easier targets for NGOs seeking to protect the orangutan. Nevertheless, it is apparent that mining companies have done more good than harm in to the cause of orangutan conservation. For example, during 1998, employees of PT Kaltim Prima Coal, operator of the Kaltim Prima project, rescued dozens of orangutan from forest fires. The company has also been cooperating with NGOs in the restoration of orangutan habitat in the area.

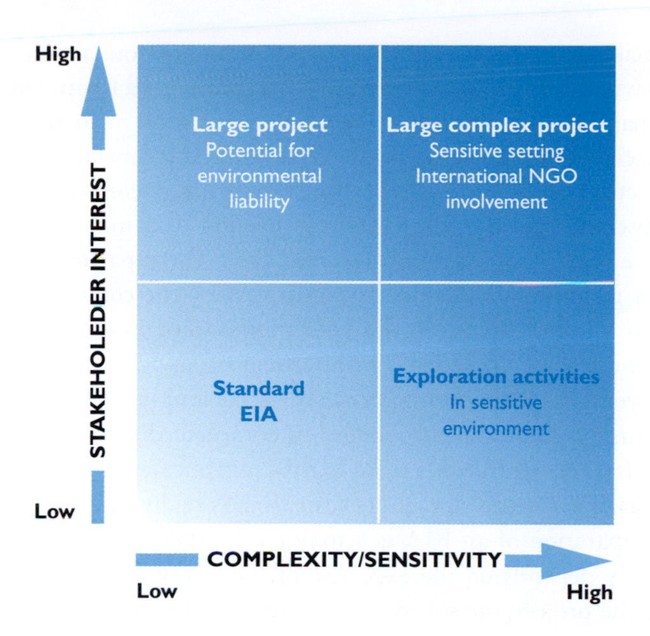

FIGURE 9.1
Resource Requirements for Environmental Assessment as a Function of Project and External Factors

Large and complex project in sensitive environmental settings always command high attention and resource commitment, including time.

Source:
Based on Walker and Johnson 1999

mining projects of different size and complexity indicates that the preparation of an EIA absorbs, on average, anywhere from US$ 100,000 to US$ 5 million. EIA preparation costs are generally lower in areas with a long history of mining where environmental conditions are well documented and impacts well understood.

By contrast, engineering feasibility and design studies may involve as much as 10% of the project capital cost while environmental mitigation measures usually account for three to five percent of the total project cost (Goodland and Mercier 1999). There are various ways to manage costs, including: (1) Integrate environmental assessment early into other project planning activities, particular into feasibility studies. (2) Seek project support from host government and approval authorities. (3) Avoid the high costs associated with exclusively hiring expatriate specialists by involving local expertise (e.g. local universities or research institutions). (4) Costs usually diminish with experience of key personnel (e.g. appointing a study team with proven EIA track record) and with the appropriate support mechanisms (e.g. translation capability).

While environmental assessment is a process that continues throughout the mine life, from exploration to mine closure, the EIA is a well-defined study constituting to a large extent the environmental permitting of the mine, if approved. The time necessary to complete an EIA varies widely and is controlled by a number of factors, predominantly by project type, the state of existing environment conditions at the mine site and the availability of relevant existing environmental information (**Figure 9.1**). Large and complex projects in sensitive environmental settings always command high attention and resource commitment, including time. Seasonal climatic cycles also influence study duration. At a minimum baseline studies should cover one complete climatic cycle, i.e. all seasons.

Other factors controlling the time to complete an EIA are statutory requirements in regard to public consultation, and the EIA review and approval process. Analyzing environmental impacts is largely in control of the study team; obtaining environmental approval is not. In fact, describing the often complex forces that drive environmental permitting deserves a textbook by itself. Project approval does not depend only on technical or regulatory considerations, but can also be influenced by which company proposes a specific mine development. State-owned companies are likely to face less scrutiny by the review team

Analyzing environmental impacts is largely in control of the study team; obtaining environmental approval is not.

than multi-national mining companies that attract attention from a wide range of advocacy groups. Locally owned companies also tend to receive favoured treatment compared to foreign-owned companies. Unfortunately there is also a common misperception in developing countries that foreign mining companies have unlimited financial resources; requests for gifts or 'financial' contributions to support approval are not uncommon.

The above illustrates that the length of time needed to complete an EIA will eventually hinge on the extent of co-operation received from third parties such as local government, the level of interest and support demonstrated by the community, the skills of the EIA team, and to a lesser extent, on the EIA methodologies employed. When all these factors are combined, the completion and approval of an EIA will vary anywhere from one year to three years for large mining projects. Accordingly, an EIA prepared at the outset of project planning should take no longer to complete than the project design. Unless the environmental assessment proves to be highly complicated (or the project proves to be highly controversial), this is probably more time than is really necessary.

Completion and approval of an EIA will vary anywhere from one year to three years for large mining projects.

Because the preparation of an EIA is a major undertaking, the success and quality of the outcome largely depends on the expertise of the study team. Based on the complexity and nature of the project, the study team needs to draw upon the skills of specialists in physical, chemical, technical, biological, social, and economic sciences, expertise that may not be readily available in the host country. Another aspect of expertise is familiarity with the environment and regulatory setting of the project area and familiarity with the proposed mining technology. As a general rule, subjective approaches and methods should not be used in environmental assessment without familiarity with local environmental conditions. Some analyses are not feasible unless extensive data sets are available. This constraint must be recognized, as well as the cost and time requirements of collecting and interpreting such data.

Objectivity of Impact Assessment

An environmental assessment is never entirely objective.

An environmental assessment is never entirely objective. In spite of the best intentions of involved stakeholders, be it the mining company, the appointed EIA consultant, the approval authority, or involved advocacy groups, stakeholders tend to be biased. And even if they were entirely objective, financial, technical, or scientific limitations probably introduce subjective view-points. This important aspect will be discussed in more detail later in this chapter. In practice, however, a wide variety of view-points is usually found among those involved – even within the EIA study team – and the presence of opposing view-points tends to mitigate against any systemic bias.

9.2 DECIDING ON A DIRECTION

A common approach used by scientists in the study of complex relationships is to break down the relationships into individual components by identifying characteristic elements and traits of the subject matter under consideration. In the study of environmental impacts, characteristic elements and traits are impact categories and significance attributes of impacts.

Since an environmental impact is independent of its categorization, categorizing impacts is a tool, not an end.

Impacts are categorized as direct, indirect, and cumulative, and as impact interactions, impact categories that overlap (**Table 9.2**). Since an environmental impact is independent of its categorization, categorizing impacts is a tool, not an end.

TABLE 9.2

Categorizing Environmental Impacts – Since an environmental impact is independent of its categorization, categorizing impacts is a tool, not an end

Direct (or primary)	Impact that results from the direct interaction between project activity and receiving environment (such as between effluent discharge and receiving water quality).
Indirect (or secondary)	Impact that follows from primary interactions between a project activity and its environment as a result of subsequent interactions within the environment (such as soil loss as a consequence of land clearing affecting down stream aquatic habitats).
Cumulative	Impacts acting together to affect a particular environmental resource or receptor. Several types of cumulative impacts are defined in the literature. Temporal: a series of impacts, in themselves not insignificant, occurring repetitively to build up to the point that they become significant (such as family stress developing due to a fly in/fly out working arrangement). Accumulative: the overall effect of different types of impact (such as air pollution and noise and traffic and visual blight) on a single receptor (such as a community or a habitat) where each single impact may not be significant, but combined they are. Additive: where impact from the planned activity occurs at the same time as impact from activities being undertaken by other parties. Interactive: where two different types of impact (which may not in themselves be significant, say overburden placement and increased infiltration) react with each other to create a new impact (that might be significant such as acid rock drainage). Synergistic: where two impacts interact together (e.g. changes in water quality with respect to two different pollutants) to create an impact that is greater then the sum of their parts.
Induced	Impact originating from other developments or activities that are encouraged to happen as a consequence of the original development (such as the mine development stimulates a requirement for improved site access leading to an increase in local population).
Non-Normal (or accidental)	Impact that results from un-planned events – incidents – within the project (such as breakdowns, failures, or human error) or in the external environment affecting the project (such as floods, seismic activity, or landslides). The probability (or likelihood) of the event becomes important.

Source:
Based on SHELL (2003)

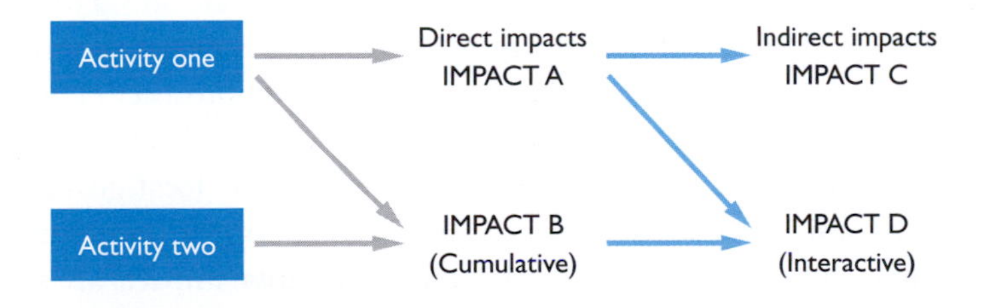

FIGURE 9.2
Categorizing Impacts

Direct project impacts are usually predictable with certainty while indirect impacts are often subtle and difficult to identify and measure.

Source:
Based on Walker and Johnson 1999

Direct impacts are defined as environmental changes directly associated with project activities (**Figure 9.2**). In mining, examples include change of topography, erosion, dust emissions, or change of water quality due to effluent discharge. Direct project impacts are usually predictable with certainty.

Indirect impacts are defined as environmental changes not directly associated with project activities, often occurring at a distance from the mine, and as a result of complex pathways. Indirect impacts are referred to as second or third level impacts, or secondary

FIGURE 9.3
Cross Media Impacts

Source:
Based on ADB 2003

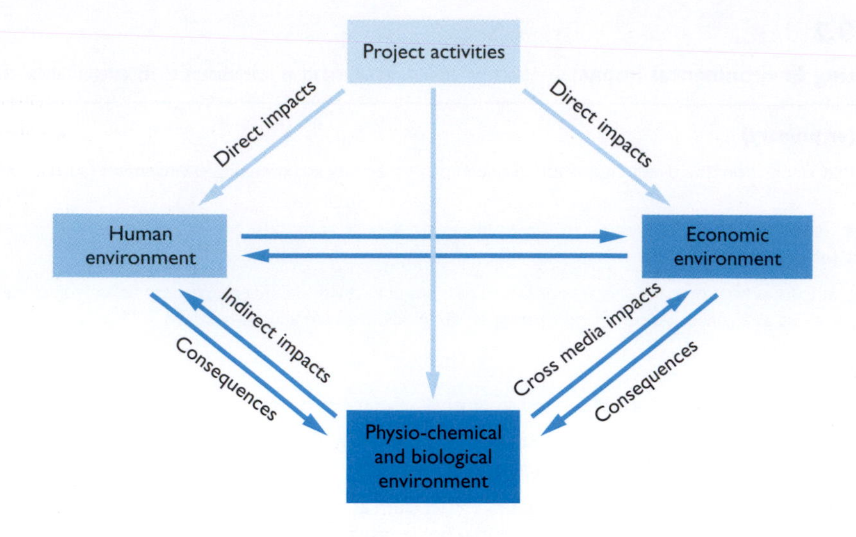

impacts. These impacts are often subtle, and difficult to identify and measure. For example mining may cause a direct change in groundwater level, thereby affecting a nearby wetland with impacts on its plants and animals. Induced development due to mining and most other social impacts are examples of indirect impacts.

Induced development does not always receive scrutiny in environmental assessment because it is accepted that it just happens.

Induced development (e.g. population increase and related environmental impacts) does not always receive scrutiny in environmental assessment because it is accepted that it just happens. Good practice, however, requires that induced development be considered in impact assessment if there is reason to believe it may occur. Ultimately, because of the uncertainty and often dispersed nature of induced development and related activities (i.e. activities may occur in many places within the host region under different jurisdictions), induced activities are best considered as part of regional development planning, involving regional administrative authorities.

Impacts on a particular environmental medium frequently affect other media (sometimes referred to as cross media impacts, **Figure 9.3**). Indirect impacts often outnumber direct impacts and tend to be more significant. An increase in noise level, for example, may have an effect on nesting birds. While the direct impact of increased noise may not qualify as significant using simple assessment methods, indirect impacts from this on the ecology may be significant. Similarly, an increase in mine production may generate only a limited number of actual employment opportunities. Doubling mine production, for example, may generate the perception of doubling working opportunities. This perceived prospect of employment will attract migrants to the area which in turn can place stress on local communities and public infrastructure, imposing social challenges on the mine, and the local government. In such cases, in the absence of indirect impact analysis, employment opportunities may be classified as presenting significant benefits, while related indirect negative impacts are severe.

Cumulative impacts are mostly indirect impacts, and are defined as incremental environmental changes caused by independent or repetitive activities. These impacts may occur as interactions between activities, between an activity and the environment, and across environment media. Cumulative effects can occur in many ways (CEAA 2003; Spalding 1994): (1) A physical or chemical constituent can be transported away from the impact source only to interact with another activity or environmental medium (e.g. soil erosion or acid rock drainage). (2) The gradual disturbance and loss of land and habitat, or nibbling loss (e.g. incremental land clearing as the mine advances or incremental road development into forest areas, alienating wildlife habitat due to sensory disturbances). (3) Spatial and temporal crowding, when too much is happening within too small an area and/or in too brief a period.

Spatial and temporal crowding, when too much is happening within too small an area and/or in too brief a period.

Spatial crowding includes overlapping effects from separate activities (e.g. downstream confluence of effluent plumes from multiple discharge points). Temporal crowding is when impacts from different actions overlap or occur before the receiving environment has had time to recover (e.g. family stress due to continuous fly-in/fly-out employment). (4) Each impact can induce further effects, sometimes called growth-inducing potential. The effects of these spin-off impacts (e.g. increase in hunting or trade of endangered species due to improved access to areas) may add to the cumulative effects in the vicinity of the mine. The magnitude of cumulative effects, or the impact propagation between causes and effects, can equal the sum of each individual effect (additive effects) or can result in an increased effect (synergistic effect).

There are many other examples of cumulative impacts – incremental noise from independent activities (either from one or several projects) or the cumulative effect of individual impacts on a particular receptor, such as human stress due to combined noise and dust exposure, or visual impacts. Impact interactions, a subset of cumulative impacts, can occur either between the impacts of one project or between the impacts of several projects. A mineral processing plant, for example, may produce two waste water streams that individually meet relevant water quality standards, but that react in combination resulting in significant pollution levels.

The opposite to cumulative impacts are cancelling, offsetting or compensating impacts. These are also quite common. Examples include water abstraction that is offset by water discharges, or increases in traffic, offset by improvements in capacity of local roads. Clearly one of the objectives of environmental management is to generate impacts that will offset the significant adverse impacts.

While categorizing impacts is helpful in identifying impacts, the fundamental question remains *'Is the impact significant?'* Defining impact significance is generally based on evaluations of a number of significance attributes. These attributes have gained common acceptance (although definitions may vary) as a means of identifying and measuring various aspects of an impact, and collectively assist in evaluating impact significance (**Table 9.3**).

'Is the impact significant?'

Direction or nature. The categorization of impacts into positive or negative (adverse) impacts is not necessarily simple as project impacts may have both positive and negative effects, for example because one group may benefit while another is disadvantaged or the impact may be positive socio-economically but not ecologically. Take household income as an example. One of the main positive impacts of mine development is the creation of significant income opportunities. Some community members however, may lose their current source of income due to displacement. In the early days of environmental assessments, study focus was mostly directed towards understanding and mitigating negative project impacts. The main question was 'How can we minimize impacts?' Even today many people may initially associate project impacts with negative project outcomes (or environmental damage). This of course does not do justice to development. If all the impacts of mining were negative, it would not exist. Fortunately the focus of environmental assessment has now shifted to maximizing positive project impacts, predominantly socioeconomic benefits for local communities and the host region, while minimizing negative impacts. The main question today is or at least should be *'How can we maximize benefits?'*

The categorization of impacts into positive or negative (adverse) impacts is not necessarily simple as project impacts may have both positive and negative effects.

The main question today is or at least should be *'How can we maximize benefits?'*

Magnitude of impact. Magnitude measures the severity of environmental effects. Effects range from minor or inconsequential with little significance to major or catastrophic with significant adverse environmental effects, which may be unacceptable. The number of project-affected people is often selected as a criterion to determine impact magnitude. ADB (1998) defines resettlement as significant if more than 200 households are displaced. When considering the magnitude of potential impacts, it is important to consider mine development within the context of regional development and the extent to which the project could

TABLE 9.3

Describing Significance Attributes – Significance attributes help to provide the answer to the following question: 'Is the impact significant?'

Nature/Direction

- Positive: Beneficial effect on environment
- Neutral: No change in environment
- Negative: Adverse effect on environment

Magnitude

- Low: Minimal or no impairment of environmental component's function or processes (e.g. for wildlife, a species' reproduction capacity, survival and habitat suitability, or for human, the number of people affected)
- Moderate: Measurable change in environmental component's function or processes in the short or medium term; however, recovery is expected at pre-project level
- Severe: Measurable change during project life or beyond (e.g. for wildlife, serious impairment of species reproduction or habitat suitability)

Reversibility

- Reversible: Environmental component recovers to pre-project level. The rate of recovery is important.
- Irreversible: Impact that causes a permanent change in the affected receptor or resource (e.g. the felling of old growth forest as a result of occupation of a site, landscape changes caused by mining).

Duration

- Short-term: Impact predicted to last only for a limited period (such as during construction, seismic studies, drilling or decommissioning) but will cease on completion of the activity, or as a result of mitigation/reinstatement measures and natural recovery. For species, impacts occur for less than one generation.
- Long-term: Impact that will continue over an extended period, (such as operation noise throughout project life or impact from operational emissions), continuous, intermittent, or repeated (such as impacts caused by repeated explosions during mining). For species, impacts occur for more than one generation.

Frequency

- Once: Occurs only once
- Continuous: Occurs on a regular basis and regular intervals
- Sporadic: Occurs rarely and at irregular intervals

Geographic extent

- Local: Impact that occurs in the vicinity of the project, and affects a locally important environmental resource (in contrast, an impact on a nearby Orangutan Conservation Area, even restricted spatially, would constitute an international impact).
- Regional: Impact that affects regionally important environmental resources or is felt at a regional scale as determined by administrative boundaries, habitat type.
- National: Impact that affects nationally important environmental resources or affects an area that is nationally important or protected.
- International: Impact that affects internationally important environmental resources such as areas protected by International Conventions.
- Trans-boundary: Impact that is experienced in one country as a result of activities in another (such as acid rain, greenhouse gases, or river pollution).

Likelihood

- Likely: high probability that impact will occur, or high certainty that impact will be significant
- Unlikely: Low probability that impact will occur, or high uncertainty in significance prediction

trigger or contribute to any cumulative effects. Often the impact of the mine is indirect change, such as induced development, which is more difficult to mitigate and which has the potential for greater environmental consequences. Perception is also important when discussing magnitude of impacts. Small positive actions and gestures, such as scholarships provided to local students, while insignificant in the larger picture of mine development, can substantially contribute to positive community perceptions towards the mine and its

Perception is important when discussing magnitude of impacts.

proponent. Similarly, small and otherwise inconsequential accidents may undermine the credibility of a proponent and create serious social tensions and negative community attitudes toward a project which may never be overcome.

Reversibility. Reversibility refers to the environmental recovery once an impact has occurred. Irreversible environmental impacts are commonly considered more significant than those that are reversible. Irreversible changes always command attention because they signal a loss of future options. Species extinction, severe soil erosion, destruction of ancient rain forest, and other habitat destruction are examples of irreversible changes. Change in land use by providing access to remote areas is also virtually impossible to reverse once land use changes commence (such as conversion of forest areas to agricultural land). If impacts are reversible, it is important to understand the rate of recovery or adaptability of an impact area. Groundwater pollution is generally reversible, but recovery may take several decades, or longer. In practice, it can be difficult to know whether the environmental effects of a mining project will be irreversible or not. It will be important to consider any planned mine closure activities that may influence the degree to which environmental effects are reversible or irreversible. Sometimes environmental change will be absolute, as in the extinction of a species, while sometimes it will be absolute for all practical purposes, as in the case of landform changes due to waste rock or tailings placement, which could only be reversed over a long period of time and with unacceptable expenditure of money and energy.

Irreversible changes always command attention because they signal a loss of future options.

Duration. This refers to the period over which an effect occurs. Long-term environmental effects may be significant. In the case of a mine, there are a number of long-term environmental effects. With few exceptions on-land waste rock and tailings storage facilities are permanent structures designed to last hundreds or thousands of years. Open pits also change the landscape permanently, except in rare cases where they are back-filled. Improvement of quality of life or education equally may last for generations. Short-term environmental effects may also be significant, especially if the short-term effects negatively affect public perception of a mine project. Community tension developed during exploration or mine construction may haunt a mining project, even though the actual impacts may have ceased long before.

Consideration should also be given to negative impacts that may develop over time. The most common examples of delayed impacts are those associated with acid rock drainage which may, in some instances, emerge after decades, sometimes after mine closure. Human health impairment associated with exposure to trace metals released by mining activities may also have latency periods of up to tens of years. Obviously, when considering future impacts, questions regarding their likelihood, latency period and duration are all important.

The most common examples of delayed impacts are those associated with acid rock drainage which may, in some instances, emerge after decades, sometimes after mine closure.

Frequency. Closely related to the duration of the effect is its frequency. The frequency of effects and the potential of the environment to recover from these effects are important. If an activity is intermittent, for example, it may allow for environmental recovery during inactive periods. A good example is turbidity. Many aquatic ecosystems have evolved in situations where surface waters are intermittently turbid. Indirect effects of mining may increase the incidences of turbidity, but if adequate intervals of clear water remain, the important ecological functions of the ecosystem may be retained.

Geographic extent. The geographic extent is defined as how far an effect propagates. Localized adverse environmental effects may not be significant, while widespread effects may. Geographic extent takes into account the extent to which adverse effects, caused by the project, may occur in areas far removed from it (e.g. the long range transportation of atmospheric pollutants), as well as how they may contribute to any cumulative environmental effects. A single stream crossing, considered separately, may represent a localized impact of small significance and magnitude; however, a number of crossings of the same stream could result in significant downstream degradation of water quality. The deterioration of fish production resulting from an access road with numerous river crossings could affect fish population in an area many kilometres away, and for months or years after construction activities have ceased. National legislation may provide guidance in determining the geographic extent of impacts. The Australian Environment Protection and Biodiversity Conservation Act (1999) defines matters of national environmental significance as (1) World Heritage properties, (2) Ramsar wetlands of international importance, (3) listed threatened species and communities, (4) migratory species protected under international agreements, (5) nuclear actions, and (6) the Commonwealth marine environment.

Likelihood. This is defined as the probability of an impact occurring. When deciding on likelihood, there are two criteria to consider: (1) Probability of occurrence – If there is a high, medium or low probability that a particular significant environmental impact (say acid rock drainage) will occur. (2) Certainty of significance – There will always be some uncertainty associated with environmental assessment, often termed as 'confidence limits'. If confidence limits are high and impacts are evaluated as significant, there is a high degree of certainty that conclusions are accurate and environmental impacts are significant. If confidence limits are low, there is a high degree of uncertainty about the accuracy of conclusions, and it will be difficult to decide whether significant environmental effects are likely or not. High certainty can lead to an unambiguous conclusion of likelihood, conversely high uncertainty cannot provide a basis for a clear-cut conclusion about likelihood. If confidence limits are low, only the probability of occurrence criterion should be used to determine likelihood.

Significance attributes, such as those in **Table 9.4**, allow impact significance to be ranked against a relative scale. Scales may be predefined by the regulatory authorities of the host country.

Hunt and Johnson (1995) propose that impacts be evaluated using a simplified risk assessment technique, as follows: (1) For each effect assign a rank order (from 1 to 5) in respect of: frequency of occurrence (F), likelihood of control loss (L) and severity of consequences (S); (2) Multiply rank orders to obtain an overall criticality factor (C): $C = F \times L \times S$; (3) Rank effects by their C values and judge the significance accordingly. Ranking criteria such as those below can be used in the calculations.

Frequency of occurrence (F)

1 = very rare e.g. infrequent production run, to
5 = continuous e.g. a treated effluent discharge

Likelihood of control loss (L)

1 = extremely unlikely e.g. complete failure of a robust process control element, to
5 = highly likely e.g. a small spillage of widely used solvent

Severity of consequences (S)

1 = very limited, localized impact e.g. local dust problem
5 = extensive and severe damage e.g. toxic spillage to a large watercourse

While objective approaches to identifying impacts and assigning significance attributes may exist, evaluating and ranking impact significance will remain largely a subjective exercise.

Table 9.5 is another example of ranking environmental impacts; in **Table 9.5** attributes such as geographic extent and duration of impact determine the impact ranking.

The discussions in this section indicate that while objective approaches to identifying impacts and assigning significance attributes may exist, evaluating and ranking impact significance will remain largely a subjective exercise. Impact evaluation should therefore remain the domain of professionals familiar with the subject matter, and, most importantly, the process of arriving at a ranking and the inputs used should be documented.

TABLE 9.4

Examples of Impact Scale Description Criteria – While objective approaches to identifying impacts and assigning significance attributes may exist, evaluating and ranking impact significance will remain largely a subjective exercise

Negligible

- Small localized impact
- Low probability of occurrence
- Impact is reversible

Minor

- Abnormal operating conditions would cause breach of legislation
- Impact and probability of occurrence both small
- Emissions are within statutory thresholds

Moderate

- Moderate impact occurring over short period
- Environment has time to recover
- Project benefits are limited to few people

Significant

- Project activity has an irreversible impact, but impact is moderate
- Project activity results in a breach of legislation under abnormal operating conditions
- Conflict with established recreational, agricultural, or other established uses of the project area
- Effect and probability of occurrence are moderate
- Tailings dam exceeds 15 m in height (World Bank 2001b)
- Project benefits entire community

Severe

- Impact is irreversible affecting a high number of people
- Impact causes resettlement of more than 500 households (ADB 1998)
- Impact exceeds legal thresholds
- Disrupts or adversely affects a property of cultural significance to a community or ethnic or social group
- Project induces substantial growth or concentration of population
- Project converts prime agricultural land to nonagricultural use

Unacceptable (applies to negative impacts only)

- High likelihood of catastrophic failure
- Loss of life
- Impact on nationally/internationally recognized environmental protection areas/heritage sites

9.3 DECIDING ON THE METHODOLOGY

There are as many methodologies to assess impacts as there are impacts. Before selecting a specific method, consideration should be given to acceptability, accuracy, relevance, confidence limits, and proportionality of efforts

Is the method *objective*? Objectivity is a prerequisite for acceptability and credibility of the environmental assessment. Objectivity diminishes the possibility that predictions automatically support preconceived notions that are usually the result of self-interest, lack of

Objectivity is a prerequisite for acceptability and credibility of environmental assessment.

TABLE 9.5
Ranking of Effects based on Effect's Attributes

Duration	Magnitude	Extent			
		Local	*Regional*	*Territorial*	*National*
Short-term	*Low*	L	L	M	M
Short-term	*Moderate/high*	L	M	M	M
Medium-term	*Low*	M	M	M	M
Medium-term	*Moderate/high*	M	M	M	H
Long-term	*Low*	M	M	H	H
Long-term	*Moderate/high*	M	H	H	H

(A ranking of L (Low), M (Moderate), or H (High) is determined based on the duration, magnitude and extent of an effect)

knowledge of local conditions, or insensitivity towards community opinions. Objectivity facilitates a fair comparison of the impacts between project alternatives. In an ideal world, impact evaluation would contain no bias.

Is the method *comprehensive*? Accuracy and completeness require a method capable of detecting the full range of potential effects, directing attention not only to obvious impacts, but also to novel or unsuspected ones (**Case 9.3**).

CASE 9.3
Ecological Risk Assessment Study of the Freeport Tailings Disposal Scheme

The Ecological Risk Assessment study to support the environmental permitting of riverine tailings disposal of the Freeport Copper and Gold Mine in Papua, East Indonesia arguably constitutes an extreme case of assessing environmental change and associated consequences in a comprehensive fashion, fully justified by the complexity of the tailings disposal scheme, and the controversy that goes with it. The study took several years to complete, engaging a wide range of task specialists, from modellers to toxicologists, and is based on several thousands of field data (Jakarta Post, April 2006).

Is the method *selective*? Often a method is preferred that focuses attention on the most relevant environmental effects. It is often desirable to eliminate unimportant impacts as early possible so they do not dissipate efforts during the detailed impact analysis. Screening, to some degree, requires a tentative predetermination of impact significance, and this may create subsequent bias. Impact assessment, however, could start with the most critical environmental concerns by formulating initial sets of impact indicators and environmental effects.

It is often desirable to eliminate unimportant impacts as early possible so they do not dissipate efforts during the detailed impact analysis.

This limited set may be analyzed, at least in part, with relatively low expenditure of resources. Analyzing only a few project impacts initially will create an appreciation, even at this preliminary stage, of the factors that are likely to be important, and those that are not. As the impact assessment progresses, impact analyses will become increasingly complex. The assessment will expand naturally as time, money, and allocated human resources permit.

Does the method estimate *confidence limits*? Subjective approaches to uncertainty are common in quite a few methods and can lead to useful impact prediction. Of course, subjective methods are generally more acceptable, if assumptions are open to critical review and, if desirable, alteration. In statistical methods standard deviation serves as measure of uncertainty. Besides representing the confidence limits, the standard deviation allows one also to consider the most likely and the least likely effect (e.g. two standard deviations from the mean). A large variation in both effects may indicate the need for further studies and/or extended monitoring. The ramifications of environmental uncertainty are considered in more detail in a later section of this chapter.

Does the method *predict impact interactions*? Environmental processes generally contain feedback mechanisms. A change in the magnitude of an environmental effect may produce unsuspected amplification or dampening in other parts of the environment. Consider populations of most large wild animals which do not increase exponentially even during the most favourable environmental conditions, due to a number of negative feedbacks such as diminishing food supplies. Methods to analyze impacts should therefore be capable of identifying impact interactions and estimating their magnitudes.

Is effort *proportional* to investigated effect? More often than not a method is selected that is either too simple to accurately assess environmental change, or too complex for a given impact. Overly simple methods do not work, since impact predictions will be neither accurate nor accepted. Overly complex methods are equally undesirable. There is little benefit in allocating substantial resources to analyzing minor effects (**Case 9.4**). These can best be dealt with using the judgements of experienced professionals.

There is little benefit in allocating substantial resources to analyzing minor effects.

9.4 LINKING CAUSE AND EFFECT

In the context of the following discussion, source is defined as the origin of an environmental impact. It is an activity that imposes change to the receiving environment. A cause

CASE 9.4
Over-complexity or not?

The BP Tangguh LNG Project in Papua, East Indonesia spent hundreds of thousands of US dollars on consultants to advise on forest clearing for the construction camp and the plant site. Direct impacts, with or without these expenditures remain the same – a cleared forest area. Could the money have been better spent on actual community development programmes or on reafforestation programmes elsewhere?

is defined as the stressor that eventually results in a change in the receiving environment. A receptor is the environmental component changed by the cause. An effect (or impact) is the change in environmental conditions traceable to a cause. An exposure pathway is defined as the physical, chemical, or biological course a stressor takes from the source to the receptors of interest (e.g. organisms or human). The pathway links the cause with the effect.

An appreciation of the Source-Pathway-Receptor (SPR) concept is helpful in identifying and evaluating environmental impacts and risks. The SPR concept is illustrated in **Figure 9.4** using the bioaccumulation of trace metals in the marine environment as an example (MSN 2005). Tailings discharged to the Java Sea are the source of impacts, and trace metals are the cause. Both seawater and sediments are primary pathways, with benthos and phyto-plankton being primary receptors. Following various pathways, trace metals will move up the food chain, eventually reaching humans, the ultimate receptor in the shown SPR model. **Figure 9.4** demonstrates that Source-Pathway-Receptor models excel in illustrating linkages between complex pathways of cause and effect, and in illustrating where data are (and are not) available. If the exposure pathway is incomplete, the stressor does not reach the receptor, and cannot cause an effect, or an environmental impact.

As illustrated in **Figure 9.4**, Source-Pathway-Receptor concepts attempt to explain how a stressor could impact an environmental component. They are conceptual models used to communicate hypotheses and assumptions about how and why effects are occurring. SPR models also indicate where different stressors may interact and where additional data collection may provide useful information. SPR models vary largely in complexity, depending on the mechanisms and ecological processes involved.

Specialist advice based on experience with similar ecological habitats is necessary when developing conceptual models.

Specialist advice based on experience with similar ecological habitats is necessary when developing conceptual models, especially when complex pathways and ecological process

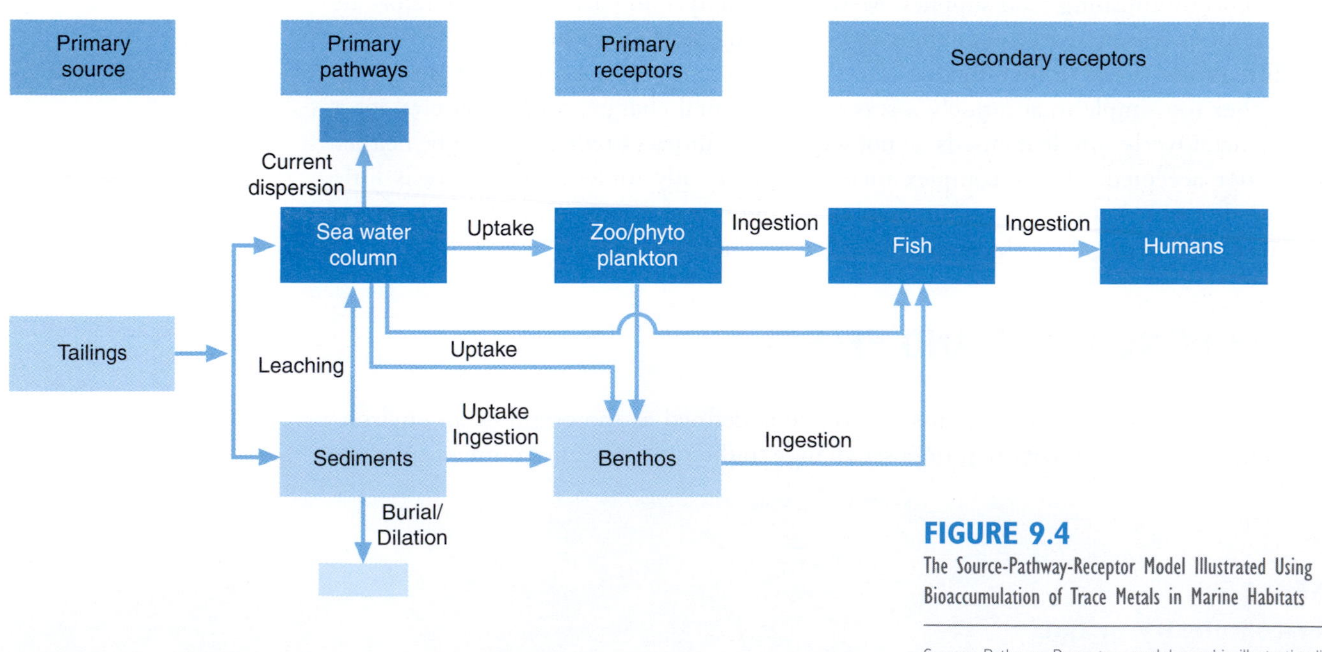

FIGURE 9.4
The Source-Pathway-Receptor Model Illustrated Using Bioaccumulation of Trace Metals in Marine Habitats

Source–Pathway–Receptor model excel in illustrating linkages between complex pathways of cause and effect, and in illustrating where data are (and are not) available.

Source:
MSN 2005

are involved. A pictorial, poster-style conceptual model is useful to introduce ecological relationships. A box and arrow diagram can follow, detailing relationships among stressors, receptors, and intermediate processes. To ensure the model does not become too complicated to be helpful, only pathways and causes relevant to a specific analysis should be considered. Separate diagrams for each stressor or pathway will help to keep the focus on the analysis steps that follow. See Jorgensen (1994), Suter (1993), Cormier *et al.* (2000), or USEPA (1998) for additional advice on developing conceptual models. As well as clarifying relationships among multiple causes and effects, the SPR concept is a powerful tool for discussing environmental change processes among the assessment team and for obtaining additional insights from external stakeholders.

9.5 IDENTIFYING PROJECT IMPACTS

In this text we differ between identifying and evaluating environmental impacts. Impact identification methods recognize how and where impacts or impact interaction may occur. Evaluation methods quantify impacts based on context and their significance attributes. Some methods are capable of both, identifying and evaluating impacts and often combinations of various techniques are used (**Figure 9.5**). Regardless of which method is selected, the method should be practical and adequate given available data, time and financial resources, and impact significance. Flexibility, data requirements, and cost are always important criteria when deciding on a particular methodology.

Regardless of which method is selected, the method should be practical and adequate given available data, time and financial resources, and impact significance.

Identifying impacts is essentially an objective exercise answering the question 'Which environmental changes are associated with mine development?' To find answers, two data sets are necessary (**Figure 9.6**).

Firstly, the receiving environment needs to be defined. Environmental assessment is a site-based approach and baseline conditions provide the context for evaluating project-induced effects. More data are required in terms of quantity, area coverage, and time horizon, when assessing indirect impacts compared to direct impacts. It is important to identify data needs early in the assessment process and to consider how data will be used before they are collected. Data collection should consider current and likely future status of potentially impacted environmental resources, so that project-induced changes can be differentiated from changes that would occur in the absence of the project. Information on historical trends, existing regulatory standards, and established regional planning are also relevant.

Secondly, information on mine development is required. Key project characteristics are ore type and reserve (such as coal, metals, or non-metals), scale of mine (annual average and

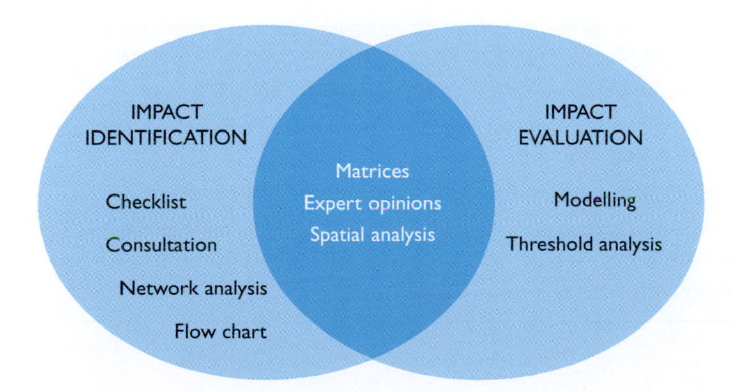

FIGURE 9.5
Methods and Tools for Identifying and Evaluating Impacts

Impacts identification methods recognize how and where impacts or impact interaction may occur. Evaluation methods quantify impacts based on context and their significance attributes.

Source:
Based on Walker and Johnson 1999

maximum ore production rate) and quantity and quality of emissions to air (gaseous emissions, noise and dust), land (waste rock and tailings), and water (process effluents, tailings decant water, and acid rock drainage). Other relevant information relates to project phasing, mine layout, auxiliary developments, labour requirement, and proposed mitigation measures. Mitigation measures themselves may result in additional environmental impacts, referred to as impact shifts. An obvious example is a waste water treatment plant that produces sludge. Identifying impacts other than the obvious ones, benefits from experience and familiarity with a wide range of projects. The ability to mentally conceptualize a wide range of possible chemical, physical, biological and human interactions is also advantageous.

Mitigation measures themselves may result in additional environmental impacts, referred to as impact shifts.

Impact assessment eventually links cause and effect, the source of impact with the impact on the receptor(s). Most impact identification methods fall in the categories of checklists, consultation, matrices, network and flowchart analysis, spatial analysis, and expert opinion.

Checklists

Checklists are built around past experience with similar projects, often adopting the Source-Pathway-Receptor concept: *'To what extent will a specific mine activity (source) affect a given environmental component (receptor)?'* Checklists can serve to remind of possible impacts caused by a proposed development. Lohani *et al.* (1997), SIDA (1998, **Table 9.6**) or World Bank (1998) provide good examples of various checklists. Standard

FIGURE 9.6
The 'Source–Pathway–Receptor' Model – Data Needs and Model Application

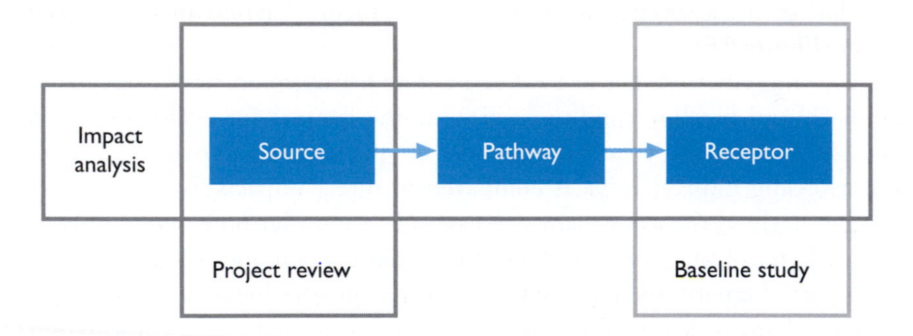

TABLE 9.6
Partial Checklist for Transport and Communications, Building, Construction, Waste Disposal and Mining Projects

People
Will the project
• result in a greater or smaller risk that diseases or other negative effects on people's health are spread as a result of pollution, poor quality building materials, poor sanitary conditions or the building of slums?
• improve or impair the living conditions of the settled population or nomadic groups?
• make it easier or more difficult, directly or indirectly, for the local population to move or to use natural resources inside or outside the project area?
• increase or reduce conflicts in respect of the present use or tenure of the land?
• damage or protect archaeological relics, places or areas of religious, cultural or historic value, and cultural monuments or make considerable changes to areas of great beauty?
• result in migration of people out of or into the project area?

Source:
SIDA (1998)

checklists exist for certain project types, commonly modified to reflect project or site-specific characteristics.

Checklists record important issues, particularly useful for the initial scoping effort. The simplest form is arguably the one with project-specific questions. One must be careful of '*yes*' or '*no*' questionnaire-type checklists, which discourage thinking and provide a false sense of assessment. Open questions such as '*to what extent?*', '*under what conditions?*' or '*in what way?*' are preferable. Checklists are often structured according to potential environmental issues, but they can also be designed to capture impacts during various project phases.

Regardless of their structure, checklists do serve as a good starting point for impact assessment, providing a simple way to identify where impacts are likely to occur. Checklists also help to focus on critical environmental issues, and to create an increased awareness of potential environmental impacts of mine development.

Regardless of their structure, checklists do serve as a good starting point for impact assessment, providing a simple way to identify where impacts are likely to occur.

The main disadvantage of generic checklists is that important site-specific factors may be omitted. They may be incomplete in their coverage, omitting important effects, or too broad in their coverage, making them difficult to manage. Checklists may also provide little help in quantifying impact significance. They tend to compartmentalize the environment, without regard for nature's complexity or the particular attributes of the subject project.

Consultation

Consultation with people that are directly or indirectly affected by the proposed mine development is particularly useful in determining their views and concerns. It also assists in defining the scope of impact assessment and identifying how and where impacts may occur, and who will be impacted. Because of this, consultation is often used at the scoping stage; however, it should be maintained throughout the project life. While consultation may identify other activities that can impact on the project, it is not always feasible during the early stages of a project when confidentiality may be an issue.

Consultation should be designed to provide answers to questions such as: What are sensitive resources/environmental receptors in the project area? What is the threshold beyond which there will be significant environmental impacts for a particular environmental component? What activities (past, present or future) may influence project impacts? Are there community members who are particularly disadvantaged by the project? Who will be unlikely to cope with the project's impacts?

It is necessary to consider carefully which respondents would be helpful in providing meaningful answers to the questions above. Questionnaires, which are particularly useful in obtaining socio-cultural and socio-economic information, should be designed to generate data suitable for use in the assessment. Practice shows, unfortunately, that while impressive quantitative socio-economic data are often collected, little, if any, subsequent interpretative analysis is carried out.

Practice shows, unfortunately, that while impressive quantitative socio-economic data are often collected, little, if any, subsequent interpretative analysis is carried out.

Matrices

In the context of evaluating environmental impacts, the term 'matrix' refers to a style of presentation rather than any mathematical implication. Matrices relate project activities to environmental components so that their intersection can be used to indicate a possible effect. Matrices are somewhat similar to checklists but present information in a tabular format, or a two-dimensional checklist. Matrices may vary according to the type and detail of information required, but all matrices are designed to present potential impacts in response to project activities.

Simple Checklist Matrix – The simple checklist matrix (Leopold 1971) can be used to identify impacts by systematically checking each project activity against each environmental component. If a particular activity is considered to have the potential to affect a particular environmental component, a mark is placed in the cell at the intersection of activity and environmental component, highlighting the need for further studies (**Figure 9.7**). Descriptive information on the nature and magnitude of impacts may replace the mark, providing information rather than just identifying whether the impact would occur or not. Patterns in completed matrices, for example columns or rows with numerous impact strikes, help to illustrate cumulative impacts on a particular environmental receptor. Likely impact interactions can also be identified.

Patterns in completed matrices, for example columns or rows with numerous impact strikes, help to illustrate cumulative impacts on a particular environmental receptor.

Weighted matrices – Some matrix applications allow the matrix to be weighted to reflect factors such as duration, frequency and extent of impacts, or to score or rank impacts.

Development	Environmental effects	Social environment									Physical environment										Biological environment													
		Recreation	Landscape/visual	Historical/cultural	Personal and social value	Risk and anxieties	Existing landuse	Land value	Employment	Public participation	Landform	Nuisance (Noise, Dust, Smell)	Climate/atmosphere	Foundation materials	Agriculture soil	Groundwater	Surface water	Sedimentation	Erosion/Land stability	River regime	Wetlands	Marine	Inter-tidal	Estuaries	Rivers	Lakes	Urban land	Crop land	Sand/shingle/rock	Herb field	Marine	Grass land	Shrub land	Forest
Prospecting	Surveys																																	
	Drilling																																	
	Sampling																								•									
Opencast mining	Overburden stripping		•	•			•	•	•		•	•		•				•	•	•	•				•									
	Blasting																																	
	Dewatering																								•									
	Crushing																																	
Under-ground mining	Methods used																																	
	Ventilating system																																	
	Dewatering																																	
Dredging	Floating plant																								•									
	Pond formation																								•									
Ore Processing	Water supply																								•									
	Washing plant																																	
	Process used																																	
	Stockpiling																																	
	Wastewater treatment																								•									
	Wastewater disposal																								•									
Tailings	Tailings dam																								•									
	Runoff control																								•									
Rehabilita-tion	Contour shaping																																	
	Planting																																	
	Overburden use																																	
General	Surface infrastructure																																	
	Access road																																	
	Energy source																																	

FIGURE 9.7
Leopold Matrix

This simple checklist matrix can be used to identify impacts by systematically checking each project activity against each environmental component.

Impact magnitudes are estimated by assigning a weight to each environmental component, indicating its importance. The impact of the project on each component is then assessed and scored.

Weighting or scoring can also provide a total score for the proposed project or for project alternatives. The most frequently used presentation of a comparison of alternatives is a matrix, in which +, 0, and – are used to indicate how each alternative affects different environmental components. This method provides a quick overview of differences between project alternatives. The presentation can be designed to illustrate the impacts of each alternative against a reference scenario (usually the existing situation); or to compare impacts of each project alternative with the preferred alternative. Using multiple signs such as + + or – – allows further differentiation.

Weighting or scoring relies heavily on professional judgement in providing ranks/weights to each project activity with respect to its environmental effects. It is important, therefore, to state assumptions made and criteria used. While illustrative, caution should be exercised in interpreting weighted matrices, because although they may provide useful indications of relative impacts, environmental impacts are often too complex to be accurately ranked by a simple numbering system.

While illustrative, caution should be exercised in interpreting weighted matrices, because although they may provide useful indications of relative impacts, environmental impacts are often too complex to be accurately ranked by a simple numbering system.

Symbolized matrices – The symbolized matrix differs from the weighted matrix by using symbols to capture significance attributes of impacts. Environmental impacts may be described with words such as 'important' or 'significant', and further classified as 'positive' or 'negative' impacts by the use of plus or minus signs (for example, + S, a positive and significant impact). Other abbreviations include 'ST' for short term and 'LT' for long term, or '10' to denote a very high impact and '1' to indicate an almost negligible one. A symbolized matrix can often be a combination of descriptive and numerical scales. Note that because these subjective, qualitative words and symbols are based on the professional judgement of the evaluator, their interpretation may differ depending on his or her cultural values, education level, or specific circumstances. Simple rating schemes are often used during early scoping exercises, before a more detailed assessment confirms or dismisses the validity of conclusions reached in the matrix.

As one would expect, a wide range of more advanced types of matrices have emerged over the years (Walker and Johnson 1999). While useful for some applications, advanced matrix types are not routinely used in environmental assessment. Whether simple or advanced, matrices do not quantify impact significance; this is only achieved by quantitative methods such as mathematical or numerical modelling; however, matrices are easy to interpret and provide a good visual summary of impacts. Matrices are especially useful when comparing project alternatives.

Network and Flowchart Analysis

Network analysis and impact flowcharts are used in most environmental assessment studies. They are based on links and interaction pathways between causes and effects, and between individual environmental components. If one component is affected, subsequent impacts will occur on those components which interact with it. Network and flowchart analyses identify pathways of an impact using a series of chains (networks) or flowcharts between a proposed activity and the receptor of the impact.

Analyzing the response of a receptor to a particular stressor and identifying where knock-on effects on other receptors or environmental components exist, enables indirect impacts and interactions between both project activities and project impacts themselves to be considered. If different activities or developments affect the same environmental receptor, cumulative impacts are identified. Feedback loops can also be incorporated into network and flowchart analyses by including a series of loops with information on existing feedback mechanisms.

The basic component of network and systems analysis is the impact chain, which illustrates the process of cause and effect including the knock-on effects to other environmental receptors (**Figure 9.8**). Linked together, complex diagrams or flowcharts are developed which include a wide range of indirect impacts and more impact interactions. There are, however, limits to the complexity that can be readily understood by reviewers. Highly complex diagrams may look impressive but may not be easily understood. Accordingly, it is preferable to produce multiple, simple flow charts or diagrams, rather than a single complex diagram that purports to identify all project interactions.

Highly complex diagrams may look impressive but may not be easily understood.

Since a direct project impact often results in a series of indirect effects, the chain becomes increasingly complex with each additional project activity. The method can be expanded to include impact interactions between environmental sub-systems. If appropriate and available data allow, it is also feasible to include some quantitative analysis in the network. The network analysis then constitutes a simple form of modelling, a quantitative impact evaluation.

FIGURE 9.8
An Example of how Impact Chains can be used to Illustrate Indirect Impacts and Impacts Interaction

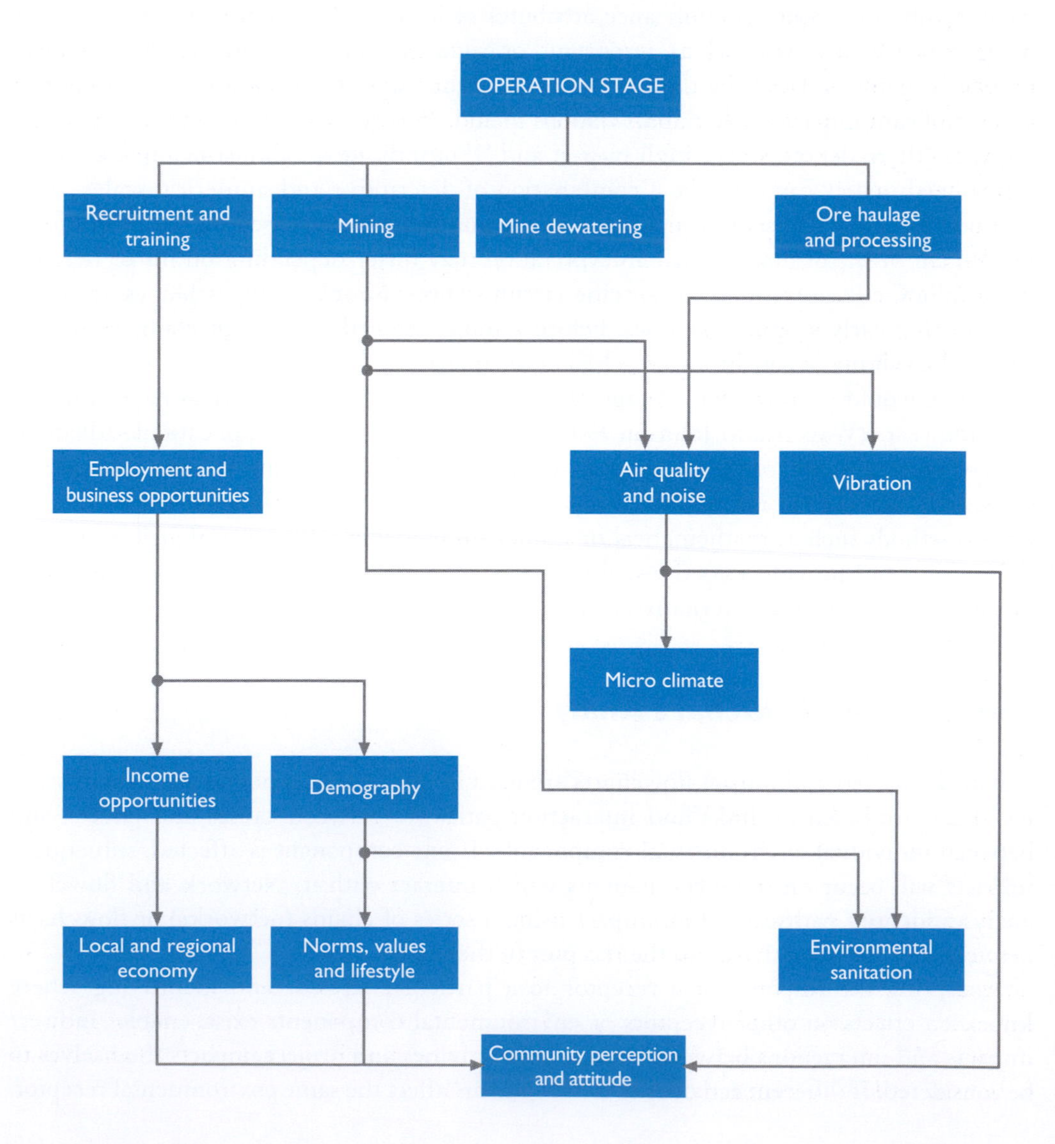

Network and flowchart analysis is a holistic approach to impact assessment, clearly illustrating the multiple and often complicated nature of project impacts, and their pathways. Unlike simpler forms of analysis, indirect impacts and impact interactions are identified using network and flowchart analysis. Processes that need further studies can also be identified.

Developing a network or flowchart analysis follows a series of individual steps: listing of all project activities; identifying environmental components and sensitive receptors; identifying primary impacts and 'knock on' effects; identifying impact interactions; and selecting either a network or flowchart approach depending on the nature of the assessment. A simple network may be appropriate at the scoping stage or when assessing project alternatives.

Large-scale mining operations affect a number of different ecosystems, and separate analyses for each ecosystem may become practical. Similarly, separate network and flowchart analyses may be established for various phases in the life of the mine.

Spatial Analysis

Overlay mapping, preferably done by Geographic Information Systems (GIS), assists in identifying the geographic extent of impacts, and can assist in identifying where cumulative impacts and impact interactions may occur. Overlay mapping is best suited for identifying physical/chemical impacts and their geographical extent. Overlay mapping involves preparing maps or layers of information which are then superimposed on one another. Overlay techniques can provide a composite picture of the baseline environment, identifying sensitive areas or resources. They help to illustrate the influences of past, present, and future activities on a project or receiving environment.

Overlay techniques can provide a composite picture of the baseline environment, identifying sensitive areas or resources.

Manual overlay mapping uses a series of transparent maps, each depicting a specific data set. When superimposed over other maps, the areas of overlapping information are highlighted. These are areas where impacts may potentially take place.

A GIS is a computerized spatial information system with a range of applications that extend well beyond the simple overlaying of information. Map production, however, remains one of the core strengths of any GIS. Spatial data are transformed in layers of information representing different environmental resources (such as watersheds or soil distribution) or impact distributions (such as simulated ground concentrations caused by project emissions). A GIS allows information to be easily overlaid, and as with manual overlay techniques, areas of potential cumulative impacts or impact interactions are identified.

Both manual and GIS overlay mapping provide valuable visual aids, and greatly assist in identifying where impacts may occur (**Figure 9.9**). Manually prepared maps or overlays are generally relatively inexpensive and quick to compile, but there are some restrictions on what can be represented manually with ease and accuracy. A GIS can be regarded as the high-tech equivalent of the manual overlay mapping method, which allows for the rapid construction of multi-layered electronic maps. A GIS is a spatial analysis system that allows the systematic compilation of a wide range of environmental data which can then be selected according a particular need. A GIS is particularly useful in dealing with large impact areas, as is the case for most mining projects.

A GIS can replace manual mapping in most applications. Because hard copy maps are static and inflexible, they are difficult and expensive to keep up-to-date. A GIS has much greater flexibility, and hard copy maps can be produced from the electronic database to meet the needs of the user. Once a site-specific GIS has been prepared, additional information can be added as and when necessary. Impacts can therefore be combined in an additive way. The main advantage of a GIS, however, is its modelling capability, allowing data to be analyzed using spatial modelling techniques.

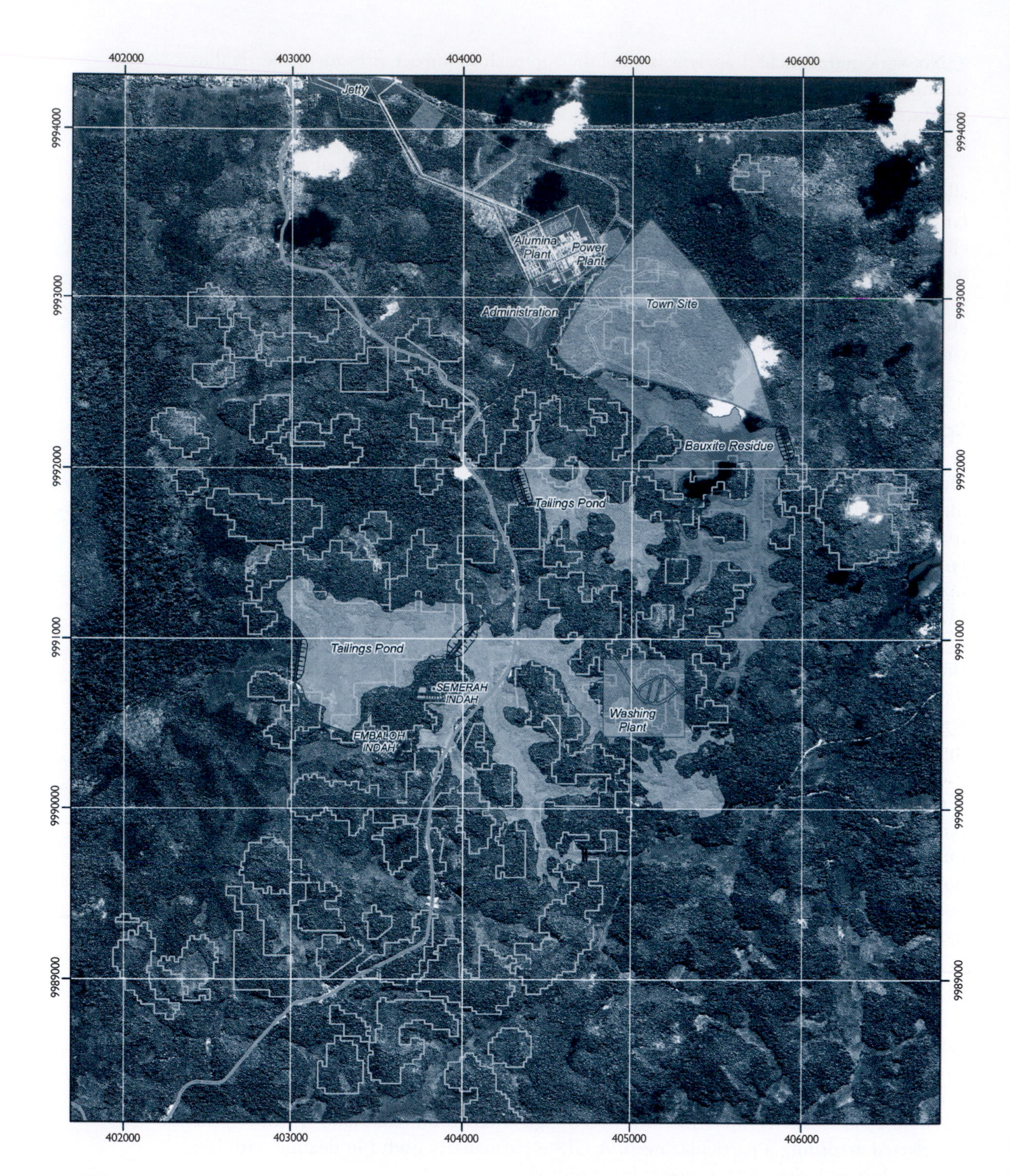

FIGURE 9.9

Illustrating Physical Project Impacts by Overlaying Baseline Information (Satellite Image) with Mine Layout (Tayan Bauxite and Alumina Refinery Project, West Kalimantan, 2006)

Expert Opinion

Professional judgements will always form an intrinsic part of environmental assessment. No matter what method is applied for identifying and evaluating impacts, it is not possible to conduct an environmental assessment without relying on expert opinions. For this reason alone, team selection is crucial to the success of any EIA. Although expert opinion is not a method, it can be considered a tool for assessing complex project impacts. The ability of standard checklists and other generic tools is limited to the accurate identification of a wide range of indirect impacts that may occur because of a project activity. Consider, for example, the range of potential impacts on an aquatic ecosystem due to mine water discharge. These include, but are not limited to (USEPA 2000): (1) death of fish, invertebrates, plants, domestic animals, or wildlife; (2) abnormalities in life forms, such as tumours, lesions, parasites, disease; (3) altered community structure, such as the absence, reduction, or dominance of a particular taxon (including increased algal blooms, loss of mussels, increase of tolerant species, etc); (4) loss of species or shifts in abundance; (5) response of indicators designed to monitor or detect biological, community, or ecological condition, such as the Index of Biotic Integrity (IBI) or the Invertebrate Community Index (ICI); (6) changes in the reproductive cycle, population structure, or genetic similarity; (7) alteration of ecosystem function, such as nutrient cycles, respiration, and photosynthetic rates; and (8) alteration of the geographic extent and pattern of different ecosystems, for example, shrinking wetlands, change in the mosaic of open water, wet meadows, sandbars, and riparian shrubs and trees. It is clear that even the most experienced aquatic biologist would be challenged to identify all potential impacts prior to further site-specific studies.

Professional judgements will always form an intrinsic part of environmental assessment.

Most impact assessments involve more than one scientific discipline, which emphasizes the need for a team approach. Sound professional judgement means engaging specialists from appropriate disciplines, with enough experience, at the right time, for a sufficient duration, and with adequate resources. Because specialists tend to work in isolation, care should be taken to ensure there is sufficient and effective discussion between them. In fact it is this interaction between members of a multi-disciplined team that provides the most effective impact assessment. While one specialist postulates an impact or evaluates its significance, other team members contribute from their own specialized view-points while others may act as 'devil's advocates'. As a result, the evaluations are likely to be well considered and the conclusions robust, well before they are subjected to scrutiny by regulators and the public. Recognizing the importance of such 'brain-storming', some jurisdictions, notably Indonesia, identify the Environmental Impact Assessment Workshop as a specific step in the Environmental Impact Analysis process.

Sound professional judgement means engaging specialists from appropriate disciplines, with enough experience, at the right time, for a sufficient duration, and with adequate resources.

The number of experts and their areas of expertise can be adapted to suit the particular mining project. Key to success of expert opinion is the appointment of an experienced team leader, identifying the requirements for specialist expertise, and ensuring cooperation between team members.

9.6 EVALUATING PROJECT IMPACTS

Once environmental impacts are identified, there is a need to evaluate their significance. Establishing significance is by no means an easy task and uncertainties will remain. Impact evaluation makes it possible to roughly rank impacts. Equal ranking of major and trivial impacts will suggest lack of judgement. If major impacts are omitted and minor ones are given too much attention that too is a warning that the environmental assessment process is flawed. Significant impacts command more attention, minor impacts less.

Significant impacts command more attention, minor impacts less.

Quantitative Assessment

Quantitative assessment methods are typically used for evaluating physical-chemical changes, for example, modelling of airborne pollutants. A wide range of models exists to quantify physical-chemical project impacts. These can range from relatively simple models, considering only one aspect of the environment, to complex models, predicting natural system responses. Although physical or analogue models do exist, mathematical models are normally used in the context of environmental impact assessment.

Mathematical Models

Mathematical models lend themselves to the spatial and temporal analysis of selected environmental aspects such as air and water quality, water volume and flows, noise levels or airborne deposition on soils and vegetation. Analytical models often provide a first estimate of the magnitude of the impact. Numerical models are applied if higher accuracy is required, but developing numerical models is generally demanding in terms of cost, expertise, time, and input data requirements. The use of proven numerical models accepted by regulatory authorities is preferable. For competent references to available mathematical models see Spitz and Moreno (1996) (groundwater), Chapra (1997) (surface water), USEPA (2005) (air), Canter (1996), and Canter and Sadler (1997). These provide excellent overviews of prediction techniques, based on US experience (**Table 9.7**).

The use of proven numerical models accepted by regulatory authorities is preferable.

Mathematical models approximate geographical and cross-media pathways from the source of the impact to its effect as shown in **Figure 9.10**. Using mathematical approximations they link inputs (x, say in terms of source and concentration) with outputs (y, say in terms of pollution concentration at the receptor). In general, the output variable (y) is a function of one or more input variables (X):

$$Y = f(X_i),\ i = 1,\ n \qquad (9.1)$$

Mathematical models range from simple analytical equations to complex numerical models. Impact evaluation can be based on one model or a combination of numerous models (**Figure 9.11**). However, evidence from case studies suggests that the use of simple models is the rule rather than the exception in most EIAs. Simple models include:

- Steady-state, single source, Gaussian plume dispersion model for air quality;
- Simple runoff model based on watershed area and rainfall;
- Water balance involving rainfall, evaporation, runoff, infiltration, and storage in a watershed or other hydrologic system;
- Steady-state dispersion model for water quality;
- Universal soil loss equation predicting erosion rates from knowledge about rainfall, slope, soil structure, vegetative cover, and management practices;
- Population dynamics, predicting the rise and fall of biological organisms and communities from knowledge of lifecycles, predator-prey relationships, food webs, and other factors affecting the lives of various species; and
- Inventory approaches for direct and higher-order effects on receptors.

Complex modelling demands accurate input data, often not available at the early stage of mine development.

Although impact evaluations based on simple models are approximates, the quality of results will depend on the particular problem and circumstances for which the model is applied. Two main factors restrict the application of complex models. First, complex modelling demands accurate input data, often not available at the early stage of mine development;

TABLE 9.7

Prediction Techniques Applicable in EIA

Air	**Source description**
	Emission inventory
	Emission factor technique
	Effect on air quality
	Box models
	Gaussian plume (analytical models)
	Single to multiple source dispersion models (numerical models)
	High order effect
	Receptor monitoring
	Pathway models for human exposure
	Others
	Urban area statistical models
	Monitoring from analogues
	Air quality indices
Surface Water	**Source description**
	Point and non-point waste loads
	Runoff models
	Effect on water quality and quantity
	Mathematical models
	Segment box models
	In situ tracer experiments
	Statistical models for selected parameters
	High order effect
	Water usage studies
	Pathway models for human exposure
	Others
	Waste load allocations
	Water quality indices
Ground Water	**Source description**
	Pollution source surveys
	Leachate testing (e.g. toxicity leachate characteristic procedure/TLCP)
	Unsaturated flow models
	Effect on water quality and quantity
	Water/solute balance
	Analytical approximation
	Numerical flow and solute transport models
	Others
	Soil and/or ground water vulnerability indices
	Pollution source indices
Noise	**Source description**
	Mobile sources
	Stationary sources
	Emission models
	Noise statistical model

(Continued)

TABLE 9.7
(Continued)

	Noise propagation
	Individual source propagation models
	Others
	Noise impact indices
Biology	Chronic toxicity testing (e.g. LD_{50}) *
	Habitat-based methods
	Species population models
	Diversity indices
	Indicators
	Biological assessments
	Ecologically based risk assessment
People	Demographic models
	Econometric models
	Descriptive checklists
	Multiplier factors based on population or economic changes
	Quality-of-life (QOL) indices
	Health-based risk assessment
Visual	Baseline inventory
	Questionnaire checklist
	Photographic or photomontage approach (still models)
	Computer simulation modelling (moving models)
	Visual impact index methods
Historical/Archaeological	Inventory of resources and effects
	Predictive modelling
	Prioritization of resources
Accidental Effects	Hazard and operability studies
	Event and fault tree analysis
	Consequence modelling

Source:
modified from Canter and Sadler (1997)

* LD50 ("Lethal Dose, 50 %"): The basic idea (and practice) of the test (and similar lethal dose tests) is to take healthy animals (usually mice or rats but sometimes dogs, monkeys or other animals) and force feed them enough poison to kill (usually slowly) approximately 50% of them. (Variations include starving the individual before testing, injecting the tested substance, or coating the animal's skin with the tested chemical.)

and second, there is a diminishing return achieved in terms of accuracy in impact prediction relative to modelling effort, compared with simple predictive models.

The number of independent variables to be considered in mathematical modelling and the nature of the relationships between them are determined by the complexity of the environmental system. The aim in mathematical modelling is to minimize the number of variables and keep the relationships as simple as possible, at the same time retaining a sufficiently accurate and workable representation of the environmental system.

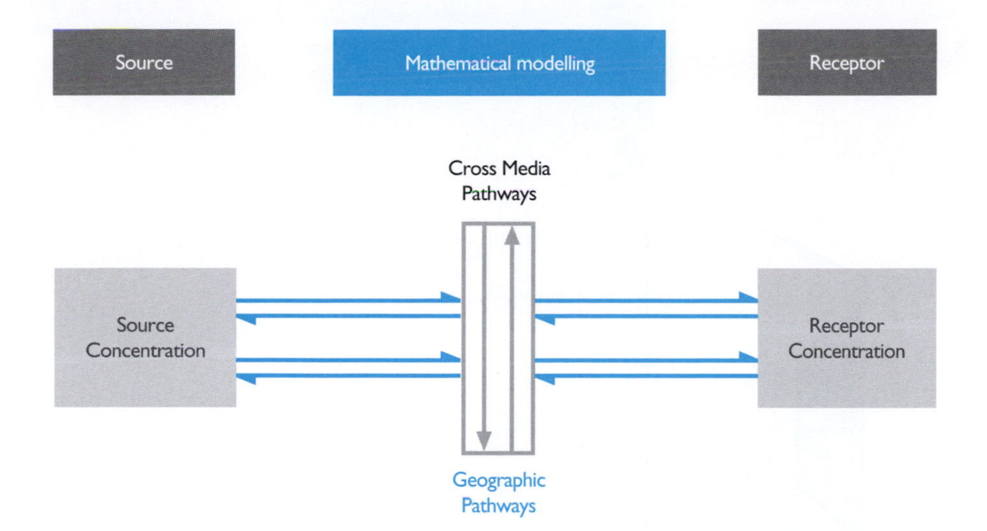

FIGURE 9.10
The Concept of Mathematical Modelling in Environmental Assessment

An example of a mathematical model found in practically all environmental assessment studies is the simple form of the Gaussian plume dispersion equation for predicting concentrations downstream of a source in uniform flow, illustrated on steady-state, single source plume dispersion modelling for air quality (total reflection at ground level):

$$c = \frac{Q}{\pi u \sigma_y \sigma_z} \exp\left(\frac{-h^2}{2{\sigma_z}^2}\right) \tag{9.2}$$

where c is the ground level concentration at a downstream distance of x metres along the symmetry axis; Q is the rate of emission; u is wind velocity; h is the height of the emission (stack height plus plume rise); and σ_y and σ_z are the lateral and vertical dispersion coefficients calculated for the required value of x from standard empirical formulae appropriate to the emission height, the roughness of the surrounding surface, and the atmospheric stability.

The above equation illustrates that the model prediction is only as good as the input data allow, a simple truth that holds true for all model applications. Using the above formula, an error of 20% in rate of emission will result directly in an error of 20% in the predicted concentration values. The accuracy of mathematical modelling is often controlled by empirical model parameters. The dispersion coefficients σ above must be defined with consideration of local conditions in terms of atmospheric stability, topography, and ground surface roughness, all of which are difficult to derive.

The universal soil loss equation (USLE) is another example of a simple equation that does not necessarily stand for a simple analysis; this is discussed further in Chapter Twenty. Empirical parameters used in the USLE are equally difficult to estimate, and long track records of site-specific historical data are essential for meaningful application of the soil loss model.

Uncertainties in model parameters will always require model calibration and validation, and analyzing model sensitivities. Model calibration is essentially done by reconciliation of predicted and observed values; then subsequent model validation compares predicted values against a series of field data from a period with changed conditions. At a minimum, validation is done by comparing predicted and observed values for a data set different to the data set used for calibration. If significant discrepancies exist, model input data should be reviewed and calibration and validation cycle need to be repeated.

Uncertainties in model parameters will always require model calibration and validation, and analyzing model sensitivities.

After calibration and validation, the mathematical model is ready for use. First, the model is applied to existing conditions without the project (for example, evaluation of runoff from a catchment area without the mine, or down stream water quality without

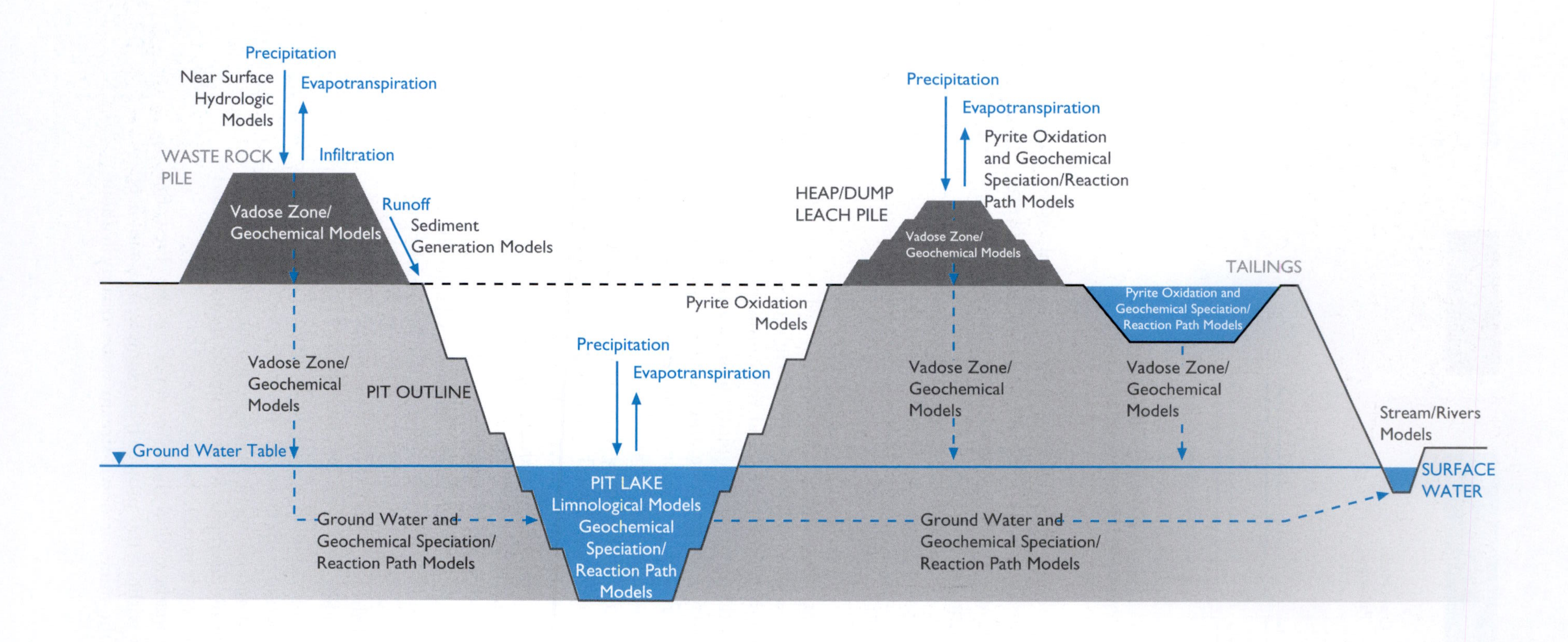

FIGURE 9.11
The Range of Models Illustrated Using Water Flow and Solute Transport as Example

Source:
Based on Kuipers *et al.* 2006

effluent discharge). Next, proposed project parameters are added (catchment area affected by mining, or predicted effluent discharge quality and quantity) and the model is applied again to predict changes in environmental quality after mine development. The difference between both predictions is equal to the project-induced environmental change.

Sensitivity analysis is an integral and, in many respects, the most important component of impact evaluation. The sensitivity analysis enables the relative importance of selected parameters to be evaluated. This is especially important considering the many uncertainties in input data in most model applications. Usually input data are selected as 'most likely' values. The relative response of outputs to changes in inputs is their sensitivity. Sensitivity analysis aims to determine the effects of individual factors and their variations on the overall results of an analysis. By changing the value of an input variable, the response of a system to new external influences is tested.

A simple yet powerful method of predicting average environmental impacts and their likely range is the Monte Carlo analysis, a form of probabilistic modelling. The Monte Carlo analysis uses randomly generated sets of model input data, representing input probability distributions. Each set of possible inputs is used to calculate a set of outputs. Calculations are repeated hundreds or thousands of times. Outputs are treated as a statistical sample, analyzed to determine the mean, standard deviation, and confidence limit. Independent of the input data, distribution outputs will follow Gaussian distribution (The Central Limit Theorem). Monte Carlo analysis provides the mean and the standard deviation of predicted outputs. The standard deviation of predicted outcomes is as important as the mean value, since it enables confidence limits of impact predictions to be defined.

The model application should reflect the quality of the model input data. Increased model complexity often closely relates to increased uncertainty in the input data. Errors in estimating the model input data are carried over to the model application, producing less accurate model outcomes. In complex environmental systems, impact predictions become difficult; at the same time, financial and scheduling constraints often emerge.

The model application should reflect the quality of model input data.

Mathematical models are best suited to assess physical-chemical changes, rather than to predict changes in the biological and human environment. Here other methods such as ecological (species habitat) and visual modelling, or threshold analysis are preferred.

Ecological Modelling

The cumulative effects of impacts on species populations or habitats can be examined using ecological models which represent component processes of natural ecosystems. They provide a simplified representation of dynamic, complex systems which often have many interacting components. As with numerical models, ecological modelling can be time consuming and often command significant commitments in establishing complex processes and related model input data. Again the accuracy of ecological modelling depends largely on the available input data and the relationships between variables within the model.

Ecological modelling can be time consuming and often command significant commitments in establishing complex processes and related model input data.

Examples of successful ecological modelling exist mainly in the forestry sector. Fragmentation of forest has a certain discernible effect on biodiversity, populations of a particular species or stream bed conditions within the area. Variables considered by such models might include the effect of habitat loss, genetic isolation, and edge effects. Ecological models are equally applicable to numerous other ecosystems. Because ecological models focus on cause and effect linkages, they can generally differentiate additive and interactive processes. As such they offer one of the best methods for analyzing cumulative and interactive environmental change, but ecological models are also some of the more difficult models to establish.

The study of ecology is a well-established and broad discipline. The reader is referred to Jorgensen (1994) for authoritative subject discussions.

Visual Modeling

Modelling of the visual impacts of a project on the surrounding area can be achieved through the creation of photomontages. These can be generated manually or on a computer using GIS software and can provide an idea of the overall visual impact on the landscape. The assessment will, however, be subjective as it is not possible to express visual impact on the landscape in numerical terms.

Threshold Analysis

Thresholds may be expressed in terms of goals or targets, standards and guidelines, carrying capacity, or limits of acceptable change, each term reflecting different combinations of scientific data and societal values.

Threshold analysis is based on the assumption that thresholds exist in most ecological systems, specifying limits in an environmental medium (predominantly but not to limited to air and water) that must be not exceeded, or levels of environmental quality that must be maintained. Thresholds may be expressed in terms of goals or targets, standards and guidelines, carrying capacity, or limits of acceptable change, each term reflecting different combinations of scientific data and societal values. Regulatory thresholds may change in time, especially in developing countries where institutional settings are still in their infancy. A threshold can be a maximum concentration of a certain pollutant beyond which health may be adversely affected, a maximum number of hectares of land cleared from its existing natural state before visual impacts become unacceptable, or a maximum number or proportion of animals lost from a habitat before the viability of the population is threatened. Other examples of environmental threshold limits are ambient environmental standards set by governments defining the degree of environmental quality that must be maintained in an environmental resource to support its continued beneficial human use. While often set to protect human health, thresholds may also be set to ensure long-term sustainability of an environmental resource or any other resource considered worthwhile to maintain. Examples are limiting the number of visitors to a natural recreation area (recreational carrying capacity), or to a mining area (operational carrying capacity). In a social context, carrying capacity could mean the limits of development that existing public infrastructure can support, or limits of newcomers into an area that can be integrated without changing the existing social fabric. Emission standards represent yet another group of threshold limits defining the maximum acceptable quantity and/or concentration of pollutants that may be discharged into the environment.

Threshold analysis is best suited to situations where regulatory thresholds exist.

Threshold analysis is best suited to situations where regulatory thresholds exist. By identifying regulatory thresholds, the mining project can be systematically assessed in terms of its environmental impacts in relation to established regulatory carrying capacities. Impacts due to project emission are said to be acceptable (no significant impact) if they meet regulatory emission standards. However, subsequent quantitative assessment may still be required to predict changes in the receiving environment which may or may not indicate a violation of regulatory ambient environmental standards for environmental resources (**Case 9.5**).

Project-specific thresholds can be developed for site-specific critical environmental parameters. The remaining forest area within a valley, for example, could constitute the controlling factor in determining whether the population of a certain bird species has reached its viable minimum. The minimum forest area that can support the bird population then becomes the threshold against which the impact of land clearing associated with the mining project is assessed.

Regulatory thresholds are generally more commonly used in identifying project impacts. If regulatory thresholds exist, they are easily compiled. The threshold analysis is then simply reduced to a checklist which compares regulatory thresholds (e.g. emission and ambient standards) with the corresponding project information (expected project emissions and expected change in environmental quality). Because pollutants affecting human health and constituents in air and water are usually regulated, these thresholds are useful for assessment purposes. Consideration of human health is often implicit in thresholds established for bio-physical components such as air quality.

In practice, however, impact assessment is often hindered by a lack of thresholds; which is particularly true for terrestrial components of ecosystems. Because of this, there is not always an objective technique to determine appropriate thresholds, and professional judgement must be used. In the course of an environmental assessment, and in the absence of defined thresholds, the study team can either (1) suggest an appropriate threshold; (2) consult various stakeholders, government agencies, and technical experts; or (3) acknowledge that there is no threshold, determine the residual effect and its significance, and let the reviewing approval authority decide if a threshold is being exceeded.

Qualitative Assessment

Often it is either impossible or impracticable to approximate nature simplistically using models. In other cases, models would require more resources for their successful application than could be justified by the project requirements.

CASE 9.5
The Importance of Considering Emission and Ambient Environment Quality Standards

Question: Is the discharge of mine dewatering water with a pH value of 7 into the aquatic environment acceptable? It depends. If the aquatic environment is swamp or peat water with a pH value of, say, 5, discharge of water of excellent quality and a neutral pH of 7 will negatively impact the natural ecosystem. It is the difference of individual chemical characteristics between discharge and ambient waters that is important, not only the absolute values.

An alternative to quantitative modelling is the less formal qualitative approach, which is based on expert advice, experience drawn from historical and scientific evidences, or conceptual models. Professional judgement based on the experience of a single expert from activities of other mining operations under similar environmental settings is the most informal method of qualitative impact assessment. From this starting point, formality can be increased by:

- substantiating the expert opinion with written or mathematical descriptions of source-pathway-receptor relationships, and/or supporting the findings by reference to historical and scientific evidence;
- relying on a group of experts for their individual opinions and taking a view of their overall conclusions; or
- consulting with a group of experts to find consensus in some formal structure and agree on their opinion of likely environmental effects. Joint workshops with mine proponent, environmental study team, and invited task specialists or community members, may provide the best approach to quantifying impacts, and to develop impact mitigation or enhancement measures.

It would be overly optimistic to expect a single environmental generalist is able to assess all impacts of a major project.

It would be overly optimistic to expect a single environmental generalist is able to assess all impacts of a major project, such as a mine development. To understand project activities that potentially interact with the environment, mining engineers should be involved. If a river is potentially impacted, inputs from an aquatic biologist or fish biologist are required. If resettlement is involved, then an anthropologist or sociologist experienced in resettlement and land use planning will be essential. Involvement of social and community development specialists is essential, commencing with project disclosure. Biodiversity specialists are needed if tropical rainforests or other sensitive habitats may be affected, and so on. It is generally accepted that a major mining project located in a relatively natural environment will require input from a number of specialists on the environmental assessment team over a period of two years or more while impacts are evaluated and management measures devised.

Prediction by analogy is where environmental impacts are predicted by direct extrapolation from similar activities at an existing mine site. Conclusions may be adjusted to accommodate different conditions at the site of the proposed activity. The analogy between an existing mine and a new project depends on the extent to which the sites are similar. Predictions based on comparable experiences are always preferred to estimates with no basis of direct observations.

Predictions based on comparable experiences are always preferred to estimates with no basis of direct observations.

Impact prediction, of course, is made easier if the proposed development is an expansion of an existing mine operation. However, extrapolation of existing impacts still needs care and understanding, since trends and correlations may not be linear and continuous. The mine expansion, for example, may reach into new and sensitive watershed areas previously unaffected by mining, with a new suite of potential changes to the environment.

Direct impacts on plants and animals can be evaluated by comparing areas of habitat destroyed with total areas of these habitats in the vicinity. In this connection the territories or ranges occupied by mammal and bird species of concern is also an important consideration. For most threatened or endangered mammal and bird species, the home ranges occupied by males, females and/or family groups are well known, as are the habitat requirements. The requirements of other, less conspicuous species may not have been studied and can only be assessed by comparison with similar organisms. Indirect impacts are more difficult to predict. A case in point is the effect of project operations and the associated noise, vibration, and lights at night, on wildlife living in nearby areas. Opponents of a project typically claim that the impacts of these activities cause damage to animal populations well beyond the boundaries of the project 'footprint', and that many of the threatened or endangered species are particularly vulnerable in this regard. There is usually little evidence to support or refute such claims, and it is difficult to obtain such evidence during the course of an EIA study. However, there are occasions where relevant observations can be made as illustrated in **Case 9.6**. In the authors' experience, mobile forms of wildlife as distinct from domesticated animals (and including both terrestrial and marine species) are capable of recognizing and avoiding danger. For example, birds do not fly through plumes of smoke issuing from chimney stacks. In addition, as evidenced by the case study, many species adapt readily to changed circumstances. Of course this is not true for all species; however, it does appear that indirect effects on wildlife are commonly exaggerated, even by biologists.

Mobile forms of wildlife as distinct from domesticated animals (and including both terrestrial and marine species) are capable of recognizing and avoiding danger.

Indirect effects on wildlife are commonly exaggerated, even by biologists.

Assessing impacts on biodiversity requires a two-part approach. Firstly, a list of rare, threatened and vulnerable species is compiled and the effects on each are assessed considering their local status, habitats affected and the ranges occupied by the subject species, and the degree of protection afforded by conservation reserves, if any. Secondly, to evaluate the potential for loss of biodiversity at a less species-specific level, the areas of each habitat that will be destroyed, damaged or significantly altered as a result of the project are calculated, and compared with the unaffected areas of each habitat in the surrounding areas. It follows from this that baseline information on habitat distribution is required for a region extending well beyond the immediate project footprint.

9.7 CULTURAL HERITAGE SITES AND MINE DEVELOPMENT

Cultural heritage sites can be defined as a human work or place that gives evidence of human activity or has spiritual, cultural, or historic value (CEAA 1996). Evaluating the significance of cultural heritage sites and predicting how mine development may impact such sites, poses particular difficulties. The challenges are threefold: (1) What makes a site a cultural heritage site? (2) How to define its significance? (3) How to objectively judge project impacts? Relevant legislation and guidelines from the host country or, if unavailable, from other jurisdictions help to answer these questions (CEAA 1996).

What makes a Site a Cultural Heritage Site?

Cultural heritage sites hold tangible and intangible attributes. They are distinguished from other resources by virtue of the historic value placed on them through their association

CASE 9.6
Bighorn Sheep near a Proposed Mine in Idaho, USA

In the early 1970s one of the authors was involved in EIA studies for an underground mine at the site of a historic silver mine at Bayhorse, in the Salmon River watershed in Idaho. State authorities identified the proposed mining area as being along the migration route for a herd of Rocky Mountain Bighorn Sheep (Ovis canadensis canadensis). Rightly or wrongly, the Bighorn Sheep is recognized as a species that is sensitive to intrusion. As a result of these considerations, a 12-month study was implemented using radio transmitters to identify the areas where the sheep travelled. This expensive study contributed little apart from confirming what was already known, that the herd did pass through the proposed mining area. Coincidentally, the author was working on another project while the Bighorn Sheep tracking study was underway. This project was for the expansion of a limestone quarry and crushing plant on the front range of the Rocky Mountains near the city of Colorado Springs in Colorado. During fieldwork for this project, a herd of Bighorn Sheep was observed on several occasions in the vicinity of the mine and on one occasion, the herd was observed grazing around the foundations of the Crushing and Screening Plant, **during operations**. So much for sensitivity.

with an aspect(s) of human history (CEAA 1996). This interpretation can be applied to a wide range of resources, including natural landscapes and landscape features, archaeological sites, structures, engineering works, artefacts, and associated records (**Table 9.8**). Frequently, cultural resources occur in complexes or assemblages, including movable and immovable resources, resources that are above and below ground, on land and in

TABLE 9.8

Defining Places of Cultural Heritage Value – Cultural heritage sites are distinguished from other resources by virtue of the historic value placed on them through their association with an aspect(s) of human history

Aboriginal Cultural Landscapes

An Aboriginal cultural landscape is a place valued by an Aboriginal group (or groups) because of their long and complex relationship with that land. It expresses their unity with the natural and spiritual environment. It embodies their traditional knowledge of spirits, places, land uses, and ecology.

Archaeological Sites

National significance can be based on one or more of the following aspects: (1) there is substantive evidence that the particular site is unique; (2) it satisfactorily represents a particular culture, or a specific phase in the development of a particular cultural sequence; or (3) it is a typical and good example.

Parks of National Significance

A park may be of national significance because of (1) the excellence of its aesthetic qualities; (2) the unique or remarkable characteristics of style(s) or type(s) which show an important period or periods in the history of the country or of its horticulture; (3) the unique or remarkable characteristics reflecting important ethno-cultural traditions which show an important period or periods in the country's history; (4) the importance of its influence over time or of a given region of the country by virtue of its age, style, type, etc.; (5) the presence of horticultural specimens of exceptional rarity or value; (6) its exceptional ecological interest or value; (7) its associations with events or individuals of national historic significance; or (8) the importance of the architect(s), designer(s), or horticulturalist(s) associated with it.

Historic Districts of National Significance

Historic districts are geographically defined areas which create a special sense of time and place through buildings, structures, and open spaces modified by human use and which are united by past events and use and/or aesthetically, by architecture and plan.

Sites Associated with Persons of National Historic Significance

The national significance of an individual should be the key to designating places associated with him or her; the nominated sites must communicate that significance effectively. For a site to be designated for its association with a nationally significant person, the nature of the association is important, and will be one or a combination of the following: (1) The site is directly and importantly associated with the person's productive life, often best representing his or her significant national contribution. (2) The birthplace, childhood home, or the site associated with the person's formative or retirement years, which should relate persuasively to the national significance of the person. (3) The site that is attributed to be the source of the inspiration for the individual's life work, which may require scholarly judgement of that relationship. (4) The site associated with a consequential event in the person's life and is demonstrably related to his national significance. (5) The site that has become a memorial (that is, that has symbolic or emotive associations with a nationally significant person) and demonstrably shows the significance of the person in the light of posterity.

Graves and Cemeteries

A cemetery (1) representing a nationally significant trend in cemetery design; (2) containing a concentration of noteworthy mausoleums, monuments, markers or horticultural specimens; (3) being an exceptional example of a landscape expressing a distinctive cultural tradition.

Historic Engineering Landmarks

To merit inclusion on the list of engineering landmarks, a site has to meet one or more of the following guidelines: (1) Embodies an outstanding engineering achievement; (2) Is of intrinsically outstanding importance by virtue of its physical properties; (3) Is a significant innovation or invention, or illustrates a highly significant technological advance; (4) Is a highly significant country adoption or adaptation; (5) Is a highly challenging feat of construction; (6) Is the largest of its kind, at the time of construction, where the scale alone constituted a major advance in engineering; (7) Has had a significant impact on the development of a major region in the country; (8) Has particularly important symbolic value as an engineering and/or technical achievement to the nation or to a particular cultural community; (9) Is an excellent and early example, or a rare or unique surviving example, of a once-common type of engineering work that played a significant role in the history of national engineering; and/or (10) Is representative of a significant class or type of engineering project, where there is no extant exceptional site to consider for inclusion.

Source:
Based on CEAA (1996)

water, and whose features are of both natural and human origin. Not all cultural heritage resources have official designation status. In developing countries, many cultural heritage sites may not even be formally recognized or documented. In developed countries, details of sites may be maintained on registers which can be accessed through criteria such as type, age, location etc.

Not all cultural heritage resources have official designation status.

How to Define its Significance?

The Newfoundland Museum (www.nfmuseum.com) defines significance as follows.

Scientific Significance

Historic resources may be scientifically significant in two respects. First, an appropriate measure of the scientific significance of a site is its potential to yield information that, if properly recovered, will substantially foster the understanding of local or national heritage. In this respect, a historic site should be evaluated in terms of its capability or potential to help resolve current heritage research problems. Scientific significance should also refer to a site's potential for making substantive contributions to other disciplines, or for providing information which may be used by industry for practical purposes. The relevance of historic resource data to private industry may also be interpreted as a particular kind of public significance.

Public Significance

A site's relative potential for enhancing the public's understanding and appreciation of the past can be considered as its public significance. In this respect, a site's interpretive, educational, and recreational potential are valid indications of its public significance. Unlike those criteria for measuring scientific significance, public significance criteria such as ease of access, land ownership, or scenic setting, are often external to the site itself.

Ethnic Significance

Ethnic significance applies to historic sites which have religious, mythological, social, or other special symbolic value to an ethnically distinct community or group of people. Archaeological, historical, and architectural sites may have some degree of ethnic significance. Determining the ethnic significance of an historic site may require consulting those groups who occupy or have occupied the site, the descendants of such groups, or people who presently own or live near the site.

Historical Significance

Historically significant sites can be readily associated with individuals or events that made an important, lasting contribution to the historic development of a particular locality or larger area. Historically important sites are also those which reflect or commemorate the historic socioeconomic character of an area. This type of significance applies to both architectural and historic sites, including those of an archaeological nature. Normally, sites having high historical significance will also have high social or public significance.

Economic Significance

The economic or monetary value of an historic site, if calculable, is also an important indication of significance. In some cases, it may be possible to project monetary benefits derived from the public's use of a historic site as a an educational or recreational facility. This may be accomplished by employing established benefit estimation methods, most of which have been developed for evaluating outdoor recreation. The objective is to determine the willingness of users, including local residents and tourists, to pay for the experiences or services the site provides even though no payment is presently being made. The calculation of user benefits will normally require some study of the visitor population.

Integrity and Condition

Both the integrity and condition of an historic site are important considerations in evaluating significance. However, an assessment of integrity and condition alone is not sufficient to establish significance. These factors are probably best viewed as specific criteria for measuring certain types of significance (e.g. public significance). Integrity refers to a site's degree of authenticity, and, in this respect, pertains chiefly to historic buildings or architectural sites. These heritage properties may possess integrity of design, workmanship, materials, and/or location or setting. Condition, on the other hand, applies to all historic sites, and refers to the degree of disturbance or dilapidation of a site.

How to Objectively Judge Project Impacts?

Impacts to sites of spiritual significance are particularly difficult to evaluate, partly because it is virtually impossible for the significance to be appreciated by anyone other than those who are spiritually involved, and partly because of the secrecy and taboos that are commonly associated with such sites.

As with most other impact assessments, direct impacts are readily evaluated, but indirect impacts may prove more difficult to assess. Construction will either damage the site or it will not. Similarly, project components will either obscure the site (from any given vantage points) or they will not. On the other hand, increased population or accessibility due to the project may or may not indirectly lead to damage to a heritage site by vandalism or by accident. Again, it may not be possible to quantify the increased risk of damage due to the project, but in most cases, recognition of the risk will lead to the adoption of protective measures, so that risks are minimized.

Impacts to sites of spiritual significance are particularly difficult to evaluate, partly because it is virtually impossible for the significance to be appreciated by anyone other than those who are spiritually involved, and partly because of the secrecy and taboos that are commonly associated with such sites. Clearly, the assessment of such sites requires close consultation with those to whom such sites are important.

9.8 THE SPECIAL NATURE OF COMMUNITY IMPACTS

The term 'community' encompasses a range of definitions, and analyzing community impacts is not an easy undertaking. Community analyses vary widely but they are usually concerned with the topics and subjects listed in **Table 9.9**. Approaches to analyzing community impacts differ from most methodologies developed to evaluate physical-chemical or biological environmental changes. This text does not attempt to capture the vast body of knowledge that exists in social sciences; the interested reader is referred to specialized literature for further reading. However, the following text briefly discusses the subjects of growth inducement, community cohesion, and relocation to illustrate the many facets of community impacts, and their evaluation.

As Table 9.9 shows, community impact analysis considers how the mining project will affect people, communities, institutions, and larger social and economic systems, sometimes referred to collectively as socioeconomic impacts. Most community impacts, of

TABLE 9.9

Community Analysis – Community impact analysis considers how the mining project will affect people, communities, institutions, and larger social and economic systems

Social impacts
Relocation
Population characteristics
Community institutions
Community stability and cohesion
Economic impacts
Change in employment
Change in land ownership
Income gain or loss
Secondary business opportunities
Land use and growth
Land use changes
Induced development
Consistency of project with existing regional planning
Shift in location where growth will occur
Public Services Impacts
Education and health systems
Sanitary systems
Police and security
Accessibility
Public utilities

course, are not mutually exclusive. Social change in local population caused by relocation may also have an effect on the local economy, and vice versa. For example, a mining project that would result in the displacement of a significant number of people would have more than just social effects. There are also economic effects because relocation will affect the local labour market and existing sources of household income at the same time.

Community impact analysis is encumbered by a lack of rigorous quantitative analytical methodologies. There are few clear thresholds, standards, or formulae for identifying potential community impacts or for evaluating their significance. The significance of a potential impact is usually determined by professional judgement on a case by case basis. Much of the information on communities is considered 'soft data', involving such areas as people's perceptions, feelings, and attitudes. Soft data typically makes the acceptance of social analyses more difficult. Credibility is improved, however, through clear and concise explanations as to study objectives, methodology, and data sources.

Community impact analysis is encumbered by a lack of rigorous quantitative analytical methodologies.

Predictive tools do exist but most, created in the 1970s, are now seldom used because of their high cost, their questionable validity, and the frequent controversy that surrounds conclusions drawn from such methodologies (Caltrans 1997). Today's community impact analyses are context-based, in line with most environmental regulations that increasingly emphasize significance and context. Context implies that project impacts should be viewed within the framework of what the project locality contains.

Community impact analyses are context-based.

Growth Inducement

The environmental assessment of a mining project has to be concerned with growth inducement, arguably one of the main consequences of mining. Such induced growth may be positive or negative depending on the point of view, and planned or unplanned, but the topic is often treated superficially. Growth inducement, defined as the relationship between mine development and future growth within the project area, is difficult to understand, to predict, and to manage. The relationship is sometimes viewed as either one of facilitating planned growth or inducing unplanned growth. Both types of growth, however, must be evaluated because they may each have varying degrees of beneficial and harmful effects.

It is not difficult to comprehend how mine development induces growth or development in a localized fashion. Development of new housing complexes will be encouraged; numerous cottage industries are likely to develop; providers of fresh produce, transportation services, and restaurants will emerge. Yet, it is commonly accepted that growth patterns depend on a whole range of economic forces that are not only local, but also regional, national, and in case of large mining projects even international in scope. As a consequence no universally accepted standard analytical methods for analyzing growth inducement have emerged. In fact, there is no single way of looking at the topic. The reality is that induced development is the product of multiple social, economic, and geographic factors, none of which is easily understood.

To analyze growth inducement, an answer has to be found to the question '*Will the mine promote future economic or population growth, and if so, to what extent?* ' Economic or population growth move forward, or are held back, mainly for economic reasons, although social, political, and environmental factors sometimes play a part. An analysis of growth and development should explain these reasons in the context of the proposed mine project and any growth that would be affected by it, and why it is so. Because growth is by its very nature a future event, conclusions require professional judgements, but these judgements should be based on a thorough consideration of economic, social, political, and environmental factors. The assessment should be more reasoned than merely a statement that induced growth is inevitable.

Local governments as well as mining companies often have limited control on growth rates and patterns.

It is important to realize that local governments as well as mining companies often have limited control on growth rates and patterns. Uncontrolled growth is a common associate of mine development, placing a heavy burden on local public infrastructure and administration. Slum areas or shanty towns developing near mines are obvious signs of uncontrolled growth, with all their associated adverse impacts, such as poor sanitation, health risks, prostitution, and crime. Secondary effects from growth should be carefully considered since often they become disproportionately significant. Growth will, for example, bring outsiders to the project area, add pupils to schools, increase sanitary waste load, add traffic to existing roads, and may eliminate habitats for plants and animals. If these indirect impacts become so great that local communities are negatively affected by eroded community cohesion, decreased access to schools, poor sanitation, deterioration of public roads, or by a decline in local plant and animal populations, then these secondary effects become important. Other indirect impacts of growth inducement include leapfrog development, pressure for the development of environmentally sensitive lands, increased energy consumption, changes in land prices, and perceived changes in quality of life. Often, these indirect impacts can be discussed only in qualitative terms, and only direct impacts can be quantified.

The discussion of growth inducement has to include governmental regional planning. A project's compatibility with governmental planning (if such plans exist) is an important consideration. Inconsistencies can become controversial, generating local opposition to the mining project. More importantly, mine development will become a nucleus for regional growth independent of what is stated in existing regional planning documents. To maximize the benefits for the region as well for the mine, mine planning and regional planning need to be carefully aligned. **Chapter 22** provides a more detailed discussion of this subject.

To maximize the benefits for the region as well for the mine, mine planning and regional planning need to be carefully aligned.

Community Cohesion

It is often argued that the social costs of mining projects are borne by those communities near the mine, while the benefits are shared by a larger population located mainly elsewhere. For this reason, analysis of social impacts should be generally directed at the community level, where the majority of negative impacts are to be felt, without neglecting regional social benefits associated with any mine project.

It is often argued that the social costs of mining projects are borne by those communities near the mine, while the benefits are shared by a larger population located mainly elsewhere.

Community cohesion is the degree to which residents have a 'sense of belonging' to their community, a level of commitment and attachment of residents to the community, and its formal and informal institutions, usually as a result of continued association over time. Cohesion depends on interactions among the individuals, groups, and institutions that make up a community. Mining projects tend to be disruptive to cohesive communities due to the influx of large numbers of outsiders, especially during the construction period when large numbers of people enter the project area for a relatively short period of time, with no particular commitment or attachment to existing communities unless it is in their interest.

Mining projects tend to be disruptive to cohesive communities due to the influx of large numbers of outsiders, especially during the construction period.

One of the traditional tools for measuring community cohesion is the stability index. The stability index, a methodological approach used for more than a quarter of a century, is based on the assumption that the longer people live in a community, the more committed they become to it and the more cohesive the community. A quick snapshot of a community may be gained by analyzing existing demographic data, providing that this includes information on how long community members have been in their current residence. The stability index may be most useful when it is viewed as just a rough indicator of neighbourhood stability. In any event, it is essential that community studies are backed up with direct field observation and other primary data collection.

Relocation

Relocation or displacement is discussed in depth in Chapter Fourteen. It has three aspects: the number and type of families and businesses displaced; the probability that comparable housing relocation sites can be found for those affected; and the psychological and economic impacts associated with relocation. Relocation impacts are among the most sensitive community-related impacts of development; they tend to modify relationships between people and their community. The relocation of families or businesses from their existing locations affects not only the relocated people themselves, but also those who are left behind and those who live in the areas where the relocated people will live.

Relocation impacts are among the most sensitive community-related impacts of development.

As with most approaches to social impact evaluation, no simple analytical formulae or engineering approaches exist to identify and evaluate the significance of relocation impacts. World Bank (2001), IFC (2006), ADB (1998), and other financial institutions have developed useful frameworks that can guide the relocation process, and assist in evaluating its significance. It is widely accepted that not all social impacts associated with displacement are offset by financial compensation or physical relocation. On the contrary, financial compensation is discouraged if relocation involves rural communities without a well-developed monetary value system. The impacts to a person's social attachment to a particular community or the loss of close proximity to customary land may not be duplicated in another community.

Financial compensation is discouraged if relocation involves rural communities without a well-developed monetary value system.

Adverse psychological and social impacts of relocation are understandably difficult to evaluate and to mitigate. Certain population groups such as Indigenous Peoples, elderly people, or low-income community members often have strong community ties and depend upon primary social relationships and important support networks that can be severed upon relocation. Effective and established techniques for assessing the severity of social and psychological impacts from displacement are not available. The severity of the impact is related to numerous factors; the effectiveness of mitigation efforts is largely related to the type and amount of compensation available and the expertise and sensitivity of approach applied to relocation. There are other situations where the social and psychological effects associated with relocating people cannot be wholly mitigated. Keep in mind, however, that while the general attitude prevails that displacement is a negative impact, this is not always the case. In many instances, individuals and families who relocate to make way for a project, do improve their quality of living because of better housing and living conditions than the ones they left behind. This of course can lead to other social problems. Practice shows that if relocation compensation is disproportionate to loss, or perceived to be so by unaffected community members, social jealousy develops. Claims for compensation by project-unaffected people are only one of the many negative consequences.

If relocation compensation is disproportionate to loss, or perceived to be so by unaffected community members, social jealousy develops.

Priorities

Obviously, not all community impacts associated with mining command the same priority or depth of analysis. The more important impacts should receive higher priority. For instance, a project-related impact that seriously affects large segments of the population for a long time (such as village relocation) is by its nature always more important than one that is not severe, affects few people, and lasts for only a short duration (such as noise impacts during construction). It would be appropriate to expend more effort and budget to analyze potential impacts and to design mitigation measures in the case of the former rather than the latter. Also, thresholds of impacts that are clearly established by legal mandates, or established social norms, usually at local or regional level, and which are often tied to social or economic concerns, should receive a high priority for analysis.

Advocacy groups may distort priorities. Projects that are highly controversial, or which are sources of substantial conflict between various peoples or groups, command a serious level of analytical effort, and hence a serious level of resource commitment. Community controversy could trigger the perception that a mine has significant impacts even when this is not the case. The reality is that advocacy groups often make it necessary to allocate significant financial and human resources in demonstrating that perceived impacts do not exist, or if impacts were to occur, that they are insignificant in the local context. The opportunity is thereby lost to direct these resources to real and more pressing issues.

Advocacy groups may distort priorities.

9.9 ENVIRONMENTAL JUSTICE

Do people count? The answer is yes, regardless of race, ethnicity, religion, gender, or social and financial status. Today, justice is more than whatever the strongest decide it to be, a definition first formulated by the great Greek philosopher Plato. In environmental science the term 'environmental justice' has risen to prominence in recent years, as briefly discussed here.

Environmental justice refers to the fair treatment of people of all races, cultures, and income with respect to the development, implementation, and enforcement of environmental laws, regulations and policies (Caltrans 1997). Environmental justice concerns may

Environmental justice refers to the fair treatment of people of all races, cultures, and income with respect to the development, implementation and enforcement of environmental laws, regulations and policies.

arise from project impacts on the living or physical-chemical environment, particularly from health, social, or economic impacts on minority and low-income groups.

To ensure that environmental justice is promoted, that is to see that the mining project is implemented in a socially equitable fashion, socioeconomic studies need to identify any ethnic and racial minorities and low-income population groups in the affected communities. Moreover they should address, as appropriate, disproportionately high and adverse human health and environmental effects of the mine development on minority and low-income groups. Central to the Equator Principles is that priority is given to mitigating impacts on identified minorities and low-income groups. It is important to determine if minority, low-income, or otherwise disadvantaged groups in the project area would be disproportionately affected by mine development and to find ways to mitigate such effects if they cannot be totally avoided.

Environmental justice is based on communication with and participation of formal and informal community leaders, in an informational outreach programme. Public outreach should not be seen as a one-time solicitation of project support, but as a continuous process. Regular community contacts are integral to community consultation and participation programmes for mining projects. Concerns of grass roots groups should be sought through consultation, and carefully documented. While this is not a new requirement, mining companies have not always actively sought out or heeded the opinions of all peoples affected by their plans and actions.

9.10 GROUP DECISION-MAKING IN ENVIRONMENTAL ASSESSMENT

In environmental assessments, task specialists typically research baseline conditions in the absence of the project, they review the mine proposal identifying activities with potential effects on the environmental component(s) of their particular expertise, and eventually they draw initial conclusions regarding impact significance, applying (or not) established methodologies of their disciplines. In this traditional approach, environmental assessments are considered as syntheses and critical appraisals of scientific information on particular environmental aspects of mining. They usually involve targeted data collection, models and analyses, result in a report, and are considered complete when they are delivered to whoever asked for the assessment, usually the mine proponent and environmental approval authorities. Recent research has developed a broader conceptualization (Moser 1999), acknowledging that environmental assessment is more than the collection of independent points of view. Assessments encompass the outcomes (models and analyses, and reports) and the social, dynamic, and iterative processes that lead to and follow them – the communicative, social interactions among assessors and project engineers, and those among other stakeholders, scientists, interested groups, and potential assessment users such as regulatory authorities, approval authorities or managers. Applied in this broader sense, environmental assessment can serve a variety of functions (Miller, Jasanoff *et al.* 1997):

- integrating disparate knowledge from many different disciplines and expertise into consensus answers;
- disseminating this consensus to, i.e. informing and educating stakeholders and the decision-maker community;
- identifying gaps in understanding;
- re-evaluating the relevance of knowledge claims; and
- providing opportunities for (new) stakeholders to interact and develop common ground with respect to choices of selected project components.

Consensus exists today on the benefits of interactive discussions between all stakeholders in a mining project and of group decision-making in environmental assessment.

Consensus exists today on the benefits of interactive discussions between all stakeholders in a mining project and of group decision-making in environmental assessment, in particular for ranking the significance of each impact and selecting and designing impact mitigation priorities that relate to the social dimension of the project.

Impact assessors think differently from decision makers.

In **Figure 9.12** the range of stakeholders involved in an assessment process is neatly organized by scale and role. Of course, they rarely fall in such clear categories. Sometimes the boundaries between categories are rather blurred. However in spite of such simplification, **Figure 9.12** helps to illustrate the wide range of interactions that can develop between different stakeholders, and the benefits of interactive workshops as part of the environmental assessment process. Impact assessors (usually scientific consultants) think differently from decision makers (e.g. mine managers, usually with an engineering background), an observation not meant as a value judgement, but to appreciate the difference in problem formulation and problem solving. The scientist tends to ask: '*How can I make my impact prediction more accurate?*' This contrasts with the thinking of the potential user of an assessment: '*What can I do about it (given the regulatory, resource, time, and political constraints that I'm working under)?*' A successful answer to one helps little in solving the other one's challenge and vice versa.

An internal workshop involving only the EIA team prevents interactions along the full spectrum of stakeholders. A workshop with only local stakeholders ignores the spatial scale of large mining projects.

Group decisions consider and weigh different perspectives about the same issue, balancing differences in importance assigned to the issue by different people.

Group decisions consider and weigh different perspectives about the same issue, balancing differences in importance assigned to the issue by different people. Group discussions also help in identification and consideration of aspects not previously taken into account in evaluating environmental impacts. Such discussions also draw on the collective experiences on a particular issue from people with different educational and practical backgrounds. Group discussion and subsequent decision on project impacts are of most benefit where the composition of group members reflects the various stakeholders involved in the project – mining engineers, task specialists, consultants, researchers, community members, and administrative authorities.

FIGURE 9.12

Participation along the Spectrum

The graph helps to illustrate the wide range of interactions that can develop between different stakeholders, and the benefits of interactive workshops as part of the environmental assessment process.

Source:
Based on Moser 1999

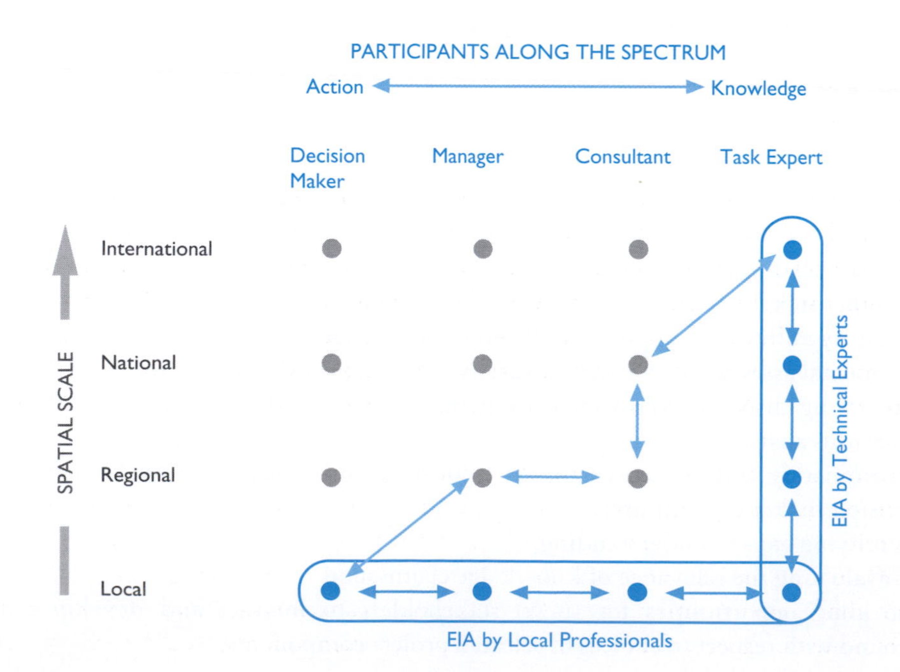

Workshops with selected participants and selected topics prove to be effective in generating useful outcomes in a limited time. At a minimum, workshops should be attended by representatives from the mining company (project manager, environmental staff members, mine operator, and government relations officer) and the full EIA study team (task specialists and document compilers). A series of workshops may be designed to discuss potential impacts and their significance, to design potential mitigation and monitoring measures, and to jointly review the final environmental management plan. Often it is of benefit to arrange different workshops for different groups of stakeholders.

Workshops with selected participants and selected topics prove to be effective in generating useful outcomes in a limited time.

A number of factors influence the successful (or unsuccessful) outcome of group decision-making. Sometimes, a few group members tend to dominate others in discussions using their authority or personality. On other occasions, group decision-making is influenced by personal biases, or by 'bandwagon' and 'halo' effects. Team members are not always accustomed to free expression of their opinions. This is particularly so in developing societies where people defer to the older, more powerful spokespersons, thus discouraging critical discussion, admission of errors, and revision of earlier judgements. Such a problem may be overcome by conducting multiple workshops, so that younger or less powerful stakeholders can provide their input without any cultural impediment.

A few rules guide successful decision-making workshops.

Preparation

Carefully select participants.
Appoint an experienced moderator.
Prepare for discussions.
Select a suitable venue.

Workshop

Allow for breaks.
Stick to the agenda.
Take notes.
Summarize outcomes.

Follow up

Act upon findings.
Keep participants informed on outcomes.

A number of more formal methods for group decision-making have been developed, which are occasionally applied in the environmental assessment of major projects, or in consensus-building for particularly controversial aspects of a specific project (such as deciding on tailings disposal schemes). The Delphi Technique (or Delphi) is a formal group decision-making tool to reach consensus on a particular issue, developed by the RAND Corporation in the late 1960s as a forecasting methodology. Delphi is particularly appropriate when decision-making is required in a political or emotional environment, or when the decisions affect strong factions with opposing preferences. It has proved to be effective in resolving seemingly-intractable site selection problems in the USA involving the 'nimby' (not in my back yard) syndrome, and is well suited for application to controversial mining projects. The main caveat is that the Delphi process requires the involvement of a skilled and experienced moderator. The tool works formally or informally, in large or small contexts, and captures the benefits of group decision-making while avoiding the limitations of group decision-making; e.g. over-dominant group members, political

The Delphi process works formally or informally, in large or small contexts, and captures the benefits of group decision-making while avoiding the limitations of group decision-making; e.g. over-dominant group members, political lobbying, or 'bandwagonism'.

TABLE 9.10

Steps Involved in Applying the Delphi Technique – Delphi uses a group of carefully selected experts who answer a series of questions. Questions are usually formulated as hypotheses. Each round of questioning (or discussions) is followed by feedback on the preceding round of replies

1. Pick a facilitation leader

Select a person who can facilitate, is a recognized mining expert, and is not a stakeholder, at least not an influential stakeholder with preconceived notions. An outsider is often chosen.

2. Select a group of experts

Group members should have an intimate knowledge of the mining project and its potential environmental impacts, or be familiar with related subject matters that would allow them to prioritize project-induced impacts effectively.

3. Generate a long list of potential impacts

In a brainstorming session, build a list of impacts and significance attributes that all think appropriate to the project at hand. At this point, there are no 'wrong' contributions from group members.

4. Rank impacts

For each impact, the group ranks it as 1 (very important), 2 (somewhat important), or 3 (not important). Each group member ranks the list individually and anonymously if the environment is charged politically or emotionally.

5. Calculate the mean and deviation

For each impact on the list, find the mean value and remove all impacts with a mean greater than or equal to 2.0. Place impacts in rank order and show the (anonymous) results to the group. Discuss reasons for impacts with high standard deviations. Group members may insert removed impacts back into the list after discussion.

6. Re-rank impacts

Repeat the ranking process among group members until the results stabilize. The ranking results do not have to have complete agreement, but a consensus such that all can live with the outcome. Two passes are often enough, but more may be performed for maximum benefit.

Source:
Based on Cline 2000

lobbying, or 'bandwagonism'. **Table 9.10** illustrates the general procedure for applying Delphi to evaluating mining impact significance. Delphi uses a group (called panel) of carefully selected experts who answer a series of questions (independently and anonymously in the traditional approach). Questions are usually formulated as hypotheses.

In a less formal approach, such as a workshop format, hypotheses or relevant subject items are openly commented on amongst group members facilitated by the moderator. Each round of questioning (or discussions) is followed by feedback on the preceding round of replies. Thus the experts are encouraged to revise their earlier answers in light of the replies of other group members. It is believed that during this process the range of answers will decrease and the group will converge towards the 'correct' answer. After several rounds the process is complete and the median scores determine final answers.

9.11 REFLECTING ON THE OBJECTIVE NATURE OF ENVIRONMENTAL ASSESSMENT

Environmental assessment can never be entirely objective. Despite the best intentions of stakeholders, be they mining company personnel, EIA consultants, approval authorities, or advocacy groups, some bias is to be expected. Even if the stakeholders were entirely objective, financial, technical, or scientific constraints would introduce subjective view-points.

Study Team and Potential Conflicts of Interest

In preparing an EIA, mining companies may sometimes rely on in-house resources, or more commonly arrange for work to be performed by external resources, usually environmental consulting firms and/or research institutions. If the EIA is prepared in-house it is indeed difficult to convince the public of the objectivity of related efforts. Moreover, many jurisdictions prohibit this approach. Consequently, most mining companies routinely rely on the services of consulting firms, recognizing that outsourcing is usually more efficient and provides access to more specialized expertise than available in-house. Outsourcing does not entirely resolve the problem of objectivity. External stakeholders may be concerned about the possibility of bias by consulting firms, selected and paid directly by the mining company. In addition mining companies often provide guidance to consultants throughout the preparation of the EIA to ensure that the EIA correctly reflects practical mining experience available in-house. Such guidance, while it may be seen as introducing the proponent's bias, is of course beneficial and in fact essential, considering that the mining operator not the consulting team owns the EIA and will have to live with all the management and monitoring commitments it contains. Experience demonstrates that environmental management and monitoring measures designed by consultants with an academic or research background are often impractical to implement. Critical review by project engineers can assist in making these measures more practicable and effective.

Outsourcing does not entirely resolve the problem of objectivity.

A point often made is that consultants are obliged or financially bound to submit an environmental assessment in favour of the project, and the objectivity of the assessment is therefore questionable. Consultants, it is argued, are hardly likely to bite the hand that feeds them. It is also argued that the mining company could attempt to exert influence over consultants in many other ways. For example, the company could select a consultant known to have a perspective likely to produce a positive assessment of the project; or attempt to influence the consultant's work by the lure of future work (or the exclusion in future work). While understandable, these perceptions seldom hold true in practice. Admittedly potential conflicts of interest do exist but are rare. Today, the environmental assessment process is too transparent to allow inappropriate collusion between the project proponent and the EIA compiler. It should be also recognized that similar conflicts of interest are unavoidable, no matter which party is responsible for preparing the environmental assessment. More and more, the interests of mine proponents, environmental consultants and other participants in the EIA process are converging. All parties seek (or should seek) a positive outcome in which serious environmental problems are avoided or mitigated, and in which sustainable benefits will accrue to local communities. Should the results be otherwise, this reflects badly, not only on the proponent, but on the consulting firm and its individual team members, and on the regulatory authorities, whose responsibility is (usually) not to prevent development but to ensure that it occurs without unacceptable impacts.

Consultants, it is argued, are hardly likely to bite the hand that feeds them.

To address concerns of real or perceived bias and quality lapses, mining companies can consider implementing a range of measures: pre-qualifying consultants based on strict requirements; selecting consultants in consultation with qualified independent third party advice; hiring a separate consulting firm or individual experts to conduct a peer review of documents; and facilitating an expanded (earlier and more accommodating) public participation process.

Regulatory Instruments to Foster Objectivity

Public consultation is generally accepted as the best instrument to objectively collect feedback from project-affected communities.

To balance potential bias in environmental assessment, regulators have introduced various instruments in the environmental assessment process, such as mandatory public consultation, governmental review, or independent third party review. *Public consultation* is generally accepted as the best instrument to objectively collect feedback from project-affected communities. In many jurisdictions, public consultation takes the form of public hearings, a prerequisite to obtaining environmental clearance. A public hearing provides a forum where community members and concerned groups come face to face with the project proponent and government authorities to voice their suggestions and concerns. A public hearing allows participation in environmental decision-making. Effective public hearings require efforts from all participants. The project proponent has to provide sufficient project information in time to prepare for the hearing. A strong moderator is necessary to allow the opportunity for all participants to express their view-points. Unfortunately, public hearings are easily disrupted by the presence of trouble-makers opposed to the project, who may seek to dominate proceedings to limit the opportunities for presenting other view-points. It is important that a public hearing be organized and conducted in accordance with local custom, which will usually effectively limit the damage that can be done by trouble-makers.

A common concern among project opponents is that the time periods available for them to review documents and appeal decisions, are too tight to effectively enable public participation. Similarly, there is a sense that few grassroots community groups have sufficient expertise and financial resources to effectively participate in the process, such as by hiring their own independent consultants.

Often as mentioned above, the spirit and intention of public hearings is sabotaged by advocacy groups who see public hearings as an opportunity to project their own general view-points on a particular subject, controlling the hearing and distracting from project-specific environmental issues.

In countries with a long mining history EIA review committees not only have the authority but also the experience to review a proposed mining project competently.

Government review committees serve as another instrument to ensure objectivity. Depending on the jurisdiction, the committee is composed of different government department representatives, scientific experts, community members, and/or other relevant parties assembled based on having the collective knowledge and experience to provide an informed review of all the significant issues. The findings of the environmental impact assessment are presented to and scrutinized by the review committee. Such committees operate successfully in jurisdictions as disparate as the Australian State of Victoria, and the Kingdom of Thailand. In countries with a long mining history EIA review committees not only have the authority but also the experience to review a proposed mining project competently. This cannot always be said for developing countries, where it cannot always be taken for granted that the review committee has a full appreciation of the environmental issues involved in a particular project. Recognizing this fact, some jurisdictions such as the national government of Papua New Guinea retain foreign experts to assist in the review process.

Independent review by financial institutions, while not a regulatory requirement, is often helpful in ensuring the objectivity of the final outcome of environmental assessment. It is said that about 80 percent of all project financing is by banks that have adopted good industry practices (The Equator Principles) in their environmental covenants. Banks investing in mining projects often appoint recognized experts to provide an independent expert opinion on the environmental assessment efforts. If past efforts fall short of expectation – that is if significant gaps exist between the existing EIA and requirements based on The Equator Principles – financial institutions will ask for additional studies or mitigation commitments, prior to committing funding to the project.

Impact Assessment Methodologies and Objectivity

When evaluating impacts, quantitative methods are usually preferable to qualitative methods as they are more precise, outcomes are reproducible, they do not contain subjective views, and often, outcomes can be linked to numerical standards. In an ideal situation, use of quantitative methods allows a baseline parameter to be measured using a numerical scale (such as ambient concentration of a trace metal in water), and impact prediction to be made using a scale that relates to the existing baseline (predicted maximum concentration of a trace metal in water), facilitating the evaluation of its significance. This is particularly ideal if the impact can be evaluated against a numerical standard (say maximum permissible concentration of the trace metal under consideration). Quantitative methods apply particularly to physical and chemical environmental aspects such as air quality, which would be hard to express in any meaningful way other than by using a quantitative approach. By expressing existing and predicted air quality levels and the relevant standard as, for example, micrograms per cubic metre, the magnitude of the predicted impact over the existing baseline, and its significance in the context of the standard, can be clearly stated.

A quantitative approach also facilitates discussions between the mining company and its stakeholders in the new mine project. It is easier to explain that effluent discharge will conform to a given discharge standard than to articulate how landscape quality will be affected by mining. In addition trade-offs between reductions in environmental impacts and associated costs are inherently easier to understand if costs relate to measurable environmental gains. The effectiveness of different abatement technologies in reducing gaseous emissions and hence in improving ambient air quality can be readily assessed and compared with associated costs. This contrasts with cost estimates for implementing mitigation measures to reduce the impact on landscape quality or visual amenity. It may be possible to assign a cost to the measure (for example rehabilitation); it is much more difficult to assess the resulting impact reduction.

It is easier to explain that effluent discharge will conform to a given discharge standard than to articulate how landscape quality will be affected by mining.

Notwithstanding their benefits, quantitative methods do not provide a total panacea for ensuring objectivity in environmental assessment.

Data collection and modelling effort can be disproportionate. Quantitative methods are often resource-intensive, involving extensive data collection and careful interpretation of model results. In contrast, expert judgement, while subjective, can be reached relatively quickly, and often fits the purpose. Some cases of acid rock drainage illustrate this point. Despite pre-mining investigations involving several hundred geochemical tests to assess acid rock drainage of waste rock, some projects have found that the actual incidence of ARD, in practice, has been much higher or lower than predicted by the quantitative pre-mining studies. Various reasons have been postulated for this, but in general, there have been incorrect classifications and correlations of different lithological units, which only became clear when mining provided large exposures, enabling the geology to be properly interpreted. In at least one such case, the mine planners would have been better advised to assume that all waste would be acid-generating, rather than attempting to implement a strategy to separate, selectively place and treat a range of different materials.

Expert judgement, while subjective, can be reached relatively quickly, and often fits the purpose.

As a completely different example, mine development may partly affect breeding grounds of commonly encountered bird species in the host region that are classified as neither threatened or endangered. A complete quantitative impact assessment would involve a wide range of studies and research, and may be disproportionate to the significance of the impact. In such cases, professional judgement may be sufficient to advise on impact significance, and to propose effective mitigation measures.

'garbage in, garbage out'.

Modelling input data can be incomplete or inaccurate. Another concern about using quantitative methods relates to the accuracy of the input data. There is the saying in modelling, 'garbage in, garbage out'. Natural systems have a high degree of natural variability. To represent nature's response to an imposed impact adequately, it is often necessary to collect a wide range of data over long time periods to represent natural data averages and their variation. In most cases, this is impractical and data collection is limited to one year at the best to capture seasonal variation. Errors are introduced. For example, the year in which the baseline was measured may have been an unusual year. This is particularly true for faunal populations which may experience large fluctuations for reasons that are not always apparent.

Model limitations do exist. For aspects such as noise or emissions to air, the science behind pollution propagation is well-understood and quantitative impact predictions can be made with confidence. However, this holds true for few aspects in nature. Most ecosystems are complex, and to quantify how they, and elements within them, will react to project impacts is difficult and uncertain. Where uncertainty exists, the causes and implications of the uncertainty need to be clearly stated.

Standards can conflict. One of the advantages of quantitative approaches is that they allow comparison with standards, if they exist, providing a transparent evaluation of significance. In the first instant, predicted impacts are compared to the relevant standards of the host region. Sometimes local standards are not established. In other cases, local standards exist but are not applicable to the specific type of industry, or discharge. Occasionally, local standards may be wrong. In all three cases reference to established international standards becomes helpful in discussions with regulatory authorities.

Model presentation can create the illusion of certainty and objectivity. As is true for all technical reports, presenting outcomes of the impact assessment study should be concise and carefully organized. Tables and charts should be prepared when needed to enhance the presentation and highlight information. Many readers of environmental documents are visually oriented while others will rely more heavily on the narrative text. Written text should accompany each table or chart to assist the reader in understanding the table or graphic. The original source for data for the compilation of charts and tables should be clearly identified. However, the danger is real that word processing and visual tools combined with creative imagination suggest an accuracy that often does not reflect reality. Care is necessary in presenting quantitative results. Limitations, assumptions, and uncertainties should be clearly stated, which, contrary to common belief, should actually strengthen the credibility of the impact assessment.

9.12 DEALING WITH UNCERTAINTIES AND RISKS

A number of project aspects introduce a degree of uncertainty related to impact prediction. Environmental Risk Assessment (ERA) is the scientific method of confronting and expressing uncertainty in predicting the future. Risk is the chance of some degree of damage in some unit of time, or the probability or frequency of occurrence of an event with a certain range of adverse consequences. These probabilistic expressions, as opposed to a single (mean or expected) value, are what distinguish ERA from mere impact assessment. The irreducible uncertainties leave uncertain the calculated absolute value of a risk, often by as much as an order of magnitude or more. Risk is therefore expressed and communicated as a mean plus its standard deviation, in addition to any upper bound or worst case calculation that may be used for purposes of conservative policy. The following summarizes the concept of an environmental risk assessment study.

Risk is therefore expressed and communicated as a mean plus its standard deviation, in addition to any upper bound or worst case calculation that may be used for purposes of conservative policy.

Impacts and Risks

Even with a final project design and an unchanging environment, impacts are difficult to predict with certainty. This is partly because impact prediction is always based on judgement, even when impacts have been assessed using quantitative tools such as mathematical modelling. For example, uncertainties are introduced when selecting the mathematical model, determining input data, or when interpreting model outcomes. Partly, this is because prediction is based on known or historical, but never certain information. Uncertainty is inherent in all geological and geotechnical investigations because the exposures and samples provided by techniques such as drilling represent but a tiny proportion of the volume of soil or rock that is the subject of investigation. Partly again, this is also because mine design is largely based on normal operating conditions although allowances for 'upset conditions' and worst case scenarios are made. Finally, accidents do happen. For example, the failure of tailings storage facilities continues to occur at mining operations around the world, often associated with significant environmental damage including loss of life. Other unplanned events are likely to occur – such as road accidents, landslides, chemical/oil spills, fires, or explosions.

How should uncertainty be allowed for in the environmental assessment process? Simple risk assessment tools help (US EPA 1998; EEA 2006).

Impact, Hazard and Risk: Problems with Terminology

In the context of an EIA, risk assessment is the evaluation of risks to the environment based on an analysis of threats to, and vulnerabilities of mining operation and activities. One of the difficulties with the concept of both risk and impact alike is that risk and impact relate to common experiences for which a language has been developed across a diverse range of disciplines and activities. This language often lacks precision, and its ambiguity can lead to confusion. The term risk assessment, as an example, may have a different meaning for the chemical engineer than it has for the biologist. The medical doctor is concerned about health risks, where as the ecologist is concerned about ecological risk and so on. Without being descriptive, the terms 'impact', 'hazard', and 'risk' used in this text are defined as follows: (1) impact is defined as an environmental change caused by the mining project (either positive or negative); (2) hazard is defined as a property (e.g. mine water quality) or situation (e.g. storm event) that in particular circumstances could lead to negative environmental impacts; and (3) risk is defined as the combination of the probability (sometimes also termed frequency or likelihood) of occurrence of a defined hazard and the magnitude of the consequences of the occurrence (that is the severity of potential negative environmental impacts). A hazard can change predicted impacts (e.g. a storm event, further increasing predicted water turbidity of mine runoff) or it can lead to the creation of a new set of environmental impacts (e.g. as it the case of an accidental break in a fuel flow line). Assessing risks to occupational health and safety is often not part of the environmental assessment; it is a well established specialized discipline, embedded in mine management rather than in the environmental impact assessment process.

The term risk assessment may have a different meaning for the chemical engineer than it has for the biologist.

The following questions help to identify hazards: (1) What can go wrong?; (2) Are all natural hazards identified?; (3) How likely is it that these hazards will occur?; (4) How frequently and where will these hazards occur?; or (5) How much confidence can be placed in existing data and information? Operating experience is most important in providing answers to these questions.

Risk Assessment Matrix

Risk is characterized in terms of the severity of consequences and likelihood of occurrence.

By definition, risk is characterized in terms of the severity of consequences and likelihood of occurrence. Combining the probability and the severity of those consequences yields an estimation of risk in the form of a simple two-dimensional diagram as schematically shown in **Figure 9.13**. As both probability and severity increase, the risk becomes less and less acceptable.

To apply the risk assessment matrix in practice, three measures are required: (1) rating of risk likelihood; (3) rating of risk consequences; and (3) rating of risk acceptability. All three measures are defined according to parameters relevant to a specific risk situation. Rating risk likelihood and consequences produces a risk assessment matrix as illustrated in **Figure 9.14**. For each hazard the risk assessment matrix provides a consistent basis for communicating risks and for deciding whether a specific risk is acceptable or whether it is not. The risk assessment matrix should of course be interpreted with caution, recognizing the oversimplification that it will normally represent. Both components – probability and consequences – are likely to be at best semi-quantitative and so each component will

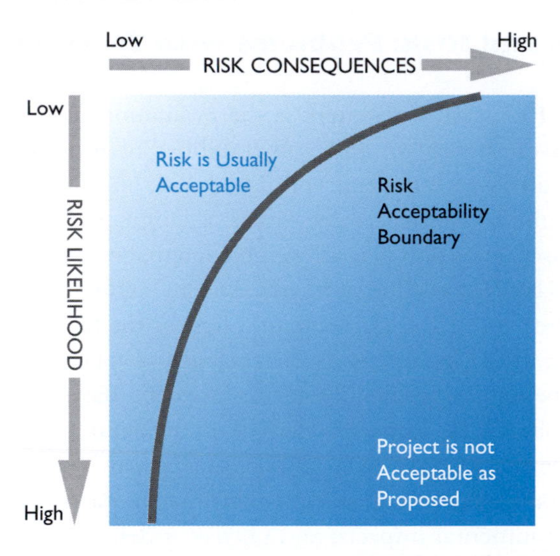

FIGURE 9.13
Schematic Two-dimensional Risk Diagram – Illustrating Risk Likelihood, Risk Consequences and Risk Acceptability

FIGURE 9.14
Schematic Risk Assessment Matrix

The risk ranking from 'minor' to 'severe' depends on the specific risk situation.

Decreasing Acceptability

LIKELIHOOD \ CONSEQUENCES	Insignificant	Moderate/ Marginal	Critical	Catastrophic
Unlikely	Minor	Minor	Moderate	High
Seldom	Minor	Moderate	High	Severe
Occasional	Moderate	Moderate	High	Severe
Will Occur/ High Certainty	Moderate	Moderate	Severe	Severe

to some extent represent judgements on the basis of the knowledge and experience available.

Descriptive ratings of hazard probability, illustrated in **Table 9.11**, range from 'unlikely' to 'will occur/high certainty'. In most risk situations it will be easier to estimate the probability of risk than to judge its consequences. In some situations, estimating the likelihood of rainfall will be based on interpretation of historical records. Statistical analysis of rain fall data to determine the design flood is one example. In others estimates will be based on operational experience, e.g. estimating the likelihood of equipment failure.

Descriptive ratings of risk consequences are illustrated in **Table 9.12** ranging from 'insignificant' to 'catastrophic'. Evaluating consequences of a risk or, respectively, evaluating the severity of negative environmental change, involves determining the broader implications of a hazard considering all three dimensions of sustainability, that is considering ecological,

TABLE 9.11
Rating of Risk Likelihood

Descriptor	Descriptive Rating
Unlikely	Very rare Not possible
Seldom	Remotely possible Not expected
Occasional	See sporadically Occurs as often as not
Will occur/high certainty	Occurs often Occurs continuously

TABLE 9.12
Example of Rating of Risk Consequences

Description	Environment Dimension	Social Dimension	Financial Dimension	Reputation Considerations
Insignificant	Limited damage to minimal areas of low significance	Limited social change in well defined communities	Financial consequences below US$ 50,000 (50k)	Occasional local grievances
Moderate/Marginal	Temporary impacts; localized, reversible damage	Social change in host communities; limited change in social cohesion	US$ 51k to 500k	Limited negative press coverage in local newspapers
Critical	Exceedance of regulatory thresholds; loss of wild life	Social tensions; uncontrolled influx of people; decline of public sanitation; potential public health impacts	US$ 501k to 2m	Frequent international news coverage; International NGO involvement; Community actions/protests
Catastrophic	Irreversible damage to sensitive habitats	Social unrest; Loss of human life; public health impacts	Above US$ 2m	Long-term international reputation loss; Violent community actions

social, and economic systems, as well as reputation. The rating of risk consequences will depend on the highest risk in each of these four aspects.

Evaluating the consequences of a risk is not very different from evaluating the significance of environmental impacts.

Evaluating the consequences of a risk is not very different from evaluating the significance of environmental impacts. As is true for assessing impacts, the complicating issue for environmental risk assessment is the lack of an easily defined measure of what constitutes harmful consequences for the environment. In some cases definitions of environmental damage are intuitive or are laid down in statute (e.g. loss of life or violation of regulatory standards); in others appropriate criteria need to be based on professional judgements.

How do environmental impacts identified in an EIA link into the risk assessment matrix? The best way to visualize the link is to understand these impacts as the consequences of risks that fall into the category 'will occur/with certainty'. Environmental impact assessment studies are concerned with environmental hazards that are certain to occur. In mining the generation of waste rock and tailings, as one example, is certain, so is the disturbance of land and water. Engineering and management efforts are designed to eliminate hazard-related environmental consequences entirely (that is moving the probability of occurrence into the category 'unlikely'), or if impacts are unavoidable, to move potential environmental risks into the lower left corner of the risk assessment matrix by reducing the magnitude of the consequences of the occurrence. As such, environmental management measures aim either to eliminate hazards, to reduce unavoidable environmental consequences to an acceptable level, or both (**Figure 9.15**). This begs the question: What is an acceptable environmental risk?

Environmental management measures aim either to eliminate hazards, to reduce unavoidable environmental consequences to an acceptable level, or both.

Risk Acceptability

The boundaries of risk acceptability are not clearly defined and are tailored to the factors influencing the significance of the risk. Clearly risks that are situated in the lower right corner of the risk assessment matrix (risks occurring frequently with catastrophic consequences) are not acceptable. Risks in the upper left corner of the matrix are more acceptable, as (1) risk occurrence is unlikely and (2) risk consequences are negligible. The acceptability of risks that fall between these two end points depends on a variety of factors; some follow.

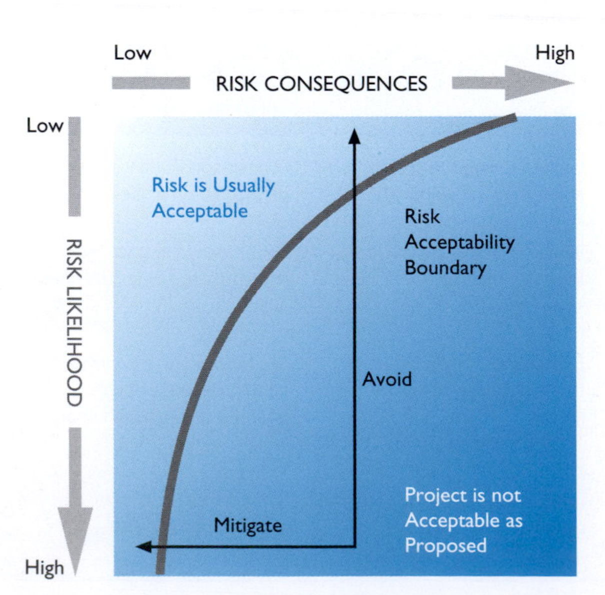

FIGURE 9.15
Using the Risk Assessment Diagram to Illustrate the Aim of Risk and Impact Mitigation Measures

Statutory and Policy Requirements

Most jurisdictions have promulgated a wide range of standards for pollution control, including exposure standards, environmental quality standards, emission standards, process or operating standards and product standards. Clearly if these are legally mandatory and a risk assessment demonstrates that an intended activity is likely to breach them, the risk is unacceptable and measures to reduce it to acceptable levels (that is ensuring regulatory compliance) should be adopted.

Loss of Life

Planners, engineers, and operators have a zero tolerance for loss of life. Particularly in the oil and gas industry, avoiding the risk of loss of life is deeply embedded into project design, so that operational risk assessment has evolved into a specialized engineering discipline. Spectacular accidents, however, such as the explosion of Space Shuttle, serve as a reminder that 100% safe design and operation do not exist even with the best of efforts. Under given circumstances the residual risk of loss of life may be acceptable, if the likelihood of the hazard occurring is remote and if the number of potential casualties is small.

Planners, engineers, and operators have a zero tolerance for loss of life.

Value Judgements

Value judgement is subjective. Defining what constitutes unacceptable harm to an ecosystem, as one example, is therefore a difficult task and ultimately depends on what values an individual or society places on ecosystems. Some may argue that the existing environment at the project site is invaluable and should be maintained at all costs. Others may hold the opinion that maintenance of ecosystem function is the main objective and that an accidental spill of tailings supernatant may not threaten this objective.

Value judgement is subjective.

Economic Considerations

Economic factors often have a significant influence on the acceptability of a given risk. An example could be the tailings storage facility for a mine. A tailings disposal site located at some distance from the mine may impose less environmental risks, but may make tailings disposal extraordinary expensive. The preferred tailings management option is likely to be the one with the greatest excess of benefits over costs. Economic considerations also extend to considering the provision of costs to mitigate a project impact over and above those that are provided. Financial consequences due to a delay in project development, or threat/disruption of operations are also important considerations when deciding on environmental project risks.

Social Aspects of Risk

The acceptability of a risk can be significantly influenced by a range of psycho-social and political factors. These may include individual risk perceptions and attitudes, cultural values, questions of trust and credibility of the mine operator, and questions of equity in risk distribution. The social aspects of risk can be reduced to some extent by constructive dialogue between project affected people and the mine operator.

Reputation Risk

Reputation risks are important (**Case 9.7**). The reputation risk includes both local and international reputation risk, as well as strained stakeholder relations. Reputation risk is of major concern for large mining companies that pursue business interest in different parts of the world. Negative publicity in one country can hinder mining development in another one. Reputation risk may also influence the capability to raise funds on the financial market.

Reputation risk may also influence the capability to raise funds on the financial market.

Residual Risk

Initial investments in risk mitigation can significantly and cost-effectively reduce the vulnerability of a mining project to identified hazards and can reduce risk consequences. After basic precautions are taken, however, considerably larger expenditures may be required to achieve an additional reduction in risk. As shown in **Figure 9.16**, at some point it may be more cost effective to transfer the remaining risk or find alternative ways to manage or to finance it than to attempt to mitigate it completely. For example, constructing a 'perfect' tailings dam to resist all seismic shocks or flood events is likely to be inordinately expensive in some situations.

Risk preparedness and emergency response planning are common tools to manage residual risks with a low probability of occurrence. In the mining sector APPEL (2000) is the accepted standard for residual risk management. The consequences of likely, irreducible environmental risks of course are the subject matter of environmental impact assessment studies.

Risk Perception and Risk Communication

Our responses to new risks are largely predictable based on our enduring values and social relationships.

Risk judgement is very responsive to verbal cues.

A myriad of characteristics other than mortality are factored into everybody's working definition of risk as illustrated in **Table 9.13.** The very same risk is likely to be understood quite differently by the EIA practitioners, community members, and mine engineers since the perceptions of risk will depend on where an individual stands on the listed dimensions. People's perception of risk is also influenced by the social context. Our responses to new risks, in fact, are largely predictable based on our enduring values and social relationships. Trust and credibility are important. Do the host communities like or dislike, trust or distrust the mining company that is putting them at risk? It is fair to say that few people trust their government or the mining industry to protect them from environmental risk. This is equally true of the passive, apparently apathetic public as it is of the activist, visibly angry public. Both publics will listen to the reassurances of the government and the mining company – if they listen at all – with considerable suspicion.

Risk communication strongly influences risk perception. For example, risk judgement is very responsive to verbal cues. Host communities are likely to be more concerned with the

CASE 9.7
Lafayette's Rapu Rapu Mine in the Philippines

Small Australian mining company Lafayette Mining established its Rapu Rapu mine on a small island located in Albay Province of the Philippines. First established in 2005 the open pit produces a polymetallic ore that is concentrated and exported. Initially, oxidised ore was mined and processed using cyanide leaching to extract gold. Early in its life, there were two unplanned releases from the tailings storage facility, the main cause of which were, reportedly, lack of sufficient freeboard to accommodate inflows from an extreme rainfall event. While both releases were relatively minor, Lafayette's reputation was seriously effected. The reputation of the mining industry as a whole had previously suffered as a result of the Marcopper incident in which large quantities of tailings had been discharged. Consequently, the general public and government regulators became highly sensitized to tailings and the potential impacts of tailings spills. The industry declined and it became more difficult to obtain environmental approvals for new projects. Ironically, the Rapu Rapu project was cited as a forerunner in a new generation of mining projects in which environmental management would be a major component. It was in this context that the Rapu Rapu releases occurred and, perhaps understandably, there was an over-reaction by the public and its political representatives. Accordingly, Lafayette was involved in extensive and extended negotiations and remedial actions before it regained its permits to operate.

Following many months when mining and processing activities were curtailed, operations were eventually permitted on a trial basis. However, soon after, the facilities suffered damage from a typhoon, causing further production delays. The combined effect of both production delays was to bankrupt the project which was placed in receivership in December 2007. In addition, the tailings incidents have further damaged the reputation of the mining industry in the Philippines, and (in the eyes of many Filipinos) the ability of the Australian mining industry to introduce good mining practices in the Philippines.

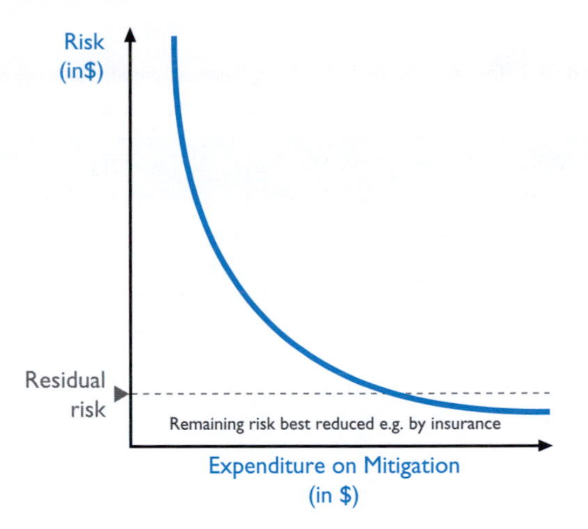

FIGURE 9.16
Decreasing Marginal Returns to Investment in Physical Mitigation

At some point it may be more cost effective to transfer the remaining risk or to find alternative ways to manage or to finance it than to attempt to mitigate it completely.

TABLE 9.13
Some Key Non-damage Attributes of Risk

Involuntary	Risks voluntarily assumed are ranked differently from those imposed by others
Uncontrollable	The inability to personally make a difference decreases a risk's acceptability
Immoral	Pollution is often viewed as a consuminate evil. And statements that hazards are `too low to worry about' can engender suspicion
Unfamiliar	Generally speaking, more familiar risks are regarded as more acceptable
Dreadful	Risks that cause highly feared or dreaded consequences are viewed as more dangerous
Uncertain	Scientific uncertainty about the effect, severity, or prevalence of a hazard tends to escalate unease
Catastrophic	Large-scale disasters such as aircraft crashes weigh more seriously in the public's mind than individual events such as exposure to radon gas in a neighbour's basement
Memorable	Risks embedded in remarkable events have greater impact that risks that arise in less prominent circumstances.
Unfair	Substantial outrage is a more likely result if people feel they are being wrongfully exposed
Untrustworthy	The level of outrage is higher if the source of the risk is not trusted.

Source:
Foundation for American Communications and National SEa Grand College Program. Reporting on Risk: A Handbook for Journalists and Citizens; The Annapolis Center, 1995. pp 84-86

fact that a tailings dam may fail within the next thousand years, than that the dam is designed to last for one thousand years. Doctors, as another example, are much more likely to prescribe a new medication that saves 50 percent of its patients than one that loses 50 percent of them.

On a final note the following rules-of-thumb should be considered when dealing with uncertainty. (1) Make conservative conclusions (i.e. assume that an effect is more, rather

TABLE 9.14

Potential Accidents Associated with Mine Sites and their Effects – The APELL for Mining Handbook provides an excellent framework for the preparation of an Emergency Response Plan that can be used by mine management, emergency response agencies, government officials, and local communities, if unplanned events do occur

Type of Incident	Typical Causes	Potential Effects
Tailings dam failure	Poor water management, overtopping, foundation failure, drainage failure, piping, erosion, earthquake.	Loss of life, contamination of water supplies, destruction of aquatic habitat and loss of crops and contamination of farmland, threat to protected habitat and biodiversity and loss of livelihood.
Failure of waste rock dump	Instability often related to presence of water (springs, poor dump drainage).	Loss of life, injuries, destruction of property, damage to ecosystems and farmland.
Pipeline failure, e.g. tailings, leach solution	Inadequate maintenance, failure of equipment, physical damage to pipeline.	Contamination of soil, water, effects on water users. May not be detected for some time.
Transport of chemicals to/ from site	Inadequate transport procedures and equipment, unsafe packaging, high-risk transportation routes.	Contamination of soil, water, effects on water users, aquatic ecosystem damage, threat to human health.
Ground subsidence	Slope failure, breakthrough to surface.	Loss of life, damage to property.
Spills of chemicals at site, e.g. fuel tank rupture, reagent store damage	Poor maintenance, inadequate containment.	Contamination of soil and water. Air pollution could have health effects.
Fire	Poor design, unsafe practices in relation to flammable materials.	Effects of air pollution on health, property damage.
Atmospheric releases	Inadequate design, failure to follow procedures, inadequate maintenance.	Community concern, possible health effects.
Explosions (plant)	Inadequate design, failure to follow procedures, inadequate maintenance.	Community concern, loss of life, destruction of property.
Blasting and explosives accidents	Poor practice, unsafe storage and handling.	Property damage, risk to life.

Source:
APELL for Mining, UNEP (2000)

than less adverse). This is referred to as the 'Precautionary Principle'. (2) Provide a record or audit trail of all assumptions, data gaps, and confidence in data quality and analyses, to justify conclusions. (3) Recommend mitigation measures to reduce adverse effects, and implement monitoring programmes, followed by evaluation and management of effects, to ensure effectiveness of these measures. (4) Implement mechanisms to evaluate the results of monitoring and provide for subsequent mitigation or project modification, as necessary.

The aim of incorporating environmental risks into the environmental assessment process is to reduce, where practical, unforeseen events with negative environmental consequences. Absolute prevention is not possible. The APELL for Mining Handbook, prepared by UNEP (2000), provides an excellent framework for the preparation of an Emergency Response Plan that can be used by mine management, emergency response agencies, government officials, and local communities, if unplanned events do occur (see **Table 9.14** for potential mining accidents and their effects). It introduces the generic objectives and organizational framework of UNEP's Awareness and Preparedness for Emergencies at Local Level (APELL) programme, covers risk factors specific to the mining industry, and describes how APELL can be applied to the mining industry.

REFERENCES

ADB (1998) Handbook on Resettlement – A Guide to Good Practice, Asian Development Bank, Manila.

ADB (2003) Environmental Assessment Guidelines, Asian Development Bank, Manila

Canter L (1996) Environmental Impact Assessment (Second Edition), McGraw Hill Publishing Company. New York, USA.

Canter L and Sadler B (1997) A Tool Kit of Effective EIA Practice – A Review of Methods and Perspectives on their Application: A Supplementary Report of the International Study of the Effectiveness of Environmental Assessment, IAIA, Environmental and Groundwater University of Oklahoma, Oklahoma, USA.

CALTRANS (1997) Community impact assessment, Environmental Handbook Volume 4, Sacramento, USA.

CEAA (1996) Assessing environmental effects on physical and cultural heritage resources; A reference guide for the CEAA (Canadian Environmental Assessment Agency), www.ceaa.gc.ca

CEAA (2003) Cumulative Effect Assessment, Practitioner Guide, Canadian Environmental Assessment Agency, www.ceaa.gc.ca

Chapra SC (1997) Surface Water-Quality Modeling, McGraw-Hill Series in Water Resources and Environmental Engineering, McGraw-Hill, New York, 844 p.

Cline A (2000) Prioritization Process Using Delphi Technique, White Paper, Carolla Development, www.carolla.com/wp-delph.htm

Cormier SM, Smith M, Norton S and Neiheisel T (2000) Assessing ecological risk in watersheds: A case study of problem formulation in the Big Darby Creek Watershed, Ohio, USA. Environ Toxicol Chem 19:1082–1096.

EEA (2006) Environmental Risk Assessment – Approaches, Experiences and Information Sources, Environmental issue report No 4, European Environmental Agency.

Goodland R and Mercier J (1999) The Evolution of Environmental Assessment in the World Bank: from “Approval” to Results. Paper No. 67, Environmental Management Series.

IFC (2006) Environmental and Social Performance Standards, International Finance Cooperation, www.ifc.org

Hunt D and Johnson C (1995) Environmental Management Systems: Principles and Practice, McGraw-Hill.

Kuipers JR, Maest AS, MacHardy KA and Lawson G (2006) Comparison of Predicted and Actual Water Quality at Hardrock Mines: The reliability of predictions in Environmental Impact Statements.

Jorgensen SE (1994) Fundamentals of Ecological Modelling, (2nd Edition), Part 19, Elsevier, Amsterdam.

Leopold LB, Clarke FE, Hanshaw BB and Balsley JR (1971) A procedure for evaluating environmental impact. Geological Survey Circular 645, Government Printing Office, Washington, D.C. 13 pp.

Lohani B, Evans, JW Ludwig, H Everitt, RR Carpenter RA and Tu SL (1997) Environmental Impact Assessment for Developing Countries in Asia. ADB, Volume 1 – Overview. 356 pp.

Miller C, Jasanoff, S *et al.* (1997) Shaping knowledge, defining uncertainty: The dynamic role of assessments. A critical evaluation of global environmental assessments. WC Clark, J McCarthy and E Shea. Cambridge, MA, Harvard University: pp. 79–113.

Moser S (1999) Global Environmental Assessment Project, Environment and Natural Resources Program, Harvard University. Impact assessments and decision-making: How can we connect the two? White paper prepared for the SLR impact assessment workshop in Charleston, SC, January 31-February 3, 1999.

MSN (2005) Ecological risk assessment for the proposed submarine tailings disposal scheme at the Toka Tindung Mine, Indonesia, prepared by ERM, confidential report.

SCOPE (Scientific Committee on Problems of the Environment) 5 (1975) Environmental Impact Assessment: Principles and Procedures; established by the International Council of Scientific Unions (ICSU) in 1969 http://www.icsuscope.org/downloadpubs/scope5 3/16/2006

Shell (2003) Environmental Impact Assessment Module, HSE Manual, Shell International Exploration and Production, EP 95-0370.

SIDA (1998) Guidelines for Environmental Impact Assessments in International Development Cooperation.

Sorensen J (1971) A framework for the identification and control of resource degradation and conflict in the multiple use of the coastal zone. Dpt. of Landscape Architecture, U. of California, Berkeley, California, USA.

Spalding H (1994) Cumulative effects assessment: concepts and principles. Impact Assessment 12 (3), pp. 231-51.

Spitz K and Moreno J (1996) A practical guide to groundwater and solute transport modeling, John Wiley & Sons.

Suter GW II ed. (1993) Exposure in Ecological Risk Assessment. Lewis Publishers, Ann Arbor, MI. 538 pp.

UNEP (2000) APELL for Mining: Guidance for the mining industry in raising awareness and preparedness for emergencies at local level.

USEPA (1998) Guidelines for Ecological Risk Assessment, EPA/630/R95/002F.

USEPA (2000) Stressor Identification Guidance Document, Office of Water Washington, DC 20460; Office of Research and Development Washington, DC 20460 EPA-822-B-00-025, December 2000.

USEPA (2005) Guideline on Air Quality Models. U.S. Federal Register, 40 CFR Part 51 (Vol. 70, No-216/Weds., Nov 9, 2005). (www.epa.gov/scram001)

Walker LJ and Johnson J (1999) Guidelines for the Assessment of Indirect and Cumulative Impacts as well as Impact Interactions, EC DG XI Environment, Nuclear Safety & Civil Protection NE80328/D1/3 May 1999.

World Bank (1989) Operational Directive 4.01 on Environmental Assessment, converted in 1999 into a new format: Operational Policy (OP) 4.01 and Bank Procedures (BP) 4.01

World Bank (2001a) Operational Policies 4.12. Involuntary Resettlement, Washington DC: The World Bank.

World Bank (2001b) Operational Policy 4.37. Safety of Dams, Washington DC: The World Bank.

10

Emphasizing Environmental Management and Monitoring

Managing what Matters

SULPHUR

Sulphur is a yellow solid non-metal with a low melting point. Although sulphur is an essential element for all organisms, many sulphur compounds such as sulphur dioxide and hydrogen sulphide are toxic to humans. Sulphur combines with many metals to form sulphides which form the major ores for copper, gold, silver, lead, zinc and many less common metals. Oxidation of sulphides generates sulphates, most of which are soluble. Native or elemental sulphur occurs in evaporates and around many volcanic vents. The most common use of sulphur is in the manufacture of sulphuric acid, one of the most common materials used in industrial processes. Sulphur is also used in vulcanizing rubber and for a variety of medicinal purposes.

10 Emphasizing Environmental Management and Monitoring

Managing what Matters

'I need only to stand in the midst of a clear-cut forest, a strip-mined hillside, a defoliated jungle, or a dammed canyon to feel uneasy with assumptions that could yield the conclusion that no human action can make any difference to the welfare of anything but sentient animals' (Rodman 1977).

In fact one can consider the EAP as the main document, with the prediction of impacts as an annex to explain why impact mitigation or enhancement is necessary.

There are many ways in which mining impacts can be reduced and collectively these are referred to as mitigation. Mitigation also refers to delivering benefit but in this context the term enhancement is more commonly used. Mitigation is essentially about answering the basic question 'What can be done about it?' What to do about recognized impacts is becoming increasingly emphasized, and the Environmental Action Plan (EAP, sometimes also referred to as Environmental Management Plan or EMP) has evolved into the single most important document of any environmental impact assessment. In fact one can consider the EAP as the main document, with the prediction of impacts as an annex to explain why impact mitigation or enhancement is necessary.

The EAP should budget for three priorities: mitigation/enhancement; monitoring; and capacity strengthening (World Bank 1991). In some jurisdictions, impact management and monitoring of management success (or failure) is developed in two separate but closely interrelated documents (referred to as environmental management (EMP) and monitoring (EMoP) plans).

Once the project is defined and significant impacts are identified, the EAP outlines protection, mitigation, and, for positive impacts, enhancement measures. Mitigation and enhancement measures, of course, vary widely, and in the case of mining projects include one or a combination of the following generic measures (see also **Figure 10.1**):

- Final changes to mine infrastructure, mine waste placement, and material management.
- Pollution controls such as wastewater or flue gas treatment plants, sediment retention ponds, encapsulation of acid forming materials, or back filling of mine wastes.

- Resource conservation through reuse, recycling, rehabilitation, or company supported environmental protection programmes (such as critical species management plans).
- Development of a protected buffer zone around the mine site, if practical and feasible (**Figure 10.2**).
- Public consultation and participation as well as community development programmes to enhance quality of community life or other aspects of the living environment.
- Training and education of workforce, and selected local community members.
- Compensatory measures for restoration of damaged resources or monetary compensation for project-affected people, including land acquisition and resettlement measures.

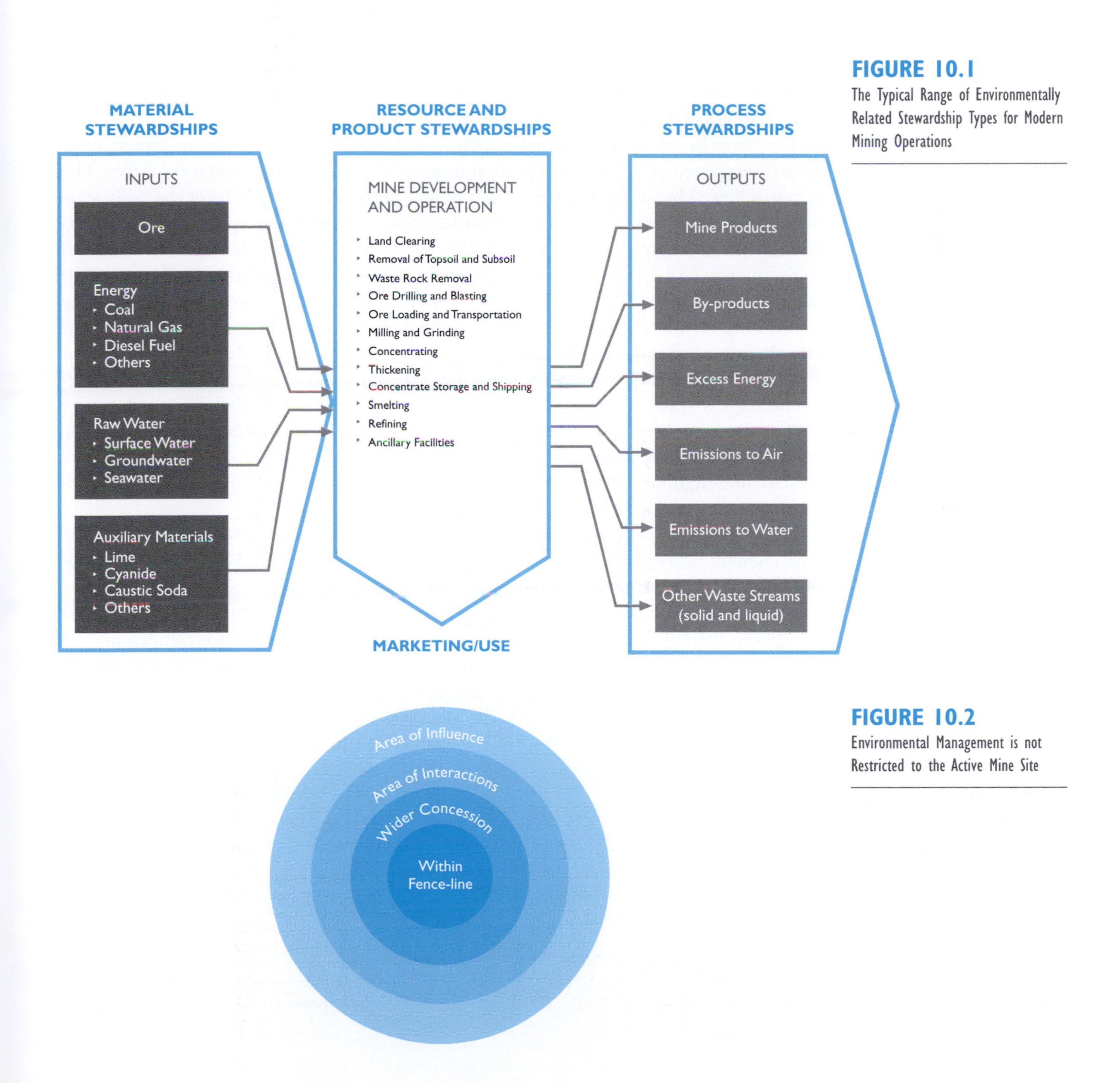

FIGURE 10.1
The Typical Range of Environmentally Related Stewardship Types for Modern Mining Operations

FIGURE 10.2
Environmental Management is not Restricted to the Active Mine Site

10.1 MANAGING WHAT MATTERS

Avoidance is better than Cure

Negative environmental impacts are best avoided.

The mitigation hierarchy of *Avoid – Minimize – Rectify – Compensate* provides a useful framework for the development of mitigation options and is illustrated in **Figure 10.3**. The mitigation hierarchy neatly summarizes common sense. Negative environmental impacts are best avoided. For this reason the study of project alternatives is so important in environmental assessment. Can the port site be moved to avoid the destruction of a mangrove forest? Is a closed cooling water cycle feasible to minimize water usage and water impacts? Is an alternative tailings disposal scheme environmentally superior?

Some impacts cannot be avoided even with the best intentions. Surface mining will change land form and land cover. Material movement consumes energy. The mine will bring changes to livelihoods in host communities.

Ideally the environment assessment team works simultaneously with the mine design team from the outset of the project until completion of the feasibility study,

The main benefit of conducting the environmental assessment early in project planning is to prevent or, if unavoidable, to minimize losses in environmental resources. This is best achieved by modifying the project design to the degree feasible, at the early stage of mine development. Ideally the environment assessment team works simultaneously with the mine design team from the outset of the project until completion of the feasibility study, often over a period of two years or more. By far the most effective and inexpensive means of environmental improvement are design changes made for environmental reasons during the design phase. At the end of the design period, environmental inputs have already improved the mine design, leaving fewer and smaller impacts to be mitigated. Incorporating mitigating measures into project design may mean that these measures are not necessarily apparent and that the associated costs are not separately recognized.

If the environmental assessment commences when the mine design is well advanced or worse, has already been completed, that should be a warning to all concerned, that deficiencies in design are likely. While still useful, much of the opportunity for zero or low cost impact mitigation may have been lost by then.

Two types of management measures apply to unavoidable negative impacts. The first type includes all remedial actions. Mine rehabilitation during mine operation is a good example. The second, used as a last resort, is to compensate for project-induced impacts. Compensation usually applies to project-affected people for their losses or negative changes to their livelihood. Environmental offsets can also be used to 'compensate' for adverse physical or ecological impacts, but more often they are designed to enhance the host environment irrespective of actual project impacts.

FIGURE 10.3
Hierarchy of Mitigation Measures

The main benefit of including the environmental assessment early in mine planning is to prevent or, if unavoidable, to minimize losses in environmental resources.

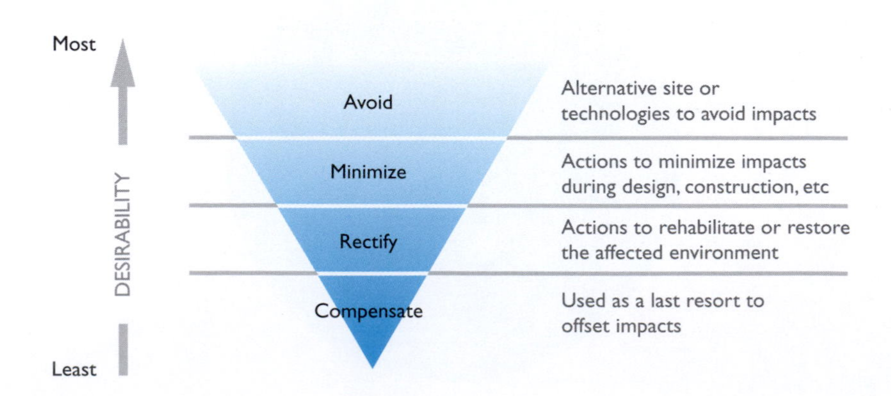

Balancing Costs and Benefits

As for most aspects of life, mitigation measures need to balance costs and benefits. As potential significant impacts are identified, mitigation measures should aim to ensure that the residual impact is 'as low as reasonably practicable'. Determining what is reasonably practicable is something that the EIA team cannot achieve in isolation. Factors such as technical feasibility, constructability, and operability are all relevant, and design input on these issues is a necessity. The second key consideration is cost. In defining 'as low as reasonably practical' for an impact/mitigation measure, the proportionality of cost to benefit should be given due consideration. Sometimes mitigation measures can save money to the project proponent, but they always improve project acceptance by the public and by approval authorities. Of course financing must always match legal obligations. Statutory emissions standards must be met irrespective of the costs involved.

Mitigation measures should aim to ensure that the residual impact is 'as low as reasonably practicable'.

Financing must always match legal obligations.

If mine development relies on external financing, related loan covenants usually specify, as a minimum, that the EAP shall be implemented on an agreed schedule with an agreed budget. The EAP itself is often an integrated part of the environmental mine permit, and stated environmental management and monitoring actions become mandatory requirements.

Focusing on the Top Ten Impacts

An EAP allocating equal attention to all predicted impacts is poorly designed. The top 10 or so impacts should capture most of the mitigation efforts, in terms of study detail and budget allocation. This is not to say that some other significant impacts do not need mitigation, but that efforts should be concentrated where most needed.

Efforts should be concentrated where most needed.

There is nothing more difficult to predict than the future. Even the best-designed mining projects cannot plan for each adjustment which is made during the project implementation. Mitigation measures should be sufficiently flexible to cope with changes during the project implementation. This usually means that there is inbuilt capacity to accommodate conditions beyond those predicted in the EIA. Flexibility also means contingency funds to finance emergencies or unforeseen changes in priorities.

Table 10.1 illustrates a commonly used summary of mitigation measures in matrix form. For each significant impact, the matrix details the proposed mitigation measures, management location, estimated costs, management schedule, and assigns responsibility. If supported

TABLE 10.1

Key Elements of an Environmental Management Matrix – For each significant impact, the matrix details proposed mitigation measures, management location, estimated costs, management schedule, and assigns responsibility

Impact	Measure	Location	Cost	Schedule	Responsibility
List impacts according to impacted environmental resources – air, land, water, fauna and flora, and people	Describe proposed measures – mitigation for negative, enhancement for positive impacts	Specify mitigation location (e.g. effluent outfall, or specific community)	Provide allocated annual budget	Define duration and frequency of management actions (e.g. quarterly, continuously)	Define institutional responsibilities – within the company and external authorities

TABLE 10.2

Key Elements of an Environmental Monitoring Matrix – A good environmental monitoring programme is based on a sound understanding of the 'what, where, when, how, and why' of monitoring

Parameters	Methodology	Location	Cost	Schedule	Responsibility
Monitoring parameters, Environmental performance indicators	Describe proposed monitoring measures – equipment, parameter, test procedure, etc.	Specify mitigation location (e.g. effluent outfall, or specific community)	Provide allocated annual budget	Define duration and frequency of management actions	Define institutional responsibilities – within the company and public authorities

All the active players in a mine development must adopt an equally rigorous stance towards environmental protection to ensure the success of the mine as a whole.

by relevant regulations, separate tables for high, medium, and low priority mitigation measures can be prepared. The Environmental Management Matrix is commonly complemented with an equivalent matrix for environmental monitoring (**Table 10.2**).

Regardless of the quality of the environmental assessment of a mining project and the design of the Environmental Action Plan, related efforts are of limited value unless recommended mitigation and enhancement measures are fully implemented (**Case 10.1**). In this context it is important that the EAP is expanded to include the EPC and all other mine contractors. All the active players in a mine development must adopt an equally rigorous stance towards environmental protection to ensure the success of the mine as a whole.

CASE 10.1
Implementation is Just as Important as Planning

When it was completed in 1996, the environment impact assessment study of the Batu Hijau Copper Mine in Sumbawa, Indonesia, set a new benchmark against which subsequent studies for other mine developments in Indonesia were measured. Shortcomings in environmental management during construction, however, caused excessive erosion, hampering the otherwise exemplary mine development and mine operation. Once the scale of this problem became apparent, however, additional resources and equipment were deployed to remedy the damage, with the result that erosion rates were dramatically reduced and, in fact, no sign of erosion damage remains.

10.2 MANAGEMENT REQUIRES MEASUREMENT

Mitigation measures, although designed in good faith, may fail to live up to their expectation or potential. Monitoring evaluates the success or failure of environmental management. Environmental monitoring is a well-researched and documented discipline, not only because of its importance, but due to the considerable costs involved. The following section captures some of the considerations that typically guide the design of a monitoring programme. For detailed reading refer to Artiola *et al*. (2004), Kim and Platt (2007), and OECC (2000).

Monitoring evaluates the success or failure of environmental management.

Monitoring Objectives

Monitoring programmes can be established for a number of purposes, typically:

- to quantify changes in physical, chemical, and biological characteristics of the environment (such as ambient water quality or vegetation);
- to ensure that results or conditions are as predicted during the planning stage, and where they are not, to pinpoint the cause and implement action to remedy the situation;
- to identify any unpredicted impacts requiring remedial measures;
- to monitor emissions and discharges to ensure that they meet established standards (such as stack emissions or decant water from tailings ponds);
- to verify the evaluations made during the planning process, in particular in risk and impact assessments and standard and target setting, and to measure operational and process efficiency;
- to measure changes in the inter-relationships of different aspects of the environment (such as metal contents in water and fish tissue);
- to determine whether environmental change is a result of project activities or a result of other activities or natural variation (such as fluctuations in water levels);
- to determine whether adopted mitigation measures are effective ('You can only manage what you measure.');
- to provide factual data that can be used to refute unfounded claims by opponents of the project;
- to monitor whether procedures, for example grievance mechanisms, emergency response plans or rehabilitation measures, are working effectively; and
- to provide the basis for mine auditing.

'You can only manage what you measure.'

Monitoring will confirm that management commitments are being met, or not. Monitoring may take the form of direct measurement and recording of quantitative information, such as amounts and concentrations of emissions, for measurement against corporate or statutory standards, consent limits or targets. It may also require measurement of ambient environmental parameters in the vicinity of the mine site using ecological or biological, physical, and chemical indicators. Equally importantly, monitoring programmes are also concerned with socioeconomic interactions, including local liaison activities and the assessment of complaints or grievances.

Monitoring programmes are also concerned with socioeconomic interactions, including local liaison activities and the assessment of complaints or grievances.

The preventative approach to management may also require the monitoring of process inputs, for example, type and stocks of chemicals used, resource consumption, and the efficiency and performance of plant and equipment. Malfunction of equipment is, in fact, one of the major contributors to environmental damage, although this is often overlooked. Where hazardous materials are to be stored, used or processed, an inventory should be maintained, and the monitoring programme should pay particular attention to the most labile and most toxic or hazardous of these.

Key Elements of an Environmental Monitoring Programme

A good environmental monitoring programme is based on a sound understanding of the 'what, where, when, how, who, and why' of monitoring. It also contains a mechanism to feed monitoring results back to decision-makers, including government authorities. References should be made to relevant legislation and standards of the host country. Monitoring methods should conform to relevant guidelines of the host country concerning monitoring techniques and equipment, and if possible and practical, certified laboratories should be used for critical tests. On more than one occasion, government authorities have rejected monitoring results from established and well-recognized international laboratories because they lacked certification by the host country (**Case 10.2**).

It is important to have a clear perspective on the purpose of the programme.

Considerable resources may be deployed in monitoring, so it is important to have a clear perspective on the purpose of the programme and what the results will be used for. When drawing up a monitoring programme, it is therefore critical to specify:

- why monitoring is being undertaken and what methods will be used;
- who is responsible for carrying out the monitoring and who should receive the results;
- what the follow-up should be (i.e. who will check and interpret the results); and
- what will happen if the monitoring indicates non-compliance with commitments.

What Should Be Measured?

The parameters to be monitored for any particular project should be selected based on the following:

- regulatory requirements;
- the chemical composition of the ore and process reagents;
- the results of baseline surveys; and
- specific concerns of local communities.

Considerable money may be wasted in measuring parameters that are not present in significant concentrations.

Regulatory requirements apply to a wide range of projects and therefore do not necessarily meet the needs of particular projects. This particularly applies in the case of metals. For example there is no requirement by the Indonesian government to monitor Silver (Ag^{+}) in surface water discharges. However, due to its potential toxicity to aquatic organisms and its high concentration in and relatively low recovery from many gold ores, this is a parameter that should be monitored on such projects.

On the other hand, the Indonesian government requires that both mercury and cyanide concentrations be monitored, regardless of whether or not mercury occurs in the ore or cyanide is used in the process. As a result, considerable money may be wasted in measuring parameters that are not present in significant concentrations.

Quality Endorsed Company

CASE 10.2
What, Where, When, How, Why, and by Whom?

As part of the environmental approval process for the Dairi Lead and Zinc Mine in North Sumatra, Indonesia, ore processing technologies were tested in Australia, with subsequent leachate tests of produced tailings by an Australian based laboratory. Test results were rejected by local government authorities since the laboratory used was not officially registered with the Indonesian government authorities.

Graphic: Registered ISO 14000 Certification Trademark in Australia.

TABLE 10.3

Typical Components of a Water Monitoring Programme (UN 1992) – Monitoring parameters should be limited to significant elements, based on experience, past monitoring results, and common sense judgement

Physical
- Temperature
- Turbidity TDS
- Water flows

Chemical
- Conductivity
- Alkalinity
- pH
- Common cations: Na^+, Ca^{++}, K^+, and Mg^+,
- Common anions: SO_4^-, Cl^-, HCO_3^-, CO_3^-
- Hardness
- Colour
- COD/BOD
- Nitrogen
- Phosphorus
- Metals – depending on those metals enriched in the ore

Biological
- Phytoplankton
- Zooplankton
- Benthic organisms
- Fish
- Water fow

The results of baseline surveys are particularly relevant. Any parameter, whose concentration is close to or exceeds ambient standards during baseline surveys, should be included in subsequent monitoring programmes.

Typical parameters to be included in water quality monitoring are listed in **Table 10.3**. It is useful to include the major cations and anions (as listed) as this enables a cation/anion balance to be calculated which may identify analytical problems or the presence of unidentified contamination.

Monitoring results may indicate that continued monitoring of a specific parameter yields no benefit, and the parameter may be excluded from future monitoring subject to approval by relevant environmental authorities.

In selecting the parameters to be monitored, particularly in relation to groundwater quality, it should be recognized that geochemical changes will take place as water flows through soil and rock. The following three examples illustrate this point:

Geochemical changes will take place as water flows through soil and rock.

1. Visual monitoring of a TSF embankment at an alumina refinery identified a seepage emerging from the embankment. Analysis of the seepage indicated a neutral pH, and virtually no hydroxide (OH^-) whereas the solution in the adjacent impoundment had a very high hydroxide concentration and a pH exceeding 13. Accordingly, the seepage was initially ascribed to rainfall infiltrating the embankment and subsequently emerging through the downstream. Subsequently it was found that the seepage had originated from within the impoundment but had been neutralized by reactions with the embankment soils. Had it been measured, the high concentration of sodium might have provided an earlier indication of this phenomenon.
2. It is common for pH levels in groundwater to remain at or about neutral levels, even when the water is draining through waste rock or tailings that are oxidizing. This is because of the buffering capacity in the system. In many project situations, maintenance of neutral conditions has led to complacency on the part of the operators in the

Maintenance of neutral conditions has led to complacency on the part of the operators in the mistaken belief that ARD was not occurring.

mistaken belief that ARD was not occurring, and the subsequent unpleasant surprise when the buffering capacity was eventually overcome, as evidenced by a rapid decline in pH. Monitoring of sulphate (SO_4) concentrations would have provided early warning that ARD was, in fact, occurring.

3. Cyanide also is readily decomposed by reactions with minerals in soil and rock, yielding carbon and nitrogen oxides. The products of these reactions may not be identified through a normal suite of parameters. Again, it is possible that the capacity for such reactions will eventually be exceeded, at which time cyanide may be detected in monitoring bores.

Such situations are discussed in more detail by Mulvey (1997). In all these cases, it may be several years or more before any direct evidence of groundwater contamination is identified.

Monitoring Locations

Monitoring locations are selected, based on statutory requirements and common sense. For water quality monitoring, any compliance points identified in the EIA documents or in government permit conditions represent the most important sampling locations. Similarly, locations are selected to enable project-related contamination to be distinguished from contamination that results from other causes. Therefore in the simple example of a project producing a single effluent discharge to a surface stream, monitoring points would be located upstream of the discharge, at the effluent source itself, and immediately downstream of the discharge. Depending on the regulatory situation, compliance points could be at the effluent discharge itself or at a prescribed distance downstream (selected to allow for thorough mixing of the effluent in the stream), or both. Since baseline data serve as a reference, monitoring locations are best linked, where possible, to sampling locations used in the initial field surveys. This applies to physical, chemical, biological, and social parameters.

Monitoring locations are best linked, where possible, to sampling locations used in the initial field surveys.

When monitoring of groundwater quality in the vicinity of tailings facilities, waste rock dumps and other locations where there is potential for groundwater contamination, consideration should be given to directions of groundwater flow. In this respect, the influence of the facility on groundwater flow directions should also be taken into account. Clearly, the majority of groundwater observation bores should be located down-gradient (in terms of the water table) from the potential source of contamination.

Monitoring Duration and Frequency

The duration and frequency of monitoring should reflect the nature of impact.

Monitoring should occur throughout the construction, operation, and de-commissioning phases of the mine. The duration and frequency of monitoring should reflect the nature of the impact. Effluent discharge often requires continuous monitoring during the entire mine operation. In contrast, progress achieving net income goals for resettled people, a requirement in line with the imperative expressed in the World Bank's Involuntary Resettlement Policy (OP 4.12), may be amenable to annual monitoring over a period of several years.

Attention should be given to seasonal and diurnal variability in designing monitoring programmes.

Attention should be given to seasonal and diurnal variability in designing monitoring programmes. This is clearly important in relation to aspects of climate, surface hydrology and groundwater. However, it is also important for many biota, the presence or population of which may differ markedly from one season to another, or in the case of marine zooplankton – diurnally. The same considerations apply to the design of social surveys. For example, door knocking with questionnaires will yield different population samples during the day compared to the evenings. Similarly, in traditional agricultural societies, surveys during the planting and harvesting periods would result in an extremely non-representative sample.

Sometimes it is necessary to experiment in order to select the most cost-effective sampling frequency. For example, measurement of groundwater levels may be made weekly or even daily at first, until the rate and magnitude of change become clear. Then it may be possible to reduce the frequency to monthly or quarterly, and still identify maximum and minimum levels.

In the absence of any compelling or statutory requirements, monthly sampling is usually selected for water quality analyses other than for effluent discharges, which may be monitored continuously or hourly as is commonly the case for cyanide, or daily for other parameters of concern. Biological monitoring is usually conducted annually, twice per year or four times per year, depending on which parameters are being monitored and the seasonal variations that apply, as revealed by the baseline surveys. Monitoring of rehabilitation success is generally conducted annually, usually near the end of the wet season, although there are benefits in conducting a second round of monitoring towards the end of the dry season.

Environmental Monitoring Plans are Living Documents

As is the case for environmental management plans, environmental monitoring plans are living documents. Monitoring locations may change with changes in mine development; monitoring results may suggest to delete (or to add) selected parameters, or to adjust monitoring frequency. As one example, there is little benefit in monitoring groundwater quality monthly if past track records demonstrate only small variations over time.

The dynamic nature of environmental monitoring plans (and management plans as well) is illustrated in **Figure 10.4**. The design of monitoring is shaped by the monitoring objectives. The objectives in turn are influenced by the outcome of monitoring results.

Because the natural environment changes with time, it is not always easy to distinguish incremental project-induced changes. But in order to understand environmental changes due to mining, it is necessary to know what environmental conditions would have been if mine development had not taken place. Hence it is important to include reference locations in the environmental monitoring programme, which will remain unaffected by mine development. For water quality monitoring these are normally monitoring locations in adjacent watersheds or upstream of the mine area.

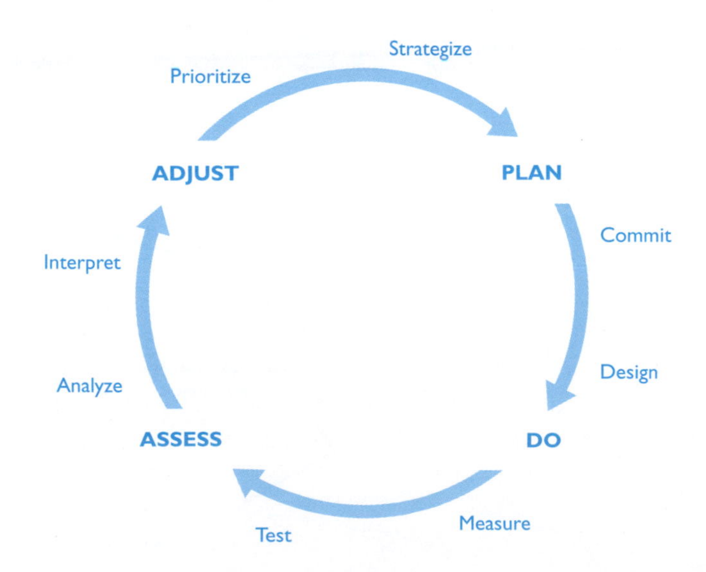

FIGURE 10.4
The 'Plan-Do-Assess-Adjust' Management Cycle

All management and monitoring systems consist of a cycle 'plan, do, assess, and adjust' process that takes learning and experiences from one cycle and uses them to improve and adjust expectations during the next cycle.

It is also valuable to include one or more reference communities, unaffected by the project, into the social monitoring programme.

While not common it is also valuable to include one or more reference communities, unaffected by the project, into the social monitoring programme. This allows more objective evaluation and demonstration of socioeconomic benefits that the mining project and its community development programmes have brought to host communities throughout the mine life. Practice shows that advocacy groups tend to take mine closure as the last opportunity to challenge the benefits of mining. A comparative study of communities within and outside the influence of the mine may help to diffuse some of the criticism.

Figure 10.5 illustrates the surface water monitoring programme for an underground the mine in Java, Indonesia. Monitoring locations include output indicators (e.g. outfall at de-toxicification plant), reference locations (upstream monitoring locations), and compliance points (outfall of polishing pond and discharge point into the river).

The collection of high-quality data depends on having adequately trained people.

Ultimately the collection of high-quality data depends on having adequately trained people. This applies both in the field and in the laboratory. One way to document satisfactory

FIGURE 10.5
Example of a Surface Water Monitoring Programme

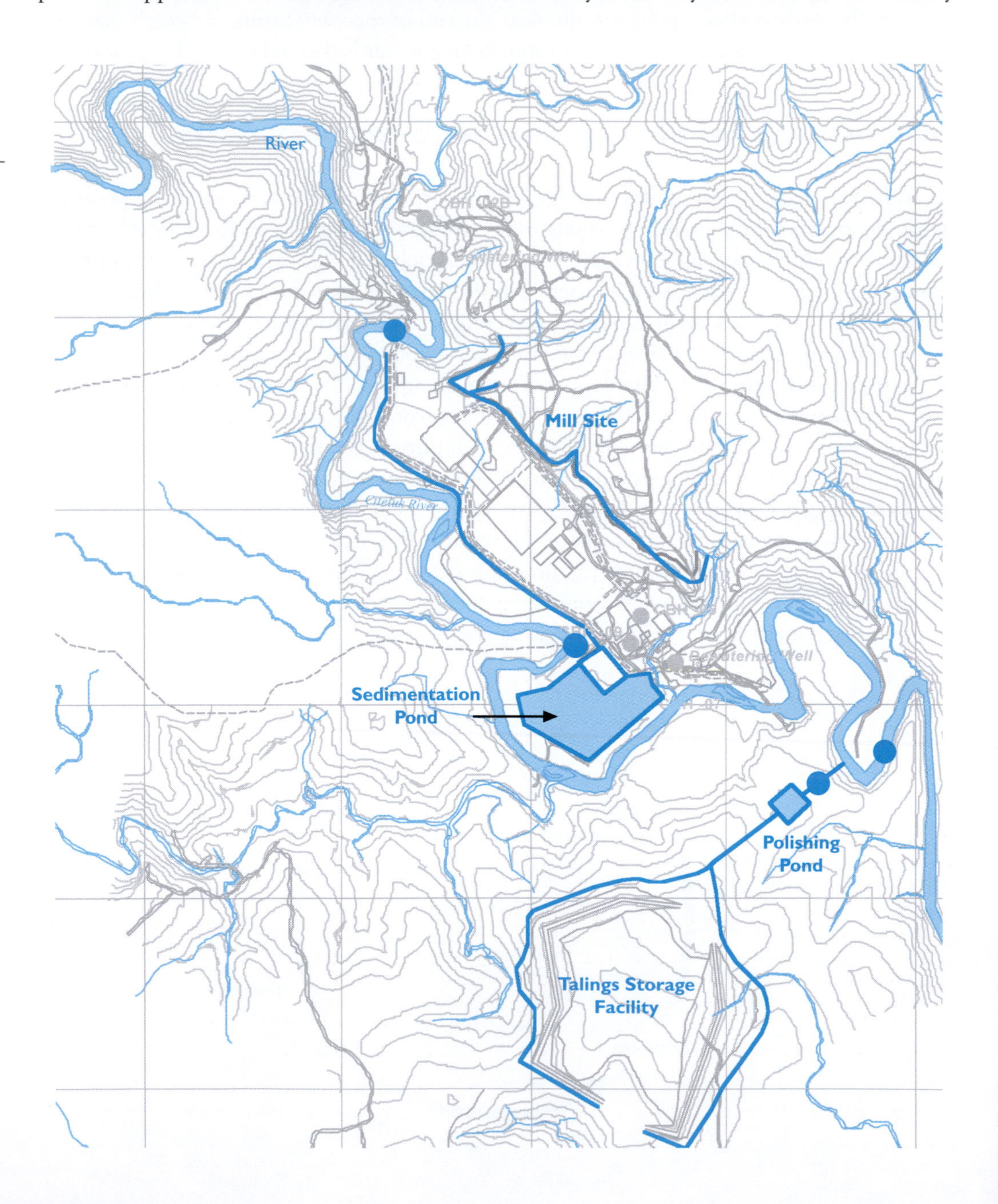

training is to have newly trained personnel collecting data using established procedures and then compare their results with those previously considered acceptable. In spite of this approach, it is difficult for a mining company to assure itself and other potential data users that different personnel collecting and analyzing samples from the same site yield comparable results.

10.3 ENVIRONMENTAL MANAGEMENT SYSTEMS

Mining operations maintain a variety of relationships, from contractor partnerships and joint ventures, to dealing with other stakeholders such as government and the public. This, together with the fact that environmental issues are now so publicized (even small incidents are communicated via SMS within minutes) and interconnected, means that an *ad hoc* approach to problem-solving is no longer effective. The need for a systematic approach to management of health, safety and environmental (HSE) issues is well recognized. Understandably mining companies mainly focus on addressing health and safety concerns through rigid management approaches. These systems are being increasingly complemented by a formal environmental management component.

An *ad hoc* approach to problem-solving is no longer effective.

The Key Components

Environmental Management Systems provide a tool for managing the environmental impacts associated with mining. EMS provides a structured approach to planning and implementing environment protection measures, and to ensure compliance with local and national laws and regulations, international treaties, and corporate requirements. As a financial management system monitors expenditure and income and enables regular checks of a company's financial performance, an EMS attempts to monitor environmental performance. An EMS integrates environmental management into a company's daily operations, long-term planning and other quality management systems.

A high-powered car does not make a good driver, and similarly EMS is a tool to support responsible management, not a substitute for good management. Environmental performance is more a reflection of the organization's culture, rather than the quality of its systems.

There is extensive literature on EMS, and this text provides only an introductory summary, based loosely on the fact sheet on EMS provided by the DEH (2006). The interested reader is directed to Hunt and Johnson (1995) for fine points on the subject, and to the excellent 'how to' implementation guide by the USEPA (1996).

To develop an EMS, an organization needs to assess its environmental impacts, set targets to reduce negative impacts, and formulate actions to achieve the targets. The environmental assessment of a mining project and its supporting documentation does just this. A well-designed environmental management and monitoring plan will go a long way towards establishing a more formalized EMS. The most important component of an EMS, however, is organizational commitment. For an EMS to be successful, commitment is required at the very top of the organization. Senior management should provide strong and visible leadership, and ensure that this commitment is translated into the necessary resources to develop, operate, and maintain the EMS, and to attain the policy and strategic objectives. Common elements of an EMS are shown in **Table 10.4.** Essentially, all management systems consists of a cycle 'plan, do, assess, and adjust' process that takes learning and experiences from one cycle and uses them to improve and adjust expectations during the next cycle (**Figure 10.4**).

The most important component of an EMS is organizational commitment.

TABLE 10.4

Components of a Formal Environmental Management System (EMS) – EMS is a tool to support responsible management, not a substitute for good management

Environmental Policy is the public statement of what an organization intends to achieve from environmental management.
Environmental impact identification is of course a prerequisite to managing actual and potential environmental impacts of an operation. Impact identification is central to the EIA. For an operating mine, impact identification is best achieved through an environmental audit.
Objectives and targets are the main mechanisms to encourage improvement of an organizations' environmental performance. Targets are regularly reviewed.
Consultation before, during and after establishment of an EMS will be necessary to ensure that staff are involved in, and committed to the EMS. Public consultation can help to improve public perception of the company, one of the benefits of committing to a formal EMS.
Operational and emergency procedures, usually in form of Standard Operating Procedures (SOP), guide operating in line with the organization's environmental objectives and targets.
Environmental management plan details the methods and procedures adopted to meet objectives and targets.
Documentation of objectives, targets, policies, responsibilities and procedures along with information on environmental performance is useful for verifying environmental performance to stakeholders, in particular regulators and the public.
Responsibilities and reporting structure need to be allocated to staff and management.
Training is necessary to familiarize staff with their responsibilities in environmental management and in achieving environmental objectives and targets.
Review audits and monitoring compliance should be undertaken regularly to ensure the EMS is achieving its objectives and to refine operational procedures to meet this goal.
Continual improvement is the ultimate aim of any EMS. Environmental objectives should be regularly reviewed to see if they can be improved or if more effective systems can be introduced.

Benefits and Limitations

An EMS should not be prescriptive; rather it should require organizations to take an active role in examining their practices.

Much has been written on the benefits of formal environmental management systems. The main benefits are arguably (1) demonstrating a good corporate image, (2) building awareness of environmental concern among employees, (3) implementing formal procedures to monitor regulatory and legislative changes affecting the organization to help ensure regulatory compliance, and (4) minimizing environmental liabilities. Admittedly (and unfortunately) documentation within an EMS requires excessive administrative efforts. That said, an EMS should not be prescriptive; rather it should require organizations to take an active

CASE 10.3
A Picture Says more than Thousand Words

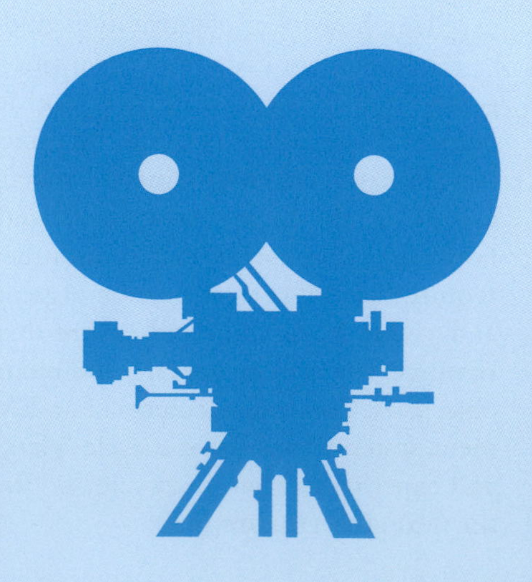

EMS consultants in particular love to indulge themselves in volumes of written Standard Operating Procedures. 'Documentation' is the buzzword, since auditing, a cornerstone of any EMS, largely depends on written evidence that mine operation adheres to agreed upon and established procedures to safeguard the environment. As a consequence the end result of a formal EMS is too often a burdensome administrative system, which may satisfy an auditor but which may consume substantial efforts to maintain.

But there are other less formal ways to implement and to maintain good environmental management. In some circumstances, less formal systems may also be more appropriate. Mining companies working in frontier regions have found it effective to replace or at least to complement written SOPs with short video clips illustrating the key message to their staff (some of whom are probably illiterate). It is also not surprising that most mining companies rely on short induction movies instead of written materials to raise the environmental, health and safety awareness of the mine visitor.

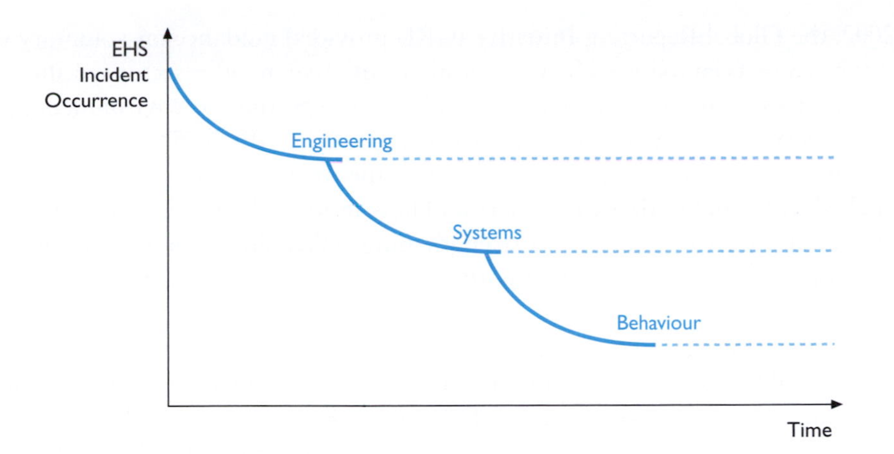

FIGURE 10.6
Ways and Means to Reduce Health, Safety, and Environmental Incidents

An environmental management system is a valuable tool to improve the environmental performance of a mine, but it is not the sole solution to environmental excellence.

role in examining their practices, and then determining how their impacts should best be managed (**Case 10.3**). This approach encourages creative and relevant solutions from the organization itself. The main limitation of any management system, however, is that human behaviour cannot be changed by systems and procedures.

Figure 10.6 illustrates that an environmental management system is a valuable tool to improve the environmental performance of a mine, but it is not the sole solution. The first step to environmental excellence is responsible mine planning and engineering; the environmental impact assessment is the vehicle to do this. The second step towards success is having appropriate management systems in place during operation. In the long-term, however, environmental excellence is based on a company's culture and on human behaviour. The point is that neither engineering nor systems can replace leadership and commitment. Buy-in from senior management is essential, eventually trickling down through the organization, and thereby developing a culture of care and environmental responsibility.

Although the implementation of an EMS is essentially a voluntary initiative, it can also become an effective tool for governments to protect the environment as it can assist regulation. For example, regulatory systems can encourage organizations to use an EMS to meet standards, by providing incentives for strong environmental performance. Likewise, organizations can use an EMS to ensure that their performance is within regulatory requirements, and to keep ahead of more stringent regulations which might be introduced in the future.

Internationally, voluntary environmental management standards have emerged to bring expectations about environmental management and environmental performance into harmony. One of these efforts, conducted by the International Organization for Standardization (also known by its French acronym, ISO) has resulted in a series of international environmental management standards (the ISO 14000 series).

Larger mining companies may find certification of their EMS more valuable considering the potential trade and market advantages of an internationally recognized and certified EMS. This was a significant factor for companies seeking certification under the ISO 9000 quality assurance standards, and is likely to be a factor in decisions regarding ISO 14001 certification.

Voluntary Sustainability Reporting

One result of the increasing emphasis on integrating non-financial operational aspects into mine management is the growing practice of reporting non-financial performance to complement the financial reports intended for investors. Often this is referred to as 'triple bottom line' reporting, or reporting of the three legs (environment, social, and economic) that are the basis of sustainable development.

In 2002, the Global Reporting Initiative (GRI) provided guidance on voluntary reporting. This has now been followed by various other initiatives to offer sector specific or alternative references to improve the quality of voluntary reporting on environmental, health and safety, social, and economic performance (e.g. IPIACA/API 2005).

Reporting of non-financial indicators has become an increasingly important requirement fuelled by the public desire to understand how industrial activities affect the environment. It can also provide a means for strengthening stakeholder relationships, improving internal values, and may become an important accountability mechanism which will eventually lead to enhanced business value (Stratos 2001).

The key to performance reporting is selecting the set of appropriate performance indicators.

The key to performance reporting is selecting the set of appropriate performance indicators. Financial key performance indicators (KPI) are well-established, but this is not true for non-financial performance indicators. Developing KPIs for a particular mine is not an easy exercise (see Chapter Eight for a discussion of environmental indicators). It is even more difficult at the corporate level where it becomes necessary to report on the collective performance of several mining operations, which may be operating in different jurisdictions and environmental settings.

10.4 COMMITMENT, FUNDS AND RESOURCES

Good design and planning are essential, but they are not a substitute for good implementation. Experience shows that more often than not, a massive gap exists between the EAP recommendations and their implementation, particularly during mine construction. The reasons for this apparent inconsistency include inadequate funding and lack of internal capacity.

Allocating Environmental Budgets

Adequate budgeting for environmental management signals the ownership, commitment, and buy-in of senior management.

Implementation of mitigation measures requires money. Not accepting environmental management costs as standard operational project costs is a warning sign of future environmental problems. In contrast, adequate budgeting for environmental management signals the ownership, commitment, and buy-in of senior management.

Environmental performance depends on the timely implementation of planned mitigation measures, and budgeting should be aligned to the environmental implementation schedule. Budgets must be time-bound. Often implementation is time-sensitive and cannot be left until later in the mine life. Environmental management of construction activities at the beginning of mine development unfortunately often serves as an example of what can go wrong. Construction activities are the cause of a number of significant adverse impacts, if not managed properly. In tropical areas, scheduling of earthworks to avoid the wet season may be the single most effective decision in avoiding environmental problems and their costly consequences. Yet only a minority of projects plan accordingly. Budgeting for and monitoring of environmental mitigation measures (of a physical and social nature) during construction, are essential to ensure that the mining project is not tainted by negative community perceptions for the rest of its life.

Adequate funding for environmental mitigation measures is now a legal requirement in most countries. Where external project funding is used, funding for environmental management is normally included in the legal covenants and loan agreements. A standard clause in loan agreements is to specify that the EAP shall be implemented in accordance with an agreed schedule and budget.

Environmental budgets, especially for mining projects, differ from normal project budgets in that environmental measures and needs often extend well beyond construction and operation into the period following the mine closure. Some environmental impacts may also only become relevant some years after the project has commenced operating. For example, separate mine pits may be developed in sequence, or new waste mine placement areas are developed years into mine operation. Then decommissioning, mine closure including final rehabilitation, and post-mining monitoring costs are needed possibly for decades after operations cease and there is no longer a positive cash flow. Budgets for priorities very late in the life of the project, specifically mine closure and post-closure environmental measures, are probably best provided for in escrow accounts or performance bonds (see Chapter Twenty One on Mine Closure).

Human Capacity – or Lack Thereof

Human capacity, that of the professionals involved in implementation of mitigation measures, is a necessary requirement for successful implementation of the EAP. However, the capacity for performing environmental management, in particular managing social aspects, is often lacking in new mining ventures, in which initial resources are primarily directed to overcoming the many technical, engineering, and logistic challenges associated with project development. Accordingly, lack of human capacity has arguably become the biggest constraint to effective implementation of environmental mitigation measures. While mining professionals readily cope with technical measures to limit physical or chemical impacts, measures to address the softer social issues remain alien to many technically oriented personnel. Even people trained in the social sciences commonly lack the sensitivity or the emotional stamina required to effectively deal with complex social issues.

The capacity for performing environmental management, in particular managing social aspects, is often lacking in new mining ventures.

Capacity strengthening requires much more time than securing budgeted funds. The process commences with the needs and capabilities analysis. These should become a routine part of the environmental assessment process and should occur early in the preparation phase of the project cycle. As preparation and implementation of environmental management plans are often the responsibilities of quite different parties (say consultant and mining company), the analysis should identify who will be doing what, and the extent to which identified professionals or institutions are suited to carry out their future tasks. In summary effective implementation of the EAP requires two main requisites, a clear budget to pay for the implementation, and adequate human capacity. Budgets are the easier of the two.

Capacity strengthening requires much more time than securing budgeted funds.

REFERENCES

Artiola J, Pepper IL, and Brusseau ML (2004) Environmental Monitoring and Characterization, Elsevier Scientific Publisher.

DEH (2006) Environmental Management Systems (EMS), Department of the Environment and Heritage (www.deh.gov.au).

GRI (2002) Sustainability Reporting Guidelines (www.globalreporting.org).

Hunt D and Johnson C (1995) Environmental Management Systems: Principles and Practice, McGraw-Hill.

IPIECA/API (2005) Oil and Gas Industry Guidance on Voluntary Sustainability Reporting (www.ipieca.org; www.ipa.org; www.ogp.org.uk.).

Kim YJ and Platt U (eds.) (2007) Advanced Environmental Monitoring 2007, XXII, 422 p., Hardcover ISBN: 978-1-4020-6363-3.

Mulvey P (1997) Conceptual Model for Groundwater Monitoring Around Tailings Dams, Evaporation Ponds and Mills. Proc. 22nd Annual Environmental Workshop. Minerals Council of Australia, Oct. 1997.

OECC (2000) Text Book for Sampling for Environmental Monitoring – Analysis Video Series for Environmental Technology Transfer, Overseas Environmental Cooperation Center, Japan.

Rodman J (1977) The Liberation of Nature, Inquiry 20: 83-145.

Stratos Inc. (2001) Stepping Forward – Corporate Sustainability Reporting in Canada.

UN (1992) United Nations, Department of Technical Co-operation for Development, and German Foundation for International Development. Mining and the Environment – The Berlin Guidelines, Mining Journal Books Ltd, pp. 41.

USEPA (1996) Environmental Management Systems: An Implementation Guide for Small and Medium-Sized Organizations. NSF International, Ann Arbor Michigan.

World Bank (1991) Environmental Assessment Sourcebook (three volumes). Technical Papers Nos. 139, 140 and 154, World Bank, Washington D.C.

World Bank (2001) Operational Policies 4.12 – Involuntary Resettlement, Washington, D.C.: The World Bank.

11

Metals, their Biological Functions and Harmful Impacts

Metals are Naturally Occurring Elements

♉ BISMUTH

Bismuth is a heavy, brittle, silver-white semi-metal with a low melting point. Most bismuth compounds are non-toxic, which has led to the use of bismuth as a replacement for lead in solders, shot and bullets. Other uses include in nuclear reactors, in alloys and fire detection devices. Commonly associated with lead and tungsten ores, the most common bismuth-bearing mineral is bismuthinite.

11 Metals, their Biological Functions and Harmful Impacts

Metals are Naturally Occurring Elements

In chemistry we learn that all known matter is made up of 109 known elements (not all of them are naturally occurring). Elements are the most basic substances consisting of atoms arranged in characteristic structures. An element cannot be broken down by chemical means into anything simpler. Elements are further subdivided into metals and nonmetals. Interestingly, most known elements are metals (86 in total) but over 80% of the Earth's crust, the area of interest for the mining industry, is made up of nonmetals. On the periodic table, metals are all the elements to the left of the diagonal line running from boron to astatine (**Figure 11.1**).

Chapter Eleven offers a brief discussion of some of these metals, their biological functions and harmful impacts. It does appear that increased global concern about the state of the environment has led to over-emphasizing the harmful impacts of metals and certainly to downplay their benefits. This may be related to the unfavourable views of society towards mining.

Metals are naturally occurring elements, and life has evolved in their presence. Industrial activities and human use of metals have liberated, transformed and accumulated metals in some geographic areas and for some metals such as lead or mercury, changes in concentrations can be detected on a global scale. It is also true that on occasions these human activities have caused harm for public health and the environment (USEPA 1995; MMSD 2001). Metals which are of most environmental concern include arsenic, cadmium, chromium, lead and mercury (listed in alphabetic order). However, the fact that methyl mercury is a serious threat to life does not support claims that mercury in all forms and concentrations is equally harmful. Similarly, while exposure to lead can result in poisoning

Metals are naturally occurring elements, and life has evolved in their presence.

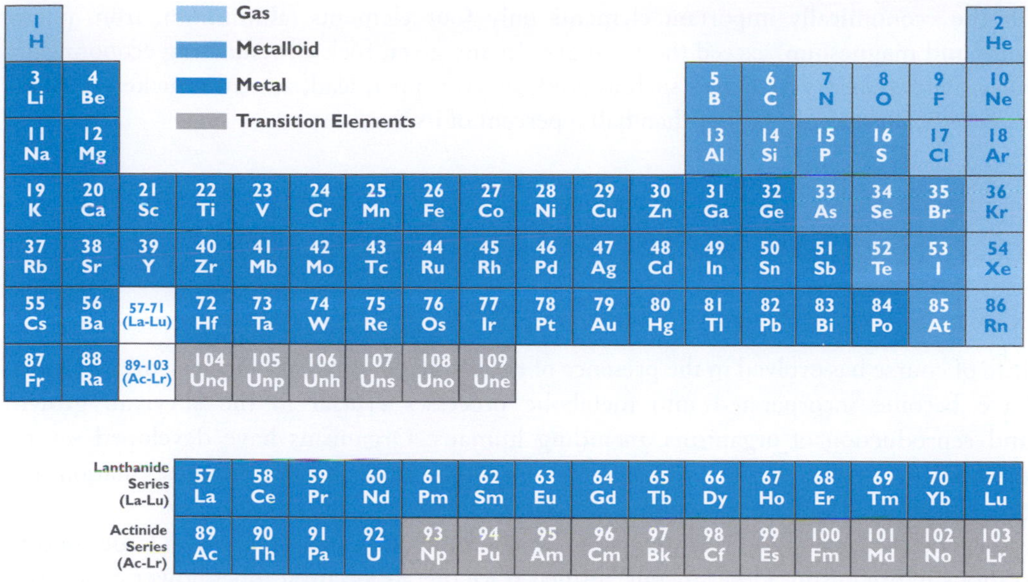

Gas
Metalloid
Metal
Transition Elements

1 H																	2 He
3 Li	4 Be											5 B	6 C	7 N	8 O	9 F	10 Ne
11 Na	12 Mg											13 Al	14 Si	15 P	16 S	17 Cl	18 Ar
19 K	20 Ca	21 Sc	22 Ti	23 V	24 Cr	25 Mn	26 Fe	27 Co	28 Ni	29 Cu	30 Zn	31 Ga	32 Ge	33 As	34 Se	35 Br	36 Kr
37 Rb	38 Sr	39 Y	40 Zr	41 Mb	42 Mo	43 Tc	44 Ru	45 Rh	46 Pd	47 Ag	48 Cd	49 In	50 Sn	51 Sb	52 Te	53 I	54 Xe
55 Cs	56 Ba	57-71 (La-Lu)	72 Hf	73 Ta	74 W	75 Re	76 Os	77 Ir	78 Pt	79 Au	80 Hg	81 Tl	82 Pb	83 Bi	84 Po	85 At	86 Rn
87 Fr	88 Ra	89-103 (Ac-Lr)	104 Unq	105 Unp	106 Unh	107 Uns	108 Uno	109 Une									

Lanthanide Series (La-Lu)	57 La	58 Ce	59 Pr	60 Nd	61 Pm	62 Sm	63 Eu	64 Gd	65 Tb	66 Dy	67 Ho	68 Er	69 Tm	70 Yb	71 Lu
Actinide Series (Ac-Lr)	89 Ac	90 Th	91 Pa	92 U	93 Np	94 Pu	95 Am	96 Cm	97 Bk	98 Cf	99 Es	100 Fm	101 Md	102 No	103 Lr

FIGURE 11.1
Periodic Table of Known Elements

From the 109 known elements most are metals, but over 80% of the Earth's crust is made up of nonmetals. From the economically important elements only 4 elements (aluminium, iron, potassium and magnesium) exceed the 1% mark. In any given rock, all remaining economically important elements combined such as gold, silver, copper, lead, zinc, tin, nickel, or platinum typically account for less than half a percent of its mass.

this does not mean that similar effects occur when being exposed to copper, nickel, zinc, or other mined metals. There are also, of course, the indiscriminate claims that all ores contain arsenic, cadmium, mercury, or even cyanide; these have become 'angst' words commonly used by anti-mining advocates to oppose mine developments.

This chapter aims to provide enough information on some mined metals and their harmful impacts to facilitate an objective environmental assessment of a mine. What follows is a highly summarized overview of what is a highly complex subject. Undoubtedly chemists and toxicologists may feel that the subject matter is dealt with in an overly simplified manner. The text, however, provides sufficient reference to the ample specialized literature on this subject to assist the reader interested in further study.

What defines a metal?

What defines a metal? In some contexts, the definition of a metal is based on physical properties. Metals are characterized by high thermal and electrical conductivity, high reflectivity and metallic lustre, strength and ductility (Masters 1991). From a chemical perspective, it is more common to use a definition that says a metal is an element that gives up one or more electrons to form a cation in an aqueous solution. With this latter definition, there are about 80 elements that can be called metals.

'Heavy metals' is an inexact term.

Among metals, heavy metals are environmentally of most concern. 'Heavy metals' is an inexact term used to describe some elements that are metals or metalloids (meaning elements that have both metal and nonmetal characteristics). Examples of heavy metals include chromium, arsenic, cadmium, lead, mercury and manganese, common associates in base metal sulphide deposits. By some definition, heavy metals are those metals that have densities above 5 g/cm^3 comprising 38 elements in total (UNEP 1994). Others define heavy metals as those metals with an atomic number 20 and greater (Eby 2004). Another, more strict definition, identifies those metals heavier than the rare Earth metals, which are at the bottom of the periodic table. More commonly, the term is simply used to denote metals that are toxic, since none of the heavy metals is an essential element in biological systems and additionally, most of the better known heavy metals are toxic in fairly low concentrations. To avoid bias, however, it is best to use the neutral term 'metal' in environmental studies.

None of the heavy metals is an essential element in biological systems.

Nature has distributed elements, including metals, in different concentrations but rather equally within the Earth's crust. Only eight elements are present in the Earth's crust in amounts exceeding 1% crustal abundance (**Figure 11.2**). These eight elements constitute more than 98% of the Earth's crust by weight. Furthermore, the two most common elements in the crust, silicon and oxygen, constitute about two-thirds of the crust's weight.

Of the economically important elements only four elements (aluminium, iron, potassium and magnesium) exceed the 1% mark. In any given rock all remaining economically important elements combined such as gold, silver, copper, lead, zinc, tin, nickel, or platinum typically account for less than half a percent of its mass.

11.1 PERSISTENCE, BIOACCUMULATION AND TOXICITY OF METALS

Life of course has evolved in the presence of metals, some of which – the essential metals – have become incorporated into metabolic processes crucial to the survival, growth and reproduction of organisms, including humans. Organisms have developed several mechanisms – with varying efficiencies – for the uptake and excretion, regulation and detoxification of both essential and non-essential metals.

Several metals (e.g. sodium, potassium, magnesium and calcium) occur in large concentrations in organisms. Other metals, termed trace metals, occur at much lower concentrations (normally <0.01%) in organisms. A number of metals, such as Fe, Mn, Zn, Cu, Co and Mo, have been identified as essential for all living organisms, while the necessity of other metals, such as Ni, V, I, Si and B, has only been established for a limited number of species. Other metals such as As, Pb, Cd, Hg, Al, Li and Sn have no known useful functions

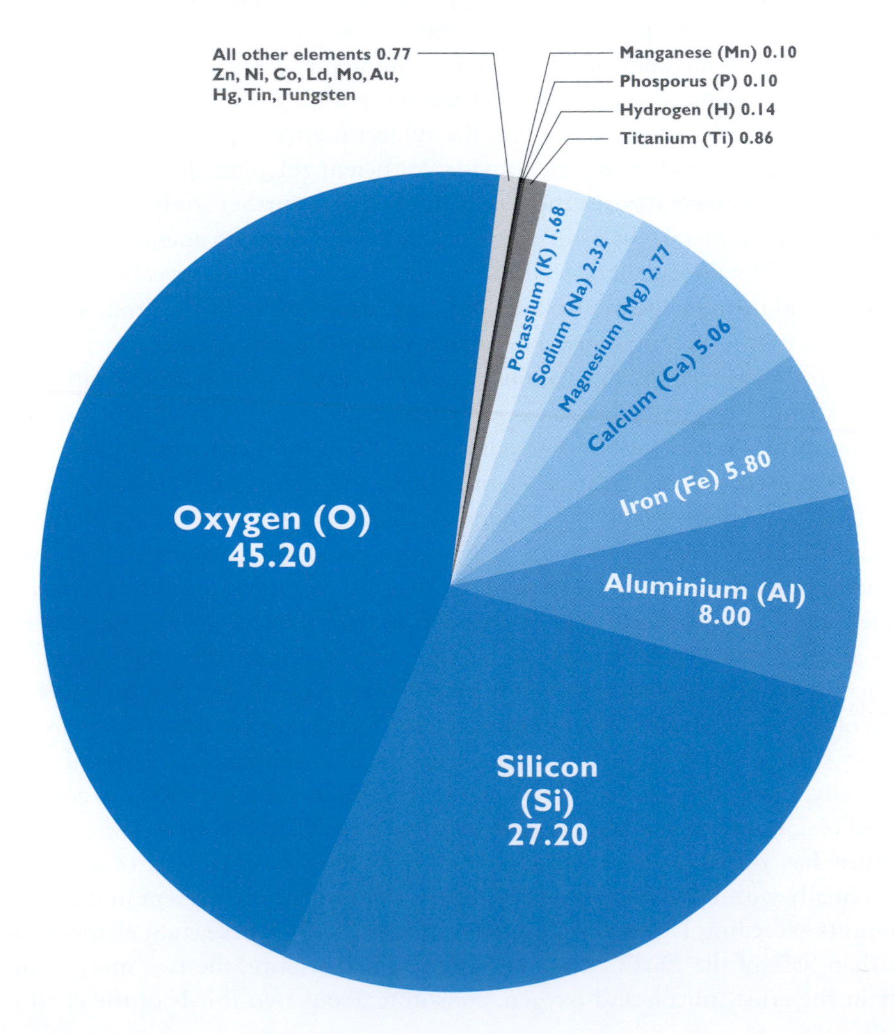

FIGURE 11.2
Composition of Earth's Crust in Percentage by Weight

The two most common elements in the crust, silicon and oxygen, constitute about two-thirds of the crust's weight. Only eight elements make up more than 98% of the Earth's crust by weight.

in biological organisms, and on the contrary, are toxic to some forms of life (Österlöf and Österlund 2003; WHO 1996). **Table 11.1** lists the functions and effects of some metals.

Some elements, as well as many compounds, can be toxic, despite the fact that they are essential. For most essential metals, either a shortage or an excess of the metal in the environment will lead to detrimental effects on organisms, populations and ecosystems. For non-essential metals, only excess environmental concentrations will cause adverse

Some elements, as well as many compounds, can be toxic, despite the fact that they are essential.

TABLE 11.1

Functions and Effects of Some Metals on Organisms (after Osterlof and Osterlund 2003; IPCS monographs WHO: 1988; 1989a, b; 1990; 1991; 1992; 1995a,b; Janssen and Muyssen 2001)

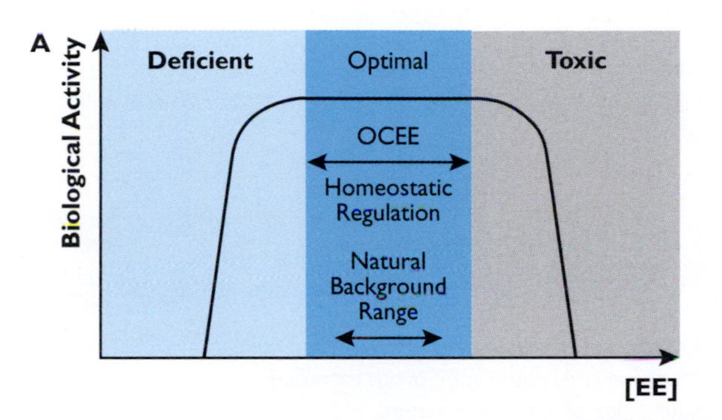

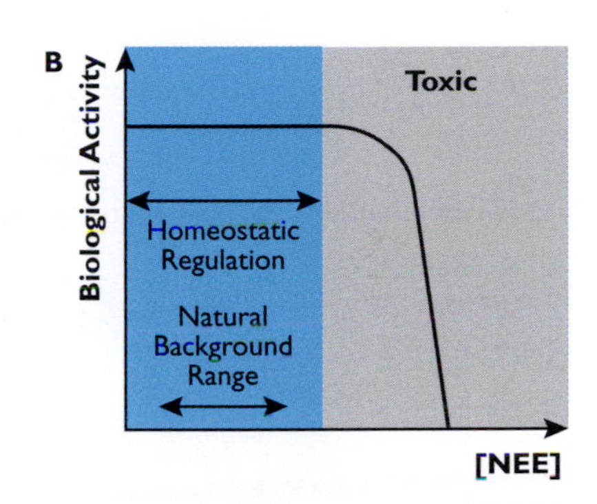

A. Biological activity as a function of the essential element concentration (EE). OCEE is the Optimal Concentration range for Essential Elements for a species in a given enviroment (Van Assche *et al.* 1997).
B. Biological activity as a function of the non-essential element (NEE).

Metal	Shortage Deficiency	Optimal Range Function	Excess Toxicity
Non-Essential Elements [NEE]			
Arsenic (As)	No known useful function	**No known useful function**	Highly poisonous and possibly carcinogenic in humans.
Cadmium (Cd)	No known useful function	**No known useful function**	Cd may be linked to renal arterial hypertension and violent nausea. Cd accumulates in liver and kidney tissue. It depresses growth of some crops and is accumulated in plant tissue.
Lead (Pb)	No known useful function	**No known useful function**	A cumulative body poison in humans and live stock. Humans may suffer acute or chronic toxicity. Young children are especially susceptible.
Mercury (Hg)	No known useful function	**No known useful function**	Hg and its compounds are highly toxic, esp. to the developing nervous system.
Essential Elements [EE]			
Cobalt (Co)		**Essential** – Present in vitamin B12, required for the formation of hemoglobin; Plays a role in biological N2-fixation	
Copper (Cu)	Anaemia, disturbed formation of bone	**Essential** – Present in cytochrome and hemocyanin, molecules; involved in (cellular) respiration; Essential in oxidases; metabolism of iron	Large doses may induce vomiting, nausea, diarrhea, cramps, or hepatic damage. Toxic to fish and aquatic life at low levels.

(Continued)

TABLE 11.1
(Continued)

Metal	Shortage Deficiency	Optimal Range Function	Excess Toxicity
Iron (Fe)	Anaemia	**Essential** – Present in hemoglobin for oxygen transport	Essentially non-toxic but causes taste problems in water; may induce vomiting, kidney damage, black stool
Manganese (Mn)	Anaemia	**Essential** – Present in pyruvate carboxylase required for the metabolism of sugars; involved in synthesis of fatty acids and glycoproteins	Affects water taste. Toxic to animals at high concentrations
Molybdenum (Mo)		**Essential** – Involved in electron transfer processes; Nitrogen fixation is also coupled to a molybdenum process	
Zinc (Zn)	Growth retardation; delayed sexual maturity	**Essential** – Essential in several enzymes catalyzing the metabolism of proteins and nucleic acids	May affect water taste at high levels. Irritation of digestive system, epithelium, vomiting. Toxic to some plants and fish
Essential Elements for Some Organisms			
Chromium (Cr)		Involved in glucose metabolism (insulin)	Cr + 6 is toxic to humans and can induce skin sensitizations. Human tolerance of Cr+3 has not been determined
Iodine (I)		**Essential for some organisms** – Present in thyroxine and related compounds for proper functioning of the thyroid system	
Nickel (Ni)		**Essential for some organisms** – Component of urease and thus a part of the CO_2 metabolism	
Selenium (Se)		**Essential for some organisms** – Activates glutathione peroxidase to scavenge free radicals	
Valadium (V)		**Essential for some organisms** – Regulation of intracellular signalling; Cofactor of enzymes involved in energy metabolism; Possible therapeutic agent in diabetes	

There is a clear concentration 'window' within which the internal metal concentration of the organism is regulated without causing detrimental effects.

effects. For both metal groups, however, there is a clear concentration 'window' within which the internal metal concentration of the organism is regulated without causing detrimental effects.

With the exceptions of mercury and gold, most metals occur rarely or not at all in metallic form under normal natural situations. Metals remaining in mine and mineral processing wastes are sooner or later transformed into other chemical or physical forms. However, unlike organic compounds, metals can not be degraded chemically or bacteriologically into simpler constituents; metals are therefore classified as persistent. Metals exist in a wide variety of physical and chemical states and several forms will coexist, depending on environmental conditions. In rocks, metals usually occur in the form of insoluble minerals such as silicates and sulphides. These minerals, when exposed to atmospheric processes, slowly decompose which may liberate some metals. In the case of sulphide minerals, as discussed elsewhere in this text, weathering may proceed rapidly with the liberation of metals in a soluble form. Once they are dissolved in surface water or groundwater, metals are likely to be bio-available. The distribution of elements into different forms is termed 'speciation' of the element. Accordingly, an element such as iron may occur in a variety of ionic forms, the simplest ones being Fe^{3+} and Fe^{2+}. Environmental conditions such as pH, redox potential, alkalinity and the occurrence of organic and inorganic compounds play an important role in speciation. Some metalloid elements such as arsenic may occur as cations (As^{3+} and As^{5+}) and as anions ($HAsO_4^{2-}$ – arsenate); arsenic may also occur in solution in various organic complexes such as methyl arsenate and di-methyl arsenate. Speciation is particularly important for the transport and the bio-availability of the metals (Parametrix 1995; ICCM 2007). Also, the toxicity of various species varies widely from highly toxic to non-toxic among different species.

The distribution of elements into different forms is termed 'speciation' of the element.

The mineral cycle as depicted in Chapter One illustrates how mining and mineral processing contribute to the release of metals into the environment. Some metals will end up unavoidably in mine wastes, particularly in tailings. Those of environmental concern are associated with mined metals; lead, mercury, cadmium and chromium in mine or mineral processing wastes that originate from the mining of other commercial metals such as zinc, copper, nickel or gold. In fact, as noted in Chapter Six, cadmium is entirely a byproduct of smelting, and there is no mining operation dedicated to its extraction.

The most important mechanisms for the release of metal from mine wastes are through leaching into ground and surface water, fugitive dust emissions and from tailings solutions. Leaching of metals depends on metal solubility. Since metals exist in a wide variety of physical and chemical states, solubility characteristics vary widely. In addition, solubility depends on many physical and chemical parameters occurring in complex patterns: e.g. pH, redox potential, presence of electron donors and acceptors, occurrence of organic and inorganic complexing agents, etc. However a decrease in the pH of water (as experienced in acid rock drainage) will cause increased leaching of most metals (e.g. Campbell and Stokes 1985; Cusimano *et al.* 1986; Evans *et al.* 1988; Kimball and Wetherbee 1989; Pagenkopf 1983; Schubauer *et al.* 1993). An increase in the acidity results in an increase in the free metal ion concentration in solution. In sum, the solubility of metals is largely pH-dependent and for most metals the solubility increases with a decrease in pH value (**Figure 11.3**). Metal hydroxides are generally more soluble than metal sulphides, and they are amphoteric, as shown in **Figure 11.3**: a metal hydroxide has one pH at which its solubility is at a minimum.

The solubility of metals is largely pH-dependent and for most metals the solubility increases with a decrease in pH value.

As discussed in Chapter Six air emissions in primary smelting also constitute a major source of metal release. Depending on the properties of the metal, such problems as volatile species, fugitive dust and stack emissions may be encountered. A small portion of metals will end up in waste or by-products from subsequent steps in metal processing, mostly in the form of slag, waste-water sludge, or dust.

Air emissions in primary smelting also constitute a major source of metal release.

Since only a small fraction of metals contained in final manufactured products are lost during use through corrosion and wear, most metals will still be present in discarded products. There are increased efforts to recycle metals, but metal containing products are often disposed of, either uncontrolled or controlled, in municipal solid waste incinerators or landfills. As a consequence, only a small fraction of total metals released to the environment originate directly from mining and mineral processing. Other than natural sources, metal usage, including the use of leaded gasoline and the final disposal of discarded metal containing products, are eventually the main sources of metal release into environmental media such as air, water, or soil.

There is ample literature on the potentially harmful effects of metals in the environment (e.g. IPCS monographs by WHO: WHO 1988; 1989a; 1989b; 1999; 1991; 1992; 1995;

FIGURE 11.3

Concentration of Dissolved Metals as a Function of pH Value

Metal hydroxides are generally more soluble than metal sulphide, and they are amphoteric: a metal hydroxide has one pH at which its solubility is at a minimum

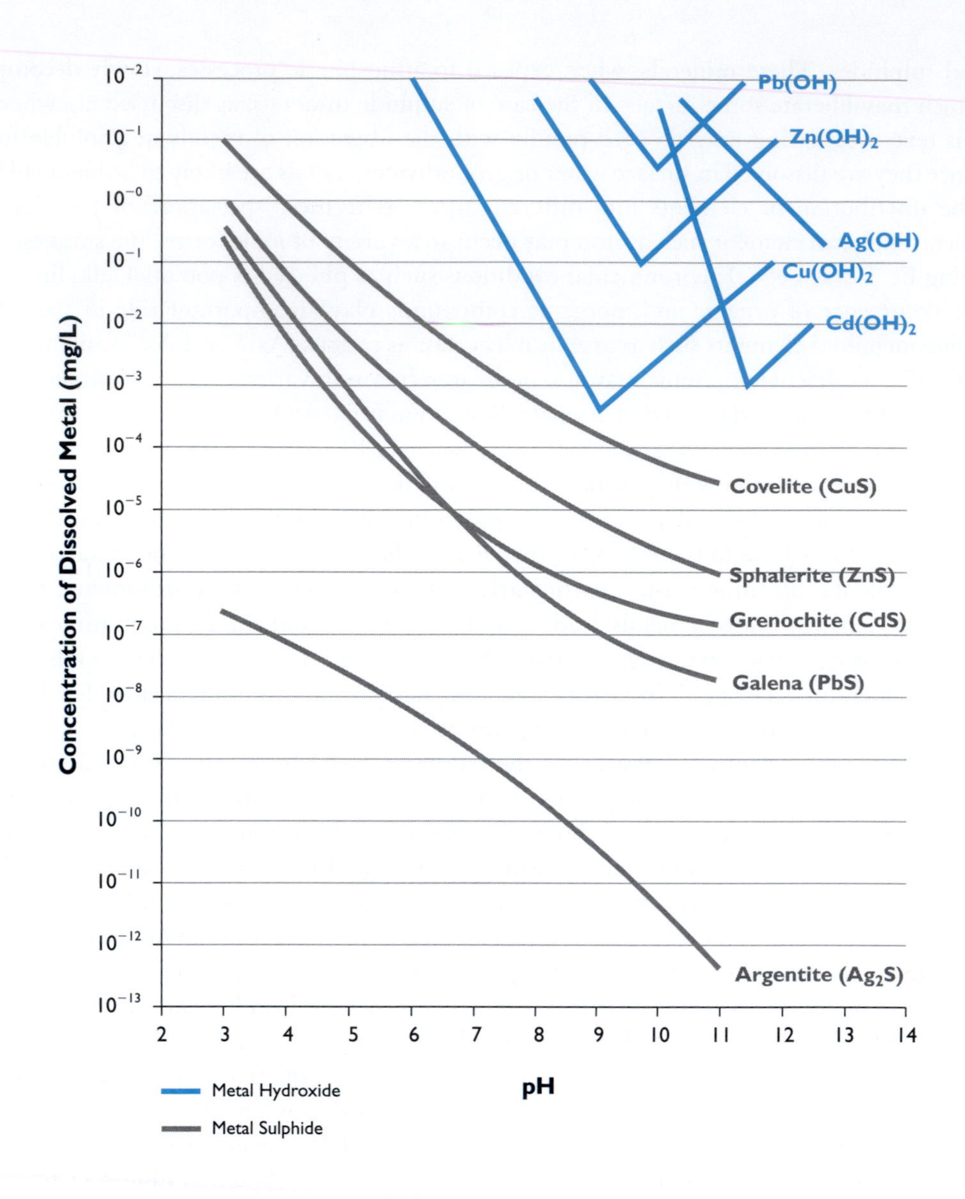

Some combinations of metals act antagonistically or even synergistically in solution.

Stephens and Ahern 2001). It should also be noted that some combinations of metals act antagonistically or even synergistically in solution: e.g. nickel and zinc, copper and zinc, and copper and cadmium in combination are more toxic than their individual toxicities would suggest (Down and Stocks 1977). The most prominent metals in the context of an environmental impact study of a metal mine are discussed in the following section.

11.2 SOME NOTES ON SELECTED METALS

By the eighteenth century 12 metals had been discovered, which doubled to 24 in the following century.

By the eighteenth century 12 metals had been discovered, which doubled to 24 in the following century. Today there are 86 known metals, of which the 7 metals of the antiquity are the metals on which civilization was based (although not classed as elements at that time): gold (5000BC), copper (4300BC), silver (4000BC), lead (3500BC), tin (1800–1600BC), iron (1400BC), and mercury (750BC) (Ball 2002). After the Renaissance the seven known metals at that time became linked with the seven known celestial bodies and the seven days of the week (**Figure 11.4**).

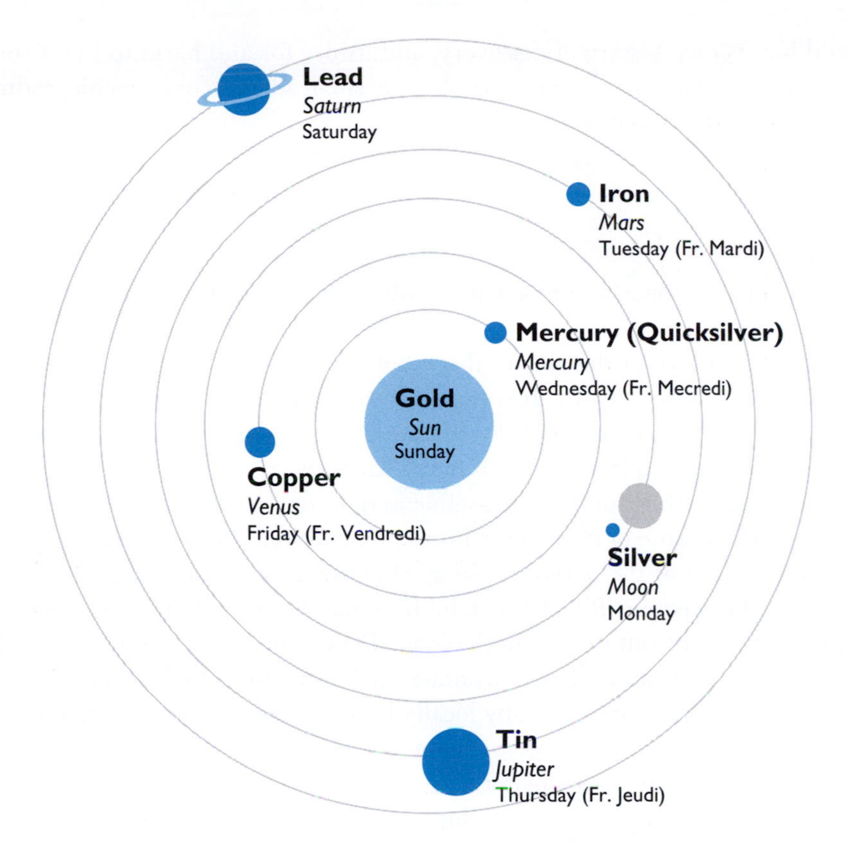

FIGURE 11.4
Metals and their Discovery

The Seven 'Classical' Metals of the Antiquity and Their Correspondences

Timeline of Metals Discovery

The Seven Metals of the Antiquity

6000 BC Gold
4200 BC Copper
4000 BC Silver
3500 BC Lead
1750 BC Tin
1500 BC Iron
750 BC Mercury

Ca 1300 Arsenic
Ca 1500 Platinum
1560 Antimony
1595 Bismuth

18th Century

1735 Cobalt
1738 Zinc
1751 Nickel
1774 Manganese
1781 Molybdenum
1782 Tellurium
1783 Tungsten
1789 Uranium
1789 Zirconium
1791 Titanium
1794 Yttrium
1797 Beryllium
1797 Chromium

19th Century

1801 Niobium
1802 Tantalum
1803 Iridium, Palladium, Rhodium
1807 Potassium, Sodium
1808 Boron, Barium, Calcium, Magnesium, Strontium
1814 Cerium
1817 Lithium, Cadmium, Selenium
1823 Silicon
1827 Aluminium
1828 Thorium
1830 Vanadium
1839 Lanthanum
1843 Erbium, Terbium
1844 Ruthenium
1860 Cesium, Rubidium
1861 Thallium
1863 Indium
1875 Gallium
1878–1885 Holmium, Thulium, Scandium, Samarium, Gadalinium, Praseodynium, Neodynium, Dysprosium
1886 Germanium
1898 Polonium, Radium
1899 Actinium

20th Century

1901 Europium
1907 Lutetium
1917 Protactinium
1923 Hafnium
1924 Rhenium
1937 Technetium
1939 Francium
1945 Promethium
1940–1961 Transuranium elements. Neptunium, Plutonium, Curium, Americum, Berkelium, Californium, Einsteinium, Fermium, Mendelevium, Nobelium, Lawrencium

Each metal has its own history of discovery; and its use for and harm to life. From the viewpoint of environmental harm, most people associate lead, mercury, arsenic, cadmium, chromium, copper and uranium with mining.

Lead

Water pipes in ancient Rome, some of which still carry water, were made of lead. The English words plumber and plumbing are derived from the Latin word for lead, *plumbum*. *Plumbum* is also the source of the chemical symbol for lead, Pb. Lead has a low melting point and volatilizes relatively easily. It is hypothesized that lead poisoning due to the use of lead piping contributed to the fall of the Roman Empire (Gilfillan 1965). Today, lead is no longer used as piping, and its main use is in car batteries.

The continued use of leaded gasoline in developing countries is a major health concern.

However the continued use of leaded gasoline in developing countries is a major health concern, particularly because of its possible impact on children. Lead influences the development and functioning of the nervous system, slowing down neural response, affecting learning ability and behaviour (RTI 1999). Children may be exposed to lead *in utero*, since embryos can receive lead from the mother's blood. They may also be exposed to lead contaminated dust and soil. Lead in the environment is known to be toxic to plants, animals and microorganisms. Effects are generally locally limited, such as metropolitan areas (use of leaded gasoline), aquatic regimes downstream of lead bearing rocks (ore bodies or mine waste), and in the vicinity of lead smelters. In a recent case in Western Australia, thousands of birds were killed and local water supplies contaminated by dust from lead concentrate which was being shipped from the Port of Esperance (see also **Case 4.6**). Although the health effects of lead have been restricted to the immediate areas of lead emissions, lead from gasoline has been widely dispersed in the environment. In fact, the oceanic distribution of lead is substantially influenced by man's activities (Bruland 1983). This is not true for any other metal.

Mercury

The term 'native mercury' is used for natural mercury associated with the mineral cinnabar. Mercury was named after the Roman god, *Mercury*. Mercury is also known as *quicksilver*, from the Greek words, *hydros* meaning *water*, and *argyros* meaning silver, because this silvery mineral occurs at room temperature as a liquid. The symbol for mercury, Hg, was derived from the name, *hydrargyrum*. Mercury is a peculiar metal, being liquid at room temperature, and naturally occurring in its pure form.

Under reducing conditions in the environment, ionic mercury changes to the uncharged elemental mercury which is volatile and may be transported over long distances by air; also it may be chemically or biologically transformed to methylmercury and dimethylmercury, of which the former is bio-accumulative and the latter is also volatile and may be transported over long distances. Mercury is not essential for plant, animal or human life. The main human exposure to mercury is via inhalation of the vapour of elemental mercury and ingestion of mercury and methylmercury compounds in food.

Mercury and its compounds are toxic to humans, with toxicity varying among the different species.

Mercury and its compounds are toxic to humans, with toxicity varying among the different species (**Case 11.1**). Methyl mercury poses the most serious threat, and can affect various organs including the brain (NRC 2000). As with lead, human embryos receive mercury across the placenta, potentially causing persistent damage to a child's mental development. Mercury toxicity first made headlines following the release of mercuric chloride (a catalyst in the production of plastics) into the bays of Minemata and Niigata, Japan in 1953 and 1960 (UNEP 2002). The source of the 'Minamata disease' (later found to be methylmercury poisoning) was the discharge of untreated effluent containing methyl mercury chloride from a polyvinyl chloride plastic plant. Once in the bay's sediments, the mercury was readily absorbed by marine species, which resulted in the contamination of the entire ecosystem. Many of the local residents consumed fish and seafood from the mercury-contaminated waters as a staple part of their diet. Thirty-nine years later, 2,252 patients had been officially diagnosed with Minamata disease, and over 1,000 had died (www.ec.gc.ca/MERCURY/EH/EN/eh-hc.cfm). Since Minemata, mercury has become an 'angst word', although people tend to associate it with mine wastes because of its earlier use in gold extraction. Mercury does bio-magnify in the food chain with the uppermost trophic level of marine organisms (sharks, barracuda, etc.) often containing levels of mercury considered too high for safe human consumption.

Mercury does bio-magnify in the food chain.

Arsenic

With the major market for arsenic being the production of wood preservatives, the demand for arsenic is closely tied to the home construction market, where timber construction materials containing arsenical preservatives have become ubiquitous.

With the major market for arsenic being the production of wood preservatives, the demand for arsenic is closely tied to the home construction market.

The name arsenic comes from the Greek word *arsenikon*, which means orpiment. Orpiment is a bright yellow mineral composed of arsenic sulphide (As_2S_3), and is the most highly-visible common arsenic mineral. Historians say that arsenic was identified in 1250 by Albertus Magnus, a German monk who spent his life studying and classifying natural materials. It is believed that he heated soap and orpiment together and isolated elemental arsenic. Pure arsenic is a grey-colored metal, but is rare in the environment. It is usually found combined with one or more other elements such as oxygen, chlorine, or sulphur,

CASE 11.1
The Almaden Mine in Spain, Home of One-third of the World's Known Reserves of Mercury

One of Almaden's most surprising aspects is its environmental situation. At first, one would suppose that a region that has been the world's principal producer of a highly contaminating substance such as mercury for more than 2,000 years would be profoundly polluted. Nevertheless mercury has had negligible environmental effects in the region, which has no degraded or unproductive lands as a result of mercury extraction.

Almaden's geological history indicates that the deposits have been close to the surface and in direct contact with subterranean aquifers during the last 250 million years. Although inorganic mercury has a low solubility, it is surprising that such high quantities of mercury over a prolonged time period have caused insignificant environmental effects.

The environmental impacts of mercury contamination in tropical climates are well known and generally attributed to the inappropriate use of mercury to recover gold by artisanal miners. Many authors maintain that under certain physical-chemical circumstances, metallic mercury can change to organic compounds with greater solubility and contaminating capacity (such as methyl mercury).

The physical-chemical conditions at Almaden are very different from those of the tropics. It is important to note that local people regularly consume fish from local rivers without a single recorded case of poisoning. This situation suggests that the mechanisms of methyl mercury generation and the cause-effect relationships linking poisoning to small-scale gold mining are more complex than previously thought.

Source: Excerpt from Girones and Viejobueno 2001: Spain's Almaden Mine: 2000 Years of Solitude

and is referred to as inorganic arsenic. Arsenic combined with carbon and hydrogen is organic arsenic. The organic forms are usually less toxic than the inorganic forms. For example, some lobsters contain sufficient arsenic to be lethal, except that the particular form is non-toxic. Large doses of inorganic arsenic can cause death. It is a notorious poison and has also been implicated as a cause of some skin and lung cancers (TRI 2007).

Cadmium

Cadmium was discovered in 1817 by the German chemist Friedrich Strohmeyer, as a by-product of the zinc refining process. The chemist noticed that some samples of zinc carbonate (calamine) changed colour when heated. Pure calamine, however, did not. Strohmeyer surmised there must be an impurity present and eventually isolated it by heating and reducing the zinc carbonate. What he isolated was cadmium metal. Strohmeyer coined the name cadmium, derived from the Latin word *cadmia* which means *calamine*.

Cadmium metal is produced as a by-product from the extraction, smelting and refining of the nonferrous metals zinc, lead and copper.

Cadmium metal is produced as a by-product from the extraction, smelting and refining of the nonferrous metals zinc, lead and copper, and industrial applications for this by-product were developed in the late nineteenth and early twentieth Centuries. Cadmium has a wide variety of unique properties and cadmium metal and cadmium compounds are used as pigments (cadmium-sulphide based pigments appear prominently in the paintings of Vincent Van Gogh), stabilizers, coatings, speciality alloys, electronic compounds, but, most of all (more than 80% of its use), in rechargeable nickel–cadmium batteries.

Cadmium tends to accumulate in the kidneys, leading to kidney dysfunction as identified by the increased secretion of proteins in urine.

Cadmium is recognized to produce toxic effects on humans. Cadmium tends to accumulate in the kidneys, leading to kidney dysfunction as identified by the increased secretion of proteins in urine. Intake of cadmium is generally through the diet, in particular vegetables, corn products and fish. Smokers are also at risk from cadmium in tobacco. Under normal conditions, however, adverse human health effects have not been encountered from general population exposure to cadmium.

Chromium

Chromium was discovered at the end of the eighteenth century. The name *chromium* is derived from the Greek word *chroma* which means *colour*, since chromium reacts with certain other materials to determine their colours. For example, the green colour of emerald is caused by the presence of very small amounts of chromium in the crystal. Chromium, in the form of Cr (III), is an essential trace element for humans and animals. Its most common harmful effect on humans is chromium allergy caused by exposure to large amounts of chromium (especially Cr (VI) compounds) in the working environment. Chromium compounds are also assumed to be carcinogenic (RTI 2007). Environmentally, Cr (VI) compounds are generally considered to be toxic.

Environmentally, Cr (VI) compounds are generally considered to be toxic.

Chromium is a hard, steel-grey metal that is highly resistant to oxidation, even at high temperatures. Chromium is used in three basic industries: metallurgical, chemical and refractory (heat-resistant applications). It is the sixth most abundant element in the Earth's crust, where it is combined with iron and oxygen in the form of chromite ore. Russia, South Africa, Albania and Zimbabwe together account for 75% of world chromite production. Chromite ore has not been mined in the United States since 1961; by 1985 the United States was completely dependent on importation for its primary chromium supply.

Chromium is released to air primarily by combustion processes and metallurgical industries.

Chromium is released to air primarily by combustion processes and metallurgical industries. Leaching from topsoil and rocks is the most important natural source of chromium entry into the hydrosphere. Electroplating, leather tanning and textile industries also release relatively large amounts of chromium in surface waters.

Copper

Copper, along with gold the first metal known to mankind, was named from the Greek word *kyprios*, that is, the Island of Cyprus, where copper deposits were mined by the ancients. The chemical symbol for copper is Cu which is derived from the Latin name for copper, *cuprium*. Of all the materials mined, copper is the most versatile and durable – it appears everywhere in our everyday lives. This miraculous metal has a number of unique properties: besides being nonmagnetic, copper is conductive, ductile, malleable, resistant and biostatic. Copper is at the heart of all technology, from telecommunications to transportation.

Copper appears everywhere in our everyday lives.

Copper is a natural element found in abundance in the Earth's crust. As a result, drinking water often contains copper, which is safe to drink, even in instances where the copper level is high enough to add a metallic taste to the water. Like chromium, copper is an essential nutrient and is required by the body in daily dietary amounts of 1 to 2 milligrams for adults. Copper deficiency can lead to illness, as will excessive copper when ingested. However, the World Health Organization has concluded that copper deficiency is much more of a global problem than copper toxicity (TRI 2007). Although those living near copper mines may be fearful of polluted air and water, acute copper poisoning is a rare event, largely restricted to the accidental drinking of solutions of copper nitrate or copper sulphate. These and organic copper salts are powerful emetics and large inadvertent doses are normally rejected by vomiting. Chronic copper poisoning is also very rare, with a few reports referring to patients with liver disease. However, the capacity for healthy human livers to excrete copper is considerable and it is primarily for this reason that no cases of chronic copper poisoning have been medically confirmed (TRI 2007). Copper is, however, much more toxic to plants and is commonly used as an herbicide and fungicide. Even at very low concentrations of less than 20 μg/L, copper is toxic to some freshwater aquatic phytoplankton species.

The World Health Organization has concluded that copper deficiency is much more of a global problem than copper toxicity.

Uranium and its Decay Products

Uranium, named after the planet Uranus, was first identified as an element in 1789 by Martin Klaproth, a German chemist (NRCC 2006, from which the following text borrows). It is generally not found in concentrated deposits, although it is more abundant than silver, cadmium or mercury. It remained little more than a curiosity for over a century, being used to manufacture vivid green glass but having little other application. Uranium occurs in nature in two main isotopic forms: U-235 and U-238. Uranium-235 was the first element in which nuclear fission was observed. An atom of U-235 will capture a neutron and become U-236. U-236 is not stable and achieves stability by splitting into two lighter atoms, releasing some energy and two or three fast-moving neutrons. If those neutrons are slowed down sufficiently they can be captured by another atom of U-235 and the process is repeated. A piece of apparatus containing uranium-235, some material for slowing neutrons down and some means to remove the heat energy produced is the basis for a nuclear reactor. Today energy from the fission of uranium produces one-sixth of the world's electricity. The same amount of energy is produced from 20 kg of uranium as from 400,000 kg of coal and nuclear power stations produce much less waste than

Today energy from the fission of uranium produces one-sixth of the world's electricity.

fossil fuel stations (NRCC 2006). Unfortunately, nuclear waste is more hazardous, and as yet no safe long-term nuclear waste treatment/disposal system has been implemented.

With the major isotope U-238 having a half-life equal to the age of the Earth, uranium is certainly not strongly radioactive. U-235 has a half-life one-sixth of this and emits gamma rays as well as alpha particles. Hence a lump of pure uranium would give off some gamma rays, but less than those from a lump of granite. Its alpha radioactivity in practical terms depends on whether it is as a lump (or in rock as ore), or as a dry powder. In the latter case the alpha radioactivity is a potential, though not major, hazard. It is also toxic chemically, being comparable with lead in this respect. Uranium metal is commonly handled with gloves as a sufficient precaution. Uranium concentrate is handled and contained to ensure that it is not inhaled or ingested (UIC 2006).

Uranium metal is commonly handled with gloves as a sufficient precaution.

Uranium ores, however, also contain over a dozen radioactive materials which are all harmful to living things. The gamma radiation detected by exploration geologists looking for uranium actually comes from associated elements such as radium and bismuth, which over geological time have resulted from the radioactive decay of uranium. The most important decay products are thorium-230, radium-226, radon-222 (radon gas) and the radon progeny, including polonium-210 (**Figure 11.5**). Having no commercial value uranium decay products are discarded together with mine wastes when uranium is mined.

Radon gas is a decay product of radium, itself a decay product of uranium.

Radon gas is a decay product of radium, itself a decay product of uranium. Scientists were baffled as to why this alpha-emitting gas, radon, was such a powerful cancer-causing agent. It seemed much more damaging than other alpha emitters such as those found in the ore dust. The mystery went unexplained for more than a decade. In the 1950s the mystery was partially dispelled when it was pointed out that the radon gas, hovering in the stagnant air of the mine, produces radioactive decay products called radon progeny (or, formerly, 'radon daughters'). These solid radioactive by products, produced a single atom at a time, hang in the air along with the radon gas. When radon gas is inhaled, the radon progeny are also inhaled, resulting in a much larger dose of alpha radiation to the lungs than would be delivered by the gas alone (**Case 11.2**). These solid radioactive materials also attach themselves to tiny dust particles and droplets of water vapour floating in the air. By itself, radon gas is exhaled as easily as it is inhaled; but when the accompanying radon progeny are inhaled, they lodge in the lining of the lung. There they bombard the delicate tissues with alpha particles, beta particles and gamma rays (**Figure 11.6**).

Three different isotopes of polonium are included among the radon progeny. They are polonium-218, polonium-214 and polonium-210. These pernicious substances are responsible for most of the biological damage attributed to radon. In particular, polonium-214 and polonium-218, when inhaled, deliver massive doses of alpha radiation to the lungs, causing fibrosis of the lungs as well as cancer. Animal studies have confirmed that polonium is extremely harmful, even in minute quantities.

Radon has a half-life of 3.8 days. This may seem short, but due to the continuous production of radon from the decay of radium-226, which has a half-life of 1,600 years, radon presents a long-term hazard. Further, because the parent product of radium-226, thorium-230 (with a half-life of 80,000 years) is also present, there is continuous production of radium-226 (WISE 2004). As such radon releases constitute a major hazard that continues long after uranium mines are shut down. USEPA estimates the lifetime excess lung cancer risk of residents living near a bare uranium tailings pile of 80 hectares at two cases per hundred. Since radon spreads quickly with the wind, many people receive small additional radiation doses. Although the excess risk for the individual is small, it cannot be neglected due to the large number of people concerned. USEPA estimates that the uranium tailings deposits existing in the United States in 1983 would cause 500 lung cancer deaths per century, if no countermeasures were taken (WISE 2004).

Radon releases constitute a major hazard that continues long after uranium mines are shut down.

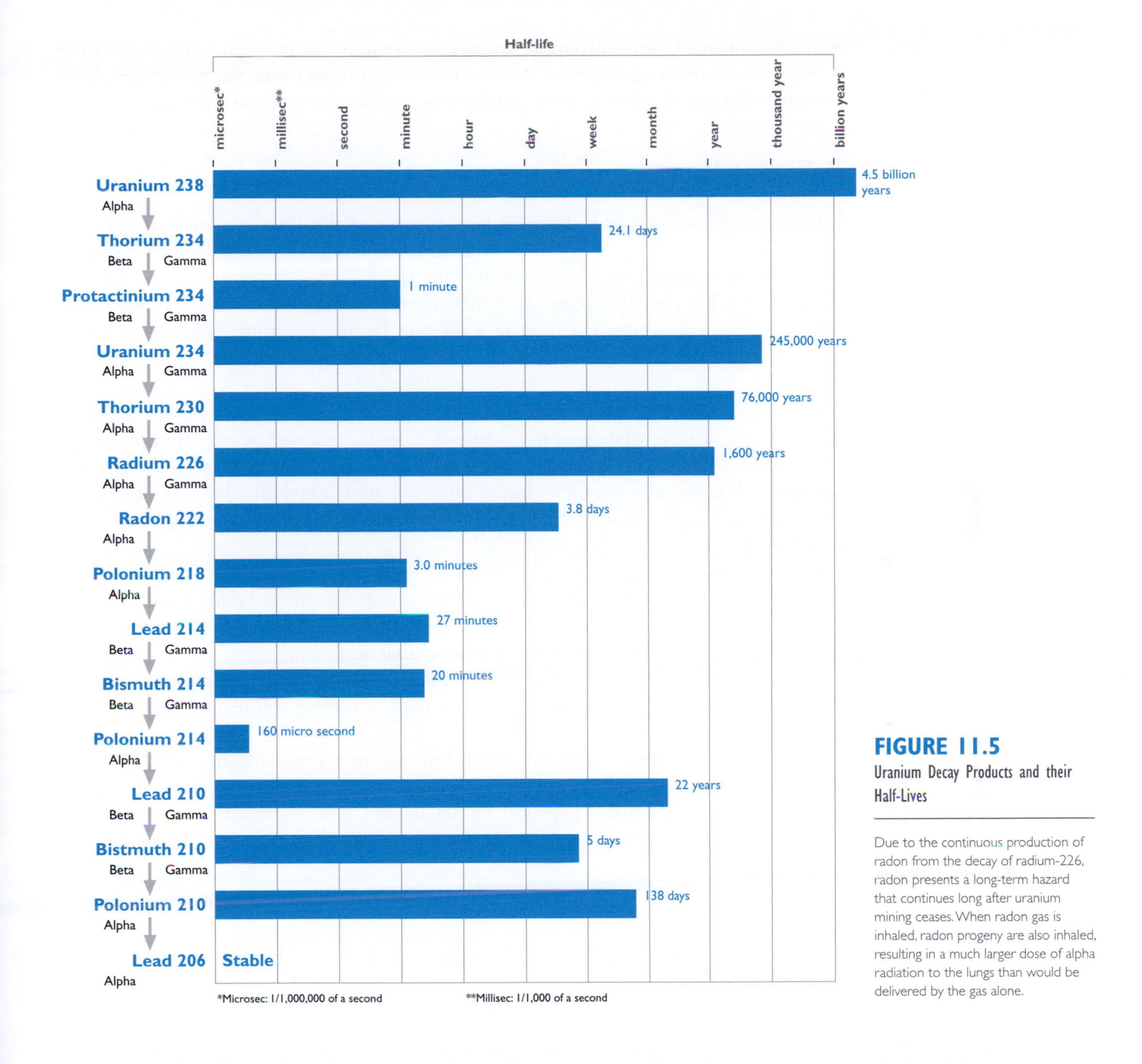

FIGURE 11.5
Uranium Decay Products and their Half-Lives

Due to the continuous production of radon from the decay of radium-226, radon presents a long-term hazard that continues long after uranium mining ceases. When radon gas is inhaled, radon progeny are also inhaled, resulting in a much larger dose of alpha radiation to the lungs than would be delivered by the gas alone.

11.3 METALS, MINERALS AND ROCK

Metals do occur as single elements in nature (termed 'native' form). Gold is an obvious example; mercury is another. More often, however, they occur in combinations of two or more elements to form a mineral (for a list of common minerals targeted for mining see Chapter Five). To be classified as a mineral scientifically, a substance generally must meet five requirements (Eby 2004): it must be naturally formed; it must be solid; it must be inorganic; it must have a specific chemical composition; and it must have a characteristic crystal structure. A mineral always contains the same elements in the same proportions by weight. Interestingly, a mineral has quite different properties from its constituents.

Minerals can only be separated into elements by chemical reactions.

Minerals can only be separated into elements by chemical reactions. This means that there is no physical metallurgical process to separate a metal from a mineral. Sometimes, however, as in the case of asbestos, the mineral not its constituents is the mined 'target'.

Asbestos Minerals

Asbestos is a commercial term used for a group of six naturally occurring fibrous silicate minerals.

While not a metal, asbestos is discussed here as many of the issues relating to metals, also relate to asbestos. Asbestos is a commercial term used for a group of six naturally occurring fibrous silicate minerals, which occur in two mineral groups – Serpentine and Amphibole (USGS 2007). The crystalline structure of these silicate minerals differ from other minerals – the Serpentine group takes the form of a sheet or layered structure, Amphiboles have a chain-like structure. While there are several different asbestos minerals, the most common is Chrysotile.

Chrysotile, known as white asbestos, is the only member of the Serpentine group of minerals. It is actually white-grey in colour, silky, flexible and very strong. Most of it is has been mined in Canada, with three-quarters of the world's chrysotile found in the province of Quebec. It is the most common form of asbestos found in buildings but it is also used in asbestos textiles and in brake linings. Beside Chrysotile two other asbestos minerals are in common use.

Amosite (brown asbestos) is a member of the amphibole group. Most of it comes from Southern Africa as the name suggests: Amosite stands for the 'Asbestos Mines of South Africa'. Some Amosite is also found in Australia. It is the second most common form of asbestos found in building materials, often in thermal insulation.

Crocidolite (blue asbestos) is another amphibole found in Southern Africa and Western Australia, less brittle than brown asbestos. It is used in asbestos textiles and high temperature applications.

Since all asbestos fibres are silicates, they exhibit common properties: incombustibility (in the past, fire was occasionally used to clean fabric made out of asbestos), thermal stability, resistance to biodegradation, chemical inertia toward most chemicals, and low electrical conductivity. Its crystalline structure also allows separation of asbestos minerals into

CASE 11.2
Polonium, a Rare and Highly Radioactive Radon Daughter

November 23, 2006 was notable for the first reported death by the deadly polonium-210 (^{210}Po), a rare and highly radioactive metalloid. The poisoning of Alexander Litvinenko was widely publicized and gave ample reason for high level conspiracy speculations: the victim, a former lieutenant colonel with the Russian Federal Security Service, being an outspoken critic of the Russian government; the use of ^{210}Po as a poison that has never been documented before; the probable first instance of testing of a person for the presence of ^{210}Po in his or her body; the fact that ^{210}Po is difficult to produce; and lastly the simple fact that ^{210}Po is deadly in minute amounts – one gram of ^{210}Po could in theory poison 2 million people of 50 kg body weight, of whom 1 million would die.

Polonium is a very rare element in nature because of the short half-life of all its isotopes. It is found in uranium ores at about 100 micrograms per metric ton (1 part in 10^{10}), which is approximately 0.2% of the abundance of radium. The amounts in the Earth's crust are not harmful. The general population is exposed to small amounts of polonium as a radon daughter in indoor air.

Polonium may be made in milligram amounts by bombardment of bismuth using high neutron fluxes found in nuclear reactors. Only about 100 grams are produced each year, practically all of it in Russia, making polonium exceedingly rare.

thin and strong but flexible fibres. Fibres that are at least 1 cm in length can further be spun into yarn. Because of these characteristics, asbestos has found numerous industrial applications: fireproof fabrics, yarn, cloth, paper, paint filler, gaskets, roofing composition, reinforcing agent in rubber and plastics, brake linings, tiles, electrical and heat insulation, cement and chemical filters. Over centuries asbestos was regarded as a miracle material, combining the strength of rock with the flexibility of silk.

Over centuries asbestos was regarded as a miracle material, combining the strength of rock with the flexibility of silk.

Like metals or coal, asbestos has to be mined. Most asbestos is mined from open pits using bench drilling techniques. The fibre extraction (milling) process is chosen to optimize recovery of the fibres in the ore, while minimizing reduction of fibre length. Dry milling operations are the most widely used taking advantage of the way asbestos crystals are formed: since in asbestos the crystals take the form of long, thin and strong fibres, fibres can be released from the rock by crushing followed by blowing.

Today mining of asbestos, either as the primary mineral or included as an unwanted material while mining for the 'target' mineral, is one of the more controversial issues facing the mining industry: asbestos fibres are dangerous when breathed. In fact the relationship between workplace exposure to airborne asbestos fibres and respiratory diseases has become one of the most widely studied subjects of modern epidemiology.

Asbestos fibres are dangerous when breathed.

The dangers of asbestos, however, have been known for more than a century, and became widely apparent soon after asbestos entered widespread commercial use. Deaths and lung problems were noticed in asbestos mining towns before 1900. The first medical diagnosis of a death from asbestos was made in England in 1906. By 1918 insurance companies in America and Canada would no longer insure workers in the asbestos industry because of suspected health hazards. In 1935, researchers in Europe and the US reported a suspected association between asbestos exposure and lung cancer.

FIGURE 11.6
Uranium Ores contain over a Dozen Radioactive Materials which are All Harmful to Living Things

Source: www.cartoonstock.com

Scientific data regarding the dangers of asbestos were, for a long time, ignored.

Unfortunately scientific data regarding the dangers of asbestos were, for a long time, ignored by many asbestos manufacturers and corporate users of asbestos materials. Companies continued to use asbestos as a building material and in manufacturing processes. Sometimes safer alternatives, such as fibreglass insulation, were ignored in favour of less expensive asbestos materials. Workers, their families and occupants of asbestos-containing buildings have been needlessly exposed to asbestos, with some suffering damage to their health or death as a consequence.

Today it is widely accepted that asbestos fibres can be associated with three diseases: asbestosis, lung cancer and mesothelioma. Workers can be exposed to asbestos during mining, milling and handling of ores containing asbestos or during the manufacture, installation, repair and removal of commercial products that contain asbestos. Local populations can also be exposed to asbestos fibres, either due to naturally occurring asbestos or due to building materials containing asbestos from which fibres may be released and subsequently inhaled.

Not surprisingly then, asbestos has led to the longest and most expensive mass tort in US history, involving more than 700,000 claimants and 8,000 defendants as of 2007 (RAND 2007). By some estimates the total costs of asbestos litigation in the USA alone will exceed US$ 250 billion (**Case 11.3**).

All asbestos-related diseases have a long latency period: it takes 20 to 40 years for the first symptoms to appear.

All asbestos-related diseases have a long latency period: it takes 20 to 40 years for the first symptoms to appear. Hence the asbestos problem today is mostly with the old, poorly

CASE 11.3
No Win No Fee

Lawyers in the US and UK frequently deal with personal injury claims with compensation on a contingency fee ('No Win No Fee') basis: if the claimant wins the case lawyers will recover their costs from the defendant; if the claimant loses lawyers will write off their costs. Appointed lawyers may also fund disbursements such as the cost of obtaining medical records, a medical report and any court fees. No wonder then that asbestos litigation has reached such levels. And this 'elephantine mass [is] still growing' states the *Chicago Tribune* in July 22, 2001.

The impact of asbestos litigation is vast and far-reaching (www.asbestossolution.org). Hundreds of thousands of claims are pending and new claims are accelerating, including ones by people who are not sick. The flood of claims and resulting settlements are forcing companies into bankruptcy and putting at risk compensation for those who are sick today or may become sick in the future.

'It's a – most of the asbestos producers are now bankrupt so that lawyers target companies once considered too small to sue, or once considered to be not really directly involved with the manufacturing of asbestos. Because there is nobody left to sue, they try to drag in people that aren't directly involved with the manufacturing of asbestos. ... This is a national problem, as the Supreme Court said, that requires a national solution. ... There are some principles which I think ought to govern Congress' actions. First, funds should be concentrated on those who are sick, not lawyers or claimants who are not ill' address by US President George W. Bush to an audience in Clinton Township, 2005.

'... [A]sbestos litigation – actually, asbestos exposure has been a national tragedy, and more than 100,000 workers have died as a consequence of asbestos exposure. But lawyers have taken this tragedy and turned it into an enormous moneymaking machine, in which ... baseless claims predominate. In the year 2003, 105,000 new claimants came into the asbestos litigation system. Each claimant will sue 40, 50, 60, 70 different companies, so we're talking about a total of 50 million, 60 million, 70 million new claims generated in the year 2003 alone. Of the 105,000 claimants, approximately 10,000 are seriously ill, some dying, some dead, because of asbestos exposure. These are the malignancies' adds Prof. Brickmann, law professor at the Benjamin N. Cardozo School of Law at Yeshiva University, at the same meeting (www.whitehouse.gov/news/releases/2005/01/20050107-8.html).

That said, the highly politicized controversy about asbestos litigation should not overshadow the quiet, sad, and directly related crisis in public health: an epidemic of asbestos-caused diseases that was for a long time ignored by many asbestos manufacturers and corporate asbestos users and that reportedly claims the life of one out of every 125 US American men who die over the age of 50 (www.ewg.org).

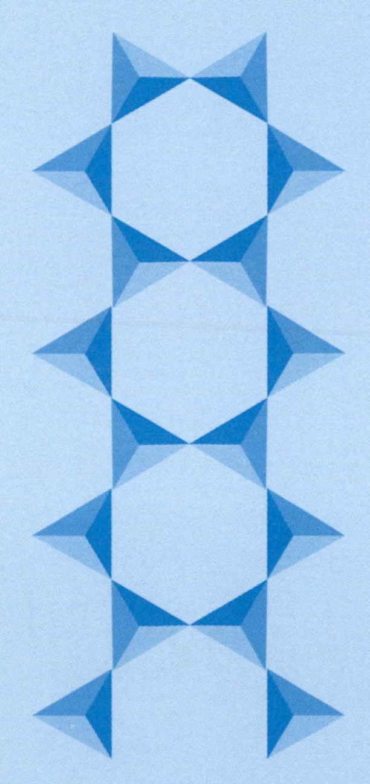

Amphiboles

Graphic: based on Kirk-Othmer Encyclopedia of Chemical Technology published by John Wiley & Sons, Inc.

controlled products of the past. Given the widespread use of asbestos minerals prior to the 1970s, poor past working conditions, and a long latency period, new cases of asbestos-related disease will probably continue to be observed. It will take many more years before the health benefits from banning of amphiboles and friable asbestos products which began in the 1970s, and the regulations which now impose strict factory controls will have a marked effect.

Asbestos materials are still widely used. Modern asbestos products, however, are different from the asbestos products used until the 1970s. Old asbestos products, principally low-density insulation materials, were very dusty and crumbled under hand pressure. Unlike today's products, they often contained amphibole fibre.

Today, only one type of asbestos is used: chrysotile. In addition, the industry now only markets dense and non-friable materials in which the chrysotile fibre is encapsulated in a matrix of either cement or resin. These modern products include chrysotile-cement building materials, friction materials, gaskets and certain plastics. Ninety percent of the world production of chrysotile, however, is used in the manufacture of chrysotile-cement, in the form of pipes, sheets and shingles.

Asbestos minerals of course represent only a minute fraction of all known 3,000 mineral species (Weiss 1985; Eby 2004), of which only 50 are rock forming (**Table 11.2**).

TABLE 11.2

Common Ore Types (Weiss 1985)

Metal	Ore Mineral	Composition	Percent Metal
Gold	Native gold	Au	100
	Calaverite	$AuTe_2$	39
	Sylvanite	$(Au,Ag)Te_2)$	
Silver	Native silver	Ag	100
	Argentite	Ag_2S	87
	Cerangyrite	$AgCl$	75
Iron	Manganiteti	$FeOFe_2O_3$	72
	Hermatite	Fe_2O_3	70
	Limonite	$Fe_2O_3H_2O$	60
	Siderite	$FeCo_3$	48
Copper	Native copper	Cu	100
	Bornite	Cu_5FeS_4	63
	Brochantite	$CuSO_4\ 3Cu(OH)_2$	62
	Chalcocite	Cu_2S	80
	Cholcopyrite	$CuFeS_2$	34
	Covelite	CuS	66
	Cuprite	Cu_2O	89
	Enarngite	$3Cu_2S\ As_2S_2$	48
	Malchite	$CuCO_3\ Cu(OH)_2$	57
	Azurite	$2CuCo_3$ 2H2O	55
	Chrysocolla	$CuSiO_3\ 2H_2O$	36
Lead	Galena	PbS	86
	Cerussite	$PbCO_3$	77
	Angiesite	$PbSO_4$	68
Zinc	Sphalerite	ZnS	67
	Smithsonite	$ZnCO_3$	52
	Calamine	$H_2Zn_2SiO_5$	54
	Zincte	ZnO	80

(Continued)

TABLE 11.2
(Continued)

Metal	Ore Mineral	Composition	Percent Metal
Tin	Cassiterite	SnO_2	78
	Stamnite	Cu_2S FeS SnS_2	27
Nickel	Pentlandite	(Fe, Ni)S	22
	Garnierite	$H_2(Ni, Mg)SiO_3$ H_2O	
Chromium	Chromite	FeO Cr_2O_3	68
Manganese	Pyrolusite	MnO_2	63
	Psilomelane	Mn_2O_3 $(OH)_2$	39
	Braunite	$3Mn_2O_3$ $MnSiO_3$	71
	Manganite	Mn_2O_3 H_2O	81
Aluminium	Bauxite	Al_2O_3 $2H_2O$	39
Antimony	Stibnite	Sb_2S_3	71
Bismuth	Bismuthianite	Bi_2S_3	81
Cobalt	Smaltite	$CoAs_2$	28
	Cobaltite	CoAsS	35
Mercury	Cinnabar	HgS	86
Molybdenum	Molybdenite	MnS_2	60
	Wulfenite	$PbMoO_4$	39
Tungsten	Wolfanite	(Fe Mn)WO_4	76

Minerals can consist of a single element such as diamond or native gold. More often minerals are in the form of combinations of two or more elements, and about 3,000 minerals species are known. Only 50 of them are rock forming.

Rocks

To the geologist, rock is any substantial portion of Earth, whether solid or loose. In the geological sense, rock also includes glaciers and sometimes even water bodies, which after all, are only liquid ice. To most of us, however, rock includes only solid materials that cannot be excavated by hand. Scientifically, rocks are a composition of mineral assemblages. Rocks are classified in various ways, the most common being according to their mode of formation. A few rocks consist of only one mineral, such as quartzite or limestone, but these are the exceptions to the rule. A mineral assemblage is a mixture of any number of minerals and elements. The composition of rock, of course, varies widely, but minerals can be separated by physical means, a key characteristic exploited in ore processing. Unlike minerals, rock retains the properties of its main constituents.

Minerals can be separated by physical means, a key characteristic exploited in ore processing.

There is the final point that to the exploration geologist, a mineral is anything that is commonly considered as such. Thus coal, limestone, or other rocks are not truly minerals, but they certainly are mineral resources, as is petroleum, which is neither mineral nor rock.

REFERENCES

Ball P (2002) The Ingredients: A guided tour to the elements, Oxford University Press.

Bruland KW (1983) Trace Elements in Sea Water. Chapter 45, Chemical Oceanography, Vol. 8, Academic Press, London.

Campbell PGC and PM Stokes (1985) Acidification and toxicity of metals to aquatic biota. Can. J. Fish. Aq. Sci. 42, 2034–2049.

Cusimano FR, Brakke DF and GA Chapman (1986) Effects of pH on the toxicities of cadmium, copper and zinc to steelhead trout (Salmo gairdneri). Can. J. Fish. Aq. Sci. 43, 1497–1503.

Down C and Stocks J (1977) Environmental Impact of Mining, Applied Science, London.

Eby GN (2004) Principles of environmental geochemistry, Brooks/Cole – Thomson Learning.

Evans RD, Andrews D and Cornett RJ (1988) Chemical fractionation and bioavailability of Co60 to benthic deposit-feeders. Can. J. Fish. Aq. Sci. 45, 228–236.

Gilfillan SC (1965) Lead Poisoning and the fall of Rome. J. Occup. Med. 1965, 7; pp. 53–60.

Girones EO and Vielobueno CD (2001) Spain's Almaden Mine: 2,000 Years of Solitude. In: Large Mines and the Community – Socioeconomic and Environmental Effects in Latin America, Canada, and Spain, International Development Research Centre (IDRC).

Kimball BA and Wetherbee GA (1989) Instream chemical reactions of acid mine water entering a neutral stream near Leadville, Colorado. US Geological Survey Toxic Substances Hydrology Program – Proceedings of the Technical Meeting, Phoenix, AR, September 26–30, 1988. US Geological Survey, Water-Resources Investigations Report 88–4220. 71–79.

ICMM (2007) MERAG: Metals Environmental Risk Assessment Guidance.

Janssen C and Muyssen B (2001) Essentiality of metals: consequences for environmental risk assessments. Fact Sheet on Environmental Risk Assessment. International Council on Metals and the Environment (ICME), Ontario, Canada.

Masters GM (1991) Introduction to Environmental Engineering and Science, Prentice-Hall International, Inc.

MMSD (2001) Worker and Community Health Impacts Related to Mining Operations Internationally; A Rapid Review of the Literature by Carolyn Stephens and Mike Ahern; London School of Hygiene & Tropical Medicine.

NRC (National Research Council) (2000) Toxicological effects of methyl mercury, National Academic Press, Washington DC, USA, 2000, http//:www.nap.edu/openbook/0309071402/html/1.html.

NRCC (2006) Periodic Table of the elements; National Research Council Canada http://www.nrc-cnrc.gc.ca/eng/education/elements/el/u.html.

Oesterloef A and Oesterlund S (2003) Environmental Impacts and Assessment of Ore Mining in a European Context; Centek, OMENTIN Report No 3.

Pagenkopf GK (1983) Gill surface interaction model for trace-metal toxicity to fish: role of complexation, pH and water hardness. Environ.Sci.Technol. 17, 342–347.

Parametrix (1995) Persistence, Bioaccumulation and Toxicity of Metals and Metals Compounds, The International Council on Metals and Environment.

RAND (2007) Asbestos Litigation Costs and Compensation – An Interim Report.

RTI (1999) Toxicological profile for lead. Research Triangle Institute for U.S. Department of Health and Human Services. Agency for Toxic Substances and Disease Registry, Atlanta.

RTI (2000) Toxicological profile for chromium. Syracuse Research Corporation for U.S. Department of Health and Human Services. Agency for Toxic Substances and Disease Registry, Atlanta.

Schubauer MK, Dierkes JR, Monson PD and Ankley GT (1993) pH-dependent toxicity of Cd, Cu, Ni, Pb and Zn to Ceriodaphnia dubia, Pimephales promelas, Hyalella azteca and Lumbriculus variegatus. Environ. Toxicol. Chem. 12, 1261–1266.

Stephens C and Ahern M (2001) Worker and Community Health Impacts Related to Mining Operations Internationally-A Rapid Review of the Literature.

TRI (2007) TRI/Right-To-Know Communications Handbook, Environmental Health Center; A Division of the National Safety Council, Washington, DC.

UIC (2006) Environmental Aspects of Uranium Mining, Uranium Information Center, Australia, Briefing Paper 10, 2006; www.uic.com.au.

UNEP (1994) Environmental management of mine sites, Training manual, Technical Report No. 30.

UNEP (2002) Global Mercury Assessment. Website: http://www.chem.unep.ch.

USEPA (1995) Human health and environmental damages from mining and mineral processing wastes; Technical Background Document Supporting the Supplemental Proposed Rule Applying Phase IV Land Disposal Restrictions to Newly Identified Mineral Processing Wastes; Office of Solid Waste.

USGS (2007) Asbestos: Geology, Mineralogy, Mining, and Uses by Robert L. Virta Open-File Report 02–149, Prepared in cooperation with Kirk-Othmer Encyclopedia of Chemical Technology, Online Edition, Wylie-Interscience, a division of John Wiley & Sons, Inc., New York, NY.

Van Assche F, Van Tilborg W and Waeterschoot H (1997) Environmental risk assessment for essential elements, Case study: Zinc. In: Langley E, Mangas S (Eds.), Zinc – Report of an International Meeting. National Environmental Health Forum Monographs, Metal Series No. 2, Adelaide, Australia, September 12–13, 1996. 33–47.

Weiss N L (Editor) (1985) SME Mineral Processing Handbook, Kingsport Press.

WISE (2004) Uranium mining and milling wastes: An introduction by Peter Diehl; www.wise-uranium.org.

WHO (1988) Chromium – Environmental Health Criteria 61. World Health Organisation, International Programme on Chemical Safety (IPCS), Geneva, Switzerland.

WHO (1989a) Lead – Environmental aspects. Environmental Health Criteria 85. World Health Organisation, International Programme on Chemical Safety (IPCS), Geneva, Switzerland.

WHO (1989b) Mercury – Environmental aspects. Environmental Health Criteria 86. World Health Organisation, International Programme on Chemical Safety (IPCS), Geneva, Switzerland.

WHO (1990) Methylmercury – Environmental Health Criteria 101. World Health Organisation, International Programme on Chemical Safety (IPCS), Geneva, Switzerland.

WHO (1991) Inorganic mercury – Environmental Health Criteria 118. World Health Organisation, International Programme on Chemical Safety (IPCS), Geneva, Switzerland.

WHO (1992) Cadmium – Environmental Health Criteria 134. World Health Organisation, International Programme on Chemical Safety (IPCS), Geneva, Switzerland.

WHO (1995a) Cadmium – Environmental aspects. Environmental Health Criteria 135. World Health Organisation, International Programme on Chemical Safety (IPCS), Geneva, Switzerland.

WHO (1995b) Inorganic lead. Environmental Health Criteria 165. World Health Organisation, International Programme on Chemical Safety (IPCS), Geneva, Switzerland

WHO (1996) Trace Elements in Human Nutrition and Health, Geneva, Switzerland.

12

Was the Environmental Assessment Adequate?

Identifying Issues, Finding Solutions

COBALT

Cobalt is a hard, magnetic, silver-gray metal. It occurs in Vitamin B12, and is an essential element for mammals. Exposure to high cobalt levels may adversely effect heart and lung function. The main uses of cobalt are in the manufacture of alloys of high strength and abrasion resistance. Cobalt compounds are widely used in inks, paints and glazes, imparting deep blue and green colours. Most cobalt is produced as a by-product from nickel or copper ores.

12 Was the Environmental Assessment Adequate?

Identifying Issues, Finding Solutions

Critical review of the Environmental Impact Assessment (EIA) is an important element of environmental assessment for a new mining development. If the EIA is prepared by external experts, as is often if not always the case, the mining company normally scrutinizes the EIA documentation internally before releasing the document to government authorities for approval. This is to ensure that the document meets the company standards, that the project description accurately represents company planning and that commitments made on behalf of the company are financially and technically achievable. Government authorities conduct their own internal review as the basis for decision-making.

The environmental audit helps to check whether environmental mitigation measures achieve the expected results.

Review of environmental assessment is not restricted to the planning stage of a new project, but may be carried out at any stage of the project. Environmental auditing assesses the environmental performance of an active mine, and as such, is best suited to compare actual with predicted impacts. The environmental audit helps to check whether environmental mitigation measures achieve the expected results. This chapter provides guidance to both reviewing an EIA as well as auditing an active mine.

12.1 REVIEWING THE ENVIRONMENTAL IMPACT STATEMENT

Defining Review Purpose and Objectives

An environmental assessment review can serve several purposes, each requiring somewhat different reviewer skills. Usually, an environmental assessment review is part of the EIA approval process. It also serves to collect public opinions on a proposed mining project. Occasionally, an independent reviewer is asked by stakeholder groups (such as potential investors in a new mine venture) to review environmental assessment in terms of its strengths and deficiencies, or more specifically, to assess its compliance with the Equator Principles. Most environmental reviews fall into one of the following categories.

Review by Approval Panel

The review of an Environmental Impact Assessment report as part of the environmental permitting process is one of the main checks built into most national EIA guidelines. Depending on national legislation the review process is done by (1) environmental government authority (e.g. Canada and most Australian states), (2) inter-agency or interdepartmental committee (e.g. USA and Indonesia), (3) commission of independent experts (e.g. Thailand, Papua New Guinea, Australian State of Victoria), (4) commission of experts within the government, or (5) planning authority using government guidelines.

The purpose of the review is to ensure conformance of the EIA with its Terms of Reference or other Scoping documentation; with applicable government regulations/guidelines; to judge the validity of impact predictions; and to check whether expected negative environmental and social impacts are within acceptable limits. This pre-decision review is an important quality control step in the EIA process. It determines whether a report meets the terms of reference, examines required or reasonable alternatives, provides a satisfactory assessment of the environmental effects of the proposed activity, adequately deals with mitigation (and, where necessary, follow-up), fairly represents public concerns and inputs, and provides the information required for decision-making (Sadler 1996). With certain exceptions, the purpose of the review, which may range from a quick check to a systematic examination depending on the nature of the EIA, is to verify that the document provides an adequate assessment and is sufficient for the purpose of decision-making. There are situations where the review panel lacks appropriate expertise and experience. In such cases the review tends to concentrate on procedures rather than content.

There are situations where the review panel lacks appropriate expertise and experience. In such cases the review tends to concentrate on procedures rather than content.

Public Review

Public review and input is accommodated in many EIA systems. At a minimum this requires sufficient information and time for the public to comment on the project. International financing institutions promote a more comprehensive and open review process, using public hearings and providing open access to documentation, for any interested party. The purpose of the EIA review is to present an opportunity for the public to comment and to involve stakeholders in environmental assessment (World Bank 1989; IFC 2006). Public review provides an insight into public perception towards the project. In most cases where there is public review, the proponent is required to respond to issues and answer questions raised by the public. In some systems the final EIA document must include a summary of public comments, and the assessment must take these comments into account.

International financing institutions promote a more comprehensive and open review process, using public hearings and providing open access to documentation, for any interested party.

Third Party Review

Environmental review by independent experts who have no vested interest in promoting or opposing the mining project is sometimes required to demonstrate technical accuracy and completeness of the environmental assessment or to identify any shortcomings. Generally the focus of the third party review is on the content rather than the procedures of the EIA.

Internal Review by Financial Institutions

International financial institutions increasingly conduct an internal environmental and social review of projects in which they are considering investments. The purpose of the review is to ensure that the project complies with applicable lender's environmental and social policies, identifies any environmental risks and meets best industry practice. In the case of Equator Banks, the review is to ensure that the project complies with the Equator Principles. Beside an internal project review, international financial institutions rely on external independent experts to conduct environmental project reviews. Each type of review has its own approach, review criteria and review documentation. The following text is predominantly concerned with the independent third party review process.

Identifying Review Criteria

An EIA review does not set out to refute findings of the environmental assessment or to develop new conclusions regarding potential impacts.

While the approach, methods and criteria differ, formal EIA reviews focus on a number of common aspects. These include the 'triple A-test' of appropriateness (coverage of key issues and impacts), adequacy (of impact analysis) and actionability (does the report provide the basis for informed decision-making?) (Sadler 1996). It is also important to realize that an EIA review does not set out to refute findings of the environmental assessment or to develop new conclusions regarding potential impacts. The focus of the review is on identifying areas of strength and weakness, omission, or inaccuracy.

An EIA review needs to be based on specified review criteria, best developed on a case-by-case basis by reference to the following questions (UNEP 2002).

Are Terms of Reference or other Guidelines Available and Applicable for the Review?

Most national EIA guidelines require a terms of reference for EIA work, which provides a useful framework for EIA review. If terms of reference do not exist, generic criteria to guide the EIA review can be derived from some or all of the following information: (1) legal EIA requirements; (2) relevant environmental guidelines and standards; (3) principles of EIA good practice as summarized in the Equator Principles; and (4) knowledge of the type of mining project and its typical impacts and their mitigation. It is clearly important that, at a minimum, the EIA must comply with the laws and regulations of the host country. In addition most international financial institutions adopt the Equator Principles as EIA good practice to complement national EIA guidelines.

Are Any Reviews of EIA Reports or Environmental Audits of Comparable Proposals in Similar Settings Available?

EIA reviews of similar mining projects in similar environmental settings provide helpful insight into the main environmental issues associated with the project under review.

EIA reviews of similar mining projects in similar environmental settings provide helpful insight to the main environmental issues associated with the project under review. It is particularly useful if environmental audits on similar existing operations are available, since environmental audits may help to identify problems experienced during actual implementation and operation.

When is a Comprehensive Review Appropriate?

The quality of the executive summary may provide a first indication of the quality of the EIA documentation.

At the outset of a review, particular attention can be directed to the executive summary, which is intended to explain the key findings accurately and in a non-technical manner. The quality of the executive summary may provide a first indication of the quality of the EIA documentation. The presentation of the documents and the information contained, may also provide some indication of the quality of the EIA. A comprehensive review of the EIA extending into an independent analysis may be necessary in situations where serious deficiencies in the information become apparent. The EIA review then becomes a mini-EIA to supplement the EIA under review.

Supporting Review Materials

Often a wide range of information exists that may facilitate an EIA review. Apart from the EIA documentation, such material may include the following:

- Contact names and telephone numbers of people, agencies, organizations and environmental information/data resource centres able to provide assistance in EIA review;
- Lists of agencies/government departments etc. responsible for review in the national jurisdiction and their specific requirements;
- Review procedures and requirements established in the applicable EIA legislation or guidelines;
- Methods and criteria that are used or could be applied locally to review the quality of EIA reports;
- International Best Practice Guidelines such as World Bank Operational Directives and guidelines;
- Mine pre-feasibility or feasibility study documents;
- EIA document for a similar project conducted in the same jurisdiction;
- Examples of reviews of EIA reports carried out locally and their results;
- Copies of research focused on the quality of EIA reports;
- Copies of project-related studies that are of environmental relevance;
- Other resources that may be available such as videos, journal articles and case studies;
- Project-related information available from the internet;
- Outline of a typical public review process and how it is related to decision-making;
- Copies of public submissions or inputs to the review of EIA reports; and
- Examples of the system of summarizing and reporting on public submissions from EIA review.

Of all review guidelines, the most pertinent are the IFC Performance Standards, which represent the quintessence of the Equator Principles.

Review Focus

The elements of EIA review and the aspects to be considered differ, depending on arrangements that are in place in a particular country and for a particular project. However, at a minimum, the review should elaborate whether the EIA has adequately covered the following issues (after IFC 2000).

Environmental Guidelines. Mining projects must meet applicable environmental guidelines as set forth by the national legislation.

International Treaties and Agreements on the Environment and Natural Resources. Various international treaties and agreements on a range of environmental and natural resource issues are relevant to particular mining projects (e.g. marine environmental protection and biodiversity). Environmental assessment should consider these treaties and agreements in the project's environmental analyses.

Public Disclosure and Participation. It is no longer acceptable to develop large-scale projects without informing affected communities on the various facets of the project and how it may affect their lives. The project documentation should demonstrate that public involvement from disclosure up to participation in decision-making is an integral part of the planning process.

It is no longer acceptable to develop large-scale projects without informing affected communities.

Indigenous Peoples. Indigenous Peoples issues are relevant for many mining projects. An Indigenous Peoples Action Plan should be developed for any mining project impacting on Indigenous Peoples. The World Bank OD 4.10, Indigenous Peoples and OP. 4.20 Indigenous Peoples provide guidance on the preparation for such action plans.

Land Acquisition and Involuntary Resettlement. A resettlement plan should be developed for every project involving involuntary resettlement and impacts from land acquisition. World Bank OP 4.12 and ADB (2000) outline the requirements for resettlement plans.

Access to land and its various primary and secondary impacts are often overlooked in the environmental assessment of mining projects.

Road Access. Access to land and its various primary and secondary impacts are often overlooked in the environmental assessment of mining projects although they may cause some of the most significant adverse impacts. Direct impacts of new access roads may include land instability and soil erosion. River crossings are of particular concern. Indirect impacts may also occur if the road facilitates third party access to natural resources, leading, in some cases to illegal logging, hunting or illegal artisanal mining.

Land Settlement. Land settlement is the managed process of opening new lands to planned permanent occupancy. Due to the complex physical, biological, socioeconomic and cultural impacts, land settlement, land use, land title and land acquisition should be carefully reviewed.

Induced Development and Other Socio-cultural Aspects. Secondary growth of settlements and infrastructure, often referred to as 'induced development' or 'boomtown' effects, have major indirect environmental and social impacts in developing countries. Often, local governments have difficulties in addressing and managing these developments. Such impacts should be taken into account in project design and in the EIA documentation.

Cultural Properties. It is in the public interest to protect non-replicable cultural property. Sites, structures and remains of archaeological, paleontological, historical, religious, cultural, aesthetic, or unique natural value should be identified and addressed in the EIA.

Natural Resources. The environmental assessment should demonstrate that exploitation of natural resources is carried out in an environmentally and socially sound manner and on a sustainable basis.

Biodiversity. Conservation of endangered plant and animal species, critical habitats and protected areas is of particular relevance for mining projects particularly for large-scale projects located in remote or pristine environments.

Coastal and Marine Resources Management. Planning and management of coastal marine resources, including coral reefs, mangrove swamps and estuaries are relevant to some mining projects, where ports are to be constructed or where submarine tailings disposal is planned.

Tropical Forest, Wetlands and Wild Lands. The environmental assessment should review to what extent areas of particular biological value are affected by the mining project. The environmental management and monitoring plans should clearly address such adverse impacts, including providing information on compensatory measures.

The environmental assessment should address to what extent, if at all, the project may be affected by natural hazards.

Natural Hazards. The environmental assessment should address to what extent, if at all, the project may be affected by natural hazards (e.g. earthquakes, floods, land slides, or volcanic activity) and should propose specific measures to address these concerns if appropriate.

Rehabilitation of Disturbed Land. A rehabilitation plan should be part of the environmental assessment for a mining project. The aim of the plan is to ensure that land is returned to a condition capable of supporting prior land use, equivalent uses or other acceptable uses (see for instance: 'Base Metal and Iron Ore Mining, Industry Sector Guidelines,' Pollution Prevention and Abatement Handbook, World Bank 1989). In most national jurisdictions the preparation of a preliminary mine rehabilitation plan is part of the project approval.

Global Warming. Methane gas emission is a recognized potent greenhouse gas. The EIA needs to address methane emissions in coal mining proposals.

Dams and Reservoirs. Large tailings dams (defined as over 15 metres in height by the World Bank OP 4.37, 2001) and dams with design complexities, require special attention in the EIA review.

Hazardous and Toxic Materials. Environmental management and monitoring plans should address the safe use, transport, storage and disposal of hazardous and toxic materials used in mining and mineral processing operations.

12.2 ENVIRONMENTAL MINE AUDITS

Environmental auditing, as the name suggests, is similar to financial auditing. It aims to evaluate the environmental performance of the mine, and to assess the effectiveness of management systems and programmes. As with financial auditors, environmental auditors are liable for their actions, and so can not afford to be other than independent and objective.

Audit History

As with environmental impact assessment, environmental auditing had its beginnings in the USA in the early 1970s. Originally applied to US multinational companies (such as the General Motors Auditing Program, 1972), it was soon extended to a wide range of industries. The initial concept of an environmental audit was a response to increased environmental awareness and liability reflected in the publication of several environmental regulations (Clean Air Act, 1970; Clean Water Act, 1972; Toxic Substances Control, 1976; Resources Conservation and Recovery Act, 1976). The environmental audit initially represented an independent and systematic process to evaluate compliance with prevailing environmental regulations.

The nature of environmental audits has changed significantly since that time. It was soon recognized that the audit process is useful well beyond merely identifying regulatory violations, by analyzing the causes of non-compliance, and environmental risks, and for evaluating environmental systems and programmes to ensure compliance. Today's audits are as much management audits as they are compliance audits, and environmental auditing has become an essential element of environmental management worldwide.

Today's audits are as much management audits as they are compliance audits.

The environmental audit offers the mining industry an opportunity to improve its environmental image in the eyes of the public. The mining industry in particular is working hard to overcome the negative environmental image under which it operates. Tailings dam failures such as in Guyana, at the Marcopper mine in the Philippines, and the widely publicized tailings dam failure in Los Frailes, Spain, do not help this image and have reverberations in the mining industry throughout the world. Adverse publicity over the Ok Tedi mine in PNG and, some years ago, the New South Wales government's refusal to grant permission for a US$ 1 billion gold project to proceed, citing environmental reasons, is representative of this attitude which is fueled by the environmental lobby groups.

Audit Type

The nature and type of environmental audit varies with the objective of the audit. Audits therefore have a number of different formats, of which the four most common are summarized here.

Environmental Compliance Audits are usually carried out internally and are used to assess compliance with existing Environmental Laws and Regulations. These audits are usually conducted annually. In mining they are also used to assess compliance with commitments made in the Environmental Impact Assessment and in the Environmental Management and Monitoring Plans. Many multinational companies measure their performance against best industry practice (Benchmark Audit) or against internal company standards which can be more stringent than most international criteria. Also auditing of mining operations for compliance with the Equator Principles is becoming increasingly important.

Environmental Performance Audits are internal audits which are designed to track environmental objectives against specific targets. These are similar to financial or production targets set by mining companies such as mill throughput per day, mill efficiency, or cost per tonne of ore mined. Typical environmental objectives could be zero release of acid rock drainage, recycling of all waste oils, water conservation by minimizing water consumption per tonne of ore mined, or a reduction of greenhouse gas emissions over time.

Environmental Liability Audits are external and often form part of the 'due diligence' process used in acquisition/divestitures and mergers in order to identify potential environmental liabilities and the cost of correction or redemption. These costs are then introduced into the terms of the financial agreement negotiated by the parties. Such audits are undertaken on behalf of mining companies; investors, including IFC and EBRD; insurance companies, including OPIC and EFIC; and government privatization agencies.

Environmental Management System Audits are internal and/or external audits designed to regularly test applied management practices for compliance with the requirements of a formal environmental management system such as ISO 14000 EMS. Environmental Management System Audits follow a strict protocol, which is very much determined by the applied management system. A common mine audit is not a complete environmental management system audit, neither should it be. A complete environmental management system audit would provide a thorough, systematic evaluation of all elements of a mine's implementation of an environmental management system, which would not necessarily provide a true reflection of the mine's environmental performance.

Mine audits differ from common auditing not only because of the magnitude of mining operations and the often unique locations in which they are sited, but also because of the controversy and agitation by NGOs.

Mine audits differ from common auditing not only because of the magnitude of mining operations and the often unique locations in which they are sited, but also because of the controversy and agitation by NGOs, whereby operations can unexpectedly find themselves in the public spotlight. Mining and mineral processing operations are particularly susceptible to this because of the scale of land disturbance, potential for large and harmful discharges to the environment, emotive land use conflicts and health risks, and the general negative perception that mining tends to receive.

Three parties participate in an audit. The auditee is the operation being audited. The client is the organization that requires the audit to be conducted. This can be the auditee itself, the auditee's holding company or its Board of Directors, a financial institution, or in some instances public authorities. Finally there is the audit team led by a lead auditor.

Audit Matters and Criteria

Mining audits need to focus on the significant issues.

Mining audits need to focus on the significant issues: e.g. resettlement and land compensation, social investment, resource conservation and recovery, tailings disposal and waste rock management, potential loss of biodiversity, rehabilitation and mine closure, applied management practice, and other controversial issues raised by NGOs and the media. Of course, assessing compliance with the host country's environmental regulations pertaining to these issues is essential. Mine audits are not the typical 'nuts and bolts' audits as applied to say manufacturing; they require a throughout understanding of the mining industry so that the significant issues are appropriately addressed.

Whatever is being audited – e.g. activities covered by legislation, EMS, mine waste management practices – is referred to as the **subject matter** of the audit. The policies, procedures, guidelines, industry standards or other requirements such as EIA commitments against which the subject matter of the audit is checked are called **audit criteria**. Other audit criteria, of course, are the Equator Principles or mining sector specific industry codes such as the cyanide code (see Chapter Seven). The audit criteria in a legislative compliance audit will be predominantly the environmental legislation of the host country plus commitments made in the EIA. The subject matters of the audit will be the environmentally relevant components and activities covered by such legislation. In an environmental management system audit, the subject matter of the audit will be the company's EMS and the audit criteria will be the requirements of EMAS or ISO 14001. Audit criteria are determined at an early stage of the audit process. They should be agreed between the lead auditor and the client and communicated to the auditee.

How does an auditor determine whether or not the subject matter of the audit conforms to the audit criteria? The auditor has to collect audit evidence defined as verifiable information, records or statements of fact, in order to determine whether or not the audit subject matter conforms to established audit criteria. The important point to make here is that the audit process involves using existing information to determine conformity; auditors do not usually generate new information. To enhance consistency and reliability, collection, documentation and presentation of audit evidence should be in accordance with documented and well-defined methodologies and systematic procedures. It should follow appropriate guidelines developed for that particular type of environmental audit. (For example ISO have published guidelines for conducting environmental management system audits – ISO 14011).

The important point to make here is that the audit process involves using existing information to determine conformity; auditors do not usually generate new information.

In spite of some explanatory words at the beginning of this section, the question may remain why voluntarily carry out or submit to an environmental audit? The simple answer is that it is always important to check that things are as they should be. Given that any mining company must comply with relevant environmental legislation and regulations, it is obviously important that management checks that it is actually in compliance. Only by auditing (not accidentally adopted from the Latin word auditare meaning hearing), whether conducted internally or externally, will management discover instances of non-compliance and be able to correct them.

It is always important to check that things are as they should be.

Developing the Audit Protocol

A mining audit is commonly based on a combination of staff interviews, pre-site visit document reviews and a site visit to the mine. Interviews are especially important. They provide the primary means of understanding the organizational relationships, roles and responsibilities, policies and systems that form the framework for environmental management. More importantly, they often reveal differences in actual versus documented practices. Document review is important to verify the formality of the system and confirm interview information. A site visit is necessary to verify the implementation and effectiveness of environmental management. An environmental mine audit typically evolves along the steps illustrated in **Figure 12.1**.

The lead auditor prepares the audit protocol (also termed audit plan) in consultation with the mine management and any audit team members. The protocol should include, if applicable: (1) audit objectives and scope; (2) audit criteria; (3) identification of the organizational units to be audited; (3) identification of those activities that are of high audit priority; (4) audit procedures to be used; (5) identification of reference documents; (5) expected time and duration for major audit activities; (6) the dates and places where the audit is to be conducted; (7) identification of audit team members; (8) schedule of meetings to be held with the mine management; (8) confidentiality requirements (an important matter although not always fully appreciated by all auditors); (9) content, format, structure, expected date of issue and distribution of the audit report; and (10) document retention requirements.

The pre-site visit information review will allow the audit team to finalize a site specific audit protocol. The audit protocol will summarize areas of environmental and social concerns on which the site visit should focus, outline the type of questions that must be asked and identify key mine personnel to be interviewed. In view of the technical, environmental and social complexities of a mine operation, audit protocols are based on a review and evaluation approach. As such the audit protocol is designed as a general list of important topics

FIGURE 12.1
The EMS Audit Process

A mining audit is commonly based on a combination of staff interviews, pre-site visit document reviews and a site visit to the mine.

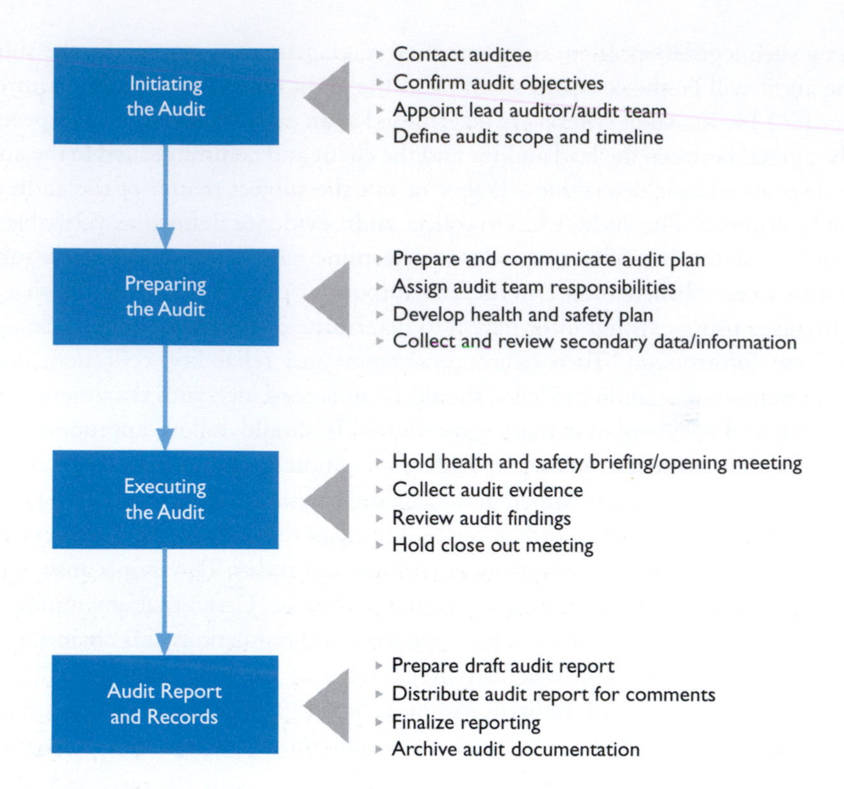

A detailed checklist would also force the site visit into a rigid structure and would most likely result in missing critical issues.

and issues to be addressed rather than a detailed audit questionnaire commonly applied to environmental due-diligence audits. There is no questionnaire that gives full respect to the complex nature of a mine operation. A detailed checklist would also force the site visit into a rigid structure and would most likely result in missing critical issues. The Australian Mineral Industry has developed a useful Code Progress Assessment Protocol Survey, 1999, to guide auditing of the mine management systems. A Rapid Assessment Method for assessing the environmental performance of a mine is presented in **Appendix 12.1**.

When implementing the audit protocol, auditors should exercise the professional care, diligence, skill and judgment expected of any auditor in similar circumstances. The audit team/client relationship should be one of confidentiality and discretion. Unless required to do so by law, the audit team should not disclose information or documents obtained during the audit or the final audit report to any third party without the approval of the client. The audit team should follow the audit protocol designed to provide quality assurance.

Selecting the Audit Team

The audit team should be selected based on auditing experience, mining expertise, regulatory expertise, working experience in the host country, familiarity with similar mining operations, proven cultural sensitivity and the ability to communicate appropriately with different levels of personnel.

The first and most important step in any audit programme is the selection of the audit team. The audit team may consist of one to five people, depending on the size of the operation. An environmental audit conducted by a single auditor is relatively rare and usually occurs only for small to medium-sized operations, or to auditing of a few selected audit segments. An audit team may comprise a mixture of external and internal personnel. External auditors bring a fresh perspective to an operational site, and if senior enough, can provide benchmark auditing, comparing practices at the subject site to practices at other mines. It is important to include professionals from the host country in any audit team.

The audit team should be selected based on auditing experience, mining expertise, regulatory expertise, working experience in the host country, familiarity with similar mining operations, proven cultural sensitivity and the ability to communicate appropriately with different levels of personnel. At least one team member needs to demonstrate a full appreciation of social issues associated with any mine operation. Each audit team needs a lead auditor who takes on the overall responsibility for the audit process and audit findings. He or she is also the main contact person for the mine management, or the auditee. The obvious question is who qualifies as the lead auditor? The lead auditor should, of course, have the knowledge, skills and experience necessary to carry out the audit. Guidance on these matters is provided in ISO 14012, the international standard on qualification criteria for environmental auditors. More importantly, the lead auditor needs to demonstrate a thorough understanding of the mining sector, and the complex web of potential environmental, social and economic issues associated with the sector. Moreover, the lead auditor must be able to communicate effectively and appropriately with a wide range of personnel, including senior management. Tact and diplomacy are also useful attributes, as it is not unusual for employees of the auditee to respond negatively to audit findings which may reflect on their performance.

Since an environmental audit should be as objective as possible, the members of the audit team should be independent of the activities they are to audit. If an internal audit is being performed then none of the audit team members should be accountable to those directly responsible for the subject matter being audited.

Agreeing on Audit Objectives and Programme

Prior to the site visit, it is essential to agree the scope and the main objectives of the audit, and to finalize and to coordinate the overall audit programme meaning the actual implementation of the audit plan with the mine management. Scope and objectives then determine the extent and the boundary of the audit. In view of the magnitude of most mining operations, it is important to narrow the audit scope to the agreed audit matters (e.g. environmental issues versus social issues, environmental management system, legal compliance versus compliance with Equator Principles, etc). A well-designed audit programme helps to avoid wasting valuable time, not only of the audit team, but even more importantly of mine personnel. It is at this time that communication systems are established, and that audit constraints if they exist, are identified. Mine auditing requires the involvement of mine personnel and such time commitments should be clearly communicated at the outset of the audit.

A well-designed audit programme helps to avoid wasting valuable time, not only of the audit team, but even more importantly of mine personnel.

Pre-site Visit Data Review

Pre-site visit data is valuable for an effective audit programme. Information needs to relates to the project operations, environmental monitoring and applied management practices. To facilitate data compilation by the mine management it is common practice that the audit team prepares an outline of information needs to facilitate information gathering by the mine management.

Site Visit

Some audits, usually in the context of investment or divestment, are carried out entirely in so called 'data rooms'. The audit team is given access to a set of information in a very controlled manner: (1) data are only available at one location which can also be a website with access restriction – a virtual data room; (2) data can only be accessed during a specified time period, usually 3 to 5 days; and (3) the number and composition of the audit team is subject to approval by the auditee. It should be realized that the findings of an audit that does not include a site visit will be qualified as it will not be possible for the auditor to verify much of the information that is provided. Only a site visit can provide a true impression of the magnitude and the environmental setting of the mine operation. The on-site component of an environmental audit is also an intensive period of collecting and evaluating specific data, and of critically reviewing applied management practices (**Case 12.1**).

An opening meeting is necessary to: (1) introduce the members of the audit team to the site management; (2) review and confirm the scope, objectives and audit plan and agree on an audit timetable; (3) provide a short summary of the methods and procedures to be used to conduct the audit; (4) establish the official communication links between the audit team and mine site management; (5) confirm that the resources and facilities needed by the audit team are available; (6) confirm the time and date for the closing meeting; (7) promote the active participation of mine personnel in the audit; and (8) review relevant site, safety and emergency procedures for the audit team, of special importance for any mine visit.

Interviews with mine personnel at various levels provide an understanding of employees environmental awareness and commitment to existing environmental objectives, employees qualifications in relation to job-specific environmental and social tasks and to obtain subjective information. Frequently, staff will take the opportunity provided by an interview to identify issues which they believe are not being adequately addressed by the mine management. An experienced auditor will ask open-ended questions and will listen attentively.

An experienced auditor will ask open-ended questions and will listen attentively.

Another critical component of the audit is the review of the environmental monitoring program. The audit should test whether the monitoring programme adequately measures the true nature and extent of all emissions to land, water and air. The mine monitoring programme, therefore, is a central focus of any audit plan, which seeks to ensure that the environmental management programme provides effective feedback on the extent of compliance and that results are systematically reviewed to identify any breach of compliance or a trend that may signal an impending breach. The monitoring programme for a project, often included in the EIA, is usually subject to approval by the regulating agency, as one of the functions of the monitoring programme is to demonstrate compliance with both the environmental conditions of the approved mine plan and with regulatory requirements. The review of past monitoring records provides the audit team with the opportunity to comment on the relevance of collected data, past trends and the adequacy of the programme. Recommendations can be made to either add or delete monitoring parameters or locations, or to modify monitoring frequency. Monitoring programmes may not necessarily be restricted to regulatory requirements, but may extend to issues which are not subject to compliance. Conformance with company internal policies is one example of beyond compliance management.

A closeout meeting allows the audit team to clarify outstanding issues while providing initial feedback to the mine management. The main purpose of this meeting is for the

CASE 12.1
Mining and Auditing

Few mine managers would question the merits of environmental auditing. However, in today's world of corporate responsibility, audit requirements may seem excessive. Depending on circumstances, the operations may be subject to (1) external and internal ISO audits; (2) corporate auditing (also external and internal); (3) government audits; (4) audits related to specific industry codes, such as the Cyanide Code; and (5) other external audits such as may be required by the Equator Principles Financial Institutions. The subject matter of these audits also varies and may range from health and safety, systems and compliance to environmental and social performance. Since these various audits are for different purposes and audiences, they are also uncoordinated, occurring at different times in the year. On the other hand, the audits cover many of the same issues and rely on much the same information. Clearly, audits require much time and attention from management and staff. However, there is scope for rationalization. As suggested by one mining executive, there may be merit in scheduling all audits to be carried out either together or sequentially during a two or three weeks 'audit season' each year.

team to present the audit findings to the site management so that they are fully understood and to enable management to correct any misunderstandings or misrepresentations. The closeout meeting thus provides an opportunity to resolve any disagreements between the auditing team and the mine management or staff. The final decision on the significance and description of the audit findings rests with the lead auditor, even if the mine management disagrees with the findings.

The closeout meeting thus provides an opportunity to resolve any disagreements between the auditing team and the mine management or staff.

Issuing the Audit Report

The client is provided with a written report of the audit findings (and/or a summary thereof). Unless the client states otherwise, the auditee should also receive a copy of the report as draft for comments. The audit report should follow an agreed content and format. Any photo-documentation should support critical issues if identified. Information that may be contained in an audit report includes (but is not limited to) the following: (1) identification of the organization audited and of the client; (2) agreed objectives and scope of the audit; (3) agreed criteria against which the audit was conducted; (4) period covered by the audit and the date(s) the audit was conducted; (5) identification of the audit-team members; (6) identification of the auditee's representatives participating in the audit; (7) statement of the confidential nature of the contents; (8) distribution list for the audit report; (9) summary of the audit process including any obstacles encountered; and (10) audit conclusions.

The lead auditor, in consultation with the client, should determine which of these items, together with any additional items, should be included in the report. Normally it is the responsibility of the auditee to determine any corrective action needed in response to the audit findings. However the auditor may provide recommendations when there has been prior agreement to do so with the client.

The evidence collected during an environmental audit will inevitably only be a sample of the information potentially available, as audits are conducted over a limited period of time and with limited resources. There will therefore be an element of uncertainty inherent in all audit reports. The audit limitations should be stated in the final audit reports so that the users of the results of environmental audits are aware of this uncertainty. The auditing process, however, should be designed to provide the client with the desired level of confidence in the reliability of the audit findings.

Finally it is fair to say that it is tempting for some auditors to present a negative and damaging audit report as the emphasis is generally on identifying non-compliance and environmental liabilities. This is believed to be a self-defeating attitude. The main thrust of the audit should be 'Here are some areas where you can improve your environmental performance' rather than 'Hey I found things at your project that are not as they should be'. Thus an audit should be viewed constructively, identifying opportunities for improving environmental performance and thereby enhancing the image of the mining company, the mining industry, and the regulators.

The main thrust of the audit should be 'Here are some areas where you can improve your environmental performance' rather than 'Hey I found things at your project that are not as they should be'.

12.3 SOMETIMES THINGS GO WRONG

Occasionally a mining project will cause serious impacts that were not predicted in the EIA. Well known examples include both the Ok Tedi (Papua New Guinea) and Grasberg (Indonesia) Copper projects, where mine wastes were discharged to rivers in the belief that the rivers would transport the wastes to the sea. In both these projects, the unplanned deposition of tailings occurred on land outside the river channels, causing widespread loss of forest. In both cases, modelling had predicted that the carrying capacity of the rivers would not be exceeded. Clearly, something was wrong with the model or, more likely, on the assumptions that were used or the input data. Any project should be aware of the possibility that something will go badly wrong, and should be prepared to respond rapidly in such an eventuality. Clearly, a well-designed audit programme is the key to identifying impending problems so they can be remedied before causing damage. However, experience has shown that nasty surprises will still occur.

The most common causes of unpredicted impacts relate to what might be termed 'system overload'.

In the experience of these authors, the most common causes of unpredicted impacts relate to what might be termed 'system overload'. For example the mill throughput may be increased beyond its design capacity, which may result in the assimilative capacity of the receiving environment being exceeded. Similarly, a tailings storage facility may be designed to contain a certain volume of tailings, but during operations, more ore is mined than anticipated and the embankment is raised to accommodate the additional tailings. In some circumstances, this could lead to failure. It is therefore most important that any changes to a project be thoroughly assessed in terms of their potential environmental consequences. This applies whether or not such additional assessments are required under host country regulations.

Other causes of unpredicted impacts are:

- Inadequate baseline data, such as insufficient number or frequency of samples to adequately represent the range of variability;
- Failure to identify all impacts. On several projects, for example, the risks of acid rock drainage (ARD) from the mine, waste rock and tailings, were adequately predicted while other sources of ARD, such as access roads exposing acid-generating material, were not considered; and
- Human error, sometimes due to complacency after many years of trouble-free operation.

Again, a well-designed audit programme should identify and highlight the potential for such impacts, before they develop.

REFERENCES

ADB (2000) Special evaluation study on the policy impact of involuntary resettlement. Manila. Available: www.adb.org/Documents.

IFC (2006) Environmental and Social Performance Standards, International Finance Cooperation, www.ifc.org.

Sadler (1996) International Study of the Effectiveness of Environmental Assessment, Final Report Environmental Assessment in a Changing World: Evaluating Practice to Improve Performance. Canadian Environmental Assessment Agency and International Association for Impact Assessment.

UNEP (2002) UNEP Environmental Impact Assessment Training Resource Manual, Second Edition, www.unep.ch/etu/publications/EIAMan_2edition_toc.htm.

World Bank (1989) Operational Directive 4.01 on Environmental Assessment, converted in 1999 into a new format: Operational Policy (OP) 4.01 and Bank Procedures (BP) 4.01.

World Bank (1989) Base Metal and Iron Ore Mining, Industry Sector Guidelines, Pollution Prevention and Abatement Handbook.

World Bank (2001) Operational Policy OP 4.37 Safety of Dams.

Appendix 12.1 Comprehensive Mine Audit Checklist (Example)

Summarize Institutional Setting

Political Dimension

- Foreign Investment Policy
- Mining Code
- Legal Agreement and Obligations
- Mine Closure Requirement
- Jurisdiction of Host Region

Legislative Setting

- Environmental Acts and Regulations
- Host Country EIA Regime
- Applicable Standards
- Environmental/Mining Licenses
- Traditional/Cultural Laws
- International Law/Industry Codes
- Governmental Spatial Planning

Review Systems and Documents

Environmental Assessment

- Baseline Data
- Environmental Impact Statement
- Environmental Management Plan
- Environmental Monitoring Plan
- Performance Indicators
- Environmental Reporting
- Equator Principles Requirements

Environmental Management System

- Senior Management Awareness and Commitment
- Internal Capacity
- Company Policies
- Environmental Handbook
- Operation Procedures/Manuals
- Previous Audit Results/Correspondence
- Management of Suppliers/Contractors
- Environmental Monitoring Program

Assess Mine Operation

Work Force

- Organization Chart
- Management
- Mine Workforce
- Environmental Department

Mineral Resources

- Mineral Resources
- Mineral Reserves
- Production Rate
- Mine Life
- Cutoff Grade
- Stripping Ratio

Mining

- Prospecting and Exploration
- Drilling, Blasting, Excavating, Crushing, Concentrate Preparation
- Mineral Processing plus Concentrate Shipping
- Mine Quarries (borrow pits)

Infrastructure

- Power Generation
- Sea/Air Port
- Hospital
- Mine Town/Industrial Park
- Recreational Facilities
- Catering
- Waste Management (land fill, waste water treatment, incineration)
- Transportation (people and goods/ equipment)

Mine Closure

- Mine Closure Planning
- Mine Closure Funding
- Timeline
- Post-Closure Commitments

Rehabilitation

- Final Landform
- Erosion Controls
- Active/Cleared Areas
- Progress of Revegetated Areas
- Type of Revegetation/Plants
- Plants Survival Success
- Plant Trials
- Fertilizer Usage
- Post-Mining Landuse

Assess Mine Induced Environmental Change

Water

Water Sources

- Ground Water
- Surface Water
- Wetlands/Lake
- Coastal Water

Water Usages

- Water Usage by Mine
- Competitive Water Users

Stormwater

- Modification of Drainage Pattern
- Stormwater Storage
- Water Erosion

Acid Rock Drainage

- Acid Generation Potential
- Mine Wastes Storage Areas
- Ore/Coal Stock Piles
- Disturbed Land Areas

Mine Effluents

- Tailings Dam Overflow
- Wastewater Treatment Plants
- Mine Run Off/Discharge Points
- Emergency Dams
- Pollution Control Dams
- Lime Dosing Facilities
- Pumps and Flow Lines

Groundwater

- Aquifer Systems
- Groundwater Usage
- Groundwater Drawdown

Environmental Water Monitoring

Air

Particulate Matter (Sources)

- Active Mine Areas
- Material Movement
- Stockpiles
- Unpaved Roads
- Dust Emission Controls
- Community Complaints

Plant Emissions

- Dust Emissions
- Gaseous Emissions
- Stationary/Fugitive Emissions

Noise

- Community Complaints
- Employee Complaints
- Occupation Health Records
- Noise levels

Greenhouse Gases

- Coal Bed Methane
- Fossil Fuel Driven Equipment
- Biomass Management
- Power Generation
- Material Movement

Odor

Environmental Air Monitoring

Land

Land Clearing

Active Mining Areas

Topsoil Management

Mine Waste Storage Areas

- Geotechnical Stability
- Long-term Integrity

Tailings Disposal

- Tailings Quantities/Qualities
- Disposal Scheme
- Environmental Monitoring System
- Overflow Points
- Pollution Control System

Waste Rock Placement

- Volume and Quantities
- Procedures for Dumping, Compacting, Contouring, Revegetation
- Acid Rock Drainage Potential

Visual Appeareance (day and night)

Vibration

- Blasting
- Material Movement
- Generators

Mine Closure

- Final Landform
- Health and Safety Aspects
- Final Landuse

Environmental Monitoring

Biological Environment

Biodiversity

- Ground Cover
- Fauna and Flora (terestrial and aquatic)
- Protected/Conservation Areas
- Forest Classification/Management
- Endangered Species
- Migratory Species
- Introduced Species
- Restricted Range Species
- Increased Hunting/Logging due to Improved Site Access

Biological Monitoring

Human Environment

Social Aspect

- Land Ownership and Resettlement
- Land Compensation
- Resettlement/ Income Restoration
- Participatory Decision-making
- Stakeholder Mapping (interest, expectation, perception)
- Community Development Programs
- Grievance Mechanism
- Code of Conduct
- Social Performance Indicators
- Complaints/Community Tensions
- Anti-Mining Activists

Economic Aspect

- Employment (local, foreign)
- Secondary Business Opportunities (multiplier effect)
- Cottage Industries
- Local, Regional and National Economy
- Distribution of Economic Benefits
- Taxes and Royalties
- Sustainability of Economic Benefits
- Economic Indicators
- Workforce Demobilization at End of Mine Life

Social Monitoring

Review Emergency Response Measures

Environment

- Emergency Response Planning
- Pollution Control Procedures
- Fire Control Procedures

Oil/Chemicals

- Oil/Non Hazardous Chemical Spillage
- Hazardous Chemical Handling
  - Procedures
  - Spill register
  - Spill atlas
  - Material Safety Data Sheets (MSDS)
  - Register

Natural Disaster Response

Register of Accidents of Environmental Relevance

13

The Range of Environmental Concerns

Separating Fact from Fantasy

ZINC

Zinc, a malleable blue-gray metal, is the fourth most commonly used metal, and has been in use for more than 3,000 years. It is commonly found in sulphide ores together with copper, or silver and lead, where its most common mineral is sphalerite. It is an essential element but may be toxic in solution, particularly to plants and aquatic organisms. It is a constituent of many alloys, notably brass, and is also used in galvanising to protect steel from oxidation. Other uses include die casting, batteries, cathodic protection and variousmedicinal products. Zinc oxide is used as a pigment and sun-block.

13 The Range of Environmental Concerns

Separating Fact from Fantasy

The intention of Chapter Thirteen is to illustrate the range of environmental concerns that are commonly associated with mining. This listing of potential impacts is not complete; it does not need to be. Environmental concerns are many and varied, since mine-induced change processes may occur in almost any environmental component (**Figure 13.1**). It is this recognition that is important, rather than any attempt to present a complete picture of all potential environmental concerns.

The main environmental and social issues, risks (and opportunities) associated with mining projects are detailed in subsequent chapters, notably: land acquisition and resettlement in Chapter Fourteen, community development in Chapter Fifteen, Indigenous Peoples issues in Chapter Sixteen, acid rock drainage in Chapter Seventeen, tailings disposal in Chapter Eighteen, waste rock placement in Chapter Nineteen, and erosion in Chapter Twenty.

The significant environmental and social risks and opportunities emerge during project construction and operation.

Although there is the potential for environmental impacts during the exploration stage (**Appendix 13.2**), the significant environmental and social risks and opportunities emerge during project construction and operation. Mine closure also has its own environmental impacts as discussed in Chapter Twenty-One. Mine closure typically results in negative consequences on the regional economy, but other impacts are positive because mine closure aims to rehabilitate the mine site for future use.

Grouping Impacts

First-time participants in an EIA are often faced with a very important question: How to organize or logically group environmental impacts? The question is not new, and there is no single correct answer. Some prefer to group impacts according to the affected environmental components (impacts to water or air are two examples). Others prefer to associate impacts with activities (e.g. impacts related to tailings management or resettlement). The difficulty of course arises when impacts cross the entire environmental spectrum. For example, tailings impact water quality, which in turn may affect fish. This in turn affects community life and community relationships. When tailings dry, the particles may be

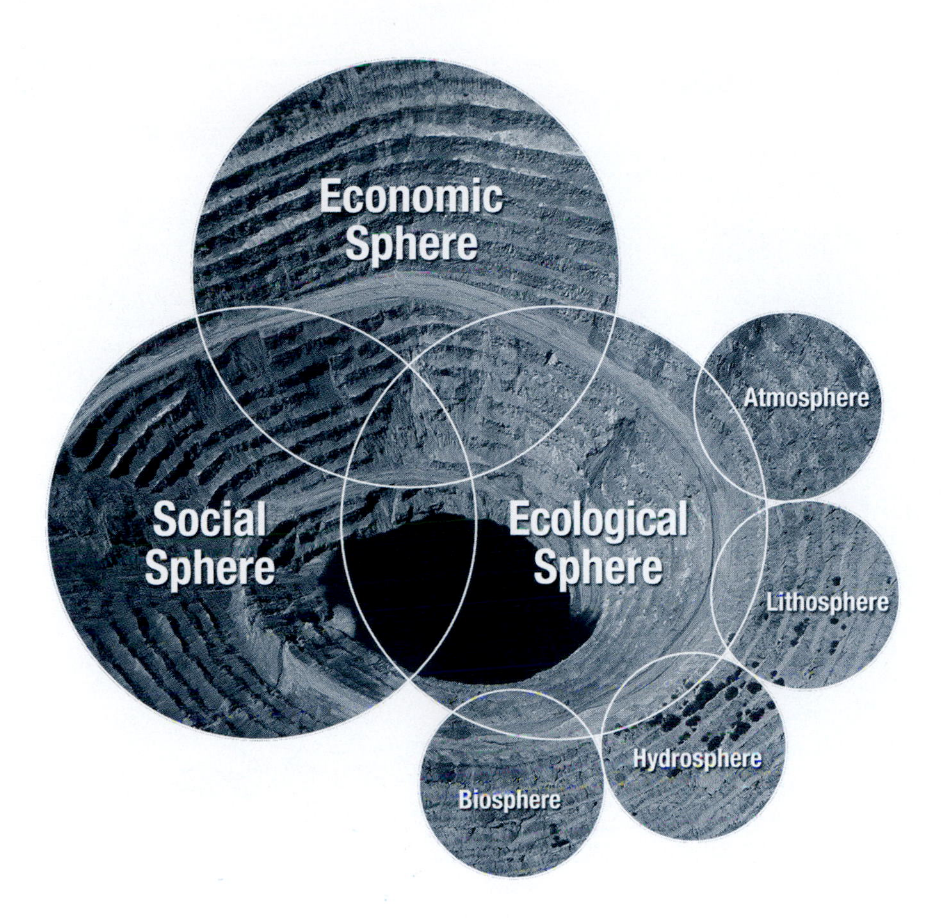

FIGURE 13.1

Environmental and Social Concerns are Many and Varied

Mining affects the physical, socio-cultural, socio-economic, and ecological spheres, and the environmental assessment of a new mine should give justice to impacts on all of them.

eroded by wind, degrading air quality. Furthermore, tailings impoundments may inundate large land areas, physically displacing or burying local fauna and flora.

Frequently the required EIA format of the host country mandates how impacts are grouped, which can lead to tedious repetition, due to the requirement to discuss each occurrence of an impact rather than as part of a significant group. In this chapter, however, impacts are grouped according to activity as well as according to the affected environmental component, depending on the most appropriate approach to discussing a given group.

CASE 13.1
The Bunker Hill Mine Complex in Idaho, USA

The Bunker Hill Mine complex is located in northwest Idaho in the Coeur d'Alene Region, and has a legacy of a century of mining-related contamination. Operations ceased in 1982, and the EPA declared much of the area a Superfund site in 1983. The complex produced lead, zinc, cadmium, silver, and gold, as well as arsenic and other minerals and materials. Much of the mining pollution was caused by the dispersal of mining wastes containing such contaminants as arsenic, cadmium, and lead into the floodplain of the Coeur d'Alene River, acid mine drainage, and a leaking tailings pond. Soils, surface water, groundwater, and air became contaminated by these metals, leading to health and environmental effects. Lead, in particular, was noted for its health effects on children in the area. EPA reports concerning lead poisoning state that experts believe blood levels as low as 10 micrograms per decilitre (μg/dL) are associated with children's learning and behavioral problems. High blood lead levels cause devastating health effects, such as seizures, coma, and death. Blood levels of children in areas near the complex ranged from about 35 to 65 μg/dL in the early 1970s to less than 5% of that figure in 1999, as remediation efforts progressed. EPA reports also state that children are at a greater risk from exposure to lead than adults because, among other reasons, children absorb and retain a larger percentage of ingested lead per unit of body weight than adults, which increases the toxic effects of the lead. Efforts by the federal government, the state of Idaho, and industry to remediate contaminated areas associated with the site are ongoing.

Source: Thomson Gale 2006

FIGURE 13.2
The Open Pit at Batu Hijau Copper Mine in Indonesia in 2006

Large mining pits can reach more than 2,000 meters in diameter and several hundred metres in depth. In fact, once completed, the bottom on the Batu Hijau Pit is expected to be at least 300 m below sea level and will be the lowest on-land point in Indonesia.

Some environmental and social impacts from mining have been raised in previous chapters, and all are well-documented in the literature (**Case 13.1**; Goodland 2004; Oesterloef and Oesterlund 2003; UNCAD 1996; UNEP 2000 and 2001; WRI 2003; Bell and Donnelly 2006; MMSD 2002). This chapter addresses key environmental and social issues, leaving detailed discussions to the remaining chapters of this book: change in land form and associated structural issues; the generation of mine wastes dominated by waste rock and tailings; mine effluents and acid rock drainage; air quality and climate change emphasizing the relationship between coal mining and its use, and greenhouse gases; mining and its perceived impact on biodiversity and habitats; and social changes such as induced development, economic growth and wealth distribution. **Appendix 13.1** provides a listing of these and other environmental issues.

13.1 CHANGES IN LANDFORM

Most mining operations change topography and landform.

Most mining operations change topography and landform. In open pit mining the pit is noticeable even to a casual passer-by (**Figure 13.2**). 'Mountain top' mining too is visually intrusive and highly controversial in the US; as the name suggests it basically means removing valuable deposits which form the top of mountains or plateaus, not at all an unusual situation. Such deposits are found in different parts of the world. Iron ore caps a series of flat-topped hills (mesas) in the Pilbara Region of Western Australia. Recent mine planning allows for a 50 to 100 m wide belt of ore to remain around the perimeters of these mesas. The original landform stays intact, at least as viewed from a distance. However as noted elsewhere in the text, mining not only comprises the removal of large amounts of rock materials; a similar amount of rock needs to be disposed of – mining not only removes mountains, it creates hills, mountains, and terraces of waste rock and tailings.

These physical changes in landform pose a unique set of problems relating to long-term structural stability of the new landforms, whether great holes in the ground or mountains of tailings and waste rock. Safety is of paramount importance, initially for workers and, subsequently, after closure, for the public. In the latter case, a range of potential downstream impacts may result from mass wasting, erosion, or structural failure and consequent uncontrolled release of contaminants.

Water Erosion

Erosion is a common problem associated with earthworks (excavation, transportation, and placement of soil and/or rock). This applies particularly in mining where large areas are cleared of vegetation, and huge volumes of soil and rock are removed and stored (environmental impacts of erosion are detailed in Chapter Twenty). Also, the most erodible material of all is topsoil, which, as discussed in Chapter Twenty One represents a valuable resource that usually warrants conserving. However, without application of stringent safeguards, topsoil from disturbed land will end up in nearby streams causing short-term or long-term degradation to aquatic habitats. Erosion of waste rock piles from runoff after heavy rainfall also transports soil and rock materials into nearby streams. Increased turbidity in natural waters will reduce the light available to aquatic plants for photosynthesis. Increased sediment loads can also smother benthic organisms in streams, eliminating important food sources for predators and decreasing available habitat for fish to migrate and spawn (Johnson 1997b). Some eroded materials are simply toxic due to their high metal content, or become so after contact with air, which causes oxidation of sulphide minerals.

Erosion is a common problem associated with earthworks.

In addition, high sediment loads can lead to aggradation, decreasing the depth of streams, with an increase in risk of flooding during times of high stream flow (Mason 1997) (**Case 13.2**). Sedimentation from mining may modify stream morphology by disrupting a channel, diverting stream flows, and changing the slope or bank stability of a stream channel. All or any of these disturbances may, and probably will, reduce water quality (Johnson 1997a).

Slope Failure and Landslides

Mining projects result in formation of two categories of slopes: cut slopes and fill slopes. Cut slopes are created by the removal of waste rock and/or ore with new, usually steeper, slopes surrounding the excavation or 'mine void' (see **Figure 13.2** as one example). During

CASE 13.2
Log Jam or Tailings – Which one is the Culprit?

From the outset of mining its Grasberg deposit, PT Freeport Indonesia adopted a riverine tailings disposal scheme. This practice and associated environmental impacts first made headlines in the early 1990 when the original tailings river overflowed its banks during a severe storm event, depositing tailings over large areas of natural rain forest. At that time environmentalists claimed that increased sediment loads due to past tailings disposal had clogged the drainage system. Company officials on the other hand ascribed the cause of bank overflow to logs transported by the river during the storm event forming a log jam. Whatever the initiating cause, the event led to the construction of extensive levees along both sides of the tailings river to reduce the risk of lateral uncontrolled spread of tailings.

During the mining period, maintenance of slope stability and minimization of related operational health and safety risks are of prime concern.

the mining period, maintenance of slope stability and minimization of related operational health and safety risks are of prime concern (**Case 13.3**). Excavated slopes may create other environmental concerns including increased erosion, rapid runoff, risks to wildlife, and the exposure of potentially reactive natural materials.

Dumping or placement of overburden soils, tailings, waste rock, or other materials creates new fill slopes. The geotechnical stability of fill slopes is of concern because, compared to most *in situ* materials, they are less dense, relatively weak, and more likely to absorb water. Water saturation of waste material can trigger slope failure. Fill materials may in some cases also be subject to rapid disintegration, or to decomposition, yielding acids and releasing potentially toxic metals. Slope failure can result in direct release or direct exposure of these materials to the surrounding environment.

Many natural slopes in mountainous terrain are only marginally stable and require only a relatively minor change to become unstable.

Mining and associated activities may also trigger landslides through natural soil and rock masses. This commonly occurs in pit walls, (Case 13.3) but may also occur along access and haulage roads, particularly where these involve excavations into steep hillsides. Many natural slopes in mountainous terrain are only marginally stable and require only a relatively minor change to become unstable. Merely clearing vegetation or excavating a pipeline trench may, in some instances, be sufficient to destabilize an entire mountain slope.

Subsidence and Earthquakes

The potential for subsidence exists for all forms of underground mining.

Mine subsidence is the movement of the surface resulting from the collapse of overlying strata into mine voids, a subject matter well-documented by Bell and Donnelly (2005). The potential for subsidence exists for all forms of underground mining. Subsidence may manifest itself in the form of sinkholes or troughs. Sinkholes are usually associated with the collapse of a portion of a mine void (such as a room in 'room and pillar' mining); the extent of the surface disturbance is usually limited in size. Troughs are formed from the subsidence of large portions of the underground void and typically occur over areas where most of the resource has been removed (**Figure 13.3**). Typically, such subsidence is not evident at the surface except along its margins.

The effects of subsidence include the development of sinkholes or depressions in the land surface that may interrupt surface water drainage patterns; ponds and streams may be drained or channels redirected. Farmland can be affected to the point that cultivation is

CASE 13.3
The Collapse of the Southern Part of the Grasberg Pit Wall

In October 2003 a landslide arrived in a torrent of 2.5 million tonnes of rock and mud at the Freeport-McMoran gold and copper mine in Papua, Indonesia. Heavy rain had fallen for five days, and water was trapped in the slope building up water pressure in the steep southern part of the pit wall. Because this section was always most at risk, the company had rigged it with more than a dozen extensometers – devices that measure the rate of movement at 20 minutes intervals. Two days before the pit wall failure, engineers noticed an accelerated rock movement and the company moved its stationary mining equipment out of the zone where it expected the slide to hit. However the engineers underestimated the magnitude of the landslide; they did not expect the slide to be so liquid. Eight workers lost their lives in this unfortunate accident.

Source: Based on The Syndey Morning Herald, November 1, 2003

disrupted; irrigation and drainage systems may be disrupted. In developed areas, subsidence has the potential to affect buildings, roads, and flow lines. However in remote areas, increasingly the location of new metalliferous mines, such structural risks are usually absent. Subsidence can contribute to increased infiltration to underground mines, potentially resulting in the flooding of underground openings, increased acid rock drainage, and a need for greater water treatment capacity in instances where mine drainage must be treated. Groundwater flow may be interrupted or accelerated as impermeable strata deform, crack and collapse, flooding mine voids. Impacts to groundwater also include changes in water quality and regional flow patterns (with subsequent impacts on groundwater recharge).

Occasionally a cave-in is felt; it may be powerful enough to be measured by seismographs. On August 6, 2007 the University of Utah seismograph station recorded a seismic event of magnitude 3.9. Scientists later realized a powerful collapse at the nearby Genval coal mine at a depth of several hundred metres had caused the disturbance. Since the mid 1990s, at least half dozen other mine collapses in the USA alone have caused similar seismic events, including a 1995 cave-in in southwestern Wyoming that caused readings as high as 5.4 on the Richter scale.

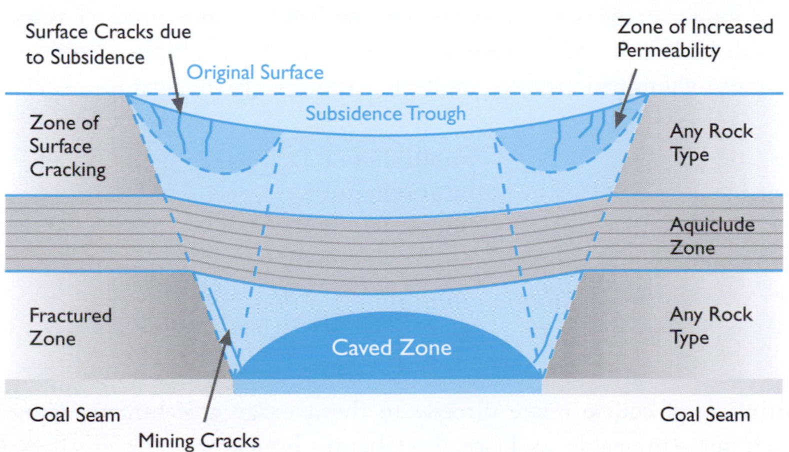

FIGURE 13.3
Land Subsidence Illustrated

Occasionally a cave-in is felt; it may be powerful enough to be measured by seismographs.

Photo: Geological Survey of Northern Ireland (www.dsi.ie) and Department of Enterprise, Trade and Investment

Loss of Visual Amenity

Reclamation and eventually mine closure aim to re-instate a stable and pleasing landform but full rehabilitation of natural vegetation is likely to require several decades, if it's possible at all.

Surface mining commonly causes scars on the Earth's surface. While the extent of such scarring is much less than that associated with rail and road construction, the scars themselves tend to be larger and more unsightly, potentially causing a loss of visual amenity. Visual impacts may diminish public enjoyment, or may impair the character or quality of a natural landform. Reclamation and eventually mine closure aim to re-instate a stable and pleasing landform but full rehabilitation of natural vegetation is likely to require several decades, if it's possible at all.

Loss of visual amenity is most noticeable during construction, when physical change occurs quickly, and at a scale that few of the local population have ever experienced. In host regions that have had little past experience with mining, socialization of the size of mining and associated visual impacts during the consultation phase with the host communities is important. This can be achieved using photos, films taken of similar mines, or 3-D computer animation of the future mine. However, the best approach is to take representatives of the prospective host community to inspect a mine of similar size and type. This will help prepare community members for the visible changes that will occur. Unprepared, an initially supportive community may be overwhelmed, potentially shifting community support to community opposition. This was a contributing factor to the local opposition that developed at the Bougainville Copper mine in Papua New Guinea, which eventually contributed to closure and abandonment of the project.

13.2 MINE WASTES

Mining operations produce a wide range of waste streams, dominated in quantity and importance by waste rock and tailings.

Mining operations produce a wide range of waste streams, dominated in quantity and importance by waste rock and tailings (**Figure 13.4**). In some cases waste rock and tailings contribute significantly to the total waste output of the host country.

The amount of mine waste produced depends on the type of mineral extracted, as well as the type and size of the mine (**Figure 13.5**). Higher market values allow mining of ore bodies with lower mineral concentrations, which in turn leads to greater generation of mine wastes. Gold and silver mines involve the highest levels of waste production in relation to the amounts of product recovered, with more than 99% of ore extracted ending up as waste, with waste rock to ore ratios that may be ten to one or greater. By contrast, mining of iron ore produces much less waste, with approximately 60% of the ore extracted processed as waste (Da Rosa and Lyon 1997; Sampat 2003). Douglas and Lawson (2000) estimated the total average material movement per unit of mine product as follows, based on actual published data and expressed in the form of a multiplier (mine waste equals multiplier times total mined product, multiplier rounded to the next number):

- Bauxite — multiplier of 3
- Coal, hard — multiplier of 5
- Iron — multiplier of 5
- Coal, brown, including lignite — multiplier of 10
- Copper — multiplier of 450
- Gold — multiplier of 950,000

These multipliers of course relate directly to the average enrichment factors needed to turn a deposit into a mineable asset (see also Chapter Four).

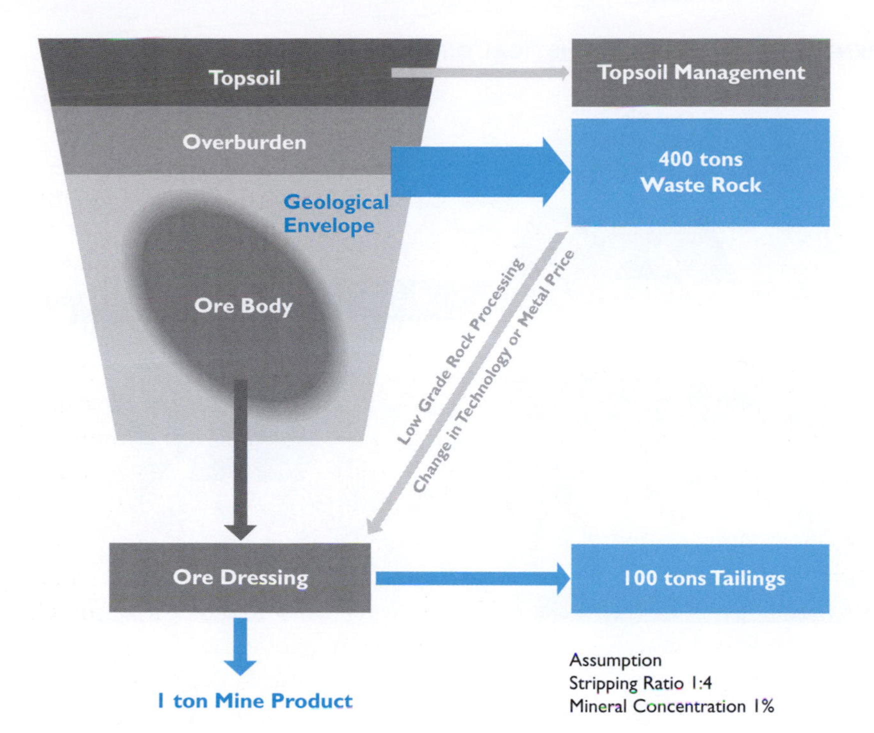

FIGURE 13.4
Waste Rock and Tailings, the Two Main Waste Streams in Mining

The amount of mine wastes depends on mineral type, stripping ratio, and mine size. Higher mineral values allow mining of ore bodies with lower mineral concentration which in turn leads to higher mine waste generation.

Disposing of such large quantities of mine wastes poses enormous challenges for the mining industry, with significant environmental impacts that are difficult if not impossible to avoid. Mine waste management is more of a challenge for open-pit than for underground mining which generally produces relatively little waste. In a sense mining companies are primarily in the waste disposal business, with the final metal merely a byproduct of excavation and processing (Navin 1978). Taking the average multiplier in copper mining as an example, 1/450th of the entire mass of material managed has to defray the costs of all the rest that is unwanted earthen materials, tailings, slag, and sulphur. Since mine waste management is also non-revenue producing, it must be conducted efficiently.

In a sense mining companies are primarily in the waste disposal business, with the final metal merely a byproduct of excavation and processing.

All mine wastes are of environmental concern. Waste rock storages and tailings impoundments stand out in this regard, not only because of their size but also because these are the storages in which toxic contaminants, if present at all, are to be found. Contaminants associated with these areas include metals, reagents, salts, and acidic solutions that may continue to degrade groundwater, surface water, and soil long after mine closure.

The issue of whether a particular material is a mine waste often depends on the specific circumstances surrounding its generation and management at the time (USEPA 1994), as well as on the jurisdiction of the host country. Mine wastes such as waste rock and tailings are generally considered to be of no value and are managed as such, typically in on-site management units. Some of these materials, however, may be useful (either on- or off-site). Waste rock and tailings, for example, may be used for construction of roads, embankments, and other engineered fills. For example, tailings embankments are often constructed, at least partially, from waste rock. And depending on economic conditions and technological improvements, today's waste rock may become tomorrow's ore resource. Research conducted to find beneficial uses for tailings, however, has been unproductive for a variety of reasons, some as in **Case 13.4**, quite unexpected. Realistically, there is little scope for finding beneficial uses for most mine wastes, because they do not contain valuable substances or properties and they are usually (although not always) sited well away from potential markets.

Depending on economic conditions and technological improvements, today's waste rock may become tomorrow's ore resource.

FIGURE 13.5

Waste Rock Management in Surface Mining

From the environmental viewpoint the advantages of open cast mining are:

1. The limited need to remove adjacent waste rock.
2. The opportunity to deposit waste rock in mined-out areas.
3. Reclamation and topsoil use can follow mining immediately.

Photo: Kennecott Bingham Copper Mine, www.bettermines.org

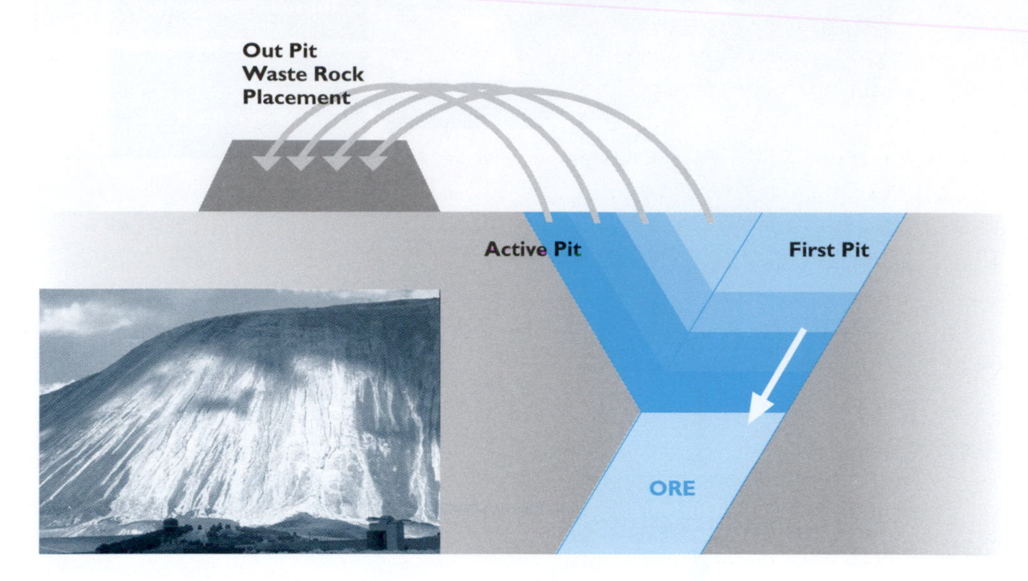

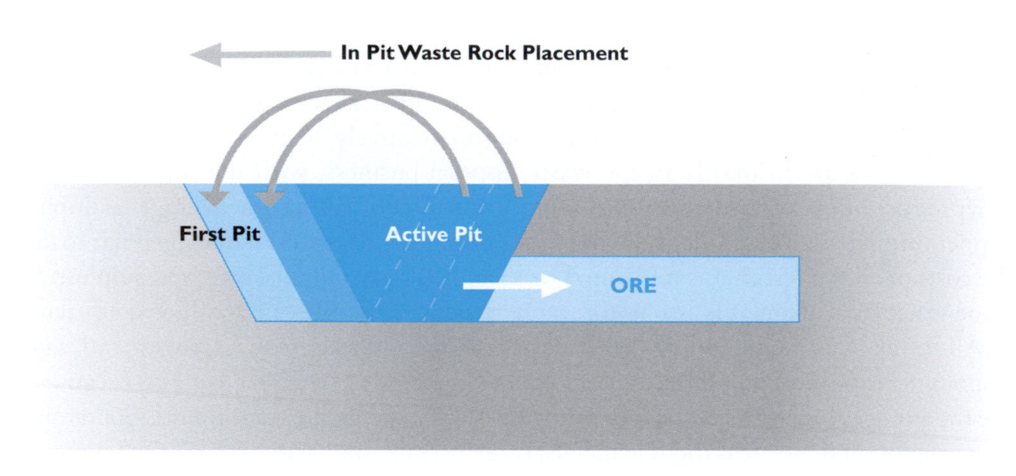

CASE 13.4
Use of Red Mud for Manufacture of Bricks

The aluminium industry has over a long period, conducted extensive research into alternative beneficial uses for its main waste product – red mud, which is the tailings remaining after extraction of alumina from bauxite, using the Bayer process. One alumina producer sponsored studies into the manufacture of bricks using red mud as the major ingredient. The results proved promising with trial batches of bricks having adequate physical properties and acceptable appearance. Plans had progressed to discussions with a potential manufacturer when someone thought to measure the level of radiation from the bricks. While not extreme, the level of radiation was sufficient that neither the mining company nor the manufacturer was prepared to proceed.

Other materials that are generated and/or used at mine sites may only occasionally or periodically be managed as wastes (USEPA 1994). These include mine water removed from underground workings or open pits, which usually is re-circulated for on-site use (e.g. as mill or leaching makeup water) but at times is discharged to surface waters. Leaching solutions are another example. They are typically regenerated and reused continuously for extended periods. On occasion, however, such as during temporary or permanent closure, the solutions may be disposed of, with or without treatment, as wastes via land application or other means.

Finally, some materials are not considered to be wastes until a particular time in their life-cycle. These include spent ore at heap or dump leaching operations: here, only when active leaching for metal recovery ends is the spent ore that comprises the heap or dump considered a waste.

Waste Rock

A number of environmental concerns are associated with the disposal of waste rock (see Chapter Nineteen for a detailed discussion). Waste rock dumps have a large footprint, and they often have a strong visual impact (**Figure 13.5**). They represent engineering structures, and geotechnical stability cannot be taken for granted. Waste rock geochemistry affects runoff and leachate, often turning water into wastewater. Finally, erosion by water and wind may occur, and rehabilitation of waste dumps may be difficult to achieve, especially in arid environments.

Waste rock removal and placement in open pit and open cast mining differ. From the environmental viewpoint the notable difference is that in open cast mining, as in coal mining in the eastern United States, the placement of overburden in mined-out areas minimizes or eliminates off-pit on-land overburden placement as illustrated in **Figure 13.5**. Thus reclamation can follow mining immediately. Another advantage is that the cut is restricted to the area of the ore deposit, and unlike open pit mines, little adjacent waste rock material is removed. However due to the large lateral extent of ore bodies typically mined by open cast mining, the total Earth movement in open cast mining often rivals or exceeds the Earth movement in open pit mining.

Waste rock removal and placement in open pit and open cast mining differ.

Tailings

After waste rock, tailings constitute the second major waste stream. Tailings consist of finely ground host rock from which valuable minerals have been removed. The host rock has essentially no commercial value. Given their unique characteristics, safe and environmentally acceptable disposal of tailings is far more challenging than the disposal of waste rock. The materials are fine-grained with high water content, the volumes are large, and depending on the chemical composition of the ore, water discharging from tailings may contain constituents which threaten the environment. Coal mining, on the other hand, may produce no tailings, or may produce relatively small amounts of tailings from coal washing operations, which remove reject materials such as stones and clay occurring within or between the coal seams.

Safe and environmentally acceptable disposal of tailings is far more challenging than the disposal of waste rock.

Many potentially severe environmental issues are related to tailings disposal. There is the risk of failure of tailings containments, and the question of their long-term integrity. Rehabilitation of tailings facilities and development of an economic land use after mining are, in many situations, difficult to achieve. Contaminated runoff and leachate from tailings storage areas pose threats to groundwater and surface water quality. Tailings may be acid generating, creating a long-term environmental liability for the mine operator. Because tailings are fine-grained they can be easily eroded by wind when they are dry. Finally, tailings disposal requires significant land areas, thereby creating undesirable visual impacts. These and other issues related to tailings management are discussed in Chapter Eighteen.

Other Wastes

Any mining project produces numerous wastes in addition to waste rock and tailings. These may be small in comparison but still substantial. Typical solid wastes include ash from coal-fired power generation, domestic waste from accommodations and administrative buildings, medical wastes from on-site medical facilities, large quantities of scrap metal including spent balls from milling operations, used batteries, used conveyor belts, and used tyres. Aqueous wastes include laboratory solutions, while non-aqueous wastes include used oils, lubricants, and solvents.

A tabular listing of solid waste streams as in **Figure 2.7** (detailing waste characteristics and intended waste management practice) helps to initiate waste management planning at the EIA stage. Waste management practices include re-use (e.g. waste tyres being used to construct retention walls or to dissipate the energy of flowing water; and waste rock or ash being used as construction material), recycling (commonly applied to scrap metal, see also **Case 13.5**) disposal on-site (e.g. tailings impoundments or the use of landfills for sanitary waste), disposal off-site (e.g. disposal of used oil using a licensed waste management contractor), return to supplier (as is sometimes the case for used tyres), and incineration (e.g. co-incineration of waste oil to generate energy or incineration of medical waste).

The best waste management practice is always waste minimization. As in the extractive industry as a whole, the mining industry increasingly aims to implement the 'Four R'

CASE 13.5
Recycling of Metal Scrap

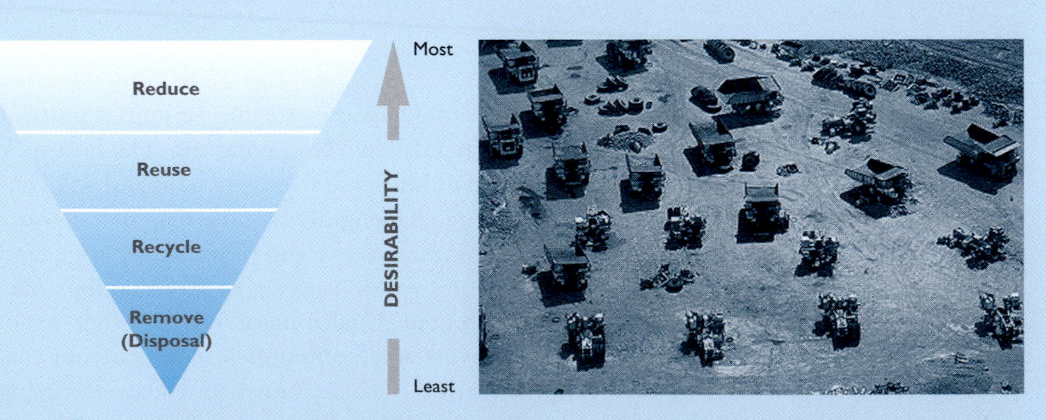

The amount of metal scrap produced by any mine can be huge. Metal recycling is the obvious solution to convert metal waste into raw material. However, the management of scrap metal is not always as simple as that. The mine operator may be exempted from import or other taxes when importing mining equipment. This in turn raises the questions of how to value scrap metals, and who should receive the earnings from the sale of the scrap metal. To give metal scrap away for free is also not an option since the host government is likely to claim at least foregone taxes. To allocate all earnings to the central government (which of course controls tax related issues) is also not necessarily an easy way out since it may conflict with claims of the regional government. The simple and practical solution to the dilemma? – Storing of valuable metal scrap at the mine site.

Photo Credit: www.airphotona.com

waste management hierarchy; i.e. reduce, reuse, recycle, and remove (disposal) (**Case 13.5**). The main goal of the waste management hierarchy is to provide an order of preference for the selection of appropriate waste management techniques. This order is based on the apparent effectiveness of each technique in conserving resources and protecting the environment against pollution.

As in the extractive industry as a whole, the mining industry increasingly aims to implement the 'Four R' waste management hierarchy; i.e. reduce, reuse, recycle, and remove (disposal).

Landfill disposal represents the most common means for disposal of solid wastes generated from a mining project (other than waste rock and tailings). In the past, these landfills were badly designed and poorly managed, often resulting in groundwater pollution and sometimes leaving an eyesore; they were a cheap method of waste removal/disposal. As environmental legislation has evolved, so has the design and management of landfills. Today's landfills at mine sites are commonly engineered to include impermeable liners and leachate collection systems to avoid groundwater contamination. Long-term monitoring is adopted to demonstrate that the landfill continues to perform over time as designed. As landfills are no longer cheap options for waste removal there is a strong incentive to shift waste management to recycling or reuse.

13.3 MINE EFFLUENTS, ACID ROCK DRAINAGE AND WATER BALANCE

Degradation of water systems, including rivers, lakes, and marine coastal waters, and their aquatic ecosystems, is one of the most significant environmental impacts of mining. Sources of water pollution vary from discharge of process effluents containing remnants of process chemicals, to acid rock drainage (ARD) from the mine or from mine waste dumps. The nature of wastewaters from mineral processing varies widely. For example, highly acidic wastes may be produced from acid leach processes as for lateritic nickel or uranium extraction; caustic soda used to dissolve alumina from bauxite, produces highly alkaline residues, while most precious metals are extracted using sodium cyanide solutions and the wastes may contain potentially toxic levels of cyanide.

The nature of wastewaters from mineral processing varies widely.

Rock containing sulphides oxidizes on contact with the air, producing sulphuric acid. ARD may originate from rainfall dissolving acidic salts from excavated mine surfaces, or as leachate emerging from oxidizing waste rock dumps or tailings disposal facilities. Leachates commonly dissolve metals from materials which they contact. In this way, metal may migrate to the groundwater or to the surrounding surface environment.

Mine effluents can also result from seepage through and below impoundment walls, percolation to the subsoil and groundwater, or overflow. A discussion of these and other common effluent streams at mine sites follows, some of them illustrated conceptually in **Figure 13.6**.

Common Effluent Streams

Process effluent can be defined as water that has come into contact with process materials or results from the production or use of any raw material, intermediate product, finished product, by-product, waste product or wastewater. It also includes blow-down water, effluent that results from plant cleaning or maintenance operations, and any effluent that comes into contact with cooling water or storm water (www.ec.gc.ca). Process effluent can be discharged from a mine site to the environment as one or a combination of the following: (1) decanted effluent from an engineered impoundment area (e.g. tailings pond); (2) decanted effluent from a polishing pond or clarification pond typically situated downstream of a tailings pond; and (3) treated effluent from an effluent treatment plant.

Cooling water is used in various mine processes for the purpose of removing heat from equipment, process operations and materials. Typically, spent cooling water is combined with other effluent streams such as a tailings slurry discharge to a tailings pond (e.g. process effluent) or mine water stream, prior to discharge to the environment.

Mine water results from the dewatering of underground or open pit mining operations. It is typically pumped from the mine and either added to the process water supply or discharged through settling ponds or combined with process effluent.

Seepage comprises water discharged to the environment by seepage from waste management areas (e.g. tailings ponds or waste rock dumps) or wastewater impoundment areas (e.g. clarification or mine water ponds).

Storm water from a mining operation comprises surface runoff from rainfall, snowmelt, and natural drainage. Storm water discharges associated with mining operations can include but are not limited to drainage from mine and mill sites; drainage collection ponds; material handling areas; raw material storage sites; and waste disposal areas, including waste rock and overburden dumps. Typically, drainage is diffuse, with a large number of discharge points to the environment. In many projects, storm water is collected and stored for use in mine processes. In other cases it is discharged to the environment after first passing through one or more settling ponds.

Typically, drainage is diffuse, with a large number of discharge points to the environment.

Other effluent streams on a mine site include sanitary wastewater discharges, emergency overflows from wastewater impoundment ponds, and backwash waters from potable water treatment plants.

Acid rock drainage also counts as mine effluent, and will be discussed in more detail in the following section.

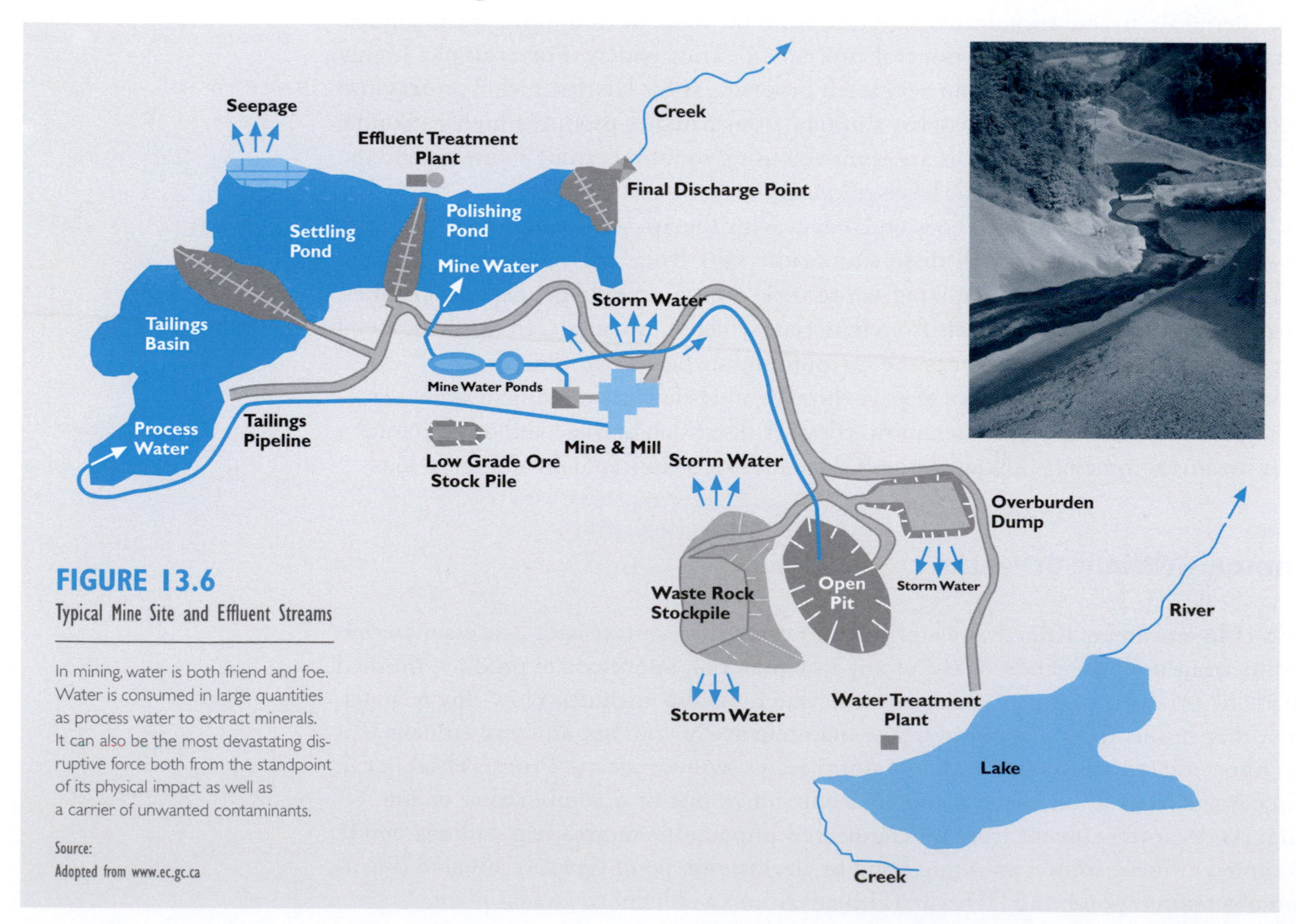

FIGURE 13.6
Typical Mine Site and Effluent Streams

In mining, water is both friend and foe. Water is consumed in large quantities as process water to extract minerals. It can also be the most devastating disruptive force both from the standpoint of its physical impact as well as a carrier of unwanted contaminants.

Source:
Adopted from www.ec.gc.ca

For compliance with regulatory effluent discharge standards, defining the final discharge point is important (**Case 13.6**).

For compliance with regulatory effluent discharge standards, defining the final discharge point is important

Acid Rock Drainage

The most serious and pervasive environmental problem related to mine waste management is arguably acid rock drainage (ARD), further detailed in Chapter Seventeen. ARD is commonly associated with gold, copper, and other metal sulphide mining projects. It is also common at coal mines where sulphide minerals may occur on partings in the coal or in interbedded shales. However, there are many mining projects where there is no potential for ARD. These include most laterite mining projects for nickel, iron and alumina, many uranium projects, diamond and placer tin mines. Acid generation occurs when mine openings, pit walls, and sulphide-rich materials in waste rock and tailings are exposed to, and react with, oxygen and water to form sulphuric acids. Sulphuric acid, which is generated easily, dissolves metals such as iron, copper, aluminium, and lead (**Figure 13.7**). This is a natural process, occurring over thousand of years as sulphide-bearing rocks are subjected

CASE 13.6
Where to Measure Compliance?

Mining differs from manufacturing and other industries in that mining activities are embedded into the natural environment. The ultimate boundary of legal responsibility for the mine operator is the boundary of the mine property or assigned mining lease. In any given mine water management system there may be several water compliance points, e.g. tailings slurry outlet at the thickener, decant outlet at the tailings pond, water outlet at the polishing pond, discharge point into the first significant surface water body, or exit point at the mining lease boundary.

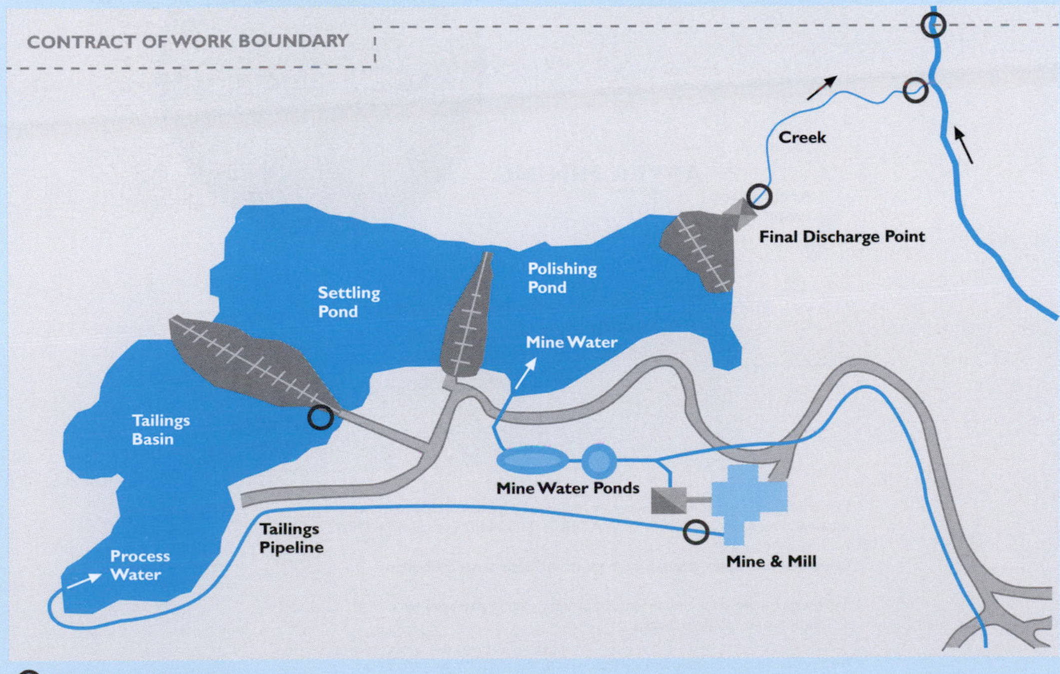

Possible Compliance Point

ARD is characterized by depressed pH values and elevated concentrations of dissolved metals: the solubility of metals is pH dependent and for most metals the solubility increases with a decrease in pH value.

to weathering. However, by exposing and fragmenting these materials, mining greatly accelerates the rate at which these reactions take place. The presence of acid-ingesting bacteria may further accelerate the process. ARD is characterized by low pH values and elevated concentrations of dissolved metals: the solubility of metals is pH dependent and for most metals the solubility increases with a decrease in pH value. Metal hydroxides are generally more soluble than metal sulphides, and they are amphoteric: a metal hydroxide has one pH at which its solubility is at a minimum (as illustrated in Chapter Ten).

Sometimes dissolved metals in ARD discolour water and cause stains on exposed rocks and streambed. If this happens, ARD is easily detectable by eye. Colours differ. 'Yellowboy'

FIGURE 13.7
Acid Rock Drainage Illustrated

Acid generation occurs when mine openings, pit walls, and sulphide-rich materials in waste rock and tailings are exposed to, and react with, oxygen and water to form sulphuric acid and acid salts. Acid rock drainage is the most pervasive environmental problem related to mine waste management.

Source:
Graphic based on "Mining Effects on Rainfall Drainage" by Philippe Rekacewicz, UNEP/Grid -Arendal Maps and Graphics Library 2004;
Photo courtesy of Jim Wark, www.airphotona.com'

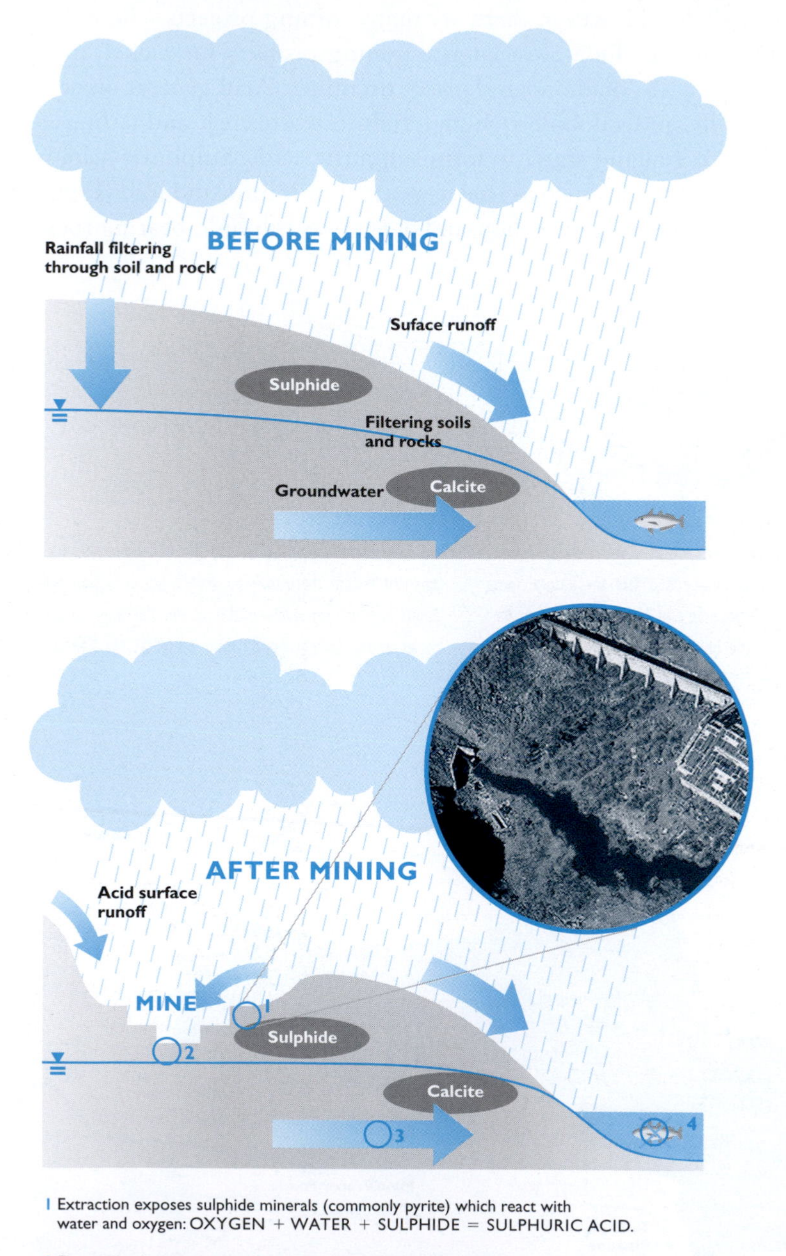

1 Extraction exposes sulphide minerals (commonly pyrite) which react with water and oxygen: OXYGEN + WATER + SULPHIDE = SULPHURIC ACID.

2 Extraction decreases groundwater depth and natural neutralization.

3 If present, calcite dissolved in water consumes acid generated by dissolution of sulphide minerals.

4 Acid water (groundwater and surface runoff) can dissolve and transport high concentrations of metals, which can harm aquatic life.

is the name for iron and aluminium compounds that stain streambeds. Copper causes a bright blue-green stain. Sometimes oxidized rock is stained black.

One of the most serious aspects of ARD is its persistence in the environment (**Case 13.7**). An acid-generating mine waste storage facility has the potential for long-term, severe impacts on surface and groundwater and aquatic life. The process will continue until there are no remaining sulphides. This can take centuries, given the large quantities of exposed rock at some mine sites. Once the process of acid generation has started, it is difficult to stop. The combination of acidity and dissolved metals may harm or even kill aquatic life, rendering streams nearly sterile and making water unfit for human consumption.

Predicting the potential for acid drainage can help determine where problems may occur. Methods vary from simple calculations involving the balance of acid-generating minerals (e.g. pyrite) against the existence of neutralizing minerals (e.g. calcium carbonate) to complex laboratory tests (e.g. kinetic testing). However, even laboratory-based tests may not always accurately predict the amount of metals that will be leached if acid drainage occurs, because of the differences in scale and composition that occur when samples are analyzed off-site (Da Rosa 1997).

CASE 13.7
The Berkeley Pit in Butte, Montana

'Our Lady of The Rockies', a 30 m statue of steel, overlooks the mining town of Butte. It is an old European tradition to place an image of the Holy Lady at the portal of underground mines to ask for heavenly protection for the men working underground.

As such the erection of the huge statue appears appropriate. For more than a century, miners, mining, and the mining industry have brought wealth and sophistication to Montana as a whole, and the town of Butte in particular. The Mining Law of 1872 still allows speculators to remove minerals from public lands for just a small fee. If not for mining, Montana would not have been settled for years after it actually was. But in those days, people did not understand the long-term impacts mining practices could have on the land.

In 1955, Butte's first large-scale open-pit mine, the Berkeley Pit, was constructed on 'the hill.' Mining ceased in 1982 after the extraction of 1.5 billion tons of ore. The Berkeley Pit reached a final depth of about 600 m.

During operation, groundwater was continually pumped from the mine so that miners could reach the ore. In 1982 the water pumps at the bottom of the pit were shut down. Acid groundwater with a pH less than 3 continues to fill the pit which has now become the Berkeley Pit Lake. The water is also toxic because large quantities of heavy metals have been dissolved in the pit lake water.

The Pit received national attention as an example of environmental damage when a flock of migrating snow geese chose to land and rest on the Berkeley Pit Lake in November of 1995. They drank the highly acidic water and close to 350 died.

Unless a way is found to safely and economically treat and remove the pit water, scientists believe that it will spill out into the Silver Bow drainage system between the years 2010 and 2020. In the meantime, Berkeley Pit Lake remains the Environmental Protection Agency's largest Superfund site in the US. That means that the US Government has deemed the Berkeley Pit the number one environmental hazard facing the USA today.

Back in 1994, after a court ruling which pressured certain mines to become a little more environmentally sensitive, the director of the Montana Environmental Information Center stated, 'The decision means you fill up a hole when you get through mining. There will be no more Berkeley Pits.'

Source: www.montankids.com; Photo Credit: www.airphotona.com

While small amounts of metals are considered essential for the survival of many organisms, large quantities are often toxic (Chapter Eleven). Few terrestrial and aquatic species are known to be naturally tolerant of heavy metals, although some have adapted over time. Many fish are highly sensitive to even mildly acidic waters and cannot breed at pH levels below 5. Some may die if the pH level is less than 6 (Ripley *et al.* 1996). In general, the number of plant and animal species decreases as the aqueous concentration of heavy metals increases. Some taxa are known to be more sensitive to the presence of heavy metals. This is particularly the case for species living in freshwater aquatic habitats. For example, salmon species are particularly sensitive to increased concentrations of copper (Kelly 1998). Furthermore, juvenile fish are more sensitive than adult fish, and the presence of heavy metals may affect the critical reproductive and growth stages of fish.

Water Balance

Mining invariably influences the local water balance and sometimes even the regional water balance. Whether these impacts are felt and become important depends on the climatic setting, the local hydrology, and the type of mine. Changes in the local water balance are the direct result of changes in landform, land cover, and from water usage by the mine. Mine dewatering also changes the water balance.

The environmental effects of groundwater rebound after shutting down pumps can be significant.

Mine closure itself can affect the water balance, a situation which may be overlooked in the environmental assessment process. The environmental effects of groundwater rebound after shutting down pumps can be significant. Furthermore, a new ecological equilibrium may have developed, in response to the changed water balance during mine operation. This is the case along the Erft River in North Germany: for more than 20 years the river received most of its water from the dewatering pumps of the large coal strip mining operations close by.

13.4 AIR QUALITY AND CLIMATE CHANGE

As with most industrial activities, mining affects air quality. The most obvious impacts are from the generation of dust and combustion gases. Considering the entire life-cycle of mine products, however, the relationship of mining and air quality is more complex than these obvious impacts.

Atmospheric and noise pollution are addressed in various parts in this text and therefore are only briefly addressed in this chapter. The emerging issue of mining's contribution to climate change is dealt with in greater detail.

Atmospheric Pollution

Mining operations can affect air quality in many ways. An air emissions inventory as illustrated in **Figure 2.7** in Chapter Two is useful in addressing atmospheric pollution issues. Material movement, stationary sources, and fugitive emissions are the three main sources of air quality degradation.

Dust generated by land transportation is often the main source of air pollution.

Land transport is one of the major sources of atmospheric emissions. Emissions from land vehicles are most often considered in terms of the exhaust from fuel combustion. However, dust generated by land transportation is often the main source of air pollution. Combustion of petrol or diesel fuel leads to the production of exhaust gas containing a

FIGURE 13.8
Particulate Matter Emitted from a Nickel Smelter (2006)

The profile of air emissions sources expands substantially if the mine operation includes downstream pyro-metallurgical mineral processing.

range of potentially harmful pollutants. Some countries still permit the use of lead additives in petrol and this generates an important air contaminant.

Stationary sources of air emissions in mining operations vary. Crushers and ore transfer points are amongst the main sources of dust emissions. So are power stations at mine sites which are typically the greatest sources of combustion emissions. The profile of air emissions sources expands substantially if the mine operation includes ore roasting, or downstream pyro-metallurgical mineral processing (**Figure 13.8**). Smelters are large-scale industrial operations, in which emissions control constitutes the main environmental challenge (Bounicore and Davis 1992, Corbitt 1993, USEPA 1996, Wirth 2000).

In principle, it is a simple matter to measure the amount of air pollutants emitted into the atmosphere from a defined source such as a stack, a crusher, or a stockpile. However, this is only relatively straightforward in the case of a stack or a vehicle exhaust; it becomes more difficult when considering diffuse sources such as wind-blown dusts. Wind-blown dust can be generated from any disturbed land area which lacks vegetation: construction sites, areas for mining, active mining areas, stockpiles, and mine waste placement areas. Tailings disposal areas, if dried, are especially prone to wind-blown dust – fine-grained tailings materials are readily eroded by wind, and may be transported over large distances.

In addressing atmospheric pollution measurements of ambient air quality can be very helpful, but they do not tell the complete story of how pollutants move and change in the air. In the case of soil pollution, for example, samples can be taken and analyzed for the presence of a pollutant. By testing enough samples at different depths and locations, a relatively accurate picture of the distribution of the pollutant emerges. The pollutant generally stays in place, making it easier to find.

In addressing atmospheric pollution measurements of ambient air quality can be very helpful, but they do not tell the complete story of how pollutants move and change in the air.

Defining atmospheric pollution is more difficult. There are additional uncertainties about how a pollutant is diluted, changed, or moved as it disperses into the air. Additional uncertainties relate to the weather – wind, sun, and temperature. Also, as the air sample is taken over a specific time period, the concentration measured is the average concentration over that interval. In sum the concentration of any air pollutant may vary widely, both temporally and spatially in three dimensions.

Air dispersion modelling is the only practical way to factor in all of these influences to predict atmospheric pollution. The model cannot provide absolute values, but it will provide a good statistical prediction about the pollutant and its movement.

Noise

The three major categories of noise sources associated with mining are (1) blasting; (2) fixed equipment or process operations such as crushers, mills or ventilation fans; and (3) mobile equipment used to load and haul ore, mine wastes, and mine products. In particular the warning sounds made by mine vehicles when reversing are, by design, quite intrusive and may disturb nearby communities.

Fixed mine installations may include a wide range of equipment including crushers, grinders, screens, generators, pumps, compressors, conveyers, storage bins, and electrical equipment. Mobile noise sources may include drilling equipment, excavation equipment, haulage trucks, pug mills, mobile treatment units, and service vehicles. Any or all of these activities may be in operation at any one time. Single or multiple effects of sound generation from these operations constitute potential sources of noise pollution.

Mitigating measures are commonly applied to reduce the effects of noise or the noise levels experienced by receptors. Reduction of sound levels at the point of generation is preferred to diminishing the effects of the noise at the point of reception. Alternative construction or operational methods, improved equipment maintenance, selection of alternative equipment, physical barriers, siting of activities, set backs, and established hours of construction or operation, are common measures that are successfully used to mitigate adverse noise effects.

Spontaneous Combustion

Spontaneous combustion is most commonly associated with coal deposits, but can occur at any deposits where chemical reactions of natural elements (such as sulphur) take place and generate heat. The chemical reactions can be aerobic (that is oxidation), or anaerobic. For example, coal seams once ignited can continue to 'burn' underground for many years, despite the absence of air.

Combustion within coal seams (underground or at the surface), in piles of stored coal, or in spoil dumps at the surface are the most common types of spontaneous combustion. Coal fires can emerge from spontaneous self-combustion or as a result of extraneous causes. Spontaneous combustion is commonly a consequence of oxidation processes in coal piles or heaps. These exothermic processes result in increasing temperatures within the rock pile which finally lead to the spontaneous ignition of the coal. Other sources of coal fires are related to careless mining operations, e.g. underground blasting, welding or grinding or natural events such as lightning strikes. In developing countries traditional 'slash and burn' practices to clear land for agricultural are common causes of uncontrolled coal and peat fires. Nevertheless, although it is illegal, this practice is commonly used to clear land for plantation development (**Case 13.8**). Uncontrolled coal and peat fires are a major contributor to air pollution, locally and globally. Of course they also contribute significantly to greenhouse gas emissions. By some estimates the uncontrolled forest, peat, and coal fires in Indonesia in 1997 contributed nearly one-third of the global CO_2 emissions in that year. Less obviously, the burning of peat increases the subsequent generation of methane from swamps.

Spontaneous combustion is commonly a consequence of oxidation processes in coal piles or heaps.

The potential for spontaneous combustion of coal depends on its aptitude for oxidization at ambient temperature. Oxidation is the absorption of oxygen at the coal surface; it is an exothermic reaction generating heat as a by-product. If the temperature reaches somewhere between 80°C and 120°C (called the 'threshold' temperature) a steady reaction commences. The temperature of the coal will almost certainly continue to increase until, somewhere between 230°C and 280°C, the reaction becomes rapid and strongly exothermic – the coal reaches 'ignition' or a 'flash' point and starts to burn.

Not all types of coal are equally susceptible to spontaneous combustion. High-ranking coals (that is, having high carbon content) are more fire prone than lower-ranking coals. Another important factor is the size of the particles; the larger the effective area of the coal (fine particles), the more rapidly the reaction can proceed. Also, external factors play a role in the oxidation reaction. Oxidation requires an adequate supply of air; cracks, fissures, and the porosity of rock and soil over the coal seams may encourage underground coal fires by allowing oxygen to reach the coal. The moisture content is also important: coal with high moisture content tends to exhibit a low oxidation rate. However, the presence of some moisture favours continued oxidation.

Micro Climate

As with urbanization, mining may change the local climate in many ways. The microclimates of urban areas differ substantially from the microclimates of surrounding natural areas due primarily to the removal of vegetation, enabling the Earth's surface to become much warmer, with the release of artificially created energy into the environment (the urban heat island effect).

At a mine site, exposed rock, roads, and buildings often replace natural vegetation, and the vertical surfaces of mine pits and buildings are added to the natural landscape. Exposed

CASE 13.8
Uncontrolled Forest, Peat, and Coal Fires in Indonesia

The combination of forest destruction, land clearance, and an exceptionally severe El Nino climatic event in 1997 led to the most severe forest and peat fires ever known in Indonesia. Between half a million and three million hectares of vegetation burned, much of it on peat and coal. The fires penetrated up to 1.5 m into the dried-out peat. It also ignited coal seams along their outcrops.

At least one billion tonnes of carbon were released into the atmosphere – more than that released by the fossil fuels burnt by the European Union in a year. It undid an estimated ten years of carbon fixation by all of the world's pristine peat bogs. By some estimates this sudden release of carbon may have added about 0.5 parts per million CO_2 to the atmosphere, a significant addition to the global greenhouse gas concentration.

Once started peat and coal fires are difficult to extinguish. They continue to burn underground during the tropical wet season only to start new forest fires in the following dry season.

land surfaces generally have greater heat conduction, and more heat storage than the natural land covers they replaced. Heat energy is also added through the operation of mining equipment and power generation. Finally, evaporation and transpiration from various natural surfaces act to cool the land surface and local atmosphere. At mine sites, drainage systems rapidly remove surface water. Thus, little water is available for cooling.

Mine areas tend to be warmer than the surrounding areas.

All factors combined, mine areas tend to be warmer than the surrounding areas. As in urban areas, there is then at least theoretically the potential for locally increased rainfall due to the combined effect of particulate air pollution and increased convectional uplift due to increased surface temperature. Particulate air pollution may enhance rainfall by increasing the number of condensation nuclei. The incremental temperature increase may add to convection currents over the mine area.

Global Climate Change

Mining also contributes to greenhouse gas emissions that are believed to contribute to global climate change. It is now widely believed that climate change has caused negative impacts on the global environment and there are many differing predictions about future impacts. Mining operations typically move, break, and grind large quantities of rock, all energy-intensive activities. Energy generation at a mine site is almost always based on the consumption of fossil fuel, releasing carbon dioxide (CO_2), a prominent greenhouse gas (GHG) to the atmosphere. Subsequent mineral processing, especially in the steel industry, adds to CO_2 emissions. Equally the use of coal as an energy source continues to be a major contributor to global CO_2 emissions.

On the other hand, the mining industry is not always acknowledged for its contribution to the reduction of vehicular emissions. Mining produces the light weight metals (aluminium and magnesium) that have enabled manufacturers to produce vehicles which are far more fuel efficient than their predecessors. Similarly, mining produces metals such as platinum and palladium which are used in catalytic converters that also improve combustion efficiency as well as reducing potentially toxic emissions. Furthermore, mining produces the materials used in the manufacture of computer chips and electronic fuel injection systems which have also improved vehicle fuel efficiency. The net effect of these technical innovations, all of which depend on products obtained from mining, is that fuel consumption in the United States has remained essentially the same over the past decade despite the increased numbers of vehicles and a large increase in the numbers of large SUVs. This also holds true for Europe and other parts of the world.

The amounts of greenhouse gases produced per tonne of concentrate shipped usually increase over the life of the mine.

Despite the best attempts of the mining industry to reduce greenhouse gas emissions (such as minimizing waiting times of mine equipment, reducing rolling resistance, and optimizing fuel efficiency), the amounts of greenhouse gases produced per tonne of concentrate shipped usually increase over the life of the mine, due to increased haulage distance, increased vertical transportation distance as the depth of mine pit increases, and sometimes also increasing stripping ratios as the mine approaches its economic mine life.

Numerous challenges are involved in addressing issues related to mining and climate change. They include:

- Developing a Greenhouse Gas Policy;
- Establishing the carbon footprint;
- Evaluating the GHG risk profile (**Case 13.9**);
- Reporting emissions; and
- Implementing measures to reduce or offset GHG emissions.

Coal mining and coal use is of course a special case. Combined coal mining and use contribute to climate change much more than metal or industrial mining. There are many reasons for this; some follow.

13.5 COAL – A SPECIAL CASE

The coal cycle differs from the general mineral cycle in that, following mining, coal is burnt or converted into other fuels (**Figure 13.9**). Coal combustion (as occurs, for example, at coal-fired power plants at many mine sites) has direct environmental impacts, with national and even international implications. Coal combustion and its environmental impacts is a major topic on its own. Thus, the purpose of this section is not so much to provide a thorough analysis of environmental impacts associated with coal use, as it is to outline the main issues relating to air pollution and the generation of greenhouse gases.

Emissions from Coal Combustion

The most significant polluting residuals from coal are the emissions from coal combustion. Flue gases contain sulphur dioxides (SO_x), nitrogen oxides (NO_x), carbon monoxide (CO), carbon dioxide (CO_2), particulates, organic compounds, and some trace metals, including mercury (OECD 1983 a,b). Each of these has been implicated in some form of undesirable environmental impact: damage to human health, corrosion of materials, damage to vegetation, and in the case of CO_2, contribution to global warming.

Acid deposition, a direct impact of SO_2 emissions, is recognized as a serious international problem in the northern hemisphere. Generally coal has a sulphur content of 0.5 to 2.5% but can be found with contents up to 5%. During combustion, in the absence of emission controls, up to 95% of the sulphur present in the coal is oxidized to SO_2 and emitted to the atmosphere. Besides petroleum use, the major portion of anthropogenic SO_2 emissions is derived from coal combustion. These problems can be minimized by using low sulphur coal and/or by installing flue gas desulphurization (FGD) as a control technology. Various FGD processes exist, generating different waste products such as calcium slurry, gypsum (a saleable by-product if the economics are favourable), elemental sulphur, or sulphuric acid. However, the simplest way to reduce sulphur emissions to the environment is through combustion of low sulphur coal. As a consequence, low sulphur coal commands

The simplest way to reduce sulphur emissions to the environment is through combustion of low sulphur coal.

CASE 13.9
Citigroup Global Mining Report 2007

While some mining industry representatives continue to debate anthropogenic contributions to climate change, a Citigroup study published in 2007 identified significant potential physical climate change impacts on mining companies including more frequent or more severe weather events, reduced supplies of fresh water, and thawing permafrost impacts on Arctic infrastructure. Of the 12 major mining companies reviewed in the study, Citigroup's research found that BHP Billiton, Rio Tinto, and Norilsk 'probably have the most significant exposures to manage.' The analysts noted in their global mining study that the potential impacts of climate change may involve 'extended operational shutdowns or removal of the license to operate due to environmental "disasters". Some changes will require adaptation and mitigation by operators.' Other potential impacts include sea ice reduction, which could open up new Arctic shipping routes, and 'food shortages and increased tropical disease in Africa', causing social and political instability.

Source: Citigroup 2008: Towards Sustainable Mining

FIGURE 13.9
The Coal Cycle – From Mining to Combustion

The coal cycle differs from the general mineral cycle in that, following mining, coal is burnt or converted into other fuels.

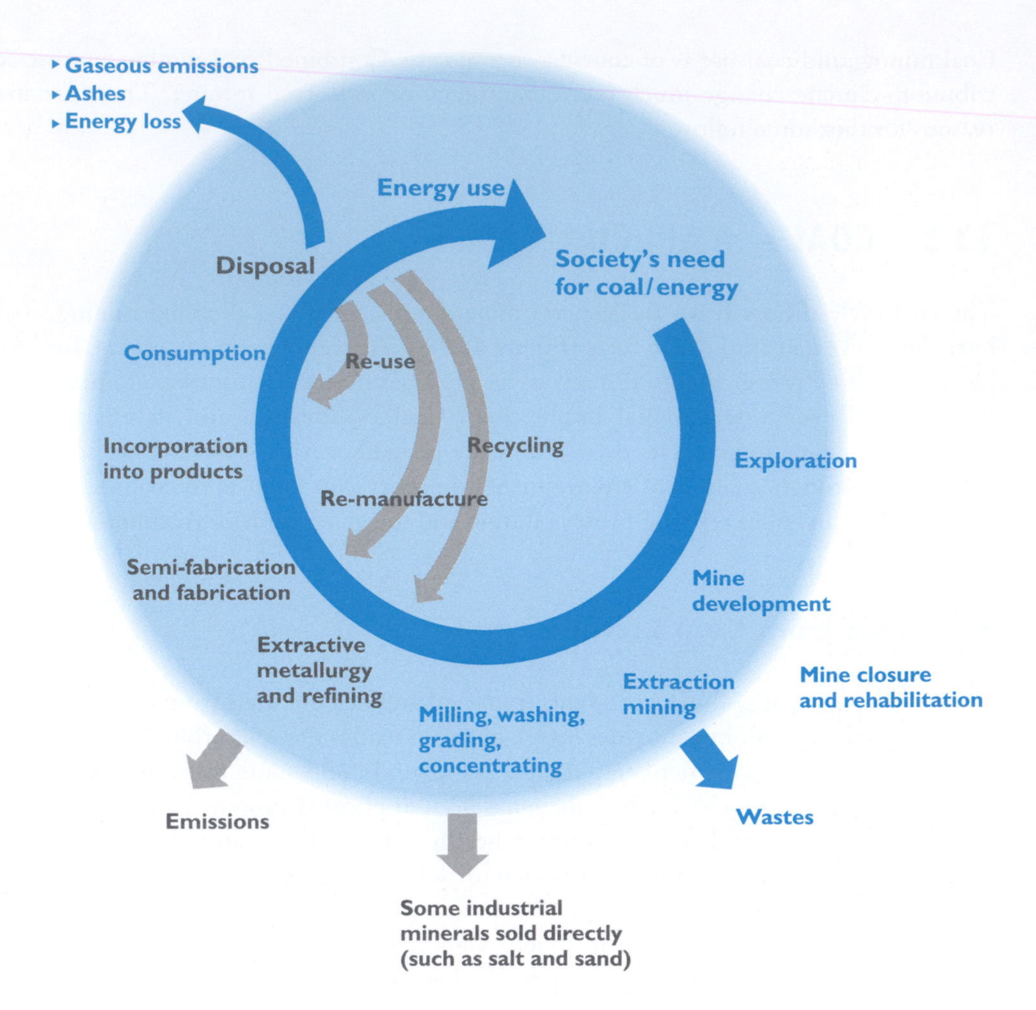

higher prices. Some countries such as Indonesia therefore encourage the export of low sulphur coal while using higher sulphur coal for domestic consumption. In the case of oil this policy leads to the interesting observation that Indonesia produces low sulphur oil for export, while it imports oil with a high sulphur content for domestic consumption.

NO_x emission control is less subject to regulatory standards; in many countries there is no NO_x emission standard for coal-fired power plants. The common method of controlling NO_x emissions is through combustion modification, and generally proves very effective in reducing emissions. The method relies on the delayed mixing of coal particles in the combustion air. The use of flue gas denitrification units involves high operating costs and is confined to a few oil-fired power plants. There is insufficient knowledge about the health effects and damage to vegetation of long-term exposure to NO_x. There is evidence that NO_x is several orders of magnitude less toxic than SO_2 (OECD 1983c).

In the mining sector, the climate change debate is very much concerned with the role of coal in the industry.

The release of carbon dioxide is of increasing concern considering that CO_2 plays a major role in the heat balance of the Earth atmosphere (**Case 13.10**). A global increase of CO_2 in the atmosphere causes an increase of the Earth's surface temperature. While much mining industry rhetoric on climate change relates to improving energy efficiency as one way of tackling greenhouse gas emissions, these future efforts promise relatively little in terms of the overall reductions required to meet GHG targets. In the mining sector, the climate change debate is very much concerned with the role of coal in the industry. And this is not only because coal mining accounts for around two-thirds of total mining output worldwide (Brendow 2006). UNFCCC estimated global CO_2 emissions in 2004 as about 16,797 million tonnes; global mining and quarrying contributed around 79 million tonnes (excluding coal) (IEA 2006) – that is less than 1%. However the picture changes dramatically if coal combustion is taken into account: based on a global coal production of about 2,620 million tonnes in 2004, the coal consumption added about 10,592 million tonnes of CO_2, or 63% of the total CO_2 emissions.

Excluding coal from the energy mix appears, at least in the foreseeable future, to be unrealistic. On the basis of energy content, coal combustion releases about 25% more CO_2 than oil and 75% more CO_2 than natural gas. However, carbon dioxide in the atmosphere is less of an issue for coal *per se* than for energy growth, and energy mix. Given the ever-increasing demand for energy, coal consumption is certain to continue growing. By some estimates in the next 20 years global coal use may grow by 60% (IEA 2002, 2006).

On the basis of energy content, coal combustion releases about 25% more CO_2 than oil and 75% more CO_2 than natural gas.

Carbon-capture storage (CCS) and 'zero carbon-emission' plants are some of the technical answers to reducing CO_2 caused by coal consumption. In the CCS process, CO_2 is captured and separated from other gaseous emissions and then injected into a suitable host rock formation. It was pioneered by the oil and gas industries to enhance oil recovery from existing reservoirs. Indeed, CCS is widely advocated as a saviour for the industry; yet the technology is still under development, its economics are uncertain and it does not completely eliminate carbon waste.

Several CCS projects are underway in the mining industry. Alcoa is operating a pilot CCS project at Kwinana in Western Australia which, as well as using underground storage, mixes CO_2 waste with bauxite residues from the refinery process. This reduces the high level of alkalinity in the bauxite residues, a beneficial side benefit to the aluminium industry. Alcoa's first carbon capture plant at Kwinana is locking-up 70,000 tonnes of CO_2 a year. The CO_2 is produced by a nearby ammonia plant and would otherwise be emitted to the atmosphere (AAC 2006).

Rio Tinto and BP are in the feasibility stage of a coal-fired power-generation project which aims to capture 4 million tonnes per year of CO_2 from 2014, also at Kwinana. Anglo American and Shell Energy Investments Australia are in the pre-feasibility stage of a project which is designed to produce ultra-clean synthetic diesel from brown coal in Victoria, Australia. Up to 50 million tonnes per year of CO_2 waste will then be injected into the ground. Completion is expected in 2016 (**Case 13.11**).

Control of particulate matters (PM) emissions is well-established. The use of electrostatic precipitators (ESP) with a removal efficiency of up to 99.5% is common practice. The major concern is that the removal efficiency of ESP for very small particles

CASE 13.10
Anvil Coal Project in NSW, Australia

The Anvil Hill Coal Project proposes an open-cut mine near Wybong in the upper Hunter Valley, to extract up to 150 million tonnes of coal over a 21-year period for both domestic electricity generation and export. The environmental assessment was prepared and exhibited during 2006. An environmental opponent successfully challenged a step in the assessment process, satisfying the court that the environmental assessment was defective because it did not contain a detailed assessment of scope 3 emissions, namely an assessment of the greenhouse gas (GHG) emissions produced by combustion of the coal. The court, however, allowed the environmental assessment process to continue, because additional information about downstream GHG emissions was in fact provided by the proponent during the assessment process. The supplemented environmental assessment and more than 2000 submissions made during the exhibition period were considered by the Director-General of the NSW Department of Planning and also by an independent panel of experts, who reported to the Minister who subsequently approved the project, subject to conditions, on 7 June 2007.

This decision sets a significant new standard for future mining developments in NSW. It requires that the global warming impacts of proposed projects be considered as part of the planning process. Indeed, its implications may extend beyond the mining sector in NSW and require all new coal mine developments to undertake a global warming impact study. What is unclear, however, is the depth of analysis of global warming impacts that might be required, and what steps to address those impacts might be deemed acceptable.

Source: Allens Arthur Robinson (2006, 2007)

(1 to 2 microns) is lower than for larger particles. Concern now focuses on the health effects of these smaller particles, since they may be enriched with more toxic substances, travel over long distances, and penetrate deeper into the human respiratory system. Also the ESP process becomes less efficient when low sulphur coal is combusted. A second control mechanism is the use of bag filters which are particularly efficient in collecting smaller particles. However, bag filters are more difficult to operate, and more costly to maintain than ESP. Removed particulates, either by ESP or bag filters, are termed fly ash. By its very nature, fly ash comprises fine particles.

Emissions of organic compounds as polycyclic aromatic hydrocarbons (PAHs), which have carcinogenic properties, have been significantly reduced by the increased use of large coal combustors with highly efficient flue gas treatment controls.

Trace metals are preferentially concentrated onto smaller ash particulates, the fraction of particulate emissions, which are the most difficult to arrest by control technology.

Another environmental concern with coal combustion is the presence of trace metals in coal. Trace metals are preferentially concentrated onto smaller ash particulates, the fraction of particulate emissions, which are the most difficult to arrest by control technology. The quantities of trace metals emitted from power plants into the atmosphere depend on the metal concentrations in the coal, boiler type, and installed flue gas emission control systems. Of particular interest are emissions of mercury from coal combustion. Coal combustion has contributed a large portion of the mercury present in the atmosphere today. While the natural component of the total atmospheric burden is difficult to estimate, a recent study (Munthe *et al.* 2001) suggests that anthropogenic activities have increased the overall levels of mercury in the atmosphere by roughly a factor of three, some 70% of it contributed by emissions from stationary combustion of fossil fuels, especially coal, and the incineration of waste materials (UNEP 2002). As combustion of fossil fuels is increasing in order to meet the growing energy demands of both developing and developed nations, mercury emissions can be expected to increase. Today's flue gas control technologies may provide some level of mercury control, but when viewed at the global level, these controls result in only a small reduction of total mercury emissions.

CASE 13.11
Monash Energy Coal-to-Liquids and Carbon Capture and Storage Project, Australia (Anglo Coal)

Anglo Coal's subsidiary Monash Energy has plans to convert brown coal mined in Australia's Latrobe Valley to ultra-clean diesel fuel. The process achieves low emissions to the atmosphere by separating a concentrated stream of carbon dioxide for transport by pipeline to injection wells for injection and secure storage in deep underground geological formations. The Monash Energy vision involves: a brown coal mine; a coal drying and gasification plant; a synthesis gas-based, hydrogen production plant; a fuel synthesis plant for converting gas to liquids; and an integrated waste heat and off-gas power plant. The plant is planned to produce more than 60,000 barrels per day of liquids – mainly ultra-clean diesel fuel. The project, currently in its pre-feasibility stage, is also acting as a catalyst for the development of local carbon capture and storage infrastructure. Monash Energy is investigating the technical and commercial potential of the adjacent offshore section of the Gippsland Basin as a site for the storage of carbon dioxide. The project includes a community engagement component to ensure the public has the necessary confidence in the technologies, and appreciates the conceptual link between a carbon-constrained future and the requirement for implementation of carbon capture and storage.

Source: OECD/IEA 2006

The combustion of coal leaves a solid, unburnable residue called bottom ash, the nature and amount of which depends on the composition of the coal and the degree of preparation. To this is added the fly ash removed from flue gases by ESP, bag filters, and scrubbers. The proportion of fly ash to bottom ash is about 80 to 20% in pulverized coal boilers. Pulverized ash is of commercial value and a significant proportion of it is used in cement making, or other construction related applications. Unused ash is disposed of in landfills. There is some concern about the health effects of residual trace metals in fly and bottom ash. This has been the subject of some controversy, and there is a marked divergence of opinion on the environmental and health risks from hazardous components of ash. In some jurisdictions such as Indonesia, waste management regulations impose regulatory challenges to the use of fly and bottom ash, and impose stringent requirements on the design of disposal sites for solid combustion wastes such as ash or scrubber wastes.

Coal Mining and the Release of Methane

In many cases, coal mining also results in emissions of methane (CH_4), a potent greenhouse gas. There are many sources of methane, usually involving the degradation of organic matter to a more thermodynamically stable form, namely methane: biological processes as a result of microbial action, as in biomass or landfills, and/or from thermal processes as a result of pressure as in case of petroleum and coal. Methane trapped in coal deposits and in the surrounding strata is released during normal mining operations in both underground and surface mines, and in handling of the coal after mining, as well as during post-mining. Methane remains in the atmosphere for approximately 9 to 15 years. Methane is more than 20 times more effective at trapping heat in the atmosphere than carbon dioxide over a 100-year period. Methane is of course also a primary constituent of natural gas and, as such, represents an important energy source.

The geological formation of coal, commonly called coalification, which involves an increase in carbon content and density as it changes from peat via lignite (65 to 72% mass C) to hard coal such as bituminous coal (76 to 90% mass C) and anthracite (93% mass C), results in methane formation together with carbon dioxide and nitrogen. The three basic stages in the formation of coal are: (1) formation of peat in a swamp; (2) production of a gel-like material, gytta, by the partial decay of the peat by bacteria and fungi (bacterial decay); and (3) thermal alteration (bituminization) of the gel to the various ranks of coal after burial under hundreds or thousands of metres of sediment. Both decay and thermal alteration increase the percent of carbon present and reduce the amount of water and other volatile gases (e.g. carbon dioxide or methane). These changes increase the heat content of the coal and hence its rank.

In summary the process of coalification converts the plant material – which consists essentially of compounds of carbon, hydrogen and oxygen – to coal which in its purest form, consists essentially of carbon. The different grades or types of coal vary between relatively unchanged plant material and pure carbon.

There are different ways in which the various types of coal can be classified (**Figure 13.10**). Some classifications rely on the use of the coal – for example coking coal or steam coal. One common classification system divides coals into various rankings based on a range of properties. The order of rank of the coals from lowest energy-value to highest energy-value is:

- Peat
- Lignite
- Sub-bituminous
- Bituminous
- Anthracite.

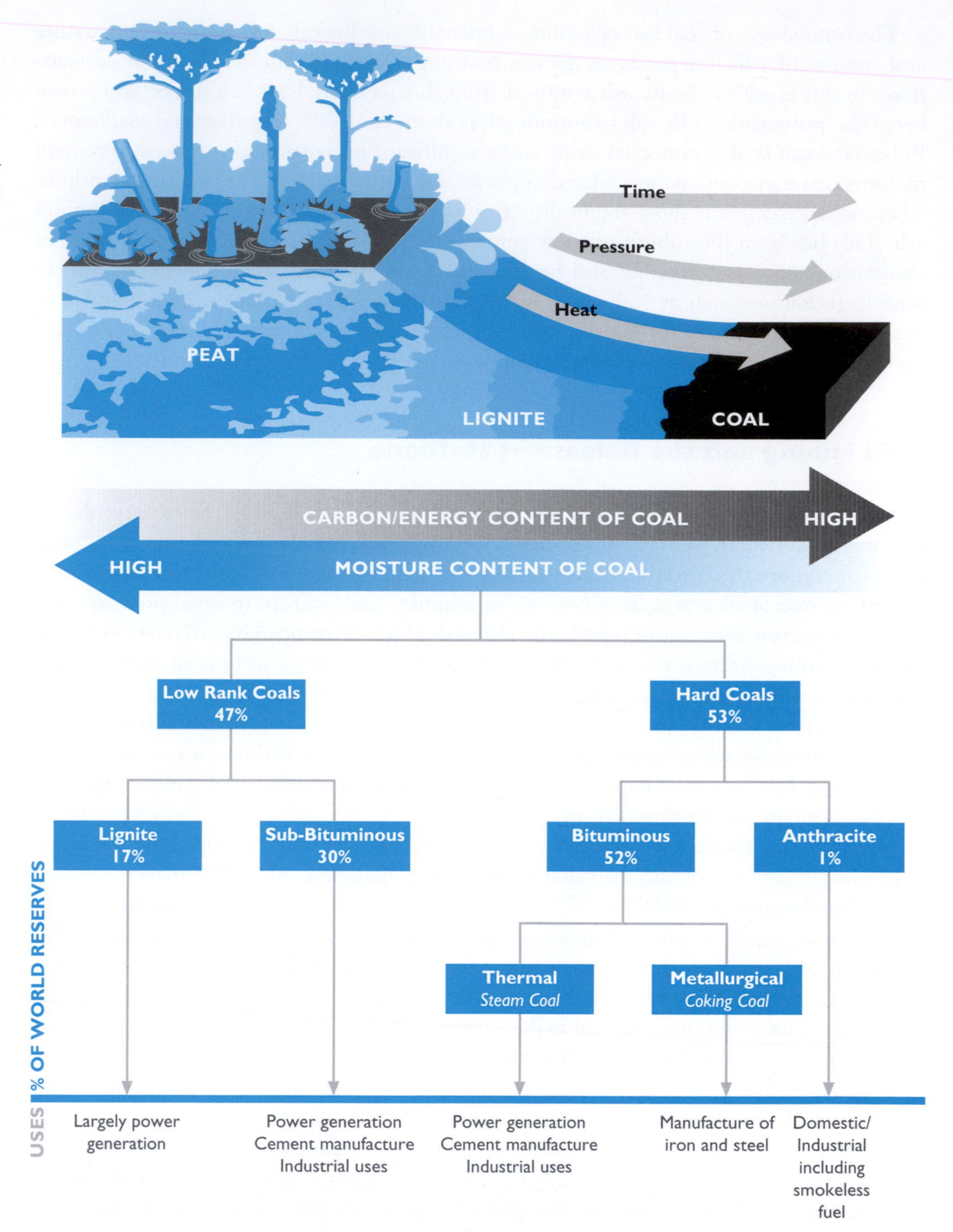

FIGURE 13.10

Coalification, the Biochemical and Geochemical Processes that Transform Organic Materials into a Combustible-Carbonaceous Solid

A one metre thick coal seam may have started out as a layer of plant material about 120 metres thick and taken over 200 million years to form.

Source:
World Coal Institute 2005

Coal bed methane (CBM) is generated microbially and/or thermogenically during coal formation. The amount of methane commonly varies between 100 and 300 cm^3/g of coal (Rice 1993). Methane may also be stored under pressure in the rock above a coal bed. Methane is adsorbed in the coal micro-pores as a function of reservoir pressure. Pressure reductions liberate these gases. Mining can liberate the gas from its pressurized environment by accessing and excavation of coal as well as subsequent fracturing of the overlying strata.

While the amount of methane gas generated is enormous, the actual quantities found today are a function of Earth movement and erosion over geological times. Coal beds and rocks above drainage store relatively little methane because coal exposed at an outcrop loses the methane during the erosional process. Thus, much methane is associated with deeper coal seams. However, because coal has such a large internal surface area (the micropores in coal provide a large surface area, approximately 200 to 300 square metres per gram of coal; OSM 2001), it can store surprisingly large volumes of methane-rich gas; six or seven times as much gas as contained in a typical natural gas reservoir of equal volume.

Because coal has such a large internal surface area, it can store surprisingly large volumes of methane-rich gas; six or seven times as much gas as contained in a typical natural gas reservoir of equal volume.

Actual methane gas quantities are, as might be expected, a function of depth, pressure, water/moisture content and the extent of coalification. Depth is important because it affects the pressure and temperature of the coal seam, which in turn determines how much methane is generated during coal formation (Williams and Mitchell 1994). If two coal seams have the same rank the deeper seam will hold larger amounts of methane because pressure increases with depth, all other things being equal. Water content is important, as when the coal and methane conversion process occurs and coal is saturated with water, methane is more easily trapped within the coal.

The translation of methane content in coal into methane gas emissions during mining is a complex matter, and so is the calculation of methane gas emissions from coal mining. The rate of methane gas release depends on the quantity and the way coal is mined, the working depths, mine drainage, the types of ventilation in case of underground coal mining, and many other factors such as density of coal or the general geology of the mining area. The OECD Expert Group recommends global average emission factors for underground coal mining between 10 and 25 M^3 CH_4/tonne (OECD 1990). The simplest estimate of annual coal mining methane (CCM) emission is then calculated as follows:

$$\text{CCM Emissions (tonnes)} = \text{Emission Factor} \times \text{Coal Production (tonnes/annum)} \times \text{Conversion Factor}$$

The conversion factor converts the volume of CH_4 to a weight measure based on the density of methane.

Little data exists on which to base emission factors from surface mining. Average emission factors ranging from 0.3 to 2 M^3 CH_4/ton are found in the literature (USEPA 1999), but these numbers are more uncertain then the corresponding factors for underground mining. However, it is clear that underground coal mines are the single largest source of coal mine methane (CMM) emissions in most countries. Emission factors do not account for post-mining methane emissions, commonly overlooked in many past studies.

In total, CMM accounts for an estimated 8% of total methane emissions resulting from human activities. In 2000, worldwide CMM emissions totalled 120 million tonnes of carbon equivalent (MMTCE), or about 30.8 billion cubic metres (BCM). By 2020, the world's coal mines are expected to produce annual emissions of 153 MMTCE (39.3 BCM) (methanetomarkets 2006). China and the United States, the world two largest producers of hard coal, are also the leading emitters of CMM, contributing about 35% (China) and 25% (USA) to global CCM emissions. Both countries also have made no commitment or obligation to reduce GHG emission under the Kyoto protocol.

CMM accounts for an estimated 8% of total methane emissions resulting from human activities.

Coal Bed Methane Capture

Coal mining does not always cause uncontrolled release of methane into the global atmosphere. Coal bed methane, previously a wasted energy resource, is being increasingly

FIGURE 13.11
The Fire Triangle

Methane is highly explosive in air concentrations between 5 and 15 mol%

Source:
OSM 2001

captured in many countries and used for heating and power generation. For example, captured coal bed methane (CBM) accounts for about 7.5% of the total natural gas production in the USA. Techniques for removing methane from deep mine workings have been developed primarily for safety reasons, because it is highly explosive in air in concentrations between 5 and 15 mol% (**Figure 13.11**). However, the same techniques are increasingly being used to capture methane gas for use as fuel, thereby reducing CMM emissions by up to 90% (Bennet *et al.* 1995).

Pre-mining degasification, often termed pre-draining, recovers methane from virgin seams, before coal is mined. Increasingly, degasification is becoming economic without subsequent coal mining. Coal bed methane is the same compound as occurs in natural gas, but is derived from a different geological situation. The gas has a wide variety of energy-related uses, and with the current energy crisis and relatively high fuel prices, increased attention is being focused on development of this resource.

Degasification is a relatively low technology enterprise, requiring limited capital expenditure. However extraction of coal bed methane is not without controversy. Increased production of coal bed methane carries with it some technological and environmental difficulties and costs. In a conventional oil or gas reservoir, for example, gas floats on oil, which, in turn, floats on water. An oil or gas well draws only from the petroleum that is extracted without producing a large volume of water. But water also permeates coal beds, and its pressure traps methane within the coal. To produce methane from coal beds, water must be drawn off first, thereby lowering the pressure so that methane can flow out of the coal and to the well or bore (**Figure 13.12**). The quantity quality, and disposal of this water are a source of much debate.

To produce methane from coal beds, water must be drawn off first, thereby lowering the pressure so that methane can flow out of the coal and to the well or bore.

CBM-produced water is pumped in large volumes from the coal seams to the surface to release the gas trapped in the coal seams. It can be discharged to the land surface, to surface water, stored in evaporation ponds, used for stock or wildlife watering, re-infiltrated, injected back into the aquifer, or treated for various uses. Without treatment, its discharge may have significant impacts on water quality. CBM-produced water can contain concentrations of sodium total dissolved solids (TDS), total suspended solids (TSS), fluoride, chloride, ammonia, and metals higher than those of the receiving waters. The sodium adsorption ratio (SAR) is the ratio of the sodium ion concentration to the combined concentrations of calcium and magnesium ions in water. Water with high SARs can cause soils to become dispersed, less permeable (resulting in reduced plant growth) and more prone to erosion; such waters may not be appropriate for irrigation. High levels of soil salinity, resulting from irrigation with some CBM-produced water can reduce crop yields. The cumulative effects of CBM-produced water on agricultural

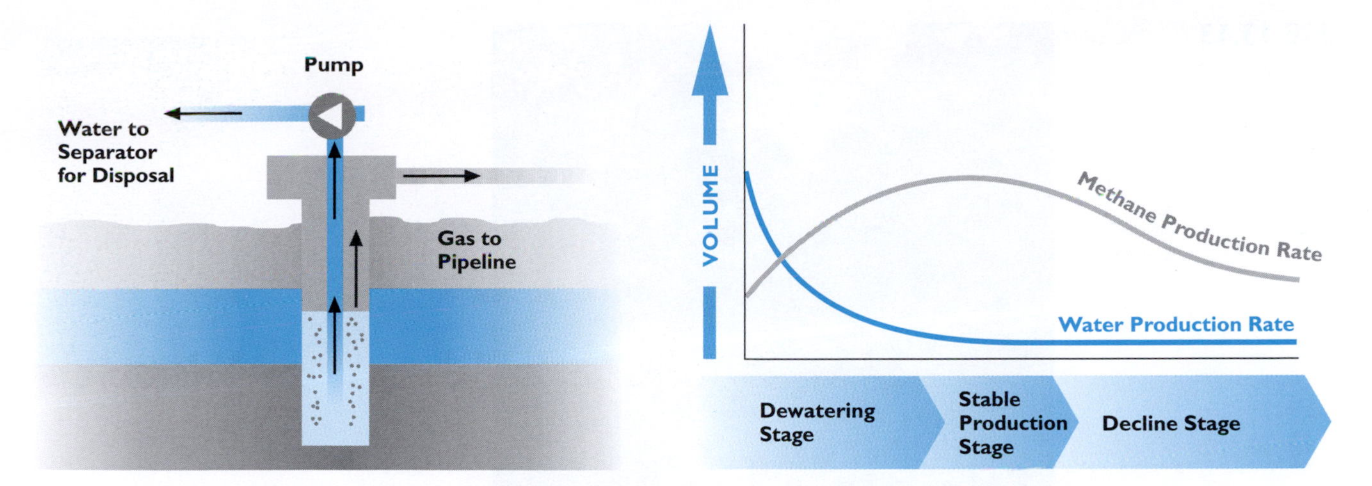

FIGURE 13.12
Relationship of Dewatering and Methane Release Over Time

CBM-produced water can contain concentrations of sodium, total dissolved solids, total suspended solids, fluoride, chloride, ammonia and metals higher than those of the receiving waters.

soils and crop yields, and the factors that influence those effects are not completely understood. Hydrologic changes resulting from CBM operations may adversely affect fisheries; the nature and extent of those effects, however, are largely unknown. Coal seam dewatering can also affect the shallow groundwater table, impacting shallow bores, springs and domestic wells.

Roads and equipment used to recover CBM may fragment the habitat. A large number of wells have to be drilled, and access to these wells needs to be provided. Vehicles operating on area roads, compressors, drilling and water-pumping equipment generate noise that can be constant and at sound levels higher than normal. The most significant sources are large compressor sites containing two or more compressors, fuelled by large diesel or natural gas engines. Noise can negatively affect wildlife and impair the quality of life for humans. Negative visual impacts can result from areas cleared for development or eroded by CBM-produced water discharges and by CBM equipment, buildings, roads and power lines.

On the regulatory side the main concern is that relatively few CBM-specific regulations have been written. Typically, CBM operators must comply with regulations applicable to other oil and gas drilling and production operations, but other regulations including mining regulations may also become relevant, potentially being in conflict with each other.

13.6 BIODIVERSITY AND HABITATS

The most obvious impact to biodiversity from mining is the removal of vegetation, which in turn alters the availability of food and shelter for wildlife. Mine waste placement areas conventionally occupy large areas, and yet almost all the world's supplies of metallic minerals to date have come from only about 4,000 square kilometres of the Earth's surface (Pearl 1973). This accounts for less than 0.003% of the total land area. This number may have changed since then but it will be small nevertheless.

Mine waste placement areas conventionally occupy large areas, and yet almost all the world's supplies of metallic minerals to date have come from only about 4,000 square kilometres of the Earth's surface.

In a wider sense mining may impact biodiversity by changing species composition and structure and, in some cases, by providing access to previously isolated wilderness areas enabling exploitation of biological resources in areas where this was previously impossible (**Figure 13.13**). Acid drainage and high metal concentrations in rivers, for example, generally result in an impoverished aquatic environment. Some species of algae and invertebrates are more tolerant of high metals and acid exposure and may, in fact, thrive in less

FIGURE 13.13
Providing Access to Hard-to-reach Areas

By providing access to previously isolated wilderness areas the construction of mining roads may enable exploitation of biological resources in areas where this was previously impossible.

competitive environments (Kelly 1998). Imported species (e.g. weedy plants and insect pests) may thrive while native species decline (Ripley *et al.* 1996). Access to wilderness areas may also result in increased hunting, logging and land development. The high rate of clearing of rain forest in the Amazon region as well as parts of Southeast Asia is a prime example of the negative consequences of such induced development.

Opponents of mining cite loss of biodiversity as one of the main reasons for their opposition. This section addresses the questions:

- Does mining pose a risk to biodiversity? and
- If so, how should the risk be minimized?

Loss of biodiversity is generally considered to be deleterious to the functioning of natural ecosystems, with many adverse ramifications to the Earth's inhabitants, including humans. The International Council on Mining and Metals (ICMM) in 2003, adopted as one of its principles '(to) *contribute to the conservation of biodiversity and integrated approaches to land use planning*'. The ICMM has published a definitive report 'Good Practice Guidance for Mining and Biodiversity' which is available through its website www.ICMM.com. The report provides a detailed account of means for evaluating impacts on biodiversity, and measures for minimizing such impacts, including the use of offsets.

Effects of Mining – Separating Fact from Fantasy

Mining has the potential to reduce biodiversity either directly, through activities which remove, damage or modify habitats, or indirectly where changes caused by project activities may create situations where species or their habitats can be damaged by activities other than those directly associated with mining (**Table 13.1**). Examples of such indirect impacts

TABLE 13.1

Sensitivity of selected ecosystems to mining, oil and gas development may pose risks to some environments due to the sensitivity and/or rarity of these ecosystems

Forests

Forests are the most biologically diverse terrestrial ecosystems. Tropical forests are particularly diverse and provide the greatest source of endemic plant species in the world. The key direct impact of mining on forest ecosystems is the removal of vegetation and canopy cover. Indirect impacts include road-building and pipeline development, which may result in habitat fragmentation and increased access to remote areas. While larger intact forest ecosystems may withstand the impacts of mining and oil development, smaller fragments are likely to be particularly sensitive to clearing.

Wetlands and Mangroves

Wetlands (including estuaries, mangroves and floodplains) act as natural pollution filters, as well as providing unique habitat for aquatic species. Mangroves act as an important interface between terrestrial and marine ecosystems, often providing food and refugia for marine organisms. Wetlands may be destroyed through direct habitat elimination or by pollution from heavy metals and oil spills upstream. Mining and oil development can also contribute to the destruction of mangroves and wetlands through altering upstream watersheds and increased sedimentation.

Mountain and Arctic Environments

Extreme northern ecosystems are characterized by cold temperatures and short growing seasons. Arctic ecosystems exhibit far fewer plant and animal species than in the tropics, but they are often highly sensitive to disturbance and the loss of one or two species has a far greater impact. Lichens and mosses are often among the first species to disappear due to pollution and human disturbance. Permafrost degradation associated with mining and oil development may extend far beyond the initial area of disturbance, due to the melting of ice, soil degradation and the impoundment of water. The arctic environment often takes longer to recover from pollution due to the slow speed of biological processes. In addition, the lack of sunlight throughout the winter months makes the management of some mining wastes (e.g. cyanide-laced tailings) more difficult.

Arid Environments

Water scarcity is the primary constraint in arid environments. Vegetation is limited, but biodiversity is high among insects, rodents and other invertebrates, especially in semi-arid regions. The main impact of mining and oil development on these ecosystems is the alteration of the water regime, especially lowering of the water table and depletion of groundwater. These impacts may result in increased salinization of the soil and erosion, which eventually lead to a decline in vegetation and wildlife species. In densely populated areas, the competition for scarce water resources makes these ecosystems especially fragile.

Coral Reefs

Coral reefs harbour the most biodiversity of any marine ecosystem. Located primarily in the Indo-Western Pacific and Caribbean regions, coral reefs are important links in maintaining healthy fisheries. Reef systems are highly sensitive to human disturbance. Sedimentation from upstream land-uses and pollution are among the greatest threats to coral reefs. Mining directly impacts coral reefs through increased sedimentation, especially in cases where wastes are dumped directly in rivers and oceans, as well as through increased pollution by heavy metals.

Source:
based on CEAA 1996

include proliferation of weeds attracted to areas disturbed by mining, and exploitation of timber in forests made accessible by mining infrastructure. While direct impacts are readily identified, some indirect impacts are more difficult to predict. Accordingly, monitoring programmes should be designed to identify indirect impacts at an early stage so that remedial action can be taken.

Figure 13.14 shows the various activities involved in mining operations and the potential intersections of these activities with environmental components that could lead to a loss of biodiversity. ICMM (2003) provides a similar summary table for the pre-operations phase.

While this figure identifies the intersections that need to be evaluated, it should not be concluded that all these intersections will reduce biodiversity. This is certainly not the case.

Despite the widespread concern that mining reduces biodiversity, there is no documented evidence that any species has become extinct as a direct or indirect result of mining. Some loss of genetic diversity has undoubtedly occurred as from any activity that causes the death of plants and/or animals before they have had the chance to reproduce. However, as mining usually involves relatively small, discrete areas, it is much less likely to cause

Despite the widespread concern that mining reduces biodiversity, there is no documented evidence that any species has become extinct as a direct or indirect result of mining.

	MINING ACTIVITIES	Exploration and Ore Extraction							Ore Processing and Plant Site								Infrastructure, Access and Energy					Decommissioning		
POTENTIAL IMPACTS		Exploration drilling	Resettlement (if necessary)	Extraction and waste rock removal/disposal	Rock blasting and ore removal	Mine dewatering	Placer and dredge mining	Small-scale artisanal mining	Plant site, material handling etc.	Stockpiling	Beneficiation	Pyrometallurgical processing	Hydrometallurgical processing	Water usage (all industrial and domestic)	Use and storage of process chemicals	Tailings containment disposal	Access roads, rail and transmision lines	Wastewater treatment and disposal	Pipelines for slurries or concentrates	Power sources and transmission lines	Construction camps, town site	Regrading and recontouring	Stabilization of waste dumps and tailings	Mine closure
Air Quality	Increased ambient particulates (TSP and PM-10)			●	●				●	●		●				●	●			●	●	●		
	Increased ambient Sulphur dioxide (SO_2)											●								●	●			
	Increased ambient Oxide of Nitrogen (NO_2)											●								●	●			
	Increased ambient heavy metals			●	●				●	●		●				●	●			●	●			
Hydrology, Hydrogeology and Water Quality	Altered hydrologic regimes			●		●	●	●	●				●	●		●	●	●		●	●	●		
	Altered hydrogeological regimes					●							●	●										
	Increased heavy metals, acidity or pollution	●		●		●	●	●	●	●			●		●	●	●	●		●	●	●	●	
	Increased turbidity	●		●			●	●	●	●			●			●	●	●	●	●	●	●	●	
	Risk of groundwater contamination	●		●	●	●		●					●		●	●		●					●	
Ecology and Biodiversity	Loss of rare natural habitats and biodiversity (OP 4.04)	●		●			●	●	●							●	●		●	●	●			
	Loss of rare and endangered species			●			●	●	●							●	●		●	●	●			
	Effects of induced development on ecology	●						●									●				●			
	Effects on riverine ecology and fish			●		●	●	●		●				●	●	●		●	●		●			
	Impacts due to effluents or emissions	●		●			●	●			●	●	●			●		●				●	●	
Social Concerns	Resettlement issue (OD 4.30)		●	●	●		●		●							●	●			●	●			
	Effects on Indigenous Peoples (OD 4.20)	●	●	●	●		●	●	●							●	●			●	●			
	Loss of cultural heritage or religious sites	●	●	●			●		●							●	●			●	●			
	Loss of livelihood		●													●				●	●			●
	Induced development issues	●	●														●			●	●			
	Effects on aesthetics and landform			●			●	●	●							●				●	●			
	Noise issues			●	●		●		●											●	●			
Occupational and Public Health Concern	Occupational health and safety concerns										●									●	●			
	Hazards from process chemicals or explosives				●			●			●				●					●				
	Potential increase in disease vectors		●					●													●			
	Increased potential for respiratory disorders				●			●			●	●			●					●				
Resource Issues	Effects of subsidence on surface resources					●																		
	Agricultural land losses	●	●	●			●		●							●	●			●	●			
	Loss of forestry resources (OP 4.36)			●			●	●	●							●	●			●	●			
	Effects surface water resources (OP 4.07)			●	●	●	●	●		●	●		●	●		●	●	●		●	●			
	Effects groundwater resources (OP 4.07)				●	●		●			●		●	●		●		●		●	●			
	Disruption to infrastructure			●			●										●		●					
	Effects on fisheries			●		●	●	●		●			●	●		●		●		●	●			

FIGURE 13.14 Example of the Intersection of Mining Operations and Biodiversity

While this figure identifies the intersections that need to be evaluated, it should not be concluded that all these intersections will reduce biodiversity. This is certainly not the case.

Source: ICMM 2003

species extinctions than land uses such as agriculture, plantation forestry or urbanization, which involve much larger areas (see also **Case 7.5**). In fact, perhaps the major concern is the cumulative impact of mining, together with other changes in land use.

Occasionally, there may be specialized habitats confined to sites where ore bodies intersect the ground surface. If all such ore bodies were to be mined, then all these sites would be destroyed, together with any species confined to these habitats. Such a situation appears to be behind a recent recommendation by the Environment Protection Authority of Western Australia to disallow mining of the Mesa 'A' and Warramboo iron deposits of the Robe River Project. Here, the concern was the perceived threat to the continued existence of 11 previously unknown species of small (4 mm) spider-like invertebrates that live in cavities and crevices in the rock at depths of 5 m to 30 m below the surface. The iron ore has a particularly fissured and honeycombed structure and is habitat to an apparently unique assemblage of subterranean organisms termed 'Troglobitic fauna'. (It should also be noted that most rock formations have never been investigated for such organisms).

Perhaps the most vulnerable ecosystems and habitats occur in freshwater aquatic environments (wetlands) downstream of major sulphide ore bodies. In these cases, there is the potential that either accidental discharges during operations, or acid rock drainage from the mine, waste dumps or tailings facility could contaminate the water downstream, killing many of the resident species. The long-term degradation of aquatic habitats in the King and Queen Rivers of western Tasmania which resulted from dumping of mine wastes from the Mt Lyell copper mine and other mines over many decades, is an example of extensive habitat damage, affecting much of a river system. However, unless the rivers involved contained endemic species, biodiversity except at the genetic level, would not have been affected.

Perhaps the most vulnerable ecosystems and habitats occur in freshwater aquatic environments (wetlands) downstream of major sulphide ore bodies.

From the foregoing, it is clear that there are situations in which the survival of particular species may potentially be threatened by mining. However, such situations are relatively uncommon; many mining operations pose no threat to any plant or animal species, although they may cause some loss of genetic diversity.

Indirect Effects – Introduction of Weeds and Pests

Introduced plants and animals pose a much greater threat to biodiversity than the direct effects of mining. While at first glance it may seem unlikely that a mining project could introduce weeds or pests, there are several mechanisms by which this does occur. These include:

1. **Ballast water disposal.** Most major mining projects involve shipping of mineral products, often from dedicated port facilities. Ships used for transport of bulk minerals must first discharge water from ballast tanks, the water having originated at another port, often located a great distance away and sometimes in a different ocean. There have been many worldwide examples of marine species introduced in this way and invading their new habitats to the detriment of native species. According to the CSIRO (1993), as one example, more than 100 marine species have been introduced in this way to Australian waters.
2. **Rehabilitation.** Weeds may be introduced deliberately or inadvertently. Some species that are not invasive in their natural habitats, prove to be highly invasive when introduced to a new area. This is the case with many of the worst weeds throughout the world. Any non-native species being considered for mine rehabilitation should therefore be carefully trialled and closely monitored prior to its introduction. The advantages and disadvantages of using non-native species for rehabilitation are discussed in more detail in Chapter Twenty-One. Inadvertent weed introduction can occur

Any non-native species being considered for mine rehabilitation should be carefully trialled and closely monitored prior to its introduction.

through the use of seed contaminated with seeds of weeds. This is quite common, even in commercial seed supplies. Suppliers should be requested to certify that seed supplied is weed free, and seedlings should be closely monitored for signs of unwanted species. Weeds can also be introduced by plant and equipment relocated from other areas.

3. **Use of Second-hand Plant and Equipment.** Many mining projects use plant and equipment that were previously used on other projects. Soil pathogens, insects and fungi can be introduced from these items unless stringent measures are taken to avoid this possibility. Similarly, it is common for construction contractors to relocate workers' accommodations from one construction site to another. In one such case numerous Redback Spiders were found when portable dongas (accommodation buildings) from Australia were being unloaded at a mine site in Indonesia. This venomous spider (*Latrodectus hasselti*) while common in Australia, does not occur in Indonesia. Fortunately, the infestation was discovered in time to eradicate the spiders before they could spread. Giant African Land Snails (*Achatina fulica*) have invaded Lihir Island, site of the Lihir Gold Project. While this pest probably arrived with timber imported by the island's residents, this may not have occurred without the wealth created by the project. The mining company is leading efforts to eradicate the snail.
4. **Household pets.** Dogs, cats and other pets kill many native birds and small animals. Feral dogs and cats are particularly damaging in this regard. Accordingly, some mining projects ban dogs and cats from their mine sites.

13.7 SOCIAL AND ECONOMIC CHANGE

Potential social changes associated with mining include land use changes; growth inducement; income generation; education and training; capacity strengthening; improvement of public infrastructure; education, health, and welfare; uncontrolled influx of people; disruption of societal organization; loss of traditional values and norms; and changes in life style and livelihood. Some of these changes are a direct consequence of project operations; most are not (**Figure 13.15**). Some change processes are amenable to management influence; others are not. All of these potential changes are discussed at length elsewhere in this text. For an overview see Chapter One. As an illustration of social change, the following discussion addresses only two of the main issues: (1) the relation between mining and induced development; and (2) mining, economic growth and wealth distribution.

Induced Development

Mining projects stimulate economic growth. A major mining development may require a workforce of several thousand people. During operation, the same project will provide hundreds of direct and many more indirect employment opportunities. Mines also project the image of wealth. As is the case with metropolitan centres, major mining operations may attract a great number of people seeking jobs, as they provide income and a better livelihood. Adding to this are the aspirations of local and central governments, seeking to use the mining project to stimulate local economic growth.

Most of the actual immigration is unplanned; induced development often outweighs expected and planned development.

Part of this economic growth is planned and is (or should be) reflected in both the EIA for the project as well as in regional land use plans. Most of the actual immigration, however, is unplanned; induced development often outweighs expected and planned development.

Induced development poses many social challenges. It disrupts existing ecosystems and local communities. Mines themselves require access to land and natural resources, such as

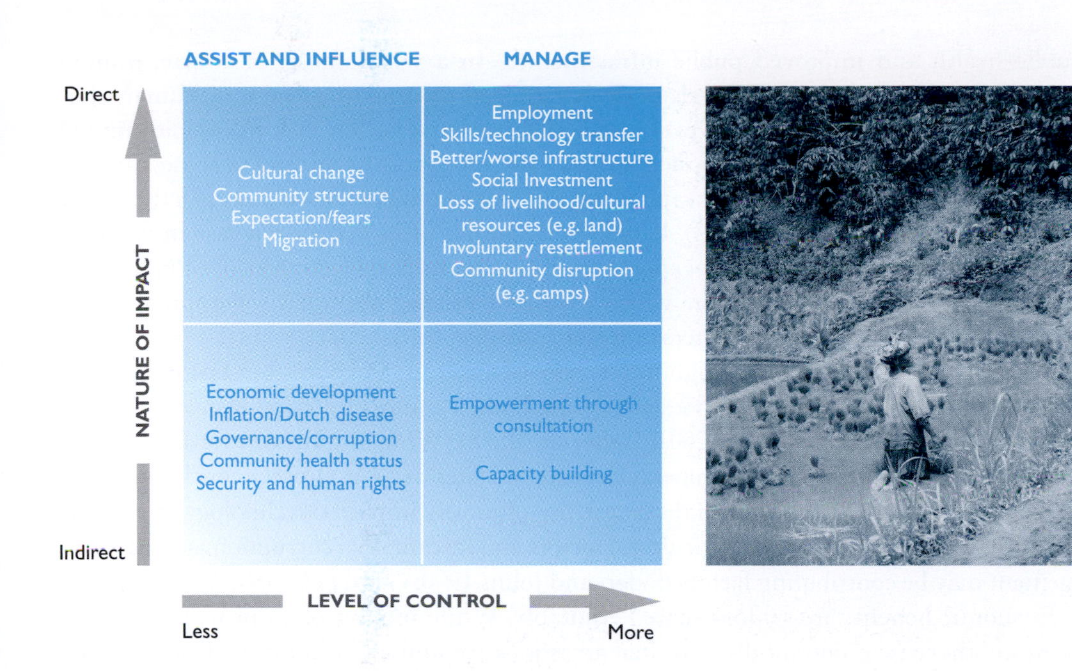

FIGURE 13.15
Mining and Social Change – Levels of Control

water, for which they often compete with other land uses (Ashton *et al.* 2001). Induced development adds to these pressures. In addition, more often than not, the existing local infrastructure is insufficient to cope with the increase in population: clean drinking water supplies may be insufficient; waste management facilities, if they exist, are inadequate; and so are existing public health and educational facilities.

It also does not help that most mine locations are fixed. The size of most mining operations may appear small compared to other land uses (e.g. industrial agriculture and forestry) but mining companies are limited in development alternatives by the location of economically viable reserves, some of which may overlap with sensitive ecosystems or traditional indigenous community lands. Induced development may easily affect communities that are least prepared for an uncontrolled influx of immigrants and the rapid community changes that may result.

Induced development illustrates that most significant social impacts from mining developments are of an indirect nature; they are unwanted consequences of providing site access to and employment opportunities in a host region that may previously have had little to offer. The fact is that mining almost always stimulates in-migration (ESSC 1999 a and b; 2003), although this may be minimized if the local community is determined that it be so, in which case the project proponent can assist by implementing appropriate workforce recruitment and accommodation policies.

Induced development illustrates that most significant social impacts from mining developments are of an indirect nature; they are unwanted consequences of providing site access to and employment opportunities in a host region that may previously have had little to offer.

In some instances concerns regarding the potential conflicts between mining and other land uses have prompted communities to pass non-binding referendums banning mineral development. As an example, in June 2002, the community of Tambogrande in Peru voted to reject mining in their community due to concerns regarding the projected displacement of half of its residents and fears regarding the potential impacts of mining on the community's traditional livelihood (Oxfam 2002).

Economic Growth and Wealth Distribution

Most direct positive impacts of mining are related to stimulation of economic growth. Growth can be economic as well as in terms of increased education levels, improved public health and improved public infrastructure. In a wider sense, of course, mining provides metals, industrial minerals or fuels required to support growth of human civilization. In the first instance, however, developing countries often seek to exploit mineral resources as a means of ensuring much needed state revenues. According to some, mineral wealth is part of a nation's natural capital and the more capital a nation possesses the richer it becomes (Davis and Tilton 2003). Papua New Guinea often serves as a case in point, as it receives almost two-thirds of its export earnings from mineral resources (GoPNG 2002). Another example is Botswana, in which diamond mining accounts for approximately one-third of GDP and three-quarters of export earnings.

Mineral exports can provide a significant share of a country's exports, but history demonstrates that mineral development does not always boost a country's economic growth (Sideri and Johns 1990; Auty 1990; Ross 2001; Gelb *et al.* 1988). While the reasons for lack of economic growth in some oil- and mineral dependent states are not entirely conclusive (Ross 1999), low levels of employment in the sector, use of mostly imported technology, high market volatility, competition with agricultural sectors and institutional corruption and mismanagement may be contributing factors (Sideri and Johns 1990; Gelb *et al.* 1988; Auty 1990).

Economic benefits are seldom shared equitably.

Economic benefits are seldom shared equitably. While mining often provides employment in otherwise economically marginal areas jobs are limited in duration. Communities that come to depend on mining to sustain their economies are especially vulnerable to negative social impacts when the mine closes. Mining tends to raise wage levels, potentially leading to the displacement of some community residents and existing businesses, and elevated expectations (Kuyek and Coumans 2003) (**Case 13.12**).

13.8 SURFACE MINING VERSUS UNDERGROUND MINING

Environmental and social impacts associated with surface and underground mining are similar in nature, but often different in magnitude.

Environmental and social impacts associated with surface and underground mining are similar in nature, but often different in magnitude. Compared to surface mining, underground mining requires relatively small openings and limited excavation. Underground mines also produce relatively little waste rock. Tailings are often used as backfill materials eliminating the need for large on-land tailings storage areas. Accordingly, two of the main sources of environmental impacts from mining (changes in landform and mine wastes) are greatly reduced. Since underground mines are generally much smaller in terms of production than surface mines, the fleet of mining equipment, and ancillary mining operations are also of smaller scale. Loosely speaking, this also leads to smaller environmental impacts. On the other hand, underground mining is invariably more labour intensive, so that there may be more employment opportunities than for a surface mine of similar production, but this may not benefit local communities as underground mining requires more highly skilled workers. Underground mines may also require more extensive mine dewatering, with the potential for more severe impacts related to water management.

Underground mining features two unique environmental features: the necessity to provide air to the workforce and the possibility of land subsidence. The mine ventilation system, while providing essential life support for miners, has relatively minor environmental impacts – noise due to the operation of ventilation fans and dust emissions from ventilation shafts. Land subsidence is frequently observed at underground mines that are relatively close to the surface. Overburden collapses into the voids created by underground mining, resulting in uneven land subsidence at the surface, surface cracks, cracks in buildings, or at the extreme, in large sink holes. In some cases holes and cracks formed by subsidence pose safety hazards.

As mentioned, higher labour skills are required in underground mining meaning that without a significant training programme, fewer employment opportunities may be available for local labour, compared to surface mining. Higher health and safety risks are also associated with underground mining. Finally, special attention is required during mine closure to avoid unsafe surface openings and to avoid public access to underground mines (**Figure 13.16**).

13.9 ACCIDENTAL ENVIRONMENTAL IMPACTS

Like all industrial sectors, the mining industry has risks specifically related to its operations. Some risks are similar to those in other industries, such as storage, handling, transportation and use of toxic chemicals (**Figure 13.17**). Hazardous materials are used at most mining and mineral processing operations, and the waste products generated by these operations and stored in specially designed facilities may also be hazardous to human health and the environment.

CASE 13.12
The Dutch Disease

Are countries with rich natural resources (minerals or oil and gas) more likely to industrialize and to develop? It may seem that having a large amount of these natural resources is a big advantage. History shows that this is not necessarily so. But why?

There is a danger when a country or region containing a large natural resource receives too much money, too fast. There is even a name for it: 'Dutch disease', a term coined in 1977 by *The Economist* to describe the decline of the manufacturing sector in the Netherlands after the discovery of natural gas in the 1960s (The Economist, November 26, 1997). *Finance and Development*, a magazine of the International Monetary Fund, defined the term as 'too much wealth managed unwisely' (Ember 2007, www.voanews.com).'

In the Netherlands, as natural gas was extracted, it increased domestic income and spending. Investment was redirected toward the natural gas sector. Dutch wages and prices began to rise. The Dutch guilder became overvalued in real terms, Dutch industrial products became uncompetitive, and the traditional manufacturing sector declined. The Netherlands experienced the phenomenon of de-industrialization despite the presence of rich natural resources.

The Dutch disease is most often associated with oil, gas and other minerals, because they tend to be discovered suddenly and in large amounts. The same phenomenon was observed in the UK some years later (see graphic above), and in many resource rich countries and regions since then.

Does this mean that having natural resources is a bad thing? Not necessarily. Natural resources are a precious earner of foreign exchange provided that these earnings are wisely spent. However, there is another complicating factor.

Logic dictates that any non-renewable natural resource will eventually be exhausted, no matter how big it is (or appears to be). There are numerous examples of what happens when the resource becomes depleted. Assume that during mine operation a complacent government enjoyed foreign revenue without redirecting these earnings in sustainable capital building. Further assume that at the same time local communities and the private sector were left unprepared for mine closure, and investments in infrastructure were predominately concentrated on the sectors supporting mine operation. What happens when natural resources start to run out? The local traditional economy has probably decreased and local communities become economically stranded. Worse, public and private spending may be out of control and unsustainable.

Today, governments and the mining sector alike, recognize that human resources count for more than natural resources.

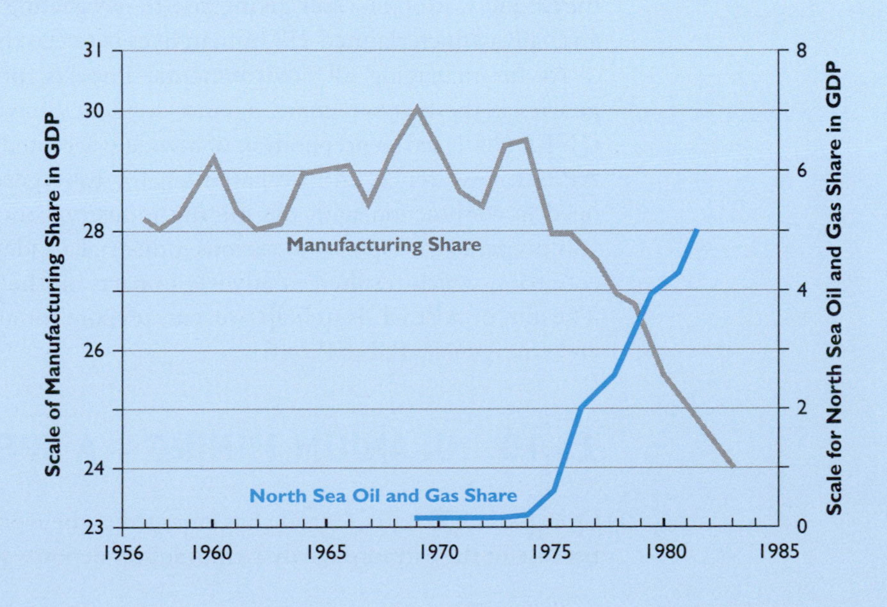

FIGURE 13.16
Safety Hazards Illustrated

Special attention is required during mine closure to avoid unsafe surface openings and access to underground voids.

While major accidents in the mining industry are relatively uncommon, given the number of global mining operations, they do occur. Any single major accident can be detrimental to public perception of a mine operation and the industry as a whole. Risks from tailings dams have drawn particular attention, following failures at mines in Europe and elsewhere (for discussion of tailings management and inherent risks see Chapter Eighteen). Accidents involving facilities or activities other than tailings disposal also occur, but they appear to be less frequent or to attract less attention. A review of incidents in the gold industry since 1975, reveals that 72% of the incidents involved tailings dams, while 14% involved pipe failure and 14% represented transport accidents (UNEP 2000). All such incidents have the potential to cause serious environmental and health effects.

Major tailings storage accidents seem to occur on average once each year and they can have serious consequences for the environment, human health and property, and the financial health of the mining company involved. The physical properties of fine-grained mine waste particles are frequently such that, when saturated with water and subjected to stress, they liquefy, in some cases giving rise to devastating mud flows, such as those at Aberfan and Stava which claimed 412 human lives between them (Panman 1998).

Preparedness and planning for emergencies is the best safeguard against accidental environmental impacts.

As for managing all environmental impacts, preparedness and planning for emergencies is the best safeguard against accidental environmental impacts. An initiative by UNEP (2001) led to preparation of a widely accepted guide to emergency response termed APELL (Awareness and Preparedness for Emergencies at Local Level), a process developed in conjunction with the mining industry, communities, and governments. APELL was prepared in response to various industrial accidents in both developed and developing countries, which resulted in adverse impacts on the environment and local communities. The aim of APELL is to help prevent, prepare for and respond to technological accidents and emergencies (UNEP 2001).

13.10 URANIUM MINING – A SPECIAL CASE

Over the last 30 years, Canada has emerged as the world's leading uranium producer, largely because of the extraordinarily rich uranium deposits found in Northern Saskatchewan, and

FIGURE 13.17
Hazardous Materials are Used at Most Mine Sites and Risks Related to Storage, Handling, Transportation and the use of these Chemicals are Similar to Those in Other Industries

a favourable legislative setting in regard of uranium mining. Australia is the second largest uranium producers. Both countries together account for over 50% of global primary production, today primarily used by the nuclear power industry to produce energy.

Since the advent of the first atomic bomb, uranium commands strong public interest. Uranium is the heaviest metal occuring in nature, unstable and gradually breaking apart or 'decaying' at the atomic level. Any such material is said to be radioactive. As uranium slowly decays, it gives off invisible bursts of penetrating energy (termed atomic radiation). It also produces more than a dozen other radioactive decay products of little or no commercial value.

In public awareness for only a few decades, impacts on human health due to radioactivity have been unknowingly reported for centuries. Beginning in 1546, it was reported that most underground miners in Schneeburg, Germany, died from mysterious lung ailments. By 1930, the same grim statistics were found among miners in Joachimsthal, Czechoslovakia, on the other side of the same mountain range. More than half of them were dying of lung cancer, at a time when lung cancer was all but unknown among the non-mining populations on both the German and Czech side of the mountains. The ores in question were particularly rich in uranium.

Much research has been conducted on uranium and its various environmental and health impacts (WHO 2001) and the following section draws on text by Edwards *et al.* (2006) and Diehl (2004), if not otherwise stated. Health impacts of uranium mining are long-term – sometimes over 20 years after end of work. Most studies find relative risks of lung cancer between 2 and 5 times higher in uranium workers who have been exposed to higher levels of radon, or to longer periods of low exposure. Some studies put these risks at levels much higher (Stephens and Ahern 2001).

Health impacts of uranium mining are long-term – sometimes over 20 years after end of work.

In most respects, conventional mining of uranium is the same as mining any other mineral ore, and the same environmental constraints apply in order to avoid off-site pollution. However uranium minerals are always associated with more radioactive elements such as radium and radon. Therefore, although uranium itself is not very radioactive, the ore which is mined, especially if it is very high-grade such as in some Canadian mines, is hazardous if handled without regard to appropriate health and safety guidelines.

In addition to the usual risks of mining, uranium miners worldwide have experienced a much higher incidence of lung cancer and other lung diseases. Several studies have also indicated an increased incidence of skin cancer, stomach cancer and kidney disease. Since uranium ore is radioactive, so are the mine wastes. The majority of mine wastes containing natural radioactivity are derived from sources of uranium and thorium, particularly in the mining and treatment of uranium ores or in the processing of beach sands containing monazite. Rare earths mining and processing may also produce radioactive wastes. Radioactive wastes occur as gases, liquids and solids. In considering the characterization of wastes, the toxicity, mobility, radioactive half-life and type of radioactive emission (alpha-particle, beta-particle, gamma ray, or neutron) are important. These characteristics govern the choice of management procedures which should be adopted.

Today, uranium mining methods, tailings and runoff management and land rehabilitation are subject to strict government regulations and inspections (e.g. Australian Code of Practice and Safety Guide: Radiation Protection and Radioactive Waste Management in Mining and Mineral Processing 2005). Mining operations are undertaken under strict health standards for exposure to gamma radiation and radon gas. In fact a recent report by the Uranium Industry Framework Group argues that the level of regulation of the Australian uranium industry 'may add to the perceived level of risk associated with uranium mining and perhaps hinders the public's understanding of the actual level of risk'. That is to say, the public sees uranium mining as more hazardous than it really is.

The public sees uranium mining as more hazardous than it really is.

In modern uranium mines acute risks from exposure to uranium and its decay products are low. Risks from uranium are carefully measured and relatively low, but the public perception of risk remains relatively high. The common opposition to uranium mining is mainly due to a significant misalignment between public concern and the objective risk, but also due to the concern of long-term management of uranium mines once operation ceases.

Environmental Concerns Related to Uranium Mining

The principle mine waste streams in uranium mining are the same as mining metal ores, but the mine waste characteristics differ. Solid wastes include overburden and barren rock in which the uranium ore is dispersed, but more importantly, tailings from the treatment of ore, in which finely ground radioactive materials are exposed to the environment. Liquid wastes are in the form of surface water runoff from waste rock dumps and ore stockpiles, water seepage through the waste rock and ore stockpiles, and water seepage from mine voids. As in metal mining, the acidity of the water increases the leaching of radionuclides, and liquid effluent generation may continue to be produced long after mining ceases. Airborne waste streams are in the form of dust, gas and fine water droplets: (1) dusts containing uranium and its decay products, (2) radon-222 released to the atmosphere during the break-up and exposure of the ore body, and (3) solid decay particles of radon gas that attach themselves to tiny dust particles and droplets of water vapour. Radon emission will continue for thousands of years.

Tailings Management

The recovery of uranium processing plants is in the order of 95%. This implies that an average of 5% of the original uranium remains in the tailings. Since the uranium extraction process selectively removes only uranium from the ore, all other chemical compounds remain in the tailings in the same concentration as in the ore. Tailings therefore represent most of the original ore and they contain most of the radioactivity in it. In particular, they contain all the radium present in the original ore, which is the most dangerous from the radiological point of view. By its decay the inert gas radon is generated. This penetrates easily through

An average of 5% percent of the original uranium remains in the tailings.

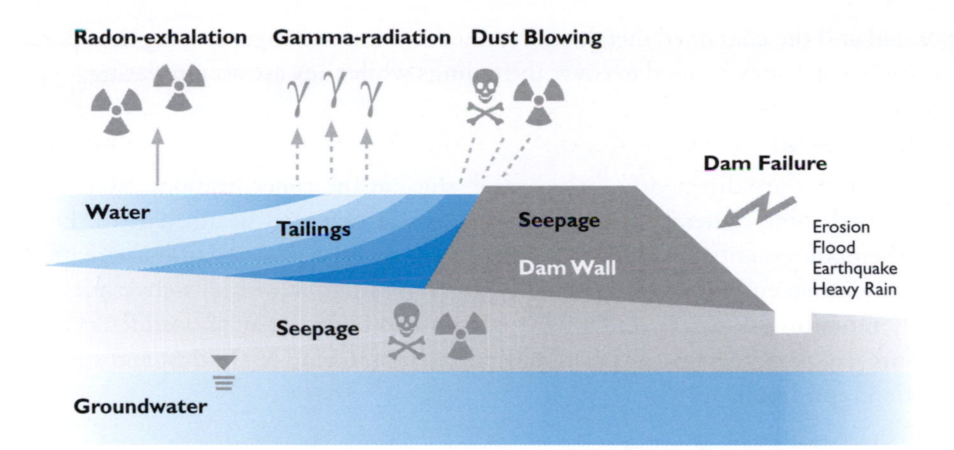

FIGURE 13.18
Uranium Mill Tailings Hazards

Source:
based on Diehl 2004

the ground rock into the atmosphere and is spread over long distances by the air. However, radon is easily soluble in water and, thus, is usually retained in any aqueous phases. Chemical reagents used in the uranium extraction together with the waste from the uranium processing are also deposited in the tailings ponds. Consequently tailings as a whole are a complex system containing inert rock matter ground to small grain sizes, radionuclides from the uranium and thorium decay series, chemical substances, mainly sulphates, carbonates, nitrates and complex salts of metals and alkali-earth elements, as well as heavy and rare metals.

During the operational life of an on-land tailings storage facility, tailings are usually covered by water to reduce surface radioactivity and radon emission (though with lower-grade ores neither pose a major hazard at these levels). On completion of the mining operation, tailings storage areas are commonly covered with some two metres of clay and topsoil to reduce radiation levels to natural background levels at the mine site. At underground mines, tailings may be used for underground fill, reducing the risk of radon exposure at the ground surface.

Figure 13.18 illustrates the various hazards associated with radioactive tailings storage facilities (Diehl 2004). The greatest risks to the environment are (1) contamination of groundwater and river systems with dissolved radioactive materials; (2) catastrophic failures of tailings containment; (3) the dispersal of radioactive dust, which finds its way into water, plants, animals, fish and humans; (4) releases of radon gas into the air, which will deposit radon progeny on the ground for hundreds of kilometres around; (5) pollution of surface and groundwater by chemical pollutants in tailings, notably heavy metals, acids, ammonia and salts. Radionuclides contained in uranium tailings also emit 20 to 100 times as much gamma-radiation as natural background levels on deposit surfaces but gamma radiation levels decrease rapidly with distance.

In the short term, chemical pollution is of most concern. Radiation hazards are more subtle and will take longer to be manifested. However, unless the tailings are properly disposed of, these hazards will continue unabated for thousands of years. Arguably tailings hazards will get worse as time goes on because of erosion, neglect and climatic change.

Water Management

Runoff from mine stockpiles and waste liquors from the milling operation are collected in secure retention ponds for isolation and recovery of any heavy metals or other contaminants. The liquid portion is disposed of either by natural evaporation or recirculation to the milling operation. Most Australian mines adopt a 'zero discharge' policy for any pollutants. Process water discharged from the mill contains traces of radium and some other metals which would be undesirable in biological systems downstream. This water is

Most Australian mines adopt a 'zero discharge' policy for any pollutants.

evaporated and the contained metals are retained in secure storage. During the operational phase, such water may be used to cover the tailings while they are accumulating.

Dust Management

Particular efforts are directed to dust control, since at the concentrations associated with uranium (and some mineral sands) mining, dust is a potential health hazard. Dust represents the main potential exposure to alpha radiation as well as a gamma radiation hazard. At any mine, employees likely to be exposed to radiation or radioactive materials are regularly monitored for alpha radiation contamination and personal dosimeters are worn to measure exposure to gamma radiation. Routine monitoring of air, dust and surface contamination is undertaken.

Mine Closure

When closing down uranium mines, large amounts of radioactively contaminated equipment and scrap are produced, which have to be disposed of in a safe manner. In most mine operation the intention is to bury radioactively contaminated scrap on-site in mined-out pits. However, no mine or mill site has yet been cleaned up in a permanently satisfactory way anywhere in the world. Since it is impossible to avoid radioactive materials once they are released from the deep rock and brought to the surface, the only remedy is 'long-term' prevention of radioactive materials escaping into human environments.

The actual length of 'long-term' for prediction of environmental and radiological impacts of abandoned uranium mines and tailings impoundments is not clearly defined.

The actual length of 'long-term' for prediction of environmental and radiological impacts of abandoned uranium mines and tailings impoundments is not clearly defined. The International Atomic Energy Agency considers long-term to be the period, beyond the design life, for which climatological and geomorphological processes are more or less predictable and would not substantially affect the integrity of tailings impoundment systems. This would be in the order of 1,000 to 10,000 years, practically forever in human terms. Once uranium mines have been abandoned, it is doubtful whether any regulations can be effective in preventing radioactive materials escaping into the environment over such long periods of time. The levels of radioactivity in the tailings, and the amount of radon gas produced by the tailings, will not noticeably diminish for more than 10,000 years. How can the natural forces of erosion, migration, dispersion and dissolution be held in abeyance? Who will monitor the wastes and take corrective action in, say two hundreds years from now? And who will pay for the future effort needed to do all this (**Case 13.13**)?

CASE 13.13
The Wismut Legacy

Immediately after the end of World War II, the Soviets started exploration and mining of uranium in the historic mining provinces in the Ore Mountains, Germany. Subsequently, the Wismut company developed the third-largest uranium mining province of the world (after the USA and Canada). With the political changes in 1989, it came to light that uranium mining in Eastern Germany had devastated large areas. Uranium production was terminated but huge shut-down uranium mines, hundreds of millions of tons of radiating waste rock and uranium mill tailings remained. The environmental legacy does not only present an immediate hazard, but also endangers future generations for tens of thousands of years. The German government estimates the clean-up to amount to more than US$ 9 billion. If this estimate is attributed to the amount of uranium produced, specific reclamation costs of US$ 43 per kg of uranium produced are obtained which compare to the current world market price for uranium of about US$ 26/kg (as of December 2006).

Uranium Leaching

In situ leaching (ISL) of uranium ore has been carried out since the early 1960s, with early production from deposits in Russia, Ukraine and Uzbekistan. After considerable research during the 1960s, commercial production of ISL uranium in the US commenced in the 1970s and has continued ever since, resulting by 2001, in a combined production of 43,000 tonnes of uranium from 28 ISL operations. Other ISL projects have been developed in China, Kazakhstan, Germany and Australia. Most ISL projects have been applied to 'roll-front' type sandstone uranium deposits which are 'uniquely amenable to ISL exploitation since ISL mining relies on physical and chemical processes similar to those that originally deposited the uranium ore bodies' (Underhill 1992).

In situ leaching is usually used for relatively low grade uranium deposits which can not be economically mined and processed by open pit or underground means. Uranium is extracted from these low grade ore bodies with either acid-leach using sulphuric acid, or alkaline-leach using sodium bicarbonate or ammonium bicarbonate. For both these processes, the uranium is subsequently recovered from the pregnant solution using either ion exchange or solvent extraction methods. Selection of the leach process is influenced by:

- Regulatory requirements;
- Composition and therefore relative solubility of ore and host rock;
- Reagent consumption rates and costs;
- Leaching rates and uranium recovery; and
- Environmental factors including aquifer quality, groundwater gradients and flow rates.

In situ leaching is usually used for relatively low grade uranium deposits which can not be economically mined and processed by open pit or underground means.

The salient features of acid and alkaline leaching are compared in **Table 13.2**.

In many cases the choice of leaching process is dictated by regulatory requirements. Groundwater in contact with uranium ore is inevitably radioactive and contaminated with dissolved uranium and its decay products, including radon and thorium. Accordingly, it is unsuitable for any form of domestic, agricultural or industrial use. US regulations require that aquifer quality after solution mining be returned to a condition approaching potability. This is more difficult to achieve in the case of acid leaching, and, as a result, alkaline leaching has been used in most uranium ISL projects in the US. Regulations in other jurisdictions have been more pragmatic, in recognition of the fact that the groundwater had no beneficial use before mining and therefore did not warrant expensive treatment after mining. In most of these countries, acid leaching has been selected, as it generally involves higher leaching rates, higher overall recoveries and consequently better economic returns. Ores containing high concentrations of calcite, however, still require alkaline leaching.

Groundwater in contact with uranium ore is inevitably radioactive and contaminated with dissolved uranium and its decay products, including radon and thorium.

A typical alkaline leach process is depicted in **Figures 13.19** and **13.20**. The acid leach process is similar except that sulphuric acid is the lixiviant. The effective location, spacing and design of injection and recovery wells requires a thorough understanding of the hydrogeology of the ore body and its surrounds. Detailed hydrogeological studies using drilling and pump testing are used to evaluate aquifer conditions. Laboratory testing is used to assess leaching rates. With these results as inputs, numerical modelling can then be used to evaluate the optimum layout of bores, injection and extraction rates, and waste (barren solution) re-injection scenarios.

Impacts

Compared to alternatives involving excavation, processing and disposal of waste rock and tailings, ISL involves much less surface disturbance and accordingly much smaller impacts on the landscape, soils, flora, fauna, land use and visual amenity. Capital and operating costs are generally also lower. On the other hand, the recoveries achieved are significantly lower than can generally be achieved by conventional mining and processing.

TABLE 13.2
Comparison of Acid and Alkaline Leaching

Acid Leaching	Alkaline Leaching
Acid leaching achieves a high rate of uranium extraction – typically 70 to 90%.	Extraction from alkaline leaching is relatively low – typically 60 to 70%.
Acid leaching yields faster dissolution of uranium – requiring 40 to 70 pore volumes.	Slower kinetics of uranium dissolution. Alkaline dissolution typically requires more pore volumes
Significant increase in concentration of dissolved solids (TDS) in groundwater – typically 10 to 25 g/L.	Insignificant increase in groundwater TDS.
High acid consumption for ores containing carbonate minerals.	Potential to treat ores containing high carbonate content.
Requires use of corrosion resistant materials and equipment.	Common materials and equipment can be used.
Addition of oxidant not always required.	Addition of oxidant always required.
Possibility of recovering by-products.	Leaching chemistry is highly selective for uranium.
Additional processing on surface may be required to produce contaminant-free product.	Product solution from ion exchange should produce required product.
Risk of reducing permeability due to chemical and gaseous plugging.	Precipitation of carbonate or sulphate minerals can also reduce permeability.
Restoration to pre-mining water quality requires an extended treatment period. Such restoration has only been demonstrated at one pilot site[1].	Restoration to pre-mining water quality has been demonstrated at several sites.
Seepage beyond borefield is unlikely, due to formation of precipitates that reduce porosity and permeability, and given natural attenuation due to reaction of contaminants with adjacent barren rock and unaffected groundwater.	Potential for residual solutions to spread beyond the limits of the areas being treated.

Source:
Taylor *et al.* 2004

[1] Note: For many acid ISL sites, restoration to pre-mining water quality has not been a requirement, as the quality of pre-mining water was already poor.

There have been examples of adverse environmental impacts as a result of ISL operations. These examples are from the former Soviet Union and resulted from the use of improper practices. There are, of course, many more and far more serious examples of environmental damage from conventional uranium mining operations in various parts of the world, mostly from operations that predated the rise of environmental consciousness that dates from about 1970.

Of primary concern are the impacts on groundwater quality. Although aquifers in contact with uranium ore are naturally contaminated with radioactive elements and therefore unsuitable for consumption by humans, livestock or wildlife, the wastewater from the process plant will contain other contaminants that could eventually migrate away from the ore body area. In this respect the alkaline leach method is preferred by US regulators as it facilitates restoration of groundwater quality to potable levels. However, in many situations such as remote, arid areas with low quality groundwater and no potential for significant groundwater interaction with the biosphere, there is no such requirement.

Closure

Given the small footprint associated with ISL operations, surface rehabilitation is relatively straightforward. The main issue is usually groundwater restoration to ensure that there are no long-term effects, particularly outside the area of the leach field. Given that

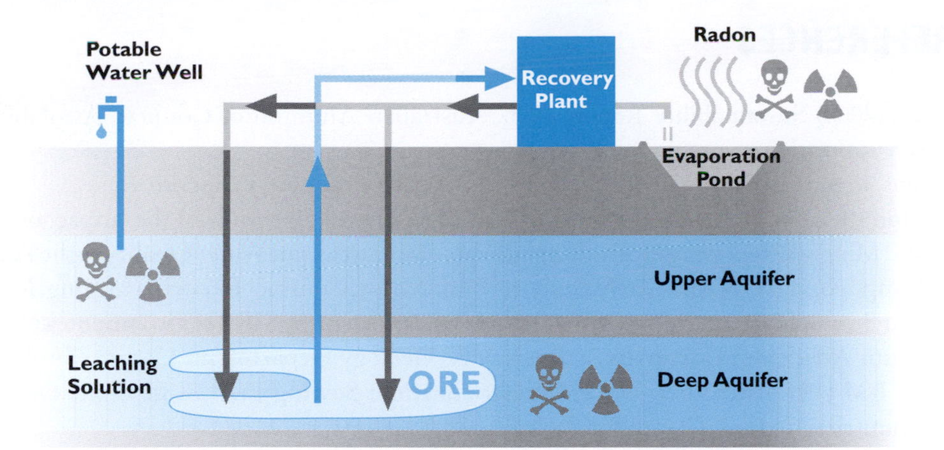

FIGURE 13.19
Scheme of Normal *In situ* Leaching Operation

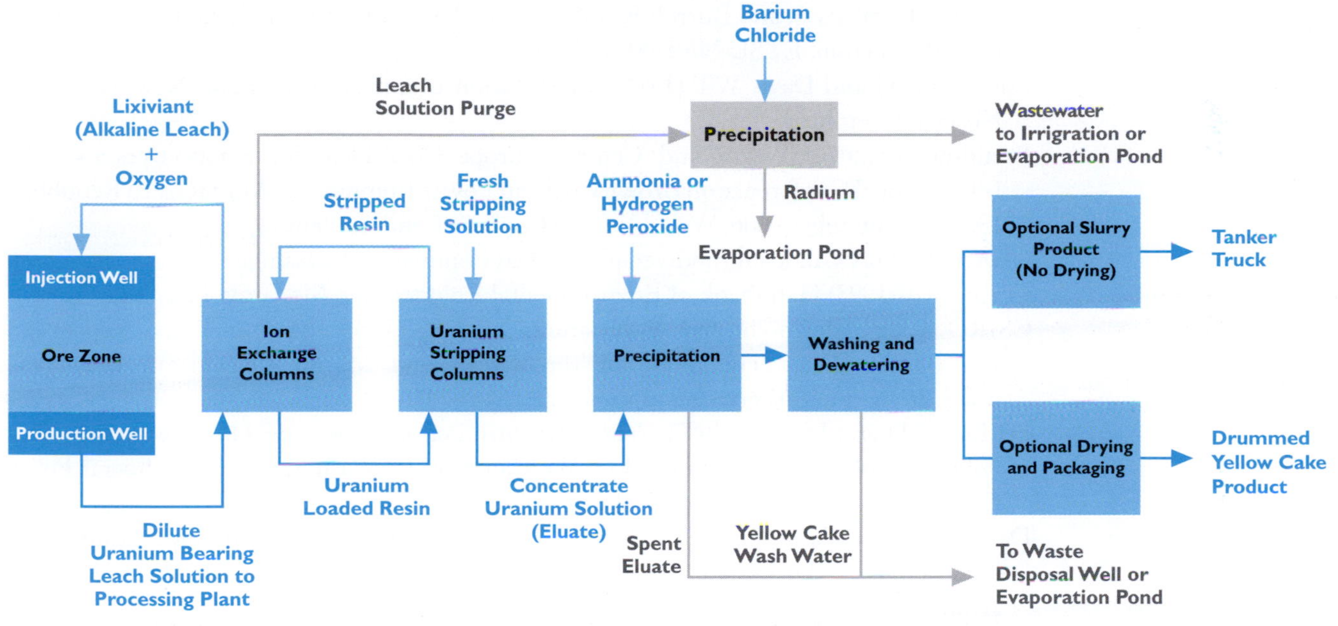

FIGURE 13.20
Process Flow Diagram for a Typical Uranium *In situ* Leach Mining Facility

Compared to alternatives involving excavation, processing and disposal of waste rock and tailings, ISL involves much less surface disturbance and accordingly much smaller impacts on the landscape, soils, flora, fauna, land use and visual amenity.

Source:
Modified from Power Resources Inc. 1991

the hydrogeology of the area is thoroughly understood as a result of both the feasibility investigations and operating experience, the groundwater flow patterns following closure can be readily predicted, enabling appropriate remediation to be carried out, if required. Usually, no remediation will be necessary. Detailed studies on groundwater restoration carried out in various countries and summarized by Taylor *et al.* (2004), indicate that natural attenuation will remove most contaminants remaining in the aquifer after completion of ISL, thereby restoring the quality of groundwater to levels comparable to adjacent unaffected groundwater over periods ranging from a few years to a few decades. Trials have shown that attenuation can be expedited if waste solutions are passed through un-oxidized rock formations.

Natural attenuation will remove most contaminants remaining in the aquifer after completion of ISL, thereby restoring the quality of groundwater to levels comparable to adjacent unaffected groundwater over periods ranging from a few years to a few decades.

REFERENCES

AAC (2006) Sustainability Report 2006, Australian Aluminium Council, Available from www.aluminium.org.au

Allens Arthur Robinson (2006, 2007) Focus, Available at: www.aar.com.au

Ashton PJ, Love D, Mahachi H and Dirks PH (2001) An Overview of the Impact of Mining and Mineral Processing Operations on Water Resources and Water Quality in the Zambezi, Limpopo, and Oilfants Catchments in South Africa, Contract Report to Mining, Minerals, and Sustainable Development Project/Southern Africa, CSIR-Environmentek: Pretoria, South Africa and University of Zimbabwe, Geology Department: Harare, Zimbabwe

Auty RM (1990) Resource-Based Industrialization: Sowing the Oil in Eight Developing Countries. Oxford: Clarendon Press.

Bell F and Donnelly L (2006) Mining and its Impact on the Environment. 541pp. Taylor & Francis

Bennet S, Kershaw S, and Burrell R (1995) Control Measures for Methane Emissions from Coal Production, ETSU N/01/00006/REP

Bounicore AJ and Davis WT (1992) Air Pollution Engineering Manual. New York: Van Nostrand Reinhold

Brendow K (2006) World and Central Europe Coal Demand Perspectives to 2030, International Conference on Brown Coal and Power Engineering Most (Czech Republic), 13. –15. September 2006, World Energy Council, London/Geneva

CEAA (1996) A Guide on Biodiversity and Environmental Assessment

Corbitt RA (1993) Handbook of Environmental Engineering, Mc Graw Hill, New York

CSIRO (1993) Research and Publications on Introduced Marine Pests; Report no 1 – Ballast Water Treatment for the Removal of Marine Organisms, June 1993 www.daff.gov.au/fisheries/invasive/research-pubs

Da Rosa CD and Lyon JS (1997) Golden Dreams, Poisoned Streams: How Reckless Mining Pollutes America's Waters and How We Can Stop It. Washington, DC: Mineral Policy Center.

Davis GA and Tilton JE (2003) Should Developing Countries Renounce Mining?: A Perspective on the Debate, Paper contributed to Extractive Industries Review (EIR). Available online at www.eireview.org.

Diehl P (2004) WISE – Uranium mining and milling wastes: An introduction; www.wise-uranium.org

Edwards G *et al.* (2006) Uranium – A discussion guide, prepared and published by The National Film Board of Canada, Canadian Coalition for Nuclear Responsibility (CCNR), www.ccnr.org

ESSC (2003) Mining and Critical Ecosystems: Philippines Case Study, Case Study commissioned by World Resources Institute. Environmental Science for Social Change: Manila, Philippines.

EESC (1999a), Decline of the Philippine Forest. Environmental Science for Social Change, Makati City, Philippines: Bookmark Inc.

EESC (1999b), Mining Revisited: Can an Understanding of Perspectives Help? Environmental Science for Social Change, Quezon City, Philippines: ESSC.

Douglas I and Lawson N (2000) An earth science approach to assessing the disturbance of the Earth's surface by mining, School of Geography, the University of Manchester.

Gelb A *et al.* (1988) Oil Windfalls: Blessing or Curse? New York: Oxford University Press.

Goodland R (2004) Examining the Social and Environmental Impacts of Oil, Gas and Mining; Sustainable Development Sourcebook for the World Bank Group's Extractive Industries Review

Government of Papua New Guinea (GoPNG) (2002) "Fourth Quarter Bulletin: Annual Bulletin, 2001" PNG Department of Mining. Port Moresby, PNG: GoPNG.

ICMM (2003) Good Practice Guidance For Mining and Biodiversity, International Council on Mining and Metals, pp 148. Author: Sally Johnson.

IEA (2002) World Energy Outlook 2002, International Energy Agency, OECD/IEA, Paris

IEA/OECD (2006) Case Studies in Sustainable Development in the Coal Industry; International Energy Agency (IEA), Head of Publications Service, 9 rue de la Fédération, 75739 Paris Cedex 15, France

Johnson SW (1997a) Hydrologic Effects, In JJ Marcus (ed.) Mining Environmental Handbook, London: Imperial College Press

Johnson SW *et al.* (1997b) Effects of Submarine Mine Tailings Disposal on Juvenile Yellowfin Sole (Pleuronectes asper): A Laboratory Study, Marine Pollution Bulletin Vol. 36 (4)

Kelly M (1998) Mining and the Freshwater Environment. London: Elsevier Applied Science/British Petroleum

Kettel B (1982) Gold, Graham and Trotman Ltd, London

Kuyek J and Coumans C (2003) No Rock Unturned: Revitilizing the Economies of Mining

Mason RP (1997) Mining Waste Impacts on Stream Ecology, In C.D. Da Rosa (ed), Golden Dreams, Poisoned Streams, How Reckless Mining Pollutes America's Waters and How We Can Stop It . Washington, DC: Mineral Policy Center.

Munthe J, Wängberg I, Iverfeldt Å, Petersen G, Ebinghaus R, Schmolke S, Bahlmann E, Lindquist O, Strömberg D, Sommar J, Gårdfeldt K, Feng X, Larjava K and Siemens V (2001) Mercury species over Europe (MOE). Relative importance of depositional methylmercury fluxes to various ecosystems. Final report for the European Commission, Directorate General XII. September 2001.

MMSD (2002) Breaking New Ground. pp420. Earthscan

Navin T (1978) Copper Mining and Management, University of Arizona Press

OECD (1983a) Coal and environmental protection: Costs and costing methods, Paris

OECD (1983b) Coal – Environmental Issues and Remedies, Organisation for Economic Co-operation and Development, Paris

OECD (1983c) Control Technology for Nitrogen Oxide Emissions from Stationary Sources, Organisation for Economic Co-operation and Development, Paris

OECD (1990) International Workshop on Methane Emissions from Natural Gas Systems, Coal Mining and Waste Management Systems, 9–13 April 1990, Washington, DC, USA

Oesterloef A and Oesterlund S (2003) Environmental Impacts and Assessment of Ore Mining in a European Context; Centek, OMENTIN Report No 3

OSM (2001) Technical Measures for the Investigation and Mitigation of Fugitive Methane Hazards in Areas of Coal Mining; The Office of Surface Mining Reclamation and Enforcement, US Department of the Interior.

Oxfam America (2002) Tambogrande Speaks Out, Oxfam America Issue Update. Available online at: www.oxfamamerica.org

Pearl R (1973) Handbook for Prospectors, McGraw-Hill, Inc.

Penman ADM (1998) The Need for Dam Safety – Case studies on tailings management, International Council on Metals and the Environment (ICME), United Nations Environment Programme (UNEP). pp. 5–6

Power Resources Inc. (1991) Company Brochure

Rice DD (1993) Hydrocarbons from Coal, Chapter 7, Composition and Origins of Coal bed Gas, AAPR Studies in Geology No 38, pp. 159–183.

Ripley EA, Redman RE and Crowder AA (1996) Environmental Effects of Mining. Delray Beach, FL: St. Lucie Press.

Ross ML (2001) Extractive Sectors and the Poor, Oxfam America: Washington, DC.

Sampat P (2003) Scrapping Mining Dependence, In C. Bright *et al.* State of the World: 2003 Washington, DC: Worldwatch Institute

Sideri S and Johns S (eds) (1990) Mining for Development in the Third World: Multinational Corporations, State Enterprises and the International Economy. New York: Pergamon Press.

Stephens C and Ahern M (2001) Worker and Community Health Impacts Related to Mining Operations Internationally-A Rapid Review of the Literature

Taylor G, Farrington V, Woods P, Ring R and Molloy R (2004) Review of the Environmental Impacts of the Acid *In Situ* Leach Uranium Mining Process, CSIRO Land and Water, Client Report, August, 2004

Thomson Gale (2006) Mining; http://www.pollutionissues.com/Li-Na/Mining.html

UNCAD (1996) Social Impacts of Mining; Workshop in Bandung, Indonesia, 14–15 October 1996

Underhill D (1992) *In-situ* leach uranium mining in the United States; Int. Energy Agency, Tech. Comm. on uranium *in-situ* leaching, Vienna, Austria, 5–8 October 1992

UNEP (2000) Mining and Sustainable Development II: Challenges and Perspectives; Industry and Environment Volume 23 Special Issue

UNEP (2001) Environmental Aspects of Phosphate and Potash Mining; United Nations Environment Programme -International Fertilizer Industry Association

UNEP (2001) APELL for Mining: Guidance for the mining industry in raising awareness and preparedness for emergencies at local level

UNEP (2002) Global Mercury Assessment. Website: http://www.chem.unep.ch

Underhill D (1992) *In Situ* Uranium Mining In The United States, Int. Atomic Energy Agency. Tech. Comm. Meeting On Uranium *In Situ* Leaching. Vienna, Austria, Oct 1992.

USEPA (1994) Technical resource document: Extraction and beneficiation of ores and metals Volume 4 Copper; EPA 530-R-94-031; NTIS PB94-200979

USEPA (1996) Emission Factors and Emission Source Information for Primary and Secondary Copper Smelters, EPA-450/3-77-051, US Environmental Pprotection Agency, Research Triangle Park, New York

USEPA (1999) Inventory of Greenhouse Gas Emissions and Sinks 1990-1997. Office of Policy, Planning, and Evaluation, U.S. Environmental Protection Agency, Washington, DC; EPA 236-R-99-003. (Available on the Internet at http://www.epa.gov/globalwarming/inventory/1999-inv.html.)

WHO (2001) Depleted uranium: sources, exposure and health effects http://www.who.int/ionizing_radiation/env/du/en/

Williams A and Mitchell C (1994) Methane emissions from coal mining, in: Mining and its environmental impact, Issues in environmental science and technology, Vol. 1, Editors: Hester R and R Harrison

Wirth JD (2000) Smelter Smoke in North America: The Politics of Transborder Pollution, Lawrence: University Press of Kansas, 264 pp. ISBN 0-7006-0984-9

World Coal Institute (2005) The coal resource: A comprehensive overview of coal, available at: www.wci.org

WRI (2003) Mining and Critical Ecosystems: Mapping the Risks. World Resources Institute

Appendix 13.1 An Overview of Environmental and Social Risks and Potential Financial Implications

Environmental Risks and Financial Implications

ENVIRONMENTAL AND SOCIAL RISKS

Direct Risks

Land requirement – Land areas for open pit excavation, dumping of mine wastes external to the pit or underground voids, and the construction of transportation corridors and infrastructure may destroy surface features of economic, cultural, or nature conservation value.

Waste rock disposal areas – Waste rock may contain heavy metals. If waste rock contains sulphur, acid rock drainage may occur. Runoff or leachate may contain elevated concentration of metals and low pH values posing a particular threat to the aquatic environment long after mine closure.

Tailings storage areas – Tailings may contain heavy metals, sulphur, or mill reagents. Acid rock drainage potentially occurs. Runoff or leachate from tailings storage areas may contain elevated concentrations of these elements and low pH values imposing a particular thread to the aquatic environment. Discharge water may qualify as effluent.

Stockpile areas – Stockpiles may contain heavy metals and mineral oxidation products. Runoff and leachate may contain elevated concentration of heavy metals and be of acid pH. Spontaneous combustion of coal stockpiles and spoil heaps may occur if coal residues are present in the heaps.

Geotechnical stability – Inadequately designed and maintained pit faces, spoil heaps, waste rock dumps, or tailings dams may become unstable leading to ground or tip collapse. Failure of tailings containment will result in the release of large volume of potentially hazardous tailings.

Ground subsidence – Surface damage due to ground subsidence is a major risk of underground mining. Structural damage to buildings, infrastructure, agricultural land and drainage system may occur, in particular if significant differential subsidence occurs.

Water erosion – Disturbed land is subject to water erosion and subsequent impacts on the aquatic environment. Land disturbance occurs due to direct mining activities as well as due to any other construction or land clearing activity (such as road construction). Runoff may be of acid pH with elevated concentrations of heavy metals.

Mine dewatering – Groundwater from mine dewatering discharged to surface water may be acid in nature and contain elevated concentrations of dissolved metals due to mineral oxidation and dissolution within the mine. Groundwater draw down may lead to drying up of well, springs, wetlands, or entire surface water bodies. Depending on geology, groundwater abstraction may lead to ground subsidence.

Groundwater rebound – Cessation of pumping at the end of mine life will result in groundwater rebound (rising groundwater levels) leading to surface discharge of potentially contaminated mine water. Groundwater rebound may also lead to ground instability and potentially flooding in areas distant from the mine. Groundwater rebound may pollute upper aquifers.

Waste generation – Increased population and workforce will generate sanitary waste. Industrial wastes originate from workshops, broken equipment, unused materials, waste tires, scrap metal, power generation and many other activities related to mine operation. Health facilities will generate medical waste. Waste rock and tailings, however, will always dominate the generated waste streams.

(Continued)

Environmental Risks and Financial Implications

Health and safety – Mining presents high risk to the health and safety of mine workers, in particular if mining is underground. This is due to accidents, geotechnical failures of walls or dumps and exposure to damaging levels of dust, gases (particular methane in underground coal mining), radioactivity, noise and vibration. The use of explosives to assist in ore fragmentation and removal creates an additional safety risk both within the mine site and in surrounding areas. High voltage electrical supply to mine equipment such as crushers, mills, screening equipment, or conveyor belts is a potential hazard. Asbestos may be present within mine buildings.

Particulate matters – Mineral extraction, crushing, materials transportation and un-vegetated surfaces may result in emission of particulate matters to the atmosphere with potential for trace metal and dust deposition over large areas.

Greenhouse Gas Emissions – Mines move a large quantity of earthen materials. With the exception of coal mining, material movement is likely is the mine cause of greenhouse gas emissions. Release of coal bed methane and coal combustions are the main sources of GHG emissions related to coal mining.

Other direct impacts – Various other direct impacts are potentially associated with mining. Noise and vibration is likely to be generated primarily by blasting, mineral extraction, mineral crushing, and vehicle movements. Spillage of fuel, oils and other chemicals particularly in workshop areas may cause contamination of both soil and water. Mines are intrusive by nature, often perceived as negative visual impacts. Lighting at night operation is equally intrusive to nearby resident and wild life fauna. Water supply to the mine may reduce water availability for other users.

Indirect impacts

Land acquisition and resettlement – Land acquisition and resettlement are closely interrelated. Resettlement per se is a social change process that places pressure on social system. Resettlement is not restricted to physical relocation of people. Any loss of assets, access to resources, or adverse impacts on livelihood due to the mine development count as resettlement.

Mine access – Access roads to previously hard to reach areas may place hunting pressures on the fauna and may give rise to illegal logging and settlement. Providing access to remote areas often constitutes one of the highest environmental impact potentials.

Induced development – Mining activities often attract people looking for direct employment or indirect economic benefits. Influx of outsiders will put considerable strain on existing social pattern and infrastructure, often causing social fabrics and existing infrastructure to fail. Increase in crime, diseases, and social tensions are direct consequences. Mine towns often wholly depend on the existence of the mine. Mine closure associated with demobilization of workforce and loss of income is challenging. Note, however, that the first round of workforce demobilization already occurs at the end of mine construction.

FINANCIAL IMPLICATIONS

Land compensation – Mine operators need to compensate landowners for the land areas occupied by the mine and supporting infrastructure. Land speculators often inflate the value of land, and encourage local residents to make increasing claims for land compensation. Landowners may sell land twice, once to the mine, and a second time to a neighbour. The mine company may end up paying the neighbour's bogus claim to avoid trouble.

Regulatory approval – Financial losses due to a delay or a refusal of environmental mine approval is deeply embedded into the history of mining. A delay of approval can easily render a financially attractive mine asset into a financial disaster.

Resource compensation – Regulatory authorities will possibly ask for compensation for loss of natural resources such as water, agricultural land, or forest. Landowners may ask for compensation for allowing temporary access to land for exploration purposes or for potential environmental damage such as vegetation clearing. Residents or squatters may cultivate in areas cleared for exploration, gambling that they would receive company compensation for lost harvest.

Fees and fines – Fees by regulatory authorities may be applied to discharges to air and water. Regulatory authorities may impose fines if discharges are above statutory levels.

Social unrest – Protest by local residents or non-government organizations to defend existing surface features, ironically often initiated by anti-mining activists with no site-specific knowledge, can lead to delays in mine permitting, in a reduction of resources that could be otherwise exploited, and in an increase in mine operational costs.

Social investment – Community development programmes are integral to any mine proposal. It is the intent of the mine operator to give something back to local communities. Funds for community development programmes need to be allocated.

Health compensation claims – Exposure of employees to occupational hazards may result in health compensation claims. Methane or radon gas release and migration is a significant hazard in coal mining.

(Continued)

Environmental Risks and Financial Implications

Investment in environmental controls – Major investment costs are necessary to provide environmental protection controls. Significant increase in operation costs could occur where outdated site facilities need to be replaced to satisfy a more stringent regulatory environment. Poor environmental performance in the past may accelerate the demands for a more stringent regulatory environment.

Claim for accidental damage – Failure of pit walls, waste rock dumps, or tailings dam has the potential to cause loss of production time and loss of life with associated financial liabilities. Failure of tailings containment imposes a high environmental and financial risk. Land subsidence caused by underground mining may have severe financial implication.

Risk related to global warming – Admittedly not everybody buys into the current dominant environment scheme of global warming. However, the potential financial consequences of a change of global climate on existing mining operations can be severe. A prolonged through can limit ore processing, and as such revenue generation.

Mine closure fund – Provisions have to be made for rehabilitation of disturbed areas and eventually for site decommissioning and final rehabilitation. Groundwater rebound if it occurs may lead to resurgence of contaminated mine water at surface increasing mine closure costs substantially. Mine closure constitutes a significant and often long-term financial commitment.

Appendix 13.2 Environmental Impacts at the Exploration Stage

Exploration activities can result in social and physical disturbance, which depend on the type of activity and at what stage it is undertaken in the exploration programme. The potential impacts of exploration activities range from low to high as illustrated in **Table 1**.

Today environmental and social considerations are integral to the planning of exploration programmes. Adequate planning commonly embraces the following philosophy:

- consultation with relevant authorities, landowners and interested parties on environmental issues both prior to and during any field activities;
- identification and documentation of potential impacts on the environment at each exploration stage;
- adoption of design and operating strategies which avoid or mitigate adverse impacts;
- to the extent practicable, restoration of all areas disturbed by exploration to an acceptable state to facilitate re-vegetation (or if appropriate, landowner sanctioned alternate, beneficial land use, e.g. market gardening);
- monitoring and auditing of environmental performance at all stages of the exploration programmes; and finally
- regular communication with and feedback to landholders and public authorities.

TABLE 1

Categorization of Exploration Activities According to Environmental Impact Potential – With the possible exception of 3D seismic exploration surveys or bulk sampling, exploration activities normally do not require the preparation of a full Hedged environmental impact assessment study; generally, good environmental management practices are sufficient

Stage	Example Activities	Impact Severity
Preliminary reconnaissance	• Air photo and land analysis • Airborne geophysical survey • Geological mapping (no clearing/ trenching required) • Stream sediment sampling (e.g. walk-in, hoisted-in) • Rock sampling (no clearing)	• Usually very low
Exploration and limited ground surveys	• Grid soil sampling; surveying • Line clearing for geophysical survey	• Usually low
Resource definition and evaluation	• Site clearing for helicopter pads, drill sites, camp sites • Operation of camp sites • Drilling • Trenching; adit development • Bulk sampling	• Low to moderate, depending on areal extent

With the possible exception of three-dimensional seismic exploration surveys or bulk sampling, exploration activities normally do not require the preparation of an environmental impact assessment study; generally, good environmental management practices apply. Consistent with safe operating practice and compliance with statutory requirements, the design and operating strategies adopted for an exploration programme should consider the following issues and opportunities.

SOCIAL FACTORS

At all stages of an exploration programme, social and ethnographic factors are important and play key roles in the planning, timing and conduct of a specific exploration activity (such as tenement access; compensation; installation and maintenance of environmental monitoring stations, e.g. weather and stream gauge stations; or land ownership surveying). It is good practice to commence a well-structured community consultation programme prior to exploration, to enhance community understanding of the exploration programme and to minimize potential social impacts.

LAND OWNERSHIP AND APPROVAL

Access to land requires consent of and approval by the landowner and land user. Exploration personnel should be familiar with the terms and conditions of the access agreement for the land on which they are working. In addition to the terms and conditions of an access agreement, areas of particular concern to individual landowners should be discussed, negotiated and, if necessary, added to the agreement. These include issues such as the location of water supplies, water lines, access roads, fences, domestic bathing and laundering sites. Regular dialogue should be maintained with landowners and land users so that potential concerns can be recognized at an early stage and resolved with general consent.

Not all land areas are open for exploration. For example, land areas designated as National Parks or Conservation Reserves, are excluded from exploration in some countries. In other cases, exploration in such areas may only be permitted under rigorous conditions determined by Government, following public consultation. Calls for the establishment of wilderness or World Heritage areas and the possibility of future native title considerations may also significantly influence areas where exploration access is impeded.

AERIAL ACCESS

Remote areas are often accessed by helicopter, or if a suitable landing strip exists, by fixed wing aircraft. Helicopters generally require no specially prepared landing areas except in forested areas. Support for personnel in remote area exploration programmes can vary widely. Some exploration managers operating major programmes consider it cost effective to access the area by helicopter on a daily basis. Field crews are based in a nearby town or base camp and are airlifted into and out of the area, eliminating the need for camping. For smaller programmes, field crews are usually flown into an area for an extended period and establish a field camp. Helicopter access is favoured by many exploration managers for exploration of remote areas. However, helicopter operations are expensive and are beyond the means of many exploration companies.

GROUND ACCESS

The type of ground access required and its environmental impact is influenced by the nature of the landscape and the exploration programme to be conducted. Initial mapping and stream sediment geochemical surveys are usually undertaken along existing roads, survey tracks, open areas or stream beds, with little disturbance to local people and the environment.

When small areas of interest are defined for more detailed geochemical or geophysical work, a system of surveyed traverse or grid lines is normally required. Grid lines in open country require no clearing. The lines are marked at regular intervals with biodegradable wooden pegs and flagging tape. Low impact geophysical, geochemical and geological surveys are carried out along these lines. Increased use of satellite-based Global Positioning Systems (GPS) has lessened the need for a grid system to be put in place. To the extent practical, grid lines across existing walking tracks should be avoided. In very sensitive areas, grid pegs can be removed after completion of the survey. Otherwise, they are normally left to degrade and disappear naturally.

In open country no special access is required for drilling rigs. In heavily timbered areas, a four wheel drive track is normally required. Permission is necessary and stringent conditions apply to the construction of any track involving clearing and Earth works. As a general rule, the need to construct tracks for vehicle access should be minimized. Clearing associated with the track and drill pads should be kept to a minimum to reduce disturbance. In timbered areas, clearing is often carried out in association with the Government to ensure that all economic timber is salvaged. Topsoil is commonly stripped from cleared areas and stockpiled for later rehabilitation.

Where a track is only to be used for a short period, it is more appropriate to leave the topsoil and root stock in place. The thorough cleaning of soil and seeds from equipment and vehicles before entering a forested area is also encouraged in order to prevent the spread of noxious weeds or pathogens. This practice is mandatory in areas infested with die-back as in Southwest Western Australia.

When drilling or other detailed ground surveys have been completed, areas that involve Earth works are rehabilitated. For example, drill sites on farmland may only require surface loosening with a ripper, re-spreading of any excavated topsoil and sowing with suitable grass or tree species. On un-cultivated land, rehabilitation of access tracks and drill sites generally involves a combination of the following steps: (1) installation of permanent drainage in wet areas; (2) deep ripping of compacted areas; (3) re-spreading of topsoil; (4) bush matting for erosion control and seed source; (5) direct seeding with appropriate local species; and (6) potentially, fertilizer application. The success of the rehabilitation should be monitored and, if necessary, the process should be repeated.

At exploration sites, walking access may be via raised walkways constructed from cleared timber above the ground to avoid disturbing the vegetation and soil. The use of constructed earthen paths in wet conditions should be avoided if possible. Ruts on paths capture runoff water which, on a well-maintained path, should flow into side drains constructed alongside. Otherwise, erosion gullies may result. Obstructions should be removed from any water drainage channels to ensure minimal scouring.

DRILLING

Many different drilling techniques are used, depending on the geographic and geological situation. Increasingly, man-portable drilling rigs are used to minimize disturbance in high quality natural environments including rainforest. Containment of drill cuttings and drilling fluids is another important issue. Following drilling, drill sites are usually restored to their pre-drilling condition, except in situations where it is certain that the area will be mined in the near future.

WATER MANAGEMENT

Like everyone else, exploration geologists depend on a continuous supply of clean water. Clean water sources in exploration areas need to be identified, if they exist. Exploration staff should be aware of existing water supplies prior to entering the exploration site. Silt generated from exploration activities should be diverted away from drawing points of clean water or washing sites.

RUNOFF AND EROSION CONTROL

Sediment control measures need to be coupled with appropriate control measures to divert water runoff away from potential erosive areas. A few principles govern the effectiveness of erosion and sediment control and should be implemented continually during the life of all exploration activities: (1) minimize disturbed areas; (2) manage site drainage (e.g. establish diversion ditch); (3) control erosion and sediment; (4) support progressive re-vegetation; and (5) closely manage field work. Managing field work includes advising all site-labor and sub-contractors of their responsibility in minimizing soil erosion and other forms of pollution.

WASTE MANAGEMENT

It is considered good industry practice to remove from the exploration area all that is taken into it, including packaging, cans, bottles, orange skins and consumables. As such, waste management during exploration depends on understanding the type and quantity of incoming materials. Materials used vary widely and can be classified into four main groups: liquid, solid, gaseous and food/putrescible materials (**Table 2**).

SURVEY MARKERS

It is good practice to use only as much flagging tape as necessary and to avoid fluorescent paints. If marking rock faces for sampling and mapping, aluminium permatags can be as effective as paint, are less visually intrusive and last longer if not physically removed. When using a hip chain with nylon thread for measuring distances, good practice is to remove the thread when walking back along the traverse. Biodegradable thread is preferred.

VEGETATION CLEARING

Vegetation protects the land surface from erosion, while foliage protects the soil from the impact of rain and reduces runoff. Tree roots bind the soil to prevent surface erosion and

TABLE 2

Waste Categories for Exploration – It is considered good industry practice to remove from an exploration area everything that has been brought in

Category	Type	Description
Liquid	Fuel Lubricants Solvents Detergent Chemicals Surfactants/drilling polymers	Diesel, Benzene and Aviation fuel Oils and Greases Biodegradable Acid, Alkaline-general multipurpose, soap/hand cleaners Paints Drilling mud, drilling additives
Solid	Containers Packaging Polyweave sheeting Drill rods PVC pipe PVC hosing Metal sheeting Timber Medical waste Sample bags	Fuel drums, lubricant drums, chemical containers, gas bottles Food packaging, etc Used to clad temporary field camp structures Unserviceable/broken drill rods Used as drill collars Used for water reticulation Used as cladding Planking imported 'Sharps', ampoules, dressing, etc Used to store samples
Gaseous	LPG Acetylene Oxygen	For cooking Used for welding Used for welding
Food	Various	Consumed, eventually leading to sanitary waste

consequent slipping and slumping. Therefore, where practical, clearing of vegetation should be kept to a minimum. Removing of rootstock or vegetation on ridges or steep slopes, or where soil development is minimal, should be avoided. Where areas are to be cleared of large timber, timber not used for construction should be stored for use at other sites where timber availability may be limited. 'Trash' vegetation (leaf litter, bark, twigs, timber off-cuts, or sawdust) can be used as a surface cover to reduce potential erosion. Vegetation declared as a 'productive plant' but which is to be disturbed or removed during exploration, needs to be tagged and numbered. Records should be kept identifying species type and diameter at breast height (DBH), and tree height. Good pre-planning will require all tagged trees to be registered on a database, with compensation to landowners to be calculated on a regular basis, as administered by the Government. Appropriate measure should be taken to minimize disturbance of rare or endangered flora and fauna. Fauna breeding or nesting areas should be avoided. Exploration drill rigs and other vehicles/machines should be thoroughly washed when moving between separate exploration areas to reduce the spread of vegetation diseases and weeds.

FIRE PREVENTION

Fire prevention is always of concern when working in remote areas. Fire extinguishers are standard items on all contractor machinery and drilling equipment. Field crew members who smoke must ensure that cigarettes and matches are extinguished after use. Local fire regulations should be observed. All field crews should be familiar with fire prevention procedures, particularly around drill sites and in field camps.

CHEMICAL STORAGE

Management procedures should be in place to ensure safety and to prevent land contamination due to poor on-site storage of chemicals. Chemicals at the exploration site are to be covered securely until used. Fuels and oils typically comprise diesel, benzene, aviation fuel and lubricating oils. A chemical inventory should be maintained at all times. Management of chemicals should follow standard industry practices: (1) fuel, oil and chemical stores should be contained, where necessary, in bonded designated safe areas, and distant from watercourses and other sensitive areas; (2) chemicals such as undiluted drilling fluids or fuels should not be discharged into water bodies (including the sea) or dumped on the ground or in excavations. They should be either incinerated in burn pits or temporarily stored prior to transportation to a suitable site for final treatment or disposal; (3) oil absorbent booms should be used where appropriate if complete containment of spills is not possible. They should not be removed until the contaminated areas are skimmed for oil; (4) generators should be stored in shipment boxes and have taps and trip trays set up to contain any spillage.

Used oil should be collected and re-used as chainsaw chain lubricant or as grease for rods. Substantial spillage (e.g. more than 10 liters) of undiluted drilling fluids, fuels, oil or other potentially harmful chemicals should be reported as soon as possible to the responsible authorities, and effective clean-up measures undertaken immediately. Closed water reticulation system should be used for drill fluids wherever possible. Where this is not practical, any necessary discharge to a water course should meet agreed standards.

HERITAGE PLACES

Exploration often occurs in hard to reach areas that may hold hidden treasures other than concealed ore deposits. If archeological materials are uncovered, works which may affect those materials should cease and the heritage branch of the Government should be notified immediately. Work should only resume following inspection by that department and implementation of appropriate recovery or protection measures.

REHABILITATION STRATEGIES AND ACTIVITIES

The general principle of minimum disturbance should apply to all exploration activities. In the event that land disturbance is necessary, the following practices should be implemented:

- Rehabilitation strategies need to be discussed with land owners and land users.
- Exploration planning should include consideration of alternative methods of testing a target to minimize land surface disturbance.
- Costeaning should be avoided where possible. Topsoil should be pre-stripped and stockpiled for top-dressing at a later date. Costeans should be aligned to assist drainage and costean length should be limited to 50 m or less.
- The smallest drilling rig capable of completing the work should be employed.
- When the project work is completed drill pads should be cleaned up. Costeans should be backfilled followed by topsoil coverage.
- Water from drill sumps should be pumped into drums and transported off-site. Sumps should then be backfilled and allowed to re-vegetate naturally.

- Other refuse resulting from exploration activities should be removed and disposed of in an acceptable manner upon completion of the exploration activities.

Sound environmental management practices require guidance from qualified professionals. Access to such expertise should be available at all stages of exploration. The nexus between safety, occupational health and pollution control should be provided for by enabling the available expertise in these areas to coordinate their activities and inputs into the design of the exploration programme. Environmental operating practices will more effectively translate into good environmental performance when field personnel are trained to recognize the potential environmental implications of their actions and are motivated to act responsibly. Exploration managers should strive to maintain a high level of environmental awareness.

Monitoring programmes as illustrated in **Table 3** help to demonstrate compliance with good environmental management practice. An internal company monitoring report, including an interpretation of all monitoring results, should be completed each year. This is to ensure successful operation of control strategies, compliance with statutory requirements and to confirm that planned management procedures have been successfully undertaken.

TABLE 3

Examples of Environmental Monitoring Programmes during Exploration – A variety of monitoring methods may be employed: visual monitoring, vegetation, water and erosion monitoring. An internal company monitoring report, including an interpretation of all monitoring results, should be completed each year

Type	Aim	Method	Frequency
Land Resources			
Erosion	To identify gully erosion on disturbed land To detect failure of contour and diversion banks, waterways and gullies To measure soil loss rates	Surveillance Surveillance Measure controls and relate to UMA	Monthly in wet season Every runoff event Annual
Revegetation	Compare vegetation growth with adjacent areas	Survey to UMA, measure vegetation cover and/or bulk	Annual to broad level, detailed when required
Land Capability	Measure tree/shrub establishment To measure land capability of rehabilitated areas (includes identification of toxic material)	Inspection/Photographic recording, Count trees/shrub >1 m in height in transects in UMA Measure of exploration activity on downstream water quality compared to upstream sites	Annual Runoff events or monthly during flows
Water Resources			
Surface Runoff	To monitor water quality	Impact of exploration activity on downstream water quality compares to upstream sites	Runoff events or monthly during flows
Groundwater (where applicable)	To monitor groundwater quality	Groundwater bores	Monthly
Air and Noises			
Proximity of dwellings from exploration activity	Mapping of any dwellings with respect to proximity to air and noise emanating from exploration activities	Investigate issues raised	On a nuisance basis
Social Issues			
Regional Infrastructure	Check for adverse impacts	Investigate issues raised	Ongoing

(Continued)

TABLE 3
(Continued)

Type	Aim	Method	Frequency
Cultural Heritage			
Cultural Sites	Check protection measures	Inspection/photo-graphic Recording/surveillance	Annual/as required
Conservation Values			
Land disturbance	Ensure distance is limited to disturbance areas	Survey	Annual/as required

* Uniform Mapping Area based on topography, soils type and vegetation type.

A variety of monitoring methods may be employed: (1) vegetation monitoring by annual paired photography, where photos of undisturbed areas are compared to rehabilitated areas; girth measurements of trees and shrubs; counting stems per hectare and/or transect species counts from ranked paired photos; (2) water monitoring (*in situ* and laboratory testing); or (3) erosion monitoring by rill counting and cross-section measurement. Actions considered appropriate to rectify causes of unacceptable results should commence as soon as possible after identification of the problem.

14

Land Acquisition and Resettlement

When Property and Development Rights Collide

NICKEL

Nickel is a hard, gray, magnetic ductile metal that is extremely corrosion resistant. It is an essential element but some nickel compounds are known to be toxic or carcinogenic. It occurs in association with iron and magnesium in laterites derived from ultra-basic and ultra-mafic igneous rocks. It also occurs in sulphide ores, the main mineral being pentlandite. The main use of nickel is in stainless steel and other alloys, including superalloys. Nickel is also used in rechargeable batteries.

14 Land Acquisition and Resettlement

When Property and Development Rights Collide

Who is entitled to acquire land from the individual for the benefit of the society-at-large?

The cultural, economic and emotional impacts of resettlement are written deeply in the history of both industrial and developing countries. In the latter, development projects and programmes that use a nation's passive natural wealth are indisputably needed; they broaden the production base, create employment, bring added revenues and improve many people's lives. However, the concept of property versus development rights is ill-defined: who is entitled to acquire land from the individual for the benefit of the society-at-large?

In developing countries, the scale of development-related resettlement has grown rapidly in the past few decades due to a combination of accelerated industrial and infrastructure development and ever-growing population densities. By some estimates more than ten million people are involuntarily displaced every year (World Bank 1994, Goodland 2004).

Dams, urban development and transportation programmes represent the main causes of development-related resettlement. The displacement toll of the 300 large dams that commence construction in an average year is estimated to exceed four million people (World Bank 1994). The displacement toll of the controversial 'Three Gorges Dam' in China alone amounts to more than one million people. Urban development and transportation programmes displace an additional six million people annually.

Providing mine access alone may not cause excessive resettlement, but associated mine waste disposal sites, mine infrastructure, roads, rails and pipeline rights-of-way often do.

Mining projects too contribute to resettlement, particularly those surface coal mines and laterite mines which involve large areas. Large-scale coal strip mining operations in Germany for example continue to force the relocation of families, and entire communities (see also Case 7.7). Providing mine access alone may not cause excessive resettlement, but associated mine waste disposal sites, mine infrastructure, roads, rails and pipeline rights-of-way often do. However there is usually some flexibility in the siting of mine facilities, access routes and infrastructure so that the resettlement of people can be minimized, if not avoided. Of course there is little flexibility with the site of the mine itself.

Project-related environmental change and pollution such as degradation of water quality, increased noise or dust, or increased traffic may also make it necessary to physically relocate people. For instance, the Skorpion mine in Namibia, South Africa is relocating an informal settlement of about 4,500 people for health and safety reasons (Rio Tinto 2005).

The argument previously voiced by developers and lending agencies that 'You can't make an omelette without breaking eggs' (Goodland 2004) is now outmoded and discredited. Globalization and the attendant global environmental awareness have forced mine developers to avoid wherever possible, or otherwise to minimize negative impacts associated with the resettlement of people. Where it is unavoidable, resettlement now constitutes a planned process involving project-affected people, host communities, mine management, local governments and other relevant stakeholders. Today the goal of resettlement is crystal clear: that displaced people will be better off after they have moved. Accordingly, where poor or disadvantaged people are involved, resettlement should be seen as a valuable project opportunity to reduce poverty.

Today the goal of resettlement is crystal clear: that displaced people will be better off after they have moved.

The involvement in resettlement of multilateral lending agencies such as the World Bank stems largely from their past involvement in financing large dam developments. These projects, of course, have become synonymous with the involuntary displacement of millions of people, often without adequate compensation or resettlement assistance. It was largely the resultant negative publicity, and criticism of the lender's failure to intervene on behalf of the displaced people that led the World Bank in 1990 to spearhead the development of policies and procedures for involuntary resettlement. The bank's current policy, applicable to private sector investment, is the IFC's Policy on Social and Environmental Sustainability and its associated Performance Standard (PS) 5 Land Acquisition and Involuntary Resettlement (IFC 2006). IFC PS 5 addresses both physical displacement (where people experience relocation or loss of shelter) and economic displacement (where people or communities experience loss of assets or access to assets that leads to loss of income or means of livelihood).

IFC PS 5 describes the World Bank Group's resettlement policy for private sector projects (**Table 14.1**) and is the most widely used international benchmark for private sector projects involving resettlement. IFC PS 5 also forms part of the Equator Principles. IFC PS 5 covers: policy objectives; types of impacts; required assistance or compensation measures; eligibility for benefits; resettlement planning, implementation and monitoring; and resettlement instruments. Required resettlement instruments may include a resettlement action plan (for projects involving physical displacement) or resettlement framework. The Guidance Notes associated with PS 5 provide additional clarification on procedural requirements to achieve the required standards.

IFC PS 5 is the most widely used international benchmark for private sector projects involving resettlement.

Dams generally displace far more people than mining, as they invariably inundate valley floor areas which are usually populated, sometimes densely. Mining, except for some surface coal mines, seldom impinges on these areas. However, the findings and lessons learnt from past dam developments are relevant and instructive in relation to resettlement due to mining. The World Commission on Dams (WCD) has published a set of guidelines for good practice, which encompass how to plan and implement involuntary resettlement. As well as covering issues related to the economic and bio-physical impact of dams, the report on 'Dams and Development' covers the experience of displaced people and their resettlement and compensation (WCD 2000). The WCD recognizes that successful resettlement relies upon supporting national legislation and development policies, as well as accountability and commitment from governments and project developers (Goodland 2004).

Successful resettlement relies upon supporting national legislation and development policies, as well as accountability and commitment from governments and project developers.

TABLE 14.1

IFC Performance Standard 5 Land Acquisition and Involuntary Resettlement

• Involuntary resettlement should be avoided or minimized wherever feasible by exploring alternative project designs.
• Resettlement action plans should be developed for cases where displacement is unavoidable.
• Land rights should be acquired through negotiated settlements wherever possible, even if the project developer has the legal means to gain access to the land without the seller's consent.
• Resettlement programmes should improve or at least restore the livelihoods and standards of living of displaced persons.
• Resettlement programmes should improve living conditions among displaced persons through provision of adequate housing with security of tenure.
• The project developer will provide opportunities to displaced persons and communities to derive appropriate development benefits from the project.
• The project developer will consult with and facilitate the informed participation of affected persons and communities, including host communities, in decision-making processes related to resettlement.
• Resettlement should be conceived and executed as a development programme.
• Adverse social and economic impacts from land acquisition or restrictions on affected persons' use of land should be mitigated by: – Providing compensation for loss of assets at full replacement cost; – Ensuring that resettlement activities are implemented with appropriate disclosure of information and the informed participation of those affected.
• Compensation and resettlement assistance should not only be limited to affected persons that hold legal title to the land. The Bank recognizes the existence of usufruct or customary rights to the land or use of resources.

The importance of managing resettlement is now recognized by a majority of bilateral and multilateral lenders, development banks and private financial institutions. Some have followed the World Bank's example and have developed their own resettlement policies and guidelines. The ADB Handbook on Resettlement 'A Guide to Good Practice' (1998) describes resettlement planning in the context of the ADB's project cycle. It elaborates on key resettlement planning concepts; explains data collection and participatory methods, and their application to resettlement planning; and reviews income restoration.

Past experiences clearly illustrate the perils of resettlement: failure will jeopardize the success of the entire mining project. Trauma, social unrest, conflict and impoverishment caused by resettlement and by the associated disruption of existing social patterns, once developed, are difficult to reverse. Their consequences may haunt the mine operation throughout its life.

14.1 SOME USEFUL DEFINITIONS

The issue of resettlement presents two opposing sides that form a whole: displacement and livelihood restoration.

The issue of resettlement presents two opposing sides that form a whole: displacement and livelihood restoration. The first side, displacement, is the process of people losing land, assets, or access to resources. The second side, livelihood restoration, is the process of assisting adversely affected people in their effort to improve, or at a minimum to restore their previous living standards. All activities related to resettlement, from land recognition, land

FIGURE 14.1
From Relocation to Restoration

The issue of resettlement presents two opposing sides that from a whole: displacement and rehabilitation.

acquisition, compensation packages, to long-term community and regional development programmes, are aligned in time with these two processes as illustrated in **Figure 14.1**.

It is also important to recognize that resettlement is a social change process that places pressure on social systems with subsequent direct and indirect social impacts. Indirect social impacts result where change processes leave people no other option but to move.

From the above the following definitions emerge:

Rights to land include many levels of land tenure and use, all of which are relevant in the determination of physical or economic displacement. These include ownership, leasehold, rights of use (such as for cultivation, grazing and gathering), customary or usufruct rights (which include grazing, hunting and fishing rights), ancestral domain (which refers to lands occupied or used by Indigenous Peoples), and third party proprietary interests such as share cropping, rental or other formal or informal agreements. It follows that 'land ownership' in the western sense is only one aspect of rights to land.

Resettlement includes any loss of assets, access to resources or adverse impact on livelihood. It follows that resettlement is not restricted to the physical relocation of people. The ADB (1998) defines the resettlement effect as the loss of physical and non-physical assets, including homes, communities, productive land, income-earning assets and sources, substance, resources, cultural sites, social structures, networks and ties, cultural identity and mutual help mechanisms.

Resettlement is not restricted to the physical relocation of people.

Trigger levels, however, do exist. For example, in relation to land loss, the ADB only requires detailed studies on the economic impacts on households if the land loss exceeds 10% of the total land owned or used by the affected household. However in many traditional rural areas, land holdings by community members are small. Assessing resettlement losses of hundreds of small land holders as is commonly required in surface mining which covers large areas, is time consuming and, at times, difficult.

Involuntary resettlement 'refers both to physical displacement (relocation or loss of shelter) and to economic displacement (loss of assets or access to assets that leads to loss of income sources or means of livelihood) as a result of project-related land acquisition. Resettlement is considered involuntary when affected individuals or communities do not have the right to refuse land acquisition that results in displacement. This occurs in cases of: (i) lawful expropriation or restrictions on land use based on eminent domain; and (ii) negotiated settlements in which the buyer can resort to expropriation or impose legal restrictions on land use if negotiations with the seller fail' (IFC PS 5).

Displacement comprises both physical and economic displacement. Physical displacement occurs where the inhabitants must be relocated in order for the development to proceed. Economic displacement occurs where a project causes loss of livelihood, regardless of whether or not there was physical displacement.

Eminent domain is the term given to the power of the State to expropriate property or restrict land use without the owner's consent. In some States, the right of eminent domain is a last resort used only after negotiations have failed. Under many jurisdictions, eminent domain procedures may be very time consuming, leading to significant project delays.

Negotiated settlements are those achieved by provision of fair and appropriate compensation, plus other incentives and benefits to affected persons and communities, and by mitigating the risks of unequal access to information and bargaining power. Included in these negotiated settlements are 'willing buyer – willing seller' transactions, which perhaps are more common in developed countries. IFC PS 5 does not apply to these voluntary transactions. However, other social impact requirements as in IFC PS 1, still apply.

Compensation refers to measures to compensate for resettlement losses and may include cash payments or compensation in kind (such as 'land for land'). Decades of international experience in resettlement suggest 'land for land' compensation alone, although preferable, is seldom fully accepted by displaced people; they often demand cash payments. As a result, most compensation schemes provide compensation in both cash and non-cash forms.

Livelihood restoration refers to measures aimed at restoring the livelihoods of affected people after they have being compensated for resettlement losses. There is a clear difference between compensation for losses and techniques for providing livelihood streams/ income replacement or community benefits. IFC PS 5 and its predecessors require compensation for land and lost assets in the first instance plus measures to restore livelihood. One is not a substitute for the other.

According to IFC PS 5, **displaced or affected people** include those '(i) who have formal legal rights to the land they occupy; (ii) who do not have formal legal rights to land, but do have a claim to land that is recognized or recognizable under national laws; or (iii) who have no recognizable or legal right to the land they occupy'. Displaced people are not the only people directly affected by the mine; host communities too may suffer. Social conflicts are likely to arise when a large population is displaced onto the land of a smaller existing population. There is always the need to investigate the capacity of the host area to sustain a resettled population without serious resource depletion. Important factors to consider include availability of clean water (in all seasons); amount and productivity of agricultural land; health services; school system; sewage disposal and water delivery systems; road network; law enforcement services; and utilities.

Displaced people are not the only people directly affected by the mine; host communities too may suffer.

A **resettlement plan** defines the policy applied to resettlement, the magnitude and severity of resettlement losses, approaches to compensation, and measures applied for livelihood restoration. It provides important protection to displaced people and to the company displacing them – where legal rights to land are poorly defined; where legislative or institutional frameworks are weak; where land markets are poorly developed or in a formative stage; where there is no institutionalized social safety net; and where displaced people may have limited or no access to legal remedies and protection in the event of a dispute.

Resettlement is an important component of the Equator Principles with the 'Land Acquisition and Resettlement Action Plan' (LARAP) document one of the key public documents to be prepared by proponents. While virtually all mining projects require the acquisition of land, physical resettlement of people from this land is not required for all or even a majority of mining projects. However, it is important to understand that the Equator Policies are applicable, even where there is no physical resettlement of people. In developing countries and in developed countries where the interests of Indigenous Peoples

The Equator Policies are applicable, even where there is no physical resettlement of people.

apply, acquisition of land for the mine and associated facilities, involves sensitive and difficult issues which must be resolved before the project can proceed. Bad feelings generated during land acquisition can remain, with adverse consequences to the project well after the acquisition process has been completed.

14.2 WHAT DETERMINES THE SEVERITY OF RESETTLEMENT LOSSES?

Resettlement effects are complex. They depend on the time and place where they occur, and on the type of people that are affected. Impacts and risks that are severe at one mining project may be of no significance for another project. Whether impacts of resettlement are severe will depend in large measure on how resettlement is managed.

Number of Affected People

Of course the number of project-affected people (PAP) is important. Affected people are defined as people who stand to lose, as a direct or indirect consequence of the project, all or part of their physical or non-physical assets. The more people affected, the more severe is the resettlement impact. The ADB Handbook on Resettlement (ADB 1998) defines resettlement of more than 200 people as significant.

The obvious cause of resettlement in a mining project is the need to provide access to ores. Mineral and coal deposits are immobile and occupation of land for ore extraction is unavoidable. Mine infrastructure such as housing, airstrip, mineral processing plants and roads also requires considerable land area, much of which may be used by communities as productive land. As infrastructure planners do, communities also avoid difficult terrain such as swamps or steep terrain. Mine waste storage and disposal facilities also occupy sizable land areas. On-land tailings disposal usually involves valleys, land that is commonly preferred for housing, farming and land traffic. That said, many mining projects are carried out in sparsely populated areas and often do not require the physical resettlement of so many people.

Increased migration to the mine area and the associated pressure on existing social systems can also contribute to resettlement losses increasing the number of affected people significantly. New settlers may migrate to the mining area with the expectation of employment opportunities or shares of other benefits. Migration will put pressure on the existing economy that has developed over generations. The existing economy may disintegrate to the disadvantage of the weakest and most vulnerable groups of the society. Increases in living costs, often associated with immigration, have most effect on the poor, and may be devastating. Other mining activities can contribute further losses. Destruction of traditional hunting grounds or spiritual places, restricted access to natural resources such as Sago trees, impacts on fishery, or disruption of traditional transportation routes, all count as resettlement effects.

Increased migration to the mine area and the associated pressure on existing social systems can also contribute to resettlement losses.

Number of Indigenous and Vulnerable Affected People

A characteristic feature of resettlement is that those most affected represent those in a society who have the least access to resources. Remote areas where large mining projects are often located, may be home to tribal and Indigenous Peoples who are disconnected from and have never learnt to represent themselves to the rest of the society, and are the least equipped to benefit from the mining project. Accepted resettlement policies specify that these and other vulnerable groups merit special attention in resettlement. Vulnerable groups include but are not restricted to Indigenous Peoples, ethnic minorities, households headed by women and the poorest without legal entitlement to assets or resources.

Magnitude of Tangible Resettlement Losses

Tangible resettlement losses include loss of land, structures, crops, trees, or any clearly defined assets. Tangible resettlement losses are relatively easy to identify and to quantify. As an example, tailings ponds have well-defined footprints, and a detailed census of affected peoples' assets will help to establish an accurate inventory of lost resources. Of course, opportunists often attempt to inflate the value of lost resources. Barren land is suddenly cultivated, crops are planted, or shelters and fences are built in the expectation that a disproportionate increase in compensation can be achieved. There have also been cases of unscrupulous landowners collecting land compensation while at the same time selling the land to a third party.

Magnitude of Intangible Resettlement Losses

Intangible resettlement losses are more difficult to identify, and even more so to quantify.

Intangible resettlement losses are more difficult to identify, and even more so to quantify. Defining intangible resettlement losses and related compensation entitlements commonly requires extensive investigations. Examples of intangible losses include user rights to access land, loss of employment opportunities, loss of cultural assets or spiritual places, or loss of traditional hunting grounds. Such losses may be intangible, but affected people have to be eligible for appropriate compensation to ensure successful resettlement programmes.

Community Participation and Integration with Host Communities

Socially responsible resettlement is guided by a partnership approach, and by transparency.

Socially responsible resettlement is guided by a partnership approach, and by transparency. Lack of participation and transparency are likely to lead to failure (**Box 14.1**). Communication with all stakeholders is central to implementing a successful resettlement strategy. Prior capacity building for displaced and host communities may be required to allow affected people to actively participate in the planning and execution of resettlement. In particular, Indigenous Peoples often lack legal recognition and formal representation; they may also lack the capacity to cope with a mining project and its implications if left alone and unsupported.

Attention to Livelihood Restoration

Resettlement is not only about the cash compensation of resettlement losses, but it is also about social and economic development to re-establish the livelihoods of affected people. Successful resettlement programmes take advantage of opportunities to enhance the economic and social conditions of vulnerable groups. Cash compensation brings only temporary relief, if at all, to affected people. Even for the most experienced resettlement specialist

Box 14.1 Some Examples of Lack of Social Preparation and Community Participation

- Consultation and participation is viewed as a 'necessary evil'.
- Mining companies disregard the diversity within communities; communities are viewed as homogeneous. Marginalized groups with no direct representation such as women, the poor or Indigenous Peoples may be overlooked. Their interests and needs are not explored; assumptions are made about their requirements.
- Failure to recognize pre-existing risks of social division or potential conflict that can be triggered when compensation and resettlement programmes benefit some more than others or lead to the monetarization and breakdown of customary modes of transaction.
- Mining companies deal only with the community leaders and/or representatives of the local government. Informal leaders are not recognized and large groups of the community are ignored. NGOs and other informal stakeholders are viewed as 'troublemakers'.
- Community consultation may not develop into community participation. Communities are excluded from planning resettlement options. They are expected to 'take it, or leave it'.
- Communication is not open and fair. Shared information may be selective and incomplete, and often issued after a delay.
- Affected people are not informed about their rights, and are thus open to abuse.

Source: These and other common issues summarized in the various boxes in this chapter are based on Sonnenberg and Muenster (2001), using involuntary mining-related resettlement in southern Africa as examples.

it remains a challenge to convert cash into sustainable measures for long-term social and economic benefits. For the inexperienced and vulnerable groups in societies affected by resettlement, this is extremely difficult. Resettlement has to involve income replacement through land-based re-establishment and re-employment.

Resettlement has to involve income replacement through land-based re-establishment and re-employment.

Allocated Resettlement Budget

Resettlement costs money. Allocation of sufficient funds is essential (**Box 14.2**). In addition, to be successful resettlement requires a multi-year commitment. Mining companies need to dedicate and use financial resources on a multi-year basis so they can respond to the needs of affected people and build trust in their commitment. Land acquisition costs represent only a fraction of the total required resettlement budget. Cost items include expenditure for planning, community consultation and participation, capacity building, relocation, income restoration and monitoring of resettlement success. Mining companies, however, must find the correct balance between the money actually spent on affected people, and the money spent on personnel involved in planning and implementing resettlement. The irony is that in more than one project, money spent on consultant fees has far outweighed the total expenditure on directly affected people.

Box 14.2 Examples of Common Budgeting Mistakes

- The mining project allocates insufficient funds for the entire resettlement process from planning, displacement and capacity building to income restoration and after care.
- Funds are insufficient for expanding benefits to host communities.
- Inflexible budgetary constraints exist, including the lack of additional compensation if the resettlement process is delayed.
- There is a shortage of skilled resources to implement capacity building for both company and affected communities.
- The mining project partly externalizes costs for resettlement: affected communities shoulder the costs, but not the benefits of the project.

Evaluation of resettlement losses eludes the purely scientific approach. The ADB (1998) and World Bank OD 4.30 (now superseded by OP 4.12 and BP 4.12, World Bank 2001 a,b) provide guidance on how to define significant resettlement losses. The ADB (1998) for example uses the following criteria to define significant resettlement: (1) 200 people or more will experience resettlement effects; (2) 100 people or more who are experiencing resettlement effects are Indigenous Peoples or who are vulnerable (for example, female-headed households, the poorest, isolated communities, including those without legal title to assets); or (3) more than 50 people experiencing resettlement effects are particularly vulnerable, for example, hunter-gatherers.

The Reliability of Data

Estimating significance and designing adequate resettlement plans requires extensive field surveys.

In most, if not all, projects associated with resettlement, estimating significance and designing adequate resettlement plans requires extensive field surveys. Initially, it is necessary to conduct a census to identify and enumerate affected people, giving special attention to identifying the poor, Indigenous Peoples, ethnic minorities and other vulnerable groups who stand to be affected by the project. If successful the census will document land ownership, land use and other rights to land, and will provide an inventory of all potential tangible resettlement losses. In conjunction with aerial mapping or high resolution satellite images, and a review of land use records, if available, a census carried out at the early stage of mine development will be the best safeguard against future land speculation. The census will help to distinguish between genuine and false claims for entitlements. It will identify those who were living in the project area prior to project approval, as distinct from those who subsequently move into the area hoping to benefit from resettlement arrangements. It will also help to verify the nature and extent of traditional activities such as harvesting of forest resources, and hunting.

First, a tabular listing of project-affected people is helpful. This table may be somewhat inexact in the early stages of project planning, but the details should become precise as project planning proceeds, enabling land requirements to be defined more clearly and as the land surveys (and ultimately the final design) are completed. The table should identify the types of people affected (e.g. as owners, tenants, squatters); the type of impact on land (e.g. farm size reduced, house or shop acquired, access limited); and the type of impact on people (e.g. reduced livelihood, reduced income, lost house).

Second, there is likely to be a need for the socio-economic survey to identify and to define intangible resettlement losses taking into account social, cultural and economic parameters such as existing markets, common land ownership or seasonal employment opportunities. Such a survey assesses sources of income and the use of existing natural resources. Generated data are used in planning for income restoration.

Third, it is necessary to understand the existing social fabric. A social-cultural survey documents existing social community fabric, formal and informal leadership structures, communication channels, conflict resolution mechanisms and the community capacity to respond to mine-induced changes. In addition the social-cultural survey supports community consultation and participation planning and the design of community development programmes.

All three surveys need to include host populations if physical displacement is unavoidable. In cases where Indigenous Peoples are involved, anthropological surveys provide a means of understanding the people and their relationships to the land and its resources.

14.3 RESETTLEMENT PRIORITIES

Two priorities should guide resettlement actions.

The first priority is to minimize the need for resettlement by means of project design, such as intelligent alignment of access roads, or siting of supporting mine infrastructure. The location of the ore body is fixed, but everything else is largely the outcome of project design. Project design should pass exceptional scrutiny if large numbers of families have to be displaced. Only if, after all alternatives have been exhausted, resettlement is still considered unavoidable, should it go ahead. Goodland (2004) cites the following example to illustrate unsustainable development involving resettlement: 'the Canadian company Manhattan Minerals Corporation has proposed relocating at least half of the 16,000 population of the Tambo Grande Township in Peru. Much of this relocation would be involuntary. The remaining citizens would then live next to two opencast gold, zinc and copper pits about a kilo-meter in diameter and 250m deep. The project would divert the main river, risking polluting the irrigation water, the main livelihood of the town. As such, this project does not fulfill the requirements of a developmentally sustainable enterprise.'

The first priority is to minimize the need for resettlement by means of project design.

The second priority is to reach a voluntary agreement on resettlement with those affected. Of course, the host government has the power to exercise eminent domain, but this should be avoided if at all possible. Invoking eminent domain indicates inadequate trust and a breakdown in the participatory process. Transparent participation and democracy in developed countries essentially provide veto power to people who may be adversely affected by a mining project.

Many societies have developed their own mechanisms for dealing with sensitive issues such as resettlement, and project proponents are advised to use these mechanisms where they exist. In Indonesia, *mufakat* and *masyarakat* refer to the processes by which social issues are discussed at length, leading to the development of consensus among participants. Failure to achieve consensus in the Indonesian situation is quite rare, although the time required may seem excessive to outsiders.

14.4 COMPENSATION FOR RESETTLEMENT LOSSES AND RESTORATION OF LIVELIHOOD – A RIGHT, NOT A NEED

Mining projects recognize land as an economic asset. This perception implies that mining should be given priority to other land uses since it yields greater economic value

(Wiriosudarmo 2002). This economic value-driven perception of land use has dominated past mine developments. However it implicitly ignores cultural and environmental land values and puts the rights of exploitation before rights of ownership; rights to use, build, lease and to develop; or the right to collect forest products. Some land rights conflict; others may be mutually exclusive. Land conflicts arise when the right to exploit is exercised without due consideration of the legitimate land rights of other people. For example, if farmers have used land over many years, they have a legitimate 'right to use' although they may not own the land.

Land conflicts arise when the right to exploit is exercised without due consideration of the legitimate land rights of other people.

In 1993, the UN Commission on Human Rights, recognizing that involuntary resettlement always causes trauma and impoverishment, ruled that forced evictions constitute a 'gross violation of human rights'. The decision was upheld in 1997. Acceptable involuntary resettlement is distinguished from forced eviction by the following conditions: (1) opportunity for genuine consultation with those affected; (2) adequate and reasonable notice for all affected persons prior to the scheduled date of eviction; (3) provision of information on the proposed evictions and where applicable, on the alternative purpose for which the land or housing is to be used, made available in reasonable time to all those affected; (4) especially where groups of people are involved, government officials or their representatives to be present during an eviction; (5) all persons carrying out the eviction to be properly identified; (6) evictions not to take place in particularly bad weather or at night unless the affected persons consent otherwise; (7) provision of legal remedies; and (8) provision, where possible, of legal aid to persons who are in need of it to seek redress from the courts.

In democratic societies people are rarely forcefully displaced, especially without just compensation, regardless of national need. There is a simple reason: moving people inevitably raises legal issues. At a minimum involuntary resettlement violates the freedom of choice. In fact the potential for violating human rights is much greater in resettlement than in any other activity associated with mining (**Box 14.3**). However when land compensation and resettlement are carried out in a lawful manner and in full respect of peoples' rights, opposition to mining projects by adversely affected people is reduced (although not eliminated) and overall mine development is likely to proceed more effectively.

Moving people inevitably raises legal issues.

Box 14.3 Some Causes Which May Contribute to Violation of Human Rights

- Lack of appropriate legislation and policy guidelines in the host country.
- Local customs and traditions conflict with state law.
- The rights of affected people are not fully taken into account.
- Intervention by government authorities (such as police or military)
- Domestic stress, alcohol abuse and domestic violence – consequence of social disarticulation.

Today no mining company will dispute that resettlement should adhere to human rights enshrined in a series of international conventions (**Table 14.2**). Forced evictions in a mine development must raise a red flag, and are rarely accepted by either the general public or financial institutions.

Poorly conceived and executed involuntary resettlement can result in violations of civil and political rights, such as the right to life, the right to security of the person, the right to non-interference with privacy, family and home, the right to the peaceful enjoyment of possessions and the right to adequate housing. The right to adequate housing implies an obligation on mine developers to make provision for adequate housing even for informal

Poorly conceived and executed involuntary resettlement can result in violations of civil and political rights.

TABLE 14.2

Conventions on International Human Rights (Office of the High Commissioner for Human Rights, Geneva)

The International Human Rights Instruments
• The United Nations Universal Declaration of Human Rights, 1948
• The UN International Covenant on Civil and Political Rights, 1966
• The International Covenant on Economic, Social and Cultural Rights, 1966
• The United Nations Convention on the Elimination of all forms of Racial Discrimination, 1963
• The United Nations Convention on the Elimination of All Forms of Discrimination against Women
• The United Nations Declaration of the Rights of the Child
• The United Nations Declaration on the Right and Responsibility of Individuals, Groups and Organs of Society to Promote and Protect Universally Recognized Human Rights and Fundamental Freedoms, 1999
The United Nations ILO 'Fundamental Conventions'
• Freedom of Association 1948
• Right to Organize and Collective Bargaining Convention 1949
• Forced Labor Convention 1930
• Abolition of Forced Labor Convention 1957
• Discrimination Convention 1958
• Equal Remuneration Convention 1951

dwellers or squatters, who may have had no legally recognizable right to the land from which they were displaced. The key instrument is the: 'Committee on Economic, Social and Cultural Rights, General Comment 7, The right to adequate housing (Art. 11 (1) of the Covenant): forced evictions, UN Doc. E/C.12/1997/4 (1997)'. Under some circumstances, violation of such rights could leave a multi-national mining company exposed to third party tort actions in European or US courts.

Local authorities often influence resettlement. They may fix the land value, regulate land compensation procedures and they may act as mediators. In practice, land values set by the local authorities may be much lower than those expected by the community. Mediation is necessary when landowner and mine developer disagree on the magnitude of compensation. If mediation fails, intervention by government authorities in some countries may escalate to direct or indirect repressive measures. Local authorities justify the involvement of the police or the military with the need to prevent public unrest. Clearly the presence of police or military personnel in the process of land compensation is threatening to landowners, and may cause them to accept unfair payments. It may also violate common human rights. Complaints of injustice in land compensation will invariably re-surface in the future. Also, people who feel aggrieved may seek to disrupt a project as an attempt at revenge. Disagreement of land acquisition may also lead to legal disputes frequently delaying mining projects at great cost to the mine developer; additionally, compensation levels may rise significantly on appeal.

In practice, land values set by the local authorities may be much lower than those expected by the community.

14.5 LAND ACQUISITION AND RELATED ISSUES

Not surprisingly, land conflicts are often related to land acquisition and compensation. No mining company will object to land compensation, especially compensation for privately owned land. However, there are two challenges. The first challenge is to ascertain the status of land ownership land use and land rights; the second challenge is to agree on the form and magnitude of compensation for resettlement losses.

The Complexity of Traditional Land Rights

International law requires equal respect of traditional property rights and titled lands.

International law (especially ILO Convention 169) requires equal respect of traditional property rights and titled lands. Formal land titles, common as they may be in the west, are seldom found elsewhere. Most Indigenous Peoples and many non-indigenous poor do not possess paper land titles to the lands they have used for generations; they may even wish to avoid land ownership in the conventional western sense.

In addition, in the context of land acquisition and compensation, the broader concept of 'Rights to Land' is perhaps more useful than 'Land ownership', especially in developing countries. In each of the following categories, those rights might be legally recognized or recognizable, or not recognized by law: (1) ancestral lands; (2) usufruct rights (such as to grazing lands, hayfields, forest lands, etc); and (3) customary use (herders, fishing families, hunters, gathers).

There may be particular difficulties associated with valuing or replacing such rights. Those reliant on natural resources are often amongst the most vulnerable and least flexible in terms of being able to adjust to new livelihoods. However, they are often not opposed to mine development or the benefits it might bring to them.

Various approaches may be used in order to protect these rights, including:

- very careful socio-economic and anthropological assessment to provide a full understanding of the dimensions of losses traditional or customary users might experience, and how these might be managed;
- extensive and culturally appropriate consultations and participative decision-making;
- concept of free informed prior consent (or consultation) as the case may be;
- royalties to traditional owners as are applicable in some jurisdictions;
- negotiating offsetting benefits; and
- additional requirements where Indigenous Peoples are involved.

Mining companies often find it difficult to deal with traditional rights to land.

That said, mining companies often find it difficult for many reasons to deal with traditional rights to land. As stated previously, land rights are not necessarily limited to ownership of agricultural land or to other economic land uses. Communities often consider themselves as the guardians of the forests, lakes, rivers, mountains and other natural features within certain geographical boundaries. The geographic boundaries themselves are defined by the communities; each land area belongs to a certain community. It is also common for a mining project to span more than one cultural area, and traditional rules and tolerance of land use are likely to differ from community to community.

Traditional land rights are not necessarily an obstacle to mine development.

Traditional land rights, however, are not necessarily an obstacle to mine development. In traditional law, land is commonly believed to be inherited by the community to sustain community life from one generation to another. Recognizing a land use such as mining in the context of collective community interests opens avenues to mine development that paper titles do not offer.

The complexity of traditional land rights, of course, also provides opportunities for anti-mining advocates to oppose a mining project. The economic value of land and the ore, they argue, cannot be the decisive factor for land compensation or mine development. To local communities, they say, land has its own 'priceless' cultural dimension, an important characteristic of community identity. Clearly, mining often does affect cultural values, which may not be exchangeable for economic or financial benefits. Undoubtedly, the feeling of loss of cultural values has caused community dissatisfaction in more than one mine development, a feeling that may reappear repeatedly. However, the indiscriminate use of cultural values as a blanket argument to oppose mining *per se,* or to create cultural values that do not exist, will ultimately hinder much needed development, and the development of civilization as a whole.

Entitlement, Form and Magnitude of Compensation

According to IFC PS 5, displaced persons and communities should be compensated for loss of assets at full replacement cost, and other assistance should be provided to help restore or preferably improve their livelihoods and standard of living. A standard approach to compensation should apply throughout the project, and it should be transparent. Where feasible, loss of land should be compensated by provision of new land of equal or better value and utility.

Where feasible, loss of land should be compensated by provision of new land of equal or better value and utility.

Physically displaced persons should be offered a choice of options for adequate housing, with security of tenure. For persons who are economically displaced, including those without legally recognized claims, loss of assets or access to assets should be compensated at full replacement cost, including the costs associated with re-establishment. Other assistance should be supplied, as necessary, to restore or improve income earning capacity which, in the case of a mining project, could include project employment. Displaced persons and communities should receive opportunities to benefit from the project.

In defining the compensation package, it is of course easier to deal with titled land than it is to deal with the uncertain boundaries of traditional or cultural land rights, uncertain both in terms of legal status and tribal ownership, or to the informal occupation of land or use of resources by squatters (**Box 14.4**). To the discredit of the mining sector as a whole, uncertainty of land status and land rights has been misused in the past, by some junior mining companies to deny compensation. Quite simply they lacked the necessary financial

It is easier to deal with titled land than it is to deal with the uncertain boundaries of traditional or cultural land rights, uncertain both in terms of legal status and tribal ownership,

Box 14.4 Some Examples of Incomplete Assessment of Entitlements and Eligibility

- Social–economic baseline data are incorrect or incomplete. Affected people with no formal legal entitlements are overlooked. Communal land tenure customs are not addressed. Traditional land rights are not recognized.
- Entitlement is solely viewed as cash entitlement. If 'land for land' compensation is planned, land available for resettlement may be marginal and unsuitable for land use compatible with previous skills.
- Emphasis of compensation schemes is on compensating for loss of physical assets such as land, vegetation or structures, not on developing sustainable value.
- Influx of people associated with mine development is not considered.
- Little support for host communities is provided. Pressures on local resources and services are not accounted for.

resources. On other occasions junior mining companies have collaborated with local authorities to avoid or minimize compensation.

Some jurisdictions restrict land compensation to private land; state land is exempted from compensation. State land, however, while unused by the government may nevertheless be claimed as cultural land by local communities. At the time of allocating the mine concession, the government (and hence the mining company) may not recognize these claims. Ignoring such land rights to deny compensation is likely to lead to long-lasting land conflicts. The same is true of delaying compensation; allegations of wrongdoing by the mining company will emerge.

The ways and conditions under which land (and land rights) is transferred are important.

The ways and conditions under which land (and land rights) is transferred are important. Land transfer should follow existing formal and traditional laws, and, of course, should respect human rights. Land transfer and compensation depend on variables such as the nature of lands themselves, prevailing land law, the way land is used, ownership, type of transaction, traditional land transfer practices, or mediation of the transfer process.

No matter what the time constraints are, compensation for traditional land rights should follow traditional procedures.

Complex traditional land ownership and land rights combined with schedule pressures in mine development too easily lead to ignoring or by-passing cultural procedures for land compensation. Mining companies are tempted to resort to government agencies or local authorities as the best practical way to deal with the host community (Wiriosudarmo 2002). This, of course, invites abuse of power by local authorities, and may also lead to corruption. No matter what the time constraints are, compensation for traditional land rights should follow traditional procedures which usually means that the process cannot be rushed. Financial compensation is necessary, but insufficient and insulting to the community if local traditions are not followed.

Compensation: Cash or Kind?

Failure in the adopted compensation scheme is one of the principal sources of project risk, especially for mining projects that require large tracts of land. It is also one of the chief ways that the poor subsidize development.

Since communities with non-monetized economies are not well accustomed to managing cash, cash compensation has many inherent risks.

Most countries have land acquisition laws that require prompt and adequate monetary compensation for people who lose land or property. However, many remote rural economies are largely based on reciprocal exchange of goods and services. Since communities with non-monetized economies are not well accustomed to managing cash, cash compensation has many inherent risks. People may change their lifestyle, start gambling and drinking or spending on luxury items, until the received money is exhausted.

'Land is like diamonds but money is like ice'.

Experience is that the best strategy for preventing landlessness and the concomitant threat to livelihoods is 'land for land' compensation. Guggenheim (1990) cited a popular saying among the *Havasupai* Apache Indians in the United States, a people displaced repeatedly by development projects: '*Land is like diamonds but money is like ice*'. Displaced people face a high risk of landlessness if measures are not taken to ensure that the land they have foregone is replaced, either by the project or by the displaced people themselves.

The goal of 'land for land' compensation is to provide land comparable to that on which the livelihoods of displaced people are based. World Bank OP 4.12 states '*Whenever replacement land is offered, re-settlers are to be provided with land for which a combination of productive potential, location advantage, and other factors are at least equivalent to the advantages of the land taken.*'

However decades of international experience in resettlement suggests that 'land for land' compensation alone is seldom fully accepted by displaced people; they often demand cash payments. As a result, most compensation schemes provide compensation in both cash and non-cash forms. A compensation package can include the following: (1) cash compensation to displaced people for land acquired by the project; (2) cash compensation to host communities for land provided to displaced people; (3) cash compensation for intangible resettlement losses; (4) construction of new homes at the resettlement site; (5) provision of formal land titles for displaced people; and (6) construction of basic public and social infrastructure at the resettlement site. In a wider sense, as discussed in Chapter Fifteen, community development programmes, regional development planning, royalties, and the like, can also become part of compensation.

Repeated land compensation claims are not uncommon.

Repeated land compensation claims are not uncommon. At first glance it may seem that communities are constantly complaining about unfair previous payments; they repeatedly request more money. Repeated compensation claims have created the distorted perception that host communities are materialistic, and it may be that they have become this way as a result of the way that compensation is administered. Sometimes, however, there are valid reasons for additional community claims; for example: (1) the local authority may have exerted strict control on the freedom of community members to negotiate fair land compensation when the original land transfer occurred; (2) loss of cultural land values may only be recognized by the community long after they have surrendered their land to mining; and (3) the community did not realize the scale of mining or the magnitude of impact it would have on their livelihood. In the case of the Bougainville Copper Mine, many young people sought compensation long after the project had been developed, claiming that while their parents had been compensated, they had not. All too frequently, of course, the root cause of demands for additional payments is the unwise use of the initial payments; meaning that at best the land compensation led to no sustainable benefit and at worst, nothing remained from the initial compensation and previous livelihoods were lost.

Inadequate compensation during exploration may influence landowners' attitudes towards the project.

In discussing land compensation it is also necessary to address land access and land use during the exploration phase (and also in post-mining land use as discussed in a separate chapter on mine closure). During exploration there is no direct gain to the mining company from acquiring land and no guarantee that there will be a future need for the land. Compensation is commonly confined to damage from exploration activities, such as compensating for trees lost or agricultural damage, and for other land disturbances and for inconveniences to land holders that may result. Experience has shown that inadequate compensation during exploration may influence landowners' attitudes towards the project: when exploration moves to the mining stage, landowners may become reluctant to allow their land to be acquired for mining. In such cases, the mineral discovery is perceived as a threat to land ownership (Wiriosudarmo 2002). On the other hand there are usually opportunities for an exploration company to generate goodwill in host communities by providing relatively modest assistance, beyond compensation for damage and inconvenience. Companies exploring in remote areas may enhance accessibility for local people to their own lands through the construction of tracks providing drilling rig access. Prior consultations with local communities will determine if the access tracks will prove useful to local people and therefore whether they should be rehabilitated or retained. Similarly, when earthmoving equipment is in the area for the purpose of establishing access, there are usually small but locally valuable tasks that can be added to assist local people.

The Timing of Land Acquisition is Important

A mining company faces two contradictory facts in regard to the timing of land acquisition. On one hand there is the simple fact that as mine development progresses, land prices increase. On the other hand, acquiring land for mineral exploitation only really makes sense when the mining project is certain to go ahead. Mine development will depend on the positive outcome of the bankable feasibility study, on securing funding and on the granting of necessary government approvals.

Legal procedures also influence the timing of land acquisition. In most countries land ownership and land use is restricted. Prevailing legislation may not allow foreign investors to acquire land prior to the formal approval of the mining project by the government. While sensible, and ultimately to the benefit of the investor and local communities, such restrictions invite abuse. Frequently, land speculators emerge hoping to profit by acquiring land from local land holders only to sell them at an inflated price to the mining company once the mining project moves into construction. Mine proponents may respond by acquiring land through creative means through decoy property purchases; using nominees or senior and loyal employees as their trustees; or through foundations established for the purpose. These strategies, of course, are contrary to the spirit of transparency. In fact, it is in relation to land acquisition that many mining companies have most difficulty with the requirement for transparency.

Frequently, land speculators emerge hoping to profit by acquiring land from local land holders only to sell them at an inflated price to the mining company once the mining project moves into construction.

In the environmental permitting process, the status of land acquisition may lead to a difficult situation. As mentioned, mining companies are reluctant to purchase land until the necessary approvals have been obtained. On the other hand, the impacts on local land holders can not be definitively assessed without detailed understanding of the compensation arrangements (and relocation arrangements if they apply). This may be overcome, at least partially, by the negotiation of options or rights to purchase, subject to receipt of project approvals. In other cases, companies proceed to purchase the required land, accepting the risk that much of the money spent will be wasted if the project is not approved.

Adequate time should be allocated to fully resolve land rights, identify ownership, plot boundaries and regularize tenure prior to commencing acquisition, and subsequently to negotiate acquisition terms and arrangements. It is a common mistake to allow insufficient time to fully resolve these matters (**Box 14.5**).

Box 14.5 Common Timeline Issues in Resettlement

- Implementation of resettlement is either too early or too late.
- Financing may require preparing resettlement plans long before mine implementation. Early plans, if not updated as new information is generated, will become obsolete and meaningless.
- Planning for resettlement is carried out as an 'afterthought', taking place too late, necessitating short cuts. The resettlement plan turns into an alibi document for resettlement that has already taken place.
- Resettlement ceases after rehabilitation and no consideration is given to post resettlement monitoring and after-care.

14.6 LIVELIHOOD RESTORATION – REALIZING SUSTAINABLE VALUE IN THE COMPENSATION OF LOST ASSESTS

Compensation for losses differs from techniques for providing livelihood streams or community benefits. IFC PS 5 and its predecessors require compensation for land and lost assets in the first instance plus measures to restore livelihood. One is not a substitute for the other: social experts agree that effective compensation for land and lost assets plus measures to restore livelihoods are essential components of a resettlement scheme.

What Compensation Scheme can Create Sustainable Value?

Monetary compensation is still the most common form of compensation; it is based on but not restricted to the present land value paid in cash. While almost universally, landowners prefer and in fact, often insist on cash compensation, there are two inherent and interrelated problems associated with cash compensation.

Landowners prefer and in fact, often insist on cash compensation.

First, if the project is located in a remote area as many mines are, land valuations are generally low. Payment for land based on such valuations will be insufficient to last for long or to sustain future livelihoods. When the money is spent, compensated community members may descend into poverty, especially when the surrendered land was the main source of livelihood.

Second, even if larger compensation payments have been made, compensated landowners often lack the knowledge and experience to use received payments wisely or to re-invest so as to achieve sustainable value. It follows that the preferred arrangement for land compensation schemes involves compensation of 'land for land', often with an additional payment or assistance to cover relocation and establishment costs.

Clearly 'land for land' compensation, if carried out appropriately offers a continuation of existing livelihoods, although in some cases adjustments or modifications may be required due to changed circumstances. But fair compensation alone is no guarantee of successful resettlement. Assistance in livelihood restoration is equally important.

What Constitutes Livelihood Restoration?

Livelihood restoration is the most difficult and critical aspect of resettlement programmes. It is also the most misunderstood and overlooked aspect. Livelihood restoration may take many forms, including project employment for affected people; equity sharing with host communities; distribution of royalties to the community (royalties are one way of compensating for lost benefits from ancestral lands that are communally held, not held by individuals, see **Case 14.1**); and sustainable community development programmes (as discussed in Chapter Fifteen). An innovative approach involving ownership of mining equipment has been successfully used at the Lihir Gold Project in Papua New Guinea (**Case 14.2**).

Royalties are one way of compensating for lost benefits from ancestral lands that are communally held, not held by individuals.

Training and education also offer opportunities for new livelihoods, and so does direct employment on the mining project as well as secondary employment through other businesses and enterprises spawned by the project. Capacity building to enable local people to benefit from small enterprise and local procurement opportunities, may be an important component of livelihood restoration programmes. Similarly, arrangements in which local enterprises receive access to finance, can be highly beneficial. At the community level, as mentioned above, equity, royalty streams and other revenue stream approaches are commonly used to provide a range of benefits including livelihood restoration and support.

Each component of land compensation and livelihood restoration is rational, but each also involves challenges. Land for land compensation appears to provide an equitable solution that overcomes many of the problems of cash compensation. However, this approach has its own problems as it is not always possible to identify and purchase land of equivalent utility. And what of the vendors; do they also require land for land compensation?

Clearly, this approach potentially multiplies the number of property transactions, increasing the scope for speculators and adding to the risk that property values will escalate.

It is fair to say that rural resettlement has rarely been totally successful. Often incomes have not been restored to pre-existing levels, or not for several years afterwards. This means that resettlement in many past projects has increased poverty instead of decreasing it.

14.7 THE SOCIAL RISKS OF RESETTLEMENT

Resettlement, unfortunately, often affects people who are least able to afford it.

Resettlement, unfortunately, often affects people who are least able to afford it. Although the land and its resources are essential to their livelihoods, vulnerable affected community groups often have no legally recognized rights to them. They may have little experience in creating a livelihood outside the community or the social system in which they live, and find it difficult to cope if they are suddenly removed from their familiar social setting. If affected people do not perceive the compensation offered as fair, they have no remedy beyond the court system that may be very distant, hopelessly slow, extremely expensive by their standards, unsympathetic to the poor, or simply corrupt (Goodland 2004).

Involuntary resettlement reveals social changes and risks that follow well-established patterns, regardless of the type of project leading to displacement. The frequently used Impoverishment Risk and Rehabilitation (IRR) Model by Cernea (1991, 1999, 2000) organizes these risk patterns into eight potential risks, or impacts: landlessness, joblessness, homelessness, marginalization, food insecurity, loss of common lands and resources, increased health risks and social disarticulation (**Figure 14.2**). The model is widely accepted, and is sometimes expanded to include two additional risks: loss of civil and human rights (Downing 1996) and loss of access to public services including disruption of formal educational services (Mathur and Marsden 1998). Combined or individually, these

CASE 14.1

The Freeport Partnership Fund for Community Development – Damned if you do, damned if you don't

Since 1996, PT Freeport Indonesia's operations have allocated 1% of revenues for the benefit of the local community through the Freeport Partnership Fund for Community Development (also dubbed the '1% Fund'). This fund is used to build schools, hospitals, places of worship, housing and community facilities in the area of the mine operation. The fund also supports a comprehensive series of health and educational programmes. All funds are disbursed to a local community governing board, the Amungme and Kamoro Community Development Organization (LPMAK), which is governed by a board consisting of local indigenous tribal, church and community leaders. LPMAK approves and funds local programmes.

By 2004, total contributions to the partnership fund since inception had reached approximately US$ 152 million. In its December 1, 2003 issue, *Business Week* magazine published the results of its survey ranking America's most philanthropic companies and listed Freeport-McMoRan Copper & Gold Inc. as America's most philanthropic company in terms of cash given as a percentage of revenues. This is largely due to the Freeport Partnership Fund for Community Development.

However, the fund does not come without controversy. Critics claim that Freeport too often intervenes in LPMAK decision-making with a 'we know best' attitude. The 1% Fund is also cited as a source of conflict, since it has created jealousy between local tribes (the 'haves' and the 'have nots'). It is also said that the purpose of the funds is not clear. Some critics perceive the fund as 'riot money' paid to settle violence in 1996. There is also the claim that too little is achieved with too much money spent.

Thom Beanal, the Amungme tribal leader and a vocal supporter of independence for Papua, has fought the company from outside and inside. He stated that the flood of money from the community fund was ruining peoples' lives.

Graphic: Logo is copyright of Freeport McMoran Company

risks are the main route to impoverishment, the central risk to which displaced people are exposed. The following paragraphs expand on these risks based on Cernea (2000).

Landlessness

Land is the foundation on which our production system is based. It is a direct or indirect source of our income. As such, land loss is a significant social impact, particularly in agricultural societies which often host new mining ventures. Land loss not only occurs to provide mine access; it is also the result of the inability of displaced people to find suitable new land. This may be either due to a lack of suitable replacement land or the inability to buy such land, since local prices may be inflated or cash compensation may have been spent to meet immediate needs or, more likely was inadequate to effect full replacement. Clearly, to minimize the risk of landlessness, resettlement plans should favour 'land for land' over cash compensation – in situations where displaced communities are substantially reliant on land for their livelihood.

Land loss not only occurs to provide mine access; it is also the result of the inability of displaced people to find suitable new land.

Joblessness

Often underestimated is the loss of pre-displacement economic opportunities that are not directly related to land ownership or that are caused by local economic cycles. Squatters, wage labourers and vendors may lose employment and sources of income. Tribal people may experience loss of traditional land rights and land use. Often overestimated are the post-displacement employment opportunities. Previous skills of affected people may become redundant, and previous markets become closed.

Often overestimated are the post-displacement employment opportunities.

The creation of new jobs is difficult. Like many other industries, mining is continually adopting less labour-intensive methods requiring a more highly skilled labour force. Without implementing extensive occupational skills development programmes, it is difficult to argue that mining brings many employment opportunities to unskilled local people. On the contrary, with each new mine development, companies compete to maintain or

CASE 14.2
Lihir Gold Project

Seeking to avoid the problems associated with cash compensation, the operator of the Lihir Gold Mine on the remote Lihir Island in Papua New Guinea developed an innovative means of providing a long-term income stream to land holders. The arrangement involved one or more land holders receiving compensation in the form of a piece of mining equipment, usually a haulage truck, which is subsequently leased by the mining company for use in operations at standard equipment leasing rates. The owner may or may not operate the equipment. The owner also receives training in concepts of depreciation, savings and investment with the objective that the owner will be able to replace the equipment at the end of its economic life. Clearly, the benefits of this approach go well beyond the income stream provided by the lease agreement; many successful entrepreneurs have developed businesses from similarly modest beginnings.

Photo: NASA Satellite Image of Lihir Island

FIGURE 14.2
The Eight Potential Risks of Involuntary Resettlement

These risks are the main route to impoverishment, the central risk to which displaced people are exposed.

Source:
based on Cernea 2000

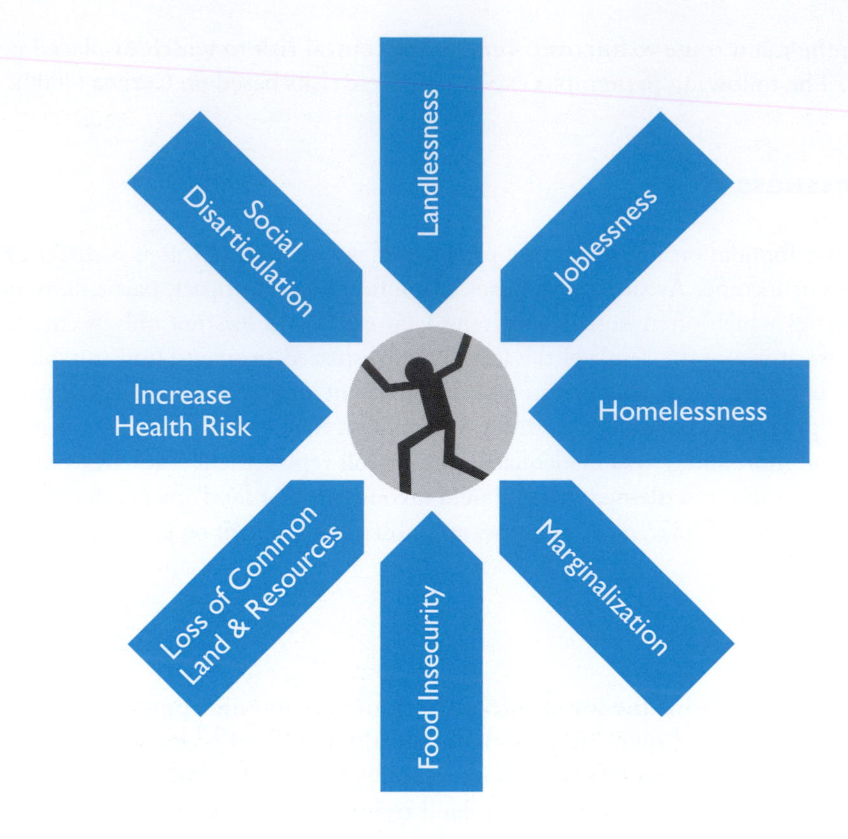

attract the few skilled workers available in a developing country. On the other hand, there are many examples of recent mining projects in which the operating company was able to recruit and train hundreds or in some cases thousands of previously unskilled or semi-skilled people from local communities. Companies that have made this investment in time and resources, invariably observe that the costs are more than justified by the more stable workforce that results, and the closer integration of the project into the local community.

Homelessness

Loss of shelter is not permanent, but for some displaced people a worsening in housing standards or loss thereof is a reality. A decline in housing conditions increases if the compensation for demolished houses is paid at assessed market value rather than replacement value.

Marginalization

Previous existing formal and informal networks are disrupted; new ones may be difficult to develop.

The risk of marginalization always exists when people are physically relocated. Previous existing formal and informal networks are disrupted; new ones may be difficult to develop. The host community may view displaced people as strangers, or even intruders. Marginalization is reflected by a decline in social status, in social isolation and in psychological depression. Substance abuse may occur, which in turn affects health and productivity, often driving affected people even further into isolation. Adverse feelings against the project that caused social change develop and grow, and are often associated with increased crime and violence.

Health Risks

Involuntary resettlement is often accompanied by trauma. Psychological depression, together with stress and anxiety experienced during resettlement, may cause a decline in physical and mental health. Substance abuse may augment such health risks. Unsanitary conditions at resettlement areas can also contribute to vector-borne diseases such as malaria or dengue fever. As might be expected, the young, old and frail disproportionately shoulder these health risks.

Food Insecurity

Food insecurity and malnourishment are symptoms of resettlement gone wrong. Food insecurity is directly related to the degree to which pre-displacement sources of income have been successfully replaced. Successful resettlement should create sustainable value for relocated people, who should be better off after relocation.

Food insecurity and malnourishment are symptoms of resettlement gone wrong.

Loss of Common Land and Resources

Common land while not owned by individuals with direct ownerships rights, often contributes significantly to a society's production system. Forest, water resources, or hunting grounds are common property assets to which displaced people potentially lose access once they are relocated. To compensate for such loss, displaced people may have to encroach on protected areas or on the common land and resources of host communities. Pressure on these resources increases, often causing environmental degradation or depletion.

Social Disarticulation

Resettlement breaks social patterns and interpersonal ties, so that affected communities risk becoming fragmented and dispersed. The risk of social disarticulation is especially real for Indigenous Peoples who do not have a long history of interacting with other cultures and societies.

Resettlement breaks social patterns and interpersonal ties.

The IRR model can be used in two ways: first to identify the main poverty risks that are involved in the economic and physical displacement associated with mine development; and second to formulate strategies to mitigate and reverse these risks. A community risk assessment and reversal matrix based on the IRR can help to concisely outline the level of each risk and the strategy that is taken to reverse that risk (**Table 14.3**).

TABLE 14.3
Outline of a Community Risk Assessment and Reversal Matrix

IRR Risk	Risk Level	Risk Reversal Strategy
Tabulate all nine risk factors, e.g. Landlessness	Quantify risk; e.g. as low, medium or high; describe risk factors	List mitigation measures

14.8 MANAGING LAND ACQUISITION AND RESETTLEMENT

Visualizing involuntary resettlement as simple, following a straight path from A to B understates the complexity of the process.

In the absence of a clear company policy, resettlement is unlikely to receive the priority it deserves and requires.

Visualizing involuntary resettlement as simple, following a straight path from A to B understates the complexity of the process. Unfortunately, the 'line from A to B approach' to resettlement all too often characterizes the mind set of technically oriented mining personnel who are better equipped to embrace the physical challenges of exploiting mineral resources in remote and often hostile environments, rather than the sustainable betterment of affected communities. The main interests, and often the only basis of financial rewards, are of a technical nature. Too often resettlement is seen as an unwelcome burden to be dealt with by imposing engineering solutions.

It follows that how resettlement is managed depends largely on the policy in place to guide and to govern planning and execution, and on the skills and commitment of the personnel responsible for its implementation (**Box 14.6**). In the absence of a clear company policy, resettlement is unlikely to receive the priority it deserves and requires. On the contrary, other company policies may even prevent appropriate resettlement actions from being implemented. Without a clear policy, resettlement will not be recognized as a priority for the project, and is unlikely to be adequately funded. There will be no clear decision-making framework; neither will there be a consensus on the scale and requirements of the resettlement action plan. Resettlement may be viewed as a public relations exercise or as a nuisance to core business. However, in recent years, resettlement has increasingly been perceived as a business risk and therefore, amongst enlightened mining groups, is being incorporated into risk management frameworks.

Box 14.6 Common Deficiencies in Resettlement Programmes

- A clear company policy on resettlement is missing and as such, resettlement is not given the priority it deserves. At worst, resettlement is viewed as a nuisance to core business.
- The complexity of resettlement is underestimated. Resettlement process is overly simplified. Mine management sees resettlement as a housing project, reduced to budgets and technicalities. Internal capacity to provide financial or technical assistance may be abundant, but social skills are missing.
- Resettlement planned and implemented by technical personnel with little if any resettlement experience, is viewed as acceptable practice.
- Assets are transferred to communities and the host government without transferring the skills to manage them. The emphasis of resettlement actions may also be on tangible assets. Infrastructure 'overkill' has occurred in more than one project and is, of course, unsustainable.
- There is a lack of capacity on the part of the host government. Institutional support is minimal. Civil servants involved in resettlement may be ill-equipped and disinterested. In addition, the host government may lack the resources to oversee the rehabilitation and post-rehabilitation phase.
- Local development opportunities and income restoration for affected people are difficult to develop, and affected people continue to depend on the mining operation after resettlement is thought to be complete.

Land compensation and resettlement policies of host countries are in many cases absent; in others they vary widely. Multi-national companies therefore commonly refer to global guidance frameworks such as the directives of the World Bank (1990; 2001a,b) (**Table 14.4**) or the ADB (1998).

Resettlement is a management process with three phases: planning, execution and monitoring/evaluation. Planning is often documented in a land acquisition and resettlement action plan (LARAP), sometimes referred to as resettlement action plan (RAP) or simply resettlement plan (RP). The World Commission on Dams guideline (WCD 2000) calls the planning document 'mitigation, resettlement, and development action plan' (MRDAP). Compliance with the Equator Principles requires preparation of a formal LARAP for Category A and B projects that involve land acquisition and displacement of individuals and communities. The LARAP should keep the community in mind, not the individual. The LARAP complements the environmental impact assessment, and has close links to the environmental mitigation and monitoring plans to be implemented during project implementation and operation. **Table 14.4** summarizes the key planning concepts to be taken into account when developing a LARAP.

The LARAP should keep the community in mind, not the individual.

Appendix 14.1 to this chapter presents an example outline of a full land acquisition and resettlement plan. The level of detail of each LARAP will vary according to the specific circumstances of a given mining project. Best resettlement practice (Goodland 2004) calls for:

- Full disclosure and informed prior consent (IFC PS 1 and 5). Essentially, this means that resettlement must be voluntary.
- Using resettlement needs as an opportunity to improve the livelihood of displaced people by compensating fully and promptly at a minimum the full replacement value for all tangible assets, any decrease in use rights, including access to common property resources and other intangible losses.

TABLE 14.4

Key planning concepts for developing a land acquisition and resettlement plan

- Policy framework – Do national policies and guidelines exist and are they adequate? Is the project proponent accountable for promised entitlements and compliance with the LARAP, if such document exists? Is the policy's focus on physical relocation and compensation rather than income restoration? Are institutional instruments in place to facilitate 'land for land' compensation alternatives?
- Defining entitlements and eligibility – Who will receive compensation and rehabilitation, and how much? How will these measures be structured to avoid cash compensation? Is recognition given to informal or customary land rights?
- Social preparation – Are the needs of women being taken into account? Will the needs of vulnerable people be met? Are neighbouring communities with a genuine interest in resettlement overlooked?
- Community participation – How best to consult with affected people. Will consultation lead to participation? Are host communities included in the consultation process?
- Organizational capacity – Do the skills, staff and organizational capacity exist to properly implement resettlement plans and to assist in income restoration? Do local government institutions have adequate resources and capability? Can affected people actively participate in resettlement?
- Budget – How will land acquisition and resettlement be financed? Are allocated resources sufficient to fully cover resettlement including reestablishment? Does the budget allow for a multi-year programme? Do the host communities have to shoulder part of the costs or are development benefits extended to host communities?
- Time line – Is resettlement planning part of the early project planning or it is more an after thought? How does resettlement fit into the project implementation schedule? When will monitoring and evaluation of resettlement success start and end?

- Sharing the benefits of mining in a transparent and participatory manner, either though community development programmes or other forms of revenue sharing.
- Accepting that the mining project is fully responsible for all resettlement-related costs. This principle needs to be incorporated in operations before a project is permitted or financed (Downing *et al.* 2002).
- Respecting basic human rights derived mainly from principles of equality, social justice and freedom of choice. Implementation of whatever human rights standards are adopted is more important than specifying exactly which standards to adopt.

The mining proponent should look on resettlement as an opportunity rather than a threat: an opportunity to achieve community development, to win the support of the local communities and to earn the social licence to operate.

Since resettlement seeks to provide restitution for assets taken for the project by improving the livelihoods of the affected people, the mining proponent should look on resettlement as an opportunity rather than a threat: an opportunity to achieve community development, to win the support of the local communities and to earn the social licence to operate. Ideally project-affected people will share project benefits over the entire life span of the project. They can be helped to develop rather than to become impoverished by their sacrifice for the good of the mining project (Goodland 2004).

There should, therefore, be no confusion over the goal of resettlement: resettlement is a valuable opportunity to reduce poverty. At least at the materialistic level, 'better off' means provision of the very standard package of compensation such as: better houses, improved water supply, equivalent agricultural plots, access to educational and health facilities and access to affordable energy. People in remote regions of developing countries are often very poor; making them modestly better off in materialistic terms is rarely expensive. However, the major risks associated with cash compensation as discussed formerly, need to be recognized; cash compensation should be the exception, rather than the rule.

Unfortunately experience demonstrates that people displaced by past mining projects have not usually benefited, even modestly, after their resettlement. Another reality is that vulnerable ethnic minority groups are not always protected. As for employment opportunities, retraining and job creation are not always offered. There is ample room for the mining sector to improve its performance in these aspects.

14.9 ARTISANAL MINING AND INVOLUNTARY RESETTLEMENT

The interests of commercial mining companies and artisanal miners will intersect from time to time as each group seeks to exploit the same resources.

Artisanal mining has previously been discussed in Chapter Five. It is inevitable that the interests of commercial mining companies and artisanal miners will intersect from time to time as each group seeks to exploit the same resources. The IFC has identified this as one of the most contentious and difficult areas they face in applying their resettlement policy on mining projects. It is perhaps one of the clearest examples of where mining rights collide with established (formal or informal) livelihoods of others. It is a particular challenge to mine development in Africa, but such conflict is also widespread in Latin America, Indonesia, Papua New Guinea and the Philippines. While the incomes of artisanal miners are usually very small, the activity commonly provides a social safety net and income of last resort for the very poor. Accordingly, loss of this livelihood can have devastating consequences. Again, there are differing degrees of 'rights' associated with artisanal mining, ranging from traditional mining spanning many centuries as in the case of the *Kankana-ey* in the Philiipines (see Chapter Five) to gold rushes that develop after a mining company has discovered a new ore body, as was the case with the Talawaan mining activities in Indonesia. There have been recent examples of cases where conflicts between artisanal miners and mining companies have caused significant losses to mining companies – in the case of Talawaan, contributing to the demise of the Australian company Aurora Gold. Diwalwal in the Philippines is another example (see Case 7.6 in Chapter Seven), where the Philippine Government continues to search for a balanced approach that allows for the opening of the area for large-scale modern mining while at the same time providing income and livelihood to thousands of existing small-scale miners.

However, there have also been examples of situations where artisanal miners and mining companies have co-existed amicably and, in some cases to the mutual benefit of both groups. A good example of the latter is in mining operations managed by the Steel Authority of India Limited, in which the larger iron ore deposits are mined and processed by modern methods while small deposits, after drilling and blasting carried out by the company, are mined by the local community using hand methods to separate ore from waste.

The main impediment to mutual cooperation between mining companies and artisanal mining communities is not the sharing of resources, but the legal ramifications, whereby companies are concerned that they may be held liable for health hazards and unsafe practices over which they have no control. Clearly, given good intentions, such concerns can be overcome with government or local government and company cooperation.

The main impediment to mutual cooperation between mining companies and artisanal mining communities is not the sharing of resources, but the legal ramifications, whereby companies are concerned that they may be held liable for health hazards and unsafe practices over which they have no control.

As discussed in Chapter Five, artisanal mining is commonly associated with lawlessness, poor sanitation, lack of safety provisions, environmental degradation and a high incidence of diseases such as HIV-AIDS. The worst examples, such as the Diwalwal gold mining area in Mindanao in the Philippines in the 1990s, represent some of the most dangerous locations in the world.

Some governments such as that of the Philippines, have developed national policy approaches seeking to formalize the artisanal/small mining sector and to facilitate engagement between these groups and the large miners. In particular, several governments including Indonesia, the Philippines and Thailand have established 'peoples' mining areas' specifically for use by artisanal miners. In some cases, the mining industry has been consulted before these areas have been established. In other cases (such as in PT Newmont Minahasa Raya's operations in North Sulawesi, Indonesia), mining companies have voluntarily made available to artisanal miners, ore bodies within their mining leases. Also common is the situation where artisanal miners continue mining of the near-surface ore while the company mines the deeper underground ore that is not amenable to small-scale mining. A good example of this arrangement is Aneka Tambang's Pongkor Gold Project in West Java, Indonesia.

Economic displacement impacts on artisanal miners can often be compensated by providing employment in the larger scale operation. As mentioned previously, incomes generated from artisanal mining are usually very low and in many cases, far exceeded by wages offered by mining companies. Added to this are the significant accompanying benefits such as training, security, health and safety provisions.

Economic displacement impacts on artisanal miners can often be compensated by providing employment in the larger scale operation.

In summary, while it is common for conflicts to develop between companies and artisanal miners, given goodwill and patient communications, mutually beneficial outcomes can usually be achieved.

REFERENCES

ADB (Asian Development Bank) (1998) Handbook on Resettlement – A Guide to Good Practice. Asian Development Bank, Manila.

Cernea MM (1991) Involuntary Resettlement: Social Research, Policy, and Planning, in MM Cernea (ed.) Putting People First: Sociological Variables in Rural Development. Oxford University Press for the World Bank, Washington DC.

Cernea MM (1999) The Economic of Involuntary Resettlement: Questions and Challenges. The World Bank, Washington DC.

Cernea MM (2000) Risks, Safeguards, and Reconstruction: A Model for Population Displacement and Resettlement, in Cernea MM and C McDowell (eds.) Risk and Reconstruction: Experiences of Resettlers and Refuges. The World Bank, Washington DC.

Downing TE (1996) Mitigating Social Impoverishment when People are Involuntarily Displaced. In Understanding Impoverishment: The Consequences of Development-Induced Displacement. In C McDowell (ed.). Oxford and Providence, RI: Berghahn Press.

Downing TE, Moles J, McIntosh I and Carmen Gracia-Downing (2002) Indigenous Peoples and Mining: Strategies and Tactics for Encounters. London: International Institute for Environment and Development, MMSD Project.

Goodland R (2004) Sustainable Development Sourcebook for the World Bank Group's Extractive Industries Review: Examining the Social and Environmental Impacts of Oil, Gas and Mining; Independent Extractive Industries Review for the World Bank Group, Washington DC.

Guggenheim S (1990) Development and the Dynamics of Displacement, Paper prepared for Workshop on Rehabilitation of Displaced Persons, sponsored by Institute for Social and Economic Change and Myrada, Bangalore, India.

IFC (2006) Environmental and Social Performance Standards, International Finance Cooperation, www.ifc.org

Mathur HM and Marsden D ed. (1998) Development Projects and Impoverishment Risks. Oxford University Press, Delhi.

Sonnenberg D and Muenster F (2001) Mining Minerals Sustainable Development. Southern Africa. Research Topic 3: Mining and Society – Involuntary Resettlement. MMSDSA Regional Research. African Institute of Corporate Citizenship, South Africa.

WCD (2000) Dams and Development – A New Framework for Decision Making, World Commission on Dams, Earthscan, London, UK.

Wiriosudarmo R (2002) Gap Analysis on Mining in Indonesia, IIED and WBCSD Publication.

World Bank (1990) The World Bank Operation Manual: Operation Directive 4.30 – Involuntary Resettlement. World Bank, Washington DC.

World Bank (1994) Resettlement and Development: The Bank Wide Review of Projects Involving Involuntary Settlement 1986–1993. Washington DC.

World Bank (2001a) The World Bank Operational Manual: Operational Policy OP 4.12 – Involuntary Resettlement.

World Bank (2001b) The World Bank Operational Manual: Bank Procedure BP 4.12 – Involuntary Resettlement.

Appendix 14.1 Full Resettlement Plan: A Recommended Outline

Topic	Contents
Scope of land acquisition and resettlement	
	Describe, with the aid of maps, scope of land acquisition and why it is necessary for main investment project.
	Describe alternative options, if any, considered to minimize land acquisition and its effects, and why remaining effects are unavoidable.
	Summarize key effects in terms of land acquired, assets loss, and people displaced from homes or livelihoods.
	Specify primary responsibilities for land acquisition and resettlement.
Socioeconomic information	
	Define, identify and enumerate people to be affected.
	Describe likely impact of land acquisition on people affected, taking into account social, cultural and economic parameters.
	Identify all losses for people affected by land acquisition.
	Provide details of any common property resources.
	Specify how project will impact on the poor, Indigenous Peoples ethnic minorities, and other vulnerable groups, including women, and any special measures needed to restore fully, or enhance, their economic and social base.
Objectives, policy framework and entitlements	
	Describe purpose and objectives of land acquisition and resettlement.
	Describe key national and local land, compensation and resettlement policies, laws, and guidelines that apply to project.
	Explain how Bank Policy on Involuntary Resettlement will be achieved.
	State principles, legal and policy commitments from borrower executing agency for different categories of project impacts.
	Prepare an eligibility policy and entitlement matrix for all categories of loss, including compensation rates.
Consultation, and grievance redress participation	
	Identify project stakeholders.
	Describe mechanisms for stakeholder participation in planning management, monitoring, and evaluation.
	Identify local institutions or organizations to support people affected.
	Review potential role of non-government organization (NGOs) and community-based organization (CBOs).
	Establish procedures for redress of grievances by people affected.
Relocation of housing and settlement	
	Identify options for relocation of housing and other structures, including replacement housing, replacement cash compensation, and self selection.
	Specify measures to assist with transfer and establishment at new sites.
	Review options for developing relocation sites, if required, in terms of location, quality of site, and development needs.
	Provide a plan for layout, design, and social infrastructure for each site.
	Specify means for safeguarding income and livelihoods.

Topic	Contents
	Specify measures for planning integration with host communities.
	Identify special measures for addressing gender issues and those related to vulnerable groups.
	Identify any environmental risks and show how these will be managed and monitored.
Income restoration strategy	
	Identify livelihoods at risk.
	Develop an income restoration strategy with options to restore all types of live hoods.
	Specify job opportunities in a job creation plan, including provisions for income substitution, retraining, self-employment and pensions, where required.
	Prepare a plan to relocate and restore businesses, including income substitution, where required.
	Identify any environmental risks and show how these will be managed and monitored.
Institutional framework	
	Identify main tasks and responsibilities in planning, negotiating, consulting, approving, coordinating, implementing, financing, monitoring and evaluating land acquisition and resettlement.
	Review the mandate of the land acquisition and resettlement agencies and their capacity to plan and manage these tasks.
	Provide for capacity building, including technical assistance, if required.
	Specify role of NGOs, if involved, and organization of affected people in resettlement planning and management.
Resettlement budget and financing	
	Identify land acquisition and resettlement costs.
	Prepare an annual budget and specify timing for release of funds.
	Specify sources of funding for all land acquisition and resettlement activities.
Implementation schedule	
	Provide a time schedule showing start and finish dates for major resettlement tasks.
	Show how people affected will be provided for before demolition begins.
Monitoring and evaluation	
	Prepare a plan for internal monitoring of resettlement targets, specifying key indicators of progress, mechanisms for reporting, and resource requirements.
	Prepare an evaluation plan, with provision for external, independent evaluation of extent to which policy objectives have been achieved.
	Specify participation for people affected in monitoring and evaluation.

Source: borrowed from the ADB's Handbook on Resettlement 2000

15

Community Development

Ensuring Long-term Benefits

MANGANESE

Manganese is a hard gray white, brittle metal. It is an essential element for all living organisms. Some of its compounds have strong oxidizing properties. Few are toxic. Manganese is essential for the manufacture of iron and steel, and is an important characteristic of many alloys, where its main contribution is to increase hardness. The oxide is used as cathode in batteries and as a pigment. Pyrolusite, an oxide of manganese, is also the main ore, from which manganese can be extracted, using acid digestion followed by electrolytic recovery.

15 Community Development

Ensuring Long-term Benefits

Most mining operations, regardless of size, either have a community development (CD) programme in place or know that they should develop one. However, there are as many definitions of community development as there are mining companies. One company may classify its support of the national Independence Day celebrations as community development, while another may use the community development budget to build a swimming pool and tennis courts for its employees. Perhaps these illustrate the two most basic approaches to development: donating to a worthy cause or building something.

While donating money or constructing facilities may represent valid components of community development, a broader definition is necessary. Frank and Smith (1999) define '*community development as the planned evolution of all aspects of community well-being (economic, social, environmental and cultural)*'. It follows that a preferred model for community development is one that is integrated with existing community structures and activities and sustainable beyond the life of the mine.

Probably no mining company has carried out CD programmes exactly as described in this chapter, and possibly no one will ever carry out programmes with all the characteristics presented here. However, the model that follows is simple, achievable, and based on practical realities. The two essential elements are consultation and collaboration. Communities are dynamic and it follows that any model must be flexible so that it can be adapted to changing circumstances.

Figure 15.1 illustrates the general steps involved in CD:

- Build support;
- Develop a plan;
- Maintain the momentum; and
- Implement and adjust the plan as the programme evolves.

Community development requires the utilization of existing capacity and, if successful, results in increased capacity.

A critical factor in community development is community capacity. Community development requires the utilization of existing capacity and, if successful, results in increased

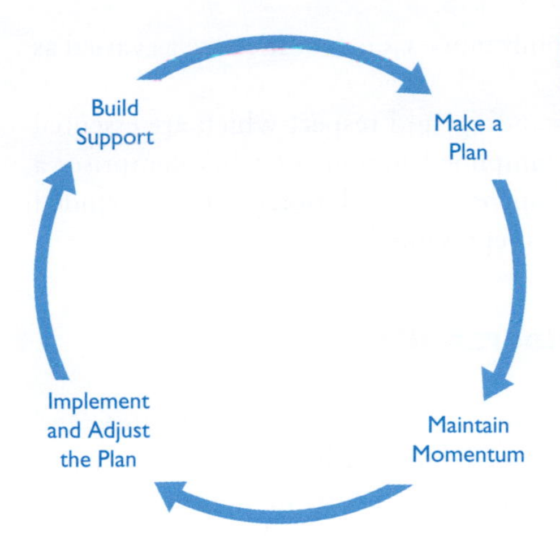

FIGURE 15.1

Main Components of a Flexible Community Development Framework

As community development is dynamic, a static or rigid approach to community development is likely to fail.

capacity. Community capacity is much broader than simply people or money; it includes leadership, commitment, resources, planning skills, and experience in community affairs. Community development may well begin with building capacity within the mining company itself, since companies may be constrained by the lack of skills available internally. Social development staff may be hired to provide the necessary knowledge, but such skills are not easily transferred to mine management and mine personnel. Similarly, not all social development experts, either employed directly by the mining project or retained as advisors, have an appropriate mind-set or relevant experience. Inexperienced or biased advisers may, (1) push the mining company to assume a role which properly belongs to government, or (2) over-commit resources, or both.

Contrary to general belief, responsible national or government owned mining companies often excel in community development initiatives. National staff are, in many cases, knowledgeable of the local context and committed to the local community, while for many public company personnel this may be 'just another project'. International expertise is generally best used to develop a company's internal capacity. The mining company's local staff can then successfully engage in local community activities.

National staff are knowledgeable of the local context and committed to the local community.

There is ample literature on community development, including the International Finance Corporation (IFC) good practice manuals and reports such as 'Doing Better Business Through Effective Public Consultation and Disclosure' (1998), 'Investing in People: Sustaining Communities through Improved Business Practice' (2000), 'Developing Value: The Business Case for Sustainability in Emerging Markets' (SustainAbility and IFC 2002), and 'Breaking New Ground' (MMSD 2002). The Community Development Toolkit (World Bank and ICMM 2005) is designed to create community development plans that help reduce conflict, promote co-operation, and enhance the contribution of private investments, such as mining, to sustainable development. Most international donor agencies provide good practice guidance (e.g. ADB 1999). Relevant reading includes UNDP (1998), USEPA (1999), Creighton and Creighton (1999), EC (2000), Wates (2000), Bruch and Filbey (2002), and Coyle *et al.* (2002).

15.1 WHAT DEFINES A COMMUNITY?

There is a tendency to define a community solely in geographic terms. However, a community can also be defined as a group of people with a common interest or a common origin. A community may consist of Indigenous Peoples (IP), or a blend of people from different origins, religions, and cultures, including IP. People may belong to more than one community. They may for example be part of a neighbourhood community, a religious

Simplified notions of what comprises a community, and expectations of a homogenous response to development are ill-founded because of the many competing and diverse interests represented.

community, an ethnic community, and a fishing community. Communities are as varied as the individuals of which they are comprised.

Each community has unique attributes, understanding, and respect which are essential for effective community development planning. Simplified notions of what comprises a community, and expectations of a homogenous response to development are ill-founded because of the many competing and diverse interests represented.

Indigenous Peoples, Culture, and Vulnerability

Indigenous Peoples are those people regarded as original inhabitants of an area, and are usually acknowledged as the traditional owners of land, which is commonly termed 'ancestral land'. World Bank (1989) proposes the following six criteria to identify IP:

1. Close attachment to ancestral territories and to natural resources in these areas;
2. Self-identification and identification by others as members of a distinct cultural group;
3. Often use a language different from the national language;
4. Presence of customary and political institutions;
5. Primarily subsistence-orientated production; and
6. Vulnerability to being disadvantaged as a social group in the development process.

Past experience from mining operations in many parts of the world, confirms that where ancestral land is involved, especially in the case of small IP communities, they are especially vulnerable to being disadvantaged – economically, socially, and politically – by a mine development. It is now widely accepted that many IP and traditional cultures have world heritage values. IP serve as custodians of the last remaining pristine territories of the Earth (ADB 2004). These territories remain largely unaffected by the environmental degradation that has occurred elsewhere as a consequence of economic growth and are now seen as resources of competitive advantage. Developed countries have few if any such areas and most are legally protected.

Insulation versus Integration

Mining companies often prefer to insulate the mine construction from the local community to the extent possible.

There are two approaches to CD for a proposed project: to insulate the local community from the project or to integrate it into the project. A decision on which approach to take is best made jointly with the affected community. Both have merits (see also **Table 14.1**). Mining companies often prefer to insulate the mine construction from the local community to the extent possible. Practical measures include fencing off the construction site and construction camps (practising the saying 'A good fence makes a good neighbour.'). Access to the construction camp is restricted so there is minimum interaction between the construction workforce and community, except for any villagers employed by the mine construction. Often the construction workforce is employed and accommodated on a single status basis. This approach is selected to minimize the worker's impact on the areas outside the camp and plant area, including hunting, and to minimize security disturbances or issues arising between local communities and the construction workforce, in particular in regard to potential sexual harassment of local women. The arrangement to insulate construction from the communities also recognizes the safety hazards that are present at any large construction site.

Insulation becomes less practical during the mine operation. The presence of a mining project will bring about many unavoidable changes in the lives of people in nearby communities. Some community members may find it difficult to accept changes that are the result of the project activities, such as the increased number of outsiders in the project area. In under-developed regions where the carrying capacity of the local area is low and not able to support a large population, there is the real risk of marginalization of the host community economically, politically, and socially. Ensuring that any project-affected community continues to thrive as an articulate, self-reliant, and mature community is an overall goal of community development programmes.

15.2 POINTERS TO SUCCESS

The maxim, 'It ain't what you do it's the way that you do it', sums up most of the lessons learned by practitioners of CD. Successful community and IP development programmes reflect the understanding that the unique needs and aspirations of individuals and communities are best identified and addressed through a process of consultation and participation. Any mining company should consciously avoid the paternalistic top-down approach to community development and the unhealthy dependency it engenders. Some pointers to success are shown in **Table 15.1**.

'It ain't what you do it's the way that you do it'.

Getting it Right from the Start

Impressions formed during the initial contact with local communities can affect relations between the mining project and the local communities for many years. Because negative initial impressions may be very difficult to overcome, project personnel need to prepare themselves well before meeting with a community. This preparation includes recognition of community rights, assessment of the company's values and its key operatives, and appropriate planning. The starting point is an assessment of whether the company and its key employees have the required cultural awareness and capacity to understand a community's needs. If not, the company must develop its internal capacity before undertaking significant interaction with an affected community.

TABLE 15.1

Core Principles of Community-Driven Development – The unique needs and aspirations of individuals and communities need to be addressed through a process of consultation and participation. Any mining company should consciously avoid the paternalistic top-down approach to community development

Core principles
Getting it right from the start
Respecting human and community rights
Minimizing direct transfers
Coordinating with mining operation
Seeking local input to design and execution
Ensuring local government is not bypassed
Focusing on project-affected people
Maintaining the natural resource base
Aiming for self-financing programmes
Benefiting both community and mining project
Maintaining Community ownership

CD planning can begin once the internal capacity has been developed. It is sensible that community development planning is well under-way prior to construction. Start of construction looms as a deadline, after which the timely implementation of community development programmes becomes a greater challenge. It is reasonable to say that the mine planning stage is the last opportunity to implement community development smoothly. Construction represents a break-point, after which the relations between the company and the community may never be the same, and good will is likely to become a rare and expensive commodity. Implementing CD programmes during construction is always problematic, as mine construction typically represents extended periods of intense confusion with continuous crisis management to correct design flaws and failed planning. Because contractors move on to new projects once construction is completed, they have little incentive to involve themselves in community development. Hence, only community development programmes planned prior to construction are likely to be implemented.

The worst time to start CD planning is the operation phase, because it is highly probable that the community has been thoroughly traumatized by the impacts of construction.

The worst time to start CD planning is the operation phase, because it is highly probable that the community has been thoroughly traumatized by the impacts of construction, and social cohesion may have broken down as a result of massive migration of workers to the area. Large numbers of demobilized construction workers, locals and immigrants, may be agitated by the perception of the company's lack of concern for their future. Local government's planning failures may have crippled its operations as well, and infrastructure inadequacies surrounding the mine may also create demands for any available funds. Implementing CD in the early years of mine operation will be a severe challenge without detailed advance planning.

Respect of Human and Community Rights

The IFC Performance Standard 2 states it clearly: the respect for human and community rights is fundamental to all interactions with the communities where the mining company operates. In preparing its human rights policy, the mining proponent should refer to international conventions and declarations, such as the United Nations Universal Declaration of Human Rights (1948), International Covenant on Civil and Political Rights (1966), the International Labour Organization (ILO) Convention Nos. 107 (1957) and 169 (1989), Agenda 21 of the United Nations Conference on Environment and Development (1992), the Vienna Declaration and Program of Action (1993), and the United Nation's Draft Declaration on the Rights of Indigenous Peoples (1993), for their implications to IP.

The ILO Convention 169 of 1989 concerning 'Indigenous and Tribal People in Independent Countries' and the Convention on Biological Diversity (CBD) of 1992 provide guidance on traditional resource rights. Many of the signatory countries have ratified the later convention, giving it the force of international law. However, an initial caveat in Article 8 limits the requirement to 'as far as possible and appropriate', and a specific caveat in section (j) empowers nations to apply the right subject to its own internal legislation. As a consequence, the CBD is only as strong as the country's national legislation in protecting traditional rights to resources.

Understanding of human rights policy must permeate the whole mining system.

Understanding of human rights policy must permeate the whole mining system. Employees and contractors need to be informed about the policy, and adherence to the established Code of Conduct must be mandatory. In particular, the mining project should respect the right of self-determination by adopting a participatory approach to CD planning. Affected communities and IP not only have the right to be substantially involved in CD programmes, but to a large extent should determine the content of those programmes. Without community participation, there is the risk that the mining company will default to an authoritarian approach, providing hand-out programmes that are either unfit for, or an insult to the community.

Minimizing Direct Transfers

A CD programme organized as a charity is hardly sustainable. Cash transfers represent a direct charge to the company's bottom line, while to the community the cash takes on the character of an entitlement. These transfers have very limited potential for initiating a higher level of development. Simply moving cash from one pocket to another is seldom the answer. The mining company, however, is morally, and often legally required to make these contributions, not least to support local government programmes. Cash transfers should be fixed, either as an absolute amount, or tied to scale of production or operation. Indexing contributions to revenues (or profits) invites mistrust about how these amounts are calculated, and more importantly, means that funds will be cut during business slowdowns, when the community will already be suffering from employment cutbacks.

Above all, cash transfer payments should be transparent – everyone should know how much is going where, before it happens. It is hardly a secret that the opportunity for corruption is a challenge faced by CD programmes. Any budget for any purpose in developing nations faces this reality. Community development money is seen as buying peoples' co-operation. Even a completely honest programme faces the perception that someone is benefiting on the side, and resentment develops from those who believe they are not getting their share. The best defence is honesty and transparency. Community relations programmes may need to push money in obscure directions at times, which is all the more reason for having clear divisions between community relations and CD programmes and budgets.

Cash transfer payments should be transparent – everyone should know how much is going where, before it happens.

It is often necessary for the mining company to finance community participation in the planning process. Most local communities in developing countries do not have the financial resources to undertake or to participate in a major development planning exercise. By assisting the community to participate, a mining project can ensure better relationships that will be extremely valuable in later discussions if there are contentious issues.

It is also no secret that mining projects often finance government participation. In some countries this is mandatory and governments prescribe standard allowances for accommodation, per diems, etc. to be provided to government participants. Financing government participation can take on many forms, e.g. providing lunch (with extra lunch boxes available to be taken home), paying travel expenses or handing out daily allowances. What is acceptable in one country may be unacceptable or even illegal in another. Cash payments of course open a window for misuse. For example money for accommodation may not be used as intended, if participants stay with relatives; money for air travel is not used as intended since a cheaper mode of transportation may be selected.

Coordinating With Mining Operations

Community development does not operate in isolation; it is part of the overall operation so it is essential to coordinate CD programmes with other mining activities. CD programmes should be coordinated with the company's community/public relations, environmental management, land acquisition, purchasing, safety and health, and security efforts. These are all areas where company actions can interact with the local community, and an inconsistent policy in any of these areas can frustrate the objectives in other areas. In practical terms, coordination of these functions means that the people who are responsible for

Community development does not operate in isolation; it is part of the overall operation so it is essential to coordinate CD programmes with other mining activities.

the day-to-day implementation should communicate closely, on a daily basis. Ideally, there should be a clear division between community relations and community development, including completely separate budgets.

Seeking Local Input to Design and Execution

Communities need to be included in the design and execution if a programme is to be accepted.

Generally speaking, the objectives of most community development programmes are simple in concept, tending to focus on improving health, public services, and education. However, communities need to be included in the design and execution if a programme is to be accepted. The process of planning and implementing a community development programme often determines its success or failure.

In outmoded CD models, community development planning is undertaken by outside experts, who then inform the communities what programmes are available for them and seek their consensus and participation. The inherent risk in this approach is that communities correctly perceive that the mining company alone prescribes and decides what will be done. Admittedly, with some effort and thought, numerous good community development programmes can be conceived without community participation. However, without the integration of the local community into the programme identification, selection, and design processes, most programmes will realize limited success or fail entirely.

Community participation in the process is not always easy, requiring at least a proactive engagement, sustained over a long period. The creation of a local institution or representative committee to conduct this process may be necessary, or participatory resource appraisal tools, described in a later section, can be applied. Some mining companies wish to be closely involved in the process of determining community development priorities. Others prefer a more 'arm's length' approach where, for example, an independent third party such as an NGO is engaged to work with the community in planning the CD programme. In this case, the company's role may be only to approve (or not) the commitments of company resources.

Community development programmes identified and developed with the people who are benefiting are more likely to be sustainable, and the people are likely to become more self-sufficient and self-reliant as a result. Participatory planning takes time, and can best be achieved by starting early. Comprehensive dialogue, consultation, and involvement of community members are essential elements of this strategy.

Local Government Authorities should not be By-passed

In many cases, CD initiatives actually address the same needs (education, health, welfare, community infrastructure) that are the clear responsibility of governments. In very remote areas or in highly impoverished societies this cannot be avoided.

While mining companies are usually in frequent contact and interaction with local and regional governments, CD programmes are seldom well-integrated with official local and regional development efforts. In many cases, CD initiatives actually address the same needs (education, health, welfare, community infrastructure) that are the clear responsibility of governments. In very remote areas or in highly impoverished societies this cannot be avoided. However, unless there are no government institutions and none can be developed, it is preferable for the mining company and government to coordinate resources, with the government in the leading role and the mining company in the background. Many companies resist this approach citing objections such as 'Only we can ensure that outcomes are achieved on time and within budgets' or 'If the government is in control, money will be wasted'. These concerns can be alleviated in several ways. For example, rather than committing funds for government programmes, the company could elect to provide personnel and equipment. In

FIGURE 15.2
The Tri-Partite Principle in the Community Development Process

The mining company should aim to support the government in delivering government services, but should not take on government responsibilities.

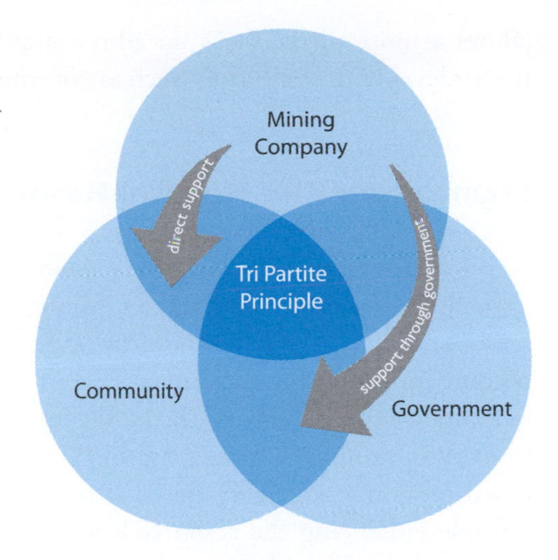

summary, mining companies may support the government in delivering government services, but CD programmes should not usurp the government's responsibilities.

Figure 15.2 shows the tri-partite principle. In an ideal world, CD efforts are planned and implemented jointly, by all three parties. During the process itself, however, individual involvement may vary. At the needs assessment stage, for example, the government authorities may not make decisions, but may contribute to issues of concern and provide guidance in terms of regulatory requirements and trade-offs that should be considered. In the implementation phase, government may take the lead, especially in community development programmes that are concerned with improving public infrastructure, with both the company and the community providing active support (funding, material, and equipment from the mining project, and labour, knowledge, or land from the community).

Focusing on Project-affected People at Various Levels

Community programmes can be applied at various levels within the community, with different end results and tradeoffs achieved through focusing on different levels of organization (Gibson 2001). However, while not discriminating on the basis of creed, gender, national origin, or broad racial affiliation, mining companies should recognize the special role in the mining project for project-affected people (PAP) (World Bank 1989; ADB 2003; Performance Standard 1, IFC 2006). The people most affected by any stage of the mining project should enjoy priority status for community development support. They are the key stakeholders even if they are not in a position to articulate their concerns and aspirations. NGOs, politicians, or business people may be more vocal or influential, but community support should be based on needs, rather than pressure applied to the company.

The people most affected by any stage of the mining project should enjoy priority status for community development support.

By focusing on individual community members with selected scholarships, for example, community programmes may ultimately help to increase knowledge and participation across the community, including those who do not traditionally hold any decision-making power. Focusing at the individual level, however, has some inherent risks including undermining the goals of formal or informal authorities, or distributing benefits subjectively.

The involvement of families (not only the family head) helps people to learn and advance together, based on a shared vision and understanding. It also allows for action focused on such immediate families needs as health and education. However, there is some danger of this undermining the traditional roles of family or community leaders. Working with formal or informal organizations such as schools, religious groups, and other community associations allows a more direct focus on administrative and project skills. Capacity building at the macro-level with institutions such as governments and companies, may also be productive.

Maintaining the Natural Resource Base

Maintenance and development of the original economy is necessary to maintain the stability of the local community, given that most residents will not be able to work in mining.

At most mining sites, the original economy is natural, resource-based primary production – agriculture, forestry, and fisheries. Mining and associated activities will constitute the new land use in some areas (temporarily or permanently), and a portion of the labour force will become mine workers. Maintenance and development of the original economy is necessary to maintain the stability of the local community, given that most residents will not be able to work in mining (see also Case 13.12). After mine closure, a return to a natural, resource-based economy is usually the only option available to the community. Sustaining and if possible enhancing the resource base is important whether mine life spans 5 years or 50 years, and this should serve as a central principle for CD programmes.

Aiming for Self-Financing Programmes

Leveraging refers to using community development funds to create projects that stimulate a higher level of economic activity than would otherwise occur. While not all community development programmes generate a profit (or break even), those that do so enable the transfer of ownership and management to a local private sector or co-operative entity. Direct transfer of funds, either as grants or as loans, provides seed money for other self-sustaining projects. A good example would be programmes in which the scrap materials and wastes produced by the mining operation are used for locally based industry and trade. Even better examples are commercial agriculture, horticulture, animal husbandry, and aquaculture projects set up to provide the mine workers with a portion of their food supply (**Case 15.1**). While neither of

CASE 15.1
Harvesting Sunshine for Biofuel

Cassava is a starch-rich tuber that grows well on degraded land and requires relatively little fertilizer and water inputs. Apart from its traditional role as a food crop, Cassava has increased its value as a fuel commodity. Its starch is already being used to produce ethanol on a large scale in Brazil, Nigeria, China, and Thailand. The technology for converting cassava into the biofuel Ethanol is being perfected and the technology is affordable.

The cassava business has caught the attention of some mining companies. Planting and harvesting cassava to produce biofuel consumed by the mine is an interesting community development concept: (1) cassava has the advantage of being relatively undemanding, and will thrive on poor and even tired soils, where few other crops will grow; (2) there is the capacity building aspect in preparing communities to enter the cassava business; (3) while the main focus is on the roots and stems, leaves are a supplementary output with a potential cash value; (4) during operation the mine provides a secure market; (5) by purchasing biofuel from host communities the mine reinvests in the local economy; and (6) the demand for biofuel is likely to remain after the mine is gone. By that time the host communities have established a solid cassava business with good organization and understanding of the specific conditions and scale of operation.

these examples would necessarily be sustainable in terms of providing employment beyond the mine lifetime, both would involve development of significant local entrepreneurial and management skills and capital formation that would almost certainly provide long-term benefits to the community. It should be expected that many ventures based on seed money will not be successful. Accordingly, investment in a wide range of ventures is preferred over a few or a narrow range of initiatives, however compelling they might seem.

Benefiting Both Community and Mining Company

A mining company exists to make profits for its shareholders. For a company to claim altruistic motives for providing benefits to the community is neither credible nor necessary. Community development programmes need to be an integral part of the mining operation, because the company needs a stable and improving socioeconomic environment in which to operate successfully. Since even the largest mining companies have limited resources to devote to community development programmes, companies should concentrate on those programmes with the potential for providing benefits to the company as well as the community.

Companies should concentrate on those programmes with the potential for providing benefits to the company as well as the community.

Community has Ownership

All sustainable community development programmes should be planned to gain community decision-making based on shared responsibility for success as well as failure. Without community endorsement, ownership is unlikely to develop and failure is likely. Some community development programmes suggest themselves without extensive consultation. A good example is where the mining company, with its own labour force, equipment, materials, and funds, builds a community water supply system where none existed before. However, if the water supply is seen as the company's system, all administration, maintenance, extensions, upgrades, and replacements will be seen as the company's responsibility, and the water will cease to flow soon after the mine closes. The preferred alternative is that before a single well is drilled or pipe is laid, a local government or community based entity is created or adapted, with responsibilities for collecting fees (however nominal) and arranging maintenance. The company's support should be phased out as soon as practicable. While this approach may lead to short-term inefficiencies, it is more likely to be sustainable.

All sustainable community development programmes should be planned to gain community decision-making based on shared responsibility for success as well as failure.

15.3 COMMUNITY DEVELOPMENT PROCESS

Community development is not a regulatory add-on, but rather a programme to allow communities within the mining operation's impact area to realize sustained improvements in standards of living and quality of life that exceed the direct effects of employment in, and higher levels of economic activity resulting from mining. The sequence for programme formulation is illustrated in **Figure 15.3**.

It is advisable to document in detail the consultations and rationale leading to the approach adopted, for the record and to meet the ever-increasing need to demonstrate compliance with the Equator Principles. This documentation is best formalized in a Community and Indigenous Peoples Development Plan (CIPDP), ideally adopting relevant guidance provided by leading international lenders such as IFC (1998, 2006) or ADB (2004). **Table 15.2** provides an example of the structure and contents of a CIPD. Subsequently as the CIPD is implemented, outcomes and changes should be documented

FIGURE 15.3
Integrated Community Development Programme Cycle

Of utmost importance in community development process is the feedback loop which enables the programmes to be modified or restructured in response to changed circumstances or past experience.

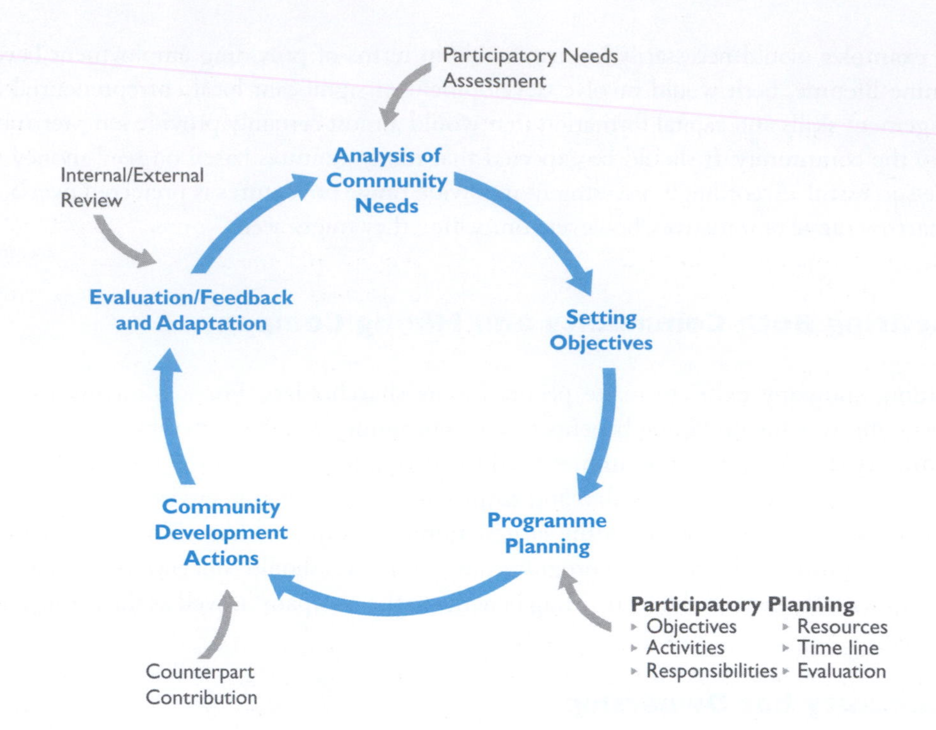

TABLE 15.2
Example Outline for a Community and Indigenous Peoples Development Plan (CIPDP) – The CIPDP is a living document which should be updated whenever significant changes occur

Introduction
Design of Mining Project
Project-affected Communities and Ethnic Context
Guiding Principles, Policy Objectives and Regulatory Requirements
Past Community Development Activities
Community Participation in Identification and Implementation of CIPDP Activities
Community Action Plans
Organizational Framework
Funding
Time Line
Monitoring and Evaluation

in regular status reports. Many companies publish annual 'Public Environmental Reports' or 'Sustainability Reports' which are ideal for this purpose.

Analysis of Current Conditions, Needs and Objectives

Every mining engineer understands that knowledge of the ore body is essential for mine planning. There is less appreciation, however, that an understanding of the nearby communities is equally important to plan for community development. Just as deficiencies in understanding of an ore body will lead to inefficiencies or incomplete recoveries, deficiencies in understanding the community will lead, at best to wasted resources and at worst to community opposition and alienation. Understanding involves collection and compilation of relevant information on the community, its history, attitudes, and aspirations and in a general sense, on the prevailing environmental, social, and political setting. While literature reviews and community surveys contribute valuable insights, the most important contributions to understanding follow from day-to-day interactions between people which, given goodwill on both sides, lead to mutual understanding and trust. This process should commence at the outset of the exploration programme.

Just as deficiencies in understanding of an ore body will lead to inefficiencies or incomplete recoveries, deficiencies in understanding the community will lead, at best to wasted resources and at worst to community opposition and alienation.

Since an exploration project is organized to develop information systematically, it should be no significant burden to add collection of information on the community to the programme. During exploration and, more particularly, from the day it becomes clear that a mine may be built, the community starts changing. Expectations build, and land speculation and speculative in-migration may soon become evident. It is important to understand the changes that happen, but this is only possible with an understanding of who and what were originally there. Also, the company on its own can do little to prevent speculation and in-migration except with the total support and co-operation of the community. Similarly, if and when NGO opponents of the project arrive, their opposition can only be effectively counteracted if the company is trusted and enjoys strong relationships in the community. However, support based on 'blind trust' is not sufficient. More important is 'informed support' wherein the community understands the project and its impacts, and the rationale behind the decisions that have been made. With such information the community is less likely to be influenced by counter arguments put forward by NGOs opposed to a mine.

From the day it becomes clear that a mine may be built, the community starts changing.

When feasibility and environmental studies begin, additional studies that will be valuable in planning for CD may be conducted. Environmental and social baseline and impact assessment studies carried out simply to satisfy regulations will not be adequate. Important activities include in-depth socioeconomic and socio-cultural studies and aptitude surveys, which also provide basic information for labour force plans. Other relevant studies include surveys of natural resource use and land capability. One of the purposes of detailed soil surveys is to serve as a base for a master land-use plan, while an agricultural census serves as the basis for an agricultural development and diversification plan. This will assist in the development of a food supply plan. An integrated community development strategy may also include a conflict analysis to assess the underlying causes of conflicts in the larger project area to ensure that mine development and CD programmes do not increase the likelihood of conflict.

Participatory Planning

In the past, mine developers as well as governments simply decided what would be in the best interests of a community which was then informed of the decision(s). Community development programmes based on this out-dated model fail even though they may be well-intended and well-suited to particular needs. Modern community development initiatives are based on a participatory, or a 'bottom-up' approach. Potential programmes emerge from discussions with community members. Community members are involved in the actual planning stages of programmes, from identifying needs, specific actions to correct problems, and resource requirements. Community development becomes community driven.

Modern community development initiatives are based on a participatory, or a 'bottom-up' approach.

This is not necessarily straightforward as communities may have completely unrealistic aspirations. The 'Cargo Cult' phenomenon of some Melanesian communities is an extreme example, but unrealistic expectations are widespread. Accordingly, as part of their involvement, community members need to be fully informed so that they develop an understanding of what can be realistically achieved and the likely time required. A CD strategy should therefore be implemented in conjunction with a broad-based Public Consultation and Disclosure Plan (PCDP) that actively solicits local input on the appropriateness and effectiveness of CD activities and approaches.

If the community's needs for information are not supplied by the Company, they will be supplied by rumours or by opponents of the project.

Experience is, however, that communities around exploration projects universally complain about a lack of information from the company. Most companies do organize information meetings, but they usually refrain from providing information other than pertaining to current activities. This is partly because exploration personnel may have little knowledge of how the ore body will ultimately be exploited, or because the spokespeople do not have the company's authority to communicate except about exploration issues. Another reason that the mining company may initially have little interest in providing specific project information is to avoid creating unrealistic expectations within the communities and even less to inform land speculators. These are valid concerns. However, if the community's needs for information are not supplied by the company, they will be supplied by rumours or by opponents of the project. It is therefore important for the company to keep the community informed, not only about site activities, but where these might lead.

The initial communication between the mining company and the communities may include general information about the project; site access issues; employment opportunities; purchasing of local goods; environmental impacts management; and health and safety issues. As the mine proposal evolves, communication should become more specific and more frequent, with increasing community participation. Progressively, the community and the mining company enter a partnership that involves community members participating in the planning stages of programmes, from identifying needs, specific actions to correct problems, and resource requirements, to deciding who in the community or what institution is most appropriate to provide resources required or undertake each activity.

Partnerships evolve in three phases (**Figure 15.4**). Initially partners explore a potential partnership by opening the dialogue, and assessing benefits and risks. To decide not to form an agreement is a possible outcome at this stage (as is often the case with anti-mining advocates). The second phase concerns the building of trust. Participants identify and formulate common interests and vision, and agree on the division of roles and resources in implementing mutually agreed actions. Measuring outcomes and adapting to change are

FIGURE 15.4
Developing Partnerships

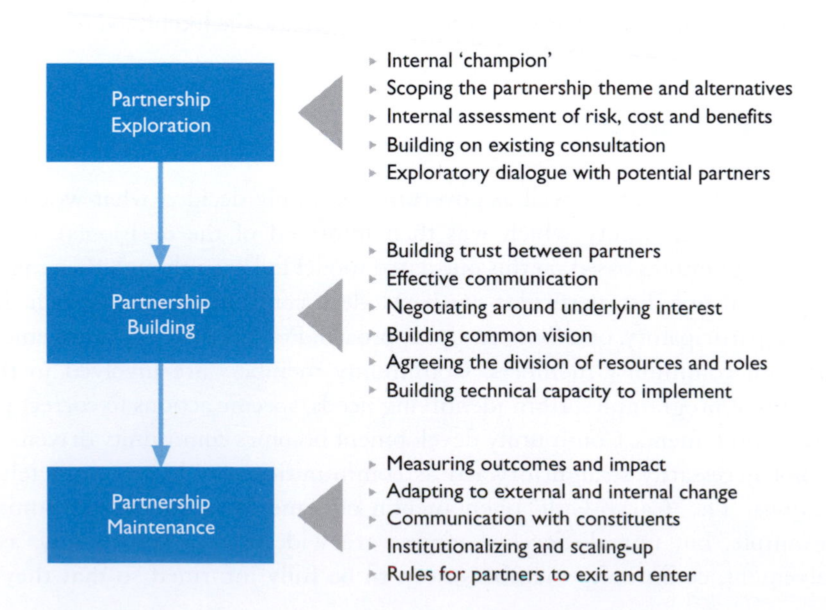

essential characteristics of maintaining and strengthening the partnership, which constitutes the third phase of partnership development.

While not mandatory, mining companies and communities often choose to formalize their partnership in a jointly developed written agreement. Names of partnership agreements vary, but common themes exist (**Table 15.3**). Especially in the early phase of community interaction, a formal partnership agreement increases the level of confidence in the success of partnering. The joint development of the partnership agreement also helps each participant to understand the other's positions and interests, and provides a common platform for future actions. Addressing the elements of a partnership agreement one by one is also a helpful educational tool: it raises awareness of the many facets of a long-term partnership relationship, and forces everybody involved to agree on some key points. As such, the joint formulation of the agreement helps build trust, and the process becomes as important as the outcome.

Mining companies and communities often choose to formalize their partnership in a jointly developed written agreement.

Planning Methods

There is no shortage of participatory planning methods (Rietbergen-McCracken and Narayan 1998). There are workshop-based methods, e.g. Appreciation Influence Control

TABLE 15.3

Common Ingredients of a Partnering Agreement (such as Letter of Intent, Memorandum of Understanding, Memorandum of Agreement, Charter, or Code of Conduct) (World Bank 2004) – Names of partnership agreements may vary, but common themes exist

Assumptions
Representatives of each partner organization and their status
Geographic boundaries and/or target population of the partnership
A (common) vision statement
The objectives of the partnership: shared by all parties; specific to each organization
Joint work plan, encompassing: activities, schedules and performance indicators; resource commitments; and responsibilities
Funding arrangements/contracts
Decision-making principles
Grievance mechanism to resolve differences
Procedures for transparency and on-going communications between partners
Measures to strengthen the capacity of partners to implement their commitments
Measures to mitigate external risks and threats to the partnership
A protocol for communicating with constituents and other parties
Procedures for monitoring and measuring the performance of the partnership against both the business and wider social objectives
Rules for individual partners to leave or join the partnership, and an exit strategy for the partnership as a whole

(AIC) (Smith 1991); Objectives ('Ziel' in German)-Oriented Project Planning (GTZ 1991); Logical Framework Analysis (LFA) (NORAD 1999), and Team Up. There are also community-based collaborative decision-making techniques – Participatory Rural Appraisal (PRA) (Chambers 1992) and Self-esteem, Associative Strengths, Resourcefulness, Action Planning, and Responsibility (SARAR) (Srinivasan 1990). For large resource development projects in isolated and under-developed areas, private sector companies have recently started to make increasing use of PRA tools (also termed participatory rapid and rural appraisal) for working with communities to design development programmes in a cooperative and participatory manner.

In addition, there are stakeholder consultation methods, Beneficiary Assessment (BA) (Salmen 1995) and Systematic Client Consultation (SCC) and supplementary techniques, Social Assessment (SA) and Gender Analysis (GA) (Finsterbursch *et al.* 1990) that are used to fill the gaps left in methods that focus on male and elite capture by highlighting the impacts on marginal groups and women. A number of organizations maintain websites to share information and experience on participatory methods, such as the Electronic Development and Environmental Information System (ELDIS) (www.nt1.ids.ac.uk/eldis/eldis.htm), the Institute of Development Studies (IDS) (www.ids.ac.uk), the International Institute for Environment and Development (IIED) (www.iied.org), the Institute for Participatory Management and Planning (IPMP), (http://www.ipmp-bleiker.com), and the International Association for Public Participation (IAP2) (www.iap2.org).

Participatory planning may require establishing a community development board (or community board) comprising formal and informal community leaders, representatives of religious groups, company representatives, and other interest groups. The CD board will probably include equal representation of all affected communities, which is important for cordial and harmonious relations between these communities. If resettlement occurs, a community board likewise contains representatives from both host and resettlement groups.

The mining company normally assists in advising and financing the community board and in strengthening the capacity of the board to carry out the assigned tasks. The key role of the community board is to identify, facilitate, and ultimately coordinate CD programmes. The role of the community board can easily be expanded to include a pivotal role in all consultation, disclosure, and negotiation processes that involve the mining company and local communities, and the implementation of decisions made through these processes. Possible tasks for the community board include: supervision of compensation arrangements, in conjunction with the mining company; confirming community membership and therefore eligibility; providing a community-based mechanism for raising and addressing community issues or grievances; and providing advice on labour recruitment.

Depending on the range of responsibilities of the community board, it may be advisable to adopt mechanisms to ensure that the board is legally constituted and to ensure adequate provisions in its charter, guaranteeing that regular and thorough auditing is conducted. Working through a CD board allows the mining company to provide community development programmes at arm's length, avoiding criticism of manipulating community interests and perceptions.

Participatory planning may take longer to develop and implement than authoritarian methods. However, the initial outlay in time and inconvenience will be repaid many times over when communities begin to assume control of their own destiny – a fundamental criterion for sustainable development.

Independent of the chosen method of community participation, community development programmes growing out of the participatory planning process enable communities to analyze their situation, gain confidence in their own ability to understand the cause of existing problems, devise solutions, and design plans for future action. Participatory planning may take longer to develop and implement than authoritarian methods. However, the initial outlay in time and inconvenience will be repaid many times over when communities begin to assume control of their own destiny – a fundamental criterion for sustainable development.

	Initiate	Plan	Implement	Maintain
Self-help Community control	Community initiate action alone			Community maintains alone
Partnership Shared working and decision-making	Company and community jointly initiate action	Company and community jointly plan	Company and community jointly implement	Mine Closure
Consultation Company asks community for opinions	Company initiate action after consulting community			
Information One way flow of information Company control	Company initiate action			

FIGURE 15.5
Participation Matrix

Ideally most of the community development programme cycle operates on a partnership model, with increasing community ownership and control towards mine closure.

Source: based on Wates 2000

A participation matrix adapted from Wates (2000), neatly demonstrates the different natures of interactions that can guide community development (**Figure 15.5**). The matrix also helps to illustrate appropriate levels of participation at different stages of the CD programme. Shared interests of community and mining company are assumed. Ideally, most of the CD programme cycle operates in the shaded areas, although either party can initiate action. Joint planning is crucial. The matrix's value is in prompting communities and mining companies alike to think about where they are in the overall process and when it is necessary to move between levels in stakeholder involvement. It also illustrates that, following mine closure the contribution of the mining project to community development ceases to exist.

Following mine closure the contribution of the mining project to community development ceases to exist.

Community Action Plans

A specific CD programme, or Community Action Plan (CAP), simplified in **Table 15.4**, is a detailed plan for implementing solutions to problems that have been identified during the needs assessment process. It will become the work plan both for the community and its development partners (mining company and/or local government) with adjustments from time to time to suit changing community priorities.

There are six principle steps in developing a CD programme, best documented in a formal manner as is normally required for most capital expenditures. These steps are:

1. Define needs and overall objectives;
2. Develop a list of actions required to achieve the objectives;
3. Assign responsibilities to the communities, the mining company, the government, and other institutions to implement each activity;
4. Identify resources required for each activity and the contributor of these resources;
5. Decide when the activities are to be done; and
6. Develop criteria to evaluate success or need for corrective actions.

A Community Action Plan is the work plan both for the community and its development partners and will be adjusted to suit changing community priorities.

A Community Action Plan is the work plan both for the community and its development partners and will be adjusted to suit changing community priorities.

Community ownership of the CAP is the most important aspect of the plan. The CAP acts as a template for future planning activities by the community and also serves as a

TABLE 15.4
Community Action Plan Illustrated

Narrative summary	Programme/ Activities	Coordinator/ Owner	Target Group/ Participants/ Beneficiaries	Resource requirements	Assigned responsibilities/ Counterpart contribution	Approximate timing

record of how much was achieved during participatory planning and implementation. Each CAP becomes part of the Community and Indigenous Peoples Development Plan (see **Table 15.2**).

CD projects should be implemented to come rapidly under community ownership and control.

There should be no such thing as a 'free lunch'.

CD projects should be implemented to come rapidly under community ownership and control. Ownership, which starts with participatory planning, requires participatory execution; therefore, CD programmes should require counterpart contributions – there should be no such thing as a 'free lunch'. The mining project, for example, may provide materials while the communities provide the labour. Through participatory execution the outcome of the CD programme is seen as community owned, with the mining project not being held responsible for administration, maintenance, extensions, upgrades, and replacements. As discussed earlier participation is a concept of sharing decision-making power and the responsibilities that accompany such power. Participation and contribution bring ownership, pride, and empowerment. Participation can be an end in itself to empower communities.

Codes of Conduct

The Code of Conduct is best established in consultation with the affected communities.

In recognition of the complex nature of dealing with communities and groups, with diverse interests, ethnicity and education, it is advisable to establish a workforce Code of Conduct prior to mine construction or any other labour intensive field activities. The Code of Conduct is best established in consultation with the affected communities, which will foster relations based on mutual respect and understanding. The Code of Conduct is a living reference, and may be revised during the life of the mining project. By consulting the local communities, matters considered important are included. Examples include:

- prohibition of narcotics;
- prohibition or limitations on possession and use of alcoholic beverages;
- limitations on access to communities, such as 'no go' areas, and curfews;
- imposition of conditions concerning relationships between non-local workers and local women;
- prohibitions regarding access to sites of religious or cultural significance; and
- limitations on exploitation of traditional resources by non-local workers.

The Code of Conduct should become part of contractual arrangements with all contractors, including consultants. A compulsory briefing on the Code of Conduct should apply to all new employees. Some provisions of the Code will be more important than others – again reflecting the importance accorded to these elements by the communities. Breach of important Code provisions may be grounds for dismissal of employees or for penalization of contractors.

TABLE 15.5

Grievances Affecting Mining Communities and the Environment – Themes (World Bank 2004)

Land-Related
Past grievances over land claims and asset compensation
Land ownership disputes ignited by speculation of mining firms
Current disputes over land ownership of project sites and related compensation payments
Disputes over land boundaries between different local land-owners
Disputes between the company and communities over access rights
Disputes over renewal arrangements for land leased by the company
Income and Employee Benefits-Related
Unfulfilled hopes for higher incomes, e.g. access to employment, employment multiplier effect of mining activities
Jealousy between community members (and between different communities) who receive higher wages or employee benefits (health care, education, etc.)
Intra-community jealousy of the way the law is perceived to favour Indigenous Peoples over more recent settlers
Cultural and Relationship-Related
Poor communication between company and communities
Communication by the company biased towards one section of a community
Lack of co-operation between different community groups on how they engage with the company
Cultural conflicts between indigenous cultural communities and more recent settlers
Political interventions at the municipal or regional level on the side of either the company or communities
Latent family and relationship disputes that fuel local grievances
Different expectations of community groups and companies, and between different community groups
Different approaches to the management of environmental and social issues between the company, communities, local government authorities, and NGOs, e.g. speed of decision-making, degree of participation and consultation, etc.
Natural Resource-Related
Competition for natural resources (e.g. wildlife, fish stocks, timber, other forest products, etc.) resulting from new technologies (e.g. motor vehicles, guns, sawmills, fishing nets)
Environmental damage by the mining company that affects local resource users
Unplanned exploitation of natural resources opened up by the project's access roads
Differential benefits for landowners (e.g. those near a road)

Grievance Mechanisms

Although a project may strive to base all activities on the concept of partnership, there is a need to recognize complaints or claims and to establish appropriate grievance identification, tracking, and resolution mechanisms. Many issues can lead to disagreement and grievance. Most grievances affecting mining communities relate to land disputes, perceived biases in recruitment, inequalities in remuneration, cultural differences, and natural resource use (**Table 15.5**).

Most grievances affecting mining communities relate to land disputes, perceived biases in recruitment, inequalities in remuneration, cultural differences, and natural resource use.

Issues are often addressed informally as they arise; however, some may require formal grievance resolution. Grievance resolution strategies range from 'do nothing' to international arbitration or dispute settlement in court (**Table 15.6**). Each strategy has its own costs and risks; 'Do nothing' has a low cost and low risk, but it is only recommended for trivial and vexatious disputes that are not likely to escalate rapidly. The judicial approach is at the other end of the scale; it is the course of last resort when all other avenues to

TABLE 15.6

Grievance Resolution Strategies (World Bank 2004) – In practical terms, disputes require a 'basket' of different grievance management approaches, for example, redirecting historical grievances to the relevant government authorities, but dealing with more readily resolvable issues through consensus-based negotiations

Do nothing – adopt this strategy when, for example, the current effect of the dispute on the business or communities is minor and not expected to escalate rapidly, and when the existing business management processes for engaging with the aggrieved parties (e.g. environmental impact assessment, on-going stakeholder consultation, social investment and partnerships) are effective. (cost – low, risk – low)
Information disclosure and consultation – adopt when specific issues arise that seem to be based on wrong perceptions about an operation, project or project impact, and where these fears can be readily allayed through the exchange of information and/or face-to-face consultation. (cost – low, risk – moderate)
Formal complaints procedures – where grieving parties seek redress through the mining company's formal complaints procedure. Adopt when there is distrust of verbal communication and when a written record of exchanges would help focus minds on 'fact' rather than 'perception' (cost – low; risk – high)
Government/regulator procedures – in cases where a grievance is clearly not the responsibility of the mining company, consider redirecting to the relevant government authority (cost – low; risk – high)
Arbitration – where a third party hears the views of the different parties and then makes a binding decision. Often applied to employer-employee issues, but may be appropriate to apply to company community grievances, especially when an agreement is potentially close but trust has broken down (cost – high; risk – moderate/high).
Courts – adopt this course as a last resort. Regardless of the merits and strength of the case of the company, community member, NGO or government agency, a judicial approach is often perceived as bullying and is therefore likely to alienate the community, including those not involved in the dispute. (cost – high; risk – high).
Consensus-based negotiation – (with or without a third-party impartial facilitator or mediator) is likely to be the best course of action where the need to maintain good stakeholder relations is paramount, where the underlying interests of the grieving parties are legitimate (i.e. 'grievance' rather than 'greed'), and where there are resources within the mining company, local communities, local government authorities and other civil societies available to contribute to a resolution – (cost – moderate; risk low)
Customary approaches – some local (particularly indigenous) communities find many of the above approaches alien and inaccessible. In these cases the customary ways in which such communities resolve their own disputes should be considered, with possible adoption of these or 'hybrid' approaches that afford confidence to all parties. (cost – moderate; risk – moderate)

dispute resolution have failed. Cost and risks are high. In practical terms, disputes require a 'basket' of different grievance management approaches, for example, redirecting historical grievances to the relevant government authorities, but dealing with more readily resolvable issues through consensus-based negotiations.

A formal grievance mechanism, a requirement of the IFC Performance Standards (IFC 2006) and the Equator Principles (EP), constitutes a partnership agreement. It enables affected community members to raise concerns and grievances about the project's social and environmental performance, and it provides a mutually agreed mechanism for resolution. EP 6 requires that communities have been informed about the grievance mechanism, and embedded procedures for conflict resolution. This is the case if communities have participated in developing the grievance mechanism. An informal practice to grievance resolution may prove to be successful, but a formal and well-communicated grievance mechanism in the form of a partnership agreement will strengthen the CD programme cycle.

The outcome of any grievance resolution is likely to be in the form of a compromise with trade-offs from both parties.

The outcome of any grievance resolution is likely to be in the form of a compromise with trade-offs from both parties (**Figure 15.6**). Preferably, the outcome falls in the upper right corner of **Figure 15.6**. If the stakes are low a party may decide to accommodate its partners' interests. However if the stakes are high, either the community or the mining company may attempt to resolve conflict by force. Community members may block mine

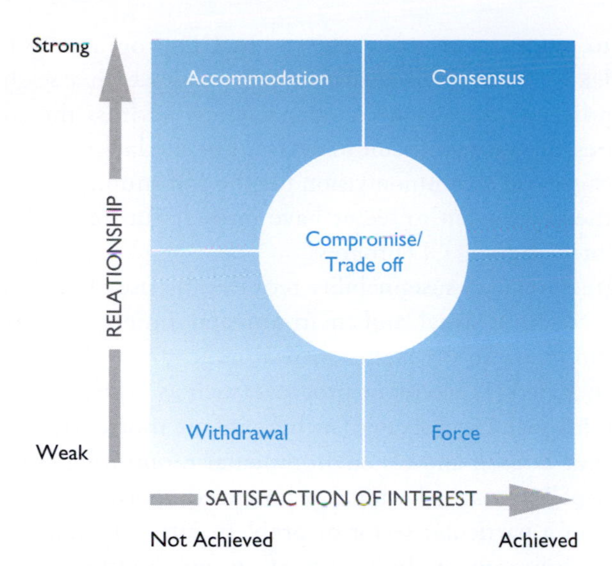

FIGURE 15.6
Plotting the Outcomes of Different Grievance Resolution Strategies

The ever-increasing duration of the project planning process is due in large measure to efforts on the part of responsible mining companies to achieve consensus-based conflict resolution.

access roads; employed community members may go on strike. Mining companies on the other hand may ask for government or legal support in conflict resolution. Damage to the long-term relationship is accepted if conflict resolution favouring one party is an essential outcome. Such a case may occur in land acquisition, when mine development can only proceed if the site is fully accessible and free from private use. The ever-increasing duration of the project planning process is due in large measure to efforts on the part of responsible mining companies to achieve consensus-based conflict resolution.

From the conflict resolution perspective, the absolute reality of a conflict situation is often less important than each party's perception of that situation (SFCG 2006). For example, while there may be no actual stated threat of violence between the mining project and its adjacent communities, the simple perception of a threat may be enough to bring one or both disputants to action. It is necessary to consider perceptions objectively in an attempt to determine how they can create misunderstandings, limit options, and hinder communication. Sensitivity to the differences between perceptions and reality is important.

The absolute reality of a conflict situation is often less important than each party's perception of that situation.

Ongoing Feedback and Adaptation

As important as it is to start the CD programme early, it is also important to evaluate initial ventures – to see if they work. Since ongoing feedback and adaptation is important, a clear system for continuously evaluating success should be incorporated into any CD programme using both objective and subjective measures. Both company and community need to be realistic about whether a CD programme is serving their respective needs, even when this requires backing down from previous positions.

In addition to the kind of feedback that allows a failing programme to be modified or terminated, there must be recognition that even successful CD programmes may cease serving their designed purpose because the community itself will change as the mining operation grows and develops. Some constituencies will develop in either the company or in the community that have vested interests in continuing CD programmes; however continuation of these programmes should be based on their inherent strengths.

Internal or external audits provide formal feedback. In the Philippines, for example, legislation requires mining companies to form multi-party working groups to monitor community programmes (**Case 15.2**). Monitoring CD progress begs the question 'How to measure success?'

Internal or external audits provide formal feedback.

Indicators help to judge the progress achieved by CD programmes (e.g. see Dow Jones Sustainability Indexes Guide, www.sustainability-index.com/assessment/criteria.html), and while some indicators are intuitive, others which address the sustainable development of communities, are more difficult to derive. For the latter, the first step in developing indicators is to agree on a common vision for the community. The indicators are then what community members want to see or have in their future world. These aspirations can then be a tool for evaluating CD progress.

Evaluating progress towards sustainability requires the use of indicators that can measure changes across economic, social, and environmental dimensions rather than just measuring changes within them. In the past, statistical indicators of how well a community or society was doing included (1) economic progress (such as GDP, growth rates, unemployment, or incomes); (2) social well-being (such as infant mortality, years of schooling, or number of people per house); and (3) environmental monitoring (such as air and water quality, emission of pollutants, or hectares of protected areas). These indicators however, give a narrow view of a particular sector or problem rather than a holistic picture of the community and its environment. Indicators of sustainability need to look at all three dimensions – environment, society, and economy. An important aspect of developing these indicators at the local level is that they need to be defined, developed, and used by the community. Ideally the community becomes the steward of the indicators, which are determined through a democratic process.

Indicators of sustainability need to look at all three dimensions – environment, society and economy.

The community's quality of life in the mining region remains the main aspect to consider when determining indicators of sustainable development. Quality of life will influence community performance, productivity, and the community's attitude and views towards the surrounding environment including mining. Four categories of key indicators help to measure quality of life (**Table 15.7**).

The human development index comprises the elements of life quality that are universal. Fulfilment of human rights including gender equality and poverty are additional indicators of quality of life.

15.4 PREPARING FOR MINE CLOSURE

CD programmes need to be designed to prevent long-term dependence on the mining company. Ideally, they initiate a process of development that will continue after mine closure.

CASE 15.2
Community Technical Working Groups in the Philippines

The Mines and Geosciences Bureau in the Philippines recommends that all mining companies establish a Community Technical Working Group (CTWG). While not directly involved in the design of Community Development (CD) programmes, the CTWG is an environmental and social assessment team that allows multi-partite monitoring of the environmental and social performance of the project including its CD initiatives. For illustration, the composition of the CTWG team for the Boyongan Copper and Gold Project in Southern Philippines is:

- MGB 13 Regional Director
- DENR-PENRO Surigao del Norte Representative
- Silangan Mindanao Mining Representative
- Tubod Municipal Mayor
- Barangay Chairmen of Timamana and San Isidro
- Principal of Timamana National High School
- President of WATSAN
- Chairman of SILKA
- Priest of Philippines Independent Church
- Church Elder of Seventh Day Adventist Church
- Deacon of Baptist Church

The CTWG provides a clear framework for continuously and objectively evaluating the success of CD programmes. Programmes that fail to deliver expected outcomes are discontinued, allowing resources to be redirected to other initiatives.

As such, CD programmes should emphasize training and capacity building rather than direct financial support. More specifically, programmes should be designed to strengthen the four crucial determinants of development as shown in **Figure 15.7** income generation, education and health improvement, infrastructure support, and capacity building.

Actual programmes rarely follow simplified theoretical models and CD design and implementation need to accommodate site-specific practicalities.

Income generation is often the linchpin of sustainable development. It depends largely on the creation of new, easily learned income-generating activities. In many cases, CD programmes must assume basic levels of skills and education amongst community members, and hence, CD programmes often commence with rudimentary activities such as agribusiness and home industries. Over time, programmes can expand into value-added activities. Income generation programmes will have a widespread economic impact because of the multiplier effect created by new economic activities and markets. A 'Family Financial Planning' component will support the CD strategy by assisting families to spend new income wisely.

Functional private and public institutions, in particular as they relate to educational and health institutions, contribute to long-term development. In developing countries the 'cycle of poverty' is real – poor health leads to low productivity and restricted incomes, leading in turn to low purchasing power, limited funds for health care, and finally back to poor health. CD programmes to improve education and health have two common components: (1) increasing the capability and effectiveness of health and educational facilities, and (2) improving educational, health and hygiene awareness among communities. NGOs as well as community-based organizations can be involved in such efforts.

Public infrastructure is critical to support any economic activity, and to improve access to educational institutions and health facilities. Since the mining company is likely to be one of the few organizations in the wider mining area with sufficient expertise in infrastructure development projects, it can make a major impact in this field. Support can also

CD programmes should emphasize training and capacity building rather than direct financial support.

Actual programmes rarely follow simplified theoretical models and CD design and implementation need to accommodate site-specific practicalities.

In developing countries the 'cycle of poverty' is real – poor health leads to low productivity and restricted incomes, leading in turn to low purchasing power, limited funds for health care, and finally back to poor health.

TABLE 15.7

The Human Development Indices – The human development index comprises the elements of life quality that are universal

Indicators directly related to people
Life expectancy
Child mortality rate
Birth rate per woman
Average life expectancy
Indicators related to health and education
People with access to clean water
Frequency and victims of contagious diseases
Ratio of people to the numbers of doctors
Indicators related to education
Participation of age-school children
Literacy rate
Percentage of people with higher education
Indicators related to economy
Rate of unemployment
Ratio of woman workers to total workers
Child workers

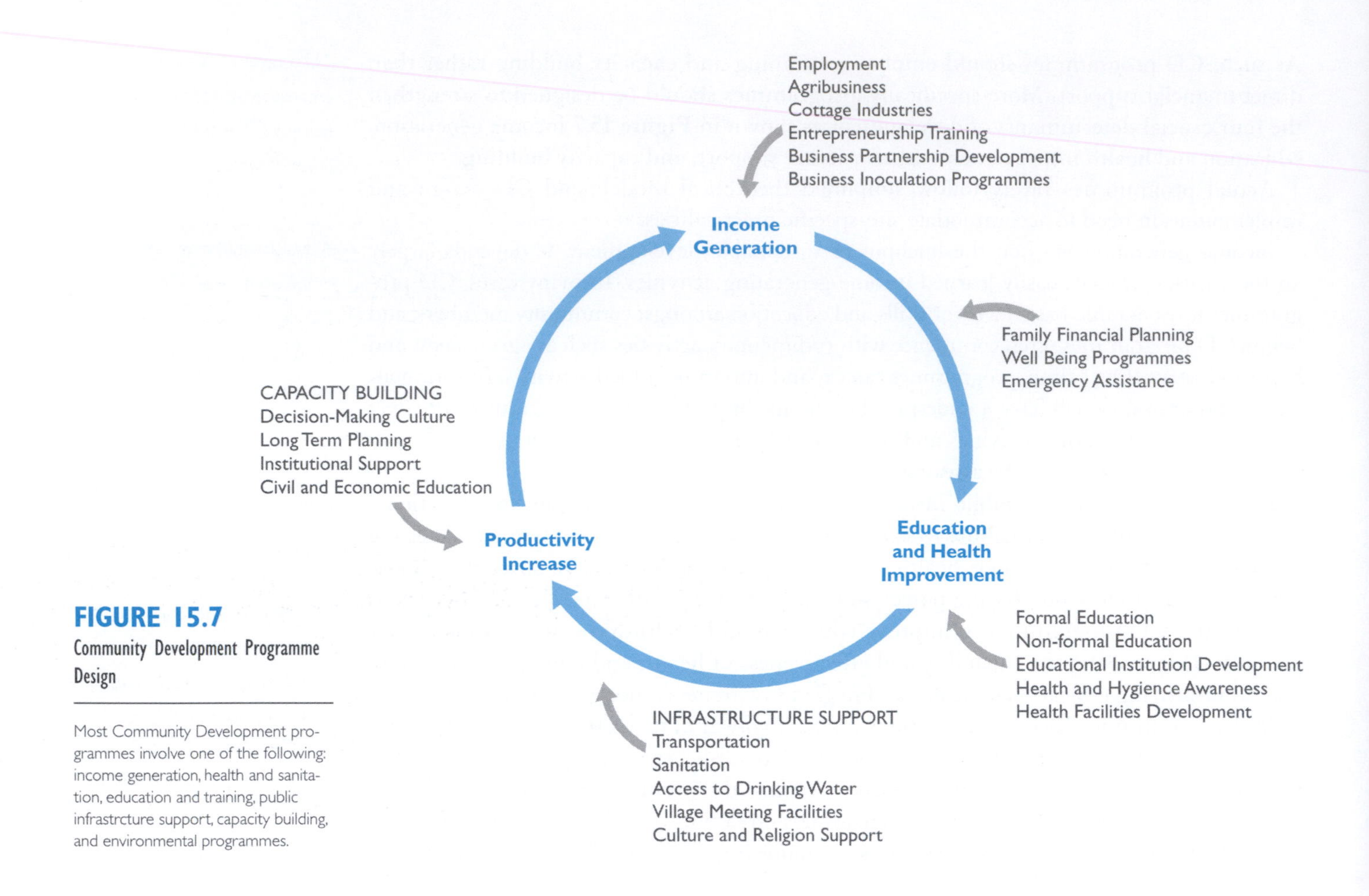

FIGURE 15.7
Community Development Programme Design

Most Community Development programmes involve one of the following: income generation, health and sanitation, education and training, public infrastrcture support, capacity building, and environmental programmes.

be given in the construction of village meeting facilities, in order to encourage greater community interaction and participation.

More often than not, communities in remote mining areas retain a subsistence mentality and are unfamiliar with long-term planning and social development activities.

The creation of a decision-making culture, which includes taking initiatives, is an essential step towards sustainable development without the support of the mining company. More often than not, communities in remote mining areas are just beginning to emerge from a subsistence lifestyle. Most retain a subsistence mentality and are unfamiliar with long-term planning and social development activities. CD efforts are best directed towards helping community members understand the benefits of long-term planning and to become more pro-active in improving their livelihood. The appreciation of 'delayed rewards' is important.

The following describes the cycle in **Figure 15.7** Increased incomes from new income generated activities will raise local purchasing power. Combined with family financial planning, higher purchasing power should lead to increased expenditure on health and education. Both increased expenditures coupled with stronger public institutions will lead to better health and education, which in turn, should result in higher productivity. Higher productivity leads to higher incomes. Improved infrastructure increases the effectiveness of health and education institutions, making them more accessible, and further boosts peoples' productivity by facilitating commercial activities.

Every programme implemented during the life of the mining operations should include implicit consideration for what happens when mining ceases.

The most crucial part of the CD programme takes place as mine closure approaches. This is the stage when the mining company scales down its input. Just as CD planning is important prior to mining, a failure to plan for this critical transition stage will have severe consequences. Every programme implemented during the life of the mining operations should include implicit consideration for what happens when mining ceases. After mine closure, the company's main responsibility should be to monitor the implementation of replacement economic activities and social-infrastructure systems. If these are not ready, creating them at the end of the mine life will be no easy task.

Although this economic transition is the responsibility of local and regional governments, the consequences of any failure of CD programmes will severely damage the public image of the company, and could potentially lead to security problems for its post-closure management and labour force. These problems are trivial when compared with those of the unprepared community, which is forced to deal with high levels of unemployment without adequate replacement opportunities, and a narrowly based, undeveloped economy. Too often when mining activities cease, the communities previously dependent on them become economically stranded.

Central to mine closure is the land use planning for rehabilitated land. This plan should be prepared in the early years of the project and, using examples and lessons learned during the mining project start-up and initial operations, should address culturally and biologically suitable vegetation, relinquishment of land to previous owners or the government, and improving the stability and fertility of top soils. Further detail on planning and achieving ultimate land use is provided in Chapter Twenty-one.

15.5 COMMUNITY PROGRAMMES – WHAT TO DO?

In the introduction to this book we discussed the various types of capital that exist in a region of a prospective mine including: human, social, natural resource, environmental, and financial capital. Mining is likely to diminish environmental resources and by its very nature it depletes natural resource capital. Thus, CD programmes are designed to (1) mitigate potential negative project impacts, and (2) stimulate social investment to increase the level of renewable resource capital to offset the loss in natural resource capital.

As illustrated in **Figure 15.7** and **Table 15.8,** most CD programmes involve one of the following: income generation, health and sanitation, education and training, public infrastructure support, capacity building, and environmental programmes.

Income Generation

Local preference in hiring construction workers, while sensible and necessary from business and ethical points of view is not a CD programme. However, the training component of employment does contribute to community development, although this may not be its primary objective. CD programmes related to employment aim to help prepare suitable local community members for project employment opportunities, by providing training where appropriate. Ongoing programmes also aim to develop emloyee skills, performance, and ability to take on more responsibility.

Any mining project is expected to provide several hundred or even thousands of temporary jobs during the construction phase and fewer permanent jobs during the mine operation. There is a big difference between manpower for operational activities and that for construction activities, because these are two different activities.

There is a big difference between manpower for operational activities and that for construction activities, because these are two different activities.

Although only temporary, the construction phase absorbs more skilled and unskilled labour than the operational phase does. It is advisable for the mining project to require its contractors to maximize the involvement of local people during construction. Attention

TABLE 15.8

Common Community Development Programmes – Community development programmes are designed to mitigate potential negative project impacts and to increase the level of renewable resource capital to offset the loss in natural resource capital

Income generation

Direct employment
Inoculation programmes such as providing micro-credits, housing or small business development loans
Business partnership programmes (Provision of fresh produce, cattle farming, bio – fuel production, etc.)
Cottage industries (e.g. clothing, repair shops, artisan entrepreneurship, etc.)
Entrepreneurship training
Management training for running family business enterprises

Health and sanitation

Baseline health survey
Support through discounted health care sponsorships
Maternal and child health training
Health and hygiene education programmes
Family planning programmes
Donation of medical equipment, computers, etc to health facilities
Malaria prevention measures
Vaccination support
Emergency response assistance (e.g. flood, fire, accidents, medical emergencies, disease epidemics)
Support in establishing adequate waste management practices

Education and training

Scholarships
On-site workplace literacy training programmes
Language training (English plus host country language)
Computer literacy programme (Project donates computer and teaches teachers and students)
Work skill development (Basic motor mechanics, electrical maintenance, welding, mining, and plant operation training)
Others, such as construction skill training, driving, cooking, trades assistants for boilermakers, electricians, etc; computer and typing training, GIS training, secretarial/administrative training
Upgrading local education centres (Electrification of schools, repair of classrooms, donations of books, etc)

Infrastructure support

Transportation (Construction and maintenance of roads, installation of waiting sheds, provision of transportation vans, etc)
Assistance in developing affordable housing schemes
Provision of electricity
Clean water supply programme
Repair of irrigation systems
Village drainage programme

Capacity building

Family financial planning assistance
Construction or repair of religious centres
Civilian security training
Establishing of Community Centre (Project provides rent, computer, books, etc.)
Community capacity building programmes including developing decision-sharing culture
Employee and community well-being programmes (such as an Employee Assistance Programme to all employees to provide employees and families with access to counselling services to help cope with personal and work-related problems; or mandatory cross-cultural training is provided to all employees and permanent contractors).
Employee family days (e.g. each year employee family members are brought to the mine for a tour).
School/ student mine tours
Assistance to local governments (e.g. provision of equipment and materials, training, assistance in regional planning, or sponsoring study tours)

Environmental programmes

Support of protection of cultural heritages sites
Sponsorship of cultural events
Support of protected area and/or endangered species
Back yard garden improvement (e.g. by providing seeds or training on bio-composting)
Environmental management and monitoring

should be paid to offering employment opportunities equally to households (or in some circumstances, to allocate an equal number of employment slots). However, specific skill requirements often mean that this ideal cannot be achieved.

In the operations phase, a smaller, highly skilled labour force is required. Many of the operating positions do not require advanced educational qualifications; rather they require training specific to the tasks involved. Priority for such positions is usually given to members from affected communities who possess the necessary aptitudes required to be trained for these jobs. Opportunities exist for both men and women from local communities to be trained in equipment operation, maintenance, HSE, administration, or security.

It is best to inform communities accurately and thoroughly of company recruitment plans as soon as possible. Relevant information on position descriptions and criteria pertaining to recruitment for all positions should be made public. The selection of potential recruits may be coordinated with community leaders, or with the Community Board, if established. Wages need to be carefully structured to reflect local practices.

Wages need to be carefully structured to reflect local practices.

Consideration needs to be given to two issues. Firstly, because the mining project may create local employee dependence, CD programmes should focus on sustainable vocational and trade programmes that can be maintained after the project ceases operation. Secondly, project employment, while in most respects highly desirable, will change family lives. Mine workers may find less time to fulfil traditional support roles for their families, shifting more responsibility and work to their partners. Whether or not this is important needs to be addressed in the planning process. Some mining companies, for example PT Newmont Nusa Tenggara at its Batu Hijau operation, actually employ mine workers on a reduced (4-day) work week. While this requires that the company trains more people, it has major community benefits, as it enables a larger pool of workers to be employed, and it enables workers, most of whom are subsistence farmers, to continue farming.

CD programmes may be established to help potential entrepreneurs from local communities develop entrepreneurial skills, choose and develop sustainable businesses, and improve their access to credit and markets so they can take advantage of the emerging opportunities associated with mine development. For example, the supply of fresh produce to the caterers serving construction and operations presents considerable opportunity for small-scale income generating projects. Supplies of meat, fish, fruit, and vegetables available locally are likely to find ready markets while a large construction workforce is on site. Successfully exploiting this market opportunity is only possible if local producers can organize their operations prior to start of construction. Therefore estimates of requirements should be made public well in advance, allowing sufficient time for local farmers to plan crops or invest in livestock. Contractors should be required to purchase locally, wherever possible. The huge increase in demand should provide opportunities even for marginal producers who meet minimum quality standards. By the end of the construction phase, when the mine workforce is reduced, the unreliable suppliers or those with unacceptable quality would have been identified. Those with good quality and reliable supplies should be readily able to meet the requirements of the smaller operational work force.

In most cases, the number of operations stage employment opportunities will be much less than the number of people seeking employment. Since the traditional economy is often based on natural resources, the implementation of long-term income-generating projects based on this are preferred for those unable to find project employment. Some programmes will generate income through increasing productivity through the introduction of new techniques, or through traditional methods, such as the use of fertilizers or superior seed strains.

In most cases, the number of operations stage employment opportunities will be much less than the number of people seeking employment.

As a cautionary measure, the price and production of fresh produce and staples of local communities should be monitored. An increase in price due to shortages may undermine the goals of any income-generating programme as the additional income from these projects may be consumed by inflated prices. Inflated prices also cause great hardship for those who, for whatever reasons, do not benefit from income-generating schemes.

Common obstacles in developing small enterprise businesses include lack of management skills and limited access to funds and markets. As part of strengthening community capability, the mining project may encourage communities to establish a local business forum to involve local business and entrepreneurs in the project. Prior to construction, and later during operations, the mining company can use this forum to inform business partners on the project's procurement requirements with a view of establishing what local business can supply, for example, hardware, stationary, electrical tools, and cloth. As the capacity of local merchants builds, mining-specific consumables such as steel balls or activated carbon may also become available locally.

Some mining companies establish 'revolving fund' credit programmes that are used to provide start up capital for small business enterprises. Repayment of the loans is redirected into other new business ventures. There is also a wide range of other community lending funding schemes that can be used. 'Affordable housing lending schemes' may be used to help house disadvantaged families and to also provide innovative social services to disadvantaged local community members; 'community development loans' are used to create jobs and to boost locally owned businesses. CD programmes can also be designed to help community members and community groups obtain affordable financing, particularly for low-income and special-needs groups. 'Micro-credit schemes' are designed to extend credit to low-income individuals seeking to start or expand their small businesses, increase their incomes, and feed their families (**Case 15.3**). Experience has shown that even very small loans, when directed to a committed group of individual borrowers, can boost local living standards. 'Small business loans' help community members, including low-income individuals, women, members of minority groups, and immigrants to start or strengthen their own businesses. Such schemes provide loans and technical assistance to small businesses including childcare businesses, rural businesses, businesses that create jobs in low-income communities, and non-profit organizations that are unable to access support from traditional sources.

Experience has shown that even very small loans, when directed to a committed group of individual borrowers, can boost local living standards.

Health and Sanitation

Mining companies have adopted differing approaches to the provision of medical facilities. All mining operations provide health and medical facilities for their workers, including ambulances and other equipment to cope with project related accidents and emergencies.

CASE 15.3
The Grameen Bank – Bank for the Poor

The Grameen Bank ('The Bank of the Villages' in Bangla language) is a community development bank started in Bangladesh by Muhammed Yunus in 1974. It provides micro-credits to the poor without requiring collaterals. The underpinning idea is that the poor have skills that are under-used because of lack of seed funds. From its modest beginnings three decades ago more than half of Grameen borrowers and their families in Bangladesh (about 6 million as of 2006) have risen out of acute poverty thanks to their loan. Most of the borrowers are women who otherwise would have no or limited access to loans. The overwhelming success of providing seed money to poor people to enable them enhance their livelihood is well recognized: The Grameen Bank and its founder were jointly awarded the Nobel Peace Price in 2006, and its system of micro-credits and self-help groups is now copied in more than 40 countries.

Photo: www.grameen.com

Arrangements range from small clinics staffed by paramedics to fully equipped hospitals with medical doctors and nursing staff, depending mainly on the size of the mining operation. Alternatively, where operations are located close to an existing town, the company may opt to sponsor additional capabilities in existing medical facilities so they can cope with mine-related cases, rather than duplicate facilities. In many circumstances, medical facilities located at the mine site are not accessible to the local villagers, although services and facilities are made available in the event of an emergency. Whatever course is adopted there is the potential for discord between the company and local community if company personnel (and their families) qualify for a much higher standard of medical care than is available to those not employed in the project. For this reason, particularly for operations in remote areas without pre-existing medical facilities, some mining companies make their facilities available to all, regardless of employment.

In many circumstances, medical facilities located at the mine site are not accessible to the local villagers, although services and facilities are made available in the event of an emergency.

It is also relatively common for a mining company to assist government medical programmes by providing funding for government health facilities, in order to provide adequate health services in the project area. Company initiatives may include identifying specific health programmes needed in the project area and considering ways of meeting those needs. Mining companies are responsible for ensuring a safe work environment for employees and for protecting the public from any direct or indirect adverse health affects from the mining activity. Direct health effects are unlikely given the attention paid to management of emissions and effluents. However, there is always potential for indirect effects such as introduction of diseases by incoming workers.

The presence of a project workforce will potentially cause changes in the profile of local health issues, so that the main health care focus during construction may be to address risks such as alcohol abuse and sexually transmitted diseases (STD). During operations, efforts should continue to enforce rules that protect public health, including the medical examination of potential recruits prior to employment, and ongoing health surveillance for the workforce including routine medical check-ups, early diagnosis, and treatment.

Most mining projects adopt a drug and alcohol policy to ensure that the work-place is drug and alcohol free. No mining project will tolerate employees under the influence of alcohol while at work, as they are a danger to themselves as well as to their colleagues.

While health care services are important to local people, and communities may anticipate support from mining projects, community health and local government's provision of health services are not the responsibility of private investors such as mining companies. A mining project should work with the relevant government agencies to improve health care through existing programmes and facilities by providing advice, information, training, and some basic supplies; however, it should not replace or take on the functions and responsibilities of the government.

Mining companies can encourage and sponsor the government in conducting health education programmes within the community, including family planning, STD awareness, general sanitation, and information regarding healthy diets. Special care should be given to advising, and as necessary, assessing project-affected people about the potential of malnourishment which may lead to the critical condition of kwashiorkor, a condition found amongst communities with an over-reliance on carbohydrates (such as rice, cassava, or sago), and who therefore do not balance protein and fat intake.

Most mining projects adopt a drug and alcohol policy.

A mining project should work with the relevant government agencies to improve health care.

Mining companies routinely implement malaria prevention where relevant. Practical measures include:

- raising general awareness;
- educating mine personnel and visitors on simple precautionary measures, such as to avoid being in the open at dusk and dawn, and to wear long-sleeve shirts and long trousers;
- eliminating mosquito-breeding habitats;

- fumigating;
- undergoing regular medical check-ups;
- monitoring of malaria cases at local health facilities;
- monitoring the types of mosquitoes in the project area; and
- selective use of malaria prophylaxis by mine personnel. (This is difficult to maintain over long periods of time without potential negative side-effects on the person's health).

Provision of emergency support is vitally important for communities.

Provision of emergency support to local communities in the form of medical assistance for life-threatening illness or accidents, or as a response to natural disasters such as landslides, floods, earthquakes, or tsunamis, is vitally important for communities. There are many examples where mining companies, such as PT Freeport Indonesia, have deployed company medical personnel and equipment in response to such crises, even when they have been remote from the company's operations. While this may be seen as providing services that are the responsibility of government, such considerations do not apply in a crisis. Although the time frame may be more compressed than under normal circumstances, given the need for emergency assistance, the mining project should still attempt to use a consultative and participatory process in an emergency response to the extent possible. CD programmes may include preparation of equipment and training programmes to enhance emergency preparedness.

In spite of their best efforts and intentions, mining companies are frequently accused of causing a deterioration of community health; historical evidence of mining supports such criticism. It is also no secret that some mines and smelters in Eastern Europe, South America, and other parts of the world, that operate with minimal regard for the environment have adversely affected public health in nearby communities. This contrasts with most new mine developments that generally contribute to improved public health in the host region. Nevertheless, opponents of mining frequently accuse companies of causing community health problems, for example, by contaminating surface waters. Rashes and other skin disorders are frequently cited as evidence of such contamination. However, such allegations rarely stand up to medical scrutiny.

Most new mine developments generally contribute to improved public health in the host region.

Education and Training

Although community development budgets during exploration are generally quite limited, some modest opportunities for community development will always arise during this phase, particularly when bulldozers and drill rigs are brought into an area without good roads. Despite these limitations, the focus should be on education, which will not only benefit the community but ultimately benefit the company too; however, rather than indiscriminately funding the local schools directly, a CD programme should be designed with careful and patient consultation as to how the available resources can best be allocated. Programmes should be devised to provide sustained benefits to the community, regardless of whether or not exploration leads to a future mining project. If a project does eventuate, improved education provides a basis for later CD efforts.

New forms of education and training are needed as the mining project evolves. The main priorities should be training and educating local community members for the:

- construction stage – short-term programmes focused on basic-skills targeting adults; and the
- operational stage – long-term programmes focused on professional and skilled tasks targeting both adults and highschool students.

The mining company should consult with government education officials on how it can assist with the redistribution of current educational resources. In addition, it may directly establish with local communities a scholarship programme for especially gifted local children.

Some host country governments apply labour quotas to encourage training with a decreasing ratio of expatriate to local personnel over the mine life. In fact the question of local labour content is often one of the more contentious issues during the environmental approval process. Once established, mining projects commonly find themselves involved in continuous training of professional and skilled workers, as employees are 'poached' by new mining ventures in the host country.

Once established, mining projects commonly find themselves involved in continuous training of professional and skilled workers, as employees are 'poached' by new mining ventures.

Infrastructure Support

Often there is very little in the way of basic or secondary infrastructure in a mining project area. As the only organization in the area with sufficient funds and expertise to develop major infrastructure, a mining company can make a major contribution to assist the local government in this field. Constructing and upgrading basic infrastructure is of widespread benefit. Although priority should be given to the immediate project area, the mining company may consider encouraging infrastructure development in surrounding areas or potential growth centres, so as to reduce the attractiveness of the mining area to further local migration. On the other hand, improvements to infrastructure may result in improved access to those seeking project employment.

Mine development will have a substantial impact on transportation, both into and within the host region, often providing an airport, seaport, as well as access roads. This requires substantial forward planning and coordination between the mining company and local government. This planning may include haul roads being used for community transportation prior to and after their use for mining operations.

If the actual workforce during operation is relatively small compared to the size of the neighbouring communities, it will be possible to seek housing for mine worker in adjacent villages rather than in an isolated mine camp. The benefits are manifold. Community members will gain financial benefits by renting accommodation to workers. Mine employees in return become members of the communities and as such act as ambassadors for the mining company. Adequate housing is not always available, either in number nor in quality. This provides the opportunity to consult with affected communities to establish adequate housing as part of community development programmes. The mining company may provide initial financial assistance in the form of loans to be repaid over time.

If the actual workforce during operation is relatively small compared to the size of the neighbouring communities, it will be possible to seek housing for mine worker in adjacent villages rather than in an isolated mine camp.

Benefits to the host region from the provision of electric power by the mining company are obvious, and it is common for companies to provide electricity to nearby communities. More distant communities may also be supplied but usually government or other assistance is required to provide the necessary transmission and distribution facilities.

The provision of a source of clean drinking water is a high priority where ever people settle. Lack of clean drinking water is often a major hardship for communities in remote areas, even in areas of high rainfall. Local people are usually aware that many of the illnesses they suffer are a result of water-borne infections. Any shortage of potable water will be exacerbated by population increases brought about by mine development. Accordingly, it is not surprising that provision of clean drinking water is one of the most common CD programmes for mining operations in developing countries. Water requirements for a particular community depend on various factors, including:

Provision of clean drinking water is one of the most common CD programmes for mining operations in developing countries.

1. size of the community and characteristics of the people;
2. climatic conditions;
3. demand from commerce and industry;
4. pressure on water;
5. quality of water;

6. sewerage facilities;
7. water rates and metering;
8. nature of supply;
9. availability of private supplies; and
10. efficiency of water management.

Finally mining companies will generally consider the merits of any requests to contribute to the community's construction of places of worship or religious education as part of its infrastructure support programme. It goes without saying that extreme care is required not to appear as favouring one religious belief over another.

Capacity Building

Experience in developing countries indicates that both local communities and local government are seldom well prepared to cope with changes that accompany mine development. To participate in the project and to leverage the many opportunities the project is able to generate, specific CD programmes should be geared towards capacity building.

Training helps to maximize local participation in a mining project, and to develop social capital.

Training helps to maximize local participation in a mining project, and to develop social capital. Training also acts as a catalyst for sustainable development. Training programmes related to a mining project typically have three components, namely:

- Training geared towards employment during construction;
- Training geared towards employment during operation; and
- Broader training and skill development for adjacent communities.

Other training opportunities involve the day-to-day activities of community members and may include management, marketing, accounting, or farming.

An assessment of skill needs and availability is central to community capacity building. In most instances, the mining company and its contractors will need to seek skilled and semi-skilled labour outside the host region, at least during the initial stages of the project. However, the company should identify semi-skilled labour requirements which can, in time, be filled through local training programmes. Non-local workers brought to the project for semi-skilled jobs should be on short-term contracts and required to provide on-the-job training to local employees. It should be anticipated that, when locals have acquired the basic knowledge and experience required to perform semi-skilled tasks, they will be able to replace their trainers. This does not pose a problem to the mining company. In fact, the workforce becomes more stable, and costs decrease as the local content increases.

Civilian security training can be an essential part of community development. A Code of Conduct for construction and mine workers, together with a well-designed construction camp system and layout of facilities will help to minimize security issues. The workforce Code of Conduct should be designed to establish a good worker-community interface and thereby minimize the potential for conflict. The construction camp system and the mine layout should be designed to minimize security disturbances or issues arising between local people and project work force by minimizing access to the camp and mine site. Safety hazards are present at any construction site and during mobilization of equipment and materials. Security measures can reduce these hazards by limiting outside exposure. Consultation with affected communities minimizes transportation hazards whenever unusual traffic is expected.

Generally, it is preferable to adopt a community-based security strategy.

Generally, it is preferable to adopt a community-based security strategy, which relies on a positive perception of local communities. The communities themselves provide the best defence and protection against robbery, unrest, or crime. Nevertheless, professional security services will be necessary. More often than not, a private security team will be established to manage security at the project site. The primary task of the security team is to protect the mine's property and personnel, and to reduce safety hazards. Apart from fenced areas, the security team patrols and guards entrances, service roads, active mine areas, explosive storages and other facilities within the work area such as auxiliary mine facilities and overburden and tailings storage facilities.

The size of the security team will increase and decrease with the scale and extent of project activities, with employment peaking at the peak of construction activities. It is good practice for the majority of the security team to comprise local people. The security team should be professionally trained, and highly disciplined. A specific, particularly strict Code of Conduct should apply to the private security team. A major consideration is the avoidance of undue use of force. It is also good practice to formally introduce security managers to the village leaders.

The mining company (and NGOs commenting on police or military involvement in mining projects) need to realize that government police and army have the ultimate legitimate role in law enforcement and protection of national resources. While the presence of increased government controlled law enforcement units does not necessarily meet with community approval, a mining company needs to co-operate with these forces and frequently to provide accommodation, subsistence, and other support. It should also be recognized that national budget allocations for security and defence in many countries requires that police and/or military forces involved in security of privately owned projects are paid for by the project proponent.

Environmental Programmes

Mining and associated activities destroy much of the environment in the immediate project area, at least temporarily. These are the direct effects. Indirect stress on existing environmental resources may result from the increased population in the project area. Improved access to previously inaccessible areas may add further stress to local fauna and flora, due to increased hunting or natural resource harvesting activities. But mining can also be beneficial for the existing environment, a fact that is unfortunately too often ignored by the media and many NGOs. Mining generates funds that, if wisely allocated, help protect existing environmental resources. Without mining, much of the existing environment is threatened by a combination of natural and human-induced factors. In most cases these factors are detrimental, and may include logging, land use changes, or uncontrolled settlement. Mining can help to protect existing natural habitats of significant value from these changes (see also **Case 9.2**).

Mining can also be beneficial for the existing environment, a fact that is unfortunately too often ignored by the media and many NGOs.

Cultural heritage sites are important, not only for the communities for which these sites are important, but for the world community as a whole. Located in remote locations, cultural sites or issues of cultural heritage often only become apparent during the project planning process. Some countries, such as Australia, maintain registers of cultural heritage sites but, even so, anthropological studies for environmental baseline purposes, often identify previously undocumented sites. There are established procedures for addressing issues of cultural significance, including maintaining the secrecy of site locations when this is required by traditional cultural lore. Contrary to common perception, a mining project, through careful planning, will support protecting cultural heritage sites, and make them accessible to a wider audience, if appropriate, or otherwise maintain their secrecy. Ore bodies located in close proximity to cultural sites have, in some instances, been left unmined to avoid damage to the sites – for example the Jabiru 3 deposit of the Ranger Uranium Project in Australia's Northern Territory. In other cases, objects of cultural significance have been salvaged and relocated, with the approval and oversight of their traditional custodians.

15.6 LOCAL BENEFITS DO NOT ALWAYS EVENTUATE

To judge the contribution of a mine to local development without time comparative data is not easy.

Mining companies, as well as most national governments' assert that mining projects contribute to local development. Other groups deny such assertions (Christian Aid 2005). To judge the contribution of a mine to local development without time comparative data is not easy. As detailed elsewhere in this text, social monitoring of a reference community unaffected by mine development is necessary. Comparisons over time between project-affected communities and reference communities may help to strengthen the case that mining spurs development, through means such as taxes received by local government and communities, generation of long-term employment opportunities, economic stimulation, and improved social services.

Taxes and the Case of Strengthening Local Governments

It is fair to state that central governments often fail to transfer even those funds that have been collected.

A mining project cannot commit to community development in isolation. Local development should always remain the primary responsibility of local governments, supported by dominant local economic players such as the mining project. To enable local governments to stimulate and direct local development, a percentage of the income the host country's government receives from the exploitation of natural resources should be transferred to the regional and local governments. Evidence from some countries indicates that such redistribution of government earnings does not always occur: 'In Peru and Indonesia, laws were created to ensure that extractive industry revenue would be returned to local communities or the government. Due to the design of the laws and the lack of transparency, however, little revenue actually reached the communities' (World Bank 2003). Unfortunately it is fair to state that central governments often fail to transfer even those funds that have been collected.

A number of parties, including local governments, mining companies and some NGOs have proposed changes to tax distribution, such as:

- Revenue distribution to include all income that the government receives as a consequence of natural resource exploitation and not just income tax;
- Communities that are affected by mining activity to be the principal and direct beneficiaries of revenue redistribution; and
- Revenue distribution to be enforced in a timely, effective manner.

It is fair to say that without support to local development from both local and national government alike, CD efforts of a mining company alone are likely to fail in the long-term.

The Perceived Fallacy of Employment Generation

Mining projects employ significant numbers of people. As a result, despite the negative impacts of mining, local people benefit from decent jobs, at least while the mine is in operation. A fair comment, however, is that the introduction of new technology has reduced the number of jobs that are now on offer compared to mining, say, a few decades ago. Moreover, critics claim that most employment opportunities require skills that local people tend not to have, so mining companies hire people from outside the region, often foreigners. It is also argued that while mining creates a certain number of jobs, its environmental impact leads to a corresponding loss of livelihoods in the agricultural sector. Furthermore, they argue that mining jobs last only as long as the mine is active, but negative effects of mining on the local agricultural economy can be permanent (Christian Aid 2005). Successful CD programmes should serve to ameliorate such criticism.

Economic Stimulation

There is little dispute that the mining industry relies on a range of goods: chemical and petroleum products, timber, iron and steel products, fresh produce, and general industrial and domestic products. Equally, each mining operation awards contracts for a wide range of services: maintenance of machines and equipment, explosives, security, accommodation, catering, engineering and consulting, administration and accounting, transportation and construction are some examples. Mining does stimulate the local economy through its demand for goods and services. Some critics argue, however, that its contribution could and should be larger. Critics often don't recognize, however, that many of the goods that a mine requires are highly specialized and are not readily available in the host country. It is also often argued that while developing countries do provide raw materials, value-added and labour intensive downstream activities remain the domain of metal consumer countries. The proper response to such critics is with the host government and in the form of appropriate investment policies, not with the mining industry. It is also clear that, as the industry grows in any particular country, local services and supplies will also grow.

Improvement of Social Services

Mining projects are often located in remote areas, far from basic services and markets. Either by law or by necessity, mining companies are required to provide a range of services for their employees, including providing clean water, education or health care, all of which also benefit neighbouring communities. Mining projects also engage in community development. Without proper planning these benefits to local communities will cease when the project ends.

Mining projects are often located in remote areas, far from basic services and markets.

It is not surprising that the main limiting factor for CD programmes is financial capacity. Sustainability requires significant community contributions (both cash and in-kind) towards initial capital outlays and, eventually, full recovery of operational and maintenance costs. The extent of community contribution is in fact a key measure of sustainability. Typically, the community is too poor to make large cash contributions. By the same token the local government often has virtually no tax revenue base and is dependent on unreliable and inadequate transfers from the central government. Other times, communities and local government authorities do not have the requisite technical expertise or the financial resources to make, say, infrastructure repairs. On yet other occasions, low degrees of social mobilization (free-rider incentives) can lead to declines in volunteer labour pledged to maintain a CD programme. As a result, without continuous funding by the mining project, CD programmes such as infrastructure support deteriorate and become of limited or no use after a few years.

Without continuous funding by the mining project, CD programmes such as infrastructure support deteriorate and become of limited or no use after a few years.

15.7 COMMON PROBLEMS AND SOLUTIONS

At least in theory, the prospects for creating sustainable, effective community development programmes can be straightforward. An integrated partnership between the company and the community's formal and informal leaders is necessary to create programmes that work and are sustainable. Intelligent planning is essential for successful community development, and it must be approached as an integral part of planning every stage of the mining project. And yet, more often than not, CD projects go wrong. Here are some of the reasons.

Most problems relate to ignoring the previously stated 'pointers to success' by doing the opposite (**Table 15.9**). However, other things can go wrong (Frank and Smith 1999).

TABLE 15.9
Common Problems in Community Development Programmes

Starting too late
Lack respect for Indigenous Peoples' rights
Use of hand-outs in community development programmes
Lack of transparency
Confusing roles and power struggles within the mining company
Top-down approach to community development
Taking on government responsibilities

Not Understanding the Community

It is often difficult to understand your own partner in life; how much more difficult is it then to understand a community with different members, traditions, culture, beliefs, religion, norms, values, education, livelihoods, language, leadership structure, etc. Our natural tendency to project our own experience, ambitions, and values to others does not help to the process of understanding communities. When undertaking a CD initiative, assuming or guessing is not good enough. Participatory needs assessment and planning provide a simple solution. The company acts as an agent to development: it listens and facilitates community self-assessment. It relies on local expertise and inputs.

The company acts as an agent to development: it listens and facilitates community self-assessment. It relies on local expertise and inputs.

Unconsciously Excluding Some Community Groups

A community comprises many sub-groups, one or more of which may easily be overlooked in the CD process. Women are an obvious example. They are often omitted from decision-making and from participation in the development of their communities. However, women may influence decisions in informal ways. Including traditional knowledge in project planning can encourage the inclusion of women by recognizing the value of their knowledge. Finding culturally sensitive ways to include women can promote their influence and contributions.

Failing to Evaluate Results

Mine production, energy consumption and most other operational inputs and outputs are monitored and evaluated on a continuous basis. If ore characteristics change, ore is blended to optimize process recovery and efficiency. If energy consumption per unit output increases, mining and mineral processing is reviewed, adjusted, or modified. Why not apply similar attention to monitoring and evaluating community development outcomes? Admittedly indicators to measure progress in the social context are more challenging to formulate than operational performance indicators, but this alone is no reason to engage in community development programmes without the mechanisms to evaluate subsequent success and performance.

Indicators to measure progress in the social context are more challenging to formulate than operational performance indicators.

Role Confusion and Power Struggles

An individual's personality or character is an easily identified cause of role confusion and power struggle. Role confusion and power struggles, however, can also emerge from the CD process itself. Community development brings about change, forges new relationships and shifts power. Some community members may perceive a loss of power or be threatened by the new relationships that they see being developed. To resist or to be threatened by this type of change is natural. How work is divided and managed can also have a large impact. With the best of intentions and in good faith, community members or organizations may be taking on work that impinges on others without realizing it. It is not possible to avoid role confusion and power struggles completely, but mutual respect, transparency, and maintaining an open dialogue do help to minimize adverse effects.

Unresolved Conflicts

Only the naïve will believe that a mining project or a CD programme can be implemented in full consensus with all involved. Conflicts and disagreement will occur. This may not be such a bad thing; it depends on how conflicts are managed. It is always best to create an atmosphere in which disagreements are clearly expressed and out in the open. This, of course, is not a universally accepted approach in all cultures. However, disagreement becomes dangerous when suppressed. If ignored, conflict will escalate. On the other hand it is clearly not possible, necessary or affordable to satisfy everybody.

Disagreement becomes dangerous when suppressed.

Throwing Money at Problems

Some problems go away with money, most won't. Throwing money at problems is treating the effect not the cause. The reality is also that community expectations rise with each dollar spent.

Community expectations rise with each dollar spent.

Not Managing Mine Contractors and Consultants

Communities do not differentiate between mining employees, contractors, or consultants; all personnel working on a project represent the project in the view of communities. Mine management must be aware of this fact, and take measures to ensure that everyone involved in the project adopts and follows corporate directives. The need for all project personnel to adopt the established project-specific Codes of Conduct, is a simple example. No formal meeting between community members and contractors or consultants should occur without adequate representation of the mining management. At no time is this more important than during the environmental permitting stage.

Making Promises that Cannot Be Kept

Great care must be taken to ensure that the project does not raise unreasonable expectations. While this is true throughout the entire mine life, the first phases of community involvement are the crucial ones. Related to un-kept promises is the tendency to over-design CD programmes. Few if any CD initiatives are cost free. Funding is often a major challenge for many community development initiatives as they do not fit into the traditional types of activities in the mining sector. What may be appropriate and in fact necessary for a large-scale mining project with an expected mine-life of 20 years or more, may represent complete overkill for a small-scale mine that may be in operation for say 10 years or less. The planning principles remain the same, but the scope and depth of CD initiatives will differ. Put simply, a small mine cannot afford to engage in community development to the same extent as governments or large mines do, or should do.

What may be appropriate and in fact necessary for a large-scale mining project with an expected mine-life of 20 years or more, may represent complete overkill for a small-scale mine that may be in operation for say 10 years or less.

Over-emphasizing NGO Involvement

NGOs themselves are rarely affected by projects; however, they tend to be more vocal than other stakeholders, often dominating the agenda.

NGOs come in many forms, with many agendas. Some are highly dedicated to the betterment of society. Others seem to be driven by opposite motives. Some cynics joke that NGO is an abbreviation of 'new government organization'. It is probably not unfair to generalize that NGOs benefit from power and influence well beyond their size or capabilities. Those who are pressing for increased NGO involvement in mining projects often tend to forget that an NGO is but one of many stakeholders. NGOs themselves are rarely affected by projects; they represent a particular interest group. However, they tend to be more vocal than other stakeholders, often dominating the agenda. This is certainly true in the case of self-proclaimed anti-mining advocacies. Typically, these groups project a hostile attitude and proclaim an extremely negative stereotyped view of mining that may evolve to the point where co-operation or even communication between the project and the NGO becomes impossible. Mining companies are well advised to engage with NGOs based on what they can add to the project. A partnership between two organizations should be based on mutual respect and mutual benefit, not on the noise that any given organization is able to generate.

In conclusion nobody can say that community development is easy. But at a minimum, a mining project should be able to shape its role in the community to that of good neighbour, rather than intruder. Good neighbours help each other.

REFERENCES

ADB (2003) Environmental Assessment Guidelines, Asian Development Bank

ADB (2004) Indigenous Peoples and the ADB. A NGO guidebook on the ADB series.

Bruch C and M Filbey (2002) Emerging Global Norms of Public Involvement, The New Public 1-15, Environmental Law Institute

Chambers R (1992) Rural Appraisal: Rapid, Relaxed and Participatory, Institute of Development Studies (University of Brighton), IDS Discussion Paper No. 311, Sussex

Christian Aid (2005) Unearthing the truth – Mining in Peru (www.christianaid.org,uk)

Coyle J *et al.* (2002) Closing the Gap: Information, Participation, and Justice in Decision-Making for the Environment, World Resources Institute, Washington DC, USA

Creighton and Creighton (1999) The Public Involvement Manual (2nd edition)

EC (2000) Public Participation and Consultation, Environmental Integration Manual: Good Practice in EIA/SEA. European Commission, Gibb Ltd., 239 (www.gibbltd.com)

Finsterbursch K, Ingersoll J and L Llewellyn (1990) Methods for social analysis in developing countries, Boulder: Westview Press.

Frank F and Smith A (1999) The Community Development Handbook, A tool to build community capacity, written for Human Resources Development Canada, Minister of Public Works and Government Services Canada (www.hrdcdrhc.gc.ca/community)

Gibson G (2001) Key elements of capacity building – An exploration of experiences with mining communities in Latin America. Mining, Minerals and Sustainable Development (MMSD) Publication September 2001.

GTZ (1991) Methods and Instruments for Project Planning and Implementation. Gesellschaft für Technische Zusammenarbeit, Eschborn: Germany.

IFC (1998) Doing Better Business through Effective Public Consultation and Disclosure, International Finance Corporation, Washington DC, USA

IFC (2000) Investing in People: Sustaining Communities through Improved Business Practice, International Finance Corporation, Washington DC, USA

IFC (2006) Environmental and Social Performance Standards, International Finance Cooperation, Washington DC, USA, available at www.ifc.org

MMSD (2002) Breaking New Ground; Mining, Minerals and Sustainable Development (MMSD) Project; published by Earthscan for IIED and WBCSD

NORAD (1999) The Logical Framework Analysis Handbook, Norwegian Agency for Development Cooperation, available at www.ccop.or.th

Rietbergen-McCracken J and Narayan D (1998) Participation and social assessment – Tools and techniques, The International Bank for Reconstruction and Development, The World Bank

Salmen L (1995) Beneficiary assessment: An approach described. World Bank, Environmental Department, Social Assessment Series Paper No. 23, Washington, DC.

Smith WE (1991) The AIC Model: Concepts and Practice. Washington, D.C.: ODII.

SFCG (2006) Washington DC., Search for Common Ground, www.sfcg.org/resources/resources_distinctions.html

Srinivasan L (1990) Tools for community participation: A manual for training trainers in participatory techniques. United Nations Development Programme, PROWWESS/UNDP Technical Series, New York. PACT Publications, 777 United Nations Plaza, New York.

SustainAbility and IFC (2002) Developing Value: The Business Case for Sustainability in Emerging Markets

UNDP (1998) Empowering People: A Guidebook on Participation, United Nations Development Program, New York (http://www.undp.org/csopp)

USEPA (1999) Community Based Environmental Protection. United States Environmental Protection Agency EPA.230-B-96

Wates N ed. (2000) The Community Planning Handbook, London, Earthscan Publications

World Bank (1989). Operational Directive 4.01 on Environmental Assessment, converted in 1999 into a new format: Operational Policy (OP) 4.01 and Bank Procedures (BP) 4.01

World Bank (2003) Striking a Better Balance, Volume 1: The World Bank Group and Extractive Industries. The Final Report of the Extractive Industries Review, December 2003.

World Bank (2004) Capacity building for local stake holder in key mining regions – awareness-raising, confidence-building and empowerment; Materials for a five day Multi-Sector Workshop on Environmental and Social Management in Mining Projects, The Philippines; published by the Philippine Business for Social Progress

World Bank and ICMM (2005) The Community Development Toolkit, published under the ESMAP Formal Report Series, Report No. 310/05, October 2005, www.esmap.org

Appendix 15.1 Evaluating Community Development Programmes

There are many ways to review CD strategies. A simple protocol follows, which can easily be adopted to serve as a regular self-assessment tool for site use. For each of the four main CD strategy steps, that is (1) need and capacity assessment, (2) planning, (3) implementation, and (4) feed back and evaluation, three performance attributes are formulated, using the following grades for evaluation:

1. No action
2. Action planned and documented
3. Systems / processes being implemented
4. Systems / processes are implemented
5. Integration into management decisions and business functions
6. Social excellence and leadership

The protocol can be further refined by assigning a number of sub-elements to each performance attribute. A percentage implementation for each of the sub-elements can be determined, then aggregated into a single score for each of the 12 performance attributes. The final results, accompanied by some descriptive passages, are best presented in the form of a wheel, or spider diagram, as depicted in **Figure 1** Key elements contributing to each performance attribute are summarized below.

Capacity/Needs Assessment

Responsible project culture – This attribute is used to summarize the various characteristics of the mining project that relate to the project's internal capacity and capability to successfully engage in CD. Does the company differentiate between public relations and community development? Do employees who are engaged in or are responsible for CD have relevant qualification and/or training? Has the project formulated a project-specific Code of Conduct and/or Code of Ethics? Are these codes known to all employees and contractors? Is there a senior management champion for CD activities? Does cross-cultural training occur? Are CD activities formally documented?

Bottom up approach – This attribute is used to describe the extent of participatory planning. Are community needs identified and articulated by community members? Does the company apply participatory planning methods? How often does the company meet the community to discuss community matters? Does the company consult with the community in all aspects of CD?

Inclusion – This attribute is used to gauge the extent to which CD addresses all community groups. Are women included in participatory planning? Are all community groups identified? Do significant differential benefits for various community groups occur? Do CD programmes

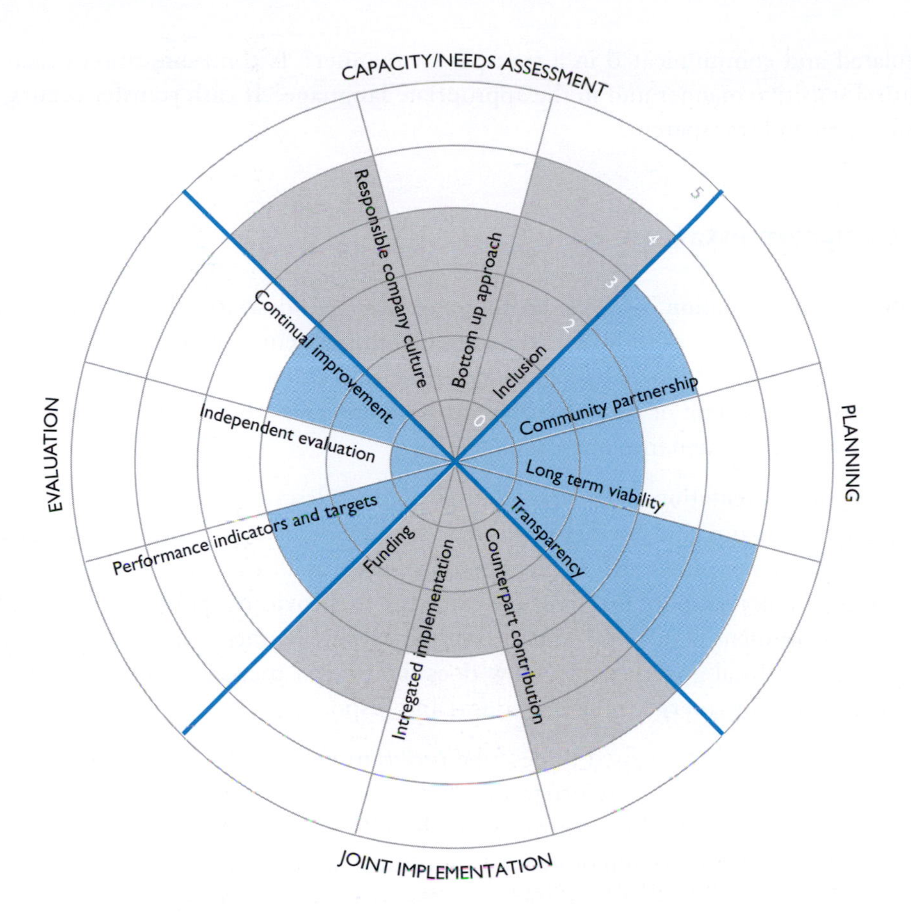

FIGURE A15.1.1
Evaluating Community Development Programmes

create jealousy? Do vocal interest groups tend to dictate the agenda? Are Indigenous Peoples present in the project area? Has an anthropological study been carried out?

Planning

Community partnership – This attribute is used to describe the strength of existing partnerships. Is partnership based on trust? Is a formal partnership agreement established? Does a conflict resolution mechanism exist? Does the mining project maintain regular communication with affected communities? Are conflicts identified at an early stage? How many grievances have occurred, and how have such grievances been resolved? Do disagreements escalate easily? Are community members proud and defensive of the project? How do community members feel about the project?

Long-term viability – This attribute is used to describe the long-term viability of CD programmes. Is the programme designed for mine closure? Has the community ownership? Is the community responsible for maintenance? Has the community the resources to maintain the programme? Does CD focus on strengthening human capital? Does the CD programme create benefits beyond the end of mine life? Does the programme generate profit? Does it become break-even? Is the focus of CD programmes on building capacity rather than on building infrastructure?

Transparency – This attribute is used to describe the transparency of decision-making. Is there a spirit of openness and frankness by all partners? Is what can be done and why, formulated and communicated in a transparent manner? Is communication practised in a cultural sensitive manner and in the appropriate language? If cash transfer occurs, is the transfer open and transparent?

Joint Implementation

Counterpart contribution – This attribute is used to describe the extent to which beneficiaries of CD programmes contribute to CD implementation. Is the CD programme designed as hand-outs? Do CD programmes require counterpart contribution? What are the forms and the extent of counterpart contributions? Does the counterpart contribution continue into the maintenance phase?

Integrated implementation – This attribute is used to describe the coordination between community, company, and local government in CD implementation. Are programmes related to improving public infrastructure coordinated with the local government? Does the mining project support the local government in providing public services? Does a regional development plan exist? What communication channels exist between company, community, and local government? How does the project manage health care? Is a programme in place for emergency preparedness and response?

Funding – This attribute is used to describe programme funding. Are funds sufficient to implement and to maintain the programme? Can the mining project maintain funding during times of economic downturns? Can the CD programme become self-financing over time (either breaking even or generating profit)? Was there ever the need to cancel a CD programme due to lack of funding?

Programme Evaluation

Performance indicators and targets – This attribute is used to describe whether performance indicators have been established. Do performance indicators and targets exist? Are they documented? Is a reference community included in social monitoring (a community with similar characteristic but being unaffected by the mining project)? Are performance indicators easy to apply?

Independent evaluation – This attribute is used to describe whether programme evaluation is objective or not. Who evaluates CD programmes, and how often? Is the review independent from design and implementation? Have past CD programmes been discontinued or changed following review findings? Are review findings formally documented? Are review findings communicated to senior company management?

Continual improvement – This attribute is used to describe whether the feedback loop is designed to allow for continual improvement. Are the priorities of CD programmes regularly reviewed and adjusted? Do CD programmes serve their purpose? Does the project receive recognition for its CD success? Is success (or failure) of CD programmes acknowledged and publicized? Does the project provide sufficient funding for CD?

16

Indigenous Peoples Issues

Respecting The Differences

 URANIUM

As the naturally occurring element with the highest atomic number, uranium is 70% more dense than lead. This silver gray metal is weakly radio-active. Three isotopes occur: U-234, U-235 and U-238, of which U-238 is the most prevalent accounting for more than 99% of uranium. As for all radioactive metals, the metallic nature is of no concern for the end user - uranium is used as nuclear fuel to produce energy, and 20 kg of uranium produces as much energy as 400,000 kg of coal. U-235, the fissile isotope used in nuclear reactions, must be enriched prior to use. Uranium is not particularly rare, being more abundant than tin, mercury or silver. Several types of uranium ore occur; the predominant uranium mineral is uraninite. Uranium is chemically reactive and oxidizes readily. It may accumulate in the bones; its toxicity is of more concern than its radioactivity. Depleted uranium, U-238, is used for armaments and other applications requiring a highly dense material.

16 Indigenous Peoples Issues

Respecting the Differences

Public attitudes towards the extractive industries continue to change rapidly. For the mining industry, this is accentuating the importance of establishing and monitoring good community relations, and emphasizing the essential role of local communities as key stakeholders in any project. Nowhere is this more important than where Indigenous Peoples are involved. This chapter aims to advance appropriate and effective practices in relation to Indigenous Peoples.

For most of modern mining history, the industry has been involved to a greater or lesser extent in dealing with aboriginal people.

For most of modern mining history, the industry has been involved to a greater or lesser extent in dealing with aboriginal people – in Canada, Australia, western USA, and parts of Africa. Over the past decade, Australian, Canadian, American, and South African companies have also been developing mines in other countries. Many of these operations involve contact and interaction with Indigenous Peoples. While there are many similarities in the issues associated with Indigenous Peoples from different countries, there are also many differences. Therefore, this chapter cannot and should not be prescriptive. Rather, it sets out some basic principles and guidance which can provide an overall approach, together with some useful insights.

This chapter looks at issues involving Indigenous Peoples (IP) and the mining industry and includes both constructive advice and warnings based on project experience. It deals with the rights of Indigenous Peoples and the responsibilities of the industry towards these peoples. Likewise, mining companies also have rights and Indigenous Peoples have responsibilities; however, these are largely beyond the scope of this publication.

Issues such as land acquisition, communication with local communities, and workforce recruitment, are likely to be far more challenging and, at the same time, more critical where indigenous communities are involved.

Most of the issues raised in this chapter also apply to varying degrees to non-indigenous societies. However, issues such as land acquisition, communication with local communities, and workforce recruitment, are likely to be far more challenging and, at the same time, more critical where indigenous communities are involved.

16.1 WHO ARE INDIGENOUS PEOPLES?

There is no one, universal definition of 'Indigenous Peoples'. The term refers sometimes to those people regarded as the 'original inhabitants' of an area, such as the American Indians of the American continents, the Dayaks who inhabit much of the island of Borneo, and the Australian Aborigines. This use of 'indigenous' is quite similar to that of the related term 'aboriginal'. The term Aborigines can be used for all original people of Australia. Aborigines in different parts of Australia have individual tribal and area names. The term Koori is such a name and is not applicable to all Australian Aborigines. Elsewhere, other terms have come to designate similar populations, such as 'ethnic minorities', 'tribals', 'minority nationalities', and even the semi-racist term, 'natives'.

This confusing welter of terms has also muddled the definition of 'indigenous'. The term is clearest when applied to the Americas, differentiating the original inhabitants from those populations that came in with the European conquests of the 1500s and later. When the term is applied elsewhere, interpretations and adjustments are made to the term, and the term's initial clarity is often lost. No wonder political arguments often rage as to who should and who should not be considered 'indigenous'.

No wonder political arguments often rage as to who should and who should not be considered 'indigenous'.

Nevertheless, over the last decade or so, a broad consensus has begun to emerge as to the broad criteria a group must have to be dubbed 'indigenous'. Beginning with the World Bank (Operational Policy 4.10), and then picked up by the other international financial and policy institutions, the following criteria are now recognized widely as indicative of indigenous status:

(a) self-identification as members of a distinct indigenous cultural group and recognition of this identity by others;
(b) collective attachment to geographically distinct habitats or ancestral territories in an area and to the natural resources in these habitats and territories;
(c) customary cultural, economic, social, or political institutions that are separate from those of the dominant society and culture;
(d) a distinct language or dialect, often different from the official language or dialect of the country or region.

Indigenous Peoples everywhere thus claim to be, and hope to be acknowledged as, the original inhabitants and *customary owners* of an area. Lands associated with particular groups may be termed *ancestral lands*. The UN recognizes over 250 million people as indigenous, with the bulk of them residing – as does 2/3 of humanity – in Asia.

16.2 REASON FOR CONCERN

Why should the mining industry be concerned about the Indigenous Peoples with whom it interacts? There are strong legal, moral, and pragmatic reasons for such engagement, not the least of which is the rising influence and presence of Indigenous Peoples on the world stage. Once completely isolated, vulnerable, and ignored, Indigenous Peoples have organized themselves globally and attained a presence internationally which was unimaginable just a decade or two ago. The United Nations Permanent Forum on Indigenous Issues, established in 2002, is one potent symbol of that new status as was the passage by the UN's General Assembly in 2007 of the Declaration of the Rights of Indigenous Peoples. Legal recognition and protection of these rights, however, differs markedly from country to country.

Once completely isolated, vulnerable, and ignored, Indigenous Peoples have organized themselves globally and attained a presence internationally which was unimaginable just a decade or two ago.

But these are all very recent developments. Until quite recently, the western model of progress prevailed. According to this model, anything that provides financial wealth is good for all concerned. It is now recognized that this does not apply universally and it certainly does not necessarily apply to Indigenous Peoples. Another prevailing view is that democracy is or should be the basis for political systems and that the countries of the western world best exemplify democratic values. This ignores the fact that each indigenous

Indigenous societies have much to teach us.

society has developed its own socio-political system to meet its own needs and that, in many cases, the system includes democratic values, some of which may be superior to those of the western world. It follows that indigenous societies have much to teach us.

World values are evolving and the world community now generally places a very high value on indigenous cultures. Increasing pressure on the world's resources, climate, and environmental conditions is focusing attention on environmental degradation and the relative harmony which apparently existed between nature and traditional livelihoods and lifestyles that are perceived as ecologically sustainable. In fact, there is evidence that the activities of some aboriginal peoples were not harmonious or environmentally sustainable (Flannery 2002); however, the popular view is otherwise. Coupled with this increased admiration for aspects of indigenous societies is increased awareness of the extremely difficult conditions under which most Indigenous Peoples live and the inherent disadvantages that they face. This reflects the harshness of the environment without modern conveniences, the way that other communities (or cultures) may have marginalized the aborigines over time, and the areas into which Indigenous Peoples have been pushed.

Many indigenous societies have been weakened by alienation from their lands, by disease, by exploitation or by misguided attempts towards integration.

While most non-indigenous populations have experienced more-or-less continuous improvement in recent generations in basic human needs such as freedom, health, education, and welfare, such benefits have generally not extended to indigenous societies, or have been slow in coming. Many indigenous societies have been weakened by alienation from their lands, by disease, by exploitation or by misguided attempts towards integration. The once close-knit relations within communities and with their land and cultural beliefs and practices, have been severely stressed as Indigenous Peoples find themselves in a cultural divide – not belonging fully to either the societal values of their own traditions or those of the larger society of which they have been made a part. Many indigenous communities that remain are therefore highly vulnerable to further social, cultural, and economic damage.

Legal and moral considerations aside, a compelling case for concern for Indigenous Peoples can be made on purely pragmatic grounds. Most, if not all, mining companies recognize the desirability (if not the absolute need) to develop and maintain good relations with the communities in which they operate. Increasingly, companies have found it impossible to operate in the face of sustained community opposition.

Confronted with the prospect of a new resource development, Indigenous Peoples generally share the same aspirations and concerns as any other communities. However, fear and mistrust born of past injustices, and deeply held concerns for the preservation of culture and identity, may overwhelm good intentions. These special concerns need to be recognized, understood, and accounted for as the basis for establishing good community relations. Unfortunately, anti-mining influences from outside, well meaning or otherwise, commonly seek to exploit these concerns, denying the benefits and exaggerating potential problems, impeding attempts to foster goodwill.

16.3 IMPORTANT CHARACTERISTICS OF INDIGENOUS SOCIETIES

Every community has unique features

It is important to recognize that no two societies or communities are the same. This is especially true of indigenous communities. Every community has unique features, and knowledge of these features is valuable for all involved as it increases understanding and respect.

What shapes a community is the blend of culture and experience. This may differ in terms of:

- Language;
- Beliefs and attitudes;
- Communication norms;

- Extended family or kinship systems, and
- Experience with outsiders.

These differences are encountered between communities from one location and/or social group to another. Similarly, these types of differences also occur within a society, particularly between generations.

However, there are certain similarities that exist both between and within many indigenous communities. These include:

- Close affiliation with the land;
- Internally-understood rules and codes of conduct;
- A community structure which gives elders a leadership role;
- Important cultural and spiritual knowledge passed from one generation to the next;
- Community memory of events, and
- Negative experience with outsiders at some time.

These aspects of a society are highly important to an understanding of how Indigenous Peoples think and view the world, both as individuals and as groups. Like people anywhere, indigenous societies generally:

- Desire improved conditions for future generations, and
- Deserve and expect respect.

Close affiliation with the land is a fundamental characteristic of most indigenous communities. This includes the concept of the Earth as 'mother' or creator of life, as well as the close identification of individuals, families or tribes with particular parts of the landscape. For Indigenous Peoples living a customary lifestyle, this identification with the land permeates their entire existence. Damage to, and, in some cases, exclusion from the land results in anguish and even grief due to the belief that the indigenous society will inevitably suffer as a consequence. Their belief system allows for a direct and consequential relationship between the land and their well being.

Close affiliation with the land is a fundamental characteristic of most indigenous communities.

The knowledge passed by Indigenous Peoples from generation to generation provides all the information required to survive and to live a customary life. Among other things, this knowledge includes how, where, and when to find different types of food and other resources, and a detailed understanding of weather patterns, seasonal changes and their effects on biological systems. It frequently provides a code of conduct for interacting with other communities, as well as treating and respecting guests.

Rules and codes of conduct differ widely from one indigenous society to another. Ignorance of these may cause amusement, disdain, or offence, and may lead to miscommunication or misunderstanding. Such rules may not be easily learned, as they are seldom available in accessible publications. This reflects the oral nature of many aboriginal traditions.

Negative experiences of outsiders are common among virtually all Indigenous Peoples, although the numbers and severity of the experiences differ from one people to another. In some cases these experiences have generated an overt hostility toward outsiders; more commonly they result in suspicion and mistrust.

Indigenous Peoples living traditional lifestyles often have values that are different to those of the non-indigenous. One example is the so-called 'work ethic'. By the standards of the non-indigenous, Indigenous Peoples are commonly seen as unmotivated, lacking application or lazy. From the indigenous viewpoint, however, the non-indigenous are seen as being pre-occupied by work and the material benefits that it generates, to the exclusion of more important social and spiritual needs. Similarly, time is valued very differently. Unused to the western concept that 'time is money', Indigenous Peoples may not subscribe to or revere the clock-driven 'virtues' of punctuality, adherence to schedules, and meeting deadlines. Accustomed

From the indigenous viewpoint the non-indigenous are seen as being pre-occupied by work and the material benefits that it generates, to the exclusion of more important social and spiritual needs.

Conflicts and misunderstandings between indigenous and the non-indigenous may result from very different perceptions of the world and the value placed on aspects of that world.

to living in a world run more by the rhythms of nature (seasons, weather patterns, tides) than that of an industrial society's timetables, Indigenous Peoples are often unlikely to appreciate the impatience or frustration of the non-indigenous faced with time delays.

Conflicts and misunderstandings between indigenous and the non-indigenous may result from very different perceptions of the world and the value placed on aspects of that world. To non-indigenous communities, time and tide wait for no one in the march of progress. To indigenous communities, the 'developed western world' – or their host nation's majority society – has gained its trappings of progress at the expense of its soul.

16.4 ISSUES AND OPPORTUNITIES

A mining development potentially offers both opportunities and threats to indigenous communities. The challenge is to realize the potential opportunities without diminishing or losing those aspects of culture or lifestyle that are important to the indigenous communities. Potential opportunities may include employment, education and training, improved infrastructure, housing, medical services, and community services. Threats include loss of land, pollution, loss of access to traditional resources, being swamped by large numbers of incoming people, introduction of new diseases, and changes in norms and values as a result of contact with outsiders. These can all lead to dependency, loss of identity, breakdown of traditional codes of conduct and ultimately spiritual and physical breakdown.

Understanding these issues is essential if mining companies are to help indigenous communities benefit from a mining development without compromising their social and cultural values and without the risk to the company of significant liabilities (including poor public relations).

Sharing of Benefits – The Chance for Betterment

Most, if not all societies seek betterment, although what constitutes betterment will differ from one society to another.

Some indigenous societies may aspire to be left alone. However most, if not all societies seek betterment, although what constitutes betterment will differ from one society to another. For most indigenous societies, aspirations for betterment would include:

- Improved health and medical facilities;
- Improved safety and security;
- Access to sufficient resources so as to be independent;
- A viable and vibrant community life that provides attraction for and helps retain young people;
- The means to deal with existing social problems such as alcohol abuse.

Other common aspirations include education and training tailored specifically to the community's circumstances, better housing, transportation, and modern amenities. For some indigenous communities, developing a mineral resource may provide the best, perhaps the only means by which the people's aspirations can be achieved.

Benefits derived from mining or petroleum production can lead to enormous differences between neighbouring indigenous groups.

Whatever the particular aspirations may be, a mining project among Indigenous Peoples should be able to contribute toward the realization of realistic goals for betterment. Benefits derived from mining or petroleum production can lead to enormous differences between neighbouring indigenous groups. One group with large mineral resources can prosper while a nearby group remains impoverished. Nowhere has this been more apparent than among the Indian tribes of North America. In Australia, this situation has been alleviated to some extent by the Land Councils through which benefits flow beyond the group that is directly involved.

In some countries the opposite occurs where the national government claims all mining revenues, and there is no special treatment of the communities affected by the mine development. Under these circumstances, the indigenous group that is directly involved feels unfairly treated because it receives the adverse effects of the development with no compensating benefits, or only a small share. This was the case in Indonesia until 1999 when the Law No. 22 Act was promulgated, ensuring that part of government revenues from a project flowed to the region involved.

Mining companies are faced with difficult choices in distributing benefits among stakeholders and would-be stakeholders (see also Chapter Fourteen on compensation). Often, the community expects the company to compensate for the inadequacies of government, but most companies would not view this as their responsibility and, in many cases, government policies can restrict the means by which this can be done. It is therefore essential that discussions, negotiations, and decisions involve the government as well as the community and the company.

Alienation of Land or Access

The availability of land and access to land and related resources are usually the most sensitive and crucial issue between Indigenous Peoples and mining companies. Most Indigenous Peoples have already lost much of their traditional lands. For some, the land remaining may already be insufficient to sustain their traditional way of life.

The availability of land and access to land and related resources are usually the most sensitive and crucial issue between Indigenous Peoples and mining companies.

Land access is culturally and historically important for Indigenous Peoples. It is considered by many a source of spiritual and physical well-being (Togolo 1997). In some cases, access to land is a matter of physical survival and, therefore, alienating land can compromise the ability of Indigenous Peoples to meet their own needs. In most cases, Indigenous Peoples have strong beliefs about land use, including understandings about the origins of land areas, special values of certain areas, and prohibitions or taboos. Some sites are considered by Indigenous Peoples as sacred. The most sacred sites are of such significance that any alienation or damage is unthinkable. There are also customary land tenure systems that dictate ownership and use of some land. Finally, aborigines generally do not regard traditional land as a tradable asset.

For these reasons, land transaction and access arrangements must be negotiated carefully and sensitively. Some areas will be strictly unavailable to mining companies; others will be accessible but only under certain circumstances. The opportunities that arise or can be created will depend on the outcome of negotiations over land and access. Most projects have sufficient flexibility that land issues can generally be satisfactorily resolved. An exception would be if the ore body was itself a sacred site. A mining company needs to be prepared to forego mining in such a case. The Ranger Uranium Mine in northern Australia provides an interesting example. Three ore bodies were discovered, but one was excluded from mining because of its close proximity to a sacred site.

There is also the reasonably common situation where a community is divided in its willingness to allow mining at or near a sensitive area. Such situations may take many years to resolve if, in fact resolution is possible.

Compensation

Compensation is, in concept, relatively straightforward. Payments are nearly always in cash, following which the compensated surrenders rights of ownership or privilege to the compensator. The negotiated price is at least sufficient to enable the compensated party to replace or renew whatever was sold.

The concepts of ownership and compensation, particularly financial compensation, are alien to some indigenous societies and may be misunderstood.

The concepts of ownership and compensation, particularly financial compensation, are alien to some indigenous societies and may be misunderstood. In the past, compensation has been negotiated with different perceptions about what the compensation is for. Traditional owners have believed that they were being compensated for allowing access to or usage of the land, while the mining company was paying for transfer of ownership with consequent denial of access to traditional owners.

Language can be a significant problem. Given their different cultures, level of education, and a tendency for non-indigenous groups to (for example) produce legalistic documents for negotiation, there is a considerable risk that neither group will succeed in communicating its concerns or aspirations clearly.

Other factors relate to the form of compensation (financial or other), the recipients (individuals or groups), and payment schedules (in one or multiple instalments). Many indigenous societies function mainly outside the cash economy. Accordingly, some Indigenous Peoples are largely unfamiliar with commerce and the 'value of money'. There are many examples from the past where cash compensation in such circumstances has led to nothing of lasting value while at the same time causing social disruption. Examples of innovative, non-cash compensation schemes are rare, as most Indigenous Peoples prefer cash payment and companies find it much easier to administer. In cases where compensation was paid in the form of housing, vehicles, and equipment, the results have often been disappointing, because the recipient communities did not have access to spare parts or the capabilities required for maintenance and repair.

Procedures for administration of compensation payments are also important to ensure that compensation goes where it is intended. As always, large amounts of money often attract opportunistic outsiders including criminals; Indigenous Peoples are particularly vulnerable to exploitation by such people.

Uncontrolled Influx of People

A common problem in developing countries is the influx of people looking for work at a new project, such as a mine under construction. If they are unable to find work, some will remain in the vicinity hoping to find casual work. Uncontrolled influxes of people can lead to 'shanty towns' characterized by poverty, poor sanitation, and high incidences of disease and crime, which are potentially damaging to local indigenous communities.

A mining company does not normally have (or want) the legal authority to control activities outside its mine site. However, companies need to anticipate where and when large influxes of people may occur and to assist the authorities in their response to the situation. Such assistance might include sponsoring and paying for planning, capacity building and strengthening of community services. In this way the mining company can cooperate with and support an indigenous community to help preserve and protect its interests.

Disruption of Societal Organization

Societal organization, or the way a society organizes itself, is determined by many factors, including population composition, activities, institutions, and interaction. Population composition is determined by factors such as gender balance, ethnicity, and age-group proportions within a community. Daily activity trends may include employment participation, schooling, and others, while institutions and practices include religions, interest groups, and union involvement, to name a few. Social interaction is a main indicator of societal organization and involves such relationships and concepts as family friendship, business relations, hierarchy, and social order. These are all important aspects of societal organization.

Disruption of societal organization due to mining activities occurs with changes to one or more of these aspects of a society. Examples of disruptions include:

- Increased number of single males of a different culture entering the community;
- An exaggerated economic gap between groups within the community – the 'haves' and the 'have nots';
- Conflicts between or within residents and newcomers;
- Introduction of outside agencies;
- Corrosion of existing values;
- Creation of additional values.

Marginalizing an indigenous community within a society also represents a disruption.

Societal change occurs naturally. However, industrialization causes accelerated societal change by introducing foreign and unfamiliar activities and opportunities. Employment opportunities, for example, may alter daily activities and population composition, while land clearing may interfere with or preclude traditional institutions or practices; e.g. hunting and harvesting of forest resources. Changes which upset the balance of interests within a community are generally not welcomed, particularly when an essential part of the community's identity is affected.

Industrialization causes accelerated societal change by introducing foreign and unfamiliar activities and opportunities.

It is clear that industrial activity presents opportunities for Indigenous Peoples. The challenge is to find ways of reducing the intensity of social disruption and maximizing the likelihood of social change associated with mining development which is a positive experience for the local people. If managed sensibly and sensitively, it should be possible for the inescapable social and cultural pressures to be offset by, for example, improved living conditions, practical employment experience, health care, etc. Social impact predictions should be undertaken early in the planning process to help forecast and develop appropriate strategies.

Loss of Customary Values and Cultural Identity

Cultural values and cultural identity are what underpin an indigenous society. Without these attributes the group's cohesion and sense of self will disintegrate.

Independent and distinct values are difficult to maintain in a world where 'modern' values are pervasive and persuasive. As a result, some indigenous societies seek to escape these influences and protect 'traditional' values by isolating themselves and avoiding contact with the outside world. An example is the 'Outstation Movement' which has been embraced by many Australian aboriginal societies. This is a clear reaction to past adverse experiences related to a wide range of issues/developments of which mining may be one.

Terms such as 'modern' can also carry an inherent bias. By common usage, 'modern' carries connotations of contemporary, progressive, desirable, and beneficial. 'Traditional' may also infer a sense of conservatism, the archaic and something that is 'past its time'. It is therefore important not to allow attitudes or the language itself to build a negative momentum in dealings between indigenous and non-indigenous groups. If we speak of indigenous culture as 'traditional' while contrasting it to 'modern' Western or national cultures, we help condemn indigenous ways and values to oblivion. It is possible to be contemporary and indigenous at the same time.

Mining on aboriginal land inevitably reduces the isolation of the indigenous community and increases exposure to outside values.

Mining on aboriginal land inevitably reduces the isolation of the indigenous community and increases exposure to outside values. Some mining projects have been designed to minimize impacts on Indigenous Peoples by largely or totally avoiding contact between the mine workforce and the local community. Such projects use the 'fly in fly out' approach. The Nabarlek Uranium Project which was located inside the Arnhem Land Aboriginal Reservation in northern Australia was an early example of this approach, which is now widespread. (There are other reasons for adopting 'fly in fly out', such as minimizing capital costs). Elsewhere, attempts have been made to control the interactions between the non-indigenous workforce and the Indigenous Peoples by the imposition of a 'Workforce Code Of Conduct' which may include such matters as:

- Limitations on visits to indigenous communities or to particular areas;
- Restrictions on possession or use of alcohol, and
- Prohibition of sexual contact with Indigenous Peoples.

Clearly, as with other company policies with potential to cause or alleviate social problems, major decisions on contact and codes of conduct should be developed in consultation with the indigenous community, whose wishes it should reflect.

Incoming workers are more likely to respect Indigenous Peoples if they have awareness, understanding and respect for their culture and values. Mining companies can assist by avoiding recruitment of biased or bigoted employees, by including cultural awareness among their goals and by providing all incoming workers with information on the importance of local indigenous culture as part of their induction.

Even more important is for the company to maintain a dialogue with the local community and to establish mechanisms through which misunderstandings, grievances, and contentious issues can be identified and resolved before harm is done.

Even more important is for the company to maintain a dialogue with the local community and to establish mechanisms through which misunderstandings, grievances, and contentious issues can be identified and resolved before harm is done. Appropriate mechanisms will vary depending on the nature of the indigenous community and the ways by which the community addresses its own social issues.

Changes in Decision-Making Authority

All societies have well-developed rules with decision-making authority clearly defined. Indigenous societies are no exception. Operating at the non-state level, authorities have tended to base their legitimacy on kinship and locality grounds. Family heads, clan or tribal chiefs, spiritual leaders, and healers all have well-understood areas of responsibility and levels of authority. There are also well-developed and generally accepted means of conciliation, conflict resolution, and dispensation of justice and punishment.

These authorities may be diminished as contact with the outside world leads to situations that are not covered by established rules or positions (i.e. rules and structures in the non-indigenous community) that do not fit with the traditional hierarchical structure. Questions such as:

- How does a society with no experience of addictive drugs respond to the sudden availability of alcohol and other drugs?
- Where does a merchant, a foreman or a person with wealth and possessions belong in a hierarchy based on bloodline, religious ritual or prowess in a traditional activity?

As the lines of authority (which include respect and leadership) become blurred, conflict increases and the indigenous society's ability to manage its own affairs decreases. A resilient community can usually adapt to these changes. However, a weakened fragile society may be unable to adapt. Traditional chiefs may also feel threatened or ineffectual because they may feel or be seen as inadequate by their community who realize they are not knowledgeable about these new issues. Suddenly a whole bunch of unknown people with more power than the local chief or leader can appear on the scene who were unknown or not interested before – national government bureaucrats, politicians, and company management, etc; these all erode the chief's status. Providing the mining company is aware of these matters, it is a relatively straightforward matter to ensure that company personnel do not undermine the authority of indigenous leaders; e.g. official business with indigenous leaders should be transacted only by the most senior mine personnel.

Official business with indigenous leaders should be transacted only by the most senior mine personnel.

Creating a Dependent Community

There is a very real risk that mining companies, in their attempts to provide benefits for Indigenous Peoples, will create a dependent community. The community may become totally dependent on the mining operation, not only for its means of subsistence but for its decision-making and management. When the operation ceases, the society, having surrendered its original means of livelihood, loses its ability to meet these basic needs. In the case of communities traditionally reliant on subsistence agriculture, ways have to be found to revive old skills or the people must move away from their traditional homelands. In mining operations where previously farmed land becomes unavailable (e.g. through burial under tailings impoundments), the old ways may simply be impossible to re-establish. Dependency on the mining operation has other disadvantages such as reducing confidence, self-reliance, and self-respect of individuals and entire communities.

There is a very real risk that mining companies, in their attempts to provide benefits for Indigenous Peoples, will create a dependent community.

To avoid creating dependency, companies should refrain from making decisions on behalf of a community, provide assistance which promotes self-help rather than handouts, and do everything possible to encourage the continuation of traditional activities and the maintenance of existing organizational structure and hierarchies.

16.5 STRATEGIES FOR INTERACTION WITH INDIGENOUS COMMUNITIES

Getting It Right from the Start

Impressions formed during initial contact can affect relations between a mining company and an indigenous community for many years. In particular, any adverse impressions may be very difficult to overcome. As the adage says, 'you never get a second chance at a first impression'. Before contacting indigenous communities a company needs to be prepared. This involves recognizing cultural differences, developing appropriate corporate values, information gathering, and planning an approach to community contact.

'You never get a second chance at a first impression'.

Partnering with Indigenous Peoples

Rather than approach relations with Indigenous Peoples as primarily a 'problem' to be dealt with, companies would do well to consider that for this group of people affected by the construction and operation of a mining project, their lives, livelihoods, and futures may hang in the balance. What Indigenous Peoples want in such a situation – more than the size of a compensation package or the dimensions of a benefits-deal – is the respect of their interlocutors from the company. By approaching the indigenous community as their partners, mining companies can prepare the psychological ground for mutual compromise and mutual benefit.

Recognizing Cultural Difference

Experience from projects around the world confirms that unless a proper participatory, consultative approach is undertaken with stakeholders through all stages, projects will encounter problems. Success requires more than just going through the right motions and employing the correct procedures. The additional requirement is a fundamental understanding by mining companies and their key personnel.

Not all people perceive the world and events the same way, yet each worldview is equally justifiable and real.

At the outset, it is absolutely critical to understand that all cultures are equally valid. Culture is a human construct by which communities make sense of reality. Cultures differ given the circumstances in which they grow and evolve. There are very many different cultures and, correspondingly, there are very many different versions of reality. Therefore, not all people perceive the world and events the same way, yet each worldview is equally justifiable and real. This must be understood and appreciated if one is to deal successfully with people of different cultural backgrounds. Of course, many people will not agree with or subscribe to these statements. The point here is that such people are unlikely to treat Indigenous Peoples with respect, and should therefore not be involved in dealing with Indigenous Peoples.

There are many definitions of culture, but all have common threads. Here are four:

- It is all the accepted and patterned ways of behaviour of a given people. It is a body of common understanding. It is the sum total and the organization or arrangement of the group's ways of thinking, feeling, and acting. 'Every people has a culture and no individual can live without culture'. (Man and Culture, Ina C. Brown).
- 'Culture is that complex whole which includes knowledge, belief, art, morals, law, customs and other capabilities or habits acquired by members of a society.' (E.B. Taylor).
- 'The sum total of ways of living built up by a group of human beings, which is transmitted from one generation to another.' (The Macquarie Encyclopedic Dictionary).
- 'The way we do things around here.' (Anon).

Finally, it is now widely accepted that the cultures and special knowledge of Indigenous Peoples are rare and valued aspects of the human tapestry. The fragility of these cultures and their relevance as a window to an appreciation of ancient human societies is reflected in the inclusion of traditional cultural values as criteria for World Heritage listing of areas (e.g. Kakadu, Australia). Also, the value of this ancient knowledge is only beginning to be appreciated in modern society, for example through recent discovery of the medicinal effectiveness of tribal herbal remedies, or by more sensitive knowledge of the ecosystems and the interactions within them.

Assess and Establish the Values of the Company

The first essential step is to evaluate the values and attitudes of the company and its key personnel towards Indigenous Peoples.

The first essential step is to evaluate the values and attitudes of the company and its key personnel towards Indigenous Peoples, identifiable cultural groups, and local communities. The company needs to ensure from the outset that it will approach stakeholders and all external parties with an appropriate attitude and value system that will allow for:

- A belief in openness and dialogue in its dealings with other parties;
- An understanding of the hopes and aspirations of Indigenous Peoples;

- A recognition and acceptance of co-existence of cultures, cultural, and ethnic differences;
- An acceptance of the validity of different socio-cultural traditions;
- A respect for customary rights of indigenous and tribal peoples to their traditional land and natural resources;
- An appreciation that the land the company wishes to mine represents irreplaceable totemic, ancestral, religious and subsistence values, and traditional practices without which the community would lose its sense of place or purpose.

Cultural Awareness Training

Non-indigenous and Indigenous Peoples may have completely different values, living patterns and perceptions of the world. When the two groups interact, these differences can lead to misunderstanding and problems. What seems 'logical' or a 'natural progression' to one is not necessarily true for the other. Indigenous Peoples may be bound by cultural obligations that take precedence over any other commitments. Many indigenous societies also observe various taboos, non-observance of which is taken extremely seriously.

An overview of the local indigenous culture will assist non-Indigenous Peoples in understanding the Indigenous Peoples with whom they will interact. Where appropriate, this overview would include the protocols for greetings, meetings, sharing food, and other social interaction, knowledge of which can help avoid awkward and embarrassing situations.

An overview of the local indigenous culture will assist non-Indigenous Peoples in understanding the Indigenous Peoples with whom they will interact.

Most Indigenous Peoples are proud of their culture and willingly share it with others. They are usually pleased to assist in the training of non-Indigenous Peoples, either through personal instruction or by assisting in the preparation of written or audio-visual materials.

Two levels of training may be appropriate. Basic induction requiring only a few minutes, could be given to all employees and visitors to a site. A more in-depth cultural training course up to one week in duration could be provided to all Non-indigenous managers.

Language training can also be invaluable in improving communication and understanding. It is common for mining personnel operating overseas to receive language training, and also for the company to sponsor English language training for locally recruited employees. However, language training for expatriates is nearly always in the national language (e.g. Bahasa Indonesia, Tagalog, Pidgin). These national languages may only be the second or third language to local Indigenous Peoples or, in some instances, may be unknown to them. Training in local indigenous languages is more difficult and requires a much greater commitment. It would however produce benefits, both by demonstrating respect and through the increased understanding that can only come from conversing with someone in their native tongue. Similarly, some basic ability in the local Aboriginal language can be effective in demonstrating commitment even though the local people may be competent in English.

Cultural awareness can also be useful in reverse. For example, Thai employees at the Chatree Gold Mine are provided with the opportunity to participate in an interactive workshop in which western and local attitudes and values are explained and compared. Feedback from Thai personnel has confirmed the benefits of this initiative in helping them understand the otherwise puzzling behaviour of expatriate personnel (Yaowanud Chandung – pers. comm.).

Build the Right Team

The starting point is, do the mining company and its key employees have the capacity to empathize with the needs of the communities with whom they will deal? If the potential for the right approach does not exist, and there is a risk that the company will take an exclusionary, confrontational or elitist approach to the communities it will be dealing with, then it is time to stop the process, and better select and prepare both the company organization and its key personnel in terms of outlook and understanding of cultural factors.

Elements of the 'Right' Approach

Assuming that the mining company is culturally aware and sensitive to Indigenous Peoples, and is committed, the adopted approach should include the following elements:

- Provide time and funding to develop company expertise, (in-house and through external experts) on local communities, traditional customs, and contemporary politics;
- Investigate the potential for conflicting objections between local communities and their higher representatives; for example Land Councils in Australia; local, regional or national governments.
- Realize that it may be necessary to manage ways of avoiding this conflict; for example by involving representatives of all tiers of stakeholders, at all stages;
- Seek to understand the different views of history that each community holds, and particularly their individual experience with governments and non-indigenous society;
- Ensure that the company is at all times represented to the community by a senior person suited to the task who has the authority to speak on behalf of the company and make firm and binding decisions;
- As far as possible continue to use this same person as the lead representative throughout the entire consultation process;
- Refrain from overstating benefits to the community or other actions that may lead to unrealistic expectations.
- Take careful steps to ensure that commitments are well thought out, and that all commitments are met;
- Develop protocols and clear lines of communication between the company and indigenous communities, and ensure that they are consistent and reliable;
- Respect the traditions and culture of the communities one is dealing with;
- Acknowledge cultural differences, commit to understanding and accommodating them as much as possible;
- Initiate contact, honest communication, and open dialogue as early as possible in the life of the project;
- Deal with all Indigenous Peoples with honesty and integrity, by developing a working relationship based on mutual respect and trust, with open and ongoing dialogue throughout the life of the project;
- Engage all stakeholder groups who represent, or may in the future profess to represent, the interests of the Indigenous Peoples (including relevant NGOs and aid agencies) and inform them throughout the negotiation process;
- Accept that mining exploration and development/production activities will have an impact on communities, minimize this impact at all times and provide compensation where justifiable and appropriate;
- Assist in community development where appropriate, and in accordance with the wishes of the community and local, regional, and national governments, and
- Employ, train, and educate Indigenous Peoples from the local communities, whenever possible, appropriate and consistent with community wishes.

Take careful steps to ensure that commitments are well thought out, and that all commitments are met.

Deal with all Indigenous Peoples with honesty and integrity.

16.6 RIGHTS OF INDIGENOUS PEOPLES

As for all human groups, Indigenous Peoples have certain well-defined rights. Moreover, in addition to these universal rights, Indigenous Peoples are protected by the provisions of various international or national laws that apply specifically to them.

First, international law is intended to guarantee to all people certain universal human rights, and Indigenous Peoples are covered by a category of general international human rights law. In many countries, the nation, or separate states within a country, also guarantee to all citizens certain civil, political, and social rights.

In looking to understand the rights of Indigenous Peoples, it is important to be aware of the origin of any statement of rights. Many statements of human rights, for example those which derive from the respected Brundtland Report, are still premised on a western perspective of what those rights should be, and both intentionally and unintentionally seek to protect non-indigenous interests in the main. Recent experience indicates that any statement of rights that does not truly derive from Indigenous Peoples, or that does not protect them, will be rejected by Indigenous Peoples – and their advocates (**Case 16.1**).

In looking to understand the rights of Indigenous Peoples, it is important to be aware of the origin of any statement of rights.

There are a number of influential and respected international non-government organizations (NGOs) concerned with the rights of Indigenous Peoples. Some of the most prominent include:

- Amnesty International;
- Cultural Survival;
- Indigenous World Association;
- International Workgroup for Indigenous Affairs;
- Survival International, and
- Oxfam and Community Aid Abroad.

The larger NGOs are well-organized, well-informed, and command respect. They have tended to act to counter-balance the inevitable bias of national governments in prescribing the rights of people, and in so doing play an important advocacy role on behalf of Indigenous Peoples. NGOs are worth listening to and are a good source of advice. It should be recognized, however, that in their efforts to protect Indigenous Peoples, NGOs

CASE 16.1
Development of an International Declaration of Rights for Indigenous Peoples

From 1957 to 1982, the International Labour Organization (ILO) was the sole international law body concerned with Indigenous Peoples rights. In 1957 it promulgated ILO Convention 107. This tended to reflect the views of settler societies and to promote the integration of indigenous populations into the changing society.

By the 1980s this was seen as unacceptable and there emerged the revised ILO Convention 169.

In 2007 the United Nations' General Assembly, after over a decade of debate and contention, passed the Declaration of the Rights of Indigenous Peoples.
The Declaration rejects the assimilationist orientations of ILO 107, and recognizes Indigenous Peoples' basic human rights and fundamental freedoms, including the right not to be dispossessed of their lands, territories, and resources.

In Part I, the document makes all previous human rights instruments applicable to Indigenous Peoples. Part II sets out guarantees of cultural rights. Part III describes property rights. Part IV deals with economic and social systems. Part V provides standards for self-determination. Part VI describes a process for resolution of conflict, defines the purpose of the declaration, and prohibits nation states from using the declaration against Indigenous Peoples.

may exaggerate problems and repeat allegations without checking their veracity. There have been many cases where NGOs have overstated social problems between mining operations and indigenous communities, and in some of these cases, the overstatements aggravated the problems.

NGOs, as well as politically aligned or unaligned indigenous communities, have increasingly well-developed national and international networks. This has been facilitated by the rapid improvement and efficiency of international communications, particularly global email and the Internet. Important and influential information including photographs of, for example, claimed pollution or substandard living conditions is now transmitted and received in real-time throughout the world. Protest action or distribution of adverse claims can be organized to take place almost globally within a matter of days.

It is therefore in the interests of mining companies or governments to avoid local disputes involving Indigenous Peoples which could generate an international issue; interest on the part of outside parties in disputes can now be taken up extremely quickly.

The key rights that are now generally accorded to Indigenous Peoples are discussed separately below:

Indigenous Peoples have the basic right to protect and maintain their traditional ways of life.

Self-determination – Indigenous Peoples have the basic right to protect and maintain their traditional ways of life. Although the right of self-determination has yet to be fully explored by many Indigenous Peoples, this is likely to be a matter of increasing conflict against the interests of nation, state, and capital interests.

The International Court of Justice has declared that self-determination is a right of people, not nation states. This is off-set by the role of nation states in imposing constraints on self-determination for Indigenous Peoples, and in determining the extent of the negotiating power of Indigenous Peoples.

In international law, self-determination implies that Indigenous Peoples have the right to independent sovereignty and hence the right of succession. In so far as Indigenous Peoples constitute identifiable communities with collective rights, they also have some rights of appeal to the United Nations Human Rights Committee.

Intellectual property rights – Indigenous Peoples' cultures invariably hold significant traditional knowledge of the natural resources on which they depend; and of specific resource values, such as medicinal values of herbs and plants, and natural processes that occur in the environment around them. Such knowledge is of great value, with significant potential commercial value. Such knowledge belongs to the traditional holders of that knowledge.

Control over and management of local resources – Clearly, if Indigenous Peoples have rights to exist and maintain their traditional culture and practices, then they must have a corresponding right to protect, maintain, and have control over those natural resources that are an integral part of their culture and support their way of life.

Preservation of cultural traditions and languages – If Indigenous Peoples are to be protected and their way of life maintained, then their cultural traditions and languages need to be preserved. Language is the most critical aspect of culture, and, throughout history, denying Indigenous Peoples the right to use their own language (or dialect) has helped destroy the independence and identity of minority cultural groups.

In Australia, the Philippines and many other countries, there is now a legislative framework which compels a serious consideration of Indigenous Peoples' rights over land and resources.

Compensation for the removal of rights to access and/or use – Under the provisions of various legislation, governments will from time to time arrange to acquire lands in the interests of achieving an over-arching benefit for the wider society. Land use planning legislation allows for this potential in most, if not all, countries. The application of such powers will at times affect the traditional lands of Indigenous Peoples.

In the past, Indigenous Peoples and their interests and affiliations to the land have been considered as separate from the concerns of regional development and resource management. In Australia, the Philippines and many other countries, there is now a legislative framework which compels a serious consideration of Indigenous Peoples' rights over land and resources.

Pursuit of betterment and prosperity – Indigenous Peoples and communities of all types have the right to pursue betterment and prosperity. If their traditional lives or livelihoods are affected by any given land development or mining operation, they have an indisputable right to some measure of benefit.

Community development – As part of self-determination, Indigenous Peoples have a right to be substantially involved in, and principally determine the content of, community development programmes, within the funding constraints that are made by the mining company.

16.7 RESPONSIBILITIES OF MINING COMPANIES IN RELATION TO INDIGENOUS PEOPLES

The responsibilities of mining companies towards Indigenous Peoples should be clear. They should abide by the moral imperatives as affirmed by responsible governments and such organizations as the United Nations and the International Court of Justice. Ten general principles are:

- Respect Indigenous Peoples and their cultural traditions;
- Deal openly and honestly with Indigenous Peoples;
- Understand the cultural differences they are dealing with, and make allowances as necessary;
- Help protect and retain indigenous communities and their traditional ways of life, so long as this is the wish of the community involved in any given situation;
- Ensure effective, ongoing, and accessible communication between the mining company and Indigenous Peoples, throughout the life of the project;
- Through pre-planning and implementation, minimize to the extent possible, adverse impacts both upon the community itself, and the environment upon which the community depends;
- Ensure that only appropriately qualified and senior experienced people represent the company and liaise with the community;
- Fund and facilitate community development programmes that assist local communities in adapting to the changes affecting the communities way of life, in so far as any given community wishes to be involved in such programmes;
- Provide assistance to the community in understanding the activities of the mining operation and its planned intentions. Allow the community the opportunity to comment on any aspect of mine operations and time to provide fair and reasonable responses;
- Do not deliberately attempt to determine the future, fate, choices or preferences of an indigenous community, with respect to any matter under a mining company's control;
- Avoid generating unrealistic expectations. Keep commitments to levels that can definitely be achieved and, above all, ensure delivery of all commitments.

Mining Companies are not Governments

It is equally important to establish what are not the responsibilities of mining companies towards Indigenous Peoples. Mining companies operating in isolated locations in underdeveloped regions often find it is practical or necessary to undertake actions that should be the responsibility of governments, particularly with respect to the supply of essential infrastructure and certain community services. Unfortunately, provision of some assistance or facilities to the community can lead to an expectation on the part of that community, of ongoing responsibility for wider support and services even after the mine closes. This is a problematic issue – remote regions of some third world countries effectively have no government, so the company ends up filling this role, in spite of what it would like to see happen.

The mining company should recognize the potential for this problem before it arises, and to ensure that the practical issues are worked through with the local, regional, and national governments at an early stage. The company should avoid providing facilities, infrastructure or services that are in practical terms unsustainable in the medium or long term. To do so can mislead the community and create a dependence that cannot be supported, and can cause greater hardship for the local community in the long run.

Check Investment Incentive Packages

Incentive packages often include provisions such as access to a region's natural resources and guaranteed rights or control over specified areas, that are potentially in direct conflict with the interests or rights (under international law) of Indigenous Peoples.

It is in the mining company's own interests as much as the Indigenous Peoples', to evaluate carefully the incentive packages offered by governments seeking to attract mining investment. Incentive packages often include provisions such as access to a region's natural resources and guaranteed rights or control over specified areas, that are potentially in direct conflict with the interests or rights (under international law) of Indigenous Peoples. In such circumstances, a mining company has the responsibility to assess the situation carefully, and not be misled into believing that the on-ground issues have been resolved in advance. Early advice from international organizations such as the UNESDC and ILO and from relevant NGOs, may be useful in defining potential longer term issues and potential liabilities for which the company could be held responsible.

Self-regulate Impacts of Mine Activities

The effects of economic development in an increasingly globalized world pose risks for most indigenous cultures and tribal peoples, particularly those who have not previously been exposed to such changes. Companies involved in developing new mining operations in previously undeveloped areas may be relatively unconstrained by laws or regulations. However, companies clearly have a moral responsibility to be as responsible and self-regulating as possible. There is also a 'self-interest' aspect to voluntarily adopting the principles outlined previously, that failure to do so, may expose the company to criticism and harassment in the international arena of environmental human rights activism, with adverse effects on shareholder relations and acceptance in other countries in which the company operates or seeks to operate.

Desired Outcomes

The desired outcomes for mining operations and effects on Indigenous Peoples, are that:

- Indigenous communities should remain intact, particularly when the communities themselves desire to retain their own values and ways of life;
- Changes to local communities affected by mining operations will have been determined to as large a degree as possible by the local communities themselves;

- Throughout the project life, interaction between the company and the community will be characterized by constant open, honest dialogue, with minimum misunderstandings, a maximum degree of free information exchange, and mutual respect;
- Local indigenous communities will receive long-term and sustained benefits from their association with a mining activity, continuing after mine closure;
- Natural resources, important land areas, and sites of critical importance to any given indigenous community remain intact, accessible to the community, and sustainable, so far as it is possible in any given situation. These must be identified and agreed in the planning and negotiation stage;
- Appropriate compensation is made to indigenous communities whenever an unavoidable and irretrievable loss has been experienced by a community as a consequence of a mining activity;
- Benefits should be dispersed widely throughout the affected communities. Often, benefits at reduced levels are also provided to adjacent communities and the surrounding region, although these communities may not be directly affected;
- Any knowledge gained by the mining company from traditional Indigenous Peoples' knowledge of the environment should be financially or otherwise rewarded. This knowledge should be incorporated into the preparation, operation, and post-operational stages of the mining activity;
- Education, training, and employment schemes for Indigenous Peoples are established and continued throughout the operating life of the mine;
- The mining company participates and assists in establishing regional social and economic development programmes for the local community, that will replace the mine following its closure;
- Arrangements are established with the national or regional governments to ensure that government community services are continued at appropriate levels after mine closure;
- Indigenous Peoples involved in any aspect of the mining activity, or the provision of external services to the mine, are left with a means of support and useful skills, once the mine has closed.

Organizational Responsibility and Authority

Mining companies need to carefully consider how their organizational structure and operational practices should be developed to best reflect and relate to the organizational structures, cultural norms, traditional activities, and concerns of Indigenous Peoples and local communities.

Mining companies need to carefully consider how their organizational structure and operational practices should be developed to best reflect and relate to the organizational structures, cultural norms, traditional activities, and concerns of Indigenous Peoples and local communities.

Many operational mines divide their basic functions into:

- mine production
- mine planning
- mine support and administration

These organizational divisions have an implied hierarchy of importance, and line responsibilities from the CEO/General Manager downward tend to reflect the relative importance of each.

An outcome of this traditional hierarchy is that some vital activities are relegated to the operational margins of the organization. In some cases this is entirely appropriate, while with others it can lead to difficulties.

Environmental management responsibilities have become so important a part of mine operations and performance compliance that in many mines the 'environmental branch' is placed with mine production, and responsibility for environmental performance directly placed with the CEO, or a senior mine director. The practice of making the company CEO directly responsible for company environmental performance was an organization innovation first adopted by Dupont in the early 1980s, and revolutionized the status and authority given to environmental management.

It is arguable that responsibility for socio-cultural performance, and the management of Indigenous Peoples issues, should follow a similar pattern. It is also arguable that as indigenous cultural values are often inextricably interwoven with land and environmental values, the two areas have equal status and should be closely coordinated in the mine management structure. It may therefore be advantageous to integrate these two functions, with the proviso that the individual with primary responsibility should be temperamentally and experientially suited to the task.

The reasons for placing responsibility for Indigenous Peoples' affairs at the highest possible level include:

- Indigenous Peoples will expect to have access to, and be able to negotiate directly with, an individual who has significant seniority in the company, who can make decisions on behalf of the company without constant referral to others;
- Indigenous Peoples will infer from the status of the designated company representative, the respect afforded to their community by the company; a low level of seniority will obviously indicate that their concerns are not considered of great importance by the company;
- Continuity of interaction with community elders through involvement of the same person over periods of at least several years.

The major debates concerning Indigenous Peoples, particularly as they are affected by the activities of international corporations and the outcomes of economic globalization, are occurring very much at international forums, even though events on the ground may be local and in remote locations. Mine CEOs and senior management are more likely to be informed of these broader international level debates and the initiatives taking place with respect to Indigenous Peoples' issues. They will also be more influential in interfacing with governments and representative groups such as Land Councils and NGOs.

Good relations with indigenous communities are the responsibility of the *entire* workforce.

Regardless of the desirability of involving the most senior managers, it is important for a company to require all workers to respect Indigenous Peoples and energetically observe the codes of conduct that apply. In other words, good relations with indigenous communities are the responsibility of the *entire* workforce.

16.8 PRESERVING OR RESTORING AUTONOMY: PARTNERING FOR THE LONG-TERM

Indigenous societies that have retained much of their traditional autonomy are at risk of losing it as a result of societal disruption, changes in values, and economic dependency. Companies interacting with such societies should consciously seek to strengthen the autonomy of these societies.

This can be achieved by:

- Communicating through indigenous leaders so that these leaders are always fully briefed and up-to-date on company plans;
- Seeking input from indigenous leaders on all matters affecting the Indigenous Peoples or their environment, and according with their wishes whenever reasonably possible;
- Avoiding actions that can lead to conflict, disrupt societal organization or weaken traditional values.

There are many indigenous societies which have already lost most or all of their former autonomy, having become completely dependent on government welfare or charity. Communities in this situation commonly have a strong interest in breaking out of the welfare cycle as a way of re-establishing independence, community pride, and self-respect (Pearce and Uren 1996). A mining project may provide an ideal opportunity to lessen dependency, and increase autonomy and self sufficiency.

In either of these situations, the company may, at the request of the community involved, assist in 'capacity building', by providing technical legal and financial advice and training, Such assistance should consciously avoid the relatively common 'pitfall' that authority is transferred from the indigenous leaders to the advisers, which defeats the purpose of capacity building.

There are many aspects of planning, operating, and closure of a mining project that are important in achieving the desired outcomes for Indigenous Peoples. The following sections address the major ones and offer ways to address them which are consistent with current effective practice. Whether or not a particular policy, procedure or activity should be implemented in a particular situation should be based on local circumstances and, most particularly, on the wishes of the Indigenous Peoples who stand to be affected.

Ethical Values and Attitudes

A clearly stated set of ethical values is the preferred basis for dealing with indigenous societies. Essential values are summarized as **HIRE** (after Cragg *et al.* 1997):

Honesty, including full disclosure of unpalatable or potentially deal-breaking information;

Integrity, including honouring all commitments;

Respect for individuals and for traditional customs, rituals, values, and rules of behaviour, however different these may be, and

Equity and fairness to all stakeholders, including involuntary stakeholders (i.e. those not party to a mining agreement established with the local community but who at some stage are affected by the operation)

Such values, communicated throughout a company and reinforced by consistent application, will go a long way towards ensuring effective practice.

It needs to be recognized that there are many people who have negative perceptions or prejudices towards Indigenous Peoples. Such people pose a real threat to relations between the company and the indigenous community. It is in the interests of all concerned that companies should: (1) develop screening programmes to minimize the chances of employing people with such prejudices in situations where they can interact with Indigenous Peoples; and (2) require dismissal or transfer of any employee who through his or her interactions with Indigenous Peoples demonstrates such prejudice.

There are many people who have negative perceptions or prejudices towards Indigenous Peoples.

A particularly difficult situation occurs in some multi-ethnic countries. The ethos within the ruling elite in some countries may be distinctly antipathetic to some minorities. The elite, socially advantaged group believes itself to be totally superior to Indigenous Peoples, who they may treat very badly and to whom they show no respect. It may prove impossible to avoid recruitment from the elite group as they may be the only qualified candidates, particularly for professional positions.

In this sort of situation, a company can only seek to minimize problems by:

- Disseminating, publicizing, and reinforcing the company's ethical values and any codes of conduct in relation to Indigenous Peoples;
- Ensuring that key individuals – those directly responsible for interaction with Indigenous Peoples – share the company's ethical values;
- Seeking to promote understanding through cultural awareness programmes;
- Monitoring workforce relations to identify unfair treatment of Indigenous Peoples and intervene accordingly;
- Taking account of ethnic and cultural differences in the organization and deployment of the workforce.

Communications, Relationships, and Trust

Communication is the key to interaction between mining companies and the community. In dealings with Indigenous Peoples, the risks of miscommunication and misunderstanding are much greater due to: (1) different languages; (2) different cultural values; and (3) different perceptions.

Most indigenous societies have an extensive oral tradition but little or no written tradition.

Communication through interpreters is essential in many cases and desirable in others, so that neither party is disadvantaged by language.

Most indigenous societies have an extensive oral tradition but little or no written tradition. Talking directly with Indigenous Peoples is preferable; however, a company will wish to maintain written records of important communications.

Communication through interpreters is essential in many cases and desirable in others, so that neither party is disadvantaged by language. The benefit of communicating complex or unfamiliar concepts such as large-scale excavation, construction, and rehabilitation by the use of images cannot be underestimated. Aerial, low level, and ground photography are readily understood by Indigenous Peoples; computer modified images are powerful in illustrating the impacts of mining, such that there is less likely to be surprise and dismay felt by the local people when construction and mining are underway.

Another important ingredient in successful interaction between a company and an indigenous community is the development of relationships between leaders and other key people from both parties. Developing such relationships, based on mutual respect and trust, usually requires multiple meetings over a considerable time. Relationships exist between individuals, not organizations. It is important for a company to recognize this fact because the regular turnover of key staff is a normal occurrence, particularly in remote and off-shore sites. It is a relatively common occurrence that social misunderstandings and other problems emerge soon after the departure of a key company representative who had developed and maintained key relationships with leaders of the local community. Relationships also need to be diligently maintained. Indigenous Peoples will lose both respect and trust if relationships are not maintained once an agreement has been negotiated.

A common scenario is for a rift to develop between a mining company and an indigenous community as a project moves from exploration to construction. The community may have been dealing with the same company representatives for several years during which strong relationships were developed. Then, suddenly, everything changes with the arrival of the construction workforce. At the time when the community needs reassurance from familiar people, management changes and the new manager is completely preoccupied with meeting construction deadlines. Such a situation should be avoided by the company, which should strive to provide continuity in key positions and to maintain regular contact with the local community. Alternatively, transition from one company representative to the next should take place over an extended period, enabling new relationships to be developed.

This 'transition sensitivity' again reflects a conscious acknowledgment of differing world-views. An indigenous community may become accustomed to the activities of a small exploration crew. To that crew, it is only natural to expect the arrival of a bustling construction crew. This transition is not necessarily expected or welcomed by an indigenous community.

Participation in Decision-Making

Some mining operations are legally required to involve Traditional Owners in decisions affecting the land. Other enlightened companies voluntarily consult Indigenous Peoples and seek their input on a wide range of issues. This is particularly important during the planning and design stages of a project when major decisions are being made. It is equally important towards the end of the mine life as various options for closure are being considered.

A major decision that should involve the indigenous community is whether to insulate from or integrate the project into the local community. **Table 16.1** lists the differences between the two approaches, either of which may be valid, depending on circumstances. The decision will be important to the company because of differing financial requirements. It may also be important to the government's plans for regional development. Above all, it will be of utmost importance to the indigenous community, and it is therefore vital that the ultimate decision is in accordance with the community's wishes.

A major decision that should involve the indigenous community is whether to insulate from or integrate the project into the local community.

Other decisions that should involve Indigenous Peoples include the siting of facilities; employment policies, particularly in relation to local recruitment; and measures for dealing with a large influx of job seekers

Insulation is usually the better option for projects with an operating life of five years or less, as there is little scope for major involvement of local, unskilled personnel in such ventures, and it is a realistic proposition to isolate the project from surrounding activities for such a relatively short period.

Longer term projects, particularly those with the potential to continue for 20 years or more, have the potential to deliver much greater and more sustainable benefits to indigenous communities. Importantly, such projects allow unskilled and semiskilled people to be

TABLE 16.1

Insulation versus Integration – Insulation is usually the better option for projects with an operating life of five years or less

Features of an Insulated Project	Features of an Integrated Project
• Fly-in/fly-out workforce	• Maximize local recruitment
• Encourage permanent, family involvement	
• Single status, barrack-style accommodation	• Open town, family accommodation
• Isolate workforce from local people to mine site	• House workforce within or near existing communities.
• Restrict access of local people to mine site	• No restriction on public access other than those required for public safety
• Separate utilities and infrastructure	• Share utilities and infrastructure
• Import goods and services	• Obtain goods and services locally
• Long working hours; little or no time for local recreation	• Normal working hours; normal time for recreation

Ideally, a mining project with the potential to continue for a generation or more, should be integrated so that the community views it as their own, regardless of whether there is any financial ownership by the community.

trained so that, over time, most of the jobs can be filled by local people. Equally importantly, such projects enable local communities to develop the capacity to supply goods and services to the project.

Ideally, a mining project with the potential to continue for a generation or more, should be integrated so that the community views it as their own, regardless of whether there is any financial ownership by the community. Such identification can be invaluable in reinforcing shared interests and, in particular, in providing political support to the company, if required. Local support is particularly important in the event of a mine accident or environmental incident.

Codes of Conduct

A mining company operating near an indigenous community should be prepared, if requested by the community, to implement a code of conduct for its non-indigenous workforce. The elements of such a code depend on the concerns of the community and include such matters as:

- Restricting access to community settlements;
- Prohibiting or restricting access to sacred sites, and
- Bans or restrictions on the possession or sale of alcohol and other drugs.

Involvement in Operations

Mining companies have five options available for involving indigenous communities in their operations:

Equity Partners – where communities own a share of the operations and benefit from operations, with dividends in proportion to that share;

Beneficiaries – where communities receive an agreed benefit from operations calculated by means of a formula which typically includes production throughput, market price, and price escalators;

Joint Venturers – where communities joint venture with the mining company or contractors to participate in operations and receive payment for their partnership in proportion to their joint venture contribution;

Contractors – where indigenous businesses contract to provide services for mine operations and receive payment for services rendered; and

Employees – where Indigenous Peoples are employed by the mine to participate in operations.

Various combinations are common. For example, a community may be a beneficiary and also provide employees. Financial arrangements between companies and Indigenous Peoples in Australia have evolved from simple royalty type arrangements to more complex arrangements as described in **Case 16.2**.

Capacity Building

Many indigenous societies lack the negotiating skills, confidence or experience to negotiate with a company on equal terms or to capitalize on opportunities that mining will bring. In some such cases governments, NGOs or churches may assist. In Australia, the Land Councils provide such assistance. Where assistance is required but not available, it is in the company's interest to provide arm's length assistance or, better yet, to assist the community in capacity building including strengthening its organization and leadership, and providing access to a range of sources of information and advice.

Recruiting and Training Programmes for Indigenous Peoples

Indigenous Peoples often have traditional and family responsibilities which may conflict with typical mine work schedules.

Because many indigenous communities are not used to the work environment and conditions experienced on a mining project, participation in operations needs to recognize a number of factors:

- Flexibility – Indigenous Peoples often have traditional and family responsibilities which may conflict with typical mine work schedules. Flexible employment conditions or the use of contract and small business service provider arrangements, with payment on completion of specific tasks, will contribute to more flexible and successful relationships between miners and indigenous communities. The Batu Hijau Copper-Gold Project provides an example of such a flexible arrangement whereby locally recruited employees work a roster of 4 days on and 4 days off. This enables local workers to continue their farming activities.
- Training – indigenous communities rarely have access to the education and work experience resources or backgrounds which other mine workers enjoy. Recruitment and training programmes need to identify community members with an interest in mine work, and assess them as being either job ready, semiskilled and ready for training, or unskilled and ready to learn work habits.
- Incentive-based – Indigenous Peoples respond to clear incentives which are structured to their capacity and aspirations. These incentives should be negotiated as part of the agreement which establishes a long-term relationship between the mining company and the indigenous community.
- Consistency – all contractors, operations managers, and corporate divisions must share a consistent approach to recruitment and training of Indigenous Peoples. All contractors

CASE 16.2
Evolution of Relationships between Companies and Australian Aborigines

In Australia, there have been three generations of relationships between mining companies and their indigenous communities. First generation relationships were those conducted to comply with regulations – such as those under the Commonwealth Northern Territory Aboriginal Land Rights Act 1976, where community members had little or no active involvement in operations, but received a share of mine profits or shared in mine returns in some other way.

Second generation relationships involve complex land transactions, enterprise agreements, and employment provisions – often negotiated with third parties such as Land Councils. In these circumstances, community organizations are often equity holders or joint venture contractors with a mine, but – like first generation relationships – community members have little or no active involvement in operations. Instead, indigenous organizations joint venture with non-indigenous companies and so Indigenous Peoples have limited active participation in operations. This is not necessarily bad since many Indigenous Peoples prefer it this way, but it makes for a different sort of relationship between a mining company and the indigenous community.

The third generation of relationships is negotiated between many stakeholder groups including indigenous organization, native title claimants, and regional community members. These relationships typically seek to include Indigenous Peoples as active participants in mine operations and the regional economy. This frequently involves multiple land use agreements and a capacity building process to ensure that indigenous communities establish some economic independence beyond closure.

Consistency from the Chief Executive Officer to the job foreman will improve the likelihood of success.

should be contractually committed to meeting corporate goals for indigenous relations. Consistency from the Chief Executive Officer to the job foreman will improve the likelihood of success.

- Staged progression – like other employees, many Indigenous Peoples aspire for management responsibility, economic independence, and self-employment. A staged recruitment and training process which clearly outlines roles and responsibilities as well as rewards, contributes to the long-term realization of goals. Again, contractors and major suppliers should also adopt this staged approach.
- Mentors – initially, indigenous employees are likely to constitute a minority of the site workforce, and the presence of indigenous mentors to support indigenous employees, will enhance the retention rate and personal development of indigenous employees.

As a result, it is common to see Indigenous Peoples restricted in their employment opportunities to jobs such as gardening, cleaning, labouring, and survey assistant.

Many, if not all, potential workforce recruits from indigenous communities will lack the immediate skills required to undertake most tasks. As a result, it is common to see Indigenous Peoples restricted in their employment opportunities to jobs such as gardening, cleaning, labouring, and survey assistant. This causes resentment, as the higher status, better paying jobs go to non-indigenous employees. This situation can be avoided, albeit at a significant cost to the company, by implementing **job readiness programmes** whereby selected individuals are trained to undertake various occupations well before these operations commence. Larger mining companies with multiple mine sites are able to organize on-the-job-training at existing operations. However, small companies may not be able to organize such training and therefore will depend on either bringing in instructors or sending recruits to existing training facilities.

16.9 PROJECT PREPARATION

Anthropological Surveys

Anthropological surveys can be carried out at any stage of a project. However, in general, the earlier a survey is undertaken, the better, as the results will provide valuable insights which should lead to improved communications. As baseline data are established, better social planning is possible and the data can serve as a future reference for measuring change. Anthropological surveys are a form of social research and an important means of gathering information about people. The results of these surveys contribute to understanding aspects of a society, whether these are general aspects or specific characteristics. Some variables include a community's previous experience with change, the beliefs and customs which prevail, specific expectations, and responses to current or past projects.

There are various components to anthropological surveying. Generally, these are simple questionnaires, observations, guided or in-depth interviews, and focused group discussions. These are all methods for obtaining data about a community or group in society. Anthropological surveys are often coordinated with other related studies such as socio-economics and archaeology. The information from these disciplines usually complements and completes the community profile obtained through anthropological surveying. Survey methodology is more than 'where', 'when', and 'how many people'. Sampling frameworks plan and account for differences in conditions and responses between groups within a community. Anthropological surveys must be representative. A full description of the survey technique as well as the data is important, as design and methodology will affect the survey's quality and usefulness.

Because surveying makes the community directly aware of a project, the way in which the survey is delivered is of special importance.

It should be recognized that anthropological surveying has an element of public involvement. Because surveying makes the community directly aware of a project, the way in which the survey is delivered is of special importance. The approach used by an anthropological surveyor can greatly affect attitudes and expectations as it is part of the experience an individual or community has with a project or company. Time is required for interested, informed, and sensitive people to successfully carry out a social survey. The process is as important as the product.

Identifying Stakeholders

The process of locating Traditional Owners or establishing Ancestral Domain, may be very difficult and time consuming, particularly if there are conflicting or overlapping claims. Accordingly, the process should be initiated as early as possible, once the presence of potentially economic mineralization has been established. This applies whether or not there is a legal requirement (traditional ownership rights of Indigenous Peoples are not recognized in all countries).

Where mining requires use of land held by Indigenous Peoples, the most important stakeholders will be the Traditional Owners. Identification of Traditional Owners may be simple or difficult depending on location and circumstances. In Australia's Northern Territory, it is the responsibility of the relevant Land Council to identify Traditional Owners.

Where mining requires use of land held by Indigenous Peoples, the most important stakeholders will be the Traditional Owners.

In some areas, identification of Traditional Owners may require extensive research. In parts of Papua New Guinea, individual land-holdings may be very small, with the result that there may be hundreds or thousands of Traditional Owners for one mining project. In these situations identifying Traditional Owners becomes one of the biggest and most difficult tasks involved in developing a mine. In some countries such as Indonesia, most land is owned by the State. However, various categories of land tenure are recognized, based on formal and informal land use.

Apart from the Traditional Owners, stakeholders will include other members of the indigenous community and, possibly, other neighbouring communities. It is important that members of communities at risk of being affected over the life of the mine such as downstream communities if there is potential for river degradation or reduced flow are included in the stakeholder identification process. There may also be non-indigenous stakeholders which, in Australia might include pastoralists and tourism operators. In developing countries of the Asia-Pacific they might include migrant groups, logging companies, missionaries, plantation companies, trading companies, artisanal miners, and government administrators.

The nature of mining developments from mineral exploration through to operation and ultimate closure means that identifying stakeholders is an ongoing process. As the project moves towards the development phase the number of stakeholders increases. Furthermore, as there is a tendency for migration to occur into the development area, these migrants also become stakeholders (notwithstanding that the local community may hold strong views regarding the level to which their rights should extend to the new migrant group).

The nature of mining developments from mineral exploration through to operation and ultimate closure means that identifying stakeholders is an ongoing process.

Informing Stakeholders

All stakeholders need to be accurately informed of company plans. This is important so that stakeholders can plan their own activities, and to avoid information gaps which will inevitably be filled by inaccurate rumours. The aim should be to avoid surprises, particularly unpleasant ones. This requires a conscious effort to avoid creating unrealistic expectations.

Mechanisms for informing stakeholders depend on circumstances. Perhaps the best method involves regular visits to individual stakeholders with information provided through one-on-one briefings. Where the number of stakeholders is too large for this approach, information can be provided through representatives of each stakeholder group. In the case of a single indigenous community this may be one or two community leaders. In Australian aboriginal communities, the Land Councils help to disseminate information.

Most Indigenous Peoples have never seen a mining operation and can have little appreciation of its nature or scale.

Many companies work through Community Liaison Groups which are established with the dual purpose of informing the stakeholders and providing a forum for feedback to the company. Each member represents a different stakeholder group. The company should help form such liaison groups and pay any expenses involved; however, the members should be selected by the stakeholders whom they will represent.

From the time when it becomes apparent that there is real potential for a mining operation, it is very important to supply accurate information and make sure that it is fully understood. Most Indigenous Peoples have never seen a mining operation and can have little appreciation of its nature or scale. Care should be taken to explain the downside of mining, as people unprepared for sudden change may otherwise react with outrage when confronted with reality. Care should also be taken not to exaggerate the benefits, as unfulfilled expectations are another source of outrage. Again, computer modified photographs are very helpful in getting the story across, including warts and all depictions.

Taking community representatives on visits to operating mines can be a valuable means of informing people on the nature and magnitude of mining.

Taking community representatives on visits to operating mines can be a valuable means of informing people on the nature and magnitude of mining. However, this applies only when operations visited are reasonably similar to the planned operations.

Negotiations

For negotiations to succeed, trust and understanding must be developed between the indigenous community and the mining company. When negotiating with an indigenous community, a mining company should identify its aspirations and the assets it is prepared to trade. It should also predict what assets and aspirations the indigenous community is likely to bring to the negotiation. A clearly defined and realistic negotiation framework for regional development, rather than an open-ended and generic settlement, enables mining companies to provide what they are prepared to deliver without compromising their long-term core interests. Maintaining the flexibility to change this framework is obviously important. However once agreed, the arrangement can only be renegotiated with full participation of both parties. A company should appreciate that during the life of the mine the community's expectations may change. This should be seen as legitimate, as community issues will obviously evolve over the typical mine life and as the inevitable cultural changes take hold and a full appreciation of the character of the mine and its effects develops in the community. Clearly, the more fully the impacts of the mine are communicated, and the more effectively the negotiations recognize and anticipate these changes, the less the probability will be of having to renegotiate an agreement during the mine life. A suggested negotiating framework is summarized in **Table 16.2** and **Figure 16.1**.

A company should appreciate that during the life of the mine the community's expectations may change.

Governments also have an interest in negotiations between mining companies and indigenous communities. These interests include:

- Political considerations relating to public and international perceptions of respect for indigenous rights and sustainable development;
- Economic considerations relating to market perceptions of investment risk and government commitment to development, and
- Institutional considerations including the relationship between state and national legislation in controlling local political and economic interests.

TABLE 16.2

Negotiating Framework for Developing Long-term Relationships between Miners and Indigenous Communities

Miners Want	Miners Prepared to Give Up
Security of access to minerals and associated resources such as water	Sole occupancy of titles
Security of right to extract and process minerals and transport them to port or market	Focus on holding underlying title as the only means of ensuring security to minerals
Surrender of indigenous intrusive rights	
Positive publicity and community support	
Security of title to minerals and essential resources such as water	Exclusive title to non- essential resources such as land
Indigenous Communities Want	**Indigenous Communities Prepared to Give Up**
Recognition of indigenous rights and obligations	Intrusive rights to delay or disrupt exploration mining activities
Economic independence	
Employment, training and enterprise	
Opportunities	
Land access	Sole occupancy of land
Preservation of culture	

Figure 16.1 shows the interrelationship among indigenous communities, government, and mining companies and sets out the mutual benefits which can accrue. **Tables 16.3** and **16.4** summarize the likely assets and aspirations between mining companies and indigenous communities in negotiations.

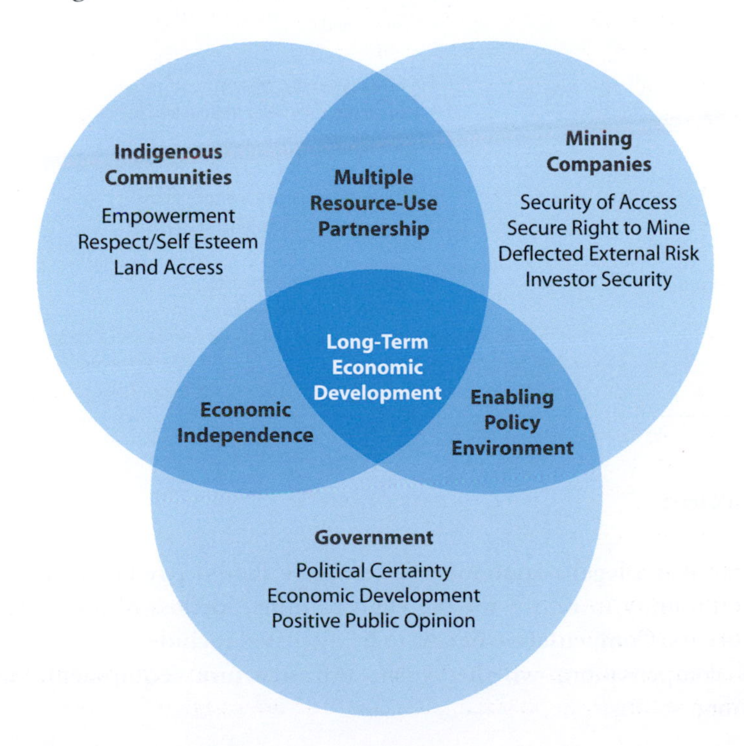

FIGURE 16.1

Negotiation Framework for Long-Term Relationship between Miners, Indigenous Communities, and Government

TABLE 16.3

Typical Assets Available for Negotiation

	Mining Company	Indigenous Communities
ACCESS	• Business knowledge • Legal tenure • Capital and equipment • Training resources • Economic stability	• Workforce with variable education levels • Knowledge of country and its ecology • Young workforce
OWNERSHIP	• Properties and other land • Financial management and corporate government knowledge	• Native Title • Indigenous rights
RIGHTS	• Consultation and representation • Inclusion in negotiations	• Consent to major land use changes • Indigenous rights
OBLIGATIONS	• Payment of resource rent • Open communication	• Open communication • Maintenance of country and culture
ECONOMIC VALUES	• Competitive tendering process • Financial credibility • Capital	• Growing community • Access to government training and employment programmes • Growing economic development organizations

TABLE 16.4

Typical Aspirations of Companies and Communities

	Mining Company	Indigenous Communities
ACCESS	• Secure access to minerals, water and associated resources	• Secure access to land and water resources
OWNERSHIP	• Confirmed rights	• Share in resource rents • Sacred Sites • Title to land
RIGHTS	• Access to resources • Protection from sovereign risk • Ability to add value	• Co-management of land • Participate in environmental and infrastructure planning and management
OBLIGATIONS	• Investor security • Risk management	• Maintenance of culture and lands • Development of young people
ECONOMIC VALUES	• Competitive access to resources	• Protection of sacred sites • Economic independence • Contract supply of goods and services to regional economy • Negotiation of socio-economic agreements to plan and manage • Corporate and community economies

Whether or not it is a legal requirement, a company should pay fair compensation to an indigenous community in return for the right to mine, for loss of amenity, and for any damage that occurs.

Compensation

Whether or not it is a legal requirement, a company should pay fair compensation to an indigenous community in return for the right to mine, for loss of amenity, and for any damage that occurs. Compensation details to be discussed include:

- Form of compensation, whether cash, infrastructure, equipment, services or a combination;

- Recipients of compensation whether individuals, communities, representative bodies or a combination, and
- Schedule of payments.

Cash payments to Indigenous Peoples can lead to undesirable social impacts particularly in communities unused to the cash economy. More lasting value is usually obtained from other forms of compensation such as the provision of equipment and infrastructure. However, it is common for Indigenous Peoples to insist that compensation be made in the form of cash. In such cases, cash compensation is usually more effective at achieving beneficial outcomes if paid progressively, e.g. royalty payments and land rental payments.

It is common for Indigenous Peoples to insist that compensation be made in the form of cash.

Many mining companies opt to contribute to the development of local indigenous communities at levels well beyond the compensation stated in the formal agreement. This is particularly so where initial commitments have been at deliberately modest levels so as to avoid the risk that they could not be afforded. In disbursing discretionary funds, the company is faced with a dilemma. On one hand it does not wish to be patronizing, to tell the community what it needs. On the other hand, it has a valid interest in successful outcomes, not only from the viewpoint of community relations but through the eyes of the critical outside world.

Some companies create foundations to manage the disbursement of discretionary funds. Others negotiate agreements that involve foundations or trusts administering royalties and other legally prescribed payments. Foundations have the following advantages:

- They enable the company to operate at 'arm's length', thereby removing the perception of self-interest.
- They can be more readily administered by people with the appropriate background and skills. Accordingly, they are less likely to lead to dependency.
- They may attract additional funding from other sources.

Another approach to the 'arm's length' disbursement of funds for community development is to use appropriately qualified and experienced NGOs. This arrangement has similar benefits to the use of a foundation.

Addressing Social Change

Indigenous societies will change whether or not there is a mining project in the vicinity. Many changes are welcomed by the communities. A mining project can accelerate the change so that the communities can no longer adapt, and therefore lose control. On the other hand, with appropriate planning and commitment, a mining project can deliver the resources that will help the communities to adapt without losing control.

Indigenous societies will change whether or not there is a mining project in the vicinity.

An impact assessment process which addresses social issues will identify likely changes and provide measures to ameliorate or control any related adverse impacts. Effective practice requires local communities to have detailed input to the environmental studies and selection of improvement and control measures. Accordingly, they need to be informed, before the event, of the impacts that may occur and the possible means of mitigation or control.

In the past, this ideal situation has been rare. More commonly the local community has not been prepared for the sudden changes that occurred when the project moved to the construction stage. In some cases this has led to outrage. In other cases, the community has been initially excited by the changes which were seen as signs of progress after years of expectation. As the excitement declines, however, the impacts become apparent. Unforeseen social impacts may emerge at this stage. It is therefore important to have an established consultative mechanism such as a Community Liaison Group where emerging issues and problems can be discussed and resolved.

To identify long-term changes it is valuable to conduct regular anthropological surveys at intervals of several years.

Some regulatory systems require monitoring and reporting of socio-economic and socio-cultural impacts as part of the environmental reporting process. Those parameters that can be readily monitored such as health and medical statistics, demographic indicators, crime rates, and workforce participation rates, while useful, do not gauge cultural impacts on indigenous societies. Major impacts such as loss of traditional values may take place over several generations. These may not be evident, even to community leaders. To identify long-term changes it is valuable to conduct regular anthropological surveys at intervals of several years.

Community Development Programmes

Successful community development programmes are based on clear and mutually agreed goals (see also Chapter Fifteen). The goals need to be supported by a set of principles which define a framework for on-going refinement of cooperation and settlement of disputes. Principles appropriate to a community development programme for indigenous communities supported by mining companies could include:

- Recognize each others rights and obligations as a basis for development;
- Use participatory processes to make decisions and identify priorities for development;
- Use negotiation to prepare and implement the development activities;
- Build on existing successes to develop a programme of development activities which form an imaginative symbol of the long-term relationship;
- Support discrete projects developed and implemented in partnership rather than hand out cash or other resources;
- Use a project investment process to maintain flexibility and communication in the development relationship;
- Make use of complementary government and aid programmes.

Community development activities need not be directly associated with the mine.

Community development activities need not be directly associated with the mine. Activities such as maternal health clinics and agricultural development may contribute more to building a relationship than mine-related activities. They are also more likely to contribute to a sustainable community with viable activities lasting beyond the life of the mine.

Table 16.5 lists some typical community development initiatives provided by mining operations to indigenous communities. For a more comprehensive listing see Chapter Fifteen.

Many of these initiatives, such as those directed at improving health and education, provide direct benefits to both the company and the community. While health and education programmes are common to most projects, the actions will differ according to circumstances, particularly the wishes of the recipient communities.

Of particular importance are those programmes that broaden the economic base and that provide ongoing employment and income-generating opportunities beyond the life of the mining operations. It is essential that practical realities are recognized in the design and implementation of such programmes. For example, there is no ongoing benefit in developing an industry to supply a particular product if, after mine closure, there is no longer a market for this product.

16.10 IN OPERATION AND CLOSING DOWN

Monitoring Effectiveness of Programmes

As discussed in relation to addressing social change, the effectiveness of some programmes can be readily monitored; others cannot. Those that can be monitored include:

- Effectiveness of training programmes;
- Workforce participation rates;

TABLE 16.5
Examples of Community Development Initiatives

Provision of materials and equipment for new education facilities – actual construction undertaken by local people.
Provision of health facilities, personnel, equipment, and medicines.
Provision of transportation equipment, providing better access to markets.
Introduction of new agricultural or horticultural crops, livestock, etc.
Establishment of fish farms.
Training in pottery or other arts and crafts, plus supply of equipment.
Training in business management and marketing.
Assistance in establishing ecotourism ventures.
Provision of sewing machines and weaving looms and training in their use.
Provision of expertise, materials, and equipment for irrigation projects.
Provision of refrigeration equipment, facilitating production of fish, meat, etc beyond subsistence needs.

- Effectiveness of community health programmes;
- Effectiveness of community development programmes;
- Status of preparations for closure;
- Socio-economic statistics relating to health, education, crime, etc.

In addition, surveys can be undertaken to evaluate community attitudes, including attitudes relating to the mining operation.

Preparing for Mine Closure

Serious adverse social effects can accompany mine closure. When a mine closes, the majority of the non-indigenous workforce usually relocates. However this may not be an acceptable option for most indigenous workers who will wish to remain on their ancestral lands within their own community. The community may then be faced with high unemployment levels at the same time that royalty payments and other sources of income cease. The lifestyle habits and expectations will also have changed so that the community cannot be sustained at this level on pre-mining economic arrangements (see also **Case 13.12**). In a worst case scenario, traditional subsistence skills will have declined during the mining period and the people may no longer be able to return to a traditional lifestyle. Preparation and planning to avoid this outcome needs to take place well before closure occurs.

The community may be faced with high unemployment levels at the same time that royalty payments and other sources of income cease.

All mines must inevitably close. In most, but not all cases, the time of closure can be reasonably well predicted, enabling the company to work together with the community, and with government, to prepare for closure.

In most, but not all cases, the time of closure can be reasonably well predicted, enabling the company to work together with the community, and with government, to prepare for closure.

For most communities there are likely to be three important requirements:

- Rehabilitation of the land to a final form that is acceptable to the Traditional Owners. This may or may not involve future use of the land for commercial purposes.
- Continuation of an income stream to enable the community to be independent. This requires that money received during operations be prudently invested to provide an ongoing return.
- Development of alternative livelihoods for the indigenous workers. As mentioned in the previous section, this is an outcome of effective community development programmes.

Directions for measures to mitigate the social impact of closure should come from the community, with government input and oversight, as appropriate. The company's role is to encourage and assist, and to provide its best available predictions of mine life and schedules for reductions in the workforce.

Clearly, where Indigenous Peoples have been involved in supervisory or management positions, they and their communities will be better equipped to undertake new ventures after mine closure. Even better is where indigenous groups organize themselves into contracting businesses, performing work at the mine. These businesses are then able to undertake contracts for other customers.

16.11 CONCLUSIONS

Mining in a region occupied, used or valued by an indigenous community will inevitably create pressures that can contribute to significant changes in that society. Early planning and effective consultation are required to ensure that this does not occur. Any 'changes' should come solely at the request of, or should be understood and accepted by, the affected community.

Mining is not necessarily incompatible with the values and aspirations of Indigenous Peoples.

Mining is not necessarily incompatible with the values and aspirations of Indigenous Peoples. On the contrary, it may provide the opportunity for indigenous societies to regain a sense of independence and a strong, or stronger, degree of economic self-reliance.

Indigenous Peoples have, or should have, the right to decide whether, and under what conditions, mining can take place on their land. They have a right not only to be fairly compensated for any losses, but to benefit from the operations.

In the end, after mining ceases, it is likely to be the Indigenous Peoples who will re-occupy the land and be the beneficiaries of its final form and condition. This may mean restoration of the land to a condition consistent with its original land use. Or, it may involve rehabilitation to suit one or more different uses as determined in consultation with the relevant stakeholders.

The challenge for the mining industry is to plan, operate and close in a way which is understood by the indigenous community, acceptable to them, offers them meaningful levels of involvement in terms of decision-making and employment, and is beneficial to the community in a collective sense throughout the project.

The challenge for the mining industry is to plan, operate and close in a way which is understood by the indigenous community, acceptable to them, offers them meaningful levels of involvement in terms of decision-making and employment, and is beneficial to the community in a collective sense throughout the project. Rather than mine closure being seen as the 'end of the good times', the company must share the responsibility with the community and its government, of establishing a sustainable socio-economic infrastructure robust enough to avoid collapse of the community's welfare after the company withdraws from the area.

Many governments require that companies provide legally binding financial guarantees or accrue funds for use in mine closure. This is to protect against the situation in which a company becomes insolvent or is wound up once operations cease – a common situation in the past, particularly among single project companies. These arrangements take various forms as discussed in Chapter Twenty-One.

As mentioned elsewhere in this book, effective environmental management is only likely to be achieved if the project is appropriately profitable so that the company involved becomes and remains financially robust. While the posting of bonds and accrual of funds for closure help to protect against sudden mine closure due to bankruptcy, this is no substitute for continued involvement by a committed company through the entire closure process.

REFERENCES

Bharat B Dhar (1996) Environment and Sustainable Development in Developing Countries with Special Reference to Mining Industry

Casagrande E (1996) Sustainable Development in the New Economic (Dis)Order: The Relationship Between Free Trade, Transnational Corporation, International Financial Institutions and Economic Miracles. Sustainable Development, Vol. 4, 121–129.

Casella FC and Syme GJ (1995) Relationships between the Mining Industry and Aboriginal Communities: A Qualitative Analysis of Stakeholder Opinions. DWR Consultancy Report No. 95–38.

Chogull C (1996) Ten Steps to Sustainable Infrastructure in Habitat International Vol. 20, No. 3, pp389–404, Elsevier Science.

Clark A (1996) Emerging Challenges and Opportunities for Minerals Industry in the 21st Century, 5th Asia Pacific Mining Conference, Jakarta, Indonesia.

Cook P and Kirkpatrick C (1997) Globalization, Regionalization and Third World Development. Regional Studies, Vol. 31.1. pp55–66.

Cragg W, Pearson D, and Cooney J (1997) Ethics, Surface Mining and the Environment, Mining Environmental Management, March 1997.

Department of Resources Development Working With Aboriginal Communities ii A Practical Approach. Department of Resources Development, Western Australia. Prepared by Ken Hayward and Michael Young, (undated).

Flannery T (2002) The Future Eaters, Grove Press.

Gana UA (1997) Problems in New Mining Regions, Outlook, No. 473/Business News 5992/4.4.1997.

Kim YW, Masser l, and Alden J (1996) Urban and Regional Development Strategies in an Era of Global Competition, Habitat International Vol. 20, No. 4, pp vii–viii, Elsevier Science Ltd.

Kepui T (1996) The Role of Culture in Development Management: Case Studies of a Public and Private Sector Organization in Papua New Guinea. Sustainable Development, Vol. 4, 111–120.

Machribie A, Murphy P, and Marsh B (1996) Environmental and Community Development Programs (Environmental and Social Audits) at P.T. Freeport Indonesia. Paper to the 5th Asia Pacific Mining Conference, Jakarta, Indonesia.

Labat-Anderson Inc (1997) Final Social Audit Report for P T Freeport, Indonesia. July 1997.

McKeon E (1997) How the Mining Industry Can Promote Community Development

In Making the Grade, The Mineral Industry and Aboriginal Community Development, Canola Forum, Nov 20–21, 1997.

Milne G (1996) Defining the Features of Sustainable Development for a Developing Country. Comments and Debates pp295–298, Journal of Environmental Planning and Management, 39(2).

Nash J (1979) We eat the mines and the mines eat us: Dependency and Exploitation in Bolivian tin mines. New York, Columbia U.P.

Noakes M (1996) RTZ-CRA in Indonesia: Building a Long Term Relationships. Paper to the 5th Asia Pacific Mining Conference, Jakarta, Indonesia.

Pearce D and Uren C (1996) Best Practice in Doing Business with Aboriginal Peoples. National Engineering Conference Darwin, April 21–24, 1996.

Robinson K (1986) Stepchildren of progress: The political economy of development in an Indonesian Mining Town. Albany: State University of New York Press.

Schumacher E (1974) Small is beautiful: A study of Economics as if People Matter edi. Abacus. London.

Togolo M P (1997) Mining Relationship Building? A Papua New Guinea Perspective.

••••

17

Acid Rock Drainage

A Widespread Problem

TITANIUM

Titanium is a light, strong, gray-white metal with notable corrosion resistant properties. It has the highest strength to weight ratio of any metal. Accordingly titanium metal has many uses including in sporting equipment, artificial joints and a variety of alloys used in the aerospace industry. Its main compound, titanium dioxide is used in the manufacture of paper and plastics and as a pigment and opacifier in most paints; titanium dioxide pigment accounts for more than 90% of the consumption of titanium minerals. Titanium is found in the minerals rutile and ilmenite which occur in igneous rocks, from which the eroded grains may be concentrated by wind or water to form heavy mineral or mineral sand ores. Both chloride and sulphate processes are used to extract titanium from ilmenite. Titanium is not an essential element for humans, and is non-toxic.

17 Acid Rock Drainage

A Widespread Problem

Many of the most difficult and expensive environmental problems faced by mining companies result from Acid Rock Drainage (ARD). Sulphide minerals such as pyrite and pyrrhotite are commonly associated with many ore types, and it is the oxidation of these minerals and subsequent leaching of their acidic oxidation products including sulphuric acid, that leads to ARD. Discharge of acid solutions from mine workings is usually known as Acid Mine Drainage (AMD). As if generation of acid itself is not of sufficient concern, the potential for widespread and possibly far-reaching environmental damage is aggravated in many situations, by the presence of 'heavy metals' such as manganese, copper, arsenic, and zinc which may dissolve in acidic solutions, thereby becoming mobilized and, potentially, 'bio-available'.

There is the potential for ARD to develop at most operations in which the ores or surrounding waste rocks contain sulphide minerals such as pyrite.

There is the potential for ARD to develop at most operations in which the ores or surrounding waste rocks contain sulphide minerals such as pyrite. Examples of such operations include most gold mines, most major copper mines and many mines producing silver, lead, zinc, or nickel. Many coal mines are also subject to ARD, due to the presence of pyrite within the coal, overburden, or interburden. ARD is not associated with projects where only oxidized ores are mined. These include lateritic mining operations for bauxite or nickel, iron and manganese mines, and many near-surface gold and copper operations. Some uranium deposits are associated with sulphide minerals (e.g. the Rum Jungle deposit in Australia's Northern Territory), while others are not.

ARD may occur in or may be derived from many parts of an operation, including:

- Runoff from surface excavations for access roads, drains, and site facilities;
- Drainage or seepage from underground excavations for mine access, exploration, development, ventilation, and extraction;
- Runoff from open pit mines – exposures in pit walls, berms, and the mine floor;
- Seepages of contaminated groundwater into surface or underground mines;
- Percolation through and drainage from rock masses fragmented by block caving or subsidence;

- Runoff and/or drainage from stockpiles of ore awaiting processing;
- Runoff and/or drainage from waste rock storages including coal rejects piles;
- Seepage or overflow from tailings storage facilities, and
- Runoff from spillages or remnants of mineral concentrates around stockpiles, bins, conveyor transfer points, etc.

It appears that nowhere in the mining world is immune from ARD. While moisture and temperature are essential for the reactions involved, sufficient moisture exists virtually everywhere and the reactions, which may be slow to initiate at low temperatures, generate their own heat so that oxidation rates can rapidly accelerate. Severe examples of ARD have been recorded in arid and temperate Australia, in the Canadian Arctic, and in equatorial Indonesia.

It appears that nowhere in the mining world is immune from ARD.

Considerable progress has been made over the past 25 years in developing of methods for predicting the nature, severity, and duration of ARD. Parallel developments in environmental management include measures to prevent, isolate, and treat ARD. Despite the considerable progress that has been made, management of ARD presents major challenges at many mining operations throughout the world and accounts for a major part of day-to-day environmental management and monitoring costs. It is also a major issue, sometimes the dominant issue, in mine closure. More exhaustive discussion of these topics can be found in Environment Australia (1997), Hutchison and Ellison (eds) (1992) and MEND (2001). The MEND Manual is a particularly comprehensive and authoritative reference, covering all aspects of ARD including its prevention and remediation.

17.1 NATURE AND SIGNIFICANCE OF ACID ROCK DRAINAGE

A Natural Phenomenon

ARD is, in fact, a natural process, which takes place wherever sulphide minerals such as pyrite are in contact with oxygen and moisture. Sulphide minerals are not only associated with economic mineral deposits. Pyrite is widespread in nature, occurring commonly in sediments such as organic clays, silts, and peats which deposit in reducing environments. These sediments lithify to become shales, claystones, siltstones, and coal, all of which commonly contain pyrite, either disseminated or concentrated in nodules. Metamorphic rocks such as schist and phyllite also frequently contain pyrite.

ARD is, in fact, a natural process, which takes place wherever sulphide minerals such as pyrite are in contact with oxygen and moisture.

Oxidation is the predominant process involved in surface weathering which, accordingly, takes place mainly in the unsaturated zone above the water table. Weathering of bedrock usually produces a profile with highly or completely weathered rock near the surface grading to fresh or unweathered rock at depth. Rock below the water table is permanently saturated and is usually unweathered except for slight weathering along fractures. As water tables tend to fluctuate seasonally, there may be a transitional zone of incomplete weathering between maximum and minimum water table levels. Ore bodies which cross these boundaries may be divided into a primary sulphide zone where sulphide minerals occur in an essentially un-oxidized condition and an oxide zone where sulphide minerals have been largely or completely decomposed by oxidation, commonly separated by a transition zone. These features are illustrated in **Figure 17.1**.

It should be noted that the existing weathering profile may reflect past water table locations, not necessarily those that occur under current conditions.

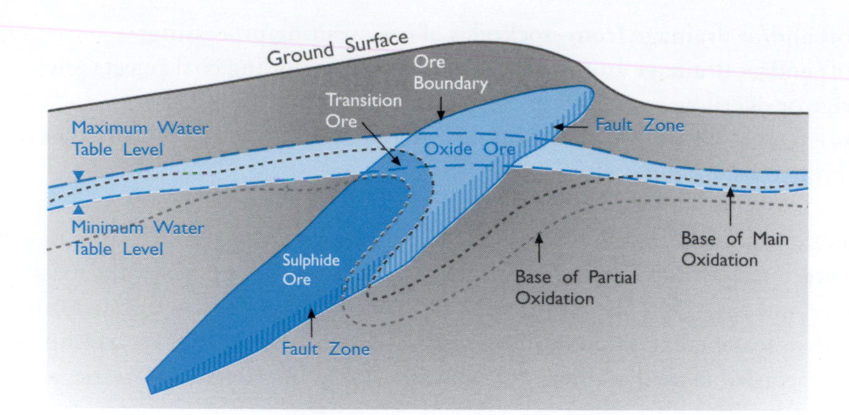

FIGURE 17.1
Typical Weathering Profile with Corresponding Ore Types

Natural groundwater in contact with sulphidic mineralization is commonly acidic, particularly in the unsaturated zone where oxygen is also present; such water may lead to acidification of surface streams. For example, near the discovery outcrop of the Batu Hijau copper porphyry ore body on the island of Sumbawa in Indonesia, there was a perennial, spring-fed stream known as Air Merah (meaning 'red water') with a pH of 4 and anomalously high concentrations of manganese and copper.

More commonly, the rate of natural oxidation is low, and the acidic groundwater may become diluted and/or neutralized by the time it reaches the surface drainage system. In such cases, the only evidence that sulphide oxidation has taken place will be the relatively high concentration of dissolved sulphate.

Development of ARD is a three stage process involving oxidation; leaching; and drainage.

Development of ARD is a three stage process involving oxidation; leaching; and drainage.

Oxidation

Pyrite, pyrrhotite, and some other sulpide minerals react with oxygen in the presence of water, according to the following reaction (for pyrite):

$$FeS_2 + 15/4\ O_2 + 7/2\ H_2O \longrightarrow Fe(OH)_3 + 2\ H_2SO_4 \qquad (17.1)$$

The reaction is complex, involving chemical, electro-chemical and, frequently, microbiological processes. The oxidation rate is highest at low pH where it is catalyzed by the bacterium *Thiobacillus ferro-oxidans*. Commonly, oxidation proceeds slowly for months or years with steady decline in pH until the pH falls declines below about 3.6, when there is a sudden acceleration in the rate of oxidation. The sulphuric acid formed is itself highly reactive and will react with adjacent minerals, dissolving metals and forming a variety of sulphate salts. For example, if calcite is present, gypsum ($Ca_2SO_4 \cdot 2H_2O$) may form. A variety of more or less acidic compounds such as jarosites are commonly formed, which may crystallize to fill cracks and pores in the rock or form coatings, encrustations or efflorescences on rock surfaces. These hydrated sulphates of iron usually combine with varying proportions of other cations, and are usually readily soluble in water.

Leaching

Oxidation of sulpide minerals in itself, does not necessarily pose a problem.

Oxidation of sulphide minerals in itself, does not necessarily pose a problem. It is only after the acidic oxidation products are flushed from their source or dissolved in water (the process known as 'leaching') – with the resulting leachate solution discharged to the environment – that adverse effects may occur. Most leaching occurs in response to rainfall, although the appearance of ARD may occur some time after the responsible event. Leaching of acid and salts in an open pit occurs as rainfall runoff flows over pit slopes.

Similarly, rain falling on a pile of oxidizing waste rock or tailings infiltrates and dissolves salts as it percolates by gravity flow through the pile. Particularly in arid and seasonally dry climates, ARD often occurs as a series of pulses, each the result of a specific rainfall event. The highest acidities and highest concentrations of salts tend to result from the first rainfall following a prolonged dry period.

Drainage

The most common drainage pathways by which oxidation products are transported to the receiving environments are:

- Dewatering of surface and underground mines to remove groundwater which commonly includes ARD derived from contact with the ore body or its surrounds. Water pumped or drained from operating mines may be used as process water, discharged to evaporation ponds (a common practice in arid Australia), or discharged to the environment. In this context it is also important to recognize that mine dewatering may expose acid generating rock to oxidation. Groundwater rebound after mining ceases may flush oxidation products into the drainage system (e.g. into the pit lake);
- Rainfall runoff into surface mines is handled similarly to groundwater, with which it may be mixed;
- Seepage from waste rock dumps or ore stockpiles, which commonly emerges from the toe at the base of the dump but may also emerge from internal percolation barriers;
- Seepage from the base of tailings storages, and
- Overflow from tailings storages or evaporation ponds.

Capillary rise represents another pathway by which oxidation products may reach the surface. This mechanism commonly occurs on the surface of tailings storages, and may also occur on waste rock dumps where fine-grained soils occur at the surface.

Effects of ARD

Much of the acidity resulting from sulphide oxidation associated with mining activities may, by design or natural circumstances, remain isolated from the biosphere, in which case it may be of no concern. Examples are found in many underground mines and some surface mines where the acidic solutions are trapped at depth with no drainage path to the surface. In other cases, however, such solutions may be in contact with aquifers which do eventually discharge to the surface through seeps or springs, or through water supply bores. In these cases, the acidic solutions may not emerge for many years following the end of mining. As in natural systems, neutralization and dilution may remove much or all of the acidity in the meantime.

Effects on Operations

Sulphide oxidation, with or without subsequent leaching and drainage, may adversely affect mining operations in numerous ways, such as:

- Corrosion of concrete foundations, culverts, metallic pipes, and walkways;
- In some open pit situations, such as the North Davao Copper Mine in the Philippines, sulphur dioxide fumes may be generated from the pit walls, which in the absence of ventilating winds, may settle in lower parts of the pit, presenting a hazard to the operations workforce. A similar situation has occurred at the Mt Newman Iron Ore Mine in Western Australia where pyrite nodules which are widespread in the black shale interburden, oxidize rapidly upon exposure;

Sulphur dioxide fumes may be generated from the pit walls, which in the absence of ventilating winds, may settle in lower parts of the pit.

- Corrosion, leading to failure of metal rock bolts, resulting in slope or roof collapse. In an underground nickel mine at Kambalda, Western Australia, oxidation of pyrrhotite in a warm, humid environment, caused corrosive failure of rock bolts within two months of installation.
- Physical disintegration of rock due to crystallization of sulphates in micro-cracks or inter-granular pores. This process has the potential to cause damage to structures such as retaining walls or rock-fill dams constructed of rock containing sulphides.

Ecological Effects

The most serious effects of ARD occur in freshwater aquatic ecosystems.

The most serious effects of ARD occur in freshwater aquatic ecosystems. While ARD may kill most terrestrial organisms in its path, the effects do not extend beyond the contact zone, which is generally confined to one or more flow paths (**Figure 17.2**) rather than spread over broad areas. In rivers, streams, and lakes, however, large areas can be affected.

In general, any increase in acidity of surface water is likely to adversely affect the resident aquatic biota. The extent to which ARD causes ecological damage depends on numerous factors, including:

- The quantity, degree of acidity, and concentrations of heavy metals in ARD solutions;
- The degree to which the ARD can be confined close to the source;
- The assimilative capacity of the receiving environment, which includes attributes such as the diluting, adsorbing, neutralizing, and buffering capacities of receiving soil or waters;
- The sensitivity of key organisms in the receiving environment, which may reflect exposure and adaptation of these organisms to acid environments during their recent evolutionary history, and
- Interaction with other substances such as cyanide.

In general, where ARD results in pH values below 5, significant effects can be expected.

Typically, where ARD occurs, there are a range of adverse impacts which are most acute close to the discharge point(s), and diminish with distance downstream. In general, where ARD results in pH values below 5, significant effects can be expected, including both reduced diversity and reduced populations of aquatic plants and aquatic invertebrates, particularly insect larvae. Fish, in such circumstances may be stunted and breeding success

FIGURE 17.2
Vegetation Absent from Path of Overland Flow of ARD

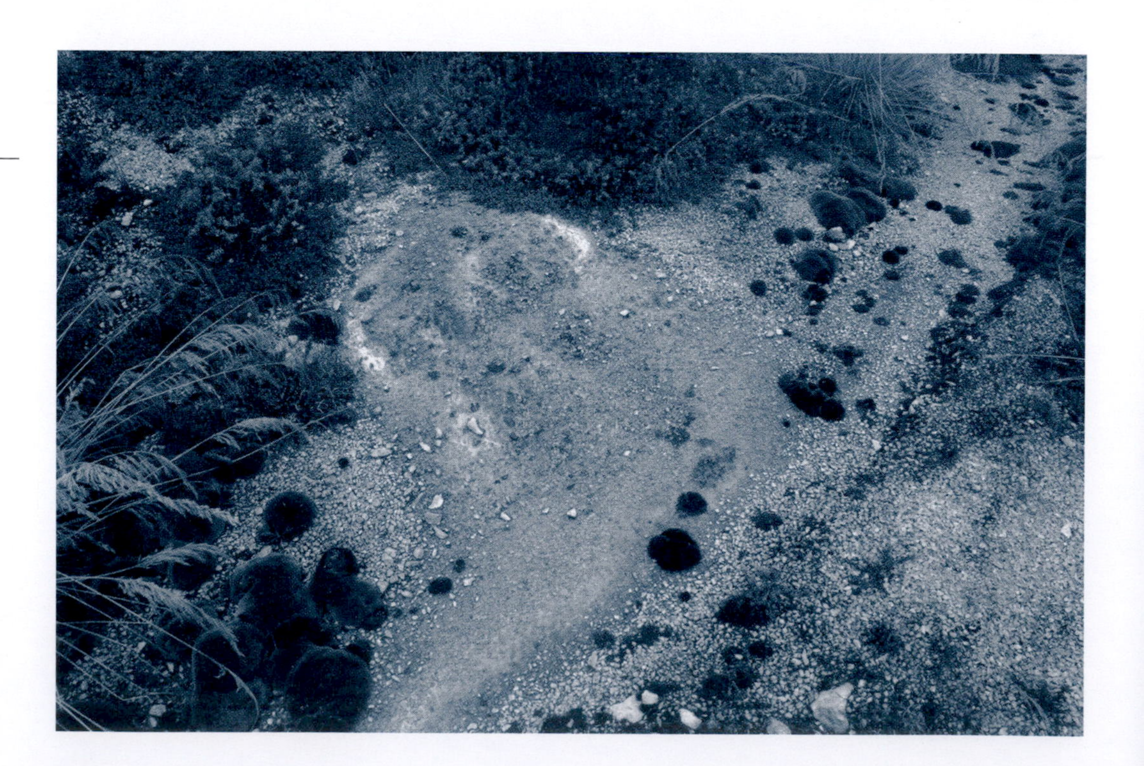

may be impaired. Once the pH falls below 4, most biota will die, although some benthic organisms may survive. At very low pH values, only specialized bacteria and algae may occur, but these will have little, if any, value in terms of the overall ecology.

In practice, the scenario described above is an over-simplification. Factors such as pH and the concentrations of cations and anions are commonly subject to considerable variation, with changes in discharge volumes in response to seasonal and specific events, such as the previously mentioned pulses which follow rainfall events. As a result, the ecosystem may be in a constant state of change, both spatially and temporally, with periodic episodes where organisms are killed, followed by partial or complete recovery, and with different parts of the ecosystem affected to different degrees at different times. Of course, for rapid recovery to occur, part of the ecosystem must remain unaffected so that it can provide supplies of organisms to re-colonize affected areas. Another significant factor is that the capacity of some environmental attributes that act to ameliorate the effects of ARD, may eventually be exceeded, so that the situation could suddenly deteriorate as, for example, when the buffering capacity of a receiving water body is overcome.

ARD discharges to the sea have been relatively rare. For many years, pyritic tailings from the large Toquepala and Cuajone Copper Mines, operated by Southern Peru Copper, accumulated along the shoreline of the Pacific Ocean. While the resulting ARD produced an unsightly scene, monitoring indicated little or no effect on the marine environment except within and immediately adjacent to the tailings deposition area which extended for a distance of 10 km along the Pacific Ocean coast. This is to be expected given the large buffering capacity of sea-water and the open ocean situation in which water is constantly circulating.

Large quantities of tailings also accumulated within Macquarie Harbour on the west coast of Tasmania, during many decades of mining at the Mt Lyell Copper Mine (see **Case 17.1**) in the Queenstown area. While ARD has occurred on exposed areas of tailings, the effects have been highly localized and Macquarie Harbour supports a productive fishery. Far more severe effects have occurred along the Queen and King Rivers, along which tailings flowed before entering the sea.

Social and Community Effects

ARD may affect surface waters or groundwater supplies used for drinking and bathing. In extreme cases, water supplies contaminated by ARD may become toxic, due to the presence of heavy metals such as arsenic at toxic concentrations. More commonly, acidic water will have an unpleasant taste, leading to its abandonment as a source of potable water.

Acidic water will have an unpleasant taste, leading to its abandonment as a source of potable water.

The ecological effects of ARD are also likely to be detrimental to local communities, particularly in developing countries where aquatic resources such as fish, shellfish and crustaceans contribute to subsistence. Again, adverse effects are more likely to result from the disappearance of organisms used for food, rather than from consumption of contaminated organisms, although the latter is possible, particularly in the event of mercury contamination (see discussion on Artisanal Mining in Chapter Five). Allegations of surface water contamination from mining and mineral processing have been made in several countries including Indonesia and the Philippines, with claims of rashes and other skin disorders ascribed to the alleged contamination. In most, if not all such cases, consequent investigations have revealed that such symptoms have not been related to ARD.

Finally, ARD commonly causes adverse visual effects such as red or brown staining of the river bed and adjacent rocks or the accumulation of unsightly gelatinous sludge (hydrated ferric hydroxides) where acidic seepage emerges or mixes with receiving waters.

ARD commonly causes adverse visual effects such as red or brown staining of the river bed and adjacent rocks.

While such adverse aesthetic effects may not, in themselves, cause damage, they certainly draw attention to ARD and potentially contribute to community concerns which may or may not be justified in terms of the real risks. There is no doubt that community concern and community outrage are aggravated by visible signs of pollution.

ARD may take many years to become apparent.

Whether its onset is rapid or follows an extended lag time, ARD is usually a long-term, persistent phenomenon.

A Long-Term Problem

In some situations, such as tailings storages associated with mines in the Tennant Creek area of northern Australia, ARD may take many years to become apparent, with signs of surface acidity appearing long after completion of mining and mineral processing activities. In other situations, signs of ARD may develop almost immediately. There is a characteristic profile of ARD behaviour, illustrated in **Figure 17.3**, in which the amount of acid generated increases slowly, then accelerates before reaching a peak, following which there is a slow decline. Understanding of the kinetics of the oxidation reactions may be important in devising appropriate management strategies and budgeting of funds for remediation, particularly if management includes treatment. Whether its onset is rapid or follows an extended lag time, ARD is usually a long-term, persistent phenomenon.

Perhaps the oldest incidence of ARD due to mining is associated with the Rio Tinto Mine in Spain, where copper has been mined for centuries. The name of the river reflects the characteristic red colour caused by iron as it precipitates from acidic solution. In the USA, where numerous cases of acidic drainage from old mining operations (many of them dating from the late nineteenth and early twentieth centuries) are being treated under the Federal CERCLA and RCRA regulations, the 'principal responsible parties' (PRPs) are committed to treating the ARD for as long as it continues. Predictions based on geochemical modelling have shown that ARD may continue for many decades or even centuries. **Case 17.2** describes the remediation of the Eagle Zinc Mine in Colorado, USA. Most of the older mines producing from sulphide ore bodies have significant acid rock drainage legacies. A notable example is the Mt Lyell Copper Mine, the subject of **Case 17.1**.

CASE 17.1
Mt Lyell Copper Mine, Tasmania, Australia

One of Australia's oldest and longest producing mines is the Mt Lyell Copper mine located in mountainous terrain at Queenstown, near the west coast of Tasmania. Copper ores have been mined at Mt Lyell for well over a century. In 1896 Mt Lyell became the world's first mine to commercially implement pyritic smelting which continued until 1969, discharging about 200,000 tonnes of sulphur dioxide to the atomosphere per year. Until the demise of the Mt Lyell Mining and Railway Company in 1994, more than 90 million tonnes of mill tailings were discharged to the Queen River from where they flowed into the King River. Tailings, together with slag from smelting operations have deposited in the bed and banks of the rivers, and also form a delta where the King River enters Macquarie Harbour. Since the mine re-opened under new ownership in 1995, tailings have been stored in a tailings impoundment. The upper part of Macquarie Harbour is included in the large Franklin-Gordon Wild Rivers National Park, which was accorded World Heritage Area status in 1982.

The Mt Lyell operations have resulted in severe impacts close to the operations and downstream along the Queen and King Rivers for about 50 km to Macquarie Harbour. Around Queenstown deforestation due to logging and fire, together with topsoil erosion and inhibition of plant growth due to atmospheric sulphur dioxide, have led to an almost completely denuded landscape, which ironically, provides a tourist attraction. There are two major sources of acid drainage from the workings at Mt Lyell: the underground workings and waste rock dumps. Rainwater percolates through the mine workings and adjacent fractured rock masses dissolving acid and metals from oxidation of pyrite. This acid mine drainage emerges from three tunnels which provide an average discharge of 148,000 ML/day into the Queen River accounting for 78% of the copper load from the mine area which ranges from 2 to 9 tonnes/day. Other significant contaminants include sulphate, zinc, iron, manganese, and aluminium. Waste rock dumps containing more than 1 million tonnes of sulphide-bearing rock account for most of the balance of copper entering the rivers. Untreated, these dumps would continue to oxidize and contribute ARD for hundreds of years. Closure of the mine would not reduce the contaminant load from the mining area.

The Mount Lyell Remediation Research and Demonstration Program was established jointly by the Tasmanian and Australian governments to investigate options for remediation of the mining area, the rivers draining the area, the delta and Macquarie Harbour itself. The final report of this initiative, issued in 1997, evaluated a wide range of options for removal and treatment of acidic streams and acid-generating wastes. Recommendations were made for a staged approach to remediation that focused initially on elimination of acid drainage from the mining area. Capital costs ranged between $ 10 million and $ 16 million for the various options with annual operating costs between $ 1.6 million and $ 10 million. The first stage of remediation removes copper by cementation from acid mine drainage, to produce a saleable product.

Sources: Koehnken, L. 1997. Mount Lyell Remediation – Final Report. Supervising Scientist Report 126, Tasmanian Government Environment Division

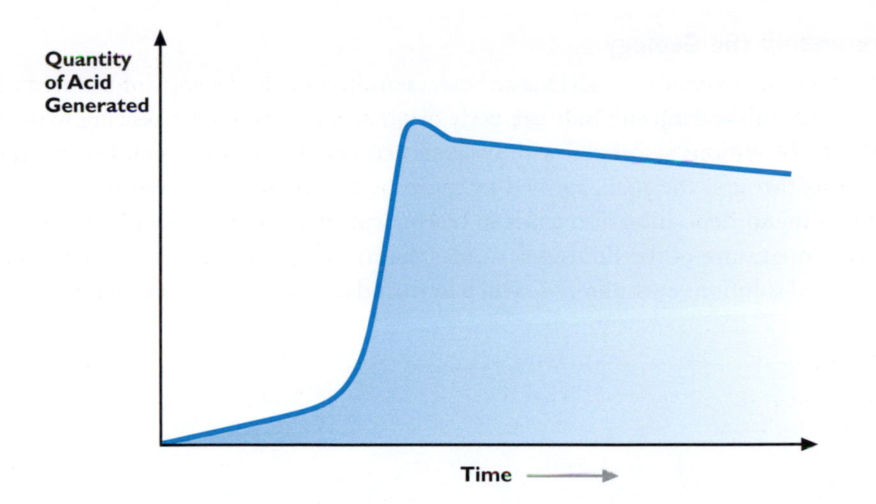

FIGURE 17.3
Typical Acid Generation Profile

17.2 EVALUATING THE OCCURRENCE OR RISK OF ARD

Identifying Areas of Active ARD

Active oxidation of sulphide minerals, the precursor to ARD, may be identified by visual observation of its common effects such as staining by hydrated oxides of iron and manganese, or by the presence of efflorescent sulphate salts, confirmed by tasting. Other, less common, indicators are emissions of steam and or oxides of sulphur with their characteristic acrid odour, and of course the presence of acidic solutions, identified by routine pH tests. Visual observations may similarly be used to identify actively oxidizing intervals in drill cores, particularly where the cores have been stored for several years in a humid environment. In some cases, acid produced by oxidation of drill cores may result in rapid corrosion and collapse of metal core trays, thereby providing an early indication of the potential for ARD.

The simplest tests used to delineate areas of developing ARD in the field, are the paste pH and electrical conductivity tests. For these tests, samples of less than 50 g are crushed to less than 1 mm and mixed with distilled water at a solid to liquid ratio of 1:2 (w/w) to form a paste. After a period of equilibration, usually overnight, the pH and conductivity are measured. A pH of less than 4 indicates an acidic sample. An electrical conductivity of more than 2 deci-siemens per metre (ds/m) indicates a high level of soluble salts, a result of oxidation with possible neutralization.

The simplest tests used to delineate areas of developing ARD in the field, are the paste pH and electrical conductivity tests.

Similar tests can be conducted on drill cores. However, usually such cores are subjected to static geochemical tests, as subsequently described.

Assessing the Potential for ARD

Theoretically, there is the potential for ARD from rocks containing a significant component of sulphide minerals. However, whether or not ARD develops depends on numerous factors including the particular sulphide minerals present; the size, shape, and texture of the sulphide grains; the permeability of the substances surrounding the sulphide grains; the oxygen concentration; pH, the presence of bacteria; and the presence of acid-neutralising substances. Generally, the presence of sulphide minerals at concentrations of 1% or more in the waste rock, tailings or pit walls, indicates the need for further investigation. The steps involved in detailed assessment are shown in **Figure 17.4**.

The presence of sulphide minerals at concentrations of 1% or more in the waste rock, tailings or pit walls, indicates the need for further investigation.

Understanding the Geology

Fundamental to any study of ARD is an understanding of the geology of the area involved. Typically a metal-bearing sulphide ore body originates from sulphur-bearing hydrothermal or epithermal solutions associated with volcanic activity. These liquids and gases follow preferred paths through the rock, formed by more permeable zones of faulting or fracturing. Sulphide mineral deposition also tends to be concentrated in these fractures. The composition and temperature of the fluids determines the metals present and the reactions with the host rock and solutions encountered, which in turn determine the resulting mineralization.

CASE 17.2
Eagle Mine, Minturn, Colorado, USA

The Eagle Mine, located 13 km southwest of the ski resort town of Vail in the Rocky Mountains of Colorado, was for many years a major producer of zinc. Mining of lead/zinc and copper/silver deposits at the Eagle Mine commenced toward the end of the nineteenth century and continued until 1984, when the extensive mine workings were allowed to flood. Subsequently, seepage from the mine began entering the Eagle River, along with contamination from tailings storage areas and waste rock dumps. The site was placed on the US EPA's Superfund list in 1986.

Remediation activities have included: raising the water level in mine workings by establishing bulk-heads in the main adits; relocation of mine wastes; capping of the consolidated tailings storage; construction and operation of a water treatment plant; and interception and diversion of uncontaminated water. The treatment plant has treated an average of 530 million litres per year, thereby removing about 20 tonnes of zinc per year that would otherwise enter the Eagle River.

A review of remediation progress carried out in 2000 concluded that health risks had been removed and that significant progress had been made in restoring the Eagle River.

The accompanying graph shows the reduction in zinc levels achieved since commencement of remediation and the corresponding increase in trout populations. Treatment is ongoing.

Source: Colorado Dept.of Public Health and Environment

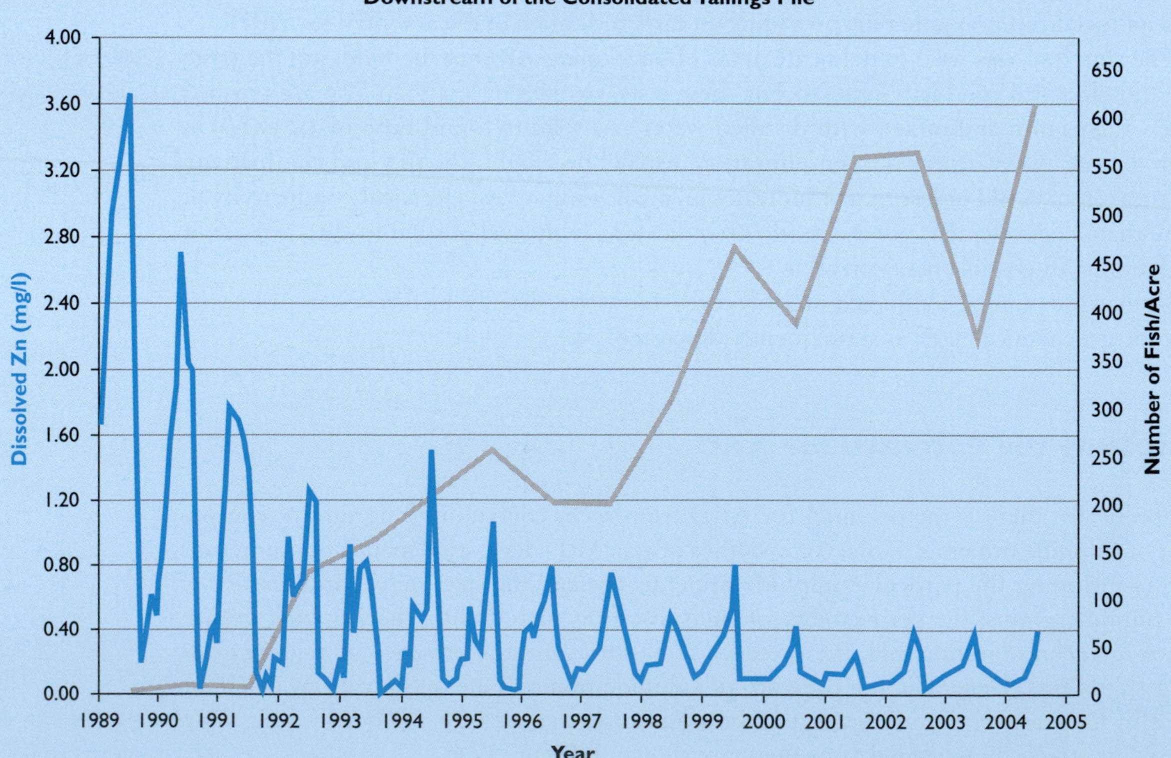

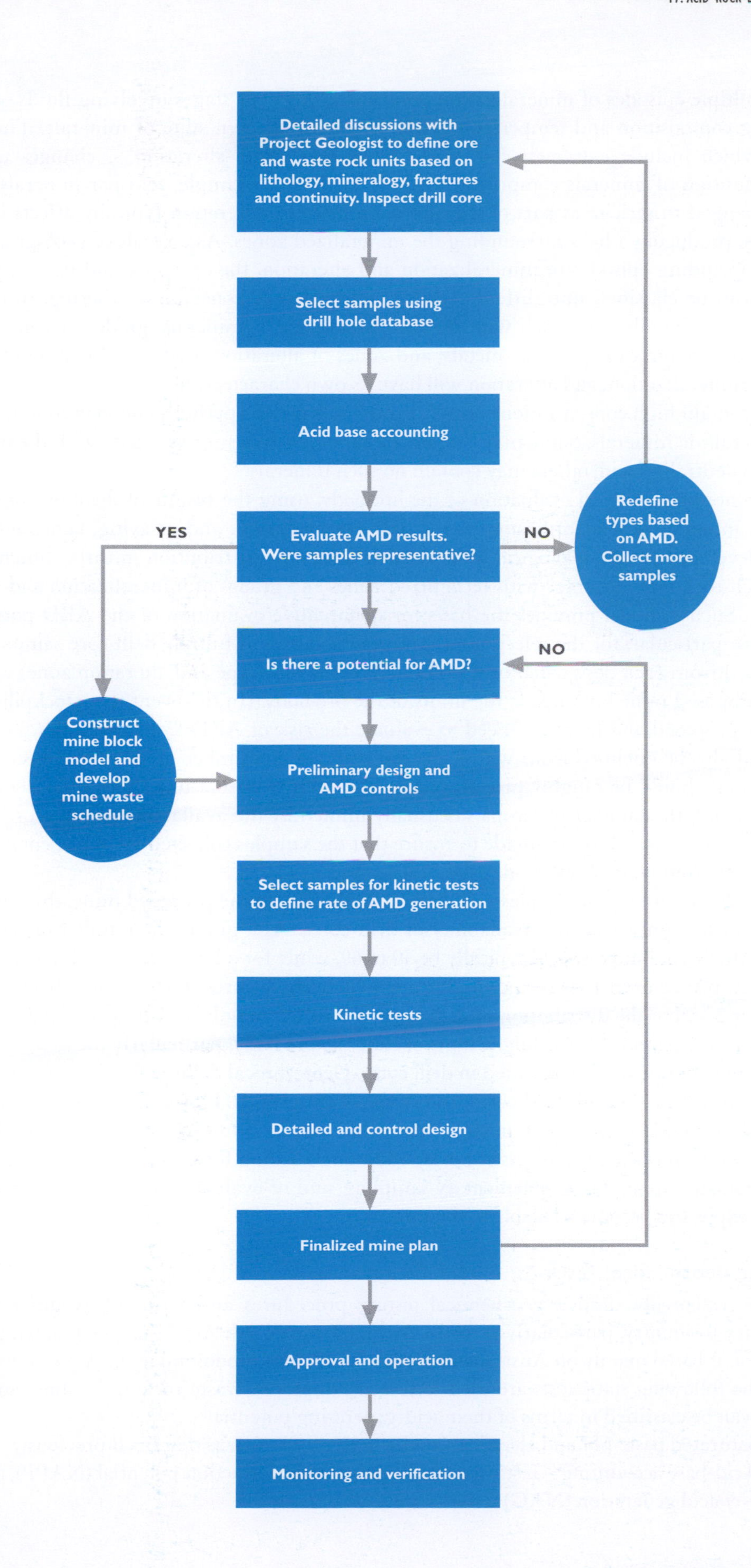

FIGURE 17.4
Evaluating Acid Rock Drainage Potential

Source:
Environment Australia 1997

Multiple episodes of mineralization occur, with different stages involving fluids of differing composition and temperature, each producing its own suite of minerals. The fluids which include extremely hot, acidic gases also cause 'alteration' – changes to the composition of minerals comprising the host rocks. For example, feldspar minerals may be changed to sericite as part of the alteration process. Alteration typically affects broad zones, producing a halo surrounding the mineralized zones. As a result of geological history, including episodes of mineralization and alteration, the ore body and its surroundings can be classified into different rock types; different zones of weathering; different stages of mineralization each with its own assemblage of minerals, grades or concentrations of economically valuable metals; and zones of alteration. Each combination of rock type, mineralization, and alteration will have its own characteristic suite of minerals. Some may contain high concentrations of sulfide minerals such as pyrite; some may contain little or no sulfide minerals. Some may contain alkaline minerals such as calcite with the potential to neutralize acid; others may contain no such minerals.

Each combination of rock type, mineralization, and alteration will have its own characteristic suite of minerals.

Economic geological evaluation of the ore body, using the results of drilling, core logging, mineralogical identification, structural interpretation, and assaying, contributes to the development of a geological model which shows the distribution in three dimensions of each rock type, together with recognized zones and grades of mineralization and alteration. Such a model provides the basis for quantitative evaluation of the ARD potential and, in particular, for the selection of samples for testing. Multiple drill core samples are selected from each designated unit (combinations of rock type and alteration zone) within the proposed mine but outside the limits of the ore body. (In the event that stockpiling of ore is proposed and there is a need to evaluate the risk of ARD from stockpiles, samples would also be obtained from within the ore body). The number of samples collected for each unit should be roughly proportional to the volume of that unit in relation to others. In practice, the number of samples is usually limited by the availability of drill core samples. In any case, efforts are made to ensure that the samples collected are representative of the units from which they are derived.

The total number of samples will depend on the size of the proposed mine, the number of identified units, and the availability of drill cores. As a guide, the number of samples for a small gold mine would typically be 40 to 60, while for a large porphyry copper mine, 400 samples or more may be required to adequately characterize the situation. It should be recognized that the distribution of drill holes is directed mainly at delineation and evaluation of the ore body. Accordingly, units outside the ore body, particularly the more distant ones, may not be well represented in drill cores. Geotechnical drilling for slope design purposes may provide additional cores from some of these units. However, it is likely that the sampling density will be low in the outer units compared to those close to the ore body. This problem is usually addressed not by additional drilling for geochemical purposes, but by geological mapping, supplementary sampling, and re-evaluation of the ARD potential, once exposures become available in the mine.

The total number of samples will depend on the size of the proposed mine, the number of identified units, and the availability of drill cores.

Static Geochemical Testing

While essentially similar, geochemical testing procedures and terminology differ from country to country, particularly between Australia and North America. The following discussion is based mainly on Australian practice, which also predominates in Asian countries.

The following static tests are used to screen samples of waste rock and tailings so that they can be classified in terms of their acid-generating potential:

- Saturated paste pH and electrical conductivity (EC) tests, as described previously;
- Acid-base accounting – tests to determine net acid production potential (NAPP), and
- Net acid generation (NAG) test.

Acid-base accounting involves measurement of the maximum potential acidity (MPA) and the acid neutralizing capacity (ANC), with the net acid producing potential (NAPP) calculated as the difference between the two values. The MPA is calculated from the total sulphur content, determined by the Leco furnace method, assuming that all the sulphur occurs as pyrite, which oxidizes to generate acid according to the following reaction:

$$FeS_2 + 15/4O_2 + 7/2H_2O \longrightarrow Fe(OH)_3 + 2H_2SO_4 \quad (17.2)$$

The MPA of a sample is calculated from the sulphur content according to the following formula:

$$MPA(kg\ H_2SO_4/t) = (Total\ \%S) \times 30.6 \quad (17.3)$$

The presence of sulphate minerals such as anhydrite and/or the presence of sulphide minerals with a lower sulphur content would result in the MPA being significantly overstated.

The ANC represents the inherent acid buffering of the sample. It is determined by the Modified Sobek method in which the sample is reacted with hydrochloric acid (HCl), followed by back-titrating with sodium hydroxide (NaCl) to quantify the unreacted HCl. The acid consumed by the reaction (HCl added minus unreacted HCl) is then expressed as kg H_2SO_4/t of sample.

The NAPP, which may be positive or negative, is the difference between MPA and ANC:

$$\text{i.e.} \quad NAPP = MPA - ANC \quad (17.4)$$

The ANP/MPA ratio provides an indication of the risk of acid generation from mine wastes. A positive NAPP equates to an ANC/MPA of less than 1, while a negative NAPP represents an ANC/MPA of more than 1. Acid-base account results are commonly presented on a plot as shown on **Figure 17.5**.

The ANP/MPA ratio provides an indication of the risk of acid generation from mine wastes.

The NAG test involves reaction of a sample with hydrogen peroxide (H_2O_2) which oxidizes the contained sulphide minerals. As both acid generation and neutralization reactions occur simultaneously in this reaction, the resultant NAG capacity is a direct measure of the amount (rather than the potential amount as in the NAPP test) of acid generated, as evaluated from the pH and acidity of the NAG liquor, after cooling. The forms of acidity may be determined by titration to pH 4.5 which determines acidity due to free acid, then continuing the titration to pH 7 which determines acidity due to soluble iron and alumina. The Sequential

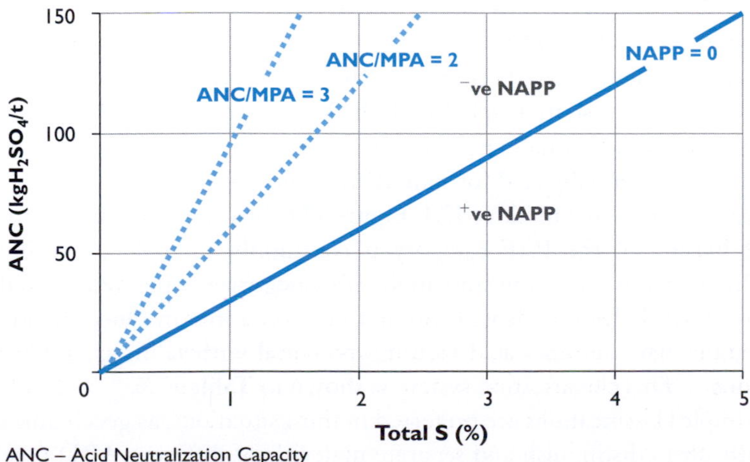

FIGURE 17.5
Acid-base Account Plot

Source:
Miller & Jeffery 1995

NAG test is a variation of this test, used for samples containing relatively high sulphur contents or high ANC which may not react completely during a single stage NAG test. In this multi-stage test, the same sample is tested repeatedly until no further reaction occurs.

Kinetic Geochemical Testing

Many geochemists consider that NAPP and NAG tests are not sufficiently conclusive to provide the sole basis for the management of mine wastes, and that selection of ARD control measures should be based on more definitive kinetic tests. This may be so in some cases, but there are also many situations where a combination of static testing and local experience enable confident predictions to be made which provide the basis for effective management.

The simplest kinetic test is a variation of the single stage NAG test – the Kinetic NAG test – in which temperature, pH, and electrical conductivity are recorded during the test, thereby providing information about reaction rates that can be related to field tests or site observations.

The most common kinetic tests are laboratory column leach tests which are used to confirm predictions of acid generation and leaching from the static testing programme, and to evaluate:

- Rates of acid generation over time;
- Neutralization, and
- Leachate water quality, including concentrations of metals over time.

As these tests are time consuming and comparatively expensive, relatively few are conducted compared to numbers of samples subjected to static tests. One approach is to undertake a column leach test on a single, representative, usually blended sample of each identified waste rock unit or tailings type. An alternative approach is to use column leach tests for those units where the static tests yielded inconclusive or contradictory results.

While laboratory leach tests attempt to simulate site conditions to the extent that is practicable, actual field conditions will be different.

While laboratory leach tests attempt to simulate site conditions to the extent that is practicable, actual field conditions will be different. If such differences are considered critical, small-scale leach pad trials may be carried out to assess acid generating and leaching behaviour of mine wastes at the mine site. In particular, such trials enable mine wastes to be placed as in operating waste dumps or tailings storages. They also enable the effectiveness of various treatment measures to be assessed. In the simplest cases, samples of leachate are collected and analyzed periodically. In more sophisticated cases the leach pads are also instrumented, with probes to measure moisture content, temperature and oxygen levels.

Characterization and Classification

Various classification schemes have been used to characterize material types in terms of their acid-forming potential. A system adapted by EGi (Miller and Jeffery 1995), and commonly applied to both waste rock and tailings, classifies materials into four categories: barren; non-acid forming (NAF); potentially acid forming (PAF); and uncertain (UC). The criteria used for this classification are shown in **Table 17.1**. **Figure 17.6** illustrates this classification scheme.

Further subdivision of the PAF category is commonly carried out with, for example, low and high or low, medium and high sub-categories being recognized. Another approach, used by BHP-Billiton at its Cannington silver mine in Queensland, Australia, uses a classification that combines acid-forming potential with salinity, based on management requirements. This classification system is shown in **Table 17.2**.

Simple classifications are preferred in most situations, as geochemical complexity makes it difficult to distinguish and separate materials of differing ARD potential.

In general, simple classifications are preferred in most situations, as geochemical complexity makes it difficult to distinguish and separate materials of differing ARD potential without first conducting large numbers of tests. As discussed later, however, the measurement of relative acid-generating potential may be required for certain ARD control strategies.

TABLE 17.1

Simple Classification of Acid-generating Potential (after EGi)

Category	Rationale	Criteria
Barren	Material is inert with respect to acid generation; includes most highly weathered materials	≤0.1%S, ANC ≤ 5 kg H_2SO_4/t
NAF	Availability of ANC is more than sufficient to neutralize all acid that could possibly be produced	NAPP is negative, NAG pH ≥ 4.5
PAF	Acid-producing potential exceeds neutralizing potential	NAPP is positive, NAGpH < 4.5
UC	Cases where there is apparent contradiction between NAPP and NAG results	NAPP is positive & NAGpH > 4.5 OR NAPP is negative & NAGpH ≤ 4.5

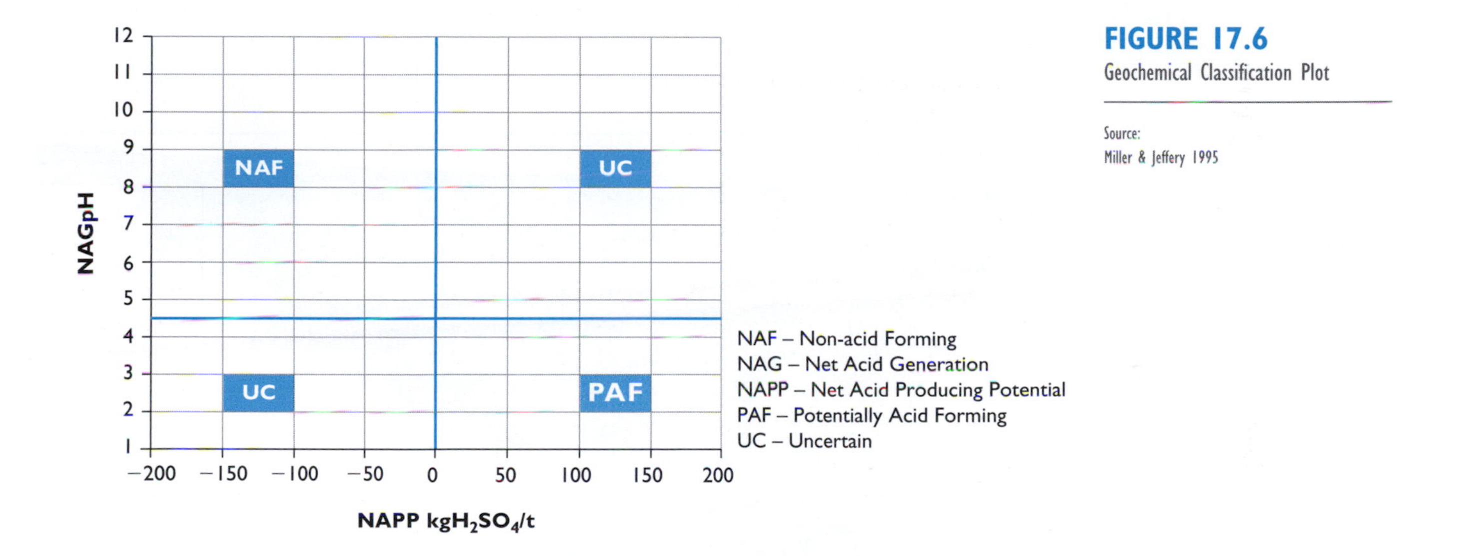

FIGURE 17.6

Geochemical Classification Plot

Source:
Miller & Jeffery 1995

TABLE 17.2

Waste Rock Classification Combining Acid-producing Potential and Salinity

Material Type	Geochemical Characteristics	Management Prescriptions
IA	Non-acid forming Nil/low/moderate salinity NAG pH > 4 & EC 1:5 < 0.8 dS/m NAG pH > 4 & EC1:2 < 1.5 dS/m	Suitable for general construction use and general fill. No specific geochemical constraints. Suitable for rehabilitation works.
IB	Non-acid forming High salinity NAG pH > 4 & EC1:5 0.8–1.3 dS/m NAG pH > 4 & EC1:2 1.5–2.5 dS/m	Suitable for general fill. Undesirable for rehabilitation use due to salinity. Avoid placing within 300 mm of final surfaces.
IC	Non-acid forming Extreme salinity NAG pH > 4 & EC1:5 >1.3 dS/m NAG pH > 4 & EC1:2 >2.5 dS/m	Can be used for general fill provided it is isolated within the core of any embankment (for example water storage). Do not place within 500 mm of rehabilitation surface.

(Continued)

TABLE 17.2
(Continued)

Material Type	Geochemical Characteristics	Management Prescriptions
II	Potentially acid-forming Low risk 3 < NAG pH < 4	Not suitable for construction use or general fill unless placed and compacted within the core of embankments and isolated from leaching. Do not place within 1 m of final surfaces or outer edge of stockpile#.
III	Potentially acid-forming High risk NAG pH < 3	Should be buried and isolated from leaching. Place and compact Type III materials in layers. Locate material towards the centre of the stockpile area. Do not place within 1 m of final surface or within 5 m of the outer edge of the stockpile. Place compacted Type IC over the Type III material before placing soil cover for rehabilitation#.

Source:
Environment Australia (1997)

EC1:2 Electrical conductivity in slurry – 2 parts water to 1 part solids
EC1:5 Electrical conductivity in slurry – 2 parts water to 1 part solids
Type II and Type III materials can be converted to Type I material by blending with limestone or other acid neutralizing materials.

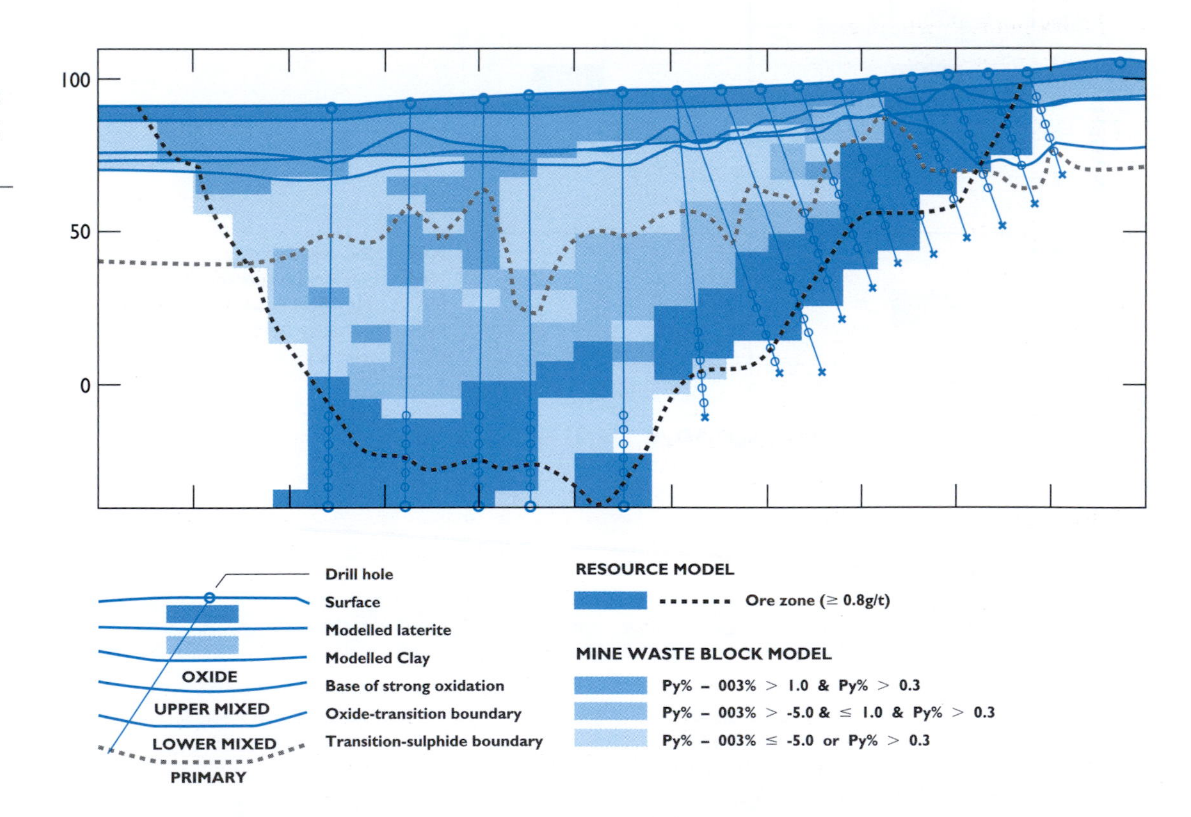

FIGURE 17.7
Cross-section Showing Block Modelling of Acid-producing Potential

Block Modelling

The results of testing and classification are stored in a 'block model', a 3-dimensional relational database which divides the rock mass to be mined into a series of small blocks, each of which receives a classification in terms of acid-producing potential. The results can be shown in various ways such as in cross-sections or in horizontal slices corresponding to individual mine benches. An example is shown on **Figure 17.7**.

Such block models are routinely used by mine geologists and mining engineers to store data on ore grades, which are then used for detailed mine planning. ARD block models, which use the same software, are similarly used for detailed planning of waste rock removal and disposal.

Typically, the ARD block model developed during the feasibility stage will lack detail, as the input data will be relatively sparse. If the geology is well defined and if there is good coincidence between geological units and acid-forming potential, there may be no need for further geochemical testing. However, more commonly, there will be a need for further testing leading to more detailed block modelling as mining proceeds. In such cases, the block model is updated progressively or periodically throughout the mine life as required to facilitate waste rock management.

Management of ARD: Prevention versus Treatment

A variety of approaches have been used for management of ARD, including:

- Toleration – no action;
- Minimization or prevention;
- Isolation;
- *In situ* neutralization, and
- Interception and treatment.

Prior to the 1970s when environmental management was introduced to mining operations, the approach to ARD was tolerance, not because the effects were not recognized, but because the means of control had not been developed. There are still examples of operations where ARD is tolerated, but these are now uncommon and are restricted mainly to arid areas where the effects of ARD are confined to the immediate vicinity of the operations. There are also many underground mines where acid-generation is tolerated because there is no potential for adverse impacts on the environment. Numerous underground mines in the Western Australian Goldfields are in this category. Drainage from these mines does not reach the surface, and the groundwaters are naturally highly saline, with a very high buffering capacity. These waters are not potable, even for wildlife, and the only beneficial use is for mineral processing.

Prior to the 1970s., the approach to ARD was tolerance.

On the other hand, underground mines situated in hilly or mountainous terrain, particularly where an access adit or shaft is located low in a valley, may act as a groundwater drain, preventing the water table from rising to its original level which would inundate much of the remaining sulphide-bearing rock. In these situations, common to numerous abandoned silver mines in the Rocky Mountains of Colorado, oxidation continues in and around the mine openings, and the adits or tunnels convey the ARD to the surface. Sealing of the mine opening does not necessarily solve the problem as, to avoid instability due to high hydrostatic pressures, the sealing plug may need to be installed well inside the slope and, in any event, seepage may continue through other pathways after mine openings have been plugged.

Some North American mining companies, conditioned to long-term management of abandoned mines, opt for the approach of interception and treatment, without any attempt to minimize or isolate ARD. However, most active mines with a potential ARD problem adopt an approach that combines minimization and isolation, as well as some interception and treatment. As mentioned previously ARD involves three steps: oxidation, leaching, and drainage. Controls can be applied to intervene at any or all of these steps. For example: (1) oxidization can be prevented by removing the means of contact with oxygen, either by encapsulation or by saturation with water; and (2) leaching and subsequent drainage can be prevented by capping to prevent the infiltration of rainfall.

Most active mines with a potential ARD problem adopt an approach that combines minimization and isolation, as well as some interception and treatment.

The best way to prevent oxidation of sulphides is to maintain the material in a saturated condition, usually by storing it below the water table.

Prevention

The best way to prevent oxidation of sulphides is to maintain the material in a saturated condition, usually by storing it below the water table. An operation involving multiple pits extending below the water table provides the potential for such an approach. After the first pit has been mined, PAF waste from the next pit is placed in the base of the first pit, up to the level of the water table. The upper part of the pit can be left unfilled, or may be filled with NAF. This approach was used towards the end of the Mt Muro project in Kalimantan, Indonesia which involved the mining of six pits, all in reasonably close proximity. A similar approach can be used in long narrow pits and in coal strip mines. Once excavation is complete at one end, PAF from elsewhere in the Pit can be placed at the base of the mined out area. This approach has been used in 'P' pit at the Chatree Gold Mine in Thailand. Whether or not the PAF can be inundated and maintained in a saturated condition, storage below grade is preferred to storage in an above ground waste rock dump, as there is much less chance of ARD reaching the surface.

Minimization and Control

Prevention, while theoretically possible, proves to be economically or practically unachievable in many circumstances. Many operating mines therefore adopt an approach which combines minimization of both oxidation and leaching with some interception and treatment (see **Case 17.3**).

CASE 17.3
Compacting Outslope of Tongaloka Waste Rock Dump, Batu Hijau

Waste rock dumps at Batu Hijau use a range of measures to minimize and treat ARD. Firstly, the dumps are formed using the bottom-up approach. This enables the formation of a low permeability skin on the surface of each layer, inhibiting both oxygen ingress and downward percolation of rainwater. The dump outslopes also incorporate compacted clay material below the final topsoil layer (see photo below) Secondly, PAF waste is placed preferentially away from the dump outslopes. Thirdly, under-drains are installed beneath each dump, draining to sumps from where any ARD that occurs is pumped to a treatment pond for neutralization.

The most common measures applied to minimize ARD from waste dumps are:

- Surface shaping and capping with low permeability material, to minimize infiltration of rainwater;
- Total encapsulation to minimize infiltration and also to restrict oxygen supply;
- Preferential placement, with high capacity PAF materials placed in central parts of the dump, and NAF materials in outer zones (Marszalek 1996);
- Compaction to reduce overall permeability, to minimize cracks that enable ready access to air and water, and to provide low permeability barriers.

These measures are facilitated where waste rock storages are formed by paddock dumping and bottom-up construction methods (see Chapter Nineteen on waste rock disposal).

Commonly, the main constraint to management of ARD in waste rock storages, is the availability of clay or other low permeability material for use in capping or encapsulation. Impermeable membranes have been used; however, the costs would be prohibitive for most operations. Another common constraint is that most of the oxidized wastes, suitable for capping or placement of the outer zones of a dump, are produced in the initial stages of mining while wastes produced in the final stages comprise mainly PAF material. This constraint can be overcome, albeit at considerable cost, by stockpiling of oxidized wastes for later use in capping.

The main constraint to management of ARD in waste rock storages, is the availability of clay or other low permeability material for use in capping or encapsulation.

One of the waste rock dumps at Mt Muro in Central Kalimantan, Indonesia, uses a low permeability zone within the dump to impound water within the dump to maintain saturation of the PAF waste as shown in **Figure 17.8**. Another approach has been to mix limestone with sulphidic waste rock so as to achieve neutralization *in situ*. (**Case 17.4**)

For tailings storage facilities with the potential for ARD, two alternative approaches may be used. These are: (1) wet cover, which means maintaining a pond on the surface of the tailings, thereby excluding atmospheric oxygen; and (2) dry cover, in which a low permeability capping is placed over the surface of the tailings, limiting both the ingress of oxygen and infiltration of rainfall.

The design of covers and encapsulating layers depends on many considerations including the properties of available materials, the physical and chemical characteristics of the wastes, climatic conditions, and potential future land uses. Usually cover design forms part of the overall dump design that also includes one or more layers for establishment of vegetation. Because of the differing, sometimes conflicting requirements, the overall design may require cooperation between geochemist, mine planner, geotechnical

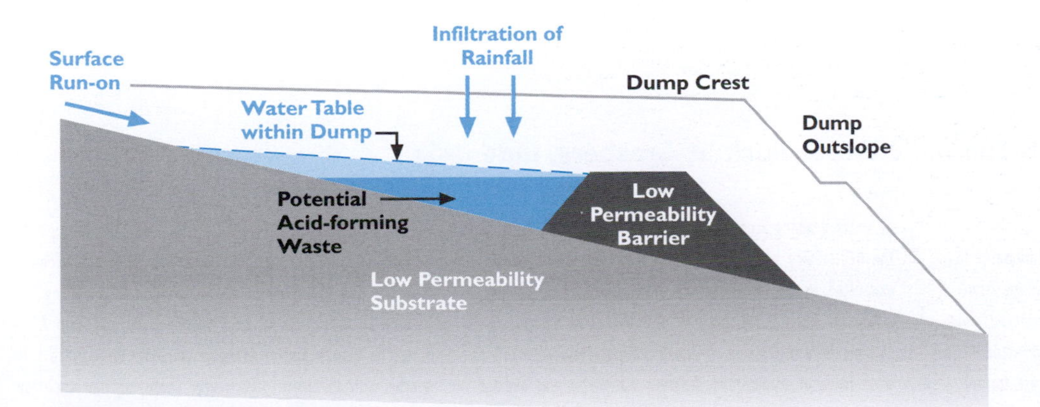

FIGURE 17.8
Schematic Profile through Mt Muro Waste Rock Dump

engineer, hydrologist, and rehabilitation specialist. Factors considered in design of covers include:

- Physical properties of capping materials, including particle size distribution, moisture content, compaction, shrinkage, and permeability;
- Surface configuration, including angle and length of slopes, presence and slope of berms to meet future land use requirements while being reasonably simple to construct;
- Permeability, which should be as low as practicable, and is minimized by conditioning and compaction;
- Potential for cracking which depends on mineralogy, climate, depth of burial, moisture content, and degree of compaction;
- Erodibility;
- Potential for capillary rise.

Treatment

Despite all the measures applied for storage of waste rock and tailings, there may still be acidic streams that require treatment before discharge to the environment. During operations, these streams are commonly collected and added to process water. In the case of most gold operations, lime is added in the process to raise the pH to 8 or more, so that this constitutes the necessary treatment. For acidic drainage streams remaining after completion of processing, ongoing treatment may be required.

For small acidic streams, this may be as simple as the addition of lime slurry to the stream. For larger discharges, the acidic drainage may be treated in a treatment plant, using lime. Various other treatments exist, to remove metals as well as to neutralize the streams. In particular there are several highly promising technologies for the extraction of metals such as copper, zinc, and cadmium, such as Biosulphide treatment process, and zeolite cementation (MEND 2001- Vol. 5) which in favourable circumstances may be amenable to metal recovery. However, conventional neutralization using limestone ($CaCO_3$) and/or lime ($Ca(OH)_2$) remains the most widely applied and most reliable treatment technology. The lime reacts with acid to produce gypsum as follows:

Conventional neutralization using limestone and/or lime remains the most widely applied and most reliable treatment technology.

$$H_2SO_4 + Ca(OH)_2 \longrightarrow CaSO_4 \cdot 2H_2O \tag{17.5}$$

As the pH rises with addition of lime, metals will precipitate from the solution, as hydroxides.

CASE 17.4
Blending Limestone with Sulphidic Waste Rock at Grasberg, Indonesia

Limestone underlies the main waste rock dumps for the Grasberg Copper Project in Papua Province, Indonesia. In the initial planning, it was assumed that drainage from the waste rock dumps would infiltrate the numerous solution cavities in the limestone foundations and would be neutralized at the same time. This was the case for the first few years of operations. However, ultimately, the pH of water draining from the dumps declined, indicating that the neutralizing capacity along the flow paths had been exhausted. In the circumstances at Grasberg, encapsulation would not be practical because of the absence of soils suitable for construction of encapsulating layers. After a wide range of studies into different options, it was decided to mix crushed limestone with potentially acid-forming wastes, particularly in near-surface zones of the dumps. A series of laboratory column tests and field trials demonstrated the effectiveness of this method in achieving *in situ* neutralization of acid produced by oxidation. A limestone quarry and crushing plant were constructed nearby, with the main objective of supplying limestone for blending with waste rock.

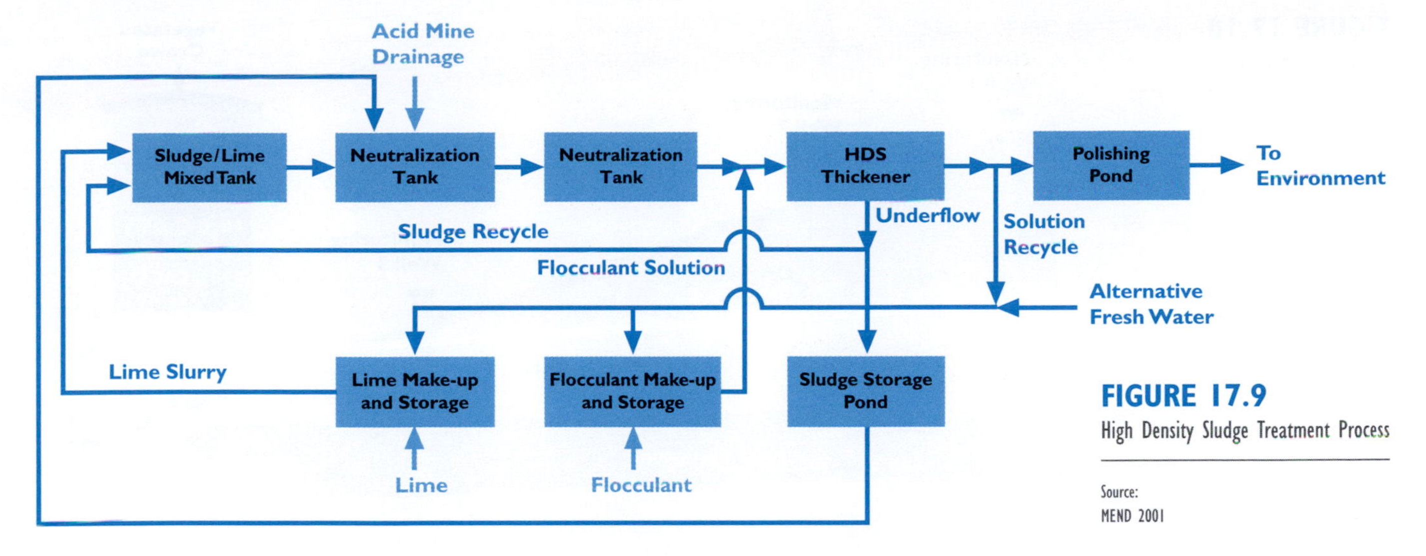

FIGURE 17.9
High Density Sludge Treatment Process

Source:
MEND 2001

For example:

$$M_{2+} + 2(OH)\_ \longrightarrow M(OH)_2 \quad (17.6)$$

where M represents a metallic cation such as Cu^{2+}. Most metals precipitate from solution as pH increases from 7 to 10.

Three types of lime-based treatment are commonly used:

- Batch treatment by the addition of lime to ponded water;
- Conventional treatment which involves continuous addition of lime to acidic streams; and
- High Density sludge (HDS) process which uses a treatment plant for continuous treatment.

Which ever treatment is used, storage and disposal of sludge is an important issue. HDS plants produce much lower volumes of sludge than conventional treatment plants. The HDS process is depicted schematically in **Figure 17.9**. These treatments are 'active' treatments in the sense that they require ongoing attention including replacement of reagents, maintenance, and process adjustments.

Which ever treatment is used, storage and disposal of sludge is an important issue.

A variety of 'passive' treatment technologies have been developed, which are intended to be self-maintaining, requiring little if any maintenance or intervention. These processes actually imitate processes that occur in nature, such as those that have led to the formation of 'bog iron ore' and the formation of pyrite in coal beds. Passive treatments include both aerobic systems where lime is used, and anaerobic systems including 'wetland treatment' where sulphur is removed, forming sulphides in an organic substrate.

Three passive treatment systems are briefly described below with much more information available from the technical literature, in particular in MEND (2001).

(1) Anoxic limestone drains (ALD)

These drains are composed of crushed limestone, through which acidic waters flow by gravity, with progressive increase in alkalinity, as shown schematically in **Figure 17.10**. Metal removal takes place downstream of the ALD. Accordingly, ALDs are used in conjunction with oxidative ponds or wetlands in which metal deposition takes place.

(2) Aerobic wetland treatment systems

A variety of complex biotic and abiotic processes take place in aerobic wetlands including oxidation, adsorption, complexation, precipitation, and filtration. As these systems are only effective at treating net-alkaline waters, they are commonly used in conjunction with ALDs.

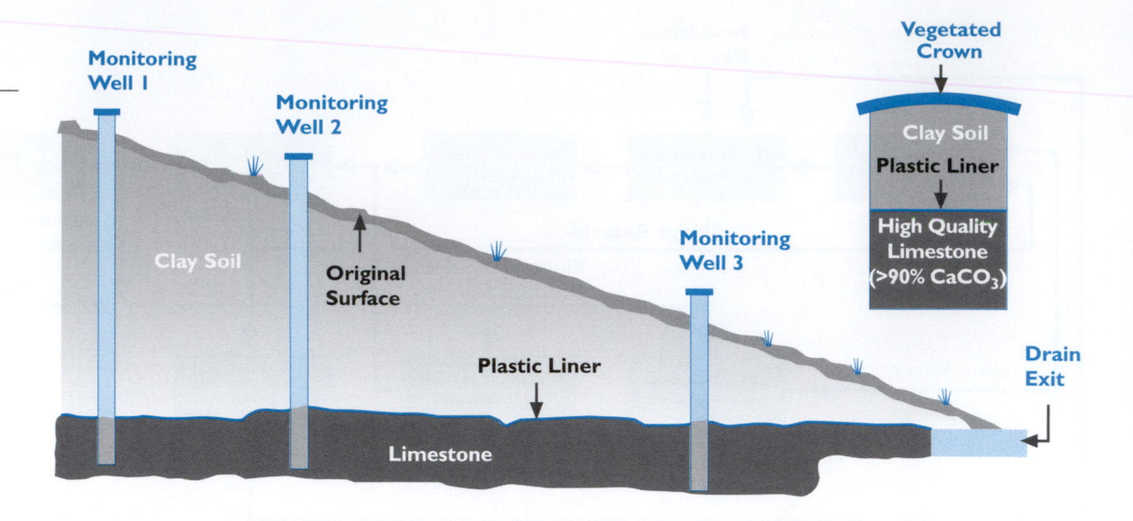

FIGURE 17.10
Anoxic Limestone Drain

Source:
Nairn *et al.* 1992

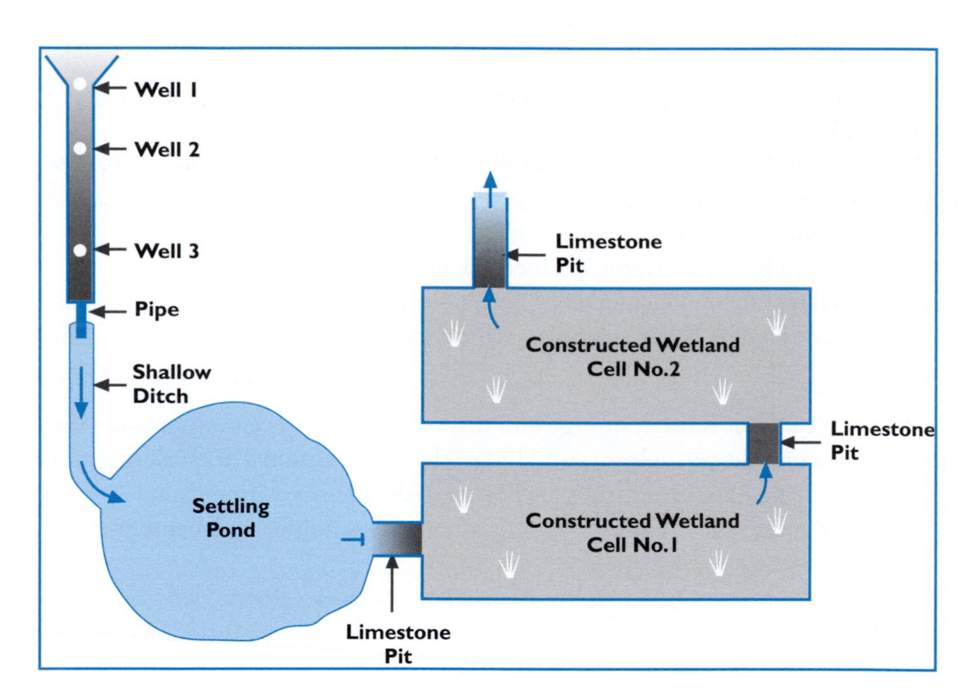

FIGURE 17.11
ALD – Wetland Passive Treatment System

Source:
Nairn *et al.* 1992

Metals such as aluminium, iron, and manganese are commonly removed in aerobic wetlands and various co-precipitation reactions also result in the removal of other metals. Aerobic wetlands are particularly well suited to provide a final 'polishing' stage of metal removal. Many factors are involved including the layout of wetlands, the selection of suitable aquatic plants and the residence time required for the processes to occur. A typical system involving anoxic limestone drains and aerobic wetlands is depicted schematically in **Figure 17.11**.

(3) Anaerobic treatment systems
These systems depend on metal removal and acid neutralization from sulphate reducing bacteria. Accordingly, they need to create the conditions in which these organisms can thrive. This requires the supply of an organic substrate which may use low cost agricultural or silvicultural wastes, and requires relatively deep water to prevent oxygen from reaching the substrate. This substrate through which the acidic water must percolate, produces the reducing environment as well as fuel for the bacteria. Bacteria reduce sulphate

ions to sulphides producing bicarbonate alkalinity and hydrogen sulphide, which reacts with metals, forming insoluble sulphides.

Remediation

In many parts of the world, there are abandoned mine sites where active ARD is occurring. In Canada, the Mine Environment Neutral Drainage Program in 1994, estimated the cost of remediation of acid generating mine wastes in Canada to be $ 3 billion. (MEND 1995) While the total quantities of potentially acid generating materials in Australia are comparable to those in Canada, the estimated remediation costs in Australia are considerably lower, amounting to less than $ 50 million for the remaining unrehabilitated abandoned sites (Harries 1997). This figure does not include the numerous sites in remote, arid parts of Australia where high evaporation rates limit the potential damage from ARD. It is also a fact that much of Australia's historic mining has taken place in the oxidized zone where sulphide minerals have been absent.

In Canada, the Mine Environment Neutral Drainage Program in 1994, estimated the cost of remediation of acid generating mine wastes in Canada to be $ 3 billion.

The issues involved and the potential control measures are essentially the same as those affecting operating mines. However, the costs may be much higher because:

- Records of construction and operation are generally unavailable or inadequate, and considerable effort is required to identify the issues and to characterize the mine wastes, and extent of contamination;
- Following abandonment, structures such as tailings embankments and retaining walls may have failed, discharging their contents which may have been transported well beyond their original storage area;
- Plumes of contaminated groundwater may have migrated well beyond the sources of contamination; and
- The capacity for treatment of problems no longer resides at the site. Investigators and designers need to be recruited from outside and require time to become familiar with the situation. Similarly, equipment for remediation may need to be mobilized from a distance. This contrasts with an operating mine situation where most of the personnel and equipment are already available.

Remediation seldom produces saleable products, although there have been numerous examples of tailings re-treatment where the income generated paid for the necessary ARD controls that had not previously been implemented. Lack of funds constrains many remediation efforts, particularly at 'orphan sites' where the previous owners no longer exist and which therefore depend, for remediation, on public funds. Nevertheless, there are many examples where ARD from abandoned mine sites has been substantially reduced (see for example **Case 17.1**), or treated to the extent that the receiving environment has been returned to a condition comparable to that which existed prior to mining.

Lack of funds constrains many remediation efforts, particularly at 'orphan sites' where the previous owners no longer exist.

REFERENCES

Environment Australia (1997) Managing Sulphidic Mine Wastes and Acid Drainage, Booklet in the series 'Best Practice Environmental Management in Mining.' Principal contributors: John Johnston and Gavin Murray, May 1997.

Harries J (1997) Acid Mine Drainage In Australia – Its Extent and Potential Future Liability. A joint initiative of the Office of the Supervising Scientist and the Australian Centre for Minesite Rehabilitation Research. Supervising Scientist Report 125. Supervising Scientist, Canberra, Australia.

Hutchison IPG and Ellison RD eds. (1992) Mine Waste Management, Pub. Lewis, London.

Koehnken L (1997) Final Report, Mount Lyell Remediation Research and Demonstration Program. Supervising Scientist Report No 126, Supervising Scientist, Canberra, Australia.

Marszalek AS (1996) Preventative and Remedial Environmental Engineering Measures to Control Acid Mine Drainage in Australia. In Preprints of Papers, Engineering tomorrow today, National Engineering Conference, Darwin, April 1996. Pub. Inst. Of Eng. Aust., Barton A.C.T., Australia.

MEND (1995) Economic Evaluation Of Acid Mine Drainage Technologies, MEND Report 5.8.1, Energy Mines and Resources, Canada, January, Ottawa.

MEND (2001) MEND Manual. Six Volumes, Sponsored by Natural Resources Canada, Northern Ontario Development Agreement, Quebec Mineral Development Agreement, Organizing Committee for the 4th Int. Conf. on Acid Rock Drainage. Eds. Gilles A Tremblay and Charlene M Hogan, March 2001.

Miller S and Jeffery J (1995) Advances In The Prediction of Acid Generating Mine Waste Materials, Proc. 2nd Acid Mine Drainage Workshop, Charters Towers, Queensland, 28–31 March, 1995.

Nairn RW, Hedin RS, and Watzlaf GR (1992) Generation of Alkalininty in an Anoxic Limestone Drain. In: Proc. From 9th Annual Meeting of Am. Soc. for Surface Mining and Reclamation, Duluth. MN, 14–18 June, 1992.

18

Tailings Disposal

Concepts and Practices

 CHROMIUM

Chromium is a hard, lustrous, steel gray metal which can sustain a high polish. It is corrosion resistant and has a high melting temperature. It is an essential trace metal; however, in its hexavalent form it is toxic and mutagenic. Chromium metal is used in stainless steel and other alloys, chrome plating and for anodizing aluminium. Chromium compounds are used in dies, paints and glazes, and in tanning of leather. It is extracted mainly from the mineral chromite which is an oxide of both chromium and iron.

18 Tailings Disposal

Concepts and Practices

Arguably, more environmental problems have resulted from tailings disposal than from any other component of mining operations. In the past, many tailings storage structures failed, some during operation, some after closure, some catastrophically causing loss of life, and others causing chronic or acute environmental damage. In the mountainous wet tropics, in particular, the historical record of tailings disposal includes more failures than successes.

Tailings are essentially finely-ground rock particles of no practical or economic value, mixed with water.

What are tailings, or 'tails' as sometimes known? In some mineral industries tailings have different names. Notable examples are 'slimes' in the phosphate industry and 'red mud' in the alumina refining industry. Despite differences in terminology, tailings are essentially finely-ground rock particles of no practical or economic value, mixed with water. They also have in common that their safe disposal constitutes a considerable environmental challenge.

The bibliography on this subject is considerable. A non-exhaustive selection of recent and most innovative titles includes Australian Government (2007), USEPA (1982, 1995), BGRM (2001), MAC (1998, 2003), Jewell and Fourie (2005), ICME and UNEP (1999), Martin *et al.* (2002) and Xenidis (2004) for an authoritative general coverage of the subject matter. The book by Vick (1983, reprinted in 1990) is the authoritative text on tailings dam planning and design. Kreft-Burman *et al.* (2005) provides a review of tailings specific regulations.

Safe tailings disposal concepts and practices comprise all the elements depicted in **Figure 18.1**, and are briefly discussed in this chapter, focusing particularly on conventional on-land tailings storage facilities (TSF) and deep sea submarine tailings placement (DSTP).

18.1 DECIDING ON THE TAILINGS DISPOSAL SCHEME

The choice of what tailings disposal scheme to adopt depends on the quantity, nature and characteristics of the tailings; potential impacts and risks associated with tailings disposal in a given the geologic, topographic and environmental setting; and on disposal objectives.

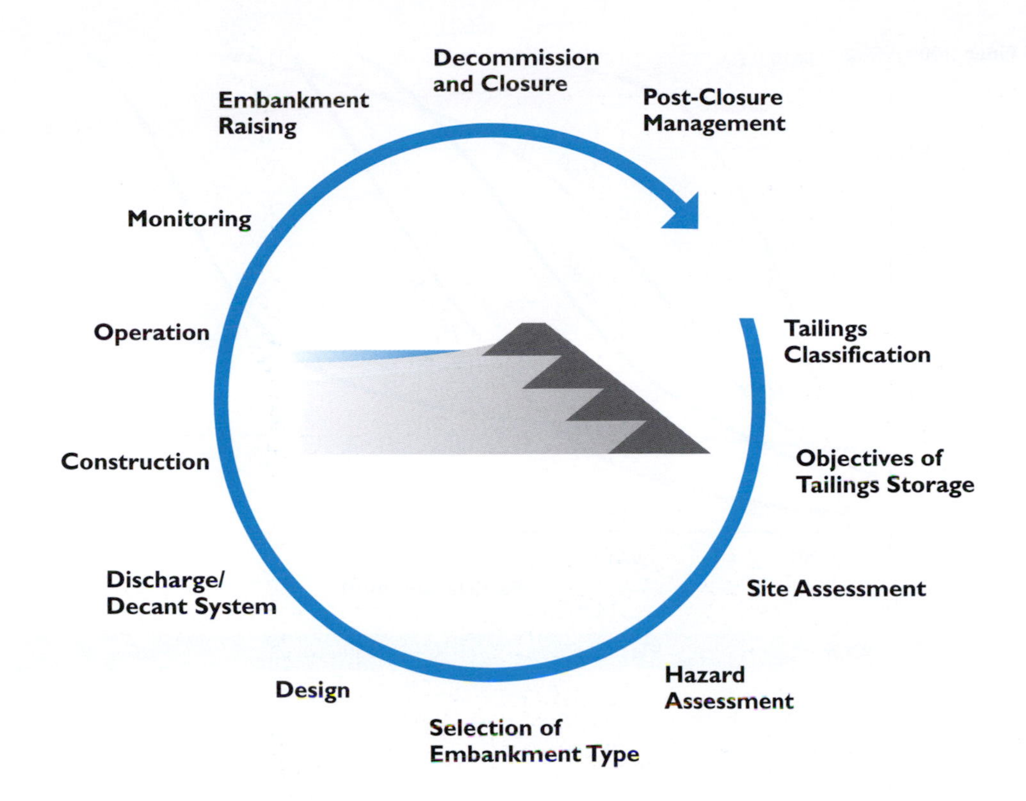

FIGURE 18.1
Dam Safety Management of Tailings Facilities

Environmental risk and risk to human life from catastrophic failure of tailings impoundments as a result of floods, earthquakes, inadequate engineering design or defective construction are of major concern in tailings management.

Quantity, Nature and Characteristics of Tailings

Tailings are the residues of ores after the valuable mineral constituents have been extracted. Hence the nature and properties of tailings vary according to the nature of the original ore, and the processes to which the ore is subjected prior to separation of the tailings. Common types of tailings include residues from the following processes:

1. Physical separation or beneficiation processes such as washing (commonly used in the coal industry); cycloning; jigging; magnetic separation; or heavy media separation using fluids of high specific gravity;
2. Froth flotation to extract sulphide minerals;
3. Leaching with cyanide solution to dissolve gold and silver, and
4. Hydrometallurgical processes such as acid leaching for some refractory gold, uranium and lateritic nickel ores; and alkaline leaching used in the production of alumina from bauxite ores and in some uranium extraction processes.

The nature and properties of tailings vary according to the nature of the original ore, and the processes to which the ore is subjected prior to separation of the tailings.

Since all of the above mineral beneficiation processes use water for transportation, leaching and separation of minerals from host rock, tailings are nearly always in the form of a slurry – a mixture of fine-grained particles suspended in water.

Other processes that may be used include roasting of refractory gold ores prior to leaching and high pressure leaching in autoclaves. Wastes produced from pyrometallurgical processes such as smelting are known as slag. These are not considered to be tailings, although sometimes they are disposed together with tailings. Other wastes from mineral processing are sludges that precipitate from various parts of the process, such as metal refining. While these are also not tailings in the strict sense, they are commonly incorporated with tailings for convenience of disposal.

Tailings emerging from mineral processing differ widely in volume, particle size distribution, pulp density and chemistry.

FIGURE 18.2
Particle Size Distributions for Typical Tailings

Tailings are commonly in the top three of the following four grain size distribution groups:

1. clay – materials less than 2 μm;
2. silt – materials lying between 2 μm to 63 μm;
3. sand – materials lying between 63μm and 2 mm.
4. gravel – More rarely, they may contain gravel-sized particles, more than 2 mm in diameter.

Source:
Sarsby 2000

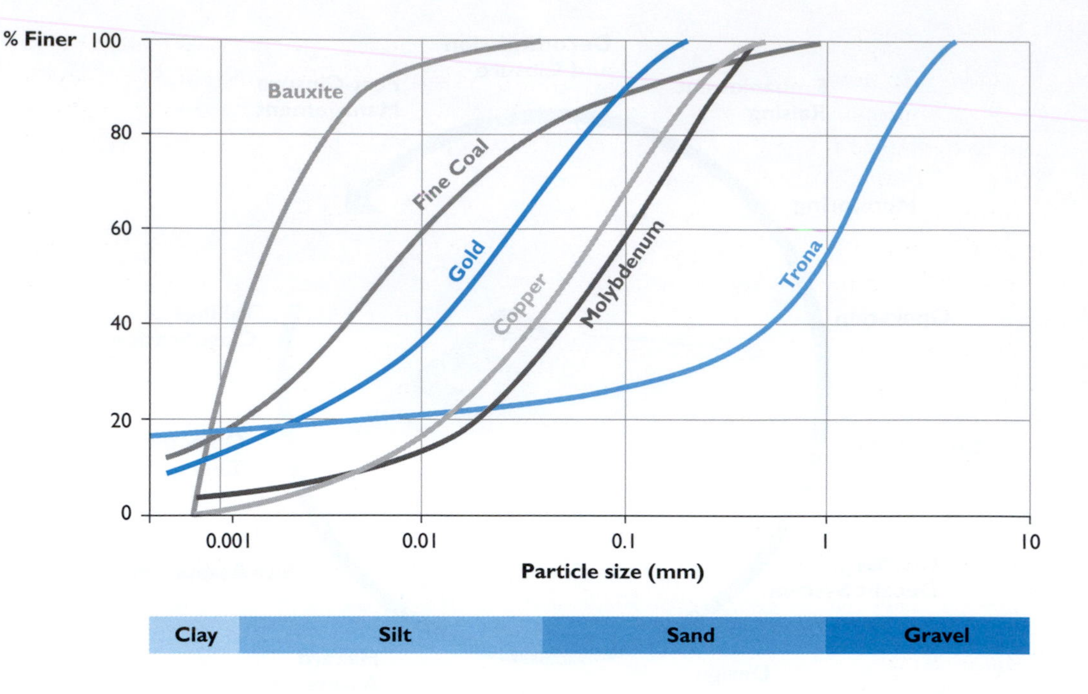

A mine producing 100,000 tons of copper ore per day also produces close to 100,000 tons of tailings.

Volume of Tailings

The sheer volume of tailings creates some of the most challenging management problems in the mining industry. The quantities of tailings produced at most precious and base metal mines are about the same as the quantity of ore that is mined because ore grades are in the order of a few percent at most. A high-grade gold ore body may contain tens of grams of gold per ton (equivalent to one thousandth of 1% gold per ton) while a high grade copper ore may contain 20 to 30 kg copper per ton (equivalent to 2 to 3 percent copper). Therefore, even if all commercial minerals are removed, most of the original ore remains as tailings. A mine producing 100,000 tons of copper ore per day also produces close to 100,000 tons of tailings. This equates to several thousand truck loads per day of tailings that need to be managed and disposed of in an environmentally acceptable manner. Where tailings are stored and how they are managed becomes critical to determining the impacts of mining: the larger the tailings volume the larger the associated environmental impacts and risks.

Particle Size Distribution

The particle size distribution is one of the most significant characteristics of tailings. Prior to ore beneficiation, most ores are first crushed and then milled or ground into fine particles. The exceptions are placer gold and tin deposits, and most mineral sand ores where the valuable minerals already occur as separate grains within essentially granular sediments. In these situations, little or no crushing or milling is required. In most other cases, the extent of grinding required depends on the particle size of the mineral grains within the ore. Most tailings consist predominantly of grains in the fine sand and silt sizes (2 μm to 0.2 mm), with engineering properties similar to those of silt (Wahler and Associates 1973), although finer and coarser gradings are not rare. **Figure 18.2** shows particle size distributions for several tailings of contrasting particle size distribution.

For on-land disposal systems, particle size distribution is the major factor controlling tailings deposition, consolidation and desiccation. In general terms, deposition occurs over days while consolidation requires years or tens of years. Tailings are difficult to drain; in general, the higher proportion of 'fines' (silt and clay fractions), the lower the permeability and hence slower drainage rates. This of course also has its benefits. Consider bauxite residues, commonly known as 'red mud'. As illustrated in **Figure 18.2** red mud predominantly consists of silt and clay-sized particles. These small particles, once they have settled, form a soil mass of very low permeability with a high moisture retention capacity. As a result of these properties, the remnants of caustic solution from the Bayer process, used to dissolve alumina from unwanted host materials, remain locked in the matrix of the mud, and therefore cannot be readily leached. These properties also mean that the mud can be used as a capping or sealing layer to prevent infiltration or drainage.

Tailings deposition occurs over days while consolidation requires years or tens of years.

If tailings contain a sufficient percentage of coarse materials (sand), the coarser fraction can be separated and used to construct embankments to impound the much weaker fine tailings materials. Coarse tailings are also used, with or without added cement, for back-filling of underground mine openings. If coarse tailings materials are absent, all construction materials for tailings embankment construction must obtained elsewhere, with the result that additional environmental impacts due to borrow pits and material movement may occur. In most mining situations, selected waste rock from the mine excavation is incorporated in tailings embankments.

Pulp Density

Pulp density refers to the water content of the solid-water mixture that forms tailings. Once the minerals of interest have been extracted from milled and processed ore, fine solids that remain are mixed in suspension with water that may also contain remnants of dissolved salts and reagents used in the process. Typical tailings slurries have solids contents of 20 to 40% by weight (**Table 18.1**). Once discharged to a tailings storage facility, water is removed by decantation, drainage and evaporation. Safe on-land storage of tailings slurry is difficult to achieve and reclamation of wet tailings storages requires long-term planning and financing. A recent trend is to reduce the water content of tailings prior to final disposal by using paste and thickened tailings technologies (Jewell and Fourie - 2005). Tailings with reduced water contents involve lower environmental risks. However, such dewatering technologies are relatively expensive and are used mainly in cases of severe water shortage, to recover expensive process reagents, or in situations where secure conventional tailings systems cannot be constructed.

Safe on-land storage of tailings slurry is difficult to achieve and reclamation of wet tailings storages requires long-term planning and financing. Tailings with reduced water contents involve lower environmental risks.

When discussing thickened tailings or paste, it is also important to recognize that no two tailings will necessarily have the same solids concentration when formed in these states, since their particle size distributions, clay content, particle shape, mineralogy, electrostatic forces and flocculants dosing vary tremendously. **Table 18.1** gives some typical slurry and paste solids concentrations for a range of tailings types (Williams 2004). The percentage of solids in the thickened state is between the slurry and paste values.

Tailings Chemistry

Tailings chemistry, not surprisingly, reflects the host rock geochemistry. Many tailings contain pyrite which may oxidize on exposure to the atmosphere, producing acid in the same way that waste rock can oxidize to give acid rock drainage. Depending on the process, reagents may be introduced that may be harmful to the environment. Since all water entering the tailings impoundment is potentially released to the environment either as decant water, run off, seepage, leachate, or by evaporation, understanding of long-term chemistry of tailings water is essential. It is important, however, to recognize that while some tailings are toxic, many tailings are not. More often than not, tailings solids are inert rock particles and their direct environmental impacts are of a physical, not chemical nature. Similarly, while some tailings liquids contain dissolved substances that may be toxic or hazardous, others contain water with no deleterious constituents.

More often than not, tailings solids are inert rock particles and their direct environmental impacts are of a physical, not chemical nature.

TABLE 18.1

Typical Solid Concentrations for Various Tailings Types – Typical tailings slurries have solids contents of 20 to 40% by weight

Tailings Type	Slurry % Solids	Paste % Solids
Bauxite red mud	25	45
Base metal tailings	40	75
Coal tailings	25–30	—
Gold tailings	45	72
Mineral sands slimes	15	24
Nickel tailings	35	45

Source:
Williams (2004)

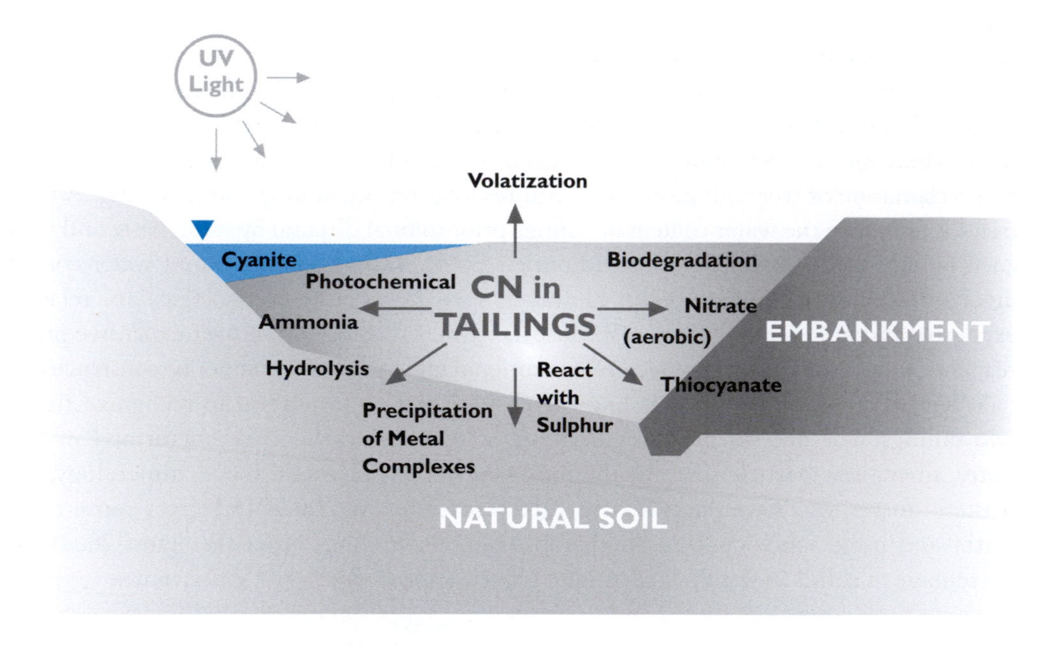

FIGURE 18.3
Simplified Diagram of Cyanide Degradation Processess in TSFs

Cyanide is generally considered as a transient pollutant: its toxic properties decrease rapidly over time due to natural degradation.

Source:
www.newmont.com

Cyanide attenuation, its decay and transformation, is generally very effective in reducing cyanide concentrations in the tailings impoundment.

Tailings from gold mines commonly contain residual cyanide, as discussed in more detail in Chapter Six in this book. Cyanide attenuation refers to the various processes that decrease the concentration of cyanide in solution, whether in the natural environment or in engineered facilities (ICME 1999). Cyanide is generally considered as a transient pollutant: its toxic properties decrease rapidly over time due to natural degradation of cyanide caused by volatilization, photodecomposition, bio-decomposition and conversion to thiocyanate as illustrated in **Figure 18.3**. Cyanide attenuation, its decay and transformation, is generally very effective in reducing cyanide concentrations in the tailings impoundment. However, there are circumstances where cyanide decomposition and attenuation occur much more slowly.

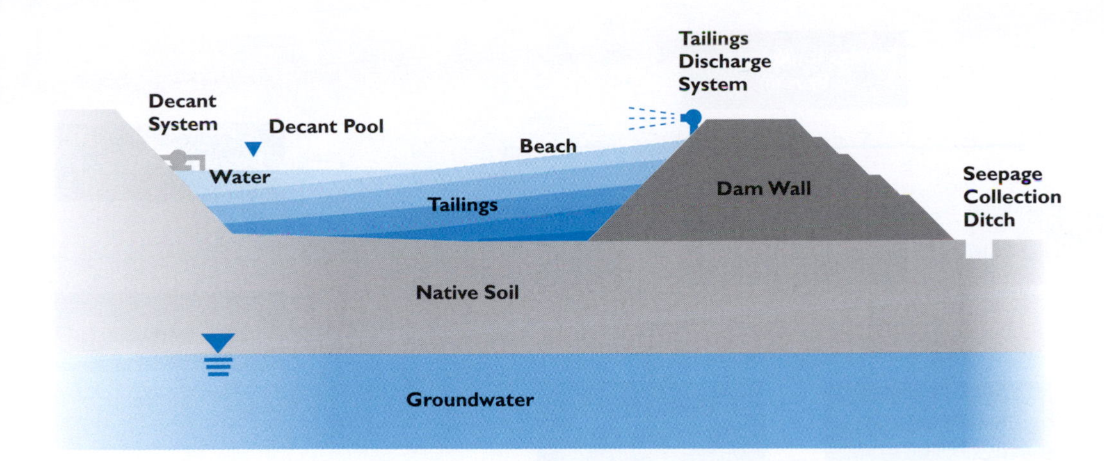

FIGURE 18.4

Typical Tailings Storage System

The solids content of the slurry, as discharged, influences the settling characteristics and properties.

Potential Impacts and Risks of Tailings Disposal

In the majority of cases, tailings are discharged into retention ponds known as tailings impoundments or tailings storage facilities, where the solids settle out, leaving a pond of supernatant water which may be recovered through a decant system and re-used. A typical tailings storage system is shown in **Figure 18.4**.

The properties of the settled tailings solids including strength, permeability, consolidation, moisture retention and potential for liquefaction under earthquake loading depend on: (1) particle size distribution; (2) mineralogy, which affects the density and cohesive properties; (3) extent of flocculation; and (3) moisture content and extent of consolidation, including that resulting from desiccation on exposure.

The solids content of the slurry, as discharged, also influences the settling characteristics and properties. Thick or viscous slurries, with high solids content, generally deposit homogeneously and form relatively steep beach slopes. Slurries with low solids contents form flat beach angles and generally exhibit particle segregation with coarser particles settling rapidly and successively finer particles settling at increasing distance from the discharge location.

Most tailings disposal operations are managed so as to maximize the extent of beach development; i.e. the area of tailings exposed to the atmosphere. This means that the size of the pool of supernatant water is minimized. The reason for this is to promote solar drying, which enables relatively high densities to be achieved relatively rapidly, thereby increasing the capacity of the storage. In some situations, the exposure of beaches is considered undesirable; for example if this would result in acid generation. In such cases, a pond is maintained over the entire TSF surface, and the tailings slurry is discharged below pond surface.

Potential Impacts

On land, tailings storage facilities are basically large landfills in structure, similar to water impoundment dams in design. By their sheer existence they have a physical impact. They permanently alter the topography of the area where they are constructed. Tailings storages are intended to exist for many centuries. Accordingly, they need to resist the same natural forces that erode and level the surrounding landscape (**Figure 18.5**). Adverse impacts may result from exposure or removal of tailings due to these natural forces. Wind erosion is often the greatest problem in arid areas and water erosion in wetter areas.

On land, tailings storage facilities are basically large landfills in structure, similar to water impoundment dams in design. Tailings storages are intended to exist for many centuries.

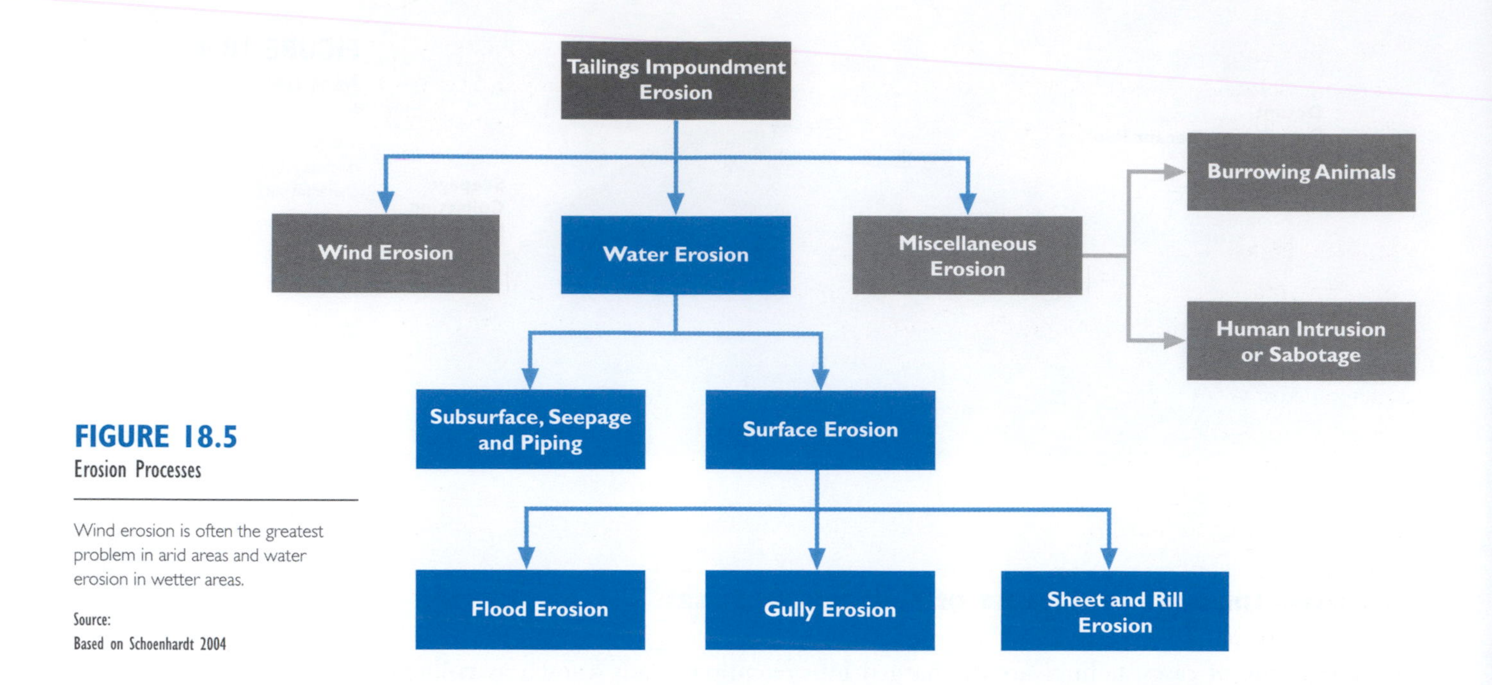

FIGURE 18.5
Erosion Processes

Wind erosion is often the greatest problem in arid areas and water erosion in wetter areas.

Source:
Based on Schoenhardt 2004

Most adverse environmental impacts of tailings are water related.

Most adverse environmental impacts of tailings are water related. As tailings consolidate, 'supernatant water' separates from the slurry, forming a pond on the surface, from where it can be decanted for disposal or, more commonly, for re-use. In tropical regions where precipitation exceeds evaporation, decant water from a TSF may carry significant loads of tailings fines that need to be managed. Once released to the environment, fine tailings will spread over long distances before eventually settling as soft sediments. Sometimes associated with fine particles are elevated concentrations of trace metals that are potentially mobilized by natural chemical or biological processes.

Even when all tailings solids are contained, discharged water may remain harmful to the environment. Commonly, large volumes of water associated with tailings placement are expelled over time. Water enters a tailings impoundment as process water, runoff from surrounding areas, seepage from abutments and precipitation. As with waste rock, the potential for tailings to impact water quality depends on the minerals involved and their geochemistry and the specific conditions at the tailings disposal site. Water infiltrating, emanating from and passing through tailings storages may become acid with elevated concentrations of dissolved trace metals. Chemicals introduced in the beneficiation process may remain in water that is ultimately released to the environment. Some reagents are toxic. Notable examples are cyanide in cyanide-based gold leaching and mercury commonly used in artisanal gold and silver mining. Leakage through the embankments or the TSF floor as a result of ongoing consolidation can continue over many decades, enabling the transport of contaminants into ground or surface water. In most cases however, natural attenuation processes will 'bind' potential contaminants, thereby limiting their travel. If tailings solids are potentially acid-generating with the potential to cause acid or heavy metal contamination of surface water or groundwater, tailings deposits present particular challenges for rehabilitation as discussed in Chapter Twenty One on Mine Closure.

Other environmental concerns relate to the mortality of wildlife or livestock alighting on or drinking from tailings ponds, or becoming bogged in unconsolidated tailings. In arid and seasonally dry areas where surface water sources are few and far between, birds tend to be attracted to tailings ponds. In such situations, bird deaths have been associated with red mud ponds adjacent to alumina refineries, or ponds containing cyanide solutions.

Where mining takes place underground, backfilling of mine openings with tailings is an option. Below ground tailings disposal schemes avoid many of the environmental issues associated with on-land storage. In addition, less tailings are usually produced from underground mining as ore grades are higher and tonnages lower. Potential impacts on water quality, however, remain.

Potential Risks

Environmental risk and risk to human life from catastrophic failure of tailings impoundments as a result of floods, earthquakes, inadequate engineering design or defective construction are of major concern in tailings management. Little imagination is required to understand how very large volumes of wet, unconsolidated tailings, contaminated sometimes with toxic reagents, may impose major risks that can induce a fearful public response. To a large extent, the negative perception of mining in our society originates from past spectacular TSF failures where fear became reality (see ICOLD 2001; USEPA 1995 and UNEP 2001 for a comprehensive listing of past TSF failures). Recent examples include Merriespruit in South Africa, 1994; Marcopper Project on Marinduque Island in the Philippines, 1996 (**Case 18.1**) resulting in the closure of the mine; Los Frailes in Spain, 1998; and Baia Mare in Romania, 2000 (**Case 18.3**). Once released, unconsolidated tailings of low density, may readily flow for large distances, inundating everything in their path. Phillips, 2000, estimates that eleven dam failures since 1965 have resulted in 863 deaths.

To a large extent, the negative perception of mining in our society originates from past spectacular TSF failures where fear became reality.

The risk of catastrophic failure of a tailings impoundment is greatest during the operational phase. Risk decreases after completion of mining due to drainage of excess water, consolidation and desiccation which increase the strength of deposited tailings, and a general aging of the tailings materials, which reduces chemical reactivity. However, abandoned tailings storage facilities continue to present considerable risks, which may be compounded by lack of attention and maintenance. Failures principally occur through over-topping of the embankment, or water finding paths of weakness through the containing structures. Foundation failure may also occur, as at the Los Frailes Mine in Spain. In seismically active areas, collapse may be due to deformation due to earthquake loading or to liquefaction of the stored tailings and/or parts of the foundation. Although relatively rare, collapse due to liquefaction is almost invariably catastrophic. ICOLD and UNEP (2001) concludes that two of the main causes of TSF failure are lack of control of water balance and construction deficiencies. Inappropriate operating procedures have also contributed to many failures.

The risk of catastrophic failure of a tailings impoundment is greatest during the operational phase.

CASE 18.1
Marcopper Tailings Storage Failure 1996

In 1996 a major tailings release occurred at the Marcopper Mine on Marinduque Island in the Philippines. The release occurred when the rock around a concrete plug in a tunnel gave way, releasing approximately 2 million dry tonnes of tailings into the local Boac River system. The tunnel was connected to the abandoned Mt Tapian pit being filled with tailings (intended as an interim storage depending on the outcome of a final sub-sea tailings disposal under study). It is believed that the rock surrounding the plug failed due to the increasing hydraulic pressure of the tailings in the pit. The bulk of tailings discharged in the first 3 days, flowing down as a 'lahar' for most of the 26 km long river into the sea. The release caused significant disruption to the lives of the people in the Boac Valley in one of the poorest province in the Philippines. Fortunately no lives were lost. The operating company, Placer Dom, committed itself to rehabilitation of the river system, the provision of infrastructure, services and compensation to the people in the valley and the permanent sealing of the tunnel, a potential financial commitment up to US$ 80 million (eventually, for various reasons, not all commitments could be fully fulfilled as initially anticipated). More importantly, however, the incident had a major impact on corporate reputation, resulted in mine closure, triggered legal proceedings by the Philippine Government, and continues to influence permitting of new mine developments in the Philippines to the current day.

Source: Based on Brehault (1997)

Objectives of Tailings Storage

The overall goal of a tailings storage facility is to provide secure containment for tailings during operation and, following closure, to retain its integrity in perpetuity. This is not easily achieved, and it is evident that many older tailings storages fall short of this goal. In planning tailings storage facilities, common objectives are to:

- Minimize the footprint of the facility, consistent with other requirements;
- Design and construct retention structures that are stable under all foreseeable conditions, with an adequate factor of safety to protect against uncertainties;
- Provide sufficient freeboard and manage the facility so as to accommodate inflow from the maximum credible operating and rainfall scenarios;
- Minimize, and in some cases intercept and recover, seepage through the base of the storage area and/or beneath the retention structure;

CASE 18.2

Riverine Tailings Disposal of Freeport's Grasberg Copper and Gold Mine in Papua, East Indonesia

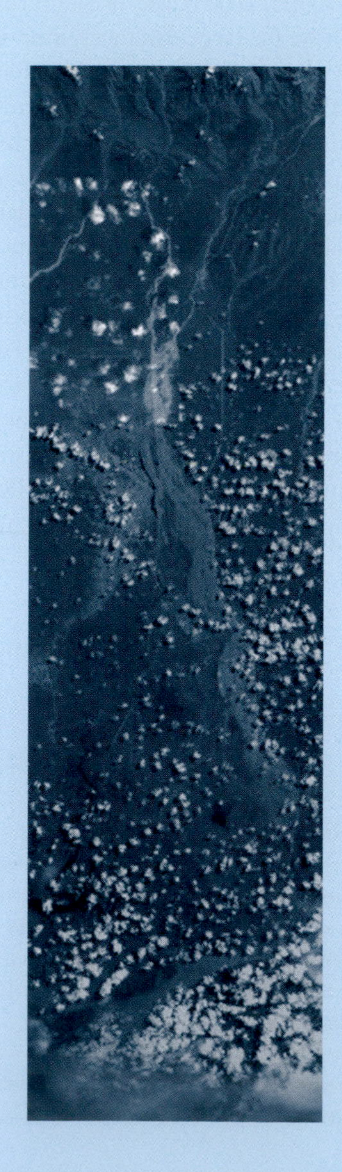

The high-mountain Grasberg mine site lies near the collision point of two tectonic plates – the Indo-Australian plate to the south and the Pacific plate to the north. The open Pit mine is located close to the highest point on the island at an elevation of more than 4,000 metres above sea level. Ore containing copper, gold and silver is sent to a mill/concentrator facility in a valley about three kilometres to the south and approximately 1,000 metres lower in elevation, where the mineral-bearing material is separated from non-economic rock material through a physical separation process by grinding and flotation methods. A pipeline sends metals bearing concentrate to the port facility near the Arafura Sea, approximately 120 kilometres to the south. Freeport uses the Aikwa river system for tailings transport to a designated area in the lowlands and coastal zone, called the 'Modified Deposition Area'. The deposition area is a portion of the floodplain of the river, confined laterally by a levee on each side, encompassing some 235 square kilometres. When mining is completed, it is envisioned that the deposition area which by then will rise up to 30 metres above the surrounding ground, will be reclaimed with natural vegetation or used for agriculture, forestry or aquaculture.

Alternative tailings storage areas were rejected due to lack of capacity or the necessity of building an extremely high dam in a seismically active area with high amounts of precipitation. The use of a pipeline to carry tailings to the deposition area was also rejected because construction and installation of a large diameter pipeline in a harsh terrain over such a long distance had never been attempted before. The integrity of the pipeline would likely be jeopardized by natural events such as landslides, floods and earthquakes.

Tailings enter the river system at rates exceeding 200,000 tonnes per day. While the majority of the tailings is captured in the designated tailings deposition area, tailings fines freely enter the Arafura Sea. The riverine tailings disposal scheme is accompanied by a comprehensive environmental monitoring programme (in 2005, water quality was monitored at over 300 locations throughout the project area, collecting over 7,000 water samples and conducting over 50,000 water quality analyses; and monitoring more than 100 sampling locations for nekton, benthos, plankton and mangrove invertebrates in the aquatic biology programme).

Freeport's riverine transport of tailings differs from the riverine disposal by the Ok Tedi mine in nearby Papua New Guinea in three main respects. First, the OK Tedi mine uses the river system for tailings and waste rock disposal. Second, at Ok Tedi, mine tailings transport occurs over 990 kilometres along the Fly River compared to river transport by the Aikwa River over 80 kilometres. Using hydraulic jargon, the transport capacity of the Ajkwa is much greater. Finally, the Freeport deposition area in the lowlands is laterally confined by a levee system. Only time will tell, however, whether such large-scale tailings management systems can be successfully managed over many decades and adequately rehabilitated.

Source: Based on Riverine Tailings Transport by PT Freeport Indonesia (2006)

- Maximize the settling rate and density of settled tailings solids, and minimize the time required to achieve this density;
- Facilitate recovery and re-use of tailings liquid;
- In cases where the facility poses a potential threat to wildlife or livestock, remove the threat or prevent access; and
- Facilitate rehabilitation of the facility after tailings discharge is complete.

As with waste rock storages, there may be a trade-off between competing objectives; e.g. minimizing the footprint of the facility involves higher retention structures and thicker tailings accumulations, which may reduce stability and make rehabilitation more difficult.

CASE 18.3
Tailings Spill at Baia Mare, Romania, 2000

The tailings storage facility at Baia Mare, designed for 'zero discharge', became probably one of the most publicized tailings spills to date. The scheme at Baia Mare involved reworking of former mine wastes including the construction of a new impoundment and a new efficient processing plant that would accept all tailings removed from the old impoundments.

At the new tailings storage facility an outer perimeter bank 2 m high was built from old tailings which also served to anchor the HDPE liner that covered the entire 90 ha storage area. About 10 m inside the perimeter, starter dams were built to heights of between 2 and about 5.5 m. Cyclones mounted along the crest of the starter dams accepted the tailings piped from the new processing plant, discharging the coarser fraction on to the downstream side to fill the space to the perimeter dam, and raise the whole dam, with the main volume of fine tailings slurry being discharged into the impoundment. Cyanide was used in the new processing plant for the extraction of gold, so that the tailings and water in the new impoundment contained considerable amounts of cyanide.

The first discharge into the impoundment was in March 1999, and during the summer everything worked well although the delivered tailings did not contain quite as much coarse material as had been envisaged and the rate of height increase of the dams was lower than intended. During the winter, however, the temperature fell below zero, freezing the cyclones. Tailings from the processing plant were warm enough to keep the operation working, but there was no further height increase for the dams because the cyclones were out of action.

Precipitation during September to January fell as rain and snow on the whole area of the impoundment. This extra water was stored in the impoundment causing the level to rise under the now thick layer of ice and snow.

On 27th January there was a marked change in the weather. The temperature rose above zero and 37 mm of rain fell. The ice and snow covering melted and the dams, where they were only starter height, were lower than the developing water level. On 30th January 2000, a section overtopped, washing out a breach 25 m long. An estimated 100,000 m^3 of mud and wastewater with a 126 mg/litre cyanide load entered through dewatering channels into the Lapus River, a tributary to the Szamos river and from there into the Tisza river and the Danube upstream of Belgrade and finally entered the Black Sea.

A very large number of fish were killed with serious consequences for the fishing industry for a time. The Hungarian authorities estimated the total fish kill to have been in excess of one thousand tonnes. Water intakes from the rivers had to be closed until the toxic plume had passed.

The concept of a closed system in which none of the process water should escape into the environment should have been excellent, with the new tailings impoundment completely lined with plastic sheeting and provision for the collection of any seepage. While in principle this was a worthy objective, both in environmental and economic terms, the design in each case contained no provision for the emergency discharge of excess waters when overflow threatened. Without specific provision for avoiding overflows, such 'zero discharge' systems are not suitable for use in meteorological conditions of heavy and intense precipitation, nor had the problems of working at low temperatures been addressed. The scheme was one that could have worked well in the hot and dry conditions found in some parts of Australia and South Africa.

Source: BMTF (2000) and BRGM (2004)
Photo: The Mineral Policy Centre

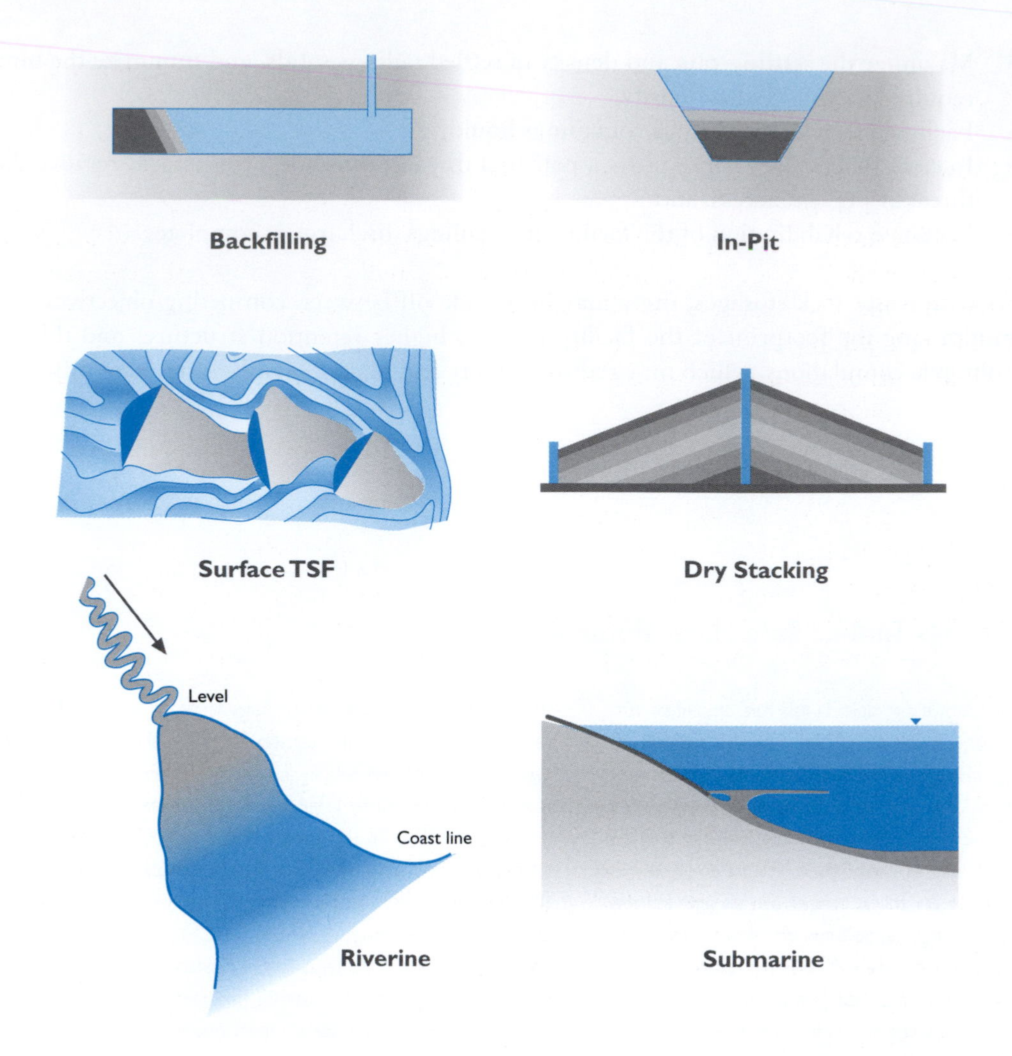

FIGURE 18.6
Alternative Approaches to Tailings Disposal

No 'One size fits all' approach to tailings management exists; various disposal schemes are common, sometime in combination.

18.2 ALTERNATIVE APPROACHES TO TAILINGS DISPOSAL

No 'One size fits all' approach to tailings management exists; various disposal schemes are common, sometimes in combination (**Figure 18.6**).

Uncontrolled Discharge

Uncontrolled discharge onto the ground surface is still common among artisanal miners.

Uncontrolled discharge onto the ground surface was relatively common practice for small operations prior to the mid-twentieth century. It is still common among artisanal miners.

Riverine Discharge

Riverine tailings discharge is an open tailings disposal scheme: tailings slurry is released to the environment without well-defined boundaries.

Riverine discharge was also common practice prior to the mid-twentieth century, even at some larger mining projects, including the Mt Lyell Copper Mine in the Australian state of Tasmania. Contrary to conventional engineered TSFs, where tailings are contained and the spatial extent of impact is well defined, riverine tailings discharge is an open tailings disposal scheme: tailings slurry is released to the environment without well-defined boundaries. The natural river system transports tailings to the final resting place which is not entirely defined. Hence, the extent of the impact remains uncertain.

While land-based tailings storage schemes are readily accepted and submarine tailings placement schemes, although controversial, have also proved to be environmentally acceptable, riverine disposal of tailings, although practised throughout mining history is now considered unacceptable by most authorities. At present, only a few mines, all located in the Asia Pacific region, rely on this practice: the OK Tedi Copper Mine, the Porgera Gold Mine and the Tol Okama Gold Mine, all located in Papua New Guinea; and Freeport's Grasberg Gold and Copper Mine in Papua, Indonesia (**Case 18.2**).

In all current cases of riverine tailings disposal, specific conditions at the mine site dictated riverine tailings disposal. At Ok Tedi, riverine disposal became the default option when a foundation failure occurred during construction of the TSF. On-land tailings storage close to the Freeport's Grasberg mine, located high up in steep mountain terrain, would be physically impossible and tailings transportation via pipeline to the lowlands over long distance and inaccessible terrain has never been attempted.

Environmental impacts on river systems due to tailings disposal are complex. The obvious impacts are related to very high suspended sediment loads and deposition of sediments in downstream areas, including the seabed. In these respects, the impacts are similar to those resulting from glacial rivers such as many in Alaska. The difference is that the ecosystems in glacial areas, have adapted to the high sediment loads over millennia. Coarse materials deposit closer to the discharge point. Fine tailings are easily transported through the entire river system, eventually reaching the coastal environment. In both the Freeport Copper Project in Indonesia and the Ok Tedi Copper Project in PNG, the quantities of discharged tailings have exceeded the transportation capacities of the rivers in their downstream reaches, causing aggradation and eventual bank overflow. In both cases, widespread loss of forest vegetation has resulted from sediment accumulation in forest areas outside the river channels. In the case of Ok Tedi, riverbed levels increased by up to 6 m. In case of the Freeport mine where levees have been constructed to confine deposition, the surface of the tailings deposition area is predicted to rise by up to 30 m. Tailings fines, however, freely enter the Arafura Sea. Trace metals are a concern in riverine tailings disposal systems, with the potential for geochemical changes by which metals may dissolve or precipitate at different distances from the discharge point, due to oxidation, reduction, dilution or other processes. Acid drainage may occur if the tailings contain sulphide minerals.

Today, riverine tailings disposal is only used for a few large mines where there is no practicable alternative and which are economically too attractive to both the host country and the mine developer, to remain undeveloped. While it can be argued that the major economic benefits provided by these mega-projects outweigh the environmental costs, it is unlikely that projects involving riverine disposal will ever be permitted in the future.

Today, riverine tailings disposal is only used for a few large mines where there is no practicable alternative and which are economically too attractive to both the host country and the mine developer, to remain undeveloped.

Surface Tailings Storage

Surface or on-land tailings storage is used by the vast majority of mining operations. In some cases tailings are stored in underground openings or surface pits but, more commonly, the tailings are deposited in purpose built containment facilities or impoundments. Most established mining companies, wherever possible and practical, prefer to store tailings in engineered tailings storage impoundments. Accordingly, tailings solids are confined within well-defined areas. In some cases the tailings of today could be the ores of tomorrow if advances in extraction technologies allow further tailings processing to be economically attractive. However, advances in recovery rates mean that such cases are likely to be less common in the future than in the past.

In some cases the tailings of today could be the ores of tomorrow.

The liquid fraction of tailings or 'tailings liquor' may in some cases contain dissolved substances which pose a potential risk to receiving waters and downstream aquatic ecosystems. In such cases decant water, runoff and seepage from tailings impoundments can be collected and treated prior to discharge or reuse.

The main alternatives in the design of surface tailings storage facilities involve:

1. Layout of retention structures, influenced mainly by site topography;
2. The retention structure itself, usually involving one or more Earth or Earth and rock embankments (concrete dams have also been used), the design of which depends mainly on the properties of materials available locally;
3. The discharge arrangement, which may involve single or multiple spigot discharge points, at fixed or movable locations, and
4. Water recovery or decant systems, which are generally selected to meet operating requirements.

Dry Stacking

Dry stacking of tailings is perhaps the most promising development in recent years as it broadens the options for tailings disposal schemes.

Dry stacking of tailings is perhaps the most promising development in recent years as it broadens the options for tailings disposal schemes. Dry stacking involves filtration of the tailings slurry to remove most of the liquid, producing a 'filter cake' that is unsaturated – typically with a moisture content of 35% or less (**Figure 18.7**). Belt or vacuum filters may be used to accomplish the dewatering. The filter cake is not pumpable, and requires transportation by truck or conveyor system to the ultimate deposition area.

The advantages of dry-stacked tailings are:

1. Most of the process water is recovered for re-use;
2. Most of the remnant process reagents are also recovered for re-use;
3. There is little or no potential for seepage losses, meaning very low potential to contaminate surface water or groundwater;
4. Dry-stacked tailings are amenable to spreading and/or compacting immediately after placement, using conventional earthmoving equipment;
5. Dry-stacked tailings are suitable for the creation of a wide range of stable landforms;
6. Dry-stacked tailings may be formed into high stacks with relatively steep slopes, involving relatively small footprints, and
7. Dry-stacked tailings are suited to progressive rehabilitation.

Low availability of water in arid areas may be the main factor in selecting dry stacking. However, while dry stacking is particularly applicable in low or medium rainfall situations, it should also prove effective in areas of high rainfall; the main management requirement in these situations will be erosion protection for active stacking areas.

FIGURE 18.7
Red Mud Dewatered to a Cake Using a Filter Press

There remains considerable controversy over the economics, feasibility, practicability and environmental trade-offs with dry tailings disposal, especially in areas of high precipitation and runoff. One significant disadvantage of dry stacked tailings is the cost, which ranges from US$ 1.00/tonne to US$ 10.00/tonne, with average cost between US$ 1.50/tonne and US$ 3.00/tonne. This is much higher than for conventional tailings disposal. However, it should be recognized that tailings storages created by dry stacking are inherently far more stable and can usually be rehabilitated more rapidly and at lower cost.

Dry stacking of tailings has not yet been attempted in the earthquake-prone, mountainous, wet tropics, where conventional tailings disposal has proved so difficult. However, *prima facie*, it provides a solution to the problems experienced in these areas. In particular, the prospect of co-disposal of tailings filter cake and waste rock appears particularly attractive. In many cases, the tailings may be suitable to create permeability barriers or encapsulating layers within the storage, assisting in control of acid rock drainage.

Tailings filter cake is well suited for back-filling of underground mine openings – to improve stability and to enable pillars to be safely extracted. Another promising application is for back-filling of shallow lateritic Pits, in which filter cake is placed, spread and compacted within each pit, prior to replacement of overburden and topsoil. Potential economies involve back-loading of ore haulage trucks with filter cake or the use of Innovative Conveyor Systems (www.innovativeconveying.com) which enable ore to be conveyed in one direction and solid waste in the other simultaneously. In such cases, disposal of tailings may be achieved without increasing the overall footprint of the project. Cementitious substances such as Portland cement have been mixed with tailings filter cake to add strength in backfill situations. Judicious use of such additives may also be used to provide erosion resistance for drystacked tailings.

Since dewatering technologies such as filtration are advancing at a rapid rate (Jewell and Fourie 2006), it is likely that costs will be reduced in comparison to alternatives, and that the use of dry stacking will become much more common in the future.

Since dewatering technologies such as filtration are advancing at a rapid rate (Jewell and Fourie 2006).The use of dry stacking will become much more common in the future.

In-pit Storage and Backfilling

As for waste rock disposal, the best sites for tailings storage may be voids remaining at the conclusion of mining, the preferred choice for final tailings placement for many regulatory authorities worldwide. Apart from the low costs involved, this means that the tailings are stored below grade, so that failure by collapse or erosion is physically impossible (USEPA 1994). Potential adverse impacts of in-pit storage of tailings may include:

1. Ore extending below the Pit becomes more difficult to extract;
2. Water discharged with the tailings, if it contains toxic or hazardous constituents, could contaminate groundwater in the vicinity of the storage;
3. Similarly, if the tailings are subject to oxidation causing acid generation, local groundwater could potentially become contaminated. On the other hand, if the storage Pit extends below the water table, this provides an ideal location for storage of sulphidic wastes so as to keep them in a saturated condition, protected from oxidation, and
4. The deep, steep-sided configuration of most mine Pits means that tailings solids will usually deposit sub-aqueously, with no scope for beach development and consequent solar drying. Tailings deposited in this way usually drain very slowly, so that decades may be required before a firm surface can be developed, suitable for rehabilitation.

While relocation of tailings into the mine Pit after completion of mining is too expensive for most mining situations, there are situations where it may be used. Three examples of in-pit tailings storage come from Australia's Northern Territory: Woodcutters Copper Project and Nabarlek Uranium Project, both of which have been closed and Ranger Uranium Project, which continues to operate. In all these cases, conventional surface tailings storage facilities were used during operations, with tailings later transferred to a nearby Pit as part of mine closure. At Nabarlek, wick drains were installed to expedite dewatering of the deposited tailings. Wick drains, usually installed in a close pattern, are commonly used to expedite consolidation of weak soils. They would normally be too expensive for dewatering of tailings. A more cost-effective approach would be to install underdrains and recovery sumps in the Pit prior to tailings discharge.

In underground mines, tailings have long been used as backfill in previously worked out voids.

In underground mines, tailings have long been used as backfill in previously worked out voids. Generally, coarse grained tailings are preferred for this purpose and, commonly, cyclones are used to separate the coarse fraction from slimes, the latter being discharged to a TSF. The tailings are commonly mixed at the surface with a binder, usually cement, and then pumped to fill voids. Using cemented backfill, a room and pillar underground mining operation is able to extract the *in situ* pillars containing ore, thereby increasing overall resource recovery; the cemented backfill provides support, preventing roof collapse and problems with subsidence.

Sub-aqueous Tailings Placement

Typically, the storage of mine tailings in a surface disposal facility is required or preferred. If the tailings are potentially acid-generating, then the tailings solids may need to be protected from exposure to the air. One option is to release tailings in a water-filled storage providing water coverage which effectively prevents oxidation (Tremblay 1998). This is referred to as sub-aqueous placement. The management of sub-aqueous tailings placement in an engineered TSF is similar to conventional on-land disposal, and the interested reader is referred to the references for further discussion.

Lake Disposal

Lake Disposal has previously been mentioned as a means of avoiding acid generation from mine tailings, and accordingly some tailings storages are maintained as lakes following mine closure. There have only been a few examples of tailings disposal in natural lakes, most of them in Canada. Conditions required, in particular, the presence of steep lake bed gradients at the discharge point and deep water where the tailings solids accumulate, are similar to those for marine disposal. Such conditions are uncommon in nature. However, they do occur in some deep glacial lakes in North America and northern Europe, some volcanic caldera lakes in Southeast Asia and deep lakes formed by recent tectonic activity (also found in Southeast Asia and elsewhere).

There have only been a few examples of tailings disposal in natural lakes, most of them in Canada.

A particular phenomenon experienced by many lakes at high latitudes and/or high altitudes (but not in the sea) is that of 'seasonal turnover'. In most parts of the sea, the presence of permanent density stratification in which each layer is denser than the one above, precludes deep water from rising to the surface. However, in lakes subject to seasonal turnover, seasonal temperature changes reverse the density difference between the surface layer and the bottom layer of the water, causing the surface layer to sink to the base of the lake as the bottom water rises. In such circumstances, any contaminated water discharged as part of the tailings stream would be brought to the surface.

In most cases, lakes are not considered as repositories for tailings, even in situations where suitable conditions occur. This may be due to the fact that lakes are generally scenic; they are usually relatively uncommon (Canada is a notable exception) and therefore particularly highly valued as aquatic habitats and also for their intrinsic values; and they often provide water supplies and recreational opportunities. Just the thought of using lakes for disposal of tailings is therefore anathema in most situations, even if a scientific evaluation was to conclude that an environmentally acceptable disposal scheme was feasible.

Just the thought of using lakes for disposal of tailings is anathema in most situations, even if a scientific evaluation was to conclude that an environmentally acceptable disposal scheme was feasible.

It has been argued that following completion of mining, the aquatic ecosystem within the lake can be re-established by raising the bottom of the lake with tailings fill, and raising the water cover over the top of the tailings with fresh or treated water (Frazer and Robertson 1994; Sly 1996; Robertson 1998). This, so it is said, effectively creates a 'natural design' at closure, which allows sunlight to penetrate to the lake bed, where previously it did not. Such 'enhanced natural design' further allows plants in the lake to produce oxygen, improving the biology and fishery of the lake. While the sub-aqueous disposal of tailings to natural water bodies is appealing, the actual circumstances where this may be acceptable are probably uncommon, and some of the potential environmental consequences are not fully understood.

The application of sub-aqueous underwater tailings placement in natural water bodies is a relatively common practice in Canada, a leading mining nation. In the past decade, six mines have been approved and fisheries compensation measures implemented at Canadian mine sites, which have been successful in restoring fisheries and wetlands habitat. Examples include: the Kemess Mine and Benson Lake in British Columbia, Buttle Lake on Vancouver Island, and Mandy and Anderson Lakes in Manitoba (www.trustingold.com). Canada has also successfully employed this technology to mitigate the potential impacts of acid rock drainage from reactive waste rock by placing the material under water.

Submarine Tailings Discharge

Marine discharge of tailings has been carried out at many sites, and has varied between shoreline disposal and discharge at great depths. The most recent and most successful examples use what is termed Deep Sea Tailings Placement (DSTP), which involves discharge of tailings slurry at the seabed in water depths of 100 m or more, with the tailings slurry flowing down the seabed slope as a density current before depositing on the seabed at depths of 400 m or more.

Unfavourable terrestrial conditions, such as a steep terrain combined with high rainfall and seismic activities, characteristic for countries such as the Philippines or Indonesia, may lead to unacceptable risks in case of on-land TSF and submarine tailings disposal may become the preferred option.

Unfavourable terrestrial conditions, such as a steep terrain combined with high rainfall and seismic activities, may lead to unacceptable risks in case of on-land TSF and submarine tailings disposal may become the preferred option.

18.3 SURFACE TAILINGS STORAGE

Traditionally, tailings management has used surface impoundments to retain tailings and mill effluent. This practice in the USA is supported by the 1982 USEPA research report: *Development Document for Effluent Limitations Guidelines and Standards for Ore Mining and Dressing – Point Source Category*. In this document USEPA profiles a variety of mining sources (gold, silver, copper and molybdenum mines, etc.) regulated under the Clean Water Act, discusses New Source Performance Standards, and outlines Best Available Treatment and effluent limitation guidelines. In the USA this document remains the basis for designating tailings ponds and settling/treatment as best available treatment technology. The factors that influence design and operation of a TSF include: (1) general design criteria, (2) tailings pulp density, (3) types of layouts, (4) embankment design, (4) tailings discharge, (5) common water discharge systems, (6) operation consideration, (7) water balance, (8) site selection, (9) site investigation, (10) risks, (11) monitoring, (12) auditing, and (13) TSF closure.

General Design Criteria

The design approach and safety requirements for tailings embankments are often adapted from the practice of conventional dam engineering and a wealth of knowledge exists.

TABLE 18.2

Target Design Criteria for Tailings Impoundments – After mining the TSF must remain as safe as during mine operations

Engineering and Construction	
Cofferdams and diversions	1:10 year < Design < 1:100 year
Factor of safety against slope failure	>1.3
No specific earthquake allowance (included in slope safety factor)	
Operation	
Spillways	1 in 1000 year flood
Earthquake	1 in 100 year event to 1 in 475 year event
Static safety factor for slopes	1.3 to 1.5
Erosion protection on slopes	1 in 1000 year rainfall
Closure	
Spillways	Probable maximum flood
Earthquake (some damage, minimal release)	Maximum credible earthquake
Static safety factor for slopes	
Trap Efficiency	
Average trap efficiency for solids	95%
Range of trap efficiencies	90–96%
Minimum size 100% trapped	0.004–0.1 mm
% trapped for fractions <0.004–0.01 mm	40% to 50%
Effluent D50	<0.004

Source:
Environmental Workshop 14–18 October 1996 Newcastle, New South Wales Australia, Proceedings Volume 2, Mineral Council of Australia p.189

Basic engineering methods such as dam stability or foundation and seepage analysis are essentially the same. Nonetheless, some aspects of tailings storage are specific such as the need to control potential environmental contamination and to design for a long closure phase. Some of the basic design criteria for TSF follow. Target design criteria commonly used for tailings impoundments are summarized in **Table 18.2**:

- Tailings Storage Facilities are designed to facilitate closure objectives.
- Embankments are engineered and constructed from local soil, waste mined rock, coarse tailings or a mixture of these materials.
- Embankments are designed to accommodate all tailings generated over the mine life, possibly allowing some contingency.
- Tailings impoundments are usually developed in stages to minimize initial capital expenditures. Embankments may be raised many times during the mine life; commonly at the rate of one raise each year.
- Allowances in the design are made for possible future expansion of the TSF to accommodate future potential increases in proven reserves.
- Diversion channels are often provided to intercept and divert clean surface water, which would otherwise flow into the TSF. This helps in overall water management and reduces the freeboard required to avoid overtopping.
- Decant water systems are designed to accommodate extreme rainfall events.
- Tailings slurry is commonly discharged from the perimeter of the impoundment above the level of previously placed tailings. This is known as aerial disposal, as distinct from sub-aqueous disposal. Near the discharge point(s), deposited tailings contain a higher proportion of coarser particles, because these settle more quickly, forming a tailings beach. Further from the discharge point(s), tailings deposits contain a higher proportion

TABLE 18.3

Comparison of Different Pulp Density and Disposal Methods – Along the continuum of slurry, thickened, paste, dry tailings and finally cemented tailings, the solids contents and costs of production increase

Pulp Density/Disposal method	Advantages	Disadvantages
Slurry (Conventional)	Low capital costs Low dewatering and operation costs Easy to pump	Large storage volume required High environmental risk Large water handling High embankment costs Difficult to rehabilitate slimes Challenging post-mine care
Thickened	Less storage volume required Low dam construction cost Reduced seepage	Increased dewatering costs High capital costs High pumping costs Large surface area
Paste	Less storage volume required Low dam construction cost Large beach area maximizes tailings density and strength Minimized seepage	Increased dewatering costs High capital cost High pumping costs
Dry	Minimum storage volume required Lower seismic risk Surface compaction possible Recycling of process chemicals Minimized seepage Co-disposal with waste rock Progressed rehabilitation	High capital cost High transport costs
Cemented	No or small dams required Erosion resistance Seismically stable	Very expensive Difficult to transport

of fine particles or slimes. Due to this segregation of particle sizes, tailings characteristics such as strength, permeability and shrinkage vary laterally across a TSF.

- Large changes in TSF operating regime typically occur during the production phase.
- After mining the TSF must remain as safe as during mine operations.
- Tailings storages are designed to contain or to control actual or potential contamination.

Tailings Pulp Density

Tailings pulp density and strength increase as water content is reduced. This can be managed at a cost: tailings come as slurry, thickened, pasted, dry and cemented tailings. **Table 18.3** summarizes how tailings pulp density influences tailings disposal.

Disposal of tailings as an un-thickened slurry (termed wet tailings) is the most common form of tailings disposal. In this method, tailings slurry is pumped from the process plant to an impoundment where the tailings undergo sedimentation, consolidation and desiccation. The embankments are designed to retain both water and solids.

Disposal of tailings as an un-thickened slurry (termed wet tailings) is the most common form of tailings disposal.

There is considerable confusion in the mining industry regarding the definition of thickened and paste tailings. Thickened tailings imply high density slurry while the latter refers to the paste-like and non-segregating nature of tailings. In these methods the tailings are thickened to high-density slurry or paste consistency using a high rate or deep thickener or similar equipment. Conventional centrifuge pumps can pump thickened tailings. Paste tailings require positive displacement pumps for their transfer. Tailings paste plants are usually located close to the tailings discharge point.

Dry stacking of tailings, a technique pioneered by Alcoa, requires that the tailings are de-watered to a 'cake' using vacuum belt filters, drum filters or similar (**Figure 18.7**). The cake is not pumpable and must be transferred to the disposal site by trucks or conveyor systems. Dewatered tailings can be disposed of in engineered piles on prepared ground. Embankments may be constructed around the perimeter of the site to collect runoff and seepage from the pile, thus preventing it from escaping into the environment. With cemented tailings, the thickened tailings are mixed with cement to generate a product that is resistant to erosion and liquefaction under seismic loading.

Along the continuum of slurry, thickened, paste, dry tailings and finally cemented tailings, the solids contents and costs of production increase. However, the requirement for large embankments reduces and the resistance to erosion and liquefaction under seismic loading increases.

FIGURE 18.8
Typical Tailings Storage Layouts

Topographic conditions usually dictate the configuration of above grade tailings storages. As with waste rock storages, additional storage capacity can be obtained by taking advantage of natural depressions in the landscape.

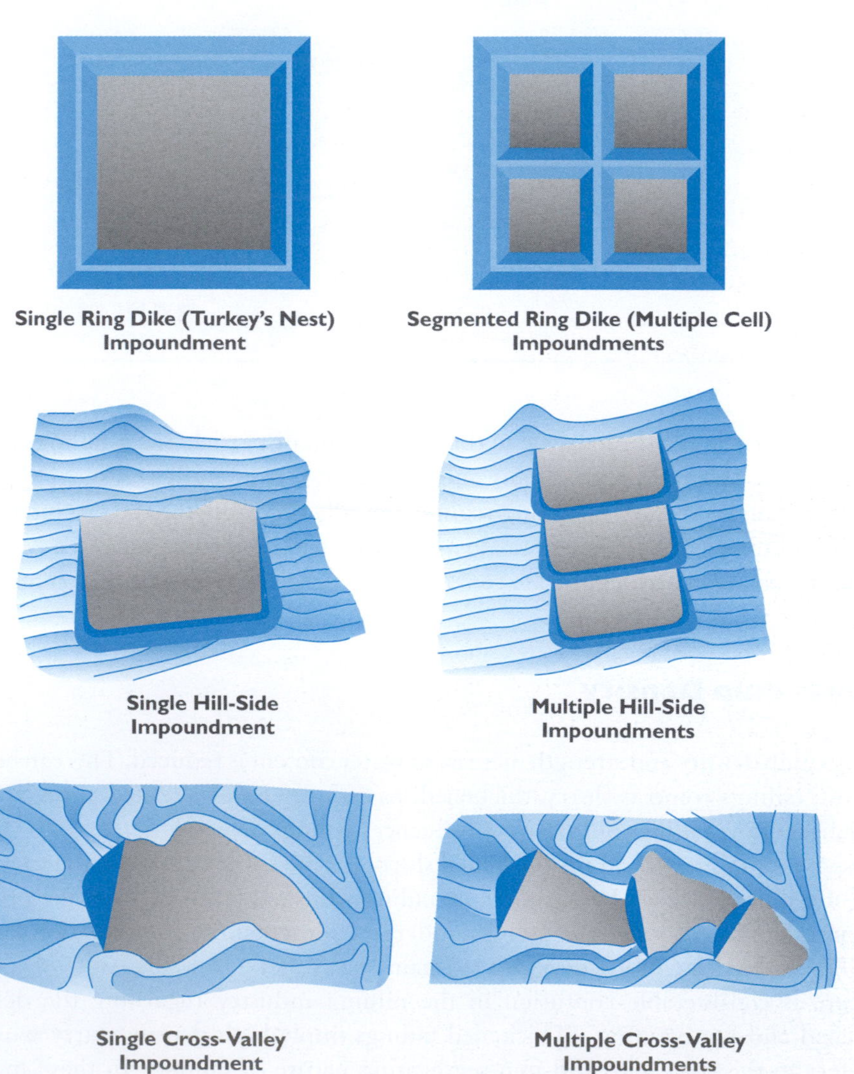

Types of Layouts

Topographic conditions usually dictate the configuration of above grade tailings storages. As with waste rock storages, optimal storage capacity can be obtained by taking advantage of natural depressions in the landscape. The main types of storage are (**Figure 18.8**):

- Turkey Nest or Ring Dyke storages, involving an embankment or dyke around the complete perimeter of the facility as shown in **Figure 18.9**. This layout provides the least storage volume per unit of area, but represents the only option on flat sites. Commonly in these situations, a series of adjacent cells are developed during the life of the project;
- Horse Shoe storage, involving a 'U'-shaped embankment, constructed on gently sloping land or at the base of a valley slope; and
- Valley storage, in which a dam is constructed across a valley (**Figure 18.10**). In mountainous areas, such sites may represent the only available storage options. Where possible, the storage is located near the head of the valley, to minimize inflow from the upstream catchment area. In other situations, it may be necessary to divert water around the storage.

Topographic conditions usually dictate the configuration of above grade tailings storages.

Embankment Design

Most tailings retention structures are Earth or Earth and rock embankments. These range from low, homogeneous dykes similar to many farm dams, to more elaborate zoned embankments incorporating sealing layers, filter zones and drainage layers. To minimize construction costs, mine waste and even tailings themselves are used in the embankments. To minimize capital costs, the embankments are usually raised in stages, with each stage typically providing additional storage capacity for one to three years of tailings production. Three approaches have been used for staged construction of tailings embankments: (1) downstream construction, in which each increase in embankment height, involves downstream extension of the embankment; (2) upstream construction, in which successive layers are placed over deposited tailings; and (3) centreline construction, in which successive

To minimize capital costs, the embankments are usually raised in stages.

FIGURE 18.9
Tailings Impoundment at Chatree Gold Mine (Thailand) soon after Commissioning

Under-drains have been installed in a 'herring-bone' pattern, to facilitate drainage and consolidation of tailings.

FIGURE 18.10
Tailings Impoundment at Mt Muro Gold Mine (Indonesia)

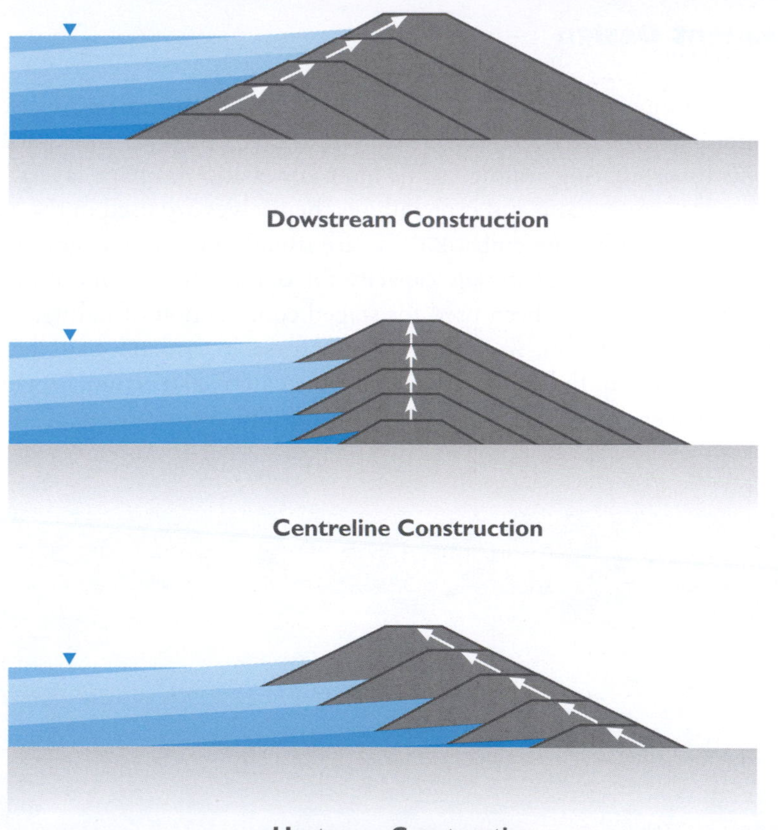

FIGURE 18.11
Alternatives Methods for Staged Embankments Construction

Risk of failure is generally highest for upstream construction and lowest for downstream construction.

layers extend both upstream over deposited tailings and downstream, retaining the same embankment centreline. These arrangements are depicted on **Figure 18.11**. The starter dam is conventionally constructed with non-acid generating borrow material.

For upstream raising using tailings, desiccated tailings are excavated from inside the wall and used for raising the existing wall and deposited tailings, with a cover of benign waste rock placed on the downstream face, crest and upstream face to prevent the erosion

TABLE 18.4

Comparison of Surface Impoundment Embankment Types – Increasingly, regulators, financial institutions and insurers are requiring that tailings retention structures be designed as water retention structures. Essentially, this means that an embankment needs to provide the mass, strength and sealing properties as if only water is stored

	Water Retention	Upstream	Downstream	Centreline
Mill Tailings	Suitable for any type of tailings	At least 40–60% sand in whole tailings. Low pulp density is desirable to promote draining-size segregation	Suitable for any type of tailings	Sands or low-plasticity slimes
Discharge Requirements	Any discharge procedure suitable	Peripheral discharge and well-controlled beach necessary	Varies according to design details	Peripheral discharge of at least nominal beach necessary
Water Storage Suitability	Good	Not suitable for significant water storage	Good	Not recommended for permanent storage. Temporary flood storage acceptable with proper design.
Seismic Resistance	Good	Poor in high seismic areas	Good	Acceptable
Raising Rate Restrictions	Entire embankment constructed initially	Less than 4.5–9 m/yr most desirable. Greater than 15 m/yr can be hazardous	None	Height restrictions for individual raises may apply
Embankment Fill Requirements	Natural soil borrow	Natural soil, sand tailings, or mine waste	Sand tailings or mine waste if production rates are sufficient, or natural soul	Sand tailings or mine waste of production rates are sufficient, or natural soul
Relative Embankment Cost	High	Low	High	Moderate

Source:
Vick (1983)

of tailings by water or wind. Upstream raising can also be carried out using waste rock or borrow material placed on top of deposited tailings that have been allowed to desiccate.

Table 18.4 highlights the very much greater volume of embankment material required for downstream raising compared to upstream raising, and the downstream advance of the toe of the containment wall in downstream raising. The schematic diagrams in **Figure 18.11** do not include details about internal drainage or clay cores within the containment walls, which may be required to ensure geotechnical stability and/or to control seepage.

Downstream construction represents the safest, most conservative approach, but also the most expensive unless waste rock suitable for the bulk of the embankment is available from the mine. Both upstream and centreline construction require that the deposited tailings, where they will be overlain by embankment fill, have achieved sufficient strength to support the imposed loads without unacceptable deformation. In practice, it is common to undertake a geotechnical investigation to evaluate *in situ* strength properties, as a basis for the decision as to which approach is used for the next embankment raise.

Due mainly to its low cost, upstream construction has been used for most tailings embankments worldwide, until recently. However, experience indicates that upstream construction involves the highest risks. Dam failure can occur if the decant pond encroaches on the embankment so that the phreatic surface of groundwater within the embankment rises above the downstream toe. While the phreatic surface location is important for all types

Downstream construction represents the safest, most conservative approach, but also the most expensive.

Due mainly to its low cost, upstream construction has been used for most tailings embankments worldwide, until recently. However, experience indicates that upstream construction involves the highest risks.

of embankment, it is more complex for upstream embankment types. Upstream dams are particularly susceptible to liquefaction under severe seismic ground motion. This motion may result from earthquakes, from nearby mine blasting, or even from nearby motion of heavy equipment. Stability also decreases with increased rates of raising the embankment. Vick (1983) considers raising rates greater than 15 m/yr as hazardous. High raising rates can produce excess pore pressure within the deposit, decreasing stability.

In the past, many tailings retention embankments were constructed largely or entirely of tailings. These designs usually depended for their long-term stability, at least in part, on strength provided by the settled tailings. In some cases the sealing properties of the tailings were required to minimize seepage from the storage. In the Western Australian Goldfields, many such structures remain intact decades after being completed and abandoned. Others have failed or have been damaged by erosion. In some cases, coarse sand used as fill to raise the embankment, has been obtained from the tailings, usually by cycloning to separate the sand from the fines that could impede drainage. Some very large tailings storage facilities have been constructed in this way. One at the Padcal Copper Mine in the Baguio district of the Philippines actually survived the Baguio earthquake of July, 1990, except that blockage of the portal of an upstream diversion tunnel caused water to flow across the storage and down the embankment face, resulting in scouring, but leaving most of the storage intact.

Increasingly, regulators, financial institutions and insurers are requiring that tailings retention structures be designed as water retention structures. Essentially, this means that an embankment needs to provide the mass, strength and sealing properties as if only water is stored; i.e. the additional strength and sealing properties of the deposited tailings cannot be included in stability and seepage calculations. This has increased the stability and safety of tailings retention structures and has also increased the costs. However, embankment design based on water retention may not provide sufficient stability in the event of an earthquake, as the dynamic forces imposed by liquefied tailings exceed those imposed by water.

A particular type of tailings storage was developed by a Canadian – Eli Robinsky (1968). This system known as the Central Discharge or Thickened Tailings Discharge (TTD) uses tailings thickened to the maximum extent consistent with being pumped, which varies considerably between different types of tailings but is typically 60% solids. The thickened tailings are discharged from a riser located in the middle of the tailings storage. Excess water including runoff from precipitation, drains to a low dyke as depicted in **Figure 18.12**. Deposition results in high beach angles, typically 1.1° to 3.4°, which assists rapid desiccation. The resulting tailings deposit is in the form of a shallow cone, which provides an attractive landform, readily accessible and amenable to rehabilitation. Other, natural-looking landforms, including fans and coalescing fans, can also be formed on gently sloping land surfaces. Tailings deposited in this way do not segregate, so that properties will be relatively consistent throughout the storage. A major benefit of this and other thickened tailings systems is the recovery of most water and process chemicals.

Perhaps the most advanced tailings storage designs in common use are those that follow the requirements of the US State of Nevada. These incorporate all or most of the following features: (1) underdrains at the base of the tailings deposit; (2) double liners with leak detection between; (3) liquid recovery systems; (4) drains beneath the embankment; and (5) secondary containment. A typical cross-section of a tailings retention system of this type is shown on **Figure 18.13**.

Clearly, tailings retention structures designed to meet these stringent requirements are expensive to build. For many tailings storage facilities which will not receive toxic or hazardous constituents, simpler and much less costly designs may be entirely appropriate.

Most surface tailings disposal systems involve settling of the tailings solids within an impoundment and recycling of the supernatant water. The exceptions are some operations

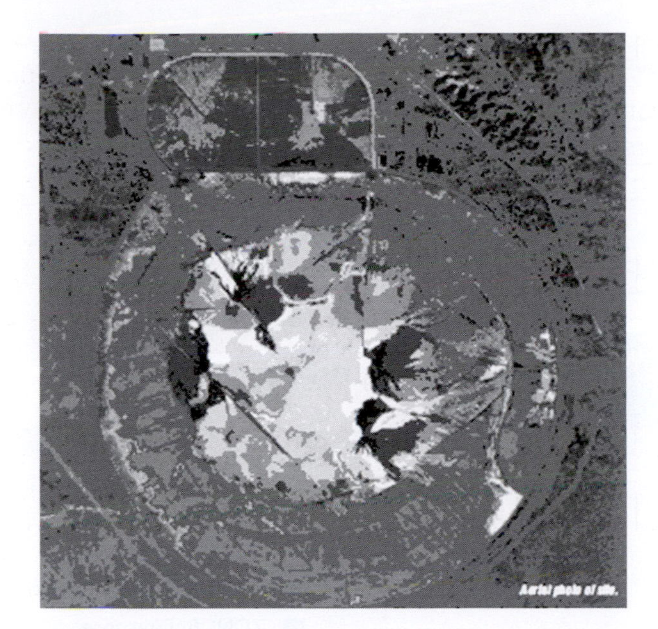

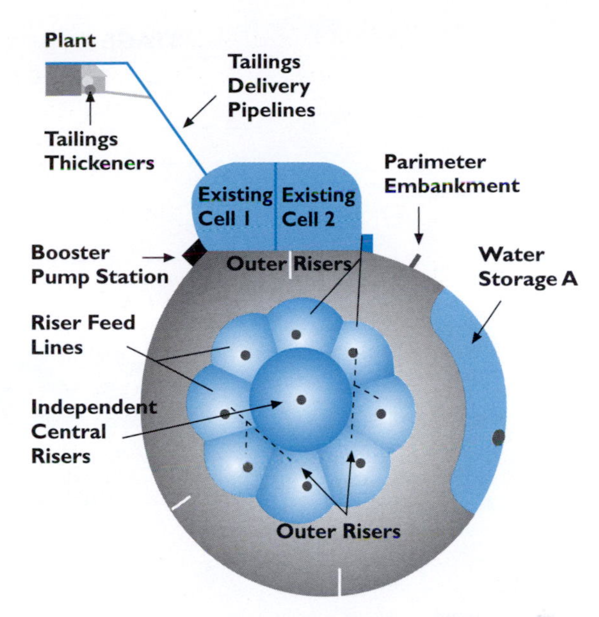

Diagram of the Centralized Discharged Tailings Storage Facility

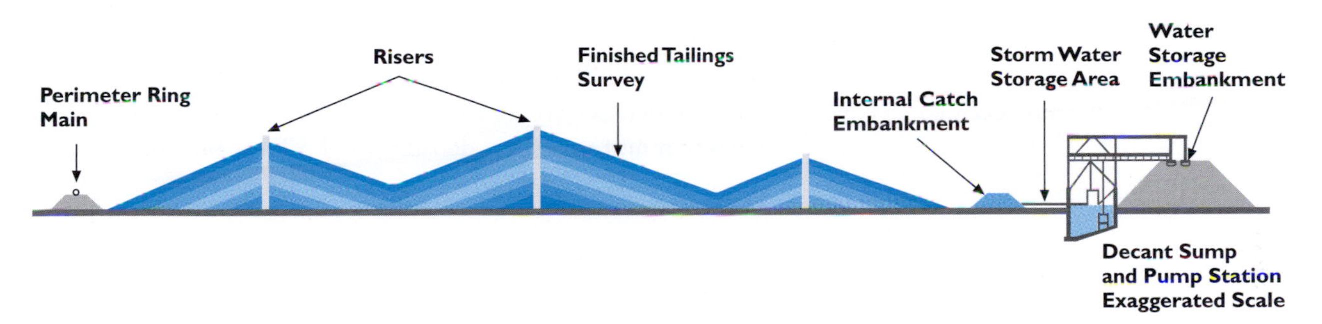

FIGURE 18.12

Schematic East/West Section through Storage Area Exaggerated Vertical Scale

Thickened Tailings Discharge uses tailings thickened to the maximum extent consistent with being pumped. The thickened tailings are discharged from a riser located in the middle of the tailings storage. The resulting tailings deposit is in the form of a shallow cone, which provides an attractive landform, readily accessible and amenable to rehabilitation.

Source:
ICME 1998

in arid areas such as central Australia where tailings supernatant is evaporated, rather than recycled. By contrast, in high rainfall areas, there is commonly an excess of liquid, so that, even with recycling there is excess water which may require treatment, prior to release.

One of the important properties of any particular tailings is its 'beach angle' which is the angle to the horizontal formed by the tailings solids as they deposit. The beach angle depends on the nature of the tailings, particularly its particle sizes and the density and viscosity of the slurry. Beach angles formed by under-water deposition are usually very low – much less than 1°. However, densities formed by deposition above the pond level may achieve 3° or more. The beach angle can be determined during feasibility studies and is an important parameter in the design and operation of tailings impoundments.

Another important property of tailings is the settled density of solids that is achieved, both initially and over time. In all cases, the higher the density that can be achieved, the better. The main reason for this is that higher densities mean that more storage can be achieved within the same volume. Higher densities also mean higher strength, which is of critical importance if upstream or centreline constructions are used. Saturated tailings may consolidate extremely slowly under their own weight. However, the surface of tailings exposed to the atmosphere dessicates rapidly leading to widespread cracking but with the intercrack soil 'slabs' having a much higher density than at the time of initial exposure. The next layer of tailings fills the cracks and adds an additional layer which, on exposure also desiccates, and so on. In this way it is possible, except in areas where

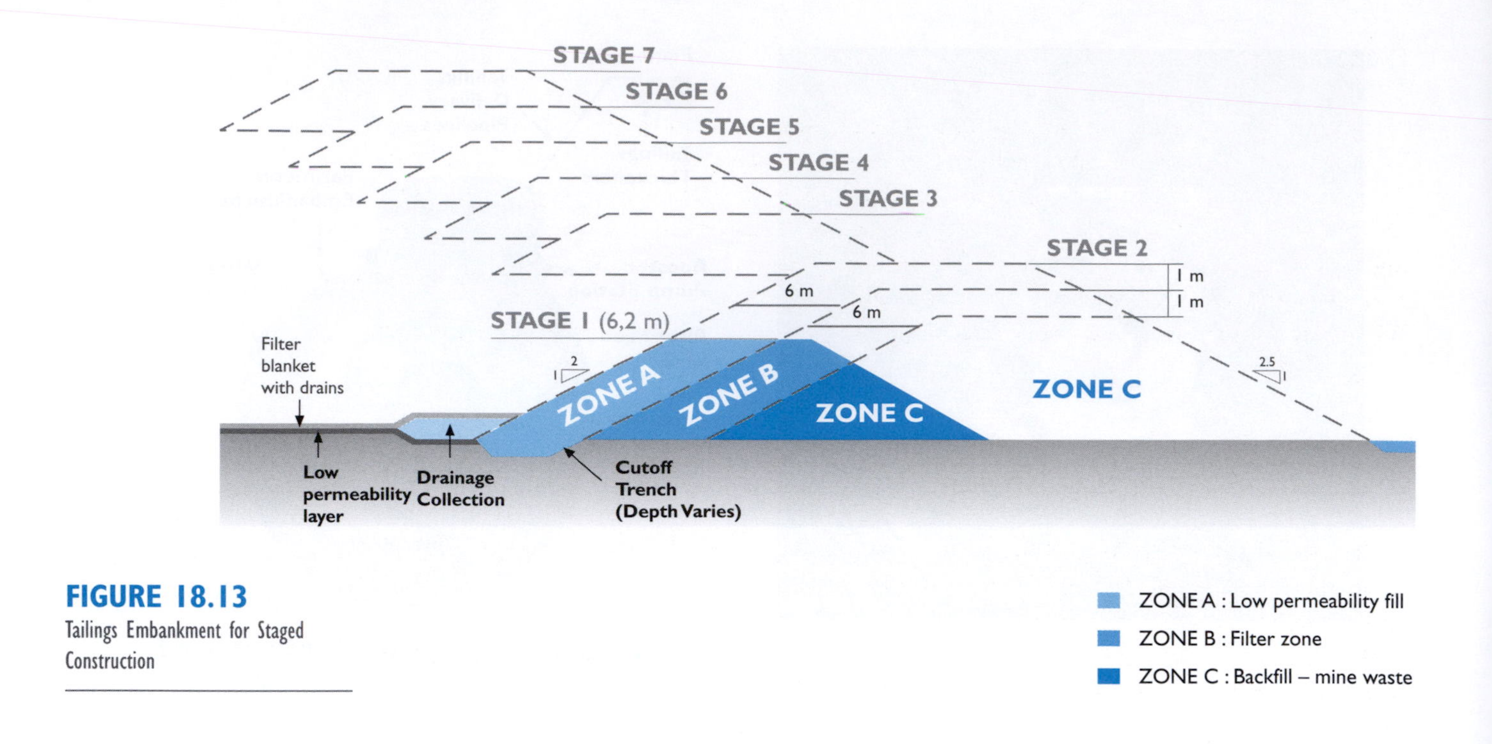

FIGURE 18.13
Tailings Embankment for Staged Construction

precipitation exceeds evaporation all year, for significant increases in density to be achieved. For example, a common initial settled density may be 0.9 to 1.0 T/m^3 (dry density), while after repeated cycles of deposition and exposure, densities of 1.3 T/m^3 or more may be achieved.

Tailings Discharge Systems

Tailings discharge systems are designed and operated to distribute the tailings so that it accumulates as planned, and to provide a pool of clear water (the 'decant pool or pond') from which the supernatant water is decanted and returned to the mill for re-use. For a long, narrow impoundment such as many valley impoundments, discharge may be from one or two locations at one end of the storage with a decant pool at the other end. For a broader impoundment, discharge is usually through multiple spigots located around the perimeter of the storage. Spigots are opened and closed to provide a more or less even accumulation of tailings within the impoundment and also to confine the decant pool in its intended location.

Tailings are sometimes segregated before being discharged to an impoundment. Cyclones may be used to extract the sand fraction from the run-of-the-mill tailings, and then used to raise the embankment or as backfill in underground workings. The remaining fines or slimes are then deposited via pipes or spigots into the impoundment.

Control of the decant pool is an important management procedure for most TSFs. Inadequate control may result in the reduction of free-board and hence increase the risk of overtopping; in the decant pool encroaching the embankment and hence increasing pore pressure; in high seepage rates and in embankment settlement (Engels and Dixon-Hardy 2004) – all of these are potential contributors to impoundment failure. Generally, the aim is to minimize the size of the decant pool, consistent with producing a clear decant. Accordingly, regular monitoring of the decant pool is required to operate a TSF safely and effectively.

Regular monitoring of the decant pool is required to operate a TSF safely and effectively.

Decant Systems

Decanting is important, not only to maximize re-use of process water, but also to maximize the area of exposed beaches subject to solar drying, and to maintain a safe free-board. The system should provide a high capacity for removing storm-water to allow for consecutive storm events, ensuring a safe free-board also under unusual but possible meteorological events. As a guide the decant system should be able to remove storm-water within 2 to 4 weeks, subject of course to prevailing climate conditions. The main types of decant systems are: (1) floating decants; (2) decant towers; (3) inclined decant: and (4) perimeter sump – for central discharge systems.

Floating Decant System

Floating decant systems are probably the most common systems in use. Essentially, this system involves a pump mounted on a floating pontoon, usually secured in place by mooring cables. The depth of intake and the pontoon location can be varied if necessary. However, in practice, the location and size of the decant pool are managed by the discharge locations and the rate of decant pumping.

Standby pumps and diesel generators are good practice to use in emergency or when the decant system cannot cope with unplanned water surge.

A floating decant system requires power to operate the pumps, adding to operating costs. A constant and reliable power source is required since loss of power will interrupt the decant system. Standby pumps and diesel generators are good practice to use in emergency or when the decant system cannot cope with unplanned water surge (e.g. in storm conditions).

Decant Tower

Decant towers are permanent installations established at the first stage and raised progressively as the tailings accumulate. Supernatant water overflows the rim of the tower, until the tailings level approaches the overflow level, at which time the tower is raised. One or more submerged pumps are located at the base of the tower, discharging to the mill via a buried pipeline. Access to the tower is usually via a causeway connected to the embankment.

Decant towers are effective at removing ponded water from a TSF but they are also problematic.

Decant towers are effective at removing ponded water from a TSF but they are also problematic (Engels *et al.* 2006). Decant towers are solid structures while tailings around them continue to settle over time. This differential movement, together with the ever-increasing weight of the tailings can crack and damage the decant system (Engels and Dixon-Hardy 2004), eventually leading to impoundment failures (ICOLD and UNEP 2001). Failure may be subtle and hence remain unnoticed over a long period over time. The Stava disaster in Italy in 1985 serves as a reminder of a decant conduit failure that led to loss of 269 lives. The rise in the phreatic water table in the embankment due to the failure of the decant system caused a rotational slip in the upper embankment. Tailings from the upper impoundment inundated the lower impoundment which eventually overtopped and failed (Penman 2001; Davies 2001). Tailings escaped down the hillside and flooded the town of Stava resulting in one of the world's worst tailings disasters in terms of loss of human life.

Inclined Decant

Inclined decants are similar to decant towers except that, instead of a vertical tower, the conduit follows the slope of the embankment or side of the impoundment. The conduit contains a series of ports into which the supernatant water flows. As the level of the settled tailings rises toward the active port, this port is plugged, and the pool level rises until overflow commences into the next highest port.

Perimeter Sump

Perimeter sumps are used in the case of central discharge system storages and a few other cases where relatively little water is available for recycling.

Operating Considerations

Operating requirements vary from case to case. However, the most common are:

- Managing the discharge and decant systems so that the decant water is clear; i.e. it contains no tailings solids.
- Consistent with the above requirement, to minimize the pool area, thereby maximizing the area of exposed tailings beaches.
- Achieving a relatively uniform accumulation of tailings around the perimeter of the impoundment.
- In some cases where the tailings embankments are not designed as water retention structures, avoidance of the decant pool encroaching on any part of the embankment.
- In other cases (e.g. the tailings/water storage for the Dizon Copper Project in Zambales, Philippines), the tailings impoundment also serves as the major water storage for the project. In such cases, water storage considerations take precedence over optimization of tailings solids storage, the solids settle sub-aqueously and remain covered by water.
- Commonly, flocculant is added to the tailings stream to assist in settling of fine tailings particles and to maintain a clear decant.

Segregation of Tailings

Coarser particles and any denser minerals will tend to deposit close to the discharge point(s), with successively finer particles depositing at increasing distances.

As previously mentioned, tailings particles discharged as slurry into conventional tailings impoundments, will segregate as they deposit, due to the higher settling rates of larger particles. As a result, coarser particles and any denser minerals will tend to deposit close to the discharge point(s), with successively finer particles depositing at increasing distances. The finest materials (slimes) will generally underlie the decant pond. A tailings deposit with distinct segregation and permeability zones develops (Kealy and Bush 1971) (**Figure 18.14**). The slimes fraction of the tailings can be retrieved and applied to areas of an impoundment that have high permeability zones that are causing seepage problems. The blending will help to lower the permeability of the deposit.

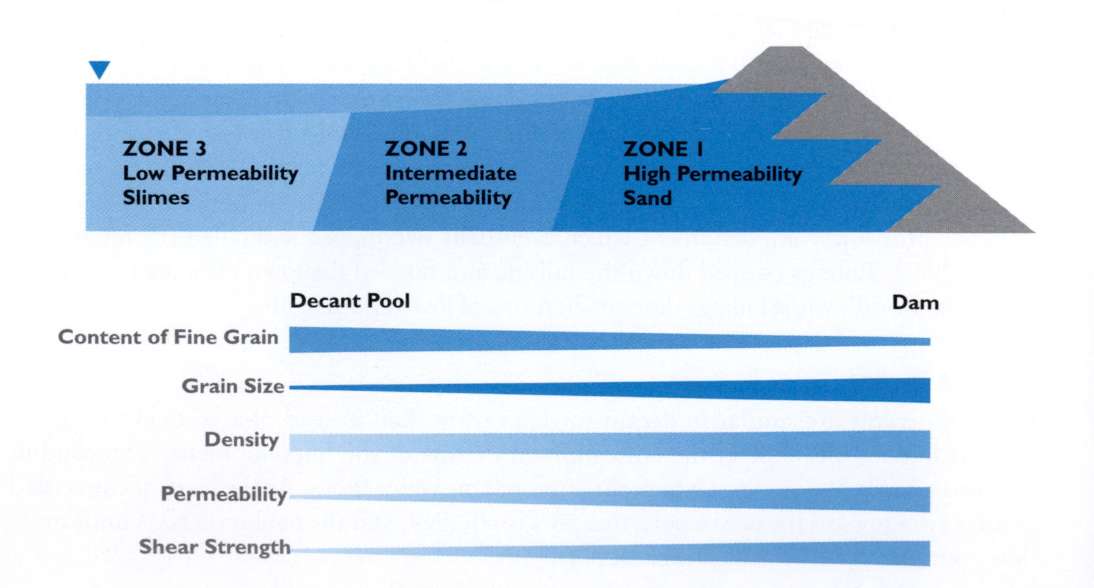

FIGURE 18.14

Tailings Segregation Zones and Soil Characteristics as a Function of Distance from Spilling Location

The coarser fraction tends to settle out closer to the tailings discharge system and the slimes travel furthest away generally towards the decant pool. A tailings deposit with distinct segregation and permeability zones develops.

Source:
Based on Schoenhardt 2004 and ATV-DVWK-M503 2001

Water Balance

As with water dams, understanding the water balance of a TSF is essential to successful operation. In fact water is cited as the underlying source of virtually all failures of tailings storage facilities between 1980 and 1996 (Fourie 2003). All these failures occurred at facilities that store their tailings by conventional impoundment methods. Fourie (2003) also notes that the presence of large quantities of stored water is the primary factor contributing to most of the recent tailings storage failures. The risk of failure of a conventional tailings facility can be reduced by having good drainage and little (if any) ponded water. Simply put, 'no water, no problem' (Engels 2006).

The risk of failure of a conventional tailings facility can be reduced by having good drainage and little (if any) ponded water.

However, as discussed, water management is not only required to ensure long-term integrity of the TSF, but also to protect the local and regional natural water regime. When developing water management measures for protecting natural water resources from the potential effects of tailings disposal, the appropriate initial response is to define the natural resource that needs protection. Accordingly, both the receiving waters and the beneficial water uses that could be affected by tailings disposal should first be identified. This includes identifying locations of water resources and potential users for both groundwater and surface water, average and extreme flow rate for both wet and dry cycles, water quality variation over time, surface water biota, depth to groundwater table and aquifer characteristics.

The next step is to assess the water balance of the TSF and to design water management measures. Several elements contribute to the water balance of a TSF (**Figure 18.15**): (1) process water that easily drains; (2) net precipitation (rainfall minus evaporation); (3) overland flow of surface water into the TSF (run-on); (4) upward directed flow of pore water which occurs as tailings consolidate (resident water content in tailings); (5) loss (or gain) of groundwater; and (6) seepage through embankment.

Process water that readily separates from tailings solids, together with net precipitation comprise the decant water. Depending on its quality, decant water is directed to silt ponds or water treatment plants for further treatment prior to final release to the environment or prior to reuse as process water. Diversion drains at the upstream end of TSF minimize overland flow of surface water into the TSF. Surface water in-flow from upstream catchment areas is commonly either channelled away from the TSF or passed under the TSF via culverts. If leachate poses a potential risk, water can be contained in several ways. First, the underlying geological formation may be of sufficiently low permeability to form a natural barrier to leachate. A second method is to use the inherent low permeability of fine tailings themselves to minimize seepage. Third, one or more liners may be installed to prevent seepage. Finally drainage systems can be installed to facilitate tailings consolidation and to

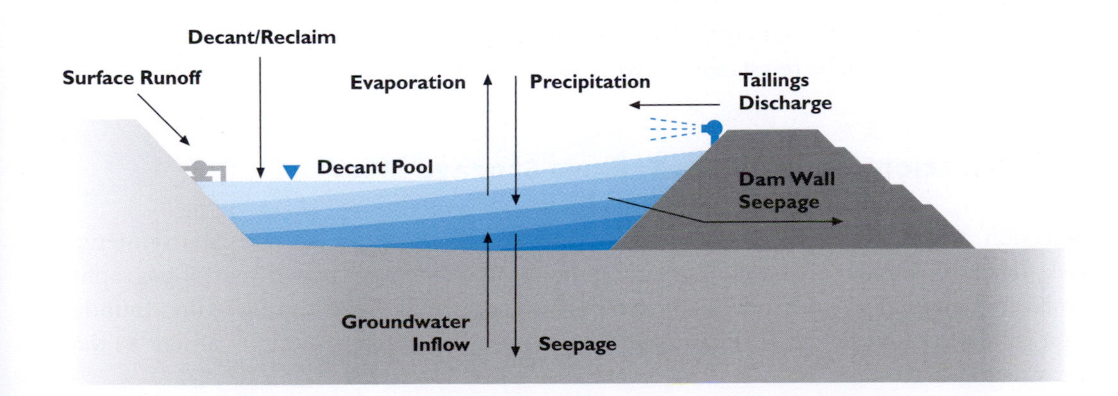

FIGURE 18.15
Water Balance of a Conventional On-land Tailings Storage Facility

It is important that the water balance not only considers average conditions, but also includes extreme operational or meteorological events.

Source:
UNEP 1996

intercept and recover leachate so that it does not enter the groundwater. Seepage can be intercepted by a subsurface drainage system such as an upstream cutoff drain along the upstream toe of the embankment. A perimeter drain around the TSF may also be installed to collect runoff and seepage water, should it occur.

It is important that the water balance not only considers average conditions and planned operating scenarios, but also includes unusual ('upset') operational scenarios and extreme meteorological events.

It is important that the water balance not only considers average conditions and planned operating scenarios, but also includes unusual ('upset') operational scenarios and extreme meteorological events. Water balances should be established for each stage of the operation, as different components may vary widely from one stage to another. For example, the quantity of water produced by mine dewatering may be zero at the initiation of mining but will increase as mining penetrates further below the water table.

Equally important are the water chemistry characteristics considering geochemistry, sources of make up water, and chemicals used in beneficiation or tailings treatment. In beneficiation processes that introduce heat, a thermal energy balance considering tailings temperature and density may also need to be included.

Given the many uncertainties and high variability of inputs and outputs, establishing the water balance is often a difficult task (Wels and Robertson 2003). As mentioned above it is important that the conceptual water balance of all inflows and outflows also includes all potential worst case scenarios (e.g. decant failure combined with snow melt and high precipitation) (DPI 2003). The simplest method to determine the water balance of a TSF is to determine average annual water inflows and outflows. Annual averages, however, are insufficient for actual design and to establish day-to-day operating conditions. Systems designed on average values may be grossly inadequate for extreme conditions. A more complex method is to use hydrological modelling (WMC 1998) which may use long-term rainfall records if they exist, or synthetic records formulated for the purpose.

Poor design of water management infrastructure and control methods increases the risk of problematic situations during operation.

Establishing a water balance as a basis for design of a TSF should prevent water management problems occurring during operation and closure, providing that upset conditions such as pump failures are also taken into account. A TSF needs to be designed to handle and control the routine inflows and outflows as well as any unusual fluctuations, either due to unusual operational or meteorological events. Poor understanding of water flow and hence, poor design of water management infrastructure and control methods increases the risk of problematic situations during operation such as uncontrolled upstream inrushes, low freeboard, or high seepage. Another consideration is that high volumes of water may require storage during the dry season to maintain plant production. Ice and snow melt, upstream inundation (valley impoundment) and severe storm recurrences (e.g. 1 in 100 year) should be considered when designing a TSF so that minimum freeboard levels, normally determined by legislation, company policy, or industry codes (**Case 18.3**) will not be reached. Freeboard is used to establish the elevation of the lowest point of the embankment crest relative to the normal or maximum operating levels of the supernatant pond (**Figure 18.16**; SANS 1998; DME 1999). As is the case with water containment, the TSF design may allow for spillways, diversion channels and/or emergency pumps as additional emergency mitigation measures. Spillways may need to be re-established with each raising of the TSF embankment.

Site Selection for Onshore Tailings Storage Sites

Selection of suitable on-land TSF sites is typically based on human and environmental risk, distance and elevation relative to the process plant, storage capacity and topography, embankment height that can be achieved, catchment area, and geology, including foundation conditions and potential for seepage. Overall, the approach to siting is similar to that for waste rock storage.

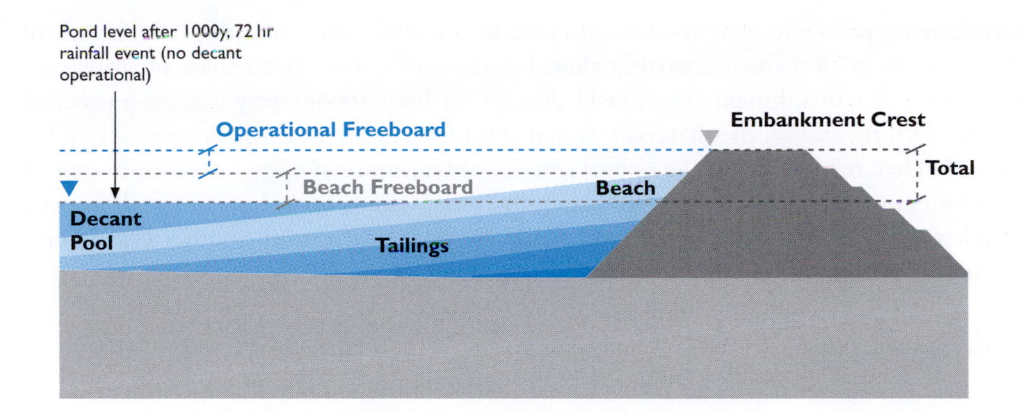

FIGURE 18.16
Freeboard of a Tailings Facility Explained

The total freeboard is defined as the vertical height between the waterline and the inside top of the embankment crest comprising beach freeboard and operational freeboard.

Source:
DME 1999

Human and Environmental Risk

Even in remote areas, human settlements or human land use occur close to the mine and the impacts of tailings disposal facilities on the public and environment must be minimized. Tailings disposal sites near villages, future townships, lakes and karst topography or limestone, where leakage may be an issue, are best avoided. Similarly, siting of tailings impoundments above underground mine workings is also best avoided.

Tailings disposal sites near villages, future townships, lakes and karst topography or limestone, where leakage may be an issue, are best avoided.

Distance and Elevation Relative to the Process Plant

Proximity to the process plant is an important factor, with a location at a lower elevation being advantageous so as to enable discharge by gravity. As such, distance from, and elevation relative to the process plant have a significant impact on the operating cost. Tailings storage facilities located a long distance from the process plant require long and costly tailings and decant water pipelines. Tailings storage facilities located well above the process plant or where the tailings discharge line traverses elevated terrain incur significant expenditure on pumping stations and pump operation costs.

Distance from, and elevation relative to, the process plant have a significant impact on the operating cost.

Storage Capacity and Tailings

An ideal tailings disposal site should have a topography which achieves the maximum storage capacity with a minimum amount of embankment fill and therefore construction cost. The fill efficiency ratio (defined as the ratio between the storage capacity and embankment fill volume) is used to quantify this aspect of the tailings storage facility. As there is usually some degree of uncertainty with respect to the required storage volume, the topography should be sufficient to permit the tailings dam embankment to be raised to meet potential extra storage capacity requirements. In many situations however, it is not possible to contain all tailings within a single impoundment. It is therefore common for mining projects to have more than one TSF.

An ideal tailings disposal site should have a topography which achieves the maximum storage capacity with a minimum amount of embankment fill.

Embankment Height

Sub-grade disposal into abandoned mine voids warrants serious consideration wherever such voids are available. However, this may be precluded if there are existing active underground workings nearby or if there is potential for future underground mining below the candidate voids. Several catastrophic events have resulted from the flow of tailings into active underground mines (e.g. the Mufulira disaster in Zambia in 1970). In most cases, however, there will be no suitable voids, and the tailings impoundment must be formed by the construction of one or more embankments. Ideally in addition to having a small volume, embankments should be kept as low as possible. In a seismically active area, high embankments pose not only design and construction difficulties but also a high risk of excessive deformation during earthquakes, leading to failure. The choice of impoundment type and embankment height will depend on local topography and its associated drainage pattern, and on the materials available for embankment construction. Generally, areas of higher relief provide the opportunity to obtain more storage volume per unit of embankment volume. However, high relief may come with its own problems such as steep unstable slopes, access difficulties and/or stream flows to be diverted or otherwise accommodated by means of spillways or alternative strategies.

Catchment Area

It is advantageous to minimize surface runoff into the impoundment.

Unless the tailings storage is also used as the main water reservoir for the project, a relatively uncommon scenario, it is advantageous to minimize surface runoff into the impoundment. This is done in two ways: (1) by minimizing the catchment area of the impoundment, usually by locating the tailings storage in the uppermost part of the drainage; and/or (2) by diverting as much surface runoff as possible around the impoundment, by means of surface drains or tunnels. In a high rainfall area, these issues are exacerbated if the catchment area of the tailings storage facility is large. Therefore tailings storage facilities with small catchment areas are preferred.

Geology

Geological factors impact upon construction costs, reservoir leakage, foundation conditions and embankment stability. Site selection needs to take account of the availability and proximity of construction materials. Clay fill for the construction of the low permeability core may be available from unmineralized, oxidized near-surface parts of the mine or from within the storage impoundment itself. Waste rock from the mine is commonly used for the construction of the outer zones of the tailings embankment, involving the bulk of the rock-fill required. Materials used for filter zones and drainage zones incorporated in the embankment, are usually required to meet tight grading specifications and are usually manufactured from crushed and screened rock. The quantities are relatively small, however, so that these materials can be transported economically from distant locations if local sources are not available.

Foundation conditions also need to be investigated, initially to identify any geological defects that could render the site unsuitable for construction, and to evaluate the need for foundation preparation, prior to embankment construction. Seepage conditions also need to be investigated, not only in the vicinity of the embankment, but from the remainder of the impoundment.

In practice, site selection involves a series of trade-offs.

In practice, site selection involves a series of trade-offs. For example, minimizing the catchment area by locating the storage high in the drainage means that the tailings impoundment will be at a higher elevation than the process plant, resulting in higher pumping costs.

A typical site selection process contains the following steps (see also Robertson 1988):

1. The topography within a reasonable radius – generally less than 20 km – of the site(s) for the process plant is examined and, using rough measurements, storage areas capable of accommodating the expected quantities of tailings, are outlined.
2. For each potential storage area, a preliminary embankment layout is made and a cumulative storage volume versus embankment height curve is established.
3. For the required storage volume, the corresponding embankment heights are determined for each site.
4. The catchment area of each site is then measured, and feasibility of diverting some or all of the upstream drainage is determined.

5. The required embankment heights are adjusted by adding freeboard to accommodate inflow from a designated storm event.
6. For each site, calculations are made of the embankment volume required to achieve the required height, including freeboard. A standard cross-section (such as 2.5:1 upstream and downstream slopes with a crest width of 8 m) is used at this stage.
7. Potential sites are then ranked according to the embankment volume as a percentage of the available storage volume, which is known as the storage efficiency, for each alternative site.
8. At this stage, comparative estimates of construction and operating costs can be made, which may confirm or modify the ranking based on storage efficiency.
9. Preliminary site screening studies are undertaken to identify any conditions that may change the ranking; these involve ground reconnaissance and include assessment of geotechnical, hydrological and environmental factors such as forest quality, wetlands, wildlife habitat values and land usage.

The ultimate selection of one or more preferred sites for more detailed investigation may be self-evident after these studies; this is a common outcome as the site with the highest storage efficiency will have the smallest footprint and, other factors being equal, the lowest environmental impact. However, cases occur where adoption of the preferred alternative from the engineering/economic view-point, would lead to environmental impacts exceeding those from alternative sites. In such cases, selection of the preferred alternative(s) may be assisted by a numerical ranking in which all factors are assigned a value, with or without weighting. In some cases, this approach does not lead to a clear identification of the preferred site(s); in such cases, more detailed field studies may be required.

As these considerations indicate, selection of a tailings storage site needs to be carried out in conjunction with early mine planning and site selection for other facilities, particularly the mill and process plant. The decisions made at this early stage may have consequences throughout the operating period. For example, supply of materials for embankment construction at each stage, may influence the scheduling of mining operations. Alternatively, it may not be possible to produce these materials as and when required, in which case, either the embankment may be raised well ahead of capacity requirements, or the construction materials may be stockpiled until required.

Selection of a tailings storage site needs to be carried out in conjunction with early mine planning and site selection for other facilities, particularly the mill and process plant.

It may seem from the foregoing discussion that environmental factors are not as important as engineering factors in the selection of tailings storage sites. In fact, this is commonly the case; when all candidate sites have very similar environmental attributes, the lowest impact will occur from the site with the smallest footprint. However, there are also cases where critical environmental factors will outweigh all other considerations.

Site Investigations

Once the preferred site(s) have been identified, site investigations are carried out to: (1) evaluate the suitability of the site for the intended purpose; (2) assess foundation conditions and the extent of foundation preparation and treatment required prior to embankment construction; (3) assess the hydrogeological situation, including the location and nature of aquifers, water table gradients, recharge/discharge relationships and water quality characteristics; and (4) to identify and characterize potential sources of material for embankment construction.

The site investigations usually include a detailed topographic survey, test pitting, drilling and sampling, laboratory testing of samples, permeability testing, plus the establishment and regular measurement of groundwater level monitoring bores (piezometers).

Other relevant data are compiled from regional or national authorities, which provide important inputs to the subsequent engineering analyses. These include climatology/hydrology data to determine the design rainfall event from which the design inflow event is computed. Similarly, seismic records are reviewed to determine seismic risk, and a design earthquake event is selected.

Once the results of site investigations are available, engineering analyses are carried out to assess the stability of various embankment layouts and designs under the entire range of operating conditions and scenarios, including 'upset events' in the process plant, extreme rainfall events and earthquake loadings. Other analyses evaluate seepage rates from the impoundment, and through, around or beneath the embankment. These are used to assess the need for sealing treatments or for cutoff walls beneath the embankment or drainage zones within the embankment. From all these studies, an embankment design is selected which satisfies stability and seepage criteria.

Stability is usually expressed in terms of factor of safety (FS). For a tailings embankment, an FS of at least 1.2 is required for steady-state operating conditions, with an FS of at least 1.1 for short-term events such as earthquakes. Acceptable rates of seepage depend mainly on the nature of the tailings solution, whether or not it contains dissolved substances that could cause environmental contamination, and the contaminant pathways which are determined by the hydrogeological situation.

While site investigations are carried out primarily for engineering design purposes, they also provide the opportunity to obtain useful environmental data with little, if any additional cost or effort. Therefore, the relevant environmental professional(s) should have the opportunity to provide input to the scoping of the site investigations and should be supplied with the relevant data as it becomes available. Environmental data generated from site investigation studies usually include: (1) water table profiles, (2) seasonal information on water table fluctuations, (3) aquifer characteristics such as storativity and transmissivity, (4) groundwater flow directions, (5) groundwater quality, (6) soil profile, (7) rock mass weathering profile information, (8) surface infiltration characteristics, (9) statistical analyses of rainfall events of different intensity and duration for different return periods, (10) rainfall/runoff relationships, and (11) seismicity. Apart from generating data for planning purposes, piezometers installed during site investigations may be incorporated in the ongoing monitoring programme.

Risks of Onshore Tailings Storage

In any discussion of tailings disposal, it is important that the risk of failure is recognized. The causes of failure are many and varied. The world-wide experience of tailings storage failures includes many incidents involving: (1) overtopping of tailings impoundments caused by unexpectedly large rainfall events; (2) collapse of tailings dams due to foundation failure; (3) internal erosion or piping, leading to the collapse of tailings embankments; (4) earthquake damage leading to overtopping or collapse; and (5) human error in terms of inadequate design, poor construction or inappropriate operating practices. In some cases, catastrophic failure involved loss of life and widespread environmental damage (ICOLD 2001, UNEP 2001).

TSF failures may be categorized as follows (Aldous 2002):

1. Structural failures generally involve the collapse, subsidence or slippage of a part of a containment structure. Structural failures often result in the discharge of large quantities of tailings with severe environmental damage, including loss of life due to inundation or damage inflicted on structures.
2. Operational failures occur where the root cause of the incident is the failure to operate or control the operation of a facility adequately. For example, failure to monitor the

water level of a TSF could result in an overflow. Lack of adequate operational control can ultimately result in a structural failure.

3. Equipment failures refer to the failure of mechanical equipment such as pipelines, pumps, or valves used for tailings management. Burst pipelines, coupling breakage and pump failure are common causes of accidental discharge. Equipment failures are usually less serious than failures due to operational or structural causes but can nevertheless cause significant harm to the environment, and reputational damage to the operator.
4. Unforeseen consequences of tailings placement may occur. Some instances of damage to the environment are the result of simple oversight in the design, operational procedures or closure of a TSF. For example, the long-term consequences of slow seepage from the base of a TSF may not have been adequately considered at the design stage and may not become apparent for some years.

Many of the past failures occurred in poorly engineered storages, constructed before tailings dam design standards were developed. However, failures continue to occur and, increasingly, it is embankments designed and constructed in accordance with internationally recognized methods and standards that are failing. According to UNEP (2001), 'the number of major incidents continues at more than one a year'. Modern tailings embankments are designed in accordance with the same principles as water storage dams, and involve comparable levels of investigation and construction control. However, it is clear that the proportion of failures in tailings storages is much higher than in water reservoirs (**Case 18.4**). Added to these failures, are many smaller incidents that cause serious environmental damage.

The number of major incidents continues at more than one a year.

CASE 18.4
How Safe are Tailings Dams?

While it appears that the mining industry has the knowledge to design dams safety, can it be said that all dams are built to the same standards, with state-of-the art technology and management? Some are better than others, and failures have occurred. Several recent studies have been conducted to compile data on tailings dam failures, isolate the causes of these failures and identify trends. No single legislative body, however, records tailings dam statistics. Furthermore, the data do not allow comparisons between the number of tailings dam failures and the total number of tailings dams built in any given area or time period.

However, comparisons have been made between tailings dam failures and incidents at hydroelectric and water-retaining structures. Although the database is incomplete, some convincing trends have emerged. The chart shows a plot of the total number of failures reported for all countries in 10-year increments for both tailings dams and water supply dams. Before the 1940s, there were very few reported failures of tailings dams, either because many of the existing dams ware not documented, or because the total number of failures was small. From the 1940s to the 1970s, the number of failures for both tailings dams and water supply dams increased substantially. The rise in the number of failures in the 1950s to 1960s may have been due to the increasing size and weight of earthmoving equipment. This trend peaked in the late 1960s for water supply dams, and in the 1970s for tailings dams. The overall behaviour of the two structure types is, in general terms very similar.

Source: UNEP 2001.

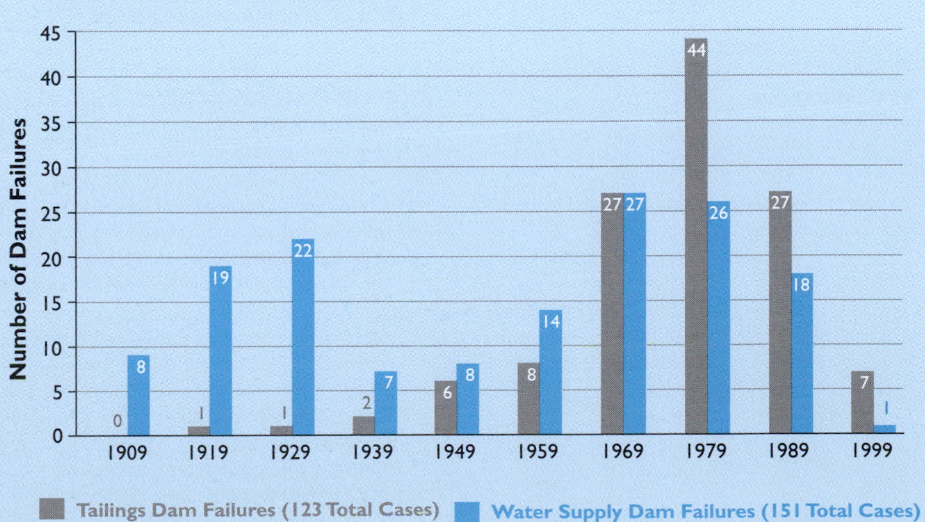

A much higher incidence of tailings storage incidents, including catastrophic events, occurs in the mountainous, wet tropics; this, despite the fact that many operations in such geographic circumstances, recognizing the risks involved, have opted for riverine or submarine disposal. Over the past two decades an increasing proportion of new mines have been developed in such areas, which are geologically active and prone to instability. The challenges involved in any development in such areas are much greater than in the traditional mining regions of Australia, Western USA, Canada, South Africa and Russia.

Table 18.5 identifies some of the more common causes and types of tailings storage failure and identifies typical measures adopted to reduce or remove the risks.

TABLE 18.5
Measures to Ensure Safety of Tailings Systems

Type and Cause of Failure	Measure(s) Incorporated in Design to Avoid Failure
Overtopping of embankment due to flood inflows, with consequent erosion. Such an event is more likely to occur where upstream flows are diverted, in which case blockage of the diversion channel or tunnel can lead to major inflow.	• Installation of interception/diversion drains outside the tailings storage perimeter. • Provision and maintenance of adequate freeboard. • Provision of capacity (and in some cases, back-up capacity) to remove water by pumping during the operating life of the impoundment, and by discharge through the spillway after operations end.
Erosion of embankment by prolonged wave action, with consequent overflow and scour.	• Development of tailings 'beach' against the embankment. • Provision of adequate freeboard. • Installation of rip rap on the upstream slope.
Sliding or collapse of downstream slope, due to inadequate strength/overly steep slope, and/or excessive pore pressures, with or without earthquake loading.	• Selection of low slope angles, and earth-fill materials of adequate strength. • Installation of drainage systems for the embankment and foundations.
Failure of upstream slope due to inadequate strength and/or excessive pore pressures, with or without earthquake loading.	• Selection of low slope angles, and earth-fill materials of adequate strength. • Installation of drainage systems for the embankment and foundations. • Avoidance of rapid drawdown during operations.
Mass downstream movement of embankment due to sliding along the foundation interface or along a weak foundation layer, caused by excess pore pressure or hydrostatic pressure, with or without earthquake loading.	• Comprehensive site investigations including engineering geological assessment to identify potential failure modes; incorporation of specific foundation treatments or structural design elements to ensure stability under all operating conditions. • Foundation drainage system ensures dissipation of pore pressures, and avoids development of uplift pressures. • Removal of weak foundation material.
Collapse, as a result of liquefaction under cyclic earthquake loading.	• Use of cohesive embankment materials and incorporation of drains to ensure a low phreatic surface within the embankment. • Avoidance of sites with saturated, granular foundation materials; alternatively removal or densification of such materials.
Internal erosion, due to 'piping' of embankment or foundation materials.	• Embankment materials must be non-dispersive. i.e. not susceptible to 'piping.' • Hydraulic gradients should be low, and toe drains should be installed, if necessary, to ensure that the phreatic surface does not intersect the downstream face of the embankment. • Foundation cutoff wall can be installed to substantially increase seepage path through foundations.
Cracking of embankment due to deep dessication or differential settlement, with consequent leakage, internal erosion, and collapse. Deep dessication can be caused by uptake of moisture by deep-rooted plants.	• Differential settlement of foundations is minimized by avoidance or removal of highly compressible foundation materials. • Differential settlement of embankment materials is avoided by placement in thin layers with adequate compaction, with construction control testing to ensure that design compaction criteria are achieved. • Cracking due to dessication is reduced by compaction and further minimized by application of a surface mulch, and by avoiding establishment of vegetation other than shallow-rooted species.

A particular concern with tailings storages is that they are generally required to maintain their integrity in perpetuity, regardless of future changes in climate, surface drainage, land use and other conditions that may not be foreseeable at present. Most onshore tailings storages are constructed above the surrounding landscape, which means that they are subject to long-term degradation by wind and water erosion. It would be preferable for long- term integrity, if tailings could be placed in below-grade storages in mined out voids or in areas subject to aggradation rather than erosion. However, opportunities for such storages are relatively uncommon.

A particular concern with tailings storages is that they are generally required to maintain their integrity in perpetuity.

To maximize the ongoing integrity of tailings storages, it is clearly beneficial if the final tailings surface can be rehabilitated to support a productive land use. Once this is achieved, the new users have an incentive to manage the area, maintain surface drainage and control erosion, as they would for any other area. Appropriate and effective rehabilitation to create a sustainable, economically productive land use, is therefore the most important means of ensuring long-term integrity. This is discussed in more detail in subsequent sections including Chapter Twentyone.

Monitoring

Monitoring of tailings disposal operations and storages has several purposes, including:

- Determining whether the tailings discharged to the storage have the same characteristics as predicted in the feasibility and environmental studies, and if not, whether the differences require changes in design or operating procedures;
- Determining whether or not the structures are performing as intended;
- Providing early warning of the development of conditions that could lead to unwanted events;
- Measuring densities achieved so that capacity calculations can be refined;
- In the case of staged construction, providing data for the design of subsequent embankment raises;
- Identifying areas or installations requiring maintenance or repairs;
- Providing feedback to improve management;
- Measuring the environmental effects of the operations, particularly the quantity and quality of seepage water.

Table 18.6 shows a typical monitoring programme for a tailings disposal operation.

Audits

For any major tailings storage, 15 m or more in height, it is advisable (and in many countries mandatory) to have the design checked by a suitably qualified and experienced independent third party. It has also become normal practice for audits of active tailings facilities to be carried out at regular intervals, usually semi-annually, annually or bi-annually, depending on the circumstances. These audits, which are best carried out by the same tailings expert who was responsible for the original design, involve a detailed inspection of all tailings facilities, discussions with operating personnel, review of all monitoring records and, in some cases, additional testing or sampling. Tailings facility audits may continue into the post-closure period until such time as closure criteria are achieved.

For any major tailings storage, 15 m or more in height, it is advisable to have the design checked by a suitably qualified and experienced independent third party.

Independent audits should commence with a review of the design and operation of the facility against the standards as set down by the regulators of the country in question and

TABLE 18.6
Tailings Monitoring Programme

Parameter	Purpose	Method	Frequency
Quantity discharged, solid: liquid ratio	Assess storage capacity Calculate water balance	Flow meter Density meter	Continuous, compiled daily
Quantity recovered through decant system	Calculate water balance	Flow meter	Continuous whenever decant system is operating
Tailings liquor quality parameters. e.g. pH, Electrical Conductivity (EC), CN_{WAD}	To assess compliance with regulations or standards	Electronic probes (pH and EC) Titration (CN_{WAD})	Depending on variability, regulations or experience. May be from 30 minute intervals to 24 hour intervals
Size and location of decant pool, Available freeboard	To help manage discharge locations and decant volumes	Visual observation or measurement, supported by completion of checklist report and photographic records	Daily
Underdrain seepage volumes	To assess effectiveness of sealing layers and drainage system	Flow meter operating during pumping	As required depending on pumping frequencies
Underdrain seepage quality; e.g. pH, EC, CN Selected cations	To assess compliance, or, possibly, to assess suitability for discharge or re-use	Electronic probes, Titration AAS	Monthly, more or less frequently depending on variability and whether toxic constituents are involved
Deformation of embankment	To assess settlement and lateral movement of embankment that might indicate impending instability	Precise survey of surface marker pins Inclinometers	Monthly More often if significant deformations are detected.
Phreatic surface within embankment	To check whether or not the system is behaving as required To provide advance warning of increasing pore pressures	Piezometers, design of which depends on nature of embankment or foundation material	Weekly or monthly
Water quality parameters: cations and anions selected as indicators, or as potential contaminants of concern, or as mandated by regulation	To check whether or not local groundwater is being contaminated by seepage from the tailings storage	Sampling from groundwater bores or piezometers, followed by laboratory analyses	Monthly or quarterly
Density and strength of settled tailings	To help assess capacity of storage To assess the feasibility of upstream or centreline construction for future embankment raises	Density testing – sand replacement or densometer methods Strength testing by static cone or vane shear methods, and/or by undisturbed sampling and laboratory triaxial testing	Annually or as required
Unusual occurrences or incidents e.g. damp areas on face of embankment, cracks or erosion gullies in embankment, springs or seeps downstream of embankment	Possible indicators of developing problems Identification of areas requiring maintenance	Reconnaissance – drive-by observation. Careful and systematic visual observations, supported by completion of checklist report and photographic records	Daily Weekly or monthly, and immediately following intense rainfall event or earthquake

the undertakings by the mine operator in their own documentation. Some standards of various countries follow.

- The Australian guidelines 'Guidelines on the Safe Design and Operating Standards for Tailings Storage' and 'Guidelines on the Development of an Operating Manual for Tailings Storage' both produced by the Department of Minerals and Energy, Western

Australia, define standards for routine inspections and operational audits. A complementary document is 'Tailings Dam HIF Audit' that describes the components of an independent audit according to the Australian standards (notesweb.mpr.wa.gov.au).
- The Canadian guidelines 'A Guide to the Management of Tailings Facilities' and 'Developing an Operation, Maintenance and Surveillance Manual for Tailings and Water Management' both produced by the Mining Association of Canada suggest that periodic inspections and reviews, audits, independent checks and comprehensive independent reviews need to be carried out as part of the surveillance programme. The documents can be found at www.mining.ca.
- The primary document controlling a mining company's tailings disposal activities in South Africa is the Department of Mineral and Energy Mandatory Code of Practice for Mine Residue Deposits (MRDs) (available on the website www.dme.gov.za). This code requires every mine to set out in writing its intended standards and procedures for the protection of the health and safety of workers, and for the reduction of the risk of damage to persons and property.
- In Sweden, generally all mining companies need to implement programmes for daily, monthly and yearly inspections/audits, but there are no requirements on independent audits.

Closure

In recent years, there have been a number of studies to consider the standards to be achieved on closure in USA, Canada, Australia, and, most recently, in Europe (Xenidis 2004.). Generally, the major issues to be considered for the reclamation of mining/milling components include the long-term (1) physical stability, (2) chemical stability and (3) land use. Based on regulations developed in Canada (Doran and McIntosh 1995), after closure, mine waste storage facilities should be physically stable under extreme events such as floods, earthquakes and perpetual disruptive forces including wind and water erosion, so that they do not impose a hazard to public health and safety or the environment. Regarding chemical stability, leaching of contaminants contained in mine wastes and migration into the environment should not endanger public health or safety, nor exceed water quality objectives in downstream watercourses. Based on the guidelines prepared by the Department of Minerals and Energy, Western Australia (DME 1999), decommissioned tailings storage facilities must be safe, stable and aesthetically acceptable. Some fundamental mine closure design criteria, based on MIRO (1999) are given in **Table 18.7**.

The ultimate land use of any tailings storage should be determined, ideally at the outset, but definitely long before tailings discharge ceases. As mentioned previously, rehabilitation to a sustainable, economically productive land use may be a critical factor in maintaining the long-term integrity of the tailings storage. However, there will be some instances where government policies or compelling ecological factors dictate that the tailings storage should be rehabilitated with local native vegetation.

The ultimate land use of any tailings storage should be determined, ideally at the outset, but definitely long before tailings discharge ceases.

Selection of the preferred land use may be obvious; more often, however, it will involve a somewhat lengthy process, including: (1) consultations with existing and potential stakeholders including land owners, ultimate land users if known, local communities and government authorities; (2) detailed assessment and consideration of land capabilities under a range of different rehabilitation strategies; (3) review of agricultural, horticultural, aquacultural and forestry products and practices in the region, with screening to assess which are potentially applicable to the tailings storage; (4) conduct of field trials to establish the feasibility of growing various crops, with or without various cappings, cultivation treatments,

TABLE 18.7

Closure Design Criteria for the Mining Planning Process – Decommissioned tailings storage facilities must be safe, stable and aesthetically acceptable.

Issue	Closure Design Criteria
Physical stability	Physical stabilization of man-made structures so that they pose no risk in terms of safety or environmental impact
Chemical stability	Chemical stabilization of physical structures, so as to provide no contamination problems to the environment or risks to public health
Biological stability	Restoration of the biological environment to a balanced, self-sustaining ecosystem typical of the area, or left in such a state that it encourages natural rehabilitation and development of a biologically diverse, stable environment
Hydrology and hydrogeology	Prevention of contaminant migration downstream (surface and ground waters)
Geographical and climatic influences	Fulfilment of the demands and compliance with the site characteristics including climatic (e.g. rainfall, storm event) and geographic factors (proximity to residential areas, topography, accessibility)
Local sensitivities and opportunities	Optimization of the opportunities for restoring the land. Upgrading of the land use, wherever appropriate and/or economically feasible
Land use	Final land use compatible with the surrounding area and the local community requirements
Financial Assurance	Adequate and appropriate financial assurance to ensure implementation of the mine closure plan
Socio-economic considerations	Encouragement of alternative opportunities for local communities. Maximization of positive socio-economic implications

Source:
MIRO (1999)

fertilization etc; and (5) in some cases, conducting bio-assays to determine whether crops grown on tailings can be safely consumed.

Once the preferred land use has been selected and its feasibility confirmed by field trials, the necessary rehabilitation plans can be formulated. These will differ substantially from one type of storage to another, so that the following discussion is necessarily general and by no means comprehensive.

Some tailings storage facilities, particularly those in high rainfall areas, are preserved as ponds. The reasons for selecting this land use may be the need to maintain tailings in a saturated condition, to ensure that they do not acidify due to oxidation; and/or the excessive costs involved in dewatering, draining and establishing a suitable medium for plant growth on the surface of the storage.

In an area of high seismicity, leaving a tailings storage facility as a pond is highly risky and cannot be endorsed, except in situations where the tailings solids have achieved sufficient strength to preclude liquefaction.

In an area of high seismicity, leaving a tailings storage facility as a pond is highly risky and cannot be endorsed, except in situations where the tailings solids have achieved sufficient strength to preclude liquefaction. While there may be the potential to establish an economically productive use for the pond, such as some form of aquaculture, the risks exceed the potential benefits. Even in a seismically inactive area, such a land use entails risks well above those that are involved in a dewatered storage. Maintenance of a pond requires that the inflow should equal or exceed evaporation; this means that water management will be an important factor – in perpetuity. A spillway is usually required to accommodate excess water, with regular maintenance required to ensure that the spillway remains functional – again, in perpetuity. Water supply reservoirs are usually considered to have operating lives of 100 to 300 years. In practice, there have been many water retention structures that have had to be substantially strengthened or re-built after periods much shorter than this. In the case of tailings storages, it is unclear who will be responsible for maintaining, strengthening or re-building them when required, or if anyone will even be responsible for assessing their condition after decades or centuries of perhaps trouble-free experience. More likely, the mining history of the land and its inherent risks will have been largely forgotten by that time.

In rare cases, tailings are returned to the mine Pit(s) as part of mine closure. Normally, this would be uneconomic unless it was factored in to the original feasibility study, and the necessary funds accrued during the operating period. Such an outcome was a condition of development of the both the Nabarlek Uranium Mine and Ranger Uranium Mine in the Northern Territory of Australia. In the case of the Woodcutters base metals project in the same region of Australia, the owner, Normandy Mining, decided during closure planning studies that the long-term risks associated with the tailings which were known to be acid-generating, could only be reduced to an acceptable level by below-grade storage in the mine Pit.

The above cases are exceptions. Rehabilitation of most tailings storages will involve a series of steps such as the following:

1. Removal of surface water from the surface, repeated as necessary until the surface has dried sufficiently to support traffic;
2. Once the surface is trafficable, there is usually a requirement for earthmoving to provide drainage and to construct a landscape consistent with the intended land use, rather than a flat basin, subject to inundation during each rainfall event;
3. Depending on the results of field trials, the selected crops may be established directly on the dewatered tailings, with whatever soil amendments are indicated by the trials;
4. In other cases where tailings are found to be an unsuitable medium for growth of the selected species, a layer of natural soil is placed over the tailings, prior to planting;
5. In yet more difficult cases, establishment of sustainable agriculture or horticulture might require capping of the tailings with clay, prior to placement of the soil growth medium, or a capillary barrier may be required between the tailings and the natural soil layer;
6. Planting of one or more leafy crops which are subsequently ploughed in to provide organic content to the soil;
7. Establishment of the selected crop(s);
8. Monitoring over at least several years to assess the sustainability of the selected land use.

It should be emphasized that in tropical areas, particularly those where rainfall exceeds evaporation, rehabilitation of tailings storages is much more difficult than in arid, semi-arid or most temperate climates. Rehabilitation in such high rainfall areas may take many years, particularly if the tailings have been very finely ground, in which case rates of consolidation will be very slow.

In tropical areas, particularly those where rainfall exceeds evaporation, rehabilitation of tailings storages is much more difficult than in arid, semi-arid or most temperate climates.

18.4 SUBMARINE TAILINGS PLACEMENT

Submarine Tailings Placement (STP) is an engineered technology developed as a disposal alternative for mines in close proximity to coastlines. Application of STP is evaluated on a case-by-case basis, and is an option where land-based disposal would have more substantial environmental impacts or high risks. A detailed set of screening criteria has been developed

TABLE 18.8

Examples of Completed Projects in which STP has been Used – While it is unlikely that such shallow outfall depths as 10 m or even 45 m will be proposed, let alone approved in the future, all these projects were more or less successful at the time, although the Minahasa Project proved to be highly controversial. In no case was there any loss of life or catastrophic environmental damage as a result of STP. This is in contrast to the history of onshore tailings disposal in the same parts of the world

Project	Years of Operation	Depth of Outfall (m)	Tailings Quantity (T)
Island Copper, Canada	1997–1997	45	350 million
Atlas Copper, Philippines	1971–1991	10	365 million
Misima Gold, PNG	1989–1999	118	80 million
Minahasa Gold, Indonesia	1995–2003	82	7 million
Kitsault Molybdenum, Canada	1979–1981	50	15 million

TABLE 18.9

Operating DSTP Systems – DSTP involves discharge of tailings slurry near the seabed in water depths of 100 m or more

Project	Year of Operation	Depth of Outfall (m)	Tailings Quantity (T)
Pechiney Alumina Refinery, Marseilles, France	Since 1967	320	>60 million
Cayeli Bakir, Black Sea, Turkey	Since 1992	350	>18 million
Lihir Gold Project, PNG	Since 1997	128	>24 million (50 million planned)
Batu Hijau, Copper Project, Indonesia	Since 1999	120	>280 million (1.5 billion planned)

to evaluate a site's suitability for STP. The screening criteria include proximity to deep-water coastlines, severe precipitation or flooding potential, seismic loading and land use considerations. STP has been employed at more than twenty sites, in the following countries: Canada, France, Denmark (Greenland), Indonesia, Norway, Papua New Guinea, Peru and Turkey (see **Tables 18.8** and **18.9** for some examples where submarine tailings placement has been implemented). The Island Copper Mine in British Columbia, Canada operated its STP system very successfully for more than twenty years. STP has not been employed in the United States, an irrelevant argument often raised by activists opposing STP, as few USA mines are located close enough to the coastline to consider the feasibility of using STP.

STP is particularly attractive if favourable sub-sea conditions exist close to the mine and if a land based TSF is difficult to realize, either due to a lack of suitable disposal areas or due to unacceptably high environmental risks (e.g. Newmont Minahasa Mine, Indonesia; Newmont Batu Hijau Mine, Indonesia; and Lihir Mine, PNG). Of course the main potential environmental impacts of STP schemes are potential negative impacts on the receiving marine environment. In all cases, STP requires intensive preparation of and consultation with affected communities to avoid negative community perception often aggravated by opponents of STP. For a mining operation, perceptions can be just as important as reality, as reality depends on what members of the community define as real.

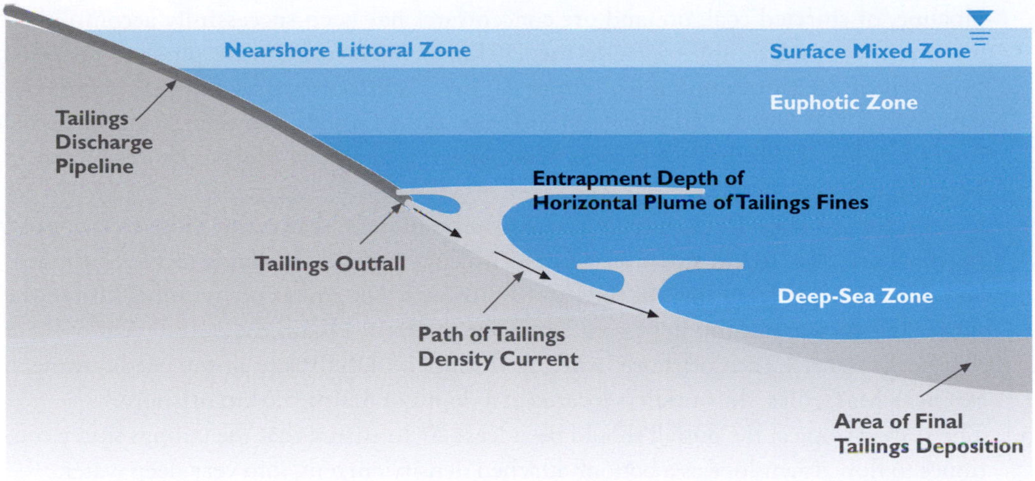

FIGURE 18.17
Deep Sea Tailings Disposal Illustrated

The risk associated with DSTP depends on the discharge depth, and is determined by three distinct layers that characterize the ocean.

It becomes appropriate at this point to mention the practice of marine tailings disposal by ship rather than by pipeline ('ocean dumping' of tailings). Examples include marine red mud disposal practised by various alumina plants (three refineries in Japan; one in France and one in Greece) or disposal of Jarosite residues from the Risdon Zinc Refinery by Pasminco in Tasmania prior to 1994. Ocean dumping of tailings is increasingly challenged by environmental groups. It is fair to assume that public pressure will eventually lead to a discontinuation of this practice, irrespective of whether or not there are serious environmental impacts.

Deep Sea Tailings Placement

Despite the controversy that invariably accompanies any proposal to dispose of mine tailings (or any other waste) in the sea, submarine tailings placement, particularly the most recent improvement known as **deep sea tailings placement** (DSTP), warrants serious consideration wherever the required environmental conditions are present. DSTP involves discharge of tailings slurry near the seabed in water depths of 100 m or more. The tailings slurry flows down slope under gravity as a coherent, bottom-attached density current. The tailings solids eventually deposit on the seabed at the base of the slope, in very deep water – typically in water depths exceeding 1,000 m, well below aquatic zones with high biological productivity (**Figure 18.17**). The risk associated with STP depends on the discharge depth and is determined by three distinct layers that characterize the ocean (Jones and Jones, 2001). The surface mixed layer in the upper ocean layer is well mixed by wind and waves and is generally of uniform temperature, density and salinity. The bottom of the surface mixed layer is marked by an abrupt change in density and temperature. The euphotic layer is defined as the depth reached by only 1% of the photosynthetically active light transmitted from the ocean surface. The euphotic zone with highest biological productivity may reach a depth of several tens of metres. The deep-sea layer lies beneath the euphotic zone. Biological productivity is low.

Despite the controversy that invariably accompanies any proposal to dispose of mine tailings in the sea, submarine tailings placement, particularly the most recent improvement known as **deep sea tailings placement** (DSTP), warrants serious consideration.

Conditions required for DSTP include:

- The processing facilities need to be located within reasonable proximity to deep sea. Of the projects undertaken to date, overland tailings slurry pipelines of up to 25 km have been used. However, studies for projects not yet constructed have indicated that pipeline lengths of 200 km or more may be feasible in some cases. (Transportation by pipeline, of slurried coal, ore and ore concentrates has been successfully accomplished for many years. Examples include the 112 km copper concentrate pipeline operated by Freeport McMoran in Papua, Indonesia, the 134 km ore slurry pipeline being constructed for the Ramu Nickel Project in Papua New Guinea and the 440 km long coal slurry pipeline from the Black Mesa Mine in Arizona to the Mojave Power Sation in Nevada, USA).

- Deep water, at least 100 m in depth, needs to be present reasonably close to the coast. In many volcanic island situations such as Misima and Lihir Islands in PNG, the seabed slopes steeply from the coastline and 100 m water depth occurs within 200 m of the shore. In other areas, the slopes are less steep. At Batu Hijau, the outfall is at a water depth of 120 m, 3.2 km offshore, while at Pechiney's DSTP site in the Mediterranean Sea near Marseilles, the outfall is located at a depth of 320 m, 5.6 km offshore.
- The seabed slope at the outfall should be at least 10° to ensure that the tailings slurry continues to flow downslope as a bottom-attached density current, into very deep water.
- Flattening of the seabed, where main tailings deposition will occur, should be in very deep water – 400 m or more.
- There should be no upwelling currents below the level of the tailings pipeline outfall.
- The surface mixed layer should not extend to the level of the tailings pipeline outfall. In many areas the surface mixed layer never exceeds 80 m in thickness. However, in some areas, particularly those that experience cyclones, hurricanes or typhoons, mixing during such events may penetrate to a depth of 120 m or more, for short periods.
- Seabed currents in the deposition areas should not be sufficient to cause resuspension of tailings solids.
- The seabed along the course of the marine pipeline route should be stable. This is a concern in the case of volcanic islands where very steep slopes of 40° or more may occur near the shore. Such slopes are frequently subject to submarine landslides. In fact, during operation of the DSTP system at Misima in PNG, a submarine landslide occurred which severed the pipeline. The new discharge point was still sufficiently deep that the tailings continued to flow downslope as before the landslide, and there were no environmental consequences from the event.

Other environmental conditions that are favourable but not essential include:

1. the presence of a submarine canyon into which the tailings slurry can be discharged (e.g. as found at the Batu Hijau DSTP site, see **Case 18.5**). The canyon serves to confine the tailings density current; and
2. the presence of sediment in the deposition area with particle sizes similar to those of the tailings. In practice this is common because both tailings and natural sediment obey the same laws of physics, with the result that similar size fractions will settle at similar locations on the seabed. This is normally the case for a slope that gradually flattens. Where deposition occurs in a more or less enclosed depression or basin, all sediment size fractions will accumulate in more or less the same area. In this case, the tailings may have quite different characteristics to the natural sediment.

From a variety of marine tailings disposal projects over the past 30 years, a considerable body of experience has been accumulated, so that it is now possible to design DSTP systems and to predict their associated environmental impacts with confidence.

From a variety of marine tailings disposal projects over the past 30 years, a considerable body of experience has been accumulated, so that it is now possible to design DSTP systems and to predict their associated environmental impacts with confidence. The main features of modern DSTP systems are shown in **Figure 18.18**.

Important components are as follows:

1. Overland Pipeline is usually constructed of welded steel, as shown in **Case 18.5**, and supported just above the ground surface on concrete stands;

2. Subsea Pipeline, also shown in **Case 18.5**, is constructed of thick-walled High Density Polyethylene (HDPE) which is delivered in 20 m or 30 m lengths and welded onsite. Through the beach and surf zones, the pipeline is buried to protect it from storm damage. Beyond the surf zone, the pipeline is placed on the seabed. Concrete saddle weights are positioned along the pipeline at intervals to anchor it in position and to resist any uplift stresses; and
3. De-aeration tank is located onshore. Here any air entrained in the tailings is released. This is important because if there was air in the tailings discharged near the seabed, it would tend to take tailings with it as it rose to the surface. In some DSTP systems, a sea water intake pipeline is also installed. The sea water is mixed with the tailings, usually in the de-aeration tank, so that the density of the liquid fraction is raised. This minimizes the tendency for fine particles to leave the tailings density current as it descends the seabed slope after discharge. At Batu Hijau, sea water is used as process water and there is no need for further sea water addition.

CASE 18.5
Newmonts' Batu Hijau DSTP Scheme

Newmont operates a large copper mine named Batu Hijau (meaning 'Green Stone') in Sumbawa, Indonesia. The tailings outfall is at a water depth of 120 m, 3.2 km offshore. The presence of a submarine canyon into which the tailings slurry is discharged serves to confine the tailings density current; eventually the tailings settle at a depth well below 2,000 m. Tailings are discharged at a daily rate of about 120,000 to 150,000 tonnes with a total by end of mine life of well over 1 billion tonnes.

The overland pipeline is constructed of welded steel and supported just above the ground surface on concrete stands. The subsea pipeline is constructed of thick-walled High Density Polyethylene (HDPE) which is delivered in 20 m or 30 m lengths and welded onsite. Through the beach and surf zones, the pipeline is buried to protect it from storm damage. Beyond the surf zone, the pipeline is placed on the seabed. Concrete saddle weights are positioned along the pipeline at intervals to anchor it in position and to resist any uplift stresses. A de-aeration tank is located onshore. Here any air entrained in the tailings is released. At Batu Hijau, sea water is used as process water and there is no need for further sea water addition.

In specific recolonization experiments undertaken using Batu Hijau tailings, the diversity and population density of benthic organisms reach natural levels within two years after deposition. Such rates of recovery are not attainable in terrestrial ecosystems, even under the most ideal circumstances.

The DSTP system is relatively simple, comprising only a deaeration tank, which is a robust structure lacking any mechanical parts, and two pipelines, involving both sub-sea and overland sections. Accordingly, the number of things that can go wrong is limited.

The Government of Indonesia congratulated the operation for its excellent environmental performance by awarding high ranking in the Governments' PROPER audit scheme. However, the Batu Hijau mine continues to be scrutinized by central government officials and by national anti-mining activists. Even many years of successful operation supported by extensive marine monitoring fails to convince 'die-hard' critics that DSTP can be a favourable option of tailings disposal in areas which impose high risks on traditional on-land tailings placement schemes.

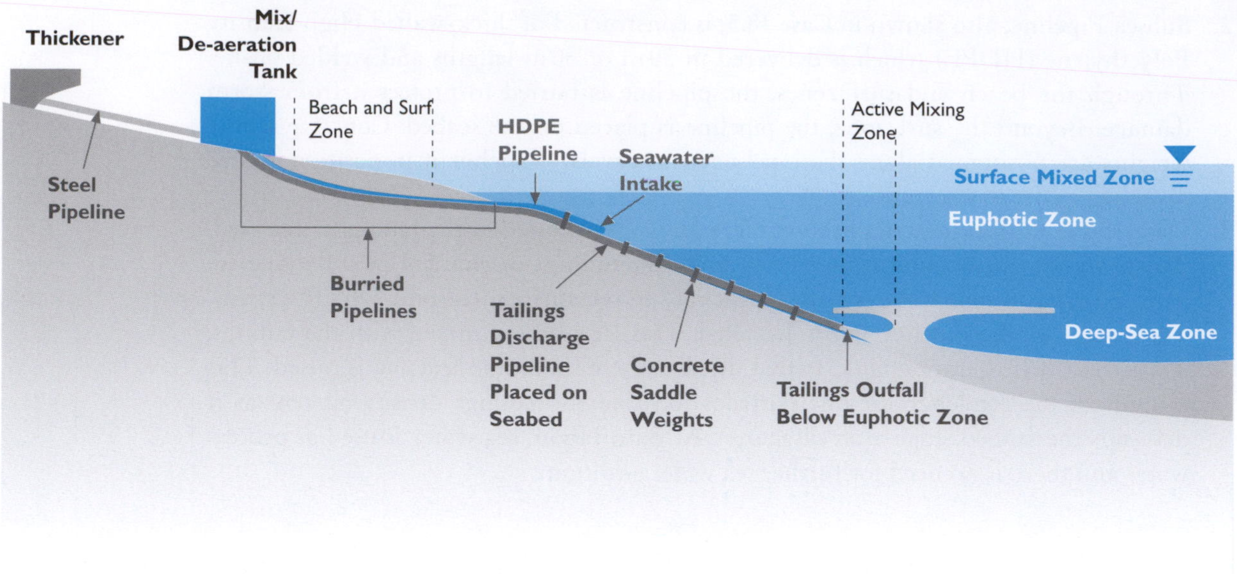

FIGURE 18.18
Typical DSTP Arrangement

The euphotic layer is defined as the depth reached by only 1% of the photosynthetically active light transmitted from the ocean surface.

Source: MMSD 2001 and Jones and Jones 2001

18.5 COMPLETED AND OPERATING STP PROJECTS

Table 18.8 lists some examples of early STD projects which are now complete. These early STD projects provided much of the design data and operating experience on which subsequent systems have been based. As in most new technologies, the first examples experienced unforeseen problems, leading to safeguards being incorporated into subsequent designs. Some examples follow.

The Island Copper Project pioneered many of the concepts that have since been incorporated in other projects. A comprehensive marine monitoring programme was conducted throughout operations and following closure, providing an invaluable long-term record of impacts and post-closure recovery. Remarkably, this project operated for a very long period and discharged hundreds of millions of tonnes of tailings, without affecting commercial salmon and crab fisheries in the vicinity, despite the fact that, by today's standards, the tailings were discharged and accumulated in shallow water.

The Atlas Copper Project was another early pioneer of STD. The sub-sea outfall was located on a tower founded on the seabed. After many years of successful operation, the tower was destroyed by a typhoon. Subsequent STD installations at other sites have used pipelines buried through the beach and surf zones and anchored to the seabed by saddle weights, thereby avoiding damage during typhoons.

Blockage of the outfall of the Minahasa STD outfall occurred during a plant shutdown. This led to the practice of maintaining water discharge through the pipeline during plant shutdowns. Subsequent STD designs have involved discharge where seabed slopes are sufficiently steep that the tailings slurry flows well away from the outfall.

The submarine landslide that severed the subsea tailings pipeline at Misima has led, in such steeply sloping areas, to implementation of geophysical investigations to help identify stable areas for the pipeline route.

Table 18.9 lists examples of DSTP systems, in operation as of 2006. These include the Lihir and Batu Hijau systems which are the most recent examples and which include safeguards based on incidents that occurred at previous operations. **Case 18.6** provides a critical comparison of the Batu Hijau DSTP scheme with the tailings disposal schemes of three other large mines developed in similar environmental settings.

Potential Impacts and Risks of Deep Sea Tailings Placement

Where bathymetric and oceanographic conditions are suitable, as in the cases listed above, DSTP systems can be designed to completely avoid impacts on shallow marine ecosystems such as mangroves, coral reefs and sea-grass meadows which constitute the most productive marine ecosystems. The deep water environments through which the tailings must flow and the very deep water environments where tailings solids accumulate, are characterized by very low productivity. There is little or no interaction between organisms in these deep habitats and those in the euphotic zone.

CASE 18.6
Comparison of Selected Tailings Disposal Schemes

It is instructive to compare the environmental outcomes from four major, geologically similar projects in similarly mountainous terrain. These are:

- Bougainville Copper Mine in PNG;
- Ok Tedi Copper Mine in PNG;
- Grasberg Copper Mine in Papua Province, Indonesia; and
- Batu Hijau Copper-Gold Mine in Sumbawa, Indonesia.

All four projects involve large open Pit mines to exploit large porphyry copper deposits containing in the order of 1 billion tonnes of ore. All reached the same conclusion prior to or during construction – that conventional on-land tailings disposal was either not feasible or entailed an unacceptable risk of catastrophic failure. (The Ok Tedi Project was designed to have an on-land tailings storage; however, a major landslide occurred during foundation preparation for the embankment, leading the owners and the government regulators to the conclusion that a secure tailings impoundment could not be developed).

The Bougainville, Ok Tedi and Grasberg operations all discharged tailings into nearby streams or rivers, while at Batu Hijau, a DSTP system was used. At Bougainville, which was developed before environmental impact assessment became established, tailings were discharged to a local stream, accumulating on parts of the flood-plain and in a large deltaic deposit beyond the mouth of the stream. A recent satellite image clearly shows the onshore tailings accumulations, and suggests that the deltaic deposits have been partly removed, presumably by wave action and currents with the remaining deposited tailings well vegetated. Prior to the unplanned closure of this project, a pipeline was being constructed to transport the tailings to a submarine outfall further offshore, and plans for rehabilitation of the deltaic area using mangrove vegetation were being implemented. For both Ok Tedi and Grasberg, the initial sediment transport modelling studies indicated that the receiving waters had sufficient capacity to transport the tailings (and, in the case of Ok Tedi, also the waste rock) to the sea, where it would deposit. In both these cases this proved not to be the case; sediment deposited in the lower reaches of the rivers where river bed gradients flattened and flow velocities decreased, leading to aggradation which eventually led to overflow of the river banks and deposition of tailings over extensive areas away from the rivers. In the case of the Grasberg project, levees were constructed to confine the area of deposition. In the case of Ok Tedi, the area of deposition has not been managed; rather the approach adopted has been to pay compensation for the resulting environmental damage. In these three cases, the environmental damage has been substantial, and at Ok Tedi and Bougainville contributed to public outrage. At Bougainville, this led to the premature closure of the project in 1989, and contributed to widespread social turmoil on the island, that continues to this day.

It may be that the benefits of both the Ok Tedi and Grasberg Projects outweigh the environmental costs. Both provide direct and indirect employment for thousands of people. The revenues from Ok Tedi comprise a major part of PNG government receipts and represent a significant portion of GNP. How history will judge these projects will depend largely on how rapidly and successfully the areas are rehabilitated, and whether or not there are secondary impacts, as yet not apparent, such as oxidation leading to mobilization of dissolved metals. It is clear that, had the magnitude of direct impacts been recognized by either proponents or regulators during initial project planning, alternative tailings disposal schemes would have been sought. In fact, after the impacts became apparent, alternative tailings disposal schemes were intensively investigated for both projects. However, the alternatives proved to be either impractical or extremely costly and the costs could not be justified, given that the environmental damage had already occurred. It is not clear whether or not the costs of implementing environmentally acceptable tailings disposal schemes at the outset, would have rendered the projects uneconomic.

Contrast this with the situation at Batu Hijau where the total terrestrial footprint (combined areas of surface disturbance) is an order of magnitude less than at the other sites. Monitoring studies have confirmed predictions made in the environmental impact assessment that:

- No tailings would be present in the water column or on the seabed above the 120 m outfall depth, which means that there can be no impact of tailings on productive shallow water ecosystems such as coral reefs;
- The tailings slurry would be largely confined within the Senunu canyon to a depth of 1,400 m; and
- The main area of tailings deposition would be on the seabed at depths exceeding 3,000 m.

In fact, the footprint of tailings on the seabed at Batu Hijau is somewhat less than the footprint of tailings on land in the cases of Ok Tedi and Grasberg. These projects also resulted in large areas of tailings deposition in shallow water marine environments, although these have not caused anything approaching the same concern as onshore deposition. The most important consideration relating to Batu Hijau is that all the impacts of tailings disposal are confined to water depths exceeding 120 m. This means that the impacts all occur in the zones of very low biological productivity, below the euphotic zone.

This comparison of four large projects in the Asia-Pacific region may lead to various conclusions; for example:

- Large porphyry copper deposits situated in the mountainous, wet tropics in earthquake prone areas, should not be developed; or
- Such deposits should not be developed unless the tailings can be discharged by means of DSTP; or
- Other methods of tailings disposal need to be developed before such deposits can be exploited.

Fine particles do become separated from the tailings density current as it descends. These particles form one or more horizontal plumes at density discontinuities in the water column, below the depth of the outfall. Sampling of these plumes indicates that the concentration of suspended sediment is generally around 50 mg/L. Similar plumes occur naturally from sediment introduced from onshore flood events, and such low sediment concentrations are not detrimental to organisms living at these depths, which are mainly zooplankton.

Another important consideration is that the marine impacts from DSTP are relatively minor and readily reversible. The main impacts relate to the burial of benthic organisms by depositing tailings. Extensive studies during and following tailings discharge at the Island Copper mine in British Columbia, Canada, have shown that deposited tailings are rapidly colonized by benthic fauna (Byrd and Ellis 1995; Ellis 2008). Similarly, studies of Misima Project tailings deposited in water depths of 300 m and more, showed rapid recolonization by benthic organisms after tailings discharge had been completed (MMSD 2002). In both these cases and in specific recolonization experiments undertaken using Batu Hijau tailings, the diversity and population density of benthic organisms reach natural levels within two years of deposition. Such rates of recovery are not attainable in terrestrial ecosystems, even under the most ideal circumstances. Commonly, decades or centuries are required before natural succession processes can restore terrestrial ecosystems of comparable diversity and productivity.

Chemical impacts are of greater concern than the physical impacts.

Clearly, chemical impacts are of greater concern than the physical impacts discussed above. Here, both the tailings liquid and solid fractions need to be considered as potential sources of contamination. In general, the same chemical substances will be of concern as for the tailings if stored onshore.

For dissolved constituents, rapid dilution occurs in the sea water mixing tank (if used) and in the 'near field' immediately downstream of the outfall where further sea water is entrained in the density current. As a result of this dilution, by the end of the near field which is about 100 m from the outfall, the concentrations of dissolved constituents within the density current itself, are close to those of natural sea water. Modelling results predict this situation and monitoring results have confirmed it at many sites. Cyanide residues in tailings from gold processing operations provide a particularly interesting case. Because of the toxicity of cyanide and public concern, some operators have opted to apply cyanide destruction to the tailings prior to discharge. For typical residual cyanide concentrations in gold tailings, however, such treatment is unnecessary, as dilution will reduce the concentration by two orders of magnitude by the end of the near field. Also, cyanide is rapidly decomposed by reactions with components of sea water. At the Misima Project, in PNG, cyanide destruction was not carried out and the tailings as discharged to the mixing tank contained cyanide at concentrations of about 80 mg/L (CN_{WAD}), which was in accordance with the water use permit. Modelling results showed that these concentrations would be rapidly reduced, and this is what happened. In fact, analyses of water samples collected close to the outfall failed to detect cyanide.

The examples of DSTP shown in **Table 18.9** have all involved tailings solutions that are more or less alkaline. As such, concentrations of dissolved metals have been quite low, even before dilution. Acidic tailings are more likely to contain high concentrations of metals. If and when such acidic tailings are discharged to the sea, it can be expected that metals will precipitate as the pH rises with the addition of sea water and as the temperature declines, adding slightly to the solids content, and reducing dissolved concentrations.

Many tailings contain one or more heavy metals in the solid fractions, at relatively high concentrations compared to natural sediment. That these metals are insoluble at the pH and temperatures used in the process, is self-evident. The remaining questions are whether or not these metals may be more soluble in sea water, and if so, whether they may reach concentrations that could adversely affect marine organisms? This is simple to ascertain by mixing tailings with sea water and analyzing for metals, before and after mixing. If this test shows that the dissolution of one or more metals is occurring, more sophisticated testing is carried out to measure the rate or flux of dissolution for each metal of concern, from which modelling can predict the maximum concentrations that could develop under operating conditions.

Finally, the fate of sulphide minerals contained in tailings should be considered. Such minerals are of concern when stored in onshore impoundments; on exposure to air they can produce acid which can leach metals from the tailings solids. One of the strategies used to avoid acidification of onshore tailings deposits is to maintain them in a permanently saturated state. Clearly, DSTP achieves the same objective. The oxygen content of deep sea water is so low that sulphides in the tailings will not oxidize. In fact, in many parts of the deep sea, metals precipitate as sulphides. According to Environment Canada (1996) 'tailings deposits on the ocean floor tend to be a sink not a source for metals'.

Tailings deposits on the ocean floor tend to be a sink not a source for metals.

DSTP systems are relatively simple, comprising only a De-aeration Tank, which is a robust structure lacking any mechanical parts, and one or two pipelines, involving both subsea and overland sections. Accordingly, the number of things that can go wrong is limited to:

- Overflow or rupture of the De-aeration Tank;
- Rupture of the onshore tailings pipeline; and
- Rupture of the submarine pipeline.

The De-aeration Tank is a robust structure which is unlikely to rupture under any circumstances. The risk of overflow is avoided in the hydraulic design.

The risk and consequences of rupture of an onshore pipeline apply equally to conventional on-land tailings storage systems. Such risks are minimized by regular inspection and maintenance programmes. Consequences may be minimized by building embankments on both sides along the pipeline corridor.

Submarine tailings pipelines have ruptured, in one instance after blocking of the outfall and subsequent over-pressurization caused the pipe to part at a joint, and in the other instance where a sea-bed landslide severed the pipe. In both these cases, temporary discharge into shallow water occurred until the ruptures were remedied. However, in neither case was there significant environmental damage. This contrasts to the situation in conventional on-land tailings disposal where many more risks need to be addressed and where the consequences of malfunction are considerably higher.

Site Selection for Deep Sea Tailings Placement

An effective approach to site selection includes the following steps.

Review of Available Bathymetric and Oceanographic Data

Although such data are regional and unlikely to provide much local information, they usually provide a good indication of whether deep water occurs close to the coast, and of ultimate offshore water depths. In addition, useful data may be available on tides and surface currents, and on the depths and strength of thermoclines in the water column. Other relevant information may be obtained from daily sea surface temperature data generated by remote sensing from satellites. In particular, a review of sea temperature records can be used to check for evidence of upwelling in which colder water from deeper parts of the water column is moved to the surface, despite its higher density. Upwelling, a natural phenomenon found in few parts of the world, can be a regional phenomenon, where submarine currents meet land masses, or localized, caused by wind shear effects, where wind-driven surface water is replaced by deeper water.

Bathymetric Survey

Assuming that the results of the initial review are favourable, the next stage is to undertake a bathymetric survey of the coastal zone to either side of the coastal location closest to the mine. The length of coastal zone to be surveyed will be selected based on information from the initial review. If a very large area is involved, the survey may be conducted in two stages – a broad survey over the entire area, followed by more detailed surveys in selected sub-areas. Sidescan sonar surveys can be conveniently conducted along with the detailed bathymetric survey. If the initial review indicates that seabed slopes are likely to be very steep (20° or more), a seismic reflection survey can also be incorporated in the programme. Data from all these surveys are used to assist in the siting of the pipeline route and outfall location and to assess seabed stability.

Seasonal Studies

Again, assuming that the results of the preceding step confirm favourable conditions for DSTP, the next stage is to collect oceanographic data required to confirm site suitability, to assist in selection of design parameters, and to provide data for input to subsequent numerical modelling studies. Typically, two areas will be selected for parallel study in case a 'fatal flaw' is identified at one or the other. Data required include current velocities and current directions, and salinity-temperature-density (STD) profiles through the water column. These data are required for a complete lunar cycle in each season of the year. Currents are measured either by deployment of current meters, of which several types are available, or by means of Acoustic Doppler Current Profiling equipment (ADCP). STD information is generated by cable-mounted sensors that measure temperature, salinity and pressure allowing temperature, salinity and density (computed from a combination of these parameters) to be plotted against depth (computed from pressure). STD probes are conducted at a variety of locations at different stages of the tidal cycle in each season, to define the variations in depth of the mixing zone throughout the year. This information is used to determine the minimum depth of the tailings outfall. The investigations outlined above should be sufficient to evaluate the feasibility of DSTP, to identify the preferred site and to provide design information such as a preferred pipeline route and outfall location. The data obtained can also be used for modelling studies to assess tailings transport and sedimentation, as part of impact assessment.

Environmental Baseline Studies

Environmental baseline studies are normally carried out in parallel with these investigations and it may be logistically convenient to conduct such studies at the same time as vessels are used to deploy or re-deploy current measuring equipment, and to carry out STD probes. The baseline studies specific to DSTP include: (1) sampling of seabed sediments for physical and chemical analyses, and for laboratory assessment of benthic organisms; (2) sea water sampling and analysis, and (3) fishing to identify species and relative populations of deep water fishes, and to obtain tissue samples for laboratory analysis of baseline metal concentrations.. In addition to these baseline studies, it is prudent to establish baseline conditions in the nearby shallow water coastal habitats. Although these habitats should not be affected by DSTP, they are likely to be of major concern to local people, and it may be necessary at some time in the future, to prove that DSTP has not affected them.

Monitoring

The impacts of DSTP are monitored using the following techniques:

- Drop cores or grab samplers to provide samples of seabed sediment for laboratory analysis;
- Transmissometer probes to identify suspended sediment in the water column, often in conjunction with STD profiling;
- Water sampling at various depths, and laboratory analysis;
- Subsea visual observations using a Remotely Operated Vessel (ROV). Such observations are used to assess the condition of the DSTP pipeline, to observe the behaviour of the tailings density current, and even to identify the behaviour of fish in the vicinity of the tailings plume;
- Fishing surveys to assess species present and to enable tissue samples to be obtained for laboratory analysis of metal concentrations.

While these monitoring procedures in water depths up to 200 m, can be undertaken at regular intervals similar to those that apply to monitoring of terrestrial systems, monitoring of sediment deposition in very deep water requires highly specialized equipment and is very expensive. The adopted approach is to undertake regular monitoring above a 200 m water depth, at monthly or quarterly intervals to identify any unexpected occurrences and to confirm the absence of impacts on the important shallow water habitats. Monitoring of the actual impacts of tailings deposition in very deep water is much less critical, and, because of its high cost is generally carried out at intervals of two to five years.

Closure

Closure of DSTP systems requires little, if any, action. There is no compelling environmental reason that the subsea pipeline needs to be retrieved, although potentially it may be used on another project or for other applications. Onshore facilities require removal, and would provide materials worth salvaging. Post-closure monitoring would logically involve a deep water sediment sampling and analysis programme two years after closure, to assess the extent of re-colonization. Should the results accord with previous studies which have shown rapid re-colonization, no further monitoring would be required.

Closure of DSTP systems requires little, if any, action.

REFERENCES

Aldous R (2002) Tailings Storage – Guidelines for Victoria (Discussion Paper), Department of Natural Resources and Environment.

ATV-DVWK-M503 (2001): Grundlagen zur Überprüfung und Ertüchtigung von Sedimentationsbecken. ISBN 3-935669-43-7.

Australian Government (2007) Tailings Management, Leading Practice Sustainable Development Program of the Mining Industry, Department for Industry Tourism and Resources.

Brehault H (1997) Tailings Management – A Placer Dome Perspective, Proceedings of the International Workshop on Managing the Risk of Tailings Disposal, Sweden 1997, International Council on Metals and the Environment (ICME).

BRGM (2001) Management of mining, quarrying and ore-processing waste in the European Union, 79 p., 7 Figs., 17 Tables, 7 annexes, 1 CD-ROM (Collected data).

Burd BJ and Ellis DV (1995) ICM Closure Plan: Review of Benthic Surveys 1970 to 1992 for Rupert/Holberg/Quatsino Inlet System.

Davies MP (2001) Tailings Impoundment Failures: Are Geotechnical Engineers Listening? Waste Geotechnics, Geotechnical News.

DME (1999) Guidelines on the safe design and operating standards for tailings storage, Department of Minerals and Energy, Western Australia, Government of Western Australia.

DPI (2003) Management of Tailings Storage Facilities – Environmental Guidelines, Department of Primary Industries, Victoria – Minerals and Petroleum Division.

Doran JR and McIntosh JA (1995) Preparation, review and approval of mine closure plans in Ontario, Canada; Proceedings, Sudbury '95, Mining and the Environment, T.P. Hynes and M.C. Blanchette (Eds.), CANMET, Ottawa, 1995, Vol. 1.

Ellis DV (1997) Year 1 Appraisal of Shoreline Biodiversity on the Beach Dump Face at Island Copper Mine, August 20/21, 1997. Report to Island Copper Mine, September, 1997.

Ellis DV (2008) The Role of Deep Submarine Tailings Placement in the Mitigation of Marine Pollution for Coastal and Island Mines, in Hofer, TN (Ed) Marine Pollution: New Research. Nova Science Publishers.

Engels J and Dixon-Hardy D (2004) Tailings disposal – Today's storage of high volumes of waste from mines. JKMRC Conference 2004, Brisbane, Australia.

Engels J *et al.* (2006) A Realistic Technique to obtain the Surface Contours of Conventional, Thickened and Paste Tailings Storage Facilities 9th International Seminar on Paste and Thickened Tailings, Limerick, Ireland, April, 2006.

Frazer W and Roberston J (1994) Subaqueous disposal of reactive mine waste: An overview and update on case studies – MEND/Canada. In: Third International Conference on the Abatement of Acidic Drainage, Pittsburg, P.A.

Fourie AB (2003) In Search of the Sustainable Tailings Dam: Do High-Density Thickened Tailings Provide the Solution, School of Civil and Environmental Engineering, University of the Witwaterstrand, South Africa.

ICME (1998) Case Studies on Tailings Management.

ICME (1999) International Council on Metals and the Environment: The management of cyanide in gold extraction, by M J Logsdon, K Hagelstein and T I Mudder.

ICME and UNEP (1999) Proceedings of the Workshop on Risk Assessment and Contingency Planning in the Management of Mine Tailings: Buenos Aires, Argentina, November 5–6, 1998. Ottawa, Ontario, Canada, International Council on Metals and the Environment (ICME) and United Nations Environment Programme (UNEP).

ICOLD and UNEP (2001) Bulletin 121: Tailings Dams – Risk of Dangerous Occurrences, Lessons learnt from practical experiences. Paris. www.icold-cigb.org

Jewell R and Fourie A (Eds), (2005) Paste and Thickened Tailings – A Guide. Second Edition. Australian Center for Geomechanics.

Jones S and Jones M (2001) Overview of Deep Sea Tailing Placement January 2001 Update, NSR Environmental Consultants Pty Ltd, Victoria, Australia for BHP Minerals.

Kealy C and Bush R (1971) Determining Seepage Characteristics of Mill – Tailings Dams by the Finite Element Method. U.S. Bureau of Mines, RI 7477.

Kreft-Burman K *et al.* (2005) Finnish Environment Institute (SYKE), Helsinki, Finland. Tailings Management Facilities – Legislation, Authorisation, Management, Monitoring and Inspection Practices Report of Workpackage 4.5 of the RTD project; Sustainable Improvement in Safety of Tailings Facilities (TAILSAFE) funded by the European Commission under contract No EVG1-CT-2002-00066; Website: http://www.tailsafe.com/

MAC (1998) A Guide to the Management of Tailings Facilities, The Mining Association of Canada, Ottawa, Canada: 54.

MAC (2003) Developing an Operation, Maintenance and Surveillance. Manual for Tailings and Water Management. Facilities, The Mining Association of Canada, Ottawa, Canada: 49.

Martin TE, Davies MP *et al.* (2002) Stewardship of Tailings Facilities, Canada, Report commissioned by Mining Minerals and Sustainable Development (MMSD) a project of Institute for Environment and Development (IIED): No. 20: 35.

MIRO (1999) A technical framework for mine closure planning, Mineral Industry Research Organisation, Technical Review Series No 20, Staffordshire, England.

MMSD (2002) Breaking New Ground; Mining, Minerals and Sustainable Development (MMSD) Project; published by Earthscan for IIED and WBCSD.

Penman ADM (2001) Risk Analyses of Tailings Dam Construction – Seminar on safe tailings dam constructions. Gallivare, Swedish Mining Association, Natur Vards Verket, European Commission.

Phillips JT (2000) Things you should not forget. Australian Center for Geomechanics Tailings – Corporate Risk and Responsibility Seminar, March 2000, Sydney.

Roberston J (1998) Subaqueous tailings disposal – The best option available; in: 3rd International and the 21st Annual Minerals Council of Australia Environmental Workshop, Proceedings Volume 2, Newcastle, New South Wales, Australia.

Robinsky EI (1979) Tailings disposal by the Thickened Central Discharge Method for improved economy and environmental control. Proc. 2nd Int. Tailings Symp. Argall G (ed). Miller Freeman, San Francisco.

Robinsky EI (2000) Sustainable development in disposal of tailings. Tailings and Mine Waste. Colorado, USA, A.A. Balkema, Rotterdam: pp. 39–48.

SANS (1998) Code of Practice for Mine Residue Deposits, South African Bureau of Standards. ISBN: 0626117003

Sarsby R (2000) Environmental Geotechnics, ISBN- 0 7277 2752 4

Sly P (1996) Review of MEND studies from 1992 to 1995; MEND Report 2.11.1e.

Tremblay G (1998) Subaqueous Tailings Disposal: Results of the MEND program – Case studies on tailings management, International Council on Metals and the Environment (ICME), UNEP.

USEPA (1982) Development Document for Effluent Limitations Guidelines and Standards for Ore Mining and Dressing – Point Source Category, Research Report.

USEPA (1995) Human health and environmental damages from mining and mineral processing wastes, Technical Background Document Supporting the Supplemental Proposed Rule Applying Phase IV Land Disposal Restrictions to Newly Identified Mineral Processing Wastes, Office of Solid Waste.

Vick SG (1990) Planning, design, and analysis of tailings dams, Vancouver, BiTech. ISBN: 0921095120. 2nd Edition, 369 p. (First edition 1983)

Wahler WA and Associates (1970) Engineering evaluation of mill tailings disposal practices and potential dam stability problems in Southwestern United States, Technical Proposal to U.S. Bureau of Mines.

Welch D (2003) Advantages of Tailings Thickening and Paste Technology, Responding to Change – Issues and Trends in Tailings Management – Golder Associates Report: 5.

Wels C and Robertson A (2003) Conceptual model for estimating water recovery in tailings impoundments, Robertson GeoConsultants, Inc., Vancouver, Canada.

Williams DA and Williams DJ (2004) Trends in tailings storage facility design and alternative disposal methods', Proceedings of ACMER Workshop on Design and Management of Tailings Storage Facilities to Minimise Environmental Impacts During Operation and Closure, Perth, Australia, Australian Centre for Minerals Extension and Research, Brisbane, Australia.

WMC Limited (1997) WMC in the Philippines. Information Paper No. 1. October 1997. In Making the Grade, The Mineral Industry and Aboriginal Community Developments. Nov 20–21, 1997.

Xenidis A ed. (2004) Tailings Management Facilities – Implementation and Improvement of Closure and Restoration Plans for Disused Tailings Facilities; Report of Work Package 4.3 of the RTD project; Sustainable Improvement in Safety of Tailings Facilities (TAILSAFE) funded by the European Commission under contract No EVG1-CT-2002-00066; Website: http://www.tailsafe.com/

19

Approaches to Waste Rock Disposal

Issues and Risks

MAGNESIUM

Magnesium is a silvery white, light weight reactive metal which tarnishes readily. The main uses of this metal are in the manufacture of light weight alloys, used in the automotive and aircraft industries, and for consumer products such as mobile phones and cameras. Alkaline compounds of magnesium are used in agriculture and for water treatment. Magnesium is an essential element for all organisms, occurring in all cells. It is the third most common element in sea water, from which it is commonly extracted. It also occurs in the minerals dolomite and magnesite. Most of its compounds are non-toxic and many have medicinal properties.

19 Approaches to Waste Rock Disposal

Issues and Risks

Waste rock, also known as spoil or mullock, is produced in most mining operations as a necessary consequence of accessing an ore body. Various terms are used for the waste rock, depending on its distribution in relation to the ore. For example, the term overburden is used for waste material overlying an ore body. Similarly in a stratified situation such as a sedimentary sequence involving multiple seams of coal, the waste rock between coal seams is known as interburden. The ratio of waste rock to ore is known as the stripping ratio. A stripping ratio of 3 means that 3 tonnes of waste is excavated for each tonne of ore.

Waste rock is produced in most mining operations as a necessary consequence of accessing an ore body.

In general, underground mining operations produce much smaller quantities of waste rock compared to surface mines. In an open Pit mine, the overall stripping ratio is determined by the dimensions and geometry of the ore body, the value of the ore and the shape of the pit, which itself may be determined by slope stability considerations. Large, low grade ore bodies, low value commodities such as limestone, or low rank coal deposits such as lignite, are usually uneconomic to mine unless stripping ratios are low – from less than 1 to 3. High rank coal seams and high grade ore bodies, particularly for valuable commodities such as precious metals, may have stripping ratios of 10 or more. **Figure 19.1** shows a cross-section through a typical open pit mine with a high stripping ratio. In this case, the ore body is a relatively narrow, steeply dipping vein, similar to many gold mining situations. The ore body continues below the base of the pit but to mine deeper would be uneconomic because the cost of removal of additional waste rock would exceed the value of additional ore. In such a situation, an increase in the price of the commodity may justify deepening the Pit, resulting in an increased stripping ratio.

19.1 NATURE AND CHARACTERISTICS OF WASTE ROCK

Waste rock from mining operations exhibits a wide range of chemical and physical properties, the understanding of which can be of critical importance in determining

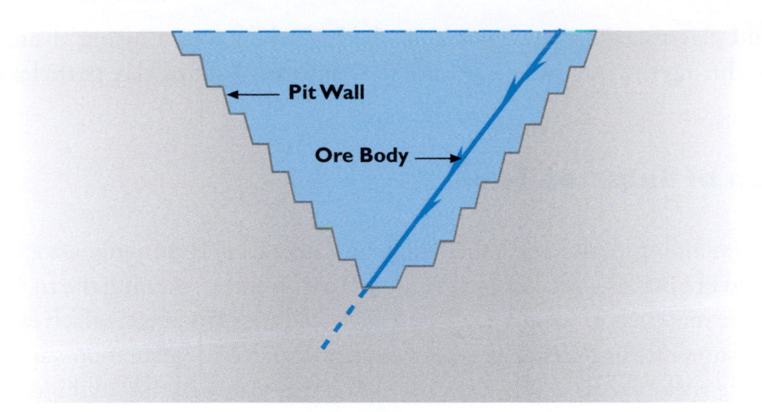

FIGURE 19.1
Profile of Open Pit with Narrow, Steeply Dipping Ore Body and High Stripping Ratio

A stripping ratio of 3 means that 3 tonnes of waste is excavated for each tonne of ore.

appropriate design and management measures. Among the most important properties are the following.

Grading or Size Range

The grading of waste rock materials depends on the nature of the rock mass prior to mining and the degree of fragmentation caused during mining. Generally, mining methods are selected to achieve sufficient fragmentation that the waste rock can be retrieved by shovel or front end loader and loaded onto trucks without the need for additional breakage. In smaller mines and in underground mines where low capacity equipment is used, a higher degree of fragmentation is required than for large mines. In the case of weak rocks the fragmentation achieved depends on the size and nature of the excavator, including bucket dimensions, teeth size and spacing, and operator method. Excavation of weak rock generally produces appreciable quantities of fines, which are particles of sand, silt and clay. In stronger rocks, explosives are used to achieve the desired fragmentation. Generally, only small quantities of fines are produced in excavation of strong rocks.

In smaller mines and in underground mines where low capacity equipment is used, a higher degree of fragmentation is required than for large mines.

Shape

The shape of particles, particularly the larger waste rock fragments, influences the way the materials behave when dumped. For example, bulky or roughly equi-dimensional fragments roll down a dump face while elongated or slabby fragments do not roll far.

Strength

Weaker waste rock particles undergo further fragmentation, and produce increasing quantities of fines during loading, transportation and dumping, whereas strong particles retain their size and shape.

Durability

Some rocks undergo rapid changes in physical properties once exposed to air and/or water. In some cases, boulders can be reduced to gravel within a few weeks of exposure. Some siltstones and claystones are susceptible to slaking when wet, causing disintegration and ultimately reducing the fragment size until the constituent silt or clay particles are released.

Oxidation of Sulphide Minerals

Waste rock containing pyrite and other sulphide minerals is commonly associated with base metal and gold ore bodies, and coal interburden. Apart from weakening the rock fabric, oxidation of sulphide materials in waste rock may lead to acid rock drainage, which can cause serious adverse environmental impacts. The rate of oxidation and acid generation varies considerably from one waste rock type to another, depending on the nature of the sulphide mineralization, the host rock and the climate. Acid drainage is discussed in detail in Chapter Seventeen.

19.2 POTENTIAL IMPACTS OF WASTE ROCK DISPOSAL

In many mining projects, more environmental damage results from waste rock disposal, than any other component of the operations. It is also usually the most visible component. The area occupied by waste rock storages commonly represents a large part of the project 'foot-print', often much more extensive than the mine itself. The most obvious impact is burial of the land occupied by the waste rock storage, with the loss of values attributable to that land, whether they be habitat values, agricultural values, or scenic values (**Figure 19.2**). This does not necessarily constitute a long-term adverse impact as there may be the opportunity to create conditions on the waste rock storage that represent an improvement over what was there before.

Storage of unconsolidated waste rock may lead to land-slips, particularly in mountainous areas, in areas subject to high rainfall, and/or in areas of high seismicity. Similarly, waste rock materials, particularly those containing weak fragments or high fines content, may be highly erodible. Erosion from waste dumps can lead to major increases in sediment in downstream streams and rivers, damaging the habitats for aquatic organisms and potentially affecting use of water for domestic purposes and irrigation (**Case 19.1**). Of all impacts caused by waste rock disposal, those involving acid rock drainage are by far the most serious. There have been many documented cases, some described in Chapter Seventeen, where large stretches of rivers or streams have been acidified by drainage from waste rock dumps, resulting in serious reductions in diversity and populations of aquatic biota and, in some cases, virtually total destruction of aquatic life.

Of all impacts caused by waste rock disposal, those involving acid rock drainage are by far the most serious.

19.3 OBJECTIVES OF WASTE ROCK DISPOSAL

The following objectives apply in planning the storage and disposal of waste rock to minimize environmental damage:

- The area occupied by the waste rock storage (i.e. its 'foot-print') should be minimized, in-so-far as this is consistent with other objectives;
- The new landform created by the waste rock storage should be consistent with its future land use(s), in terms of slopes, accessibility, surface roughness;
- The new landform should be stable;
- The new landform should not be any more erodible than comparable natural landforms in the vicinity;

A. Kennecott-Bingham Copper Mine, the world largest man-made excavation
Photo: www.bettermines.org

B. Waste Rock Iron Ore Mines at Wabush, Labrador
Photo: Courtesy of Jim Wark, www.airphotona.com

FIGURE 19.2
Waste Rock Disposal

The area occupied by waste rock storages commonly represents a large part of the project 'foot-print', often much more extensive than the mine itself.

- If the waste rock includes materials that have the potential to generate acid drainage, the storage should be designed to prevent or control this occurrence, as discussed in Chapter Seventeen;
- The surface of the waste rock storage should be amenable to rehabilitation, consistent with its future land use, and
- If possible, the new landform should be congruent with its surrounds, meaning that it should not be visually intrusive.

In practice, the planning of waste rock disposal involves trade-offs between different considerations. For example, the smallest possible foot-print for an above-grade storage would involve the highest and steepest slopes that could be achieved, which would also represent the least stable, least accessible and most difficult conditions for rehabilitation.

In practice, the planning of waste rock disposal involves trade-offs between different considerations.

19.4 SITE SELECTION FOR WASTE ROCK STORAGES

The main factors influencing the cost of waste rock disposal are:

- Haulage costs which depend on the distance of the storage site from the mine, and whether or not the site is elevated with respect to the mine;

- The design of the storage; in particular the need for encapsulation of sulphidic materials, or incorporation of elaborate drainage provisions;
- The method of construction, including foundation preparation; whether 'top-down' or 'bottom-up' (**Figure 19.3**); compacted or uncompacted; and the extent of regrading required to achieve the final surface configuration.

From an environmental view-point, the best possible sites for storage of waste rock are mined-out areas.

From an environmental view-point, the best possible sites for storage of waste rock are mined-out areas. Burial of mine wastes below the original ground surface is known as below-grade storage, as opposed to above-grade storage as is the case where wastes are stored above the natural ground surface. Below grade storage of waste rock is achieved routinely in the case of strip mines for coal, and is also common in extensive shallow surface mines such as those for the extraction of bauxite and lateritic nickel ores. It is also being used increasingly in open Pit gold mines where multiple but discrete ore bodies occur in close proximity. The advantages of using waste rock to fill mine voids may include:

- Minimizing the foot-print and, in particular, the 'residual foot-print', which is the land disturbance that remains after the completion of rehabilitation;
- Stabilizing the void. Many surface mines have unstable slopes or slopes that may become unstable over time;
- Isolation of potential acid-generating wastes. The best possible site for such wastes is below the water table, and many mine voids provide such sites. However, even if the void does not extend below the water table, the storage of potential acid-generating waste below grade is far preferable to storage above grade, as the chances of acid drainage reaching the surface are much reduced.

If below grade storage is not feasible, the next best option, in terms of minimizing the foot-print, is to utilize one or more natural depressions in the landscape. In mountainous terrain, filling or partially filling a valley, may represent the only realistic option available

CASE 19.1
Adverse Agricultural Effects from Mines in Northern Luzon, Philippines

The Cordillera region of Northern Luzon in the Philippines has long been the location of gold mining activity. During a period of high mineral production in the Philippines in the 1970s and early 1980s, prior to the introduction of environmental controls, uncontrolled dumping of waste rock and tailings, contributed large quantities of sediment to the Abra River which drains from the mountains across the coastal plain to the South China Sea (see photo). Accumulation of this sediment in the river channels led to bank overflow during the wet seasons, so that sediment composed largely of mine waste became widely distributed over the flood plain, clogging irrigation ditches and depositing on paddy fields. Adverse impacts included lower crop yields due to the low fertility of the mine-derived sediment and high costs for maintenance of irrigation systems. Consequent unrest among the agricultural communities contributed to the eventual strengthening of environmental legislation relating to mining in the Philippines.

Top-down in which the waste rock is dumped over an advancing phase.
Photo: Courtesy of Peter Essick, National Geographics

Bottom-up storage in which the waste rock is dumped in a series of piles, and then spread to form a relatively thin layer (copper mining waste rock piles, Pinal County, Arizona).
Photo: Courtesy of Jim Wark, www.airphotona.com

FIGURE 19.3
The Top-down and Bottom-up Construction of Waste Rock Dumps

for waste rock disposal. The disadvantage is that valleys contain streams, and filling a valley requires that surface drainage be accommodated in the design. The best storage sites in these circumstances are at the heads of valleys where there is little or no upstream catchment area. However, such sites may be remote from the mine.

On flat or gently sloping plains, where conditions are similar over wide areas, the main criterion for site selection will be proximity to the mine. It is also relevant to the site selection process to consider how the storage will be created, whether using a top-down or bottom-up approach. The comparative benefits of these approaches are discussed in the next section.

The foregoing discussion addresses the vast majority of waste rock storages used by the mining industry. However, two other approaches warrant consideration, namely riverine disposal and marine disposal. Riverine disposal of waste rock is practised at only a few sites, including the Ok Tedi Mine in Papua New Guinea (PNG). It has proved to be problematic and controversial (**Case 19.2**). It is unlikely that riverine disposal would have been proposed or permitted, had the outcomes been predicted in advance.

CASE 19.2
Ok Tedi Copper Project, Papua New Guinea

The Ok Tedi Copper-gold ore body is one of the world's largest porphyry copper deposits. The Panguna Mine which has been exploiting this ore body since 1984, is located at an elevation of 2000 m in the Star Mountains of western PNG. Waste rock from the open Pit mine amounting to 40 million tonnes each year, is discharged together with tailings, to the Ok Tedi River which flows into the Fly River, one of PNG's major river systems, before entering the Arafura sea about 1000 km by river downstream of the mine. The Ok Tedi River is a high velocity river which transports the bulk of mine derived sediment plus natural sediment and erosion products, without appreciable deposition. However, from where the Ok Tedi meets the Fly River at an elevation of only 18 m above sea level, the bed gradient and flow velocities are much reduced, resulting in deposition of much of the sediment load.

River bed levels became elevated as a result of sediment deposition. Eventually, the river overflowed its banks leading to widespread deposition within the adjacent lowland forest areas, causing loss of or damage to more than 1300 square kilometres of forest. Social effects included loss of livelihoods for residents of up to 130 villages.

Following social unrest and court action, the operating consortium which included the PNG government, agreed in 1996, to provide compensation of US$ 28.6 million. In addition, one-third of profits from operations are now directed to development projects in western PNG.

Waste Rock Dump by Top-down Dumping, Chhattisgarh, India

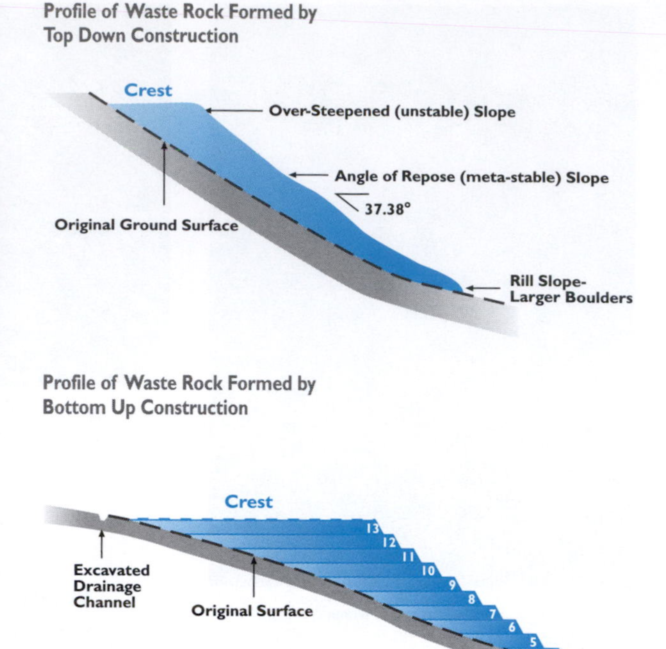

Paddock Dumping of Waste Rock. Three Layers are Visible – Lowest Layer at Left

Profile of Waste Rock Formed by Bottom Up Construction

Crest

Excavated Drainage Channel

Original Surface

13 12 11 10 9 8 7 6 5 4

Toe

FIGURE 19.4
Profile of Waste Rock Storage Formed by Top Down and Bottom Up Construction

While both cases of marine disposal have been widely criticized by environmental NGOs, monitoring has confirmed that impacts on the marine environment have been relatively minor, temporary, and in accordance with predictions made in the relevant EIAs.

Marine disposal of waste rock was used at the Misima Gold Project in PNG, and is currently used at the Lihir Gold Project (also in PNG). In both cases, the reasons for adopting marine disposal were the lack of land suitable for onshore disposal and the predicted low impacts of marine disposal. While both cases of marine disposal have been widely criticized by environmental NGOs, monitoring has confirmed that impacts on the marine environment have been relatively minor, temporary, and in accordance with predictions made in the relevant EIAs.

19.5 ALTERNATIVE DESIGN AND CONSTRUCTION APPROACHES

There are two main approaches to construction of waste rock storages:

- Top-down dumps, in which the waste rock is dumped over an advancing face, known as the angle of repose, sloping at approximately 38° from the horizontal, as shown in profile on the upper part of **Figure 19.4**. After dumping is complete, the dump is re-shaped to its intended configuration, usually using bulldozers. In many older mining operations, no re-shaping was carried out, the angle of repose slope being the final outslope.
- Bottom-up storages in which the waste rock is dumped in a series of piles, and then spread to form a relatively thin layer. This is sometimes referred to as paddock dumping (**Figure 19.4**). Subsequently, the process is repeated until the ultimate storage configuration is achieved.

Hybrid or intermediate approaches are also used, whereby top-down dumping is used to produce relatively thick, (e.g. 10 m or 15 m) layers, which are then overlain by subsequent equally thick layers. This approach is safer and requires less re-shaping than the full height top-down approach.

The advantages and disadvantages of top down versus bottom up construction are compared in **Table 19.1**.

TABLE 19.1

Comparison of Waste Rock Storages – Top Down and Bottom Up

Characteristic	Comparison
Density	Much higher densities are achieved by Bottom Up Construction (BUC), due to the compaction achieved as each layer is spread. If achieving high density is important, this can readily be accomplished using compaction equipment. TDC dumping does not provide any control over compaction.
Homogeneity 	Top Down Construction (TDC) results in particle size segregation with larger, bulky fragments rolling to the foot of the slope, and fines staying near the dump crest. Little segregation occurs during paddock dumping or subsequent spreading so that BUC storages are generally much more homogeneous than TDC storages.
Stability and Safety During Construction 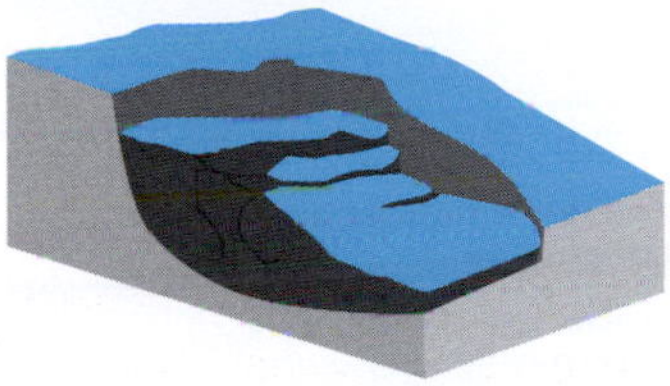	Angle of repose slopes formed in TDC dumps are marginally stable or 'meta-stable', and commonly result in land-slips, posing a risk to dumping equipment and personnel. Regrading operations may be even more hazardous to operators. BUC enables stable slopes to be achieved from start to finish.
Potential for Settlement Profile of Waste Rock Storage Formed by Top Down Construction Showing Development of Cracks 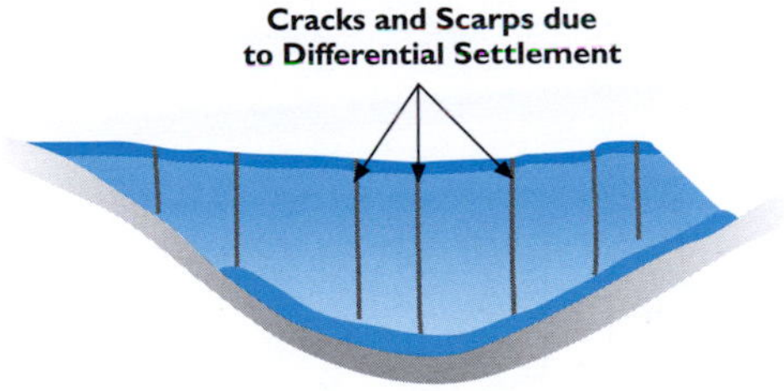 Cracks and Scarps due to Differential Settlement	TDC dumps settle substantially over long periods of time as a consequence of low initial density and gradual readjustment of particles. Differing thicknesses in TDC dumps lead to differential settlement causing cracks which provide percolation paths for rainfall, leading to internal erosion, further settlement and potential instability. Settlement and consequent cracking is much less likely in BUC storages, due to higher densities achieved.
Permeability 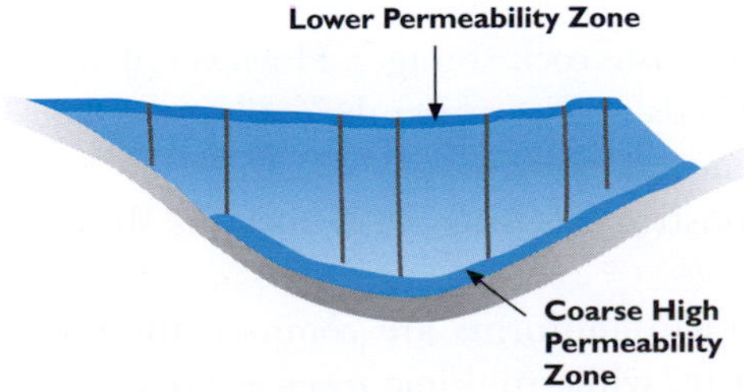 Lower Permeability Zone Coarse High Permeability Zone	TDC dumps typically have a highly permeable zone at the base and a low permeability zone at the crest. Initially the TDC dump will be resistant to infiltration by rainfall; however, as settlement cracks develop, infiltration will be facilitated. BUC storages have a relatively constant permeability throughout, except that a low permeability 'skin' tends to be present at the top of each layer.

TABLE 19.1
(Continued)

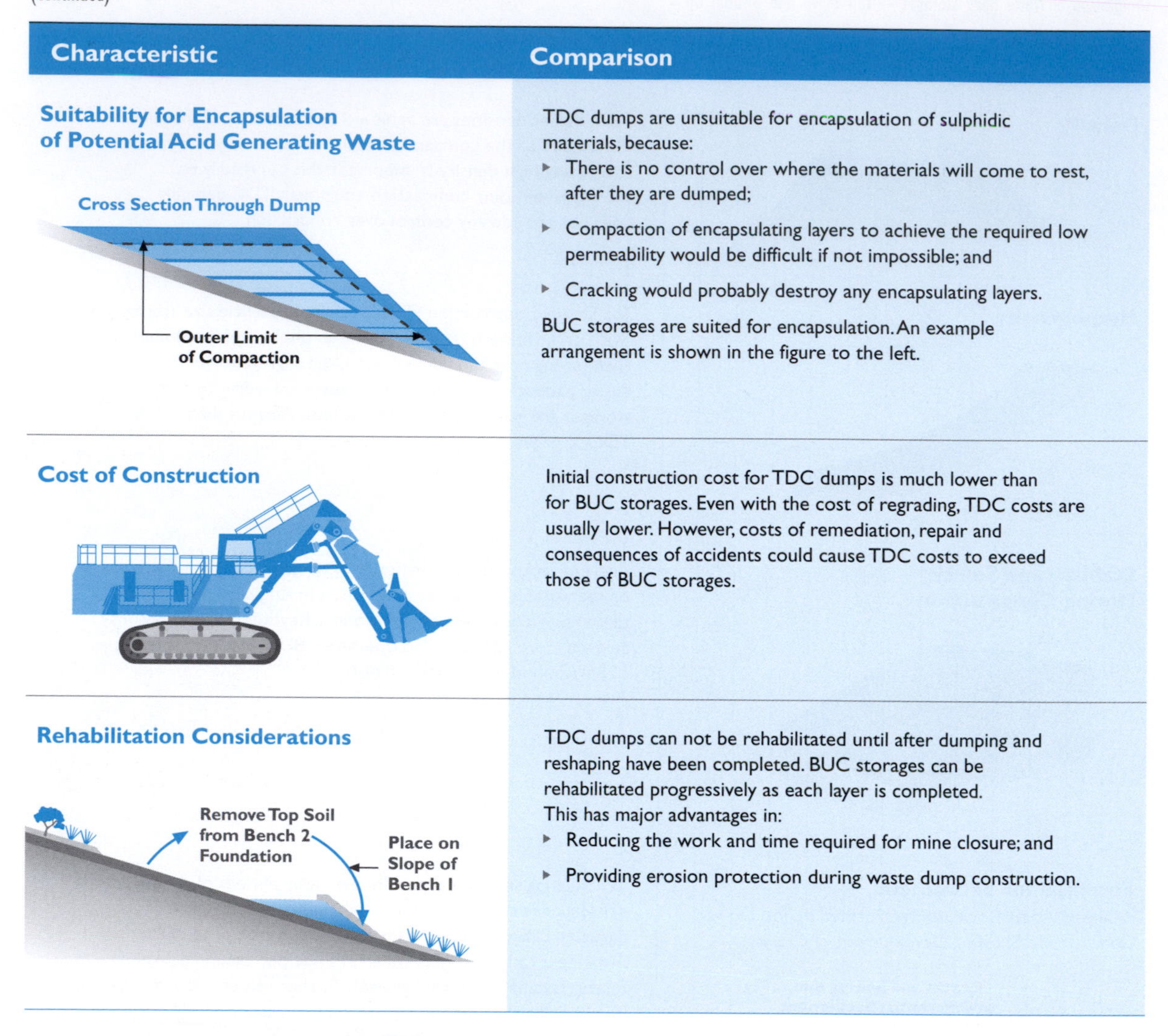

Characteristic	Comparison
Suitability for Encapsulation of Potential Acid Generating Waste	TDC dumps are unsuitable for encapsulation of sulphidic materials, because: ▸ There is no control over where the materials will come to rest, after they are dumped; ▸ Compaction of encapsulating layers to achieve the required low permeability would be difficult if not impossible; and ▸ Cracking would probably destroy any encapsulating layers. BUC storages are suited for encapsulation. An example arrangement is shown in the figure to the left.
Cost of Construction	Initial construction cost for TDC dumps is much lower than for BUC storages. Even with the cost of regrading, TDC costs are usually lower. However, costs of remediation, repair and consequences of accidents could cause TDC costs to exceed those of BUC storages.
Rehabilitation Considerations	TDC dumps can not be rehabilitated until after dumping and reshaping have been completed. BUC storages can be rehabilitated progressively as each layer is completed. This has major advantages in: ▸ Reducing the work and time required for mine closure; and ▸ Providing erosion protection during waste dump construction.

The sequence involved in bottom-up waste rock dump construction, including topsoil placement and vegetation establishment is shown on **Figure 19.5**.

19.6 LANDFORM DESIGN

Here, discussion of landform design is focused on waste rock storages. However, it should be born in mind that planning for mine closure needs to encompass the entire project area of which the waste rock storages represent only one component.

A range of different landforms that may be constructed using waste rock are shown in **Figure 19.6.**

In mountainous terrain, single bench or terrace landforms are common for underground mines producing small quantities of waste, while multiple terraces are common for more extensive waste rock storages. In flat or gently sloping terrain, the mesa, breakaway (or hog-back) and mound landforms are used. Ideally, the selection of landform is

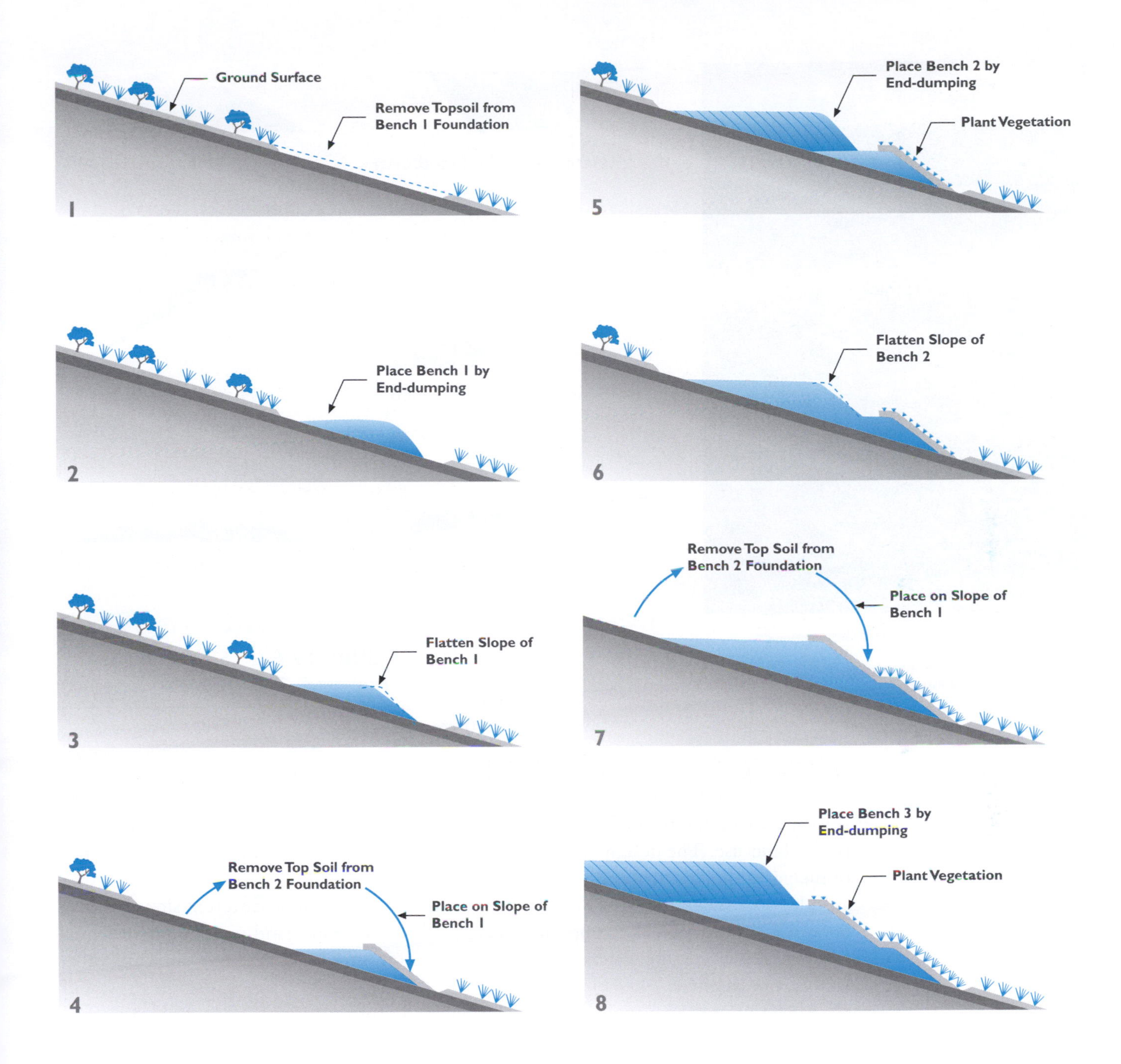

FIGURE 19.5
Stages in Construction of Top-down Waste Rock Dump

based on the characteristics of the natural local landscape, so that the waste rock storage will ultimately blend in to its surrounds. In strip mining situations, where large contiguous areas are mined and partially back-filled, there is the opportunity to create an extensive landscape incorporating a range of landforms, including undulating hills, broad valleys and ponds or wetlands of varying character and dimensions. Again, the final surface configuration should be congruent with surrounding areas, and should meet the needs of future land uses.

Once the site and proposed landform(s) have been selected, a geotechnical investigation should be carried out to assess foundation characteristics and the need for foundation treatment prior to dumping. Typically, foundation treatment involves the removal of soft, weak or highly compressible materials from the foundation area, particularly near the toe

Photo: Courtesy of Jim Wark, www.airphotona.com

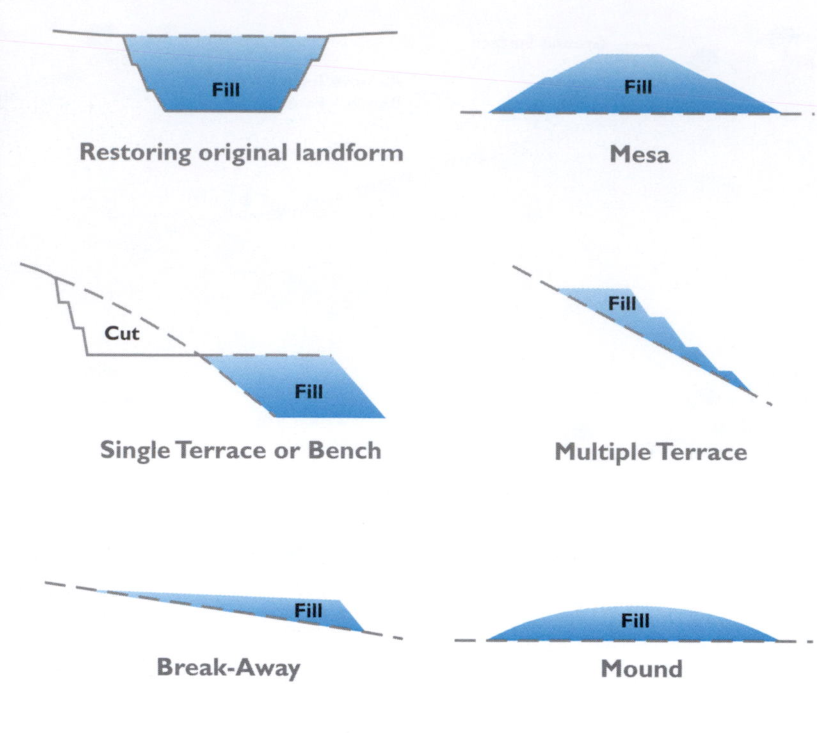

FIGURE 19.6
Types of Landforms Created by Waste Rock Storage

A geotechnical investigation should be carried out to assess foundation characteristics and the need for foundation treatment prior to dumping.

of the storage, if located on a steep slope. In some cases, it may be necessary to excavate one or more keys so that the waste rock is in contact with sound foundations. Detailed design is then undertaken to configure the shape of the storage in accordance with its intended future land use. The design also needs to provide the required characteristics to facilitate drainage, erosion control and rehabilitation.

Table 19.2 shows the types of land use that can be carried out on different slopes. As can be seen from this table, low slope angles are preferred for most land uses and are essential for some. However, the selection of slope angle requires tradeoffs between slope and other factors. For example:

- Lowering the slope angle increases the overall footprint;
- Lowering the slope angle usually increases the construction effort, increasing the overall cost;
- Lower slope angles mean longer slopes which increase runoff volumes, possibly increasing the need for erosion controls, and
- Lowering the slope angle increases the area to be revegetated.

If infiltration of rainfall is to be encouraged as is often the case in arid situations, concave surface configurations, pitting or 'moon-scaping', and berms sloping back into the overall slope are used (Lindbeck and Hannan 1998). If, on the other hand, infiltration is to be minimized as may be required in wet tropical areas, convex surfaces and berms sloping outwards may be used to promote runoff. Whichever approach is adopted, it needs to be supported by an appropriate drainage network. The drainage density and drain designs are based on calculations of runoff, based on rainfall intensities, catchment areas, slopes and infiltration capacity. As the drains are located on more or less erodible materials, lining may be required to provide protection against scouring.

Whichever approach is adopted, it needs to be supported by an appropriate drainage network.

TABLE 19.2
Land Capability in Relation to Slope

Slope	Suitable Land Uses
Up to 6°	All including residential*, industrial, intensive agriculture, sporting
7° to 9°	As above except that more site preparation may be required
9° to 13°	Horticulture, grazing, forestry, vehicular access possible
14° to 17°	Horticulture, grazing, forestry, off-road vehicular access difficult
18° to 24°	Horticulture, grazing, forestry, vehicular access requires roads
25° to 32°	Grazing, natural vegetation, forestry, roads can be established
More than 32°	Marginal for grazing except for goats. Natural vegetation, forestry. Generally too steep for road access. Not recommended for Waste Rock Storage slopes

* Assumes that ongoing consolidation/settlement is not a problem

19.7 SHORT-TERM AND LONG-TERM EROSION CONTROL

The potential for soil erosion depends on soil properties, slope angle and slope length, among other factors (see also Chapter Twenty). Slope angle and slope length are determined in the design of the overall landform, but further erosion protection may be provided by various measures designed to protect against raindrop impact or to interrupt surface flow. Protection from raindrop impact may be provided using gravel or vegetative mulch, or by spreading a degradable fabric such as jute netting. Examples of treatments to interrupt surface flow include (1) hay bales; (2) micro-terraces; (3) furrows created by deep ripping; (4) low fences such as the bamboo fences supported by stakes; (5) living hedges, such as formed by planting strips of Vetiver Grass (see Chapter Twenty-One on Mine Closure); and (6) stone walls or catch dams.

These erosion control treatments should be oriented along the contours of the waste storage. Clearly, some of these measures, such as furrows or hay bales, provide only temporary erosion protection while others such as living hedges and stone walls provide long-term protection. The choice will depend on local availability of suitable materials, labour costs and the time required to establish a cover of vegetation sufficient for longer term erosion control.

The distance between treatments to interrupt surface flow will depend mainly on the slope angle. **Table 19.3** provides a guide for the spacing of treatments, for different slopes. The spacing may be increased if the surface soil is very gravelly or stony.

As previously mentioned, long-term erosion control is best achieved by the establishment of appropriate sustainable vegetation. This is discussed in detail in Chapter Twenty-One under the topic of Rehabilitation.

Long-term erosion control is best achieved by the establishment of appropriate vegetation.

19.8 MONITORING

Monitoring of waste rock storages involves inspections to identify any problems so that appropriate precautions, repairs or maintenance can be undertaken. For large top-down

TABLE 19.3
Spacing between Erosion Control Treatments for Different Slope Angles

Slope Angle (°)	Spacing – Slope Distance (m)
Up to 6	500
7 to 9	100
10 to 13	60
14 to 17	30
18 to 24	15
25 to 32	6

For large top-down waste rock dumps, it is usually necessary to undertake frequent detailed inspections for cracks.

waste rock dumps, it is usually necessary to undertake frequent (at least daily in areas of active dumping), detailed inspections for cracks that may be due to instability. Unfortunately, it may be difficult to distinguish surface cracking due to settlement, from cracking associated with developing slope failure. In some cases, instrumentation may be used to identify and monitor slope deformations. This requires detailed geotechnical investigation and analysis. Surface cracks should be filled and covered, to prevent them becoming internal conduits for rainfall runoff which could lead to or aggravate slope instability.

For waste rock storages constructed from the bottom up, inspections may be much less frequent; weekly or even monthly inspections may be sufficient, except that additional inspections should always be carried out following significant rainfall events. Inspections are focused on identifying surface cracks, erosion gullies or damage to erosion control measures, so that appropriate repairs can be carried out before major damage occurs. Inspections should also check for signs of seepage or springs emanating from the face of the storage, and if such occur, additional drainage and/or erosion control measures may be required. Eventually additional monitoring programmes for acid drainage and for rehabilitation may be required as discussed elsewhere in this book.

REFERENCE

Lindbeck K and Hannan J (1998) Landform Design For Rehabilitation. Booklet in the series Best Practice Environmental Management in Mining, produced by Environment Australia, May, 1998.

20

Erosion

The Perpetual Disruptive Forces of Water and Wind

ALUMINIUM

Although it was not discovered until the 18th century, aluminium is the third most abundant element in the Earth's crust. It is light weight but strong and ductile, with a low melting point and silver-white colour. It is highly reactive but difficult to extract from most of the minerals in which it occurs. Virtually all aluminium is extracted from bauxite, a lateritic ore containing gibbsite-a hydrated aluminium oxide. The process involves digestion in caustic soda followed by calcination to produce alumina; subsequently aluminium is obtained from the alumina by electrolytic reduction. Accordingly, production of aluminium is highly energy intensive. Aluminium is used in a variety of light weight alloys, particularly in transportation where light weight equates with reduced energy consumption, thus offsetting the energy used in its extraction. Aluminium compounds are widely used in water treatment, paper manufacture, in medicine and as refractory materials.

20 Erosion

The Perpetual Disruptive Forces of Water and Wind

Chapter Twenty is concerned with erosion. While neither difficult in concept or involving complicated physical processes, erosion presents significant environmental management challenges, comparable to those involving acid rock drainage and mine waste management. Erosion is the dislodgement and transportation of soil materials through the action of water and wind, with potentially significant direct and indirect adverse impacts, both onsite and offsite.

Direct onsite impacts include the degradation and possible failure of mining infrastructure constructed with Earth materials, such as roads, embankments and dams. Failure of a tailings impoundment due to erosion is potentially catastrophic in terms of the environmental contamination that would result. Such a failure, or, in fact, the failure of any form of slope or embankment at a mine site, also potentially endangers property and human life in the immediate area or further down-slope.

All slopes at mine sites which are composed of soil or highly weathered rock are potentially at risk of significant erosion, whether they are natural or man-made. Significant erosion can occur as a result of a discrete event, or from a slow cumulative process over time. An example of the former is the overtopping of a dam during or after a severe rainfall event, possibly causing severe gouging, leading to failure of the dam. More gradual erosion can slowly steepen an Earth slope, leading to disrepair, decreased stability and possible eventual collapse. In the case of natural slopes, the clearing of vegetation commonly leads to accelerated erosion and instability.

All slopes at mine sites which are composed of soil or highly weathered rock are potentially at risk of significant erosion.

An indirect consequence of erosion at a mine site is acid rock drainage. This occurs when erosion exposes underlying material containing metal sulphides which, when exposed to air and water, generate acidity as discussed in Chapter Seventeen. Another indirect impact is the reduction of soil fertility in the eroded area. As mentioned in Chapter Twenty-One, topsoil is a valuable resource. It is this topsoil that is most erodible. Erosion of course may also destroy any achievement in site rehabilitation.

Where eroded Earth material is transported and deposited downstream or down-wind of the erosion site, such deposits may: block streams and raise river bed levels increasing the risk of flooding and interfering with navigation; block drains and irrigation channels; cause siltation of reservoirs resulting in a potential loss of scarce irrigation and drinking water resources; and cause silting of stream-beds, thereby destroying the habitat for many invertebrates and generally degrading aquatic habitats. If eroded particles contain metals, or other contaminants subsequently dissolve, adverse impacts on the downstream aquatic environment can be exacerbated. Increased turbidity may turn a clear stream into muddy water with significant negative visual impacts, leading to adverse perceptions in local communities, whether or not there is significant contamination.

The deposition of eroded material on land areas can negatively impact flora, fauna and ecosystems, as well as agricultural land. The impact on agricultural land may also be positive, depending on the nature of the deposits. Increased fertility in deposition areas is possible.

Subsurface erosion also occurs, although less commonly. The movement of water underground results in seepage forces that may dislodge soil particles from within or underneath an Earth dam, road embankment or constructed slope. The progressive erosion of particles in an area of concentrated leakage extending into the soil body is known as piping. There are several causes of piping, such as: (1) poor construction control, resulting in inadequate compaction within the soil embankment or slope; (2) leakage through soil tension cracks or cracked pipes that result from dam or embankment settlement; (3) leakage under dams where there is a natural variation in foundation soils; or (4) the presence of dispersive soils that disaggregate when saturated. The design and construction of dams, slopes and embankments is complex and requires the guidance of experienced geotechnical engineers. In addition, adequate supervision during construction and ongoing monitoring and maintenance are needed in order to ensure adequate stability.

The remainder of this chapter will concentrate on ground surface water erosion and wind erosion, including a description of the physical processes involved, and the principles and techniques for effective erosion management. While water erosion is one of the most devastating perpetual disruptive forces in mining both from the standpoint of its physical impact and as a carrier of unwanted materials, wind erosion can also present a formidable management challenge. Wind erosion is commonly associated with most Earth-moving activities, and significant fugitive dust emissions occur as a direct consequence. Airborne particles can cause nuisance and potentially negative health effects for communities downwind of the mine. In fact, fugitive dust is one of the most common sources of complaints from communities located near surface mining operations.

Fugitive dust is one of the most common sources of complaints from communities located near surface mining operations.

20.1 SURFACE WATER EROSION

Surface water erosion can result from the impact of raindrops, as well as from soil particles being picked up and carried by water flowing across the ground surface. Such surface flow can be classified according to the concentration and depth of flow (**Figure 20.1**). If surface water erosion occurs uniformly as a consequence of overland flow across slopes, it is termed sheet erosion, and results from slopewash that moves as a thin and relatively uniform film. With time the flow may concentrate into pathways, resulting in erosion paths classified with increasing pathway depth as rill erosion, gully erosion and stream-channel erosion.

Surface water erosion is an ongoing natural phenomenon. It can be accelerated or intensified if either the natural slope angle is increased, if the vegetative ground cover is removed or the surface is disturbed. Mining activities do all these.

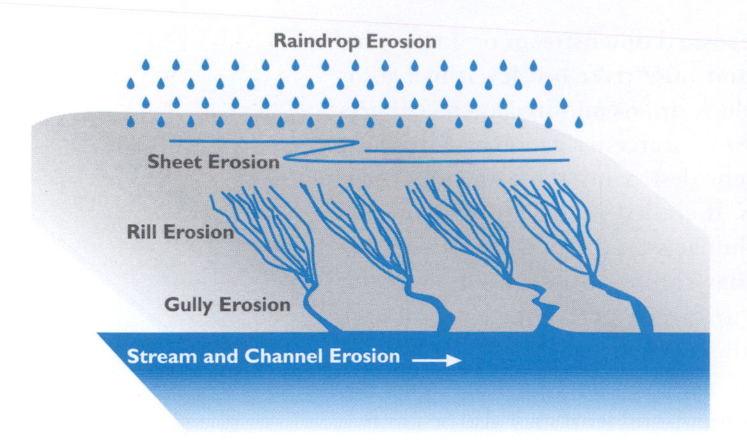

FIGURE 20.1
The Various Forms of Surface Water Erosion Illustrated

Source:
based on SCS 1980

Vegetative cover shields the soil surface from the impact of falling rain, slows the velocity of runoff, holds soil in place and maintains the capacity of the soil to absorb water (Day 2000; Goldman *et al.* 1986). It also helps to protect the soil from drying out. Therefore, the loss of protective vegetative cover increases the erosive impact of both raindrops and runoff.

Potential targets of significant erosion are any areas with changes to natural topography, and any cleared land surface, including exploration tracks. Access and haul roads, including road embankments and road cuts and fills, spoil piles, active mine faces, high wall surfaces, waste rock or overburden dumps, and tailings storage areas, including impoundment structures are especially susceptible to erosion.

The Controlling Factors

The widely accepted universal soil loss equation (USLE, US Department of Agriculture 1980) relates average soil loss to five factors: rainfall and runoff (R), soil erodibility (K), a combined slope length and steepness factor (LS), vegetative ground cover and cover management (C), and support practices (P) (**Figure 20.2**). It is an empirical equation that calculates an estimate of soil loss as the numerical product of these five empirical factors. The resulting estimate of soil loss in tonnes per hectare per year is based on average rainfall conditions. It was developed to predict soil loss for sheet and rill erosion, not for more concentrated flow in gullies or channels, which may result in larger volumes of eroded soil than predicted by the USLE.

While the USLE has limitations, it provides a good basis for considering the factors that are important for understanding and minimizing erosion. These factors are discussed below.

Rainfall and Runoff (R)

The force of impact of raindrops on unprotected soil initiates erosion by dislodging particles, while overland flow or runoff can both dislodge and carry soil particles down-slope.

Technically, the rainfall and runoff factor, R, combines both a rainfall energy component and a rainfall intensity factor into one number which is calculated by first selecting an average rainfall year. Then, for each storm in a particular area during the chosen year, the total energy is multiplied by the maximum 30-minute rainfall intensity, and the resulting values for each storm are added together (Day 2000). The intensity of a rainfall event is arguably the single largest contributing factor to soil loss. Rainfall events of more than 200 mm in

The intensity of a rainfall event is arguably the single largest contributing factor to soil loss.

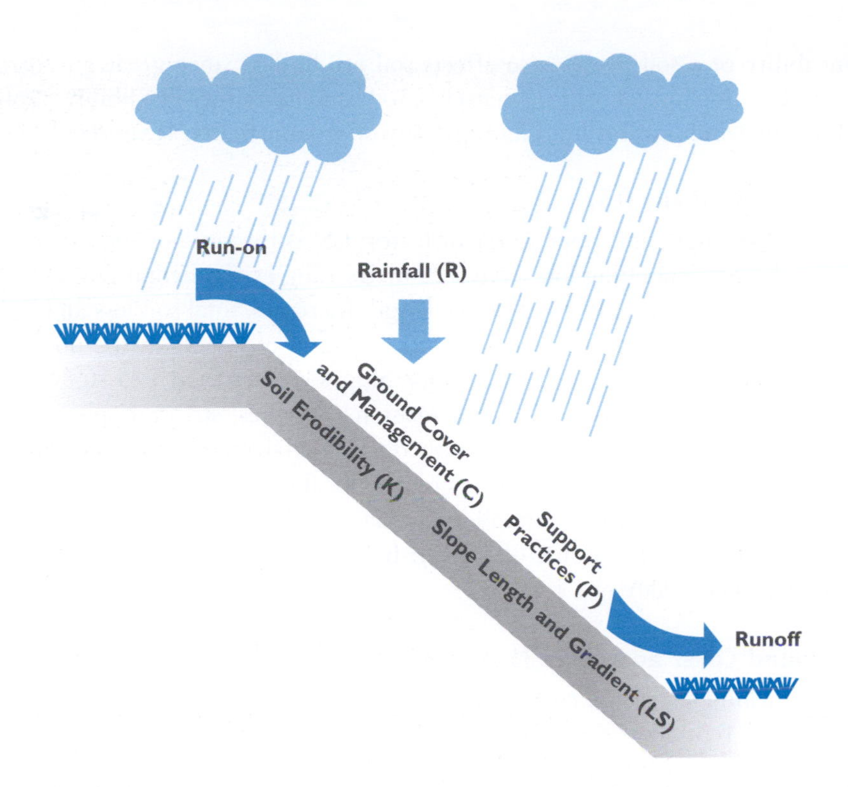

FIGURE 20.2
Factors Controlling Water Erosion

Soil Loss = R · K · LS · C · P

R = Rainfall frequency, intensity, storm kinetic energy, etc.

K = Particle size, organic matter, permeability, rock fragment, etc.

LS = Length, average gradient, geometry, etc.

C = Vegetation, mulch, surface roughness, cover management, etc.

P = Supporting mechanical practices such as diversion ditch, silt fences, etc.

less than 30 minutes are common in the tropics, and no erosion controls exist to completely eliminate soil loss in such circumstances. Analyses of data indicate that when factors other than rainfall are held constant, soil loss is directly proportional to rainfall intensity.

Soil Erodibility (K)

Soil erodibility, K, is an empirical measure of the ability of rainfall and runoff to detach and transport soil particles. It correlates with inherent physical soil properties, such as particle size, organic matter content, soil plasticity and cohesion, the dispersiveness of clay soils, and the presence of any particle-cementing agents. The two main factors are particle size and plasticity (Day 2000). Soil plasticity refers to the ability of fine-grained silt and clay soils to be rolled and moulded without breaking apart.

Cohesive, highly plastic clay soils are usually resistant to detachment from the soil matrix and soil erodibility is low. However, some types of clay, known as dispersive clay, may have high plasticity, but are more susceptible to erosion because of deflocculation of clay particles. Deflocculation occurs when the repulsive forces between clay particles exceed the attractive forces in the presence of water, forming colloidal suspensions. It is related to the chemical composition of the clay, and various tests can be used to identify dispersive clays, such as the laboratory pin hole test.

In contrast to non-dispersive cohesive clay, soils with a relatively high content of cohesionless particles, such as silt and sand, are often highly susceptible to erosion. Silt-size particles are the most susceptible, as they are non-plastic and easily detached, producing high erosion rates and large volumes of runoff. However, as the particle size of cohesionless soils increases, erosion resistance increases due to the increased weight of the individual particles. The scale of cohesionless soil particle sizes ranges from silt, for which the individual particles are indistinguishable to the naked eye, through fine to course sand, then gravel, cobbles and, finally, boulders.

Soils with a relatively high content of cohesionless particles, such as silt and sand, are often highly susceptible to erosion.

The permeability of a soil profile also affects soil erodibility, since greater infiltration reduces runoff. The presence of organic matter in soil can also reduce erodibility, probably due to the absorptive and possibly binding capability of organic matter (Day 2000).

Slope Length and Gradient (LS)

The combined slope length and slope gradient factor, LS, is the ratio of soil loss per unit area on a site to the corresponding loss from a 20 metre-long experimental plot with a 9% slope (Day 2000; Goldman *et al.* 1986). As slope length increases, total soil loss and soil loss per unit area increase due to the progressive accumulation of runoff in the down-slope direction. Slope gradient determines flow velocity. Slopes greater than 5 to 10% will usually require special controls to prevent water erosion. Terraces, strip cropping, contour ploughing, and similar techniques are often applied to retard overland flow. The effectiveness of such erosion-control practices depends on the local conditions. For example, contouring is far more effective in low-rainfall areas than the high-rainfall areas commonly encountered in the tropics. The value of LS can have a wider variation than any other factor in the USLE (Day 2000).

Slopes greater than 5 to 10% will usually require special controls to prevent water erosion.

Vegetative Ground Cover and Cover Management (C)

The 'vegetative ground cover and cover management factor' in the USLE equation is an empirical factor that represents the effects of the type and extent of vegetation at a site, as well as soil cover management, such as the use of protective mulch cover. Clearly, highly erodible conditions exist during site preparation, construction and mining periods when the soil is bare and highly disturbed. Day (2000) reports a value for C of 0.01 for an undisturbed area with native vegetation, and a value of 1.0 for bare ground. This suggests that clearing an undisturbed area can increase erosion by a factor of 100. Exposed cut and fill slopes and the slope of mine waste placements are expected to erode unless protected, particularly in the tropics.

Exposed cut and fill slopes and the slope of mine waste placements are expected to erode unless protected.

Support Practices (P)

Support practices account for control measures that reduce the erosion potential of runoff by their influence on drainage and on hydraulic forces exerted by runoff on soil. The value for P reflects the effects of practices on both the amount and velocity of water runoff. For example, higher P values apply to sites where the graded surface is smooth and uniform, while lower P values are applicable for prepared surfaces that are rough and irregular (Day 2000). This is because increased surface roughness reduces surface water flow velocity and, thereby, reduces erosion.

Practices that control the runoff from adjacent areas are essential in erosion control. Such runoff to disturbed land is sometimes termed run-on, and can create major problems. Unless run-on resulting from rainfall on adjacent tributary slopes is properly controlled, severe erosion of exposed soil will occur. When run-on to disturbed areas is concentrated at distinct locations rather than as sheet flow, erosion due to run-on often outweighs erosion due to direct rainfall on the disturbed area itself.

Unless run-on resulting from rainfall on adjacent tributary slopes is properly controlled, severe erosion of exposed soil will occur.

Assessing the Impact of Water Erosion

The USLE is not rocket science: the factors are intuitive, and the equation is simple algebra. However, soil loss equations such as USLE are based on empirical factors that in turn depend on numerous and often unknown parameters. The USLE equation, originally developed for croplands, was revised in 1997 to account for soil losses from lands disturbed by other human activities, including surface mining and activities at construction sites (RULSE; Renard *et al*. 1997, Office for Surface Mining 1998). The structure of the original USLE remains unchanged, but the estimation of the five factors has been refined to characterize the special site conditions resulting from mining, construction and reclamation activities.

RUSLE is a very powerful tool that can be used to estimate soil loss under a wide variety of site-specific conditions. The equation can include the full spectrum of land manipulation, including rock cover, mulches, random surface roughness, effects of mechanical equipment on soil roughness, terraces and vegetation types. Its successful application is directly related to the experience of the model user in estimating site-specific empirical control factors as model input data.

RUSLE originated in the USA. The extensive field data that are available to guide the selection of the empirical factors apply to weather and soil conditions in the USA, but not necessarily to other areas. Model users outside the USA are rarely able to rely on a similar set of supporting data; erroneously they often transfer US based data blindly to their host region. However, some countries have generated their own set of empirical input data. In Australia, for example, the SOILOSS model, while based on RUSLE, specifically reflects Australian weather and soil conditions (Rosewell 2005). More often than not, however, experienced professional judgement is needed to allow the empirical factors to provide adequate predictive value.

Model users outside the USA are rarely able to rely on a similar set of supporting data; erroneously they often transfer US based data blindly to their host region.

It is also important to recognize that water erosion related to Earth moving activities such as mining often depends on small-scale local features that escape generic assessment methods (**Figure 20.3**). The degree to which erosion will occur is directly related to applied management efforts and installed sediment retention measures. In these circumstances, estimating water erosion will remain qualitative, in the form of an expert opinion.

Empirical equations, however, provide useful estimates of erosion on various surface covers for mine waste placements. As with any quantitative approach, estimates should be used in conjunction with engineering judgement. Empirical equations are particularly

As with any quantitative approach, estimates should be used in conjunction with engineering judgement.

FIGURE 20.3
Localized Water Erosion Illustrated

(a) Gully erosion along a site access road.

(b) Multiple catch dams constructed across steep gully to stabilize debris from limestone mine. (Sirmour area, Lesser Himalayas, Himachal Pradesh, India) useful in comparing relative levels of erosion susceptibility under different conditions, such as different cover alternatives, rather than estimating absolute sediment yields.

As mentioned above in the discussion of controlling factors, the equation assumes average rainfall conditions and predicts soil loss for sheet and rill erosion only. Greater than average rainfall or the presence of concentrated rainfall in gullies or channels (Figure 20.3) will result in more erosion than predicted by RUSLE. It should also be noted that the equation predicts soil loss and not soil deposition. Coarser particles may only be transported to the toe of a slope and may not reach areas further downstream.

Water erosion is one of the most prominent and visible negative impacts of a mining operation. The preceding discussion illustrates the basis of the powerful methods available to quantify erosion potential, but their application is often beyond the scope of an environmental impact assessment. Professional judgement with a sound understanding of erosion management practices is key to predicting erosion significance during the EIA process. Erosion control and sediment retardation during mine construction and operation are the least expensive and the most effective means to reduce water-borne contamination. These should be regarded as major priorities in preparing the environmental management plan for a mine during the environmental impact assessment process.

Managing Water Erosion

The two key principles of water erosion management are to sustain soil cover and hence to fence off erosion (erosion controls), and to control water quality through the reduction of sediment loads if erosion occurs (sediment controls). The first principle is a priority, since prevention is always better than cure.

Erosion management is part of mine design and mine water management. Consideration is given to the hydrologic, geomorphologic and geological attributes of the mining area; to mining technology and mine waste placement; and to the combined effects of these factors on quantity and quality of runoff throughout the mine life. Finally, and most importantly, the design of the post-mining landform commands special attention: each piece of land is different (in terms of soil type, gradients, exposure to rainfall, runoff from adjacent areas, etc), and the final landform needs to take these differences into account.

The most effective erosion and sediment controls are those built into mine design and into the rehabilitation sequence.

The most effective erosion and sediment controls are those built into mine design and into the rehabilitation sequence. Key erosion and sediment controls are then part of the mining operation and do not require additional engineering design or construction. For example, mining in an up-slope direction where feasible may allow use of worked-out pits to act as runoff and sediment retention basins. Backfilling and revegetation of pits or strips as soon as possible minimizes the extent of disturbed land at any one time, significantly reducing total erosion.

Some erosion and sediment controls are only needed temporarily during construction and during active mining phases, whereas mine rehabilitation requires the provision of permanent solutions. It is important that erosion and sediment controls, even if only temporary, are stable, robust, and designed to accommodate the significant peak flows common at the location of the mine, particularly if the mine is in the tropics. Emergency spillways should be constructed, where appropriate in sediment ponds and other impoundments.

The choice of erosion and sediment controls, particularly the design of retention basins, depends on the consequences of failure.

The choice of erosion and sediment controls, particularly the design of retention basins, depends on the consequences of failure. It is important to understand the quality and significance of waters receiving eroded material, from the perspectives of safety, community use, economic value and environmental value. Where failure will not endanger life and will not cause significant economic or environmental damage, controls may be established according to standard engineering practices. If failure potentially leads to loss of life or to significant environmental damage, retention basins should be designed more conservatively in order to withstand the 'Probable Maximum Rainfall Event'. Erosion and sediment controls should be assessed for economic and environmental feasibility, considering all costs, including those of environmental impacts and the costs for ongoing, long-term maintenance.

Erosion and sediment controls should be planned and designed as a sequence of components that incorporates the following five potential measures (**Figure 20.4**): (1) minimizing land disturbance; (2) managing run-on to disturbed areas; (3) managing drainage within disturbed areas; (4) managing ground cover; and (5) managing runoff and sediment exiting disturbed areas. These are listed in priority order, as the aim is to favour preventative measures, minimizing the need for corrective measures and repairs.

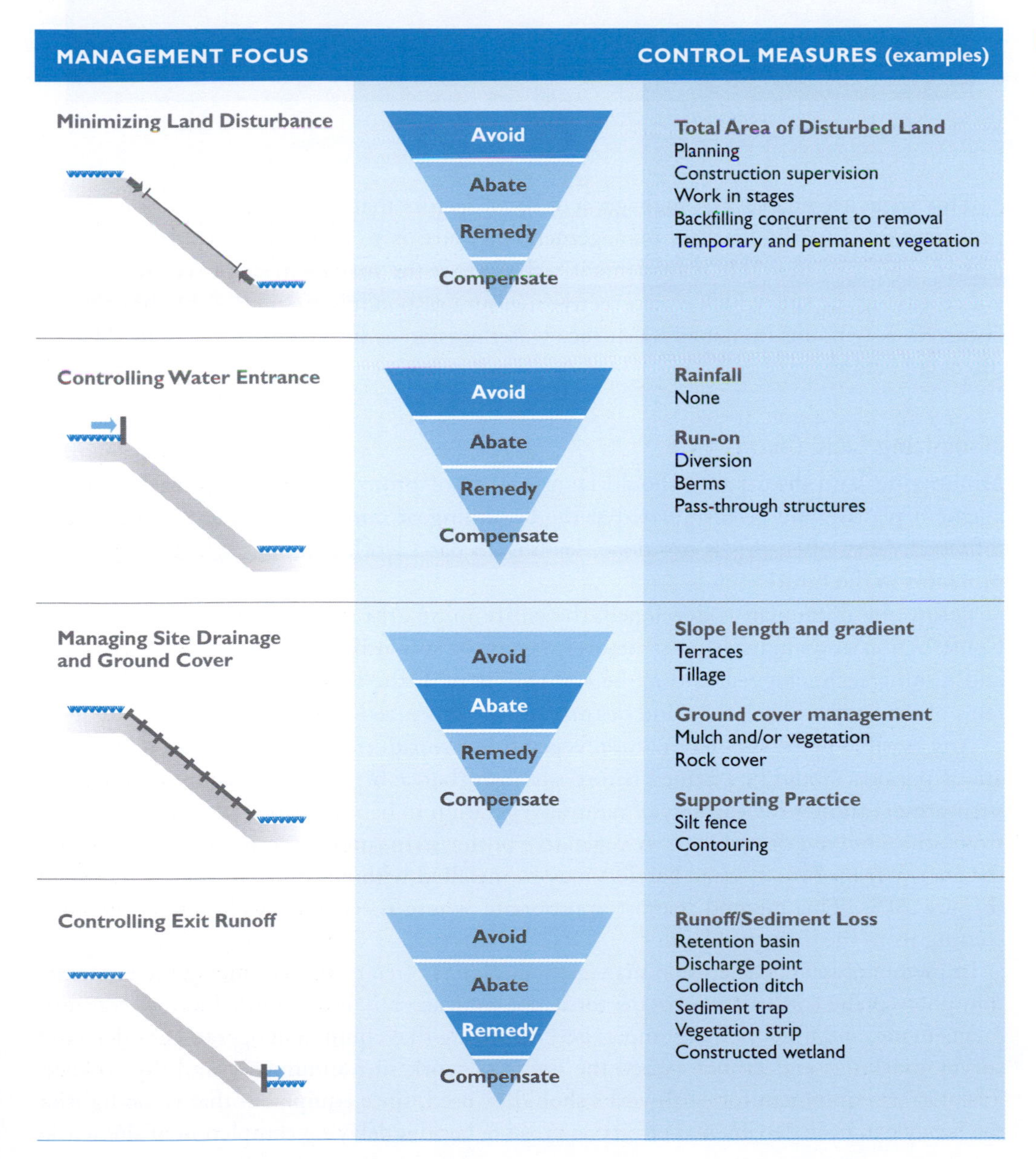

FIGURE 20.4
The Water Erosion Prevention and Treatment Train

FIGURE 20.5
Using Down-slope Vegetation Cover to Dissipate Energy from Runoff and to Collect Eroded Soil Materials

This sequence enables the management of all factors that contribute to erosion, except precipitation. Clearly, the best management measure is to avoid land disturbance altogether. The least desirable management technique is the management of sediment-laden water leaving the site, which is a corrective measure responding to soil loss that has already occurred. Key points in managing surface water erosion as summarized in **Figure 20.4** are discussed below.

Minimizing Land Disturbance

An erosion management plan should be formulated at the beginning of mine development and carefully followed.

Minimizing land disturbance should be a priority of project planning. An erosion management plan should be formulated at the beginning of mine development and carefully followed. Very often this is not done, leading to patchwork solutions which may create problems in the future.

Before the mine plan is developed, the entire mine area should be carefully surveyed. Construction and engineering structures, layout of supporting facilities, transport routes and mining technology should all be compatible with the terrain, landscape and ecosystems of the site area such that land disturbance is minimized.

Any feasible preservation of natural vegetation cover during the construction and operations periods should be planned before site disturbance begins. One major advantage of such preservation is the capacity of natural vegetation to handle increased runoff resulting from adjacent land disturbance. A vegetative buffer strip running across a slope dissipates the energy from runoff; small barriers may be installed within the strip to collect sediment (**Figure 20.5**). This method is most appropriate where there is no well-defined channel coming from the disturbed land.

Implementation of the erosion management plan is often neglected during the construction phase if the construction contractor is not contractually obliged to adhere to commitments made, and if not properly supervised. Ensuring that equipment operators understand survey markings and do not exceed the limits of work sites minimizes land disturbance. The correct equipment for earthworks should be used, since equipment that is too light or too heavy can have detrimental effects on erosion, besides delaying completion of site work.

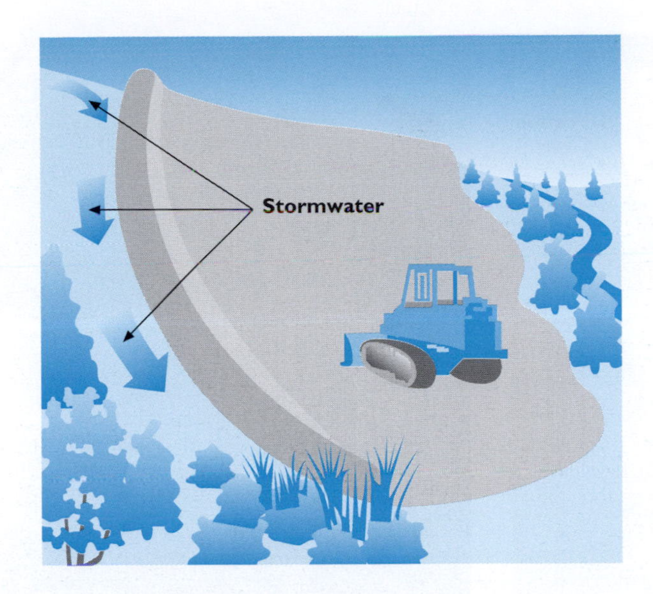

FIGURE 20.6

Diversion should be Used to Divert Stormwater away from Disturbed Area

- Make sure the diversion discharges to a stable outlet or channel.
- If practical, seed the diversion ditches and berms.
- Avoid erosion of the diversion (Channel grades should be relatively flat).

Backfilling as soon as feasible minimizes the area of disturbed land at any one time. Once final grade is achieved, areas should be mulched and vegetated as soon as possible.

Controlling Water Entrance

Prior to disturbing land, sites should be carefully studied to: verify existing drainage patterns and topography; classify runoff from adjacent areas, either as sheet or stream flow; and determine the effects of storm runoff. Runoff from undisturbed land should be diverted away from or around disturbed areas, or directed through the disturbed area. This can be engineered using berms, diversion ditches, protected channels and pipes (**Figure 20.6**). It is important that drainage does not collect in new cuts, initiating the development of gullies.

Runoff from undisturbed land should be diverted away from or around disturbed areas, or directed through the disturbed area.

Managing Site Drainage and Disturbed Area

Mats, mulches and blankets can be used for temporary stabilization and establishing vegetation on disturbed soils. Control blankets can help abate raindrop and sheet erosion (**Figure 20.7**). Mats and blankets are typically used on slopes or channels while mulches are effective in helping to protect the soil surface and foster the growth of vegetation.

Terraces, check dams, tillage and other grading control measures can be utilized to reduce erosion and to ensure slope stability. Such measures are designed to minimize both the quantity of flow and its velocity. Site drainage should be designed to control rill erosion and to prevent the deepening or enlargement of channels, which can eventually lead to the development of gullies.

Gradient terraces can be used to control slope lengths and slope gradient, and to address particular erosion problem spots. Terraces are typically Earth embankments or ridge-and-channel forms constructed along the contour of a slope at regular intervals (**Figure 20.8**). Terraces should be designed with adequate outlets, such as grassed waterways or vegetated areas, to direct runoff to a point where it will not cause additional erosion. Disturbed areas should be reclaimed to an appropriate grade to achieve long-term stability. Although the natural topography of surrounding areas can help determine achievable slopes, construction and mining activities can modify the engineering properties and the behaviour of Earth materials (for example, soil strength properties and permeability can change). Slope design should be conducted by a qualified geotechnical engineer. All final grading should be completed along contours.

Terraces should be designed with adequate outlets to direct runoff to a point where it will not cause additional erosion.

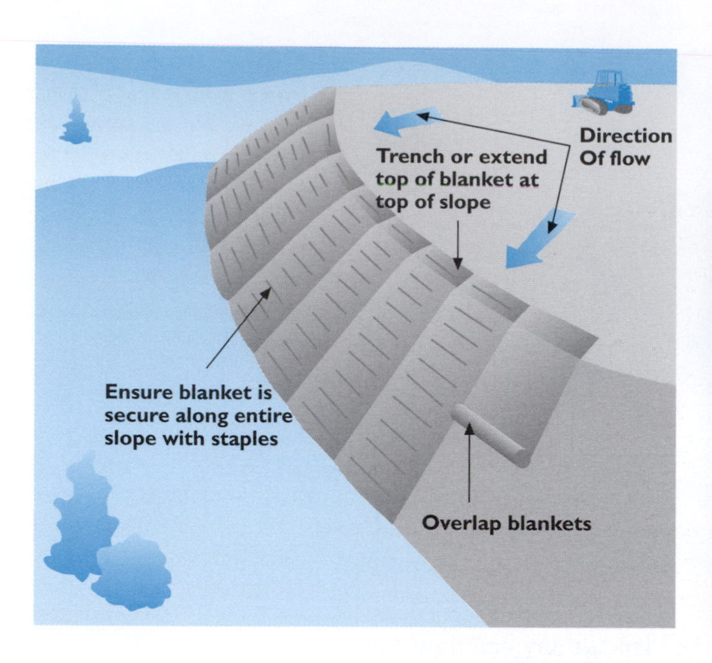

FIGURE 20.7
Erosion Control Blanket

- The blanket or mat should come into complete contact with the soil.
- The top of the blanket should be trenched-in (there should be no evidence of water flowing under the blanket or mat).
- Mulch should not be placed in concentrated flow areas.
- Check to see if erosion is occurring in mulched areas (more mulch may need to be applied).
- Install blankets and mats so that sections are overlapped about 10 cm and staples are about 30 cm apart on tops and about 60 cm apart down the sides and in the middle.

Water erosion of tailings or waste-rock management facilities during the operational phase can be avoided by using the following techniques (EC 2004): (1) covering the sloping surfaces of the impoundment with a protective layer such as gravel, a soil and grass cover, a geo-fabric and grass cover, or some form of synthetic coating; (2) impregnating the surface layer of the tailings with a chemical which can repel water or result in particle binding, such as a silica compound, cement, porcelain, bitumen or bentonite; and (3) using the chemical properties of the tailings, to assist in particle binding.

Managing Ground Cover

Preserving existing vegetation and re-vegetating disturbed land as soon as possible after disturbance are the most effective ways to control erosion (US EPA 1992). Vegetative stabilization includes temporary or permanent seeding and sodding. Vegetative stabilization helps prevent erosion at exposed mining or construction sites by re-establishing vegetation on exposed soils. Native and non-invasive species are highly preferred to introduced grasses (**Figure 20.9**).

The steps involved in revegetation typically include seedbed preparation, fertilizing, liming (to neutralize soil acidity), seeding, mulching and maintenance. Bio-solids (the residual solid fraction, primarily organic material, of processed sewage sludge or of other sludge derived from organic waste sources) are a low-cost alternative to the use of commercially available lime and fertilizer. Bio-solids are available in various forms, such as sewage sludge or pulp mill or palm oil mill wastes. They are beneficial in creating a soil substitute and in developing the soil structure through the addition of significant quantities of organic matter.

A vegetative cover shields the ground surface from the impact forces of sunlight and rainfall, attenuates surface water runoff velocity, holds soil particles in place and maintains the capacity of soil to absorb water. The combined effect is to stabilize disturbed land against erosion.

Trees are especially effective in areas with unstable soil conditions or on steep slopes

Vegetative cover can be grass, trees, legumes, or shrubs. In most cases, it will be necessary to first provide a ground cover with grass or other quick growing species to provide erosion protection within a short time. Permanent seeding and planting follow, where a long-living plant cover is desired. Trees are especially effective in areas with unstable soil conditions or on steep slopes, and a closed canopy forest cover is effective in reducing the impact of precipitation. **Table 20.1** outlines a sample revegetation plan.

FIGURE 20.8

Construction of Terraces and Channels to Manage Runoff and Slopewash, to Reduce Erosion and to Ensure Slope Stability

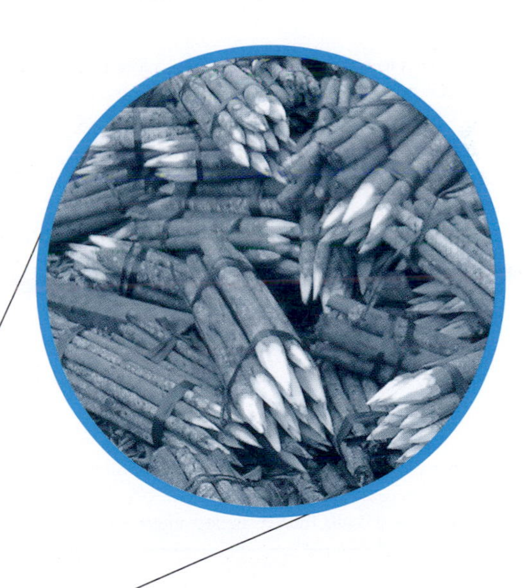

FIGURE 20.9

Using Living Fences to Control Runoff and to Establish a Vegetative Cover

An interesting approach to stabilizing slopes and help establishing a ground-cover is the concept of a living fence. Fast growing bushes or trees that are able to be duplicated by stakes or, are planted close to each other in a row along to the contour. They serve two purposes:

1. Initially they support a silt fence;
2. After rooting they develop into a firm vegetation belt.

TABLE 20.1

Example Revegetation Practice and Maintenance Plan (US EPA 1999)

Revegetation Plan

- Systematic sample collection and analysis of topsoil, subsoil and overburden materials is to be conducted to determine the type and amount of soil amendments necessary to maintain vegetation growth.
- Topsoil placement and seeding is to occur no later than the first period of favourable planting after backfilling and grading. Backfilled areas prepared for seeding during adverse climatic conditions are to be seeded with an appropriate temporary cover until permanent cover is established (a cover of small grain, grasses, or legumes can be installed until a permanent cover is established).
- Disturbed areas are to be seeded in such a manner as to stabilize the surface and establish a diverse, effective and permanent vegetation cover, preferably with a native seasonal variety or species that support the approved post-mining land use.
- Re-graded areas are to be disced (tillage using disk rippers) prior to application of fertilizer, lime and seed mixture. Fertilizer mixture is to be applied as determined necessary by soil sample analyses. Treatment to neutralize soil acidity is to be performed by adding agricultural grade lime at a rate determined by soil tests. Neutralizers are to be applied immediately after regrading. A minimum pH of 5.5 is to be maintained.
- Mulch is applied to promote germination, control erosion, increase moisture retention, insulate against solar heat and supply additional organic matter. Straw, hay, or wood fibre mulch are to be applied in an amount in the order of 2.5 to 6.25 tonnes/hectare. Small cereal grains can be used in lieu of mulch (small grains absorb moisture and act as a soil stabilizer and protective cover until a suitable growing season occurs that will allow permanent revegetation to establish).
- Conventional equipment and methods are to be used, such as broadcast spreaders, hay blowers, hydroseeders, discs, cyclone spreaders, grain drills, or hand broadcasting methods. Excess compaction is to be prevented by using only tracked equipment. Rubber tyred vehicles are to be kept off reconstructed seedbeds.

Maintenance

- Vegetative cover is to be inspected regularly. Areas are to be checked and maintained until permanent cover is satisfactory. Bare spots are to be reseeded, and nutrients added to improve growth and coverage. Areas that are damaged due to abnormal weather conditions or pests are to be repaired.
- Unwanted rills and gullies are to be repaired with soil material. If necessary, the area is to be scarified and (in severe cases) back-bladed before reseeding and mulching.
- Revegetation success is to be determined by systematic sampling, typically at a minimum sampling rate of 1% of the area. Aerial photography can be used to determine success (typically at the 1% level–or higher if necessary). Standard of Success (SOS) for revegetation is based on percent of existing ground cover or achievement of vegetation growth adequate to control erosion.
- Periodic mowing is to be performed to allow grasses and legumes a greater chance of growth and survival. Plants are not to be grazed or harvested until well-established.
- Previously seeded areas are to be reseeded as necessary, or on an annual basis, until covered with an adequate vegetal cover to prevent accelerated erosion. Areas where herbaceous cover is bare or sparsely covered after 6 to 12 months are to be re-limed and/or re-fertilized as necessary to promote vegetative growth, then reseeded and mulched.

Controlling Exit Runoff and Sediment Load

A silt fence or sediment filter is relatively simple to install and is a down-gradient barrier intended to intercept sheet flow runoff and settle out sediment upslope while allowing runoff to filter through. A silt fence or sediment filter can be used to control exit runoff and sediment load from small disturbed areas. For larger areas a series of filters may need to be installed at appropriate intervals (**Figure 20.10**). A silt fence works particularly well in conjunction with an existing vegetative buffer strip running across a slope – the vegetation cover helps dissipate the energy of runoff, while silt fences collect sediment. However,

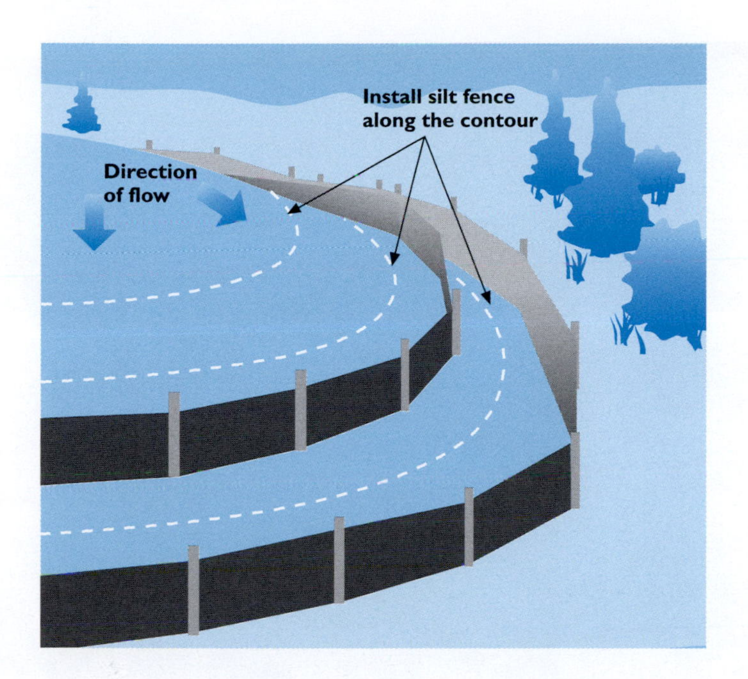

FIGURE 20.10
Illustration of Silt Fence Installed along the Contour

- Silt fence should be installed along the contour (on a level horizontal plane)
- Turn up the ends (J-hooks) to help pond the water behind the filter
- Ensure that the filter is trenched in with the stakes on the downhill side (trench should be 15 cm deep by 15 cm wide).
- Remove sediment when it reaches 1/3 the height of the barrier
- Filters should not be installed where concentrated flow is expected

as mentioned in an earlier section, a silt fence or sediment filter is less effective if there are well-defined channels coming from the disturbed land. Concentrated water flow requires other measures.

The basic principle in designing control measures for concentrated water flow leaving disturbed land is to ensure that the viability of ecosystems, aesthetic values and down-slope areas are not significantly adversely affected. Managed water, whether diverted or leaving the site, usually has an increased flow rate and sediment load. Poorly designed exits from disturbed land can result in off-site erosion near discharge points. Water and sediment retention basins, sediment traps and constructed wetlands are common erosion control techniques for managing this problem.

Managed water, whether diverted or leaving the site, usually has an increased flow rate and sediment load.

Sediment retention basins are impoundments designed to intercept sediment-laden runoff and to retain a significant portion of the sediment. Their design and operation depend primarily on the nature of eroded materials and thus must be site-specific. Coarse-grained sediment settles faster than fine-grained sediments. Clay particles may form a colloid that settles in a more loosely packed structure, taking up more volume, reducing the effectiveness of the settling pond.

Three design criteria are used in sizing the settling zone of a sediment basin, namely 'surface area' (defined as design volumetric flow rate divided by the settling velocity of the design particle), depth and the ratio of length to width. The final basin design must allow for sediment storage for either the entire duration of soil loss or until the basin is cleaned out.

Sediment retention basins also attenuate runoff, regulating water flow leaving disturbed land and hence, reducing flow velocities and stream-channel erosion downstream. In effect, they act as run-on controls for the next area down-slope.

An integrated part of water and sediment retention basin design and function is the design of water discharge points. Care must be taken to ensure that the energy generated by the outflow is safely dissipated. Techniques for energy dissipation vary with the anticipated flow rates. They range from a relatively simple protective rip rap lining of the stream-channel close to the outlet to permanent structures comprising heavy gabions (baskets of rocks), stones or concrete structures (**Figure 20.11**). Another technique to dissipate energy is to direct the discharge into a suitably large body of water that has the capacity to absorb the discharge without damage.

An integrated part of water and sediment retention basin design and function is the design of water discharge points.

FIGURE 20.11
Techniques for Energy Dissipation Vary

They range from a relatively simple protective rip rap lining of the stream-channel close to the outlet to permanent structures comprising heavy gabions (baskets of rocks), stones or concrete structures along an entire stream section.

20.2 WIND EROSION

Wind erosion is the detachment, transportation and re-deposition of soil particles by wind. The risk of wind erosion is increased by loss of vegetative cover, a loose, dry and smooth soil surface, large barren areas, equipment movement and strong winds.

In farming, the most familiar result of wind erosion is the loss of topsoil and nutrients, which reduces the soil's ability to support crop production. Wind removes the smaller dry silt and clay particles and dry organic matter from the soil surface, while coarser materials are left behind. The continued loss of fine particles reduces soil quality. In mining, the most familiar result of wind erosion is the receipt of complaints from community members living near the mining operation. For them, air borne dust is not merely a visual nuisance; long-term exposure to mine dust may impact health.

In mining, the most familiar result of wind erosion is the receipt of complaints from community members living near the mining operation.

Clearly, dust is more an issue for mining projects in arid and seasonally dry areas. Particularly in the early stages of a project when weaker, weathered materials are being mined, large quantities of dust can be produced by blasting as well as by loading and trucking of ore and overburden. Dust due to blasting may be quite dense (**Figure 20.12**), but particles are relatively coarse. The dust cloud is generally short-lived and most particles settle relatively close to the source. On the other hand, dust due to loading and haulage operations, while not so dense, is generally finer. Dust clouds produced by traffic can persist all day and may drift over long distances (**Figure 20.12**).

Dust clouds produced by traffic can persist all day and may drift over long distances.

Wind erosion is more likely to occur during dry weather periods with or followed by high winds. Given the fine-grained nature of deposited materials, tailings impoundments can be a major source of wind blown dust. Dust may not be generated from tailings areas until after final completion of the tailings disposal operations, when the tailings material dries out. Only then may elevated levels of metals in wind blown particles (such as cadmium or arsenic) start triggering health concerns. As discussed elsewhere, mine wastes at metal mines commonly contain metals that may be released as fugitive dust, contaminating areas downwind, as particles settle out of suspension in the air.

FIGURE 20.12
Mining Dust Illustrated

Dust from blasting at an early stage (1974) in the Mt Newman Iron Mine in the Pilbara Region of Australia.

Distant view of dust cloud from haulage of laterite ore at the Hinatuan Nickel Mine, near Surigao, Philippines.

Table 20.2 lists ways in which mine waste particles can be dispersed from dams or storage heaps and offers some prevention options. To minimize wind erosion from tailings beaches, the surface is usually kept wet. For example, water spraying on dried tailings beaches is applied when dusting conditions are imminent. This is generally more cost effective than placing decaying vegetation, such as hay, on the tailings surface. Sprinkling of the beach in combination with continuous management of the discharge point of the tailings onto the beach is normally satisfactory for managing wind erosion. Although generally less cost effective, a sometimes feasible alternative method to avoid dust generation is to cover tailings beaches with non-dusting material, such as topsoil, straw or bitumen. This method is only practical when the beaches are raised in distinctly separate campaigns, and not continuously. The beach must be stable enough for machinery to work on it in order to spread out the material. The application of a vegetative cover can be very effective,

TABLE 20.2

Dispersion by Wind Erosion of Mine Waste Particles from Storage Facilities and Common Prevention Options (EC 2004)

Solid may be dispersed by	Prevention
Wind erosion of the surfaces of the impoundment structure, such as: • crest of dam/heap • slopes of dam/heap • surface of tailings beaches	• dam crest and slopes may be treated in the same manner as for the prevention of water erosion • surfaces may need wind breaks, water spray, application of binding material, i.e. spraying with bituminous emulsion, surface mulch, lime slurry • in extreme cases tailings may have to be deposited under water • surface vegetation, either floating or on inactive areas • frequent change of tailings discharge points around perimeter to achieve a continuous and constantly wet surface

but the cover inhibits the maturation of the tailings deposits. The technology to apply these on very soft but maturing tailings is expensive to develop and to operate (EC 2004).

Besides tailings storage areas there are many other sources of fugitive dust emissions that may be carried away with prevailing winds. Fugitive dust can originate from site preparation (e.g. land clearing, soil stripping and topsoil management), mining operations (e.g. drilling and blasting), transportation – probably the most common source of continuous dust emissions (e.g. from loading equipment, haul vehicles, conveyors), comminution (crushing and grinding), and mine waste management operations (i.e. waste rock dumping).

During day-to-day mining operations, wind erosion control or dust control consists mostly of applying water or other dust palliatives as necessary to prevent or alleviate dust nuisance generated by mining, transport and storage activities. Covering small stockpiles or areas is an alternative to applying water or other dust palliatives (**Case 20.1**).

Dust controls generally focus on (1) stabilizing exposed surfaces and (2) minimizing activities that potentially suspend dust particles (**Table 20.3**). For heavily trafficked areas such as haul roads and for disturbed areas (e.g. land clearing areas), wet suppression (watering), chemical dust suppression, gravel or asphalt surfacing, and equipment washdown areas can be employed as dust control measures. Permanent or temporary vegetation

CASE 20.1
Judicious Application of Water for Dust Suppression

Dust control may not be the main objective for covering stockpiles. Managing moisture content in the mined product may be more important. The cost of a 1% moisture increase in shipped ore can amount to several hundred thousand dollars per shipment, due to the increased cost of shipping the additional mass plus the reduced value of product to the purchaser. The composition of the shipped mine product is carefully analyzed by both mining operator and receiving mineral processing facility. This makes for interesting discussions if port facilities apply water for dust control, or for quarantine purposes as is common at Australian ports. Water spraying commonly affects the same surface materials from which control samples are taken by the processing facility.

TABLE 20.3

Dust Control Practices (CASQA 2003)

Site Condition	Dust Control Practices							
	Permanent Vegetation	Mulching	Wet Suppression (Watering)	Chemical Dust Suppression	Gravel or Asphalt	Silt Fences	Temporary Gravel Construction Entrances/ Equipment Wash Down	Haul Truck Covers	Minimize Extent of Disturbed Area
Disturbed Area not Subject to Traffic	x	x	x	x	x				x
Disturbed Area Subject to Traffic			x	x	x		x		x
Material Stockpile Stabilization			x	x		x			x
Demolition			x				x	x	
Clearing/ Excavation			x	x		x			x
Truck Traffic on unpaved roads			x	x	x		x	x	
Mud/Dirt carry out					x		x		

and mulching can be employed for areas of occasional or no construction traffic (e.g. reclamation areas). Preventive measures include minimizing surface areas to be disturbed, timing the sequence of site clearing activities to coincide with actual mining activities in each area, and controlling the numbers and activities of vehicles on a site at any given time. Phasing of land clearing is considered to be especially critical for cleared sites greater than 40 hectares in size (Cheminfo Services 2005). Where feasible, clearing should be started at the location that is upwind with reference to the prevailing wind direction.

All in all, however, the application of water is typically the most common dust control method employed by mining companies across the world. Practically all mining operations apply water to mitigate dust generation from haul roads and mine waste storage areas be it by trucks, water pulls, water cannons, hoses or sprinklers.

The application of water is typically the most common dust control method employed by mining companies across the world.

A variety of chemicals are available to help suppress fugitive dust emissions. While being more expensive than water, chemical dust suppressants are also more effective and need to be applied much less frequently. Examples of dust suppressants include (1) liquid polymer emulsions; (2) agglomerating chemicals (e.g. lignosulphonates and polyacrylamides); (3) cementitious products (e.g. lime-based products and calcium sulphate); (4) petroleum based products (e.g. petroleum emulsions); and (5) chloride salts (e.g. calcium chloride and magnesium chloride) (Cheminfo Services 2005).

While the application of water and chemical dust suppressants have proved to be effective options for mitigating dust, they have to be applied judiciously. Their use, while mitigating dust, can trigger other environmental consequences. For instance, in areas where water is a scarce resource, water for dust suppression competes with other needs for water. There are other potential environmental consequences resulting from the over-application of water that must be considered. These include runoff problems, soil instability, spreading of contaminants in the environment (for example, oil from engines), and water erosion. The over-application of water can also lead to transport problems if the wetted surface becomes slippery or boggy. Chloride salts also need to be used with care. Magnesium chloride used for dust control in the USA has been implicated in the death of pine trees located immediately downslope of the sites of application. It is important to keep these environmental consequences in mind when deciding on the extent to which water and chemical dust suppressants are to be used.

Stabilization of disturbed sites upon completion of earthworks is important. Disturbed soil may be compacted with rollers or other similar equipment in order to reduce the erosion potential of the area. With time, however, disturbed land areas need to be stabilized through revegetation. Although the main objective of revegetation in wet climates is to reduce water erosion, reducing wind erosion is also important.

As a rule of thumb, surfaces of completed earthworks (including landscaping) should be re-vegetated (i.e. seeded and mulched) within one or two weeks after active operations have ceased.

As a rule of thumb, surfaces of completed earthworks (including landscaping) should be re-vegetated (i.e. seeded and mulched) within one or two weeks after active operations have ceased. Groundcover should be of sufficient density to expose less than one-third of un-stabilized ground within two to three months of planting, and all times thereafter (Cheminfo Services 2005). Restoration control measures should be maintained and reapplied, if necessary. The area should be restored such that the vegetative groundcover and soil characteristics are similar to adjacent or nearby undisturbed native conditions. Reseeding using native grasses is prudent. Care must be taken to avoid introducing or promoting the spread of noxious weeds and plants or those foreign to the area (see also Chapter Twenty-One on mine closure).

Temporary seeding and mulching may be applied to cover bare soil and to prevent wind erosion. The soil must be kept moist to establish cover. Mulch can protect the soil surface until newly seeded vegetation can take over and improves the chance of rapidly establishing a dense grass stand. Some types of mulch require tilling to integrate them into the upper layer of soil, if they are to be effective in dust control. Light mulches, such as straw, should be tacked in place, either mechanically or by application of a chemical tacking agent. Areas may need to be reseeded, so that a stabilized surface is formed within eight months of the initial application.

Vegetation belts can provide extra protection against wind erosion reducing the wind velocity for distances up to 30 times the height of the vegetation. Vegetation grown on and around strategic locations to mitigate fugitive dust emissions may consist of scrubs, bushes, or trees in one to ten rows. One, two, three and five-row barriers of trees are found to be the most effective arrangement for planting to control wind erosion (Cheminfo Services 2005). The type of tree species planted also has considerable influence on the effectiveness of a windbreak. In arid and semi-arid regions where rainfall is insufficient to establish vegetative cover, mulching may be used to conserve moisture, prevent surface crusting, reduce runoff and erosion, and help establish vegetation. Storage piles can also be sited in order to take advantage of existing landscape features and vegetation, which can act as a windbreak.

Site rehabilitation including providing for adequate landforms and revegetation of disturbed areas and mine waste placement areas provides the only feasible long-term preventive erosion measures after mining operations cease. Mine waste placement areas are designed to last for centuries, and relying on nature to prevent both water and wind erosion is the only feasible option.

REFERENCES

CASQA (2003) California Stormwater BMP Handbook Construction; available at www.cabmphandbooks.com

Cheminfo Services (2005) Best Practices for the Reduction of Air Emissions From Construction and Demolition Activities; prepared by: Cheminfo Services Inc. in conjunction with the Construction and Demolition Multi-Stakeholder Working Group; prepared for Environment Canada, Transboundary Issues Branch

Day RW (2000) Geotechnical Engineer's Portable Handbook, McGraw-Hill, New York, ISBN 0-07-135111-6

EC (2004) Draft Reference Document on Best Available Techniques for Management of Tailings and Waste-Rock in Mining Activities. European Commission, Edificio EXPO, Seville, Spain:.563

Goldman SJ, Jackson K, and Bursztynsky TA (1986) Erosion and Sediment Control Handbook, McGraw-Hill, New York, 454pp

Office of Surface Mining (1998) Guidelines for the Use of the Revised Universal Soil Loss Equation (RUSLE), Version 1.06 on Mined Land, Construction Sites and Reclaimed Lands, August 1998

Renard KG, Foster GR, Weesies GA, McDool DK and Yoder DC (1997) Predicting Soil Erosion by Water: Guide to Conservation Planning with the Revised Universal Soil Loss Equation (RUSLE). USDA, Agricultural Handbook Number 703

Rosewell CJ (2005) Soiloss: A Program to Assist in the Selection of Management Practices to Reduce Erosion, by Colin J Rosewell. 2nd ed. Soil Conservation Service of NSW, Sydney

US Department of Agriculture (1980) Soil Conversation Service – Predicting Soil Loss Using the Universal Soil Loss Equation

USEPA (1999) Remining Database

• • • •

21

Mine Closure

It is Not Over When It is Over

Carbon, a non-metal, is the fourth most abundant element, and occurs as three distinct allotropes: amorphous carbon (as found in charcoal), graphite and diamond. The contrast is extreme with graphite, black in colour, among the softest substances while diamond, colourless, is the hardest. Carbon provides the chemical basis for life on Earth, forming more than 10 million known compounds, more than any other element. Coal, the predominant fuel of the industrial revolution and still the major fuel for power generation, is a major source of carbon as are the hydrocarbons – natural gas and petroleum. However, most of the carbon in the Earth's crust occurs as carbonates in limestone, dolomite and magnesite deposits, which are predominantly biological in origin. Less than 1% of the world's carbon occurs in the biosphere, with the atmosphere accounting for a very much smaller fraction. The carbon cycle within the biosphere is complex and imperfectly understood. Diamonds, because of their unique physical properties, have numerous industrial uses. However, their main value is as precious stones, used in jewellery.

21 Mine Closure

It is Not Over When it is Over

Mine closure, the subject matter of this chapter, is the term applied to the activities undertaken after completion of mining and mineral processing. It includes decommissioning, dismantling and removal of plant and equipment, and the rehabilitation of areas disturbed by mining and associated activities. Rehabilitation and reclamation are the most commonly used terms for the procedures and practices applied to return land disturbed by mining to a functional condition. Rehabilitation is the term most commonly used in the Australian mining industry, while reclamation is the term favoured in the USA. The terms are essentially interchangeable, but the term rehabilitation is used throughout this book. Revegetation is another commonly used term, its meaning being confined to that part of the rehabilitation process involving establishment and maintenance of plants. Restoration is a more specific term which refers to rehabilitation that returns the land surface to its pre-existing function and condition.

Many environmentalists object to use of the term 'restoration', arguing that mining causes such drastic changes to the landscape, soils, habitats and hydrologic conditions, that it is impossible to re-establish pre-existing conditions. This negative view-point is not supported by the evidence. In the relatively short period since the advent of environmental awareness in the mining industry in the 1970s, there have been many examples of successful restoration in the functionality and productivity of agricultural lands, including rangelands and a variety of annual and perennial cropland systems. Substantial progress has also been made in the restoration of natural ecosystems; however, the time required for natural succession means that many areas rehabilitated with native vegetation, do not yet match the condition of adjacent unmined areas in terms of biodiversity and habitat diversity. On the other hand, evidence of natural succession and progressively increasing biodiversity indicate that full restoration will be achieved in time.

There are also many situations where mining takes place in areas that are relatively barren and depauperate in biota due to natural infertility, physical impediments or a prior history of man-made degradation. In many such cases, rehabilitation provides the opportunity for substantial functional improvement in terms of productivity, biodiversity or both. Common examples include lateritic areas where shallow caprock provides a physical barrier to root penetration for many plants. Removal or fragmentation of the caprock improves the situation, enabling establishment and maintenance of a more diverse range of plant and animal species. Again, some environmental purists will argue that depauperate natural areas are preferable to 'unnatural' areas of higher productivity and diversity. Naturally infertile or physically constrained systems may indeed have valuable or unique attributes which warrant preservation. If so, such attributes can be preserved by establishing conservation areas. In fact, such areas are commonly over-represented in the conservation estate compared to areas of high productivity, because they tend to be in public ownership, whereas productive areas have always been in demand for agriculture and hence tend to be privately owned.

This said, prior to 1985, the issue of mine closure had a low priority for most governments (and hence for most companies), as evidenced by the large numbers of abandoned mines that exist in virtually every major mining country. Very few countries had in place, and the majority still does not have, mineral policy and legislation that provide for comprehensive mine closure (Clark and Cook Clark 2000). The basic components of a comprehensive mine closure policy and associated legislation, as further detailed in this chapter, would (1) have a comprehensive mine closure financing programme; (2) provide specific provisions for abandonment and post-closure activities; (3) provide specific provisions for rehabilitation; (4) require an Environmental and Social Impact Assessment; and (5) have specific monitoring and enforcement procedures to ensure compliance (Clark and Cook Clark 2000). Most if not all countries with a history of mining have legacies of problems from abandoned mines. To date none have devised appropriate and cost-effective means of dealing with the issue, although some countries are well advanced in tackling the problems.

Prior to 1985, the issue of mine closure had a low priority for most governments.

By some estimates there are more than 500,000 abandoned mine sites in the USA alone (USEPA 2000; UNEP and Chilean Copper Commission 2001). Today a number of countries (e.g. Australia, Bhutan, Bolivia, Burkina Faso, Canada, Indonesia, the Lao People's Democratic Republic, Mongolia, the Philippines, UK, USA, Vietnam and Wales) have established comprehensive policy and legislation that provides both for mine closure and for post-mining sustainable development. With the exceptions of these countries, whose policies and legislation have been implemented to varying degrees, the majority of the world's largest mining countries have inadequate policies and legislation for comprehensive mine closure in place, and still fewer provide for post-mining activities and sustainable development (Clark and Cook Clark 2000).

21.1 REASONS FOR MINE CLOSURE

According to Laurence (2006) in a survey of Australian mine closures since 1980, only 25% of the 800 mine closures examined occurred following resource exhaustion or depletion. The remaining closures were due to a variety of causes including economic, geotechnical, safety and hydrological causes. This means that up to 75% of closure were either unplanned or were hurriedly planned over a very short period. These figures, however, may be somewhat misleading in terms of the total mining industry, as the vast majority of closures involved small or medium operations. The larger mines which are mainly operated by large mining companies, tend to operate until the resource is depleted, although there may be significant fluctuations in production or even interruptions, as a result of fluctuating demand.

These larger mines account for most of the production of most commodities except, possibly gold, where the numerous small producers make a substantial contribution. In fact gold mines accounted for about 50% of the 800 mine closures in the survey.

Many mine closures are either unplanned or inadequately planned.

It follows from this survey that many mine closures are either unplanned or inadequately planned. It is in this context that most host countries now require some form of financial assurance to ensure that funds are available for decommissioning and rehabilitation of a site in the event that the operator does not fulfil its obligations and commitments.

21.2 OBJECTIVES OF MINE CLOSURE

The main objective of mine closure and land rehabilitation is to return as much as possible of the land disturbed by mining to a condition in which it can sustainably support one or more pre-determined land uses. Within this overall objective there will be a number of secondary goals, including:

- Restoration of public safety;
- Stabilization to avoid landslide, serious erosion or surface collapse;
- Removal of components which are incompatible with the intended land use (these include plant, equipment and buildings);
- Removal, immobilization or safe burial of toxic or hazardous materials from mining operations;
- Restoration of hydrologic functions, involving the storage and/or release of rainfall;
- Restoration of visual amenity.

Sometimes, mineral resources will remain within the area, in which case there may be another goal – to facilitate access to these resources should they become economically mineable in the future.

Arguably the most important factor in mine land rehabilitation is the establishment and maintenance of public safety.

Arguably the most important factor in mine land rehabilitation is the establishment and maintenance of public safety. Risks associated with abandoned mine sites vary widely and are many. **Table 21.1** lists components of mining operations, the risks or hazards that they might present and the appropriate management measures to protect public safety.

21.3 FINANCING MINE CLOSURE – THE 'POLLUTER PAYS' PRINCIPLE

Much, if not all the costs involved, are incurred after revenues from the project have ceased.

A major disconnect in relation to mine closure is that much, if not all the costs involved, are incurred after revenues from the project have ceased. Not surprisingly then governments have moved increasingly over the past decade towards securing adequate financial provisions to deal with closure liabilities associated with mining operations. Estimation of and provision for closure liabilities, of course, is also often a requirement for publicly listed companies (e.g. to conform to the Sarbanes Oxely Act).

Closure liabilities vary widely and depend on type, size, complexity, and environmental and regulatory setting of the mine. There is tremendous variation in mine deposits and their ecological setting that may result in a wide range of complexities in mine and mill design. This means that closure liabilities may range from US$ 1 million or less for a small mine to US$ 100 million and more for old and large-scale mining operations (IFC 2002). Governments frequently apply the 'polluter pays' principle and, to ensure that adequate funding is available for closure, have applied a variety of different instruments of financial

TABLE 21.1

Post-Mining Risks and Hazard Reduction Measures – Risks associated with abandoned mine sites vary widely and are many

Mine Component	Type of Risk	Management Measures
Open pit	Collapse or slides due to unstable walls Accumulation of deep water, possibly of poor quality	Usually a combination of: Warning signs, Fences and/or abandonment berms, Flattening of batters and highwalls, and less often, Back-filling – partial or complete.
Shaft, tunnel or adit	Collapse Entry point to underground workings that may be unstable, flooded or poorly ventilated	Usually a combination of: Warning signs, Fences and/or abandonment berms, Capping of shafts, Back-filling of shafts – partial or complete. Installation of barricades, locked gates or bulkheads at or inside the portals of tunnels or adits.
Tailings impoundment	Collapse or overflow due to flooding or blockage of spillway, Boggy surface could trap human visitors or livestock,	Combination of: Warning signs, Fences, Removal of water and establishment of surface drainage system, Establishment and maintenance of adequate spillway capacity.
Waste rock dumps	Landslips	Shaping to form stable profile.
Mill and processing plant, plus ancillary facilities	Becoming unsafe over time, if abandoned.	Decommissioning, Dis-assembly, and Removal

Source:
Khanna 2000

assurance – either imposed on operators by legislation, negotiated in project-specific agreements or as voluntary commitments. These instruments include: third party agreements, cash deposits, letters of credit, trust funds, insurance policies and corporate guarantees based on a pledge of assets. The advantages and disadvantages of each are listed in **Table 21.2**, which has been reproduced from Fleury and Parsons (2006).

It is common for the amount covered by a financial assurance instrument to change regularly in response to revised calculations of closure liabilities. In some cases involving cash deposits, trust funds or balance sheet entries, the funds are accrued progressively in accordance with an agreed schedule, for example per unit of product or per unit of area disturbed. Similarly, accrued funds can be reduced as various closure tasks are complete as confirmed by the achievement of pre-arranged performance criteria.

The amount covered by a financial assurance instrument is changed regularly in response to revised calculations of closure liabilities.

It is evident that performance guarantees have been devised mainly to protect against default by small and medium operators, who have been responsible for the vast majority of inadequate closure efforts in the past. Companies with strong balance sheets and, in particular, companies with multiple operations, are much less likely to default. A good example of this is the unplanned closure of the Marcopper Mine in the Philippines following a major tailings discharge (Case 18.1). Here, the part owner and operator, Placer Pacific, accepted responsibility for remediation, reportedly spending more than $ 40 million in clean-up costs. It follows that balance sheet provisions may be sufficient in the case of such companies, whereas for a small company, a balance sheet entry may provide little protection for the host country.

TABLE 21.2

Financial Assurance Options – Governments frequently apply the 'polluter pays' principles

Options	Description	Advantages	Disadvantages
Third-party guarantee	Include unconditional bank guarantee and insurance bonds. All are required to be unconditional and/or irrevocable	• Relatively inexpensive (usually between 1 and 1.5% of amount) for the operator establish • Has full backing of financial institution (funds available 'on demand') • Transparent and operation-specific • Cannot normally be unilaterally withdrawn by the issuer • Can be altered as requirements change	Often considered by financial institution to be part of working capital, thereby reducing available operating funds
Cash deposit	Normally deposited direct with government and only usually accepted for 'small' operations	• Provides an advantage to the government which has direct control over funds and has sole responsibility for making funds available if required • The cash is returned to the company, normally on completion of closure works	• Providing cash 'upfront' is a financial impediment to the operator and potential loss of income through interest • If operator goes bankrupt cash may be classed as a company asset and available to all creditors • Government must have a system to ensure segregation of funds for their intended use
Letter of credit	A form of third party guarantee which normally has a one year term, usually extended following review by the issuer. If not extended the beneficiary (government) is notified and has the option of drawing down the full value	Relatively inexpensive for the operator to establish	• Can be unilaterally withdrawn by the issuer at the end of the credit term • May restrict company access to other credit
Trust fund	Administered by a third party trustee with a defined investment policy. Intended to cover the costs of a specific closure plan through a structured series of contributions. Surplus funds are returned to the operator	• The fund is visible to government (and the public) • Any surplus after completion of the closure/decommissioning plan are returned to the operator	• A transition period is required to allow the operator to build up the fund • Administrative requirements (similar to a pension fund) can be cumbersome
Insurance Policy	Several jurisdictions nominate this as an acceptable method of providing financial assurance. No example has been located of this being implemented	• Relatively inexpensive for the operator to establish • Less administration required than with a cash trust fund	• Only valid if annual premium paid • Recourse to financial assurance often takes place some years after operator becomes inactive and is unable to pay the premium
'Soft' options	Examples of soft options include: Financial strength rating (where a company is rated as investment grade); Self-funding; Financial test (e.g. balance sheet test); Corporate guarantee based on financial grade; Parent company guarantees; Pledge of assets	• Does not involve direct costs • Relatively inexpensive for the operator to establish	Does not provide the save level of security as hard forms of assurance

Source:
Fleury and Parsons 2006

The type of instrument applied has been and continues to be a source of concern to governments and mining companies alike. From the point of view of governments, it is critical that funds be available for use in case the company defaults due to insolvency, liquidation, take-over or other cause. From the point of view of companies, there is general opposition to paying money into a fund when such money could be used to retire debt or for other productive purpose. In many developing countries, companies are also concerned about the security of payments made to funds to which government agencies may have access.

Experience is that rehabilitation requirements and final closure costs exceed initial estimates, particularly in the following areas (USDA 2004):

- Interim management of process fluids;
- The need for and cost of water treatment;
- Detoxification and rinsing of spent ore from heap leach piles;
- Closure of tailings impoundments;
- Removal, isolation (liners/covers) or treatment of hazardous materials (chemicals, spent ore, waste rock);
- Site drainage, interim and long-term;
- Monitoring and maintenance of the mine site during and after closure;
- Indirect costs of closure/reclamation; and
- Lack of site-specific information at the time of mine planning.

Experience is that rehabilitation requirements and final closure costs exceed initial estimates.

By necessity, most financial mine closure instruments are initially estimated based on conceptual mine plans. Actual mine operation, however, will differ from these initial plans as the mine evolves and new information becomes available. Actual plans should be used to revise mine closure estimates, and mine closure funds, at regular intervals. Finally, estimating mine closure costs requires considerable judgement that can only be gained through experience. It is fair to say that mine closure costs are often underestimated if developed by persons with inadequate education, training or experience.

Estimating mine closure costs require considerable judgement that can only be gained through experience.

Mining is dynamic. New ore deposits may be discovered during the mine operation, and mine rehabilitation often occurs in parallel with operations. This raises the issue of the period covered by the financial mine closure provisions. For example, mine closure bonds may cover a single year, multiple years or the entire life of a mine. Selection of the bond period may be based upon some logical stage of mine development, e.g. construction or reclamation of major mine facilities like tailings ponds, heaps and mine facilities. Of course the bond period needs to reflect regulatory requirements.

What are the mine closure costs? No single formula can be used to arrive at exact estimates. Each mine is a unique situation. However, common direct and indirect costs exist.

Direct costs that relate to rehabilitation tasks can be grouped into the following eight categories (USDA 2004):

- Interim operation and maintenance;
- Hazardous materials;
- Water treatment;
- Demolition, removal and disposal of structures, equipment and materials;
- Earthwork;
- Revegetation;
- Mitigation; and
- Long-term operation, maintenance and monitoring.

Clearly the occurrence of acid rock drainage adds significant costs in terms of mine wastes rehabilitation and water treatment (IFC 2002). In the case of some mines, substantial annual expenditures may be needed, virtually in perpetuity, to neutralize acid rock drainage from inactive workings.

The occurrence of acid rock drainage adds significant costs in terms of mine wastes rehabilitation and water treatment.

Indirect costs commonly fall into the following categories (USDA 2004):

- Engineering redesign;
- Mobilization and demobilization;
- Contractor's costs;
- Government project management;

- Contingencies; and
- Inflation.

Contingency costs are necessary to address the errors that exist in every cost estimate that is based on assumptions and conceptual information rather than actual data.

Finally, while they may be considered separately, 'social' costs related to redundancy payments or contributions toward future maintenance and operation of social assets also need consideration and funding.

21.4 REHABILITATION

Rehabilitation and revegetation planning is an essential element of mine closure planning, and commences at the mine planning stage. As far as possible, the mine operator aims to progress rehabilitation in parallel with mining. In open cast mining, rehabilitation follows as mining progresses, and final rehabilitation efforts at the end of mine life are minimal. The post-mining land use for the mine area, and hence rehabilitation and revegetation should be agreed upon with local government, traditional landowners, and other interested parties from the beginning. This can be a challenging process since opinions on post-closure land use differ. The central government, for example, may insist on reinstalling native vegetation, while the local government may favour productive agricultural use.

The primary focus of post-closure activities is monitoring the progress of required activities, and when necessary providing additional management inputs until landform, vegetation, water quality and social infrastructure meet agreed targets, or until management of the former mining area is integrated into existing local or regional management systems.

A check-list of issues, relevant rehabilitation objectives and rehabilitation measures that apply to each component of a mining operation is provided in **Table 21.3,** which has been reproduced from Khanna (2000).

Mine closure measures aim to leave a stable and productive post-mining environment behind.

Mine closure measures aim to leave a stable and productive post-mining environment behind. Post-closure activities provide additional management efforts until landform, vegetation, water quality and social infrastructure are self-sustaining and meet the requirements of their users, or until management is integrated into the management of the surrounding area.

Ultimate Land Use and Design of Final Landform

Selection of the ultimate land use, and related to it, the final landform is an important initial step in rehabilitation planning.

Selection of the ultimate land use, and related to it, the final landform is an important initial step in rehabilitation planning. In the USA, it is a commonly stated requirement that the ultimate land use should be of the same or higher value than that which existed prior to mining. This presupposes a hierarchy of land use values such as the following, in order of increasing value:

- Unused land, unsuitable for any type of farming;
- Range land – natural;
- Range land – improved pasture;
- Crop land;
- Land used for intensive agriculture; and
- Residential/commercial/industrial land.

TABLE 21.3
Environmental Aspects of Mine Closure (based on Khanna 2000)

Issues	Objectives	Control
Underground mine workings		
Physical Stability • Shafts • Adits • Declines • Subsidence	• Prevent access • Seal • Safety • Stability	• Backfill • Plug openings • Vent water and gas • Infill underground and surface spaces • Re-contour surface
Chemical Stability • Mineral leaching • Acid drainage • Contaminants • Methane	• Clear water • Meet water • Quality regulations • Prevent release	• Flood workings • Plug openings • Remove contaminants • Treat water discharge • Collect and use gas
Land Use • Productivity • Aesthetics • Drainage	• Restore to original or accepted alternative use • Re-establish drainage patterns	• Backfill disrupted areas • Contour surfaces • Flood workings
Open pit mine workings		
Physical Stability • Steep slopes • Unstable faces • Erosion • Hydrology • Safety	• Stable surfaces • Remove hazards • Control erosion • Clean water	• Re-contour • Establish vegetation • Fence and erect signs • Install embankments
Chemical Stability • Metal leaching • Acid drainage	• Install drainage • Meet water • Quality regulations	• Seal surface • Flood pit • Control hydrology • Treat discharge
Land Use • Productivity • Aesthetics • Drainage	• Restore to original or accepted alternative use • Re-establish drainage patterns	• Backfill disrupted areas • Re-contour slopes • Establish vegetation • Flood
Waste rock and spent ore		
Physical Stability • Steep slopes • Unstable faces • Erosion • Drainage • Dust	• Stable surfaces • Avoid failures, slumps and sediment release	• Site selection • Internal drains • Gentle slopes • Contour surfaces • Cap • Water ditches • Settling ponds • Establish vegetation • Monitor

TABLE 21.3
(Continued)

Issues	Objectives	Control
Chemical Stability • Metal leaching • Acid drainage • Mill reagents • Contaminants	• Clean water	• Dump design • Isolate of reaction material • Cap and revegetate • Control drainage • Collect and treat effluent • Monitor
Land Use • Productivity • Aesthetics • Drainage	• Restore to original or accepted alternative use • Establish drainage patterns	• Re-contour slopes • Establish vegetation
Tailings impoundments		
Physical Stability • Dust • Erosion • Dam wall • Drainage	• Stable surfaces • Avoid failure and slumps • Control sediment	• Site selection • Dam design • Tailings disposal method • Cap and revegetate • Control drainage

Chemical Stability
• Metal leaching
• Acid drainage
• Mill reagents
• Dam structure
	• Clean water by:
	  • Control reactions
	  • Control migration
	  • Collect and treat
		• Use chemically stable material in dam wall construction
		• Pre-treatment of tailings
		• Cover to control reactions
		• Form wetland
		• Divert runoff
		• Collect and treat effluent
		• Monitor

Land Use • Productivity • Visual impacts	• Restore to appropriate land use	• Re-contour • Cap and establish vegetation • Flood and form wetland
Water Management		
Physical Stability • Dam walls • Structure • Pipelines • Ditches • Settling ponds • Culverts • Erosion	• Long-term • Safety of • Flood capacity • Prevent blockage • Prevent erosion • stability • structures • Free passage of water	• Breach dam • Remove structure • Plug intakes and decants • Upgrade flood design • Remove pipes • Fill in ditches • Provide for long-term maintenance • Monitor
Chemical Stability • Contaminant of surface and/or groundwater	• Clean water	• Remove or prevent contamination • Drain, treat and discharge • Install barriers • Establish vegetation • Monitor

(Continued)

TABLE 21.3
(Continued)

Issues	Objectives	Control
Land Use • Interruption of water supply • Productivity of land drainage	• Restore drainage patterns or establish alternative • Return to appropriate land use	• Stabilize and maintain dam or breach and establish erosion resistant drainage • Establish vegetation
Mine structures		
Physical Stability • Steep slopes • Unstable faces • Erosion • Drainage • Dust	• Stable surfaces • Avoid failures, slumps and sediment release	• Site selection • Internal drains • Gentle slopes • Contour surfaces • Cap • Water ditches • Settling ponds • Establish vegetation • Monitor
Chemical Stability • Metal leaching • Acid drainage • Mill reagents • Contaminants	• Clean water	• Dump design • Isolation of reaction material • Cap and revegetate • Control drainage • Collect and treat effluent • Monitor
Land Use • Productivity • Aesthetics • Drainage	• Restore to original or accepted alternative use • Establish drainage patterns	• Re-contour slopes • Establish vegetation
Mine infrastructure		
Physical Stability • Building • Equipment • Roads • Airstrips • Services	• Make area safe • Control access	• Disassemble and remove all buildings, equipment and other services • Excavate buried tanks and backfill • Restore drainage • Revegetate
Chemical Stability • Fuel and chemical storage areas • PCBs and insulation • Explosives • Fuel or oil spill	• Make secure and safe • Clean water	• Remove all unwanted materials • Treat contaminated soil or dispose of in an approved site • Control and treat drainage
Land Use • Alternative uses • Productivity • Visual impact	• Return to appropriate land use	• Remove foundations and re-contour • Restore natural drainage • Revegetate

TABLE 21.3
(Continued)

Issues	Objectives	Control
Socio-economic mitigation		
Workforce	• Re-employment • Relocation	• Assistance with looking for other work and moving • Financial assistance • Counselling
Local communities	• Stable economy • Good health • Education facilities	• Regional development plan • Develop local self-sustainable enterprises • Establish foundation or trust fund for essential services • Relocate in-migrants

Notwithstanding the foregoing, in the majority of cases, the most logical and practically achievable land use will be the same as that which existed prior to mining. Before this can be confirmed, however, it is necessary to consider the condition of the site as it will be at the conclusion of the project, and the measures that will be required to return the surface to a condition that can support and sustain the intended land use. It follows that selection of the ultimate land use should take place during the pre-construction (mine feasibility/ environmental permitting) stage so that environmental management measures required to establish the intended land use can be at least conceptually designed, quantified and costed at the outset. This is also the best time to initiate consultations, in relation to post-mining land use, with local communities.

Local communities, beside the host government and the mining company, are the main stakeholders in mine closure planning, but other groups and individuals may also show interest in mining closure issues as illustrated in **Figure 21.1**. Closure of the Kelian Gold Mine in Indonesia (**Case 21.1**), provides a compelling case for the use of a forum-based participatory approach to achieve 'buy in' of stakeholders to closure plans.

The importance of consulting host communities is intuitive – they have to live with mined out areas.

The importance of consulting host communities is intuitive – they have to live with mined out areas. However, the main responsibility for mine closure rests with the host government and the mining company. The host government is responsible for setting roles and responsibilities, and for enforcing regulatory requirements. Eventually the ultimate responsibility for the mine site also remains with the government once the mining company has released the area. The mining company is responsible for planning and implementation of mine closure, its financing, and initial post-closure management until the mine site is declared safe and environmentally stable.

Two distinct types of land use are recognized: (1) economically productive land uses such as agriculture, forestry and tourism; and (2) natural ecosystems which provide habitats for flora and fauna that are native to the area. Natural ecosystems may also support other activities such as hunting and recreation which may have an economic component.

Landform design involves planning the final configuration of the land surface. The landforms that may be retained or created are constrained by topography and natural barriers such as rivers; property boundaries; type and size of mine void; the extent of site preparation

Final Responsibilty | Monitor | Support/Co-operate

ACTORS	Framework	Exploration	Construction	Operation	Closure	Post-Closure
Government	Set roles and Responsibles	Monitor, enforce, inform				Monitor and inform
Mining Company	Support	Consult – design site with closure in view – partnership				Initial monitoring then support
Community	Support	Integrate closure planning in business processes form and sustain partnerships with company and others				
Local Government	Support	Begin regional planning processes with closure in mind early on set-up and sustain partnerships – sustainable economic activities				
NGOs/Civil Society Org.	Support	Links to international NGOs – capacity building for local communities – monitor and inform				
International Agencies	Support	Disseminate best practice – develop and propagate standards and guidelines – work with government, companies and communities				

FIGURE 21.1
Who Does What and When in Mine Closure

The host government sets roles and responsibilities, and enforces regulatory requirements. The mining company plans and implements mine closure, and provides the necessary funding.

Source:
IFC 2002

carried out during construction; and the nature, dimensions and physical properties of waste rock storages and tailings impoundments. As a consequence of these constraints the design of landforms usually involves compromises between what is desirable and what is practically achievable. The overriding objective of landform design is to produce landforms that are compatible with the intended land use(s), and that landforms are stable. A subsidiary objective is that new landforms should be similar in appearance and functionality to natural

The design of landforms usually involves compromises between what is desirable and what is practical.

CASE 21.1
Closure of the Kelian Gold Mine in Indonesia

Until now, the largest mine in Indonesia to carry out the closure process is the Kelian Gold Mine in Kalimantan. At the time that closure was being planned (the early 2000s), the Indonesian environmental regulatory system was in transition from central to regional autonomy, and there were no official regulations or standards on which to base mine closure plans. In order to achieve 'buy in' of stakeholders, the operator – Kelian Equatorial Mining – decided to use a forum-based, participatory approach. To this end, a Mine Closure Steering Committee was established, comprising representatives of key stakeholder groups from central, regional and local government; local communities; traditional authorities; and company representatives. The aim of this committee, which met quarterly over a two-year period, was to achieve responsible closure, through an accountable and transparent process. Outcomes included:

- Agreement of environmental and social standards, including water quality standards for ongoing discharges, rehabilitation standards, and social mitigation programmes;
- Identification of appropriate authorities and transitional provisions for company assets and community programmes; and
- Establishment of environmental trust funds for maintenance and contingency purposes.

Kelian Equatorial Mining saw this process as an investment for sustainable mine closure. In the words of its Managing Director (Charles Lenegan), *'We could have done it ourselves and we could have told people what we were doing, and we could have then tried to persuade them to buy in. We could have tried to do it bilaterally with our various stakeholders, but there is too much conflict and competition between the various stakeholders. The most important thing ... is the buy in, and we would not have got it! So ... I don't think there was another way that it would work. The only way you get buy in is to involve the people and you have to do it in a forum.'*

Source: Kunanayagam, R. (2006). Sustainable Mine Closure – Issues and Lessons Learnt. *In*: Proc. 1st Int. Seminar on Mine Closure, Perth, Australia, 13–15 September, 2006 *Eds*: Fourie, A. and Tibbett, M.

landforms in the vicinity. This may be quite a challenge, for example, if the project site is a flat plain and the operations produce a deep pit and several large hills.

To support economically productive land uses, landforms need to provide the following:

- Accessibility – provision and maintenance of access to and access within the site;
- Usability – including factors such as all weather trafficability, ease of cultivation;
- Physical stability against slope movement or erosion;
- Provision of soils suitable for establishment and survival of the proposed vegetation; and
- Potential productivity and economic yield

To support natural ecosystems, landforms need to provide:

- Physical stability;
- Surface relief, including micro-relief, to provide a range of habitat types to support a wide range of flora and fauna;
- Provision of soils suited for growth of natural vegetation; (soils suitable for establishment of native vegetation may be more or less easily provided than soils suitable for agriculture); and
- Visual amenity.

For both types of land use, aspect can be very important, especially at higher latitudes. Aspect, in this sense refers to the direction of a slope. It is an important but commonly overlooked determinant of exposure to solar radiation and exposure to prevailing winds which, in turn have a major effect on soil moisture. Landform design can be used in the case of intended agricultural or forestry land use, to maximize formation of the most favourable aspects. In the case of establishing natural ecosystems, the landform design may be used to create a variety of different aspects, each of which will ultimately develop different edaphic and habitat characteristics. Guidance in either case may be obtained by observation of surrounding areas (**Case 21.2**).

For surface mines, waste rock provides the most readily available medium for creation of landforms. Placement of waste rock can be arranged to form the required shape, minimizing the need for subsequent re-shaping. Of course, there may be constraints such as acid rock drainage, which preclude certain shapes or arrangements of waste rock storages. However, it is usually possible to create a final profile that, when vegetated, will blend in with the surrounding landscape. As discussed elsewhere, the creation of steep, 'angle-of-repose' final dump slopes is inappropriate from several view-points, including stability, practicability of rehabilitation and long-term accessibility.

The creation of steep, 'angle-of-repose' final dump slopes is inappropriate from several view-points, including stability, practicability of rehabilitation and long-term accessibility.

Conventional tailings impoundments are less versatile in terms of landforms that can be developed, as they unavoidably produce a large, flat to very gently sloping, concave crest. Thus a drainage system needs to be installed to avoid periodic inundation. However, these flat or gently concave surfaces may prove to be well suited to a variety of land uses ranging

CASE 21.2
Limestone Mining in Northern India

In the limestone mining areas of the Lesser Himalayas in India, trees are naturally restricted to the north-facing slopes, the more exposed south-facing slopes supporting only shrubs and grasses. Clearly, it is important that, if forestry is the intended land use, landform design in this situation should maximize the formation of slopes with a northerly component.

from aquaculture to irrigated rice cultivation. The disposal of thickened tailings enables somewhat steeper beach gradients, and a gradual transition between the tailings storage and its surrounds, to be achieved. Filtration of tailings promises to provide much greater versatility, as the filter cake can be placed in virtually any configuration to produce a variety of shapes and profiles.

The most problematic landform is usually the mine void itself, particularly in the case of large, open Pit mines. While it may be economically feasible to fill or re-shape small and shallow mines so that they blend with the surrounds, this is not the case for large, deep mines. Typically, these pits have steep, rocky, sometimes unstable walls which are not amenable to the establishment of vegetation and will at least partially fill with water. Often, they have no potential for economic productivity, and provide only very limited ecological function. While the final void or voids represent only a fraction of the original 'footprint' of each operation, they usually comprise the 'residual footprint', all other areas being totally rehabilitated.

While the final void or voids represent only a fraction of the original 'footprint' of each operation, they usually comprise the 'residual footprint', all other areas being totally rehabilitated.

The main elements of landscape design are:

- Shape – usually complex, involving trapezoids, wedges, and benches, forming landforms such as mesas, hogbacks, spinal ridges and terraces, usually with abrupt slope changes which may be rounded by regrading;
- Size – length, width, height and plan area (footprint);
- Slopes (including pit inslopes and dump outslopes) – gradient, aspect, length and height;
- Drainage – pattern and density, with each drain defined by its length, depth, width, gradient and cross-section, and
- Micro-relief – rip-lines, furrows, banks, micro-terraces and pitting (moonscaping).

There are at least eleven significant factors to be considered in design, as discussed below:

1. Stability of Slopes – which depends on:
   - Material properties – internal friction, cohesion, moisture content, permeability;
   - Drainage conditions – rainfall, run-on, runoff, internal drainage, groundwater level (If internal drainage is impeded, pore pressures develop, reducing stability);
   - External loading, most commonly due to Earthquakes.
2. Susceptibility to Erosion – which depends on:
   - Soil characteristics – particle size distribution, texture, permeability, cohesion, dispersivity;
   - Slope angle and slope length;
   - Rainfall intensity and duration, and as it develops
   - Vegetation – the above ground foliage that absorbs raindrop impact, the leaf litter, and the root systems that bind soil particles. Depending on susceptibility, a wide range of temporary or permanent erosion control measures can be considered in the design.
3. Settlement and Consolidation – which can result from the collapse of underground workings, or from consolidation of waste rock or tailings deposits. The amount of settlement at the surface and the rates at which settlement proceeds depend on:
   - Nature of materials;
   - In the case of underground workings – depth to and size of openings and strength of 'roof';
   - In the case of waste rock – total thickness and differential thickness, degree of compaction achieved during placement, and
   - In the case of tailings, the amount of dessication achieved following placement.
4. Requirements of Vegetation – which include:
   - A stable ground surface;
   - Soil of adequate depth and strength to anchor the plants;

- Adequate infiltration and moisture retention capacities to supply plant water needs;
- Other soil characteristics such as suitable texture, fertility and absence of growth inhibitors. Plants adapted to arid and semi-arid areas are usually deep-rooted, while tropical plants usually have relatively shallow root systems.

5. Accessibility and Usability – this includes both access to and access within the rehabilitated area and depends on the intended land use and what will be involved. For example, will vehicular access be required for planting, cultivation, harvesting, etc? Maintenance of accessibility commonly involves compromises in selection or design of erosion controls.
6. Drainage Capacity – needs to be provided:
   - To collect and accommodate runoff from the most intense rainfall events likely to occur, including any run-on from surrounding areas;
   - To incorporate measures such as stone-pitching, lining, and drop-structures to avoid scouring;
   - To convey runoff beyond the rehabilitated areas into the natural drainage system.

   It should be recognized that, if adequate drainage is not provided, it will develop by surface erosion, the results of which may be highly damaging.

   If adequate drainage is not provided, it will develop by surface erosion, the results of which may be highly damaging.

   There are usually trade-offs to be made between infiltration and runoff. Generally, in tropical, high rainfall areas, the priority is to 'shed' as much water as possible. Accordingly, drains are relatively closely spaced, with moderate to steep gradients. In arid areas, infiltration is encouraged by broad inwardly-sloping terraces and, in some cases by surface pitting or moon-scaping, with the result that drains may be few and far between. Whichever approach is adopted, it should be consistent with the hydrologic management of the overall watershed.
7. Phyto-toxicity – the most common cause of which is acid rock drainage (ARD), which can severely restrict or prevent plant growth. ARD is discussed in detail in Chapter Seventeen. In tailings impoundments, ARD can be prevented by means of a 'dry cover' or capping beneath the plant growth zone, or by a 'wet cover' involving the maintenance of a surface pond. In waste rock storages, ARD is managed by:
   - Deep burial of potentially acid-forming material beneath the water table;
   - Encapsulation within compacted clay, or
   - Capping alone, with compacted clay or artificial membranes, which may be sufficient in arid areas.

   All these cases involve careful design to ensure that the systems are effective at managing ARD and, at the same time, provide conditions adequate for establishment of sustainable agricultural or natural vegetation communities.
8. Appearance – to achieve congruity with the surrounds, designs should:
   - Avoid sharp edges and straight lines;
   - Blend gradually between the interfaces of natural and artificial landforms;
   - Provide screening vegetation in areas of unavoidable scarring, and
   - When planting native vegetation, straight lines, regular patterns and repetitive spacing should be avoided in favour of more random arrangements.
9. Final Void Hydrology and Chemistry – hydrological and chemical studies may be required to evaluate the extent to which the pit will fill with water, the rate of filling, the likely seasonal fluctuation and whether or not water in the pit will 'turn over' as a result of seasonal temperature differences. The water quality of the final pit lake is also of environmental concern (**Case 21.3**). This information is required to assess the need for treatment of pit water and whether additional structures such as overflow spillway and drains are required.

   The water quality of the final pit lake is of environmental concern

10. Pit Wall Acid Rock Drainage – ARD from the pit wall prevents vegetation from establishing and is usually the main source of poor quality water draining into the pit. If necessary, steps can be taken to minimize this source by covering or inundating areas of potential acid generation.
11. Final Void Abandonment Bund – this is to prevent inadvertent access to the pit area and should be designed in accordance with the following:
    - The location is based on an evaluation of the long-term stability of the pit walls. In the absence of definitive slope stability studies, it should be assumed that, in the long-term, the slopes will fail back to an angle of 37° from the toe. The bund should be located beyond this notional final crest.
    - The bund, in conjunction with the vegetation grown on it, will provide a screen, shielding the pit walls from view, and
    - The bund profile should be such that it prevents vehicular access and deters pedestrian access.

Soil Preparation and Use of Topsoil

Much has been written about the salvage, storage and re-use of topsoil in mine land rehabilitation. Some advocates claim that the benefits of using natural topsoil are so overwhelming that its use should be mandatory. Others claim that the benefits are not justified by the costs, and that artificial soils can be produced that provide the same or better conditions for plant growth. Examples can be found to support either case. However, it is probably true that increasing recognition of topsoil as a valuable resource (a concept that resonates with the mining industry) has resulted in its use on most projects developed during the past decade. However, there are examples of successful projects where

It is probably true that increasing recognition of topsoil as a valuable resource (a concept that resonates with the mining industry) has resulted in its use on most projects developed during the past decade.

CASE 21.3
Flooding of the Island Copper Mine Pit

The Island Copper Mine Pit near Port Hardy, Vancouver Island, BC, Canada, was flooded in 1996 with seawater and capped with fresh water to form a meromictic (permanently stratified) pit lake of maximum depth 350 m and surface area of about 1.7 km^2. The pit lake is being developed as a passive treatment system for acid rock drainage (ARD). The physical structure and water quality has developed into three distinct layers: a brackish and well mixed upper layer; a plume stirred intermediate layer; and a thermally convecting lower layer. To date the upper halocline has risen due to the injection of buoyant acid rock drainage (ARD) into the base of the intermediate layer. An upper layer 'equilibrium' depth is expected in the near future reflecting a balance between meteorological forcing from above and rising ARD plumes from below. The lower halocline depth fluctuates seasonally: deepening in the winter due to high ARD flows, which erode the lower halocline; and rising in the summer when thermal convection in the lower layer dominates over reduced intermediate layer stirring. The proposed treatment of acid rock drainage by metal-sulphide precipitation using sulphate-reducing bacteria is dependent on anoxia developing in the intermediate and lower layers. The major issues facing the passive treatment system are developing and maintaining anoxia in the intermediate layer and maintaining meromixis.

Source: Fisher and Lawrence 2006

Photo Credit: www.minemosaic.com

it has not been used. One such example is the Ranger Uranium Mine in Australia's Northern Territory, where native vegetation is being established directly on the surface of waste rock dumps. Another example is the large tailings deposition area of the Freeport Gold and Copper Mine in Papua where reclamation is proceeding without the use of topsoil.

The advantages of using topsoil are:

- It usually contains a 'seed bank', representing all or most of the plant species that previously grew on it, including both pioneer and climax species;
- By providing seeds of the same provenance, application of topsoil also helps to conserve genetic diversity;
- It contains a variety of micro-organisms, many of them beneficial to the growth of the plant species with which they are associated;
- It is a proven growth medium in terms of texture, moisture absorption, moisture retention and lack of phyto-toxins or growth inhibitors.

Application of topsoil is particularly beneficial for the restoration of natural vegetation communities. In a rehabilitation programme, it is usually impractical to plant more than about 10 species by seeding or planting of seedlings. However, topsoil may contain 50 or more species, leading to the rapid re-establishment of diversity without the need to wait for the slow spread of species from unmined areas.

Extended stockpiling of topsoil results in the progressive loss of viable seeds as well as a reduction in populations of micro-organisms.

It has been established that extended stockpiling of topsoil results in the progressive loss of viable seeds as well as a reduction in populations of micro-organisms. Much of the benefit may therefore be lost if the topsoil is not used soon after it has been excavated. 'Bottom up' waste rock dump construction and other approaches that enable progressive rehabilitation, overcome this problem by providing sites for placement of freshly excavated topsoil as shown in Figure 19.7 in Chapter Nineteen.

It is not always practical or necessary to remove and use only the topsoil layer per se, as its thickness may vary considerably over short distances, particularly in high relief areas. The approach here is usually to establish a 'stripping depth' that removes most of the topsoil, and to excavate to this depth, resulting in some sub-soil being mixed with the topsoil. Unless the sub-soil has a particularly difficult texture or other deleterious constituent, this approach is usually successful, and helps to augment limited topsoil supplies. Sometimes, to meet specific requirements of particular plant species, up to three different soil horizons are separately excavated and replaced on the rehabilitation area, in their original configuration. Similarly, soil profiles may be artificially constructed to suit plant requirements using, for example, tailings sand for topsoil and slimes to provide sub-soil. Additives such as compost or sewage sludge may be mixed with the sand to provide organic content.

Tailings are themselves soil materials which in many cases provide excellent substrates for plant establishment, without the need for a topsoil capping. The main requirement is that they be free of phyto-toxic substances. They may benefit from amendments such as pH adjustment, fertilizing, and/or addition of organic matter as part of the rehabilitation prescription. Simple nursery or field trials can be carried out to assess the need for and effectiveness of such amendments. Husin *et al.* (2005a and 2005b) describe successful establishment on tailings from the Grasberg operations, of a variety of natural forest species as well as horticultural species, with and without soil amendment.

In summary, while not always essential in mine land rehabilitation, application of topsoil usually offers benefits that more than justify the effort and costs involved. This is particularly so for cases where mined areas are to be rehabilitated with the same natural vegetation communities that occurred prior to mining.

Soil Amendments and Fertilization

Soil characteristics can be improved by a variety of practices, including:
- Addition of lime to raise pH and to improve texture of clayey soils;
- Addition of gypsum or sulphur to lower pH;
- Addition of organic matter such as manure, sewage sludge, compost or green manure to improve the texture and moisture retention properties;
- Fertilization, involving the addition of nutrients and elements essential for the growth of most plants.

Simple agronomic tests are used to assess the need for soil amendments. Many handbooks are available that provide detailed instructions for assessing soil amendment requirements and for their practical application. See for example, Minerals Council of Australia (1998). In agricultural and production forestry situations, repeated harvesting will progressively remove nutrients and deplete some elements, requiring repeated fertilizing. In natural vegetation communities, however, the nutrients are recycled through natural processes as vegetation dies, decomposes and is incorporated in the soil for subsequent uptake.

Simple agronomic tests are used to assess the need for soil amendments.

Mulching

Mulching is another practice that can be highly beneficial to the rehabilitation process. Land clearing usually produces plenty of organic matter and, even if the timber is harvested, the branches and leaves can be shredded, chipped or trittered to form a useful mulch which, when spread over the ground surface, has the following benefits:
- Absorption of raindrop impact, a major initiator of soil erosion;
- Shading and cooling of soil layer;
- Assistance in retaining soil moisture;
- As it decomposes, release of nutrients, and
- Providing food and shelter for a range of organisms; once the appropriate invertebrates colonize the area, additional organic matter will be incorporated in the topsoil.

It should be recognized, however, that nutrients, particularly nitrogen, are consumed by the mulch as it decomposes. It may therefore be important to add sufficient fertilizer that nutrients are not depleted to levels that would limit the growth of vegetation.

Nutrients, particularly nitrogen, are consumed by the mulch as it decomposes.

Establishment of Vegetation

Cover crops are commonly planted to provide groundcover, pending the establishment of the vegetation required for the intended land use(s). The purposes are (Hypers *et al.* 1987):
- To protect the soil against the full force of the rainfall, and the retention of the soil (short-term erosion control);
- To protect the soil against strong heat so that the humus is broken down less quickly;
- To eliminate weeds by water and light deficiency;
- To increase the humus level and structure of the soil, by working in the crop and litter, and
- To increase soil fertility.

To protect the soil against the full force of the rainfall, and the retention of the soil.

Usually, cover crops are planted by broadcasting seed or by hydro-seeding. Where seed is in short supply, it may be sown in rows – in furrows located along the contours. In this case, vegetation forms dense strips in the first year, spreading between the strips in subsequent years.

Where agricultural land use is planned, cereal grasses and legumes such as clover are commonly used as cover crops. Where natural vegetation is to be re-established, local pioneer species are sought, particularly low, ground covering species with a scrambling or sprawling habit. However, it is sometimes difficult to find suitable species with plenty of easily harvested seed, or for which the germination requirements are understood. In such cases, introduced species are sometimes used.

The main requirements of a cover crop are:

- Seed should be readily available;
- Seed should germinate rapidly;
- Species must be capable of growing in the prevailing climate and soil conditions;
- Following germination, the plants should be capable of providing an extensive groundcover within the initial growing season;
- Species used should not have the potential to become 'weeds';
- Preferably, the plants should become established without the need for supplementary watering.

Species used should not have the potential to become 'weeds'.

Pioneer species are preferred because they are naturally suppressed as other species become established by plant succession. Many are annuals and will continue to reproduce each year until perennial species become established. Legumes are favoured because they add nitrogen.

The requirement to avoid establishing weeds does not relate solely to exotic species. One of the most widespread plants in South East Asia is *Imperata cylindrica*, known in Indonesia as *alang alang*. This species is easily established in disturbed areas where it rapidly achieves total coverage. However, it is allelopathic and therefore patently unsuitable for use in mine land rehabilitation. Once established, it will preclude the establishment of all other species. Not only should the species not be used but in many areas it is necessary to prevent it from invading areas being reclaimed.

In contrast to *alang alang*, there are several perennial legumes that are commonly used for mine land rehabilitation in Indonesia and elsewhere. A list of widely used legumes, their main characteristics and requirements, is provided in **Figure 21.2**. Commonly used legumes in SE Asia are Centro (*Centrosema pubescens*), Bejuco (*Calopogenium caeraria*) and Puero (*Pueraria phaseoloides*), which are commonly combined in the same seed mixture. Centro and Bejuca are both native to South America but have become naturalized in many parts of SE Asia, where they have been used for erosion control following logging operations. Puero is native to East and SE Asia. All germinate rapidly following rainfall and grow vigorously to form dense, leafy thickets totally covering the surface in the first year if planted at relatively high rates (>4 kg/ha), or in the second year if sown at lower rates. Centro has the additional advantage of being drought tolerant while both Bejuco and Puero can tolerate acidic soils. Both Bejuco and Puero have a climbing, twining habit, while Centro has a creeping, sprawling habit. Bejuco has the potential to become a weed species in a grazing situation, because it is generally unpalatable. However, in reclamation where forest is being established, all three species tend to disappear over time as other herbs (particularly ferns), shrubs and trees become established (**Case 21.4**). Apart from their use as groundcover for erosion control and nitrogen addition, these plants are also suited to the provision of green manure, which is usually achieved by ploughing to bury the foliage in the surface soil layer. This can help to build and improve the topsoil layer.

	Plant Characteristics									Adaptation to the Climate								Adaptation to the Soil								Use				
										Temperate Zone				Tropical and Sub-tropical Region																
	Annual	Landscape/Visual	Perennial	Upright habit	Creeping	Winding, Climbing	Poisonous	Slightly poisonous	Drought resistant	Artic zone	Excessive rainfall	Only winter rains	Only summer rains	Excessive rainfall	Moderate rainfall	Irrigation	Arid and semi-srid regions	Light soils	Heavy soils	Limey soils	Shallow soils	Well drained soils	Wet soils	Alkaline soils	Acid soils	Fodder crop	Pasture crop	Human food	Green manuring, groundcover	Soil conservation (erosion control)
Arachis hypogaea	●			●	●									●	●	●		●				●				●	●	●	●	
Cajanus cajun			●	●					●					●	●		●	●								●	●	●	●	
Calopogonium mucunoides			●	●	●									●	●								●				●	●	●	
Cavanalis ensiformis	●			●		●		●	●					●	●	●										●		●	●	
Casia hirsula			●	●			●							●	●														●	
Centrosema pubescens			●		●	●			●					●	●		●					●				●	●		●	●
Cicer arietinum	●							●	●			●	●	●	●	●			●			●				●		●	●	
Crotalaria anagyroides			●	●					●					●	●										●				●	
Crotalaria juncea	●			●				●	●					●	●	●									●	●	●	●	●	
Crotalaria mucronata			●	●										●	●										●	●	●		●	
Crotalaria spectabilis			●	●			●		●					●	●			●							●				●	
Crotalaria usaromoensis			●	●										●	●										●				●	
Desmodium adscendens			●											●	●							●					●		●	
Dalichos lablab			●		●	●		●						●	●											●	●	●	●	
Glycine javanica			●			●								●	●											●	●		●	
Lens esculenta	●			●							●	●	●	●	●	●		●	●									●	●	
Leucaena glauca			●	●			●							●	●				●			●	●			●	●	●	●	●
Medigaco arabica	●			●	●							●								●		●		●			●		●	
Medigaco sativa			●	●					●		●	●	●		●	●			●	●		●		●		●	●	●	●	
Mimosa invisa	●	●	●	●	●	●								●	●														●	
Pueraria phaseoloides			●		●	●			●					●	●				●				●			●	●		●	●
Sesbania aculaeta	●			●										●	●				●			●	●		●				●	
Sesbania exaltata	●			●								●	●	●	●	●													●	
Sesbania macrocarpa	●			●					●			●	●	●	●	●			●			●	●			●	●	●	●	
Stizobolium aterrimum	●				●				●					●	●			●	●		●					●	●	●	●	
Stizobolium deeringianum	●	●	●		●	●			●			●	●	●	●			●	●		●					●	●	●	●	
Stylosanthes gracilis			●	●	●				●					●	●			●	●		●	●	●		●	●	●		●	
Tephrosia candida			●	●			●							●	●								●					●	●	●
Vigna oligosperm	●	●	●		●	●			●			●	●	●	●											●	●	●	●	

FIGURE 21.2

Widely Used Legumes, their Main Characteristics and Requirements

Source: Hypers *et al.* 1987

For effective erosion control, groundcover at or close to 100% is preferred.

The ground coverage that can be expected from cover crops depends largely on climatic factors, particularly rainfall. For effective erosion control, groundcover at or close to 100% is preferred. In tropical areas and in wetter parts of the temperate climatic zones, coverage of 100% in the first wet season can usually be achieved. In arid areas, such coverage can only be achieved by means of regular watering. If practicable, such watering in the first year can certainly be most beneficial. However, for economic reasons, watering is seldom continued beyond the first year.

Species Selection

For rehabilitation programmes aimed at restoring natural vegetation, most if not all the species planted will be those that occur naturally in the immediate vicinity. For many of these species, seeds can be harvested manually and either planted directly or first established in a nursery, prior to planting (**Figure 21.3**). For those species from which seed cannot be readily harvested or that defy attempts at germination, natural seedlings are sometimes excavated and grown in pots in a nursery before being planted out. As mentioned previously, there is a practical limit to the number of species that can be planted in any given situation. Many more species will appear over time, either from seeds stored in the topsoil (if applied), by invasion from nearby areas, or introduction by birds.

Outside their natural habitat where populations are controlled by natural mechanisms, many species proliferate at the expense of more vulnerable native plants.

Considerable controversy attends the use of introduced or exotic species in situations where natural vegetation is to be established. This is because of the severe and widespread ecological damage that has been caused by many exotic species introduced for agricultural or decorative purposes. Outside their natural habitat where populations are controlled by natural mechanisms, many species proliferate at the expense of more vulnerable native plants. However, there are situations where plants suitable for rapid revegetation of a given site, are not readily available within the local natural plant community. In such cases, the use of introduced species may be considered. Any introduction of exotic species should only be made after careful consideration of the characteristics and requirements of both the candidate introduced species and the species to be established in the longer term.

CASE 21.4
Rehabilitation at Mt Muro Gold Project in Indonesia

The Mt Muro Gold project in Central Kalimantan, Indonesia involved mining of multiple small pits. Waste rock was used partially as pit back-fill and was also placed in six waste rock dumps which were constructed primarily using bottom-up methods, and were rehabilitated progressively. Accordingly, by the time that operations ceased in 2002, considerable experience had been gained, and rehabilitation procedures had been well established. The target land use for rehabilitated land, as required by the Ministry of Forests was for Production Forest, and revegetation prescriptions were directed to this end. Following placement of topsoil, a legume seed mixture was applied as initial groundcover. Subsequently, in the second year, seedlings of fast growing tree species, including an Albizzia sp and a Eucalypt, were planted. Monitoring indicated that 100% groundcover was achieved in the first year and maintained thereafter. After five years, (see photo) the fast growing timber trees had reached more than 10 m in height and provided a canopy cover of more than 50%. Furthermore, the initial cover crop legumes had all but disappeared and volunteer species, including ferns substantially outnumbered the planted species. Local people had already commenced harvesting of the most advanced trees (*Albizzia*).

FIGURE 21.3
Nurseries are Surprisingly Small, Even for Large Mines

Legumes and Grasses

Legumes are commonly used in rehabilitation; some as discussed above make good cover crops. Others can be used to make hedges, or as individual shrubs or trees. Most vegetation communities benefit from inclusion of one or more legume species because of their nitrogen fixing properties. Grasses are also used as components of many rehabilitation vegetation communities. Most grasses have good soil binding properties while their seed provides food for a variety of insects, birds and small mammals. Clearly, if local grasses are available, these should receive preference. However, there may be difficulties in establishing local grasses in the short-term, in which case consideration may be given to temporary use of carefully selected exotic grasses. Hybrid grasses which are sterile, are obvious candidates for consideration. A readily available example is Regreen, a sterile wheat/wheatgrass hybrid.

If local grasses are available, these should receive preference.

Vetiver Grass

Vetiver grass (*Vetiveria zizanoides*) is another widely used species in mine land rehabilitation, which has proved to be both highly effective and non-invasive (World Bank 1993). This densely tufted grass has been used for erosion control in many tropical and semi-tropical situations, particularly on moderate to steep slopes where it is planted along the contours to form dense hedges (**Figure 21.4**) which interrupt overland flow and trap suspended sediment. The species is extremely versatile, being tolerant of extended flooding as well as drought conditions.

Vetiver grass (*Vetiveria zizanoides*) is widely used species in mine land rehabilitation, which has proved to be both highly effective and non-invasive.

This is not a short-term solution to erosion control as it commonly requires two or three years to establish dense, continuous hedges. However, as this is a perennial grass which is sterile outside its natural swampy habitat, it is suitable for retention as part of the long-term vegetation community in both agricultural and natural vegetation applications. In a handbook entitled 'Vetiver Grass – The Hedge Against Erosion', The World Bank (1993) compares the use of Vetiver Grass hedges to constructed Earth bunds or banks. A particular

FIGURE 21.4
Vetiver Grass Hedges Established on Slope of Waste Rock Storage

advantage is the much lower requirement for surface drains where Vetiver Grass is used. Further, while surface banks fill over time, the hedges are self-maintaining.

Plant Succession

Plant succession is the natural progression in which different plant communities appear over time on the same area. Each time that the vegetation is removed, for example by fire, logging or landslip, the succession process commences with the appearance of the pioneer community. Subsequently, several different communities may come and go before the climax community becomes established.

Succession is a long, slow process requiring decades, even centuries to complete each cycle.

Attempts are often made to 'short circuit' the succession process by transplanting seedlings of climax tree species early in the rehabilitation process. These attempts are usually unsuccessful as the climax species have particular requirements that preclude their establishment until the final stage of succession. In particular, many climax canopy species in the humid tropics require dense shade for the first few years of their lives. Succession is a long, slow process requiring decades, even centuries to complete each cycle. Accordingly, it is not possible for natural vegetation to approach climax conditions within a few years of rehabilitation. It is important that this is recognized in the ongoing management procedures, in interpretation of monitoring data and in the selection of acceptance criteria.

Biodiversity of Rehabilitated Areas

A common goal of rehabilitation is to produce biodiversity comparable to that which occurred prior to mining. However, for some of the reasons already discussed this can only be a long-term goal. In a tropical forest situation, primary forest communities at or close to climax condition, have a relatively low diversity of plants. Secondary forests undergoing succession following logging, commonly have much higher diversity. The maximum diversity results from a mixture of forest communities, with different aspects and soil types, and with areas at different stages of succession.

In assessing the success of rehabilitation, the diversity of plant species is not particularly important in the first few years. What is important is that the number of species increases from one year to the next. The same applies to fauna. Some insect species will rapidly colonize even newly established vegetation, particularly if natural vegetation occurs nearby. However, it will be many years before the faunal community on reclaimed sites will be comparable with undisturbed natural areas. Again, the important thing to establish is whether or not there is a trend toward increasing diversity.

The important thing to establish is whether or not there is a trend toward increasing diversity.

Culturally Important Plants

Indigenous Peoples throughout the world have used specific plants for a variety of purposes including sustenance, medicines, decoration and ritual observances. Tropical forests, in particular, contain many plant species that are used by the local Indigenous Peoples. Where mining takes place in these situations, it is important that culturally important plants are included in the rehabilitation programmes. To this end, an ethno-botanical survey is undertaken as part of the original baseline study. From this study, a list of culturally important species is compiled with details of the nature and occurrence of each. Depending on the species, seed can then be harvested or, in some cases seedlings can be removed from areas to be mined and transplanted, for use in rehabilitation.

Progressive Rehabilitation versus Post-Closure Rehabilitation

Strip mining for coal, shallow open Pit mining for bauxite and 'bottom-up' waste rock dumping practices all provide the opportunity for progressive rehabilitation, rather than waiting until mine closure. There are several significant advantages to progressive rehabilitation, including:

- Alternative practices and treatments can be tried and compared, with experience gained from each rehabilitated area leading to improvements in subsequent efforts;
- Unsatisfactory results can be identified and readily rectified;
- Topsoil can be used immediately, rather than after a period of stockpiling, with the result that more of the contained seeds and micro-organisms will be viable.
- In many cases – such as where Vetiver Grass is used – newly rehabilitated areas may provide propagules for subsequent rehabilitation;
- The full complement of mine staff and equipment are available to be involved in both implementation and evaluation of rehabilitation efforts and in any supplementary treatments;
- In the case of 'bottom-up' waste rock dump rehabilitation, vegetation on the lower benches assists in erosion control;
- Less funds need to be accumulated for closure, and
- Once successful results have been achieved, costs for the balance of areas to be rehabilitated after closure can be estimated with confidence.

Alternative practices and treatments can be tried and compared, with experience gained from each rehabilitated area leading to improvements in subsequent efforts.

There is usually no potential for progressive rehabilitation of tailings storages, except where multiple cells are filled sequentially. Accordingly, tailings rehabilitation may be the major task remaining following mine closure. Filtration of tailings to produce filter cake, promises to offer considerably more options in tailings management, including the possibility of progressive rehabilitation.

FIGURE 21.5
Surface of Tailings Storage, Showing Regrowth of Acacia sp. Following a Wild-fire, Tennant Creek Area, N.T., Australia

Sustainability

Rehabilitation of natural vegetation requires that the resulting vegetation community is self-sustaining. This depends on many factors, including the following:

- The total assemblage of species including micro-organisms, and fauna;
- The ability of species to survive, at least to first reproduction, under the prevailing conditions;
- The ability of seed to germinate under the prevailing conditions;
- Genetic diversity within each species;
- Absence of invasive weeds and serious pathogens;
- Resilience to disturbance – the ability of the vegetation community to recover following natural or man-made perturbations such as flooding or fire (**Figure 21.5**), and
- Recycling of nutrients.

Nutrients may be absent or at very low levels, requiring at least an initial input of essential nutrients at the commencement of rehabilitation, particularly if mulch has been applied.

The sustainability of agricultural or production forestry communities depends on many of the same factors. However, the substantial harvesting of biomass means that nutrient recycling is relatively limited, so that on-going additions of fertilizer are required. In other words, these artificial systems are sustained by repeated human intervention; there is no requirement that they be self-sustaining.

Acceptance (or Completion) Criteria

Acceptance criteria are the pre-determined conditions to be met to demonstrate that the overall long-term objective of rehabilitation has been achieved. In the case of natural vegetation which may take decades to develop, the acceptance criteria are selected to demonstrate

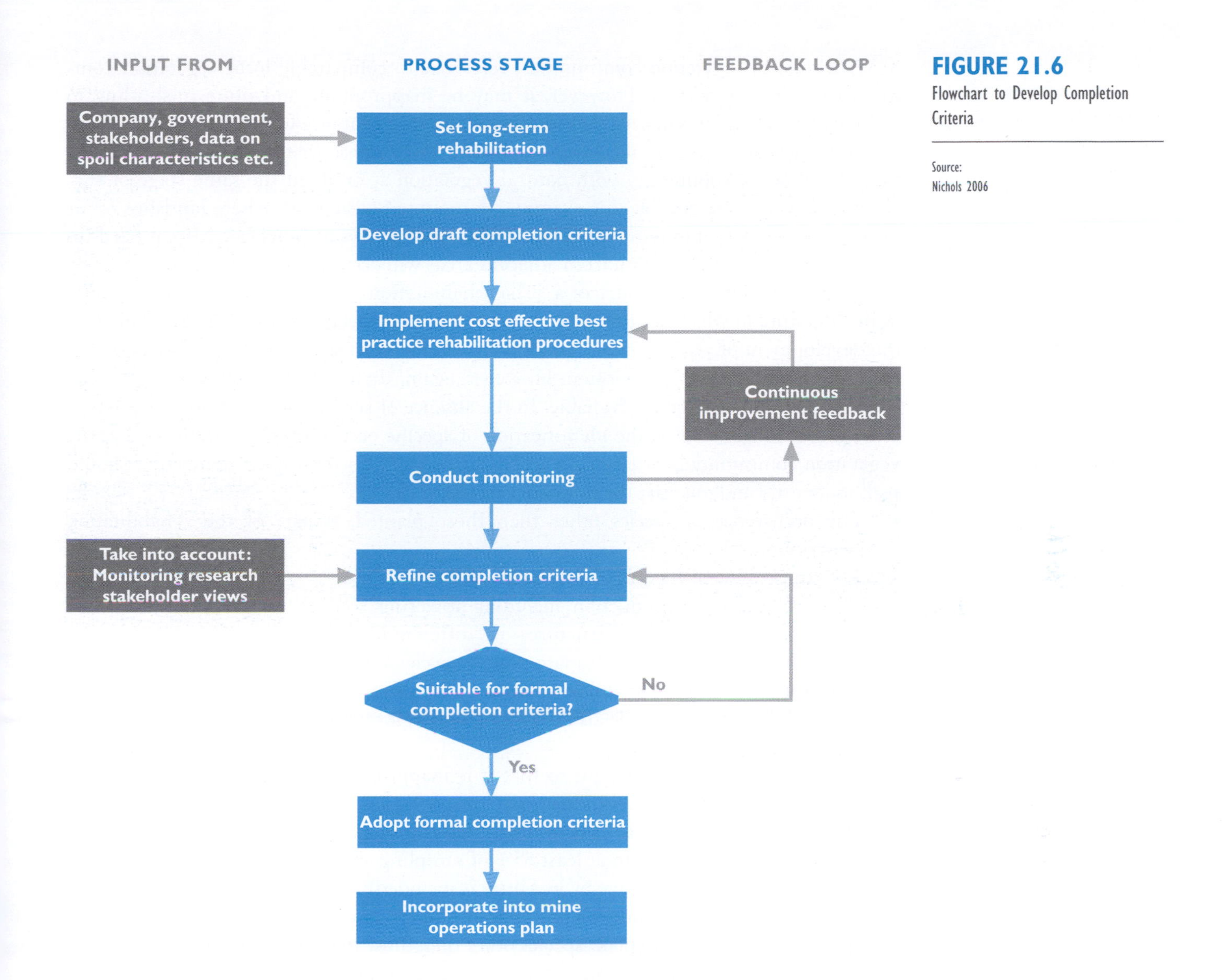

FIGURE 21.6

Flowchart to Develop Completion Criteria

Source:
Nichols 2006

that the succession processes are underway, that will lead to a sustainable ecosystem. In some cases, the attainment of one or more specific acceptance criteria, represents a regulatory milestone, usually signalling the conclusion of the maintenance period for rehabilitation and/or the time for transfer of ownership. Acceptance criteria should be objective, readily measured and reproducible. Because of the inherent variability in any natural system, statistical acceptance measures are preferred to absolute (pass or fail) criteria. Stages involved in developing acceptance or completion criteria are shown on **Figure 21.6**.

In some cases, the attainment of one or more specific acceptance criteria, represents a regulatory milestone.

The selection of acceptance criteria warrants careful consideration as criteria that are inappropriate or insufficient may mean that the final rehabilitation does not achieve its objective(s). On the other hand, overly onerous acceptance criteria may be impossible to achieve, even if the overall objectives are achieved.

With agricultural systems, acceptance criteria should be based on measures of production and quality compared to local agricultural practice. Year to year variation should be taken into account. For example, a crop yield should be compared with local crop yields in the same year. An acceptance criterion might therefore be 'a yield of 90% of the average yield (for a particular crop) in (a specified local) locality in the same season'.

It may be inappropriate and quite misleading to compare the parameters in a 4-year old rehabilitation community with nearby natural vegetation that is at or close to its climax condition.

With natural vegetation communities, reference to comparable local vegetation communities is also important. However, it may be inappropriate and quite misleading to compare the parameters in a 4-year old rehabilitation community with nearby natural vegetation that is at or close to its climax condition. More appropriate would be to compare the rehabilitation community with natural vegetation at or about the same stage of succession. This may be possible, for example, in mountainous terrain where landslips occur during the wet season in most years, but this will not always be practical. Also, it needs to be recognized that colonization from adjacent areas will occur much more rapidly in a 1 ha forest clearing than in the centre of a 50 ha rehabilitation area. There are many examples in the literature in which the details of recolonization have been documented. An example is the development of vegetation on Krakatoa Kecil, the new island that emerged following the catastrophic Krakatoa explosion. However, again, there will be many cases where no specific quantitative data are available. In the absence of such information, the acceptance criteria should be based on the identification of specific occurrences that indicate that the vegetation community is progressing according to plan. For example, monitoring should seek to identify and quantify the following trends:

- The occurrence of species other than those planted as part of the rehabilitation prescription;
- The suppression of cover crop species, if used;
- Signs of successful reproduction, including flowering, seed set and germination;
- Development of vegetation structure – i.e. different layers;
- Appearance of seedlings of ultimate canopy species, and
- Appearance of fauna, particularly insects such as ants and termites that are important in recycling nutrients, while also attracting higher level predators.

Clearly, each situation is different, so that it is inappropriate to suggest a generic set of acceptance criteria. However, for illustrative purposes, a hypothetical set of acceptance criteria for rehabilitation of a min esite in the humid tropics could be:

- No bare ground exposed in at least 85% of sampling sites, at end of wet season (access tracks are excluded, coverage by leaf litter is included);
- Progressive increases in floristic diversity over at least four years;
- Occurrence of at least 20 plant species other than those in initial cover crop;
- Surface coverage by original cover crop species less than 5%;
- Occurrence of a canopy involving at least four species;
- At least 60% tree canopy cover (these need not necessarily be final canopy tree species); and
- Occurrence of at least two climax species as seedlings or saplings.

No Rehabilitation without Maintenance

Rehabilitation is at its most vulnerable before, during or immediately following planting, germination and emergence.

It is rare for rehabilitation to be totally successful after a single preparation/planting programme. This is particularly true of initial rehabilitation efforts. More commonly, supplementary plantings and various maintenance treatments are required in order to achieve acceptable groundcover in the early stages, and to ensure the development of appropriate vegetation structure and diversity in later stages. Rehabilitation is at its most vulnerable before, during or immediately following planting, germination and emergence (**Figure 21.7**). Heavy rain at such times can erode the seed bed soil, together with most of the seed, leaving no choice but to repeat the programme. Mulch, jute netting and other temporary soil covers can reduce but not eliminate the risk of such occurrences.

FIGURE 21.7
Development of Gully Erosion at the Early Stages of Rehabilitation

Adequate design of land form and drainage followed by years of maintenance is a prerequisite for successful rehabilitation.

Supplementary plantings are commonly made in the case of canopy trees, including some climax species, which need to be shaded during the first few years of establishment. These species would not survive if planted in the initial programme.

Frequently, routine monitoring of rehabilitated areas reveals bare patches requiring follow-up treatment. It helps if the reason for bare patches can first be deduced so that follow-up treatments can be selected which overcome whatever factor prevented the vegetation from establishing. Common causes of bare areas include: uneven distribution of seed, gaps in topsoil application, the presence of acidic material at or near the surface, or localized erosion. Follow-up treatment may also be required in cases where vegetation dies after the initial establishment. This may be caused by a phyto-toxic agent including acidic soil or seepage, or by dessication due to dry conditions. Lime can be applied to neutralize small acidic areas. In seasonally dry areas, it is common for water to be applied during the first dry season after the initial establishment.

Other types of maintenance that may be applied include weed control, and fencing to exclude grazing by wildlife or livestock during the period that vegetation is becoming established.

For agricultural or silvicultural land uses, repeated cycles of planting and harvesting with appropriate maintenance, are normal, and apply whether or not these systems are established on rehabilitated land. However, where natural vegetation is being established, it is important that it becomes self-sustaining, which means that maintenance should not be required beyond the first few years.

Natural Rehabilitation

Natural processes of weathering, erosion, deposition, plant succession and faunal invasions will eventually result in the development of sustainable vegetation communities over all areas disturbed by mining, although the time required may be centuries or millennia.

Rehabilitation of areas disturbed by mining occurs naturally, without any intervention from man. Natural processes of weathering, erosion, deposition, plant succession and faunal invasions will eventually result in the development of sustainable vegetation communities over all areas disturbed by mining, although the time required may be centuries or millennia. The purpose of rehabilitation procedures and measures described in this chapter is essentially to accelerate these natural processes and to avoid the damage, particularly offsite damage that could occur in the meantime.

There have been many examples of historic mining operations where natural processes restored the affected areas. For example, in the gold fields of central Victoria, Australia, where extensive mining operations were carried out in the mid-nineteenth century, little can now be seen to indicate that such activities took place. However, these mines were very small in comparison to many modern mines. In this context, Bougainville Copper Mine in Papua New Guinea provides a unique and fascinating recent example of a large mining operation that was terminated abruptly with no subsequent rehabilitation measures. As shown in **Case 21.5**, natural rehabilitation has occurred since mining and tailings discharge ceased in 1989.

CASE 21.5
Bougainville Copper Mine, PNG

The Panguna Mine on the island of Bougainville is interesting due to the fact that dissatisfaction with environmental damage due to the riverine tailings disposal and the social impacts of mining caused the mine to be closed in 1989. For several years afterwards, Bougainvilleans were embroiled in a civil war with the Papua New Guinean state. Even now with peace restored, the prospects of the Panguna mine re-opening appear very slim. At its height, the Panguna mine drove much of Papua New Guinea's economy.

The right-hand satellite image of 1989 shows the mine at its greatest extent, just before it was closed. Tailings affecting the Jaba River can be clearly seen, as can the delta formed by deposition of sediments in the sea.

The image from 2001 shows changes in the extent and shape of the delta of the Jaba River, after twelve years. The image also shows the natural revegetation of tailings deposited along parts of the river valley. Today, all mines in Papua New Guinea have closure plans which require the rehabilitation of areas impacted by mining; but on Bougainville, due to civil war, this process could only happen naturally.

Notwithstanding evidence of promising recovery, large barren areas remain. It is clear from comparison of the images that vegetation is advancing over the disturbed areas, from the nearby undisturbed areas. Narrow areas have already been vegetated while the remaining barren areas, require more time to be vegetated.

May 2001

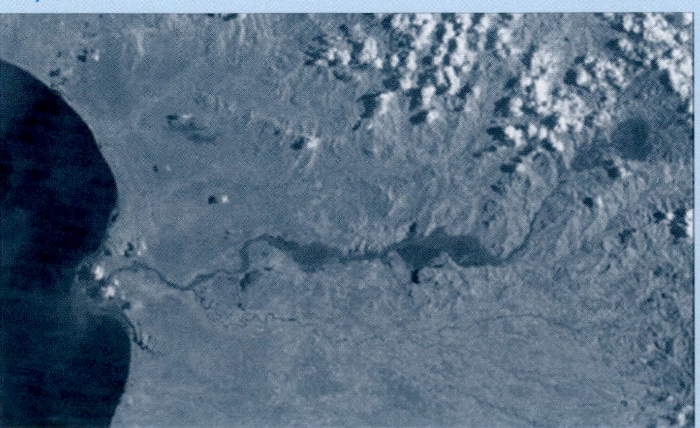

March 1989

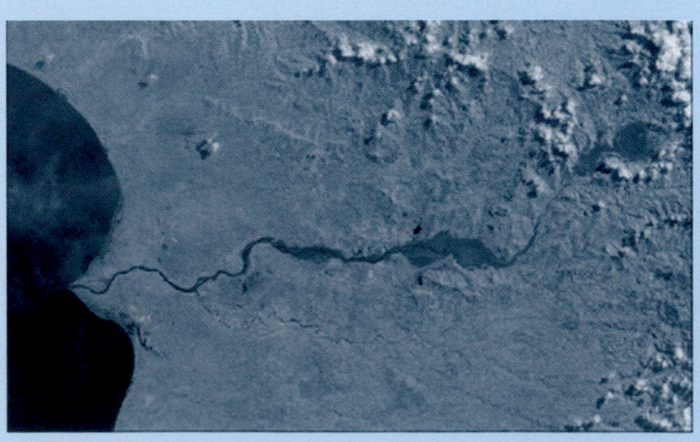

Source: Jacka 2007

21.5 PIT LAKES

The chemistry of a pit lake is, in part, a reflection of the groundwater that existed prior to mining as well as the reactions of the surrounding rock that host the pit lake. Since most pit lakes do not have much surface water flowing directly into them, groundwater comprises most of the influent water. Groundwater surrounding the mine provides the appropriate comparison for water quality in the pit lake.

Three main factors influence the chemistry of the final pit lake (the following discussion is partly drawn from Miller 2002).

First, an important factor of pit lake water quality is the interaction of groundwater with the wall rock around the pit. In an open pit, oxidation reactions on the exposed walls may release sulphate, acid and metals into the lake. Additionally, when a pit is excavated below the regional water table, the aquifer in the host rock is dewatered. If present sulphide-rich host rock will oxidize when exposed to air that is pulled into the evacuated pore spaces, generating reaction products on the exposed surfaces. As the aquifer recovers following mining, those oxidation products will be flushed into the pit lake by groundwater flowing into the cone of depression, initially as acid rock drainage.

Second, lake chemistry is influenced by the pH value. In fact the pH of the pit lake is the single greatest determinant of the water quality, or toxicity, of a pit lake. As oxidation of the pit wall rock releases sulphuric acid, calcium carbonate that may be present in the wall rock also dissolves and neutralizes the acid. Where sufficient carbonate is present, the water that enters the pit lake will be near-neutral; if there is a lack of calcium carbonate, the pit lake will be acidic. Metal solubility, particularly of divalent problematic metals such as cadmium, nickel, zinc and copper, is much higher at low, acidic pH and can render water quality poor.

Third, pit lake stability is important. The stability of the pit lake depends on the comparison of the stabilizing mechanisms and the disruptive forces, commonly expressed in terms of the densimetric Froude Number. Fully mixed lakes have a Froude Number greater than 1.

Forces that may disrupt stratification of a pit lake include wind and wave action, flow through due to runoff into the lake pit, seasonal temperature fluctuation, or potentially thermal sources at deep pits that potentially could warm the water. Seasonal atmospheric temperature fluctuation, the cause of mixing of lake waters at higher latitudes, is not a significant factor for tropical lakes. Mechanisms that enhance stable stratification of the pit lake include density stratification, relative constant temperature throughout the year and a small surface area relative to volume and depth. As a result of this stratification, higher salinity water containing most of the heavy metal ions may occur in an anoxic layer at the base of the pit where reduction reactions may cause metals to precipitate.

The environmental problems related to pit lakes have been recognized in the United States and elsewhere for only 20 years. PHREEQC, a model developed by the United States Geological Survey (USGS) finds wide application in simulating the lake hydraulics and chemistry. This model has the ability to simulate the pertinent processes occurring in a pit lake, such as mixing of multiple solutions, precipitation/dissolution of selected solids, redox reactions, evaporation, atmospheric interaction and adsorption of metals.

It is now recognized that pit lakes represent a commitment of groundwater resources in perpetuity. Pit lake management is important, particularly in arid climates where water is the limiting resource for agriculture, domestic watersupplies and wildlife needs. However, pit lake management is still very much the topic of research. Lime dosing of the pit water is one measure to improve short-term water chemistry. Accelerating the filling of the mine pit (pumping or directing surface water flow to the opening) as in the case of the Island Copper Pit in Canada in 1996 (**Case 21.3**) is another measure.

21.6 SOCIAL ASPECTS OF MINE CLOSURE

The most important aspects of mine closure from the point of view of local communities are:

- Managing workforce reductions to minimize adverse economic impacts;
- The success of Community Development programmes in providing alternative, post-mining livelihoods and income sources for those directly and indirectly affected by mine closure;
- The success of environmental management measures in avoiding or minimizing off-site impacts such as reduced water quality, as discussed in Chapter 8 and other parts of this book, and
- The ultimate return of affected land to sustainable beneficial usage, as discussed earlier in this chapter.

It is clear that all of these aspects, if they are to be handled adequately, require substantial planning and investment, well before closure takes place. Workforce reductions, while unavoidable, pose particular difficulties and require detailed planning. Elements of a workforce reduction plan that have proved beneficial in minimizing adverse impacts include:

- Preparation and communication of employee redundancy schedules, well in advance of their implementation;
- Scheduling the workforce reduction process over as long a period as practicable;
- Development, sponsorship or promotion of alternative employment schemes as part of the Community Development programmes;
- Re-training of employees, as appropriate to meet the needs of alternative employment schemes, and
- Accrual of funds for use in future, as yet unidentified initiatives, possibly administered by a reliable NGO or other third party.

Community development programmes, as discussed in Chapter Fifteen, are perhaps the most important contributors to the minimization of adverse social impacts. This applies particularly where Indigenous Peoples are involved.

Community development programmes are perhaps the most important contributors to the minimization of adverse social impacts.

As mine closure issues have come to the attention of mining companies relatively recently, there are very few case studies involving mine closure with successful social outcomes, from which to learn. There are, of course, many examples of poor outcomes from past projects that closed without adequate environmental management, rehabilitation, or Community Development programmes. The approaches suggested above have been based on critical analysis of the outcomes of past projects, combined with a common sense evaluation of practical options for improvement. Clearly, as for other aspects of environmental management with a longer history, lessons will be learnt from each future mine closure, enabling social aspects of mine closure to be improved and also enabling social outcomes to be predicted with more confidence.

REFERENCES

Clark AL and Cook Clark J (2000) An international overview of legal frameworks for mine closure.

Fisher T and Lawrence GA (2006) Development of the Island Copper Mine Pit Lake for the Treatment of Acid Rock Drainage; Fisher T, Cornet Contracting Ltd, London; and Lawrence, Greg A., Department of Civil Engineering, University of British Columbia, Vancouver, Canada.

Fleury A and Parsons AS (2006) Financial Assurance for Mine Closure and Reclamation, in Proc. 1st Int. Seminar on Mine Closure, Perth, Australia, 13-15 September, 2006 Eds: Fourie A and M Tibbett.

Husin Y, Susetyo W, Puradyatmika P, Sarwom R, Macpherson J, and Chamberlain D (2005a) Reclamation and Natural Succession In PTFI Tailings Deposition Area, Indonesian Mining Conference & Exhibition, September 21 – 28, 2005, Jakarta.

Husin Y, Chamberlain D, Macpherson J, Puradyatmika P, and Sarwom R (2005b) Potential of PTFI Tailings Deposition Area For Non-Conventional Estate Crops, Seminar Sehari Kebijakan Fiscal Unutuk Meningkatkan Kinerja Industri Perkebunan, Jakarta, 24 November, 2005.

Hypers H, Mollema A and Egger T (1987) Erosion Control In The Tropics, Agrodok 11, Agromisa foundation, Waginingen, the Netherlands, April 1987.

IFC (2002) It's Not Over When It's Over: Mine closure around the world, International Finance Cooperation, Washington DC, USA.

Jacka J (2007) Environmental Impacts of Mining at Panguna, Bougainville, Department of Sociology and Anthropology, North Carolina State University, available at www4.ncsu.edu/~jkjacka/panguna.htm

Khanna (2000) Mine closure and sustainable development, Mining Journal Books Ltd, The World Bank Group, Mining Department and Metal Mining Agency of Japan, Edited by T Khanna, p. 116–112.

Laurence DC (2006) Why Do Mines Close? In: Proc. 1st Int. Seminar on Mine Closure, Perth, Australia, 13-15 September, 2006 Eds: Fourie A and M Tibbett.

Nichols OG (2006) Developing Completion Criteria for Native Ecosystem Reconstruction – A Challenge for the Mining Industry, In: Proc. 1st Int. Seminar on Mine Closure, Perth, Australia, 13-15 September, 2006 Eds: Fourie A and M Tibbett.

Miller GC (2002) Precious Metals Pit Lakes: Controls on Eventual Water Quality; September/October 2002, Southwest Hydrology; Department of Environmental and Resource Sciences, University of Nevada, Reno.

UNEP and Chilean Copper Commission (2001) Abandoned Mines Problems, Issues and Policy, Challenges for Decision Makers, Santiago, Chile, 18 June 2001, Summary Report.

USEPA (2000) The Abandoned Mine Site Characterization and Cleanup Handbook, EPA 910-B-00-001.

World Bank (1993) Vetiver Grass - The Hedge against Erosion, The World Bank, Washington, DC. Fourth Edition.

• • • •

22

Looking Ahead

SILICON

Silicon, a brittle, gray metalloid, is the most abundant metal and the second most abundant element in the Earth's crust. Relatively unreactive, it occurs in nature as the oxide form, silica, of which quartz is the most common mineral, and in numerous silicate minerals which include many of the most common constituents of igneous and metamorphic rocks. Silicon is essential to many living organisms, particularly plants and many marine invertebrates. It is not toxic to humans. The majority of silicon is used in specialized alloys. However, the most important applications are in electronics where silicon is used in transistors, computer chips and photovoltaic cells. Silica is used to make glass, cement and refractory materials. Silicon metal is produced from high purity silica in an electric arc furnace.

22 Looking Ahead

Given the massive capital involved, the mining industry is not given to frivolous or fashionable change. However, when new and proven technology becomes available, technology that will increase productivity, the industry will move as quickly as financing systems allow to put it in place. This closing chapter will consider key changes that are possible in the mining industry, not only in technology but in structure, in the near to medium term future, as well as discussing the environmental consequences of these changes. The chapter will also reflect on the role the environmental impact assessment practice may play as the industry moves to re-establish itself in the eyes of the public, not only as an essential element in modern industrial society but as an environmentally responsible, international citizen. Finally consideration is given to current global economic change and its ramification for the mining industry and for the expected environmental performance of new mines.

22.1 EXISTING TRENDS IN THE MINING SECTOR

Bigger and Better?

The large mining companies have grown large, generally, because they have successfully operated multiple mining projects over long periods and, in the process, have become financially strong. This implies that they have been at least moderately well managed and that their successful projects have outnumbered their unsuccessful ones. With few exceptions, large projects, although not necessarily discovered by large companies, are operated by large companies. Exceptions do occur, such as the Grasberg Copper and Gold Project in Indonesia which was discovered and developed by a medium sized company – Freeport McMoran – that on the strength of this large project has become a major producer. More to the point, only large companies have the financial capacity to undertake such multi-billion dollar developments. A major benefit of their financial strength is that large companies are unlikely to renege on their environmental obligations and commitments, even for those projects that, for one reason or another, are prematurely closed as unprofitable.

Only large companies have the financial capacity to undertake multi-billion dollar developments.

Another strength of large companies is the research and development that they sponsor, and the internal training they provide, much of which subsequently benefits others in the industry. Similarly, large companies contribute substantially by providing funding and personnel resources to professional societies and technical symposia, conferences and workshops. In many cases, they set an example for other companies to follow.

In developing countries, large mining companies are often seen as flag-bearers for their home countries. It is common for them to be accompanied in their ventures by service and equipment suppliers from their home countries, thus contributing to broader trade links.

Given the stakes, large companies are naturally cautious, tending to 'overkill' so as to leave nothing to chance. In the environmental impact process this may mean that the breadth, depth and duration of studies is much more than is required by local regulators or international practice, and also more than is required for adequate assessment of impacts. While this may be commendable in principle, it can also have the effect of 'raising the bar', making it difficult for the proponents of projects that follow, especially for small companies with limited resources.

Given the stakes, large companies are naturally cautious, tending to 'overkill' so as to leave nothing to chance.

Large companies also attract close attention from international environmental NGOs. This does not lead inevitably to confrontation, but on occasion may result in productive dialogue. There have been instances where NGOs have successfully influenced corporate policy. For example, BHP Billiton undertook not to use submarine tailings disposal after campaigns by NGOs opposed to this disposal method.

As noted in Chapter Four, junior companies often play an important role in exploration. Often staffed by professionals who may have received much of their training and experience in the employ of major companies, junior companies are in many ways complementary to the majors. Large ore bodies discovered by juniors are usually developed by the majors, while the majors frequently sell their smaller discoveries to junior companies. Formal and informal cooperative agreements between the two sectors are not uncommon.

Junior companies usually have relatively limited financial resources. When developing projects in their own right, they are closely controlled by their lenders, which can impose less than optimal operating constraints in terms of environmental outcomes. From an environmental perspective, lack of financial strength presupposes a risk that the company will be unable or unwilling to meet its environmental commitments. There have been many past examples of small mining companies going bankrupt and abandoning their mines with little or no warning. Such 'orphan' sites are widespread in the USA, Canada and elsewhere. The accrual of funds to pay for closure is therefore much more critical in the case of junior companies. In countries requiring project approval based on a formal feasibility study, it would be appropriate for host country regulators to grant approval only to those projects where the feasibility study demonstrates robust profitability under all credible assumptions. Under such a regulatory regime, junior companies would be unlikely to receive approval to develop marginal projects, thus avoiding future 'orphan' sites.

Junior companies. are closely controlled by their lenders, which can impose less than optimal operating constraints in terms of environmental outcomes.

Future Supply and Demand

Demand for most mineral commodities has increased substantially throughout the twentieth and twenty-first centuries, growing steadily except for temporary reductions during recession periods. As shown in **Table 22.1**, the worldwide production of commonly used metals increased by between 420% and 4,380% between 1900 and 2004.

Despite improvements in recycling, particularly during periods of high commodity prices, demand for most metals and other mineral commodities is expected to continue increasing for the foreseeable future. The current pace of industrialization in China and India, if continued, will ensure that demand increases at even higher rates than those experienced to date. This means that known resources will be extracted at increasing rates,

Demand for most metals and other mineral commodities is expected to continue increasing for the foreseeable future.

TABLE 22.1

Worldwide Production of Common Metals 1900–2004 – Demand for most mineral commodities has increased substantially over the last 100 years.

Metal	Production in 1900 (tonnes)	Production in 2004 (tonnes)
Aluminium (as alumina)	6,800	29,800,000
Copper	495,000	14,600,000
Gold	386	2,430
Iron (as iron ore)	95,500,000	1,340,000,000
Lead	749,000	3,110,000
Nickel	9,290	1,390,000
Zinc	479,000	9,600,000

Source: USGS Statistics

requiring that new resources be discovered and developed at an ever increasing rate to meet future demand.

In the past, there have been dire predictions that the world's resources would be depleted to the extent that particular minerals would become unavailable in the relatively near future. For example, in 1972 the Club of Rome forecast that the world would run out of gold by 1981, mercury by 1985 and tin by 1987. In fact, the world has not run out of any significant mineral commodity and shows no sign of doing so. Resources are closely related to price so that any price increase leads to the identification of increased resources. Given that even after use (nuclear reactions excluded) the inventory of metals in the world is constant, this situation is never going to change. Only the technologies and economics of extraction will change and, if past history is any indication, the cost of extraction for most mineral commodities will continue to decline, in real terms.

If past history is any indication, the cost of extraction for most mineral commodities will continue to decline, in real terms.

The current (2008) high prices for most mineral commodities, which are well above long-term trend lines, are due mainly to the rapid increase in demand, largely from China. This followed many years of oversupply and has caught the industry unprepared. That the supply is taking longer than usual to catch up with demand is due to several factors, including:

- Relatively little exploration took place over the decade prior to the increase in demand, so that there were fewer than usual projects in the planning stages;
- The lead time required to develop a new project has more than doubled over the past 20 years, due in part to environmental permitting requirements, and
- Major mining companies have shown a reluctance to develop large new projects that could, in combination, lead to a return to an oversupply situation, with reduced prices as a consequence. Rather, they have acquired other companies with existing or developing operations, particularly those with long-term strategic value.

Economies of Scale

For most major mineral commodities there has been a trend over the past few decades for an increasing percentage of the world's production to be sourced from a few very large

FIGURE 22.1
The Trend of Ever Larger Haulage Trucks

Despite the lower grades and higher stripping ratios the lower haulage costs and similar economies of scale in milling and treatment processes have reduced the cost in real terms of production of many mineral commodities.

Photo Credit: Jerilee Bennett, The Gazette

mines. For some metals, 90% of production is sourced from 5% of the mines. Much of this phenomenon relates to economies of scale which apply particularly to surface mines. Most of this is the result of using ever larger haulage trucks (**Figure 22.1**). In the early 1970s a haulage truck of 50-tonne capacity was considered large, while 30 years later 350-tonne trucks are widely used in larger mining operations; and it may not be long before 500-tonne trucks are common. Lower bulk haulage costs have had two marked effects:

For some metals, 90% of production is sourced from 5% of the mines.

- They enable much lower grade deposits to be economically mined. (Average grades of copper ores mined today are, in many cases, lower than cutoff grades 40 years ago).
- They allow much higher stripping ratios, which means that deeper mineral deposits which were previously uneconomic, can now be economically mined by open pit methods.

Despite the lower grades and higher stripping ratios, however, the lower haulage costs and similar economies of scale in milling and treatment processes have reduced the cost, in real terms, of production of many mineral commodities. This is the main reason that, recent price rises notwithstanding, the price of commodities such as copper has declined progressively over the past century, as shown in **Figure 22.2.**

As a consequence of this trend, ore that would previously have been mined underground, if at all, is now being extracted in surface mines. At the same time, much more of the resource is now being used; previously, most underground mining extracted only the high-grade ore, leaving much more extensive lower grade mineralization which, in many cases, contained more minerals than the high-grade ores. A good example of this is the Super Pit near Kalgoorlie in the Western Australian Goldfields, where high-grade gold ores were mined by underground methods for many decades. Subsequently, a large open pit was developed to extract the low-grade ores that surrounded the high-grade mineralization.

One of the consequences of this trend toward very large mining operations is that production is being concentrated among a few, very large mining companies, as these are the only entities that can access the huge capital required to develop such projects. The largest operations are often jointly owned by two or more companies.

One of the consequences of this trend toward very large mining operations is that production is being concentrated among a few, very large mining companies.

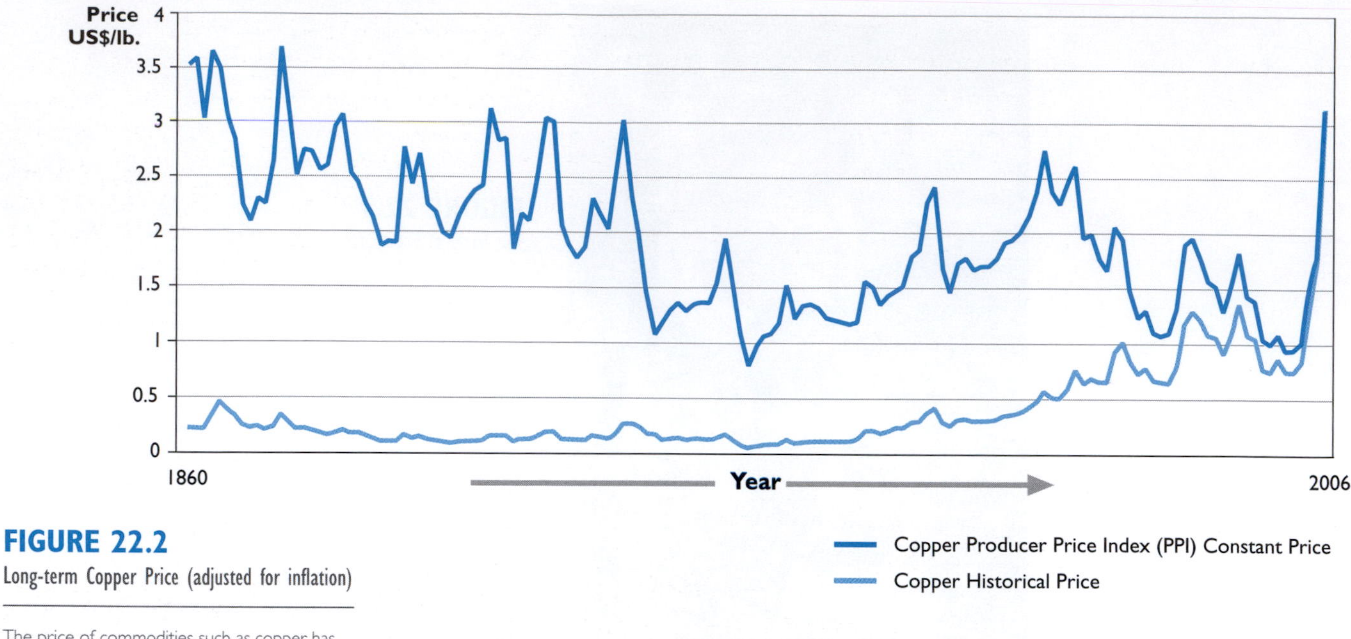

FIGURE 22.2

Long-term Copper Price (adjusted for inflation)

The price of commodities such as copper has declined progressively over the past century.

Source:
International Nuclear INC.

Accordingly, although future demand may lead to the establishment of more mining companies, more and more of the world's production is likely to be from a small number of large mines operated by large companies.

Reduced Mine Life

Despite the fact that many mines are now much larger, the total mine life may be much shorter.

Another consequence of economies of scale is that, despite the fact that many mines are now much larger, the total mine life may be much shorter. The Net Present Value (NPV) of a mineral deposit will vary depending on the rate of production; commonly within practical limits, the higher the production rate, the higher the NPV. This means that mineral deposits which may have previously been mined over 25 years are now being mined over 10 years or less. As discussed in Chapters One and Fifteen of this book, this may have important ramifications in terms of social impacts on local communities.

It is quite common for a mining project to expand its production rate during its operating life. This may be in response to one or more of the following factors:

- Discovery of additional resources;
- Commodity price increase enabling deepening of the mine and/or lowering of the cutoff grade, with consequent increase in tonnage;
- Commodity price decrease reducing or eliminating profitability at the original production rate, or
- The availability of new technology or equipment which changes the economic balance in favour of higher production rates.

Many other factors may also influence the selection of production rate and hence mine life. One is supply and demand. Oversupply will inevitably lead to lower prices, which means that some operations – usually not the ones responsible for the increased supply – will become uneconomic. There is no point in producing more of a commodity than can be sold at a profit. Large companies, in particular, will curtail production in response to declining prices. Other considerations concern the conditions that apply to new mining investments. For example, incentives that provide tax breaks for a number of years and regulations that require divestment of ownership after a period, both encourage maximizing production as quickly as possible. Finally, there are political situations which may influence a company to mine a deposit in the minimum possible time. For example, in politically unstable countries, operators will be concerned about the policies of future regimes, providing a compelling incentive to maximize production in the near term.

In politically unstable countries, operators will be concerned about the policies of future regimes, providing a compelling incentive to maximize production in the near term.

While it would seem that the limits of 'economies of scale' will eventually be reached, the trend toward larger mines and faster extraction is likely to continue in the near term.

The Role of Junior Companies

Junior companies are generally more entrepreneurial and less risk averse than their larger counterparts. Accordingly, they are prepared to explore in areas that, for reasons of perceived low prospectivity or high risk, are avoided by larger companies. Many of the recent mining projects in developing countries are the result of discoveries made by junior companies. It is also true that small companies can profitably operate small projects, something that large companies either cannot do or prefer not to do.

Canada is home to a large number of junior companies, many of which are active outside Canada. Australia also hosts many junior companies, although the majority confine their activities to their home country. The USA produces relatively few junior companies and these are generally not active elsewhere. An increasing number of small companies are being established in the United Kingdom. In addition to these junior public companies, there are a large number of small private companies throughout the world, many of them involved in single local projects.

Small companies may be categorized as those involved purely in exploration and those that aspire to develop their own mining projects. The first of these approaches has been used largely by Canadian exploration companies, which set out to identify and evaluate an ore body for subsequent sale to a larger company. Companies from Australia, the United Kingdom and elsewhere have generally favoured the second approach, which may involve one or both of the following:

- Exploration to identify and evaluate one or more ore bodies, which are then developed by the company itself, or
- Purchase and development of small projects identified by larger companies, but not meeting their internal development criteria.

While exploration activities were once dominated by larger companies, this is no longer the case. In recent years, most exploration, particularly 'green fields' exploration, has been carried out by junior companies, sometimes with prior agreement for a large company to become involved in the event that a substantial ore body is located. Thus, exploration is a major role undertaken by junior companies. There is no indication that large companies will again dominate exploration. However, there is intense competition between companies to increase reserves, and if it appears that grass roots exploration is more effective in this regard than acquisitions, then the situation could easily change.

Environmentally, junior companies exhibit the entire spectrum of performance – from the worst to the best. On the one hand there are small companies that operate without any consideration for the impacts that are generated by their projects. While the numbers of such companies have been diminishing for the past three decades, there are signs that

Environmentally, junior companies exhibit the entire spectrum of performance – from the worst to the best.

the recent escalation of commodity prices is generating a large number of small, privately owned operations, some of which do not subscribe to good environmental practice **(Case 22.1)**. On the other hand, there are many junior companies that are at the forefront of environmental practice and achievement. While large companies tend to develop standards and procedures that apply worldwide, small companies target their practices to the specific circumstances of the location where they operate. This, and the fact that smaller companies tend to be more adaptable and to make more decisions locally, means that the best small companies may perform better environmentally than some of their larger counterparts.

The Emergence of New Technologies

New technologies are continually being introduced into mining and processing operations. Most of these produce incremental improvements in efficiencies and economies. Where they follow established trends, such as in the case of larger capacity haulage or milling equipment, these improvements are readily accepted by the industry. From time-to-time, more innovative technologies are introduced. The mining industry, being essentially conservative, is usually slow to adopt such technologies until their effectiveness is established. Large international companies, due to their broad spread across the world, are exposed to more new technologies than smaller companies. On the other hand small companies tend to adopt new technology more readily than large companies.

Small companies tend to adopt new technology more readily than large companies.

In-pit crushing with conveyor transportation is becoming more common as open Pit mines become larger and deeper. To date, this approach has been applied only to ore; the cost savings achieved from conveyor transport of waste rock do not justify the cost of crushing. In-pit crushing is likely to become even more economically attractive with the availability of far more versatile enclosed (**Figure 22.3**), high angle, tight radius conveying systems such as recently developed by Innovative Conveying Systems International (www.innovativeconveying.com).

The focus of the mining industry has shifted in recent years from merely increasing the tonnes of mined materials to dramatically increasing the amount of product generated per tonne of material removed.

To increase production, the focus of the mining industry has shifted in recent years from merely increasing the tonnes of mined materials to dramatically increasing the amount of product generated per tonne of material removed (National Mining Association

CASE 22.1
Modern 'Gold Rush'

The mining sector is accustomed to business cycles – years of high commodity prices are followed by years of oversupply and depressed market prices. However with commodity prices as high as in 2008, mining does not only attract established mining companies: entrepreneurs with money to spare and the necessary political connections engage in mining. Direct shipment operations (that is mining of ore to be shipped to distant metal producers or energy consumers) often convert large areas to large Earth moving sites. Some operators adopt good industry practices and plan for long-term rehabilitation, other do not. In many respect these mining operations show similar characteristics than the 'gold rushes' of the past: enjoy as long as it lasts.

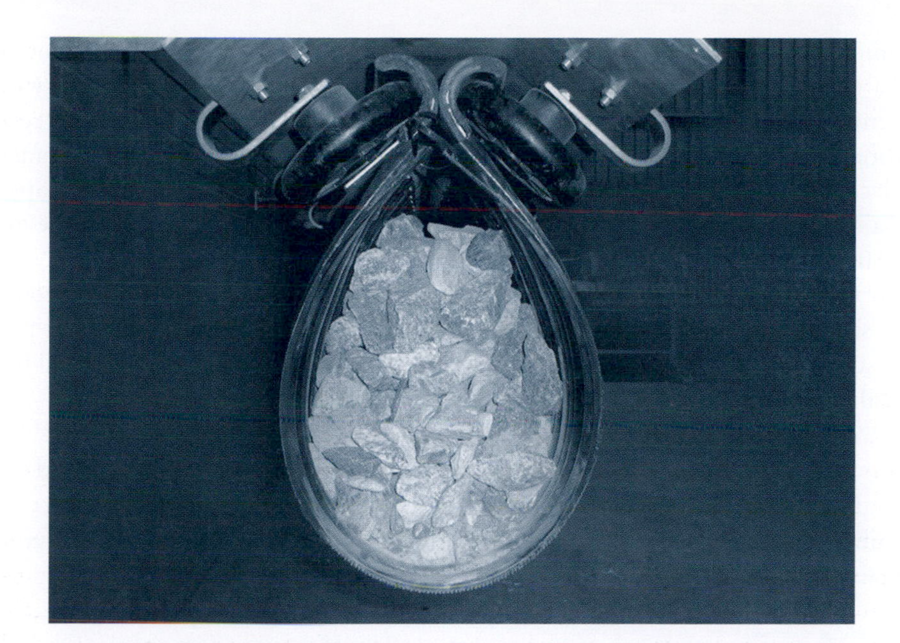

FIGURE 22.3

Cross-section through Enclosed Conveyor System Developed by ICSI

In-pit crushing with conveyor transportation is becoming more common as open Pit mines become larger and deeper, and with the availability of far more versatile conveying system.

1998). This strategy produces less waste and thereby reduces waste treatment and disposal costs. For example, sensing technologies which allow a mining machine to follow a seam of ore may result in less waste being removed. An *in situ* separation process in the mine may result in less waste material handled by a processing plant. Although these advances result in increased productivity, they may also significantly reduce waste generation. Other innovations, in instrumentation and computer hardware and software – particularly Expert Systems – have resulted in progressive improvement across many areas of mining and mineral processing: these include optimizing fuel injection, detecting and diagnosing impending malfunctions, automatically adjusting throughput and the delivery of reagents in response to real-time monitoring of process variables, and responding to upset conditions or emergencies. Such technologies have, in particular, seen the enhancement of recovery rates. In many cases, there have also been improvements in product quality and the minimization of waste. Similar advances are likely to continue, although perhaps at declining rates, as process instrumentation and computing capability continue to improve.

Another major focus of mining research is 'automated mining' including remote control mining using robotics. Mining by remote control has been mainly used in circumstances that are dangerous for miners. These include high-grade underground uranium mines where exposure to radiation is a concern, and very deep mines where 'rock bursts' may occur. Continuous mining machines have been used for many years in coal mining. Recent research has sought to develop continuous mining machines for other applications, including mining of lateritic ores.

More recently, the impetus for remote control of mining and associated operations is being driven by a shortage of experienced people willing to work in remote locations. **Case 22.2** highlights research and development sponsored by Rio Tinto which aims at automating a wide range of activities from blast hole drilling to ore transportation and port loading.

Tailings Management

Non-conventional approaches to tailings disposal are becoming more common in response to otherwise intractable problems associated with conventional tailings disposal. Pulp thickening and filtration to produce a 'filter cake' are being used increasingly, not only in alumina refining where the costs of tailings dewatering are met by the recovery of process chemicals, or in arid areas where scarcity of water requires maximum water recovery, but in many other operations. The flexibility provided by filtered tailings in transportation, placement and tailings storage configuration is only beginning to become apparent. The availability of more versatile conveying systems will further advances in the area. In particular, in the case of strip mining and shallow surface mining, such as many bauxite and lateritic nickel operations, tailings filtration in combination with versatile conveying systems should enable tailings to be returned to mined-out areas, eliminating the need for tailings storage facilities at these operations. Underground disposal of tailings will be similarly facilitated.

There is also a trend toward co-disposal of mine wastes: waste rock and tailings. In most cases this involves constructing large waste rock dumps that also serve to impound tailings. Co-mingling of waste rock and tailings has generally been avoided, as has the placement of alternating layers of waste rock and tailings; to date, such approaches have involved more risks than rewards.

CASE 22.2
Automating Mining Activities from Blast Hole Drilling to Ore Transportation and Port Loading

Key building blocks for automated mine-to-port iron ore operations are being commissioned by Rio Tinto. These include:

- Mine operations in Pilbara to be controlled 1,300 kilometres away at a new centre in Perth;
- Driverless trains to carry iron ore on most of the 1,200 km of track;
- Driverless 'intelligent' truck fleet; and
- Remote control 'intelligent' drills

Major components of the 'mine of the future' are being commissioned in Rio Tinto Iron Ore operations in 2008 and 2009 including establishing a Remote Operations Centre (ROC) in Perth to manage operations in the Pilbara mines hundreds of kilometres away. This allows operators overseeing Rio Tinto Iron Ore mines and process plant facilities to be physically located in Perth, Western Australia. Remote control 'intelligent' trains, drills and trucks will be operational within Rio Tinto Iron Ore during 2008. Humans will no longer need to be hands on as all this equipment will be 'autonomous' – able to make decisions on what to do based on their environment and interaction with other machines. Operators will oversee the equipment from the ROC. The centre will feature an operational control room, office block and supporting infrastructure, and allow for potential significant expansions beyond its initial scale.

Driverless trains

Studies are being finalized on the application of Autonomous Train Operations technology in a heavy haul capacity and are expected to lead to significant efficiency benefits. Mainline trials conducted with the Western Australia Office of Rail Safety have progressed well and a decision on the next stage of the project is expected in mid-2008. Automated rail management is the first major operation scheduled to be run from the Remote Operations Centre.

'Intelligent' driverless trucks

Rio Tinto will introduce the industry-leading Komatsu Autonomous Haulage System into Pilbara, which will allow for a fleet of 320-tonne off-highway trucks to be operated without drivers. The system will be commissioned before the end of 2008 and is expected to be more widely deployed in new and existing Rio Tinto Iron Ore operations by 2010.

Remote control 'intelligent' drills

Rio Tinto is already using bespoke autonomous drill technology in Pilbara to support the 'mine of the future' strategy.

A pathway to fully automated mine-to-port operations

Rio Tinto began work on defining building blocks for the 'mine of the future' over a decade ago and the key components required for an integrated mine-to-port operating system are being assembled and tested by Rio Tinto Iron Ore. A number of key technologies have been introduced on a staged basis, beginning in 2006 with the development of autonomous drilling rigs for the Pilbara mines. In early 2007, Rio Tinto established and funded on a long-term basis the Rio Tinto Centre for Mine Automation in partnership with The University of Sydney. Under this partnership Rio Tinto has secured exclusive access to world renowned robotics experts dedicated to addressing Rio Tinto's 'mine of the future' opportunities.

This year (2008) Rio Tinto Iron Ore will start running extensive trials at dedicated mine test site. Trials will combine the world leading Komatsu Autonomous Driverless Haulage System with a range of other advanced remote control and autonomous technologies to provide an industrial-scale proving ground and template. Experience gained by the business will allow for further deployments in Pilbara in 2010 and will also to be applied at other Rio Tinto mining operations. The 'mine of the future' programme will provide opportunities for technology driven performance improvements to support Rio Tinto Iron Ore's announced plans to take annual global iron ore production beyond 600 million tonnes.

Source: Rio Tinto Press Release 18th January, 2008.

The environmental impacts of mining are a dominant driver for the development of technology in the industry. Although much of this effort is focused on waste treatment and disposal, a significant amount of waste prevention has also occurred. Development of new technologies to extract useful products from mine wastes (solid wastes and mine effluents) has long been a goal of the industry. In the future, the industry would like to find constructive use for all material removed in the mining process, either as input materials for other industrial activities or as backfill material once mining finished. To achieve this vision there is a need for mining countries to provide a beneficial environment for the development of new technologies; currently some regulatory statutes discourage companies from trying unproven and innovative processes (National Mining Association 1998). And in some cases, government regulations actually discourage beneficial use of mine wastes, as noted in **Case 22.3.**

The environmental impacts of mining are a dominant driver for the development of technology in the industry.

Mining in the Last Frontiers

Recently there has been a trend for more countries to become involved in the export of mineral commodities. While the majority of minerals are still supplied by the historical producers (USA, Canada, Australia, South Africa, Russia, Brazil and Chile), an increasing share of worldwide production now originates in other countries. Resource-rich countries

Recently there has been a trend for more countries to become involved in export of mineral commodities.

CASE 22.3
Limestone Mining in the Sirmour Area, Himachal Pradesh, India

Numerous small limestone quarries situated in the Sirmour area of the Lesser Himalayas, produce limestone for the cement industry. Waste materials from these operations comprise limestone which is not of suitable quality for cement production. These wastes are well suited to a range of other uses including aggregates and road surfacing materials. However, the producing companies are discouraged from selling their wastes for such purposes because, if they did, then the wastes would become products, subject to government royalties, which would render the transactions uneconomic.

such as Zambia and the Philippines, which were once major exporters, are once again developing new projects and are being joined by many countries that have not previously had significant mining industries. These include Ireland, Argentina, Laos, Madagascar, Mongolia and Tanzania. Countries such as China, Mexico and India which have always had many, mainly small mines, are adopting modern technologies for exploration, project definition, production and environmental management, with the result that large-scale mines are beginning to replace relatively inefficient small operations.

There appear to be at least four main reasons for the redistribution of mining away from historical mining countries. These are:

- Most of the relatively easy to find, near-surface deposits in the historical mining countries have already been found and, in many cases, have already been mined;
- Many countries, particularly those in the Pacific Rim, are geologically highly prospective for very large, although relatively low-grade deposits that have proven to be highly profitable, using large-scale equipment;
- In several cases, pioneering exploration in new regions without a history of mining, has been successful (for example Kingsgate's Chatree Project in Thailand and Oxiana's Sepon Project in Laos), encouraging other companies to follow, and
- Many geologically prospective countries have, belatedly in some cases, recognized the economic benefits that mining can bring, and have consequently introduced policies to attract exploration and encourage mine development. The World Bank and other multi-lateral funding agencies have been influential in assisting many developing countries to rewrite their mining laws so that they encourage exploration while, at the same time, ensuring that the host countries will benefit from the resulting mineral developments.

Many geologically prospective countries have, belatedly in some cases, recognized the economic benefits that mining can bring.

Anti-mining sentiment, now common in many western countries, may also be a factor influencing mining companies to explore elsewhere, although strong anti-mining sentiment is not confined to western countries. It has also been claimed, particularly by anti-mining activists, that lower environmental standards have been a factor in attracting mineral exploration to developing countries. There does not appear to be much, if any, truth in this assertion, as:

- Many developing countries have environmental standards that are equal to or higher than those of many western countries, although the capacity for enforcement may not yet be commensurate;
- Most international mining companies have stringent internal standards which apply to all their projects, regardless of location, and
- Financing institutions have increasingly adopted the Equator Principles which require adherence to high standards, particularly in relation to social impacts and community development.

It appears that government regulatory efforts target the international companies, while local companies are able to get away with lower standards.

It is clear, however, that there are small, home-grown operations in developing countries that have lower standards of performance than those of international companies. In many of these countries, it appears that government regulatory efforts target the international companies, while local companies are able to get away with lower standards.

No mining has ever been carried out on the continent of Antarctica. There are international treaties that preclude mining there, which would probably deter any systematic exploration. Added to this, the costs and risks associated with exploration, development, operations and shipping of the product would be prohibitive. While it is possible to imagine geo-political circumstances that could change this situation, such circumstances are unlikely in the near term.

Another new frontier being posited as a source of minerals is the deep sea. It has long been known that high-grade nodules containing metals such as manganese are present on the sea-bed in various parts of the world. Similarly high-grade deposits of base metals are associated with sub-sea volcanic activity in many parts of the ocean. Technologies for economic extraction of such minerals are only in the conceptual or prototype stages. However, if commodity prices continue to rise, the possibility of deep sea mining may improve.

Another new frontier being posited as a source of minerals is the deep sea.

Mining outside the Earth remains, for the foreseeable future, the domain of science fiction. The financial disincentives that apply to Antarctica appear minor compared to those that would apply to even our nearest celestial neighbour.

Based on the above considerations, the factors that have led to increased mining in previously non-mining countries are likely to continue and to intensify in the future.

22.2 TRENDS IN ENVIRONMENTAL PRACTICE

End to Paternalism

Until quite recently, mining companies believed without question that what was good for them must also be good for the communities in which they operated. Opposition tended to be dismissed as uninformed, or the work of a small minority of trouble-makers. Companies and governments co-operated in land acquisition and in many cases, land was acquired compulsorily, sometimes forcibly. This situation has changed significantly over the past two or three decades. Now, in most countries mining companies have to negotiate land acquisition with landholders, and there are recent examples of projects remaining undeveloped because proponents were unable to acquire the necessary land. Previously, the only approvals needed for a new mining development were from the appropriate government(s). Now, equally important is the consent of the local community. Although it is still possible in some countries to develop a mine without local consent, few companies and none of the major mining companies, would be foolish enough to do so.

Until quite recently, mining companies believed without question that what was good for them must also be good for the communities in which they operated.

There is another type of paternalism, which could be termed 'environmental paternalism', which is common among expatriate staff of some mining companies and their consultants. Typically, this is the attitude that 'best practice' is Australian (or American or Canadian) and that anything else is inferior. The worst aspect of environmental paternalism, apart from the implied insult to local practitioners, is that local knowledge and experience are ignored, to the detriment of the project.

Environmental paternalism is the attitude that 'best practice' is Australian (or American or Canadian) and that anything else is inferior.

Although NGOs have long criticized the real or perceived paternalism shown by mining companies, many NGOs have been similarly paternalistic. Of course, as discussed in previous chapters, there is a broad spectrum of NGOs and therefore an equally broad spectrum of attitudes and views. There are many, mainly community-based NGOs, that are well informed and enjoy strong community support. On the other hand, some of the most well-known international NGOs are as paternalistic as the most paternalistic mining company. Again, this attitude can best be summarized as 'We know what's best for you'. Rarely, it appears, do international NGOs spend sufficient time in local communities that would enable them to speak on behalf of these communities, and yet this is what they do. Frequently, one cannot help getting the impression that they will embrace the cause of a disaffected party without investigating the source of disaffection, or the views of the community at large. There are few signs that NGO paternalism is diminishing. However, there are signs of a growing anti-NGO backlash which seeks to hold NGOs accountable for their public statements. To the extent that this anti-NGO trend is effective, paternalism is likely to diminish.

There are signs of a growing anti-NGO backlash which seeks to hold NGOs accountable for their public statements.

Government Regulation of the Mining Industry

Quite different trends are apparent in government regulation from one country to another. In democratic countries, government regulations generally reflect the perceptions and aspirations of the population. When the mining industry is perceived to be a source of major pollution, more stringent environmental regulations are likely. Where the industry is seen to be overly profiteering or exploitative, higher payments to government in the form of taxes or royalties are likely. In countries without well-developed democratic institutions, government regulations may not reflect community attitudes. This can cause problems for companies seeking to satisfy conflicting requirements.

In countries with a long history of mining such as Australia, the rapid evolution of environmental regulation that commenced in the early 1970s has now ceased.

In countries with a long history of mining such as Australia, the rapid evolution of environmental regulation that commenced in the early 1970s has now ceased. This is partly due to the maturity of the industry and the confidence, particularly among regulators, that any environmental problems will be identified and overcome. In fact, for most mining operations in Australia, environmental management has become routine. Another important factor in the Australian industry has been the Minerals Council of Australia's Code For Environmental Management. This Code, to which the majority of Australian mining companies subscribe, involves a voluntary commitment to high standards of performance, including 'continuing improvement' and 'public reporting'. As a result, the industry in Australia generally performs at standards considerably higher than those mandated by regulations. The success of 'self regulation' in Australia's mining industry does not appear to have significantly influenced other countries.

In countries with a shorter mining history, the industry is less mature and the public is likely to be less informed and more concerned about environmental impacts, particularly in the presence of adversarial environmental NGOs. In these situations, environmental regulations are still evolving. Indonesia is a good example of this situation, with major new environmental regulations appearing regularly. Another feature of Indonesia's regulatory system is the 'stick and carrot' approach wherein the government can penalize breaches of regulations, but also provides awards for good performance (**Case 22.4**). Eventually, however, effective environmental regulation is dependent on government monitoring and enforcement capacity, and the ability of the mining sector to finance the costs of compliance.

A particularly difficult situation occurs where past mining operations have produced catastrophic outcomes that have generated public outrage.

A particularly difficult situation occurs where past mining operations have produced catastrophic outcomes that have generated public outrage. The Philippines is a case in point. This was one of the last countries to require sound environmental management of mining operations and, by the time it did, the industry had almost disappeared because of political interference, financial mismanagement and numerous incidents of environmental

CASE 22.4
The PROPER System (Program for Pollution Control Evaluation and Rating) in Indonesia

In the 1990s Indonesia introduced the PROPER system to evaluate the environmental performance of industrial operations. Each year a government team audits selected operations, and publicly announces the outcome of its finding: 'gold' ranking is assigned to operations that go beyond environmental compliance in all operational aspects (a ranking that no company has achieved up to year 2008); a 'black' ranking may lead to the suspension of operations if the company cannot demonstrate significant improvement within 6 months. While participation in PROPER is in theory voluntary, in practice it is not. The PROPER system is yet another tool by the central government to put pressure on multinational companies, admittedly for a good cause.

damage, some serious. Despite a major overhaul of the regulations and their enforcement, public outrage toward mining in the Philippines continues, fuelled by a strong anti-mining lobby. In the Philippines, it would appear that the public and its elected representatives have become over-sensitized to the impacts of mining so that even minor spills, such as those that occurred at Rapu Rapu in 2005, cause a disproportionate level of outrage.

Another current trend is the increasing influence exerted on governments by NGOs. There has long been entrenched opposition to mining from a variety of environmental groups, some of which are radical and outspoken, with an influence that, on occasions, seems to be disproportionate to their size and scientific credentials. This influence seems to reflect a perception that the NGOs speak on behalf of the host communities (see also Case Studies 3.6 and 8.11). While this may be the case in some circumstances, there are other cases where local people are exploited by NGOs to support their own ideologies. In general, the popular press finds the views of these groups more newsworthy than the views of those involved in the industry. As a result, the industry continues to be portrayed in a less than complimentary light. The situation is unlikely to change in the foreseeable future; however, this does not mean that biased reporting should not be challenged.

Evolution of Government Capacity

One of the most serious problems in developing countries, particularly those without a significant history of mining, is the lack of capacity among government personnel, to evaluate the merits and demerits of new projects, as well as to adequately supervise the performance at existing operations. Too often there are insufficient numbers of adequately trained personnel, lack of relevant experience and a chronic shortage of funds and equipment. It is easy to understand why delays occur at the EIA stage as regulators, under constant pressure from the proponent to approve a project, struggle to understand large amounts of highly technical information. Their situation may be compounded by outspoken opponents generating daily press coverage claiming that the project will have disastrous impacts. This situation is not unique to developing countries: in the 1970s the environmental approval system in Australia, the US and other newly environmentally conscious countries, was administered largely by new graduates, many of whom had never seen a mine. Fortunately, experience has shown that this situation does improve with time. Co-operation and assistance on the part of industry can expedite capacity building to the benefit of all.

One of the most serious problems in developing countries, particularly those without a significant history of mining, is the lack of capacity among government personnel.

An effective approach followed in many countries to augment government capacity, is through the use of 'expert committees' to review new mining proposals, particularly large or controversial projects. Commonly, such committees are assembled from academia and industry as well as senior government experts. The main criterion is that the committee as a whole is able to understand all issues involved in the project. Clearly, the success of this approach depends on the quality and objectivity of the people involved.

One major impediment to capacity building in developing countries concerns the low salaries paid to government personnel and also to academics. This undoubtedly contributes to corruption in some countries. It may also lead to a situation where the best technical people obtain employment in industry, with less capable people in government. As mentioned previously, capacity generally improves over time as government personnel become more knowledgeable and experienced. However, this is not inevitable. One effect of the current mining boom is that key government personnel, including environmental specialists, are being recruited by an industry that is itself short of qualified personnel, leading to

One major impediment to capacity building in developing countries concerns the low salaries paid to government personnel and also to academics.

a rapid reduction in government capacity. Again, it is in the interests of all for the industry to help overcome capacity constraints in government. One effective method is for industry to second experienced personnel who can transfer knowledge and experience to government staff. Even better are exchange arrangements in which government personnel work in industry while industry personnel work in government.

Public Perception and its Influence on Regulation

The critical role of the mining industry in the world's economy is often overlooked.

The visual impact of a large mining project is obvious; however, its benefits are widely distributed geographically and economically and are therefore difficult to discern. As a result the critical role of the mining industry in the world's economy is often overlooked. Clearly the industry needs to show that improved mining technologies increasingly being put in place have not only created a cleaner, safer and more technologically advanced industry than is commonly perceived, but one that is vital to meet not just the needs of today's industrial world, but also tomorrow's.

Negative public perceptions, however unrealistic, can lead to increased regulation and supervision of the industry, although the incidence and magnitude of environmentally damaging events has declined to low levels, much lower than in many other industries. This suggests that the industry needs to widen its approach to the public, not only to show the massive contribution mining makes to domestic and international economic health, but also the degree to which mining practice has evolved in response to environmental concerns over the last few decades.

Unfortunately, the current heightened concern about climate change is likely to increase adverse public perception of the mining industry, largely because of the coal industry, a primary target of environmental activists. As a counter to the massive CO_2 output of coal-powered electricity generation, some in government and industry have seized on so-called 'clean coal' technology to reduce global warming. This label, which originally applied to the removal of sulphur from coal, now refers to stratigraphic sequestration of carbon dioxide recovered from the emissions of coal combustion. While undoubtedly feasible from a technical viewpoint, the economics of such an approach are unproven and it is probable that the same result in terms of carbon dioxide reduction could be achieved much more economically by other means.

If the industry is ever to improve its image, it will need to attend to the visual impacts of its major operations.

Perhaps the single factor that most helps to maintain negative perceptions of mining is the sheer visibility of many mining operations, many of which are now much larger than in the past. The remnants of historical mining operations, by contrast, are small with elements that are generally in scale with the surrounding landscape. It is apparent that visibility is one of the major factors, perhaps *the* major factor influencing public perceptions. There are many other examples of visible activities that cause public outrage although they may cause little or no actual environmental damage, such as visible steam emissions from chimney stacks, or the appearance of wind generators in rural landscapes. Furthermore, unlike many other industries, mining's effects on the landscape will often remain as a permanent legacy of the industry. If the industry is ever to improve its image, it will need to attend to the visual impacts of its major operations. A radical suggestion as to how this might be achieved is presented in the next section. It is true that the total area subject to mining will never represent more than a tiny fraction of the Earth's surface and that the total 'residual footprint' of mining will be much less again. However, that does not diminish the visual impact of close proximity to large mines.

A Strategy for Eliminating Adverse Visual Effects of Mining

The aim of this initiative would be to enable all areas affected by future mining activities to be totally rehabilitated, meaning that mining would have a 'zero residual footprint'. This would generally require that there be no more 'final voids'. Many operations such as underground mines, mineral sands, and open cast bauxite and other shallow laterite mines, already achieve this objective and would therefore be unaffected. Most coal strip mines, although leaving final voids under existing mine plans, could modify their spoil placement operations to eliminate final voids, without enormous financial detriment. Other mines would either need to change their methods or would need to accrue funds for the back-filling of mined out areas, following the completion of mining. Imposition of such requirements would bankrupt existing operations involving large open pits. Such operations currently supply most of the world's copper, iron ore and lignite. Clearly, if such a strategy was to be implemented, existing projects including those under development at the time of agreement, would need to be 'grandfathered', which means that they would be permitted to continue until closure using the original approach.

It is likely that any agreement to eliminate final voids would mean that some types of deposit (such as lignites) would not be mined, others would be only partly mined, while yet others would be mined using a different approach. For example, a long narrow ore body such as the Mt Newman iron deposit in Western Australia could be mined sequentially, with hind-casting of waste rock, once full depth is achieved at one end. The large copper porphyries could be partly mined by open cast methods, particularly where the ore body outcrops on the crest of a mountain. However, the suggested strategy would tip the financial balance in favour of underground mining, so that methods such as block caving would become more common.

If adopted, this approach would lead to increased costs of some commodities, particularly, copper, gold and iron ore. However, as the industry adjusts and develops new, lower cost technologies for backfilling, those incremental costs would reduce.

To have any chance of successful implementation, such a strategy would require overwhelming acceptance by the mining industry, followed by regulatory support from countries throughout the world. The key to acceptance would be that there be no exceptions – all companies would be subject to the same conditions. There would also need to be agreement as to what constitutes acceptable rehabilitation.

Building Capacity for Improved Environmental Performance

In the recent past, there have been important initiatives aimed at improving environmental impact assessment practice, most notably the IFC Environmental and Social Performance Standards of 2006. Arguably, the main change is the shift of attention from planning to implementation. In the past an environmental impact assessment study was seen, in isolation, as a planning tool; today it is the foundation of an environmental management programme, addressing environmental business risks and opportunities from the beginning of the mine development up to the mine closure and beyond.

Arguably, the main change is the shift of attention from planning to implementation.

But the process is increasingly complex, and it would be unrealistic to assume that the necessary expertise is always accessible. All stakeholders, including financial institutions, local government officials, communities that are now increasingly asked to participate in the EIA process, and the mining companies themselves, face a bewildering maze of international standards and local laws and practices, highlighting the need for capacity building at all levels of the approval and implementation processes. This is a challenge the mining sector must confront, and respond to positively over the foreseeable future.

Environmental Assessment and the Sustainability of Mining at a Micro-economic Level

Mines come in all forms and sizes, but modern mining operations may be of such a scale that the associated capital expenditure can make a measurable impact on the economy of the host country. Because of the relatively small size of the economies of most developing countries, the economic impacts of mining tend to be relatively large, particularly in the region that hosts the mine. The challenge for a mineral-rich developing region is the appropriate use of the revenues arising from resource development. Part of this revenue needs to be put aside to replace the capital being used up and, indeed, add to it (Humphreys 2000). Natural capital, that is, the valuable minerals, needs to be converted into other forms: built capital, or social capital. As economists understood long before the contemporary notion of sustainability, what counts in the creation of wealth is not the particular form of available capital, but the net stock of capital in all its forms. Sustainability is about maintaining and augmenting that stock.

The challenge for a mineral-rich developing region is the appropriate use of the revenues arising from resource development.

More recently, the focus of debate over mining's sustainability has shifted from the macro-economic to the micro-economic level. Specifically, it has moved to address the issue of the longer term impacts of individual mining operations on the local and regional economies in which they are located. There is no guarantee that the host communities of a mine will necessarily be beneficiaries, even during project operation. Companies have increasingly come to recognize that returning benefits to the host region is not a matter which can be left to government authorities alone. Many companies have determined that they need to take an active part in ensuring that the regions in which they are operating receive a reasonable share of the economic benefits from those operations. Moreover, they realize the need to prepare the basis for sustainable economic activities after they have gone. While this is difficult, it can be done. The slow-moving nature of the mining industry means that there is often, at least in principle, enough time to work out how mining can leave lasting economic benefits where management has specifically set out to create them. The environmental impact assessment process is the appropriate vehicle to commence such planning. It is fair to say, however, that current EIA practice does not always provide enough understanding of the host region's micro-economy to provide a sound platform for necessary planning and subsequent implementation. There is a need in the environmental assessment process to add to the traditional focus on the physical-chemical and, more recently, from the socio-cultural, to include a balanced analysis of the economic impacts of a new mine development (**Figure 22.4**). These impacts certainly include economic risks, but should also identify the opportunities that would ultimately help to justify permission for a new mine to proceed.

The slow-moving nature of the mining industry means that there is often, at least in principle, enough time to work out how mining can leave lasting economic benefits where management has specifically set out to create them.

Mining as Part of Integrated Regional Development Planning

It is better to address the need to optimize a mine's potential for integrating with regional economic development at the outset of a new project rather than part way through. A carefully prepared EIA that addresses all economic dimensions of the project will go a long way to optimizing regional development, while mitigating costs to the affected communities. However, as a result of the tendency of governments, supported by public opinion, to shift more non-operational responsibilities onto the mine operator, good intentions and good ideas may turn into major cost commitments. The host government retains primary responsibility for identifying, coordinating and financing regional development, particularly in view of the often vast government revenues generated by the project. This, however, is an ideal not always encountered in the real world.

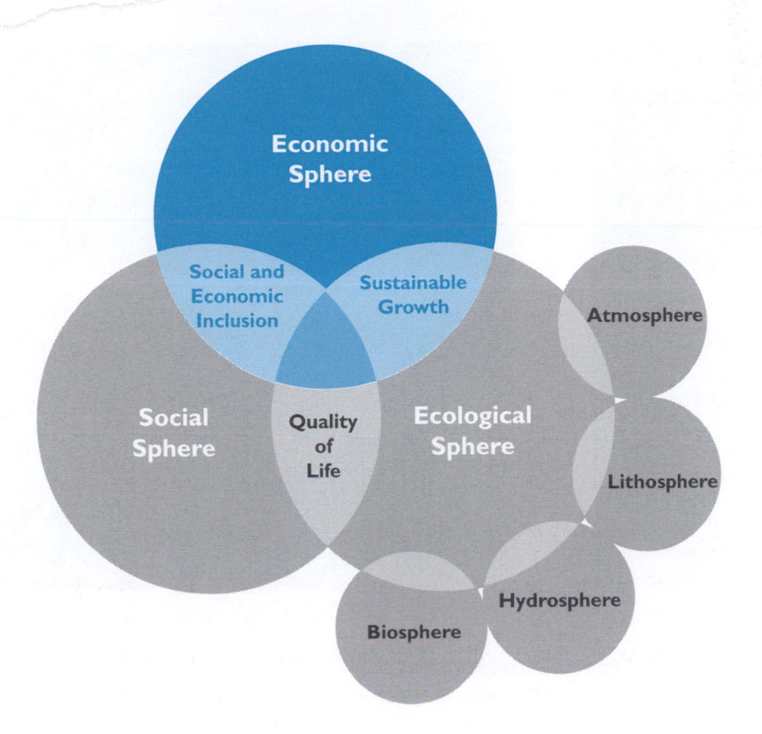

FIGURE 22.4

The Need for Balanced Consideration of all Three Spheres of Sustainability

There is a need in the environmental assessment process to shift from the traditional focus on the physical-chemical and, more recently, from the socio-cultural, to include a balanced analysis of the economic impacts of a new mine.

For a new mining development, rather than considering only the economic effects of the project, it is preferable to consider the project as one element of a broadly defined, integrated regional development plan (**Figures 22.5** and **22.12**) (Humphreys 2000, on which the following section is based). In this context, the objective becomes to develop mining in conjunction with other economic activities, such as training centres, transportation hubs, supporting industries, manufacturing, forestry and tourism, to complement the mining development. The mining would be an essential element in unlocking the region's potential and providing significant injections of capital to initiate development activities. Other sectors would provide support during mine operations, and continue as the basis for sustained development after mining ceases.

For a new mining development, rather than considering only the economic effects of the project, it is preferable to consider the project as one element plan.

The interactions of mining operations and the local economies in which they are located are often incompletely understood. Also, there have been relatively few well-documented case studies on this subject, and there are few other industrial sectors which can provide useful lessons (**Case 22.5**). Rigorous regional economic analysis and planning requires detailed information on the flow of goods and services between the various economic sectors. Such information is rarely readily available; neither can it be assumed that the local government capacity exists (in term of human resources as well as funding) to develop such information. This is perhaps an area where international agencies could assist. It is also reasonable to imagine that the mine developer assists the local government in establishing an integrated regional development plan. The mining industry is not short of relevant ideas and has a growing experience of dealing with regional development. However, the design of new mechanisms to promote the economic sustainability of mining by companies and regional authorities alike, would undoubtedly require new company outlooks and improved government policies.

Looking Beyond the Fence

The cumulative effect of environmental disasters and adverse social impacts of mining and other industrial developments of the last century, has resulted in an international awareness of environmental issues, and acknowledgement of the fact that even local developments can

	Global	National	Regional	Local
Ecological				
Social				
Economic				
Institutional/ Government				

Life Cycle Analysis e.g. GHG Emissions
Regional Economic Development
Improved Governance

GHG: Greenhouse Gases

FIGURE 22.5
Possible Trends in the Environmental Impacts Assessment Practice

(1) Addressing the entire life-cycle of the mine and its products (e.g. in regard of greenhouse gas emission);
(2) Increasing its geographical boundaries to encompass regional, if not global effects and interrelationship, and
(3) Focusing on a match of planning and implementing.

have regional or even global consequences; global environmental protection and resources development have now become inseparable.

As a consequence some commentators have observed that the environmental impact assessment should (1) be expanded to address the entire life-cycle of the mine and its products, and (2) increase its geographical boundaries to encompass regional, if not global effects and interrelationships (**Figure 22.5**). Both suggestions have their merits; both have their challenges.

Life Cycle Assessment (LCA) is a technique for assessing the environmental impacts associated with a product over its life-cycle. Especially in the case of coal, the contemporary growing concern for climate change has prompted legislators and environmentalists

CASE 22.5
The Tangguh LNG Development in Papua

'The Tangguh LNG Project will not only meet best international practice, but it will set the benchmark against which future resources development projects will be measured', claims BP in its foreword to the environmental impact assessment study for the project. In its project development, BP went beyond the classic EIA concept by helping the Government of Indonesia to develop a regional development plan for the Bird's Head region of Papua, known as the 'Diversified Growth Strategy'. The project is scheduled to start in 2008, and it is too early to evaluate the success of the strategy. However, once the outcomes become clear, the Tangguh project will provide useful lessons for future large grass roots resource developments.

to study how both coal mining and coal use will affect the global environment (see also the Case 13.10 on Anvil coal mining in Chapter Thirteen). In the mining industry there is now increasingly an internal requirement for the estimation of energy consumption and greenhouse gas emissions for each project and the triggering mechanism for implementing measures to reduce GHG emissions. LCA provides a quantitative and scientific basis for these new concepts. In simple terms, LCA is a method of quantifying and balancing the environmental impacts of a product in relation to its production, transport (to end user), use (consumption) and eventual disposal. This can be a complex and daunting task. Fortunately, there is flexibility in the level of detail required for an LCA, depending on its objectives and intents.

For mining projects, undertaking an LCA for products of a specific mine is difficult, if not impossible. Mine products are part of the global pool of raw materials, and singling out one product stream is impossible. An LCA does provide the basis for developing new policies to influence producer and consumer behaviour as well as preferences. However, because environmental degradation is seldom contained by political boundaries, the ultimate challenge is to coordinate policy and legal responses across broad regions. Even in the European Union, this can be difficult.

For mining projects, undertaking an LCA for products of a specific mine is difficult, if not impossible.

But there are other reasons why it makes good sense to look beyond the geographic study boundary when conducting an environmental impact assessment. As there is a link between upstream and downstream activities, there is a strong link between projects of similar types. These links may be easily detected as in the case of two mining projects in the same host region, or they may be more subtle as in the case of new developments in different regions (**Case 22.6**).

As there is a link between upstream and downstream activities, there is a strong link between projects of similar types.

CASE 22.6
Developing Palm Oil Plantations in Papua

While not related to the mining sector, the recent push by the Indonesian government to develop palm oil plantations in Papua illustrates the complex environmental and political links between similar activities in the same jurisdiction. Question: Is it environmentally acceptable to convert pristine rainforest on the island of Papua into oil palm plantations while, by some estimates, some 2 to 3 million hectares on the island of Kalimantan are readily available, previously deforested for the same purpose but never developed as intended? This example illustrates the need for strategic environmental assessments (SEA) to complement project-specific EIAs. The purpose of an SEA is to integrate environmental and sustainability factors in policy and national programme making.

22.3 ON AND BEYOND THE HORIZON – GLOBAL CHANGE AND CHALLENGES

This final section attends to the future and the impact of some aspects of a rapidly changing world already bearing down on the international mining industry, heavily drawing on personal communication with and on a forthcoming paper by Jim and Helen Singleton (2008 – in press). A critical factor underlying the following discussion, an underlying theme in this text, is the increasing trend towards the exploration and development of exploitable mineral resources in sometimes difficult parts of the world.

As discussed earlier in this chapter, this inevitably means undertaking operations in more challenging physical environments, e.g. deep ocean sea-bed sites such as currently planned north of the coastline of PNG; extreme climatic zones such as the case of the Ekati diamond mine in Northern Canada; or more isolated geographic locations such as West Papua, home of Freeport's Grasberg Copper and Gold Mine. It also means operating in more unstable and difficult nations and political settings. Mining companies are probably better equipped than most other industry sectors to meet these types of challenges, but difficult geography is only part of the range of changes and challenges facing the mining sector. Increasingly, challenges of a different nature are emerging.

An Increasingly Challenging Global Operating Environment

Global shifts in the world economy are evident. Until recently, industry, including mining, has evolved to some extent in the perception of a simplistic 'bi-polar' model of the global economy where western business management practices and diplomacy dominate, and are assumed as the template for development in the rest of the world. This is not happening. Clearly, the global economy is not so simple; rather, it is fast becoming 'multi-polar', with massive wealth streaming in from the OECD economies, USA and Europe in particular, to the leading performers of the developing world, particularly China and India, but also to many other rising nations. In addition, increasing numbers of developing economies are shifting from being passive recipients within the global economy to becoming active shapers.

Along with these shifts, three powerful trends have emerged: (1) the power of information and communication technologies (or 'infosphere' as discussed in a later section); (2) the rising economic openness; and (3) the expanding size and reach of global corporations.

The rising power of the BRIC nations of Brazil, Russia, India and China is well recognized, but this is now extending rapidly to other countries particularly in North and South East Asia, and Latin America. Recently, the Middle Eastern and North African (MENA) economic group of nations has emerged as a growing force. With rising wealth comes increasing influence on the way in which business occurs. This diversification is already having, and will increasingly have, bearing on the western economy-based mining companies. Mining investors from different nations and cultures will play to different rules, thus changing the traditional competitive landscape. The cultural assumptions underlying the ethics, value systems and best practice notions of western-based companies will not necessarily constrain new operators from new countries. Different assumptions imply doing business on a different and negotiated basis to achieve satisfactory business outcomes with new national governments in rising nations.

Hence, a global business scene is emerging that involves rising economic interdependencies in different key spheres, all of which are directly relevant to international mining, and all of which are likely to influence its environmental performance. Some follow:

- Winning talent – professional and intellectual talent is becoming a highly competitive global resource open to all bidders, and international mining companies from the traditionally dominant economies must increasingly compete for that pool of talent. As one example, mining professionals from the Philippines are highly sought after throughout the Asia Pacific region. In fact, recruitment has been so successful that local mining expertise in the Philippines is now becoming a rare commodity.
- Sources and flow of capital – are shifting from foreign direct investment (FDI) sourced solely from developed economies to increasing FDI from developing economies, and hence there are new competitors joining the field and playing to new roles with different strategic objectives in mind. The opening of the Indonesian economy to Chinese investors is a point in case. While practically non-existent less than 10 years ago, Chinese investment in the Indonesian mining sector (and elsewhere) now represents a formidable economic force.
- Battle for resources and commodities – the developing world is increasingly competing for commodities; for example, the developing world accounted for 85% of additional energy demand in recent years (as of 2008).
- Emerging consumers in the emerging economies – introducing new supply chains, preferred sites for down-stream processing and strategic natural advantages in different parts of the world.
- New map of global innovation – new economies and markets are becoming innovators and proving they are capable of catching up very rapidly on technical innovation with the traditional leaders from Europe and North America. This capacity shift will change the traditional leadership and locations of research and development, innovation, and 'early adopter' beneficiaries.

A global business scene is emerging that involves rising economic interdependencies in different key spheres, all of which are directly relevant to international mining.

For mining companies, the outcomes of growing competition across the above five spheres are the heightened risk of increased volatility in all respects of their operating environments. It means that more effort will be needed to attract and retain the talent mining companies require for their operations. This challenge will involve dealing with different health risks, more ideological challenges, diverse cultural differences and preferences, all of which can translate to higher staff turnover, higher operating costs, and eventually also new environmental and social issues. As in other industry sectors, the mining sector will increasingly face the heightened imperative to recognize, respond and adapt quickly to a broad range of economic, political, environmental and technological risks. Accordingly, mining companies need to build new capabilities and acquire new skills from professional disciplines not necessarily regarded as part of the traditional mining industry skill-set.

Mining companies need to build new capabilities and acquire new skills from professional disciplines not necessarily regarded as part of the traditional mining industry skill-set.

Emerging New Geo-Political Risk Factors

Operating in a changing global economy is likely to be both potentially more rewarding and more difficult than previously. In certain respects, a robust global economy has been achieved by the World Trade Organization (WTO) through the increased openness in trade and banking, with the removal of trade barriers and deregulation of the global financial system. However, other forces are contributing to new challenges and risks.

Geo-political risk and regional/global security is one. The management of operations in unstable or post-conflict affected countries and failing nation states; addressing local and regional terrorism; and assessing the implications of potential shifts in political strategic alliances within and between governments; are all more difficult than previously. International development agencies (e.g. Multilateral Investment Guarantee Agency (MIGA), part of the World Bank Group) and international business, including the mining industry, are all challenged to develop new strategies and techniques to deal with these realities.

For example, recent practice seems to have gone beyond conventional political risk insurance to the adoption of proactive, flexible and adaptive business management regimes that attempt to mitigate potential risk factors, or allow rapid response to risk issues before they have an effect. This concern seems to be steering large mining investments towards the following response mechanisms:

- Good corporate governance (e.g. in line with macro-level agency principles and guidelines, such as the World Bank, OECD, IFC and EPFI) conceived to minimize risks arising from lax operational management and insufficient attention to multifarious risk factors.
- Increasingly sophisticated approaches to political risk management in response to the above, with a shift away from traditional solutions (e.g. diversified portfolios, balancing risk and return, and weathering the worst with political risk insurance) to more flexible, holistic and proactive strategies.
- The application of more systematic and whole-of-operation approaches, i.e. Comprehensive Enterprise Risk strategies, which by definition incorporate an integrated approach to all risk issues, responses and overall risk management programmes.
- Support for Technical Assistance Programs (i.e. MIGA principles) and 'Turn Around Strategies' conducted by appropriate international agencies, governments and NGOs, to strengthen institutions, legal systems and community cohesion in regional economies facing trauma and instability, seeking thereby to contribute to a more stable operating environment.

Given the nature of its business, mining is at the forefront of these broader development risk issues. Each mining operation needs to address its viability not only in strict commercial profitability terms, but also increasingly in terms of the potential impacts of these external risk factors.

Transparency and Corruption

A particular challenge for international mining companies, as with other industry sectors operating globally, is dealing with corruption in their operating environments. Corruption is a huge drawback in doing business. While most governments have strict anti-corruption policies in place, the difference between will and action is often profound. 'It's like you look at a mirror and want to lose weight. But then you sit at the table and eat.' comments Mussemeli, US Ambassador to Laos, a country that increasingly attracts mining (Forbes Asia February 11, 2008).

'It's like you look at a mirror and want to lose weight. But then you sit at the table and eat.'

The trend for more geographically dispersed mining ventures increases the likelihood of operating in a 'corruption challenged' setting. In a rising effort to restrict or eradicate corrupt practices associated with corrupt governments and their officials, NGOs, such as Transparency International, and international watch-dog agencies increasingly monitor and report on company performance. This trend will lead to more stringent operational requirements on legitimate mining companies. For example, the OECD has agreed to international laws addressing transparency and corruption, and the United States Federal Corruption Prevention Agency (FCPA) has developed a surprising global reach in pursuing corrupt practice. An important influence for mining companies is the Extractive Industry Transparency Initiative (EITI) programmes, currently undergoing further endorsement by governments and international agencies, such as the World Bank.

Mining companies operating internationally are in a difficult position with regard to the issue of corruption. Short of a pre-emptive decision not to operate in a given context in the first place, it is very difficult to cut across locally normative values and behaviour in a specific country or cultural context where low-level corruption is a normal part community and government practice. There are also difficulties with western definitions of corruption, which sometimes fail to recognize the culturally specific legitimacy of local practices (e.g. speed-up or facilitation money; normalized remunerative practices for public servants, which consequently do not constitute bribes). Another example is the common practice in many developing countries to pay honorariums or travel allowances to the members of governmental review committees (e.g. the review committee of an environmental impact assessment). While common practice, and often underpinned by published directives of local or national government authorities in the host country, such payments may conflict with the FCPA; they are also easily perceived as 'bribery' by the inexperienced and naive observer.

Mining companies operating internationally are in a difficult position with regard to the issue of corruption.

The changing pitch for mining companies is that they will have to accommodate increasingly stringent safeguards and developments in anti-corruption techniques, although the companies themselves may have no such tendencies. In future, mining companies can expect to have to absorb costs associated with new corporate governance requirements relating to various aspects of international business, including obtaining environmental approvals, for example:

- International policing and tests for determining bribery and kick-back payments;
- Due diligence and additional documentation for all overseas payments and currency transactions;
- Guarding against corporation reputation risk outcomes due to unanticipated prosecutions and legal challenges;
- Retaining greater detail of all royalty payments as governments agree to report and publish royalties and other moneys paid by mining companies;
- Increased disclosure requirements by the companies themselves, particularly payments to governments and officials, and
- Increased pressure on international tax haven countries to co-operate with international policing agencies as part of new global initiatives to trace and allow restitution of plundered assets by corrupt governments and dictatorships, financial assets and pathways for global terrorist organizations, etc.

The unpalatable truth for all international business including mining is that coming enforcement procedures in this arena will probably involve effective national investigative agencies, including arms of the police forces, intelligence agencies and tax offices. Already, corporations are held as liable for corrupt practices overseas with which they may become unintendedly involved, including the risk of being judged as complicit. As an example of the challenge, 'facilitation payments' to speed up normal government approvals processes (a normal practice in many countries) has been legitimized under international law, but the requirement is that specific records of such payments must be held for seven years. Given the massive complexity of international mining operations, this one requirement alone will create a significant extra burden of internal reporting and record keeping. This example, however, also illustrates the best defence against the perception of corruption: transparency and openness. The second line of defence is to have well-documented and well-communicated procedures in place that deal with the issue.

Corporations are held as liable for corrupt practices overseas with which they may become unintendedly involved, including the risk of being judged as complicit.

Performance Requirements and Accountabilities Driven by Lenders

Many of the challenges arising from global economic change have been translated into proactive industry-wide business frameworks. This particularly applies to the resource industry sector. International agencies, such as the World Trade Organization (WTO), World Bank, United Nations, and industry organizations and associations, such as the World Business Council, all contribute to the rise and evolution of international 'leading practice' guidelines and compliance requirements. However, the pressure for additional performance accountabilities has also come from a range of global lending institutions such as the International Finance Agency (IFC), Asian Development Bank (ADB) and the Japanese Bank for International Cooperation (JBIC), as well as many private sector commercial banks.

In 2003, private financial institutions (The Equator Principles Financial Institutions – EPFIs) started to adopt a set of 'Equator Principles' that represent a financial industry benchmark to determine, assess and manage social and environmental risks in project financing.

As discussed at the outset of this text, in 2003, private financial institutions (The Equator Principles Financial Institutions – EPFIs) started to adopt a set of 'Equator Principles' that represent a financial industry benchmark to determine, assess and manage social and environmental risks in project financing. There is a close connection between the derivation of various recent and evolving international performance standards and the Equator Principles, which explicitly use and refer to various IFC and World Bank Group policy guidelines.

Further examples of international standards increasingly applicable to all major international projects include the Global Reporting Initiative of the World Business Council, established in the late 1990s (with a first mining-specific guide published in 2005 – 'GRI Mining and Metals Sector Supplement, Pilot Version 1.0', incorporating an abridged version of the GRI 2002 Sustainability Reporting Guidelines) and the OECD Principles of Corporate Governance published in 2004. These two connect to and are cognisant of various other international standards and compliance or reporting requirements.

Despite the laudable intents, these new performance requirements may result in overkill.

In sum, governments, international development agencies and lenders have introduced an extensive range of incentives, frameworks and obligatory requirements to improve business performance, including social and environmental performance, and to reduce risks. International mining is, and will increasingly be, subject to these multi-factorial and interrelated performance requirements. Unfortunately, it is also clear that despite the laudable intents, these new performance requirements may result in overkill, demand substantial efforts on the part of proponents, and increase the time required for developments, which may jeopardize well-conceived projects. After all mining is about providing mineral resources to the world at affordable prices, in monetary as well as environmental terms.

Mining and Sustainability – Conventional Perceptions versus Facts

Clearly many of the pressures and incentives for improved performance align with the sustainability agenda. Sustainability has been a rising issue for business over the last two decades. However, since 2006 with the release of former US Vice-President Al Gore's film, 'An Inconvenient Truth' and UK economist Nicholas Stern's report 'The Stern Review', public awareness has sky-rocketed. Leading business sectors of course embraced sustainability well before this, with support in the late 1990s for initiatives such as the Global Reporting Initiative of the World Business Council (GRI 2000) and, more recently, The Global Environmental Management Initiative (GEMI) in 2002. However, there is now an avalanche of more current initiatives, for example the establishment of the World Business Council for Sustainable Development by the World Business Council (2006).

The fundamental issue facing business including mining, in light of the global environmental challenges in general, and climate change, in particular, is the growing requirement for business to address and respond to challenges of the external world in which it operates. Sustainability is a holistic concept embracing not only the environment, but the other dimensions of the 'triple bottom line' that include economics and social elements, but also many other domains including culture, governance and the interrelationships between all of these.

Mining often receives bad press in sustainability debates. The industry is often perceived as an unsustainable activity; as underpinned by the view that after all, 'mining is engaged in the exploitation of non-renewable resources'. As elaborated throughout this text this view is neither fair nor valid. The mining industry has contributed significantly to leading practice in the many technical areas in which it is involved, including, notably, the broader societal concept and practice of 'sustainable development'.

The industry is often perceived as an unsustainable activity; this view is neither fair nor valid.

All industries experience an industrial 'life-cycle' evolving from origins and establishment, through to obsolescence and closure. Not only do all individual industrial plants eventually close down, move on and become replaced by other activities; so do entire industrial regions. This evolving life-cycle process is a common characteristic of industrial manufacturing business systems. Mining is no different. Admittedly it can leave a more pronounced regional footprint of disturbance through the mining voids that remain. Nevertheless, if at the appropriate stage in the project, sufficient thought and technology is committed, many of these mining voids can be re-used, put to beneficial subsequent use, or left in an environmentally benign state.

The sustainability imperative aligns mining activities with broader development challenges and opportunities. Mining ventures consume water and energy; create a carbon footprint; build infrastructure including roads, rail and settlements; impact upon local communities; create economic multipliers; modify the landscape; and manage the environment. These soft and hard mining industry infrastructure issues mean that mining ventures are critical in laying the development foundations of new regional economies, particularly in remote locations. The various strategies applied have considerable implications for the very sustainability of these emergent and evolving regional economies. These and many other aspects of mining ventures are areas where there is significant opportunity for innovative approaches to improve multi-faceted sustainability performance. These strategies range across the environmental, social and cultural, economic, technical, energy, building transport aspects of mining, to mention just some.

The traditional mining engineer may not be equipped to deal with such issues. There may be as increasing need for mining companies to equip themselves to undertake 'change management' and training, to facilitate more sustainable development approaches to mining ventures. At a macro-organizational level, corporate commitment to organizational change contributes to corporate sustainability. At a micro-employee level, change management and training can establish 'sustainability awareness' within the project workforce. A strategic corporate approach towards sustainable development requires an organizational transformation process where the issues of innovation and culture play key roles. This change can only come about through a process that enables managers to understand how the internal and external business drivers and opportunities influence an organization's long-term sustainability. Only when an organization understands its sustainability impacts and considers the drivers affecting it over the long term, can opportunities be sought for innovative practices that mitigate negative impacts and enhance positive impacts.

In any mining project there are inevitably internal and external constraints (such as timeliness and critical path delivery, or budgetary limitations) that affect the immediate extent to which sustainability can be incorporated into project design, execution and commissioning. It is therefore often necessary to adopt a tiered approach to sustainability.

In any mining project there are inevitably internal and external constraints. That affect the immediate extent to which sustainability can be incorporated into project design, execution and commissioning.

One approach (pioneered by the Australian engineering firm GHD) follows for illustration. The first step addresses social, economic and environmental impacts, but acknowledges that there may be external constraints (such as time critical path delivery, or budgetary considerations) that can limit the extent to which sustainability can be incorporated into project design, execution and commissioning. Hence, a three-level approach to sustainability engagement is provided to cater for the varying possibilities, as indicated in **Figure 22.6**.

At level 1, project impacts are identified and evaluated. When the sustainability impacts of projects are understood, opportunities can be sought to minimize the negative impacts, maximize the positive impacts and enhance the overall sustainability outcome projects. At level 2, a 'critical review' selection identifies the measures easily adopted within the context of the project's critical budget and timeline requirements; such 'low hanging fruit' can be easily incorporated into mine design. Examples may include: eco-efficiency measures, green design, fuel switching, use of alternative energy (e.g. solar panels) and grey water systems. Opportunities that have been identified, but are considered too cumbersome within the time and budget constraints can then be seeded into the pre-feasibility phase of subsequent activities or projects. Finally, level 3 provides for integrated strategic sustainability approaches for adoption through long-term partnerships.

The importance of the logic of this framework is that a realistic approach is applied first to acknowledge the significant constraints of a sanctioned project operating on critical path and where innovation cannot be allowed to obstruct the immediate operational objectives and performance requirements. However, it also provides for lessons learnt from immediate activities and projects to be incorporated into subsequent projects or project phases.

While the above model indicates a way in which specific sustainability considerations can be introduced into mining projects, the challenge remains for mining corporations to achieve corporate sustainability. Eventually to become truly sustainable in all its actions, products and outcomes, an organization must become internally sustainable in all its structures, processes and functions. A key finding within the literature is that it involves, as Porter and Kaplin (2006) note, fundamental attention to the integration of inside-out and outside-in practices. In other words, both the internal and the external environment are crucial domains with which business must engage equally.

To become truly sustainable in all its actions, products and outcomes, an organization must become internally sustainable in all its structures, processes and functions.

Traditional Mining Industry Culture and its Inherent Limitations

The fundamental issue for many international mining companies as for other multinational corporations is that there is often a characteristic 'prevailing company' culture which dominates the approach to mining operations no matter in which region they occur.

FIGURE 22.6
Developing Sustainability in Tiers

Source:
GHD 2008

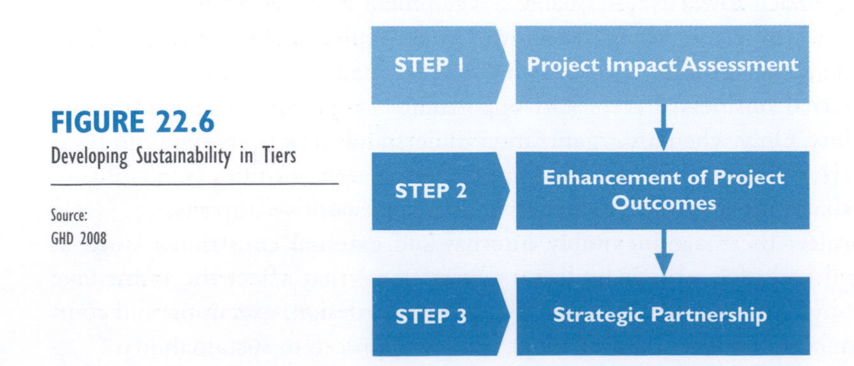

First, mining corporate culture is often dominated by assumptions that derive from the distinctive historical and cultural legacy of 'European' males conventionally from a variety of education and training disciplines, notably mining engineering, engineering and geology. Furthermore, because mining exploration is a 'rugged' activity often conducted in challenging physical settings, a tough-minded approach is embedded in mining culture. Mining exploration camps in developing countries are often the personification of 'frontier' culture (Singleton *et al*. 1997).

Second, mining professionals in the field are both adaptive and receptive to their surrounds. They have to be. Alternatively, that very same 'tough mindedness' can be an Achilles heal where a more complex notion of adaptation is required.

Following the exploration and feasibility stages, the project moves to construction and then operation. During construction, the focus of management attention is to meet budgets and schedules. The personnel recruited to manage construction, whether on the proponent's staff or the contractors, are those with a successful track record of delivering projects on time and on budget. At this stage, it is common for other aspects of the project including environmental and social aspects, to be down played, if not totally ignored.

Following construction and commissioning, the focus changes again. Early issues involve recruitment and training of the operations workforce, and fine tuning the operations in an attempt to achieve target production levels. In fact, for the remainder of the operating period, the achievement of production targets and the cost of production per unit of product, are the overriding focuses of traditional mining operations. Miners then traditionally had a rigid sense of what constitutes the core business of mining. The tough mindedness so characteristic of miners can become intolerance for different views and value sets, and a strong capacity to resist external pressures. In essence, this conventional mindset translates to business models that clearly articulate mining as concerned predominantly with the sequence of mining activities. The danger here is that other critical aspects may be under-estimated or overlooked.

There may be merits in reframing the nature of mining in the minds of both mine managers and their corporate employers. This may involve revising the mining industry's notion of the mining value chain and indeed the value proposition of international mining. This would involve a basic shift from the understanding of mining as an extraction and processing industry, to recognizing it additionally as a complex logistic operations industry with pronounced global and local ethical and sustainability obligations taking place in a world of diverse interests. A revised value proposition would also mean accepting the wider responsibilities of mining ventures and accepting non-mining principles, such as:

This would involve a basic shift from the understanding of mining as an extraction and processing industry, to recognizing it additionally as a complex logistic operations industry with pronounced global and local ethical and sustainability obligations taking place in a world of diverse interests.

- A new mining venture in a region characterized by traditional livelihoods and long established communities will be transformed by some measure as a result of the mining activity;
- International mining operators therefore have a moral obligation to set in place plans, processes and mechanisms that will guide that transformation in a way that is beneficial to local and regional communities;
- This requires mining operators to go beyond limited programmes of community engagement such as employment and training schemes for local communities, providing assistance to establish local business enterprises and purchasing materials from local suppliers, to more comprehensive partnering of regional stakeholders, and
- In essence, international mining operators now take responsibility for assisting in the sustainable transformation of local communities and traditional economies, over the entire life of the mining activity, and in such ways that the beneficial legacies of sustainable outcomes extend long after the mining activity itself has ceased.

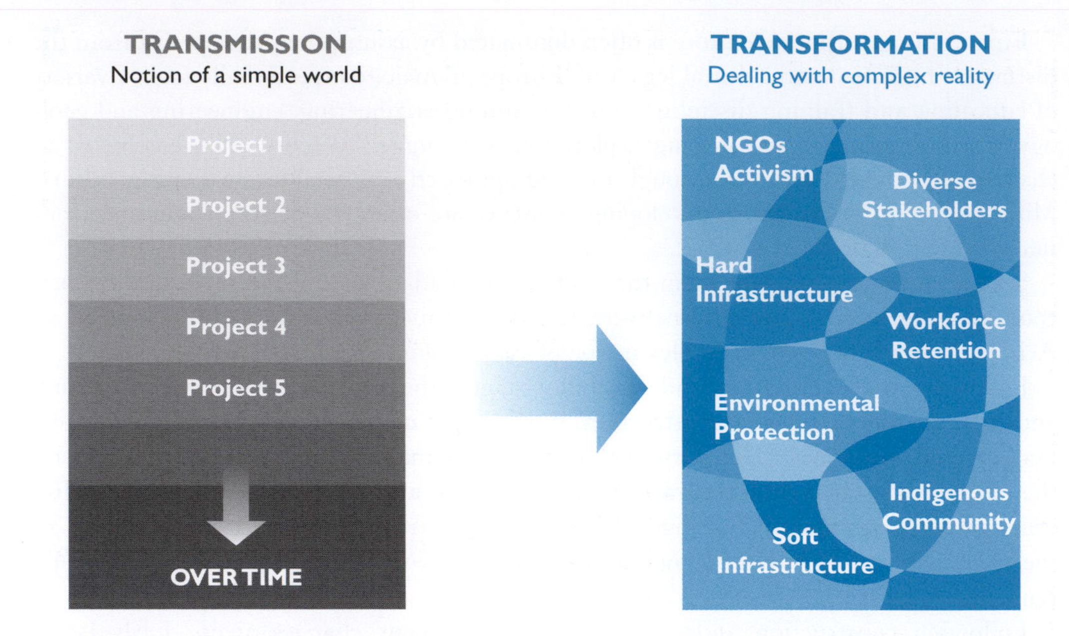

FIGURE 22.7
The Transmission and Transformation Logic Underpinning Mining Projects

Clearly this cannot and should not be achieved by mining companies alone. It requires the commitment and understanding of host governments and all other significant stakeholders involved, directly or indirectly, in the mine development.

The Notions of a Simplistic World and Complex Realities

Conventional business practice is relying heavily on science and technology to solve problems and provide answers. As effective as this world-view has been, it has limitations in dealing with an increasingly interconnected world.

A recent thinking world order framework is characterized by a more complex and organic world-view. Underlying notions of heterarchy, open systems, a holistic interconnected world with interdependencies at all levels, where complex relationships between the biotic and abiotic world exist, and where humans are part of nature, not separate from it, guide emergent interpretations from this world order thinking. This living systems approach may offer a better framework for understanding and dealing with the world as it really is (Cosby and Bryson 2005). What does this mean to individual mining companies and individual mining ventures?

Figure 22.7 illustrates the conventional representation of the two distinct worldviews towards mining projects.

The left 'transmission' side represents the conventional view of mining projects perceived as a reductionist and linear sequence of project developments, dominated by internal business concerns. In this view, the dominant mode of operation is termed 'transmission', to refer to the ordered implementation of standardized and expert-based practice, as 'transmitted' by the conventional project management process. The right 'transformation' side represents the complex realities – projects are confronted with diverse and interrelated stakeholders and issues (NGOs, regional infrastructure, governance and politics, indigenous communities, etc). This is the world of the highly responsive 'learning organization' operating with the reality of endless adaptation and change, requiring a transformation approach to the management of projects.

To operate effectively and deal with complex reality, a mining company may need to adapt in line with changes in perception of the world in which it operates.

To operate effectively and deal with complex reality, a mining company (or any corporation for that matter) may need to adapt in line with changes in perception of the world in which it operates. **Figure 22.8** illustrates the transmission to transformation change that

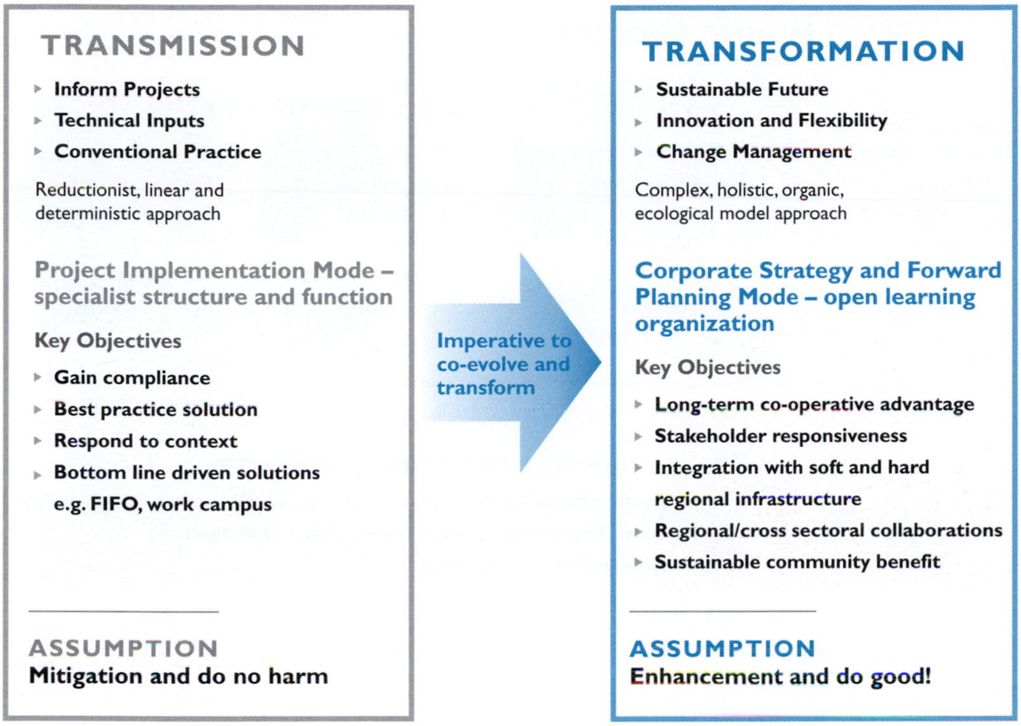

FIGURE 22.8
The Transmission to Transformation Shift for Improved Project and Corporate Performance

may be needed both at the project level, and in the overall organization. Applied to a mining company, conventional practice on the left side, with conventional specialist structure and functions applied to both company and projects, will evolve to a more complex open learning organization capable of innovation and flexibility, learning and adaptation, and the ability to engage change management. Transformation in structure and function allows more progressive and wider project objectives to be understood as 'value adding', and hence, to reframe the value proposition of the company and its broader and longer-term business objectives. It can then become an early adopter to meet the challenges of changing operating environment.

How practical is this? Clearly corporate innovation and change cannot occur quickly. The answer is that such change is likely to evolve progressively over time and at a rate that is practical and feasible. **Figure 22.9** suggests that sequential mining projects over time allow plenty of opportunity for progressive change and innovation at all levels – at the corporate level and at the project operations level. **Figure 22.9** illustrates this evolving 'transformational' change and innovation process. In practice, actual change may differ. The point however is to acknowledge that the environmental scene is continually changing, and so should industrial and societal response.

The point is to acknowledge that the environmental scene is continually changing, and so should industrial and societal response.

New Voices and New Power

The revolutionary development of digitized information communication technologies heralds the emergence of a new and important global power and global voice – in this text termed 'infosphere'. The infosphere is relevant for mining companies beyond the conventional stakeholder relationships with communities, society organizations and routine dealings with national, regional and local governments. It represents something entirely new, and its open-sourced nature and fantastic growth rate indicate its growing relevance for the mining sector and for civilization as a whole.

FIGURE 22.9
The Evolving Transformational Corporate Change and Innovation Process

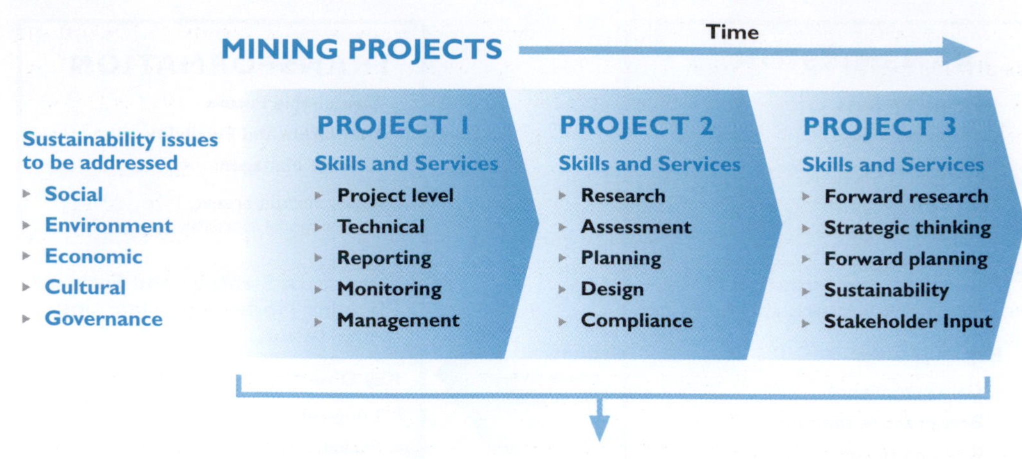

The threat of exposure of mistakes and impacts is now constant.

Catapulting off the 'up-loading' and 'down-loading' technologies of the second generation of the worldwide web its emerging impact indicates that there is nowhere to hide from the keen attention of activists and NGOs. The threat of exposure of mistakes and impacts is now constant. For example an Amazon forest-dweller can take a photo on a mobile telephone, of a mine or tailings pond spill and have it on the internet within minutes – and sent to Paris, New York, London or anywhere else. Consider also that in the case of the BP Tangguh LNG project in remote Papua, Indonesia, a 'Summary Social and Environmental Assessment Summary Report' posted to the Asian Development Bank website in 2006, achieved four million downloads within two months of posting.

The rapid expansion and open-source capabilities of the internet contribute to the creation of a participatory platform with a phenomenal capacity to mobilize global communication and engagement. Its impact is unfolding with impacts that both individuals and businesses alike, are still grasping to fully understand (**Case 22.7**). With limited control, information and ideas now flow across the globe far faster than conventional modes of communication, fostering new and cross-boundary collaboration, communication and community engagement on an unimaginable scale.

CASE 22.7
New Terminology

As with any new technology, the infosphere has coined new terms. Consider some of the diverse terms and names of Web 2.0-based resources, such as:

- Wiki (Wikipedia, wiki sites)
- Aggregator (retrieve syndicated Web content)
- Ajax (for interactive sites)
- Blog (Weblog – chronicles info)
- Folksonomy (categorize and retrieve – user generated taxonomy)
- Mash-up (mix or overlay different services from disparate Web sites)
- Open source software (source code available under license)
- Podcast
- RSS (Rich Site Summary, Really Simple Syndication)
- Tag (key word or term for searching)
- Web 2.0 Tools (applications, methods, tools, technologies – as part of Web2)
- Web feed (auto site accessing)
- Web Syndication (portion of site available to other sites or subscribers)
- Webcast (sending audio, video live)
- XML (Extensible Markup Language – W3C general-purpose mark-up language. To facilitate sharing of data across different information systems, e.g. RSS, MathML, GraphML, XHMTL, Scalable Vector Graphics, MusicXML
- Webopedia

So what is the relevance of all this for mining? The answer quite simply is that mining has always been a visible target for controversy, and hence, for activists of all types. Mines are spatially fixed, physically obvious and highly visible. They attract interest, controversy and strong opposition. The immediacy and global reach of the new global communication phenomena give stakeholders and activists a presence that mining companies and international mining ventures must deal with. The reactionary and collective power of diverse voices and the speed of information exchange by hostile or obstructive stakeholders will before long take on an entirely new dimension. Conventional approaches to compliance and accountability in all aspects of mining projects will not cope unless the mining industry responds to this challenge.

Conventional approaches to compliance and accountability in all aspects of mining projects will not cope unless the mining industry responds to this challenge.

Take something that all mining ventures must deal with – project impact assessment, the primary subject matter of this text. While science is commonly used within impact assessment, in itself impact assessment processes are not science. They are a political process. At best, they are a systematic process for society to mediate acceptable levels of change arising from human action. At worst, they are really a process of negotiating through compliance trade-offs for what was intended anyway. The issues facing impact assessment with the global voice of the internet are many. They range from such things as ethical paradoxes between individuals and organizations to preferred reading of issues by powerful organizations with a privileged world purview.

Take something else that all mining ventures probably must deal with at least once during mine life – environmental incidents. Until recently, the mining company involved was among the first to be informed of an environmental incident. The response was immediately directed at investigating the situation and implementing measures to minimize the damage, including potential damage to the company image. Before the outside world became aware of the incident, the company had assembled the facts and was well underway in its response, including possibly the retention of public relations firms to advise on appropriate responses. The company remained in control. Now, the situation has changed completely. An employee or even a passer-by, witnessing an incident, is able to take a digital photograph and transmit it anywhere in the world, before the operating company is notified. Accordingly, the New York investment community, not to mention NGOs opposed to a project, may become aware of an environmental incident at the same time or even before local management is informed. There may be no time to investigate the incident or to devise a considered response before the world press starts demanding information. In this scenario, unlike previously, the company is not in control of the agenda. This means that, to be effective, the new breed of company managers must respond in different ways to their predecessors. An additional skill set is required; that is to communicate and to respond correctly to a crisis scenario, under pressure and with little if any lead time. Clearly, in this situation, transparency and honesty are likely to be more effective than obfuscation, 'spin' or denial, and it is no coincidence that this is the way in which company responses to adverse events are moving.

Transparency and honesty are likely to be more effective than obfuscation, 'spin' or denial.

Another factor in this scenario is that employees of foreign companies operating in developing countries may find that their primary allegiance is to their country rather than their employer. This may mean – and there have already been instances of this – that an employee witnessing an environmental incident such as violation of a statutory requirement, may transmit evidence of the violation, by email directly to the government regulatory authority, without even informing the operating company. Such situations, of course, are not confined to developing countries. 'Whistle blowers' have also been active in the developed world and in some cases have received legal protection from retaliatory action from their employers. Clearly, such actions are much more likely in a situation where the employee does not respect or trust his or her employer. Accordingly, the best protection for an employer is to earn the trust of its employees through honest and transparent behaviour.

The best protection for an employer is to earn the trust of its employees through honest and transparent behaviour.

'Bottom of the Pyramid' Theory and the Notion of Regional Development

The term Bottom of the Pyramid (BOP) refers to the theory proposed by Prahalad (2005) calling for global business engagement with the world's four billion plus poorest people. Prahalad observes that the world economic pyramid consists of four tiers of relative wealth or consumption potential of the global community, measured in terms of annual per capital income. Tier 1 includes the 100 million or so middle and upper income earners on more than US$ 20,000 per year, Tiers 2 and 3 include the 1.5 to 1.7 billion poor customers of the developed countries and the aspiring middle class of the developing countries earning between US$ 1,500 and US$ 20,000 per year. Tier 4 comprises the poor at the bottom, overwhelmingly in the developing or underdeveloped countries and earning less than US$ 1,500 per year ('purchasing power parity' equivalent) which is the threshold income considered necessary to sustain a minimalist standard of living. Of this four billion over one billion are estimated to earn less than US$ 1 per day. On current projections Tier 4 will grow to more than six billion over the next 40 years.

The BOP theory rests on the principle that the Tier 4 population, who are currently not able to participate in the global market economy, represents a global ethical dilemma of immense proportions yet also a potential multi-trillion-dollar market.

The BOP theory rests on the principle that the Tier 4 population, who are currently not able to participate in the global market economy, represents a global ethical dilemma of immense proportions yet also a potential multi-trillion-dollar market. Critics argue that business mistakenly overlooks the potential to engage Tier 4 successfully and profitably. The global sustainability agenda, and the potential for continuing global economic prosperity and political stability necessitate that this challenge be met (Prahalad and Hart 2005). These are, of course, relevant arguments for the mining sector, particularly as it becomes more active in developing countries.

In fact mining is an interesting case study because it differs in operational and functional terms from many of the case studies illustrated by Prahalad (2005) and other authors in discussing the BOP theory. Specifically, mining activities have fixed spatial characteristics, rather than geographically dispersed or non-spatial product markets or services. Next, mining proponents have a very rigid sense of what constitutes the core business of mining. Hence, the $100 million or so spent by such mining companies as PT Freeport Indonesia on community programmes is seen principally as a 'cost of doing business', and not a potential investment in 'non-core' but legitimate business development. Essentially, this conventional mindset translates to business models that clearly articulate mining as being concerned with sequential mining activities.

The core business of mining, however, does engage with local populations and communities in the limited phases of the mining venture. Contact during mining exploration is usually limited, but sets up expectations in the local region at both the political and administrative level, and this interest streams down to the level of the local community. Opportunists quickly respond, including the poor who gravitate towards mining in search of employment and a better life. The scale and range of stakeholders that can be involved is illustrated in **Figure 22.10**, derived from a proposed mine in Indonesia. This indicates stakeholders ranging from local village communities, regional populations and four tiers of government. Even this figure does not include diverse business interests or local, national and international NGOs.

Mining companies have the opportunity to turn something previously viewed as 'a cost of doing business', to a business opportunity.

By embracing BOP theory and principles, mining companies have the opportunity to turn something previously viewed as 'a cost of doing business', to a business opportunity. The motivations for doing so differ from those driving existing practice to address the issues of social impact management and net community benefit. The latter are statutorily required, whilst the former are more related to the expanding sustainability agenda which is becoming increasingly embedded in the corporate charter of a company. The former

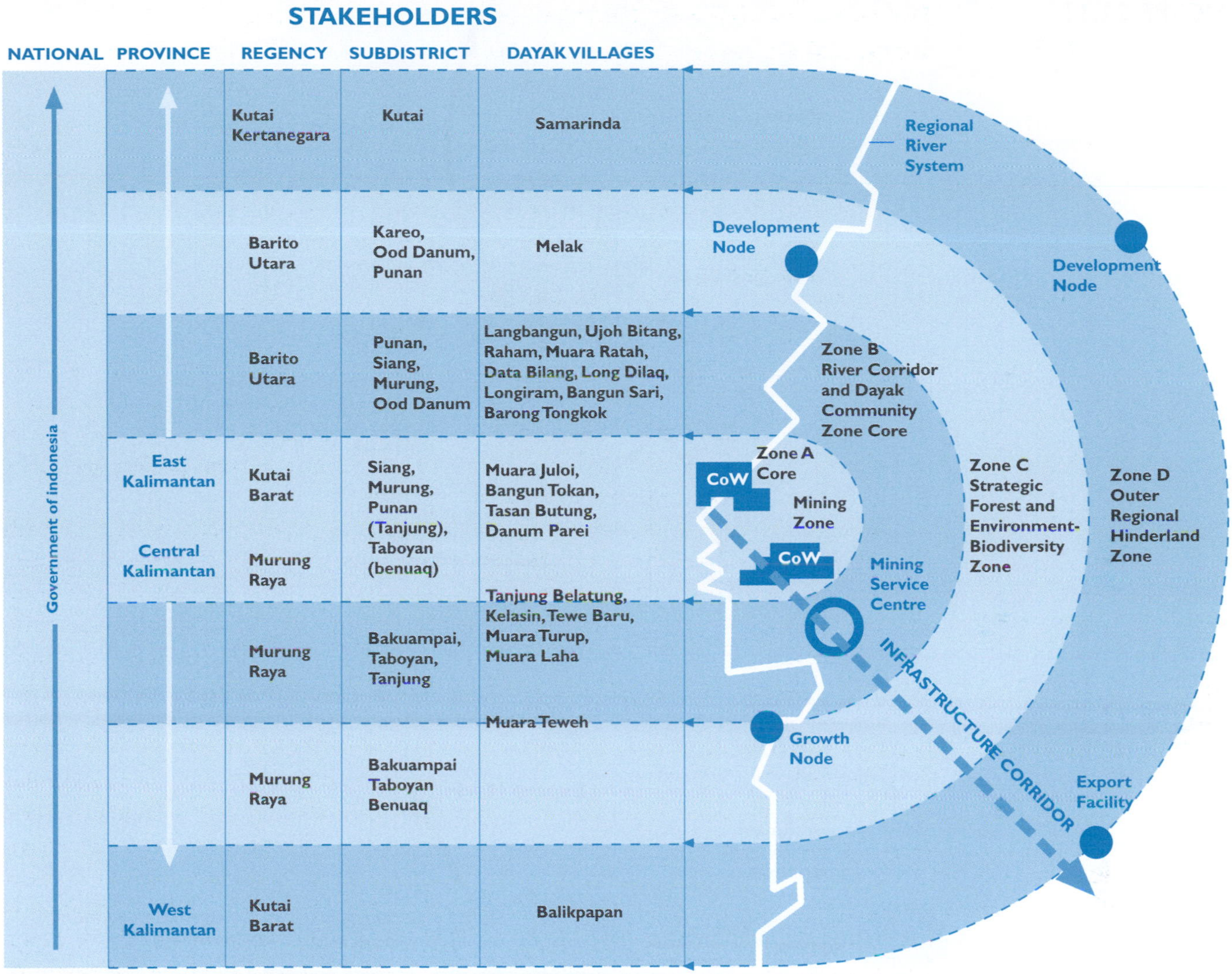

FIGURE 22.10

The Scale and Range of Stakeholders in a Large Mine

Stakeholders range from local village communities, regional populations, to numerous tiers of government. Note that the figure does not include diverse business interests or local, national and international NGOs.

idea is also a logical extension of the increasing recognition of the governance issues facing mining companies operating in the poor regions of developing countries in particular. The challenge is for mining to go beyond the traditional and current practice of activities dependent on mining that result in a single sector regional economy, to the notion of creating a diversified regional economy. In a diversified economy, certain key aspects may be linked to the mining sector, but are certainly devised to have separate and external economic links outside the mining sector, and hence create far more resilient and independent businesses and economy. **Figures 22.11** and **22.12** illustrate these two economies.

The thinking here is for mining companies to broker or negotiate with national and regional governments, a mechanism for doing so and one which can consciously engage not just the local communities, but the wider 'bottom of the pyramid' in the region in which they are operating.

FIGURE 22.11
A Single Sector Economy Illustrated

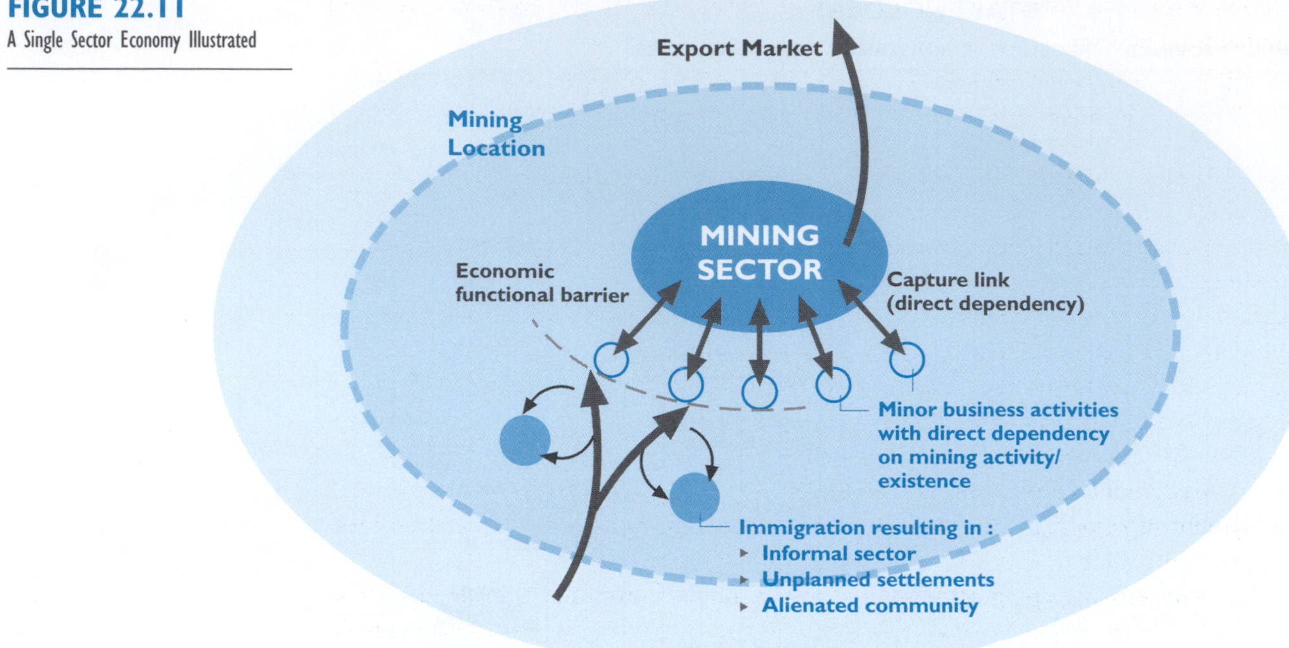

FIGURE 22.12
A Diversified Regional Economy Illustrated

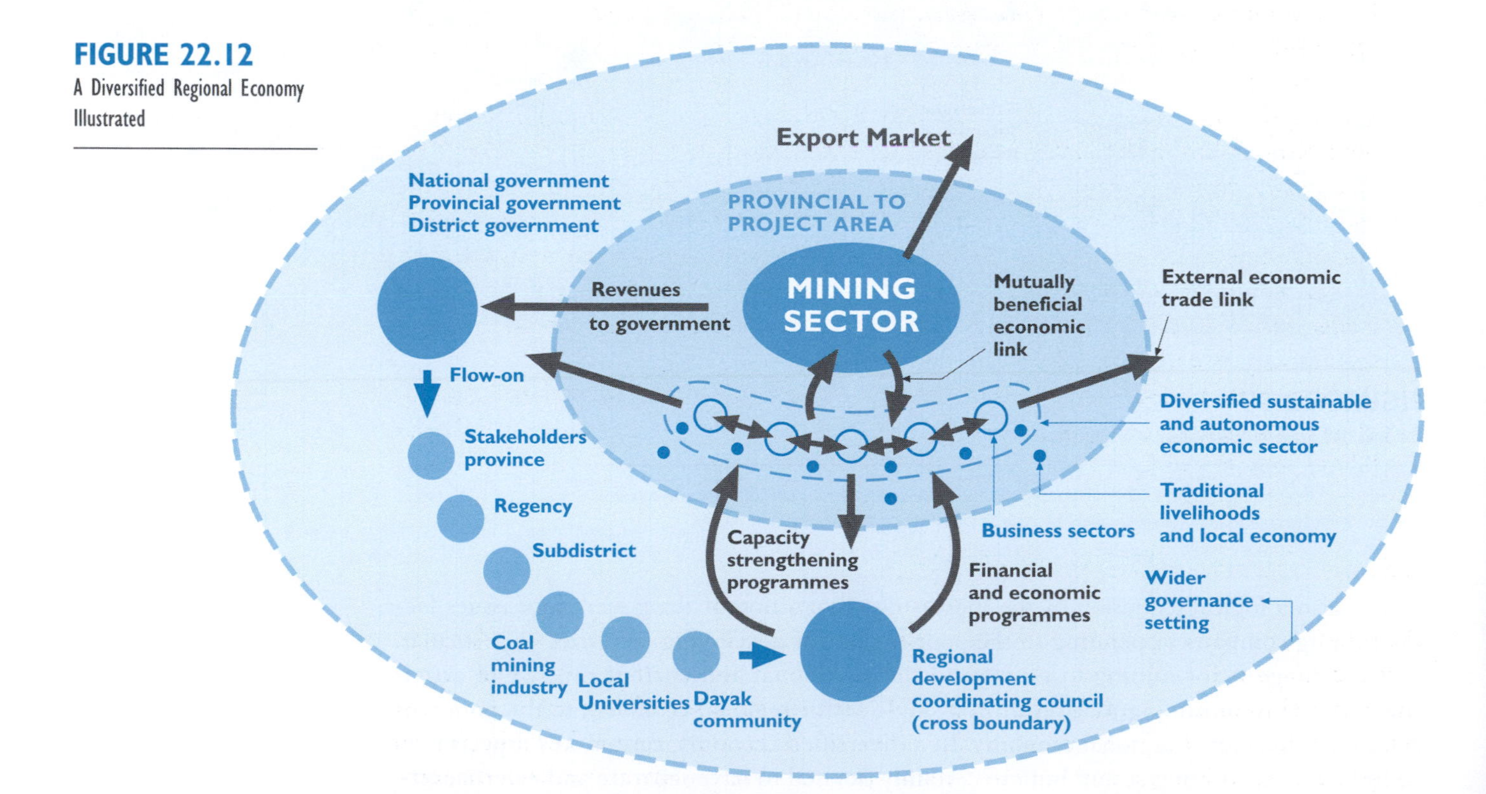

The 'Diversified Regional Economy' involves the following processes:

- Providing funding to the region from the mining revenues, whether from tax to the central government, or in the case of very large mining ventures direct funding from the mining company;
- Establishing a legal entity to receive and handle such funds;

- Creating a regional development coordination mechanism that is representative of all regional stakeholders, and
- Developing and implementing programmes to strengthen the regional non-mining sector capacity through the provision of financing and economic development programmes.

This model is now at the cutting edge of more innovative international mining and natural resource industry activity in developing countries (**Case 22.5**). However, there are good reasons to argue that this model for sustainable regional development and mining companies should go further.

Essentially within the framework of this model lies the potential for mining companies to explore, identify and develop a variety of BOP business enterprises. These may be partnerships with local interests and communities. The novel idea promoted here is that they should be treated as part of the core business of mining and natural resource exploitation. This requires the following changes in approach:

- A more fundamental and deeper commitment to regional partnerships;
- Deployment of resources into research and development to identify regional business enterprise potential;
- Employment within mining companies of a new suite of professional skills to address diverse BOP business potential (e.g. economists, micro-business analysts, organizational change experts, international cross-cultural management experts, economic anthropologists, sustainability scientists, etc);
- Change management and business analyses to develop new and more appropriate business models that allow concurrent financial differentials to operate within a corporation;
- Culture change within the corporation that allows acceptance of the legitimacy of opportunistic business initiatives outside the conventional 'core business' mindset;
- Acceptance that 'cost-neutral' low return or nil-return business activities are entirely legitimate within a business model, if it can be demonstrated that such activities derive tangible qualitative or quantitative benefits to either the corporation itself (i.e. corporate reputation, good regional politics and relationships) or the wider community (stable society, social harmony, growing regional economy, progress toward regional self-sufficiency and sustainable local economy), and
- Realization that small beginnings in BOP micro-enterprises can lead to regional, national or global business growth, and hence to a valid economic engagement of the world's four billion poorest people.

22.4 CONCLUDING REMARKS

As noted in the opening chapter, mining is a human activity that long predates civilization. It was a fundamental element in the growth of most historical civilizations, and is one of the essential foundations of the industrial and post-industrial societies of the twenty-first century. However, in its modern industrial phase mining can have a profound impact on the physical environment, and environmental consequences may extend far from the mine and may last for many years even decades after the cessation of mining. For years, this was considered the price of extracting the minerals necessary for building and maintaining the vast consumer societies of Europe, North American and Japan, whose products are now found around the globe. But in the 1970s, concerns about the health and aesthetic impacts of increasing air and water pollution led to the passage of laws which established the basis for what is now the nearly universal process of EIA or environmental impact assessment.

Although the requirements of the process can be onerous and expensive, very few mining projects are immune, perhaps only a few small, local projects in developing countries.

Artisanal mining, a simple affair with pick, shovel and pan, persists in many developing countries, but modern, large-scale mining is a much more complex affair, even before considering the EIA. And that process, as described at length in the foregoing chapters, adds layer upon layer of complexity that makes the engineering phases of planning and implementation seem almost simple by comparison. Because the EIA now goes far beyond the physical impacts on local environment to address social and political realities, it represents a particularly sensitive management challenge, demanding continuing attention to the human environment around, and often far beyond the mine site.

Because the EIA now goes far beyond the physical impacts on local environment to address social and political realities, it represents a particularly sensitive management challenge, demanding continuing attention to the human environment around, and often far beyond the mine site.

One need only look again at the Ok Tedi copper and gold mine in remote Papua New Guinea, as referred to more than once in this text, to grasp the complexities. Although PNG has comprehensive environmental laws, and the mine itself was planned to meet high standards of environmental management, when during the construction stage, a major landslide occurred on the abutment of the embankment planned to store tailings from the project, the mining consortium and the PNG government were confronted with a difficult decision. Given the cost and the unstable geology of the region, rebuilding the dam was out of the question. With a major stake in the project, and with an eye to long-term cash flow, the government would have been loath to cancel the project. The consortium, likewise, was not keen to write off its hundreds of millions of invested capital. Therefore, it was agreed that mine wastes would be discharged into the Fly River system. The long-term consequences of this political/economic decision are now well known and widely publicized: heavy siltation of the river, adverse effects on lifestyles of 30,000 local inhabitants, loss of lowland forest vegetation over hundreds of square kilometres, reputational damage to involved consortium members and the mining industry as a whole, and several hundred million dollars in legal costs and compensation. However, given the massive investment, and the revenue stream essential to pay it off, would any other decision have been likely?

Such are the complexities of large mining projects.

Clearly, the mining industry has learnt from projects such as Ok Tedi. The industry has now developed a suite of technologies to effectively manage environmental impacts under most circumstances. And, more recently the industry has tackled the more difficult issues involved with the impacts of mining on the communities involved. It is hoped that this review of all aspects of the EIA process associated with modern mining as presented in this text, has provided a comprehensive coverage of the subject, which will enable the reader to understand and participate actively and intelligently in the complexities of environmental assessment.

The mining industry has encountered many difficult challenges over its history, none more serious than the environmental and social problems that emerged in the second half of the twentieth century. The industry has responded to the challenges, initially in response to public criticism but increasingly in a proactive way so that, in many respects, performance now exceeds legislative obligations and public perceptions. Challenges remain, particularly in relation to social impacts and opportunities for social benefits in developing countries. However, the industry is well placed to meet these challenges and to achieve beneficial social outcomes. Today the industry has shifted the emphasis of environmental assessment from understanding, avoidance and mitigation of harm to the generation of environmental, social and micro-economic benefits, thereby creating a favourable net impact from mine development. Creating these benefits, rather than questions of legal and regulatory compliance, and sharing benefits equitably, will continue to dominate discussions about mining in developing countries over the next decades.

The industry has responded to the challenges, initially in response to public criticism but increasingly in a proactive way so that, in many respects, performance now exceeds legislative obligations and public perceptions.

REFERENCES

Crosby B and Bryson J (2005) Leadership for the Common Good: Tackling public problems in a shared world. Jossey-Bass, A Wiley Imprint. USA.

Humphreys D (2000) Mining as a Sustainable Economic Activity; Paper prepared for presentation to an informal seminar on the mining and metals industry at the OECD, Paris, 9 February 2000

GHD (2008) Sustainability Awareness Training Program. In-house technical report; Unpublished

National Mining Association (1998) The Future Begins with Mining: A Vision of the Mining Industry of the Future; (www.oit.doe.gov/mining/vision.shtml).

Singleton J and Singleton H (2008) 'Impact Assessment and the New World Order'. Paper presented to the International Association for Impact Assessment Conference, 'The Art and Science of Impact Assessment, Perth, Western Australia (2008) (In Press)

Prahalad CK (2005) The fortune at the Bottom of the Pyramid: Eradicating Poverty Through Profits, Wharton School of Publishing, USA

Prahalad CK and Hart SL (2005) The Fortune at the Bottom of the Pyramid. Retrieved 8th October 2005 from http://www.changemakers.net/library./temp/forutnepyramid.cfm

• • • •

Index

A

Acceptance criteria 107, 834–6
Acid *see also* pH
 mine drainage (AMD) 52, 55, 65–6, 216, 527, 696, 702, 705, 715, 717–18
 neutralizing capacity (ANC) 707, 709
 producing potential *see* NAPP (net acid production potential)
 rock drainage (ARD) 24, 216, 230, 403, 409–10, 413–15, 420, 451, 454, 456–7, 471–2, 487, 510, 516, 526, 528, 537–42, 559, 575, 695–718, 723, 776, 788, 815, 824–5, 839–40
Affected communities 33, 66, 75, 85–7, 124, 128, 137, 143, 148–9, 156, 362, 445, 596, 610, 620–1, 632, 634, 643, 647–8, 657, 677, 692, 760, 860
Air pollutant 299–300, 543
Alternatives *see* Project alternatives
Amalgamation 192, 210, 227, 245, 247–50, 267
Anoxic limestone drains (ALDs) 715–16
Anthropological surveys 597, 684
APELL (Awareness and Preparedness for Emergencies at Local Level) 56, 111, 460, 462, 564, 574
Aquatic environment 249, 435, 575
Argyle Diamond Mine, Australia 60
Arsenic (As) 27, 166, 196, 227, 347, 389, 482–3, 485, 487, 490–2, 527, 567, 696, 701, 802
Artisanal mining 166, 210, 222, 243–5, 247, 249–51, 409, 612–13, 790–2, 794, 796, 801, 880
Asbestos 496–9, 502, 576
Assets, loss of 576, 590–1, 601, 615
Atlas Copper Mine, Philippines 244, 760, 764
Atmosphere 23, 177, 282–3, 295–300, 302, 307–8, 314–15, 386, 422, 543, 545–51, 566–7, 576, 653, 723, 725, 861
Audit
 criteria 510–11
 findings 513, 515
 protocol 511–12
 team 510–14
Audit Checklist 517, 519–23
Auditing, environmental 39, 97, 107–8, 164, 504, 509–15, 755, 757
Audits, environmental management system 510–11

B

Backfill 730, 733, 744, 795
Bacterial leaching 260, 263, 269
Baia Mare Mine, Romania 60, 138, 275, 727, 729
Bankable Feasibility Study (BFS) 190, 197
Base metals 138, 175, 195, 222–4, 235, 237, 268, 303, 776, 855
Baseline 84, 94, 97, 347–9, 353, 356–7, 365–6, 371, 378–9, 404–5, 451–2
 surveys 73, 315, 324, 349, 360–1, 364–9, 371, 373, 375, 377–9, 381, 383, 437, 470–1, 473
Batu Hijau Copper Gold Mine, Indonesia 60, 181, 322, 528, 698, 712, 760, 762–5
Berkeley Pit 61, 541
Best available technology 42
Bingham Canyon Copper Mine, USA 60
Biodiversity 296, 310–15, 326, 331, 339, 344, 351, 353, 364, 366–7, 378, 392, 396, 433, 460, 507–8, 510, 528, 555–9, 572–3, 811, 832
Biogeography 311, 314
Bio-oxidation 269
Bottom of the Pyramid (BOP) 876–7, 881
Bottom Up Construction (BUC) of mine waste rock storages 780–2
Bougainville Copper Mine, Papua New Guinea 5, 14, 49, 61, 196, 217, 532, 603, 765, 838, 841

C

Cadmium (Cd) 166, 227, 482–3, 485, 487–90, 492–3, 501–2, 527, 714, 802, 839
Calcination 91–2, 279–81
Cancer, lung 492, 497–8, 565–6
Capacity building 71, 75, 125–6, 129, 131, 156, 315, 561, 595–6, 605, 616, 619, 626, 639–42, 648, 654–5, 657, 666, 679, 682–3, 821, 857, 859
Capital
 natural 17–18, 562, 860
 natural resource 8, 46, 86, 396, 641–2
Carbon dioxide (CO2) 242, 273, 545–9
Carbon-in-Leach (CIL) 267
Carbon-in-Pulp (CIP) 249, 267, 269
Carcinogen, carcinogenic 271, 485, 492, 550, 587

Cash compensation (*see also* Compensation, financial) 594, 602–3, 605, 607, 611–12, 666, 689
Catchment area 431, 433, 748, 750, 784
Chatree Gold Mine (Thailand) 214, 671, 712, 739, 854
Checklists 78, 419–21, 435, 512
Chromium (Cr) 166, 223–4, 482–3, 486–7, 489–90, 492–3, 501–2, 719
Coal 4, 7, 11, 52, 138, 179, 191, 204, 207, 211–13, 218, 222, 227, 231–4, 243, 282–3, 300, 302, 323–4, 343, 406, 419, 493, 544–54, 573–4, 696–7
  combustion, fires 544–5, 547, 549–51, 576, 858
  mining 31, 80, 203–4, 206, 211, 214–15, 230–2, 528, 535, 547, 551, 553–4, 573–4, 576, 851, 863
Coal bed methane (CBM) 95, 223, 552–4, 576
Code of conduct 100, 622, 631, 634, 648–9, 656, 663, 668, 682
Combustion, spontaneous 544–5
Communities
  human 359, 369
  neighbouring 207, 611, 647, 651, 685
  project-affected 450, 621, 650
  reference 369, 474, 650, 658
  regional 142, 871
  rural 14, 364, 443
Community
  consultation (*see also* Public consultation) 52, 76, 134, 157, 343, 364, 445, 579, 595, 597
  development (CD) 71, 74, 126, 131, 148, 526, 591, 606, 612, 618–24, 626–30, 632–4, 636, 638, 640–2, 644, 646, 648, 650–4, 656, 658, 672, 675, 689, 840, 854
  development programmes 22, 96, 107, 147, 215–16, 321, 465, 474, 576, 597, 603, 612, 621–2, 624–7, 629, 632, 640, 642, 647, 652, 656–7, 675, 690–1, 840
  health (*see also* Public health) 38, 161, 296, 321, 450, 501–2, 505, 574, 645–6
Community and Indigenous Peoples Development Plan (CIPCP) 627–8, 634
Compensation 85, 99, 148, 231, 246, 335, 370, 444, 466, 498, 501, 576, 579, 582, 589–92, 594–608, 611–12, 615, 665–6, 672, 674, 677, 688–9, 727, 765, 779
  financial 148, 443, 666
Control sites 369, 378
Conservation, environmental 2, 20, 47, 328–9
Conservation reserves 315, 437, 579
Construction workforce 620, 680
Copper (Cu) 1, 3, 4, 6, 9, 21, 54, 60, 63, 65–6, 69, 123, 138, 166–7, 221, 223–4, 254–6, 258–65, 290–1, 483–5, 487–90, 492–3, 539, 541–2, 702, 722, 846–8
Costs of delay 71, 103–5
Cultural heritage sites 397, 437–9, 649
Cultural sensitivity 132, 155, 512
Cultural norms and values *see* Norms and values
Cultures 20, 45, 48, 121, 135, 143, 296, 314, 316, 318–20, 341, 398, 403, 438, 444, 477, 609, 619–20, 652–3, 661–2, 664, 666–8, 670–2, 674, 687–8, 869
Cumulative impacts 409–11, 422, 424–5, 559
Customary law *see* Traditional law
Cutoff Grade (COG) 114, 169–71, 188, 190, 848
Cyanate (CNO) 271–3, 275
Cyanide (CN) 54–5, 210, 235, 254, 263, 267–8, 271–2, 274–8, 290–1, 343–4, 357, 458, 470, 472–3, 483, 514, 537, 700, 724, 726, 729, 766, 770
  destruction 275, 766
  free 271–2, 275–7
  total 272, 277
  weak acid dissociable (CNWAD) 272, 277
Cyanide Amenable to Chlorination (CATC) 272, 277
Cyanide Code 276–8, 510

D

Darling Range Bauxite, Australia 62, 171
Decant
  systems 721, 725, 732, 745–6, 756
  water 469, 723, 726, 732, 746–7
Deep Sea Tailings Placement (DSTP) 60, 720, 735, 760–9
Delphi technique 447–8, 461
Depositing tailings 60, 529, 766
Diamonds 53, 60, 62, 64, 167, 175, 181, 194, 211, 222–3, 227, 231–2, 323, 385, 392, 500, 539, 602, 809
Disclosure 37–8, 95, 124–5, 127, 129–30, 143, 153, 155, 278, 507, 590, 611, 619, 632, 654, 679
  plan 100, 129, 131, 630
  public 35, 89, 93
Displaced people 589, 590, 592, 601–3, 607–9, 611, 614
Disposal, riverine (tailings) 728, 731, 779
Dry stacking of tailings 230, 732–3, 738
Dump leaching 194–5, 239–41, 261–2, 287–8, 535
Dust control 568, 804, 806
Dutch disease 561, 563

E

Eagle Mine, Colorado 704
Earth Summit 32, 52, 124, 126, 309, 326, 339–41, 343
Ecological
  function 313, 332, 413, 823
  modelling 433, 461
  risk assessment 35, 462
Ecological model 386, 433
Ecosystems 2, 41, 81, 83, 275, 296, 305, 310, 313–14, 324–5, 329–34, 336, 356–8, 364, 366–7, 377, 403–4, 413, 425, 427, 433, 435, 457, 460, 557, 701
Electrostatic precipitator (ESP) 549–51
Electrowinning 192, 254–5, 262, 264–5, 269–70
Emission factors 288–90, 553
Encapsulation 464, 711, 713–14, 782
Endangered Species 52, 79, 95, 189, 315, 342, 366–8, 411, 437, 558, 642
Environmental Action Plan (EAP) 73–6, 83, 96, 98, 464, 467–8, 478–9
Environmental approvals 97, 191, 198, 200, 458
  obtaining 407, 867
Environmental assessment and impact assessment (EA and EIA) 2–3, 15–16, 23, 33–9, 50, 70–3, 75–93, 97–121, 126, 128, 137, 151, 157–8, 164, 166, 171, 197, 201, 362–4, 405–8, 410–11, 414–15, 427–8, 445–51, 461–2, 466, 504–10, 514–16, 765, 780
Environmental assessment guidelines 40, 48, 461, 654
Environmental assessment process 40, 79, 90, 93, 135, 326, 330, 360–1, 364, 427, 446, 449–50, 453, 460, 479, 549, 860, 861
Environmental audits 107, 344, 476, 504, 506, 509, 511–15
Environmental Impact Statement (EIS) 7, 34–5, 74, 98, 99, 126, 160, 200, 461
Environmental impacts
  direct 409–10, 417, 419, 424, 436, 440, 442, 508, 547, 557, 576, 723, 765
  indirect 81, 134, 208, 409–10, 419, 423–5, 427, 436, 440, 442, 508, 556–7, 576, 788
Environmental indicators 348, 351–7, 359, 391–2, 478
Environmental justice 348, 351–7, 359, 391–2, 478
Environmental laws/legislation 28, 31, 34, 52, 354, 397, 444, 509–11, 537, 778
Environmental management 6, 25, 30, 42–3, 45, 49, 52, 74–5, 96, 98, 100, 105–7, 110, 171, 252, 290, 321, 325, 343, 463–8, 474–8, 508–9, 511, 697, 840, 856
  effective 96, 212, 692
Environmental Management and Monitoring 105, 107, 464–5
Environmental management programme 74, 514, 859
Environmental Management Systems (EMSs) 11–12, 25, 36, 50, 53, 96, 105, 206, 216, 229, 254, 281, 286, 461, 475–7, 479–80, 510–11, 513
Environmental mitigation 98, 99, 357, 611
  measures 72, 74, 85, 150, 407, 478–9, 504
Environmental monitoring 75, 96, 107, 116, 359, 369, 391, 468–9, 480, 513, 638
Environmental Monitoring Programmes 468, 470, 473, 584
Environmental performance indicators (EPI) 355, 392, 468
Environmental Protection Agency 61, 219, 252, 541, 574
Environmental regulations 55, 441, 509–10, 856
Environmental Risk Assessment 452, 456, 461, 501–2
Environmental risks 28, 29, 36, 197, 284, 456–8, 505, 509, 575–7, 616, 721, 727, 748–9, 868
Environmental and Social Performance Standards 37–8, 49, 113, 461, 516, 614, 655, 859
Environmental standards 37–8, 101, 324, 434, 854
Equator Principles 3, 36–7, 39–40, 53, 76, 78, 93–6, 113, 124, 133–5, 137, 198, 397, 445, 450, 505–7, 513–14, 589, 611, 627, 636, 868
Equator Principles Financial Institutions (EPFIs) 36, 39–40, 88, 94, 124, 133–5, 137, 505, 514, 866, 868
Erosion 118, 168, 230, 282–3, 302, 304–5, 397, 409, 422, 460, 526, 529, 535, 552, 554, 557, 567–8, 581, 738, 754–5, 787–8, 790–8, 806–7, 817–19, 822–3, 838
  control 238, 336, 580–1, 784, 792, 794, 828–9, 831, 833, 841
  management plan 796
EW 262, 264
  process 264–5
Exotic species (*see also* Introduced plants, pests) 313, 394, 828, 830–1
Exploration
  activities 178, 188, 386, 407, 578–9, 581, 583–4, 603, 849
  areas 581–2
  drilling 173–4, 180–1, 198, 199, 558
  mineral 12, 14, 28, 183, 218, 246, 685, 854
  programme 173–4, 183, 185, 196, 384, 578–80, 584, 629

F
Failures, tailings storage 106, 747, 752, 754
Feasibility study (FS) 15, 72, 85, 104, 108–10, 188–90, 196–8, 466, 752, 759, 845
Flotation 192, 194, 209–10, 230
Food and Agricultural Organization (FAO) 149, 157, 306–7, 315, 344
Forests, protected 113, 305, 338
Fugitive emissions 281, 287, 542

G
Genetic diversity 313, 366, 394, 557, 559, 826
Geographic Information System (GIS) 349, 390–1, 425
Geomembranes 94, 235–6
Globalization 29, 589, 678, 693
Global Mining Initiative (GMI) 48, 53, 56
Global warming 8, 31, 42–3, 49, 57, 101, 283, 308, 328, 342, 508, 547, 549, 577, 858
Gold (Au) 167, 223–4, 268, 277, 483–4, 499
Gold rushes 3, 169, 243–4, 251, 612, 850
Google Earth 21, 136, 229, 363, 376
Grasberg Copper Mine, Indonesia 3, 21–2, 61, 63, 84, 96, 108, 136, 140, 173, 191, 244, 300–1, 309, 339, 345, 381, 385, 412–14, 416, 22, 452, 472–3, 515–16, 529–30, 606, 646, 693, 714, 728, 731, 762, 765, 826, 844, 864
Global Reporting Initiative (GRI) 33, 49, 53, 106, 111, 478–9, 868
Greenhouse gases (GHG) 52, 81, 290, 343, 397, 508, 510, 544–6, 548–9, 551, 574, 576, 862–3
Grievances, grievance mechanisms 149, 469, 615, 632, 635–6, 657, 668
Gross domestic product (GDP) 21–2, 59, 63, 108, 136, 140, 173, 191, 322, 335–6, 351, 386, 396, 405, 562, 638
Groundwater 217, 232, 236, 241, 244, 271, 274, 278, 285, 295, 307–9, 367, 370, 385–6, 422, 428, 471–2, 537, 540–1, 567, 569–71, 575, 699, 725–6, 747–8, 839
- contamination 274, 472, 537, 558, 567
- quality 118, 278, 367, 397, 471–2, 570–1, 752
- rebound 575, 577, 699
- resources 181, 307, 839

Groups, vulnerable 95, 128, 251, 593–6, 615–16
Growth inducement 440, 442, 560

H
Habitats 41, 119–20, 205–6, 312, 314–15, 326–7, 338, 363–4, 366–7, 377–8, 390, 394, 398, 399, 403–4, 406, 409–10, 412–13, 418, 433–4, 436–7, 460, 528–9, 555–9, 661, 768–9, 830–1
Health and safety (*see also* Public health) 28, 157, 278, 355, 371, 453, 475–6, 478, 512, 514, 530, 558, 563, 565, 576, 589, 613, 630, 757
Heap leaching 192, 194–5, 222, 235–41, 261–3, 267, 271, 287–8, 432, 535, 544, 575, 804, 815
Heavy metals 66, 283, 385, 483, 542, 557–8, 567, 575, 696, 700–1, 766
Heritage sites, cultural 437–9, 649
High density sludge (HDS) treatment process 715
High Pressure Acid Leaching (HPAL) 265–6
Historic sites 79, 420, 437–40
HPAL (High Pressure Acid Leaching) 265–6
Human rights 31, 71, 79, 126, 140, 340, 345, 561, 598, 599, 602, 606, 612, 622, 638, 673–4, 676
Hydrogen cyanide (HCN) 268, 271, 273, 275
Hydrometallurgical processes 255, 261–7, 283, 721
Hydrometallurgy 265–7, 269, 271, 273, 275, 279, 286–7, 290

I
IAIA (International Association for Impact Assessment) 34, 49, 54, 314, 344, 363, 461
ICME (International Council on Metals and the Environment) 501, 573, 720, 724, 743, 769–71
ICMM (International Council on Mining and Metals) 54, 363, 501, 556–8, 573
IEA (International Energy Agency) 52, 549–50, 573
IIED (International Institute for Environment and Development) 49–50, 54, 56, 111, 240, 614, 632, 655, 771
ILO (International Labour Organization) 622, 673, 676
INCO SO2 Air process 270–1, 273, 275
Indicator Frameworks 352–3, 378–9
Indicators 74, 90, 96, 99, 121, 130, 229, 326, 336, 345, 348–9, 351–60, 379, 387, 391, 427, 430, 443, 638–9, 652, 657, 667, 703, 756
- alarm 353–5

Indigenous
- Communities/societies 660, 662–4, 666–72, 674–5, 677–90, 692, 872
- leaders 669, 678–9
- Peoples 20, 38, 128, 133, 315, 350, 364, 367, 377, 444, 507, 558, 591–7, 600, 609, 619–20, 622, 652, 654, 660–85, 689, 692, 833, 840

Indonesian government 21, 140, 389, 470, 863
Indonesian Government Regulations 49, 94
Induced development 19, 95, 410, 412, 441–2, 508, 528, 556, 558, 560–1, 576
In situ leaching (ISL) 192, 234, 241–2, 262, 569, 571
Industrial minerals 8, 53, 165, 228, 230–1, 234, 548, 562
International Association for Impact Assessment *see* IAIA
International Council
  on Metals 501, 573, 769–71
    and the Environment *see* ICME
  on Mining and Metals *see* ICMM
International Cyanide Management Code 54, 271, 343
International Energy Agency *see* IEA
International Institute for Environment and Development *see* IIED
International Labour Organization *see* ILO
International law 336–8, 345, 600, 622, 673–4, 676, 866–7
Introduced plants 559
Involuntary resettlement 38, 57, 367, 462, 480, 508, 516, 561, 589–91, 598, 606, 608–10, 612, 614–15
Island Copper Mine, Canada 760, 766, 770, 825, 840

K

Kankana-ey traditional miners 247, 249, 251, 612
Kinetic geochemical tests 708

L

Laboratories 176–7, 187, 195, 233, 285, 382–5, 470, 474, 569, 585, 751
  environmental 384–5
Land acquisition 74, 100, 148, 174, 200, 211, 370–1, 465, 508, 526, 576, 587, 590–4, 596, 598–604, 606, 608, 610–12, 614–16, 623, 637, 660, 855
Land Acquisition and Resettlement Action Plan (LARAP) 592, 611
Land compensation 148, 510, 576, 594, 598–603, 605, 611
Landform design 782–3, 786, 820–2
Landforms 236, 240, 301, 303–5, 422, 528–9, 531–2, 542, 558, 562, 742–3, 782–3, 785, 806, 816, 821–3
Land ownership 100, 132, 150, 189, 399, 439, 441, 579, 635
Land use 76, 79, 83, 100, 108, 120, 148, 300–1, 306–7, 330–1, 398, 413, 441, 508, 559–60, 591–2, 596, 600–1, 603–4, 641, 755, 757–60, 776–7, 784, 816–22, 827–8
Leaching 7, 192, 194–5, 216, 222, 230, 235–7, 239–42, 254–5, 260–8, 274–5, 287–8, 302, 307, 418, 458, 487, 492, 535, 566, 569–70, 698, 710–12, 721, 817–19
Lead (Pb) 7, 223, 265, 299, 485, 488–9, 502, 539, 862–3
Legumes 798, 800, 828, 830–1
Leopold matrix 422, 461
Life Cycle Assessment (LCA) 7, 862–3
Lihir Gold Project 560, 607, 760, 780
Limestone 216, 223, 228–9, 257, 288, 302–3, 500, 710, 714, 716, 749, 774, 793, 809, 853
Livelihood restoration 590, 592, 594, 597, 599, 604–5
Los Frailes Mine, Spain 509, 727

M

Macquarie Harbour, Australia 65, 701–2
Management
  plans, environmental 341, 447, 464, 473, 476, 479, 794
  practices
    applied 510, 513–14
    environmental 32, 37, 70, 137, 337, 578–9, 584
  systems 37–9, 75, 106, 475, 477, 509, 512
    environmental 96, 105, 477, 510, 513
Manganese (Mn) 166, 223–4, 483–4, 486, 489, 617, 696, 698, 702–3, 716, 855
Mass balances 279, 288–9
Montana Environmental Information Center (MEIC) 274, 541
MEND 697, 714–15, 717–18, 770–1
Mercury 27, 60, 64, 166, 170, 172, 195–6, 210, 223, 227, 244–6, 249–51, 267, 293, 389, 470, 482–3, 485–91, 493, 495, 502, 547, 550, 659, 726, 846
Merrill Crowe Process 267
Metal
  recovery 195, 208, 235, 237, 241, 307, 535, 714
  release 487
Metallurgy 7, 8, 27, 49, 165–6, 187, 190, 196, 218, 254–5, 260, 290, 492, 548, 552
Methane 57, 342, 544, 551–5, 576, 817
Mine access, access roads 6, 84, 331, 336, 338, 576, 588, 607, 696
Mine closure 216, 413
  costs 577, 815
  design criteria 757–8
  liabilities 812
  planning 43, 96, 100, 108, 216–17, 226, 758, 770–1, 814, 816, 820–1, 838
Mineral Resources Forum (MRF) 55, 158

Minerals Council of Australia (MCA) 56, 277, 480, 827, 856
Mining accidents 13, 148, 460
Mining Association of Canada 757, 770
Mining life-cycle 164–5
Mitigation hierarchy 466
Misima Gold Mine, Papua New Guinea 760, 762, 764, 766, 780
Model input data 431, 433, 793
Models, mathematical 428, 431, 433, 453
Monitoring
locations 389, 472–4
programmes 75, 96, 107, 278, 352, 369, 378, 381, 460, 469–71, 473, 514, 557, 584, 752, 755, 786
Montana Environmental Information Center (MEIC) 274, 541
Mineral Resources Forum (MRF) 55, 158
Mt Lyell Copper Mine, Tasmania, Australia 65–6, 559, 701–2, 730
Mulching 798, 800, 805–6, 827
Multiplier effect 170, 532, 635, 639

N

Nabarlek Uranium Mine, Australia 668, 733–4, 759
NAF (non-acid forming) 708–9, 712
NAG pH 709–10
NAPP (net acid production potential) 706–7, 709
National Environmental Policy Act *see* NEPA
NEPA (National Environmental Policy Act) 5, 31, 34–5, 126
Net acid production potential *see* NAPP
Net Present Value (NPV) 189, 196, 848
Nickel 11, 55–6, 166, 199, 223–4, 253, 265, 272, 303, 483–4, 486–9, 503, 587, 696, 839, 846
Noise 113–15, 119, 145, 200, 207, 211, 213, 231, 296, 300–1, 335, 345, 367, 393, 398, 403–4, 409–12, 420, 424, 428–30, 436, 444, 544, 555, 576, 584
Non-acid forming *see* NAF
Non-government organizations (NGOs) 28, 29, 38, 79, 87–8, 112, 126, 128–9, 131–2, 141, 150–2, 154, 156–8, 339–40, 406, 510, 615–16, 624–5, 635–6, 649–50, 654, 673–4, 845, 855, 857, 866, 874–5
Norms and values 296, 319, 376–7, 395, 664
Nutrient recycling 834, 836

O

Office of Technology Assessment 264–5
Ok Tedi Copper Gold Mine, Papua New Guinea 14, 27, 66, 136, 138, 309, 509, 515, 728, 731, 765, 779, 880
Output Indicators 353, 378–9, 474

P

PAF (potentially acid forming) 708–9, 712
PAP *see* Project-affected people
Participation 12, 35, 73–4, 76, 89–90, 93, 121, 124–7, 129–32, 137, 143, 146–7, 152–4, 156, 159–60, 199, 392, 403, 445–6, 450, 507, 594–5, 621, 624–5, 633–5, 654–5
Participatory planning 624, 628–9, 632, 634, 656
Periodic table 226, 351, 482–3, 501
Performance Indicator 354, 478, 631, 658
environmental 355, 392, 468
Performance Standards (PS) 20, 37–8, 40, 289, 321, 589, 625
Periodic table 226, 351, 482–3, 501
Pests 314, 403, 556, 559–60, 572, 800
pH 24, 61, 210, 263, 268, 272–3, 275, 281, 286, 382, 385, 387, 394, 471–2, 487–8, 501, 540–1, 698, 701, 703, 707–8, 714–15, 756, 766, 827, 839
Pioneer species 828
Pit
lake 61, 432, 699, 825, 839–40
walls 530, 539–40, 577, 696, 699, 703, 775, 825
Plant succession 828, 832, 838
Political risk 27, 49, 138–40, 196–7, 865–6
Political Risk Insurance (PRI) 138–40, 866
Polonium 489, 494–6
Potentially acid forming (PAF) 708–9, 712
Poverty reduction 13, 405, 589, 612, 644, 881
Pre-feasibility study (PFS) 85, 188–9
Pressure-state-impact-response (PSIR) model 21, 252
Project-affected people (PAP) 87, 93, 133–4, 145, 349, 411, 465–6, 589, 593, 596, 612, 621, 625, 645
Project alternatives 73, 79, 89–92, 113, 416, 423, 425, 466
Public consultation 74, 76, 90, 95, 99–100, 124, 126–9, 150, 158, 278, 362, 381, 407, 450, 476, 579
Public Consultation and Disclosure Plan (PCDP) 130, 134, 630
Public health 42, 54, 99, 109, 124, 126–7, 129–30, 150, 157–9, 161, 345, 350, 361, 396, 398, 422, 450, 482, 498, 561–2, 632, 645–6, 757–8

Public involvement and public participation 125–7, 129, 131–6, 147, 149–51, 156–7, 350, 361, 507, 654, 684
Pyrometallurgy 255, 257, 259, 281, 283, 285

Q
Quality Assurance (QA) 77–8, 382–4, 387, 390
Quality control 76–7, 106, 277–8, 382
Quality standards 38, 113

R
Radioactive materials 226, 494, 566, 568
Radium 223, 489, 494–6, 565–7, 571
Radon 177, 459, 494–5, 565–9, 571
Ranger Uranium Mine, N.T., Australia 665, 759, 826
RAP *see* Resettlement action plan
Rapu Rapu Mine, Philippines 458, 857
Rare earth elements 223, 226–7, 252, 483
Receptors, environmental 421–4
Reclamation (*see also* Rehabilitation) 117, 206, 238, 274, 288, 307, 532, 534–5, 568, 573, 718, 723, 757, 793, 805, 810, 815, 826, 828, 841
Recycling 8, 9, 27, 169, 197, 217
Red mud (bauxite residues) 281, 534, 720, 723–4
Reference sites/locations *see* Control sites
Refining, refinery 7, 8, 11, 165, 193, 254–6, 258–60, 265–6, 268–70, 279–80, 465, 548
Regional development plans (*see also* Spatial plans) 100, 861–2
Rehabilitation 8, 24, 43, 45, 100, 165–6, 179, 184, 202–3, 205–6, 217, 233–6, 371, 465–6, 559, 580, 584–5, 610–11, 726–7, 729, 742–3, 777, 784–6, 810–12, 815–17, 823–38
Remediation 66, 217, 571, 697, 702, 704, 717, 782, 813
Remote sensing 173–4, 176, 178–9, 364, 390–1, 767
Resettlement 74, 132, 134, 403, 415, 436, 461, 510, 526, 558, 576, 587–98, 600–2, 604–16, 632
    action plan (RAP) 100, 589–90, 592, 605, 610–11
    planning 508, 589–90, 592–3, 596, 604, 607, 610–11, 615–16
Residues 244, 283–4, 286, 721
Resource rent 16–17, 688
Review Committee 112, 450, 867
Rights of Indigenous Peoples 622, 660–1, 673
Risks
    acceptability 454, 456
    assessment matrix 454, 456
    consequences 454–6, 458
    country 27–8, 189, 640–2, 649
    economic 197, 860
    geological 196
    highest 197, 456, 741
    real 150, 621, 669, 701
    unacceptable 735, 765
Roasting, roaster 95, 192, 255–7, 269, 389, 543, 721
RP *see* Resettlement action plan

S
Sacred sites (*see also* Cultural heritage sites) 665, 682, 688
Scoping 72–3, 89–90, 142, 161, 185, 188, 349, 360–4, 399, 630, 752
Seabed mining 232, 243, 252, 342
Shuttle Radar Topography Mission (SRTM) 56, 305, 363
Silver (Ag) 1, 3, 9, 21, 27, 55, 59, 61, 63, 65–6, 123, 166–7, 194–5, 222–4, 227, 235, 243–4, 249, 260, 263, 267, 269–71, 483–4, 488–90, 499, 659
Site selection 72,85, 447, 735, 748, 750–1, 767, 777, 779, 817–19
Small-scale mining 3, 222, 243–7, 249–51, 317, 613
Social impacts 410, 441, 443–4, 505, 508, 528, 561–2, 607, 689, 692, 838, 840, 848, 854, 861, 880
Social investment 86–7, 104–5, 107, 131, 510
Solution mining 5, 222, 233–4, 241–2, 271, 569
Solvent extraction-electrowinning (SX-EW) 256, 262, 264–5, 287–8, 290
Sound pressure level (SPL) 300–1
Source-pathway-receptor (SPR) concept/model 418, 420, 436
Spatial plans (*see also* Regional development plans) 120, 306, 398
Species
    diversity 313, 326, 345, 394
    exotic 394, 828, 830
    indicator 356–7
    sentinel 356, 358–9
Stakeholders 16, 33, 35–6, 47–8, 51–3, 70–1, 75–6, 78, 79, 87–9, 93, 95, 125–6, 129–35, 137–9, 141–3, 146–7, 154, 278, 342–3, 348–50, 445–9, 475–6, 654, 685–6, 820–1, 875–7
Standard Operating Procedures (SOPs) 176, 357, 382–4, 476
Static geochemical tests 703, 706
Storage capacity 738–9, 748–9, 756

Stripping ratio 18, 207, 533, 774–5, 847
Study boundaries 80–2, 863
Submarine Tailings Placement/Disposal (STP) 64, 230, 389, 720, 731, 759–61, 763
Sulphide minerals/sulphides 194, 210, 254–7, 260–3, 282, 286, 487, 529, 539–41, 587, 696–8, 700, 703, 707, 712, 715, 717, 731, 767–8, 776
Sulphur 67, 463
Sulphur dioxide (SO2) 257, 271, 273, 275, 282, 332, 547–8, 558
Super Pit, Kalgoorlie, Australia 63, 847
Superfund sites 61, 527, 541, 704
Sustainability 42–8, 52, 236, 240, 247, 295, 359, 619, 638, 651, 655, 759, 834, 860, 861, 868–70, 874
Sustainable Development (SD) 2, 3, 16, 20, 43–7, 49–57, 71–2, 86, 96, 105, 126, 141, 236, 239–40, 252, 340–1, 344–5, 572–4, 604–5, 614, 632, 638–40, 654–5, 692–3, 771, 811, 868–9

T

Tailings 6, 7, 18, 19, 21–2, 64–6, 115–16, 195, 210–11, 230–1, 244, 247–9, 275–6, 389, 516, 528–30, 532–7, 566–9, 701–3, 713–14, 720–47, 749–50, 754–5, 758–9, 761–7, 770–1, 798, 852
  disposal 106, 239, 249, 457, 510, 526, 529, 535–6, 564, 719–20, 722, 724–6, 728, 730–4, 736–8, 740, 744, 746–8, 750, 752, 754, 756, 758, 762–6, 768–72, 851
  operations 755, 802
  disposal, riverine 728, 731, 779
  freeboard 236, 458, 728, 736, 748–9, 751, 754, 756
  management 195, 202, 205–6, 210, 526, 536, 564, 566, 573, 721, 727, 730, 735, 753, 769–71, 833, 851
  pipeline 211, 538–9
  ponds 91, 116, 210, 271, 274, 281, 287, 354, 368, 469, 537–9, 567, 594, 726, 735, 815
  storage facility (TSF) 57, 60, 66, 84, 106, 210, 214, 239, 274–5, 370, 457–9, 535–6, 567–8, 575, 699, 702, 720–1, 723–9, 731–7, 739–40, 742–53, 755–9, 765, 769–71, 821–4, 833–4
Tailings dam failures 460, 509, 753
Thickened tailings 211, 723, 737–8, 742–3, 770, 823
Thorium (Th) 223, 225–6, 489, 495, 566, 569
Threshold analysis 419, 433–5
Thresholds 90, 169, 331, 365, 421, 434–5, 444
Titanium (Ti) 223, 225, 259, 484, 489, 695
Top down construction (TDC) of waste rock dumps 781–2
Topsoil 18, 19, 100, 115–16, 165, 203–7, 240, 465, 492, 529, 533–4, 567, 580, 583, 702, 712, 733, 782–3, 800, 802–3, 825–8, 830, 837
Toxicity of Metals 485, 487, 501
Trace metals 84, 284, 290, 300, 324, 357–8, 394, 403, 413, 418, 451, 547, 550, 576, 719, 726, 731
Traditional/customary law 148, 397, 600, 629

U

UN Conference on the Environment and Development *see* UNCED
UNCED (UN Conference on the Environment and Development) 126, 306, 309, 340
UNCLOS (United Nations Convention on the Law of the Sea) 57, 342
UNEP (United Nations Environment Program) 55, 57, 80, 106, 111, 158, 165, 217–18, 229–30, 311, 325, 339, 345, 391, 460, 462, 483, 491, 502, 506, 516, 564, 573–4, 727, 752–3, 770–1
United Nations Convention on the Law of the Sea (UNCLOS) 57, 342
United Nations Environment Program *see* UNEP
Universal soil loss equation (USLE) 428, 431, 457–8, 790, 792–3, 807
Uranium 15, 21, 67, 223–6, 234, 264, 310, 489–90, 493–5, 565–70, 572, 659, 721
  mining 67, 494–5, 502, 564–9, 571–2

V

Vetiver grass 785, 831–3, 841
Vibration 115, 119, 207, 231, 424, 436, 576
Visual impacts 411, 434, 532, 536, 818–19, 858

W

Waste
  management 46, 165–6, 212, 215, 252, 404, 533, 536, 539–40, 581, 718, 788
  rock
    disposal 535, 713, 728, 733, 773–4, 776–80, 782, 784, 786
    storages 216, 533, 697, 713, 729, 738–9, 748, 776–86, 821–2, 824, 832
Water
  balance 236, 275, 298, 305, 309, 370, 537, 539, 542, 727, 735, 747–8, 756
  erosion 95, 529, 575, 725–6, 755, 757, 789, 791–4, 806

management 52, 116, 309–10, 343, 345, 562, 567, 581, 648, 736, 747, 757–8, 770, 794, 818
quality 31, 33, 79, 117, 134, 309, 329, 332, 357–9, 365, 367, 376, 384, 394, 397, 404, 409, 411, 413, 428–9, 453, 529, 727–8, 816, 824–5, 839
monitoring 386, 471–3
retention structures 741–2, 746, 759
storages 306–7, 370, 386, 709, 743, 746
table 246, 472, 557, 697, 711–13, 733, 748, 778, 824
treatment 91, 114–15, 225, 272, 274, 283, 285, 420, 531, 538, 572, 704, 747, 773, 787, 815
plants 115, 272, 704, 747
Weeds 557, 559–60, 582, 827–8
Wind erosion 725–6, 789, 802–6
Wismut 67, 568
Workforce, employment 2, 17, 19, 31, 134, 142, 216–17, 239, 353, 367, 410–11, 560–2, 593–5, 605, 607, 612–13, 627, 635, 640–3, 645, 647–50, 666–7, 683–4, 687–8, 692, 840
Workforce recruitment 148, 561, 660
World Health Organisation 502

Z

Zinc 9, 41, 55, 59, 61, 65, 84, 123, 138, 166–7, 172, 194, 210, 223–4, 227, 267, 269, 272, 483–4, 486–9, 492, 501–2, 525, 696, 702, 704